INTERSECTIONS BETWEEN PARTICLE AND NUCLEAR PHYSICS

INTERSECTIONS BETWEEN PARTICLE AND NUCLEAR PHYSICS

6th Conference

Big Sky, MT May 1997

EDITOR
T. W. Donnelly
Massachusetts Institute of Technology

American Institute of Physics

AIP CONFERENCE
PROCEEDINGS 412

Woodbury, New York

Authorization to photocopy items for internal or personal use, beyond the free copying permitted under the 1978 U.S. Copyright Law (see statement below), is granted by the American Institute of Physics for users registered with the Copyright Clearance Center (CCC) Transactional Reporting Service, provided that the base fee of $10.00 per copy is paid directly to CCC, 222 Rosewood Drive, Danvers, MA 01923. For those organizations that have been granted a photocopy license by CCC, a separate system of payment has been arranged. The fee code for users of the Transactional Reporting Service is: 1-56396-712-X/ 97 /$10.00.

© 1997 American Institute of Physics

Individual readers of this volume and nonprofit libraries, acting for them, are permitted to make fair use of the material in it, such as copying an article for use in teaching or research. Permission is granted to quote from this volume in scientific work with the customary acknowledgment of the source. To reprint a figure, table, or other excerpt requires the consent of one of the original authors and notification to AIP. Republication or systematic or multiple reproduction of any material in this volume is permitted only under license from AIP. Address inquiries to Office of Rights and Permissions, 500 Sunnyside Boulevard, Woodbury, NY 11797-2999; phone: 516-576-2268; fax: 516-576-2499; e-mail: rights@aip.org.

L.C. Catalog Card No. 97-77178
ISBN 1-56396-712-X
ISSN 0094-243X
DOE CONF- 970564

Printed in the United States of America

CONTENTS

Introduction xix
Organizing Committee and Parallel Session Coordinators xx
Conference Participants xxi

PLENARY SESSIONS

Insight into Hadron Structure from Lattice QCD 3
    J. Negele
High Energy Tests of QCD 18
    G. C. Blazey
Lepton Flavor Violation 34
    M. Cooper, M. Brooks, G. E. Hogan, V. D. Laptev, R. E. Mischke,
    P. S. Cooper, Y. Chen, M. Dzemidzic, A. Empl, E. V. Hungerford III,
    K. Lan, W. von Witsch, K. Stanz, J. Szymanski, C. Gagliardi,
    R. E. Tribble, D. D. Koetke, R. Manweiler, S. Stanislaus,
    K. O. H. Ziock, and L. E. Piilonen
CP Violation in K and B Decays 49
    G. Buchalla
Recent Results from Relativistic Heavy Ion Collisions 67
    T. K. Hemmick
Signatures and Status of the Quark-Gluon Plasma 82
    B. Müller
Meson Spectroscopy and the Search for Exotics 91
    C. A. Meyer
An Incomplete Review of Experimental Charm Physics 101
    G. Blaylock
Chiral Perturbation Theory in Few-Nucleon Systems 110
    U. van Kolck
Low Energy Tests of the Standard Model
    S. Freedman (No manuscript provided by speaker.)
Quark and Gluon Structure of the Nucleon 125
    U. Straumann
Structure functions
    A. Vogt (No manuscript provided by speaker.)
Prospects for a Neutrino Oscillation Experiment at the National
Spallation Neutron Source 141
    B. Bugg, N. Cohn, L. Chatterjee, Y. Efemenko, A. Fazely, T. Gabriel,
    Y. Kamyshkov, F. Plasil, and R. Svoboda
Topics in Neutrino Physics Including New Results
from Super-Kamiokande 146
    T. Kajita
Supernova Heavy Element Nucleosynthesis: Can it Tell Us
about Neutrino Masses? 160
    G. Fuller

Neutron Stars 181
    F. Weber

Supersymmetry and Flavor 197
    L. J. Hall

Results from CEBAF Experimental E89-012: Measurements of Deuteron Photo-Disintegration up to 4 GeV 213
    M. A. Miller, D. J. Abbott, A. Ahmidouch, C. S. Armstrong, J. Arrington,
    K. A. Assamagan, O. K. Baker, S. P. Barrow, D. P. Beatty, D. H. Beck,
    S. Y. Beedoe, E. J. Beise, J. E. Belz, C. W. Bochna, P. E. Bosted,
    E. J. Brash, H. Breuer, R. V. Cadman, L. Cardman, R. D. Carlini, J. Cha,
    N. S. Chant, G. Collins, C. Cothran, W. J. Cummings, S. Danagoulian,
    F. A. Duncan, J. A. Dunne, D. Dutta, T. Eden, R. Ent, B. W. Filippone,
    T. A. Forest, H. T. Fortune, V. V. Frolov, H. Gao, D. F. Geesaman,
    R. Gilman, P. L. J. Gueye, K. K. Gustafsson, J-O. Hansen, M. Harvey,
    W. Hinton, R. J. Holt, H. E. Jackson, C. E. Keppel, M. A. Khandaker,
    E. R. Kinney, A. Klein, D. M. Koltenuk, G. Kumbartzki, A. F. Lung,
    D. J. Mack, R. Madey, P. Markowitz, K. W. McFarlane, R. D. McKeown,
    D. G. Meekins, Z-E. Meziani, J. H. Mitchell, H. G. Mkrtchyan,
    R. M. Mohring, J. Napolitano, A. M. Nathan, G. Niculescu, I. Niculescu,
    T. G. O'Neill, B. R. Owen, S. Pate, D. H. Potterveld, J. W. Price,
    G. L. Rakness, R. Ransome, J. Reinhold, P. M. Rutt, G. Savage, R. E. Segel,
    N. Simicevic, P. Stoler, R. Suleiman, L. Tang, B. P. Terburg, D. van Westrum,
    W. F. Vulcan, S. E. Williamson, M. T. Witkowski, S. A. Wood, C. Yan,
    and B. Zeidman

PARALLEL SESSIONS

Relativistic Heavy Ions

Dilepton Production in Relativistic Heavy Ion Collisions 227
    V. Koch

nomalous J/Ψ Suppression in 158 GeV/c Pb-Pb Collisions at the CERN SPS 233
    M. C. Abreu, B. Alessandro, C. Alexa, J. Astruc, C. Baglin, A. Baldit,
    F. Bellaich, M. Bedjidian, S. Beole, A. Borhani, V. Boldea, G. Bonazzola,
    P. Bordalo, A. Bussère, V. Capony, J. Castor, M. Cerú, T. Chambon,
    B. Chaurand, I. Chevrot, B. Cheynis, E. Chiavassa, C. Cicalo,
    S. Constantinescu, W. Dabrowski, A. De Falco, G. Dellacasa, N. De Marco,
    A. Devaux, S. Dita, O. Drapier, B. Espagnon, J. Fargeix, F. Fleuret, P. Force,
    M. Gallio, Y. K. Gavrilov, C. Gerschel, P. Giubellino, M. B. Golybeva,
    M. Gonin, P. Gorodetzky, J. Y. Grossiord, P. Guaita, F. F. Guber,
    A. Guichard, R. Haroutunian, M. Idzik, D. Jouan, T. L. Karavitcheva,
    R. Kossakowski, L. Kluberg, A. B. Kurepin, Y. Le Bornec, G. Landaud,
    C. Lourenco, L. Luquin, P. Macciotta, F. Ohlsson-Malek, A. Marzari-Chiesa,
    M. Masera, A. Masoni, S. Mourgues, A. Musso, P. Petiau, W. L. Prado Da Silva,
    A. Piccotti, J. R. Pizzi, G. Puddu, C. Racca, L. Ramello, S. Ramos,
    P. Rato-Mendes, L. Riccati, A. Romana, S. Sartori, P. Saturnini, E. Scomparin,
    R. Shaoian, S. Silva, S. Serci, P. Sonderegger, X. Tarrago, P. Temnikov,
    N. S. Topilskaya, G. Usai, E. Vercellin, and N. Willis

Non-Strange and Strange Antibaryons in Au+Pb Collisions at 11.5 A GeV/c 241
    J. G. Lajoie for the E864 Collaboration

Directed Flow in Au+Au Collisions at the AGS 247
    W. Chang for the E877 Collaboration

Multiplicities and Angular Distributions of Nucleus-Nucleus Interactions at SPS Energies: Protons to Lead 253
    M. L. Cherry, P. Deines-Jones, A. Dabrowska, J. Dugas,
    R. Holynski, W. V. Jones, D. Kudzia, B. S. Nilsen, A. Olszewski,
    M. Szarska, A. Trzupek, C. J. Waddington, J. P. Wefel, B. Wilczynska,
    H. Wilczynski, W. Wolter, B. Wosiek, and K. Wozniak

An Excitation Function of Particle Production at the AGS 259
    J. C. Dunlop for the E866 and E917 Collaborations

LEXUS 264
    S. Jeon

Peculiarities of Secondary Particle Generation Process in Pb-Pb Interactions at 158 A GeV 269
    N. M. Astafyeva, N. A. Dobrotin, I. M. Dremin, E. L. Feinberg,
    L. A. Gonchanova, K. A. Kotelnikov, A. G. Martynov,
    and N. G. Polukhina

Coherent Photons and Pomerons in Heavy Ion Collisions 274
    S. Klein and E. Scannapieco

Future Prospectives at RHIC 279
    T. S. Ullrich

Facilities and Detectors

The BNL AGS Accelerator Complex Status and Future Plans 297
    M. Tanaka

The Fermilab Long-Baseline Neutrino Program 300
    M. Goodman for the MINOS Collaboration

First Commissioning Results from Hall A at TJNAF 307
    R. Michaels for the Hall A Collaboration

K2K: KEK to Super-Kamiokande Long-Baseline Neutrino Oscillation Experiment 311
    R. J. Wilkes for the K2K Collaboration

The BaBar Detector at the SLAC B-Factory 316
    D. H. Fujino

OPPIS Development at TRIUMF 322
    A. N. Zelenski, G. Dutto, C. D. P. Levy, P. W. Schmor, W. T. H. van Oers,
    and G. W. Wight

Polarized Beam for the TRIUMF Parity Violation Experiment 328
    A. N. Zelenski, A. R. Berdoz, J. Birchall, J. D. Bowman, J. R. Campbell,
    C. A. Davis, A. A. Green, P. W. Green, A. A. Hanien, D. C. Healey,
    R. Helmer, S. Kadantser, Y. Kuznetsov, R. Laxdal, L. Lee, C. D. P. Levy,
    R. E. Mischke, S. A. Page, W. D. Ramsay, S. D. Reitzner, G. Roy,
    P. W. Schmor, J. Soukup, G. M. Stinson, V. Sun, T. S. Stocki, N. A. Titov,
    W. T. H. van Oers, and G. W. Wight

Neutron-Antineutron Transition Search at HFIR Reactor 335
    Y. A. Kamyshkov
Status of the Focal Plane Polarimeter for Hall A at TJNAF 342
    M. K. Jones, F. T. Baker, L. Bimbot, E. J. Brash, R. Gilman,
    C. Glashausser, G. Kumbartzki, J. McIntyre, C. F. Perdrisat, V. Punjabi,
    G. Quéméner, R. Ransome, P. M. Rutt, K. Wijesooriya, G. D. Zainea,
    and the TJNAF Hall A Collaboration

Tests of Fundamental Symmetries

The TRIUMF Parity Violation Experiment 351
    A. R. Berdoz, J. Birchall, J. D. Bowman, J. R. Campbell, C. A. Davis,
    A. A. Green, P. W. Green, A. A. Hamian, D. C. Healey, R. Helmer,
    S. Kadantsev, Y. Kuznetsov, R. Laxdal, L. Lee, C. D. P. Levy,
    R. E. Mischke, S. A. Page, W. D. Ramsay, S. D. Reitzner, G. Roy,
    P. Schmor, A. M. Sekulovich, J. Soukup, G. M. Stinson, T. J. Stocki,
    V. Sum, N. Titov, W. T. H. van Oers, R. J. Woo, and A. Zelenski
**Tests of Chiral Perturbation Theory in Threshold Pion Production
at SAL** 359
    E. Korkmaz
Isospin Splittings in the Pion-Nucleon Couplings from QCD Sum Rules 365
    T. Meissner and E. Henley
**Time Ordered Products, Vector-Meson Exchange, and Nuclear
Charge Asymmetry** 368
    S. A. Coon, B. H. J. McKellar, and A. A. Rawlinson
**CPLEAR Experiment at CERN: Measurement of CP, T, and CPT
in the Neutral Kaon System** 372
    R. Adler, A. Angelopoulos, A. Apostolakis, E. Aslanides, G. Backenstoss,
    P. Bargassa, C. P. Bee, O. Behnke, A. Benelli, V. Bertin, F. Blanc, P. Bloch,
    P. Carlson, M. Carroll, J. Carvalho, E. Cawley, S. Charalambous,
    G. Chardin, M. B. Chertok, A. Cody, M. Danielsson, M. Dejardin,
    J. Derre, A. Ealet, B. Eckart, C. Eleftheriadis, I. Evangelou, L. Faravel,
    P. Fassnacht, C. Felder, R. Ferreira-Marques, W. Fetscher, M. Fidecaro,
    A. Filipčič, D. Francis, J. Fry, E. Gabathuler, R. Gamet, D. Garreta,
    H.-J. Gerber, A. Go, A. Guyot, A. Haselden, P. J. Hayman,
    F. Henry-Couannier, R. W. Hollander, E. Hubert, K. Jon-And, P.-R. Kettle,
    C. Kochowski, P. Kokkas, R. Kreuger, R. Le Gac, F. Leimgruber, A. Liolios,
    E. Machado, I. Mandić, N. Manthos, G. Marel, M. Mikuž, J. Miller,
    F. Montanet, A. Muller, T. Nakada, B. Pagels, I. Papadopoulos, P. Pavlopoulos,
    J. Pinto da Cunha, A. Policarpo, G. Polivka, R. Rickenbach, B. L. Roberts,
    T. Ruf, L. Sakeliou, P. Sanders, C. Santoni, M. Schäfer, L. A. Schaller,
    T. Schietinger, A. Schopper, P. Schune, A. Soares, L. Tauscher, C. Thibault,
    F. Touchard, C. Touramanis, F. Triantis, E. Van Beveren, C. W. E. Van Eijk,
    S. Vlachos, P. Weber, O. Wigger, M. Wolter, C. Yeche, D. Zavrtanik,
    and D. Zimmerman

Direct CP Violation in $B^{\pm} \to \rho^{\pm} \pi^{+} \pi^{-}$ in the ρ^0-ω Interference Region 383
  S. Gardner, H. B. O'Connell, and A. W. Thomas
A New Experiment to Measure the Electric Dipole Moment
of the Neutron? 387
  R. Krause, J. Doyle, and R. Golub
Searching for Time Reversal Invariance Violation in
Polarized Neutron Decay 399
  L. J. Lising, J. M. Adams, J. M. Anaya, T. J. Bowles, T. E. Chupp,
  K. P. Coulter, M. S. Dewey, S. R. Elliot, S. J. Freedman, B. K. Fujikawa,
  A. Garcia, G. L. Green, S.-R. Hwang, G. L. Jones, J. S. Nico,
  H. G. R. Robertson, T. D. Steiger, W. A. Teasdale, A. K. Thompson,
  E. G. Wasserman, F. E. Wietfeldt, and J. F. Wilkerson
T-Odd Correlations in Z^0 Decay into Three Jets and in
Nuclear Beta Decay 403
  H. E. Conzett
Search for Physics Beyond the Standard Model via a
Polarization-Asymmetry Correlation Experiment on ^{107}In 408
  P. Schuurmans, J. Camps, N. Severijns, P. DeMoor, J. Deutsch, T. Otto,
  J. Govaerts, B. A. Brown, B. Holstein, R. Kirchner, O. Naviliat-Cunic,
  R. Prieels, P. A. Quin, E. Thomas, A. Van Geert, B. Vereecke,
  and L. Vanneste
The Mass-8 Experiment—Measuring the $\beta - \alpha$ Angular Correlations 416
  M. Beck, J. F. Amsbaugh, L. DeBraeckeler, D. W. Storm, E. Swanson,
  K. B. Swartz, J. P. S. van Schager, D. C. Wright, and Z. Zhao
A Search for Antiproton Decay at the Fermilab Antiproton Accumulator 419
  T. Armstrong, C. Buchanan, B. Corbin, S. Geer, R. Gustafson, M. Hu,
  M. Lindgren, J. Marriner, M. Martens, T. Müller, R. Ray, G. Snow,
  J. Streets, and W. Wester
The SAMPLE Experiment 423
  R. D. McKeown
Increased Sensitivity to Possible Muonium to Antimuonium Conversion 429
  V. Meyer, A. Grossmann, K. Jungmann, J. Merkel, G. zu Putlitz,
  I. Reinhard, K. Träger, P. V. Schmidt, L. Willmann, R. Engler, H. P. Wirtz,
  R. Abela, W. Bertl, D. Renker, H. K. Walter, V. Karpuchin, I. Kisel,
  A. Korenchenko, N. Kravchuk, N. Kuchinsky, A. Moiseenko, J. Bagaturia,
  D. Mzavia, T. Sakhelashvili, and V. W. Hughes
An Update on Cosmological Anisotrophy in Electromagnetic Propagation 432
  J. P. Ralston and B. Nodland
Tests of Fundamental Symmetries 438
  S. A. Page and B. R. Holstein

Hadron Spectroscopy

Interplay between the $f_0(980)$ and $a_0(980)$ Mesons 447
  J. Gunter
Search for the Pentaquark in Fermilab E791 451
  D. Ashery for the E791 Collaboration

Searches for H Dibaryons at the AGS................................. 457
    B. Bassalleck
Search for Strange Quark Matter with AGS E864 465
    J. C. Hill for the E864 Collaboration
Exotic Meson Signal in the $\eta\pi^-$ System in π^-p Interactions
at 18 GeV/c 471
    N. M. Cason for the E852 Collaboration
PWA Analysis of BES Data on $J/\Psi \to \gamma\pi^+\pi^-\pi^+\pi^-$ 476
    Y. C. Zhu
In-Flight Hadron Spectroscopy at LEAR 481
    D. W. Hertzog
Search for Two-Photon Production of $f_J(2220)/\xi(2230)$ at CLEO 489
    R. S. Galik for the CLEO Collaboration
Our Present Understanding of the N and Δ Baryons 494
    D. M. Manley
Spectroscopy of Nonstrange Baryons: Recent Results from PNPI 499
    I. V. Lopatin
Pentaquark Phenomenology 504
    H. J. Lipkin
The Signature of Glueballs in J/Ψ Radiative Decays 510
    Z. Li
Glueball Spectroscopy in a Many Body QCD Hamiltonian Approach 515
    S. R. Cotanch, A. Szczepaniak, E. S. Swanson, and C.-R. Ji
Electromagnetic Form Factors of the Nucleon............................ 519
    R. Bijker and A. Leviatan
Strange Skyrmion Molecules 524
    V. B. Kopeliovich and B. E. Stern
Hadron Spectroscopy Summary...................................... 530
    J. Napolitano and A. Schwartz

Lepton Probes of Hadron Structure

Measurement of the Neutron's Electric Formfactor $G_{E,n}$
in $D(e,e'n)p$ and $^3He(e,e'n)pp$ 541
    M. Ostrick for the A3 Collaboration
The E2/M1 Ratio in Δ Photoproduction 547
    S. Hoblit, G. Blanpied, M. Blecher, A. Caracappa, C. Djalali,
    G. Giordano, K. Hicks, M. Khandaker, O. C. Kistner, A. Kuczewski,
    M. Lowry, M. Lucas, G. Matone, L. Miceli, B. Preedom, D. Rebreyend,
    A. M. Sandorfi, C. Schaerf, R. M. Sealock, H. Ströher, C. E. Thorn,
    S. T. Thornton, J. Tonnison, C. S. Whisnant, H. Zhang, and X. Zhao
Excited Baryon Form Factors at High Q^2 552
    P. Stoler for the E94-014 Collaboration
Polarized Parity Violating Electron Scattering in ^3He and ^3H 558
    S. L. Mintz, G. M. Gerstner, M. A. Barnett, and M. Pourkaviani

**Proton Propagation Through Nuclei and the Quasi-Free
Reaction Mechanism Studied with (e,e′p) Reactions** 561
    D. Dutta, D. Abbott, T. S. Amatuni, A. Ahmidouch, C. Armstrong,
    J. Arrington, K. A. Assamagan, O. K. Baker, S. Barrow, K. Beard,
    D. Beatty, S. Beedoe, E. Beise, E. Belz, C. Bochna, H. Breuer, E. Bruins,
    R. Carlini, J. Cha, N. Chant, C. Cothran, W. J. Cummings, S. Danagoulian,
    D. Day, D. DeSchepper, J.-E. Ducret, F. Duncan, J. Dunne, T. Eden,
    R. Ent, J. Fedchak, H. T. Fortune, V. Frolov, D. F. Geesaman, H. Gao,
    R. Gilman, P. Gueye, J. O. Hansen, W. Hinton, R. Holt, C. Jackson,
    H. E. Jackson, C. Jones, S. Kaufman, J. J. Kelly, C. Keppel,
    M. Khandaker, W. Kim, E. Kinney, A. Klein, D. Koltenuk, L. Kramer,
    W. Lorenzon, A. Lung, K. McFarlane, D. Mack, R. Madey, P. Markowitz,
    J. Martin, A. Mateos, D. Meekins, M. Miller, R. Milner, J. Mitchell,
    H. Mkrtchyan, R. Mohring, G. Niculescu, I. Niculescu, T. G. O'Neill,
    D. Potterveld, J. W. Price, J. Reinhold, C. Salgado, J. P. Schiffer, R. E. Segel,
    P. Stoler, R. Suleiman, V. Taderosyan, L. Tang, B. Terburg, W. Turchinetz,
    D. van Westrum, P. Welch, C. Williamson, S. Wood, C. Yan, J.-C. Yang,
    J. Yu, B. Zeidman, W. Zhao, and B. Zihlmann

η-Meson Photoproduction Dynamics and Missing Resonances 567
    B. Saghai, F. Tabakin, J. Ajaka, and P. Hoffman-Rothe

Measurements of the Reactions $^{12}C(\nu_\mu, \mu^-)^{12}N_{g.s.}$ and $^{12}C(\nu_\mu, \mu^-)X$ 570
    E. D. Church

New Results from the SMC 573
    A. Sandacz

**Measurement of the Neutron Spin Structure Function g_1^n
and a pQCD NLO Analysis** 579
    S. Rock for the E154 Collaboration

**Inclusive Nucleon Resonance Electroproduction: Recent Results
from Jefferson Lab** 585
    C. Keppel

Measurements of $R_d = \sigma_L/\sigma_T$ for $0.03 < x < 0.1$ 588
    P. Bosted and J. Fellbaum for the E143 Collaboration

Diffractive Deep Inelastic Scattering at HERA 591
    R. Todenhagen for the H1 and ZEUS Collaborations

Vector Meson and Heavy Flavor Production at HERA 599
    S. M. Wang for the H1 and ZEUS Collaborations

Jets in DIS and Photoproduction at HERA 605
    L. Jönsson

Exotic Searches at HERA 612
    M. Kuze for the H1 and ZEUS Collaboration

Hadron Dynamics

Color Transparency—Color Coherent Effects in Nuclear Physics 621
    G. A. Miller

A-Dependent Effects in Jet Production 628
    M. D. Corcoran

High P_T Production of Neutral Mesons and Direct Photons 632
    G. Ginther

Quasi-Elastic Hadronic Scattering at High Momentum Transfer 637
    Y. Mardor

High p_t Quasi-Exclusive Scattering with Resonance Production 640
    I. Mardor

A Preliminary Measurement if the $\bar{u}/\bar{d}$ Asymmetry in the Proton Sea 643
    P. E. Reimer, T. C. Awes, M. E. Beddo, C. N. Brown, J. D. Bush,
    T. A. Carey, T. H. Chang, W. E. Cooper, C. A. Gagliardi, G. T. Garvey,
    D. F. Geesaman, E. A. Hawker, X. C. He, L. D. Isenhower, S. B. Kaufman,
    D. M. Kaplan, P. N. Kirk, D. D. Koetke, G. Kyle, D. M. Lee, W. M. Lee,
    M. J. Leitch, N. Makins, P. L. McGaughey, J. M. Moss, P. M. Nord,
    B. K. Park, V. Papavassiliou, J. C. Peng, G. Petitt, M. E. Sadler, J. Selden,
    P. W. Stankus, W. E. Sondheim, T. N. Thompson, R. S. Towell,
    R. E. Tribble, M. A. Vasiliev, Y. C. Wang, Z. F. Wang, J. C. Webb,
    J. L. Willis, D. Wise, G. R. Young, and B. Zeidman

Hadroproduction of Charm in FNAL E769 and E791 648
    J. Slaughter

Dynamics of Open Charm Production 654
    R. W. Gardner

Beauty Hadroproduction at Fixed Target in WA92 Experiment 659
    C. Gemme

Structure Functions from Chiral Soliton Models 664
    H. Weigel, L. Ganberg, and H. Reinhardt

Quark and Proton Spin Structure in the Instanton Liquid Model of QCD 670
    A. Blotz and E. Shuryak

Nucleon Strangeness and Spin Crisis? 673
    M. D. Scadron

Hunting the d' (2065) with Various Probes 679
    G. J. Wagner

The d'-Dibaryon in a Colored Cluster Model 685
    A. J. Buchmann, G. Wagner, and A. Faessler

Dibaryons from Hadron Supersymmetry and a Diquark Model 689
    D. B. Lichtenberg

Search for Dibaryons in Reactions of K^- Mesons with Deuterium and ^{3}He Nuclei at 0.87 GeV/c 694
    H. Piekarz

Diffractive Phenomena at Tevatron 699
    A. Santoro for the D0 Collaboration

Exclusive Near Threshold Two-Pion Production with the MOMO Experiment at COSY 704
    S. Bavink, F. Bellemann, A. Berg, J. Bisplinghoff, G. Bohlscheid,
    J. Ernst, C. Henrich, F. Hinterberger, R. Ibald, R. Jahn, L. Jarczyk,
    R. Joosten, A. Kozela, H. Machner, A. Magiera, R. Maschuw,
    T. Mayer-Kuckuk, G. Mertler, J. Munkel, P. V. Neumann-Cosel,
    D. Rosendaal, P. V. Rossen, H. Schnitker, K. Scho, J. Smyrski,
    A. Strzalkowski, R. Tölle, and R. Wurzinger

Microscopic In-Medium NN Cross Sections up to 2 GeV 707
    R. Machleidt and F. Sammarruca

The N-N Interaction Inside Nuclei: Evidence for Partial
Chiral Restoration? 708
    E. J. Stephenson and F. Sammarruca

Elastic pp Scattering Excitation Functions at Intermediate Energies 713
    F. Hinterberger for the EDDA Collaboration

New Perspectives in Multi-Nucleon Pion Absorption on Light Nuclei 717
    A. Lehmann, D. Androić, G. Backenstoss, D. Bosnar, H. Breuer,
    H. Döbbeling, T. Dooling, M. Furić, P. A. M. Gram, N. K. Gregory,
    A. Hoffart, C. H. Q. Ingram, A. Klein, K. Koch, J. Köhler, B. Kotliński,
    M. Kroedel, G. Kyle, A. O. Mateos, K. Michaelian, T. Petković,
    M. Planinić, R. P. Redwine, D. Rowntree, U. Sennhauser, N. Šimičević,
    R. Trezeciak, H. Ullrich, M. Wang, M. H. Wang, H. J. Weyer, M. Wildi,
    and K. E. Wilson

Pion Double Charge Exchange on Nucleus and the
Two-Pion Intermediate States 720
    A. B. Kaidalov and A. P. Krutenkova

Study of Hyperonic Atoms at the FNAL Main Injector 722
    Y. M. Ivanov and A. A. Petrunin

On the Structure of the First Excited Nucleon States 725
    C. Schütz, J. Haidenbauer, and J. Speth

ϕ Meson Couplings to the Nucleon and Strange Vector Currents 730
    U. G. Meissner, V. Mull, J. Speth, and J. W. Van Orden

Nucleon Strangeness Content through Vector Meson Dominance 733
    S. R. Cotanch and R. A. Williams

Sum Rule Analyses for the Light Quark Masses Revisited 736
    K. Maltman, R. Gupta, and T. Bhattacharya

The Three Nucleon System in the Skyrme Model 739
    N. R. Walet

Baryon Magnetic Moments and Axial Coupling Constants
with Relativistic and Exchange Current Effects 743
    C. Helminen, K. Dannbam, L. Y. Glozman, and D. O. Riska

A Dynamical η'-Mass from an Infrared Enhanced Gluon Exchange 746
    L. von Smekal, A. Mecke, and R. Alkofer

Strong $U_A(1)$ Breaking in Radiative η Decays 750
    M. Takizawa, Y. Nemoto, and M. Oka

Low Energy QCD from an Effective Quark-Quark Interaction 753
    T. Meissner and M. Frank

Instanton-Monopole Correlations in Lattice QCD 756
    M. Feurstein, H. Markum, and S. Thurner

Meson and Lepton Decays

First Physics from KTeV 763
    J. Belz

Status of BNL E787: Search for the Decay $K^+ \to \pi^+ \nu \bar{\nu}$ 769
    T. Numao
Experimental Studies of Rare K^+ and π^0 Decays 774
    S. Eilerts for the E865 Collaboration
Status Report of the NA48 Experiment at the CERN SPS 779
    C. Talamonti
Radiative ϕ Decays at Jefferson Lab 786
    R. W. Gardner
Status of the BNL Muon (g−2) Experiment 792
    J. P. Miller, L. M. Barkov, J. Benante, D. H. Brown, H. N. Brown,
    G. Bunce, R. M. Carey, A. Chertovskikh, J. Cullen, P. Cushman,
    G. T. Danby, P. T. Debevec, H. Deng, W. Deninger, S. K. Dhawan,
    A. Disco, V. P. Druzhinin, L. Duong, W. Earle, K. Endo, E. Efstathiadis,
    F. J. M. Farley, G. V. Fedotovich, X. Fei, J. Geller, J. Gerhaeuser,
    S. Giron, D. N. Grigorev, V. B. Golubev, M. Grosse Perdekamp,
    A. Grossmann, U. Haeberlen, E. S. Hazen, D. W. Hertzog, H. Hirabayashi,
    H. Hseuh, B. J. Hughes, V. W. Hughes, S. Ichii, K. Ishida, J. W. Jackson,
    L. Jia, K. Jungmann, D. Kawall, B. I. Khazin, J. Kindem, T. Kinoshita,
    F. Kriener, S. Kurokawa, R. Larson, Y. Y. Lee, I. Logashenko, M. Mapes,
    R. McNabb, W. Meng, Y. Merzliakov, D. Miller, Y. Mizumachi, V. Monich,
    W. M. Morse, Y. Orlov, J. Ouyang, C. Pai, C. Pearson, I. Polk, C. Polly,
    R. Prige, G. zu Putlitz, S. Rankowitz, S. I. Redin, O. Rind, B. L. Roberts,
    N. Ryskulov, J. Sandberg, T. Sato, S. Sedykh, Y. K. Semertzidis,
    S. Serednyakov, Y. M. Shatunov, R. Shutt, L. Snydstrup, E. Solodov,
    A. Soukas, A. Stillman, L. R. Sulak, T. Tallerico, M. Tanaka, F. Toldo,
    C. Timmermans, A. Trofimov, D. Urner, P. von Walter, D. Winn,
    K. Woodle, W. A. Worstell, A. Yamamoto, and D. Zimmerman
Charmless B Decays at CLEO 801
    P. A. Pomianowski for the CLEO Collaboration
Search for the CP-Violating Decay $K_L^0 \to \pi^0 \nu \bar{\nu}$ at BNL E926 807
    T. Numao
CP and CPT Violation Searches with K_S Mesons 812
    T. Alexopoulos, C. Bhat, D. Bergman, T. J. Devlin, J. Doornbos,
    A. Eichenbaum, A. Erwin, P. Martin, S. R. Schnetzer, S. V. Somalwar,
    R. Stone, M. Thompson, G. B. Thomson, and H. White
Detecting K Mesons Leptonic Decays with KLOE 819
    P. de Simone for the KLOE Collaboration
Measurement of the Michel Rho Parameter in Direct Muon Decay........... 826
    L. Piilonen, J. F. Amann, R. D. Bolton, Y. Chen, M. D. Cooper,
    P. S. Cooper, M. Dzemidzic, W. Foreman, C. A. Gagliardi, D. Haim,
    R. Harrison, G. Hart, G. E. Hogan, E. V. Hungerford, C. C. H. Jui,
    J. E. Knott, D. D. Koetke, T. Kozlowski, M. A. Kroupa, K. Lan,
    F. S. Lee, F. Liu, R. Manweiler, B. W. Mayes, R. E. Mischke,
    C. Pillai, L. Pinsky, S. Schilling, T. D. S. Stanislaus, K. M. Stantz,
    J. J. Szymanski, R. E. Tribble, X. L. Tu, L. A. Van Ausdeln,
    W. von Witsch, D. Whitehouse, B. K. Wright, S. C. Wright, Y. Zhang,
    and K. O. H. Ziock
Tau Decays at LEP.. 832
    R. J. Sobie

A Precision Measurement of Muon Decay 838
    D. H. Wright for the E614 Collaboration
Search for T-Violation in $K^+ \to \mu^+ \pi^0 \nu_\mu$ Decay 842
    M. D. Hasinoff for the KEK-246 Collaboration

Physics with Strangeness and Charm

Strangeness Photoproduction with the SAPHIR Detector 849
    D. Menze, J. Barth, M. Bockhorst, W. Braun, R. Burgwinkler,
    K. H. Glander, S. Goers, J. Hannappel, N. Jöper, U. Kirch,
    F. Klein, F. J. Klein, W. Neuerburg, E. Paul, W. J. Schwille,
    M.-Q. Tran, R. Wedemeyer, F. Wehnes, B. Wiegers, F. W. Wieland,
    J. Wisskircher, J. Ernst, H. J. Jüngst, H. Kalinowsky, E. Klempt,
    J. Link, H. V. Pee, R. Plötzke, M. Schumacher, F. Smend, T. Mzt,
    and C. Bennhold
Kaon Productions Off Nucleons and the Structure of
Baryon Resonances. 852
    Z. Li
Off-Shell Effects in Electromagnetic Production of Strangeness 855
    C. Fayard, G. H. Lamot, T. Mizutani, and B. Saghi
Hyperon-Production with Anti-protons at LEAR 858
    J. Franz for the PS 185 Collaboration
Antihyperon-Hyperon Production in a Quark Model 862
    M. A. Alberg, E. M. Henley, P. D. Kunz, and L. Wilets
Recent Charm Physics Results from CLEO 865
    D. H. Fujino for the CLEO Collaboration
Charm- & Strangeness-Production in Σ^--Nucleus-Interactions 875
    E. B. Wittmann for the WA89 Collaboration
Physics Goals and Experimental Status of SELEX: Fermilab E781 883
    M. Procario for the SELEX Collaboration
Recent Results from Experiment E835 at Fermilab 887
    M. Ambrogiani, S. Argiro, S. Bagnasco, W. Baldini, F. Bertini, D. Bettoni,
    M. Bombonati, D. Bonsi, G. Borreani, A. Buzzo, R. Calabrese, M. Cardarelli,
    A. Ceccucci, R. Cester, P. Dalpiaz, S. Frabetti, X. Fan, G. Garzoglio,
    K. E. Gollwitzer, A. Hahn, S. Jin, J. Kasper, G. Lasio, M. Lovetere, E. Luppi,
    P. Maas, M. Macri, M. Mandelkern, F. Marchetto, M. Marinelli, W. Marsh,
    M. Martini, R. McTaggart, E. Menichetti, R. Mussa, M. Obertino,
    M. Pallavicini, N. Pastrone, C. Patrignani, T. Pedlar, J. Peoples Jr., S. Pordes,
    E. Robutti, J. Rosen, L. Rossetto, P. Rumerio, A. Santroni, M. Savrie,
    J. Schultz, K. K. Seth, J. Streets, G. Stancari, M. Thompson, L. Tomassetti,
    S. Werkema, and G. Zioulas
Recent Results from FNAL E791 892
    A. J. Schwartz for the E791 Collaboration
Recent Results from Fermilab E687 on Charm Spectroscopy 901
    P. Lebrun for the E867 Collaboration
Hypernuclear Spectroscopy and Weak Decays at KEK 907
    H. Noumi

**The Parity-Violating Asymmetry in the Weak Decay
of Polarized Hypernuclei**..912
    C. Bennhold, A. Parreño, and A. Ramos
Nonmesonic Weak Decays of Light Hypernuclei..........................916
    T. Inoue
Systems with Strangeness -2 BNL..919
    G. B. Franklin
Enhanced Production of $\Lambda\Lambda$ Pairs near Threshold in (K^-,K^+) Reaction......923
    J. K. Ahn, S. Aoki, K. S. Chung, M. S. Chung, H. En'yo, T. Fukuda,
    H. Funahashi, Y. Goto, A. Higashi, M. Ieiri, T. Iijima, M. Iinuma, K. Imai,
    Y. Itow, J. M. Lee, S. Makino, A. Masaike, Y. Matsuda, Y. Matsuyama,
    S. Mihara, C. Nagoshi, I. Nomura, I. S. Park, N. Saito, M. Sekimoto,
    Y. M. Shin, K. S. Sim, R. Susukita, R. Takashima, F. Takeutchi, P. Tlustý,
    S. Weibe, S. Yokkaichi, K. Yoshida, M. Yoshida, T. Yoshida, and S. Yamashita
$S=-1$ and $S=-2$ Few Body Hypernuclei................................927
    B. F. Gibson
Search for Strange Matter via Heavy Ion Activation.....................931
    M. C. Perillo Isaac, Y. D. Chan, R. Clark, M. A. Deleplanque,
    M. R. Dragowsky, P. Fallon, I. D. Goldman, K. Nishiizumi,
    R.-M. Larimer, I. Y. Lee, A. O. Macchiavelli, R. W. MacLead,
    E. B. Norman, L. S. Schroeder, and F. S. Stephens
Strangeness Physics Sessions Summary...................................935
    R. A. Schumacher

Neutrino and Non-Accelerator Physics

Recent Double Beta Decay Experimental Results..........................941
    C. S. Sutton for the NEMO Collaboration
**Super-Kamiokande Solar Neutrino Analysis:
The First 201.6 days' Results**..946
    R. E. Sanford for the Super-Kamiokande Collaboration
Toward Future Solar Neutrino Experiments...............................951
    R. E. Lanou
The Atmospheric Neutrino Muon-Like Fraction Above 1 GeV................958
    R. Clark for the IMB Collaboration
The Atmospheric Neutrino Flavor Ratio in Soudan 2......................962
    M. Goodman for the Soudan 2 Collaboration
**The 1000 Ton Liquid Scintillation Detector Project at Kamioka
(Kam-LAND)**..969
    F. Suekane

Particle and Nuclear Astrophysics

Very High Energy Gamma-Ray Astronomy...................................979
    M. Catanese
The rp-Process in X-Ray Bursts...987
    H. Schatz, L. Bildsten, J. Görres, M. Wiescher, and F.-K. Thielemann

Neutrino Capture and r-Process Nucleosynthesis 992
    B. S. Meyer
Strange Stars... 999
    J. Madsen
The Equation of State in Nucleon and Strange Stars...................... 1007
    M. Prakash
Composition and Energy Spectra of Cosmic Rays—Implications for Cosmic Ray Origins 1022
    M. L. Cherry
Cosmic Ray H and He Spectra from 2 to 800 TeV/nucleon from the JACEE Experiments... 1031
    B. S. Nilsen, K. Asakimori, T. H. Burnett, M. L. Cherry, K. Cherli,
    M. J. Chrisl, S. Drake, J. H. Derrickson, W. F. Fountain, M. Fuki,
    J. C. Gregory, T. Hayashi, A. Iyono, J. Iwai, J. Johnson,
    M. Kobayashi, J. Lord, O. Miyamura, K. H. Moon, H. Oda, T. Ogata,
    E. D. Olson, T. A. Parnell, F. E. Roberts, K. Sengupta, T. Shiina,
    S. C. Strausz, T. Sugitate, Y. Takahashi, T. Tominaga, J. W. Watts,
    J. P. Wetel, B. Wilczynska, H. Wilczynski, R. J. Wilkes, W. Wolter,
    H. Yokomi, and E. Zager

Author Index 1035

INTRODUCTION

This volume contains written versions of most of the talks presented at the Sixth Conference on the Intersections of Particle and Nuclear Physics, which was held from 27 May to 2 June, 1997 at the Big Sky Ski and Summer Resort in Big Sky, Montana. The participants were nuclear, particle, and astro physicists from 21 countries who gathered in a rather remote, but beautiful location to discuss their overlapping areas of research.

The topics discussed at the conference covered many subjects, from relativistic heavy ion physics and high-energy tests of QCD, which require the highest energy accelerators, to low-energy tests of the electroweak standard model. The energy range extended from physics with ultracold neutrons involving kinetic energies below 10^{-6} eV to cosmic ray physics studying primaries at an energy of 10^{20} eV. There were many connections between topics, such as the role of strange quarks in neutron stars or as a possible signal of a quark-gluon plasma. There were new results including neutrino physics from Super-Kamiokande and evidence for a glueball mixed into the scalar mesons. There were reviews of topics such as CP violation, lepton-flavor violation, and charm physics. These are just examples; the full volume has all the details.

The conference format followed the model that has been established in the previous conferences in this series. In the mornings, plenary talks were given by speakers chosen by the Organizing Committee. In the afternoons, parallel sessions were arranged into ten subject areas. Each of these sessions were organized by a pair of coordinators who invested considerable time to arrange interesting sessions.

Not all the time was spent in meetings. During an afternoon's excursions away from the conference facilities, the participants were able to enjoy activities such as whitewater rafting, a trip to Lewis and Clark Caverns, or a visit to see the wonders of Yellowstone National Park.

The success of the conference was dependent on the efforts of many people. The financial assistance provided by the sponsors (ANL, Bates, BNL, FNAL, IUCF, LANL, LBNL, TJNAF, and TRIUMF) was essential to help cover the costs of the conference. The Conference staff was lead by Susan Ramsay (LANL) who devoted many hours over many months to preparations for the conference. During the conference she was assisted by Jeanne Bowles (LANL), Heidi Demers (MIT), Anne MacInnis (MIT/Bates), Lucy Maestas (LANL), and Laura Zaharatos (BNL). All of the staff worked hard to keep the conference running smoothly. Tony Carter (MIT/Bates) brought in computers and was very resourceful in establishing our link to the outside world from what seemed to be the middle of nowhere. Carol Kuc from Complete Conference Coordinators (CCC) provided her expertise from initial site selection through all the arrangements with the hotel and Conference staff. Susan Kuc of CCC handled the hotel and conference registrations. The staff at Big Sky worked hard to make our stay pleasant and productive. Following the Conference Anne MacInnis coordinated the assembly of the proceedings - her tireless efforts during this editorial phase are much appreciated. Our thanks to everyone who helped for again we see that the intersections of particle and nuclear physics are many and deserve to be explored.

Richard E. Mischke

ORGANIZING COMMITTEE

R. Mischke (Chairman)
S. Kowalski (Co-Chairman)
J. Appel
R. G. Arnold
A. Astbury
P. D. Barnes
E. L. Berger
J. Cameron
E. M. Henley
R. L. Jaffe

K. Kilian
A. Masaike
B. Mecking
F. Plasil
M. Schwartz
J. A. Thompson
W.T.H. van Oers
H-C. Walter
M. E. Zeller

SESSION COORDINATORS

Hadron Dynamics	H. Reinhardt
	A. Zieminski
Hadron Spectroscopy	J. Napolitano
	A. Schwartz
Leptonic Probes of Hadronic Structure	D. Beck
	P. Truoel
Facilities and Detectors	B. Baller
	S. Majewski
Relativistic Heavy Ions	T. Hallman
	J. Kapusta
Physics with Strangeness and Charm	P. Cooper
	R. Schumacher
Meson and Lepton Decays	J. Macdonald
	G. Thomson
Neutrino and Non-accelerator Physics	E. Gates
	J. Learned
Particle and Nuclear Astrophysics	T. Haines
	M. Wiescher
Tests of Fundamental Symmetries	B. Holstein
	S. Page

SPONSORS

Argonne National Laboratory
Indiana University Cyclotron Facility
MIT-Bates Linear Accelerator Center
Los Alamos National Laboratory
Brookhaven National Laboratory
Fermi National Accelerator Laboratory
TJNAF
TRIUMF

CONFERENCE PARTICIPANTS

Adelberger, Eric G.	University of Washington
Agnello, Michelangelo	I.N.F.N. sez. di Torino
Ahn, Jung Keun	Kyoto University
Alberg, Mary	Seattle University
Arnold, Ray,G.	The American University/SLAC
Arrington, John R.	California Institute of Technology
Ashery, Daniel	Tel Aviv University
Bachman, Mark G.	University of California at Irvine
Baglin, Christian C.	LAPP Annecy/France
Baker, Keith	Hampton University and Jefferson Laboratory
Baller, Bruce	Fermilab
Barber, Robin E.	University of Houston
Barnes, Peter D.	Los Alamos National Laboratory
Bassalleck, Bernd	University of New Mexico
Bearden, Ian G.	Niels Bohr Institute
Beck, Marcus	University of Washington
Beck, Doug H.	University of Illinois at Urbana
Bellemann, Frank	ISKP University of Bonn
Belz, John W.	Rutgers University
Benesh, Charles	Indiana University Nuclear Theory Center
Bennhold, Cornelius	George Washington University
Bijker, Roelof	ICN-UNAM, Mexico
Bildsten, Lars	University of California at Berkeley
Blaylock, Guy	University of Massachusetts
Blazey, Jerry Charles	Northern Illinois University
Blotz, Andree	Los Alamos National Laboratory
Bochna, Christopher W.	University of Illinois at Urbana
Bock, Greg	Fermilab
Bosted, Peter E.	The American University/SLAC
Bowles, Jeanne M.	Los Alamos National Laboratory
Bressani, T.	INFN - Sezione di Torino
Buchalla, Gerhard	Stanford Linear Accelerator Center
Buchmann, Alfons	University of Tübingen
Carnahan, Bryan F.	Catholic University of America
Carroll, Alan	Brookhaven National Laboratory
Carroll, Jim B.	University of California/LBL
Carter, Tony	MIT-Bates Linear Accelerator Center
Cason, Neal M.	University of Notre Dame
Catanese, Michael A.	Iowa State University/Whipple Observatory
Cates, Gordon D.	Princeton University
Chang, Wen-Chen	University of California at Riverside
Cherry, Mike L.	Louisiana State University
Church, Eric D.	University of California at Riverside
Clark, Russell J.	Louisiana State University
Cole, Brian A.	Columbia University

Cole, Phil L.	George Washington University
Conzett, Homer E.	University of California at Berkeley/LBL
Coon, Sidney A.	New Mexico State University
Cooper, Martin D.	Los Alamos National Laboratory
Cooper, Peter	Fermilab
Corbin, Brent A.	University of California at Los Angeles
Corcoran, Marjorie D.	Rice University
Cotanch, Stephen R.	North Carolina State University
Crannell, Hall	Catholic University of America
Crawford, Henry J.	University of California at Berkeley
Dasgupta, Sudebsankar	Burdwan University, Burdwan, India
de Simone, Patrizia	L.N.F. Infn
Demers, Heidi M.	Massachusetts Institute of Technology
Denissov, Oleg	INFN-Torino, Dubna
Donnelly, T. William	Massachusetts Institute of Technology
Drees, Axel	Universitat Heidelberg
Dunlop, James C.	Massachusetts Institute of Technology
Durrant, Simon C.	Brookhaven National Laboratory
Dutta, Dipangkar	Northwestern University
Efremenko, Yuri V.	University of Tennessee
Eichblatt, Steve	Fermilab
Eilerts, Scott	University of New Mexico
Empl, Anton	University of Houston
Falk, Willie R.	University of Manitoba
Fazely, Ali R.	Southern University
Fenimore, Ed E.	Los Alamos National Laboratory
Fleming, Sean P.	University of Wisconsin at Madison
Franklin, Gregg B.	Carnegie Mellon University
Franz, Juergen	University of Freiburg/Germany
Freedman, Stuart J.	University of California at Berkeley
Fujikawa, Brian K.	Lawrence Berkeley National Laboratory
Fujino, Don	Lawrence Livermore National Laboratory
Fuller, George M.	University of California at San Diego
Gagliardi, Carl A.	Cyclotron Institute, Texas A & M Univ.
Gai, Moshe	University of Connecticut
Galik, Richard S.	Cornell University
Gamberg, Leonard	University of Oklahoma
Gardner, Rob	Indiana University
Gardner, Susan V.	University of Kentucky
Gates, Evalyn I.	University of Chicago and Adler Planetarium
Gemme, Claudia	Universita' di Genova
Gates, Gordon	Princeton University
Gibson, Ben F.	Los Alamos National Laboratory
Ginther, George	University of Rochester
Goodman, Maury	Argonne National Laboratory
Gordeev, V. A.	PNPI, Russia
Greene, Geoffrey L.	Los Alamos National Laboratory
Gunter, Jeff L.	Indiana University
Haines, Todd J.	Los Alamos National Laboratory

Hall, Lawrence J.	University of California at Berkeley
Hallman, Timothy	Brookhaven National Laboratory
Hasinoff, Michael D.	University of British Columbia
Haxton, Wick C.	INT, University of Washington
Helminen, Christina I.	University of Helsinki
Hemmick, Thomas K.	State University of New York at Stony Brook
Henley, Ernest M.	University of Washington
Heppelmann, Steven F.	Penn State University
Herczeg, Peter	Los Alamos National Laboratory
Hertzog, David W.	University of Illinois at Urbana
Hill, John C.	Iowa State University
Hinterberger, Frank	ISKP, Bonn University
Hoblit, Sam D.	Brookhaven National Laboratory
Hungerford, Ed V.	University of Houston
Inoue, Takashi	University of Tokyo
Ivanov, Yuri Michailovich	Petersburg Nuclear Physics Institute
Jacobs, Peter	Lawrence Berkeley Laboratory
Jahn, Rainer	ISKP, University of Bonn
Jeon, Sangyong	University of Minnesota
Joensson, Leif S.	Lund University, Lund, Sweden
Johnson, Denis P.	I.I.H.E.-Vrije Universiteit Brussels
Jones, Mark K.	College of William & Mary
Jones, Billy D.	The Ohio State University
Kahana, David E.	State University of New York at Stony Brook
Kajita, Takaaki	Kamioka Observatory, University of Tokyo
Kalelkar, Mohan S.	Rutgers University
Kamyshkov, Yuri A.	ORNL/University of Tennessee
Keppel, Cynthia E.	Hampton University/Jefferson Laboratory
Kharzeev, Dmitri	University of Bielefeld
Kiel, Bert	University of Erlangen - Germany
Kinney, Edward R.	University of Colorado
Klein, Spencer R.	Lawrence Berkeley Laboratory
Kleinfeller, Jonny E.A.	Forschungszentrum Karlsruhe (KARMEN)
Koch, Volker	Lawrence Berkeley National Laboratory
Koetke, Donald D.	Valparaiso University
Kopeliovich, Vladimir B.	Inst. for Nucl. Research of Russian Acad. of Sciences
Kopeliovich, Boris Z.	Max-Planck-Institut fuer Kernphysik, Heidelberg
Korkmaz, Elie	University of Northern British Columbia
Korsch, Wolfgang K-H.	University of Kentucky
Kowalski, Stanley	MIT-Bates Linear Accelerator Center
Krutenkova, Anna	ITEP Moscow, Russia
Kuze, Masahiro	KEK/IPNS, Japan
Laget, Jean-Marc	CEA-Saclay
Lajoie, John G.	Yale University
Lanou, Bob E.	Brown University
Learned, John G.	University of Hawaii
Lebrun, Paul	Fermilab
Lechanoine-Leluc, Catherine	D.P.N.C. University of Geneva
Lee, Fei-sheng	Virginia Polytechnic Inst. and State Univ./LANL

Lee, Henry W.	Sudbury Neutrino Observatory
Lehmann, Albert A.	Paul Scherrer Institute
Li, Zhenping	Peking University
Li-Scholz, Angela	Atomic Data and Nuclear Data Tables
Lichtenberg, Don	Indiana University
Lipkin, Harry	Weizman Institute
Lisa, Michael A.	Ohio State University
Londergan, J. Tim	Indiana University
Lopatin, Igor	Petersburg Nuclear Physics Institute
Maas, Peter A.	Northwestern University
Macdonald, John A.	TRIUMF
Machleidt, Ruprecht	University of Idaho
MacInnis, Anne	MIT-Bates Linear Accelerator Center
Madsen, Jes	University of Aarhus, Denmark
Maestas, Lucy P.	Los Alamos National Laboratory
Maggiora, Angelo	INFN-Torino
Maltman, Kim R.	York University
Manley, D. Mark	Kent State University
Manweiler, Robert W.	Valparaiso University
Mardor, Yael	Tel-Aviv University
Mardor, Israel	Tel-Aviv University
Masaike, Akira	Kyoto University
McGrory, Joseph B.	U.S. Department of Energy
McKee, Paul M.	University of Virginia
McKeown, Robert D.	California Institute of Technology
Meissner, Thomas	Carnegie Mellon University
Menze, Dietmar Wolf	Physikalisches Institut - Universitat Bonn
Meyer, Brad S.	Clemson University
Meyer, Volker	University of Heidelberg
Meyer, Curtis A.	Carnegie Mellon University
Michaels, Robert W.	Thomas Jefferson National Accel. Facility
Michel, Thilo	University of Erlangen - Germany
Mikuz, Marko	University of Ljubljana - Slovenia
Miller, Jerry A.	University of Washington
Miller, James P.	Boston University
Miller, Michael A.	University of Illinois at Urbana
Mills, Geoffrey B.	Los Alamos National Laboratory
Mintz, Stephan L.	Florida International University
Mischke, Dick E.	Los Alamos National Laboratory
Molzon, William R.	University of California at Irvine
Mueller, Jim A.	University of Pittsburgh
Mueller, Berndt	Duke University
Murdock, David P.	Tennessee Technological University
Napolitano, Jim J.	Rensselaer Polytechnic Institute
Negele, John W.	Massachusetts Institute of Technology
Nilsen, Bjorn S.	Louisiana State University
Noumi, Hiroyuki	IPNS KEK
Nozar, Mina	Rensselaer Polytechnic Institute

Numao, Toshio	TRIUMF
Ostrick, Michael	Universität Mainz
Page, Philip R.	Thomas Jefferson National Accel. Facility
Page, Shelley A.	University of Manitoba
Perillo Isaac, Maria Celia	Lawrence Berkeley National Laboratory
Piasetzky, Eli	Tel Aviv University
Piekarz, Henryk	Florida State University
Piilonen, Leo E.	Virginia Tech
Ping, Jia-Lun	Nanjing Normal University
Polukhina, Natal'ya G.	Lebedev Phys. Inst. of Russian Academy of Sciences
Pomianowski, Paula A.	Virginia Tech/CLEO
Prakash, Madappa	State University New York at Stony Brook
Preedom, Barry M.	University of South Carolina
Procario, Michael P.	Carnegie Mellon University
Quinn, Brian D.	Carnegie Mellon University
Ralston, John P.	University of Kansas
Ramsay, W. Des	University of Manitoba/TRIUMF
Ramsay, Susan M.	Los Alamos National Laboratory
Redwine, Robert P.	Massachusetts Institute of Technology
Reimer, Paul E.	Los Alamos National Laboratory
Riepenhausen, Frank	University of Zurich
Roberts, B. Lee	Boston University
Rock, Stephen	The American University
Roos, Philip G.	University of Maryland
Ryan, James	University of New Hampshire
Saghai, Bijan	CEA - Saclay
Sammarruca, Francesca	University of Idaho
Sandacz, Andrzej	Soltan Inst. for Nuclear Studies, Warsaw, Poland
Sanford, Robert E.	Louisiana State University
Santoro, Alberto	F.S.Lafex/Cbpf/Fnal
Sato, Yoshihiro	Utsunomiya University
Scadron, Michael D.	University of Arizona
Schaefer, Thomas M.	University of Washington
Schatz, Hendrik	University of Notre Dame
Scholz, Wilfried W.	State University of New York-Albany
Schroder, Bent	Lund University
Schumacher, Reinhard A.	Carnegie Mellon University
Schuurmans, Paul	I.K.S., K. U. Leuven
Schwartz, Alan	Princeton University
Seunarine, Surujhdeo	University of Kansas
Slaughter, Jean	Yale University
Sobie, Randall J.	Institute of Particle Physics & Univ. of Victoria
Sokolsky, Pierre	University of Utah
Somalwar, Sunil V.	Rutgers University
Speth, Josef	Juelich, Germany
Stanislaus, Shirvel	Valparaiso University
Stankus, Paul W.	Oak Ridge National Laboratory
Stephenson, Edward J.	Indiana University Cyclotron Facility
Stoler, Paul	Rensselaer Polytechnic Institute

Straumann, Ulrich D.	University of Heidelberg
Suekane, Fumihiko	Tohoku University
Sutton, Sean	Mount Holyoke College
Svoboda, Robert C.	Louisiana State University
Takizawa, Makoto	Showa College of Pharmaceutical Sciences
Talamonti, Cinzia	University of Edinburgh
Tanaka, Mitsuyoshi	Brookhaven National Laboratory
Thomson, Gordon	Rutgers University
Thurner, Stefan	Technische Universitaet Wien
Todenhagen, Ralf	MPI-Kernphysik, Heidelberg
Truoel, Peter	Physik-Institut, Univ. of Zurich, Switzerland
Ullrich, Thomas S.	Yale University
van Kolck, Bira L.	University of Washington
Vogt, Andreas	Wuerzburg Univ., Inst. for Theoretical Physics
von Smekal, Lorenz J.M.	Argonne National Laboratory
von Witsch, Wolfram H.	University of Bonn, Germany
Vrana, Thomas P.	University of Pittsburgh
Wagner, Gerhard J.	Physikalisches Inst. - U. of Tübingen
Walet, Niels R.	UMIST
Walter, Hans-Christian	Paul Scherrer Institut
Wang, Song Ming	University of Iowa
Weber, Fridolin	University of Munich
Weigel, Herbert	Inst. of Theor. Physics, Tuebingen University
Weygand, Dennis P.	Brookhaven National Laboratory
Whisnant, Steve	University of South Carolina
Wilkerson, John F.	University of Washington
Wilkes, R. Jeffrey	University of Washington
Wittmann, Eva B.	MPI-Kernphysik, Heidelberg
Wolfe, Carl E.	York University
Wright, Dennis H.	TRIUMF
Zaharatos, Laura A.	Brookhaven National Laboratory
Zelenski, Anatoli	INR, Moscow/TRIUMF
Zhang, Yangling	University of Puerto Rico, Mayaguez
Zhu, Yucan	IHEP, China
Zieminski, Andrzej T.	Indiana University
Zoeller, Michael M.	Ohio State University - CLEO

PLENARY SESSIONS

Insight into Hadron Structure from Lattice QCD

J.W. Negele

Center for Theoretical Physics[1]
Laboratory for Nuclear Science and Department of Physics
Massachusetts Institute of Technology, Cambridge MA 02139

Abstract. A variety of evidence from lattice QCD is presented revealing the dominant role of instantons in the propagation of light quarks in the QCD vacuum and in light hadron structure. The instanton content of lattice gluon configurations is extracted, and observables calculated from the instantons alone are shown to agree well with those calculated using all gluons. The lowest 128 eigenfunctions of the Dirac operator are calculated and shown to exhibit zero modes localized at the instantons. Finally, the zero mode contributions to the quark propagator alone are shown to account for essentially the full strength of the rho and pion resonances in the vector and pseudoscalar correlation functions.

INTRODUCTION

For the nearly three decades since the experimental discovery of quarks and the formulation of QCD, understanding the essential physics of light quarks in QCD and the structure of light hadrons has remained an elusive goal. Since analytic theoretical techniques are as yet inadequate to solve QCD, a number of very different QCD-inspired models have been developed that present quite disparate physical pictures. For example, non-relativistic quark models focus on constituent quarks interacting via an adiabatic potential. Bag models postulate a region in which relativistic current quarks are confined and interact by gluon exchange. Motivated by large N_c arguments, Skyrme models describe the nucleon as a topological soliton built out of $q\bar{q}$ pairs. Finally instanton models emphasize the role of topological structures in the vacuum corresponding in the semiclassical limit to instantons and of the quark zero modes associated with these topological excitations.

---

[1] This work is supported in part by funds provided by the US Department of Energy (DOE) under cooperative research agreement #DF-FC02-94ER40818.

How can one understand which, if any, of these fundamentally different pictures describes the essential physics of light hadrons? Phenomenology has proven inconclusive, with each of the models being rich enough that with sufficient embellishment it can be made to fit the data. Whereas perturbative QCD has proven extremely useful in extracting quark and gluon structure functions from high energy scattering experiments, it is inadequate to understand their origin. Hence, it is necessary to turn to nonperturbative methods, and the only known techniques to solve, rather than model QCD, is lattice field theory. Our strategy, then, will be to use the fact that lattice calculations numerically evaluate the path integral for QCD as a tool to identify the paths that dominate the path integral and thereby identify the physics that dominates hadron structure.

The lattice results described below indicate that gluons play an extremely important dynamical role in light hadrons. Thus, QCD with light quarks is unique among the many-body systems with which we are familiar in the sense that the quanta generating the interactions cannot be subsumed into a potential but rather participate as essential dynamical degrees of freedom. In atoms, for example, photons play a negligible dynamical role, and to an excellent approximation may be subsumed into the static Coulomb potential. In nuclei, mesons play a minor dynamical role, and to a good approximation nuclear structure maybe described in terms of two- and three-body nucleon forces. Indeed, experimentalists need to work very hard and pick their cases carefully to observe any effects of meson exchange currents. And in heavy quark systems, much of the physics of $c\bar{c}$ and $b\bar{b}$ bound states may be understood by subsuming the gluons into an adiabatic potential with Coulombic and confining behavior. It turns out that nucleons, however, are completely different in that gluons are crucial dynamical degrees of freedom. This result is not entirely unexpected, since from perturbative QCD, we already know by the work of Gross and Wilczek [1] and Hoodbhoy, Ji and Tang [2] that approximately half ($16/3n_f$ to be precise, where n_f is the number of active flavors and equals 5 below the top quark mass) of the momentum and angular momentum comes from glue in the limit of high Q^2. Furthermore, experiment tells us that this behavior continues down to non-perturbative scales of the order of several GeV^2.

The physical picture that arises from this work corresponds closely to the physical arguments and instanton models of Shuryak and others [3–5] in which the zero modes associated with instantons produce localized quark states, and quark propagation proceeds primarily by hopping between these states. The support that lattice calculations provide for this picture includes quantitative determination of the instanton content of the QCD vacuum, a comparison of the effects of all gluon contributions versus those of instantons alone, direct calculation of the quark zero modes, and demonstration that these modes dominate the rho and pion contributions to vector and pseudoscalar correlation functions.

BACKGROUND

Lattice QCD

A QCD observable is evaluated by defining quark and gluon variables on the sites and links of a space-time lattice, writing a Euclidean path integral of the generic form [6]

$$\langle T e^{-B\hat{H}} \hat{\bar\psi}\hat{\psi}\hat{\bar\psi}\hat{\psi}\rangle = Z^{-1}\int \mathcal{D}(U)\mathcal{D}(\bar\psi\psi) e^{-\bar\psi M(U)\psi - S(U)} \bar\psi\psi\bar\psi\psi \qquad (1)$$

$$= Z^{-1}\int \mathcal{D}(U) e^{\ln\det M(U) - S(U)} M^{-1}(U) M^{-1}(U)$$

and evaluating the final integral over gluon link variables U using the Monte Carlo method. The link variable is $U = e^{iagA_\mu(x)}$, the Wilson gluon action is $S(U) = \frac{2n}{g^2}\sum_\Box (1 - \frac{1}{N}\operatorname{Re}\operatorname{Tr} U_\Box)$ where $U_\Box$ denotes the product of link variables around a single plaquette, and $M(U)$ denotes the discrete Wilson approximation to the inverse propagator $M(U) \to m + \not{\partial} + ig\not{A}$. Evolution in Euclidean time is required to assure a dominantly positive integral and thus to obtain a statistically accurate Monte Carlo result, and we will utilize the fact that the Euclidean evolution operator projects out the lowest energy state having a specified set of quantum numbers:

$$e^{-BH}\psi = \sum_n e^{-BE_n} C_n \psi_n \xrightarrow[B \gg (E_1-E_0)^{-1}]{} e^{-BE_0} C_0 \psi_0 . \qquad (2)$$

A physical way of understanding the final integral in (1) is to expand $M \equiv (1 + \kappa u)^{-1}$ in powers of the so-called hopping parameter κ which couples neighboring sites with gauge fields, and thereby generate all the quark time-histories. In the case of propagation of a meson from x to y, the integral for $\langle T e^{-BH} \bar\psi(y)\psi(y)\bar\psi(x)\psi(x)\rangle$ has the following three contributions. Expansion of the two $M^{-1}(U)$ terms generates all valence quark and antiquark trajectories that begin at the source x and terminate at the sink y. Expansion of $\ln \operatorname{Det} M(U)$ generates all disconnected quark loops corresponding to excitation of quark-antiquark pairs from the Dirac Sea. Omission of the determinant, which is very expensive computationally, yields the so-called quenched approximation in which the quark-antiquark pairs excited from the sea are neglected. Finally, when the sum over all plaquettes in $S(U)$ is expanded out of the exponent, the lattice is tiled in all possible ways by any number of plaquettes. After integration over $\int \mathcal{D}(U)$, only those combinations of link variables from expanding $M(U)$ and $S(U)$ survive that correspond to color singlets. The simplest non-vanishing tiling for a meson corresponds to completely filling in the region between the valence quark and antiquark with gluon plaquettes. If one imagines cutting this and more and more complicated tilings on a single time slice, one obtains the physical picture of a quark and antiquark (from cutting the two valence quark lines generated by M^{-1}) connected by a gluon flux tube (from cutting all the gluon surfaces that connect the quark and antiquark). Typical lattices range from $16^3 \times 32$ to $32^3 \times 64$ and thus involve numerical integrations over $\sim 10^7$ to 10^8 real variables.

Correlation functions

As in the case of other strongly interacting many-body systems, to understand the structure of the vacuum and light hadrons in nonperturbative QCD, it is instructive to study appropriately selected ground state correlation functions, to calculate their properties quantitatively, and to understand their behavior physically.

The vacuum correlation functions we consider are the point-to-point equal time correlation functions of hadronic currents

$$R(x) = \langle \Omega | T J(x) \bar{J}(0) | \Omega \rangle \tag{3}$$

discussed in detail by Shuryak [3] and recently calculated in quenched lattice QCD [7]. The motivation for supplementing knowledge of hadron bound state properties by these correlation functions is clear if one considers the deuteron. Simply knowing the binding energy, rms radius, quadruple moment and other ground state properties yields very little information about the nucleon-nucleon interaction in each spin, isospin and angular momentum channel as a function of spatial separation. To understand the nuclear interaction in detail, one inevitably would be led to study nucleon-nucleon scattering phase shifts. Although, regrettably, our experimental colleagues have been most inept in providing us with quark-antiquark phase shifts, the same physical information is contained in the vacuum hadron current correlation functions $R(x)$. As shown by Shuryak [3], in many channels these correlators may be determined or significantly constrained from experimental data using dispersion relations. Since numerical calculations on the lattice agree with empirical results where available, we regard the lattice results as valid solutions of QCD in all channels and thus use them to obtain information comparable to scattering phase shifts.

The correlation functions we calculate in the pseudoscalar, vector, nucleon and Delta channels are

$$R(x) = \langle \Omega | T J^P(x) \bar{J}^P(0) | \Omega \rangle ,$$
$$R(x) = \langle \Omega | T J_\mu(x) \bar{J}_\mu(0) | \Omega \rangle ,$$
$$R(x) = \tfrac{1}{4} \text{Tr} \left(\langle \Omega | T J^N(x) \bar{J}^N(0) | \Omega \rangle x_\nu \gamma_\nu \right) ,$$

and

$$R(x) = \tfrac{1}{4} \text{Tr} \left(\langle \Omega | T J_\mu^\Delta(x) \bar{J}_\mu^\Delta(0) | \Omega \rangle x_\nu \gamma_\nu \right) ,$$

where

$$J^P = \bar{u}\gamma_5 d ,$$
$$J_\mu = \bar{u}\gamma_\mu\gamma_5 d ,$$
$$J^N = \epsilon_{abc}[u^a C\gamma_\mu u^b]\gamma_\mu\gamma_5 d^c ,$$

and

$$J^\Delta_\mu = \epsilon_{abc}[u^a C\gamma_\mu u^b]u^c .$$

As in Refs. [3] and [7], we consider the ratio of the correlation function in QCD to the correlation function for non-interacting massless quarks, $R(x)/R_0(x)$, which approaches one as $x \to 0$ and displays a broad range of non-perturbative effects for x of the order of 1 fm. Typical results of lattice calculations of ratios of vacuum correlation functions are shown in Fig. 1.

Note that the lattice results (solid line) agree well with phenomenological results from dispersion analysis of data (long dashed curves). Also, observe that the vector and pseudoscalar correlation functions are strongly dominated by the rho and pion contributions (dotted lines) in the region of 0.5 to 1.5 fm. We will subsequently show that these rho and pion contributions in turn arise from the zero mode contributions associated with instantons.

As discussed in Refs. [3,7], these vacuum correlators show strong indications of instanton dominated physics. As shown by 't Hooft [8], the instanton induced interaction couples quarks and antiquarks of opposite chirality leading to strong attractive and repulsive forces in the pseudoscalar and scalar channels respectively and no interaction to leading order in the vector channel. Just this qualitative behavior is observed at short distance in all the channels we computed. Furthermore, as shown by the open circles with error bars in Fig. 1, the random instanton model of Shuryak et al. [4] reproduces the main features of the correlation functions at large distance as well.

Instantons

The QCD vacuum is understood as a superposition of an infinite number of states of different winding number, where the winding number characterizes the number of times the group manifold is covered when one covers the physical space. Just as there is a stationary point in the action of the Euclidean Feynman path integral for a double well potential corresponding to the tunneling between the two degenerate minima, so also there is a classical solution to the QCD equations in Euclidean time, known as an instanton [9], which describes tunneling between two vacuum states of differing winding number. The action associated with an instanton is

$$S_0 = \frac{1}{4}\int d^4x\, F^a_{\mu\nu}F^a_{\mu\nu} = \frac{48}{g^2\rho^4}\int d^4x\, \Big(\frac{\rho^2}{x^2+\rho^2}\Big)^4 = \frac{8\pi^2}{g^2} . \qquad (4)$$

Note that the action density has a universal shape characterized by a size ρ, and that the action is independent of ρ. Furthermore, the instanton field strength is self-dual, *i.e.* $\tilde{F}^a_{\mu\nu} \equiv \epsilon_{\mu\nu\alpha\beta}F^a_{\alpha\beta} = \pm F^a_{\mu\nu}$, so that the topological change of an instanton is

$$Q \equiv \frac{g^2}{8\pi^2}\frac{1}{4}\int d^4x \tilde{F}^a_{\mu\nu}F^a_{\mu\nu} = \pm 1 \ .$$

Two features of instantons are particularly relevant to light hadron physics. The first is the fact that although the fermion spectrum is identical at each minimum of the vacuum, quarks of opposite chirality are raised or lowered

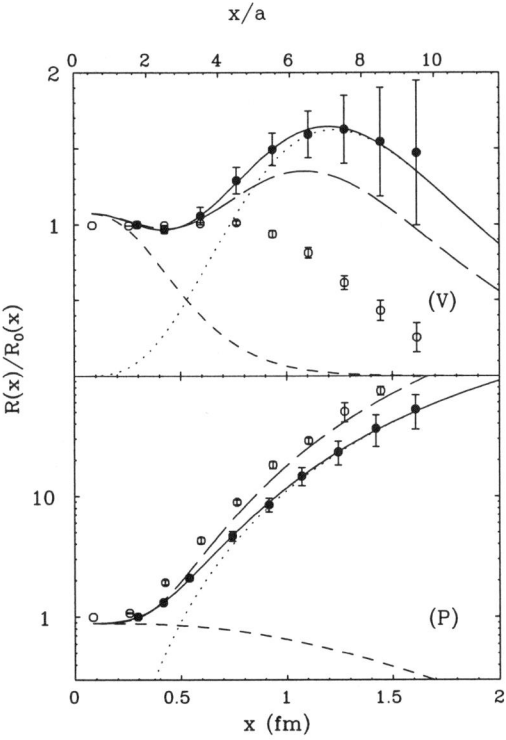

FIGURE 1. Vector (V) and Pseudoscalar (P) correlation functions are shown in the upper and lower panels respectively. Lattice results [7] are denoted by the solid points with error bars and fit by the solid curves, which may be decomposed into continuum and resonance components denoted by short dashed and dotted curves respectively. Phenomenological results determined by dispersion analysis of experimental data in Ref. [3] are shown by long dashed curves, and the open circles denote the results of the random instanton model of Ref. [4].

one level between adjacent minima. Thus, an instanton absorbs a left-handed quark of each flavor and emits a right-handed quark of each flavor, and an anti-instanton absorbs right-handed quarks and emits left-handed quarks. Omitting heavier quarks for simplicity, the resulting 't Hooft interaction involving the operator $\bar{u}_R u_L \bar{d}_R d_L \bar{s}_R s_L$ is the natural mechanism to describe otherwise puzzling aspects of light hadrons. It is the natural mechanism to flip the helicity of a valence quark and transmit this helicity to the glue and quark-antiquark pairs, thereby explaining the so-called "spin crisis." It also explains why the two valence u quarks in the proton would induce twice as many $\bar{d}d$ pairs as the $\bar{u}u$ pairs induced by the single valence d quark. The second feature is that each instanton gives rise to a localized zero mode of the Dirac operator $D_\mu \gamma_\mu \phi_0(x) = 0$. Hence, considering a spectral representation of the quark propagator, it is natural that the propagator for the light quarks is dominated by these zero modes at low energy. This gives rise to a physical picture in which $\bar{q}q$ pairs propagate by "hopping" between localized modes associated with instantons.

INSTANTON CONTENT OF LATTICE GLUON CONFIGURATIONS

Identifying instantons by cooling

The Feynman path integral for a quantum mechanical problem with degenerate minima is dominated by paths that fluctuate around stationary solutions to the classical Euclidean action connecting these minima [10]. In the case of the double well potential, a typical Feynman path is composed of segments fluctuating around the left and right minima joined by segments crossing the barrier. If one had such a trajectory as an initial condition, one could find the nearest stationary solution to the classical action numerically by using an iterative local relaxation algorithm. In this method, which has come to be known as cooling, one sequentially minimizes the action locally as a function of the coordinate on each time slice and iteratively approaches a stationary solution. In the case of the double well, the trajectory approaches straight lines in the two minima joined by kinks and anti-kinks crossing the barrier and the structure of the trajectory can be characterized by the number and positions of the kinks and anti-kinks.

In QCD, the corresponding classical stationary solutions to the Euclidean action for the gauge field connecting degenerate minima of the vacuum are instantons, and we apply the analogous cooling technique [11] to identify the instantons corresponding to each gauge field configuration.

The results of using 25 cooling steps as a filter to extract the instanton content of a typical gluon configuration are shown in Fig. 2, taken from Ref. [12] using the Wilson action on a $16^3 \times 24$ lattice at $6/g^2 = 5.7$. As one can see, there is no recognizable structure before cooling. Large, short wavelength fluctuations of the order of the lattice spacing dominate both the action and topological charge density. After 25 cooling steps, three instantons and two anti-instantons can be identified clearly. The action density peaks are completely correlated in position and shape with the topological charge density peaks for instantons and with the topological charge density valleys for anti-instantons. Note that both the action and topological charge densities are reduced by more than two orders of magnitude so that the fluctuations removed by cooling are several orders of magnitude larger than the topological excitations that are retained.

Setting the coupling constant, or equivalently, the lattice spacing, and quark

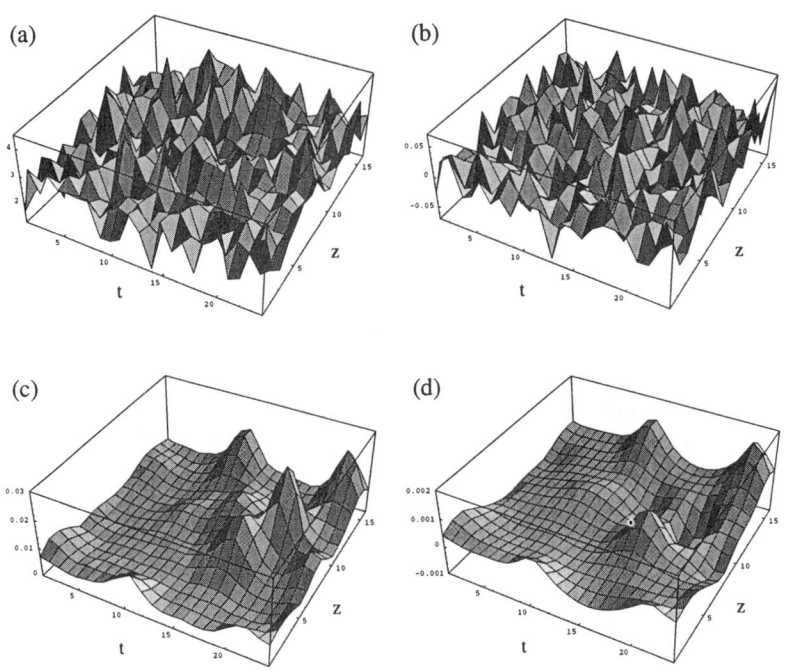

FIGURE 2. Instanton content of a typical slice of a gluon configuration at fixed x and y as a function of z and t. The left column shows the action density $S(1,1,z,t)$ before cooling (a) and after cooling for 25 steps (c). The right column shows the topological charge density $Q(1,1,z,t)$ before cooling (b) and after cooling for 25 steps.

mass by the nucleon and pion masses in the usual way, it turns out that the characteristic size of the instantons identified by cooling is 0.36 fm and the density is 1.6 fm^{-4}, in reasonable agreement with the value of 0.33 fm and 1.0 fm^{-4} in the liquid instanton model [4].

Comparison of results with all gluons and with only instantons

One dramatic indication of the role of instantons in light hadrons is to compare observables calculated using all gluon contributions with those obtained using only the instantons remaining after cooling. Note that there are truly dramatic differences in the gluon content before and after cooling. Not only has the action density decreased by two orders of magnitude, but also the string tension has decreased to 27% of its original value and the Coulombic and magnetic hyperfine components of the quark-quark potential are essentially zero. Hence, for example, the energies and wave functions of charmed and B mesons would be drastically changed.

As shown in Fig. 3, however, the properties of the rho meson are virtually unchanged. The vacuum correlation function in the rho (vector) channel and the spatial distribution of the quarks in the rho ground state, given by the ground state density-density correlation function [13] $\langle \rho | \bar{q}\gamma_0 q(x) \bar{q}\gamma_0 q(0) | \rho \rangle$, are statistically indistinguishable before and after cooling. Also, as shown in Ref. [11], the rho mass is unchanged within its 10% statistical error. In addition, the pseudoscalar, nucleon, and delta vacuum correlation functions and nucleon and pion density-density correlation functions are also qualitatively unchanged after cooling, except for the removal of the small Coulomb induced cusp at the origin of the pion.

Although these cooling studies strongly indicate that instantons play an essential role in light quark physics, cooling has the disadvantage of modifying the instanton content of the original gluon configuration. It is possible to avoid the gradual shrinkage of a single instanton until it eventually falls through the lattice by using an improved action that is sufficiently scale independent [14]. However, pairs of instantons and anti-instantons will eventually attract each other and annihilate, thereby continually eroding the original distribution. Hence, it is valuable to complement these cooling calculations by studies of the zero modes associated with instantons, which, as we show in the next section, can be carried out successfully on the original uncooled gluon configurations.

QUARK ZERO MODES AND THEIR CONTRIBUTIONS TO LIGHT HADRONS

Eigenmodes of the Dirac operator

In the continuum limit, the Dirac operator for Wilson fermions approaches the familiar continuum result

$$D\psi_x = \psi_x - \kappa \sum_\mu \left[(r-\gamma_\mu)u_{x,\mu}\psi_{x+\mu} + (r+\gamma_\mu)u^\dagger_{x-\mu,\mu}\psi_{x-\mu}\right] \to \frac{1}{m}\left[m + i(\slashed{p} + g\slashed{A})\right]\psi.$$

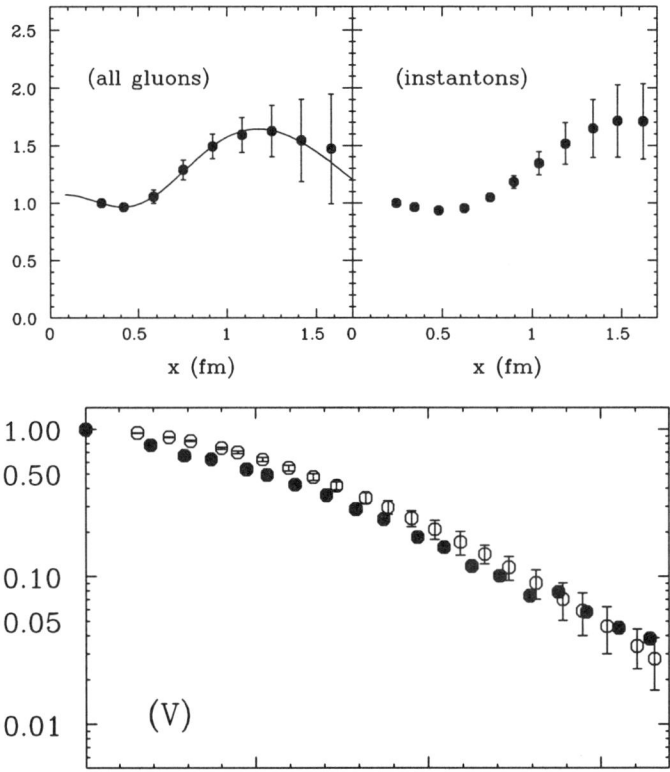

FIGURE 3. Comparison of rho observables calculated with all gluon configurations and only instantons. The upper left-hand plot shows the vacuum correlator in the rho channel calculated with all gluons as in Fig. 1 and the upper right-hand plot shows the analogous result with only instantons. The lower plot shows the ground state density-density correlation function for the rho with all gluons (solid circles) and with only instantons (open circles). Error bars for the solid circles are comparable to the open circles and have been suppressed for clarity.

In the free case, the continuum spectrum is $\frac{1}{m}[m+i|\vec{p}|]$ and the Wilson lattice operator approximates this spectrum in the physical regime and pushes the unphysical fermion modes to very large (real) masses. In the presence of an instanton of size ρ at $x = 0$, it is shown in Ref. [15] that the lattice operator produces a mode with zero imaginary part that approaches the continuum result

$$\psi_0(x)_{s,\alpha} = u_{s,\alpha} \frac{\sqrt{2}}{\pi} \frac{\rho}{(x^2+\rho^2)^{3/2}}$$

and whose mixing with other modes goes to zero as the lattice volume goes to infinity. In addition, instanton-anti-instanton pairs that interact sufficiently form complex conjugate pairs of eigenvalues that move slightly off the real axis. Thus, by observing the Dirac spectrum for a lattice gluon configuration containing a collection of instantons and anti-instantons, it is possible to identify zero modes directly in the spectrum.

Fig. 4 shows the lowest 64 complex eigenvalues of the Dirac operator on a 16^4 unquenched gluon configuration for $6/g^2 = 5.5$ and $\kappa = 0.16$, both before and after cooling (where 100 relaxation steps with a parallel algorithm are comparable to 25 cooling steps). The lower, cooled, plot has just the structure we expect with a number of isolated instantons with modes on the real axis and pairs of interacting instantons slightly off the real axis. However, even though the uncooled case shown in the upper plot also contains fluctuations several orders of magnitude larger than the instantons (as seen in Fig. 2), it shows the same structure of isolated instantons and interacting pairs. To set the scale, note that if we had antiperiodic boundary conditions in time, the lowest Matsubara mode ($ip = i\frac{\pi}{L}$) would occur at 0.06 on the imaginary axis, so all the modes below this value are presumably the results of zero modes.

Zero mode expansion

The Wilson–Dirac operator has the property that $D = \gamma_5 D^\dagger \gamma_5$, which implies that $\langle \psi_j | \gamma_5 | \psi_i \rangle = 0$ unless $\lambda_i = \lambda_j^*$ and we may write the spectral representation of the propagator

$$\langle x | D^{-1} | y \rangle = \sum_i \frac{\langle x | \psi_i \rangle \langle \psi_{\bar{i}} | \gamma_5 | y \rangle}{\langle \psi_{\bar{i}} | \gamma_5 | \psi_i \rangle \lambda_i}$$

where $\lambda_i = \lambda_{\bar{i}}^*$. A clear indication of the role of zero modes in light hadron observables is the degree to which truncation of the expansion to the zero mode zone reproduces the result with the complete propagator.

Fig. 5 shows the result of truncating the vacuum correlation functions for the vector and pseudoscalar channels to include only low eigenmodes [15]. On a 16^4 lattice, the full propagator contains 786,432 modes. The top plot of Fig. 5 shows the result of including the lowest 16, 32, 64, 96, and finally 128 modes. Note that the first 64 modes reproduce most of the strength in the rho resonance peak pointed out in Fig. 1, and by the time we include the first 128 modes, all the strength is accounted for. Similarly, the lower plot in

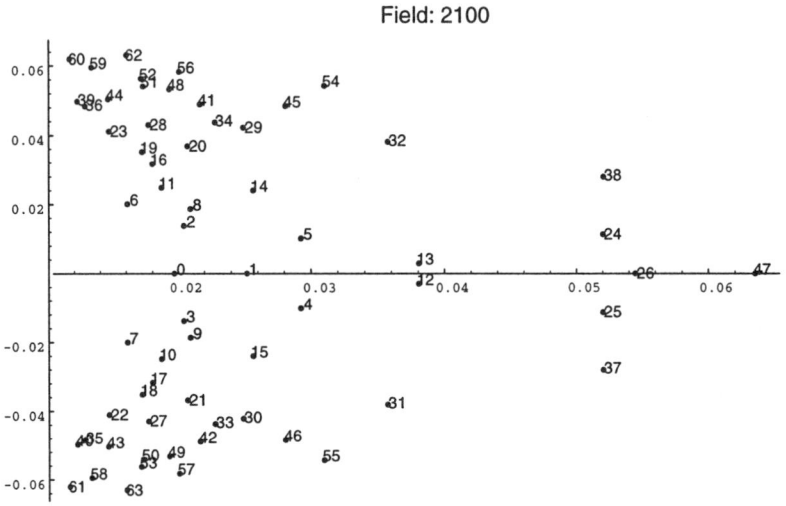

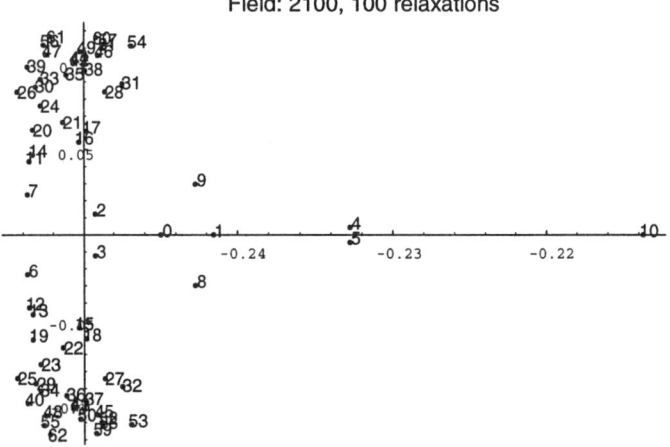

FIGURE 4. Lowest 64 complex eigenvalues of the Wilson–Dirac operator for an unquenched gluon configuration both before (upper plot) and after cooling (lower plot). The scale is such that 0.06 on the imaginary axis roughly corresponds to the lowest Matsubara frequency, 380 MeV.

Fig. 5 shows that the lowest 128 modes also account for the analogous pion contribution to the pseudoscalar vacuum correlation function. Thus, without having to resort to cooling, by looking directly at the contribution the lowest eigenfunctions, we have shown that the zero modes associated with instantons dominate the propagation of rho and pi mesons in the QCD vacuum.

Localization

Finally, it is interesting to ask whether the lattice zero mode eigenfunctions are localized on instantons. This was studied by plotting the quark density distribution for individual eigenmodes in the x-z plane for all values of y and t, and comparing with analogous plots of the action density. As expected, for a cooled configuration the eigenmodes correspond to linear combinations of localized zero modes at each of the instantons. (Because there are no symmetries, the coefficients are much more complicated than the even and odd combinations in a double well or the Bloch waves in a periodic potential.) What is truly remarkable, however, is that the eigenfunctions of the uncooled configurations also exhibit localized peaks at locations at which instantons are identified by cooling. Thus, in spite of the fluctuations several orders of magnitude larger than the instanton fields themselves, the light quarks essentially average out these fluctuations and produce localized peaks at the topological excitations. When one analyzes a number of eigenfunctions, one finds that all the instantons remaining after cooling correspond to localized quark fermion peaks in some eigenfunctions. However, some fermion peaks are present for the initial gluon configurations that do not correspond to instantons that survive cooling. These presumably correspond to instanton–anti-instanton pairs that were annihilated during cooling.

CONCLUSION

Altogether, the lattice calculations reported here provide strong evidence that instantons play a dominant role in quark propagation in the vacuum and in light hadron structure. We have shown that the instanton content of gluon configurations can be extracted by cooling, and that the instanton size and density is consistent with the instanton liquid model. We obtain striking agreement between vacuum correlation functions, ground state density-density correlation functions, and masses calculated with all gluons and with only instantons. Zero modes associated with instantons are clearly evident in the Dirac spectrum, and account for the rho and pi contributions to vector and pseudoscalar vacuum correlation functions. Finally, we have observed directly quark localization at instantons in uncooled configurations.

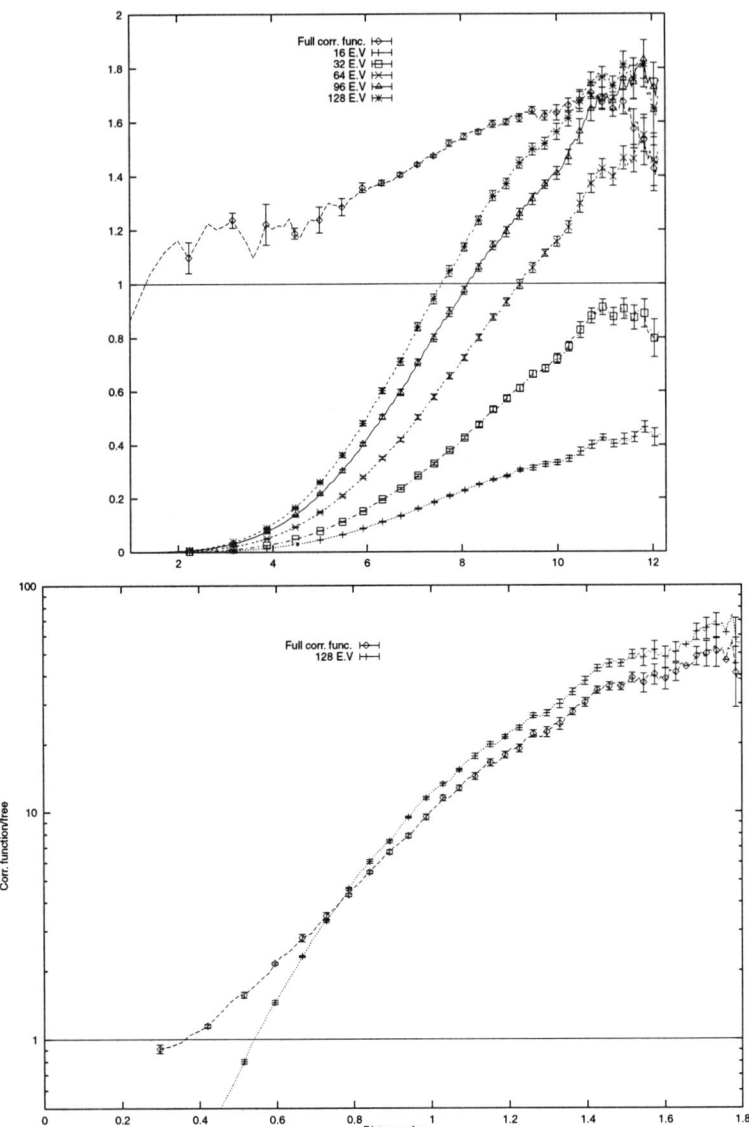

FIGURE 5. Contributions of low Dirac eigenmodes to the vector (upper graph) and pseudoscalar (lower graph) vacuum correlation functions. The upper graph shows the contributions of 16, 32, 64, 96, and 128 eigenmodes compared with the full correlation function for an unquenched configuration with a 63 MeV valence quark mass. The lower graph compares 128 eigenmodes with the full correlation function for a quenched configuration with a 23 MeV quark mass.

Acknowledgments

It is a pleasure to acknowledge the essential role of Richard Brower, Ming Chu, Jeff Grandy, Suzhou Huang, Taras Ivananko, Kostas Orginos, and Andrew Pochinsky who collaborated in various aspects of this work. We are also grateful for the donation by Sun Microsystems of the 24 Gflops E5000 SMP cluster on which the most recent calculations were performed and the computer resources provided by NERSC with which this work was begun.

REFERENCES

1. Gross D., and Wilczek F., *Phys. Rev.* **D9**, 980 (1974).
2. Ji X., Tang J., and Hoodbhoy P., *Phys. Rev. Lett.* **76**, 740 (1996).
3. Shuryak E.V., *Rev. Mod. Phys.* **65**, 1 (1993), *Nucl. Phys.* **B** (Proc. Suppl.) **34**, 107 (1994), and Schäffer T., and Shuryak E.V., hep-ph/9610451v2.
4. Shuryak E.V., and Verbaarschot J.J.M., *Nucl. Phys.* **B410**, 55 (1993); Schäffer T., Shuryak E.V., and Verbaarschot J.J.M., *Nucl. Phys.* **B412**, 143 (1994).
5. Dyakonov D.I., and Petrov V. Yu, *Nucl. Phys.* **B245**, 259 (1984); **B272**, 457 (1986).
6. See for example, Creutz M, *Quarks, Gluons and Lattices*, Cambridge: Cambridge Univ. Pr., 1983.
7. Chu M.-C., Grandy J.M., Huang S., and Negele J.W., *Phys. Rev. Lett.* **70**, 225 (1993); *Phys. Rev.* **D48**, 3340 (1993).
8. 't Hooft G. *Phys. Rev.* **14D**, 3432 (1976).
9. Belavin A.A., Polyakov A.M., Schwartz A.P., and Tyupkin Y.S. *Phys. Lett.* **59B**, 85 (1975).
10. Negele J.W., and Orland H., *Quantum Many-Particle Systems*, New York: Addison-Wesley, 1987.
11. Berg B. *Phys. Lett.* **104B**, 475 (1981); Teper M., *Nucl. Phys.* **B** (Proc. Suppl.) **20**, 159 (1991).
12. Chu M.-C., Grandy J.M., Huang S., and Negele J.W., *Phys. Rev.* **D49**, 6039 (1994).
13. Chu M.-C., Lissia M., and Negele J.N., *Nucl. Phys.* **B360**, 31 (1991); Lissia M., Chu M.-C., Negele J.W., and Grandy J.M., *Nucl. Phys.* **A555**, 272 (1993).
14. de Forcrand P., Garcia Pérez M., and Stamatescu I.-O., hep-lat/9701012 (1997).
15. Ivanenko T., MIT Ph.D. dissertation 1997; Ivanenko T., and Negele J.W., *Proceedings of Lattice '97*, to be published.

High Energy Tests of QCD

Gerald C. Blazey

Northern Illinois University
DeKalb, Illinois 60115

Abstract. Selected, recent, high energy results, primarily from collider experiments but including some fixed target experiments, are presented as illustrations of the status of Quantum ChromoDynamics (QCD). The concepts of leading order (LO) and next–to–leading order (NLO) QCD are introduced. Inclusive $\bar{p}p$ jet production as a function of jet transverse energy and cone size and di–jet angular distributions are shown to be in reasonable agreement wiht QCD. Jet shapes from $\bar{p}p$, ep, and e^+e^- colliders are also compared to NLO QCD. Discrepancies between NLO calculations and data are apparent for measurements which involve a clearly identified final state parton. This is demonstrated by an unexpectedly large ratio of W+1–jet production to W+0–jet production and an unexpectedly large b–quark cross section in $\bar{p}p$ scattering. Similarly, a compilation of $\bar{p}p$, pp, and fixed target scattering results shows prompt photon production to be systematically greater than NLO predictions. Measurements of the Casimir color factors and the strong coupling constant (α_s) are presented and are in good agreement with theoretical expectations. The decrease of α_s with momentum transfer is apparent in a compilation of world results. New measurements of α_s from e^+e^-, ep, and fixed target experiments do not affect the world average $\alpha_s = 0.118 \pm 0.003$.

INTRODUCTION

The field of high energy quantum chromodynamics (QCD) is in a period of great excitement and rapid advance. Results from three complementary high energy colliders ($\bar{p}p$, ep, e^+e^-) as well as fixed target experiments are providing a broad and comprehensive view of experimental QCD. This, in turn, has stimulated a great deal of theoretical work and progress in the field. For example, measurements of jet production and semi–inclusive W+jet, photon, and heavy quark production illustrate the breath of experimental progress. Although theoretical calculations adequately describe most aspects of jet production, they fall short of describing the semi–inclusive measurements. On a more fundamental level, there has been great improvement in determination of basic QCD parameters involving the quark and gluon couplings. These increasingly more accurate and complete measurements are testing the limits and applications of QCD.

The proton–antiproton interaction, a fairly general scattering process, nicely introduces the concepts of leading order and next–to-leading order QCD. As shown in Fig. 1, hadron–hadron scattering can be considered a convolution of

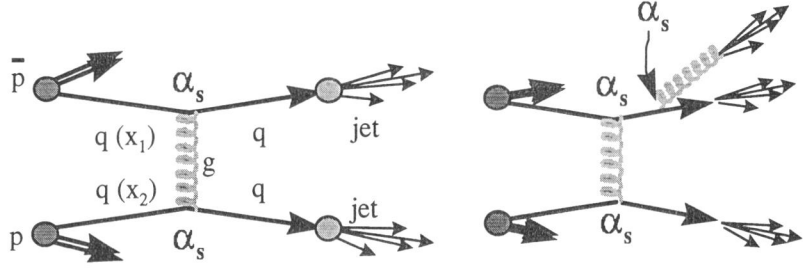

FIGURE 1. Factorization of the scattering process: Incoming quarks with momentum fraction x_i of the incident hadrons scatter through gluon exchange and fragment into final state jets.

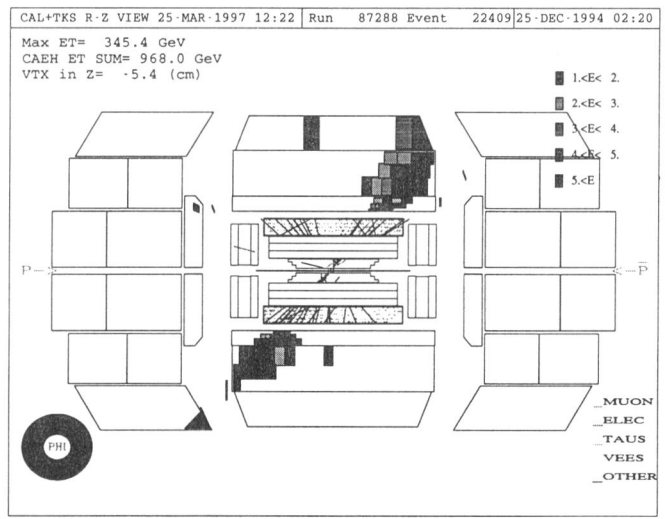

FIGURE 2. A high energy two jet event.

parton distribution functions, a hard two–body interaction, and nonperturbative fragmentation or hadronization functions. The nonperturbative parton distribution function (pdf) for each incoming particle $q(x_i)$, describes the momentum fraction, x_i, carried by each constituent or parton of the parent hadron. The hard two–body interaction, $\hat{\sigma}$, between quarks and gluons is described by perturbative QCD. The figure depicts $qq \to qq$ elastic scattering. In general, pdf's describe the gluon, quark, and antiquark contents of protons, and the hard scattering describes interactions between the partons. The nonpeturbative fragmentation or hadronization function, d, represents the manifestation of the final state quarks and gluons as observed particles. The cross section for the process in Fig. 1 can be written succinctly as $\sigma = \int q(x_1)q(x_2)\hat{\sigma}_{12}d_1d_2$.

The factorized scattering has been illustrated with a leading order graph; that is, a graph proportional to two powers of the strong coupling constant, α_s. A good experimental example of such a leading order process is shown in Fig. 2. This represents a proton–antiproton collision recorded by the DØ [1] detector at the Fermi National Laboratory Tevatron Collider in Batavia, Illinois. The incoming beam energies are 900 GeV/c (center of mass energy 1800 GeV/c). At leading order, the scattering can be understood as a quark from the proton scattering elastically off an antiquark from the antiproton. The final state partons then fragment and hadronize as showers or jets of energy seen in the detector. The two final state jets have energy transverse to the beam axis, E_T, of 480 GeV/c and are the highest transverse energy jet pair observed. Note that the final state jets are considered equivalent to the final state partons.

Although useful the leading order picture is too simple. Current next–to-leading order (NLO) or O(α_s^3) calculations include radiative corrections or additional gluon emission. This is shown in the second illustration of Fig. 1, here a final state parton has radiated an additional gluon and so the entire scattering process is proportional to α_s^3. The final state radiation in the hard scatter can be thought of as a replacement for the fragmentation functions so that $\sigma = \int q(x_1)q(x_2)\hat{\sigma}_{12}$. At this point a technical note is warranted: The perturbative expansion of $\hat{\sigma}$ must be evaluated at some momentum transfer which is typically taken to be the momentum transfer between the partons, errors incurred in the theoretical prediction due to this choice are of O(α_s^4). To a large degree current high energy QCD is simply a study of the additional radiation. As shown in the next section this is well illustrated by jet production in the high energy regime. (Measurement and determination of the parton distribution functions and hadronization functions are each vast and interesting topics of discussion. See the talk by U. Straumann in these proceedings for details on parton distributions.)

JET PRODUCTION

The Inclusive Jet Cross Section

A complete theoretical description of inclusive jet production, $p\bar{p} \to j + X$, requires proper treatment of the final state radiation and accurate measurements of the parton distribution functions, pdf's. Thus the inclusive cross section is a basic test of perturbative NLO QCD as well as the pdf's. Perhaps most interestingly, with current data sets, the inclusive jet cross section constitutes a search for new physics at a distance scale of 10^{-17} cm. In a manner completely analogous to Rutherford scattering, excess jet production at very large transverse energies signals the presence of quark compositeness.

The cross section is typically reported as $d^2\eta/dE_T d\eta$. $E_T = E \sin\theta$ where E is jet energy and θ the angle between the proton direction and the jet. The pseudorapidity, η, is defined as $-ln(tan(\theta/2))$. Kinematically, an individual jet is characterized by E_T, η, and ϕ where ϕ is the azimuthal direction of the jets. At ninety degrees to the beam line the jet energy is equal to E_T and the pseudorapidity is zero. As the jet nears the beam axis in the forward or backward direction the magnitude of η grows. Jets are found by clustering energy in a cone of radius $R = 0.7$ in $\eta - \phi$ space.

Figure 3 shows the central inclusive jet cross section as measured by the DØ collaboration at beam energies of 900 GeV/c for $|\eta| < 0.5$ [2]. The data

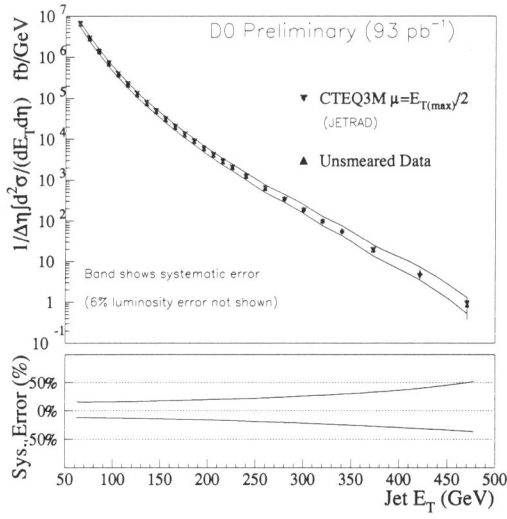

FIGURE 3. The central inclusive jet cross section.

spans seven orders of magnitude for jet energies between 50 and 450 GeV. Data points include statistical errors only and the inset indicates the magnitude of the systematic errors. The systematic error is dominated by uncertainties in the jet energy scale. The figure includes a NLO prediction due to Giele, Glover and Kosower [3]. The cross section has also been calculated by Ellis, Kuntz and Soper [4]. The percentage difference as a function of E_T between the data and theory is shown in Fig. 4. Note there is excellent agreement at all E_T and no indication of new physics.

As shown in Fig. 5 the CDF experiment [5] shows a similar result below 200 GeV for jets in the region $0.1 < |\eta| < 0.7$. However, there is a clear discrepancy at high energy. The open symbols represent published data from a 1992-1993 data run [6]. The closed circles represent a high statistics 1994-1995 data run [7]. The CDF systematic errors (slightly better than those shown in Fig. 3) cannot explain the high energy discrepancy with NLO QCD.

A direct comparison of CDF data to DØ data in the region $0.1 < |\eta| < 0.7$ shows the two experiments to be consistent within systematic errors. The CDF result is roughly 10–20% above the DØ result with a 10% dependence on E_T. Despite the experimental agreement, the two theoretical comparisons suggest differing levels of agreement with NLO QCD. The apparent discrepancy can be attributed to different parameter selections for the NLO predictions. Each theoretical prediction requires a choice of pdf, renormaliza-

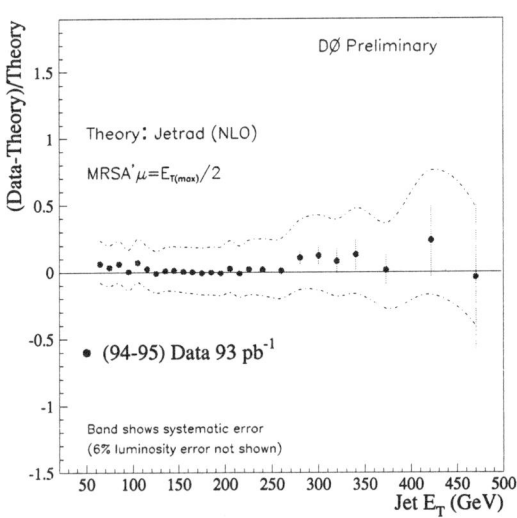

FIGURE 4. Difference between data and NLO QCD for the central inclusive jet cross section as measured by DØ.

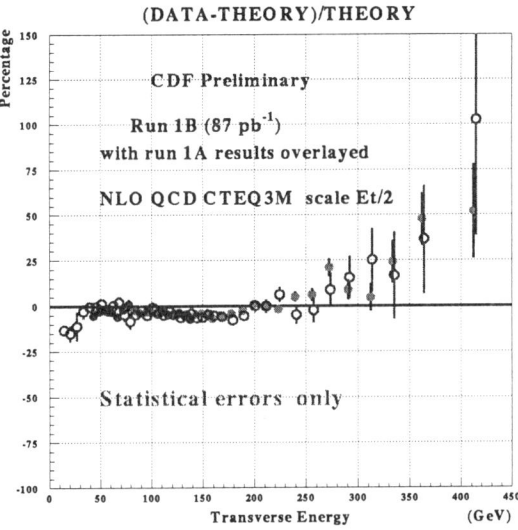

FIGURE 5. Difference between data and NLO QCD for the central inclusive jet cross section as measured by CDF.

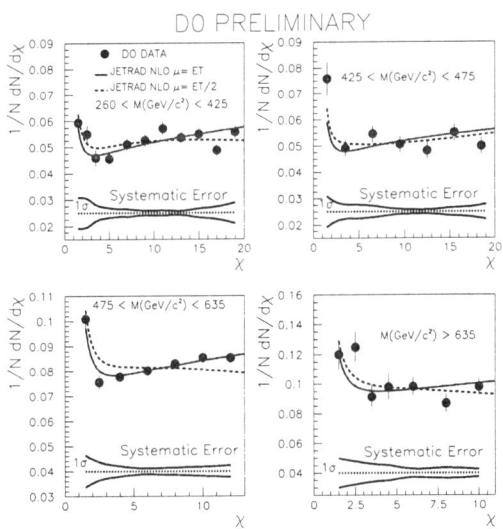

FIGURE 6. Dijet angular distributions.

tion scale, and parton clustering algorithms. The pdf, scale, and clustering choice can engender 15%, 10%, and 3% variations in the NLO prediction. The DØ comparison in Fig. 4 incorporates the MRSA' pdf [8] and a renormalization scale of Et(max)/2 where Et(max) represents the maximum jet E_T in an event. The CDF comparison incorporates the MRSA' pdf and a renormalization scale of Et/2 where Et represents jet E_T. Further clarification of the theoretical comparison awaits more accurate measurements of the pdf's, higher order calculations to reduce renormalization and clustering uncertainties, and a reduction of systematic errors. Nonetheless within 15-20% NLO QCD describes the inclusive jet cross section over eight orders of magnitude.

Dijet Angular Distributions

Angular distributions between the leading two jets of a hard $\bar{p}p$ event also constitute a rigorous test of NLO QCD. In the lab frame, the two jets are produced at rapidities η_1, η_2. When boosted into the center–of–mass the jets or partons will be at opposite rapidity $\eta_1' = -\eta_2'$ so that the hard scatter is characterized by the angle, θ^*, between the leading jet and the direction of the proton beam. Typically, the angular distribution is plotted in the normalized form $(1/N_{jets}) \times dN/d\chi$ where $\chi = (1 + cos\theta^*)/(1 - cos\theta^*)$. Since $\chi \sim 1/sin^4(\theta^*/2)$, $dN/d\chi$ is flat for Rutherford scattering. Fig. 6 illustrates the remarkable agreement between measured angular distributions and NLO QCD for four dijet mass bins [2]. The data shows a preference for NLO over LO QCD. This indicates proper treatment of the radiative corrections is required to describe the angular distributions.

The highest mass region is also quite sensitive to new physics in the form of a contact interaction or constituent exchange between scattering quarks. As with the inclusive cross section, additional contributions will enhance production at ninety degrees to the beam line or at $\chi \sim 1$. Because gg, gq, and qq scattering processes all have very similar angular distributions, the dijet angular distribution is quite insensitive to input pdf's. Thus, deviations from QCD expectations will not be obscured by pdf uncertainties. This is in contrast to a compositeness search with the inclusive jet cross section. The high mass χ distributions can be compared to NLO QCD predictions with and without composite interactions to set lower limits on quark compositeness [2], [9]. The current best limit of 2.3 TeV indicates that for the case of dijet angular distributions there is no indication of new physics.

Jet Shape

The inclusive $\bar{p}p$ cross section can also be measured as a function of cone size [10], [11]. Of particular interest is the ratio of the cross section for various cone sizes as a function of jet E_T [10]. The ratio of the 1.0 cone cross section to the

0.7 cross section and the 0.5 to the 0.7 cross section, for example, shows the cross section to increase with cone size. Above $E_T = 80$ GeV/c the ratios are in good agreement with NLO QCD. Since the theoretical calculations indicate that there is little dependence of the ratios on the pdf's, NLO QCD accurately predicts gluon emission characteristics.

Closely related to the variance of the inclusive cross section with cone size is jet shape or energy flow profile. Quantitatively this may be defined as $\Psi(r) = \Sigma(E_T(<r)/E_T(<R=1))/N_{jets}$. The function $\Psi(r)$ reflects the sum of the transverse energy within a subcone r normalized to the total energy within the cone $R = 1$. Qualitatively, Ψ will rise to unity rapidly for a thin jet and slowly for a fat jet. To describe jet shape a new parameter must be added to NLO theory. The parameter, D, sets the maximum distance between partons clustered into a final state jet. For example, partons within a distance D=1 in $\eta - \phi$ space may be clustered into a single jet and those more than D=1 apart remain unclustered. Thus, at NLO a single event may include one, two, or three jets.

Jets shapes or transverse energy flow for jets produced at the Tevatron collider and at LEP, the e^+e^- collider in Geneva, Switzerland, can be found in the literature [12]. Jet production at the Tevatron is well described by the single additional gluon of NLO QCD with D $\sim$1 . At comparable jet energies, since Tevatron jets are dominated by gluons (from gg elastic scattering) and LEP jets by quarks (from pair creation) and since gluons radiate more than quarks, QCD predicts that Tevatron jets will be "fatter" than LEP jets. In fact, very beautiful analyses of LEP three jet events clearly demonstrate that u, d, and s quark jets are narrower than gluon jets [13].

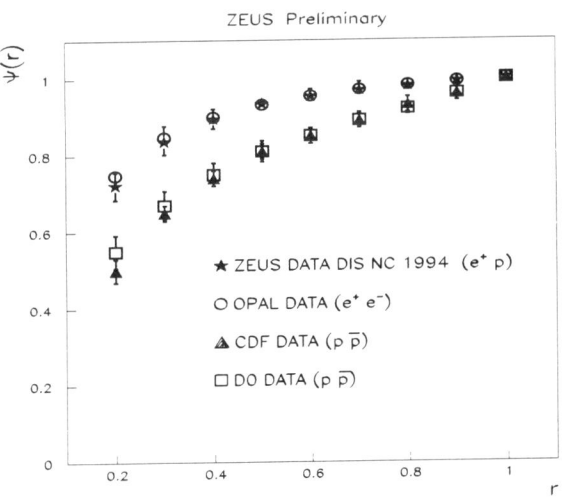

FIGURE 7. Jet shapes from various colliders.

With recent results from HERA, the electron–proton collider in Hamburg, Germany, jet shapes can be compared for all three colliders. For momentum transfer, Q^2, greater than 100 GeV^2, ep scattering is dominated by photon exchange between the incident electron and struck quark. As a result, the final state partons always include a quark. At comparable energies then, QCD predicts that LEP and HERA jets will have similar profiles. The similarity is clearly demonstrated in Fig. 7 [14]. The stars show the jet energy profile as measured by the ZUES collaboration at HERA for a positron beam energy of 28 GeV and a proton energy of 820 GeV. The jet E_T is between 37 and 45 GeV. The results track neatly the LEP profile as measured by the OPAL collaboration for jet energy less than 35 GeV and beam energy of 45 GeV.

Although not shown, the HERA jets shapes can be described by NLO QCD in all kinematic regions if the parton distance variable, D, is allowed to vary between 1 and 1.5 [15]. The other two data curves in Fig. 7 correspond to jets as measured by CDF ($40 < E_T < 60$ GeV) and DØ ($45 < E_T < 75$ GeV) at the Tevatron. As expected because of gluon dominance the Tevatron jets are "fatter" than LEP and HERA jets. Apparently, jet structure seems to be universal or independent of environment and surprisingly well described by single gluon emission or $O(\alpha_s)$ calculations.

SEMI–INCLUSIVE PRODUCTION

W+jet

To lowest order, W production at the Tevatron $\bar{p}p$ collider involves only an electroweak vertex and is unaccompanied by partons or jets. When an additional parton or jet is produced through annihilation ($q\bar{q}' \to Wg$) or Compton scattering ($qg \to W\bar{q}'$), an additional factor of α_s enters the picture. Current QCD calculations are exact to order α_s^2 and describe the emission of an additional gluon. In the cases of annihilation or Compton scattering, the final state gluon or quark may radiate.

Since systematic uncertainties cancel, a particularly useful test of NLO QCD is given by the ratio of W+1–jet production to W+0–jet production, R^{10}. Both the lowest order and $O(\alpha_s)$ processes contribute to W+0–jet production and the $O(\alpha_s)$ and $O(\alpha_s^2)$ processes to the W+1–jet production. The ratio has been calculated to $O(\alpha_s^2)$ by Giele, Glover and Kosower [16]. Figure 8 shows the theoretical calculation to be a factor of two or more below the measurement at all jet E_T [17]. Jets are counted as a function of a minimum threshold E_{Tmin}. The two theoretical curves differ only by the choice of pdf and demonstrate that pdf uncertainties cannot explain the discrepancy. As with the NLO jet comparisons, uncertainties due to the choice of the renormalization scale and parton clustering are small ($\sim 10\%$). The R^{10} discrepancy represents the first of several inconsistencies between NLO QCD and semi–inclusive processes (processes in which at least one of the final state partons is clearly identified).

Heavy Quarks

Heavy quark production ($\sigma_{t\bar{t}}, \sigma_{b\bar{b}}$) in hadron–hadron scattering provides another important test of NLO QCD. At leading order, heavy quarks are produced primarily through valence quark annihilation and subsequent pair creation. In the case of top production at Tevatron energies, NLO corrections are dominated by gluon radiation from the final state t or $\bar{t}$ [18]. Current NLO predictions for the top cross section, $4.7 - 5.5$ pb for a mass of 175 GeV are in good agreement with measurements from the CDF and DØ collaborations, $7.5^{+1.9}_{-1.6}$ pb for a mass of 175 GeV and 5.5 ± 1.8 pb for a mass of 173 GeV, respectively [19]. The top cross section will not be a precise test of QCD until Tevatron Run II when statistical samples increase tenfold.

Interestingly, NLO corrections for $b\bar{b}$ production are of the same order as LO contributions. This can be attributed to "gluon splitting" in which valence quark annihilation leads to two final state gluons, one of which splits into a $b\bar{b}$ pair. The b quark cross section has been measured in many semi–inclusive channels. These include observation of the muon p_T spectrum from the semi–leptonic decays of the b quarks with or without jets, from dimuon final states,

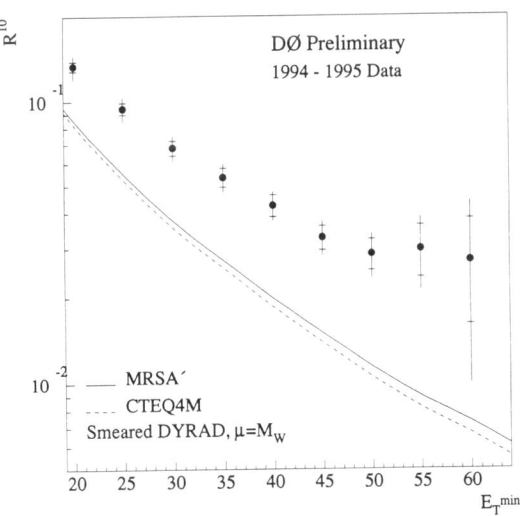

FIGURE 8. Ratio between W+1jet and W+0jet production.

and from meson resonances. Experimental measures of $\sigma_{b\bar{b}}$ as a function of the b quark p_T are well above the nominal NLO predictions [20].

The Photon Cross Section and "k_T"

At leading order direct photon production can occur by annihilation, $q\bar{q} \to g\gamma$, and by Compton scattering, $qg \to q\gamma$ or $\bar{q}g \to \bar{q}\gamma$, and is therefore quite sensitive to the gluonic content of the proton. Higher order bremsstrahlung graphs, for instance $qg \to qg$ where the final state quark radiates a photon, also provide beyond–leading–order tests of QCD. Inclusive photon measurements are free of uncertainties due to jet energy scale and reconstruction. Likewise, theoretical calculations are not hampered by parton clustering algorithms. The measurement does suffer one important drawback: the signal must be extracted from the large π^0 and η meson decay background. There are several methods employed to reduce or estimate the background and details can be found in the references [21]. Nevertheless, because the photon and jet based measurements have very different systematic errors, they provide important complementary tests of QCD.

The CTEQ collaboration has noted an excess of photon production at low x in nearly all the direct photon data accumulated over the last ten years [23]. Figure 9 plots the percentage difference between the NLO predictions for inclusive production and the data for an impressive array of results. The

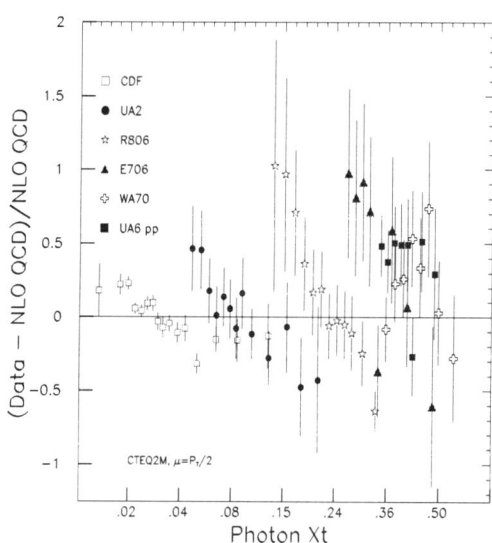

FIGURE 9. Compilation of direct photon measurements compare to NLO QCD.

curve is plotted versus photon $x_t = 2p_T/\sqrt{s}$ where p_T is the photon transverse momentum and s the center of mass energy. At the lowest x accessible by each experiment there appears to be excess production. A NLO prediction using all the photon data to determine the pdf shows nearly identical behavior.

A possible origin of the excess, as suggested by CTEQ, could be higher order processes which impart transverse momentum or "k_T" to the initial partons. The presence of such transverse momenta on the observed cross section would be profound. Since the k_T would be misinterpreted as p_T, the observed cross section as a function of p_T would be a smeared version of the true cross section as a function of p_T. The high cross section at lower p_T would make a large contribution at higher p_T's.

Supporting evidence for nonzero k_T can be found in diphoton production data in hadron colliders [24]. In all cases, the average p_T of the diphoton system is nonzero. At the Tevatron the average is ~ 4 GeV/c. Detailed simulations show this to be the correct magnitude required to explain the excess production at low p_T. Finally, theoretical predictions for prompt photon production from pBe fixed target scattering at $\sqrt{s} = 31.6$ GeV/c are a factor two low unless an average k_T of 1.3 GeV/c is included in the calculation [22]. The "k_T" requirement is most assuredly an interesting observation; higher order calculations may be required to explain these results.

QCD PARAMETERS

Casimir Factors

The only free parameter of QCD is the strong coupling constant α_s. However, the relative strengths of the three distinct quark and gluon vertices, $\alpha_s C_F$ for $q \to qg$, $\alpha_s C_A$ for $g \to gg$, and $\alpha_s T_F$ for $g \to q\bar{q}$ are completely determined by the structure of the gauge group describing the strong force. For $SU(N_c)$ where N_c is the number of colors $C_A = N_c$, $C_F = (N_C^2 - 1)/2N_C$, and $T_F = 1/2$. The probability for a gluon to radiate a gluon is roughly twice the probability for a quark to radiate a gluon. (This is reflected in the narrowness of quark jets discussed earlier.) Four jet production in e^+e^- scattering provides a beautiful experimental measurement of these color or Casimir factors and so a direct test of the gauge couplings [25].

All three of the basic vertices are present in four jet production: the two final state quarks may each radiate a gluon, one quark may radiate two successive gluons, a single quark may radiate one gluon which splits into two gluons, or a single quark may radiate one gluon which splits into a quark–antiquark pair. Since each diagram involves spin–1 and spin–1/2 particles in different configurations each graph results in different angular distributions for the final states. Thus, the observed four jet angular distributions can be fit to theoretical predictions with C_F, C_A, and T_F as free variables.

Figure 10 shows the results of a recent analysis from data taken at LEP (beam energy 45 GeV) by the ALEPH collaboration [25]. Ratios of the Casimir factors are taken to remove sensitivity to the strong coupling constant. Note the 65% contour eliminates many possible gauge groups and comfortably encompasses SU(3). The most probable values of $C_A/C_F = 2.20 \pm 0.16$ and $T_F/C_F = 0.29 \pm 0.08$ compare quite well with SU(3) expectations of 2.25 and 0.375, respectively. Also shown are results from similar analyses by OPAL and DELPHI.

The Strong Coupling Constant

An enormous body of research has been dedicated to the study of the strong coupling constant, α_s since it is the only free parameter of QCD and must be

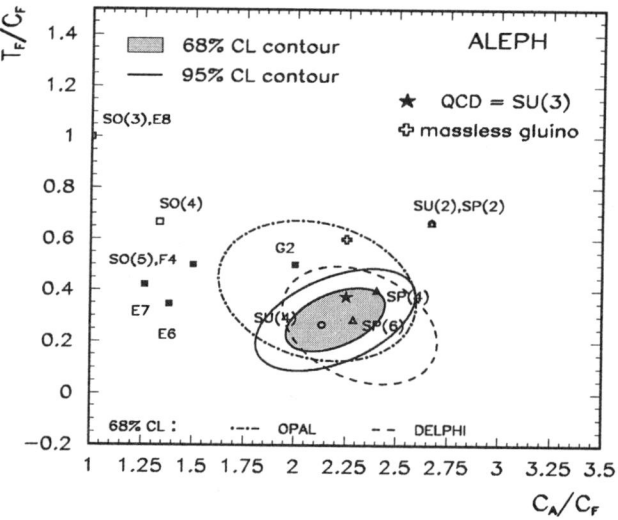

FIGURE 10. Contour plot for Casimir factors measured with 4–jet events.

determined experimentally. Of equal interest is the dependence of the coupling constant on momentum transfer Q^2, $\alpha_s(Q^2) = 12\pi/(33-n_f)log(Q^2/\Lambda^2)$ where n_f is the number of quark flavors and Λ is experimentally determined. Notice α_s decreases or "runs" with momentum transfer. This is, in fact, the basis for the perturbative NLO QCD calculations described earlier. By convention α_s is reported at momentum transfer equal to the Z mass.

The strong coupling constant can be derived in a myriad of ways, from absolute decay rates of the Z boson and τ lepton, energy levels of bound heavy quarks, jet event shapes, jet production rates and angular distributions, and scaling violations in deep inelastic scattering. Details of these derivations can be found in the many references on the strong coupling constant. Figure 11, taken from the excellent, recent review by M.Schmelling, is a compilation of some of the many derivations of α_s and is a beautiful demonstration of running [26]. The curves are derived from the evolution of α_s using the running equation and the world average, $\alpha_s(M_Z) = 0.118 \pm 0.003$. Note that the strong coupling constant is known to three percent accuracy.

The plot does not include some of the most recent results from LEP, the CCFR neutrino scattering experiment, and HERA. In particular, LEP recently ran at center-of-mass energies of 133, 161, and 172 GeV and from event shape measurements determined $\alpha_s(133 GeV) = 0.113 \pm 0.003 \pm 0.007$, $\alpha_s(161 GeV) = 0.105 \pm 0.003 \pm 0.006$, and $\alpha_s(172 GeV) = 0.103 \pm 0.003 \pm 0.006$ where the first error is statistical and the second theoretical [27]. Recent

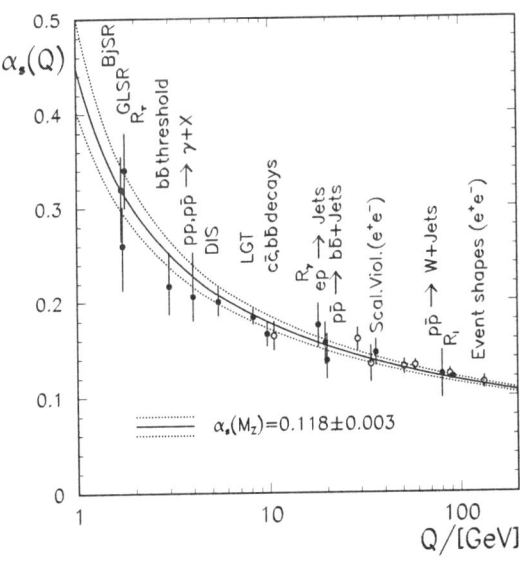

FIGURE 11. A compilation of α_s measurements.

jet event shape results from the H1 collaboration experiment at HERA yield $\alpha_s(M_Z) = 0.118 \pm 0.001 \pm 0.007$ [28] and from the CCFR neutrino experiment at Fermilab $\alpha_s(M_Z) = 0.119 \pm 0.002 \pm 0.004$ [29]. The last measurement is notable in that it represents the single most accurate measurement of α_s to date and that a long–standing discrepancy between deep inelastic scattering derivations and LEP derivations of α_s has vanished. (Previously deep inelastic scattering measurements where about three standard deviations below LEP measurements.) These latest measurements do not change the world average.

CONCLUSIONS

The vitality of QCD studies has never been greater. Recent results from the three collider and many fixed target environments have stimulated great experimental and theoretical progress. Jet production and jet characteristics are well described by perturbative QCD and are setting ever higher limits on "new" physics. Semi–inclusive measurements are providing important new clues as to the nature of higher order corrections to QCD calculations. And perhaps of a more fundamental nature, beautiful measurements of the QCD coupling constants have confirmed the correctness of SU(3) as the gauge group of strong interactions. Many other ongoing analyses will continue to explore QCD and the future remains bright with commissioning of the high luminosity Main Injector at Fermilab and high energy LHC at CERN.

REFERENCES

1. S. Abachi et al. (DØCollaboration), "The DØDetector", NIM A**338** 185 (1994).
2. M.Bhattacharjee, "Inclusive Jet Cross Section and Dijet Angular Distribution at DØ", Proc. of Hadron Collider Physics Conf., Stony Brook, NY (1997)
3. W.T.Giele, E.W.N.Glover, and D.A.Kosower, Nucl. Phys. B**403**, 633 (1993). Calculations performed with their program JETRAD.
4. S.Ellis et al. Phys. Rev. Lett **64** 2121 (1900).
5. F.Abe et al. (CDF Collaboration) NIM Res., Sect. **A272**, 376 (1988) and references therein.
6. F.Abe et al. (CDF Collaboration), Phys. Rev. Lett., **77** 438 (1996).
7. P. Melese, (CDF Collaboration), Proc. 11th Les Rencontres de Physique de la Vallee D'Aosta: Results and Perspectives in Particle Physics, La Thuile, Italy, 1997. FERMILAB-CONF-97/167-E.
8. A.D.MArtin, R.G.Roberts, W.J.Stirling, Phys. Lett. **B354** 155 (1995).
9. F. Abe et al. (CDF Collaboration), Phys. Rev. Lett., **77** 5336 (1996).
10. M.Bhattacharjee, "Inclusive Jet Cross Sections", Proc. of DPF96, Minneapolis, MN, 1996 and FERMILAB–CONF–96/304–E.
11. F.Abe et al. (CDF Collaboration), Phys. Rev. Lett. **68** 1104 (1992).

12. S.Abachi et al. (DØCollaboration), Phys. Lett. **B357**, 500 (1995). F. Abe et al. (CDF Collaboration), Phys. Rev. Lett. **70** 713 1993., R.Akers et al. (OPAL Collaboration) Zeit. Phys. **C63** 197 (1994).
13. G.Alexander et al. (Opal Collaboration), Ziet. Phy. **C69** 543, (1996).
14. M.Martinez, "Jet Shapes at HERA", Proc. of 5th Int. Workshop on Deep Inelastic Scattering and QCD, Chicago, (1997), See URL: http://www.hep.anl.gov/dis97/.
15. M.Klasen and G.Kramer, "Jet Shapes in ep and $p\bar{p}$ Collisions in NLO QCD", DESY 97-002 and hep-ph/9701247 preprints (1997).
16. W.T.Geile, E.W.N.Glover, and D.A.Kosower, Nucl,Phys. B403, 633 (1993).
17. T.Joffe–Minor "$W/Z + jets$ Production at the Tevatron", Proc. of 5th Int. Workshop on Deep Inelastic Scattering and QCD, Chicago, (1997), See URL: http://www.hep.anl.gov/dis97/.
18. E.Laenen, J.Smith, and W.van Neerven, Phys.Lett **321**, 254 (1994). E.Berger and H.Contopangaos, Phys. Rev. D **54** 3085 (1996). S.Catani, M.L.Mangano, P.Nason, and L.Trentadue, Phys. Lett. **B378**, 329 (1996)
19. S. Krzywdzinski, "Top Cross Sections at the Tevatron", Frontiers in Contemporary Physics, Vanderbilt University, May 1997.
20. F.Stichelbaut, "Properties of $b\bar{b}$ Production at the Tevatron", Rencontres de Moriond, Les Arcs, France 1997, FERMILAB-Conf/159–E preprint.
21. F.Abe et al. (CDF Collaboration), Phys. Rev. Lett. **73**, 2662 (1994). S. Abachi et al. (DØCollaboration), Phys. Rev. Lett. **77**, 5011 (1996).
22. H.Montgomery, "Overview of Results from the Fermilab Fixed Target and Collider Experiments", Proc. of 5th Int. Workshop on Deep Inelastic Scattering and QCD, Chicago, (1997), See URL: http://www.hep.anl.gov/dis97/ and FERMILAB-Conf-97/193 preprint.
23. J.Huston et al. (CTEQ Collaboration), "A Global QCD Study of Direct Photon Production", MSU–HEP–41027 and CTEQ-407 preprints, (1995).
24. F.Abe et al. (CDF Collaboration) Phys. Rev. Lett. **70**, 2232 (1993).
25. R.Barate et al., (ALEPH Collaboration), "A Measurement of the QCD Colour Factors and a Limit on the Light Gluino", CERN-PPE/97–002 preprint (1997), Submitted to Zeit. Phys. C.
26. M.Schmelling, "Status of the Strong Coupling Constant", Proc. of the XXVIII International Conf. on High Energy Physics, Warsaw, (1996).
27. S.Marti i Garcia, "A Review of α_s Measurements at LEP", Proc. of 5th Int. Workshop on Deep Inelastic Scattering and QCD, Chicago, (1997), See URL: http://www.hep.anl.gov/dis97/.
28. H.U.Martyn and K.Rabbertz, "Results on Event Shapes in DIS", Proc. of 5th Int. Workshop on Deep Inelastic Scattering and QCD, Chicago, (1997), See URL: http://www.hep.anl.gov/dis97/. M.Martinez,
29. W.G.Seligman et al. (CCFR Collaboration), NEVIS–REPORT–292, (1997), Accepted for publication in Phys. Rev. Lett.

Lepton Flavor Violation

M. D. Cooper, M. Brooks, G. E. Hogan, V. D. Laptev,
and R. E. Mischke

Los Alamos National Laboratory
Los Alamos NM 87545

P. S. Cooper

Fermi National Accelerator Laboratory
Batavia, IL 60510

Y. Chen, M. Dzemidzic, A. Empl, E. V. Hungerford III,
K. Lan, and W. von Witsch

University of Houston
Houston, TX 77004

K. Stantz and J. Szymanski

Indiana University
Bloomington, IN 47405

C. Gagliardi and R. E. Tribble

Texas A&M University
College Station, TX 77843

D. D. Koetke, R. Manweiler, and S. Stanislaus

Valparaiso University
Valparaiso, IN 46383

K. O. H. Ziock

University of Virginia
Charlottesville, VA 22901

L. E. Piilonen

Virginia Polytechnic Institute and State University
Blacksburg, VA 24061

Abstract. The connection of rare decays to supersymmetric grand unification is highlighted, and a review of the status of rare decay experiments is given. Plans for future investigations of processes that violate lepton flavor are discussed. A new result from the MEGA experiment, a search for $\mu^+ \to e^+\gamma$, is reported to be B.R. $< 3.8 \times 10^{-11}$ with 90% confidence.

INTRODUCTION

The Standard Model of electroweak interactions gives a type of periodic table of the elementary fermions, where the periodicity is labeled by the family of the particle. The repetition of families is not understood, and neutral current transitions between the families appear to be forbidden by experiment. The Standard Model of electroweak interactions is a remarkably robust phenomenological theory that encompasses all current measurements and tempts us to look for process outside its sphere of applicability. As it is generally accepted that the Standard Model is not likely to be a complete description of nature, many extensions have been proposed.

Searching for decays that change total lepton family number is an excellent method to explore potential physics beyond the Standard Model because those processes are predicted to be zero except when new physics is present. Even the addition of neutrino oscillations would produce only an immeasurably small rate. Essentially all extensions of the Standard Model that introduce new, heavy particles predict the existence of these rare decays, though the most probable channel is highly model dependent. If a lepton-violating process is observed, measuring many related decays will be important in uncovering the underlying physics. This paper will review the status of searches for rare processes.

There have been many reviews of possible extensions of the Standard Model and their implications for the observation of rare decays, e.g., refer to the one by Melese (1). Recently, the prejudice has grown within the physics community that supersymmetry is an extension that is likely to be related to nature. Barbieri, Hall, and Strumia (2) show that rare decays are signatures for grand unified supersymmetry and calculate the rates for $\mu^+ \to e^+\gamma$ and related processes for a wide range of parameters of these models. They conclude that $\mu^+ \to e^+\gamma$ has the largest rate by more than two orders of magnitude, and it ranges between the current experimental limit and 10^{-14}. Figure 1 was prepared by Barbieri (3) for a minimal SO(10) supersymmetry and plots the predicted branching ratio for $\mu^+ \to e^+\gamma$ and $\mu^-\text{Ti} \to e^-\text{Ti}$ as a function of the top Yukawa coupling at the grand unification scale. These theories have many free parameters, e.g., the mass of the supersymmetric intermediate-vector-boson, for which there is no preferred values, and each point in the figure is the calculated result for sampling these parameters randomly over their reasonable range. The points above the heavy, horizontal line are already excluded by experiment. A tantalizing feature of this plot is that all points appear to be above roughly 10^{-14}, leaving open the possibility that a new generation of experiments might actually push the parameters into an awkward corner of their allowed region. In fact, Hall (4) states that the most natural parameters would lead to values above 10^{-13}.

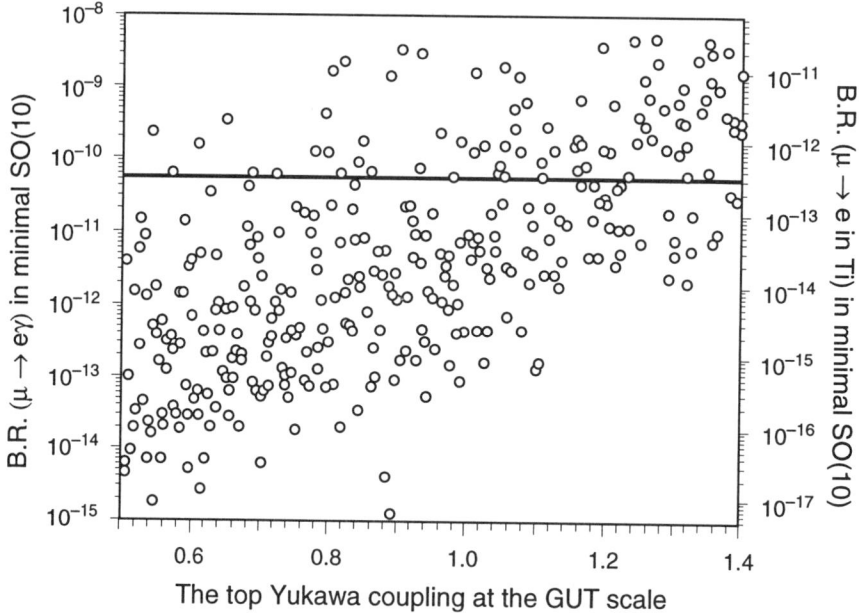

FIGURE 1. Predicted branching ratios for $\mu^+ \to e^+\gamma$ and $\mu^-\text{Ti} \to e^-\text{Ti}$ as a function of the top Yukawa coupling at the grand unification scale of a minimal SO(10) model. Each point is the result when the other parameters of the model are chosen randomly over their likely range.

ONGOING EXPERIMENTS

The smallest limits on branching ratios for rare decays come from muon and kaon decays. However, there are interesting limits from other mesons that can provide useful constraints on specific models. Therefore, this overview will try to touch on all the recent results.

A characteristic aspect of all muon experiments is that the muon lifetime is sufficiently long to require stopping the muons in a detector. A typical detector is that of SINDRUM II, shown in Fig. 2 as it will look after upgrades in 1998. The main features include the pion-muon converter (PMC) (A) designed to provide 10^8 muons to stop in the target (D), a superconducting magnet (L) to analyze the momentum of the decay products, a set of tracking chambers (I,J) to visualize the electron trajectories, and a set of trigger counters (F). The most important upgrade is the PMC, which will pass a very high rate of muons without allowing pions into the detector to very high precision. Decays of pions in the apparatus would produce unacceptable backgrounds. The goal of this detector system is a sensitivity of 2×10^{-14} for the muon-electron conversion process from a nucleus by 1999.

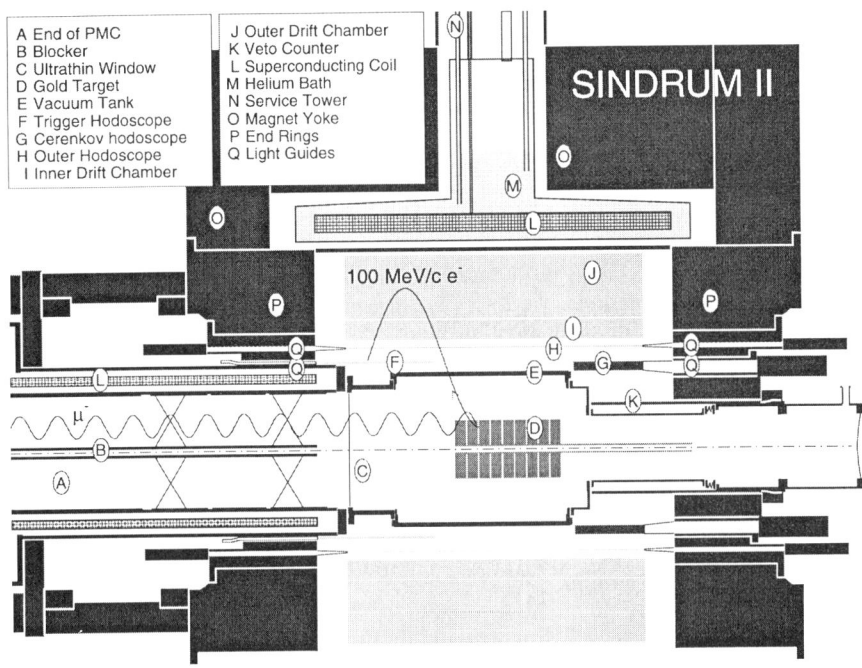

FIGURE 2. The SINDRUM II detector as it will look after addition of the PMC.

Even without the PMC, the SINDRUM II detector has produced the best limits on muon-electron conversion. Figure 3 shows their data when the beam was prepared with degraders and a beam scintillator as a method to eliminate pions. For this data set, the most serious backgrounds are a photon from a cosmic-ray shower that converts to a high-energy electron in the target and prompt decays associated with residual pions. The three curves are the electrons from the target, the spectrum after the

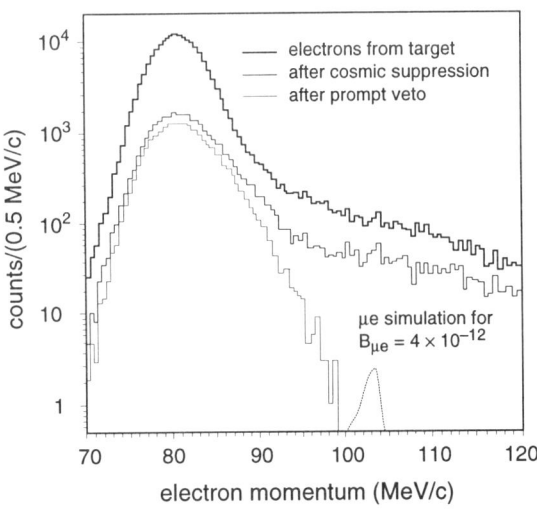

FIGURE 3. Electron spectra from SINDRUM II searching for muon-electron conversion in Ti.

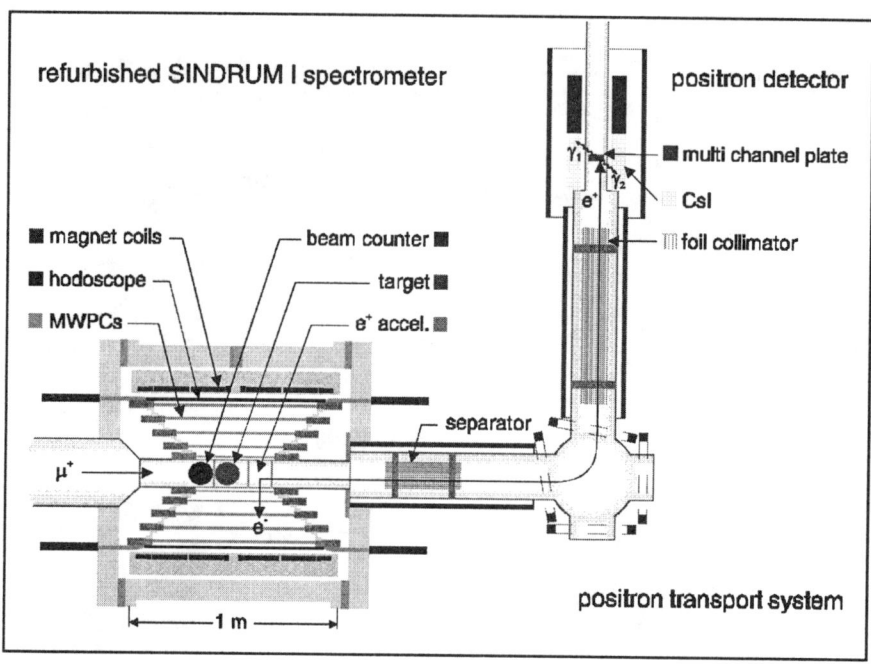

FIGURE 4. The refurbished SINDRUM I detector used to look for muonium, anti-muonium conversion.

suppression of cosmic rays, and the spectrum after the elimination of the prompt signals. Also shown is the expected signal if the branching ration were 4×10^{-12}. They see no signal events and set limits of 7×10^{-13} for $\mu^-\text{Ti} \to e^-\text{Ti}$ and 1.7×10^{-12} for $\mu^-\text{Ti} \to e^+\text{Ca(g.s.)}$ (5). The peeling away of backgrounds as more stringent requirements are placed on events is quite typical of how a potential signal is isolated in these types of experiments.

The SINDRUM I spectrometer, originally used to search for $\mu^+ \to e^+e^+e^-$, has been refurbished to seek the spontaneous conversion of muonium to anti-muonium. The new detector is shown in Fig. 4. It features a target where the muons are brought to atomic velocities to produce muonium that subsequently drifts into a decay region outside the target. If a conversion to anti-muonium were to occur, a high energy electron from μ^- decay would be observed in the magnet spectrometer and a very low energy positron would be left behind. To suppress backgrounds, it is necessary to observe the positron. The positron is seen by accelerating it and analyzing it in the transport system. Finally, its position is determined by a multi-channel plate and its sign is verified by observing the annihilation products. Figure 5 shows the events that might be candidates for muonium, anti-muonium conversion. The ordinate is the quality of the vertex

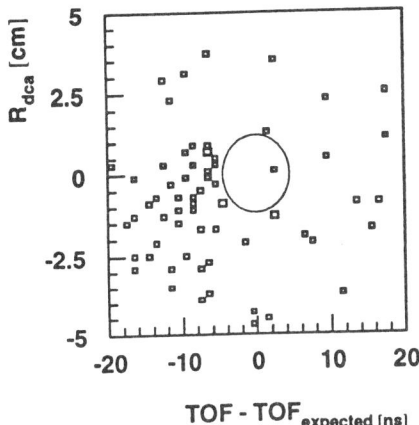

FIGURE 5. Candidate events for muonium, anti-muonium conversion characterized by the quality of their vertex and their relative timing.

between the high-energy electron and the low-energy positron, and the abscissa is their relative time. The circle is a 3σ-ellipse around the signal location. The one event inside is consistent with background, and there are no events in the region of 90% confidence. The result is an upper limit for the size of the coupling constant that could induce such a transition of $< 3 \times 10^{-3} \, G_F$, which is a factor of 2500 better than the previous measurement (6).

Rare decays of the kaon must be handled differently in experiments from those of the muon because of the much shorter lifetime. A typical kaon experiment is Brookhaven experiment 871, a search for $K_L \to \mu e$, whose apparatus is sketched in Fig. 6. The detector system features a decay region and high-rate, position-sensitive elements for tracing the decay products back to a vertex. The decay products are momentum analyzed and then particle-identified in Čerenkov detectors and muon range-finders. Results from this search are expected in the near future. The collaboration has observed more than 5000 $\mu\mu$ events, a remarkable example of the improvement in experimental technique considering that the $K_L \to \mu\mu$ had not been seen convincingly roughly a decade earlier (7). Eventually, their sensitivity for $K_L \to \mu e$ should be a few times 10^{-12}.

A related process is sought after by Brookhaven experiment 865, the decay $K^+ \to \pi^+ \mu^+ e^-$. The group is reporting a result of B.R. (90% C.L.) $< 2 \times 10^{-10}$, and when combined with their previous value gives B.R. (90% C.L.) $< 1 \times 10^{-10}$ (8). A second related decay is $K_L \to \pi^0 \mu e$ that has been studied at Fermilab. The CP-group has a result of B.R. (90% C.L.) $< 3.2 \times 10^{-9}$, but expects the eventual results from KTEV to be better than 10^{-10} (9).

A number of other rare processes have had limits set on them at around the 10^{-6} level. Usually, they are not too restrictive on extensions to the Standard Model, though they can press special models. The new results for τ, D^0, B^0 are summarized in Table 1, which is discussed at the end (10). One somewhat different set of decays are those of the Z^0, where the mass of an external line is more comparable to the masses of the internal lines. Some new limits for purely leptonic processes are B.R. ($Z^0 \to e\mu$, 95% C.L) $< 2 \times 10^{-6}$; B.R. ($Z^0 \to e\tau$, 95% C.L) $< 7.3 \times 10^{-6}$; and B.R. ($Z^0 \to \mu\tau$, 95% C.L) $< 10 \times 10^{-6}$ (11).

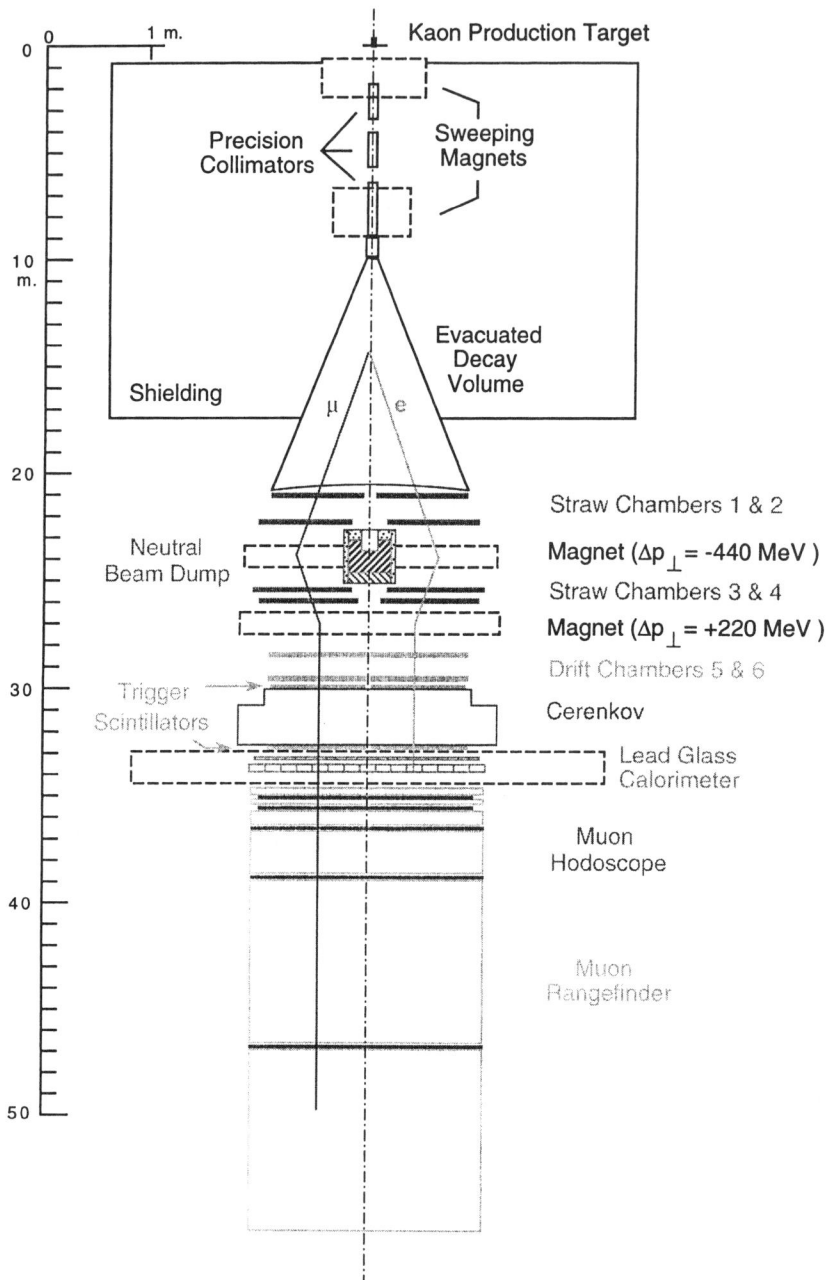

FIGURE 6. A sketch of the apparatus for Brookhaven experiment 871, a search for $K_L \to \mu e$.

TABLE 1. Results on Rare Decays since the last CIPANP Conference

Process	New Limit	Reference
$\mu^+ \to e^+\gamma$	3.8×10^{-11}	This report
$\mu^-\text{Ti} \to e^-\text{Ti}$	7.0×10^{-13}	(5)
$\mu^-\text{Ti} \to e^-\text{Ca}$	1.7×10^{-12}	(5)
$\mu^+ e^- \to \mu^- e^+$	$3.0 \times 10^{-3}\ G_F$	(6)
$K^+ \to \pi^+\mu^+ e^-$	1.0×10^{-10}	(8)
$K_L \to \pi^0 \mu e$	3.2×10^{-9}	(9)
$\tau \to e\gamma$	2.7×10^{-6}	(10)
$\tau \to \mu\gamma$	2.9×10^{-6}	(10)
$\tau^- \to e^-\pi^0$	3.7×10^{-6}	(10)
$\tau^- \to \mu^-\pi^0$	4.0×10^{-6}	(10)
$D^0 \to \mu e$	1.9×10^{-5}	(10)
$D^0 \to \pi^0 \mu e$	8.6×10^{-5}	(10)
$D^0 \to \phi \mu e$	3.4×10^{-5}	(10)
$B^0 \to K^- \mu e$	1.2×10^{-5}	(10)
$B^0 \to K^{*0} \mu e$	2.7×10^{-5}	(10)
$Z^0 \to \mu e$	2.0×10^{-6}	(11)
$Z^0 \to \tau e$	7.3×10^{-6}	(11)
$Z^0 \to \tau \mu$	1.0×10^{-5}	(11)

IDEAS FOR NEW EXPERIMENTS

The excitement being generated by the possibility of observing a signal associated with supersymmetry has initiated designs for future experiments that may span the full range of predictions in Fig. 1. The most promising arrangements aim for sensitivities of 10^{-14} and 10^{-16} for $\mu^+ \to e^+\gamma$ and $\mu^-\text{Ti} \to e^-\text{Ti}$, respectively. The MECO detector (12) would search for the latter process by utilizing a very intense source of μ^- (10^{11}/s) from a superconducting-solenoid bottle (13). Pion contamination in the beam is suppressed sufficiently by pulsing the proton beam and by the momentum acceptance of the transition region between the source and the detector. The detector, which is designed to see high energy electrons only, is shown in Fig. 7. It has three important regions. To the left is a multi-layer target designed to minimize contributions to the energy resolution due to straggling in the stopping material. The central element is a set of straw tube detectors designed for tracking the high energy positrons with a resolution of about 0.8-MeV FWHM. Finally, there is a cylindrical barrel of scintillators for the trigger. The detection elements have a hole in the center to allow the products of normal muon decay pass harmlessly through the detector. Estimates of background rates in this very high rate environment are under study.

In March of 1997, there was a workshop held at the Paul Scherrer Institute in Switzerland on "A New $\mu^+ \to e^+\gamma$ Experiment" (14). Many configurations were studied, but a final design is unsettled. In general, the problem is to keep the

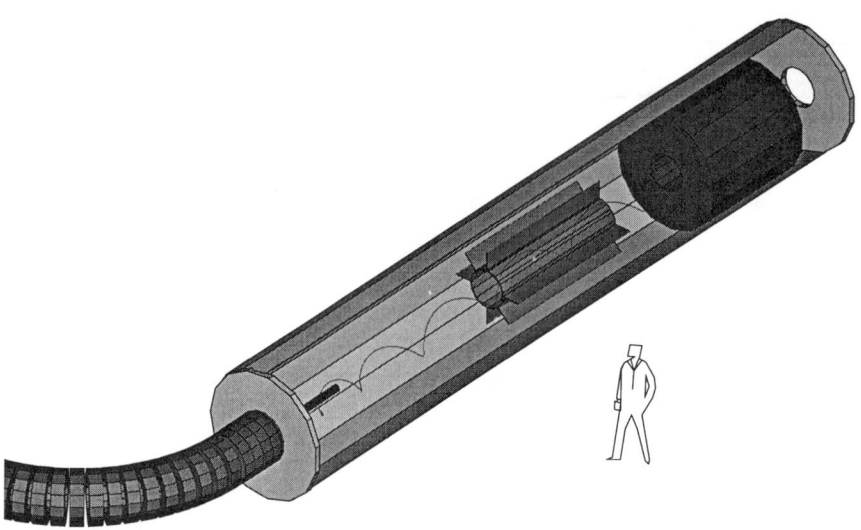

FIGURE 7. The MECO detector under design for a new search for $\mu^-\text{Ti} \to e^-\text{Ti}$ with a sensitivity of 10^{-16}.

acceptance high while suppressing accidental coincidences. However, the conclusion of the workshop was that a 10^{-14} experiment looks feasible.

One interesting idea for suppressing accidental backgrounds has been developed for stopped, polarized muons (15). If the muons are polarized along the beam direction, then the angular distribution of the positrons and photons is given by

$\mu^+ \to e^+ \nu\nu$ $d\Gamma/d\theta_e \sim 1 + P_\mu \bullet k_e$ for $E_e \sim 53$ MeV,

$\mu^+ \to e^+ \gamma\nu\nu$ $d\Gamma/d\theta_\gamma \sim 1 + P_\mu \bullet k_\gamma$ for $E_\gamma \sim 53$ MeV,

$\mu^+ \to e^+ \gamma$ $d\Gamma/d\theta_e \sim$ unknown   for $E_e \sim 53$ MeV.

As the angular correlation of the $\mu^+ \to e^+\gamma$ process is unknown, it is necessary to search in both the forward and backward hemispheres. At backward angles, either the high-energy positron or photon is suppressed. The suppression factor is large and crudely $(1-\cos\theta)/(1+\cos\theta) \sim 0.05$ for $\theta \sim 25°$. To realize this factor, two back-to-back apparatuses are needed, one with the photon detector at back angles and the other with the electron detector at back angles. With a large solid angle detector, the suppression factor is considerably worse but still worth incorporating into a design.

A large solid-angle detector is needed for beam intensities of 10^8/s. However, for intensities of 10^{10}/s, a small solid angle detector would be practical and well matched to the use of polarized muons (16). The idea would be to use a beam similar to that planned for MECO. It is unknown whether the rate, about 1/10 of the total possible, can be achieved with a high polarization. The idea is based on the fact that the sensitivity formula depends on the product of the solid angle and the rate:

$$S\,(90\%\text{ C.L.}) = 2.3/M,$$

where

$$M = (\Omega_0/4\pi) \bullet \varepsilon_\gamma \bullet \varepsilon_p \bullet E_c \bullet R \bullet T,$$

and Ω_0 is the overlap solid angle, ε_γ is the gamma-ray detection efficiency, ε_p is the positron detection efficiency, E_c is the cut efficiency, R is the average stop rate, and T is the live time. If the rate is as high as suggested above, then the solid angle can be small. Hence, small solid-angle, special-purpose spectrometers can be used that solve the problems of high singles rates and costs. The sensitivity is estimated to be 10^{-14}, and the result would be free of background.

STATUS REPORT ON MEGA

The experimental signature for an at-rest $\mu^+ \to e^+\gamma$ is a 52.8-MeV positron that is back-to-back and in time coincidence with a 52.8-MeV photon. The MEGA experiment, designed to search for it, has been described several times (17). Briefly, it consists of a magnetic spectrometer for the positron and three pair spectrometers for the photon. The apparatus has been optimized for high rates and for good resolution to suppress backgrounds; the principal background is random coincidences. MEGA had three period when it took beam, one during each of 1993, 1994, and 1995. The data samples have a ratio of sizes of roughly 1:2:3. The apparatus is mothballed and scheduled to be dismantled unless the analysis shows something surprising.

The total number of muons stopped in the apparatus was 1.5×10^{14} in roughly 10^7 s. There are 4.5×10^8 events on magnetic tape awaiting analysis. The analysis is proceeding in three stages. The first reconstructs the kinematic parameters of the particles; the second refines the reconstruction, and the last cuts away kinematically uninteresting events. At the time of this report, the data from 1993, about 1/6 of the total, has been processed through all three steps.

In general, the reconstruction algorithms trade improving the resolution of the particles for maximizing the efficiency and suppressing backgrounds. The three easiest response functions to measure are the photon energy resolution, the positron-photon timing, and the positron energy resolution. Each is done with a different technique.

The primary beam conditions with stopping muons do not contain any sharp photon lines. In order to get a sharp photon line, negative pions are stopped in polyethylene. They charge exchange roughly 50% of the time and produce a slowly moving π^0 that, in turn, decays into two photons. If one selects those photons that happen to be nearly back-to-back, one gets a narrow line at 55 MeV from the lower energy photon, quite near the endpoint of the location of any possible photon from $\mu^+ \to e^+\gamma$. The spectrum of such events is shown in Fig. 8. The energy resolution is near that predicted.

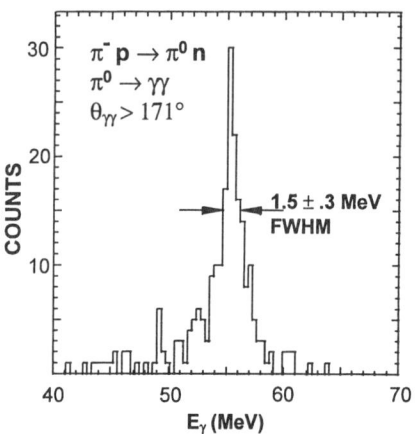

FIGURE 8. Photon energy response for 55-MeV gamma rays from π^0 decays.

The relative time resolution can be measured by looking for the allowed process $\mu^+ \to e^+\gamma\nu\nu$. This internal bremsstrahlung correction to ordinary muon decay can only be seen easily at low rates where the random backgrounds are greatly reduced. The timing spectrum is shown in Fig. 9. Improvements in calibration constants are expected to improve the timing to be nearly 1 ns FWHM. Observation of this decay is reassuring because it is the proof that the detector sees some events that it should.

The positron energy spectrum has a kinematic edge at 52.8 MeV. The resolution is given accurately by the energy difference between the 10 and 90% points on the edge and is about 500 keV FWHM. At high rates, the positron spectrum acquires a high-energy tail due to the improper reconstruction of unphysical events made from random hits in the detector; these are shown in the upper left panel of Fig. 10.

The panels of Fig. 10 show the shapes of the random backgrounds for the energies, times, and directions of the positron and photon. The Monte-Carlo response to a signal is also shown. The shapes of each curve is quite distinct between signal and background for all the variables even on these expanded scales. The deviation from a constant time spectrum for accidental coincidences is understood in terms of the acceptance of the on-line filtering software. Some of the contribution to the photon spectrum in the unphysical region above 55 MeV has been identified as originating from two separate photons.

Figure 11 is a plot of photon versus positron energy for those events with cuts on all other variables around the signal region. The box shows the signal region. It contains no events. The absence of signal corresponds to a branching-

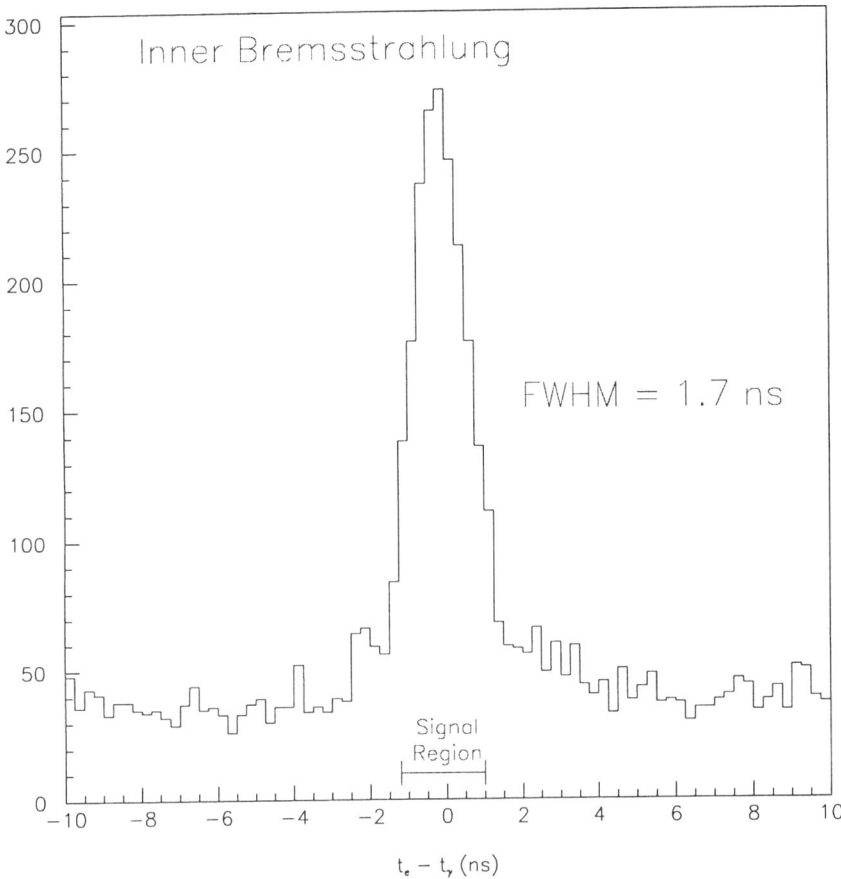

FIGURE 9. Positron-photon timing for the process $\mu^+ \to e^+\gamma\nu\nu$ process at low rates.

ratio limit (90% C.L.) of $< 3.8 \times 10^{-11}$ for $\mu^+ \to e^+\gamma$, a small improvement over the published value but background free.

It is to be expected that Fig. 11 contains no events for positrons above 53 MeV because the probability of getting such an unphysical particle is small. However, as noted above, a background from two separate photons has been identified that gives high energy events that are equally probable near 50-MeV positron energy as near 52.8-MeV positron energy. Hence, it would be nice to eliminate the five events with photon energies above 50 MeV. It is work still in progress, but indications are that four of the five photons are dubious; eliminating such events will cost about 10% of the acceptance. Hence, there is a reasonable probability that the box will remain empty if there is no signal as the balance of

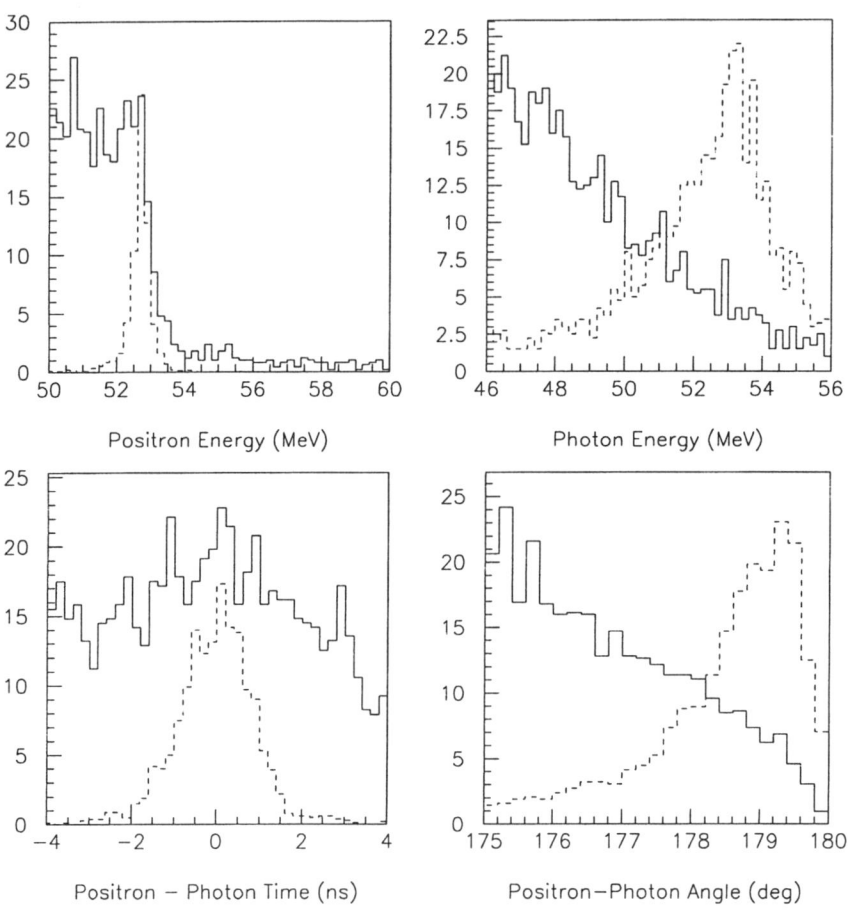

FIGURE 10. Solid curves are the data for random events near the signal region for the positron and photon energies as well as the relative timing and angle. Dashed curves are the Monte-Carlo simulated events for the $\mu^+ \to e^+\gamma$ signal.

the data are analyzed, and it is expected that a background-free sensitivity can be obtained at a level between $3-5 \times 10^{-12}$ for the full data set.

SUMMARY

The search for lepton-number violation has been a very active field since the last Conference on the Intersections of Particle and Nuclear Physics in St. Petersburg. Table 1 shows all of the new results since that meeting. Many more

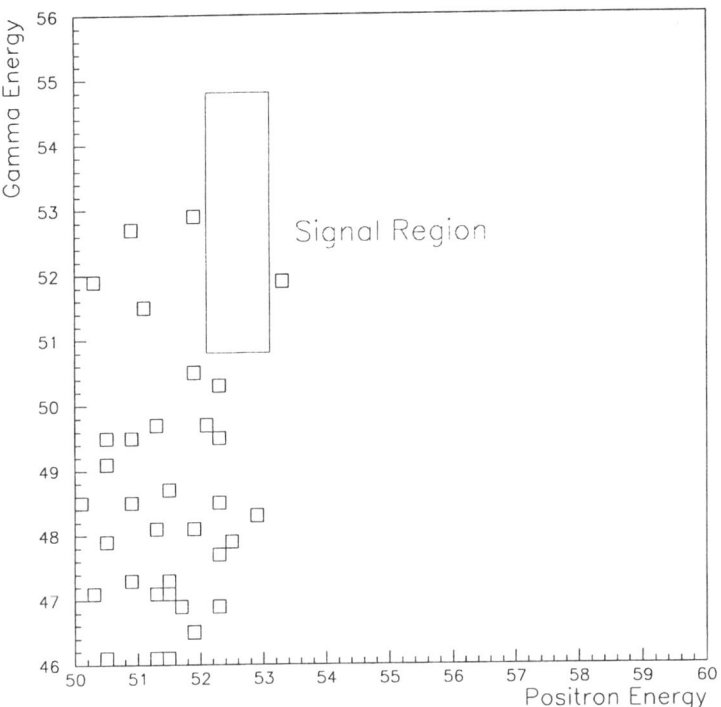

FIGURE 11. Photon energy versus positron energy with cuts on timing and back-to-back angle from high-rate data.

experiments are taking data, and plans are underway for even more ambitious goals. The popularity of supersymmetric extensions to the Standard Model is currently driving this area of research, and rare decays may provide evidence for both supersymmetry and grand unification. The field appears both vibrant and compelling.

REFERENCES

1. Melese, P., *Comments Nucl. Part. Phys.* **19**, 117 (1989).
2. Barbieri, R., Hall, L., and Strumia, A., *Nucl. Phys.* B **449**, 437 (1995).
3. Barbieri, R., in *Workshop on a New $\mu^+ \to e^+\gamma$ Experiment*, Paul Scherrer Institute, 1997.
4. Hall, L., in *Conference on the Intersections of Particle and Nuclear Physics*, Big Sky, Montana, 1997.
5. Riepenhausen, F., in *Conference on the Intersections of Particle and Nuclear Physics*, Big Sky, Montana, 1997.
6. Meyer, V., in *Conference on the Intersections of Particle and Nuclear Physics*, Big Sky, Montana, 1997.

7. Bachman, M., in *Conference on the Intersections of Particle and Nuclear Physics*, Big Sky, Montana, 1997.
8. Eilerts, S., in *Conference on the Intersections of Particle and Nuclear Physics*, Big Sky, Montana, 1997.
9. Ansaka, K. et al., University of Chicago Preprint EFI 95-08; Corcoran, M., in *Conference on the Intersections of Particle and Nuclear Physics*, Big Sky, Montana, 1997.
10. Edwards, K. W. et al., *Phys. Rev. D* **55**, 3919 (1997); hep ex/9704010; Freyberger, A. et al., *Phys. Rev. Lett.* **76**, 3065 (1996); Cornell University CLEO94-4.
11. Adriani, O. et al., *Phys. Lett. B* **316**, 427 (1993); *Lepton-Photon Conference*, Brussels, 1995.,
12. Bachman, M. et al., UC Irvine Phys. Tech. Report 96-30; Molzon, R., in *Conference on the Intersections of Particle and Nuclear Physics*, Big Sky, Montana, 1997.
13. Djilkibaev, R., and Lobashev, V., *Sov. J. Nucl. Phys.* **49(2)**, 384 (1989).
14. Walter, H. K., in *Conference on the Intersections of Particle and Nuclear Physics*, Big Sky, Montana, 1997.
15. Kuno, Y. et al., *Phys. Rev. D* **55**, 2517 (1997); Kuno, Y., and Okada, Y., *Phys. Rev. Lett.* **77**, 434 (1996).
16. Cooper, M., in *Conference on Flavor Physics*, Tsukuba, Japan, 1996.
17. Hogan, G. E. et al., in *International Conference on High Energy Physics*, Warsaw, 1996; Los Alamos National Laboratory document LA-UR-96-3749.

CP Violation in K and B Decays

Gerhard Buchalla*

*Stanford Linear Accelerator Center[1]
Stanford University, Stanford, California 94309

Abstract. We review basic aspects of the phenomenology of CP violation in the decays of K and B mesons. In particular we discuss the commonly used classification of CP violation – CP violation in the mass matrix, in the interference of mixing with decay, and in the decay amplitude itself – and the related notions of direct and indirect CP violation. These concepts are illustrated with explicit examples. We also emphasize the highlights of this field including the clean observables $B(K_L \to \pi^0 \nu \bar\nu)$ and $\mathcal{A}_{CP}(B \to J/\Psi K_S)$. The latter quantity serves to demonstrate the general features of large, mixing induced CP violation in B decays.

INTRODUCTION

Until today CP violation has only been observed in a few decay modes of the long-lived neutral kaon, where it appears as a very small ($\mathcal{O}(10^{-3})$) effect. Despite continuing efforts since the first observation of this phenomenon in 1964 and respectable progress in both experiment and theory, our understanding of CP violation has so far remained rather limited. Upcoming new experiments with K and B mesons are likely to improve this situation substantially. The great effort being invested into these studies is motivated by the fundamental implications that CP violation has for our understanding of nature: CP violation defines an absolute, physical distinction between matter and antimatter. It is also one of the necessary conditions for the dynamical generation of the observed baryon asymmetry in the universe. In addition CP violation provides a testing ground for Standard Model flavor dynamics – the physics of quark masses and mixing.

The source of CP violation in the Standard Model (SM) is the Cabibbo-Kobayashi-Maskawa (CKM) matrix V entering the charged-current weak interaction Lagrangian

$$\mathcal{L}_{CC} = \frac{g_W}{2\sqrt{2}} V_{ij} \bar{u}_i \gamma^\mu (1 - \gamma_5) d_j W_\mu^+ + h.c. \qquad (1)$$

---

[1] Work supported by the Department of Energy under contract DE-AC03-76SF00515.

where $(u_1, u_2, u_3) \equiv (u, c, t)$, $(d_1, d_2, d_3) \equiv (d, s, b)$ are the mass eigenstates of the six quark flavors and a summation over $i, j = 1, 2, 3$ is understood. The unitary CKM matrix ($V^\dagger V = 1$) arises from diagonalizing the quark mass matrix and relating the original weak eigenstates of quark flavor to the physical mass eigenstates. The off-diagonal elements of V describe the strength of weak, charged current transitions between different generations of quarks.

In general, a $n \times n$ unitary matrix has n^2 free (real) parameters. Not all of them are physical quantities in the present case since one has the freedom of redefining the $2n$ fields u_i and d_j ($i, j = 1, \ldots, n$) by arbitrary phases α_i and β_j, respectively. From (1) one sees that only the differences $\alpha_i - \beta_j$ can affect V in this redefinition. There are $2n - 1$ independent $\alpha_i - \beta_j$. The number of independent, physical parameters that characterize V is therefore $n^2 - (2n - 1) = (n - 1)^2$. Out of these $(n - 1)^2$, $n(n - 1)/2$, the number of parameters of a real, orthogonal $n \times n$ matrix, represent rotation angles. The remaining $(n - 1)(n - 2)/2$ are complex phases. Obviously, then, for one or two generations of quarks the matrix V can be chosen to be real. For the realistic case of three generations, however, a physical complex phase is in general present in V [1]. As a consequence, if this phase $\delta \neq 0, \pi$, the weak interaction Lagrangian (1) is not invariant under CP. (A further requirement for this to be true is that all three up-type quark masses must be different from each other and the same must hold for the down-type quarks. Otherwise an arbitrary unitary rotation may be performed on the degenerate quark fields and the complex phase be removed. Also, none of the rotation angles must be 0 or $\pi/2$.) In the sections following this Introduction we will discuss how this violation of CP symmetry at the level of the fundamental Lagrangian manifests itself in observable CP asymmetries occuring in the weak decays of K and B mesons.

The CKM matrix has the following explicit form

$$V = \begin{pmatrix} V_{ud} & V_{us} & V_{ub} \\ V_{cd} & V_{cs} & V_{cb} \\ V_{td} & V_{ts} & V_{tb} \end{pmatrix} \simeq \begin{pmatrix} 1 - \lambda^2/2 & \lambda & A\lambda^3(\varrho - i\eta) \\ -\lambda & 1 - \lambda^2/2 & A\lambda^2 \\ A\lambda^3(1 - \varrho - i\eta) & -A\lambda^2 & 1 \end{pmatrix} \quad (2)$$

where the second expression is a convenient parametrization in terms of λ, A, ϱ and η due to Wolfenstein. It is organized as a series expansion in powers of $\lambda = 0.22$ (the sine of the Cabibbo angle) to exhibit the hierarchy among the transitions between generations. Ordering transitions $i \to j$ according to decreasing strength, this hierarchy reads $i \to i > 1 \to 2 > 2 \to 3 > 1 \to 3$, as is manifest in (2). The explicit parametrization shown in (2) is valid through order $\mathcal{O}(\lambda^3)$, an approxiamtion that is sufficient for most practical applications. Higher order terms can be taken into account if necessary [2].

The unitarity structure of the CKM matrix is conventionally displayed in the so-called unitarity triangle (Fig. 1). This triangle is a graphical representation of the unitarity relation $V_{ud}V_{ub}^* + V_{cd}V_{cb}^* + V_{td}V_{tb}^* = 0$ (normalized by $-V_{cd}V_{cb}^*$) in the complex plane of Wolfenstein parameters (ϱ, η). The angles α, β and γ of the unitarity triangle are phase convention independent and can be determined in CP violation experiments. The area of the unitarity triangle, which is proportional to η, is a measure of CP nonconservation in the Standard Model.

The framework for a theoretical treatment of weak decays in general, and CP violating processes in particular, is provided by low energy effective Hamiltonians, which have the generic form

$$\mathcal{H}_{eff} = \frac{G_F}{\sqrt{2}} V_{CKM} \sum_i C_i(m_t, M_W/\mu, \alpha_s) Q_i \qquad (3)$$

Here G_F is the Fermi constant, V_{CKM} the appropriate combination of CKM elements, C_i are Wilson coefficients, which include also strong interaction effects, and the Q_i are local four-fermion operators. (3) provides a systematic approximation that applies to processes where the relevant energy scale is much smaller than the W-boson or the top quark mass, such as for instance K and B meson decays. An example for a typical operator is $Q_2 = (\bar{s}u)_{V-A}(\bar{u}d)_{V-A}$, which appears in the analysis of nonleptonic kaon decays. In essence the operators Q_i are nothing else than (effective) interaction vertices and the coefficients C_i the corresponding coupling constants. The Hamiltonians (3) can be derived from the fundamental Standard Model Lagrangian using operator product expansion and renormalization group techniques. They may be viewed as the modern generalization of the original Fermi-theory of weak interactions. To calculate decay amplitudes, matrix elements of the operators have to be evaluated between the initial and final states under consideration. This is a problem that involves nonperturbative QCD dynamics – in general a difficult task not yet satisfactorily solved in many cases. The coefficients C_i on the other hand are calculable in perturbation theory, as they incorporate the short distance contributions to the decay amplitude. The factorization of short distance and long distance contributions (Wilson coefficients and op-

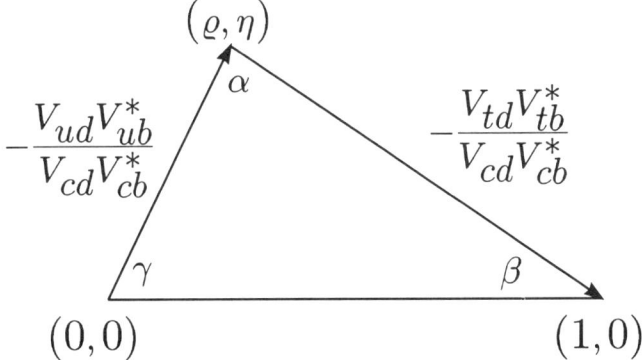

FIGURE 1. The normalized unitarity triangle in the (ϱ, η) plane

erator matrix elements, respectively), inherent in the effective Hamiltonian approach, is a key feature of this framework. Although we will not further elaborate on these issues here, the effective Hamiltonian picture should be kept in mind as the theoretical basis for weak decay phenomenology. A review of the current status of this subject as well as an introduction to the basic concepts may be found in [2]. For a general introduction to CP violation see [3].

The outline of this talk is as follows. After this Introduction we briefly recall the physics of particle-antiparticle mixing, which is crucial for the discussion of CP violation in neutral K and B meson decays. We then describe a classification of CP violating phenomena in K and B decays. To illustrate the concepts we will here use kaon processes as specific examples. Subsequently we discuss the rare decay mode $K_L \to \pi^0 \nu \bar{\nu}$ and some of the basic issues of CP violating asymmetries in B decays. A short summary concludes our presentation.

PARTICLE-ANTIPARTICLE MIXING

Neutral K and B mesons can mix with their antiparticles through second order weak interactions. They form two-state systems ($K^0 - \bar{K}^0$, $B_d - \bar{B}_d$, $B_s - \bar{B}_s$) that are described by Hamiltonian matrices $\hat{H}$ of the form

$$\hat{H} = \begin{pmatrix} M_{11} & M_{12} \\ M_{12}^* & M_{11} \end{pmatrix} - \frac{i}{2} \begin{pmatrix} \Gamma_{11} & \Gamma_{12} \\ \Gamma_{12}^* & \Gamma_{11} \end{pmatrix} \qquad (4)$$

where CPT invariance has been assumed. The absorptive part Γ_{ij} of $\hat{H}$ accounts for the weak decay of the neutral meson $F = K$, B_d, B_s. In Fig. 2 we show typical diagrams that give rise to the off-diagonal elements of $\hat{H}$ for the example of the kaon system. Diagonalizing the Hamiltonian $\hat{H}$ yields the physical eigenstates $F_{H,L}$. They are linear combinations of the strong interaction eigenstates F and $\bar{F}$ and can be written as

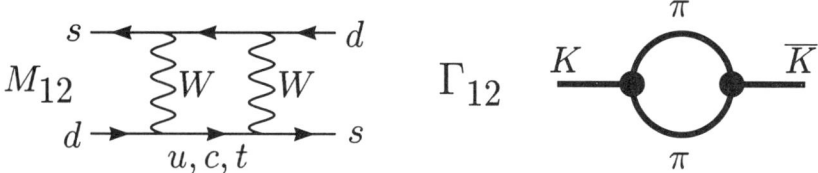

FIGURE 2. Diagrams contributing to M_{12} and Γ_{12} in the neutral kaon system.

TABLE 1. Important properties of neutral K and B meson systems. Here $\Gamma \equiv (\Gamma_H + \Gamma_L)/2$. The kaon entries and ΔM for B_d are experimental results, the remaining numbers theoretical expectations.

	K^0	B_d	B_s
$\Delta\Gamma/\Gamma$	$2.0, \Gamma_L = 579 \cdot \Gamma_H$	~ 0	$\sim 0.16 \pm 0.10$
$\Delta M/\Gamma$	0.95	0.73 ± 0.05	$\sim 25 \pm 15$

$$F_H = \mathcal{N}_{\bar{\varepsilon}}\left[(1+\bar{\varepsilon})F + (1-\bar{\varepsilon})\bar{F}\right] \equiv pF + q\bar{F} \tag{5}$$

$$F_L = \mathcal{N}_{\bar{\varepsilon}}\left[(1+\bar{\varepsilon})F - (1-\bar{\varepsilon})\bar{F}\right] \equiv pF - q\bar{F} \tag{6}$$

with the normalization factor $\mathcal{N}_{\bar{\varepsilon}} = 1/\sqrt{2(1+|\bar{\varepsilon}|^2)}$. Here $\bar{\varepsilon}$ is determined by

$$\frac{1-\bar{\varepsilon}}{1+\bar{\varepsilon}} \equiv \frac{q}{p} = \frac{M_{12}^* - \frac{i}{2}\Gamma_{12}^*}{\left(\Delta M + \frac{i}{2}\Delta\Gamma\right)/2} \tag{7}$$

where ΔM and $\Delta\Gamma$ are the differences of the eigenvalues $M_{H,L} - i\Gamma_{H,L}/2$ corresponding to the eigenstates $F_{H,L}$

$$\Delta M \equiv M_H - M_L > 0 \qquad \Delta\Gamma \equiv \Gamma_L - \Gamma_H \tag{8}$$

The labels H and L denote, respectively, the heavier and the lighter eigenstate so that ΔM is positive by definition. We employ here the CP phase convention $CP \cdot F = -\bar{F}$. Using the SM results for M_{12}, Γ_{12} and standard phase conventions for the CKM matrix (see (2)), one finds in the limit of CP conservation ($\eta = 0$) that $\bar{\varepsilon} = 0$. With (5), (6) it follows that F_H is CP odd and F_L is CP even in this limit, which is close to realistic since CP violation is a small effect. As we shall see explicitly later on, the real part of $\bar{\varepsilon}$ is a physical observable, while the imaginary part is not. In particular $(1-\bar{\varepsilon})/(1+\bar{\varepsilon})$ is a phase convention dependent, unphysical quantity.

Important characteristics of the three cases $F = K^0$, B_d, B_s are collected in Table 1. A crucial feature of the kaon system is the very large difference in decay rates between the two eigenstates, the lighter eigenstate decaying much more rapidly than the heavier one. For the kaon system the states F_L and F_H are therefore commonly denoted as short-lived (K_S) and long-lived (K_L) eigenstates, respectively. The same hierarchy in decay rates is expected for the B_s mesons, although far less pronounced as $\Gamma_H/\Gamma_L = \mathcal{O}(1)$. In the case of B_d $\Delta\Gamma/\Gamma$ is essentially negligible. The labeling of eigenstates as heavy/light is therefore more common for B mesons. The basic reason for this pattern is the small number of decay channels for the neutral kaons. Decay into the predominant CP even two-pion final states $\pi^+\pi^-$, $\pi^0\pi^0$ is only available for

K_S, but not (to first approximation) for the (almost) CP odd state K_L. The latter can decay into three pions, which however is kinematically strongly suppressed, leading to a much longer K_L lifetime. This somewhat accidental feature is absent for B mesons, which have many more decay modes due to their larger mass. We may summarize this discussion by noting that in general the following correspondence holds for the eigenstates of the neutral K and B systems. One has Heavy=Long-lived≈CP odd, and Light=Short-lived≈CP even, where the CP assignments are only approximate due to CP violation.

CLASSIFICATION OF CP VIOLATION

The CP noninvariance of the fundamental weak interaction Lagrangian leads to a violation of CP symmetry at the phenomenological level, in particular in decays of K and B mesons. For instance, processes forbidden by CP symmetry may occur or transitions related to each other by CP conjugation may have a different rate. The phenomenology of CP violating decays is very rich, already for kaons and even more so for B mesons. In this situation it is certainly helpful to have a classification of the various possible mechanisms at hand. One that is commonly used in the literature on this subject employs the following terminology.

a) *CP violation in the mixing matrix.* This type of effect is based on CP violation in the two-state mixing Hamiltonian $\hat{H}$ (4) itself and is measured by the observable quantity $\text{Im}(\Gamma_{12}/M_{12})$. It is related to a change in flavor by two units, $\Delta S(\Delta B) = 2$.

b) *CP violation in the decay amplitude.* This class of phenomena is characterized by CP violation originating directly in the amplitude for a given decay. It is entirely independent of particle-antiparticle mixing and can therefore occur for charged mesons ($K^{\pm}$, $B^{\pm}$) as well. Here the transitions have $\Delta S(\Delta B) = 1$.

c) *CP violation in the interference of mixing and decay.* In this case the interference of two amplitudes, necessary in general to induce observable CP violation, takes place between the mixing amplitude and the decay amplitude in decays of neutral K and B mesons. This very important class is sometimes also refered to as *mixing-induced* CP violation, a terminology not to be confused with a).

Complementary to this classification is the widely used notion of *direct* versus *indirect* CP violation. It is motivated historically by the hypothesis of a new superweak interaction [4,5], that was proposed as early as 1964 by Wolfenstein to account for the CP violation observed in $K_L \to \pi^+\pi^-$ decay. This new CP violating interaction would lead to a local four-quark vertex that changes flavor quantum number (strangeness or beauty) by two units. Its only effect would be a CP violating contribution to M_{12}, so that all observed CP violation could be attributed to particle-antiparticle mixing alone. Today, after the advent of the three generation SM, the CKM mechanism of CP violation appears more natural. In principal the superweak scenario represents a logical possibility, leading to a different pattern of observable CP violation effects. In fact, all experimental measurements available to date are still consistent with the superweak hypothesis.

Now, any CP violating effect that can be entirely assigned to CP violation in M_{12} (as for the superweak case) is termed *indirect CP violation*. Conversely, any effect that can not be described in this way and explicitly requires CP violating phases in the decay amplitude itself is called *direct CP violation*. It follows that class a) represents indirect, class b) direct CP violation. Class c) contains aspects of both. In this latter case the magnitude of CP violation observed in any one decay mode (within the neutral kaon system, say) could by itself be ascribed to mixing, thus corresponding to an indirect effect. On the other hand, a difference in the degree of CP violation between two different modes would reveal a direct effect.

The classification a) – c) is especially common in the context of B physics but it applies to kaon physics as well. To emphasize this point and to provide concrete examples for the above general concepts, we will next illustrate these classes by important applications in kaon decays. We will also use this opportunity to discuss several aspects of kaon CP violation in more detail. After all K physics is the area from which our entire present experimental knowledge of CP violation derives. For a general review of CP violation in kaon decays see [6].

a) – Lepton Charge Asymmetry

The lepton charge asymmetry in semileptonic K_L decay is an example for CP violation in the mixing matrix. It is probably the most obvious manifestation of CP nonconservation in kaon decays. The observable considered here reads ($l = e$ or μ)

$$\Delta = \frac{\Gamma(K_L \to \pi^- l^+ \nu) - \Gamma(K_L \to \pi^+ l^- \bar\nu)}{\Gamma(K_L \to \pi^- l^+ \nu) + \Gamma(K_L \to \pi^+ l^- \bar\nu)} = \frac{|1+\bar\varepsilon|^2 - |1-\bar\varepsilon|^2}{|1+\bar\varepsilon|^2 + |1-\bar\varepsilon|^2}$$
$$\approx 2\mathrm{Re}\,\bar\varepsilon \approx \frac{1}{4}\mathrm{Im}\frac{\Gamma_{12}}{M_{12}} \qquad (9)$$

If CP was a good symmetry of nature, K_L would be a CP eigenstate and the two processes compared in (9) were related by a CP transformation. The rate difference Δ should vanish. Experimentally one finds however [7]

$$\Delta_{exp} = (3.27 \pm 0.12) \cdot 10^{-3} \qquad (10)$$

a clear signal of CP violation. The second equality in (9) follows from (5), as applied to K_L, noting that the positive lepton l^+ can only originate from

$K \sim (\bar{s}d)$, l^- only from $\bar{K} \sim (\bar{d}s)$. This is true to leading order in SM weak interactions and holds to sufficient accuracy for our purpose. The charge of the lepton essentially serves to tag the strangeness of the K, thus picking out either only the K or only the $\bar{K}$ component. Any phase in the semileptonic amplitudes is irrelevant and the CP violation effect is purely in the mixing matrix itself. In fact, as indicated in (9), Δ is determined by $\text{Im}(\Gamma_{12}/M_{12})$, the physical measure of CP violation in the mixing matrix.

From (10) we see that $\Delta > 0$. This empirical fact can be used to define positive electric charge in an absolute, physical sense. Positive charge is the charge of the lepton more copiously produced in semileptonic K_L decay. This definition is unambiguous and would even hold in an antimatter world. Also, using some parity violation experiment, this result implies in addition an absolute definition of left and right. These are quite remarkable facts. They clearly provide part of the motivation to try to learn more about the origin of CP violation.

b) – CP Violation in the Decay Amplitude

Observable CP violation may also occur through interference effects in the decay amplitudes themselves (pure direct CP violation). This case is conceptually perhaps the simplest mechanism for CP violation and the basic features are here particularly transparent. Consider a situation where two different components contribute to the amplitude of a K meson decaying into a final state f

$$A \equiv A(K \to f) = A_1 e^{i\delta_1} e^{i\phi_1} + A_2 e^{i\delta_2} e^{i\phi_2} \tag{11}$$

Here A_i ($i = 1, 2$) are real amplitudes and δ_i are complex phases from CP conserving interactions. The δ_i are usually strong interaction rescattering phases. Finally the ϕ_i are weak phases, that is phases coming from the CKM matrix in the SM. The corresponding amplitude for the CP conjugated process $\bar{K} \to \bar{f}$ then reads (the explicit minus signs are due to our convention $CP \cdot K = -\bar{K}$, $(CP \cdot f = \bar{f})$)

$$\bar{A} \equiv A(\bar{K} \to \bar{f}) = -A_1 e^{i\delta_1} e^{-i\phi_1} - A_2 e^{i\delta_2} e^{-i\phi_2} \tag{12}$$

Since now all quarks are replaced by antiquarks (and vice versa) compared to (11), the weak phases change sign. The CP invariant strong phases remain the same. From (11) and (12) one finds immediately

$$|A|^2 - |\bar{A}|^2 \sim A_1 A_2 \sin(\delta_1 - \delta_2) \sin(\phi_1 - \phi_2) \tag{13}$$

The conditions for a nonvanishing difference between the decay rates of $K \to f$ and the CP conjugate $\bar{K} \to \bar{f}$, that is direct CP violation, can be read off from (13). There need to be two interfering amplitudes A_1, A_2 and these amplitudes must simultaneously have both different weak (ϕ_i) and different strong phases (δ_i). Although the strong interaction phases can of course not generate CP violation by themselves, they are still a necessary requirement for the weak phase differences to show up as observable CP asymmetries. It is obvious from (11) and (12) that in the absence of strong phases A and $\bar{A}$ would have the same absolute value despite their different weak phases, since then $A = -\bar{A}^*$.

A specific example is given by the decays $K(\bar{K}) \to \pi^+\pi^-$ (here $f = \pi^+\pi^- = \bar{f}$). The amplitudes can be written as

$$A_{+-} = \sqrt{\frac{2}{3}} A_0 e^{i\delta_0} + \frac{1}{\sqrt{3}} A_2 e^{i\delta_2}$$

$$\bar{A}_{+-} = -\sqrt{\frac{2}{3}} A_0^* e^{i\delta_0} - \frac{1}{\sqrt{3}} A_2^* e^{i\delta_2} \quad (14)$$

where $A_{0,2} = \langle \pi\pi(I=0,2)|\mathcal{H}_W|K\rangle$ are the transition amplitudes of K to the isospin-0 and isospin-2 components of the $\pi^+\pi^-$ final state. They still include the weak phases, but the strong phases have been factored out and written explicitly in (14). Taking the modulus squared of the amplitudes we get

$$\frac{\Gamma(K \to \pi^+\pi^-) - \Gamma(\bar{K} \to \pi^+\pi^-)}{\Gamma(K \to \pi^+\pi^-) + \Gamma(\bar{K} \to \pi^+\pi^-)} = \sqrt{2}\sin(\delta_0 - \delta_2)\frac{\text{Re}A_2}{\text{Re}A_0}\left(\frac{\text{Im}A_2}{\text{Re}A_2} - \frac{\text{Im}A_0}{\text{Re}A_0}\right)$$
$$= 2\,\text{Re}\,\varepsilon' \quad (15)$$

The quantity so defined is just twice the real part of the famous parameter ε', the measure of direct CP violation in $K \to \pi\pi$ decays. The real parts of $A_{0,2}$ can be extracted from experiment. The imaginary parts have to be calculated using the effective Hamiltonian formalism briefly sketched in the Introduction. Ultimately the amplitudes derive from quark level diagrams. The most important contributions, the gluon penguin and the electroweak penguin, are depicted in Fig. 3. The importance of the electroweak penguin graph might be surprising at first sight; after all it is a contribution suppressed by small electroweak couplings compared to the strong interaction effect represented by the gluon penguin diagram. However, there are several circumstances that actually conspire so as to enhance the impact of the electroweak sector substantially. First of all, the electroweak diagrams contribute to ImA_2, in contrast to the gluon penguins, which correspond to pure $\Delta I = 1/2$ operators (the gluon coupling conserves isospin) and can only lead to an isospin-0 final state, starting from a kaon with isospin $1/2$. Furthermore, the suppression $\sim \alpha/\alpha_s$ from coupling constants is largely compensated by the fact that Re$A_0 \gg$ ReA_2, reflecting the empirical $\Delta I = 1/2$ rule in nonleptonic kaon decays. In addition the electroweak contribution grows strongly with the top quark mass [8,9] and turns out to be quite substantial for the actual value

$\bar{m}_t(m_t) = 167\ GeV$ ($\overline{MS}$-mass). Entering with sign opposite to the (positive) gluon penguin contribution, the electroweak penguin contribution tends to cancel the latter. This feature makes a precise theoretical prediction of ε', which anyhow suffers from large hadronic uncertainties, even more difficult. The typical order of magnitude of ε' can however be understood from (15). The size of $\mathrm{Im} A_i/\mathrm{Re} A_i$ is essentially determined by the small CKM parameters that carry the complex phase and which are related to the top quark in the loop diagrams from Fig. 3. Roughly speaking $\mathrm{Im} A_i/\mathrm{Re} A_i \sim \mathrm{Im} V_{ts}^* V_{td} \sim 10^{-4}$. Empirically we have, from the $\Delta I = 1/2$ rule, $\mathrm{Re} A_2/\mathrm{Re} A_0 \sim 10^{-2}$. This leads to a natural size of ε' of $\sim 10^{-6}$, or possibly even smaller due to the cancellations mentioned before.

We should stress that the quantity in (15) is not the observable actually used to determine ε' experimentally. We have discussed it here because it is of conceptual interest as the simplest manifestation of ε'. The realistic analysis requires a more general consideration of $K_L, K_S \to \pi\pi$ decays to which we will turn in the following paragraph.

c) – Mixing Induced CP Violation in $K \to \pi\pi$: ε, ε'

In this section we will illustrate the concept of mixing-induced CP violation with the example of $K \to \pi\pi$ decays. These are important processes, since CP violation has first been seen in $K_L \to \pi^+\pi^-$ and as of today our most precise experimental knowledge about this phenomenon still comes from the study of $K \to \pi\pi$ transitions. There are two distinct final states and in a strong interaction eigenbasis the transitions are $K^0, \bar{K}^0 \to \pi\pi(I=0), \pi\pi(I=2)$, with definite isospin for $\pi\pi$. Alternatively, using the physical eigenbasis for both initial and final states, one has $K_L, K_S \to \pi^+\pi^-, \pi^0\pi^0$.

Consider next the amplitude for K_L going into the CP even state $\pi\pi(I=0)$, which can proceed via K $(\sim (1+\bar{\varepsilon})A_0)$ or via $\bar{K}$ $(\sim (1-\bar{\varepsilon})A_0^*)$. Hence (to first order in small quantities)

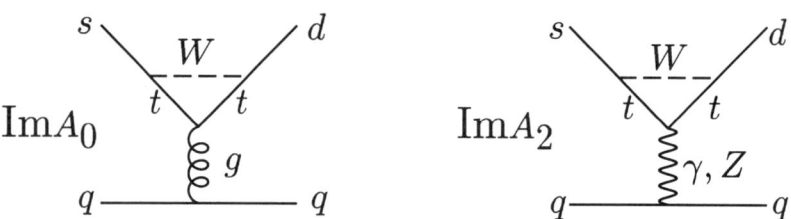

FIGURE 3. Gluon penguin and electroweak penguin diagram contributions to the parameter ε'.

$$A(K_L \to \pi\pi(I=0)) \sim (1+\bar{\varepsilon})A_0 e^{i\delta_0} - (1-\bar{\varepsilon})A_0^* e^{i\delta_0} \sim \bar{\varepsilon} + i\frac{\text{Im}A_0}{\text{Re}A_0} = \varepsilon \quad (16)$$

This defines the parameter ε, characterizing mixing-induced CP violation. Note that ε involves a component from mixing ($\bar{\varepsilon}$) as well as from the decay amplitude ($\text{Im}A_0/\text{Re}A_0$). Neither of those is physical separately, but ε is. Note also that the physical quantity $\text{Re}\bar{\varepsilon}$ discussed above satisfies $\text{Re}\bar{\varepsilon} = \text{Re}\varepsilon$. More generally one can form the following two CP violating observables

$$\eta_{+-} = \frac{A(K_L \to \pi^+\pi^-)}{A(K_S \to \pi^+\pi^-)} \qquad \eta_{00} = \frac{A(K_L \to \pi^0\pi^0)}{A(K_S \to \pi^0\pi^0)} \quad (17)$$

These amplitude ratios involve the physical initial and final states and are directly measurable in experiment. They are related to ε and ε' through

$$\eta_{+-} = \varepsilon + \varepsilon' \qquad \eta_{00} = \varepsilon - 2\varepsilon' \quad (18)$$

The phase of ε is given by $\varepsilon = |\varepsilon|\exp(i\pi/4)$. The relative phase between ε' and ε can be determined theoretically. It is close to zero so that to very good approximation $\varepsilon'/\varepsilon = \text{Re}\varepsilon'/\varepsilon$.

Both η_{+-} and η_{00} measure mixing-induced CP violation (interference between mixing and decay). Each of them considered separately could be attributed to CP violation in $K - \bar{K}$ mixing and would therefore represent indirect CP violation. On the other hand, a nonvanishing difference $\eta_{+-} - \eta_{00} = 3\varepsilon' \neq 0$ is a signal of direct CP violation. Experimentally one has [7]

$$|\varepsilon| = (2.282 \pm 0.019) \cdot 10^{-3} \quad (19)$$

Theoretically ε is related to the first diagram shown in Fig. 2. Comparison of the theoretical expression [2] with the experimental result yields an important constraint on the CKM phase δ (this is the phase of the CKM matrix in standard parametrization [7]; it coincides with the phase γ of the unitarity triangle). The quantity ε' can be measured as the ratio $\text{Re}\varepsilon'/\varepsilon \approx \varepsilon'/\varepsilon$ using the double ratio of rates

$$\left|\frac{\eta_{+-}}{\eta_{00}}\right|^2 \doteq 1 + 6\,\text{Re}\frac{\varepsilon'}{\varepsilon} \quad (20)$$

Currently the following measurements are available

$$\text{Re}\frac{\varepsilon'}{\varepsilon} = \begin{cases} (23 \pm 7) \cdot 10^{-4} & \text{CERN NA31} \\ (7.4 \pm 5.9) \cdot 10^{-4} & \text{FNAL E731} \end{cases} \quad (21)$$

These results are somewhat inconclusive and it remains presently still open whether or not a direct CP violation effect exists in $K \to \pi\pi$ decays. As mentioned before, the theoretical predictions suffer from large hadronic uncertainties. A representative range from a recent analysis of Buras et al. [10] is

$$2 \cdot 10^{-4} \leq \varepsilon'/\varepsilon \leq 19 \cdot 10^{-4} \tag{22}$$

Similar results have been obtained by other groups [11–13]. Currently running or future experiments at CERN, FNAL and Frascati aim at an improved sensitivity of $\Delta\varepsilon'/\varepsilon \approx 10^{-4}$. If ε'/ε is not too small, the new round of measurements has a good chance to finally resolve the question of direct CP violation in $K \to \pi\pi$ experimentally.

THE RARE DECAY $K_L \to \pi^0 \nu\bar{\nu}$

One of the most promising opportunities for future studies of flavor physics and CP violation is the rare decay $K_L \to \pi^0 \nu\bar{\nu}$. This process combines strongly motivated phenomenological interest (sensitivity to high energy scales, top quark mass and CKM couplings) with a situation where all theoretical uncertainties are exceedingly well under control. With these features $K_L \to \pi^0 \nu\bar{\nu}$ is unparalleled in the phenomenology of weak decays.

$K_L \to \pi^0 \nu\bar{\nu}$ is a flavor-changing neutral current process, induced at one-loop order in the SM. It proceeds entirely through short distance weak interactions because the neutrinos can couple only to heavy gauge bosons (W, Z). The transition can be effectively described by a local $(\bar{s}d)_{V-A}(\bar{\nu}\nu)_{V-A}$ interaction (and $h.c.$), whose coupling strength is calculable from the SM. This interaction is semileptonic and the required hadronic matrix element $\langle \pi^0 | (\bar{s}d)_V | K^0 \rangle$ can be extracted from the well measured decay $K^+ \to \pi^0 e^+ \nu$ using isospin symmetry. The knowledge of short distance QCD effects at next-to-leading order ($\mathcal{O}(\alpha_s)$) [14]; essentially eliminates the dominant theoretical uncertainty in this decay mode from scale dependence. The process is theoretically under control to an accuracy of better than $\pm 3\%$.

In the limit of conserved CP, the relevant hadronic matrix element would be $\langle \pi^0 | (\bar{s}d)_V + (\bar{d}s)_V | K_L \rangle$. Because of the CP properties of K_L, π^0 and the transition current this matrix element is zero in this limit. In the SM $K_L \to \pi^0 \nu\bar{\nu}$ therefore measures a violation of CP symmetry. It belongs to the class of mixing-induced CP violation. Considering the amplitude ratio $\eta_{\pi^0 \nu\bar{\nu}} = A(K_L \to \pi^0 \nu\bar{\nu})/A(K_S \to \pi^0 \nu\bar{\nu})$, which is analogous to η_{+-} for $K \to \pi^+ \pi^-$ (17), one finds $\eta_{\pi^0 \nu\bar{\nu}} = \mathcal{O}(1)$ in the SM, essentially because $K \to \pi^0 \nu\bar{\nu}$ is a rare decay. Thus we have $\eta_{\pi^0 \nu\bar{\nu}} \gg \eta_{+-} = \mathcal{O}(10^{-3})$, which means that $K_L \to \pi^0 \nu\bar{\nu}$ is a signal of very large direct CP violation within the SM. The branching ratio $B(K_L \to \pi^0 \nu\bar{\nu})$ is proportional to $(\text{Im} V_{ts}^* V_{td})^2$, which makes it an ideal measure of $\text{Im} V_{ts}^* V_{td}$ or the parameter η.

The current SM prediction for the branching ratio is $B(K_L \to \pi^0 \nu\bar{\nu}) = (2.8 \pm 1.7) \cdot 10^{-11}$ [15], where the sizable range reflects our presently still quite limited knowledge of CKM parameters, but not intrinsic theoretical uncertainties, which are negligible. Using the experimental limit on $K^+ \to \pi^+ \nu \bar\nu$, a model independent upper bound can be set at $B(K_L \to \pi^0 \nu \bar\nu) < 1.1 \cdot 10^{-8}$ [16]. Current experimental searches, not optimized for this process, have yielded a (published [7]) upper bound of $5.8 \cdot 10^{-5}$ (Fermilab E799). Dedicated experiments will aim at an actual measurement of $K_L \to \pi^0 \nu \bar\nu$ in the future. A proposal already exists at Brookhaven (BNL E926) and there are further plans at Fermilab and KEK.

CP VIOLATION IN B DECAYS

Decays of B mesons offer a wide range of possibilities to expand our knowledge of CP violation and to test further what we have learned from the kaon system. Among those are truly superb opportunities with esssentially no theoretical uncertainty and predicted large CP asymmetries. The prototype observable is the time-dependent CP asymmetry in $B_d(\bar B_d) \to J/\Psi K_S$, which is without doubt the highlight of this field. We will first focus on this case in the following because of its importance and because it exhibits the characteristic features of a large class of CP violating observables in B physics. We will briefly mention further possibilities later on.

$$B_d \to J/\Psi K_S$$

The CP asymmetry in $B_d \to J/\Psi K_S$ belongs to the class of mixing-induced CP violation, that is CP violation in the interference of mixing and decay. In the kaon system an essentially pure beam of a definite eigenstate, the K_L, can easily be produced due to the vast difference in lifetimes between K_L and K_S, which is ideal for CP violation studies. Since the lifetime difference between eigenstates is negligibly small for the $B_d - \bar B_d$ system, the same method can not be applied in this case. Instead explicit flavor tagging (determination of the flavor of one of the B mesons (produced in pairs), for instance by means of the lepton charge in the semileptonic decay of the other) is required and one has to consider the time dependence of $B - \bar B$ mixing.[2] Solving the time dependent Schrödinger equation with the mixing Hamiltonian $\hat H$ (4), and neglecting $\Delta\Gamma$, one has

$$B(t) = e^{-iMt - \frac{1}{2}\Gamma t} \left[\cos\frac{\Delta M t}{2} B - \frac{q}{p} i \sin\frac{\Delta M t}{2} \bar B \right] \qquad (23)$$

$$\bar B(t) = e^{-iMt - \frac{1}{2}\Gamma t} \left[\cos\frac{\Delta M t}{2} \bar B - \frac{p}{q} i \sin\frac{\Delta M t}{2} B \right] \qquad (24)$$

---

[2] This latter strategy can in principle also be used for neutral kaons and is in fact the method realized in the CPLEAR experiment at CERN (see M. Mikuz, these proceedings).

$B(t)$ and $\bar{B}(t)$ are the time evolved states that started out as flavor eigenstates B and $\bar{B}$, respectively, at time $t = 0$.

The CP asymmetry in $B_d \to J/\Psi K_S$ is the prime example of the important class of asymmetries in neutral B mesons decaying into a CP eigenstate, in this case $f = J/\Psi K_S$, which is CP odd. There are two basic contributions to the decay amplitude, distinguished by the combination of CKM parameters $V_{cb}^* V_{cs}$ or $V_{tb}^* V_{ts}$. Representative diagrams are shown in Fig. 4. The third possible factor $V_{ub}^* V_{us}$ can be expressed in terms of the above two by CKM unitarity, $V_{ub}^* V_{us} = -V_{cb}^* V_{cs} - V_{tb}^* V_{ts}$. Choosing the latter two as independent parameters is useful in the present case, since $V_{ub}^* V_{us}$ is Cabibbo suppressed. A crucial feature of the $B_d \to J/\Psi K_S$ mode is that the relative weak phase between $V_{cb}^* V_{cs}$ and $V_{tb}^* V_{ts}$ is negligibly small. Consequently the $B_d \to J/\Psi K_S$ amplitude can to excellent approximation be represented as $A(B_d \to J/\Psi K_S) = V_{cb}^* V_{cs} \cdot A_{red}$ as far as the weak phase structure is concerned. The quantity A_{red} involves nontrivial hadronic dynamics, but it will drop out when forming the ratio that defines the asymmetry (see (25) below). This fact lies at the bottom of the theoretically clean nature of the $B_d \to J/\Psi K_S$ asymmetry.

Using this property of the amplitude we can now see how the mixing-induced asymmetry comes about. As illustrated in Fig. 5, an initial B state can decay to the CP self-conjugate final state f via two different paths: directly $(B \to f)$, or through mixing $(B \to \bar{B} \to f)$, since the same final state can be reached by both B and $\bar{B}$. The mixing phase phase $(B \to \bar{B})$ is determined by the box diagram, similar to the first graph in Fig. 2, and reads $(V_{tb}^* V_{td})^2/|V_{tb}^* V_{td}|^2 \equiv V_{tb}^* V_{td}/(V_{tb} V_{td}^*)$. The two different decay paths therefore have a relative phase of $V_{tb}^* V_{td} V_{cb} V_{cs}^*/(V_{tb} V_{td}^* V_{cb}^* V_{cs}) = \exp(-2i\beta)$. The CP conjugate situation (starting out with $\bar{B}$) has the opposite phase. Putting everything together (using (23), (24)) one finds the time-dependent asymmetry

$$\mathcal{A}_{CP}(B_d \to J/\Psi K_S) \equiv \frac{\Gamma(B(t) \to \Psi K_S) - \Gamma(\bar{B}(t) \to \Psi K_S)}{\Gamma(B(t) \to \Psi K_S) + \Gamma(\bar{B}(t) \to \Psi K_S)}$$
$$= -\sin 2\beta \cdot \sin \Delta M t \qquad (25)$$

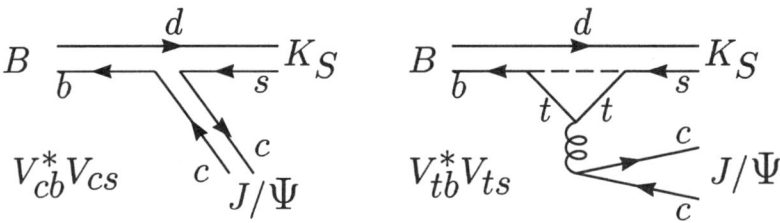

FIGURE 4. Representative diagrams contributing to $B_d \to J/\psi K_S$.

A few points about this result are worth emphasizing, the first two of which summarize the basic reasons why $\mathcal{A}_{CP}(B_d \to J/\Psi K_S)$ plays such an important role in flavor physics.

- As mentioned before, the part of the amplitude containing the dependence on the uncalculable hadronic dynamics has canceled out in the asymmetry. The asymmetry depends only on the CKM quantity $\sin 2\beta$. This result holds to within a theoretical uncertainty of less than 1%.

- The effect is quite large in the SM, where one expects approximately $\sin 2\beta \approx 0.6 \pm 0.2$. This information comes from the observed CP violation in the kaon system, which implies that the CP phase η must not be too small. It follows (see Fig. 1) that also $\sin 2\beta$ has to be sizable.

- Two a priori unrelated features of the fundamental SM parameters are very helpful to make a measurement of (25) feasible. First, the B_d lifetime (about $1.5ps$) is relatively large due to the smallness of $V_{cb} \approx 0.04$, which is crucial for being able to resolve the time dependence. Furthermore, ΔM is sizable due to the large top quark mass such that ΔM turns out to be of almost the same size as Γ (see Table 1), which is almost perfect for optimizing the effect of mixing.

Other Possibilities

A case very similar to $B_d(\bar{B}_d) \to J/\Psi K_S$ is the CP asymmetry for $B_d(\bar{B}_d) \to \pi^+\pi^-$. Here the dominant contribution to the amplitude has CKM factor $V_{ub}^* V_{ud}$ ($V_{ub} V_{ud}^*$) and the relative phase between the mixed decay $B \to \bar{B} \to f$ and the direct decay $B \to f$ is $V_{tb}^* V_{td} V_{ub} V_{ud}^* / (V_{tb} V_{td}^* V_{ub}^* V_{ud}) = \exp(-2i(\beta + \gamma)) = \exp(2i\alpha)$. Accordingly the CP asymmetry is a measure of $\sin 2\alpha$. The situation is, however, somewhat complicated by the second, non-negligible contribution to the decay amplitude from penguin

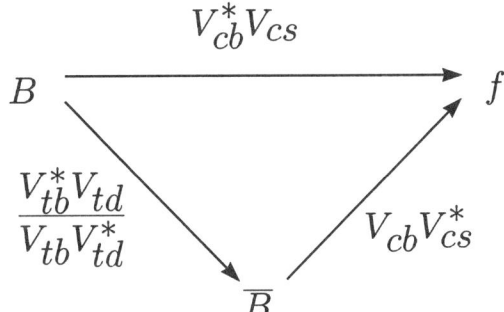

FIGURE 5. Possible decay paths for an initial B meson to decay into the final state $f = J/\Psi K_S$ that is common to B and $\bar{B}$.

graphs. This contribution comes with CKM factor $V_{tb}^*V_{td}$, which, unlike the case of $B \to J/\Psi K_S$, has a different phase than the leading contribution ($\sim V_{ub}^*V_{ud}$). Consequently, the amplitude no longer has the simple structure of the $B \to J/\Psi K_S$ amplitude with its single weak phase where all hadronic uncertainties cancel, and some poorly calculable hadronic dynamics will invariably enter the CP asymmetry $\mathcal{A}_{CP}(B_d \to \pi^+\pi^-)$ ('penguin pollution'). Strategies have been devised to eliminate this uncertainty, for instance using additional information from related modes as $B_d(\bar{B}_d) \to \pi^0\pi^0$ and $B^{\pm} \to \pi^{\pm}\pi^0$ together with isospin symmetry [17]. Assuming this has been achieved, $B \to \pi\pi$ determines $\sin 2\alpha$, an example of mixing-induced CP violation just as the case of $B \to J/\Psi K_S$ and $\sin 2\beta$. As explained before, each of these cases considered separately represents indirect CP violation. However any deviation from the equality $\sin 2\beta = -\sin 2\alpha$ would reveal a direct CP violation effect [18] (the minus sign appears here due to the opposite CP parities of $J/\Psi K_S$ (CP odd) and $\pi^+\pi^-$ (CP even)).

A good example of direct CP violation is provided by the decays $B^{\pm} \to D^0_{(CP+)}K^{\pm}$. No flavor-tagging or time-dependent measurements are required here and the asymmetries can be used to extract the angle γ in a clean way [19,20].

Also B_s mesons offer opportunities for interesting CP violation studies, although they are more challenging experimentally because of the very large oscillation frequency $\Delta M/\Gamma > 10$. For instance, $B_s \to J/\Psi \phi$ is the B_s analog of $B_d \to J/\Psi K_S$ decay. The asymmetry is Cabibbo suppressed in this case but would allow, in principle, a clean determination of η. A measurement of γ is possible with $B_s \to D_s^+ K^-$ [21].

There are many more strategies and scenarios discussed in the literature. In our brief account we have focused on those cases that can yield insight into the mechanisms of CP violation with exceptionally small theoretical uncertainties. For general reviews see e.g. [22,23].

CONCLUSIONS

The violation of CP symmetry has so far been observed in just five decay modes of the long-lived neutral kaon, namely $K_L \to \pi^+\pi^-$, $\pi^0\pi^0$, $\pi e \nu$, $\pi \mu \nu$, $\pi^+\pi^-\gamma$. All asymmetries can be described by a single complex parameter ε. The question of direct CP violation in $K \to \pi\pi$, measured by ε'/ε, is still open and currently further pursued by ongoing projects. Although our knowledge of this phenomenon is rather limited, the established pattern of CP violation with kaons, $\varepsilon \sim 10^{-3}$ and $\varepsilon' \lesssim 10^{-6}$, is well accounted for by the three generation Standard Model. The smallness of ε and ε' is related to the size of $\text{Im} V_{ts}^* V_{td} = A^2 \lambda^5 \eta \sim 10^{-4}$. This quantity is small due to suppressed intergenerational quark mixing ($\sim \lambda^5$), but not due to smallness of the CP violating phase ($\sim \eta$), which in fact is quite substantial (typically $\eta \approx 0.3 - 0.4$). As a consequence, large asymmetries are predicted in the B meson sector, which has many decay channels and a very rich phenomenology. The highlight of this latter area is $\mathcal{A}_{CP}(B_d \to J/\Psi K_S) \sim \sin 2\beta \approx 0.6 \pm 0.2$, exhibiting a large effect with essentially no theoretical uncertainties and good experimental feasibility.

Theoretical progress during recent years that is of relevance for this type of physics includes heavy quark effective theory, the calculation of higher order QCD effects and improvements in lattice QCD computations. In many cases a serious remaining problem is the nonperturbative strong dynamics governing weak decay matrix elements (ε'/ε, $B \to \pi\pi$). Exceptions are clean observables with practically negligible theoretical error. The prime examples of this class are $\mathcal{A}_{CP}(B_d \to J/\Psi K_S)$ ($\sim \sin 2\beta$) and $B(K_L \to \pi^0 \nu \bar{\nu})$ ($\sim \eta^2$). Also important for a further understanding of CP violation is the study of flavor-changing neutral current rare decays with small theoretical ambiguities such as $K^+ \to \pi^+ \nu \bar{\nu}$, $B \to X_s \gamma$, $B \to X_s e^+ e^-$, $B \to l^+ l^-$, $B \to X_s \nu \bar{\nu}$ or $\Delta M_{B_s}/\Delta M_{B_d}$. Since the number of clean processes is very limited, and much complementary information is needed, all of them should be pursued as far as possible.

CP violation is intimately connected with flavor dynamics, the least understood sector of our current Standard Model. It is therefore also closely related to the question of electroweak symmetry breaking, presently one of the most urgent open problems in fundamental physics. The coming years hold great promise for decisive progress in our knowledge about CP violation and for obtaining a clearer picture of what may still lie behind this remarkable phenomenon.

REFERENCES

1. M. Kobayashi and K. Maskawa, Prog. Theor. Phys. **49** 652 (1973).
2. G. Buchalla, A.J. Buras and M.E. Lautenbacher, Rev. Mod. Phys. **68**, 1125 (1996).
3. C. Jarlskog, ed., CP Violation, World Scientific, Singapore, (1989).
4. L. Wolfenstein, Phys. Rev. Lett. **13**, 562 (1964).
5. J. Liu and L. Wolfenstein, Phys. Lett. **B197**, 536 (1987).
6. G. D'Ambrosio and G. Isidori, hep-ph/9611284.
7. R.M. Barnett et al., Particle Data Group, Phys. Rev. **D54**, 1 (1996)
8. J.M. Flynn and L. Randall, Phys. Lett. **B224**, 221 (1989); erratum ibid. **B235**, 412 (1990).
9. G. Buchalla, A.J. Buras and M.K. Harlander, Nucl. Phys. **B337**, 313 (1990).
10. A.J. Buras, M. Jamin and M.E. Lautenbacher, Phys. Lett. **B389**, 749 (1996).
11. M. Ciuchini et al., Z. Phys. **C68**, 239 (1995).
12. S. Bertolini, J.O. Eeg and M. Fabbrichesi, Nucl. Phys. **B476**, 225 (1996).
13. J. Heinrich et al., Phys. Lett. **B279**, 140 (1992); E.A. Paschos, DO-TH 96/01, talk presented at the 27th Lepton-Photon Symposium, Beijing, China (1995).
14. G. Buchalla and A.J. Buras, Nucl. Phys. **B400**, 225 (1993).

15. A.J. Buras, M. Jamin and M.E. Lautenbacher, to appear.
16. Y. Grossman and Y. Nir, Phys. Lett. **B398**, 163 (1997).
17. M. Gronau and D. London, Phys. Rev. Lett. **65**, 3381 (1990).
18. B. Winstein and L. Wolfenstein, Rev. Mod. Phys. **65**, 1113 (1993).
19. M. Gronau and D. Wyler, Phys. Lett. **B265**, 172 (1991).
20. D. Atwood, I. Dunietz and A. Soni, Phys. Rev. Lett. **78**, 3257 (1997).
21. R. Aleksan, I. Dunietz and B. Kayser, Z. Phys. **C54**, 653 (1992).
22. Y. Nir and H. Quinn, Ann. Rev. Nucl. Part. Sci. **42** 211 (1992).
23. A.J. Buras and R. Fleischer, hep-ph/9704376.

Recent Results from Relativistic Heavy Ion Collisions

Thomas K. Hemmick

Department of Physics and Astronomy
University at Stony Brook
Stony Brook, New York 11790-3800

Abstract. The Standard Model predicts that at sufficiently high temperature and/or baryon density, nuclear matter will undergo a deconfinement and chiral phase transition to a state of nearly free quarks and gluons, the Quark-Gluon Plasma (QGP). Access to this state in the laboratory is possible only through high energy collisions of large nuclei. Studies of relativistic heavy ion collisions using fixed targets and heavy beams of 10 GeV/c and 160 GeV/c have been ongoing for several years. I will present a brief summary of recent experimental results.

INTRODUCTION

Although the Standard Model's description of electromagnetic, weak, and strong forces has survived unmodified and unviolated for many years, the properties of the strong force and the particles which experience it (quarks and gluons) remain somewhat elusive. The reason for this is that although standard calculational techniques such as perturbation theory have been tremendously successful in describing electromagnetic and weak processes, the large coupling constant for low momentum transfer strong interactions prevents similarly broad success for QCD interactions.

Numerous theoretical and experimental studies are underway to gain access to the low q domain of QCD. On the theoretical side Lattice QCD calculations [1] are the straightforward approach, however, they are computationally intensive. More recently approaches such as that of instanton liquids [2,3,1], strive to distill the essence of the strong interaction via its apparent leading term, instanton tunneling. The experimental side also hosts a traditional approach and one that is more exotic. The traditional approach involves the detailed investigation of nucleon structure functions, while the more exotic approach is via relativistic heavy ion collisions and studies of the Quark-Gluon Plasma.

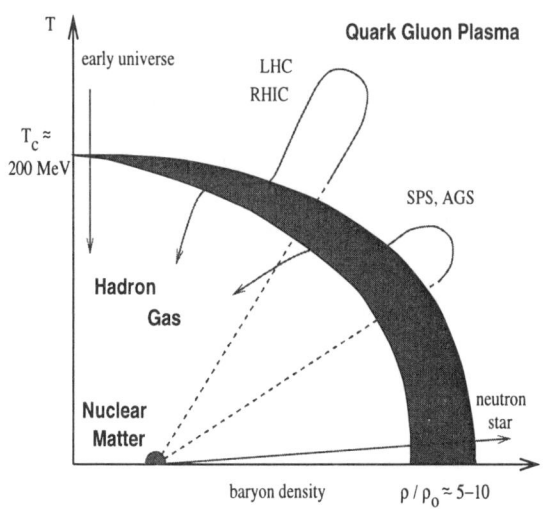

FIGURE 1. The phase diagram of nuclear matter.

Lattice QCD calculations provide the most fundamental predictor of the critical temperature and density necessary for plasma formation, however, simple considerations yield similar results. One can imagine that since nucleons have finite size, at some density nucleons will overlap and dissolve. Since the volume of a nucleus is roughly four times as large as the sum of the volumes of its constituent nucleons, a phase transition would be expected at densities somewhat higher than four times nuclear matter density. Similarly, at temperatures high enough to produce many secondary particles ($T > m_\pi$) the volume of a normal nucleus would be filled with secondaries, again leading to hadron overlap and a deconfinement phase transition to Quark-Gluon Plasma.

Such considerations lead to a cartoon phase diagram of nuclear matter such as the one shown in Figure 1. At sufficiently high temperature and/or density hadronic degrees of freedom melt away yielding a system whose thermodynamic properties (latent heat, heat capacity) are governed by the essential features of strong interaction physics. Such an approach to learning about the strong interaction is analogous to learning quantum mechanics from blackbody radiation rather than atomic spectra.

During a relativistic heavy ion collision nuclear matter will be compressed and heated hopefully forming a QGP state for a short time as indicated by the arrows in Figure 1. This state will be transitory as the system expands and cools passing back through the phase of hadronic matter prior to breakup. The challenge to the experimentalist is to determine the properties of the exotic matter formed during each stage of the collision (Pre-equilibrium, QGP, hot

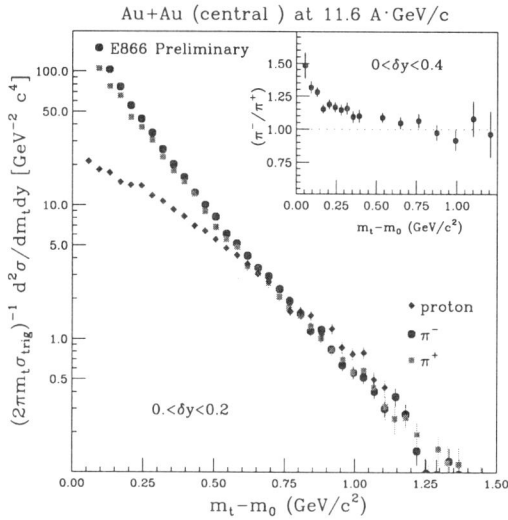

FIGURE 2. Proton and Pion spectra at central rapidity for AGS Au+Au collisions.

hadron gas, freezeout).

HADROCHEMISTRY

A necessary precondition to the discussion of "the phase" of nuclear matter is thermalization. Although the earliest experiments showed marked disagreement with a simplistic isotropic thermal fireball picture, there has been a recent resurgence of activity attempting to describe the final state of RHI collisions as a rapidly expanding or "flowing" thermal source. Here we discuss the ways in which a static fireball is seen to fail and how the superposition of expansion resolves many of the discrepancies between the data and thermal models.

Shown in Figure 2 are preliminary results for pion and proton spectra from 11 GeV/c Au+Au collisions as measured by the E866 collaboration [4]. In a thermal picture, these spectra would be expected to be exponential with a slope constant of $1/T$. Clearly this is not the case. Protons exhibit a flattening near the origin while the overall steeper pion spectra show a clear upturn at low transverse mass. As can be been in the right hand panel of Figure 3 [4], the mean transverse momentum rises monotonically with ejectile mass. This mass-driven trend in slope constants is further supported by the close match of the slope constants of nucleons and the nearly equal mass Λ baryons (see Figure 4) as measured at the AGS by the E891 collaboration [5]. A similar trend of temperature parameter with mass is observed at SPS energies (Figure 5) by the NA44 and NA49 collaborations [6,7]. Such behavior alone

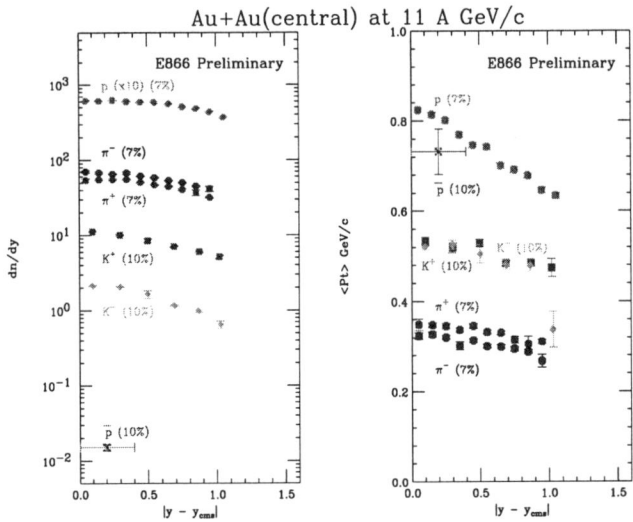

FIGURE 3. Rapidity spectra and mean transverse momentum of hadrons produced at AGS energies.

is sufficient to rule out the isotropic fireball picture, however, the systematic trends of the mass dependence of the apparent temperature hint that a simple picture could suffice to explain the data.

From model calculations and simple considerations of interaction cross sections vs. particle density, it is apparent that system freezeout must occur at significantly larger volume than the initial overlap zone (the numerous produced particles must be spatially separated to cease interacting!). The system's inevitable expansion can be included in a thermal model as an underlying common "flow" velocity superimposed upon the otherwise random particle motions. Indeed, such a picture would lead to a linear dependence of apparent temperature with ejectile mass [6]. Fits to a flowing thermal model [8] describe the transverse momentum spectra surprisingly well (Figure 6) using temperatures in the range 120-140 MeV and flow velocities near 1/3 c. A single common expansion velocity has also been shown to match the trend of increasing rapidity width with ejectile mass [8].

Another interesting puzzle is the production of strange mesons. It has long been observed [9] that Kaon production in heavy ion collisions is enhanced with respect to pion production. Shown in Figure 7 is the integrated positive kaon yield near mid rapidity as a function of the number of participating nucleons in AGS Au+Au collisions measured by the E866 collaboration [10]. This plot reveals how the production character evolves in going from essentially nucleon-nucleon to nucleus-nucleus collisions. A distinct quadratic trend is observed in the data, characteristic of production in secondary collisions. This

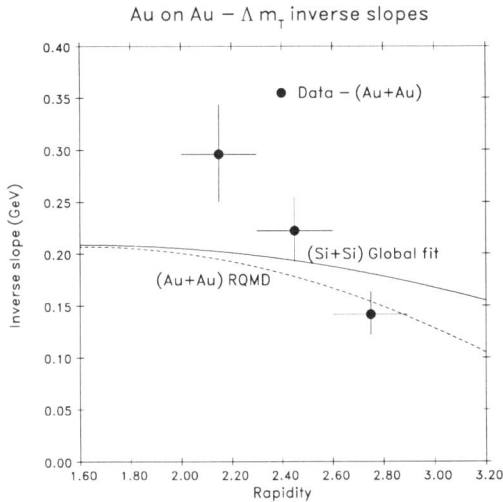

FIGURE 4. Temperature parameters for Λ production.

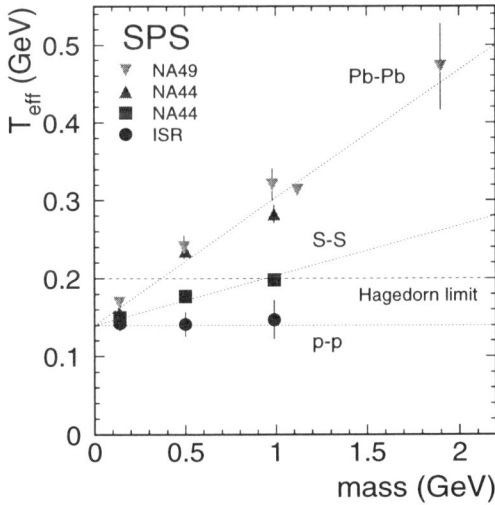

FIGURE 5. Temperature parameters as a function of particle mass from SPS collisions.

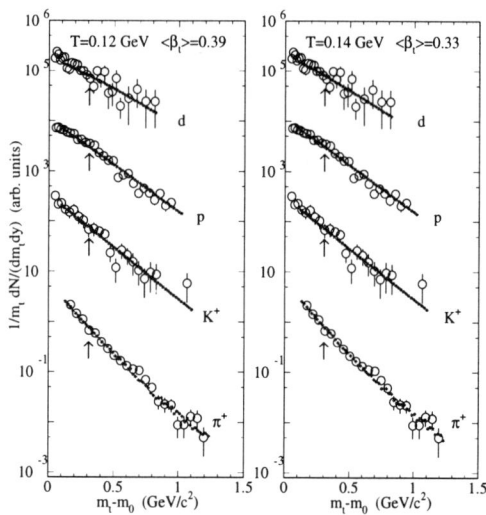

FIGURE 6. Fits to AGS hadron using an expanding thermal source.

behavior is at first sight counterintuitive since it implies that secondary, low center of mass collisions are the primary mechanism for strange ("heavy") quark production. The data are nonetheless reproduced quite well using the RQMD [11] model, a simple cascading Monte Carlo simulation code without "new physics".

Within the framework of the Monte Carlo, one can deduce the mechanism for enhanced strangeness production. There, incident nucleons are frequently left in excited resonant states (Δ, N^*) following their initial collisions. The excitation energy stored in the resonant states is available to secondary collisions, thereby allowing for a later collision to access an "accumulated" or collective energy reserve, which opens a new and dominant channel for strange quark production. This effect has become known as the *Resonance Mechanism* for strangeness production. In the RQMD model, fully 1/3 of all nucleons are found in excited states at freezeout. Interestingly, this excitation fraction matches well the expectation for a system in equilibrium at 140 MeV temperature as do the relative yields of most all hadronic species (Figure 8) [8].

ANISOTROPIC FLOW

The relative success of the flowing thermal hadron gas model could indicate that hadron gas (driven by large, resonant reinteraction cross sections) is relatively easily thermalized. If this is indeed the case, signatures of the plasma present in hadronic channels could be thermalized away prior to freezeout.

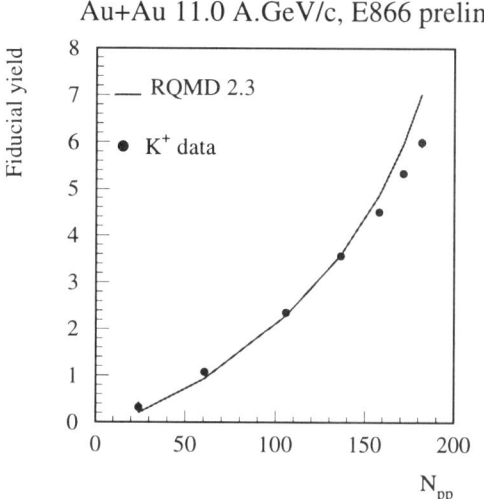

FIGURE 7. Kaon production as a function of number of participating nucleons.

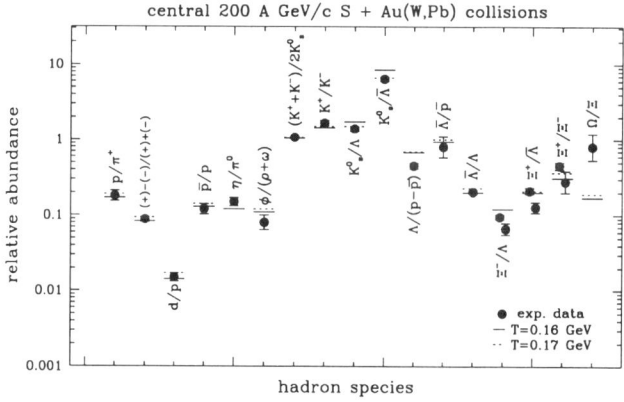

FIGURE 8. Particle abundance ratios from the thermal model.

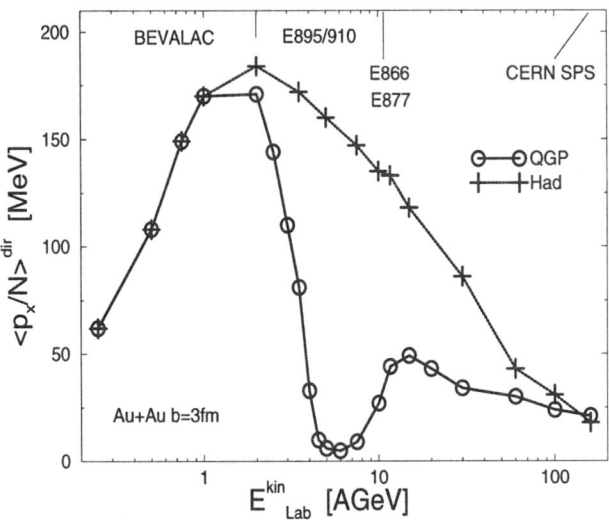

FIGURE 9. Hydrodynamical calculations predict a minimum in azimuthally anisotropic flow if a plasma were created at any stage in nuclear collisions.

For this reason recent studies have focused upon observables which inherently include a memory of an earlier stage. One such observable is anisotropic flow.

Anisotropic flow has been studied extensively at lower beam energies [12]. The signature of this phenomenon is an azimuthal asymmetry in particle production with respect the the reaction plane of incomplete overlap nuclear collisions. This momentum space asymmetry is a consequence of and therefore a memory of the initial state's spatial asymmetry. If the asymmetry were lost in the "soft" plasma phase, it could not be restored during hadronization. Theoretical studies predict a so-called "softest point" at which flow would nearly cease. This would be visible experimentally [13] as a minimum in the mean transverse momentum of ejectiles projected along the reaction plane (Figure 9) for collisions run at the beam energy corresponding to the softest point.

Flow studies have been carried out by AGS experiment E877 [14] for a variety of particle species at 10.8 GeV/nucleon. Results for the mean transverse momentum into the reaction place as a function of rapidity are shown in Figure 10 and compared with model calculations. The data indicate that the flow at AGS energies is not small. Indeed, RQMD calculations produce only 1/2 of the flow signal when run in cascade mode and require the extra boost of a repulsive mean field to reproduce the data. Flow studies as a function of beam energy from 2 - 10 GeV/nucleon will soon be available from the E895 and E866 collaborations and will directly answer whether a local flow minimum exists just below the AGS standard energy.

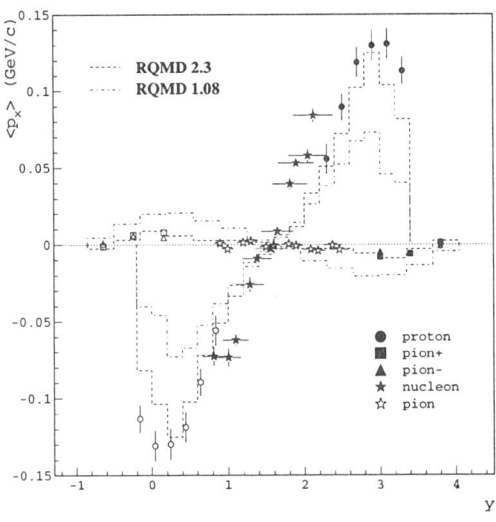

FIGURE 10. Flow measurements of charged pions and nucleons.

ELECTROMAGNETIC RADIATION

Another more direct approach to learning about the early stages of the collision is to study electromagnetic radiation and decays. Electrons and photons are expected to have means free paths in QGP of more than 100 fm, and would thus escape the collision zone during all stages of development including the plasma phase. The CERES experiment [15], uses a pair of Ring Imaging Cerenkov detectors to track only the electrons emitted by the S+Pb and Pb+Pb collisions at the CERN SPS.

Results from the Ceres experiment for p+Au collisions are shown in Figure 11. The vertical axis shows the ratio of the electron multiplicity to the overall charged particle multiplicity directed into the experimental aperture. The many curves denote calculations of electron yield from a variety of hadronic decay sources. The intensities of the hadron sources are known on an absolute scale from previously measured data. The gray band indicates the level of uncertainty in the hadronic source predictions. Excellent agreement is seen between the data and the prediction indicating that e^+e^- production in p+Au collisions follows the same mechanisms as in p+p collisions.

Shown in Figure 12 are preliminary results for S+Au and Pb+Au collisions along with the same cocktail prediction [15]. A clear excess in the yield of electron pairs is seen to begin within the data at or near twice the pion mass. One can immediately conclude that the electron production mechanism for heavy ion collisions differs significantly from that of p+p collisions. This is in fact no surprise. All nucleus-nucleus measurements differ significantly (*e.g.*

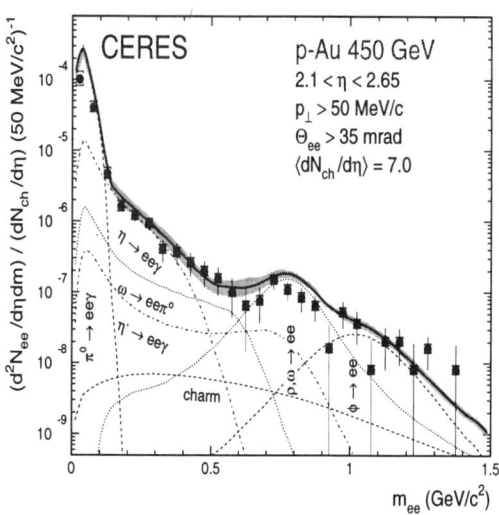

FIGURE 11. Electron mass spectra for p+Au collisions.

K^+ production) from p+p collisions due to the vitally important production steps unique to secondary collisions (Resonance Mechanism).

A more interesting analysis of the data comes from comparison to flowing thermal models similar to those which were highly successful in describing the hadronic data [16]. Numerous such calculations have been performed. A brief summary is shown in Figure 13. Although each of these calculations produces a greater electron yield than the simple p+p extrapolation, none of them are able to reproduce the Ceres data adequately. Comparisons are also made to muon spectra measured by the Helios 3 collaboration. Again, all the thermal model calculations fail to reproduce the lepton yield.

The failure of flowing thermal models to reproduce lepton spectra in heavy ion collisions has lead to the development more exotic models. Long before the data were taken, Brown and Rho [17] predicted that as the density of a nuclear system increased, a gradual restoration of chiral symmetry would take place. One effect of this restoration would be a drop in meson masses (particularly the ρ) inside the nuclear medium. The mass drop and the subsequent increase in available phase space would result in the production of numerous anomalously low mass ρ mesons. Since the ρ is quite short-lived, it would decay in medium and produce an enhancement in lepton yields in the region between $2m_\pi$ and m_ρ. Comparisons of two such calculations [17,18] to the data (Figure 14) show remarkable agreement, perhaps providing preliminary evidence for the partial restoration of chiral symmetry within nuclear collisions.

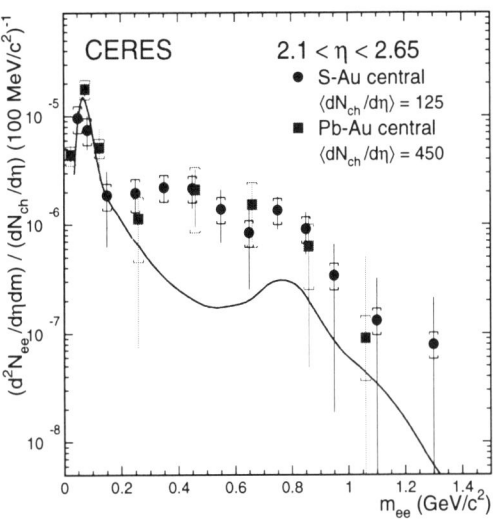

FIGURE 12. Electron mass spectra for nucleus+nucleus collisions.

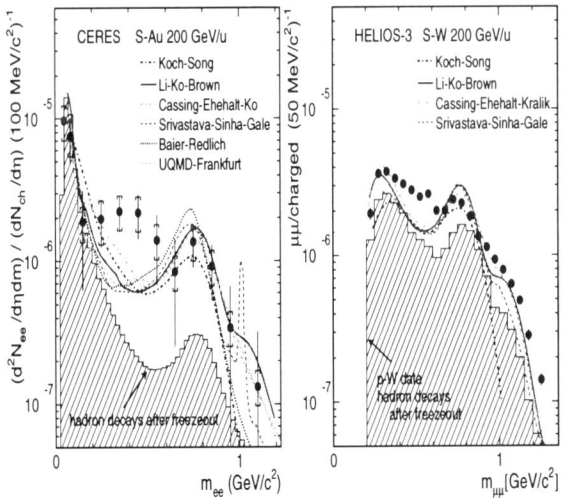

FIGURE 13. Comparison of the Ceres and Helios 3 lepton data to various hydrodynamical thermal models.

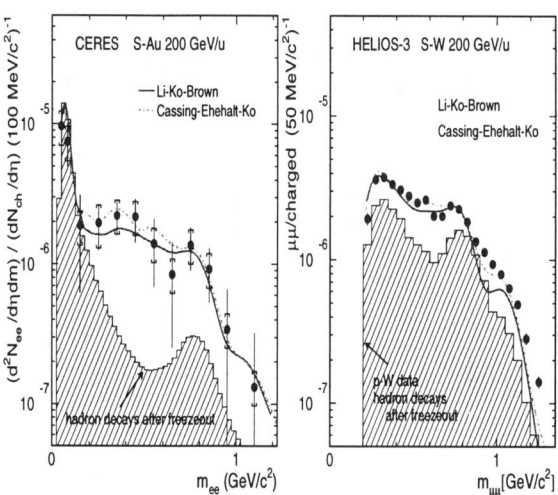

FIGURE 14. Comparison of the Ceres and Helios 3 data to models which allow for medium modification of meson masses.

CHARMONIUM

One of the earliest proposed signatures of the QGP was the dissolution and subsequent disappearance or suppression of heavy quark states [19]. Most quark bound states will be dissolved by definition in the plasma, however, heavy quarkonium states (charmonium, bottomonium) are unique due to the rarity of their valence quarks. States such as the J/Ψ vector meson are formed from a correlated pair of charm quarks produced at a single binary interaction point. Should this pair dissolve, it is unlikely they would find each other or any other charms quark during hadronization due to the fact that charm quarks are rare. The loss of J/Ψ production would thus be a permanent and lasting effect of the Quark-Gluon Plasma regardless of the hadronization scenario.

The NA50 experiment measures the decay of J/Ψ into muon pairs using a high resolution muon spectrometer and has studied a variety of colliding systems. A summary of their data is shown in Figure 15 as a function of the product of the projectile atomic number (A) with the target atomic number (B) [20]. Overall a systematic trend of an exponentially decreasing production cross section is seen. This loss is parameterized as a reinteraction of the produced J/Ψ with the surrounding nuclear medium. All measured points follow a single systematic except for the Pb+Pb colliding system. In that case, the yield is dramatically lower than would be anticipated from the reinteraction systematics.

In the case of nuclear collisions, one can roughly select the impact parame-

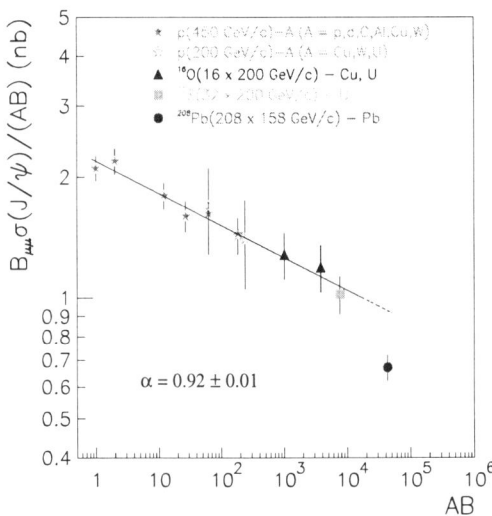

FIGURE 15. Concentration profile climatology and climate model situation.

ter of a given collision by measuring one of a variety of global observables such as transverse energy or charged particle multiplicity. If the anomalous suppression were truly due to plasma formation, one would expect that the effect would be greatest for central collisions. Since the effect is only noticeable as a deviation from an understood trend (interactions with nuclear matter) one must find a way to characterize interactions with nuclear matter as a function of centrality to interpret the data.

The NA50 collaboration has performed an analysis of their data in comparison to a simple Monte Carlo [20]. The Monte Carlo is used to measure, at each level of centrality, the typical "length" of normal nuclear matter which would wash over top of any produced J/Ψ prior to it escaping the collision and subsequently being detected. This parameter would have analogous systematics to the AB parameter, an exponential suppression as a function of L.

Shown in Figure 16 is the measured ratio of J/Ψ to Drell-Yan as a function of the L parameter. For small systems (p+A, S+A) the exponential trend of the suppression due to simple reinteractions of J/Ψ is clearly seen. However, for the Pb+Pb system the data deviate dramatically from the exponential trend, now by a factor of two at the highest centrality. There is considerable debate as to whether this suppression could be explained as being due to interactions with produced particles [21] or whether it is uniquely a signature of QGP formation [22].

79

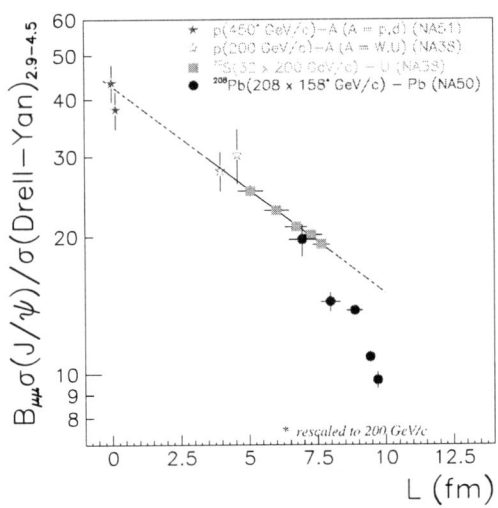

FIGURE 16. Concentration profile climatology and climate model situation.

SUMMARY

Much progress has been made in formulating accurate and simple descriptions of the final state of Relativistic Heavy Ion Collisions. The internal dynamics of the hadronic phase are dominated by the Resonance Mechanism, a collective energy storage process unique to heavy ion collisions. Thermal parameterizations indicate a system which freezes with large size, significant flow velocities, a temperature near the pion mass, and a large population of internally excited hadronic states with many more degrees of freedom than the classic pion gas. Lepton studies show a dramatic enhancement in yield between twice the pion mass and the Rho mass. So far, only models which rely upon partial restoration of chiral symmetry have succeeded in reproducing the data. A dramatic suppression of J/Ψ and Ψ' beyond that explicable by reinteraction with nuclear matter is observed.

The author would like to thank all his colleagues around the world who helped to collect and interpret data for this presentation. He is especially indebted to Y. Akiba, C. Gerschel, A. Drees, C. Ogilvie, J. Dunlop, and W.C. Chang. Support of the United States Department of Energy is gratefully acknowledged.

REFERENCES

1. J. Nagle, these proceedings.

2. E.V. Shuryak and J.J.M. Verbaarschot, *Nucl. Phys.* **B410**, 55 (1993).; T. Schaefer, E.V. Shuryak and J.J.M. Verbaarschot, *Nucl. Phys.* **B412**, 143 (1994).; T. Schaefer and E.V. Shuryak, *Phys. Rev.* **D54**, 1099 (1996).;E.V. Shuryak, *Rev. Mod. Phys.* **65**, 1 (1993).
3. T. Schaefer, these proceedings.
4. Y. Akiba for the E802 Collaboration, *Nucl. Phys.* **A610**, 139c (1996); J. Dunlop, these proceedings.
5. S. Ahmad, et al., *Phys. Lett.* **B382**, 35 (1996).
6. N. Xu for the NA44 Collaboration, *Nucl. Phys.* **A610**, 175c (1996).
7. P.G. Jones Akiba for the NA49 Collaboration, *Nucl. Phys.* **A610**, 188c (1996).
8. P. Braun-Munzinger, J. Stachel, J.P. Wessels, and N. Xu, *Phys. Lett.* **B344**, 43 (1995).
9. T. Abbott et al., *Phys. Rev. Lett.* **64**, 847 (1990); T. Abbott et al., *Phys. Rev.* **C50**, 1024 (1994).
10. C. Ogilvie and J. Dunlop, Private Communication.
11. H. Sorge, H. Stocker and W. Greiner, *Ann. Phys.* **192**, 266 (1989).
12. M. Lisa, these proceedings.
13. D. Rischke, *Nucl. Phys.* **A610**, 88c (1996).
14. J. Barrette et al., *Phys. Rev.* **C55**, 1420 (1997); W.C. Chang, these proceedings.
15. T. Ullrich for the Ceres Collaboration, *Nucl. Phys.* **A610**, 317c (1996); A. Drees, *Nucl. Phys.* **A610**, 536c (1996); A. Drees, these proceedings.
16. V. Koch and C. Song, preprint LBL-38619, nucl-th/9606028; G.Q. Li, C.M. Ko, and G. Brown, *Phys. Rev. Lett.* **75**, 4007 (1995); W. Cassing, W. Ehehalt, and C.M. Ko, *Phys. Lett.*, **B363** 35 (1995); D.K. Srivastava, B. Sinha, and C. Gale, *Phys. Rev.* **C53** R567 (1996).
17. G.E. Brown and M. Rho, *Phys. Rev. Lett.* **66**, 2720 (1991).
18. W. Cassing, W. Ehehalt, and C.M. Ko, *Phys. Lett.*, **B363** 35 (1995)
19. T. Matsui and H, Satz, *Phys. Lett.* **B178**, 416 (1986).
20. M. Gonin for the NA50 Collaboration, *Nucl. Phys.* **A610**, 404c (1996); C. Gerschel, private communication; C. Baglin, these proceedings.
21. S. Gavin and R. Vogt, *Nucl. Phys.* **A610**, 442c (1996).
22. D. Kharzeev, these proceedings.

Signatures and Status of the Quark-Gluon Plasma

Berndt Müller

Department of Physics, Duke University, Durham, NC 27708-0305

Abstract. I review new theoretical results concerning the QCD equation of state and the formation of a dense parton plasma in high energy collisions of nuclei. I also survey experimental signatures of a quark-gluon plasma in the light of present and future experiments.

I INTRODUCTION

In the preceding talk, Tom Hemmick [1] surveyed the status of relativistic heavy ion experiments with fixed targets at the Brookhaven AGS and the CERN-SPS. The observed hadron yields and spectra indicate that the final state—at the moment of break-up—is approximately thermal, with temperatures in the range of 120–160 MeV. More interesting dynamical information about earlier stages of the reaction is provided by three phenomena: Anisotropics of the transverse momentum spectra of emitted particles are evidence of directed and radial collective flow caused by the high pressure of heated and compressed nuclear matter. Enhanced emission of lepton pairs with invariant masses below and above the vector meson resonance region may indicate medium modifications of the rho meson in dense matter. Anomalously large suppression of charmonium production in Pb + Pb collisions appears to reveal a new medium effect, maybe even the formation of color deconfined matter. Further experiments will hopefully tell us whether a thermalized quark-gluon plasma is formed in these reactions.

The theoretical challenge appears quite different at the higher energies where the future heavy ion colliders, RHIC and LHC, will operate. New theoretical developments over the past few years have raised the hope that reliable calculations of the initial phase of nuclear reactions at high energies will eventually be possible. Numerical simulations of lattice QCD are approaching the stage where quantitatively reliable predictions of the equation of state of strongly interacting matter and its dynamical properties is coming forward. The better understanding of parton cascades and nuclear structure functions at small x opens up the prospect of a seamless description of the reaction dynamics within the framework of QCD from the initial nuclear structure up to a thermalized quark-gluon plasma.

In this talk I will briefly review the current status of lattice Monte-Carlo computations of the equation of state of QCD and discuss our current understanding of the thermalization process at collider energies, including recent new theoretical developments. After that, I will review the status of quark-gluon plasma signatures.

II THE QCD PHASE DIAGRAM

Rigorous results about the phase diagram of QCD are based on numerical Monte-Carlo simulations of lattice-QCD at finite temperature. For the pure SU(3)-gauge theory these simulations have reached an impressive level of precision. Calculations by the Bielefeld group [2] on lattices up to $32^3 \times 8$ yield a critical temperature extrapolated to infinite volume of

$$T_c/\sqrt{\sigma} = 0.629 \pm 0.003, \tag{1}$$

or $T_c = 260$ MeV if the value of the string tension σ derived from the Regge slope is used. The phase transition is of first order with a latent heat of slightly less than $2T_c^4$.

The simulations also provide information about quantities of dynamical relevance, such as the speed of sound, which is found rising from $c_s^2 < 0.1$ near T_c to $c_s^2 \approx 1/3$ at $T = 5T_c$.

Simulations of full lattice-QCD with light dynamical quarks have not yet reached this advanced stage, but have also made significant progress. State-of-the-art calculations [3] show a steep rise in the quantity ϵ/T^4 between 150 and 160 MeV, indicating a rapid unthawing of new degrees of freedom. It still remains unclear whether thermodynamic quantities will show a singularity at T_c in the thermodynamic limit and what order phase transition it may be. The simulations also clearly demonstrate that the quark condensate $\langle \bar{\psi}\psi \rangle$ drops steeply at exactly the same point: color deconfinement and chiral symmetry restoration occur at the same temperature. The results are expected to reach a similar quality as those presently existing for the pure gauge theory when the next generation of parallel computers with close to teraflops performance will become available.

Perturbative approaches, valid far above T_c, have also made significant progress. The equation of state is now known [4,5] up to order g^5, and a scheme yielding a complete perturbative result with only minimal lattice input has been worked out [6]. It is now understood that the convergence of the perturbative series (even in the sense of an asymptotic series) requires $g \leq 1$ or $\alpha_s(T) \leq 0.1$, which only holds for temperatures above the electroweak unification scale [7]. Nevertheless, thermal perturbation theory provides important insight into the dynamics of the high-temperature phase of QCD.

III INITIAL CONDITIONS AT RHIC AND LHC

Most recent theoretical predictions for the initial conditions at which a thermalized quark-gluon plasma will be produced at heavy ion colliders are based on the concept of perturbative partonic cascades. The parton cascade model [8,9] starts from a set of relativistic transport equations of the form [10]

$$p^\mu \frac{\partial}{\partial x^\mu} F_i(x,p) = C_i(x,p|F_k) \qquad (2)$$

$$p^2 \frac{\partial}{\partial p^2} F_i(x,p) = S_i(x,p|F_k) \qquad (3)$$

where $F_i(x,p)$ denote the phase space distributions of QCD quanta. The collision terms C_i are obtained in the framework of perturbative QCD from elementary $2 \to 2$ scattering amplitudes. The splitting terms S_i describe additional initial- and final state radiation due to scale evolution of the perturbative quanta, as well as saturation effects at high density due to parton fusion. To regulate infrared divergences, the original parton cascade model required a cut-off for the $2 \to 2$ scattering (usually $p_T^{\min} = 1.5 - 2$ GeV/c) and a cut-off for time-like branchings ($\mu_0^2 = 0.5 - 1$ GeV$^2/c^2$).

Numerical simulations of such cascades for heavy nuclei reveal a scenario where a dense plasma of (predominantly) gluons develops in the central rapidity region between the two colliding nuclei shortly after the impact. Detailed studies [11,12] show that the momentum spectrum of partons becomes isotropic and exponential, i.e. practically thermal, within a time $\tau \ll 1$ fm/c measured as proper time in the comoving rest frame. For the very high initial energy densities predicted by the parton cascade model one concludes that a thermal hydrodynamical picture makes sense after $\tau_i \approx 0.3$ fm/c.

The high density of scattered partons in $A+A$ collisions allows to replace the arbitrary infrared cut-off parameters $p_T^{\min}$ and μ_0^2 by dynamically calculated medium-induced cut-offs [12]. The dynamical density-dependent screening of color forces eliminates the need for $p_T^{\min}$, and the suppression of radiative processes provided by the Landau-Pomeranchuk-Migdal (LPM) effect renders the virtuality cut-off μ_0^2 unnecessary. The viability of this concept depends on the presence of a dense parton medium, because the dynamical cut-off parameters must lie in the range of applicability of perturbative QCD. Since the density of initially scattered partons grows as $(A_1 A_2)^{1/3} (\ln s)^2$, this condition requires both large nuclei and high collision energy. The calculations indicate that this criterion will be met at RHIC and LHC, but becomes questionable at the presently accessible energies of the SPS. The framework is not applicable to

pp or $p\bar{p}$ collisions at current energies because the parton density remains too low.

The dynamic screening of parton cascades has been implemented in several different ways. In one approach [13] the "hard" parton interactions with large momentum transfer p_T are considered as effectively screening the softer ones. The screening mass applicable to parton collisions of scale p_T is then determined self-consistently from the screening effect of the more violently scattered partons. For Au + Au collisions at RHIC energy, the screening mass saturates somewhat below 1 GeV at small p_T, and around 1.5 GeV for Pb + Pb collisions at the LHC. Both these values are comfortably within the range of applicability of perturbative QCD, demonstrating that there may be no need for artificial infrared regulators.

The self-screened parton cascade model predicts initial conditions achieved at RHIC and LHC that safely above the QCD phase transition. For the heaviest nuclei one expects thermalization to occur at $T = 730$ MeV (RHIC) or $T_c = 1150$ MeV (LHC) and initial energy densities of order 60 GeV/fm^3 (RHIC) or 430 GeV/fm^3 (LHC). At the initial instant ($\tau_i \approx 0.25$ fm/c) the parton plasma is very hot but tenuous. It takes another several fm/c to achieve full chemical equilibrium. Following the evolution with the hydrodynamical model including self-consistent screening, the quark-gluon plasma is expected to last for about 5 fm/c at RHIC and about 10 fm/c at the LHC [14].

Another approach to consider medium effects has been implemented in a numerical realization of the parton cascade model [15]. Here soft collisions between partons are suppressed due to coherence effects. This consideration, similar to the LPM effect, sets an effective lower limit on the momentum exchange between partons in a dense medium. The model has recently been applied to nuclear collisions at the SPS [16]. In the case of Pb + Pb collisions about half-the particle production at midrapidity is found to originate from parton scatterings, the balance coming from the fragmentation of the beam remnants (unscattered partons). Combined with a hadronization model [17] developed to describe jet fragmentation in e^+e^- annihilation, a complete description of the space-time evolution of the nuclear collision smoothly interpolating between partonic and hadronic phases can be given. Results [16] show that the partonic phase reaches an energy density of 5 GeV/fm^3 (corresponding to $T \approx 210$ MeV in a fully equilibrated quark-gluon plasma) in Pb + Pb, but only 2 GeV/fm^3 (corresponding to $T \approx 170$ MeV) in S + Au.

While representing a major advance in the application of perturbative QCD to relativisitic nuclear collisions, self-screened parton cascade models still rely on the input of experimentally measured parton structure functions of the colliding nuclei. A new approach [18] promises to make these structure functions themselves calculable. The basic idea underlying this approach is that, as seen by partons with $x \leq 10^{-2}$, the valence quarks in a heavy nucleus constitute a very dense, sheet-like random color source. The large area density $\rho \approx 3A/\pi R^2$ defines a large scale parameter $\mu^2 = \rho$ at which the QCD coupling $\alpha_s(\mu^2)$ is weak. It is then possible to formulate a systematic program for calculating gluon and quark (sea) structure functions at small x as generated by classical random color fields and their quantum fluctuations. Loop corrections will introduce $(\ln x)/x$ contributions, possibly leading to a power-like behavior at small x upon resummation [19].

The random light-cone source model can be extended to the collision between two nuclei, viewed as the interaction among two counter-propagating sheets of valence quarks [20,21]. Since screening effects are already included in this approach, it may provide a complementary description of the initial state produced in a nuclear collision at RHIC or LHC.

A completely different approach to the thermalization problem in non-Abelian gauge theories emerges from solutions of the classical Yang-Mills equations. Numerical calculations have shown that these equations form a strongly chaotic, infinite-dimensional dynamical system [22,23]. The evidence is derived from the study of the divergence between two initially neighboring field configurations as a function of time:

$$\|A_1^\mu(t) - A_2^\mu(t)\| \sim e^{\lambda t}. \tag{4}$$

The Lyapunov exponent λ can be obtained from numerical solutions of the Yang-Mills equations on a lattice in Minkowski space. These calculations also show that the lattice fields thermalize in the microcanonical sense.

The Lyapunov exponents provide a measure of the rate of entropy growth and, hence, of the thermal equilibration time. Using the largest Lyapunov exponent for the SU(3) gauge theory [23], one finds a characteristic time of order 0.2 - 0.3 fm/c in the range of temperatures relevant for RHIC. This agrees nicely with the results from parton cascade simulations.

IV QUARK-GLUON PLASMA SIGNATURES

A wide variety of signatures for the formation of a quark-gluon plasma in heavy-ion collisions has been proposed. Considerable progress has been made in recent years in understanding the background to many of these signals [24,25]. Here, I restrict my discussion to four types of signatures: the time delay caused by a first-order phase transition, charmonium suppression, jet quenching, and disoriented chiral condensates.

The steep rise in the active number of degrees of freedom predicted by the QCD lattice simulations will cause a significant delay in the expansion of the hot matter when it reaches the critical temperature. A true first-order phase transition would easily increase the total lifetime of the fireball threefold, and even a smooth cross-over as indicated by recent numerical results would still double the total lifetime. Such a delay would be revealed by differences in the identical particle correlation functions in the outward direction (toward the detector) and in the direction sideways to both beam and detector [26]. A detailed analysis [27] of the prospects of using this effect, called HBT interferometry, to identify a phase transition or rapid cross-over confirms that the ratio of measured outward/sideways radius parameters faithfully tracks the expected time delay as a function of initial energy density. The effect would be best visible if the initial energy density would exceed the critical density by a least a factor 10, as predicted for RHIC.

Recent progress in the theory of heavy quarkonium production at the Tevatron [28–30], has clarified many puzzling aspects of charmonium production on nuclear targets. The suppression of charmonium states observed in $p + A$ collisions is now understood as being due to the absorption of the dominant color octet component $[c\bar{c}]^{(8)}$ in the nuclear target [31]. This provides a well-defined benchmark against which additional suppression effects can be identified.

The experimental results [32] for nucleus-nucleus collision at the SPS from NA38/50 show no evidence of additional absorption of the J/ψ in S + U collision, but clear evidence in Pb + Pb. On the other hand, additional suppression effects are clearly seen in both collision systems for the weakly bound ψ' state. This fits nearly into a picture where the energy density reached in even the most central S + U events is below the threshold ($T \approx 1.2 T_c$) [33] required for the dissociation of the J/ψ due to color screening, whereas this critical energy density is surpassed in central and semiperipheral Pb + Pb collisions [34,35]. It also matches well with the recent results from the parton cascade model [16] showing a much higher parton density created in Pb + Pb than in S + U (see previous section).

On the other hand, the possibility of additional absorption by comoving hadrons (rather than partons) must be seriously considered. Leaving the J/ψ-hadron cross section as free parameter, some theorists [36,37] have found it possible to explain the full set of data from NA38/50, while others [35] have not. Although further experimental data may shed more light on this issue, the problem of charmonium interaction with mesons ($\pi, \rho, \ldots$) is interesting from a theoretical point of view. Exploratory calculations within an effective theory of J/ψ interactions with π and ρ mesons indicate that these mesons remain ineffective as absorbers of J/ψ particles unless the hadron gas temperature reaches at least 300 MeV [38]. This is different for the easily absorbed ψ' [39].

As of late, there has been significant new insight into the mechanism of energy loss of hard partons in a quark-gluon plasma [40]. The decisive difference between radiative energy loss of a fast charged particle in QED and the analogous process in QCD is that radiated gluons rescatter in the medium whereas photons, in first approximation, do not. The gluon rescattering effectively reduces the coherence length for the emission of radiation by increasing the transverse momentum of the radiated gluon. As the average k_T^2 of the radiation increases with the target thickness, this leads to the prediction of an energy loss that grows like the square of the length of the traversed medium. Estimates for the effective rate of radiative energy loss by a fast quark traversing a layer of quark-gluon plasma 10 fm thick lie in the range 2-3 Gev/fm.

This would result in a dramatic "quenching" of jets seen in nuclear collisions at RHIC which could be observed by its effect on leading particle spectra or in tagged photon events [41].

The formation of chirally disorientated domains of the quark condensate in the vacuum is a new, intriguing signature of the chiral phase transition [42]. Such domains correspond to coherent excitations of the pion field that would decay by producing large fluctuations away from 1/3 in the π^0/π ratio. Detailed numerical studies [43] of the linear sigma model have shown how disorientated domains can grow if the chiral transition occurs rapidly [44], producing a temporarily unstable state of the chiral order parameter $\langle\bar\psi\psi\rangle$. Sophisticated techniques for the identification of domain structures, such as wavelet analysis [45], may help observing these domains if they are produced experimentally.

An amusing aspect peculiar to collisions between heavy nuclei is due to the simultaneous presence of strong electric and magnetic fields in semipheripheral collisions. These fields intereact with the neutral pion field through the axial anomaly

$$\mathcal{L}_a = -\frac{\alpha}{\pi f_\pi}\vec{E}\cdot\vec{H}\pi_3. \qquad (5)$$

The anomalous interaction causes an off-set in the chiral condensate with alternating sign above and below the collision plane [46]. At RHIC, the anomaly induces a momentum kick in the neutral pion field when the nuclei collide. Although the effect is small, it may noticeably influence the DCC domain formation because it is spatially coherent over distance much larger than m_π^{-1} [47].

V SUMMARY

As RHIC proceeds toward completion, we are getting a much clearer view of the physics to be expected in heavy ion collisions at high energy. Lattice-QCD simulations have unambiguously established the rapid "unthawing" of color and the restoration of chiral symmetry at a temperature around 150 MeV. Calculations of parton transport processes based on perturbative QCD predict very high initial temperatures, above 500 MeV at RHIC and 1 GeV at the LHC. New approaches to low-x parton structure and the thermalization problem hold the promise of an ab-initio description of the formation and evolution of a quark-gluon plasma at RHIC energy and beyond. Finally, the various plasma signatures are becoming much better understood, providing valuable guidance for the experimental program at RHIC which will start in the year 1999.

ACKNOWLEDGMENTS

This work was supported in part by a grant from the U.S. Department of Energy, Office of Energy Research (DE-FG02-96ER40945).

REFERENCES

1. T. Hemmick, these Proceedings.
2. G. Boyd, et al., *Nucl. Phys.* **B469**, 419 (1996).
3. C. Bernard. et al., *Phys. Rev.* **D54**, 4585 (1996).
4. P. Arnold and C.X. Zhai, *Phys. Rev.* **51**, 1906 (1995).
5. C.X. Zhai and B. Kastening, *Phys. Rev.* **D52**, 7232 (1995).
6. E. Braaten and A. Nieto, *Phys. Rev.* **D51**, 6990 (1995).
7. E. Braaten and A. Nieto, *Phys. Rev. Lett.* **76**, 1417 (1996).
8. K. Geiger and B. Müller, *Nucl. Phys.* **B369**, 600 (1992).
9. K. Geiger, *Phys. Rep.* **258**, 237 (1995).
10. K. Geiger, *Phys. Rev.* **D54**, 949 (1996).
11. K.J. Eskola and X.N. Wang, *Phys. Rev.* **D49**, 1284 (1994).
12. T.S. Biró, et al., *Phys. Rev.* **C48**, 1275 (1993).
13. K.J. Eskola, B. Müller, and X.N. Wang, *Phys. Lett.* **B374**, 20 (1996).
14. D.K. Srivastava, M.G. Mustafa, and B. Müller, *Phys. Lett.* **B396**, 45 (1997).
15. K. Geiger, VNI3.1, ⟨hep-ph/9701226⟩.
16. K. Geiger and D.K. Srivastava, ⟨nucl-th/9706002⟩.
17. J. Ellis and K. Geiger, *Phys. Rev.* **D52**, 1500 (1995).
18. L. McLerran and R. Venugopalan, *Phys. Rev.* **D49**, 2233 and 3352 (1994).
19. J. Jalilian-Marian, A. Kovner, L. McLerran, and H. Weigert, *Phys. Rev.* **D55**, 5414 (1997).
20. A. Kovner, L. McLerran, and H. Weigert, *Phys. Rev.* **D52**, 3809 and 6231 (1995).
21. Yu. Kovchegov and D. Rischke, ⟨hep-ph/9704201⟩, *Phys. Rev. D* (in print).
22. B. Müller and A. Trayanov, *Phys. Rev. Lett.* **68**, 3387 (1992).
23. C. Gong, *Phys. Lett.* **B298**, 257 (1993).
24. C.P. Singh, *Phys. Rep.* **236**, 147 (1993); B. Müller, *Rep. Prog. Phys.* **58**, 611 (1995).
25. J.W. Harris and B. Müller, *Annu. Rev. Nucl. Part. Sci.* **46**, 71 (1996).
26. S. Pratt, *Phys. Rev. Lett.* **53**, 1219 (1994).
27. D.H. Rischke and M. Gyulassy, *Nucl. Phys.* **A608**, 479 (1996).
28. E. Braaten and S. Fleming, *Phys. Rev. Lett.* **74**, 3327 (1995).
29. P. Cho and A.K. Leibovich, *Phys. Rev.* **D53**, 150 and 6203 (1996).
30. M. Beneke and M. Kramer, *Phys. Rev.* **D55**, 5269 (1997).
31. D. Kharzeev and H. Satz, *Phys. Lett.* **B366**, 316 (1996).
32. M.C. Abreu, et al. [NA50 collaboration], *Nucl. Phys.* **A610**, 404c (1996).
33. F. Karsch and H. Satz, *Z. Phys.* **C51**, 209 (1991).
34. J.P. Blaizot and J.Y. Ollitrault, *Phys. Rev. Lett.* **77**, 1703 (1996).

35. D. Kharzeev, C. Lourenco, M. Nardi, and H. Satz, ⟨hep-ph/9612217⟩.
36. N. Armesto and A. Caprella, ⟨hep-ph/9705275⟩.
37. W. Cassing and E.L. Bratkovskaya, ⟨hep-ph/9705257⟩.
38. S.G. Matinyan and B. Müller, work in progress.
39. C.Y. Wong, *Phys. Rev. Lett.* **76**, 196 (1996).
40. R. Baier, et al., *Phys. Lett.* **B345**, 277 (1995); *Nucl. Phys.* **B483**, 291 (1997).
41. X.N. Wang, Z. Huang, and I. Sarcevic, *Phys. Rev. Lett.* **77**, 231 (1996).
42. K. Rajagopal and F. Wilczek, *Nucl. Phys.* **B404**, 577 (1993).
43. M. Asakawa, Z. Huang, and X.N. Wang, *Phys. Rev. Lett.* **74**, 3126 (1995).
44. J. Randrup, *Phys. Rev. Lett.* **77**, 1226 (1996).
45. Z. Huang, I. Sarcevic, R. Thews, and X.N. Wang, *Phys. Rev.* **D54**, 750 (1996).
46. H. Minakata and B. Müller, *Phys. Lett.* **B377**, 135 (1996).
47. M. Asakawa, H. Minakata, and B. Müller, work in progress.

Meson Spectroscopy and the Search for Exotics

Curtis A. Meyer

Carnegie Mellon University, Pittsburgh, Pennsylvania 15213

Abstract. I will review the current status of exotic hadrons in the meson sector. There is currently strong evidence that a scalar glueball mixed into the normal scalar mesons has been found. There is also an interesting candidate for the tensor glueball state. Finally, I will discuss hybrid mesons with explicitly exotic quantum numbers.

INTRODUCTION

In the context of this paper I will review the current status of exotic mesons; glueballs and hybrid mesons. Our current best evidence for a glueball comes from the scalar meson sector, $J^{PC} = 0^{++}$. In this sector there appears to be two states, the $f_0(1500)$ and $f_0(1750)$ which are produced in several glue–rich mechanisms. The decay rates of the $f_0(1500)$ have been studied in detail by the Crystal Barrel experiment which finds that this state cannot be explained as either a pure glueball nor a pure meson. The $f_0(1500)$ combined with the $f_0(1750)$ leads us to the interpretation that these states are mixtures of both glueball and normal mesons. In the tensor sector, there is currently a very interesting candidate state, the $\xi(2230)$. Its best evidence comes from radiative J/ψ decays, and appears to have several properties suggestive of glueballs. Lastly, there is evidence from E852 at Brookhaven for a state with explicitly exotic quantum numbers. However several properties of this state seem to disagree with expectations for a hybrid meson.

GLUEBALLS

The Theoretical Situation

A pure glueball is a bound state of only gluons. Due to the fact that gluons carry the color charges of QCD, it is theoretically possible for them to form bound states devoid of any quark content. The quantum numbers of these states are derived by considering them to contain either 2 or 3 *valence* gluons. While we are still unable to solve QCD exactly in the nonperturbative regime, most models which are able to explain observed phenomena predict that glueballs should exist, and most predict the lightest will be the scalar, $J^{PC} = 0^{++}$. At the moment, we believe that lattice calculations come closest to actually solving non–perturbative QCD. A calculation of the entire glueball spectrum [1] finds that the scalar glueball has a mass of 1550 ± 50 MeV/c^2 while the next lightest state is the tensor at a mass of 2270 ± 100 MeV/c^2. A recent calculation with a factor of 10 improvement in lattice density predicts that the scalar glueball has a mass of 1740 ± 71 MeV/c^2, and for the first time makes predictions for decay rates into flavorless pseudoscalar meson pairs as a function of mass [2]. Given that the two groups have a different procedures to extrapolate to the continuum limit, we take the average of these, $1610\pm70\pm130$ MeV/c^2, as the prediction for the mass of a pure scalar glueball.

Next we consider where we should search for glueballs. There are several reactions which are considered as glue–rich. Radiative J/ψ decays, $\psi \to \gamma X$ are generally considered the best source of glueballs simply because the c and $\bar{c}$ quark have to annihilate in order for the decay to proceed. Defining $b(R_J \to xx) = \Gamma(R_J \to xx)/\Gamma_{\text{tot}}$, expectations from [4] give:

$$b(R(\bar{q}q) \to gg) \simeq 0.1 \sim 0.2 \qquad (1)$$
$$b(R(G) \to gg) \simeq 0.5 \sim 1. \qquad (2)$$

Where equation 1 is for a normal meson and equation 2 is for a glueball. These quantities can be related to the radiative decay rate via equations 3 and 4.

$$(10^3)B(J/\psi \to \gamma R(0^{++})) = \left(\frac{m}{1500 MeV}\right)\left(\frac{\Gamma_{R\to gg}}{96 MeV}\right)\frac{x \mid H_T \mid^2}{35} \qquad (3)$$

$$(10^3)B(J/\psi \to \gamma R(2^{++})) = \left(\frac{m}{1500 MeV}\right)\left(\frac{\Gamma_{R\to gg}}{26 MeV}\right)\frac{x \mid H_T \mid^2}{34} \qquad (4)$$

In a similar argument, $\bar{p}p$ annihilations are also considered a likely source of glueballs simply because there are so many quarks and antiquarks. It is unfortunately difficult to predict rates, and any signal will be mixed into a background of normal mesons. Finally double pomeron exchange in central production, $pp \to p_f \mathcal{G} p_s$ is believed to be a likely source of glueballs as the pomeron seems to involve glue. In addition to glue rich sources, there are glue–poor source such as 2γ and photoproduction. There is no direct coupling between the photon and the electrically neutral gluons, so production of glueballs should be suppressed.

Finally, the glueballs are expected to have a decay pattern which in some sense is flavour blind. The gluon couples equally to all flavours of quarks, so the production of u, d and s quarks are expected to be more or less the same. One calculation [3] yields the values in table 1 as the expected relative strengths of for two-pseudoscalar decays. In addition, the 4π decay involving two pairs of $I = 0$ s-wave dipions, $(\pi\pi)_s$ is expected to be a significant and large decay mode of a glueball [5]. This combined with the two-pseudoscalar decays yields the following expectations.

Decay	$\pi\pi$	$\bar{K}K$	$\eta\eta$	$\eta'\eta$	$(\pi\pi)_s(\pi\pi)_s$
Rate	(3)	(4)	(1)	(0)	Large

TABLE 1. Predicted glueball decay rates.

In particular with the scalar glueball there is the problem of the nearby scalar mesons. Excluding the $a_0(980)$ and $f_0(980)$ from consideration, the scalar nonet is presumably made up of the $f_0(1370)$, $a_0(1450)$, $K_0^*(1430)$ and an as yet unidentified f_0' state. The $a_0(1450)$ is a recently identified state with a mass of 1450 ± 40 and width of 270 ± 40 observed in $\bar{p}p$ annihilation at rest and decaying into $\eta\pi$, $\eta'\pi$ and $\bar{K}K$ [6], [7], [8]. Enough of this nonet is known to allow us to make predictions both on the mass and decay rates of the missing f_0' state as a function of the nonet mixing angle. If we find additional scalar states, we should be able to identify they are pure meson or pure glueball.

The Scalar Sector

The Crystal Barrel experiment at LEAR has done a high statistics study of $\bar{p}p$ annihilations at rest into both charged and neutral final states using a nearly 4π solid angle detector for both charged particles and photons. These data have been analyzed in a consistent coupled channel analysis. The analysis is done within the framework of the isobar model using a K-matrix formulation to maintain unitarity and handle multiple decay modes of a given meson [7], [9], [10]. Dalitz plots for several of these final states are shown in figures 1 and 2. In particular, $\bar{p}p \to \pi^\circ\pi^\circ\pi^\circ$ ($\sim 700,000$ events), $\bar{p}p \to \eta\eta\pi^\circ$ ($\sim 200,000$ events) and $\bar{p}p \to K_L K_L \pi^\circ$ ($\sim 50,000$ events) show a new isoscalar scalar state, $((I^G)J^{PC} = (0^+)0^{++})$, the $f_0(1500)$. Its mass and width are found to be $m = 1500 \pm 15 \text{MeV}/c^2$ and $\Gamma = 120 \pm 20 \text{MeV}/c^2$. In addition to 3-pseudoscalar final states, $\bar{p}p \to \pi^\circ\pi^\circ\pi^\circ\pi^\circ\pi^\circ$ has also been studied [11]. These data show evidence for $f_0(1500) \to (\pi\pi)_s(\pi\pi)_s$. These consistent analyses yield the phase space corrected decay rates for the $f_0(1500)$ as given in table 2.

Using an SU(3) calculation which accurately predicts the relative decay rates for the tensor mesons, we can see if the Crystal Barrel decay rates of the $f_0(1500)$ are consistent with the f_0'. Figure 3 shows the predicted phase space corrected decay rates normalized to the $\eta\eta$ decay rate as a function of the scalar meson mixing angle. The Crystal Barrel data are shown as a horizontal band, and the allowed mixing angles are shown as the shaded

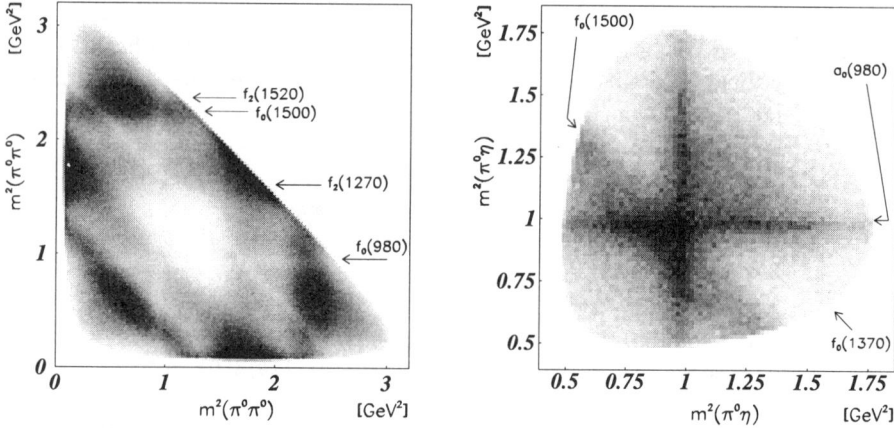

FIGURE 1. The Dalitz plots for $\bar{p}p \to \pi^0\pi^0\pi^0$ (left) and $\pi^0\eta\eta$ (right). The $f_0(1500)$ is seen clearly as the band near 2.25 GeV^2 in the $3\pi^0$ Dalitz plot. It is also seen as the lower diagonal band in the $\pi^0\eta\eta$ Dalitz plot.

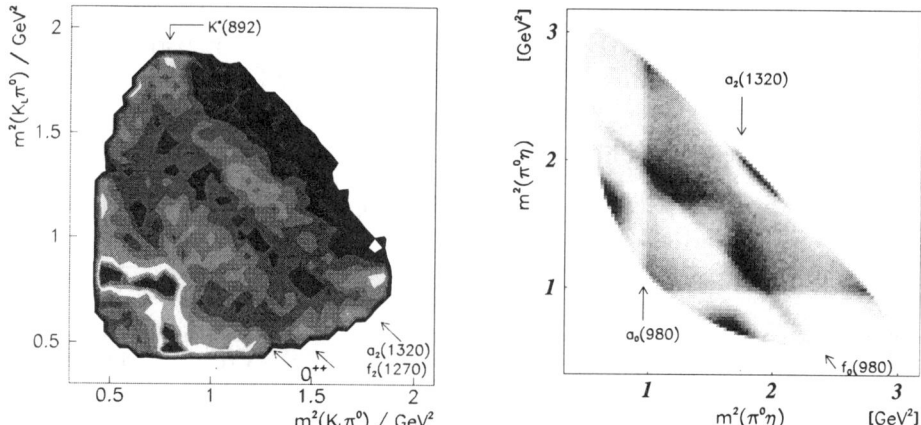

FIGURE 2. The Dalitz plots for $\bar{p}p \to K_L K_L \pi^0$ (left) and $\pi^0\pi^0\eta$ (right). The $f_0(1500)$ is seen labeled as the 0^{++} band in the $K_L K_L \pi^0$ Dalitz plot. The $\eta\pi^0\pi^0$ Dalitz plot is used to constrain amplitudes in other channels.

boxes along the mixing angle axes. From the $\pi\pi$ and $\eta'\eta$ plots, a consistent value of $\theta_s = (68.5 \pm 1.5)°$ is found. Using this value to predict $\bar{K}K$, we find the region shown by the shaded cross, and expect a decay ratio of about 10, nearly 9 times larger than the measured Crystal Barrel Rate. The $f_0(1500)$ cannot be the missing f'_0 state. In addition, in table 2 we compare this with the pure glueball. This is also a poor explanation for this object, the $f_0(1500)$ appears to be neither a pure meson nor a pure glueball.

Decay	$\pi\pi$	$\bar{K}K$	$\eta\eta$	$\eta'\eta$	$(\pi\pi)_s(\pi\pi)_s$
$f_0(1500)$	$(4.39 \pm .16)$	$(1.1 \pm .4)$	(1)	$(1.42 \pm .96)$	(14.9 ± 3.2)
GlueBall	(3)	(4)	(1)	(0)	Large
f'_0	(4.4)	(10)	(1)	(2)	

TABLE 2. Measured $f_0(1500)$ decay rates compared to a pure glueball and the expected f'_0 state.

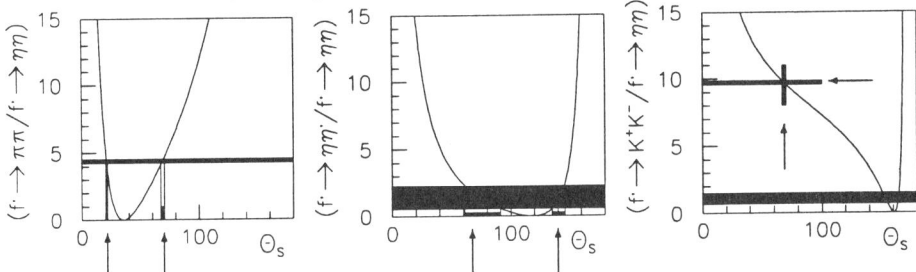

FIGURE 3. Predicted decay rates of the f'_0 normalized to $\eta\eta$ as a function of the scalar mixing angle θ_s. The horizontal shaded bands are the Crystal Barrel limits. For $\pi\pi$ and $\eta'\eta$ the allowed values of θ_s are indicated by the arrows. In $\bar{K}K$ the consistent values of θ_s from the previous two (upward pointing arrow) are used to predict a ratio. This is indicated by the left pointing arrow near a ratio of 9.5.

Finally a third scalar state, the $f_0(1750) \to (\pi\pi)_s(\pi\pi)_s$ is hinted at in the $\bar{p}p \to \pi°\pi°\pi°\pi°\pi°$ final state [11]. However, this state is near the edge of available phase space and is so far only observed in the one decay mode. However as we examine radiative J/ψ decays we find that this $f_0(1750)$ is probably the scalar component of the $f_J(1710)$. A reanalysis of MKIII data [12] on $J\psi \to \gamma\pi^+\pi^-\pi^+\pi^-$ finds two scalar states both decaying to $(\pi\pi)_s(\pi\pi)_s$. The $f_0(1500)$ and the $f_0(1750)$, ($m = 1750 \pm 15$, $\Gamma = 160 \pm 40$). In addition they find a 2^{++} state at a mass of $1620 \pm 16 \text{MeV}/c^2$ and a width $\Gamma = 140^{+60}_{-20}$. These latter two states both come from the $f_J(1710)$ region. They find the radiative decay rates to the scalars as given in 5 and 6.

$$B(J\psi \to \gamma f_0(1500) \to \gamma 4\pi) = (5.7 \pm 0.8) \times 10^{-4} \quad (5)$$
$$B(J\psi \to \gamma f_0(1750) \to \gamma 4\pi) = (9.0 \pm 1.3) \times 10^{-4} \quad (6)$$

Recently, BES has reported both a tensor and scalar state in the $f_J(1710)$ region, $f_2(1690)$ ($m = 1696 \pm 5^{+9}_{-34}$, $\Gamma = 103 \pm 8^{+10}_{-31}$) and $f_0(1780)$ ($m =$

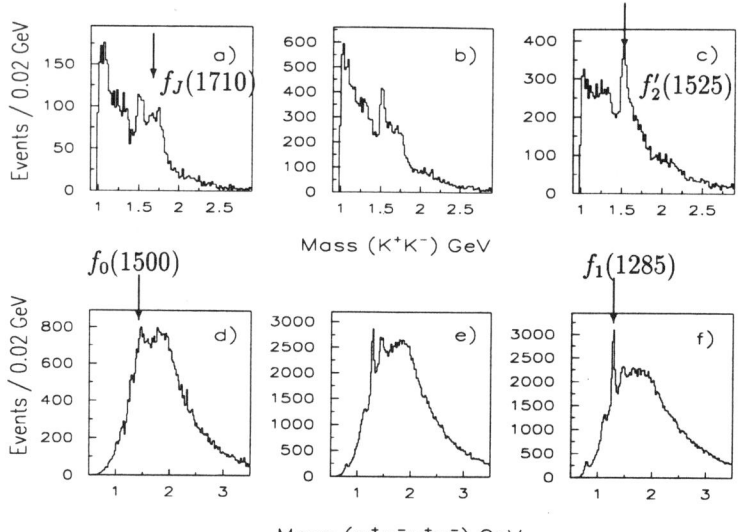

FIGURE 4. (a,d) $dP_T < 0.2$, (b,e) $0.2 < dP_T 0.5$ and (c,f) $dP_T > 0.5$, (see text).

$1781 \pm 8^{+10}_{-31}$, $\Gamma = 85 \pm 24^{+22}_{-19}$), both decaying to $\bar{K}K$ [13]. These are presumably the same states as seen in MKIII, and we will adopt the name $f_0(1750)$ for the scalar state. At this conference, BES reported on radiative decays to four pions [14]. They find the results; and 8.

$$B(J\psi \to \gamma f_0(1500) \to \gamma 4\pi) = (4.0 \pm 0.6) \times 10^{-4} \quad (7)$$
$$B(J\psi \to \gamma f_0(1750) \to \gamma 4\pi) = (5.5 \pm 0.8) \times 10^{-4} \quad (8)$$

Taking these together with the fact that 4π is on order of 50% of the decay rate, one finds using equations 3 and 4 that

$$b(f_0(1500) \to gg) \simeq 0.5 \sim 0.8$$
$$b(f_0(1750) \to gg) \simeq 0.5.$$

Both of which are highly suggestive of a large gluonic content in the states [5].

Finally, we examine data from central production — $pp \to p_f(X)p_s$. Both the $f_0(1500)$ and the $f_J(1710)$ have been reported in central production. However a recent observation made by the WA91 and WA102 collaborations [15] yields an interesting separation of normal mesons from the exotic candidates. In central production, double pomeron exchange is believed to be responsible for formation of gluonic states. However, normal mesons are also produced, and separation of these has always been difficult. If we consider the transverse momentum transfer from each proton as P_{T1} and P_{T2}, and the beam axis along z, one can define a variable

$$dP_T = \sqrt{(P_{x1} - P_{x2})^2 + (P_{y1} - P_{y2})^2}. \tag{9}$$

In a sense, this is the difference in transverse momentum between the two pomerons. When this quantity is small one could visualize the two pomerons as traveling together, while when this is large, they are moving apart. Data for $(X) = (K^+K^-)$ and $(\pi^+\pi^-\pi^+\pi^-)$ are shown in figure 4. The interesting feature is that for large dP_T, (**c** and **f**), K^+K^- shows a clear signal for the $f'_2(1525)$ and 4π shows a clear signal for the $f_1(1285)$ — both of which are normal mesons. However in the small dP_T region, (**a** and **d**) both of the previous states are gone and the $f_0(1500)$ and $f_J(1710)$ appear. Having the pomerons moving *together* enhances the exotic candidates, while having them move *apart* enhances the normal mesons.

These three production mechanisms taken together lead one to the conclusion that both the $f_0(1500)$ and the $f_0(1750)$ have a large gluonic component. Given that these are near the scalar mesons, one interpretation is that the pure f_0, f'_0 and $\mathcal{G}$ states have mixed and become the observed $f_0(1370)$, $f_0(1500)$ and $f_0(1750)$ states. Two mixing schemes have been proposed to explain this. A lattice inspired mixing based on computed masses [16] and a more data inspired approach [5]. Both of these claim to explain the measured decay rates of the $f_0(1500)$ state and are quite similar in their predictions. At this point it seems we have very strong evidence for a scalar glueball state mixed into the scalar nonet, and the question is now down to details of mixing.

The Tensor Sector

Assuming we have found the scalar glueball near the lattice predictions, we also can ask about the next state, the tensor glueball. Here the information is not nearly as clear, but a new candidate, the $\xi(2230)$ or $f_J(2230)$ has reemerged in recent years due to new measurements from BES [17]. This state has a mass of 2230 and an extremely narrow width of 20. In addition, its decays appear nearly flavour-blind over $\pi\pi$, $\bar{K}K$ and $\bar{p}p$. It also appears to have a rather large rate for $J/\psi \to \gamma\xi$ which would be indicative of a large gluonic component. However it's spin is currently unclear — being either 2^{++} or 4^{++}. Even though it has been reported by BES to decay to $\bar{p}p$, this state has not been observed in $\bar{p}p$ direct production. Dave Hertzog [18] has given a nice summary of the results and consequences of this at this conference. The current situation is that these measurements are just barely compatible, but as the $\bar{p}p$ limits improve, this may change. In addition, this state has been searched for and not observed in two-photon production [19]. Its nonobservation lends support to the glueball interpretation if this state is indeed a tensor. Finally, BES has recently reported the observation of 4π decays of a state $f_2(2220)$ [14]. They find a spin 2 state at a mass of $m = 2220$ and a width of $\Gamma = 105^{+70}_{-50}$ decaying to $f_2(1270)(\pi\pi)_s$. While the width is larger than that observed for the $\xi(2230)$, if they are the same state it is a measurement of the spin. In addition, they find a radiative rate of $B(J/\psi \to \gamma f_2 \to \gamma 4\pi) = (2.6 \times 10^{-4})$.

HYBRIDS

Hybrids are meson–like states to which a valence gluon has been added. The quantum numbers of these states are given by those of the underlying meson plus the quantum numers of the gluon. These states are predicted in several models, and are expected to appear in nonets. In the *fluxtube* model of Isgur and Paton [20] one expects 8 nonets with quantum numbers $\underline{0^{+-}}$, 0^{-+}, 1^{++}, 1^{+-}, $\underline{1^{-+}}$, 1^{--}, $\underline{2^{+-}}$ and 2^{-+}. What makes these states intriguing is that the underlined quantum numbers are not possible for a $\bar{q}q$ pair. Observation of a state with these quantum numbers would be a *smoking gun* for a non-$\bar{q}q$ state. The masses of the lightest hybrid mesons are expected to be in the 1700 to 1900 MeV/c^2 mass range, and an additional signature is that they are expected to decay into a pair of P and S wave mesons, e.g. $f_1(1285)\pi$, $a_2(1320)\pi$, etc. whereas final states like $\eta\pi$ and $\pi\pi$ should be suppressed. In addition, the hybrids with $\bar{q}q$ quantum numbers might very well mix with their normal meson counterpart. A recent calculation of the decays of all mesons and hybrids [21] withing the framework of the fluxtube model will provide a guide to interpreting these states, but we need to observe a state with non-$\bar{q}q$ quantum numbers to prove the existence of hybrids.

At this conference, Brookhaven E852 has presented evidence for a 1^{-+} state decaying into $\eta\pi$ [22]. This state, the $\pi_1(1370)$ is observed in the reaction $\pi^- p \to \eta\pi^- p$ at 18GeV/c. The state appears from a detailed partial wave analysis, and is seen via its interference with the much stronger $a_2(1320)$ state. The data are most easily explained by the introduction of a resonant state with a mass $m = 1370 \pm 16^{+50}_{-30}$ and a width of $\Gamma = 385 \pm 40^{+65}_{-105}$. In figure 5 are shown the results of their partial wave analysis. The results yields 8 possible solutions, and the ranges of each value are shown as solid vertical bars for the fits. **a** shows the D_+ wave which corresponds to the strong $a_2(1320)$, while in **b** is the resulting amplitude of the P_+ partial wave as a function of $\pi\eta$ mass. The amplitude peaks near 1370, and is about 3% of the strength of the D_+ wave. In **c** is shown the fit phase difference between the two waves. In order to explain this, they postulate that a resonant state, the $\pi_1(1370)$ is produced in addition to the $a_2(1320)$. They allow a relative production phase between these and treat both states as Breit–Wigner resonances. This hypothesis is able to reproduce the data, and yields the 4 phases shown in **d** with the relative production constant over the $\pi\eta$ invariant mass. Their data agree quite well with earlier data from VES in the reaction $\pi^- N \to \eta\pi^- N$ at 37Gev/c. In particular the same phase difference is observed in both data sets.

While there does appear to be something here, it's interpretation is not so

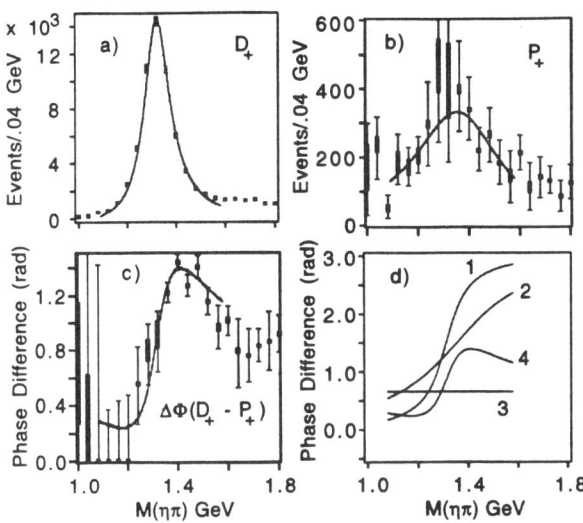

FIGURE 5. a Data and fit for the D_+ wave showing a strong $a_2(1320)$. b Data and fits for the P_+ partial wave. c Relative phase between the D_+ and P_+ partial waves. d (1) Phase of the D_+, (2) P_+, (3) difference and (4) production.

clear. The mass seems too low to be the hybrid meson predicted by the flux tube model, and the decay mode is also not favored. If it is a hybrid meson, then we expect it to be a member of a nonet. In particular we expect to find both an η_1 and an η'_1 state. Depending on their masses, likely decay modes would be either $\eta\eta'$ or $a_1(1260)\pi$. The latter would be particularly difficult to extract from data, but its observation may be quite important as establishing this state as a hybrid meson.

CONCLUSIONS

The scalar meson sector shows strong evidence for a glueball mixed into the normal mesons. Two states, $f_0(1500)$ and $f_0(1750)$ have both been observed in three glue-rich mechanisms. In radiative J/ψ decays, their large rates are indicative of a large gluonic content. In central production, both states stand out from normal mesons via the dP_T in equation 9. Finally, in $\bar{p}p$ annihilation, the relative decay rates of the $f_0(1500)$ have been measured and lead one to the conclusion that the state is neither a pure meson nor a pure glueball. The simplest explanation for what we observe is that the three observed scalars, $f_0(1370)$, $f_0(1500)$ and $f_0(1750)$ are the result of mixing between the pure f_0 and f'_0 with the scalar glueball $\mathcal{G}$. In the tensor sector, we have a rather interesting candidate, the $\xi(2230)$. This state is observed in radiative J/ψ decays with a rate suggestive of a large gluonic content, has a very narrow width $\Gamma \sim 20$MeV and seems to have flavor blind decays. While more details are needed, this does appear as our best candidate for the tensor glueball.

Finally, we have evidence for a state with exotic quantum numbers, $\pi_1(1370)$ observed in $\pi^- p \to \eta \pi^- p$. The state is consistent between two different $\pi^- N$ experiments, but its interpretation is murky. Its mass is much lower than one expects for a hybrid meson, but observation of one of its partners could help clarify this.

ACKNOWLEDGMENTS

Discussions with Z. Li, D. Hertzog, and N. Cason are gratefully acknowledged. This work was supported in part by the U.S. Department of Energy (contract No. DE-FG02-87ER40315).

REFERENCES

1. G. S. Bali, et al., *Phys. Lett. B* **309**, 378 (1993).
2. J. Sexton, A. Vaccarino and D. Weingarten, *Phys. Rev. Lett.* **75**, 4563 (1995).
3. F. E. Close, *Rep. Prog. Phys.* **51**, 833 (1988).
4. Z. Li, *The Signatures of Glueballs in J/ψ Radiative Decays*, this conference.
5. F. E. Close, G. R. Farrar and Z. Li, *Phys. Rev.* **D55**, 5749(1997).
6. C. Amsler et al., *Phys. Lett.* **B333**, 277, (1994).
7. A. Abele et al., *Phys. Lett.* **B385**, 425, (1996).
8. A. Abele et al., Accepted for publication in *Phys. Lett. B* (1997).
9. C. Amsler et al., *Phys. Lett. B* **355**, 425, (1995).
10. C. Amsler et al., *Phys. Lett. B* **340**, 259, (1994).
11. A. Abele et al., *Phys. Lett. B* **380**, 453, (1996).
12. D. V. Bugg, et al., *Phys. Lett. B*—bf 353, 378, (1995).
13. J. Z. Bai, et al., *Phys. Rev. Lett.* **77**, 3959, (1996).
14. Y. Zhu, *Recent Results on Meson Production in Radiative J/ψ Decay*, thise conference.
15. D. Barberis, et al., *CERN/PPE 96-128*, submitted to *Phys. Lett.*, (1996) and D. Barberis, et al., *CERN/PPE 96-197*, submitted to *Phys. Lett.*, (1996).
16. Don Weingarten, *hep-lat/9608070*, (1996).
17. Z. Z. Bai, et al., *Phys. Rev. Lett.* **76**, 3502, (1996).
18. D. Hertzog, *The $\xi(2230)$ and Other States Studied at LEAR*, this conference.
19. R. Galik, *Limits on Two-Photon Production of the $\xi(2230)$ at CLEO*, this conference.
20. N. Isgur and J. Paton, *Phys. Rev.* **D31**, 2910, (1985).
21. T. Barnes, et al., *Phys. Rev.* **D55**, 4157, (1997).
22. N. Cason, *Exotic Meson Signatures from BNL E852*, this conference.

An Incomplete Review of Experimental Charm Physics

Guy Blaylock

Dept. of Physics and Astronomy; LGRT
University of Massachusetts
Amherst, MA 01003

Abstract. This article reviews a selection of recent topics in experimental charm physics, focussing mostly on the rare processes of doubly-Cabbibo-suppressed decays, D^0 mixing, FCNC and lepton number violating decays. A brief discussion of current fixed target charm experiments at FNAL is included.

INTRODUCTION

In the last decade, charm physics has undergone a profound change of focus. Experiments have shifted from basic discovery experiments with a few hundred reconstructed charm decays to experiments with tens or hundreds of thousands of reconstructed charm. These new experiments now concentrate on precision tests and searches for rare phenomena that were previously impossible to explore. Many of the searches for rare phenomena are especially interesting as they relate to physics beyond the standard model. For processes where standard model rates are very low, discovery of a signal may indicate a contribution from new physics. Many of the rare charm processes now being studied fall into this category, and provide sensitive probes of new physics models. The newest charm experiments currently underway expect another order of magnitude increase in statistics, and will pursue these studies even further.

In this context, I will try to summarize a portion of charm physics, emphasizing the comprehensive nature of charm studies with an example from charm lifetime studies, and the intriguing searches for rare phenomena, including doubly-Cabibbo suppressed (DCS) decays, D^0 mixing, and rare and forbidden decays. I will close with a brief discussion of some of the new experiments we should look for in the immediate future.

CHARM LIFETIMES

For an example of the comprehensive work now being done in charm experiments it is useful to look back on the history of charm lifetime studies. Shortly after the charmed D mesons were discovered, it was ascertained that the charged and neutral D's had surprisingly different lifetimes. The D^+ lifetime (1.057 ps) is about 2.5 times longer than the D^0 lifetime (.415 ps). This was surprising in lieu of the expectation that the decay of both particles is dominated by spectator diagrams (figures 1a and b).

Several possible explanations were discussed. The fact that the dominant D^+ final state has two identical $\bar{d}$ antiquarks suggested that interference might play a role. If the interference is destructive, that could explain why the D^+ lifetime is longer. It was also known that there are additional weak decay diagrams which contribute to the decay of one or the other particle, but these

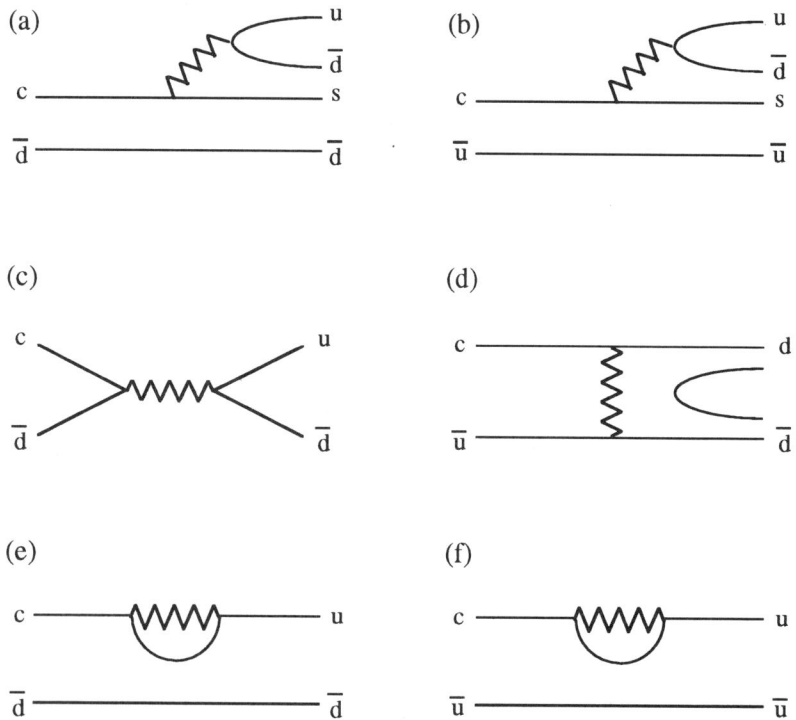

FIGURE 1. Weak decay diagrams which are expected to contribute to the decay of the D^+ (left) and the D^0 (right).

were expected to be suppressed. For the D^+, there is the annihilation diagram (figure 1c) and for the D^0 there is the W exchange diagram (figure 1d), both of which were expected to be helicity and color suppressed. The D^+ annihilation diagram is also Cabibbo suppressed. Finally, there are penguin diagrams (helicity, color and Cabibbo suppressed) which contribute to both decays (figures 1e and f).

To sort out the possible effects, information on weak decays of other charm particles is a clear advantage. Interference effects are expected to follow different patterns for baryons, especially for the Ω_c^0, which has two identical quarks in the initial state. W-exchange diagrams might also be better studied in baryon decays, where they are neither helicity nor color suppressed. The annihilation diagram is not Cabibbo suppressed for the D_s^+, and its contribution might be more readily studied there. Clearly, a comprehensive study of charm decays could be a big help in disentangling the many contributions to the weak decay dynamics.

It is noteworthy that the lifetimes of all seven weakly decaying charm particles have now been measured, including the most recently observed Ω_c^0. Experiment E687 stands out as having dominated these measurements in the last four years. Table 1 shows the lifetime measurements from that experiment, together with the reconstruction yields for the various decays. Yields for the mesons and the Λ_c^+ are in the hundreds or thousands, while yields for the charm-strange baryons remain under a hundred. The most elusive of these particles, the Ω_c^0, has been observed even more convincingly by the WA89 collaboration, who has also measured the Ω_c^0 lifetime [1]. In most cases, these lifetime measurements demonstrate a dramatic improvement in precision over what was previously available.

Current belief states that the lifetime difference of the D^0 and D^+ is largely due to interference effects in D^+ decays. However, some contribution from QCD final state interactions must increase the hadronic rate for both particles to explain the measured semileptonic branching fractions. More precise measurements of the charm-strange baryon lifetimes, as well as measurement of some branching fractions, should help establish this hypothesis.

DCS DECAYS

Doubly-Cabibbo-suppressed (DCS) decays offer another window on the study of weak decay dynamics. For DCS decays of the D^+, all the final state quarks have different flavors, thus eliminating the chance for interference. Recently, both the E791 [2] and E687 [3] experiments have observed the DCS decay $D^+ \to K^+\pi^+\pi^-$. Figure 2 shows the mass plot from E791 data, which contains evidence for both the DCS decay of the D^+ and the singly-suppressed decay of the D_s^+. Both E791 and E687 measure the ratio of DCS to Cabibbo-favored (CF) modes to be $\Gamma_{DCS}(D^+)/\Gamma_{CF}(D^+) \approx 3 \times \tan^4(\theta_c)$, in

TABLE 1. E687 results on lifetimes of the seven weakly decaying charm particles.

Decay Mode	no. events	lifetime (psec)
$D^+ \to K^-\pi^+\pi^+$	≈ 9000	$1.048 \pm 0.015 \pm 0.011$
$D^0 \to K^-\pi^+, K^-\pi^-\pi^+\pi^+$	≈ 16000	$0.413 \pm 0.004 \pm 0.003$
$D_s^+ \to \phi\pi^+$	900 ± 43	$0.475 \pm 0.020 \pm 0.007$
$\Lambda_c^+ \to pK^-\pi^+$	≈ 700	$0.215 \pm 0.016 \pm 0.008$
$\Xi_c^+ \to \Xi^-\pi^+\pi^+$	29.7 ± 7.0	$0.41^{+0.11}_{-0.08} \pm 0.02$
$\Xi_c^0 \to \Xi^-\pi^+$	42 ± 10	$0.101^{+0.025}_{-0.017} \pm 0.005$
$\Omega_c^0 \to \Sigma^+ K^- K^- \pi^+$	≈ 50	$0.089^{+0.027}_{-0.020} \pm 0.028$

line with expectations from spectator diagram dominance.

Evidence also exists for the DCS decay $D^0 \to K^+\pi^-$ from Cleo [4] and E791 [9]. In these studies, the DCS decay appears as a background to the more rare and intriguing possibility of D^0 mixing, discussed next.

D^0 MIXING

To understand why D^0 mixing is such a rare phenomenon, it is useful to compare with the case of kaon mixing. In the standard model, both particles are expected to receive contributions to mixing from box diagrams (figure 3). In the case of the K^0, the internal quark propagators can be u, c or t quarks, although the t quark is greatly suppressed by small CKM couplings and is usually ignored. Taking all of the u and c quark box diagrams into account, one may calculate the amplitude for kaon mixing to be proportional to $(m_c^2 - m_u^2)/m_W^2$. This factor expresses the familiar GIM suppression, whereby the u and c diagrams cancel perfectly in the limit $m_u = m_c$. The fact that the c quark is so much heavier than the u quark produces the kaon mixing that we observe.

In the case of D^0 mixing, the internal propagators of the box diagrams can be d, s or b quarks, with the b quarks once again suppressed by small couplings. In this case, the amplitude for mixing is proportional to $[(m_s^2 - m_d^2)/m_W^2] \times [(m_s^2 - m_d^2)/m_c^2]$. The first factor shows the GIM suppression which is much more severe in this case since the s and d quarks are closer in mass. In addition, there is an additional suppression, given by the second term, which results from the necessity of transfering the heavy c quark four-momentum through the light quark internal propagators. This pulls the internal quarks

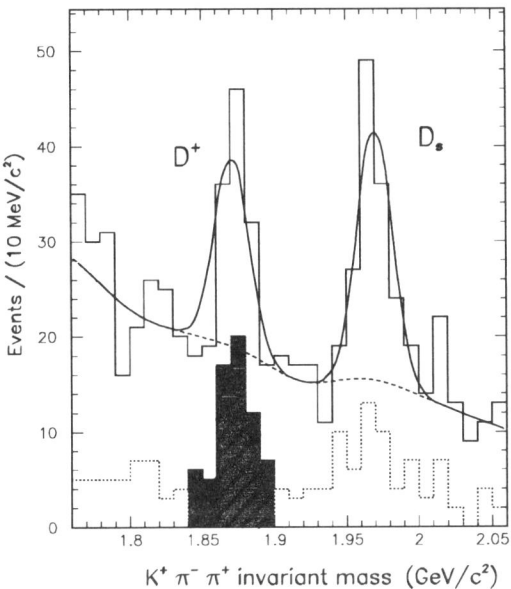

FIGURE 2. E791 data showing the signals for the DCS decay $D^+ \to K^+\pi^+\pi^-$ and the SCS decay of the D_s^+ to the same final state. There are about 60 events in the DCS signal peak.

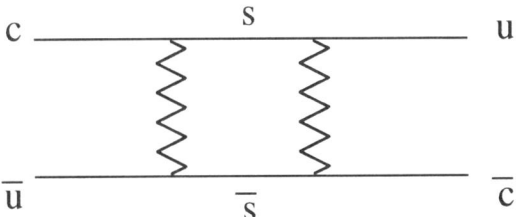

FIGURE 3. A typical box diagram showing the transition from D^0 to $\overline{D^0}$.

off-shell and reduces the amplitude.

When all is said and done, the D^0 mixing rate from box diagrams, expressed as $r_{mix} = \Gamma(D^0 \to \overline{D^0} \to f)/\Gamma(\overline{D^0} \to f)$ is expected to be on the order of $r_{mix} \approx 10^{-10}$. This is a terrifically small number when compared to the current experimental sensitivities of order $r_{mix} \approx 10^{-3}$. A recent calculation by Petrov [5] has determined that dipenguin contributions are of the same order of magnitude. Dispersive contributions have been examined in general terms [6] and also in the context of heavy quark effective theory [7]. Although large SU(3) flavor breaking in charm decays suggest that long-range dispersive contributions might play an important role in D^0 mixing, the range of predictions covers $r_{mix} \approx 10^{-10}$ to 10^{-7}, still many orders of magnitude below experimental sensitivity.

In contrast to the SM predictions for mixing, there is a large menu of theories beyond the standard model which allow for substantial charm mixing. These include a fourth generation b quark, left-right symmetric models, supersymmetry, leptoquarks and extended Higgs models. In most of these cases, certain combinations of new particle masses and couplings allow for D^0 mixing up to the current experimental sensitivities. The possibility of discovering a signature of one of these models is the primary reason for studying D^0 mixing.

The most recent experimental results on D^0 mixing come from the E791 collaboration [8,9]. In this experiment, the search for mixing involves reconstructing the decay chain $D^{*+} \to D^0 \pi^+$ with $D^0 \to K\pi, K3\pi, Ke\nu$ or $K\mu\nu$. The charge of the pion from the D^{*+} decay determines whether the neutral D is produced as a D^0 or a $\overline{D^0}$. The D decay products then determine if the D decays according to the same assignment. If the kaon and the pion from the D^* decay have the same charge, the decay is said to be "wrong-sign", and is a possible signal for mixing. For hadronic final states, an additional source of wrong-sign decays can come from DCS decays, which are expected to contribute branching ratios of order $r_{DCS} = \Gamma_{DCS}/\Gamma_{CF} \approx 10^{-2}$. In this case, decay time information can be used to discriminate statistically between DCS and mixed decays.

Figure 4 shows the E791 data for $D^0 \to K\pi$, both right-sign and wrong-sign decays. One axis shows the $K\pi$ mass, where we expect a D^0 signal at 1.86 GeV, while the other axis shows the kinetic energy from the D^* decay ($Q = m(K2\pi) - m(K\pi) - m(\pi)$) which should peak at 0.006 GeV for real D^* decays. A substantial signal is evident in the right-sign decays, with about 5000 events in the peak. The wrong-sign plot shows no striking evidence for a mixing signal, though there is about a 2σ excess in the signal region (about 40 events), consistent in its decay time distribution with DCS decays. Figure 5 shows the kinetic energy distribution for the $D^0 \to Ke\nu$ final state for both right-sign and wrong-sign decays. Once again there is no evidence of a mixing signal in the wrong-sign plot, while the right-sign plot shows a signal of about 1000 events. Similar data for the $K3\pi$ and $K\mu\nu$ final states are not shown here. Although event yields are lower in the semileptonic final state, and although the mass resolution is much worse due to the unmeasured neutrino, there is also no contribution from DCS decays to confuse the issue. This fact compensates for the worse yield and resolution so that, in the end, both final states have roughly the same sensitivity to r_{mix} of a few times 10^{-3}.

RARE DECAYS

With the recent increases in charm data samples, there has been considerable interest in looking for rare charm decays. These include flavor-changing neutral current (FCNC) decays such as $D^0 \to \mu^+\mu^-$ and $D^+ \to \pi^+\mu^+\mu^-$, lepton family violating (LFV) decays such as $D^+ \to \pi^+e^+\mu^-$ and lepton number violating (LNV) decays such as $D^+ \to \pi^-\mu^+\mu^+$. Figure 6 gives examples of the Standard Model diagrams which might contribute to $D^0 \to \mu^+\mu^-$ and $D^+ \to \pi^+\mu^+\mu^-$. The box diagram of figure 6a makes clear the relationship between FCNC processes and D^0 mixing. Not surprisingly, FCNC decays are expected to be very small in the Standard Model ($B(D^0 \to \mu^+\mu^-) \approx 10^{-19}$ to 10^{-15} and $B(D^+ \to \pi^+\mu^+\mu^-) \approx 10^{-11}$) for the same reasons that mixing is small. Also not surprisingly, many of the extensions of the Standard Model which could contribute to mixing are also expected to contribute to FCNC decays. LFV and LNV decays are forbidden in the standard model, and ob-

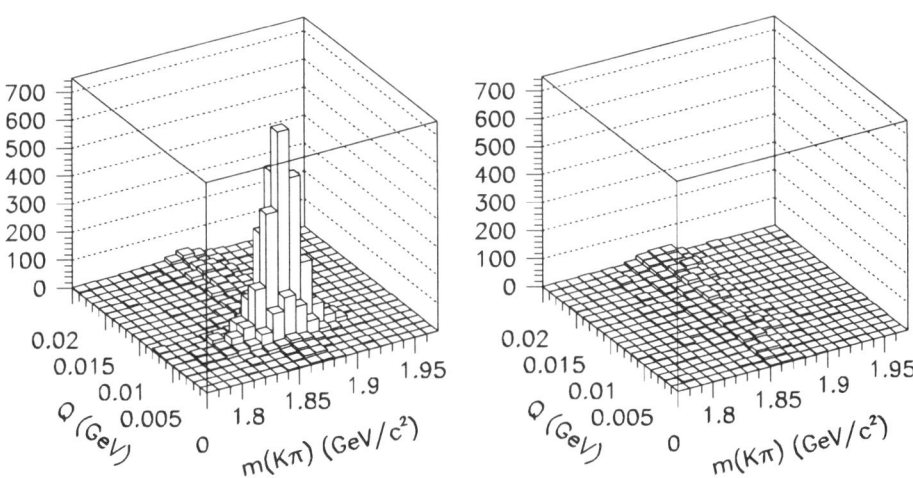

FIGURE 4. E791 data showing the right-sign signal for $D^0 \to K^-\pi^+ + c.c.$ (left) and the wrong-sign $D^0 \to K^+\pi^- + c.c.$ (right). About 5000 signal events are apparent in the right-sign plot, with no evidence for mixing in the wrong-sign data.

servation of a signal for any of these decays is an immediate indication of new physics.

Many experiments are contributing to the search for rare decays. Recently, CleoII [10], E687 [11] and E791 [12] have all contributed to limits on rare charm decay modes. A wide variety of FCNC, LFNV and LNV decay modes of D mesons (and some of Λ_c^+) have been explored, with branching fraction sensitivities in the range of 10^{-5} to 10^{-6}. So far, no hint of new physics is evident, but this area will be fertile ground for the search in the years to come.

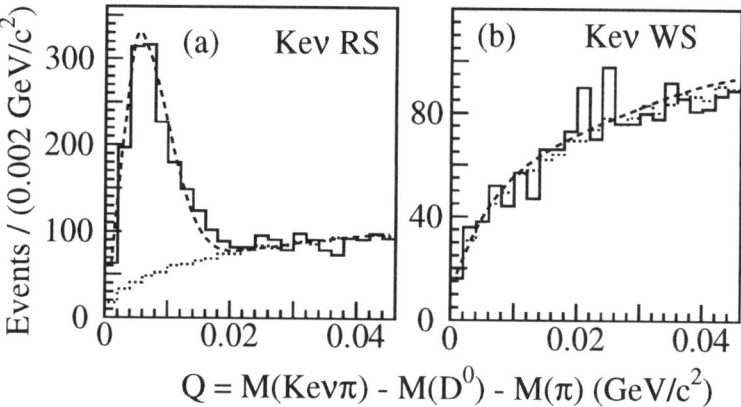

FIGURE 5. E791 data showing the right-sign signal for $D^0 \to K^- e^+ \nu + c.c.$ (a) and the wrong-sign $D^0 \to K^+ e^- \nu + c.c.$ (b). There are about 1000 signal events in the right-sign plot, with no evidence for mixing in the wrong-sign data.

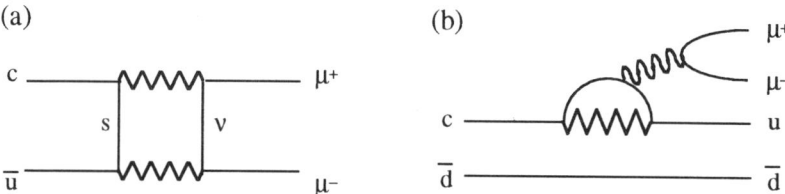

FIGURE 6. Weak decay diagrams for the FCNC decays $D^0 \to \mu^+ \mu^-$ (a) and $D^+ \to \pi^+ \mu^+ \mu^-$ (b).

NEW EXPERIMENTS

In the next few years, the most interesting work on charm physics is likely to come from three new charm experiments now operating at Fermilab, taking data during the current fixed target run from Sept. '96 through Sept. '97. Experiment E831 (the FOCUS collaboration) is a charm photoproduction experiment which is an upgrade of the previous E867 experiment. They are expecting a data sample of roughly 10^6 reconstructed charm, about an order of magnitude beyond the previous generation of experiments. They will concentrate on rare and forbidden decays of charm, D^0 mixing, CP violation, leptonic decays of the D^+ and D^{*+}, and excited charm states, among other topics. Experiment E781 (the SELEX experiment) uses a Σ^- beam to preferentially produce charm baryons, especially charm-strange baryons. In this respect it follows the lead of the successful WA89 experiment at CERN. SELEX also expects to reconstruct about 10^6 charm decays, half of them baryons. This collaboration will naturally focus on charm baryon spectroscopy and production systematics, providing a nice complement to the work by the E831 group. Finally, experiment E835 (the CHARMONIUM collaboration), an upgrade of the previous E760 experiment, uses the FNAL $\bar{p}$ beam on a hydrogen gas jet target to produce charmonium at resonance through $\bar{p}p$ annihilation. They are hoping for a total integrated luminosity of about 200 pb^{-1}, seven times that of the E760 experiment. The charmonium masses are determined from knowledge of the beam energy, providing excellent mass resolution. E835 hopes to discover the η_c' and measure its mass and decay width, measure the masses of the $^{3,1}D_2$ charmonimum states, study the $^1P_1$ state seen by E760, and improve the measurements of the η_c and χ^0 parameters. All told, the prospects for charm physics in the next few years look very exciting.

REFERENCES

1. M. I. Adamovich et al., *Phys. Lett.* **B358** (1995) 151.
2. E. M. Aitala et al., hep-ex/9706025, June 1997. To be published in *Phys. Lett.*
3. P. L. Frabetti et al., *Phys. Lett.* **B359** (1995) 403.
4. D. Cinabro et al., *Phys. Rev. Lett.* **72** (1994) 1406.
5. Alexei A. Petrov, hep-ph/9703335, To be published in *Phys. Rev.* **D**.
6. Donoghue et al., *Phys. Rev.* **D33** (1986) 179; E. Golowich, *Proceedings of the Conference on B Physics and CP Violation*, Honolulu, March 1997.
7. Howard Georgi, *Phys. Lett.* **B297** (1992) 353; Thorsten Ohl et al., *Nucl. Phys.* **B403** (1993) 605.
8. E. M. Aitala et al., hep-ex/9606016, *Phys. Rev. Lett.* **77** (1996) 2384.
9. E. M. Aitala et al., hep-ex/9608018, Submitted to *Phys. Rev.* D.
10. A. Freyberger et al., *Phys. Rev. Lett.* **76** (1996) 3065.
11. P. L. Frabetti et al., *Phys. Lett.* **B398** (1997) 239.
12. E. M. Aitala et al., *Phys. Rev. Lett.* **76** (1996) 364.

Chiral Perturbation Theory in Few-Nucleon Systems

U. van Kolck[1]

*Department of Physics,
University of Washington,
Seattle, WA 98195-1560*

Abstract. The low-energy effective theory of nuclear physics based on chiral symmetry is reviewed. Topics discussed include the nucleon-nucleon force, few-body potentials, isospin violation, pion-deuteron scattering, proton-neutron radiative capture, pion photoproduction on the deuteron, and pion production in proton-proton collisions.

INTRODUCTION

Although studies of its perturbative regime show that QCD is the theory of strong interactions, most of the structures of nuclear physics are apparent only at low energies where the QCD coupling constant is not small. Another expansion parameter is necessary in this energy regime, and an obvious candidate is energy itself.

The idea of a low-energy expansion is as old as nuclear physics itself. Already in the 30's, Bethe and Peierls [1] considered the two-nucleon system in this light. They based their approach on a previous argument about saturation due to Wigner, that the nuclear potential is of order 100 MeV and thus much larger than the deuteron binding energy of 2.2 MeV, but has a range R ($\simeq 1.4$ fm) much smaller than the size λ of the deuteron ($\simeq 4.4$ fm). Bethe and Peierls reasoned that as a consequence, for distances r such that $R \lesssim r \lesssim \lambda$, only s-waves are important and the sole effect of the potential is to provide an energy independent boundary condition at $r \sim R$. Up to an error of $O(R/\lambda)$, then, the system could be described by a free Schrödinger equation with the boundary condition that the logarithmic derivative of the radial wavefunction is a constant at $r = 0$. This constant cannot be calculated without detailed knowledge of the potential, so it was fitted to the deuteron binding energy. If that was all, nothing would have been learned. Bethe and Peierls' point,

---

[1] After Jan 1 1998: W.K. Kellogg Radiation Laboratory, Caltech, Pasadena, CA 91125

however, was that they could then predict (with a 30% uncertainty) other processes —such as $pn \to pn$, $\gamma d \to pn$, $\gamma d \to \gamma d$, and $ed \to e'pn$— at energies comparable to the deuteron binding energy.

This approach can be rephrased in an effective field theory (EFT) language. At typical momenta Q much smaller than the pion mass m_π, the relevant degree of freedom is the nucleon, the important symmetries are parity, time-reversal and Galilean symmetry, and the appropriate expansion parameter is $Q/m_\pi \sim R/\lambda$. (Electromagnetic processes can also be considered by adding the photon, $U(1)_{em}$ gauge invariance, and α_{em} to this list.) The most general Lagrangian involving nucleons only consists of an infinite number of terms, which are quadratic, quartic, ..., in the nucleon fields with increasing number of derivatives. By dimensional reasons, derivatives come associated with inverse factors of a mass scale m_π or greater. Nucleons are non-relativistic and the corresponding field theory has nucleon number conservation. The T-matrix for the two-nucleon system is simply a sum of bubble graphs, whose vertices are the four-nucleon contact terms that appear in the Lagrangian. Formally, this is equivalent to solving a Schrödinger equation with a low-energy, effective potential consisting, schematically, of a sum $\bar{C}_0 \delta(\vec{r}) + \bar{C}_2 \delta''(\vec{r}) + ...$, where the $\bar{C}_n$'s are the coefficients of the contact terms, expected to scale as $\bar{C}_n \sim \bar{C}_0/m_\pi^n$. The net effect is thus to replace the "true", possibly complicated potential of range $\sim 1/m_\pi$ by a multipole expansion with moments $\bar{C}_n$. Life is somewhat more complicated, however, because this field theory, like any other, has to be renormalized. The bubbles are actually ultraviolet divergent, requiring regularization and absorption of the regulator dependence in renormalized parameters. It is not difficult to show [2] that the effect of renormalization is to turn the effective potential into a generalized pseudo-potential, or equivalently, turn the problem into a free one with boundary conditions at the origin which are analytic in the energy. The first, energy-independent term, parametrized by $\bar{C}_0^R$, is just the one considered by Bethe and Peierls. Much effort has been spent during the last year in trying to understand issues related to regularization and fine-tuning of this "pionless" theory [3].

To the extent that there are no assumptions about the detailed dynamics of the "underlying" theory, the effective theory cannot be wrong and it is useful as long as $Q \ll m_\pi$. One may ask how strong this restriction is, however, when we consider physics of more than two nucleons. Clearly, we can always apply the pionless theory to sufficiently low-energy scattering situations, where we can control the momenta of the initial and final nucleons. But one noble goal of nuclear physics is to understand nuclei themselves. Let me use $2m_N B/A$ as a measure of a typical momentum Q of a nucleon of mass m_N in a nucleus with A nucleons and binding energy B. (Other quantities such as charge radii give similar estimates.) Q/m_π is then about 0.3 for ^{2}H, 0.5 for ^{3}H, 0.8 for ^{4}He, ..., and 1.2 for symmetric nuclear matter in equilibrium. The same argument that justified the use of a pionless theory for the deuteron now suggests that understanding the binding of typical nuclei (^{4}He and heavier) requires *explicit* inclusion of pions, but *not* of heavier mesons such as the rho.

Now, we are in luck because QCD does explain the special role of the pion, and in the process, provides a rationale to treat pion effects systematically. This procedure goes by the name of Chiral Perturbation Theory (χPT); it will be described briefly in the next section, and exemplified even more briefly in the simplest case of at most one nucleon in the following section. I then come to the main portion of this review, where we tackle nuclear forces and external probes of light nuclei.

EFFECTIVE CHIRAL LAGRANGIAN

Why is the pion special? It is one of the nicest features of QCD that it provides a scenario where the lightness of the pion results from its (pseudo-)Goldstone boson nature. Here for simplicity I will limit myself to the case of two quark flavors.

If the quark masses were zero ("chiral limit"), the QCD Lagrangian would be invariant under transformations of the group $SU(2)_L \times SU(2)_R$ of independent rotations of the quarks' left- and right-handed components. When acting on quark bilinears, this chiral symmetry is equivalent to $SO(4)$. A quick look at the hadronic spectrum convinces us that the chiral limit can only be phenomenologically relevant if there is spontaneous breakdown of chiral symmetry down to its diagonal subgroup, the $SU(2)_{L+R}$ ($\sim SO(3)$ for bilinears) of isospin. Although the mechanism of spontaneous breaking is not sufficiently understood at the present for a detailed derivation, we know that the effective QCD potential as a function of four quark bilinears ($\bar{q}\gamma_5\tau_i q, \bar{q}q$) has to have roughly a mexican hat shape, i.e. $SO(4)$ symmetry with minima in a "chiral circle" away from the origin.

Goldstone's theorem assures us there is in the spectrum, as a consequence, a (pseudo-)scalar boson, which corresponds to excitations in the coset space $SO(4)/SO(3) \sim S^3$. We call the radius of this "circle" f_π, which is a function of Λ_{QCD} that ends up being the pion decay constant $\simeq 92$ MeV. At sufficiently low energies it is convenient to assign a field π to the pion in the effective Lagrangian. An infinitesimal chiral transformation is of the form $\pi \to \pi + f_\pi \epsilon + ...$. $SO(4)$ symmetry of the dynamics implies that the Lagrangian will have a piece that is a function of π only through derivatives of π on the circle, which are non-linear. The Lagrangian, in principle completely determined by Λ_{QCD}, has an infinite number of terms with arbitrarily high pion self-interactions, but without a pion mass term.

We know, however, that not all quark masses are zero. Quark masses generate two terms in the QCD Lagrangian. One term, $\bar{m}\bar{q}q$ with $\bar{m} = (m_u+m_d)/2$, is the fourth component of an $SO(4)$ vector and therefore breaks $SO(4)$ explicitly down to $SO(3)$ of isospin. It causes a tilt of the effective potential in the $\bar{q}q$ direction determined by the small parameter $\eta = \bar{m}/\Lambda_{QCD}$. The effective Lagrangian will acquire then a piece that breaks $SO(4)$ explicitly in the same way as $\bar{q}q$. This piece is another infinite set of terms that do depend on π in an isospin invariant way, but are all proportional to powers of η. In particular, $m_\pi^2 \propto \eta \Lambda_{QCD}^2$. The other quark mass term, $\epsilon \bar{m} \bar{q} \tau_3 q$ with $\epsilon = (m_u - m_d)/(m_u + m_d) \simeq 1/3$, is the third component of another $SO(4)$ vector and further breaks $SO(3)$ down to $U(1) \times U(1)$. Likewise, the effective Lagrangian will inherit yet another infinite set of terms, this time that break isospin as $\bar{q}\tau_3 q$ and are, in principle, of order ϵ relative to the isospin conserving chiral breaking effects. Why isospin breaking is in fact much smaller in most observables will be explained later.

QCD therefore has all the ingredients to provide a rationale not only for the special role of the pion, but also for a systematic treatment of its effects: pion interactions are weak at low energies due to (approximate) chiral symmetry.

We can now formulate an EFT for momenta $Q \sim m_\pi \ll M_{QCD} \sim m_\rho \sim m_N \sim 4\pi f_\pi$ along the same lines of the pionless case. The extra degrees of freedom —besides non-relativistic nucleons and photons— are obviously pions and also non-relativistic delta isobars, since the delta-nucleon mass difference $m_\Delta - m_N \sim 2m_\pi$ is of the order of the momenta Q we want to consider. The new and very important symmetry is approximate $SU(2)_L \times SU(2)_R$. The expansion parameter is expected to be Q/M_{QCD} —besides α_{em}. The most general Lagrangian with these ingredients has schematically the form

$$\mathcal{L} = \sum_{\{dqnpf\}=1}^{\infty} C_{dqnpf} \left(\frac{\mathcal{D}}{M_{QCD}}\right)^d \left(\frac{m_\Delta - m_N}{M_{QCD}}\right)^q \left(\frac{m_\pi}{M_{QCD}}\right)^n \left(\frac{\boldsymbol{\pi}}{f_\pi}\right)^p \left(\frac{\psi^\dagger \psi}{f_\pi^2 M_{QCD}}\right)^{\frac{f}{2}} f_\pi^2 M_{QCD}^2$$

$$= \sum_{\Delta=0}^{\infty} \mathcal{L}_{(\Delta)}. \qquad (1)$$

Here the C's are parameters assumed to be natural, i.e. of $O(1)$ —or $O(\epsilon^\#)$ with # a positive integer, in the case of isospin breaking operators originating in the quark mass difference. ψ stands for a nucleon or delta isobar, and $\mathcal{D}$ for a covariant derivative. The interactions are naturally grouped in sets $\mathcal{L}_{(\Delta)}$ of common index $\Delta \equiv d + q + n + \frac{f}{2} - 2$ but arbitrary values of p. For non-electromagnetic interactions, we find that $\Delta \geq 0$ only because of chiral symmetry.

Consider an arbitrary irreducible contribution to a process involving A nucleons and an arbitrary number of pions and photons, all with momenta of order Q. It can be represented by a Feynman diagram with A continuous nucleon lines, L loops, C separately connected pieces, and V_Δ vertices from $\mathcal{L}_{(\Delta)}$, whose connected pieces cannot be all split by cutting only nucleon lines. ($C = 1$ for $A = 0, 1$; $C = 1, ..., A - 1$ for $A \geq 2$. The reason to consider irreducible diagrams and $C > 1$ will be discussed in the section on few-nucleon systems.) It is easy to show [4] that this contribution is typically of $O((Q/M_{QCD})^\nu)$, where

$$\nu = 4 - A + 2(L - C) + \sum_{\Delta} V_\Delta \Delta. \tag{2}$$

Since L is bounded from below (0) and C from above (C_{max}), the chiral symmetry constraint $\Delta \geq 0$ implies that $\nu \geq \nu_{min} = 4 - A - 2C_{max}$ for strong interactions. Leading contributions come from tree diagrams built out of $\mathcal{L}_{(0)}$ and coincide with current algebra. Perturbation theory in Q/M_{QCD} can be carried out by considering contributions from ever increasing ν.

Note that this approach is:

(i) systematic. It is a *perturbation* in the number of loops, derivatives/fermion fields, and many-nucleon effects.

(ii) consistent with QCD. The only (very important) QCD inputs are confinement (color singlet fields), symmetries (*chiral*, ...), and naturalness. QCD can be represented by a point in the space of renormalized parameters (at some renormalization scale) of the effective theory. An explicit solution of QCD (such as from simulations on a lattice) would provide knowledge of the exact position of this point, and then the effective theory would be completely predictive. Until such a solution is found —or as a test of QCD after it is found— we can recourse to fitting low-energy experiments in order to determine the region in parameter space allowed on phenomenological grounds. Even in this case the theory is predictive, because to any given order the space of parameters is finite-dimensional. After a finite number of experimental results are used, an infinite number of others can be predicted up to an accuracy depending on the order of the expansion. In practice, because the number of parameters grows rapidly with the order, model-dependent estimates of parameters based on specific dynamic ideas —such as saturation by tree-level resonance exchange— are sometimes used.

(iii) a *theory*. It is applicable in principle to all low-energy phenomena. I will in the next section mention some of the highlights of this program for processes with at most one nucleon, and then most of the rest of the paper is devoted to nuclear physics *per se*.

MESON AND ONE-NUCLEON SECTORS

In the case of strong mesonic processes, $\nu = 2 + 2L + \sum_\Delta V_\Delta \Delta \geq 2$ with $\Delta = d + n - 2$ increasing in steps of two. Since Weinberg discovered this systematic generalization of current algebra [5] and Gasser and Leutwyler implemented it [6], many processes (including weak interactions) involving pions (and kaons) have been examined, typically to $\nu = 4$. The most thoroughly studied process has been $\pi\pi$ scattering, where a $\nu = 6$ calculation was carried out [7]; fits to other data plus resonance saturation show convergence and good agreement with phase shifts through energies of more than 100 MeV above threshold. Many good reviews exist on the mesonic sector; see for example Ref. [8].

For processes where one nucleon is present, $\nu = 1+2L+\sum_\Delta V_\Delta \Delta \geq 1$, where $\Delta = d+q+n+\frac{f}{2}-2$ (with $f = 0, 2$) increases in steps of one. (Convergence can be expected to be slower compared to purely mesonic interactions.) Thanks to this power counting, a low-energy nucleon can in a very definite sense be pictured as a static, point-like object (up to corrections in powers of Q/m_N), surrounded by: i) an inner cloud which is dense but of short range $\sim 1/m_\rho$, so that we can expand in its relative size Q/m_ρ; ii) an outer cloud of long range $\sim 1/m_\pi$ but sparse, so that we can expand in its relative strength $Q/4\pi f_\pi$.

The first attempt at including the nucleon in χPT [9] had limited success because it did not fully explore the non-relativistic nature of the nucleon and did not consider the delta isobar explicitly. This was remedied in the work of Jenkins and Manohar [10], while a more complete analysis of the role of the delta in one-nucleon process has been carried out recently [11]. However, most calculations have been limited to the threshold region where the delta contribution is relatively unimportant and a "deltaless" theory useful; see for example Ref. [12] for an extensive review.

Here, for illustration, I mention explicitly the case of pion photoproduction at threshold which has received a lot of attention lately. At threshold the amplitude for a photon of polarization $\vec{\epsilon}$ incident on a nucleon of spin $\vec{\sigma}$ is $T \sim i\vec{\sigma}\cdot\vec{\epsilon} E_{0+}$. This process has been studied up to $\nu = 4$ in the deltaless theory in Ref. [13] (where references to the experimental papers can be found; see also Ref. [14]). Results for E_{0+} from a fit constrained by resonance saturation are presented in Table 1.

We can observe that the big size of the charged pion channels result from a non-vanishing $\nu = 1$ contribution (the Kroll-Ruderman term). Convergence and agreement with experimental values are pretty good. For the neutral pion channels convergence is less apparent, but the absolute values much smaller. The $\nu = 4$ result for $\gamma p \to \pi^0 p$ is in relatively good agreement with the recent results from Mainz and Saskatoon. To the same order there is a prediction for the $\gamma n \to \pi^0 n$ reaction, which would be important to test. If isospin symmetry breaking is neglected, there are only three independent amplitudes; if we use the three measured amplitudes, their uncertainties limit extraction

TABLE 1. Values for E_{0+} in units of $10^{-3}/m_{\pi^+}$ at various orders ν in χPT and in experiments, for the different channels.

$E_{0+}(10^{-3}/m_{\pi^+})$	$\nu = 1$	$\nu = 2$	$\nu = 3$	$\nu = 4$	Experiment
$\gamma p \to \pi^+ n$	34.0	26.4	28.9	28.2	27.9 ± 0.5.
					28.8 ± 0.7.
$\gamma n \to \pi^- p$	-34.0	-31.5	-32.9	-32.7	-31.4 ± 1.3.
					-32.2 ± 1.2.
$\gamma p \to \pi^0 p$	0	-3.58	0.96	-1.16	-1.31 ± 0.08.
					-1.32 ± 0.08.
$\gamma n \to \pi^0 n$	0	0	3.7	2.13	?

of $E_{0+}(\pi^0 n)$ to the range $-0.5 \mapsto +2.5$. Newer, more accurate data for the charged channels were presented at this conference by Korkmaz [14]. Isospin breaking entered the calculation of Table 1, albeit in an incomplete form. A direct measurement of $E_{0+}(\pi^0 n)$ could not only check the consistency of the calculation, but also provide a new isospin violating observable. This however requires a deuteron target, and a description of $\gamma d \to \pi^0 d$ in the same framework. So, even from the point of view of nucleon properties we are led naturally into the study of light nuclei.

NUCLEAR PHYSICS

A non-trivial new element enters the theory when we consider systems of more than one nucleon [4]. Because nucleons are heavy, contributions from intermediate states that differ from the initial state only in the energy of nucleons are enhanced by infrared quasi-divergences. These are linked to the existence of small energy denominators of $O(Q^2/m_N)$, which generate contributions $O(m_N/Q)$ larger than what would be expected from Eq. (2). The latter is still correct for the class of sub-diagrams —called irreducible— that do not contain intermediate states with small energy denominators. For an A-nucleon system these are A-nucleon irreducible diagrams, the sum of which we call the potential V. When we consider external probes all with $Q \sim m_\pi$, the sum of irreducible diagrams forms the kernel K to which all external particles are attached. A generic diagram contributing to a full amplitude will consist of irreducible diagrams sewed together by states of small energy denominators. These irreducible diagrams might have more than one connected piece, hence the introduction of C in Eq. (2). The infrared enhancement requires that we sum diagrams to all orders in the amplitude, creating the possibility of the existence of shallow bound states (nuclei). For an A-nucleon system, this is equivalent to solving the Schrödinger equation with the potential V. The amplitude for a process with external probes is then $T \sim \langle \psi' | K | \psi \rangle$ where $|\psi\rangle$ ($|\psi'\rangle$) is the wavefunction of the initial (final) nuclear state calculated with the potential V.

Because our Q/M_{QCD} expansion is still valid for the potential and the kernel, the picture of a nucleon as a mostly static object surrounded by an inner and an outer cloud leads to remarkable nuclear physics properties that we are used to, but would remain otherwise not understood from the viewpoint of QCD.

Nuclear Forces

If we put a few non-relativistic nucleons together, each nucleon will not be able to distinguish details of the others' inner clouds. The region of the potential associated with distances of $O(1/m_\rho)$ can be expanded in delta-functions and their derivatives as Bethe and Peierls did. The outer cloud of range $O(1/m_\pi)$ yields non-analytic contributions to the potential, but being sparse, it mostly produces the exchange of one pion, with progressively smaller two-, three-, ...- pion exchange contributions.

For the two-nucleon system, $\nu = 2L+\sum_\Delta V_\Delta \Delta$, with Δ as in the one-nucleon case. A calculation of all the contributions up to $\nu = 3$ was carried out in Ref. [15]. In leading order, $\nu = 0$, the potential is simply static one-pion exchange and momentum-independent contact terms [4]. $\nu = 1$ corrections vanish due to parity and time-reversal invariance. $\nu = 2$ corrections include several two-pion exchange diagrams (including virtual delta isobar contributions), recoil in one-pion exchange, and several contact terms that are quadratic in momenta. At $\nu = 3$ a few more two-pion exchange diagrams have to be considered. As in the pionless case, regularization and renormalization are necessary. It is not straightforward to implement dimensional regularization in this non-perturbative context, so we used an overall gaussian cut-off, and performed calculations with the cut-off parameter Λ taking values 500, 780 and 1000 MeV. Cut-off independence means that for each cut-off value a set of (bare) parameters can be found that fits low-energy data. A sample of the results for the lower, more important partial waves is presented in Fig. 1 and for the deuteron quantities in Table 2. (See [15] for more details and reference to experiments and phase shift analyses.)

The fair agreement of this first calculation and data up to laboratory energies of 100 MeV or so suggests that this may become an alternative to other, more model-dependent approaches to the two-nucleon problem. Further examination of regularization effects, fine-tuning in the $^1S_0$ channel, and different aspects of two-pion exchange in this context can be found in Refs. [3,16].

Perhaps more impressive is that we can get some insight into other aspects of nuclear forces. Let us look for the new forces that appear in systems with more than two nucleons. The dominant potential, at $\nu = 6-3A = \nu_{min}$, is the two-nucleon potential of lowest order that appeared in the two-nucleon case. We can easily verify that a three-body potential will arise at $\nu = \nu_{min} + 2$,

TABLE 2. Effective chiral Lagrangian fits for various cut-offs Λ and experimental values for the deuteron binding energy (B), magnetic moment (μ_d), electric quadrupole moment (Q_E), asymptotic d/s ratio (η), and d-state probability (P_D).

Deuteron	Λ (MeV)			
quantities	500	780	1000	Experiment
B (MeV)	2.15	2.24	2.18	2.224579(9)
μ_d (μ_N)	0.863	0.863	0.866	0.857406(1)
Q_E (fm^2)	0.246	0.249	0.237	0.2859(3)
η	0.0229	0.0244	0.0230	0.0271(4)
P_D (%)	2.98	2.86	2.40	

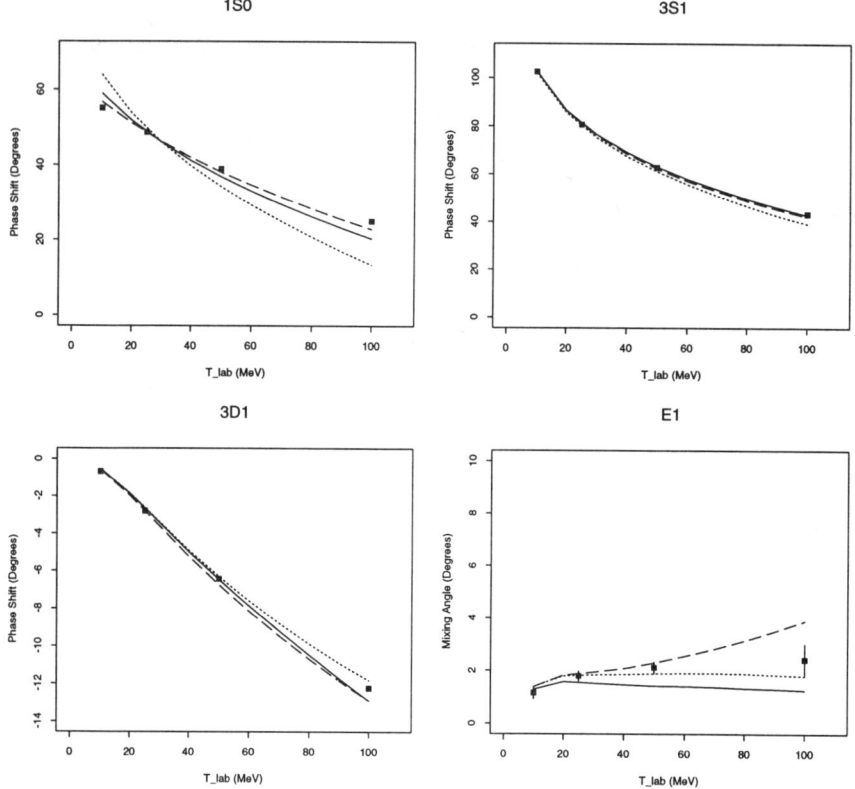

FIGURE 1. $^1S_0$, $^3S_1$, and $^3D_1$ NN phase shifts and ϵ_1 mixing angle in degrees as functions of the laboratory energy in MeV: chiral expansion up to $\nu = 3$ for cut-offs of 500 (dotted), 780 (dashed), and 1000 MeV (solid line); and Nijmegen phase shift analysis (squares).

a four-body potential at $\nu = \nu_{min} + 4$, and so on. It is (approximate) chiral symmetry therefore that implies that n-nucleon forces V_{nN} are expected to obey a hierarchy of the type $\langle V_{(n+1)N}\rangle/\langle V_{nN}\rangle \sim O((Q/M_{QCD})^2)$, with $\langle V_{nN}\rangle$ denoting the contribution per n-plet. If we estimate $\langle V_{2N}\rangle \sim \frac{g_A^2}{16\pi f_\pi^2} m_\pi^3 \simeq 10$ MeV, we can guess $\langle V_{3N}\rangle \sim .5$ MeV, $\langle V_{4N}\rangle \sim .02$ MeV, and so on. This is in accord with detailed few-nucleon calculations using more phenomenological potentials. The explicit three-body potential at $\nu = \nu_{min} + 2$ (from the delta isobar) and $\nu_{min} + 3$ was derived in Ref. [17].

We can also look at isospin-dependent effects. Within the context of χPT it can be shown [18] that isospin is an accidental symmetry, in the sense that it does not appear in the low-energy EFT in lowest order, and therefore is typically not an $O(\epsilon)$ effect, but $O(\epsilon(Q/M_{QCD})^n), n \geq 1$. In the case of nuclear forces, we find moreover a hierarchy among different types of components. It is standard to call class I the strongest forces that are isospin symmetric, class II weaker forces that are isospin violating but charge symmetric, class III even weaker forces that are charge symmetry breaking but symmetric under permutation of particles in isospin space, and class IV the weakest, remaining forces. In the chiral expansion, one indeed finds [18] that higher class forces appear at higher orders: $\langle V_{M+1}\rangle/\langle V_M\rangle \sim O(Q/M_{QCD})$, where $\langle V_M\rangle$ denotes the contribution of the leading class M potential. This qualitatively explains, for example, the observed isospin structure of the two-nucleon Coulomb-corrected scattering lengths, $a_{np} \simeq 4 \times ((a_{nn} + a_{pp})/2 - a_{np}) \simeq 4^2 \times (a_{pp} - a_{nn})$. Precise calculations of simultaneous electromagnetic and strong isospin violation in the nuclear potential have also been carried out [19].

Despite these successful fits and insights, the main advantage of the method of EFT lies in its concomitant application to many other processes, which might yield more predictive statements. I now discuss some of these.

Nuclear Probes

Before plunging into hard results, let me point out another generic result of the chiral expansion. As a result of the factor $-2C$ in Eq. (2), we see immediately —in an effect similar to few-nucleon forces— that external low-energy probes (π's, γ's) will tend to interact predominantly with a single nucleon, simultaneous interactions with more than one nucleon being suppressed by powers of $(Q/M_{QCD})^2$. Again, this is a well-known result that arises naturally here.

This is of course what allows extraction, to a certain accuracy, of one-nucleon parameters from nuclear experiments. More interesting from the nuclear dynamics perspective are, however, those processes where the leading single-nucleon contribution vanishes by a particular choice of experimental conditions, for example the threshold region. In this case the two-nucleon contributions, especially in the relatively large deuteron, can become important.

$\pi d \to \pi d$ *at threshold.* This is perhaps the most direct way to check the consistency of χPT in few-nucleon systems and in pion-nucleon scattering. Here the lowest-order, $\nu = -2$ contributions to the kernel vanish because the pion is in an s-wave and the target is isoscalar. The $\nu = -1$ term comes from the (small) isoscalar pion-nucleon seagull, related in lowest-order to the pion-nucleon isoscalar amplitude b_0. $\nu = 0, +1$ contributions come from corrections to pion-nucleon scattering and two-nucleon diagrams, which involve besides b_0 also the much larger isovector amplitude b_1. Weinberg [20] has estimated these various contributions to the pion-deuteron scattering length, finding agreement with previous, more phenomenological calculations, which have been used to extract b_0.

$np \to \gamma d$ at threshold. This offers a chance of a precise postdiction. Here it is the transverse nature of the real outgoing photon that is responsible for the vanishing of the lowest-order, $\nu = -2$ contribution to the kernel. The single-nucleon magnetic contributions come at $\nu = -1$ (tree level), $\nu = +1$ (one loop), etc. The first two-nucleon term is an one-pion exchange at $\nu = 0$ long discovered to give a smaller but non-negligible contribution. There has been a longstanding discrepancy of a few percent between these contributions and experiment. At $\nu = +2$ there are further one-pion exchange, two-pion exchange, and short-range terms. Park, Min and Rho [21] calculated the two-pion exchange diagrams in the deltaless theory and used resonance saturation to estimate the other $\nu = +2$ terms. With wavefunctions from the Argonne V18 potential and a cut-off $\Lambda = 1000$ MeV, they found the excellent agreement with experiment shown in Table 3. The total cross-section changes by .3 % if the cut-off is decreased to 500 MeV. (See [21] where reference to experiment can be found.)

$\gamma d \to \pi^0 d$ at threshold. As emphasized earlier, this reaction offers the possibility to test a prediction arising from a combination of two-nucleon contributions and the neutral pion single-neutron amplitude. Here, it is the neutrality of the outgoing s-wave pion that ensures that the leading $\nu = -2$ terms vanish. The single-scattering contribution is given by the same $\nu = -1, 0, +1, \ldots$ mechanisms described earlier, with due account of p-waves and Fermi motion inside the deuteron. The first two-nucleon term enters at $\nu = 0$, a correction appears at $\nu = +1$, and so on. At threshold the amplitude for a photon of polarization $\vec{\epsilon}$ incident on the deuteron of spin $\vec{J}$ is $T \sim 2i\vec{J} \cdot \vec{\epsilon} E_d$. Results for E_d up to $\nu = +1$ have been obtained [22] and are shown in Table 4. They correspond to the Argonne V18 potential and a cut-off $\Lambda = 1000$ MeV. Other realistic potentials and cut-offs from 650 to 1500 MeV give the same result within 5%, while a model-dependent estimate [23] of some $\nu = +2$ terms suggests a 10% or larger error from the neglected higher orders in the kernel itself. The single-scattering amplitude depends on $E_{0+}(\pi^0 n)$ in such a way that $E_d \sim -1.79 - 0.38(2.13 - E_{0+}(\pi^0 n))$ in units of $10^{-3}/m_{\pi^+}$. Thus, some sensitivity to $E_{0+}(\pi^0 n)$ survives the large two-nucleon

TABLE 3. Values for various contributions to the total cross-section σ for radiative neutron-proton capture in mb: impulse approximation to $\nu = 2$ (^{imp}), impulse plus two-nucleon diagrams at $\nu = 0$ ($^{imp+tn0}$), impulse plus two-nucleon diagrams up to $\nu = 2$ ($^{imp+tn}$), and experiment (^{expt}).

σ^{imp}	$\sigma^{imp+tn0}$	σ^{imp+tn}	σ^{expt}
305.6	321.7	336.0	334±0.5

TABLE 4. Values for E_d in units of $10^{-3}/m_{\pi^+}$ from single scattering up to $\nu = 1$ (^{ss}), two-nucleon diagrams at $\nu = 0$ (^{tn0}), two-nucleon diagrams at $\nu = 1$ (^{tn1}), and their sum ($^{ss+tn}$).

E_d^{ss}	E_d^{tn0}	E_d^{tn1}	E_d^{ss+tn}
0.36	-1.90	-0.25	-1.79

contribution at $\nu = 0$.

Some old Saclay data gave $E_d = -(1.7 \pm 0.2) \cdot 10^{-3}/m_{\pi^+}$ on its latest reanalysis, but a new precise measurement is called for. A test of the above prediction will come from new Saskatoon data, currently under analysis [14]. An electroproduction experiment will also be carried out in Mainz [24].

$pp \to pp\pi^0$ close to threshold. This reaction has attracted a lot of interest because of the failure of standard phenomenological mechanisms in reproducing the small cross-section near threshold. It involves larger momenta of $O(\sqrt{m_\pi m_N})$, so the the relevant small parameter here is the not so small $(m_\pi/m_N)^{\frac{1}{2}}$. It is therefore not a good testing ground for the above ideas. But $(m_\pi/m_N)^{\frac{1}{2}}$ is still < 1, so at least in some formal sense we can perform a low-energy expansion. In Ref. [25] the chiral expansion was adapted to this reaction and the first few contributions estimated. Again, the lowest order terms all vanish. The formally leading non-vanishing terms —an impulse term and a similar diagram from the delta isobar— are anomalously small and partly cancel. The bulk of the cross-section must then arise from contributions that are relatively unimportant in other processes. One is isoscalar pion rescattering for which two sets of χPT parameters were used: "ste" from a sub-threshold expansion of the πN amplitude and "cl" from an one-loop analysis of threshold parameters. Others are two-pion exchange and short-range $\pi NNNN$ terms, which were modeled by heavier-meson exchange: pair diagrams with σ and ω exchange, and a $\pi\rho\omega$ coupling, among other, smaller terms. Two potentials —Argonne V18 and Reid93— were used. Results are shown in Fig. 2 together with IUCF and Uppsala data. Other χPT studies of this reaction can be found in Ref. [26], while Ref. [27] presents a related analysis of the axial-vector current.

The situation here is clearly unsatisfactory, and presents therefore a unique window to the nuclear dynamics. Work is in progress, for example, on a similar analysis for the other, not so suppressed channels $\to d\pi^+, \to pn\pi^+$ [28].

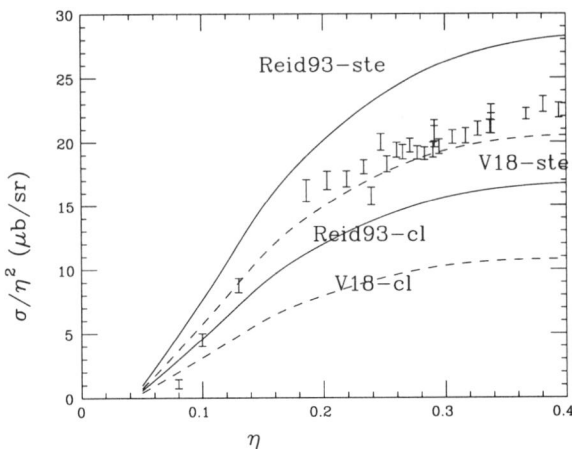

FIGURE 2. Cross-section for $pp \to pp\pi^0$ in μb/sr as function of the pion momentum η in units of m_π for two NN potentials (Argonne V18 and Reid93) and two parametrizations of the isoscalar pion-nucleon amplitude (ste and cl).

CONCLUSIONS

Mesonic χPT is now a mature subject, where the validity of the approach from both phenomenological and internal consistency standpoints has been demonstrated. χPT has also passed several tests in systems with one nucleon, even though some issues —such as delta isobar effects— remain to be fully investigated.

In nuclear physics, only the very initial steps of a systematic chiral expansion have been attempted so far, despite the amount of information available. I have tried to argue that the first results are very auspicious. The chiral expansion has the basic ingredients of nuclear forces, as evidenced by a quantitative fit to two-nucleon data and by the qualitative insights into the size of few-body and isospin-violating forces. It provides a consistent framework for scattering on the nucleon and on light nuclei, which in turn offers a handle on nucleon parameters (as for pion-deuteron scattering and pion photoproduction), successful quantitative postdictions (such as in radiative neutron-proton capture), and quantitative predictions (such as in pion photoproduction). And best of all, it has open problems such as pion production in the pp reaction. There is still a lot to be done: consistent potential/kernel calculations, the above processes away from threshold, many other processes, extension to $SU(3)$ and nuclear matter, to mention just a few topics. Perhaps χPT will then fulfill the role of a long-lacking theory for nuclear physics based on QCD.

Acknowledgements. I am grateful to my collaborators for helping making this research program possible. This manuscript benefitted from communications with E. Korkmaz and T.-S. Park, and criticism by P. Bedaque. This research was supported by the DOE grant DE-FG03-97ER41014.

REFERENCES

1. H. Bethe and R. Peierls, *Proc. Roy. Soc.* **A148** (1935) 146; **A149** (1935) 176.
2. U. van Kolck, in preparation.
3. D.B. Kaplan, M.J. Savage, and M.B. Wise, *Nucl. Phys.* **B478** (1996) 629; T.D. Cohen, *Phys. Rev.* **C55** (1997) 67; D.R. Phillips and T.D. Cohen, *Phys. Lett.* **B390** (1997) 7; K.A. Scaldeferri, D.R. Phillips, C.-W. Kao, and T.D. Cohen, Maryland preprint UMD-PP-97-053 (nucl-th/9610049); D.B. Kaplan, *Nucl. Phys.* **B494** (1997) 471; M. Luke and A. Manohar, *Phys. Rev.* **D55** (1997) 4129; D.R. Phillips, S.R. Beane, and T.D. Cohen, Maryland preprint UMD-PP-97-119 (hep-th/9706070).
4. S. Weinberg, *Phys. Lett.* **B251** (1990) 288; *Nucl. Phys.* **B363** (1991) 3.
5. S. Weinberg, *Physica* **96A** (1979) 327.
6. J. Gasser and H. Leutwyler, *Ann. Phys.* **158** (1984) 142; *Nucl. Phys.* **B250** (1985) 465.
7. J. Bijnens, G. Colangelo, G. Ecker, J. Gasser, and M.E. Sainio, *Phys. Lett.* **B374** (1996) 210.
8. G. Ecker, "Chiral Perturbation Theory", in *Quantitative Particle Physics, Cargèse 1992*, edited by M. Lévy, J.-L. Basdevant, M. Jacob, J. Iliopoulos, R. Gastmans and J.-M. Gérard, NATO ASI Series B, Vol. 311, New York:Plenum, 1993, pp. 101-148.
9. J. Gasser, M.E. Sainio, and A. Švarc, *Nucl. Phys.* **B307** (1988) 779.
10. E. Jenkins and A. Manohar, *Phys. Lett.* **B255** (1991) 558; *Phys. Lett.* **B259** (1991) 353.
11. T.R. Hemmert, B.R. Holstein, and J. Kambor, *Phys. Lett.* **B395** (1997) 89.
12. V. Bernard, N. Kaiser, and U.-G. Meißner, *Int. J. Mod. Phys.* **E4** (1995) 193.
13. V. Bernard, N. Kaiser, and U.-G. Meißner, *Phys. Lett.* **B378** (1996) 337; *Phys. Lett.* **B383** (1996) 116.
14. E. Korkmaz, "Tests of Chiral Perturbation Theory in Threshold Photoproduction of Pions at SAL", these proceedings.
15. C. Ordóñez, L. Ray, and U. van Kolck, *Phys. Rev. Lett.* **72** (1994) 1982; *Phys. Rev.* **C53** (1996) 2086; C. Ordóñez and U. van Kolck, *Phys. Lett.* **B291** (1992) 459.
16. L.S. Celenza, A. Pantziris, and C.M. Shakin, *Phys. Rev.* **C46** (1992) 2213; J. Friar and S.A. Coon, *Phys. Rev.* **C49** (1994) 1272; C.A. da Rocha and M.R. Robilotta, *Phys. Rev.* **C49** (1994) 1818; *Phys. Rev.* **C52** (1995) 531; *Nucl. Phys.* **A615** (1997) 391; J.-L. Ballot, M.R. Robilotta, and C.A. da Rocha, *Int. J. Mod. Phys.* **E6** (1997) 83; M.J. Savage, *Phys. Rev.* **C55** (1997) 2185;

G.P. Lepage, nucl-th/9706029; N. Kaiser, R. Brockmann, and W. Weise, nucl-th/9706045.
17. U. van Kolck, *Phys. Rev.* **C49** (1994) 2932.
18. U. van Kolck, U. of Texas Ph.D. dissertation (1993); Washington preprint DOE/ER/40427-13-N94 in preparation; *Few-Body Syst. Suppl.* **9** (1995) 444.
19. U. van Kolck, J.L. Friar, and T. Goldman, *Phys.Lett.* **B371** (1996) 169; and in preparation.
20. S. Weinberg, *Phys. Lett.* **B295** (1992) 114.
21. T.-S. Park, D.-P. Min, and M. Rho, *Phys. Rev. Lett.* **74** (1995) 4153; *Nucl. Phys.* **A596** (1996) 515.
22. S.R. Beane, C.Y. Lee, and U. van Kolck, *Phys. Rev.* **C52** (1995) 2914; S.R. Beane, V. Bernard, T.S.H. Lee, U.-G. Meißner, and U. van Kolck, *Nucl. Phys.* **A618** (1997) 381.
23. P. Wilhelm, Mainz preprint MKPH-T-97-8 (nucl-th/9703037).
24. MAMI Proposal A1/1-96, Spokesperson: R. Neuhausen.
25. T.D. Cohen, J.L. Friar, G.A. Miller, and U. van Kolck, *Phys. Rev.* **C53** (1996) 2661; U. van Kolck, G.A. Miller, and D.O. Riska, *Phys. Lett.* **B388** (1996) 679.
26. B.Y. Park, F. Myhrer, J.R. Morones, T. Meissner, and K. Kubodera, *Phys. Rev.* **C53** (1996) 1519; E. Gedalin, A. Moalem, and L. Razdolskaya, Negev preprint BGU-PH-97-07 (hep-ph/9702406); T. Sato, T.S.H. Lee, F. Myhrer, and K. Kubodera, South Carolina preprint USC-NT-97-01 (nucl-th/9704003).
27. T.-S. Park, D.-P. Min, and M. Rho, *Phys. Rept.* **233** (1993) 341.
28. C.A. da Rocha, G.A. Miller, T.-S. Park, and U. van Kolck, in progress.

Quark and Gluon Structure of the Nucleon

U. Straumann

University of Heidelberg, Germany

Abstract. Recently deep inelastic physics in the very low Bjorken x region has made significant progress due to the large kinematical region covered by the HERA experiments by now. Stringent tests of perturbative QCD predictions on the evolution of the parton density functions in the nucleon are possible over 5 (4) orders of magnitude in x (Q^2). Newer results from structure function studies on scaling violations, the charm contribution and the effect of longitudinally polarized virtual photons give a **consistent picture of a strongly rising gluon density distribution towards low** x. Some aspects of the search for an indication of BFKL like behaviour in the low x regime are presented and data covering the region at low Q^2 are shown, where perturbative QCD is believed to be no longer valid. - This contribution concentrates on HERA results, but other data are also taken into account where appropriate.

In a second part the events observed at HERA in excess over the standard model prediction at very high Q^2 are discussed.

I STRUCTURE FUNCTION DATA

Deep inelastic lepton proton scattering (DIS) is generally desbribed by two lorentz invariant quantities, for instance $Q^2 = -q^2 = -(k-k')^2$, the invariant mass of the virtual boson exchanged between the scattered lepton and the proton, and $x = Q^2/2pq$ the Bjorken scaling variable (k, k', p are the 4-momenta of the incoming and outgoing lepton and the incoming proton respectively). Also often used variables are $y = pq/pk \approx Q^2/xs$ and $W = (p+q)^2 \approx ys$, the invariant mass of the final state hadronic system.

In the one photon exchange approximation the DIS cross section is given by

$$\frac{\partial^2 \sigma}{\partial x \partial Q^2} = \frac{2\pi \alpha^2}{xQ^4}((1+(1-y)^2)F_2(x,Q^2) - y^2 F_L(x,Q^2)) \qquad (1)$$

$$F_2(x,Q^2) = \sum_{flavours} (e \cdot (f(x,Q^2) + \bar{f}(x,Q^2))) \qquad (2)$$

By measuring the DIS cross section we can derive the structure function F_2 and therefore measure directly the singulett quark parton density functions $f + \bar{f}$. At high y the contribution F_L from longitudinally polarized virtual photons to the cross section can not be neglected and usually predictions by perturbative QCD are used [13], which depend however on the gluon densities themselves (see section D).

Since the first year of HERA operation in 1992 we have seen every year a *significant decrease of errorbars* due to better and better understanding of the detector systematics, improved analysis methods and of course higher luminosities available. It is fair to say that H1 and ZEUS at least w.r.t. the inclusive measurements have become precision experiments. A closer look at the 1994 data [1,2] shows, that the data from HERA have overlap at its lower kinematical boundaries with fixed target data for instance from E665. In the overlap region the errorbars are comparable and the relative normalisation is compatible [3]. In addition global QCD fits to the data with unconstraint normalisation factors for the various experiments show good agreement between fixed target and ep collision data.

Also the kinematical range in the variables x and Q^2 used for the analysis has been increased every year due to several reasons: Apart from higher statistics, a better understanding of the detectors allows to include higher y data (relevant for F_L). Furthermore to study the transition region to the nonpertubative regime at very low Q^2, data sets with the ep vertex shifted along the beam axis are now available to allow the detection of electrons with smaller scattering angles.

Structure function data from HERA cover more than 5 orders in magnitude in x and more than 4 magnitudes in Q^2. Figure 1 shows combined F_2 data from 1994.

In the upgrade of 1995 ZEUS has installed a beam pipe calorimeter BPC, which covers even lower electron scattering angles, while H1 has replaced its full "backward" detector system (located in the direction of scattered electrons) by a new lead - scintillator calorimeter in spaghetti technique and a new tracking system, which allows significantly higher precision for the inclusive measurements. Apart from a higher accuracy in the kinematical quantities of the scattered electron, this allows also a more complete hadronic final state measurement. Some data from the upgraded detectors will be shown in section E.

A Scaling violations in F_2 at low x

The standard QCD analysis of F_2 to derive parton densities proceeds in three steps:

1. First a parametrisation for the parton density functions at a given Q_0^2 is chosen, for example

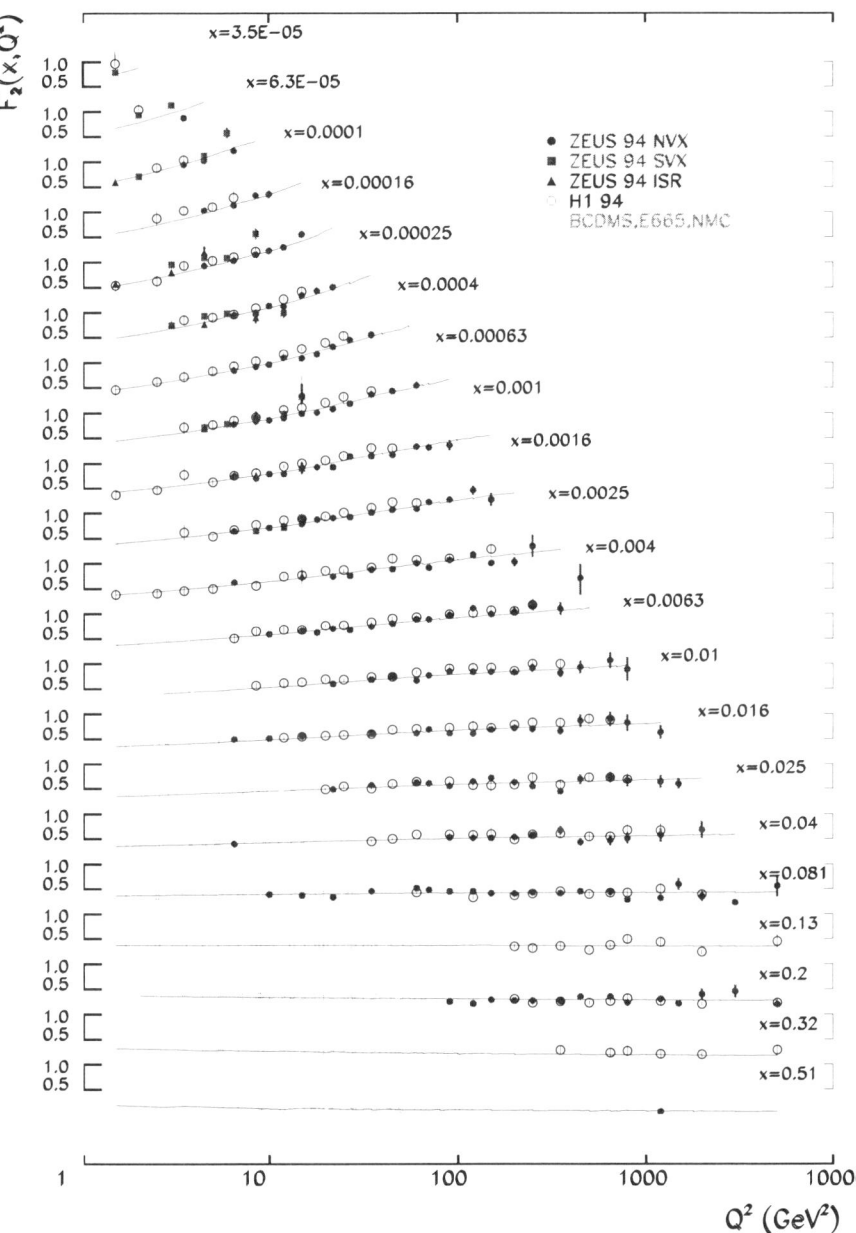

FIGURE 1. F_2 structure function data from H1 and Zeus (taken 1994) and various fixed target experiments as a function of Q^2. Nominal (NVX) and shifted vertex (SVX) data as well as an analysis of initial state QED radiation (ISR) events are combined.

$$x(f(x, Q_0^2) + \bar{f}(x, Q_0^2)) = A_s x^{B_s}(1-x)^{C_s}(1 + D_s x + E_s \sqrt{x}) \qquad (3)$$

$$xg(x, Q_0^2) = A_g x^{B_g}(1-x)^{C_g} \qquad (4)$$

where g is the gluon density function, A, B, C, D, E are parameters to be fitted to the data. For the low x region the term x^B is dominating, its form is inspired by theoretical thoughts in the "double leading log approximation".

2. Next the QCD evolution equation (DGLAP) are used to derive the parton density functions $f(x, Q^2), g(x, Q^2)$ at all Q^2, where data are existing. This is nowadays usually done in next to leading order.

3. Finally the parameters A, B, C, D, E are fitted to all data simultanously.

This relatively simple parametrisation fits the data over the full kinematical range extremely well. At very low x ($x < 10^{-3}$) the valence quarks are neglectable and DGLAP tell us, that the quark singulett distributions $f + \bar{f}$ are directly connected to the gluon density: *At low x the gluon drives F_2.* DIS inclusive measurements at such low x measure the gluon density function and thus provides us with a direct and accurate probe for the QCD dynamics in a large kinematical range. Figure 2 shows the gluon density as a function of x, which results from such a fit. The rise towards low x can be clearly seen. It is also instructive to observe, that the steepness of the rise increases with increasing Q^2, as required by the QCD evolution equations.

B QCD predictions for low x and BFKL

Apparently the DGLAP equations are able to describe the Q^2 dependance of the rising F_2 very well. The question arises, whether QCD is also able to describe this rise of the gluon density intrinsically. At very low x the higher order terms $ln^n(1/x)$ become significant. The BFKL [6] formalism tries to evolve the parton densities into the direction of $ln(1/x)$, keeping $\alpha_s(Q^2)$ constant. This results in a prediction for the gluon density:

$$xg(x) \approx x^{-0.5} \qquad (5)$$

Although this is qualitatively the right direction, the rise is too steep compared with the data, which has been shown recently in [9].

Another observation has been made a long time ago by several authors: Looking at the DGLAP equations at $x < 10^{-3}$ shows, that the splitting function P_{gg} (gluon couples to gluon) can be simplified to

$$P_{gg}(x) \approx \frac{6}{x}. \qquad (6)$$

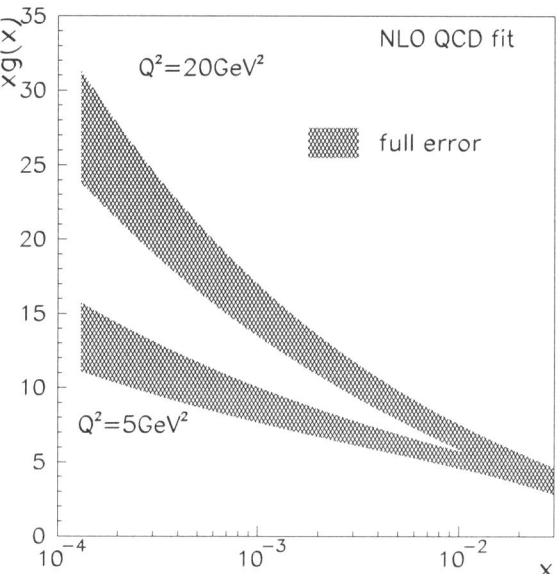

FIGURE 2. The gluon density as a function of x, obtained from a NLO QCD fit to the H1 - F_2 data from 1994

Assuming further, that the quark densities are small compare to the gluon density, the DGLAP equation for the gluon can now be solved analytically, using the leading order term for $\alpha_s(Q^2)$ only [4]:

$$xg(x, Q^2) = e^{\sqrt{\gamma \cdot ln(T) \cdot ln(\frac{1}{x})}} \qquad (7)$$

where T is only a function of Q^2 and the QCD - Parameter Λ and γ is a well defined number, depending on the number of flavours involved. This is sometimes called *double asymptotic scaling* (DAS) and it has been shown also by Yndurain and others, that this expression is numerically equivalent to

$$xg = x^{-\lambda} \qquad \lambda \approx 0.3..0.4 \qquad (8)$$

which is not incompatible with the results of the QCD fits. However to make predictions with a better accuracy - in order to make full use of the experimental information - next to leading order terms were included and a very nice agreement with the 1994 data has been achieved [5].

The question arises now, whether the BFKL mechanism is still needed to explain any feature of the data. A. Muller et al. have proposed an experimental test, which basically selects events at HERA which have simultanously a low x_{Bj} value determined from the scattered electron and a distinct "forward" jet (arising from a parton with low scattering angle) with a high value of x_q determined from the jet quantities ($\theta_{jet} < 20°$). Requiring now in addition, that the transverse momentum of the jet p_T^{jet} is comparable with Q^2 determined by the scattered electron, selects a situation, where a gluon ladder is formed for the momentum transfer between the high x_q struck quark and the low x_{Bj} quark pair scattered by the intermediate photon of the DIS process without having any transverse momentum left. The standard DGLAP formalism would suppress such events, while the BFKL due to the large x range would allow for a significantly larger rate in this kinematical region. Figure 3 shows data from H1 together with BFKL and standard DGLAP predictions. Although the data points are significantly above the DGLAP prediction, they are not as high as BFKL would like it. (For more details on the various calculations see in reference [7]).

Another test for BFKL predictions has been performed by the D0 experiment at the TEVATRON. They looked for the angular correlation in Φ for events with two jets. In leading order (no QCD radiation) the two jets should be directly opposite, while gluon radiation will allow for deviations from this correlation. In the BFKL scenario the amount of gluon radiation and therefore this angular correlation should now decrease strongly with increasing ladder length, i.e. with increasing rapidity difference between the two jets. In [8] the average angular correlations for two jets events from D0 as a function of the rapidity difference between the two jets are compared with a standard prediction and a BFKL calculation. It shows, that the data points are definitely much closer to the standard prediction than to the BFKL calculation.

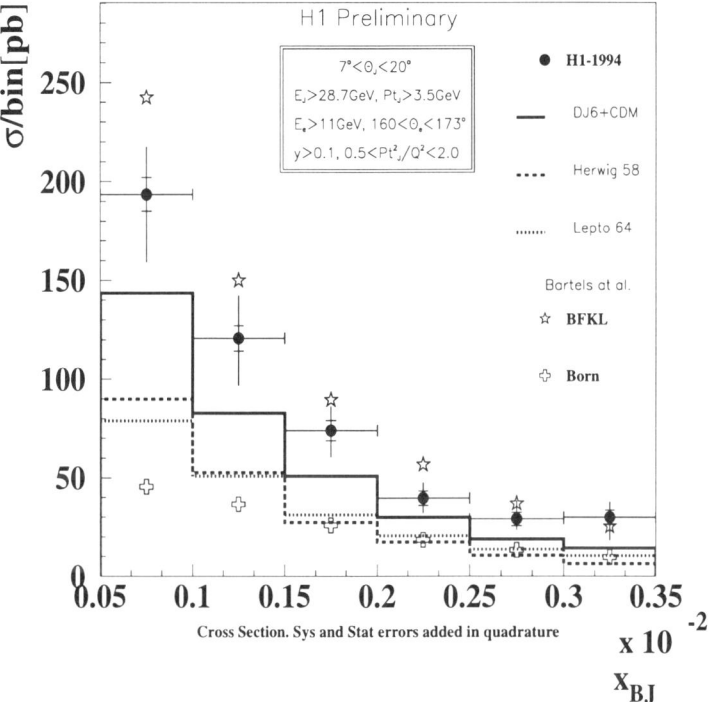

FIGURE 3. The rate of forward jets as a function of x. LEPTO and HERWIG denote standard Monte Carlo programs, while CDM (color dipole model) is a Monte Carlo prediction, which also enlarges the gluon radiation. See text for more details.

So the situation with BFKL is unclear. Both the steepness of the structure functions at low x and the D0 study seemed to indicate, that present BFKL calculations are not in any way close to the data. However the forward jet study at HERA shows, that the standard DGLAP predictions are not very good as well.

C The charm contribution

Both HERA experiments have analysed the D^* production in DIS [10], [11]. Charmed Mesons can be generated in DIS either by an intrinsic charm (IC) component of the quark sea or in a photon - gluon fusion (PGF) process, where a $c\bar{c}$ pair of quarks is generated. D^* are identified by their decay

$$D^* \to D^0 \pi_s^+ \to (K^-\pi^+)\pi_s^+, \qquad (9)$$

looking at the mass difference $\Delta M = M(K\pi\pi) - M(K\pi_s)$, and the cross section is then calculated using the branching ratios from the PDG. An analysis of the kinematics shows, that the distribution of the fractional momentum x_{D^*} of the D^* in the photon proton system allows to distinguish between intrinsic charm and the photon gluon fusion process. While for IC this distribution would be peaked at large x_{D^*}, in the case of PGF it is centered at $x_{D^*} < 0.5$. Both collaborations have used this distribution to show, that PGF is clearly dominating the charm production in DIS in this kinematical range, and gives therefore a sensitive handle to study the gluon density distribution in the nucleon.

From this data the structure function F_2^c has been determined in the standard way, using LEP results for the probablity, that a D^* is formed from a charm quark. The results are shown in figure 4. The total fraction of the charmed part of the structure function at this low x region ($x \approx 10^{-3}$ is as high as $0.237 \pm 0.021 \pm 0.041$ [10], much higher than at high x, where EMC had shown, that this fraction is in the percent region. The data compare nicely to a calculation using the gluon density function, which resulted from the QCD fit to the structure function described in section A, and they thus confirm the gluon dominance at low x.

D The longitudinal structure function

The two structure functions F_2 and F_L in formula (1) are related to the transversely (σ_T) and longitudinally (σ_L) polarized virtual photon proton cross section:

$$F_2 \sim \sigma_T + \sigma_L \qquad F_L \sim \sigma_L \qquad (10)$$

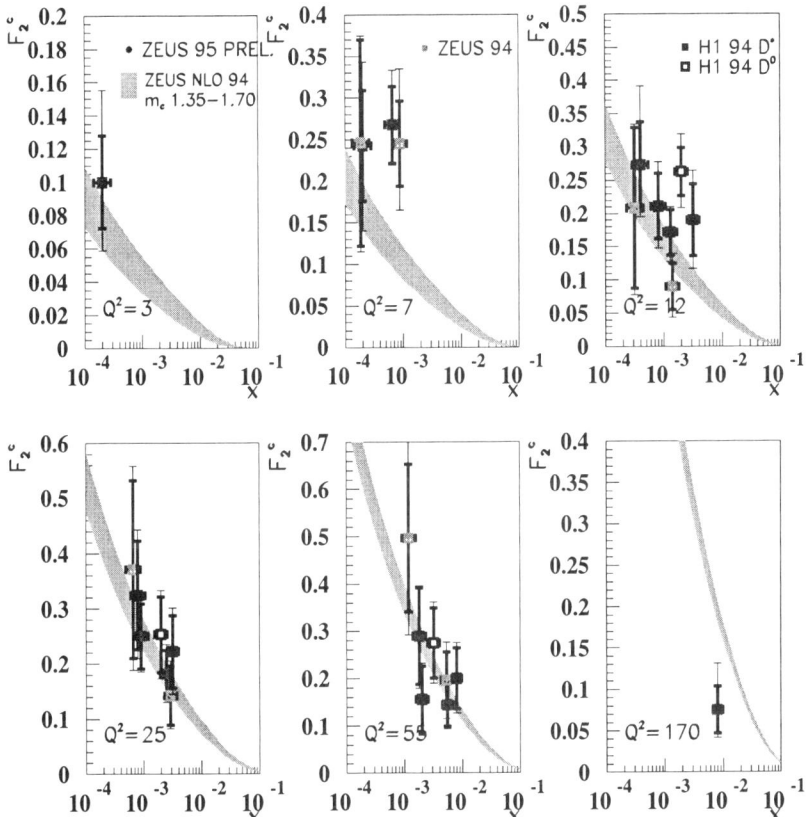

FIGURE 4. The combined results of H1 and ZEUS on F_2^c, compared to the prediction using a next-to-leading order QCD fit to the 1994 structure function data of ZEUS.

Since longitudinally polarised photons can only interact with quarks, which have non zero transverse momentum, F_L is zero in the naive quark parton model. In QCD F_L acquires a non zero value due to gluon pair production or gluon radiation. Next-to-leading order predictions [13] show, that at low x F_L is generated to over 90% from the gluons. A measurement of F_L is therefore another interesting probe for the gluon density distribution apart of representing a significant QCD test by itself.

Inserting (10) into formula (1) results in

$$\sigma^{eff}(x,Q^2) \sim \sigma_T(x,Q^2) + \epsilon \cdot \sigma_L(x,Q^2) \tag{11}$$

$$\epsilon = \frac{2(1-y)}{1+(1-y)^2} \tag{12}$$

At low y ϵ is close to one and we just measure F_2. However approaching $y = 1$ we become sensitive to the relative strength of the transverse and the longitudinal part of the cross section. Analysing the y dependance of the inclusive DIS cross section in principle allows therefore a measurement of F_L.

H1 has published such an analysis with their 1994 data. A full NLO QCD fit was performed to determine F_2 using the evolution equation to all data with $y < 0.35$, where the influence of F_L to the data is small. The difference between using F_L from the QCD prediction and $F_L = 0$ in this low y region is below 1.9% on the F_2 extrapolation and is included in the systematical error. Then F_2 is extrapolated to $y = 0.7$ and compared with the data. Figure 5 shows the double differential cross section in 6 different Q^2 bins as a function of x together with the extrapolated cross section assuming $F_L = 0$, $F_L = F_L^{QCD}$, $F_L = F_2$ respectively. The data clearly allows to exclude the two extreme assumptions $F_L = 0$ and $F_L = F_2$.

Obviously there is some model dependance in this method, since QCD knowledge of how to extrapolate F_2 into the high y region enters the analysis. A more model independant way of measuring F_L would be to change the accelerator energy (the proton or electron beam momentum, or both). This would allow to vary y and therefore ϵ keeping x and Q^2 constant (remember: $Q^2 = xys$). Such measurements are being discussed presently at HERA, studies show however, that a large amount of statistics would be needed.

Figure 6 shows F_L as a function of x for $y = 0.7$, obtained by subtracting the contribution from the extrapolated F_2 from the measured cross section. The data is fully compatible with the QCD prediction and thus confirms again the rising gluon distribution, although the errorbars are still very large.

This result can be compared with a recent analysis of NMC data together with older results from BCDMS ad CDHS, which all are at much higher x, but they extrapolate smoothly to the data presented here [14].

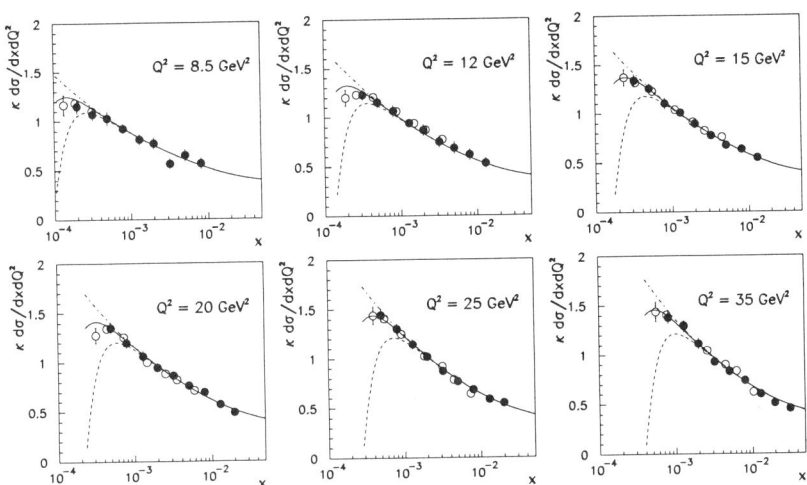

FIGURE 5. The double differential cross section as a function of x for different Q^2. The open points are from a newer analysis, the black points are from the published 1994 data. The solid line corresponds to the fit, assuming $F_L = F_L^{QCD}$, dashed-dotted line is for $F_L = 0$, dashed line is for $F_L = F_2$

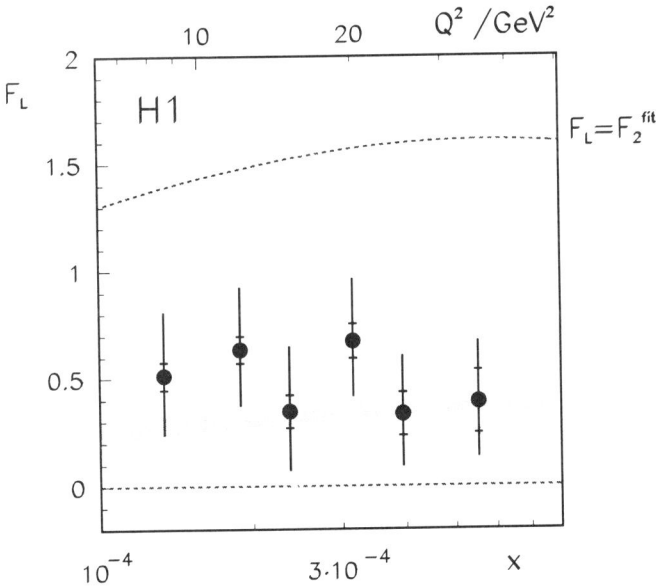

FIGURE 6. The resulting F_L as a function of x at $y = 0.7$. The dashed lines indicate the two extreme solutions described in the text, the band gives the predicted value for F_L using QCD extrapolation from lower y.

E Transition to the photoproduction limit

As we have seen, QCD predicts a form of $x^{-\lambda}$ for F_2 at low x. Expressing this behaviour in the virtual photon proton mass dependance ($W^2 = Q^2/x$) gives us for the virtual photon proton cross section

$$\sigma^{\gamma^*p} \sim W^{2\lambda} \qquad \lambda \approx 0.3..0.4 \qquad (13)$$

However in the photoproduction limit ($Q^2 = 0$) the total cross section has been measured at HERA as well and is compatible with a universal exponent of $\lambda = 0.08$, also being observed in hadron hadron total cross section. This universality is explained within the Regge formalism by a soft pomeron exchange by Donnachie and Landshoff and others.

At some Q^2 the transition between the two regimes must take place. It is often assumed that this transition is the place, where perturbative QCD breaks down. However it should be noted, that already the QCD evolution equation (DGLAP) predict, that the exponent becomes smaller and smaller, when you go to lower Q^2.

Both experiments at HERA have published data down to Q^2 significantly lower than 1 GeV^2, using shifted vertex HERA runs and the new beam pipe calorimeter of ZEUS [15], [16] (see figure 7). Several ways of parametrically combining perturbative QCD with Regge inspired models are available, but none of them describes the data accurately yet.

II THE HIGH Q^2 EXCESS EVENTS

Recently both HERA experiments reported the observation of an excess of events over the standard model prediction at $Q^2 > 15'000$ GeV^2 [17,18].

At lower Q^2 there is perfect agreement with F_2 calculated from the present parton density functions. For instance H1 observes 443 neutral current events for $Q^2 > 2500$ GeV^2, where 427 ± 38 are expected. However at $Q^2 > 15'000$ GeV^2 12 events were observed, where 4.71 ± 0.76 are expected. Zeus observes 4 events in a similar kinematical region ($x > 0.55, y > 0.25$), where they expect 0.91 ± 0.08. Numerous other cuts were tried also, and the statistical probability for a fluctuation is in most cases around 1%.

The H1 events seemed to be gathered around an invariant mass of the electron quark system of around 200 GeV (equivalent to $x \approx 0.45$), indicating a possible s-channel resonance. The ZEUS data on the other hand is more scattered in x, in average a sharp resonance is therefore unlikely. (Further statistics taken after this conference showed a broader scattering of events also in H1, while the statistical significance of any excess stayed about the same).

There were numerous tests, to find out, what could have gone wrong. First of all the systematical errors of the analysis had been checked again, confirming

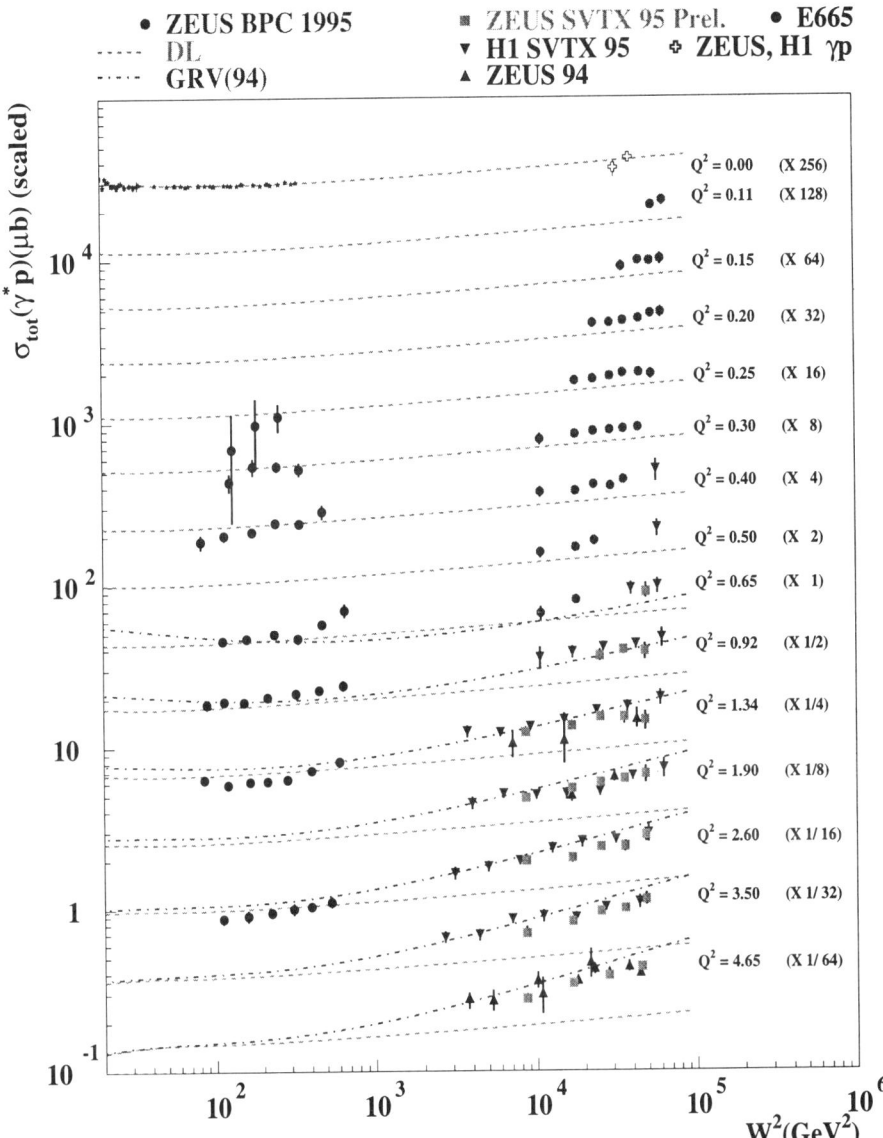

FIGURE 7. The cross section as a function of $W^2 = \frac{Q^2}{x}$ for different Q^2. Data taken in 1995 with the ZEUS beam pipe calorimeter (BPC), and with shifted vertex running (SVTX) together with data from the Fermilab collaboration E665 are shown. DL denotes a prediction using the model of Donnachie and Landshoff (soft pomeron). GRV(94) is a prediction by Glueck, Reya and Vogt, based on a pure perturbative QCD fit to the 1994 data.

2.3% (H1 and Zeus) for the luminosity measurement, 3% for the electromagnetic energy scale (H1 and Zeus), 4% for the hadronic energy scale (H1 only, Zeus does not use this for this analysis). Both collaborations and various other people have checked the parton density functions, all coming up with possible variations between 5 and 8% on the predicted event rate. Similar the value of α_s and higher order QED corrections were checked, no effects of more than 4% and 2% respectively could be explained.

In the first three months after the publication of the observed effect, 42 papers with possible explanations were published in "hep-ph". They can be grouped in 4 categories:

1. *Non standard QCD effects:* It turns out, that the parton density functions would have to be modified only relatively slightly at $x \approx 1$, however they need to have very strange shapes.

2. *Contact interactions* of new non standard model interaction: Several such analyses have been done, the mass of the exchange particle needs to be in the order of 2 TeV, depending on the assumptions of the coupling.

3. *Leptoquarks*, s-channel production of a new resonance with simultanously leptonic and baryonic quantum numbers: Here exist severe limits from TEVATRON, the newest mass limit is 230 GeV, assuming the resonance decays 100% back to electron and quark. If we assume a lower branching for this decay, the limit becomes less constraint, and an effect at HERA would be possible.

4. *R-parity violating production of a supersymmetric squark:* This is the most appealing explanation. Constraints of the double beta decay and atomic parity violation experiments could not allow such a strong coubling to a $\tilde{u}$ quark, and the Brookhaven $K^+ \to \pi^+ \nu \nu$ experiment makes $\tilde{c}$ unlikely, such that the best candidate would be $\tilde{t}$ production.

The last possibility would also be good for the explanation of a few strange events, which have been observed by H1: There are so far 3 events with a high momentum muon and no electron in the final state. The kinematics and the cross section is such, that a W production is unlikely [20]. These events could be explained for instance by $\tilde{t}$ gauge decays [19]. Unfortunately Zeus has not reported yet any such event.

Despite the relatively high statistical probability for a new effect we have to be very carefull in interpreting this as a sign for new physics beyond the standard model. Specially the discrepancies between the two experiments have to be clarified, and most of this can probably only be done, when more statistics is available.

REFERENCES

1. H1 Collaboration, S. Aid et al., Nucl. Phys. **B470** (1996) 3.
2. ZEUS Collaboration, M. Derrick et al., Z. Phys. **C72** (1996) 399.
3. Fermilab E665 Collaboration, M.R. Adams et al. Phys. Rev. **D54** (1996) 3006.
4. A. DeRujula et al., Phys. Rev. **D10**, 1649 (1974).
   R.D. Ball and S. Forte, Phys. Lett. **B336**, 77 (1994).
5. R.D. Ball and S. Forte, *Double Scaling Violations*, Workshop on Deep Inelastic Scattering, Rome 1996.
6. V.S. Fadin, E.A. Kuraev and L.N. Lipatov, Phys. Lett. **B60** (1975) 50.
7. H1 Collaboration, *Forward Jet Production at HERA*, 28th International Conference on High Energy Physics, Warsaw 1996.
8. D0 Collaboration, PRL **77** (1996) 595.
9. I. Bojak and M. Ernst, CERN preprint hep-ph/9609378.
10. H1 Collaboration, C. Adloff et al., Z.Phys. **C72** (1996) 593.
11. ZEUS Collaboration; J.Breitweg et al., DESY preprint 97-089.
12. H1 Collaboration, C. Adloff et al., Phys. Lett. **393B** (1997) 452.
13. G. Altarelli and G. Martinelli, Phys. Lett. **B76** (1978) 89; M. Glück and E.Reya, Nucl. Phys. **B145** (1978) 24.
14. E. Kabuss, *Final Results from NMC*, 5th International Workshop on Deep Inelastic Scattering and QCD, Chicago 1997.
15. H1 Collaboration, C. Adloff et al., DESY preprint 97-042.
16. ZEUS Collaboration; J.Breitweg et. al. DESY preprint 97-135.
17. H1 Collaboration, C. Adloff et al., Z. Phys. **C74** (1997) 191.
18. Zeus Collaboration, J. Breitweg et al., Z. Phys. **C74** (1997) 207.
19. T. Kon and T. Kobayashi, CERN preprint hep-ph/9704221.
20. H1 Collaboration, T. Ahmed et al., DESY preprint 94-248.

Prospects for a Neutrino Oscillation Experiment at the National Spallation Neutron Source

B.Bugg[*], N.Cohn[*], L.Chatterjee[*], Y.Efremenko[†], A.Fazely[‡],
T.Gabriel[†], Y.Kamyshkov[†], F.Plasil[†], and R.Svoboda[**]

[*] *University of Tennessee - Knoxville, Knoxville, TN 37996-1200*
[†] *Oak Ridge National Laboratory, Oak Ridge, TN 37831-6123*
[‡] *Southern University, Baton Rouge, LA 70813*
[**] *Louisiana State University, Baton Rouge, LA 70803-4001*

Abstract. Plans for the construction of a neutron spallation source with high intensity (4 MW and 1 MW) and short extraction time (1 μs and 0.5 μs) are presented. It is shown that such a facility would produce a neutrino pulse 20 times more intense than ISIS and 6 times more intense than LAMPF, and that a 1.5 kton detector would allow performing a neutrino oscillation experiment ($\bar{\nu}_\mu \to \bar{\nu}_e$) roughly 20-50 times more sensitive than existing results from KARMEN and LSND. Preliminary parameters of a typical detector are presented, including background estimates.

INTRODUCTION

Conservation of lepton flavor is a relatively *ad hoc* part of the Standard Model of particle physics, even though no exceptions to this conservation law have ever been found. The implications of the discovery of lepton flavor violations will not be discussed here. Suffice it to say that experiments to search for neutrino flavor oscillations (and hence a violation to the lepton flavor conservation law) now span almost twenty years. Many different neutrino sources have been used, both artificial (accelerators, reactors) and natural (the sun, cosmic rays). To date, there is no conclusive evidence for the existence of such oscillations, though several claims exist from solar, cosmic ray, and low-energy stopping muon source experiments. This paper will concentrate on the latter.

Stopping muon neutrino sources have several advantages over other types: (1) a well-understood ν flavor spectrum, (2) no contamination from kaons, (3) no contamination from $\nu - N$ pion production, and (4) a well-understood, stable energy spectrum. Such experiments have typically been limited by low intensity (e.g. KARMEN), long beam extraction times (e.g. LAMPF), and single-distance running (e.g. KARMEN and LAMPF). In this paper we discuss the construction of a stopping muon facility at the National Spallation Neutron Source (NSNS) that would address all three of these problems.

THE NATIONAL SPALLATION NEUTRON SOURCE

A project to develop a conceptual design for a 4 MW spallation neutron source was initiated in February, 1995. The new project, entitled the National Spallation Neutron Source (NSNS), is a collaboration of several national labs (LBL,LANL,BNL,ANL,ORNL). It is likely (but not certain) that the facility will be located at Oak Ridge. It will consist of a 1 GeV proton LINAC, two proton storage rings, two target stations, and several beam dumps for testing and tuning.

Spallation neutron sources produce copious amounts of neutrinos. Neutrino experiments exist at two "beam stop" sources, LANL and ISIS. The NSNS would be significantly more intense than LANL and ISIS, and would have a short pulse structure (similar to ISIS) that would reduce non-beam associated backgrounds significantly compared to LANL. In addition, plans call for two targets with a separation of 143 meters. This would be extremely attractive for neutrino experiments, since it would allow measurements from two different distances simultaneously with the same detector.

Existing accelerator-based neutron sources around the world are summarized in Table 1. Both target stations of the NSNS are also listed for comparison.

Each target station will have a target whose size is designed to match that of the beam, which will be roughly $70mm \times 10mm$ at this point. The targets will consist mostly of bulk mercury which is circulated for cooling. Protons from the 1 GeV LINAC will interact in the target and produce π^+ and π^- along with the spallation neutrons. Many π^+ will survive long enough to decay, producing a pulse of monoenergetic ν_μ with the characteristic decay time of 26 ns. The μ^+ from the π^+ decay will decay to produce 3-body continuous spectra of $\overline{\nu}_\mu$ and ν_e with the characteristics decay time of 2.2 μs. Since the

TABLE 1. Existing Accelerator-Based Neutron Sources.

Facility	Location	Energy (MeV)	Power (MW)	Pulse Structure
IPNS	Argonne	500	0.014	
LAMPF	Los Alamos	800	0.80	0.07 sec
ISIS	Rutherford	800	0.20	100 ns
SINQ	PSI	590	0.89	continuous
NSNS (target 1)	Oak Ridge	1000	1.0,2.0,4.0	500,1000 ns
NSNS (target 2)	Oak Ridge	1000	1.0	500 ns

ν_μ are prompt, they can be eliminated by a time cut. In addition, if a target with hydrogen is used, then $\overline{\nu}_\mu \to \overline{\nu}_e$ have a charged-current interaction that produces a double coincidence between a prompt positron and a gamma from the absorption of the recoil neutron. This can be used to distinguish them from the unoscillated $\overline{\nu}_\mu$, which are below the charged-current threshold. One can then search for $\overline{\nu}_\mu \to \overline{\nu}_e$ appearance.

Of course, some π^- and μ^- *will* survive long enough to produce the C-conjugate chain to the above, and hence background to such a search. This will be discussed in more detail in the next section.

DETECTOR CHARACTERISTICS

The desired characteristics for an NSNS detector, tentatively given the name ORLANDO, for Oak Ridge Large Area Neutrino DetectOr, can be roughly scaled from the LSND detector numbers, and are generally consistent with preliminary numbers presented here.

The LSND experiment at LAMPF measured 22 $\overline{\nu}_e$-like events, with an expected background of 4.6 ± 0.6 events. [1]. LSND is located 30 meters from the LAMPF beam stop where their 167-ton detector is illuminated by both decay-at-rest and decay-in-flight neutrinos. Calculations of the neutrino beam are believed to be good to 7% based on earlier calibration measurements and checks with the $\nu -^{12}C$ cross-section. With these measurements and experimental parameters, they deduce an oscillation probability of $0.31 \pm 0.12 \pm 0.05\%$. Figure 1 shows the region in oscillation parameter space allowed by their results. Also shown are 90% c.l. exclusion regions from Bugey and BNL E776.

ORLANDO will look for $\overline{\nu}_e$ appearance from the oscillation of $\overline{\nu}_\mu$. Sensitivity is therefore determined by the number of signal events to be expected with given oscillation parameters as compared to the number of background events expected.

The short spill time reduces non-beam associated backgrounds to essentially zero compared to other backgrounds. We propose to locate ORLANDO between the two target stations, at a 90 degree angle to both targets. Preliminary studies show that decay-in-flight backgrounds should be very small in this position. The main background will then be $\overline{\nu}_e$ created in the beam stops themselves.

Because we have a proton beam, the production of π^+ is somewhat favored over the production of π^-. Our calculations, based on the pion production measurements of Lemaire, et al. [2] and using GEANT/FLUKA transport, give a π^+/π^- ratio of 1.4. This is roughly consistent with the model of Berman and Plische [3] who give the ratio of the cross-sections $(\sigma_{\pi^+}/\sigma_{\pi^-})$ at 1 GeV as $(152.9\text{mb}/97.5\text{mb})=1.6$. Here their values have been exprapolated from Pb

(Z=82) to Hg (Z=80) using the suggested $Z^{1/3}(N^{2/3})$ dependence for π^+ (π^-).

After production, most of the π^- are efficiently absorbed in the Hg target, but the beam stop also has helium and water filled cooling channels and moderators, lead reflectors, and structural aluminum and iron. π^- that get into the low-Z material regions have a better chance of decaying. Preliminary calculations show that about 0.56% of the π^- survive long enough to decay, as compared to essentially all of the π^+. About 68% of these occur in the aforementioned light-material regions. Subsequently, about about 17% of the daughter μ^- survive long enough to decay. Thus the preliminary calculation gives an expected $\bar{\nu}_e/\bar{\nu}_\mu$ ratio of 6.8×10^{-4}. If one makes a timing cut to reject μ^- decay in high-Z materials (it must occur quickly to avoid capture), then this ratio would be roughly 68% lower.

The fluxes per year are expected to be roughly 15 (3.8) times those realized at LAMPF [4] for a 4 MW (1 MW) beam. This factor is due to: (1) enhanced pion production at 1000 MeV versus 800 MeV, and (2) higher power, (3) 70% facility duty factor for the NSNS. Thus a 20 event signal for $\bar{\nu}_\mu \to \bar{\nu}_e$ at LAMPF becomes a 300 (75) event signal over three years for a 4 MW (1 MW) target and an LSND-size detector at the same 30 m distance. However, a larger detector (1-2 kilotons) is desirable to allow sufficient statistics from the far beam stop. This would raise the signal to $(1500/167)*300 = 2700$ events.

Given the above signal and background, figure 1 shows the sensitivity of ORLANDO for three years running at the close and far beam stops. It is clear

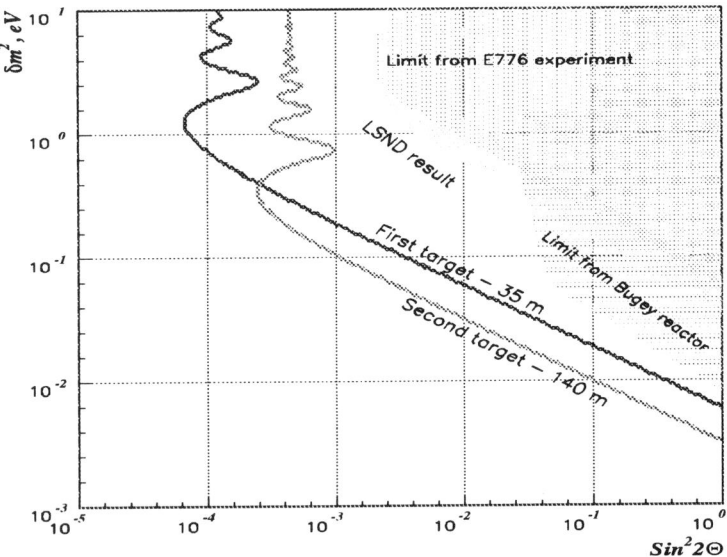

FIGURE 1. Sensitivity of ORLANDO to $\bar{\nu}_\mu \to \bar{\nu}_e$ oscillations.

that ORLANDO would push a factor of 20-50 beyond current experiments (including the upgraded KARMEN detector) in sensitivity. Of course, in addition to the above, multiple beam stops and a 500 ns extraction time would also improve reliability of the result and reduce systematic backgrounds.

CONCLUSIONS

NSNS is a high priority for a broad section of the U.S. scientific community due to the lack of high-power, modern accelerator-based neutron sources in the U.S., the cancellation of the ANS, and the shutdown of several reactor-based neutron sources due to age and safety considerations. Because of this, construction is planned to start in FY1999 after detailed designs are completed in FY1997 and FY1998. It is anticipated that Target 1 would become available at 1 MW in 2004, with an upgrade to 2 MW in 2006. Target 2 would also come on-line at that time. Final upgrade to 4 MW would be completed for Target 1 in 2010.

Since it is anticipated that ORLANDO might have to located underground to reduce dead time due to stopping muons, a neutrino "cave" will have to be included in NSNS plans before the middle of 1998 to be ready for the start of construction. Construction of ORLANDO would then be geared to coincide with Target 1 availablility in 2004, meaning that detector design would have be to done in FY1999 and FY2000, with construction to begin in FY2001.

In conclusion, the NSNS offers the opportunity to advance the research programs pioneered at Los Alamos and Rutherford, essentially combining the background-reducing short pulse structure of ISIS with significantly higher power than LAMPF. Neutrino ($\overline{\nu}_\mu \rightarrow \overline{\nu}_e$) oscillation sensitivities 20-50 times current limits can be expected, with measurements at two distances providing an essential check of the results.

REFERENCES

1. C.Athanassopoulos, at al., Phys. Rev. Lett. **77**, 3082 (1996).
2. M.-C.Lemaire et al., Phys. Rev. C **43**, 2711 (1991).
3. R.L.Berman and P.Plischke, submitted to NIM (1997).
4. C.Athanassopoulos, et al., Phys. Rev. C **54**, 2685 (1996).

Topics in Neutrino Physics Including New Results from Super-Kamiokande

Takaaki Kajita

Kamioka Observatory, Institute for Cosmic Ray Research, Univ. of Tokyo
Higashi-Mozumi, Kamioka, Gifu, 506-12, Japan

Abstract. Experiments related neutrino oscillations are reviewed. Special emphases on solar neutrinos, atmospheric neutrinos and LSND and related experiments are made. The results from these experiments suggest neutrino oscillations. New data from Super-Kamiokande on solar and atmospheric neutrinos are shown.

INTRODUCTION

It is generally believed that the measurement of neutrino mass and mixing could be one of a few ways to explore the physics beyond the standard model of particle physics.

At present, there are three "evidence" for neutrino oscillations. Two of them are from solar and atmospheric neutrino data. Solar neutrino problem has been with us for more than 25 years. There is also a problem in the (ν_μ/ν_e) ratio of the atmospheric neutrino data for about 10 years. With the increasing amount of data from various underground experiments, it is getting more and more difficult to explain these results based only on known physics. There is also positive oscillation signal in the LSND experiment at LANL. The favored oscillation parameters are just outside of the excluded regions by the other accelerator and reactor experiments.

Super-Kamiokande will provide important data for the understanding of the first two topics. It started data taking in April 1996. The Super-Kamiokande detector is a 50 kton water Cerenkov detector located at a depth of 2700 meters water equivalent in the Kamioka mine in Japan. See Fig 1. It consists of two layers of detector. 11,146 50 cm ϕ photomultiplier tubes (PMTs), instrumented in all surfaces of the inner detector, detect Cerenkov photons radiated by relativistic charged particles. 1,885 20 cm ϕ PMTs are instrumented in the

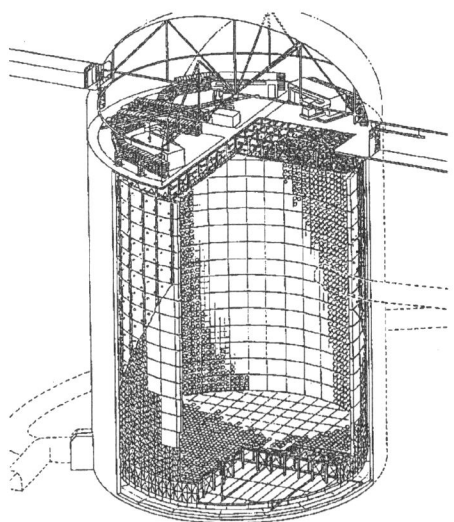

FIGURE 1. The Super-Kamiokande detector. The height is 42 meters and the diameter of 39 meters containing 50 kton of pure water.

outer anti-counter. The anti-counter is useful for identifying entering cosmic-ray muons and measuring energy leakage of the neutrino interactions occurring in the inner detector. Pulse-height and timing information from each PMT are recorded and used in the data analysis.

In the following sections, I will describe the experimental status of these three topics including new results from Super-Kamiokande.

SOLAR NEUTRINO EXPERIMENTS

The Sun is an intense source of low-energy ($E_\nu \lesssim 15\,\text{MeV}$) electron neutrinos ($\nu_e$) that are products of the nuclear processes in its central region. Solar neutrinos directly prove the interior of the Sun, unlike the photons emitted from its surface layer. Also, they provide a means of searching for as yet undetected intrinsic properties of neutrinos through the wide range of matter density and the very long distance from the Sun to the earth.

Our knowledge of the Sun is formulated in a quantitative description known as the standard solar model (SSM) [6]. The SSM predicts the flux and spectra of solar neutrinos, together with many other physical quantities of the Sun. Fig. 2 shows the energy spectra of the solar neutrino flux on the earth.

There have been five solar neutrino experiments: the Homestake experiment [1], Kamiokande [2], SAGE [3], GALLEX [4] and Super-Kamiokande [5]. There are two types of solar neutrino experiments. One is the radio chemical experiments which count the number of atoms which are the products of

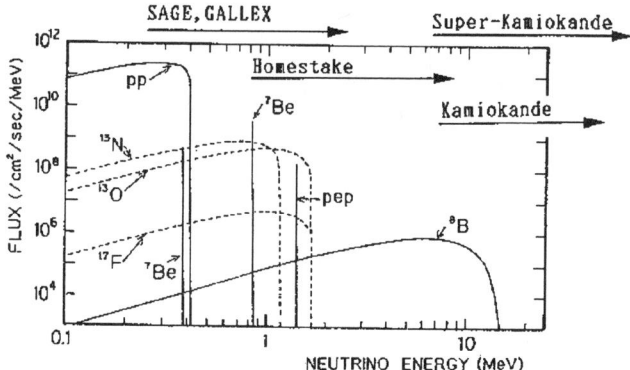

FIGURE 2. The energy spectra of the solar neutrino flux. Also shown are the energy threshold of the solar neutrino experiments.

the solar neutrino interactions. The gallium experiments and chlorine experiments use this technique. In these experiments, the energy threshold of the experiments are determined uniquely, i.e., 233keV for gallium and 814keV for chlorine.

The other type of the experiments is real-time counter experiments which detect radiation (electrons for example) from the solar neutrino interactions. The energy threshold depends on a experiment. For example, the energy threshold was 7.0MeV for Kamiokande.

These two types of experiments have advantages and disadvantages and are complementary. For example, the energy threshold could be lower in radio chemical experiments. On the other hand, it is easy, for counter experiments to study energy spectrum or short time flux variation, such as the day-night effect.

Table 1 shows the summary of the solar neutrino flux measurements and compares with the the SSM prediction [6]. It is a quite distinctive feature that all of the results from the solar neutrino experiments are significantly less than the SSM prediction. It should be mentioned that both the GALLEX and SAGE experiments carried out neutrino source experiments using 51Cr [4] [3]. ^{51}Cr provide mono-energetic neutrinos of 746keV(81%), 751keV(9%), 426keV(10%). The counting rates from the GALLEX and SAGE source experiments were 0.92±0.08 and 0.95±0.12 of the expectations, respectively. It can be concluded that the deficit of the solar neutrinos observed by gallium experiments is not due to unknown experimental effects.

Is is also seen that the (Data/SSM) value measured by the Homestake experiments is substantially smaller than the results from the other experiments. The Kamiokande and Super-Kamiokande results on the flux of ^{8}B solar neutrinos and the Homestake result (mainly) on the ^{8}B and ^{7}Be solar neutrinos suggest that the flux of ^{7}Be solar ν_e should be significantly lower than the SSM prediction. This observation, together with the data from the gallium

TABLE 1. Experimental results and the SSM predictions on the solar neutrino flux. Here the unit is SNU for ^{37}Cl and ^{71}Ga and $10^6 \nu/\text{cm}^2/\text{sec}$ for the $e^-(\text{H}_2\text{O})$ target, respectively, where SNU means "solar neutrino unit" defined by 1SNU=$10^{-36}\nu$ captures per atom per second.

Target(Exp.)	Data	SSM prediction	Data/SSM$^{(*)}$
^{71}Ga(SAGE)	$69 \pm 10^{+5}_{-7}$	137^{+8}_{-7}	$0.504^{0.082}_{-0.089}$
^{71}Ga(GALLEX)	$69.7 \pm 6.7^{+3.9}_{-4.5}$	137^{+8}_{-7}	$0.509^{0.057}_{-0.059}$
^{37}Cl(Homestake)	$2.55 \pm 0.14 \pm 0.14$	$9.3^{+1.2}_{-1.4}$	0.273 ± 0.021
e^-(H$_2$O,Kamiokande)$^{(**)}$	$2.80 \pm 0.19 \pm 0.33$	$6.62^{+0.93}_{-1.12}$	0.423 ± 0.058
e^-(H$_2$O,Super-Kamiokande)	$2.65^{+0.09+0.14}_{-0.08-0.10}$	$6.62^{+0.93}_{-1.12}$	$0.400^{+2.5}_{-1.9}$

(*) Experimental errors only. Statistical and systematic errors are added in quadrature. SSM of Ref. [6] are used.
(**) Preliminary.

experiments, makes it extremely difficult to consistently explain the data in the context of solar physics [7].

The MSW mechanism [8] is very attractive for explaining all of the existing solar neutrino data. Fig. 3, taken from Ref. [7] shows the 95% C.L. allowed regions in the neutrino-oscillation parameter space. There are two allowed regions, so called "small mixing solution" and "large mixing solution". Fig. 4 shows the probability of $\nu_e \to \nu_e$ for two sets of Δm^2 and $\sin^2 2\theta$. These two sets are selected from the allowed regions in Fig. 3. It is clear that, in both cases, the flux of ^{7}Be solar ν_e is significantly suppressed relative to the SSM value.

In the above discussion, we observed that the MSW mechanism could explain the solar neutrino data well. However it should be noted that the above discussion and the resultant allowed parameter regions of neutrino oscillations rely heavily on all the experiments. If one ignores one of the three results, i.e., (Kamiokande+Super-Kamiokande), Homestake and (GALLEX+SAGE), the allowed regions get much bigger, see Fig. 3. Much stronger evidence for the MSW effect might be necessary before settling the "solar neutrino problem". In Fig. 4, we saw that the spectrum of the ^{8}B solar neutrinos could be distorted for the case of the "small angle solution", or the day-night effect could be observed for the case of the "large mixing solution". Super-Kamiokande can provide the information for both the energy spectrum and the day-night effect.

Fig. 5 shows the day-night data from Super-Kamiokande. Within the statistics, there is no evidence for the day-night flux difference. Fig. 6 shows the energy spectrum of the electrons from Super-Kamiokande. The thick bars show the correlated systematic error. Obviously, the Super-Kamiokande experiment needs much more study to reduce the systematic error before mentioning anything on the spectrum distortion.

The present data and the MSW solution have several predictions to be

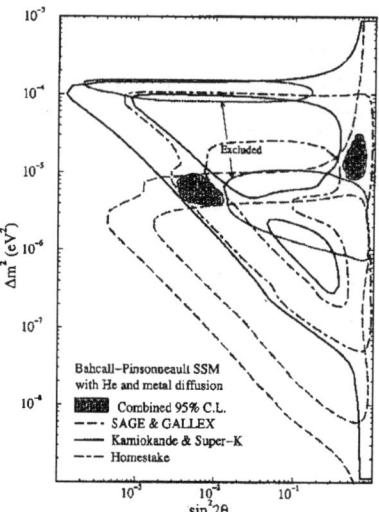

FIGURE 3. The allowed regions of the neutrino oscillation parameters. This figure was taken from Ref. [7]. The shaded regions show the allowed regions obtained by combining all the solar neutrino experiments and assuming the SSM of Ref. [6].

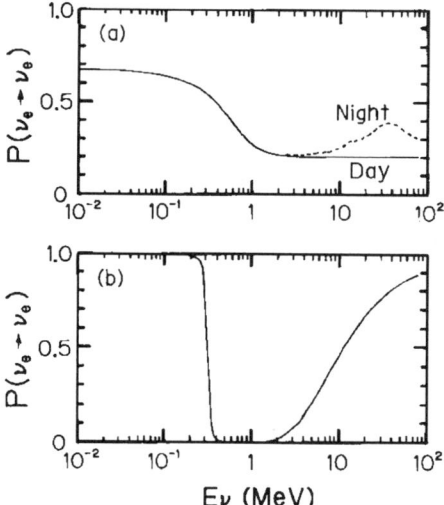

FIGURE 4. The probability that a solar ν_e remains ν_e for (a)$(\Delta m^2, \sin^2 2\theta)=(1\times 10^{-5}, 0.6)$ and (b)$(\Delta m^2, \sin^2 2\theta)=(0.5\times 10^{-5}, 0.6\times 10^{-2})$. In (a), the oscillation probabilities for both day and night are shown.

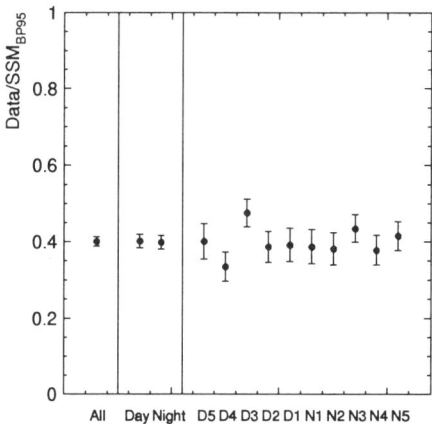

FIGURE 5. Preliminary day-night data from Super-Kamiokande. The day time and night time data are further subdivided into five bins according to the angle between the direction to the Sun and the nadir at the detector.

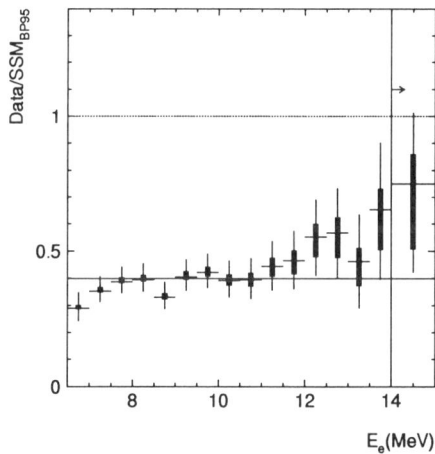

FIGURE 6. Very preliminary plot of Data/SSM as a function of electron energy from Super-Kamiokande. The thick error bars show the correlated systematic errors which are dominated by the energy scale and energy resolution uncertainties.

observed in future solar neutrino experiments. In the next generation solar neutrino experiments (including Super-Kamiokande), it would be necessary to to obtain a conclusive evidence for the solar neutrino oscillations. Super-Kamiokande will improve the data substantially in the near future and will be able to provide a high quality data on the day-night effect and on the energy spectrum of the ^{8}B solar neutrinos.

SNO [9] is a 1000 ton heavy water Cerenkov detector designed to study the flux of ^{8}B solar neutrinos independent of the neutrino oscillation effect. ^{8}B solar neutrinos are detected through the reactions (a)$\nu_e + d \to e^- + p + p$ and (b)$\nu_x + d \to \nu_x + p + n$(neutrons are detected as a signature of ν_x interactions.). Since the reaction (b) is equally sensitive to all neutrinos, the total ^{8}B solar neutrino flux can be measured by this reaction. The reaction (a) is useful for measuring the energy spectrum of ^{8}B solar neutrinos. The SNO experiment will start early 1998.

BOREXINO [10] is a liquid-scintillator detector with its fiducial mass of 100 tons. The use of liquid-scintillator with low-radioactive contamination would enable the detection of ^{7}Be solar neutrinos with a manageable background contamination. The BOREXINO detector is under construction. It will start experiment in 1999.

The data from Super-Kamiokande, SNO and BOREXINO should provide conclusive tests of solar neutrino oscillations.

ATMOSPHERIC NEUTRINO EXPERIMENTS

Atmospheric neutrinos arise from the decay of $\pi(K)$ and μ produced by primary cosmic-ray interactions in the atmosphere. In the energy range below $E_\nu \sim 1$ GeV, where all muons decay, we roughly expect that $(\nu_\mu + \overline{\nu}_\mu)/(\nu_e + \overline{\nu}_e) \simeq 2$. This is because a π-decay produces a ν_μ and a μ, and the μ, when it decays, produces another ν_μ and a ν_e. These neutrinos enter into a detector from all directions. In the energy range above $E_\nu \sim 1$ GeV, a fraction of muons do not decay before reaching the ground. Accordingly, the $(\nu_\mu + \overline{\nu}_\mu)/(\nu_e + \overline{\nu}_e)$ ratio increases as increasing E_ν. Since the $(\nu_\mu + \overline{\nu}_\mu)/(\nu_e + \overline{\nu}_e)$ ratio is determined by such a simple kinematics, the ratio can be calculated accurately. This implies that useful information on neutrino oscillations can be obtained if the $(\nu_\mu + \overline{\nu}_\mu)/(\nu_e + \overline{\nu}_e)$ ratio was measured. Also, since the path length of these neutrinos ranges from less than 10 km to 13000km, the diameter of the earth, additional information on neutrinos oscillation can be obtained by measuring the zenith-angle and energy dependence of the above ratio.

There have been several experimental results. Table 2 summarizes the results from various atmospheric neutrino experiments. The first three experiments, NUSEX [15], Frejus [16] and Soudan-2 [17], in Table 2 are tracking detectors, and the last three, IMB-3 [13] [14], Kamiokande [11] and Super-Kamiokande [12], are water Cerenkov detectors.

TABLE 2. Experimental results on the atmospheric $(\nu_\mu + \bar{\nu}_\mu)/(\nu_e + \bar{\nu}_e)$ ratio.

Exp.	Data		MC		$(\mu/e)_{data}/(\mu/e)_{MC}$
	e-like	μ-like	e-like	μ-like	
NUSEX	18	32	20.5	36.8	$0.99^{+0.35}_{-0.25}\pm?$
Frejus	75	125	81.4	136.2	$1.00\pm0.15\pm0.08$
Soudan-2	79.1	54.6			$0.67\pm0.15^{+0.04}_{-0.06}$
IMB-3(sub-GeV)	325	182	257.3	268.0	$0.54\pm0.05\pm0.12$
(multi-GeV)	25	47	30.8	41.2	$1.40\pm0.66\pm0.21$
Kamiokande(sub-GeV)	248	234	227.6	356.8	$0.60^{+0.06}_{-0.05}\pm0.05$
(multi-GeV)	98	135	66.5	162.2	$0.57^{+0.08}_{-0.07}\pm0.07$
Super-Kam.(sub-GeV)[*]	460	466	385.5	609.6	$0.64^{+0.043}_{-0.041}\pm0.059$
(multi-GeV)[*]	105	169	88.9	263.9	$0.54^{+0.072}_{-0.063}\pm0.071$

(*)Preliminary

The sub-GeV $(\mu/e)_{data}/(\mu/e)_{MC}$ data from the three water Cerenkov detector dominate the whole statistics of the atmospheric neutrino data. They agree each other well and the $(\mu/e)_{data}/(\mu/e)_{MC}$ values are substantially smaller than unity. Older data from NUSEX and Frejus show the values closer to 1. However, their statistics are limited and these data may not repudiate conclusively the results from the data from the water Cerenkov detectors. Furthermore, recent data from Soudan-2 agree with the data from the water Cerenkov detectors, although the value of unity is still allowed at 2σ level.

The multi-GeV data from Kamiokande and Super-Kamiokande agree well. The recent IMB-3 multi-GeV data [14] have higher $(\mu/e)_{data}/(\mu/e)_{MC}$ value but the statistics is limited.

In the next few years, the Soudan-2 experiment will increase the statistics and it will allow a further check of the small $(\mu/e)_{data}/(\mu/e)_{MC}$ ratio with a different detector technology and systematics. The data from Super-Kamiokande is based on about 200 days of data. Super-Kamiokande will rapidly increase the statistics in the near future.

It is generally believed that the zenith-angle dependence of the $(\mu/e)_{data}/(\mu/e)_{MC}$ ratio observed in the Kamiokande multi-GeV data, if confirmed, could be strong evidence for neutrino oscillations, since the systematic dependence of $(\mu/e)_{data}/(\mu/e)_{MC}$ on zenith-angle could not be explained by any other conventional physics.

Fig. 7 shows the zenith angle dependence of $(\mu/e)_{data}/(\mu/e)_{MC}$ for the sub- and multi-GeV data from Kamiokande and Super-Kamiokande. Data from these two experiments agrees well for the sub-GeV data. The Super-Kamiokande data on the multi-GeV region seems to have less zenith angle dependence. But since the error bars are large, the Super-Kamiokande data are consistent with both assumptions; flat and Kamiokande zenith-angle distributions. Within 1~2 more years of Super-Kamiokande data, it would be possible to get a definite conclusion on the possible zenith-angle dependent

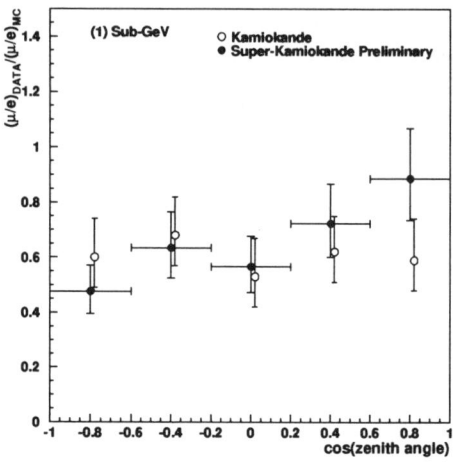
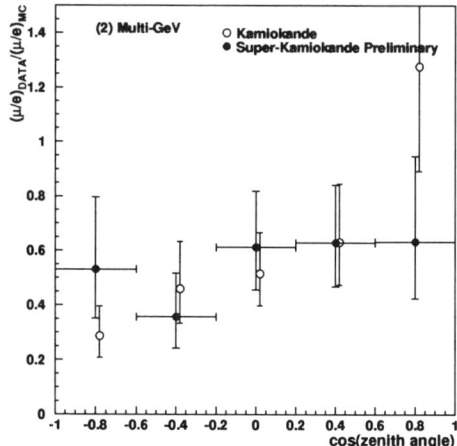

FIGURE 7. The $(\mu/e)_{data}/(\mu/e)_{MC}$ ratio as a function of zenith angle for (1) the sub-GeV energy region and (2) the multi-GeV energy region. White circles show the Kamiokande data and the black circles show the preliminary Super-Kamiokande data. $\cos\Theta = 1$ corresponds to down going particles.

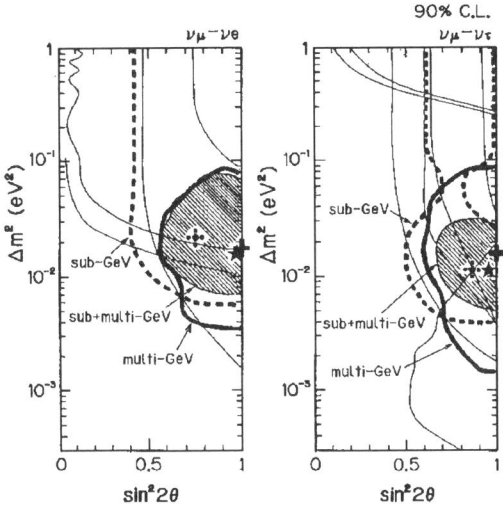

FIGURE 8. 90% C.L. allowed neutrino oscillation parameters as obtained from the Kamiokande atmospheric neutrino data (Shaded region). Allowed regions obtained separately by the Kamiokande sub- and multi-GeV are also shown. 90% C.L. excluded regions from the other experiments are also shown by thin curves.

$(\mu/e)_{data}/(\mu/e)_{MC}$ ratio.

Fig. 8 shows the allowed parameter regions of neutrino oscillations from the Kamiokande data. Since the prediction in the absolute number of events has large (20~30%) uncertainty, mainly due to the uncertainty in the absolute flux values of the primary cosmic ray particles, both $\nu_\mu \to \nu_e$ and $\nu_\mu \to \nu_\tau$ oscillations are allowed. The high Δm^2 regions $\Delta m^2 \gtrsim O(0.1) eV^2$ are excluded due to the observed zenith-angle dependence in the multi-GeV region. One sees that the most probable allowed regions are around $\Delta m^2 \sim 10^{-2} eV^2$ for both oscillation channels.[1]

The suggested Δm^2 region from the Kamiokande data could be studied by long-baseline neutrino-oscillation experiments. There are two types of experiments, reactor experiments and accelerator experiments. Reactor experiments are only sensitive to $\overline{\nu}_e \to \overline{\nu}_x$ oscillations. At present, there are two such experiments, Chooz [18] and Palo Verde [19]. The typical parameters of these experiments are; 1 km of baseline, 10 tons of liquid-scintillator with 0.1% of Gd for detecting neutrons and the expected event rate of 30~50/day. The Chooz experiments are taking data. The Palo Verde will start the experiment in the summer of 1997. We will be able to hear interesting results from these

---

[1] The question of oscillation channels can be settled by the future Super-Kamiokande data. For the $\nu_\mu \to \nu_\tau$ channel the zenith angle distribution for the multi-GeV electrons should be as expected, on the other hand, for the $\nu_\mu \to \nu_e$ channel, there should be more upward-going electrons than the downward-going ones.

experiments soon.

There are three accelerator long-baseline projects, KEK to Super-Kamiokande (K2K) [20], Fermi lab. to the Soudan site (MINOS) [21] and CERN to Gran Sasso [22]. The typical parameters of these experiments are; (250(K2K) to 730(MINOS))km of baseline, (1.5(K2K) to 25(CERN))GeV of average neutrino energy and the expected event rate of (200(K2K) to 40,000(MINOS))/yr. The sensitivity of these experiments clearly cover the suggested allowed regions for both of the oscillation modes. The MINOS and the experiment at CERN to Gran Sasso uses higher energy neutrinos which enable to produce and detect τ's if the oscillation is $\nu_\mu \to \nu_\tau$. K2K will start data taking in 1999. MINOS is at the R&D stage now, and will start the experiment in 2001 with 1/3 of the full MINOS. These experiments could be the most convincing ones to settle the current atmospheric neutrino problem.

We expect that the Super-Kamiokande data and the data from the long-baseline experiments will settle the atmospheric neutrino problem in the near future.

LSND AND RELATED EXPERIMENTS

The LSND experiment detects neutrinos from pion and muon decay at rest. Since the primary neutrino beam contains ν_μ, $\overline{\nu}_\mu$ and ν_e and the contamination of $\overline{\nu}_e$ is estimated to be 7.8×10^{-4}, a sensitive search for $\overline{\nu}_\mu \to \overline{\nu}_e$ is possible by searching for $\overline{\nu}_e$ interactions. The $\overline{\nu}_e$ are detected via the reaction $\overline{\nu}_e p \to e^+ n$, correlated with a γ from $np \to d\gamma(2.2 MeV)$. They observed 22 such events with the estimated background of 4.6±0.6 events. The probability that these events could be due to a statistical fluctuation of the background events was 4.1×10^{-8}. This result could be interpreted as evidence for neutrino oscillations with the oscillation probability of (0.31±0.12±0.05)%. The allowed parameter regions are shown in Fig. 9.

An experiment with a similar energy spectrum of neutrinos, KARMEN [24], did not observe any evidence for neutrino oscillations. But this experiment could not exclude the whole allowed regions of LSND. There were also results from accelerator and reactor oscillation experiments. Non of these experiments observed positive signal of neutrino oscillations. Excluded regions from these experiments are shown in Fig. 9. The high Δm^2 regions allowed by LSND are excluded by many experiments, but the low Δm^2 and small mixing region is still allowed.

The LSND allowed regions must be further studied.[2] KARMEN [24] finished the upgrade and the background was substantially reduced. Within a few years, KARMEN will cover the whole allowed regions of LSND.

---

[2]) In this conference, we heard that the analysis of the decay in flight neutrino events in LSND [25] also shows the oscillation signal.

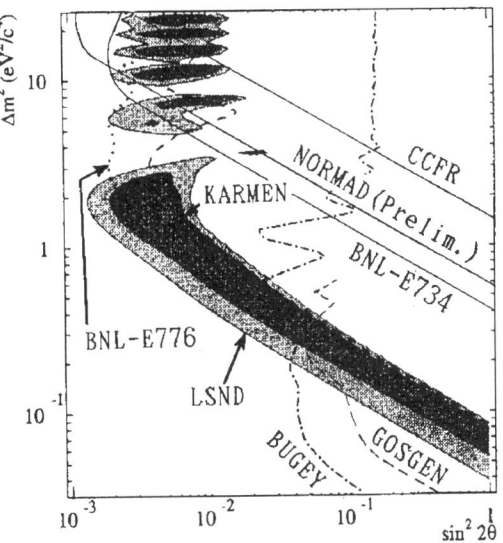

FIGURE 9. The allowed parameter regions by the LSND experiments are shown by gray(90%C.L.) and black(99%C.L.). Also shown are excluded regions [24] [26] by accelerator and reactor experiments.

SUMMARY

At present, there are three sets of data, solar neutrinos, atmospheric neutrinos and LSND, which suggest neutrino oscillations.

All the solar neutrino data show deficit of solar neutrinos relative to the SSM predictions. It is very unlikely that the data could be explained by any modifications of the solar models. The solar neutrino data could be explained by the MSW mechanism and there are two allowed regions, "small mixing region" and "large mixing region". The MSW solutions have three important predictions, low ^{7}Be solar ν_e flux, the distortion of the ^{8}B solar ν_e spectrum or the day-night effect. Super-Kamiokande have already improved significantly the day-night data. These predictions should be studied by the second generation solar neutrino experiments such as Super-Kamiokande, SNO and BOREXINO.

The small atmospheric (ν_μ/ν_e) ratio was confirmed by Super-Kamiokande. The most convincing evidence for neutrino oscillations could be the confirmation of the zenith-angle dependence of the $(\mu/e)_{data}/(\mu/e)_{MC}$ ratio observed in the Kamiokande multi-GeV data. The present Super-kamiokande data are not conclusive, but within (1~2) more years, Super-Kamiokande will produce a definite conclusion on this issue. Finally a special emphasis must be made on the long baseline experiments.

The LSND results could be evidence for $\overline{\nu}_\mu \rightarrow \overline{\nu}_e$ oscillations with the oscillation probability of $(0.31\pm0.12\pm0.05)$%. Unfortunately, so far, there has been no experiment which confirmed the LSND result. Confirmation by independent experiments are highly desired.

In conclusion, we expect really important results to appear in the field of neutrino oscillations in the near future.

REFERENCES

1. K.Lande, to be published in *Neutrino* 96, Proceedings of the 17th International Conference on Neutrino Physics and Astrophysics, Helsinki, Finland, June 1996. See also B.T.Cleveland et al., Nucl. Phys. B (Proc. Suppl.) 38 (1995) 47.
2. Y.Fukuda et al., Phys. Rev. Lett. 77 (1996) 1683.
3. J.N.Abdurashitov et al., Phys. Rev. Lett. 77 (1996) 4708.
4. W.Hampel et al., Phys. Lett. B 388 (1996) 384.
5. The Super-Kamiokande collaboration, draft in preparation; see also R.Sanford, for the Super-kamiokande collaboration, in these proceedings.
6. See, for example, J.N.Bahcall and M.H.Pinsonneault, Rev. Mod. Phys. 67 (1995) 781.
7. See, for example, N.Hata and P.Langacker, Preprint LASSNS-AST 97/29 (May1997), and references therein.
8. S.P.Mikheyev and A.Yu.Smirnov, Sov. J. Nucl. Phys. 42 (1985) 913; L.Wolfenstein, Phys. Rev. D 17 (1978) 2369; ibid. 20 (1979) 2634.
9. H.Lee, for the SNO collaboration, in these proceedings.
10. "BOREXINO at Gran Sasso − proposal for a real time detector for low energy solar neutrinos" (1991).
11. Y.Fukuda et al., Phys. Lett. B 335 (1994) 237.
12. The Super-Kamiokande collaboration, draft in preparation; see also T.Haines, for the Super-kamiokande collaboration, in these proceedings.
13. Becker-Szendy et al., Phys. Rev. D 46 (1992) 3720.
14. R.Clark et al., to be published in Phys. Rev. Lett., R.Clark, for the IMB-3 collaboration, in these proceedings, R.Clark, private communication for the $(\mu/e)_{data}/(\mu/e)_{MC}$ value.
15. M.Aglietta et al., Europhys. Lett. 8, (1989) 611.
16. K.Daum et al., Z. Phys. C 66 (1995) 417.
17. W.W.M.Allison et al., Phys. Lett. B 391 (1997) 491; M.C.Goodman, for the Soudan-2 collaboration, in these proceedings.
18. H.de Kerret et al., The Chooz experiment, Proposal, LAPP Report (1993).
19. D.Lawrence et al., Preprint Stanford-HEP-97-01, to appear in the proceedings of the 1997 Lake Louise Winter Institute.
20. K.Nishikawa INS-Rep.-924, (1992); see also, J.Wilkes, for the K2K collaboration, in these proceedings.
21. E.Ables et al., The MINOS collaboration, P-875 proposal, (1995).
22. See for example, C.Rubbia, Nucl. Phys. B (proc. Suppl.) 48 (1996) 172.
23. C.Athanassopoulos et al., Phys. Rev. Lett. 77 (1996) 3082.

24. J.Kleinfeller, for the KARMEN collaboration, in these proceedings.
25. G.Mills, for the LSND collaboration, in these proceedings.
26. L.A.Ahrens et al., Phys. Rev. D 36 (1987) 702; L.Borodovsky et al., Phys. Rev. Lett. 68 (1992) 274; Z.Zacek et al., Phys. Rev. D 34 (1986) 2621; B.Achkar et al., Nucl Phys. B 434 (1995) 503; A.Romosan et al., Phys. Rev. Lett. 78 (1997) 2912; A.De Santo, for the NORMAD collaboration, talk presented at the XXXIInd Rencontres de Moriond "Electroweak Interactions and Unified Theories", Les Arcs, France, March 1997.

Supernova Heavy Element Nucleosynthesis: Can It Tell Us About Neutrino Masses?

George M. Fuller

Department of Physics
University of California, San Diego
La Jolla, CA 92093-0319

Abstract. Here we describe a new probe of neutrino properties based on heavy element nucleosynthesis. This technique is in many ways akin to the familiar light element Primordial Nucleosynthesis probe of conditions in the early universe. Our new probe is based on the fact that neutrino masses and vacuum mixings can engender matter-enhanced neutrino flavor transformation in the post core bounce supernova environment. Transformations of the type $\nu_{\mu(\tau)} \rightleftharpoons \nu_e$ in this site will have significant effects on the synthesis of the rapid neutron capture (r-Process) elements and the light p-nuclei. We suggest that an understanding of the origin of these nuclides, combined with the measured abundances of these species, may provide a "Rosetta Stone" for neutrino properties. Heavy element nucleosynthesis abundance considerations give either constraints/evidence for neutrino masses and flavor mixings, or strong constraints on the site of origin of r-Process nucleosynthesis. The putative limits on neutrino characteristics are complimentary to those derived from laboratory neutrino oscillation studies and solar and atmospheric neutrino experiments. Preliminary studies show that the existence of r-Process nuclei in the abundances observed in the Galaxy cannot be understood unless neutrinos have small masses (possibly in the cosmologically significant range).

INTRODUCTION

In this paper we connect the question of whether or not the neutrinos have rest masses to another mystery, the origin of the neutron-rich heavy nuclei. The thesis which we advance here is as follows: the observed abundance pattern of the rapid neutron capture (r-Process) nuclei [1,2] may be able to serve as a "fossil record" which can tell us either about neutrino masses and flavor mixings and/or about the site(s) of origin of the r-Process elements. As we will outline, the most likely sites of r-Process nucleosytnesis are in environments with intense neutrino fluxes. Since the neutrino fluxes can set the conditions required for r-Process nucleosynthesis, and the characteristics (e.g., energy spectra) of these fluxes may depend on neutrino properties, we may be able to use the observed nucleosynthetic abundance pattern to inform us about basic neutrino properties. In fact, it is even possible that the r-Process nuclei cannot have been produced in the abundances observed in the Galaxy today unless (some) neutrinos have masses $\sim 1\,\mathrm{eV}$!

Though a connection between neutrino mass and the heavy elements may seem bizarre on first encounter, it becomes less fantastic when one realizes that these neutron-rich species have to have been produced in a *neutron-rich* environment. Since free neutrons are unstable they must either be "mined" from nuclei or from neutron stars, or produced directly via the weak interaction (e.g., $\bar{\nu}_e + p \to n + e^+$). The former neutron production channels are disfavored since neutrons are bound in nuclei by $\sim 8\,\mathrm{MeV}$ and in neutron stars by $\sim 100\,\mathrm{MeV}$.

Mining neutrons from these sites in sufficient quantities to provide for r-Process nucleosynthesis therefore would require very energetic particles. The direct weak interaction production channel for free neutrons is disfavored because the weak coupling G_F is indeed weak, and so very energetic and intense fluxes of neutrinos would be required to effect the right conditions for r-Process nucleosynthesis.

Nevertheless, all of these neutron production channels have been proposed at one time or another to engineer the synthesis of the r-Process nuclei [3,4]. As we shall see, only two proposed sites have a chance of producing r-Process material in the requisite quantities to explain the observed abundances of these species in the galaxy. And, of these, only one proposed site can account for the correct rate of synthesis of r-Process nuclei: the neutrino-heated supernova ejecta site. However, we will argue that this site cannot make the heaviest r-Process nuclei unless new neutrino physics is introduced.

In what follows we will describe briefly the favored r-Process nucleosynthesis sites. We then will outline how nucleosynthesis in a freeze-out from nuclear statistical equilibrium proceeds, both in the Big Bang and in a neutrino-driven wind. We will outline the fundamental problem of low neutron-to-seed nucleus ratio which prevents the neutrino-driven wind from being an ideal site for r-Process nucleosynthesis. Finally, we will discuss ways in which new neutrino physics can alleviate this and other problems.

WHERE DOES RAPID NEUTRON CAPTURE NUCLEOSYNTHESIS TAKE PLACE?

The synthesis of the r-Process nuclei takes place in conditions where the ratio of free neutrons to "seed" nuclei is large. The basic idea [1,2] is that the free neutron number density should be large enough that neutron capture rates on the seed nuclei will be fast compared to nuclear beta decay rates. Very neutron-rich species can be built-up in this way. When the neutrons are exhausted, these beta-unstable neutron-rich nuclides will undergo a series of charge-changing beta decays as the whole population of nuclei shifts back toward the valley of beta stability. The three characteristic abundance peaks of the r-Process nuclei at masses $A \approx 80$, 130, and 195 are then understood to be the result (after the decay back toward stability) of the "bottlenecks" in the neutron capture flow caused by the small neutron capture cross sections of the species with closed neutron shells at neutron numbers $N = 50$, 82, and 126, respectively.

The key to understanding the synthesis of the r-Process nuclei lies in identifying an astrophysical environment where the neutron-to-seed nucleus ratio is high enough. Since the typical seed nuclei we will be dealing with have masses somewhat less than 100, and we want to build species like Uranium with more than 200 nucleons, we must have a neutron-to-seed ratio of order 100 or more. Additionally, the candidate astrophysical event must occur often enough, with enough r-Process production, to explain the observed amount of r-Process material in the Galaxy $\sim 10^4$ $M_\odot$, yet not overproduce this material.

At the present time the two best candidate sites for r-Process nucleosynthesis are (1) binary neutron star merger events [5] and (2) neutrino-heated supernova ejecta [6–8]. Unfortunately, both of these proposed sites are flawed.

Binary neutron star mergers are expected to occur on the order of once per galaxy per million years (give or take an order of magnitude or so). That would imply that each such event would have to produce more than a few tenths of a solar mass of r-Process material, while numerical simulations of these events suggest that the actual yield is at least an order of magnitude smaller than this. While neutron star mergers are unlikely to have produced all of the r-Process material in the Galaxy, they could have produced some species. The neutrino-heated supernova ejecta models can neatly explain the synthesis rate and total r-Process abundance in the Galaxy [6]. However, these models are also problematic. With the expected standard supernova and neutrino physics, there will not be a high enough neutron-to-seed nucleus ratio in the neutrino-driven supernova ejecta to get an acceptable r-Process abundance pattern.

However, it is interesting to note that both of these proposed sites are expected to be accompanied by intense neutrino fluxes. But large neutrino fluxes may imply that beta decay is not the only weak charge-changing process relevant in r-Process nucleosynthesis. Studies [9] of neutrino capture on heavy nuclei have revealed the way in which this process can affect conditions where the r-Process nuclides are synthesized and the manner in which the neutron capture nuclear flow is determined. In fact, arguments have been advanced that the measured abundance pattern of r-Process nuclides cannot be understood unless these species were synthesized in an environment possessing sizeable neutrino fluxes and fluences.

McLaughlin and Fuller [10] suggested that the best fit to weak steady flow would include a neutrino capture contribution to the total charge-changing rate. Unfortunately, we do not know for certain that we must have staedy flow equilibrium during the r-Process. Though Kratz et al. [11] have argued from the measured solar system isotopic abundances that weak steady flow must obtain, systematic errors inherent in these studies may be large.

In a potentially more compelling argument, Qian, Haxton, Langanke, and Vogel [12,13] have examined both neutral current and charged current neutrino-nucleus interaction neutron spallation processes in r-Process models. They claim that these spallation processes are *necessary* to understand the structure in the measured r-Process abundance peaks, most importantly, the lack of pronounced "holes" on the low-mass-number sides of these peaks. These authors then suggest that there is a minimum neutrino fluence to which the r-Process nuclides must be exposed. Though this argument is attractive, neutrino-induced neutron spallation may not be the only process which can move nucleons from the peaks to lower mass nuclei [6].

We can conclude that we have a *hint* that wherever the r-Process nuclei are synthesized, there must be an intense neutrino flux. Again, that singles out the two sites under consideration. All other proposed [3,4] sites lack a significant neutrino flux and, futhermore, are even more deeply flawed than neutron star mergers or neutrino-heated supernova ejecta. This suggests that we must find a way to make one (or both) of these candidate sites work for the production of the r-Process. Since, as argued above, there has to be a fairly frequent synthesis event, supernovae must produce at least some of the r-Process material in the Galaxy.

In this spirit, let us now examine the physical environment around hot proto neutron stars and how r-Process nucleosynthesis would proceed in neutrino-heated supernova ejecta.

NEUTRINOS AND TYPE II SUPERNOVAE

It is Type II supernovae that involve neutronization and core collapse and so will be of interest for r-Process nucleosynthesis. The progenitors of Type II supernova explosions are massive stars (mass larger than about $10\,M_\odot$). These stars evolve very quickly ($\sim 10^7$ yrs) through successive burning stages until they form a core consisting of iron peak material in nuclear statistical equilibrium (hereafter, NSE). This "iron core" is supported by electron degeneracy pressure and so closely resembles a white dwarf. It will have essentially a Chandrasekar mass, $\approx 1.4\,M_\odot$. As it loses entropy through neutrino cooling, the iron core eventually will be destabilized by a combination of general relativistic effects, electron capture, and a "phase transition" as NSE shifts from predominantly heavy nuclei toward nuclei plus a significant number of alpha particles.

Once the collapse of the core begins, it will proceed at near the free-fall rate until the nucleons begin to overlap at nuclear density. This will take roughly one second. The high electron fermi energies characterizing the collapse imply a large amount of neutronization through the reaction $e^- + p \to n + \nu_e$. Since the entropy-per-baryon s in units of Boltzmann's constant is low, most of the protons which capture electrons reside inside large nuclei. Neutrinos are also produced thermally as pairs. At first any neutrinos produced will stream freely out of the core. However, once the central density of the core reaches about 1% of nuclear saturation density, the mean free path for the neutrinos will become smaller than the size of the core. Once this neutrino trapping has occurred, neutrinos are thermalized and are well described as diffusing out of the core.

The collapse is violently halted ("core bounce") when nuclear density is obtained and the large nuclei merge. The reason for the rapid ($\sim$ 1 millisecond) halt to the collapse is the tremendous increase in pressure from degenerate nonrelativistic nucleons. A shock wave is generated at the outer boundary of the halted, near-hydrostatic inner core. The shock wave begins to propagate outwards with an initial energy of about 10^{51} ergs. This is roughly the total amount of energy in optical radiation and kinetic energy seen by astronomers. Therefore, if this prompt shock were to propagate into the envelope, the Type II supernova display would be explained. However, the material behind the shock has a much higher entropy, $s \sim 10$, than does the material in the outer core which is flowing through it.

The material always remains in NSE, so that there is a shift in equilibrium from heavy nuclei ahead of the shock where the entropy is $s \approx 1$, to free nucleons and alpha particles behind the shock where $s \approx 10$. This saps energy from the propagating shock, since it costs $\sim 8\,\mathrm{MeV}$ to pull a nucleon out of a nucleus. Put another way, 10^{51} ergs are lost from the shock for every $0.1\,\mathrm{M}_\odot$ of material traversed by the shock. Since the outer core itself may comprise $\sim 0.6\,\mathrm{M}_\odot$, the shock quickly stalls, ceases to move out, and becomes an accretion shock. All of this happens by $t_\mathrm{PB} \approx 0.1\,\mathrm{s}$, where t_PB denotes time *post core bounce*.

To understand where the supernova explosion originates we must examine where the gravitational binding energy has gone during the collapse of the core. Roughly $\sim 10^{52}$ ergs of gravitational binding energy is released when the core collapses from an iron white dwarf configuration to a hot proto neutron star with a radius $\approx 45\,\mathrm{km}$. Of order 99% of this energy is radiated as neutrinos of all six species ν_e, $\bar{\nu}_e$, ν_μ, $\bar{\nu}_\mu$, ν_τ, and $\bar{\nu}_\tau$. A further $\sim 10^{53}$ ergs of gravitational binding energy is released in neutrinos as the proto neutrino star shrinks down to a radius $\sim 10\,\mathrm{km}$. This Kelvin-Helmholtz contraction phase lasts for of order a neutrino diffusion timescale $\tau_\nu \sim 10\,\mathrm{s}$.

It is now thought that the intense flux of neutrinos from the proto neutron star deposit energy behind the shock and "re-energize" it, causing the supernova explosion. This shock re-heating epoch occurs prior to about $t_{PB} \approx 1$ s. We must honestly admit that the supernova explosion process is still not understood in detail. However, for the purposes of the study in this paper it will be sufficient to note that nature somehow finds a means to make these objects explode! Once the shock is re-energized, it sweeps out the overlaying layers and ejects them. We will then be left with a still hot proto neutron star and a very low density, tenuous gas of material left above it.

This tenuous envelope, or "hot bubble," is heated by the intense flux of neutrinos coming from the surface of the neutron star. We can get a simple estimate of the neutrino energy luminosity in each species by making the reasonable assumption of equipartition of energy among all 6 neutrino species, and by assuming that all of the gravitational binding energy $E_{\rm GRAV}$ is released over a neutrino diffusion timescale τ_ν. In terms of the gravitational mass of the neutron star $M_{\rm NS}$ and the radius $R_{\rm NS}$ the gravitational binding energy is,

$$E_{\rm GRAV} \approx \frac{3}{5}\frac{GM_{\rm NS}^2}{R_{\rm NS}} \approx 3 \times 10^{53}\,{\rm ergs}\left(\frac{M_{\rm NS}}{1.4\,{\rm M}_\odot}\right)^2\left(\frac{10\,{\rm km}}{R_{\rm NS}}\right), \quad (1)$$

from which we can derive the energy luminosity of *each* neutrino species:

$$L_\nu \sim \frac{1}{6}\frac{E_{\rm GRAV}}{\tau_\nu} \sim 4 \times 10^{51}\,{\rm ergs\,s}^{-1} \quad (2)$$

The neutrinos are thermalized and diffusing inside the proto neutron star, but decouple from a "neutrino sphere(s)" near its edge. The energy spectra and distribution functions characterizing the freely-streaming neutrinos above the neutron star surface can be approximated as Fermi-Dirac black bodies. The normalized neutrino distribution functions can be cast in the form,

$$f(E_\nu) \approx \frac{1}{T_\nu^3 {\rm F}_2(\eta_\nu)} \cdot \frac{E_\nu^2}{\exp(E_\nu/T_\nu - \eta_\nu) + 1}, \quad (3)$$

where E_ν is the neutrino energy, T_ν is the temperature of the neutrino distribution function, and $\eta_\nu \equiv \mu_\nu/T_\nu$ is the degeneracy parameter, with μ_ν the chemical potential of the neutrino distribution function. In Eq. (3), ${\rm F}_2(\eta)$ is the standard Fermi integral of order 2. Fermi integrals of order k are defined as,

$${\rm F}_k(\eta_\nu) \equiv \int_0^\infty \frac{x^k dx}{\exp(x - \eta_\nu) + 1}. \quad (4)$$

The differential number flux above the neutron star for a neutrino species with a neutrino sphere temperature T_ν and degeneracy parameter η_ν in energy interval dE_ν and in a pencil of solid angle directions $d\Omega_\nu$ is,

$$d\phi_\nu \approx \frac{c}{2\pi^2(\hbar c)^3} \cdot \frac{E_\nu^2 dE_\nu}{\exp(E_\nu/T_\nu - \eta_\nu) + 1} \cdot \frac{d\Omega_\nu}{4\pi}. \quad (5)$$

At any point at distance r from the center of a neutron star with radius R_ν (roughly, the radius of the neutrino sphere(s)), the angle of the limb of the neutron star θ_0 is given by $\sin\theta_0 = R_\nu/r$, so that

$$\frac{1}{4\pi}\int d\Omega_\nu = \frac{1}{2}(1-\cos\theta_0) \approx \frac{1}{2}\left(1-\sqrt{1-R_\nu^2/r^2}\right) \approx \frac{R_\nu^2}{4r^2}, \quad (6)$$

where the last approximation is valid where $r \gg R_\nu$. Multiplying the expression in Eq. (5) by E_ν and integrating over neutrino energy, and over all allowed solid angle as above, yields the total energy flux $\Phi_E = (L_\nu/4\pi r^2)$ of a neutrino species at position r:

$$\left(\frac{L_\nu}{4\pi r^2}\right) \approx \frac{c}{2\pi^2(\hbar c)^3}\left(\frac{\int d\Omega_\nu}{4\pi}\right)T_\nu^4 F_3(\eta_\nu). \quad (7)$$

The total (integrated over all contributing solid angle) differential number flux at position r in energy interval dE_ν for a neutrino species can be written by employing Eqs. (3, 5, 7),

$$d\Phi_\nu \approx \left(\frac{L_\nu}{4\pi r^2}\right)\frac{1}{\langle E_\nu\rangle}f(E_\nu)\,dE_\nu, \quad (8)$$

where the average energy of the neutrino species is defined as the distribution function-weighted mean and is a function of T_ν and η_ν:

$$\langle E_\nu\rangle \equiv \int_0^\infty E_\nu f(E_\nu)\,dE_\nu = \frac{F_3(\eta_\nu)}{F_2(\eta_\nu)}\cdot T_\nu. \quad (9)$$

Numerical calculations [14] show that the degeneracy parameters of all six neutrino species are reasonably fit by $\eta_\nu \approx 3$. In this case the proportionality constant between T_ν and average neutrino energy will be $F_3(3)/F_2(3) \approx 3.992$.

The average energies of the energy-distribution functions corresponding to the ν_μ, $\bar{\nu}_\mu$, ν_τ, $\bar{\nu}_\tau$, $\bar{\nu}_e$, and ν_e are $\langle E_{\nu_\mu}\rangle$, $\langle E_{\bar{\nu}_\mu}\rangle$, $\langle E_{\nu_\tau}\rangle$, $\langle E_{\bar{\nu}_\tau}\rangle$, $\langle E_{\bar{\nu}_e}\rangle$, and $\langle E_{\nu_e}\rangle$, respectively. In the absence of neutrino flavor mixing effects, these average energies always satisfy the generic hierarchy:

$$\langle E_{\nu_\mu}\rangle \approx \langle E_{\bar{\nu}_\mu}\rangle \approx \langle E_{\nu_\tau}\rangle \approx \langle E_{\bar{\nu}_\tau}\rangle > \langle E_{\bar{\nu}_e}\rangle > \langle E_{\nu_e}\rangle. \quad (10)$$

This hierarchy is easy to understand. The mu and tau neutrinos and their antiparticles have nearly identical energy spectra because, at the neutrino energy scales relevant for the supernova environment, these species have only neutral current interactions and opacity sources in the neutron star. Neutral current interactions are identical for all active neutrinos. By contrast, the ν_e and $\bar{\nu}_e$ have charged current interactions and opacity sources in addition to the neutral current interactions. The result is that the mu and tau neutrinos decouple deeper in the neutron star (where it is hotter) than do the $\bar{\nu}_e$ and

ν_e. Since there are more neutrons than protons in the neutron star, the ν_e's have the largest charged current opacity contribution and therefore decouple furthest out and have the lowest average energies. A particular numerical calculation [14] suggests that at the times relevant for r-Process nucleosynthesis $\langle E_{\nu_\mu}\rangle \approx \langle E_{\nu_\tau}\rangle \approx 27\,\mathrm{MeV}$, $\langle E_{\bar\nu_e}\rangle \approx 16\,\mathrm{MeV}$, and $\langle E_{\nu_e}\rangle \approx 10\,\mathrm{MeV}$. While other numerical neutrino transport schemes may give different numbers, they will always yield energies which follow the hierarchy in Eq. (10). This will be very important for the connection between neutrino oscillations and r-Process nucleosynthesis.

Unlike the complicated shock re-heating epoch, the thermodynamic and outflow conditions of the hot bubble r-Process nucleosynthesis epoch ($t_{\mathrm{PB}} \gtrsim 3\,\mathrm{s}$) are relatively simple and straightforward to treat. A neutrino-driven "wind" can arise at these late epochs as a result of the intense neutrino fluxes driving mass loss from the neutron star surface and due to heating of the tenuous, near-hydrostatic envelope which sits on top of the neutron star [15–17]. This envelope has very little mass ($\sim 10^{-4}\,\mathrm{M}_\odot$), so that the gravitational field in the region above the neutrino sphere is dominated by the neutron star.

In steady state wind solutions the entropy per baryon, s, is roughly constant throughout the hot bubble envelope, and the enthalpy per baryon (product of temperature T and s) of the material at a given distance from the neutron star is approximately the gravitational binding energy of a nucleon at that position [15–17],

$$Ts \approx \frac{GM_{\mathrm{NS}}m_b}{r}. \qquad (11)$$

Here $m_b \approx 938.26\,\mathrm{MeV}$ is the mass of a proton and r is the distance from the neutron star center. From this expression we can easily find the radius above the neutron star corresponding to a given temperature ($T_9 \equiv T/10^9\,\mathrm{K}$) and entropy per baryon ($S_{100} \equiv s/100$),

$$r_7 \approx 2.25\left(\frac{M_{\mathrm{NS}}}{1.4\,\mathrm{M}_\odot}\right)T_9^{-1}S_{100}^{-1}. \qquad (12)$$

Here r_7 is the distance from the neutron star center expressed in units of $10^7\,\mathrm{cm} = 100\,\mathrm{km}$.

The hot bubble material above the neutron star is everywhere radiation dominated, so that the entropy per unit proper volume is $S_p = (2\pi^2/45)\,g_s T^3$, with $g_s = g_b + 7/8 g_f$, where g_b and g_f are the total statistical weights of all relativistic bosons and fermions, respectively. In the hot bubble and wind, generally $g_s = 11/2$ whenever $T_9 \gtrsim 5$ and we can be assured of relativistic electron/positron pairs as well as photons, while for $T_9 \lesssim 1$ only photons contribute to the statistical weight and $g_s = 2$. The entropy per baryon is $s = S_p/(\rho N_A)$, where ρ is the density in units of $\mathrm{g\,cm}^{-3}$ and N_A is Avogadro's number. We can then give the density $\rho_3 \equiv \rho/10^3$ corresponding to a location with temperature T_9 in a steady state isentropic wind model with entropy S_{100}:

$$\rho_3 = C(T_9) T_9^3 S_{100}^{-1}. \tag{13}$$

In this expression $C(T_9) \approx 0.607 g_s \approx 3.339$ for $T_9 \gtrsim 5$ and $C(T_9) \approx 0.607 g_s \approx 1.214$ for $T_9 \lesssim 1$.

If the entropy per baryon in the hot bubble is constant with radius as expected, then we can combine Eqs. (13,12) and find that the $\rho \propto 1/r^3$,

$$\rho \approx \left[\frac{2\pi^2}{45} g_s \frac{1}{N_A} \left(\frac{M_{NS} m_b}{m_{pl}^2} \right)^3 \frac{1}{s^4} \right] \frac{1}{r^3}. \tag{14}$$

Here $m_{pl} \approx 1.2211 \times 10^{22}$ MeV is the Planck mass (the gravitational constant is then $G = 1/m_{pl}^2$) and we can numerically evaluate Eq. (14) to find,

$$\rho_3 \approx 38 \left(\frac{g_s}{11/2} \right) \left(\frac{M_{NS}}{1.4\, M_\odot} \right)^3 S_{100}^{-4} r_7^{-3}. \tag{15}$$

Note that the whole radial distribution of the hot bubble material is proportional to the inverse fourth power of the entropy per baryon. Simple estimates [16] lead us to believe that $S_{100} \approx 1$. For numerical supernova calculations which do not employ exotic phase transitions or allow for a very relativistic core, the most extreme value given for the entropy in the hot bubble is $S_{100} \approx 5$ [6]. The value of the entropy remains a key parameter. The entropy s must be obtained through a detailed numerical neutrino transport calulation which follows neutrino heating and cooling processes in the region above the neutron star.

If we assume a constant dynamic expansion timescale $\tau_{dyn} = r/v$, where v is the fluid outflow velocity, then the position of a fluid element will increase exponentially with time $r \propto \exp(t/\tau_{dyn})$. Though this relation clearly cannot obtain for very long, it is a reasonable approximation to the flow as seen in some numerical calculations over the length scales and timescales relevant for r-Process nucleosynthesis. The mass outflow rate can then be written as $\dot{M} = 4\pi r^2 \rho v$, from which we can derive,

$$\dot{M} \approx 1.2 \times 10^{-6}\, M_\odot\, s^{-1} \left(\frac{g_s}{11/2} \right) \left(\frac{M_{NS}}{1.4\, M_\odot} \right)^3 \left(\frac{0.2\, s}{\tau_{dyn}} \right) S_{100}^{-4}. \tag{16}$$

Alternatively, we can turn this around and define the *expansion rate* of the hot bubble material as $\lambda_{exp} \equiv v/r = 1/\tau_{dyn}$. This is sometimes useful, as the the mass outflow rate can be roughly constant for some relevant timescales. Using the above arguments, we can see that $\lambda_{exp} \propto s^4 M_{NS}^{-3}$,

$$\lambda_{\text{exp}} \approx \left(\frac{45}{44\pi^2}\right)\left(\frac{11/2}{g_s}\right) N_A \left(\frac{m_{\text{pl}}^2}{M_{\text{NS}} m_b}\right)^3 s^4 \dot{M}. \qquad (17)$$

We have shown that the general thermodynamic and hydrodynamic conditions around the hot proto neutron star can be understood very simply in terms of a few key parameters like S_{100} and $\dot{M}$. Though these quantities themselves are difficult to extract with great precision and confidence from the various supernova calculations, we will see that such precision is not necessary for nucleosynthesis considerations.

When viewed in this context, the physical environment around the neutron star is really determined by just a few well defined quantities: the Chandrasekhar mass; the strength of the weak interaction G_F; and the saturation density of nuclear matter. Though the reader may protest that in reality the outflow from the neutron star is essentially multi-dimensional with convection and the like, we will argue that for nucleosynthesis it is really only S_{100}, τ_{dyn}, and the ratio of neutrons-to-protons that matter. In turn, all of these quantities are set by the neutrinos!

HEAVY ELEMENT NUCLEOSYNTHESIS IN THE NEUTRINO-DRIVEN WIND

It is useful to consider the outflow of one representative fluid element. This element will begin to move out from the "gain radius," which is in the vicinity of the neutron star surface. Here the temperature of the plasma will be very high and all nuclear reactions will be in NSE. As the fluid element moves further out, it will cool (everywhere the relation between temperature and radius is as in Eq. (12)). Note that sufficiently far above the neutron star it will be true that $T \ll T_\nu$. This is why neutrino interactions with particles in the fluid element tend to heat it. Eventually the fluid element will have moved so far out that the temperature falls below $T \approx 0.75 \text{ MeV}$, and the charged particle nuclear reaction rates, which are strongly dependent on T, can no longer match the material expansion rate λ_{exp}. At this point, the nuclear ensemble begins to fall out of NSE. This freeze-out from NSE results in the build-up of a relatively small number of seed nuclei of mass $A \sim 50$ to 100. Subsequently, when the fluid element has moved far enough from the neutron star that $T_9 \lesssim 3$ or 4, neutrons are captured on these seed nuclei and, under the right circumstances of high enough neutron-to-seed ratio, the r-process elements can be synthesized.

This scenario bears more than a superficial resemblance to the standard picture of Big Bang Nucleosynthesis (BBN). It is easy to see this parallel by identifying the radius or temperature of an outgoing fluid element in the wind with the temperature or age of the universe in the Big Bang. Both BBN and r-Process nucleosynthesis in the neutrino-driven wind are freeze-outs from NSE at high entropy. In both cases the nucleosynthesis is determined by three key parameters: the entropy per baryon; the expansion rate; and the neutron-to-proton ratio [18–20].

The entropy is $S_{100} \sim 1$ in the wind, while it is $S_{100} \approx 2.53 \times 10^6 \Omega_b^{-1} h^{-2} \sim 10^8$ in BBN, where the baryonic contribution to the closure density is determined from the observed deuterium abundance in Lyman limit systems to be $\Omega_b h^2 \approx 0.02$ [21]. The expansion rate of the wind material is given by Eq. (17), as compared to the quantity relevant for BBN, the expansion rate of the universe $\lambda_{\rm BBN} \sim T^2/m_{\rm pl}$ during the epoch $30 \gtrsim T_9 \gtrsim 10^{-2}$. The net ratio of neutrons-to-protons, n/p, can be parametrized through the electron fraction, or net number of electrons per baryon, $Y_e = 1/(1 + n/p)$. (Note that Y_e = number density of electrons minus the number density of positrons, all divided by the baryon number density.) It is in this latter parameter that BBN and the neutrino-driven wind fundamentally differ.

In the absence of neutrino flavor transformation, the neutrino-driven wind will be neutron-rich, with $Y_e < 0.5$ or $n/p > 1$. Such neutron-rich conditions are a *required* and a *necessary*, though not a sufficient condition to obtain r-Process nucleosynthesis. In stark contrast, the freeze-out from NSE in BBN is characterized by proton-rich conditions, $Y_e > 0.5$, or $n/p < 1$. This is one of the reasons that BBN makes little more than helium and trace amounts of deuterium and ^{7}Li.

In both BBN and the neutrino-driven wind it is the neutrino interactions with particles in the plasma that set the physical scale for the thermodynamic quantities and determine Y_e at NSE freeze-out. In the Big Bang we can describe an epoch of "Weak Decoupling" where the rates of neutrino scattering on relativistic particles fall below the expansion rate of the universe. This occurs roughly at $T \approx 2\,\text{MeV}$ or $3\,\text{MeV}$ and is where the neutrinos no longer rapidly share energy with the electrons and photons in the plasma. The analogous event in the neutrino-driven wind occurs at the neutrino sphere (more strictly, the chromo sphere), where the neutrinos decouple and freely stream. The corresponding temperature of this event is where $T_\nu \approx T$ for each neutrino species - approximately coincident with a position just inside the surface of the neutron star.

We also can talk about "Weak Freeze-Out" in BBN, where the electron neutrino (electron antineutrino) and positron (electron) capture rates on neutrons (protons) fall below the expansion rate. This is the epoch $T = T_{\rm WFO} \approx 0.7\,\text{MeV}$, after which neutrons and protons are no longer rapidly inter-converted one to the other. In effect, the equilibrium value of of n/p (and so Y_e) freezes out at this epoch, assuming whatever value it had at the last time it was in equilibrium ($T_{\rm WFO}$) corrected for free neutron decay. In the Big Bang, in the absence of active/sterile neutrino transformations, all neutrino species will have essentially identical spectra (so long as the three net lepton numbers are of order the baryon-to-photon number $\eta \approx 2.79 \times 10^{-8} \Omega_b h^2$) with all $\eta_\nu \approx 0$. As a result, in the competition between $\nu_e + n \rightleftharpoons p + e^-$

and $\bar{\nu}_e + p \rightleftharpoons n + e^+$, the fact that the neutron is heavier than the proton means that the reactions creating protons will win out over those that create neutrons. The equilibrium neutron-to-proton ratio at Weak Freeze-out will be $n/p \approx \exp(-\delta m_{rmnp}/T_{\text{WFO}}) \approx 1/7$, where $\delta m_{\text{np}} \approx 1.293 \, \text{MeV}$ is the neutron/proton mass difference.

In a direct analogy to the Big Bang case, we can define a Weak Freeze-Out radius in the neutrino-driven wind. This is the location where the rates of $\nu_e + n \rightarrow p + e^-$ and $\bar{\nu}_e + p \rightarrow n + e^+$ fall below the expansion rate of the material λ_{exp} as given in Eq. (17). Typically, the position of the Weak Freeze-Out radius will be where $T_9 \approx 10$ [14]. (We alternatively could refer to the corresponding Weak Freeze-Out temperature T_{WFO}, related to the radius through Eq. (12).) This is very near (usually just inside) the NSE freeze-out position. For a fluid element still inside of this radius, the neutron-to-proton ratio will be in steady state equilibrium with the fluxes of $\bar{\nu}_e$ and ν_e coming from the neutrino sphere(s). For a fluid element which has passed beyond this radius, into a temperature regime where $T < T_{\text{WFO}}$, the n/p ratio will be whatever the last equilibrium value was at T_{WFO}, corrected for free neutron decay and other residual weak reactions.

The neutrino processes principally responsible for setting the equilibrium n/p ratio inside the Weak Freeze Out radius involve the free nucleons [14,22]:

$$\nu_e + n \rightarrow p + e^- \tag{18}$$

$$\bar{\nu}_e + p \rightarrow n + e^+. \tag{19}$$

The rates of the reverse processes are generally negligible in the region well above the neutron star, as the matter temperature T is small compared to the energy scale characterizing the freely streaming neutrinos. The competition between the rates of the processes in Eq. (18), $\lambda_{\nu_e n}$, and Eq. (19), $\lambda_{\bar{\nu}_e p}$, will determine n/p and hence Y_e at the weak freeze-out position. These rates can be cast in the form $\lambda = $(neutrino number flux)·(cross section),

$$\lambda_{\nu_e n} \approx \left(\frac{L_{\nu_e}}{4\pi r^2}\right) \frac{1}{\langle E_{\nu_e}\rangle} \langle \sigma_{\nu_e n}\rangle, \tag{20}$$

$$\lambda_{\bar{\nu}_e p} \approx \left(\frac{L_{\bar{\nu}_e}}{4\pi r^2}\right) \frac{1}{\langle E_{\bar{\nu}_e}\rangle} \langle \sigma_{\bar{\nu}_e n}\rangle. \tag{21}$$

The neutrino distribution function-weighted cross section is defined as, for example,

$$\langle \sigma_{\nu_e n}\rangle \equiv \int_0^\infty \sigma_{\nu_e}(E_\nu) f_{\nu_e}(E_\nu) dE_\nu, \tag{22}$$

where f_{ν_e} is the ν_e distribution function and,

$$\sigma_{\nu_e n}(E_{\nu_e}) \approx 9.72 \times 10^{-44} \, \text{cm}^2 ((E_{\nu_e} + \delta m_{np} - m_e)/\text{MeV})^2, \qquad (23)$$

$$\sigma_{\bar{\nu}_e p}(E_{\bar{\nu}_e}) \approx 9.72 \times 10^{-44} \, \text{cm}^2 ((E_{\bar{\nu}_e} - \delta m_{np} - m_e)/\text{MeV})^2, \qquad (24)$$

and where $m_e \approx 0.511$ MeV is the electron/positron rest mass.

Integration of the rate equations corresponding to the processes in Eqs. (18,19) yields [14],

$$n/p \approx \frac{\lambda_{\bar{\nu}_e p}}{\lambda_{\nu_e n}} \approx \frac{L_{\bar{\nu}_e} \langle E_{\bar{\nu}_e} \rangle}{L_{\nu_e} \langle E_{\nu_e} \rangle} \approx \frac{\langle E_{\bar{\nu}_e} \rangle}{\langle E_{\nu_e} \rangle}. \qquad (25)$$

Here all quantities are evaluated at the position (temperature) of the Weak Freeze Out radius. The second approximation in the above equation is obtained by assuming that the energies of the neutrinos are large compared to δm_{np} and m_e. The last approximation follows on noting that the numerical calulations show that the luminosities of all neutrino species can be approximately the same at late enough epochs.

From Eq. (25) it is clear that neutron-rich conditions will obtain in the wind. As long as there is no neutrino flavor transformation in the region between the neutron star surface and the Weak Freeze-Out radius, the average neutrino energy hierarchy in Eq. (10) *guarantees* that $n/p > 1$. In fact Eq. (25) shows that with the expected neutrino average energies, there will be about a 40% excess of neutrons over protons. However, this apparently will not produce a high enough neutron-to-seed nucleus ratio to give the $A \approx 195$ peak in the r-Process.

After a fluid element passes the Weak Freeze Out position and the NSE Freeze Out position, the n/p ratio has been set and the charged particle nuclear reactions can no longer keep up with the demand of NSE. As the fluid element moves to regions of lower temperature, NSE would suggest that the equilibrium shift from free nucleons to more tightly bound alpha particles and heavier nuclei. In fact, nearly every nucleon that can be incorporated into an alpha particle does so, leaving a sea of free neutrons, very many alphas, and only a small number of seed nuclei. Not many seed nuclei can be formed, as these must be built up from alpha particles, neutrons, and a very small number of free protons.

As in BBN, the high entropy mitigates against the assembly of alphas into heavier seed nuclei: heavy nuclei are inherently low entropy configurations (many nucleons moving around as a unit). Another way to see this is that higher entropies imply a higher rate of photo-disintegration of heavy species. Also as in BBN, the assembly of heavy seed nuclei from alpha particles and free nucleons is hampered by the lack of stable nuclear species at masses 5 and 8. The pathway to heavy seed nuclei must then go through three body reactions: $3\alpha \to {}^{12}\text{C} + \gamma$; and, in the neutron-rich conditions of the wind, $\alpha + \alpha + n \to {}^9\text{Be} + \gamma$. These three body reactions introduce a "bottleneck" in the nuclear flow toward heavy seed nuclei which introduces dependence on the expansion timescale $\tau_{\rm dyn} = 1/\lambda_{\rm exp}$. The smaller is $\tau_{\rm dyn}$, the less time for assembly of heavy seeds and so the fewer of them there will be [18,19].

Numerical studies [18,19] show that there is not nearly a high enough neutron to seed-nucleus ratio in the neutrino-driven wind to produce the $A \approx 195$ peak in the r-Process. It is clear that if we are to "fix" this problem and produce a respectable n/p ratio, we must figure out how nature turns the "knobs" to raise the neutron to seed nucleus ratio. These are either or all of: (1) turning up the entropy to reduce the assembly of seeds and, hence, increase the neutron to seed ratio; (2) reducing $\tau_{\rm dyn}$ to suppress assembly of seeds; and (3) decreasing Y_e directly by altertering the weak interaction or neutrino physics. There are three general and generic ways to turn these knobs and fix the r-Process in the neutrino driven wind.

Number one, we could envision a multi-dimensional hydrodynamic fix, the "fast-slow-fast" outflow scenario. This scenario would have an initially small $\tau_{\rm dyn}$ to suppress the formation of seed nuclei, followed by a larger $\tau_{\rm dyn}$ to allow neutron capture in near (n,γ)-(γ,n) equilibrium conditions. One could imagine that this would be caused by material in the outflow being caught in in turbulent eddies, and then subsequently re-accelerated and ejected. Unfortunately, none of the various numerical calculations on the market give anything which looks like this scenario. Instead, they tend to give an always vigorous outflow [23–25].

Another fix involving general relativity has been proposed [16,17]. The deeper gravitational potential well caused by general relativistic effects will decrease $\tau_{\rm dyn}$ and increase s in the outflow (see Eq. 11). These general relativistic effects are plausible, since a phase transition to a softer equation of state in the neutron star during the neutrino Kelvin-Helmholtz epoch is a real possibility [26]. Though these effects can be engineered to give a high enough neutron to seed nucleus ratio to get a good r-Process, they must be quite finely tuned to avoid either under producing the r-Process material or to avoid problems with the differential redshift of $\bar{\nu}_e$ and ν_e [17,27].

Finally, we are left with alterations in the neutrino and/or weak interaction physics as a means to save models of r-Process nucleosynthesis sited in the neutrino-driven wind. Neutrino flavor and/or type (e.g., active $\rightleftharpoons$ sterile) transformation has great leverage on Y_e and, hence, on the neutron to seed nucleus ratio.

NEUTRINO FLAVOR/TYPE CONVERSION AND IMPLICATIONS FOR NEUTRINO MASS AND MIXING SCHEMES

If neutrinos have vacuum rest masses then there arises the possibility that their energy (or, mass) eigenstates ν_i will not be coincident with their flavor eigenstates ν_α. We can relate the amplitudes of these representations to one another through a unitary transformation,

$$\nu_\alpha = \sum_i U_{\alpha i} \nu_i. \tag{26}$$

In the particular case of two neutrino mixing between, for example, a ν_μ and a ν_e, there is a simple parametrization which serves to define a vacuum mixing angle θ. The conjugate transpose of the unitary operator $U_{\alpha i}$ relates the flavor kets to the mass kets in this case:

$$|\nu_e\rangle = \cos\theta |\nu_1\rangle + \sin\theta |\nu_2\rangle \tag{27}$$

$$|\nu_\mu\rangle = -\sin\theta |\nu_1\rangle + \cos\theta |\nu_2\rangle. \tag{28}$$

These vacuum mixings can be *matter-enhanced* in astrophysical environments, leading to partial or even complete transformation of a neutrino's flavor label as it propagates down a matter gradient. Studies have been carried out on the effects of neutrino flavor transformation in the region above the neutron star on the shock re-heating epoch [28], and on r-Process/neutrino-driven wind models [14,29].

In matter, freely streaming neutrinos will acquire effective masses as a result of forward scattering on particles which carry weak charge. We can then write a general neutrino state as $|\nu(t)\rangle = a_e(t) |\nu_e\rangle + a_\mu(t) |\nu_\mu\rangle$. Here t can be any parameter along the world line of a neutrino, including time or radial distance above the neutron star. If we know the (forward scattering) interactions of the neutrinos along their trajectories, then we can solve a Schroedinger-like equation for the amplitudes $a_e(t)$ and $a_\mu(t)$:

$$i\frac{d}{dt}\begin{bmatrix} a_e(t) \\ a_\mu(t) \end{bmatrix} = H \begin{bmatrix} a_e(t) \\ a_\mu(t) \end{bmatrix}, \tag{29}$$

where the neutrino propagation Hamiltonian H includes a contribution from neutrino-electron and neutrino-positron charged current forward exchange scattering H_e, and an analagous contribution from neutrino/neutrino neutral current forward exchange scattering $H_{\nu\nu}$.

We can then write the different contributions as a sum, $H = H_e + H_{\nu\nu}$. A particularly convenient flavor-basis representation for the electron/positron background contribution is,

$$H_e = \frac{1}{2}\begin{pmatrix} -\Delta\cos 2\theta + A & \Delta\sin 2\theta \\ \Delta\sin 2\theta & \Delta\cos 2\theta - A \end{pmatrix}, \tag{30}$$

where $\Delta = \delta m^2/(2E_\nu)$ and $A = \sqrt{2}G_F(n_{e^-} - n_{e^+})$, with n_{e^-} and n_{e^+} being the electron and positron number densities, respectively. Here the difference of the squares of the vacuum neutrino mass eigenvlues is defined to be,

$$\delta m^2 \equiv m_2^2 - m_1^2. \tag{31}$$

Similarly, a convenient representation for the neutrino background is [29],

$$H_{\nu\nu} = \begin{pmatrix} B & B_{e\mu} \\ B_{\mu e} & B \end{pmatrix}. \tag{32}$$

In this expression we follow the notation of Qian & Fuller 1995 [29]. Here B is a term analagous to A but involving, in this example, the net number of ν_e neutrinos and the net number of ν_μ neutrinos. The off-diagonal term is a result of neutrinos not being in flavor eigenstates. Both B and the off diagonal terms, $B_{e\mu}$ and $B_{\mu e}$, are given in Qian & Fuller 1995 [29].

A neutrino mass level crossing, or Mikheyev-Smirnov-Wolfenstein [30–32] "resonance," will occur when,

$$\Delta \cos 2\theta = A + B. \tag{33}$$

The degree of flavor conversion of a neutrino on propagating through resonance depends on a comparion of the neutrino oscillation length at resonanace, $L_{\text{res}} = 4\pi E_\nu / (\delta m^2 \sin 2\theta)$ to the local scale height of weak charges at resonance, $L_{\text{H}} \equiv |d\ln(A+B)/dt|^{-1}$. (Here we could employ radial distance r, as well as t, in evaluating the density scale height.) Adiabaticity obtains, and complete flavor conversion occurs, when $L_{\text{res}} \ll L_{\text{H}}$. We can define an adiabaticity parameter $\gamma \equiv 2\pi L_{\text{H}}/L_{\text{res}}$. In this case, the Landau-Zener approximation can be applied [33] and the survival probability for a neutrino of flavor α propagating through resonance is,

$$P_{\nu_\alpha \to \nu_\alpha} \approx 1 - \exp\left(-\frac{\pi}{2}\gamma\right). \tag{34}$$

Though the neutrino background $H_{\nu\nu}$ is quite important for determining the nucleosynthesis effects caused by neutrino transformation at (especially) small δm^2 [29], we can get a fair idea of the range of δm^2 required to get a mass level crossing in the region of interest for the r-Process in the neutrino-driven wind by simply employing the baryon distribution (Eq. 14) and neutrino energy spectra (Eq. 3) discussed in the last section. The range of δm^2 which will give resonances in the region above the neutron star surface, but inside the Weak Freeze Out radius, and so can affect the n/p ratio, is

$$0.1\,\text{eV}^2 \lesssim \delta m^2 \lesssim 10^4\,\text{eV}^2. \tag{35}$$

Remarkably, and fortuitously, this range encompasses the neutrino mass scales which are of potential interest for cosmology and neutrino dark matter considerations. Significant cosmological effects could result if the sum of the light neutrino masses is $\sum m_\nu \sim 1\,\text{eV}$ to $100\,\text{eV}$. Though, like laboratory neutrino oscillation experiments, r-Process nucleosynthesis considerations essentially "measure" δm^2 and not neutrino mass directly, we can get a crude idea of the mass scale probed by examining $\sqrt{\delta m^2}$. If we can read the "fossil record" left by whatever site has produced the r-Process species, then we may be able to obtain evidence/constraints on neutrino transformation in the hot bubble when δm^2 is in the range given by Eq. (35), or take what we know about neutrino masses and mixings from laboratory experiments and constrain or potentially rule out neutrino-driven supernova ejecta as a site for r-Process nucleosynthesis.

We can also employ the baryon density distribution in an isentropic wind to get an idea of the degree of flavor transformation for neutrinos of various masses and energies in the region between the neutron star surface and the Weak Freeze Out position. First, note from Eq.s (14,15) that the density scale height of the baryons and, hence, the density scale height of the net electron number, is

$$L_{\rm H} \sim |d\ln(A)/dt|^{-1} \approx \frac{1}{3}r \approx L_{\rm H0}\left(\frac{M_{\rm NS}}{1.4\,M_\odot}\right)T_9^{-1}S_{100}^{-1}, \qquad (36)$$

where $L_{\rm H0} \approx 75.0\,{\rm km}$. This implies that for 90% or higher probability of conversion at resonance (i.e., $\lesssim 10\%$ survival probability), the vacuum mixing angle should satisfy,

$$\sin^2 2\theta \gtrsim 1.54 \times 10^{-3} \left(\frac{75\,{\rm km}}{L_{\rm H0}}\right)\left(\frac{1.4\,M_\odot}{M_{\rm NS}}\right)\left(\frac{E_\nu}{20\,{\rm MeV}}\right)\left(\frac{1\,{\rm eV}^2}{\delta m^2}\right)\left(\frac{10}{T_9}\right) S_{100}. \quad (37)$$

But in fact, we do not require complete adiabatic neutrino flavor transformation to produce significant effects on r-Process nucleosynthesis in the neutrino-driven wind/hot bubble. Qian et al. 1993 [14] have shown that as little as 20% to 30% conversion in the channel,

$$\nu_{\mu,\tau} \rightleftharpoons \nu_e \qquad (38)$$

will cause sufficient increase in the neutron-to-proton ratio to drive the material in the neutrino-driven wind proton-rich! This would, of course, completely preclude r-Process nucleosynthesis in this site. This result is a consequence of high energy mu or tau neutrinos changing their flavor labels to become ν_e neutrinos in a region sufficiently close to the neutron star that the n/p ratio of an outgoing fluid element can still be affected by the reactions in Eqs. (18, 19). In the hot bubble, because their energy spectra start out nearly the same, ν_μ or ν_τ mixing with the ν_e produce essentially identical effects on the n/p ratio and r-Process nucleosynthesis.

Given the neutrino energy hierarchy extant at the neutrino sphere (neutron star surface) as in Eq. (10), it is clear that conversion of $\sim 25\,{\rm MeV}$ ν_μ or ν_τ neutrinos into ν_e's inside the Weak Freeze Out radius will drive the $n/p < 1$. This can be seen from Eq. (25). In this case, we will have $\langle E_{\bar\nu_e}\rangle < \langle E_{\nu_e}\rangle$ at the Weak Freeze Out radius, so long as $\delta m^2 \gtrsim 2\,{\rm eV}^2$.

Not only must we have $n/p > 1$ for r-Process nucleosynthesis to be *possible*, but in fact even a $\sim 40\%$ excess of neutrons is not enough to get an *acceptable* r-Process abundance pattern that will match that of the solar system! In a sense then, if we *knew* for certain that the r-Process material originated in neutrino-heated supernova ejecta, then we could place powerful limits on the mixing angles and masses of neutrinos in the oscillation channel given by Eq. (38). (This will be the case for matter-enhancement in the neutrino channel, which has as a requirement that the vacuum neutrino mass eigenvalue most closely associated with the ν_μ or ν_τ be larger than that most closely associated with the ν_e. If the opposite is true, then the corresponding antineutrino oscillation channel is matter-enhanced.)

If the r-Process, even just some of it, comes from this site, then we can conclude that the vacuum mixing parameter ranges $\delta m^2 \gtrsim 2\,\text{eV}^2$ and/or vacuum mixing angles given by $\sin^2 2\theta \gtrsim 10^{-4}$ are excluded. Note that, in principle, r-Process nucleosynthesis can probe neutrino mixing angles on scales at, or even far better, than present or future experiments can, and it can do so for a significant range of δm^2 values.

If future neutrino experiments find vacuum neutrino mixing parameters in this restricted range, then we will have established for certain that the neutrino-heated supernova ejecta *is not* the production site of even a small component of r-Process nucleosynthesis. This would probably imply that the neutron star merger scenario accounts for the r-process and, therefore, the rate of such mergers in the galaxy is much higher than current thinking indicates. In turn, this would be of great interest to schemes for gravitational wave detection such as LIGO, for which the neutron star merger phenomenon is the most promising source.

Present experiments are beginning to probe some of these mass and mixing ranges. The Los Alamos LSND experiment does claim a $\bar\nu_e$ excess in their detector which could be interpreted as stemming from vacuum neutrino oscillations in the channel $\bar\nu_\mu \rightleftharpoons \bar\nu_e$ [34,35]. Their signal can come from a fair range of δm^2 and vacuum mixing angle, some of which conflicts with the r-Process "disallowed" region.

There are several ways to resolve this conflict, all of which involve interesting astrophysical or nuclear and particle physics possibilities. First, the r-Process may be coming from a site other than neutrino-heated supernova ejecta as discussed above. Second, there may be another explanantion for the $\bar\nu_e$ excess observed in the LSND experiment, or the experiment may be simply in error. Third, perhaps the vacuum mass of the ν_μ is *smaller* than that of the ν_e, so that the r-Process limits do not apply. Fourth, there could be mixing between active neutrino species and one or more SU(2) singlet sterile neutrino species in the supernova which help evade the putative r-Process limits. Finally, it could be simply that the mixing parameters required to explain the LSND excess events are in their parameter region which is not in conflict with our r-Process considerations, i.e., $\delta m^2 \lesssim 2\,\text{eV}^2$.

On this latter possibility it is interesting that recent work [36] indicates that neutrino flavor conversion as in Eq. (38) has quite beneficial effects on the neutron to seed nucleus ratio when the vacuum mixing parameters satisfy $0.1\,\text{eV}^2 \lesssim \delta m^2 \lesssim 1\,\text{eV}^2$ and $\sin^2 2\theta \gtrsim 10^{-4}$. This is because neutrino flavor conversion with mixing parameters in this range would not entail the highest energy neutrinos transforming in a region where they could do the most damgae to the n/p ratio. That is, the region below the Weak Freeze Out radius.

It is interesting to note that neutrino mixing parameters in this beneficial range are consistent with those picked out by attempts to fit with only three active neutrino species (i.e., only two independent values of δm^2) all of the hints we have on neutrino mass and oscillations from solar and atmospheric neutrino considerations as well as from experiment. For example, the three-neutrino mixing schemes of Cardall and Fuller [37] and Acker and Pakvasa [38] purport to account for all accelerator and reactor data on vacuum neutrino mixing, as well as the data from atmospheric neutrinos and solar neutrinos. Both of these schemes would allow for neutrinos to have mass differences and mixing angles consistent with, and in the above-described range potentially beneficial for, r-Process nucleosynthesis in neutrino-heated supernova ejecta. The proposed three-neutrino mixing schemes will rise or fall in the next few years on the results of future long and intermediate baseline neutrino oscillation experiments and on whether or not neutrino oscillations are at the root of the solution to the atmospheric muon neutrino deficit and whether or not this deficit shows a zenith angle dependence.

As mentioned above, yet another scenario for helping to increase the n/p ratio and so facilitate r-Process nucleosynthesis involves active $\rightleftharpoons$ sterile neutrino transformation. For example, Caldwell, Fuller, & Qian 1997 invoke $\nu_{\mu,\tau} \rightleftharpoons \nu_s$ (where ν_s is a sterile SU(2) singlet neutrino) as occurring near the surface of the neutron star, followed by $\nu_{\mu,\tau} \rightleftharpoons \nu_e$ further out, but still in a region below the Weak Freeze Out radius, where Y_e could be affected. This would have the effect of unbalancing the competition inherent in Eqs. (18, 19) in favor of the proton destruction reaction. This could in principle drive the material in the neutrino-driven wind quite neutron-rich [39].

CONCLUSION

Understanding of the nucleosynthesis of some of the heavy neutron-rich nuclei like the r-Process and light p-Process [9,40] species may provide us ultimately with a new neutrino physics "laboratory." As we have seen, the potential constraints/hints of neutrino masses and mixings are complimentary to those obtained from terrestrial laboratory experiments and solar and atmospheric neutrino considerations.

This connection between the synthesis of the heavy nuclei in neutrino-heated ejecta and neutrino flavor transformation is a completely new way in which astrophysics and laboratory nuclear and particle physics overlap. It is especially promising that these considerations may allow us new insights into the issue of whether or not neutrinos have cosmologically significant masses. Alternatively this connection between neutrino masses and mixings and nucleosynthesis means that laboratory neutrino oscillation experiments may hvave a new relevance for astrophysics. By unambiguously measuring neutrino mass and mixing parameters in the r-Process "excluded" region, we may gain insight into the possible allowed sites of r-Process nucleosynthesis as well as new information on neutrino physics. Who would have guessed that a neutrino mass-squared difference of $\delta m^2 \sim 1\,\mathrm{eV}^2$ might be of great interest to gravity wave detection schemes? It may even be that human life itself might depend on this datum. Why? Because iodine, which is made in the r-Process, is necessary for advanced life [41]!

ACKNOWLEDGMENTS

I would like to acknowledge D. O. Caldwell, C. Y. Cardall, and Y.-Z. Qian for many useful discussions. I also would like to acknowledge partial support from NSF grant PHY-9503384 at UCSD.

REFERENCES

1. Burbidge,E.M.,Burbidge,G.R., Fowler, and Hoyle,F., *Rev. Mod. Phys.* **29**, 624 (1957).
2. Cameron, A.G.W., *PASP* **69**, 201 (1957).
3. Mathews, G.J., and Cowan, J., *Nature* **345**, 491 (1990).
4. Cowan, J.J., Thielemann, F.-K., and Truran, J.W., *Phys. Rep.* **208**, 267 (1991).
5. Lattimer, J.M., Mackie, F., Ravenhall, D.G., and Schramm, D.N., *Astrophys. J.* **213**, 225 (1977).
6. Meyer, B.S., Howard, W.M., Mathews, G.J., Woosley, S.E., and Hoffman, R.D., *Astrophys. J.* **399**, 656 (1992).
7. Woosley, S.E., Wilson, J.R., Mathews, G.J., Hoffman, R.D., and Meyer, B.S., *Astrophys. J.* **433**, 229 (1994).
8. Takahashi, K., Witti, J., and Janka, H.-Th., *Astron. and Astrophys.* **286**, 857 (1994).
9. Fuller, G.M., and Meyer, B.S., *Astrophys. J.* **453**, 792 (1995).
10. McLaughlin, G.C., and Fuller, G.M., *Astrophys. J.* **464**, L143 (1996).
11. Kratz, K.-L., Bitouzet, J.P., Thielemann, F.-K., Moller, P., and Pfieffer, B., *Astrophys. J.* **403**, 216 (1993).
12. Haxton, W.C., Langanke, K., Qian, Y.-Z., and Vogel, P., *Phys. Rev. Lett.* **78**, 2694 (1997).

13. Qian, Y.-Z., Haxton, W.C., Langanke, K., and Vogel, P., *Phys. Rev.* C **55**, 1532 (1997).
14. Y.-Z. Qian, G.M. Fuller, G.J. Mathews, R.W. Mayle, J.R. Wilson, and S. E. Woosley, *Phys. Rev. Lett.* **71**, 1965 (1993).
15. Duncan, R.C., Shapiro, S.L., and Wasserman, I., *Astrophys. J.* **309**, 141 (1986).
16. Qian, Y.-Z., and Woosley, S.E. *Astrophys. J.* **471**, 331 (1996).
17. Cardall, C.Y., and Fuller, G.M., *Astrophys. J.* **486**, L111 (1997).
18. Hoffman, R.D., Woosley, S.E., and Qian, Y.-Z., *Astrophys. J.* **482**, 951 (1996).
19. Meyer, B.S., and Brown, J.S., *Astrophys. J. Suppl.* **112**, 199 (1997).
20. Fuller, G.M., and Cardall, C.Y., *Nucl. Phys. B (Proc. Suppl.)* **51B**, 71 (1996).
21. Cardal, C.Y., and Fuller, G.M., *Astrophys. J.* **472**, 435 (1996).
22. McLaughlin, G.C., and Fuller, G.M., *Astrophys. J.* **472**, 440 (1996).
23. Burrows, A., Hayes, J., and Fryxell, B.A., *Astrophys. J.* **450**, 830 (1995).
24. Herant, M., Benz, W., Hix, W.R., Fryer, C.L., and Colgate, S.A., *Astrophys. J.* **435**, 339 (1994).
25. Miller, D.S., Wilson, J.R., and Mayle, R.W., *Astrophys. J.* **415**, 278 (1993).
26. Bethe, H.A., and Brown, G.E., *Astrophys. J.* **445**, L129 (1995).
27. Fuller, G.M., and Qian, Y.-Z., *Nucl. Phys. A* **606**, 167 (1996).
28. Fuller, G.M., Mayle, R.W., Meyer, B,S., and Wilson, J.R., *Astrophys. J.* **389**, 517 (1992).
29. Qian, Y.-Z., and Fuller, G.M., *Phys. Rev.* D **51**, 1479 (1995).
30. Mikheyev, S.P., and Smirnov, A.Yu., *Nuovo Cimento* **9C**, 17 (1986).
31. Wolfenstein, L., *Phys. Rev.* D **17**, 2369 (1978).
32. Bethe, H.A., *Phys. Rev. Lett.* **56**, 1305 (1986).
33. Haxton, W.C., *Phys. Rev.* D **35**, 2352 (1987).
34. Anthanassopoulos, C, et al., *Phys. Rev. Lett.* **75**, 2650 (1995).
35. Anthanassopoulos, C, et al., *Phys. Rev. Lett.* **77**, 3082 (1996).
36. Fuller, G.M., and Qian, Y.-Z., "Can Matter-Enhanced Neutrino Oscillations Aid r-Process Nucleosynthesis?," *preprint* (1997).
37. Cardall, C.Y., and Fuller, G.M., *Phys. Rev.* D **53**, 4421 (1996).
38. Acker, A., and Pakvasa, S., *Phys. Lett.* B **397**, 209 (1997).
39. Caldwell, D.O., Fuller, G.M., and Qian, Y.-Z., "Active/Sterile Neutrino Transformation and Supernova Heavy Element Nucleosynthesis," *preprint* (1997).
40. Hoffman, R.D., Woosley, S.E., Fuller, G.M., and Meyer, B.S., *Astrophys. J.* **460**, 478 (1996).
41. Fuller, M.O., private communication (1997).

Neutron Stars

Fridolin Weber

Institute for Theoretical Physics
Ludwig-Maximilians University
Theresienstr. 37, 80333 Munich, Germany

Abstract. Containing matter compressed to densities ranging from a few times the density of normal nuclear matter to about an order of magnitude higher, depending on the star's mass, neutron stars contain matter in one of the densest forms found in the universe. Such objects appear as natural stellar probes whose properties may tell us about the behavior of superdense matter too, complementary to relativistic heavy-ion physics. This paper gives a brief overview of the present status of such research.

INTRODUCTION

One of the most challenging but also complicated problems of modern physics consists in exploring the behavior of matter under extreme conditions of temperature and/or density. Knowledge of its behavior is of key importance for our understanding of the physics of the early universe, its evolution in time to the present day, neutron stars, various astrophysical phenomena, and laboratory physics. On the earth relativistic heavy-ion colliders provide the only tool by means of which such matter can be created and its properties studied. What is not widely appreciated is the fact that from the study of the properties of neutron stars, one too gains knowledge about the behavior of superdense matter.

Neutron stars are associated with two classes of astrophysical objects – pulsars and compact X-ray sources. Matter in their cores possess densities ranging from a few times the density of normal nuclear matter (2.5×10^{14} g/cm^3) to about an order of magnitude higher, depending on mass. They thus contain matter in one of the densest forms found in the universe, which makes them nearly ideal probes for a wide range of physical studies [1]. The equation of state of the stellar matter decisively links neutron stars with nuclear and particle physics plus various other branches of physics, as schematically illustrated in Fig. 1. It is the basic input quantity whose knowledge over a broad range of densities, from the density of iron at the star's surface up to ~ 15

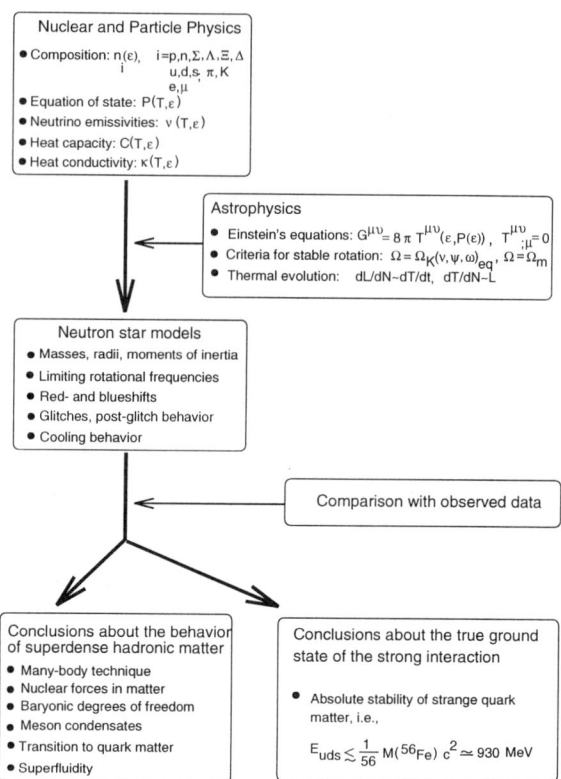

FIGURE 1. The nuclear equation of state is the fundamental input quantity for the construction of neutron star models. By means of comparing their theoretically determined properties with the observed ones conclusions about the behavior of the equation of state at supernuclear densities can be drawn.

times the density of normal nuclear matter reached in the cores of massive stars, is necessary when solving Einstein's equations for the properties of neutron stars. In this paper, I shall begin with reviewing the present status of modern neutron star matter calculations, which result in a broad competitive collection of equations of state that will be applied to the construction of non-rotating as well as rotating neutron star models. By means of comparing their theoretically determined properties with the observed ones conclusions about the behavior of the equation of state at supernuclear densities can be drawn, which addresses the open fundamental problem of modern physics concerning the behavior of the high-density equation of state.

DENSE NEUTRON STAR MATTER

Theoretical framework

Non-relativistic approach

For non-relativistic models, the starting point is a phenomenological nucleon-nucleon interaction. In the case of the equations of state reported here, different two-nucleon potentials (denoted V_{ij}) which fit nucleon-nucleon scattering data and deuteron properties have been employed. Most of these two-nucleon potentials are supplemented with three-nucleon interactions (denoted V_{ijk}). The hamiltonian is of the form

$$H = \sum_i \left(\frac{-\hbar^2}{2m}\right) \nabla_i^2 + \sum_{i<j} V_{ij} + \sum_{i<j<k} V_{ijk} . \quad (1)$$

The many-body method adopted to solve the Schroedinger equation is based on the variational approach, as outlined, for instance, in [2–5].

Relativistic fieldtheoretical approach

As discussed elsewhere [6,7], the lagrangian governing the dynamics of relativistic neutron-star matter has the form

$$\mathcal{L}(x) = \sum_{B=p,n,\Sigma^{\pm,0},\Lambda,\Xi^{0,-},\Delta^{++,+,0,-}} \mathcal{L}_B^0(x) \quad (2)$$

$$+ \sum_{M=\sigma,\omega,\pi,\varrho,\eta,\delta,\phi} \left\{ \mathcal{L}_M^0(x) + \sum_{B=p,n,\ldots,\Delta^{++,+,0,-}} \mathcal{L}_{B,M}^{\text{Int}}(x) \right\} + \sum_{\lambda=e^-,\mu^-} \mathcal{L}_\lambda(x) .$$

The subscript B runs over all baryon species that become populated in the dense stellar matter. The nuclear forces are mediated by that collection of scalar, vector, and isovector mesons (M) that is used for the construction of relativistic one-boson-exchange potentials [8,9]. The equations of motion resulting from Eq. (2) for the various baryon and meson field operators [6] are to be solved subject to the conditions of electric charge neutrality,

$$\rho_{\text{tot}}^{\text{el}} = \rho_{\text{Bary}}^{\text{el}} + \rho_{\text{Lep}}^{\text{el}} \equiv 0 , \quad (3)$$

and β (chemical) equilibrium [7,10],

$$\mu^B = \mu^n - q_B \mu^e . \quad (4)$$

μ^n and μ^e denote the chemical potentials of neutrons and electrons, q_B denotes the electric charge of baryon B. These equations have been solved elsewhere in the framework of the relativistic Greens function technique [6,10,11]. One ends up with a set of three independent equations, of which the first one determines the *effective scattering matrix* T in matter,

$$T = v - v^{\text{ex}} + \int v \Lambda T. \quad (5)$$

In the simplest case, which is the so-called Λ^{00} approximation, both nucleons propagate freely in the intermediate particle states. This is different for the relativistic Brueckner-Hartree-Fock (RBHF) approach where both intermediate baryons feel the nuclear many-body background and so scatter only into states outside occupied Fermi spheres [12–14]. The basic input quantity in Eq. (5) is the free nucleon-nucleon interaction as derived, for example, in the Bonn meson-exchange model [9]. We have adopted the latest version of this interaction together with Brockmann's potentials A–C to compute the T-matrix in neutron star matter up to several times nuclear matter density [11–14]. The important feature of such meson-exchange models is that the potential parameters are adjusted to the two-body nucleon-nucleon scattering data and the properties of the deuteron, whereby (in this sense) a parameter-free treatment of the many-body problem is achieved. The "Born" approximation of T sums the various meson potentials of the nucleon-nucleon interaction in free space, i.e.,

$$<12\mid v\mid 1'2'> = \sum_{M=\sigma,\omega,\pi,\rho,\eta,\delta,\phi} \delta^4_{11'}\, \Gamma^M_{11'}\, \Delta^M_{12}\, \Gamma^M_{22'}\, \delta^4_{22'}, \quad (6)$$

and thus neglects dynamical nucleon-nucleon correlations. It is this approximation which leads to the relativistic Hartree and Hartree-Fock approximations [6,10,15,16], whose parameters are adjusted to the properties of bulk nuclear matter. The symbol Γ^M in Eq. (6) stands for the various meson-nucleon vertices, and Δ^M denotes the free meson propagator of a meson of type M.

The second independent equation specifies the *baryon self-energy*, which is obtained from the T matrix as [6,17]

$$\Sigma^B = i \sum_{B'} \int \left[\operatorname{tr}\left(T^{BB'}\, g_1^{B'}\right) - T^{BB'}\, g_1^{B'}\right]. \quad (7)$$

Finally the two-point Green function, which describes the propagation of a baryon in matter, is given as the solution of *Dyson's equation*,

$$g_1^B = g_1^{0B} + g_1^{0B}\, \Sigma^B(\{g_1^{B'}\})\, g_1^B, \quad (8)$$

which terminates the set of self-consistent matter equations.

The equation of state follows from the stress-energy density tensor $\mathcal{T}_{\mu\nu}$ of the system as

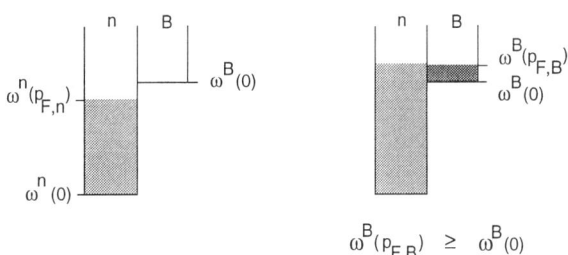

FIGURE 2. Condition for the onset of hyperon population in chemically equilibrated matter. Left: the single-particle energy is not high enough for the neutrons to transform to baryons of type B. Right: high-energy neutrons overcome the baryon threshold, that is, $\mu^B \equiv \omega^B(p_{F_B}) \geq \omega^B(0)$, and populate particle species B.

$$E(\rho) = <\mathcal{T}_{00}>/\rho - m, \qquad \text{where} \qquad (9)$$

$$\mathcal{T}_{\mu\nu}(x) = \sum_{\chi=B,\lambda} \partial_\nu \Psi_\chi(x) \frac{\partial \mathcal{L}(x)}{\partial(\partial^\mu \Psi_\chi(x))} - g_{\mu\nu}\mathcal{L}(x). \qquad (10)$$

The sum in the latter equation sums the contributions coming from the baryons and leptons ($\lambda = e, \mu$) [10]. Energy density $\epsilon(\rho)$ and pressure $P(\rho)$ are obtained from Eq. (10) as

$$\epsilon(\rho) = \rho[E(\rho) + m], \qquad P(\rho) = \rho^2 \frac{\partial}{\partial \rho} E(\rho), \qquad (11)$$

which leads to the equation of state in the form $P(\epsilon)$ that enters in the stellar structure calculations.

Baryon-lepton composition

Solving equations (5)–(8) in combination with the conditions for electric charge neutrality and chemical equilibrium (3) and (4) at nuclear matter densities ϵ_0 ($\epsilon_0 = 140$ MeV/fm^3, which corresponds to a baryon number density of $\rho = 0.16$ fm^{-3}) shows that neutron star matter at such densities consists primarily of p, n, e^-, and μ^-. At higher densities the more massive baryon states become populated as soon as the neutron chemical potential μ^n exceeds their masses (modified by interactions), Fig. 2. From a physical point of view, the particle population can be understood by remembering that the energy per baryon of neutron star matter lies way above the energy per baryon of isospin-symmetric nuclear matter, which amounts -16 MeV. So as soon as there are additional degrees of freedom accessible to the system which allow neutron star matter to escape from this energetically unfavorable high energy state by becoming more isopin symmetric, neutron star matter will do so. That is,

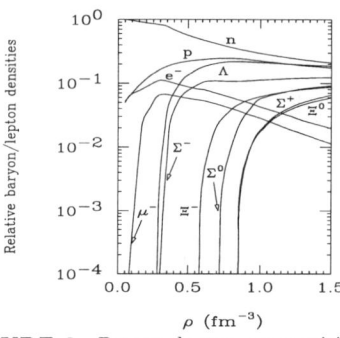 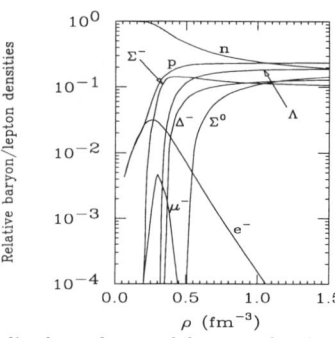

FIGURE 3. Baryon–lepton composition, normalized to the total baryon density ρ, of neutron star matter computed for representative relativistic Hartree (left) and relativistic Hartree-Fock (right) approximations.

the population of baryon states whose isospin orientation is opposite to the one of the neutron is *isospin* favored. In addition to that, negatively charged baryons are *charge* favored over the positive ones, since these will replace electrons with high Fermi momenta. Besides these two guiding principles (isospin and electric charge), the nuclear many-body background influences the threshold densities of the various baryon states too. This arises, for example, from a modification of the mass of a baryon in matter, which leads to an effective, density-dependent baryon mass to be determined self-consistently [6].

The results of two representative neutron star matter calculations, performed for the relativistic Hartree (HV) and relativistic Hartree-Fock (HFV) approximation, are shown in Fig. 3. One sees that for HV the Λ hyperon, which remains unaffected by the above two guiding principles, since its electric charge $q_\Lambda = 0$ and the third component of isospin $I_{3\Lambda} = 0$, has the lowest threshold of all hyperons. In the case of HFV, the charge-favored (but isospin-unfavored) Σ^- possesses the lowest threshold, followed by the relatively heavy Δ^-. The important net effect of these new baryonic degrees of freedom is a softening of the equation of state, as we shall see below, which is important for the maximum possible mass of a neutron star.

The difference in particle population between HV and HFV has its origin in the different descriptions of the nuclear forces in both treatments. These are described for HV via σ, ω, and ρ-meson exchange, while for HFV the π meson contributes too. Moreover HFV contains the exchange terms, which are absent in HV. Thereby the large attractive and repulsive self-energies, which are typical for the Hartree approximation, are somewhat weakened for the Hartree-Fock treatment. This has an important impact on the baryon thresholds, such that the (charge-favored) Σ^- is no longer prevented from becoming populated by reasons of isospin orientation arguments. Similarly, the isospin of the Δ is no longer a liability, and the most favored charge-state of the Δ, i.e., the Δ^-, becomes populated first. Another striking difference between HV and HFV, which is important for the population of the Δ^- too, consists in the extremely different values of the coupling constant of the ρ meson, $g_{\rho N}$. It is well known that Fock terms are essential in obtaining the correct value for the symmetry energy coefficient ($a_4 \simeq 32$ MeV). There is therefore no need in Hartree-Fock calculations to use large values of $g_{\rho N}$ to get a_4 right, as it is the case in Hartree calculations [18]. For the present parametrizations the difference amounts a factor of about 13! The presence of the Δ^- manifests itself in a considerable reduction of the number of electrons. Hence, neutron star matter described within the Hartree-Fock approximation is less electrically conductive than for the Hartree case, which may register itself in the decay time of the magnetic field of a neutron star and thus in the active lifetime of a pulsar. Besides that, the chemical potential of electrons is considerably smaller for the HFV case ($\mu^e \lesssim 120$ MeV as compared to about 220 MeV for HV) which is of great relevance for the question of whether or not K^- mesons condense in the cores of neutron stars.

Models for the nuclear equation of state

A representative collection of nuclear equations of state that are determined in the framework of non-relativistic Schroedinger theory and relativistic nuclear field theory is listed in Table 1. This collection of equations of state will be applied below for the construction of models of nonrotating as well as rotating neutron stars. The specific properties of these equations of state are described in Table 1, where the following abbreviations are used: N = pure neutron; NP = n, p, leptons; π = pion condensation; H = composed of n, p, hyperons ($\Sigma^{\pm,0}$, Λ, $\Xi^{0,-}$) and leptons; $\Delta = \Delta_{1232}$-resonance; Q = quark hybrid composition, i.e., n, p, hyperons in equilibrium with u, d, s-quarks, leptons; K = incompressibility (in MeV); $B^{1/4}$ = bag constant (in MeV). Not all equations of state of our collection account for neutron matter in full β equilibrium (i.e., entries 13–17). These models treat neutron star matter as being composed of only neutrons, or neutrons and protons in equilibrium with leptons, which is however not the ground-state of neutron star matter predicted by theory [7,10]. As an example of such an equation of state, we exhibit the FP(V_{14} + TNI) model in Fig. 4. The relativistic equations of state account for *all* baryon states that become populated in dense star models constructed from them. As representative examples for the relativistic equations of state, we show HV, HFV, G_{300}, and G_{B180}^{DCM1} in Fig. 4. A special feature of the latter equation of state is that it also accounts for the possible transition of baryonic matter to quark matter. The population of hyperon states and/or the possible transition of confined hadronic matter to quark matter (see below) sets in typically at $\epsilon \gtrsim (2-3)\,\epsilon_0$, which causes a softening of the equation of state.

TABLE 1. Nuclear equations of state applied for the construction of models of general relativistic neutron star models.

Label	EOS	Description (see text)	Reference
	Relativistic field theoretical equations of state		
1	G_{300}	H, K=300	[19]
2	HV	H, K=285	[7,10]
3	G_{B180}^{DCM2}	Q, K=265, $B^{1/4}=180$	[20,21]
4	G_{265}^{DCM2}	H, K=265	[22]
5	G_{300}^{π}	H, π, K=300	[19]
6	G_{200}^{π}	H, π, K=200	[23]
7	Λ_{Bonn}^{00} + HV	H, K=186	[11]
8	G_{225}^{DCM1}	H, K=225	[22]
9	G_{B180}^{DCM1}	Q, K=225, $B^{1/4}=180$	[20,21]
10	HFV	H, Δ, K=376	[10]
11	Λ_{Bro}^{RBHF} + HFV	H, Δ, K=264	[12–14]
	Non-relativistic potential model equations of state		
12	BJ(I)	H, Δ	[24]
13	WFF(UV$_{14}$+TNI)	NP, K=261	[5]
14	FP(V$_{14}$+TNI)	N, K=240	[25]
15	WFF(UV$_{14}$+UVII)	NP, K=202	[5]
16	WFF(AV$_{14}$+UVII)	NP, K=209	[5]
17	MS94	NP, K=234	[26]

The stiffer behavior of HFV in comparison with HV at high densities has its origin in the exchange (Fock) contribution, which is absent for HV.

An inherent feature of the relativistic equations of state is that they do not violate causality, i.e., the velocity of sound, given by $v_s = c\sqrt{dP/d\epsilon}$, is smaller than the velocity of light at all densities, which is a problem for the non-relativistic equations of state. Among the latter only WFF(UV$_{14}$ + TNI) and MS94 do not violate causality up to the highest densities encountered in the heaviest neutron star models constructed for these equations of state.

With the exception of Λ_{Bonn}^{00} + HV and Λ_{Bro}^{RBHF} + HFV, the coupling constants of all relativistic equations of state are adjusted to the bulk properties of symmetric nuclear matter. Λ_{Bonn}^{00} + HV and Λ_{Bro}^{RBHF} + HFV make use of the relativistic Bonn and Brockmann meson-exchange models for the nucleon-nucleon interaction, respectively, whose parameters are determined by the free nucleon-nucleon scattering problem and the properties of the deuteron. The influence of dynamical two-particle correlations calculated from the scattering matrix leads to a relatively soft behavior of these equations of state in the vicinity of the saturation density of nuclear matter. This is indicated by the rather small compression moduli in the range between 190 and 260 MeV. The non-relativistic equations of state of our collection account for dynamical two-particle correlations too. These are calculated for different hamiltonians.

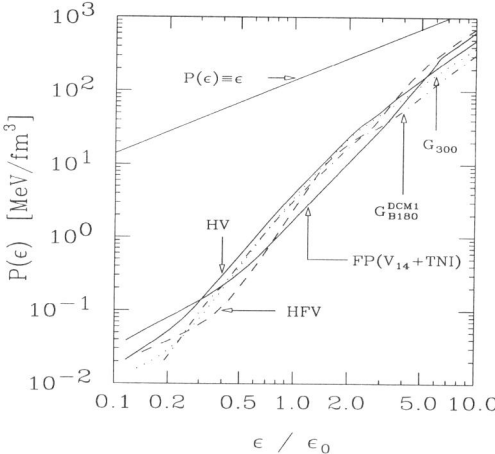

FIGURE 4. Graphical illustration of some of the equations of state, that is, pressure versus total energy density (in units of nuclear matter density, $\epsilon_0 = 140$ MeV/fm^3) listed in Table 1. $P(\epsilon) = \epsilon$ constitutes the stiffest possible model for the equation of state.

With the exception of BJ(I) and MS94, the calculations are performed for the Urbana and Argonne two-nucleon potentials V_{14}, UV_{14} [27] and AV_{14} [28], respectively, supplemented by different models for the three-nucleon interaction. These are the density-dependent three-nucleon interaction of Lagaris and Pandharipande, TNI [29], and the Urbana three-nucleon model, UVII [27].

The possible transition of confined hadronic matter into quark matter is taken into account in equations of state denoted G_{B180}^{DCM1} and G_{B180}^{DCM2}. Here a bag constant of $B^{1/4} = 180$ MeV has been used for the determination of the transition of baryon matter into quark matter, which places the energy per baryon of strange matter at 1100 MeV, well above the energy per nucleon in ^{56}Fe (≈ 930 MeV). As already mentioned above, the transition to quark matter sets in already between two and three times ϵ_0 [20,21,30]. This is rather different from what has been found in earlier investigations. The reason for this lies in the realization that the transition between confined hadronic matter and quark matter takes place subject to the conservation of baryon and electric charge. Correspondingly, there are two chemical potentials, and the transition of baryon matter to quark matter is to be determined in three-space spanned by pressure and the chemical potentials of the electrons and neutrons [20,21,31].

PROPERTIES OF NEUTRON STAR MODELS

Non-rotating star models

The structure of spherical neutron stars is determined by the Oppenheimer-Volkoff equations [32],

$$\frac{dP}{dr} = -\frac{\epsilon(r)\,m(r)}{r^2} \frac{[1 + P(r)/\epsilon(r)]\,[1 + 4\pi r^3 P(r)/m(r)]}{[1 - 2\,m(r)/r]}, \quad (12)$$

which describe a compact stellar configuration in hydrostatic equilibrium. (Here we use units for which the gravitational constant and velocity of light are $G = c = 1$. Hence $M_\odot = 1.5$ km.) The boundary condition reads $P(r = 0) \equiv P_c = P(\epsilon_c)$, where ϵ_c denotes the energy density at the star's center, which constitutes the free parameter that is to be specified when solving the above differential equation for a given equation of state. The latter determines P_c and the energy density for all pressure values $P < P_c$. The pressure profile is to be computed out to that radial distance where $P(r) = 0$, which determines the star's radius R, i.e., $P(r = R) = 0$. The mass contained in a sphere of radius r ($\leq R$), denoted by $m(r)$, follows from $\epsilon(r)$ as $m(r) = 4\pi \int_0^r dr'\, r'^2 \epsilon(r')$. The star's total (gravitational) mass is given by $M \equiv m(R)$.

Figure 5 exhibits the gravitational mass of non-rotating neutron stars as a function of central energy density for a sample of equations of state of Table 1. Each star sequence is shown up to densities that are slightly larger than those of the maximum-mass star (indicated by tick marks) of each sequence. Stars beyond the mass peak are unstable against radial oscillations and thus cannot exist stably (collapse to black holes) in nature. One sees that all equations of state are able to support non-rotating neutron star models of gravitational masses $M \geq M(\text{PSR } 1913 + 16)$. On the other hand, rather massive stars of say $M \sim 2\,M_\odot$ can only be obtained for equations of state that exhibit a rather *stiff* behavior at very high densities. Lars Bildsten (see his contribution elsewhere in this volume) reported that neutron stars that heavy may indeed exist in low-mass X-ray binaries, as indicated by the observation of quasi-periodic oscillations in luminosity from such stellar systems.

Knowledge of the maximum-possible mass of a neutron star sequence is of great importance for two reasons. Firstly, quite a few neutron star masses are known, and the largest of these impose a lower bound on the maximum-mass of a theoretical model. The current lower bound is about $1.56\,M_\odot$ [neutron star 4U 0900–40 ($\equiv$ Vela X-1)], which, as we have just seen, does not set too stringent a constraint on the nuclear equation of state. The situation could easily change if an accurate future determination of the mass of 4U 0900–40 (see Fig. 5) should result in a value that is close to its present upper bound of $1.98\,M_\odot$. In this case most of the equations of state of our collection would

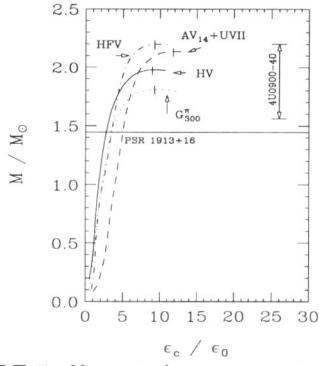 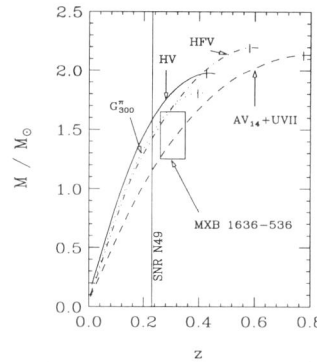

FIGURE 5. Non-rotating neutron star mass as a function of central density (left) and redshift (right).

be ruled out. The second reason is that the maximum mass can be useful in identifying black hole candidates [33–35]. For example, if the mass of a compact companion of an optical star is determined to exceed the maximum mass of a neutron star it must be a black hole. Since the maximum mass of stable neutron stars in our theory is $2.2\,M_\odot$, compact companions being more massive than that value are predicted to be black holes.

The connection between neutron star mass and gravitational redshift is given by

$$z = \frac{1}{\sqrt{1-2M/R}} - 1 \,. \qquad (13)$$

The right panel of Fig. 5 shows that the maximum-mass stars have redshifts in the range $0.4 \lesssim z \lesssim 0.8$, depending on the softness (stiffness) of the equation of state. Neutron stars with masses of typically $M \approx 1.5\,M_\odot$ (e.g., PSR 1913+16) are predicted to have redshifts in the considerably narrower range $0.2 \leq z \leq 0.32$. The solid rectangle covers masses and redshifts in the ranges of $1.30 \leq M/M_\odot \leq 1.65$ and $0.25 \leq z \leq 0.35$, respectively. The former range has been determined from observational data of X-ray burst source MXB 1636–536 [36], while the latter is based on the neutron star redshift data base provided by measurements of gamma-ray burst pair annihilation lines [37]. From the redshift value of SNR N49 (if correct) we predict a neutron mass star of $1.1 \lesssim M/M_\odot \lesssim 1.6$, which is consistent with the observed mass range. The relativistic equations of state set a narrower mass limit for SNR N49 given by $1.4 \lesssim M/M_\odot \lesssim 1.6$.

The left panel of Fig. 6 displays the radius as a function of gravitational redshift. The solid dots refer to the maximum-mass star of each sequence. Under the assumption that the interpretation of γ-ray bust redshifted annihilation

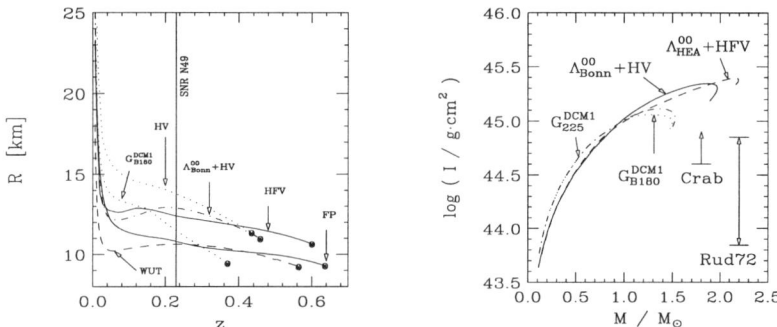

FIGURE 6. Left: neutron star radius as a function of redshift. Right: moment of inertia as a function of neutron star mass, for a sample of equations of state of Table 1.

lines as gravitationally redshifted $e^{\pm}$ pair annihilation lines from the surfaces of neutron stars is correct, SNR N49 is predicted to have a radius in the range of 10 to 14 km. The relativistic equations of state lead to a narrower radii range, i.e., 12.5 to 14 km. Madappa Prakash (see his contribution elsewhere in this volume) reported the radius measurement of neutron star RXJ 185635–3754 which may be as small as possibly 8 km! If this measurement is correct, then the underlying equation of state surely must be extremely soft up to several times nuclear matter density to get such a small neutron star. The possible condensation of K-mesons may be the cause for such an extreme softening, as Prakash pointed out. We note that the possibly very soft behavior of the equation of state required by this radius measurement is not in contradiction to the possible need for a rather stiff behavior at very high densities, as required by the possible existence of heavy neutron stars of masses around $\sim 2\,M_\odot$ (see above). Interestingly such a trend is obtained for many-body approximations that account for dynamical two-baryon correlations, arising from baryon-baryon potentials.

The right panel of Fig. 6 shows the moment of inertia of neutron stars, given by [38]

$$I(\Omega) = 4\pi \int_0^{\pi/2} d\theta \int_0^{R(\theta)} dr\, e^{\lambda+\mu+\nu+\psi} \frac{\epsilon + P(\epsilon)}{e^{2\nu-2\psi} - (\Omega-\omega)^2} \frac{\Omega-\omega}{\Omega}, \qquad (14)$$

as a function of gravitational mass. In general the incorporation of baryonic degrees of freedom as well as the possible transition of confined hadronic matter to quark matter causes a softening of the equation of state, which leads to somewhat smaller star masses and radii. From the functional dependence of I on radius and mass, which is of the form $I \propto R^2\, M$, one expects a relative decrease of the moment of inertia of star models constructed for such equation of state. Of course, the general relativistic expression for the moment of inertia, given in Eq. (14), is much more complicated. It accounts for the dragging effect of the local inertial frames (frequency dependence $\omega(r,\theta)$) and the curvature of space-time [38]. Nevertheless the qualitative dependence of I on mass and radius as expressed in the classical expression remains valid [39]. Estimates for the upper and lower bounds on the moment of inertia of the Crab pulsar derived from the pulsar's energy loss rate (labeled Rud72), and the lower bound on the moment of inertia derived from the luminosity of the Crab nebula (labeled Crab) [40–42] are shown in Fig. 6. (The arrows refer only to the value of I_{Crab} and not to the mass of the crab pulsar, which is not known.)

Rotating star models

Minimal rotational periods

Figure 7 exhibits the limiting rotational periods of neutron stars as set by the gravitational radiation-reaction driven instability [43–45]. It originates from counter-rotating surface vibrational modes, which at sufficiently high rotational star frequencies are dragged forward. In this case, gravitational radiation which inevitably accompanies the aspherical transport of matter does not damp the modes, but rather drives them [46,47]. Viscosity plays the important role of damping such gravitational-wave radiation-reaction instabilities at a sufficiently reduced rotational frequency such that the viscous damping rate and power in gravity waves are comparable [48]. The instability modes are taken to have the dependence $\exp[i\omega_m(\Omega)t + im\phi - t/\tau_m(\Omega)]$, where ω_m is the frequency of the surface mode which depends on the angular velocity Ω of the star, ϕ denotes the azimuthal angle, and τ_m is the time scale for the mode which determines its growth or damping. The rotation frequency Ω at which it changes sign is the critical frequency for the particular mode, m (=2,3,4,...). (For details we refer to [6,45].)

The critical rotational periods at which emission of gravity waves sets in in *hot* ($T = 10^{10}$ K) newly born pulsars is shown in the left panel of Fig. 7. The right panel refers to old and therefore cold stars of temperature $T = 10^6$ K, like neutron stars in binary systems that are being spun up by mass accretion from a companion. The temperature dependence arises from the dependence of the instability modes on viscosity. One sees that the instability periods are shifted toward smaller values the colder the star due to the larger viscosity in such objects. Consequently, the instability modes of neutron stars in binary systems are excited at smaller rotational periods than is the case for hot and newly born pulsars in supernovae. The rectangles in Fig. 7 denoted "observed" cover both the range of observed neutron star masses, $1.1 \lesssim M/M_\odot \lesssim 1.8$ as well as observed pulsar periods, i.e., $P \geq 1.6$ ms. One sees that even the most rapidly rotating pulsars observed to date have rotational periods

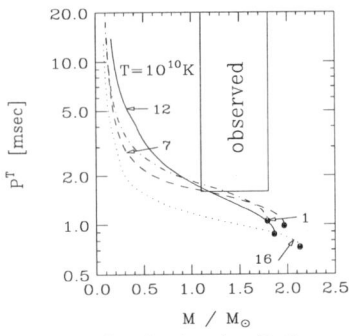 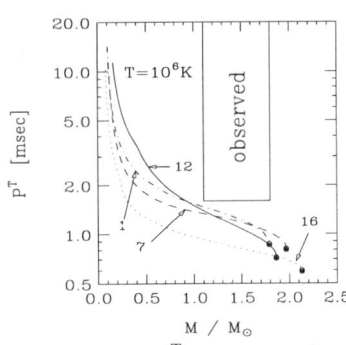

FIGURE 7. Gravitational radiation-reaction instability period P^T versus mass for newly born stars of temperature $T = 10^{10}$ K (left) and old and therefore cold stars of temperature $T = 10^6$ K.

larger that those at which the gravity-wave instability is excited and therefore can be understood as rotating neutron stars made up of baryonic matter, which may be in equilibrium with quark matter. The observation of a pulsar with a canonical mass around $1.4\, M_\odot$ but a rotational period considerably smaller than say 1 ms would be in clear contradiction to these calculations. To get such small pulsar periods seems to be very hard (if not impossible) for stars made up of baryonic matter. Therefore the interpretation of possibly existing submillisecond pulsars as rapidly rotating neutron stars is likely to fail. Such objects are plausibly made up of *self-bound* matter, of which absolutely stable 3-flavor strange quark matter is the most plausible form that fits most comfortably with our present understanding of the true nature of superdense matter [38,49]. This conclusion is strengthened by the construction of neutron star models that are rotating at their Kepler periods, P_K, at which mass shedding at the star's equator sets in. This period sets an *absolute* limit on rapid rotation, which cannot be overcome by any rotating star. On the basis of neutron star models constructed from the selection of equations of state studied here, the smallest Kepler periods are found to be in the range of $0.7 \lesssim P_K \lesssim 1$ ms, depending on the softness of the equation of state. P_K is given by [6,50,51]

$$P_K \equiv \frac{2\pi}{\Omega_K}, \text{ with } \Omega_K = \omega + \frac{\omega'}{2\psi'} + e^{\nu-\psi}\sqrt{\frac{\nu'}{\psi'} + \left(\frac{\omega'}{2\psi'}e^{\psi-\nu}\right)^2}. \quad (15)$$

Note that P_K can only be obtained by means of solving Eq. (15) self-consistently in combination with Einstein's equation,

$$\mathcal{R}^{\kappa\lambda} - \frac{1}{2}g^{\kappa\lambda}\mathcal{R} = 8\pi\, \mathcal{T}^{\kappa\lambda}(\epsilon, P(\epsilon))\,, \quad (16)$$

and the equation of energy-momentum conservation, $\mathcal{T}^{\kappa\lambda}{}_{;\lambda} = 0$, which renders the problem extremely complicated [50,52]. The quantities $\mathcal{R}^{\kappa\lambda}$, $g^{\kappa\lambda}$, and $\mathcal{R}$ denote the Ricci tensor, metric tensor, and Ricci scalar (scalar curvature), respectively. The dependence of the energy-momentum tensor $\mathcal{T}^{\kappa\lambda}$ on pressure and energy density, P and ϵ respectively, is indicated in Eq. (16). The quantities ω, ν, and ψ in Eq. (15) denote the frame dragging frequency of local inertial frames, and time- and space-like metric functions, respectively.

Finally, the fact that any successful model for the nuclear equation of state must accommodate pulsars with rotational periods of (at least) 1.6 msec and masses larger than typically 1.6 $M_\odot$ leads to an overall constraint on its density dependence (double constraint of fast rotation and a large enough neutron star mass): it must behave soft in the vicinity of the density of normal nuclear matter and intermediate nuclear densities in order to lead to small enough rotational pulsar periods, but rather stiff at high nuclear densities to account for large enough masses [11,53]!

REFERENCES

1. R. D. Blandford, A. Hewish, A. G. Lyne and L. Mestel (eds.), *Pulsars as physics laboratories*, Phil. Trans. of the Royal Soc. of London, Series A, Vol. **341**, Number 1660, (1992), 1–192.
2. D. W. L. Sprung, Adv. Nucl. Phys. **5** (1972) 225.
3. B. D. Day, Rev. Mod. Phys. **51** (1979) 821.
4. V. R. Pandharipande and R. B. Wiringa, Rev. Mod. Phys. **51** (1979) 821.
5. R. B. Wiringa, V. Fiks, and A. Fabrocini, Phys. Rev. C **38** (1988) 1010.
6. F. Weber and N. K. Glendenning, *Hadronic Matter and Rotating Relativistic Neutron Stars*, Proceedings of the Nankai Summer School, "Astrophysics and Neutrino Physics", ed. by D. H. Feng, G. Z. He, and X. Q. Li, World Scientific, Singapore, 1993, p. 64–183.
7. N. K. Glendenning, Astrophys. J. **293** (1985) 470.
8. K. Holinde, K. Erkelenz, and R. Alzetta, Nucl. Phys. **A194** (1972) 161; **A198** (1972) 598.
9. R. Machleidt, K. Holinde, and Ch. Elster, Phys. Rep. **149** (1987) 1.
10. F. Weber and M. K. Weigel, Nucl. Phys. **A505** (1989) 779.
11. F. Weber, N. K. Glendenning, and M. K. Weigel, Astrophys. J. **373** (1991) 579.
12. H. Huber, F. Weber, and M. K. Weigel, Phys. Lett. **317B** (1993) 485.
13. H. Huber, F. Weber, and M. K. Weigel, Phys. Rev. C **50** (1994) R1287.
14. H. Huber, F. Weber, and M. K. Weigel, Phys. Rev. C **51** (1995) 1790.
15. P. Poschenrieder and M. K. Weigel, Phys. Lett. **200B** (1988) 231.
16. F. Weber and M. K. Weigel, Z. Phys. **A330** (1988) 249.
17. F. Weber and M. K. Weigel, J. Phys. G **15** (1989) 765.
18. A. Bouyssy, J. F. Mathiot, and N. Van Giai, Phys. Rev. C **30** (1987) 380.
19. N. K. Glendenning, Nucl. Phys. **A493** (1989) 521.

20. N. K. Glendenning, Nucl. Phys. B (Proc. Suppl.) **24B** (1991) 110.
21. N. K. Glendenning, Phys. Rev. D **46** (1992) 1274.
22. N. K. Glendenning, F. Weber, and S. A. Moszkowski, Phys. Rev. C **45** (1992) 844.
23. N. K. Glendenning, Phys. Rev. Lett. **57** (1986) 1120.
24. H. A. Bethe and M. Johnson, Nucl. Phys. **A230** (1974) 1.
25. B. Friedman and V. R. Pandharipande, Nucl. Phys. **A361** (1981) 502.
26. K. Strobel, F. Weber, M. K. Weigel, and Ch. Schaab, "Neutron Star Properties in the Thomas-Fermi Model", to appear in the International Journal of Modern Pysics E, Nuclear Physics.
27. V. R. Pandharipande and R. B. Wiringa, Nucl. Phys. **A449** (1986) 219.
28. R. B. Wiringa, R.A. Smith, and T. L. Ainsworth, Phys. Rev. C **29** (1984) 1207.
29. I. E. Lagaris and V. R. Pandharipande, Nucl. Phys. **A359** (1981) 349.
30. B. Hermann, Master thesis, University of Munich, 1996 (unpublished).
31. F. Weber, N. K. Glendenning, and S. Pei, "Signal for the Quark-Hadron Phase Transition in Rotating Hybrid Stars", (LBNL–40217, astro-ph/9705202), to appear in the proceedings of the 3rd International Conference on Physics and Astrophysics of Quark-Gluon Plasma, March 17–21, 1997, Jaipur, India.
32. J. R. Oppenheimer and G. M. Volkoff, Phys. Rev. **55** (1939) 374.
33. R. Ruffini, in *Physics and Astrophysics of Neutron Stars and Black Holes* (North Holland, Amsterdam, 1978) p. 287.
34. G. E. Brown and H. A. Bethe, Astrophys. J. **423** (1994) 659.
35. H. A. Bethe and G. E. Brown, Astrophys. J. **445** (1995) L129.
36. M. Y. Fujimoto and R. E. Taam, Astrophys. J. **305** (1986) 246.
37. E. P. Liang, Astrophys. J. **304** (1986) 682.
38. N. K. Glendenning and F. Weber, Astrophys. J. **400** (1992) 647.
39. W. D. Arnett and R. L. Bowers, Astrophys. J. Suppl. **33** (1977) 415.
40. G. Baym and C. Pethick, Ann. Rev. Nucl. Sci. **25** (1975) 27.
41. V. Trimble and M. Rees, Astrophys. Lett. **5** (1970) 93.
42. G. Borner and J. M. Cohen, Astrophys. J. **185** (1973) 959.
43. L. Lindblom, Astrophys. J. **303** (1986) 146.
44. L. Lindblom, *Instabilities in Rotating Neutron Stars*, in The Structure and Evolution of Neutron Stars, Proceedings, ed. by D. Pines, R. Tamagaki, and S. Tsuruta, Addison-Wesley, 1992.
45. F. Weber and N. K. Glendenning, Z. Phys. **A339** (1991) 211.
46. S. Chandrasekhar, Phys. Rev. Lett. **24** (1970) 611.
47. J. L. Friedman, Phys. Rev. Lett. **51** (1983) 11.
48. L. Lindblom and S. L. Detweiler, Astrophys. J. **211** (1977) 565.
49. N. K. Glendenning, Mod. Phys. Lett. **A5** (1990) 2197.
50. J. L. Friedman, J. R. Ipser, and L. Parker, Astrophys. J. **304** (1986) 115.
51. N. K. Glendenning and F. Weber, Phys. Rev. D **50** (1994) 3836.
52. J. L. Friedman, J. R. Ipser, and L. Parker, Phys. Rev. Lett. **62** (1989) 3015.
53. J. M. Lattimer, M. Prakash, D. Masak, and A. Yahil, Astrophys. J. **355** (1990) 241.

Supersymmetry and Flavor

Lawrence J. Hall

Physics Dept., University of California, Berkeley, CA 94720
and
Lawrence Berkeley National Laboratory, Berkeley, CA 94720.

Abstract. The motivation for extending spacetime to incorporate supersymmetry is presented in Part I, together with arguments for why the superpartner masses should be close to the weak scale. In Part II the new flavor mixing matrices of supersymmetry are discussed. They offer the prospect of a variety of new experimental signals of flavor and CP violation, and provide a new window to the underlying physics of flavor. A $U(2)$ theory of flavor and its predictions are presented in Part III.

I THREE SYMMETRY AXES AND SYMMETRY BREAKINGS

A Vertical and Horizontal Symmetries

Symmetries play a central role in our understanding of the elementary particles and their interactions. The symmetry structure of the known matter is shown in Figure 1.

As we move in the vertical direction we move amongst particles with different charges and gauge interactions – this is also known as the gauge direction. The non-Abelian gauge bosons act in this direction – for example the W^+ connects u_L to d_L, and the gluons change the quark colors, which are understood to be part of this direction. There are 15 states in the vertical direction, forming a generation. Not all of them are connected by the known gauge interactions – although this could occur if the $SU(3) \times SU(2) \times U(1)$ gauge group of the standard model was embedded in a larger vertical group. Such a grand unified theory could provide an understanding for the gauge quantum numbers of the quarks and leptons.

The horizontal direction of Figure 1 shows a threefold repetition of the basic vertical structure, corresponding to the three generations. The horizontal symmetry group acting on these three generations is $U(3)$. In fact, the horizontal symmetry group of the known gauge interactions is much larger: there

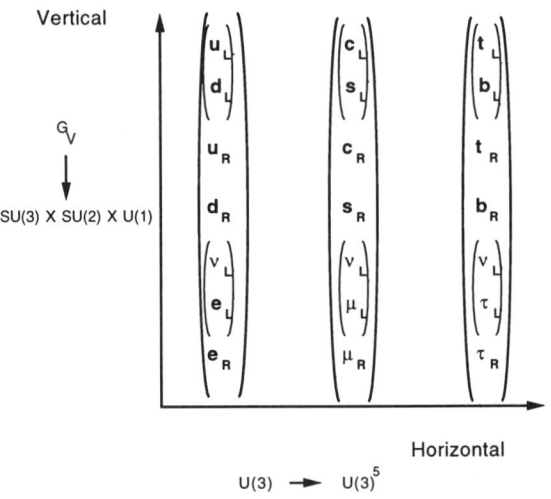

FIGURE 1. Vertical and horizontal symmetries of matter.

is a $U(3)$ factor for each of the 5 multiplets within each generation. The horizontal, or flavor, symmetry group of the standard model gauge interactions is therefore $U(3)^5$, with one $U(3)$ factor for each of $(u,d)_L, u_R, d_R, (\nu,e)_L$ and e_R.

The vertical and horizontal symmetry groups of the quarks and leptons are both broken. The electroweak symmetry is broken at the weak scale, which I label by M_Z, to the $U(1)$ of electromagnetism:

$$SU(3) \times SU(2) \times U(1)_Y \xrightarrow{M_Z} SU(3) \times U(1)_{EM} \qquad (1)$$

Experiments give very tight constraints on any breaking of color or electromagnetism. The flavor group is broken by the quark and lepton masses and by the CKM mixing to baryon number, B, and the three individual lepton numbers $L_{e,\mu,\tau}$:

$$U(3)^5 \xrightarrow{M_F,\epsilon} B \times L_{e,\mu,\tau} \qquad (2)$$

While the experimental constraints on B and $L_{e,\mu,\tau}$ breaking are strong, it is far from clear that they are exact symmetries of nature.

There is a crucial difference between the flavor and vertical symmetry directions: while we know that at least some of the vertical symmetry is gauged, we do not know that any of the horizontal direction is gauged. This means that the breaking of equation (1) necessarily occured spontaneously – the underlying theory must have interactions which possess the $SU(3) \times SU(2) \times U(1)$ symmetry. On the other hand, we cannot be sure that there is an underlying theory with an exact flavor symmetry. It is logically possible that, at the most fundamental level, the theory possesses interactions which contain small dimensionless parameters which explicitly break the flavor group. This would mean that, even in principle, it is not possible to understand the origin of the quark and lepton masses. This seems unreasonable to me. Instead, I will assume that there is an underlying theory which possesses an exact flavor symmetry group, G_F. This theory must contain new interactions which lead to a breaking of G_F at a scale M_F – the fundamental scale of flavor physics. Furthermore, these interactions must break G_F in a way that generates a set of small dimensionless G_F-breaking parameters, ϵ, which are the origin of the small quark and lepton mass ratios and mixing angles. The flavor group G_F may not be as large as $U(3)^5$. The maximal flavor group gets smaller as the vertical gauge group is extended, and we cannot be sure that nature has chosen the maximal possible flavor group.

B Mass Scales

There is a mass scale associated with each of the known forces: $m_\gamma, \Lambda_{QCD}, M_Z$ and M_{Pl} for the electromagnetic, strong, weak and gravitational forces respectively. The masses of the quarks and charged leptons are all proportional to the weak scale, M_Z, while the large hierarchies in these masses is due to the set of small G_F-breaking parameters, ϵ, discussed above. Since short distance physics is the most fundamental, we can use the Planck scale, M_{Pl}, to define our units of mass. It is convenient to set $M_{Pl} = 1$, which is analogous to defining units by setting $\hbar$ and c equal to unity. Having defined the unit of mass, our experience with quantum field theory tells us that all further mass scales should have a symmetry description. This is not a mathematical requirement – it is simply a requirement that we have a physical understanding of the new mass scales. The massless photon is understood because the $U(1)_{EM}$ symmetry is unbroken, while the scale of strong interactions arises from the dynamics of QCD, which is based on the symmetry group $SU(3)$. However *the standard model provides no symmetry description for the weak scale*. To date, the standard model provides a successful mathematical description of particle phenomena, but I believe that it is not tenable as the ultimate theory of the weak interactions. These thoughts are summarized in Table 1.

There are several possible extensions of the standard model which allow a symmetry description of the weak scale. The standard model, with an elementary Higgs boson certainly provides an economical description of weak symmetry breaking. If we require that our theory include this picture, then

TABLE 1. Symmetry descriptions of mass scales (Planckian units)

Mass	Symmetry description
$m_\gamma = 0$	$U(1)_{EM}$ unbroken
$\Lambda_{QCD} = 10^{-20}$	$SU(3)$ dynamics
$M_Z = 10^{-17}$	????

the *only* known symmetry description of the weak scale is provided by supersymmetry.

C Supersymmetry

Supersymmetry is an extension of the continuous symmetries of spacetime. The familiar generators of translations (P), rotations (J) and Lorentz boosts (K) are augmented by those of supersymmetry (Q), and the Poincare algebra of commutation relations amongst (P, J, K) is augmented by the anticommutation relation $\{Q, Q\} \sim P$. It is this relation which implies that the structure of spacetime has changed.

In Figure 1 the vertical and horizontal symmetries of matter were shown, but the spacetime symmetry axis was ignored. This is rectified in Figure 2, where the third axis represents increasing spacetime symmetry. The Lorentz symmetry implies that there is a repetition of the entire sheet of three generations of matter, giving particles of the same spin but opposite gauge charges – this is the antimatter. The particles of this replicated sheet could aptly be refered to as "Lorentz partners" of the original matter particles. Supersymmetry represents a further step along the spacetime axis, and generates an entirely new sheet, once again replicating the three generations. This time the particles have the same gauge quantum numbers as the original matter, but have spin 0 instead of spin 1/2, and are denoted by a tilde above the particle symbol.

The spacetime structure of the electron is shown in Table 2. When first discovered 100 years ago, it was believed to be a particle with just two attributes: mass and electric charge. However, the electron has non-trivial spacetime properties also; the discovery of its nature due to rotations and Lorentz boosts each led to a doubling of states. It is not surprising that a further increase in spacetime would result in one more such doubling, leading to the selectron $\tilde{e}$.

In Figure 2 we know that the underlying fundamental symmetries of the world in both the vertical and horizontal directions are broken. Understanding these breakings, shown in eqs. (1) and (2), are perhaps the most important task of particle physics. If supersymmetry is also a fundamental symmetry of

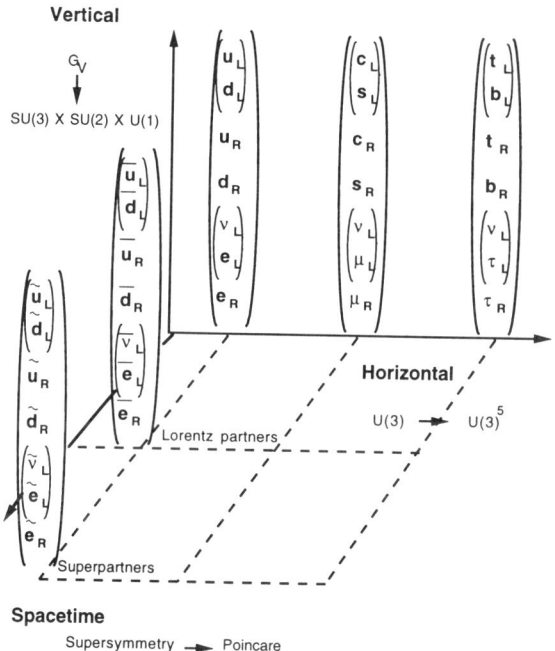

FIGURE 2. The three symmetry axes. For clarity, the Lorentz partners and the superpartners of the second and third generations are not shown.

TABLE 2. Spacetime structure of the electron

Symmetry Generator	Electron states	Discovered
Momentum	$e(m)$	1897 J.J. Thompson
Angular Momentum	$(e^\uparrow, e^\downarrow)$	1922 Stern-Gerlach
Lorentz Boost	$(e, \bar{e})$	1932 Anderson
Supersymmetry	$(e, \widetilde{e})$	??

nature, then there is also symmetry breaking along the third axis of Figure 2:

$$\text{Supersymmetry} \xrightarrow{\widetilde{m}} \text{Poincare} \qquad (3)$$

In this case the Lorentz and translation symmetries are simply unbroken remnants of a larger fundamental symmetry, in analogy with electromagnetism, QCD, baryon number and lepton numbers.

The discovery of the Lorentz-partners led to a profound puzzle: why does the universe predominantly contain baryons and leptons rather than the corresponding anti-matter? If supersymmetry is correct there is no such analogous puzzle: the superleptons and superbaryons are not present in the universe because the supersymmetry breaking of eq. (3) gives them a mass of order $\widetilde{m}$, so that they are unstable and decay into ordinary matter. In fact, the lightest superpartner may be stable – this could be the dark matter in the universe, thus solving a puzzle rather than creating one.

D   The masses of the superpartners

I have argued that a spacetime with broken supersymmetry allows a symmetry understanding of the weak scale. For this to occur, it is perhaps not surprising that the scale of the superpartner masses, $\widetilde{m}$, should be of order the weak scale, M_Z. Three questions come to mind:

• *Has any progress really been made? What is the advantage of replacing electroweak symmetry breaking at scale M_Z by supersymmetry breaking at scale $\widetilde{m}$?*

Progress has been made because there are many quantum field theories known which allow an understanding of supersymmetry breaking, including the generation of $\widetilde{m}$ at a scale much beneath M_{Pl}.

• *What is the mechanism which connects electroweak symmetry breaking to supersymmetry breaking?*

In supersymmetric extensions of the standard model there are many spin zero particles: in addition to the Higgs boson there are all the superpartners of the quarks and sleptons. When supersymmetry is broken, each of these scalars receives a squared mass parameter, m^2, with a size governed by the scale $\widetilde{m}$. These parameters, like all others in quantum field theories, are running parameters depending on the renormalization scale μ, $m^2(\mu)$, and the connection with physical measurements has to be considered carefully. At the scale of the superpartner masses, if an m^2 is positive, then m really does represent the physical mass of the corresponding spin zero particle. However, if it is negative then it implies spontaneous symmetry breaking with the corresponding scalar field acquiring a vacuum expectation value. Hence the crucial question is: why does the Higgs mass squared evolve to negative values, triggering electroweak symmetry breaking, while all the squark and slepton squared masses remain positive, giving mass to these superpartners and not leading to the spontaneous breaking of color and electric charge? Apparently the theory could only correspond to physical reality if we make one particular choice from a large discrete set of possibilities.

The dynamics of the theory solves this puzzle in a very convincing way. When supersymmetry is first broken the m^2 parameters for all the scalars are found to be positive at a large "messenger" scale, M_{mess}. However, on running the parameters down in energy to the weak scale, the large top quark Yukawa coupling, λ_t, leads to a term in the renormalization group equation which decreases the value of m^2 only for the Higgs boson and the top squarks, since they alone are directly affected by this interaction. Thus the theory guarantees that all the m^2 are positive, except possibly those for the Higgs boson and the top squarks. On further study, it is found that the gauge quantum numbers of the Higgs and top are such that the m^2 for the top squark is prevented from becoming negative. On the other hand, the correction to m_H^2 for the Higgs is given approximately by

$$\Delta m_H^2 \simeq -\frac{3}{8\pi^2}\lambda_t^2 \ln\left(\frac{M_{mess}}{\widetilde{m}}\right)\widetilde{m}^2. \tag{4}$$

The observed value of the top quark mass implies that λ_t is sufficiently large that it is very difficult to prevent m_H^2 from becoming negative and triggering electroweak symmetry breaking. *Electroweak symmetry breaking is hard to avoid in supersymmetric theories, as it is triggered by the same interaction which generates the large top quark mass.*

• *Can the superpartner masses be predicted quantitatively in terms of M_Z?* Unfortunately not, one can only make naturalness arguments, which are not precise. In any supersymmetric theory in which electroweak symmetry breaking is triggered as a heavy top quark effect, as shown in eq. (4), it is possible to derive a formula for the weak scale which has the form:

$$M_Z^2 = \sum_i c_i \widetilde{m}_i^2 \tag{5}$$

where $\widetilde{m}_i$ represent the several possible fundamental parameters of order the scale $\widetilde{m}$ which control the superpartner masses, and the c_i are constants. Some

c_i are of order unity, while others may be small, but none are large. One sees that the $\widetilde{m}_i$ that have c_i of order unity cannot become much larger than M_Z, unless several terms in the sum conspire to cancel against each other. As these $\widetilde{m}_i$ become larger, such a fine tuning renders the theory artificial and physically unacceptable. Hence, there are always some $\widetilde{m}_i$, and some superpartner masses, which cannot be made much larger than M_Z. In the minimal supersymmetric standard model, all the $\widetilde{m}_i$ parameters have c_i of order unity, and hence all of the superpartners must be close in mass to the weak scale. How "close" depends on the particle and the degree of cancellation allowed in eq. (5), and is illustrated in Figure 3.

While not predictions, the expectations of Figure 3 are certainly exciting. The particles of the standard model were found one at a time – some new facilities discovered one, some none. With weak scale supersymmetry, we expect that all the superpartners will show up at 1 or 2 new facilities. Finding 21 supermatter particles and 7 supergauge/superHiggs particles at once will present unusual experimental challenges! The expectations of Figure 3 are broadly applicable to many more models than the so-called "minimal" one. On the other hand, it is possible that many superpartners are much heavier, with multi-TeV masses which do not substantially contribute to the right-hand side of eq. (5) due to certain c_i being small. Even in this case, there are always some colored superpartners, and some non-colored superpartners, for which the expectations of Figure 3 are still broadly correct.

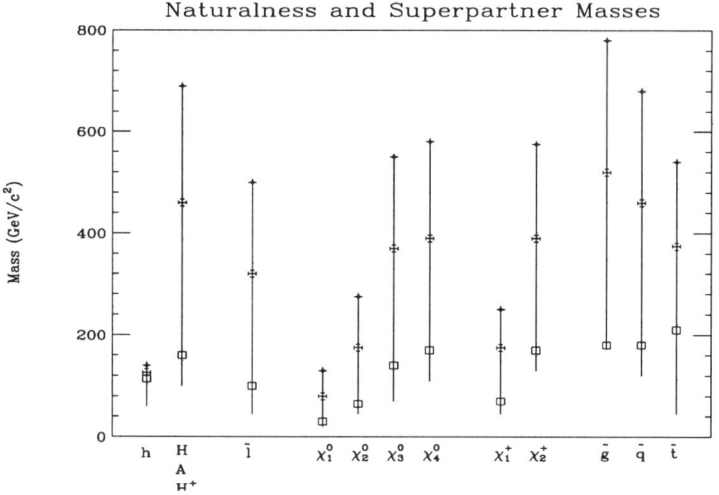

FIGURE 3. Expectations for superpartner masses in the minimal supersymmetric extension of the standard model. For each particle, the line extends up to a point where the fine tuning of parameters is 1 part in 10. (Figure supplied by G. Anderson.)

Reprinted from *Physics Letters B*, **401** 47 R. Barbieri et al. pg. 47 ©1997 with kind permission of Elsevier Science - NL, Sara Burgerhartstreet 25, 1055 KV Amsterdam, The Netherlands.

II FLAVOR MIXING IN SUPERSYMMETRY

Flavor and CP violation arises in the standard model when the quark mass matrices are diagonalized – the CKM matrix, V, results as a relative rotation between u_L and d_L type quarks in the 3-dimensional generation space. The W gauge boson couples to $\bar{u}_{Li} V_{ij} d_{Lj}$. If either the three up quarks or the three down quarks were degenerate, V could be rotated to the unit matrix by rotating the basis for the u_L or d_L quarks. Thus the standard model CP and flavor violation is a consequence of the non-trivial flavor structure of the quark mass matrices.

In any supersymmetric extension of the standard model, there are 3×3 mass matrices for the squarks and for the sleptons, in addition to those for the quarks and leptons. The m^2 parameters discussed above actually form 3×3 matrices in each charge sector. If these scalar mass matrices are non-trivial, then the supersymmetric gauge interactions will involve new flavor mixing matrices, W, analogous to the CKM matrix. For example, a relative rotation between the up quarks and the up squarks leads to a matrix, W^u, appearing at the gluino vertex: $\tilde{u}_i^\dagger W^u_{ij} u_j \tilde{g}$. Similarly if there is a relative rotation between the charged leptons and charged sleptons necessary to diagonalize the mass matrices, then the photino vertex allows CP and flavor violation via a matrix W^e: $\tilde{e}_i^\dagger W^e_{ij} e_j \tilde{\gamma}$. These W mixing matrices, one each for $u_{L,R} d_{L,R}$ and $e_{L,R}$ sectors, are a generic feature of supersymmetric theories.

In a superfield basis where the fermion masses are diagonal, it is the structure of the squark and slepton mass matrices which determines the flavor and CP violation from the W matrices. Scalar mass matrices in field theory are quite unlike fermion mass matrices, since they have different symmetry properties. If all entries of the scalar mass matrices are of order the weak scale, and there are no precise degeneracies amongst the entries, Feynman diagrams with virtual superpartners and vertices involving the W matrices, give contributions to processes which violate flavor (e.g. $\mu \to e\gamma$) and CP (e.g. ϵ_K) which are several orders of magnitude larger than allowed by experiment. This is no longer viewed as a problem for supersymmetry, rather it provides information about how supersymmetry should be broken, and, broadly speaking, has led to two classes of theories. In the first, the scalar masses are generated by some new dynamics which is generation independent, so that the mass matrices are proportional to the unit matrix. In this case a mass basis for the scalars can be found where the W matrices are also the unit matrix, and hence do not violate flavor or CP. I will not discuss this case further. The second class of theories has a non-trivial structure for the scalar mass matrices. However, this structure is not arbitrary, but is controlled by the same flavor symmetries that control the quark and lepton masses and mixings.

The approximate U(3) flavor symmetry of the standard model, which treats the three generations as a triplet, is badly broken to U(2) by the large top quark mass, while the U(2) symmetry of the lighter generations is broken only weakly by the masses of the second generation fermions. These symmetries give scalar mass matrices with eigenvalues $(m_1, m_1 + \delta, m_3)$. There is only a small mass splitting, δ, between the masses of the scalars of the first two generations, while the scalar of the third generation can have a very different mass. Hence, for supersymmetric theories with flavor symmetries, the W mixing matrices and scalar mass splittings are non-trivial and controlled — leading to what we call "1—3" flavor and CP violating signals.

Predictions for the sizes of these supersymmetric effects require a supersymmetric theory of flavor. A simple such theory of flavor based on the approximate flavor group U(2) has recently been constructed [1], and here we summarize the results presented in more depth in Ref. 2.

III THE U(2) THEORY OF FLAVOR

A Structure of the Theory

Under the U(2) flavor symmetry group, the three generations of matter fields ψ transform as $\mathbf{2} \oplus \mathbf{1}$, $\psi = \psi_a \oplus \psi_3$. Since the group has rank 2, it can have two stages of symmetry breaking

$$\mathrm{U}(2) \xrightarrow{\epsilon} \mathrm{U}(1) \xrightarrow{\epsilon'} 0, \tag{6}$$

controlled by two small parameters ϵ and ϵ', which are the origin of the generation mass hierarchies $m_3 \gg m_2 \gg m_1$ in the fermion spectrum.

Taking the Higgs bosons to be flavor singlets, the Yukawa interactions transform as: $(\psi_3\psi_3)$, $(\psi_3\psi_a)$, $(\psi_a\psi_b)$. Hence the only relevant U(2) representations for the fermion mass matrices are 1, ϕ^a, S^{ab} and A^{ab}, which can be viewed as "flavon" fields. S and A are symmetric and antisymmetric tensors, and the upper indices denote a U(1) charge opposite to that of ψ_a.

We make the simplifying assumption that each of these fields participate in only one stage of the symmetry breaking in (6). Since A^{ab} alone would break U(2) down to SU(2), whereas it would break U(2) completely in association with ϕ^a and/or S^{ab}, it can only participate in the last stage of breaking in (6): U(1) → 0. Therefore, $A^{12} = -A^{21} = \mathcal{O}(\epsilon')$. On the other hand, to account for $|V_{cb}| \simeq m_s/m_b$ in term of a unique parameter ϵ, both ϕ^a and S^{ab} must participate in the first stage of breaking in (6): U(2) → U(1). Hence, in the basis where $\phi^2 = \mathcal{O}(\epsilon)$ and $\phi^1 = 0$, $S^{22} = \mathcal{O}(\epsilon)$ and all other components of S vanish – if they were non-zero they would break U(1) at order ϵ, which is excluded by (6). We are thus led to Yukawa matrices in up, down and charged lepton sectors of the form:

$$\lambda = \begin{pmatrix} 0 & \epsilon' & 0 \\ -\epsilon' & \epsilon & \epsilon \\ 0 & \epsilon & 1 \end{pmatrix}. \tag{7}$$

All non-vanishing entries have unknown coefficients of order unity, while still keeping $\lambda_{12} = -\lambda_{21}$. With $\epsilon \simeq 0.02$ and $\epsilon' \simeq 0.004$, such a pattern agrees qualitatively well with the observed quark and lepton masses and mixings, with a few exceptions which can be understood in terms of the composition of the Higgs which couple to the D/E sectors and of the intra-generation structure of the Yukawa couplings [3].

The mass and interaction matrices for the scalars arising from supersymmetry breaking can be discussed along similar lines. We assume that the same representations, ϕ^a, S^{ab} and A^{ab}, which play a role in the fermion Yukawa sector, are also relevant in the description of flavor breaking in the scalar sector. The scalar trilinear "A-term" interactions have the same structure as the Yukawa matrices in (7), whereas the only terms of numerical relevance in the scalar mass matrices are those linear in ϕ^a, $\phi_a^\dagger$, $\phi^a\phi_b^\dagger$ and $S^{ab}S_{bc}^\dagger$. Hence, the resulting scalar mass matrices have the form

$$m^2 = \begin{pmatrix} m_1^2 & 0 & 0 \\ 0 & m_1^2(1+\epsilon^2) & \epsilon m_4^{2*} \\ 0 & \epsilon m_4^2 & m_3^2 \end{pmatrix}, \tag{8}$$

where m_1, m_3 and m_4 are supersymmetry breaking masses.

B   Predictions for the CKM matrix

This simple form of U(2) breaking, giving the pattern of zeros in (7), leads to a CKM matrix

$$V \approx \begin{pmatrix} 1 & s_c & -s_{12}^U s \\ -s_c & 1 & s e^{-i(\alpha+\beta)} \\ s_{12}^D s & -s e^{i\beta} & e^{-i\alpha} \end{pmatrix}, \tag{9}$$

where

$$s_{12}^D = \sqrt{\tfrac{m_d}{m_s}}\left(1 - \tfrac{m_d}{2m_s}\right), \quad s_{12}^U = \sqrt{\tfrac{m_u}{m_c}}, \quad s_c e^{-i\beta} = s_{12}^D - s_{12}^U e^{i\alpha}, \tag{10}$$

with m_u and m_c (m_d and m_s) renormalized at the same scale. The four independent parameters can be taken to be m_d/m_s, m_u/m_c, $s = |V_{cb}|$ and α. The unitarity triangle is given by the last expression of (10), and has angles α, β and $\gamma = \pi - \alpha - \beta$. We have made a fit to this form of the CKM matrix by inputting values of $|V_{us}|$, $|V_{cb}|$, $|V_{ub}/V_{cb}|$, m_c and m_s listed in Table 3, and values for the light quark mass ratios, m_u/m_d and m_d/m_s, discussed below.

Second order chiral perturbation theory for the pseudoscalar meson masses determines, to a remarkable accuracy, the combination [4]

$$Q = \frac{m_s/m_d}{\sqrt{1 - m_u^2/m_d^2}} = 22.7 \pm 0.08. \tag{11}$$

TABLE 3. Values of the parameters used in the text.

$(m_s)_1\text{GeV}$	$(175 \pm 55)\,\text{MeV}$	$\|V_{us}\|$	0.221 ± 0.002
$(m_c)_{m_c}$	$(1.27 \pm 0.05)\,\text{GeV}$	$\|V_{cb}\|$	0.040 ± 0.003
$(m_b)_{m_b}$	$(4.25 \pm 0.15)\,\text{GeV}$	$\|V_{ub}/V_{cb}\|$	0.08 ± 0.02
$(m_t)_{m_t}$	$(165 \pm 10)\,\text{GeV}$	$\alpha_s(M_Z)$	0.117 ± 0.006

Additional assumptions, plausible but not following from pure QCD, lead to [4]

$$m_u/m_d = 0.553 \pm 0.043. \tag{12}$$

Finally, a better determination of the scale of the light quark masses is possible if use is made of the SU(5) relations, valid at the unification scale,

$$m_b = m_\tau, \qquad m_d m_s = m_e m_\mu \tag{13}$$

as illustrated in Ref. 3. Given the different level of uncertainty and/or assumptions in these equations, we describe the results of 4 "combined fits" with different inputs for the light quark mass ratios: i) (11) only; ii) (11) and (12); iii) (11) and (13); iv) (11), (12) and (13).

Treating all errors as gaussian, at 90% C.L., we obtain the results shown in Figs. 4. In Fig. 4a, the boundary obtained from the unitarity constraints on a general parametrization of the V_{CKM} is also shown. No similar boundary is given in Figs. 4b,4c, since unitarity alone does not limit the CKM phase ϕ. Note that in these fits there is no input from B mixing or from CP-violation, so that $|V_{td}/V_{ts}|$ and the angles α, β and γ of the unitarity triangle, are predictions.

For comparison, the fit of Fig. 4d results when the CP violating parameter in K physics, ϵ_K, and the B_d mixing mass, are added to the inputs. The standard quantities which parameterize QCD uncertainties are taken as $B_K = 0.8 \pm 0.2$ and $\sqrt{B}f_B = (200 \pm 40)\,\text{MeV}$. This last plot does not have general validity because of the "1—3 signal"; loop diagrams involving internal gluinos make important contributions to $\epsilon_K, \Delta M_{B_d}$ and ΔM_{B_s}. Such strong corrections are absent from the decays $K^+ \to \pi^+ \nu \bar{\nu}$ and $K^0 \to \pi^0 \nu \bar{\nu}$ which have branching ratios proportional to $|V_{td}|^2$ and $|V_{td}|^2 \sin^2 \beta$, respectively. Predictions for these rare K branching ratios, in the four fits i) – iv), are given in Table 4.

C Supersymmetric contributions to flavor and CP violation

The mass matrix (8) leads to highly degenerate scalars of the first two generations, $(m_2^2 - m_1^2)/m^2 \approx \epsilon^2 \approx 10^{-3}$, with m^2 an average scalar mass

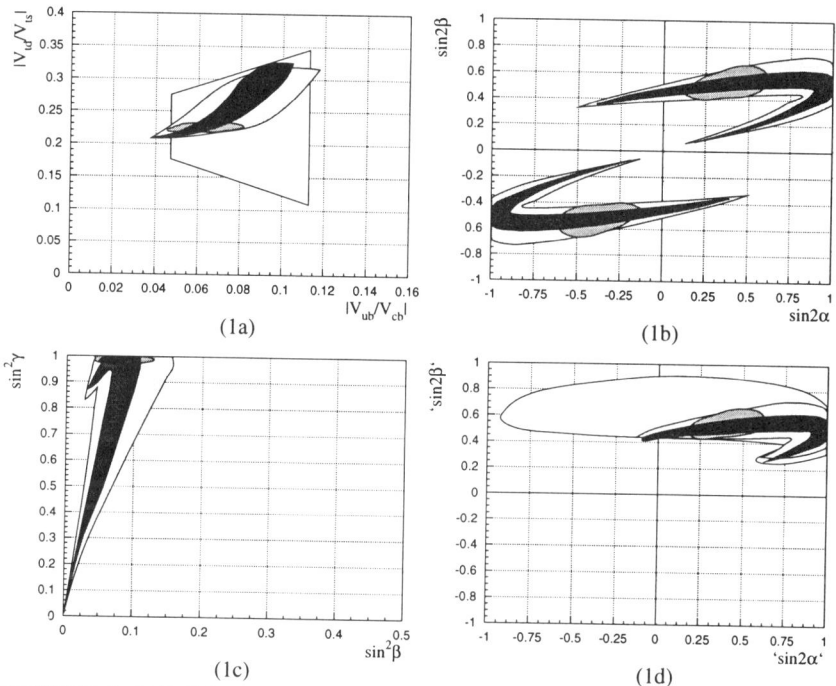

FIGURE 4. 90% C.L. contours from the four combined fits defined in the text: i) larger, lighter area; ii) smaller, lighter area; iii) larger, darker area; iv) smaller, darker area. In Figs. 1a and 1d also shown is the contour for a general parametrization of the CKM matrix. Figure 1d is the only one to include ϵ_K and Δm_{B_D} among the inputs. This figure is taken from ref. [2].

TABLE 4. Predictions for $K \to \pi \nu \bar{\nu}$ decays in the four fits described in the text.

	i	ii	iii	iv
BR($K^+ \to \pi^+ \nu \bar{\nu}$)/$10^{-10}$	$0.98^{+0.40}_{-0.30}$	0.83 ± 0.20	$0.97^{+0.50}_{-0.30}$	$0.84^{+0.17}_{-0.22}$
BR($K_L \to \pi^0 \nu \bar{\nu}$)/$10^{-10}$	0.22 ± 0.13	0.17 ± 0.07	$0.20^{+0.08}_{-0.10}$	0.14 ± 0.05

squared. On the contrary, the scalars of the third generation are likely to have masses very different from their first and second generation partners, $(m_3^2 - m_{1,2}^2) \approx m^2$, leading to flavor and CP signals from this "1—3" effect.

By going to a diagonal basis both for (7) and (8), mixing matrices are generated in the gaugino interactions: $(\bar{f}_{L,R} W_{L,R} \tilde{f}_{L,R})\tilde{g}$ for the fermions of given chirality, L or R, and their superpartners. By appropriate phase redefinitions of the fermion and scalar fields, while keeping the form (9) of V_{CKM} and the mass eigenvalues real and positive, it is possible to write the six matrices $W_{L,R}^{U,D,E}$ for U-quarks, D-quarks and charged leptons in terms of two new parameters each, $s_{L,R}^{U,D,E}$ and $\gamma_{L,R}^{U,D,E}$

$$W_{L,R}^{U,D,E} = \begin{pmatrix} c_{12} & -s_{12} & s_{12} s_{L,R} \\ s_{12} & c_{12} & -s_{L,R} \\ 0 & s_{L,R} e^{i\gamma_{L,R}} & e^{i\gamma_{L,R}} \end{pmatrix}^{U,D,E} \quad (14)$$

where $s_{12}^{U,D}$ have already been defined and, in analogy with them, $s_{12}^E = \sqrt{m_e/m_\mu}$. The parameters $s_{L,R}^{U,D,E}$ are all of order ϵ. In this basis, the $L - R$ mixings induced by the A-terms are still non diagonal and complex. They can, however, be treated as a perturbation except, maybe, for the $\tilde{t}_L - \tilde{t}_R$ mixing, which, in any event, does not alter the mixing matrices (14).

The W-matrices give rise to new FCNC and CP-violating phenomena via loop diagrams with internal supersymmetric particles. A close inspection shows that the most important effects occur in ϵ_K, $B - \bar{B}$ mixing, electric dipole moments of the electron (d_e) and the up and down quarks (d_u, d_d) and, finally, in $\mu \to e\gamma$ and $\mu \to e$ conversion in atoms. In the case of ϵ_K and $B - \bar{B}$ mixing one obtains effects comparable to those present in the Standard Model. On the other hand, the effects in the dipole moments and in the lepton flavor violating processes are at the level of the present experimental limits. The calculation of some typical, although partial, contributions to these observables gives: [1]

$$\epsilon_K \approx 2 \cdot 10^{-3} \left(\frac{500 \text{ GeV}}{m_{\tilde{q}}}\right)^2 (\omega^D)^2 \sin 2\beta \quad (15)$$

---

[1] In Equations (17) and (18) we consider the photino contribution. For the EDM of the u-quark one has $d_u \approx 8 d_d (v_1/v_2)^2 (\omega^U/\omega^D)(\sin(\gamma_L^U - \gamma_R^U)/\sin(\gamma_L^D - \gamma_R^D))$.

$$|\Delta m_{B_d}| \approx 0.1 \,\mathrm{ps}^{-1} \left(\frac{500\,\mathrm{GeV}}{m_{\tilde{q}}}\right)^2 \omega^D \qquad (16)$$

$$\mathrm{BR}(\mu \to e\gamma) \approx 2 \cdot 10^{-11} \left(\frac{100\,\mathrm{GeV}}{m_{\tilde{l}}}\right)^4 (\omega^E)^2 \left(\frac{v_2}{v_1}\right)^2 \qquad (17)$$

$$d_e \approx 6 \cdot 10^{-27} e\,\mathrm{cm} \left(\frac{100\,\mathrm{GeV}}{m_{\tilde{l}}}\right)^2 \omega^E \sin(\gamma_L^E - \gamma_R^E) \left(\frac{v_2}{v_1}\right) \qquad (18)$$

$$d_d \approx 1 \cdot 10^{-26} e\,\mathrm{cm} \left(\frac{500\,\mathrm{GeV}}{m_{\tilde{q}}}\right)^2 \omega^D \sin(\gamma_L^D - \gamma_R^D) \left(\frac{v_2}{v_1}\right) \qquad (19)$$

where $\omega^{U,D,E} = (s_L^{U,D,E} s_R^{U,D,E})/V_{cb}^2 = \mathcal{O}(1)$ and v_2/v_1 is the ratio of the vacuum expectation values of the Higgs doublets.

In these equations, the effect of the splitting between the first two generations of scalars is neglected. This is completely justified for all the observables except ϵ_K, in which case an additional effect of the same order of Eq. 15 is present, still proportional to $\sin 2\beta$ as in Eq. 15 and the SM contribution themselves. The effects are all due to the splitting $(m_3^2 - m_{1,2}^2)$, taken large enough that the GIM suppression can be safely neglected. Therefore, $m_{\tilde{q},\tilde{l}}$ denotes the mass of the lightest ($Q = 1/3$ squark, charged slepton), neglecting in both cases the difference between L- and R-states. The gluino mass in Eqs. 15, 16, 19 and the photino mass in Eqs. 17, 18 are taken equal to $m_{\tilde{q}}$ and $m_{\tilde{l}}$ respectively. Equation 16 only includes the contribution of the $(V-A)(V+A)$ 4-quark operator. The A-terms are neglected in Eqs. 17, 18, 19, whereas the "μ-parameter" is taken equal to the relevant sfermion mass.

The "1–3" contribution to B meson mixing leads to a modification of the phase in the neutral B meson eigenstates: $(q/p)_{B_{d,s}}$ acquire an extra phase $e^{-2i\phi_{B_{d,s}}}$. This modifies the CP asymmetries in neutral B meson decays: for example, the asymmetry in B_d decay to ψK_S ($\pi\pi$) becomes $-\sin 2(\beta + \phi_{Bd})$ ($\sin 2(\alpha - \phi_{Bd})$). [6]

IV CONCLUSIONS

In the first part of this talk I discussed the vertical, horizontal and spacetime symmetries of nature. A symmetry description of the weak scale is possible if spacetime is extended to incorporate supersymmetry. In this case, information about horizontal or flavor physics is contained in scalar m^2 matrices as well as in the quark and lepton mass matrices. This leads to an exciting prospect both for new experimental discoveries of flavor and CP violation, and for new insights into the fundamental origin of flavor physics.

In the second half of the talk, I presented the predictions of a simple, supersymmetric U(2) theory of flavor. While the detailed predictions of the CKM matrix, shown in Figure 4, is special to this model, we believe that the order of magnitude flavor and CP signals of Eqs. 15 — 19 are a generic consequence of a flavor symmetry solution to the supersymmetric flavor-changing problem. As the flavor symmetry dictates the structure of both scalar and fermion mass matrices, the large breaking of U(3), necessary for generating the top quark mass, will generically lead to a large scalar mass splitting, $m_3^2 - m_{1,2}^2$, and to values of W_{ij} set by the scale of the corresponding CKM matrix elements V_{ij}, resulting in the "1 — 3" flavor and CP signals discussed above. While these signals are a generic feature, they do not follow strictly as a necessary consequence of the flavor symmetry; for example, lepton flavor violation could be inhibited by a suitable extension of the U(2) flavor symmetry. However, this is not compatible with quark-lepton gauge unification: [5] these signals are a generic feature of grand unified supersymmetric theories where the hardness scale of supersymmetry breaking is higher than, or close to, the flavor mass scale.

The supersymmetric flavor contributions of Eqs. 15 — 19 could all be observed in experiments planned for the next decade. The effect of Eq. 16 could be observed in CP asymmetries in B meson decays. We look with particular interest at improvements of the experimental sensitivity in the search for $\mu \to e\gamma$ (or $\mu \to e$ conversion) and for the electron and neutron electric dipole moments.

REFERENCES

1. R. Barbieri, G. Dvali, and L.J. Hall, *Phys. Lett. B* **377**, 76 (1996); R. Barbieri and L.J. Hall, hep-ph/9605224, *Nuovo Cimento A* **110A**, 1 (1997).
2. R. Barbieri, L.J. Hall, and A. Romanino, *Phys. Lett.* B*401* 47 (1997).
3. R. Barbieri, L.J. Hall, S. Raby, and A. Romanino, *Nucl. Phys.* B*493* 3 (1997).
4. H. Leutwyler, CERN–TH/96–25, hep-ph/9602255.
5. R. Barbieri and L.J. Hall, *Phys. Lett. B* **338**, 212 (1994).
6. I. Bigi, *Perspectives in Particle Physics*, La Thuile, March 1994, p. 137.
7. G. Buchalla, A. Buras, and M. Lautenbacher, *Rev. Mod. Phys.* **68**, 1125 (1996).
8. I. Dunietz, *Phys. Lett. B* **270**, 75 (1991); M. Gronau and D. Wyler, *Phys. Lett. B* **265**,172 (1991).
9. N. Deshpande, B. Dutta and S. Oh, *Phys. Rev. Lett.* **77**, 4494 (1996); Y. Grossman and M. Worah, *Phys.Lett. B* **395**, 241 (1997).

Results from CEBAF Experiment E89-012: Measurements of Deuteron Photo-disintegration up to 4 GeV

M. A. Miller,[1] D. J. Abbott,[2] A. Ahmidouch,[3]
C. S. Armstrong,[4] J. Arrington,[5] K. A. Assamagan,[6]
O. K. Baker,[6] S. P. Barrow,[7] D. P. Beatty,[7] D. H. Beck,[1]
S. Y. Beedoe,[8] E. J. Beise,[9] J. E. Belz,[10] C. W. Bochna,[1]
P. E. Bosted,[11] E. J. Brash,[12,17] H. Breuer,[9] R. V. Cadman,[1]
L. Cardman,[2] R. D. Carlini,[2] J. Cha,[6] N. S. Chant,[9] G. Collins,[9]
C. Cothran,[13] W. J. Cummings,[14] S. Danagoulian,[8]
F. A. Duncan,[9] J. A. Dunne,[2] D. Dutta,[15] T. Eden,[6] R. Ent,[2]
B. W. Filippone,[5] T. A. Forest,[1] H. T. Fortune,[7] V. V. Frolov,[16]
H. Gao,[1] D. F. Geesaman,[14] R. Gilman,[17] P. L. J. Gueye,[6]
K. K. Gustafsson,[9] J-O. Hansen,[14] M. Harvey,[6] W. Hinton,[6]
R. J. Holt,[1] H. E. Jackson,[14] C. E. Keppel,[6] M. A. Khandaker,[18]
E. R. Kinney,[19] A. Klein,[20] D. M. Koltenuk,[7] G. Kumbartzki,[17]
A. F. Lung,[9] D. J. Mack,[2] R. Madey,[2,5] P. Markowitz,[21]
K. W. McFarlane,[18] R. D. McKeown,[5] D. G. Meekins,[4]
Z-E. Meziani,[22] J. H. Mitchell,[2] H. G. Mkrtchyan,[23]
R. M. Mohring,[9] J. Napolitano,[16] A. M. Nathan,[1] G. Niculescu,[6]
I. Niculescu,[6] T. G. O'Neill,[14] B. R. Owen,[1] S. Pate,[24]
D. H. Potterveld,[14] J. W. Price,[16] G. L. Rakness,[19]
R. Ransome,[17] J. Reinhold,[14] P. M. Rutt,[17] G. Savage,[6]
R. E. Segel,[15] N. Simicevic,[1] P. Stoler,[16] R. Suleiman,[3] L. Tang,[6]
B. P. Terburg,[1] D. Van Westrum,[19] W. F. Vulcan,[2]
S. E. Williamson,[1] M. T. Witkowski,[16] S. A. Wood,[2] C. Yan,[2]
B. Zeidman[14]

[7] *University of Illinois at Urbana-Champaign.* [1] *Thomas Jefferson National Accelerator Facility.* [2] *Kent State University.* [3] *College of William and Mary.* [4] *California Institute of Technology.* [5] *Hampton University.* [6] *University of Pennsylvania.* [8] *North Carolina A&T*

State University. [9] University of Maryland. [10] TRIUMF. [11] American University.
[12] University of Regina. [13] University of Virginia. [14] Argonne National Laboratory.
[15] Northwestern University. [16] Rensselaer Polytechnic Institute. [17] Rutgers University.
[18] Norfolk State University. [19] University of Colorado at Boulder. [20] Old Dominion University. [21] Florida International University. [22] Temple University. [23] Yerevan Physics Institute. [24] New Mexico State University.

Abstract. The first measurements of differential cross sections for deuteron photo-disintegration at photon energies up to 4 GeV were performed at the Thomas Jefferson National Accelerator Facility early in 1996. Cross section results for D(γ,p)n at proton center of mass angles of 35°, 53° and 90° will be presented. These results are in good agreement with previous measurements at low energy and extend to higher energies where data were previously unavailable. The 90° degree data show behavior consistent with the constituent counting rules up to 4 GeV and are also in fair agreement with the asymptotic meson exchange model. The 37° and 53° data do not show clear signs of counting rule behavior, although a threshold in transverse momentum for the onset of scaling cannot be excluded.

INTRODUCTION

An important question in nuclear physics is whether the effects of quarks in nuclei can be seen at high energy or large transverse momentum. Quarks in particle physics manifest themselves as a rather abrupt change in the momentum transfer dependence of the cross section, as in Bjorken scaling for example. Observation of such a scale change in photo-nuclear reactions may indicate that quarks play a role in the reaction and as such would be an important step towards understanding nuclear physics in terms of the underlying quark degrees of freedom.

Deuteron photo-disintegration has several characteristics that make it a likely place to search for such scaling. Theoretically, the deuteron is perhaps the simplest and most well understood nucleus. The photon is a well understood probe. In addition, the kinematics of $\gamma + d \rightarrow p + n$ are such that a relatively large momentum transfer to the constituents can be obtained in exclusive photo-nuclear reactions at photon energies of a few GeV [1]. This is because the absorbed photon transfers all of its energy to the constituents, in contrast to, say, e + d scattering where the scattered electron retains some energy.

Dimensional analysis can be used to arrive at scaling behavior for exclusive processes [2-4]. For A + B → C + D at high energy and large transverse momentum, dimensional analysis results in the constituent counting rule for the differential cross section:

$$\frac{d\sigma}{dt} \propto s^{-(n-2)}, \tag{1}$$

where s and t are the Mandelstam variables, and n is the total number of elementary fields (photons and quarks) in the initial and final states. High energy exclusive photo-reactions on nucleons [5] as well as other hadron reactions involving nucleons [6] exhibit an energy dependence consistent with the constituent counting rule. Thus the question at hand is whether such behavior can be observed in nuclear reactions. For deuteron photo-disintegration, this counting rule is

$$\frac{d\sigma}{dt} \propto s^{-11}. \tag{2}$$

Therefore, an observation of an s^{-11} dependence of $d\sigma/dt$ would be an indication of scaling.

Measurements of $d\sigma/dt$ for $\gamma + d \rightarrow p + n$ carried out at SLAC [7-9] showed that the energy dependence of the cross section was consistent with the constituent counting rule $\theta_{cm} = 90$ and $53°$, but not at $37°$. The SLAC results for $s^{11}d\sigma/dt$ at $\theta_{cm} = 90°$ are shown in Fig. 1, along with previous low energy data [10-13]. The onset of scaling at photon energies above 1 GeV is quite apparent. The solid and and dotted lines represent meson-exchange calculations of Lee [14] and Laget [15], respectively. Both are traditional calculations that reproduce measured NN phase shifts up to 2.0 GeV and are constrained by photo-meson production data. They are tuned to match the data in the region of the delta peak and do so well. Above 0.5 GeV they are not expected to be valid. The reduced nuclear amplitude calculation of Brodsky and Hiller [16] is shown by the dashed line. It has been normalized at $E_\gamma \simeq 2$ GeV.

The primary goals of this experiment are to see if the s^{-11} dependence continues at photon energies up to 4 GeV at $\theta_{cm} = 90°$ and to measure cross sections over a relatively large kinematic range in order to see if scaling occurs at other angles and energies. This large coverage will allow us the chance to investigate various other scaling variables, such as transverse momentum transfer, and to look for systematic behavior in the location of any observed scaling threshold. For example, the photon energy at which scaling appears in Fig. 1 corresponds to a transverse momentum of approximately 1 GeV/c per nucleon. If transverse momentum is an appropriate scaling variable, we might expect to see scaling appear at 1 GeV/c at other scattering angles. For example, 1 GeV/c transverse momentum corresponds to approximately $E_\gamma = 3$ GeV at $\theta_{cm} = 37°$.

EXPERIMENT

This experiment, E89-012, was carried out at the Thomas Jefferson National Accelerator Facility using the Continuous Electron Beam Accelerator Facility (CEBAF) with bremsstrahlung photons incident on a liquid deuterium target in CEBAF's Hall C. The photons were produced with the CEBAF electron

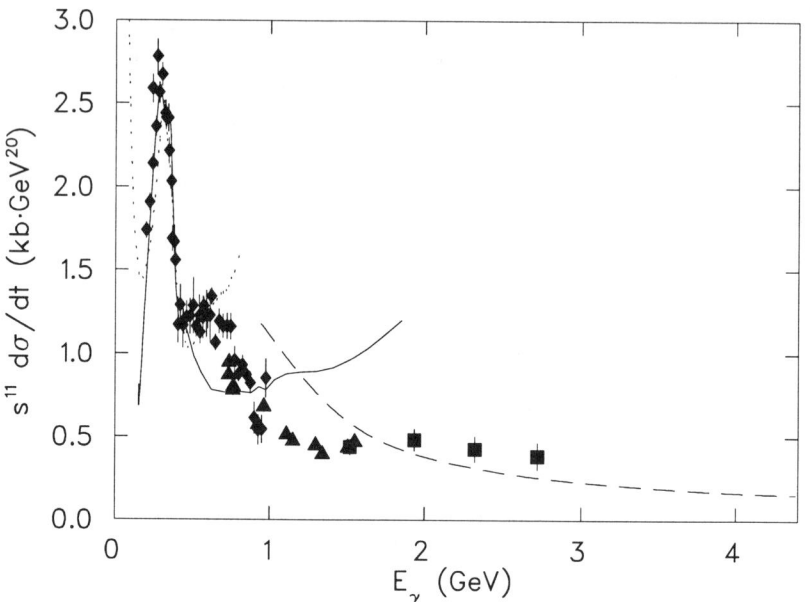

FIGURE 1. Existing deuteron photo-disintegration cross section data multiplied by s^{11} and plotted versus photon energy at $\theta_{cm} = 90°$. The squares are the SLAC NE17 data, the triangles are the SLAC NE8 data, and the diamonds are all other existing data [10–13].. The solid and dotted lines are meson-exchange calculations of Lee [14] and Laget [15], respectively. The dashed line is the reduced nuclear amplitude calculation of Brodsky and Hiller [16].

beam and a copper radiator. Photo-produced protons were detected with a magnetic spectrometer. The measured proton momentum and scattering angle were used to calculate the energy of the bremsstrahlung photons. Reactions which included a pion in the final state were excluded by accepting data with only the highest energy photons such that photo-pion production was kinematically excluded. Overall, the experiment was quite similar the SLAC experiments described in detail in references [8,17].

The electron beam was a continuous (CW) beam at beam energies of 0.8 to 4.0 GeV, with typical currents of 20 μA. A 6% copper radiator was used to produce the bremsstrahlung photons which were in turn incident on a 15 cm long liquid deuterium target. Both the Hall C high momentum spectrometer (HMS) and short orbit spectrometer (SOS) were used: the HMS for forward angle protons (in the center-of-mass system) and SOS for the backward angle proton measurements. The two spectrometer detector packages are quite similar. Triggering and time-of-flight were provided by plastic scintillator hodoscopes. Particle momenta and scattering angles were measured with drift chambers. A gas Čerenkov detector and a lead glass shower counter were also used for particle identification. In addition to the 15 cm deuterium target cell, a 4 cm long liquid deuterium cell and 4 and 15 cm long empty target cells were used for background measurements and calibrations.

Measurements were made at 0.8, 1.6, 2.4, 3.2 and 4.0 GeV and at center of mass angles of 37, 53, 70, 90, 114, 130 and 143°. The preliminary results presented here represent only the 37, 53 and 90° measurements, all of which were made with the HMS.

ANALYSIS

The detected proton momenta and angles were used to calculate photon energies. This requires that the protons be cleanly identified and that only two-body final states be allowed in the data. For protons with momenta below 2.7 GeV/c, time-of-flight between scintillator planes in the spectrometer provided clean separation of protons, deuterons and pions. Above 2.7 GeV/c, the gas Čerenkov detector was used to exclude deuterons and to separate protons from pions. Cuts on reconstructed spectrometer quantities were also applied, both at the focal plane and target.

Background for the target cell was eliminated by applying cuts on the reconstructed vertex distribution and by subtracting empty target measurements. Protons from deuteron electro-disintegration were removed by measuring yields without the bremsstrahlung radiator and subtracting. A correction factor was applied to account for the modification of the electron beam distribution due to the radiator. Background subtracted spectra showing the bremsstrahlung endpoint are shown in Fig. 2.

PRELIMINARY RESULTS

Cross sections were calculated from yields in an E_γ window below the bremsstrahlung endpoint. This window was chosen to exclude protons from photo-pion production and to avoid the region near the tip of the bremsstrahlung spectrum where the photon flux varies rapidly with energy. Photon fluxes were calculated with the code of Matthews and Owens [19] and checked against an independent code by Belz [17]. The estimated uncertainty in the calculated bremsstrahlung flux is 3%. Electron beam currents were continuously monitored with radio frequency cavity monitors calibrated to 2% for 20 μA beams. The beam energy was measured to 0.1%. Density variations in the liquid deuterium target were less than 1%. The spectrometer solid angle for the extended target was calculated for each kinematic setting with a monte carlo simulation. Acceptance measurements agree with the simulation to better than 5% for central momenta and point-like targets. A correction for proton attenuation in the detectors was applied to the measured yields. This

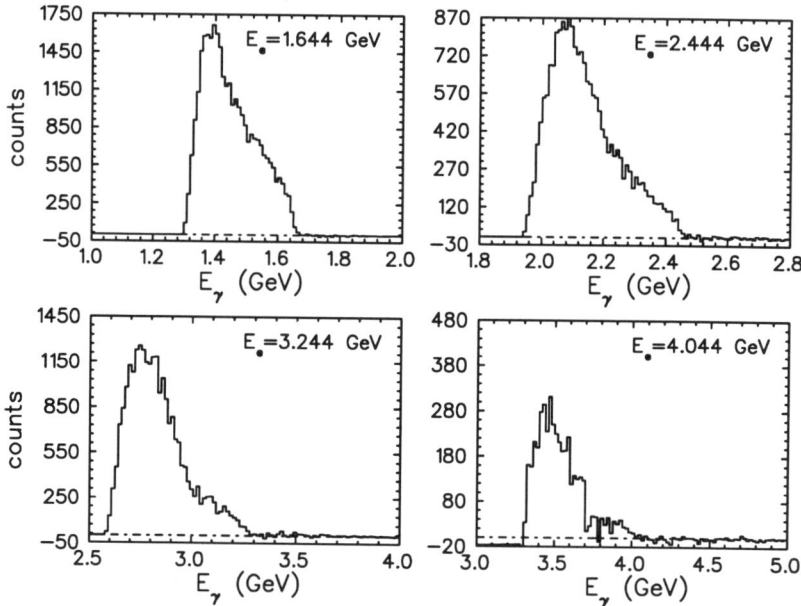

FIGURE 2. Background subtracted yields at $\theta_{cm} = 90°$. The cutoff on the high energy side is the bremsstrahlung edge, determined by the electron beam energy E_e. The low energy cutoff is the edge of the spectrometer acceptance.

correction factor was measured to be 8% within 1% absolute. An additional correction on the order of 15% was necessary due to the presence of a valve plate in the spectrometer acceptance The uncertainty in this correction is still under investigation. Detector tracking efficiency corrections and dead-time corrections were also applied.

The overall systematic uncertainty in the measurements is < 19%. This is dominated by the uncertainties in the background subtractions (10%) and in the spectrometer acceptance (~ 15%). Uncertainties in target thickness contribute < 3% while beam energy and current uncertainties contribute < 3%. After further analysis it is expected that the overall systematic uncertainty will be ~ 5-10%.

Preliminary cross section results at $\theta_{cm} = 90°$ are shown in Fig. 3 where the cross sections have been multiplied by s^{11} as before. The circles are data points from from the present measurement and are drawn with statistical uncertainties only. The squares are from SLAC experiment NE17 data, the triangles from SLAC NE8 and the diamonds are all other existing data. Where the kinematics overlap, the present results are in good agreement with the SLAC data. The new data continue to show clear scaling behavior to 4.0 GeV. Fig. 4 shows the preliminary CEBAF results at $\theta_{cm} = 37°$ and $53°$. Again the

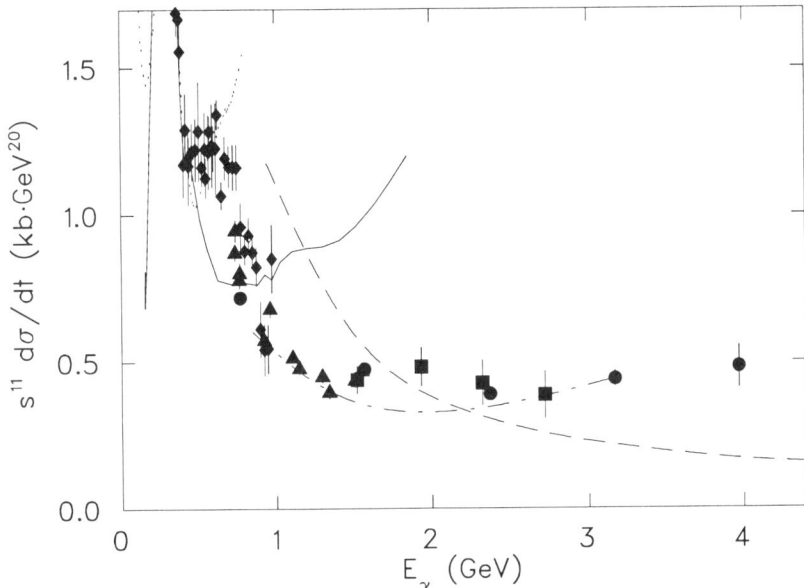

FIGURE 3. Preliminary E89-012 deuteron photo-disintegration cross section data multiplied by s^{11} and plotted versus photon energy at $\theta_{cm} = 90°$. The circles are from the present measurement. The squares are the SLAC NE17 data, the triangles are the SLAC NE8 data, and the diamonds are all other existing data [10–13].. The solid and dotted lines are meson-exchange calculations of Lee [14] and Laget [15], respectively. The dashed line is the reduced nuclear amplitude calculation of Brodsky and Hiller [16]. The dash-dot line is the asymptotic meson-exchange calculation of Nagornyĭ et al. [18].

circles are preliminary data from the present experiment data points, shown with statistical uncertainties only, and are in good agreement with the SLAC NE17 data [9], show as diamonds.

Neither the 53° data nor 37° data exhibit the strong scaling behavior seen at 90°. As mentioned above, a scaling threshold in transverse momentum of 1 GeV/c corresponds to $E_\gamma = 3$ GeV at 37°, however, no scaling threshold is readily apparent at that energy. On the other hand, the onset of scaling at $E_\gamma = 3$-4 GeV cannot be excluded in the 37° and 53° data.

SUMMARY

Preliminary results from the CEBAF deuteron photo-disintegration experiment are in agreement with previous measurements available at lower energies. The new results continue to exhibit scaling behavior at $\theta_{cm} = 90°$ but do not show clear signs of scaling at 37° or 53°, although scaling cannot be ruled out.

Further measurements at higher energies, as well as complete analysis of all the E89-012 data are needed to help shed light on the source of the scaling behavior at 90°. Measurements to determine whether hadron helicity conservation (an important requirement for a rigorous derivation of the counting rule) holds at these energies would also be very interesting. Measurements of proton polarization in the $D(\gamma, \vec{p})n$ reaction are planned at CEBAF.

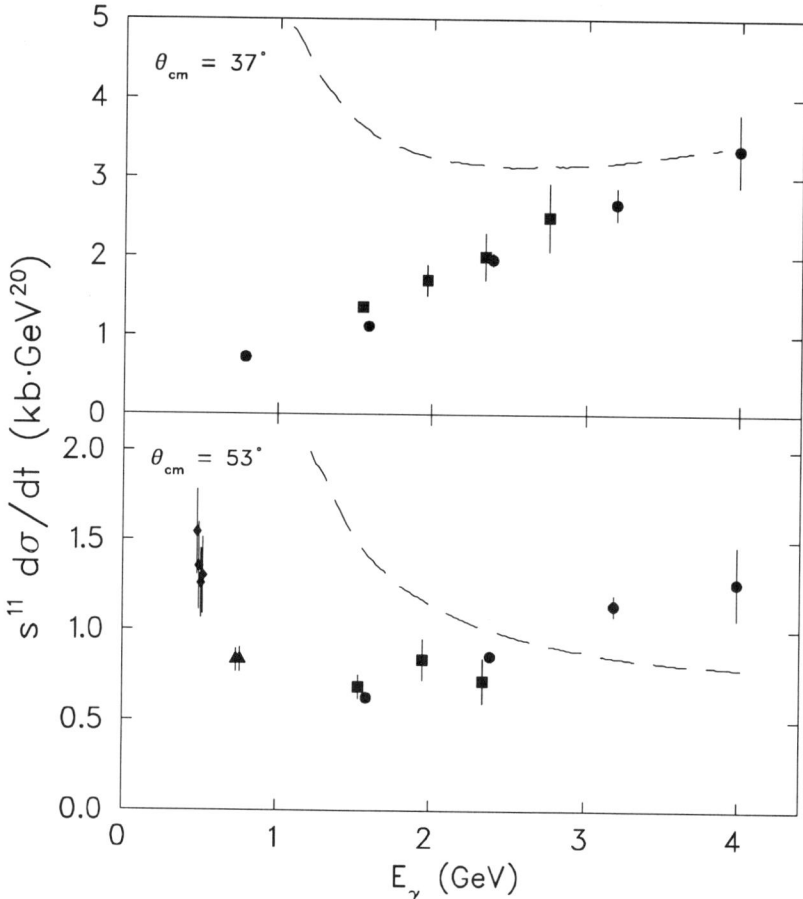

FIGURE 4. E89-012 deuteron photo-disintegration cross section data multiplied by s^{11} and plotted versus photon energy at $\theta_{cm} = 37$ (top) and 53° (bottom). The circles are from the present measurement and the squares and triangles are the SLAC NE17 and NE8 data, respectively. The diamonds are previous data from Ref. [10–13]. The dashed lines are reduced nuclear amplitude calculations of Brodsky and Hiller.

ACKNOWLEDGEMENTS

This work has been supported in part by research grants from the U.S. Department of Energy and the National Science Foundation.

REFERENCES

1. R. J. Holt, Phys. Rev. C **41**, 2400 (1990).
2. S. J. Brodsky and G. R. Farrar, Phys. Rev. Lett. **31**, 1153 (1973).
3. V. Matveev *et al.*, Nuovo Cimento Lett. **7**, 719 (1973).
4. G. P. LePage and S. J. Brodsky, Phys. Rev. Lett. **31**, 1153 (1973).
5. R. L. Anderson *et al.*, Phys. Rev. D **14**, 679 (1976).
6. G. White *et al.*, Phys. Rev. D **49**, 58 (1994).
7. J. Napolitano *et al.*, Phys. Rev. Lett. **61**, 2530 (1988).
8. S. J. Freedman *et al.*, Phys. Rev. C **48**, 1864 (1993).
9. J. E. Belz *et al.*, Phys. Rev. Lett. **74**, 646 (1995).
10. H. Myers *et al.*, Phys. Rev. **121**, 630 (1961).
11. R. Ching and C. Schaerf, Phys. Rev **141**, 1320 (1966).
12. P. Dougan *et al.*, Z. Phys. A **276**, 55 (1976).
13. J. Arends *et al.*, Nucl. Phys. **A412**, 509 (1984).
14. T.-S. H. Lee, Argonne National Laboratory Report No. PHY-5253-TH-88; T.-S. H. Lee, in *Proceedings of the International Conference on Medium and High Energy Nuclear Physics, Taipai, Taiwan, 1988* (World Scientific, Singapore, 1988), p.563.
15. J. M. Laget, Nucl. Phys. **A312**, 265 (1978).
16. S. J. Brodsky and J. R. Hiller, Phys. Rev. C **28**, 475 (1983).
17. J. E. Belz, PhD thesis, California Institute of Technology, 1994 (unpublished).
18. S. I. Nagornyĭ, YU. A. Kasatkin, and I. K. Kirichenko, Sov. J. Nucl. Phys. **55**, 189 (1992).
19. J. L. Matthews and R. O. Owens, Nucl. Instr. and Meth. **111**, 157 (1973).

PARALLEL SESSIONS

Relativistic Heavy Ions

Dilepton Production in Relativistic Heavy Ion Collisions

Volker Koch

Lawrence Berkeley National Laboratory
Berkeley, CA 94720

Abstract. The present status of our understanding of low mass dilepton production in relativistic heavy ion collisions is discussed. The focus of the discussion will be the sensitivity of dilepton measurements to in medium changes of hadrons and the restoration of chiral symmetry. We will finally discuss how the presence of strong long wavelength pion modes, i.e. disoriented chiral condensates can be seen in the dilepton spectrum.

INTRODUCTION

Electromagnetic probes, such as photons and dileptons are especially useful to investigate the early stage of an ultrarelativistic heavy ion collision, since they leave the system without any final state interaction. However, as measurements by the DLS collaboration at the BEVALAC, and more recently by the CERES collaboration at the CERN-SPS have shown, a large fraction of the dilepton yield arises from the decay of long lived states, such as the π_0, eta, or the omega. These resonances decay well outside the hot and compressed region and thus careful analysis of the dilepton spectra is needed in order to extract the information about the properties of the hot and dense matter, such as e.g. possible in medium changes of hadrons. In the low mass region, below the phi-meson, the most important production channels are: (i) Dalitz decays of η, Δ, ω, a_1. (ii) Direct decays of the vector mesons, such as ρ, ω and Φ. Unique to heavy-ion collisions are rescattering channels such as pion annihilation and bremsstrahlung due to secondary collisions. These latter channels certainly carry information about the hot and dense region and due to the vector dominance formfactor pion annihilation may reveal possible in medium changes of the ρ-meson. (For an review of in medium changes of hadronic properties see e.g. [1]).

DILEPTON PRODUCTION AT SPS-ENERGIES

As discussed in detail in the contribution by A. Dress [2] dilepton measurements for nucleus-nucleus collisions at 200 GeV per nucleon by the CERES collaboration show a considerable enhancement over the expected yield from hadronic decays. For p+Be as well as p+Au collisions, on the other hand, the data are consistent with the hadronic decays only. Certainly a large fraction of this enhancement is due pion annihilation, which is unique to heavy-ion experiments since they create a dense system of pions which then can annihilate. Thus the CERES data are proof that heavy ion collisions are more then the simple superposition of individual nucleon-nucleon collisions and that indeed an interacting hadronic system is formed (similar evidence is also derived from the measurement strange particle production).

Aside from measuring in medium modifications of hadrons, dilepton measurements may provide complementary information about the reaction dynamics and may thus help to further specify the properties of the hadronic system generated in these collisions. This question has been addressed in [3], where the dilepton spectrum has been calculated for a large variety of initial conditions under the constraint the the final hadronic spectra are in agreement with experiment. Within the CERES acceptance, the variation of the resulting dilepton mass spectra is rather small (see fig. 1). Thus the measurement of an dilepton invariant mass spectrum is unlikely to further specify the configuration of the hadronic phase. On the positive side this result shows that large deviations of the data from the hadronic calculation cannot simply be attributed to the lack of knowledge of the specific configuration of the hadronic phase. And indeed, as compared with the central points of the CERES measurement for $S + Au$, there is a considerable deviation at invariant masses of about 400 MeV. However, one should also point out, that within the systematic and statistical error quoted by the CERES collaboration, the data can be understood by simply including pion annihilation without any further in medium modifications. However, if one seeks to reproduce the central values of the CERES data, certain in medium modifications have to be included. One is follow the conjecture of Brown and Rho [4] that the mass of the ρ meson scales with the quark condensate, which is expected to be reduced at the densities and temperatures reached in these collisions. As a consequence the mass of the ρ drops, providing more strength in the low mass region of the dilepton spectrum. Following this prescription, Li et al. can reproduce the central points of the CERES measurements [5]. However, at least at low temperatures, the Brown-Rho scaling hypothesis can be ruled out by simple current algebra arguments [6], which shows that to order T^2 the mass of the ρ does not change while to this order the quark condensate drops.

More conventional calculations of in medium effects on the dilepton production determine the properties of the ρ-meson or more precisely the current-current correlator in a system of pions and nucleons/deltas. So far essentially

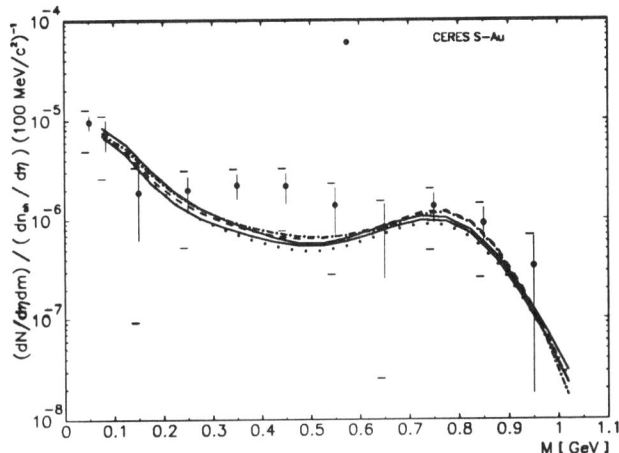

FIGURE 1. CERES data of S+Au in comparison with calculation based on different initial hadronic configurations.

three different in medium correction have been considered, which are schematically depicted in fig. 2.

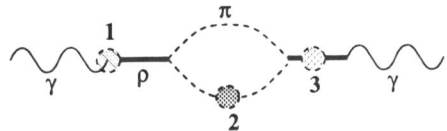

(1) The $\rho\gamma$ coupling is screened due to pion loops [7,8] This effect is a direct consequence of the partial restoration of chiral symmetry and it reduces the strength below the ρ peak.
(2) If one understands the ρ as a $\pi - \pi$ resonance, its properties are changed due to in medium modifications of the pions. These include a change of the pion dispersion relation due to thermal pions [3,9] and due to the coupling to delta-hole and nucleon-hole states [10–12] once baryons are taken into account. These states also give rise to additional inelasticities at low invariant resulting in an increased strength around 400 MeV in the imaginary part of the current-current correlator. One should note, however, that to leading order in the density, these contributions are nothing else but typical bremsstrahlung diagrams. Furthermore, these additional inelasticities at low masses seem to be sufficient to saturate the QCD-sum rules [11], without an explicit change in the mass of the ρ meson.
(3) The ρ can also couple to $N^*(1720)$-hole states. This effect, originally proposed by Friman and Pirner [13] leads to a softening of the dispersion relation of the ρ meson and provides additional strength in the dilepton spectrum at

low invariant masses. Since the coupling to the $N^*(1720)$-hole state is p-wave, only dileptons with finite momentum with respect to the matter restframe are enhanced.

A combination of effects (2) and (3) seems leads to an improved description of the CERES data [10].

DILEPTONS FROM DCC-STATES

The restoration of chiral symmetry in relativistic heavy ion collisions can, under certain circumstances, lead to a strong enhancement of low momentum pion modes which form a so called disoriented chiral condensate (DCC) [14,15]. So far, proposed observables which are sensitive to these DCC states have been in the pion sector only, where strong final state interactions may destroy them. However, the presence of a DCC state also leads to a strong and unique signal in the dilepton channel. Assume a thermal pion annihilates with a pion from a DCC. Since the DCC represents a large phase space density localized at small momenta, one would expect that this phase space distribution is reflected in the dilepton invariant mass as well as momentum spectrum. This is indeed the case as one can see in figures 3 and 4. (for details see [16]). In fig. 3 we show the resulting invariant mass distribution for thermal initial conditions and for so called quench initial conditions; the latter lead to the formation of DCC-states. The resulting momentum spectrum for an invariant mass of $M = 300\,\text{MeV}$ is shown in fig. 4. Clearly a strong enhancement (about a factor of 100) can be seen close to twice the pion mass. The enhancement is localized in invariant mass as well as in momentum, reflecting the localized phase space distribution of the DCC-state. Since this enhancement is confined to momenta below 300 MeV, it does not

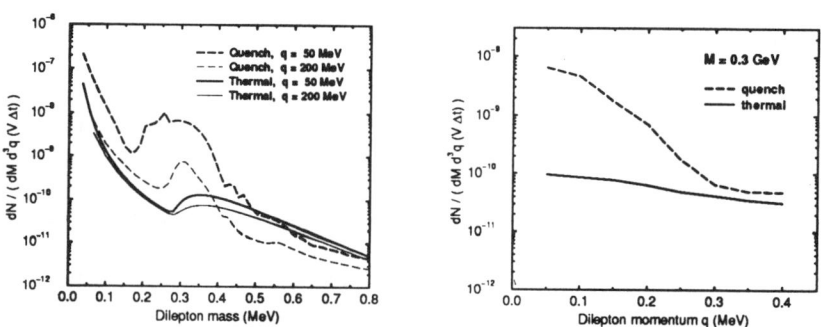

FIGURE 2. Dilepton invariant mass (left) and momentum (right) spectra for thermal (full lines) and quench (dashed lines) initial conditions. The mometum spectrum is for an invariant mass of $M = 300\,\text{MeV}$.

affect the CERES measurement, where an acceptance cut of $p_t \geq 200\,\text{MeV}$

for each individual dilepton is imposed. However, if this cut could be relaxed to $p_t \geq 100$ MeV a factor of 10 enhancement in the invariant mass spectrum should be visible.

Acknowledgments: I would like to thank Y. Kluger, J. Randrup, C. Song, and X.N. Wang, whom I have collaborated with on the topics discussed here. This work was supported by the Director, Office of Energy Research, Office of High Energy and Nuclear Physics, Division of Nuclear Physics, and by the Office of Basic Energy Sciences, Division of Nuclear Sciences, of the U.S. Department of Energy under Contract No. DE-AC03-76SF00098.

REFERENCES

1. C.M. Ko, V. Koch and G.Q. Li, nucl-th/9702016, to appear in Ann. Rev. Nucl. Part. Sci, Vol 47.
2. A. Drees, these proceedings; G. Agakichiev et al., Phys. Rev. Lett. **75**, 1272 (1995)
3. V. Koch and C. Song, Phys. Rev. C54 1903 (1996).
4. G.E. Brown and M. Rho, Phys. Rev. Lett. **66** 2720 (1991).
5. G.Q. Li, C.M. Ko and G.E. Brown, Nucl. Phys. **A 606** 568 (1996)
6. M. Dey, V.L. Eletzky and B.L. Ioffe, Phys. Lett. **B252** 620 (1990).
7. C. Song, S.H. Lee and C.M. Ko, Phys. Rev. **C52** R476 (1995)
8. C. Song and V. Koch, Phys. Rev. C54 3218 (1996).
9. C. Song, V. Koch, S.H. Lee and C.M. Ko, Phys. Lett. **B366** 379 (1996).
10. R. Rapp, G. Chanfray and W. Wambach, Nucl. Phys. **A 617** 472 (1990).
11. F. Klinkl, N. Kaiser and W. Weise, hep-ph/9704398.
12. J. Steele, H. Yamagishi and I. Zahed, hep-ph/9704414.
13. B. Friman and H. Pirner, Nucl. Phys. **A617** 496 (1997).
14. J.D. Bjorken, K.L. Kowalski, and C.C. Taylor, SLAC preprint SLAC-PUB-6109, Proc. of Les Rencontres de la Vallée D' Aoste, La Thuile, 1993, ed. M. Greco, Editions Frontier, p. 507 (1993).
15. K. Rajagopal and F. Wilczek, Nucl. Phys. B404, 577 (1993).
16. Y. Kluger, V. Koch, J. Randrup and X.N. Wang, nucl-th/9704018.

ANOMALOUS J/ψ SUPPRESSION IN 158 GeV/c Pb-Pb COLLISIONS AT THE CERN SPS

presented by C. Baglin for the NA50 collaboration:

M.C. Abreu[6], B. Alessandro[11], C. Alexa[2], J. Astruc[8], C. Baglin[1], A. Baldit[4], F. Bellaiche[12], M. Bedjidian[12], S. Beole[11], A. Borhani[9], V. Boldea[2], G. Bonazzola[11], P. Bordalo[6], A. Bussière[1], V. Capony[1], J. Castor[4], M. Cerú[3], T. Chambon[4], B. Chaurand[9], I. Chevrot[4], B. Cheynis[12], E. Chiavassa[11], C. Cicalo[3], S. Constantinescu[2], W. Dabrowski[11], A. De Falco[3], G. Dellacasa[11], N. De Marco[11], A. Devaux[4], S. Dita[2], O. Drapier[12], B. Espagnon[4], J. Fargeix[4], F. Fleuret[9], P. Force[4], M. Gallio[11], Y.K. Gavrilov[7], C. Gerschel[8], P. Giubellino[11], M.B. Golubeva[7], M. Gonin[9], P. Gorodetzky[10], J.Y. Grossiord[12], P. Guaita[11], F.F. Guber[7], A. Guichard[12], R. Haroutunian[12], M. Idzik[11], D. Jouan[8], T.L. Karavitcheva[7], R. Kossakowski[1], L. Kluberg[9], A.B. Kurepin[7], Y. Le Bornec[8], G. Landaud[4], C. Lourenço[5], L. Luquin[4], P. Macciotta[3], F. Ohlsson-Malek[12], A. Marzari-Chiesa[11], M. Masera[11], A. Masoni[3], S. Mourgues[4], A. Musso[11], P. Petiau[9], W.L. Prado Da Silva[11], A. Piccotti[11], J.R. Pizzi[12], G. Puddu[3], C. Racca[10], L. Ramello[11], S. Ramos[6], P. Rato-Mendes[11], L. Riccati[11], A. Romana[9], S. Sartori[11], P. Saturnini[4], E. Scomparin[11], R. Shaoian[5], S. Silva[6], S. Serci[3], P. Sonderegger[5], X. Tarrago[8], P.Temnikov[3], N.S. Topilskaya[7], G. Usai[3], E. Vercellin[11], and N. Willis[8].

[1] Laboratoire de Physique des Particules (LAPP), IN2P3-CNRS, Annecy-le-Vieux, France; [2] Institute of Atomic Physics (IFA), Bucharest, Romania; [3] Università di Cagliari/INFN, Cagliari, Italy; [4] Laboratoire de Physique Corpusculaire, Université Blaise Pascal et IN2P3-CNRS, Clermont-Ferrand, France; [5] CERN, Geneva, Switzerland; [6] Laboratorio de Instrumentaçao e Física Experimental de Partículas (LIP), Lisboa, Portugal; [7] Institute for Nuclear Research (INR), Moscow, Russia; [8] Institut de Physique Nucléaire, Université Paris-Sud et IN2P3-CNRS, Orsay, France; [9] Ecole Polytechnique et IN2P3-CNRS, Laboratoire de Physique des Hautes Energies, Palaiseau, France; [10] Centre de Recherches Nucléaires, Université Louis Pasteur et IN2P3-CNRS, Strasbourg, France; [11] Dipartimento di Fisica Sperimentale, Università di Torino et INFN, Torino, Italy; [12] Institut de Physique Nucléaire de Lyon, Université Claude Bernard et IN2P3-CNRS, Villeurbanne, France.

Abstract

Heavy ion production of J/ψ mesons has been studied at CERN for a decade, in search for the Quark-Gluon Plasma predicted by lattice-QCD. The suppression observed very early in O-Cu,O-U and S-U reactions has been shown to be eventually explained by nuclear absorption, and follow the same exponential behavior as the proton-nucleus reactions. The new Pb-Pb data deviate from this behavior very strongly, indicating the onset of a new regime in J/ψ suppression.

1 Introduction

Quarks and gluons are known to be confined within hadronic matter. In the frame of the Standard Model the early universe was a deconfined quark and gluon plasma (QGP). As it expanded and cooled down to its present configuration, a phase transition from QGP to confined matter occured at some stage. Within the QCD theory, lattice-calculations predict the existence of such a phase transition [1], which can possibly occur in heavy-nucleon collisions at the SPS energies. If a QGP were produced in these collisions, the resulting color screening would dissociate the $c\bar{c}$ pairs, therefore preventing their hadronization to a J/ψ meson. J/ψ-suppression has thus been proposed [2] as a signature of the QGP.

From 1986 to 1992 the CERN experiment NA38 has studied dimuon decays of charmonium produced in various $A_{projectile}B_{target}$ (nucleon-nucleus and/or nucleus-nucleus) reactions. Dimuons originating from the Drell-Yan (DY) process were naturally recorded at the same time; they were used as reference sample since this primary electromagnetic process cannot be affected by the QGP. From the early days of NA38, important decreases of the ratio J/ψ/DY were observed in all central ion-collisions [5]. A long series of subsequent theoretical efforts converged to a consistent interpretation of these results in term of nuclear absorption of a pre-resonance $c\bar{c}g$ state [4]

2 The Pb-Pb experiment: setup and data analysis

The muon spectrometer was that of NA38 [5] whose electromagnetic calorimeter was supplemented with a Zero Degree Calorimeter (ZDC), made of tantalum-imbedded Čerenkov quartz fibers and a silicon-strips multiplicity detector, for centrality measurements. To cope with the high level of radiation damage from the Pb-beam, the new beam hodoscope (BH), (for luminosity measurement and beam pile-up detection), was made of Čerenkov quartz blades. The system of 7 active Pb-subtargets (17% int. length) also used quartz Čerenkov blades, whose output signals were recorded for off-line determination of the interaction-subtarget and identification of possible secondary and/or pile-up interactions.

The data were taken in two successive november-runs 1995 and 1996 with the CERN Pb-beam (158 GeV/c per nucleon) at an average intensity of $3.5\ 10^7$ ions per burst. This article reports on the analysis of the 1995 sample (about 20% of the total statistics). The opposite-sign mass spectrum is shown in figure 1 together with the Monte-Carlo determined individual contributions from J/ψ, ψ', DY and

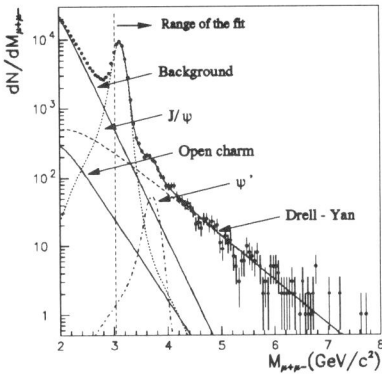

Figure 1: Opposite-sign dimuon mass spectrum.

open-charm decays as seen by the spectrometer. The same-sign dimuon spectra resulting from pure π and K decays were used to determine the background in opposite-sign dimuons. To minimize the effect of uncertainties in the open-charm contribution, we limited our analysis to dimuon masses above 2.9 GeV/c: the charm contamination is negligible in the J/ψ peak and less than 10% under the ψ'. The fit gives 49000 J/ψ, 350 ψ' and 630 DY events in the kinematical domain of the spectrometer: rapidity $3 < y < 4$ ($0 < y* < 1$ in the CM system) and $|cos\theta| < 0.5$ for the muon polar angle in the Collin-Soper frame.

From BH luminosity measurements, spectrometer acceptance and target detection efficiency calculations, the following cross-sections for the dimuon events have been obtained:

- $B_{\mu\mu}\sigma_{J/\psi} = 21.9 \pm 0.2 \pm 1.6 \mu$b
- $\sigma_{DY} = 1.49 \pm 0.02 \pm 0.11 \mu$b (for mass > 2.9 Gev)
- $B_{\mu\mu}\sigma_{\psi'}/B_{\mu\mu}\sigma_{J/\psi} = (0.59 \pm 0.09)\%$.

See ref [3] for details on the experimental procedure.

3 Comparison with previous experiments

p-p, p-nucleus and nucleus-nucleus collisions do not allow straightforward comparison of cross-sections. It is natural to divide the measured cross-sections by

$A_{projectile}B_{target}$, so as to obtain an average cross-section "per nucleon-nucleon collision" for each reaction.

First of all, as an experimental check, we have measured the well known K factor which relates the experimental value of the DY cross-section to the lowest order theoretical prediction calculated from a standard set of parton distribution functions [6]. This K factor is experimentally known to be independent from the reaction and from the incident momentum. The result $K = 2.56 \pm 0.04 \pm 0.18$ obtained from our data is in agreement with the other experiments.

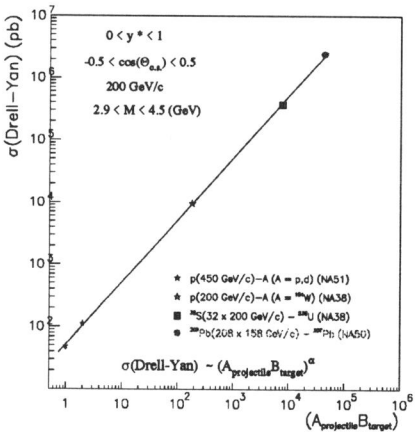

Figure 2: Drell-Yan cross-section vs. $A_{projectile}B_{target}$.

Figure 2 shows our DY cross-sections measured for various reactions with the same spectrometer. The data are observed to scale as $(A_{projectile}B_{target})^\alpha$ with $\alpha = 1.001 \pm 0.010$.

The NA38 p-nucleus J/ψ cross-sections have also been checked to follow the scaling law: $(A_{projectile}B_{target})^\alpha$ for each incident momentum separately and the exponent values are the same within experimental errors. The global fit to the whole set of NA38 p-nucleus data gives $\alpha = 0.92 \pm 0.015$, in good agreement with the other CERN and FNAL experiments. Our 450 GeV/c data can thus be rescaled to 200 GeV/c, and the rescaling factor obtained from the exponential fit is in agreement with the value given by the theoretical parametrization of J/ψ cross-sections [8].

Fig 3 shows $B_{\mu\mu}\sigma_{J/\psi}$ vs. $A_{projectile}B_{target}$: all NA38 data, including O-Cu, O-U and S-U results, fit very nicely with the same exponential behavior. The J/ψ

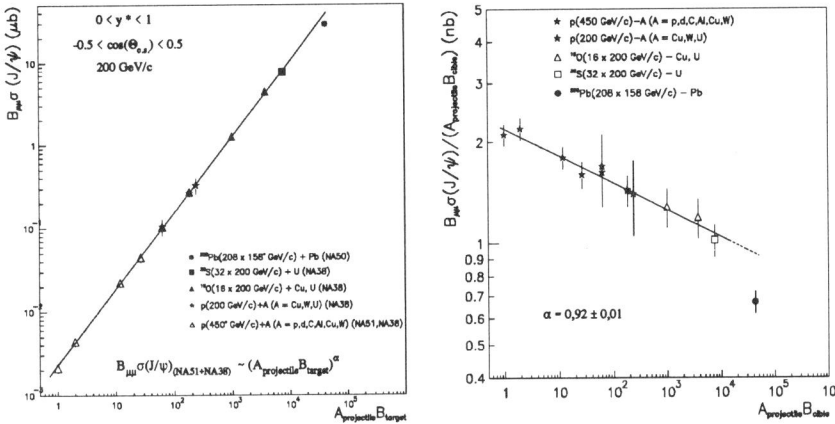

Figure 3: Left: J/ψ cross-sections as exponential function of $A_{projectile}B_{target}$
Right: same for the cross-sections "per nucleon-nucleon collision" (see the text).

suppression observed in NA38 can thus be regarded as the result of an unique absorption mechanism common to all reactions. (In the sequel this behavior is referred to as "normal" absorption.) Several authors [4] attribute the phenomenon to the absorption of a colored preresonant $c\bar{c}g$ state in the nuclear matter.

In order to extend the comparison to the new Pb-Pb data at 158 GeV/c, we have rescaled the J/ψ cross-section to 200GeV/c with the theoretical parametrization [8] before display in figure 3. The Pb-Pb J/ψ cross-section stands below the expected exponential line by a factor 0.74 ± 0.06 : the J/ψ suppression is significantly stronger for Pb-Pb than the "normal" suppression due to absorption in nuclear matter observed for lighter ions.

In the framework of [4] the normal J/ψ suppression can be parametrized as $\sigma(A_{projectile}B_{target} \rightarrow J/\psi) \propto A_{projectile}B_{target} \cdot exp(-\rho_0 \sigma_{abs} \bar{L})$: where ρ_0 is the standard nuclear density, σ_{abs} is $\approx$ 6mb, and $\bar{L}$ is the path length of the $c\bar{c}g$ state in nuclear matter, averaged over the whole range of impact parameter. The straigth line in figure 3 corresponds to this parametrization. Since the DY cross-sections scale according to $A_{projectile}B_{target}$, the ratio J/ψ / DY, which is free from systematic errors related to the normalization procedure, can be used to compare the experimentally observed J/ψ suppression with the "normal" exponential behavior.

To pursue the analysis we subdivided our data sample, in equipopulated bins, as a function of the collision centrality. For this purpose an average impact pa-

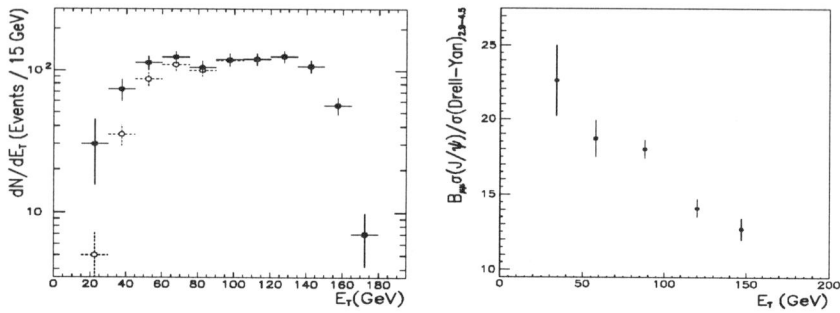

Figure 4: Left: Drell-Yan E_T-distribution, the dotted symbols are prior to correction for target detection efficiency. Right: Ratio J/ψ / DY vs. E_T.

rameter b has been determined from the transverse energy E_T of each event, as measured from the electromagnetic calorimeter. E_T is strongly (anti)correlated with the energy deposited in the ZDC. Figure 4 shows the E_T-distributions for DY events and the ratio J/ψ / DY versus E_T.

Figure 5 shows the ratio J/ψ / DY vs L(b), i.e. the path length in nuclear matter at impact parameter b. We see that for low L the Pb-Pb data agree with the S-U data; but the larger L values correspond to larger deviation from the "normal" exponential behavior. The average ratio of observed to expected values of $\sigma_{J/\psi}/\sigma_{DY}$ is 0.71 ± 0.03, about 10 standard deviations below "normal" suppression.

Figure 6 shows the ratio $\sigma_{\psi'}/\sigma_{J/\psi}$ vs. $A_{projectile}B_{target}$ for proton and ion-induced collisions. The proton data remain almost independent of incident momentum and target mass. The S-U cross-sections are significantly smaller. The new Pb-Pb data stand at least as low as the S-U data.

It also shows that $B_{\mu\mu}\sigma_{\psi'}/\sigma_{DY}$ has the same centrality(E_T)-dependence for Pb-Pb and S-U interactions. These remarks indicate that the ψ' undergoes additional suppression with respect to J/ψ for ion-induced reactions; several authors [4] attribute this effect to ψ' absorption by the comovers, which do not affect the J/ψ.

4 Summary and Conclusions

J/ψ, ψ' and DY production cross-sections have been measured for Pb-Pb and compared to p-p, p-nucleus, O-Cu, O-U, and S-U collisions measured with the same spectrometer. The Pb-Pb data presented in this report represent only 20% of the

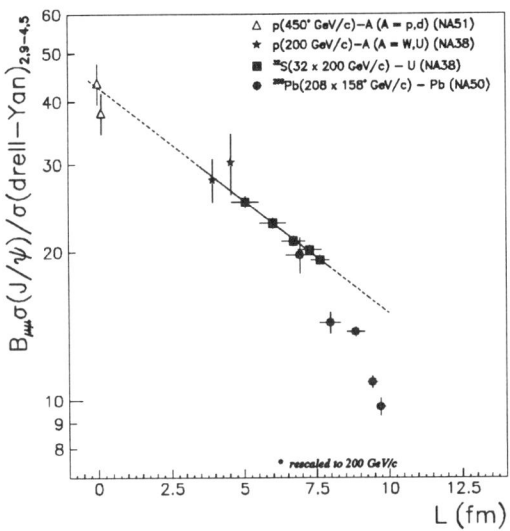

Figure 5: J/ψ/DY vs. centrality.

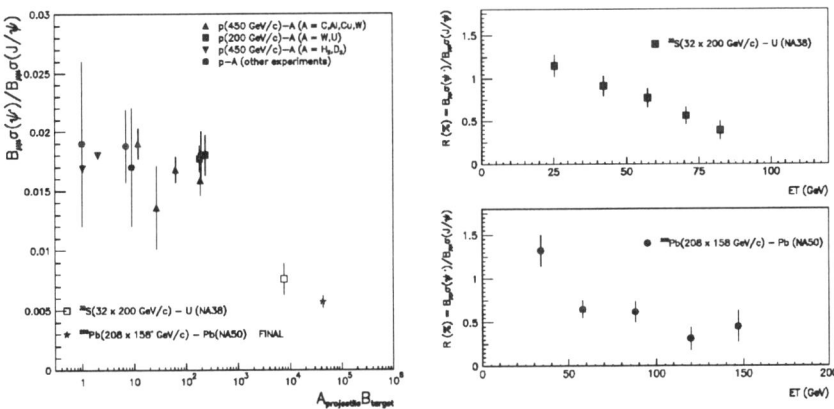

Figure 6: Ratio of ψ' to J/ψ integrated cross-sections (left) and versus $E_{\rm T}$ in S-U and Pb-Pb collisions (right).

total statistics. Nevertheless some important points can be stressed:

- DY cross-sections scale as $(A_{projectile} B_{target})^\alpha$ with $\alpha = 1$.
- J/ψ cross-sections exhibit the same exponential behavior with $\alpha = 0.92 \pm 0.015$ for all collisions, with the noticeable exception of Pb-Pb:
    - $\sigma_{J/\psi}/\sigma_{DY}$ is a factor 0.71 ± 0.03 below the value extrapoled from the other reactions.
    - Study of event samples corresponding to different centrality bins show that the deviation from previous reactions increases with centrality.
    - For lowest centrality the Pb-Pb data are in agreement with the other ion-collisions.
- $\sigma_{\psi'}/\sigma_{J/\psi}$ is constant for proton-induced reactions but considerably smaller for S-U and Pb-Pb collisions. Both ion-collisions show the same centrality dependence.

After the first presentation of our Pb-Pb results, some theorists have attributed the anomalous J/ψ suppression to the onset of a new absorption mechanism [8]. Others attempted to attribute it to the normal interaction with hadronic comovers [9]; up to now they can roughly reproduce the general trend, but miss the detailed distribution of experimental data.

I am indebted to my colleagues M. Gonin and C. Lourenço for valuable discussions and help.

References

[1] H. Satz, Ann. Rev. Nucl. Part. Sci. 35 (1985) 245.
[2] T. Matsui and H. Satz, Phys. Lett. B178 (1986) 416.
[3] "J/ψ and Drell-Yan cross-sections in Pb-Pb interactions at 158 GeV/c per nucleon", M.C. Abreu et al., Submitted to Phys. Lett. B
"Anomalous J/ψ suppression in Pb-Pb interactions at 158 GeV/c per nucleon", M.C. Abreu et al., Submitted to Phys. Lett. B.
[4] D. Kharzeev and H. Satz, Phys. Lett. B366 (1996) 316.
[5] C. Baglin et al., Phys.Lett. B220 (1989) 471.
[6] M. Gluck, E. Reya and A. Vogt, Phys. Lett. B306 (1993) 391.
[7] G.A. Schuler, CERN-TH.7170/94.
[8] see e.g., J.P. Blaizot and J.Y. Ollitrault, Phys. Rev. Lett. 77 (1996) 1703.
[9] see e.g., A. Capella et al., Phys. Lett. B393 (1997) 431.

Non-Strange and Strange Antibaryons in Au+Pb Collisions at 11.5 A GeV/c

John G. Lajoie* for the E864 Collaboration[†]

*Department of Physics
Yale University, New Haven, CT 06520*

[†] *Ames Lab-Bari-BNL-Iowa State-U. Mass-MIT-Penn. State-Purdue
UCLA-USMA-Vanderbilt-Wayne State-Yale*

Abstract. In this paper we discuss preliminary measurements of antiproton production in 11.5 A GeV/c Au+Pb nucleus collisions in Experiment 864 at the Brookhaven AGS. By comparing the E864 measurements with those of Experiment 878, we can infer the production of antihyperons in these collisions via the "feed-down" contribution of antiprotons from the decay of the antihyperon. This comparison indicates that the production of strange antibaryons is greatly enhanced, and presents significant challenges for thermal and transport models of heavy-ion collisions.

INTRODUCTION

Antiproton production in heavy ion collisions has been proposed as a sensitive probe of the produced matter due to the large annihilation of the $\bar{p}$ in matter and the possibility of enhanced antimatter production in a QGP scenario [1]. In addition, $\bar{p}$ measurements at the AGS may also contain a large feed-down contribution from antihyperon ($\overline{Y}$) decays. By combining experimental results we may be able to infer $\overline{Y}$ production in Au+Pb collisions.

Due to space constraints the discussion here is necessarily brief. A full publication of these results is in preparation [2].

EXPERIMENT E864

E864 is high rate, open geometry spectrometer designed to search for novel forms of matter that could be created in heavy ion collisions [3]. The spectrometer consists of two dipole bending magnets, with time-of-flight (TOF)

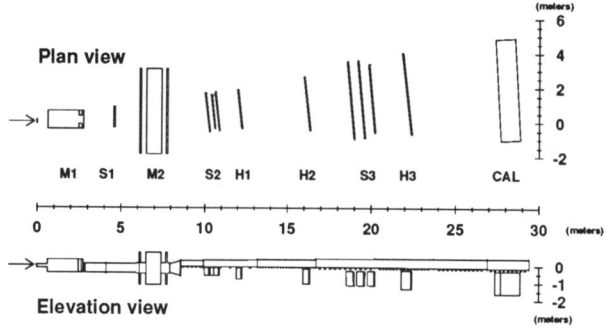

FIGURE 1. The E864 spectrometer. The Au beam is incident from the left in this diagram.

hodoscopes and straw tube tracking chambers downstream of the second magnet (see Figure 1).

The hodoscopes provide space points for tracking, as well as redundant charge and TOF measurements with a resolution of $\sim$130 ps. The straw tube chambers provide high precision space points; the mass resolution of the spectrometer at the -0.45T field setting is $\sim$3.5%. At the end of the apparatus is a lead/scintillating fiber hadronic calorimeter, which is used to confirm the energy of the particle determined by the tracking detectors. The uninteracted beam is carried away above the experiment in a large vacuum chamber. The centrality of the collision is measured by a charged particle multiplicity counter near the target.

A more detailed description of the apparatus will be available in the near future [4].

ANALYSIS

The data presented in this paper are derived from 20.1×10^6 10% central Au+Pb interactions taken during the 1994 run at the -0.45T field setting, and 85.6×10^6 10% interactions sampled at the -0.75T field setting during the 1995 run.

For the 1994 run, the experimental apparatus was not complete. The calorimeter was $\frac{1}{4}$ complete, two layers of the S3 straw array were only $\frac{1}{3}$ complete, and S1 was not in place. The calorimeter was stacked to have optimal acceptance for neutral particles, and thus was not used in the analysis.

For the 1995 run, the complete detector was in place, including the a "late energy" trigger that correlates energy and time in each calorimeter tower. This trigger is used by E864 in searches for high-mass composite objects (such as

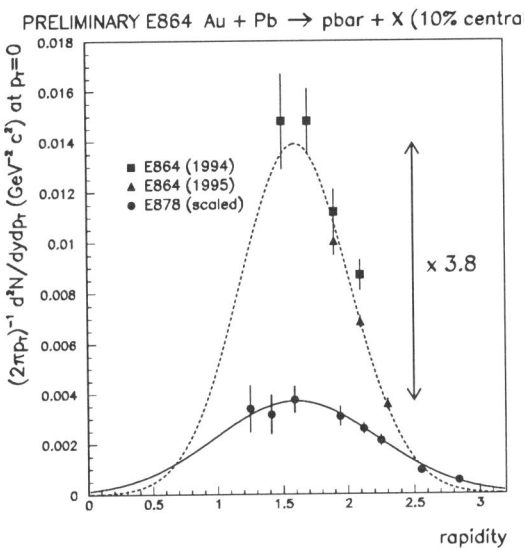

FIGURE 2. This figure shows $\bar{p}$ invariant multiplicities (in units of $\text{GeV}^{-2}c^2$) as measured by E864 and E878 at $p_T = 0$. The errors shown are statistical only. The E878 data points are scaled up by a factor of 1.5 as described in the text.

strangelets) and is effective in enhancing the $\bar{p}$'s in the data sample due to their additional annihilation energy in the calorimeter.

The mass of a particle in the E864 spectrometer is reconstructed as $m = Z\frac{r}{\beta\gamma}$ where the rigidity r is reconstructed from the downstream slope and intercept of the track in the bend plane, and the charge Z and velocity β are measured by the hodoscopes.

RESULTS

Figure 2 shows the preliminary invariant multiplicities for $\bar{p}$'s produced in 10% central Au+Pb collisions at 11.5 A GeV/c as measured in E864. For comparison, measurements made by the E878 collaboration in 10.8 A GeV/c Au+Au collisons [5] are also shown. The E878 points have been scaled up by a factor of 1.5 to account for the different beam energies between the experiments [6]. We estimate the systematic errors in our measurements to be 20%, dominated by our understanding of the experimental acceptance and track quality cut efficiencies. E878 reports a systematic error of 30% on their measurements. Figure 2 shows that the E864 measurements are consistently higher than their E878 counterparts.

Due to its large acceptance, the E864 spectrometer will detect $\bar{p}$'s from $\overline{Y}$

decay (where $\overline{Y} = \overline{\Lambda}, \overline{\Sigma^0}$, and $\overline{\Sigma^+}$). Therefore, the $\overline{p}$'s detected in E864 are a combination of primary $\overline{p}$'s and $\overline{p}$'s from $\overline{Y}$ decay, in a ratio that reflects their production ratio. The E878 collaboration have also evaluated the acceptance of their spectrometer for feed-down from $\overline{Y}$ decay [7]. At midrapidity the acceptance for $\overline{p}$'s from $\overline{\Lambda}$ and $\overline{\Sigma^0}$ decay is 14% of the spectrometer acceptance for primordial $\overline{p}$'s, and 10% of the $\overline{p}$ acceptance for $\overline{\Sigma^+}$ decays.

We can in principle separate the $\overline{p}$ and $\overline{Y}$ components if we make two explicit assumptions: both E864 and E878 understand their systematic errors, and the entire difference between the two experiments can be attributed to antihyperon feed-down. It is important to note that in energy scaling the E878 results we have implicitly assumed that the $\overline{Y}$'s scale with energy by the same factor as the $\overline{p}$'s. A detailed statistical analysis of the $\overline{Y}/\overline{p}$ ratio as a function of the E864 and E878 measurements (see Figure 3 for details), and the relevant statistical and systematic errors involved, shows that

$$\left(\frac{\overline{Y}}{\overline{p}}\right)_{\substack{y=1.6 \\ p_T=0}} \approx \left(\frac{\overline{\Lambda} + \overline{\Sigma^0} + 1.1\overline{\Sigma^+}}{\overline{p}}\right) > 2.8 \ (98\%\text{C.L.}) \qquad (1)$$

while the most probable value of this ratio is ~ 5. The factor of 1.1 multiplying the $\overline{\Sigma^+}$ arises due to the different branching ratio and acceptance for the $\overline{\Sigma^+}$ compared to the $\overline{\Lambda}$. This indicates an $\overline{Y}/\overline{p}$ ratio in Au+Pb collisions that is significantly greater than one at midrapidity and $p_T = 0$. It should be noted that if the $\overline{Y}$'s and the $\overline{p}$ are produced with different distributions in y and p_T, then the ratio of integrated yields of these particles will differ from the ratio at central rapidity and $p_T = 0$. Preliminary results from Si+Au collisions based on direct measurements of $\overline{p}$ and $\overline{\Lambda}$ production by the E859 collaboration also indicate a dn/dy ratio of p_T integrated yields greater than one [8].

In comparing the results of two experiments the potential exists for differences in the overall normalization. We note that preliminary measurements of protons, K^-, deuterons, and He^3 in E864 are consistent with preliminary E878 measurements within the quoted statistical and systematic errors of both experiments [9].

SUMMARY

In summary, E864 has made preliminary measurements $\overline{p}$ production about midrapidity in Au+Pb collisions at 11.5 A GeV/c. If we interpret the difference between E864 and E878 as a measure of $\overline{Y}$ production, we can infer that $\overline{Y}/\overline{p}$ is much greater than one at midrapidity and $p_T = 0$.

This work was supported by grants from the Department of Energy (DOE) High Energy Physics Division, DOE Nuclear Division, and the National Science Foundation.

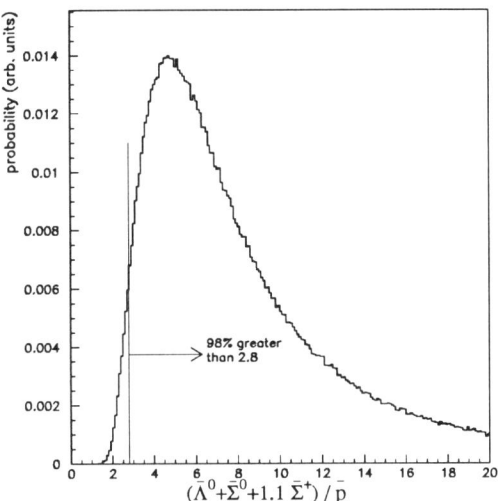

FIGURE 3. The probability distribution for the ratio $(\overline{\Lambda} + \overline{\Sigma^0} + 1.1\overline{\Sigma^+})/\overline{p}$ at midrapidity and $p_T = 0$ extracted from the E878 and E864 measurements. This distribution is generated by varying the E878 and E864 measurements within their systematic and statistical errors. Statitiscal errors are treated as Gaussian, while systematic errors are treated as indicating a flat range within which the measurements may vary.

REFERENCES

1. P. Koch, B. Müller, H. Stöcker and W. Greiner, Modern Physics Letters A **8** 737 (1988); J. Ellis, U. Heinz and Henry Kowalski, Phys. Lett. B **233** 223 (1989); U. Heinz, P. R. Subramanian, H. Stöcker and W. Greiner, Journal of Physics G: Nuclear Physics **12**, 1237 (1986).
2. T. Armstrong et al., submitted to PRL.
3. J. Sandweiss et al., *E864: Proposal For Funding* (1991)
4. T. Armstrong et al., to be submitted to NIM.
5. D. Beavis et al., *Phys. Rev. Lett.* **75** 3633 (1995)
6. S. Kahana et al., *Phys. Rev.* **C47** 1356 (1993)
7. M. Bennett et al., Yale preprint 40609-1148, to be submitted to Phys. Rev. C.
8. The E859 Collaboration, Y. Wu et al., in *HIPAGS 96*, in *Heavy Ion Physics at the AGS (HIPAGS 96)*, C. A. Pruneau, G. Welke, R. Bellweid, S. J. Bennett, J. R. Hall and W. K. Wilson Eds., WSU-NP-96-16, Wayne State University, Dec. 1996.
9. K. Pope, Ph.D. Thesis, Yale University, 1997; The E864 Collaboration, J. Lajoie et al., in *HIPAGS 96*, see ref [8].

Directed Flow in Au + Au Collisions at the AGS

Wen-Chen Chang[1] for the E877 collaboration[2]

*BNL, GSI, INEL, McGill Univ., Univ. of Pittsburgh,
SUNY at Stony Brook, Univ. of São Paulo, Wayne State Univ.*

Abstract. Studies of anisotropic transverse flow at AGS are presented. With the reconstruction of the reaction plane event by event, flow signals of the global variables transverse energy (E_T) and charged multiplicities (N_{ch}), and as well as, protons and charged pions are quantified by Fourier moments of the azimuthal distributions. The analysis shows that protons (or nucleons) exhibit a strong directed flow while pions collectively move in the opposite direction. The excitation functions of the directed flow and elliptical flow shows an interesting change of flow behavior as a function of beam energy.

INTRODUCTION

The measurement of collective motion driven by the pressure gradient provides a diagnostic tool to study the transient pressure built up through all the stages of compression, heating and subsequent expansion. There are two different patterns of collective flow [1] in the transverse direction to be distinguished, "azimuthally isotropic transverse flow" and "directed flow" where the latter one is the focus of this paper.

E877 has reported the first observation of directed flow in Au+Au at AGS energy [2,3]. The existence of this collective effect enables us to reconstruct the reaction plane event by event so that one extra degree of freedom ϕ, the azimuthal angle with respect to the reaction plane, is introduced into the analysis. Note that is practically all other experiments the ϕ dependence is neglected altogether. This paper will summarize the results of this flow analysis.

The experimental setup of the E877 apparatus and the way to determine the reaction plane angle and the associated resolution are described in detail

---

[1] Present address: University of California, Riverside, California 92507
[2] Experiment 877 is supported in part by the US DoE, the NSF, the Canadian NSERC and CNPq Brasil.

elsewhere [4–6]. We present in the following: (i) the azimuthal anisotropy of the global variables N_{ch} and E_T ; (ii) the flow signals of identified particles, protons and charged pions at forward rapidities, by utilizing the tracking information; (iii) the excitation functions of the azimuthally anisotropic flow. A brief summary will be given at the end.

I ANISOTROPY OF N_{ch}, E_T

After determination of the reaction plane angle determined in pseudorapidity window which doesn't overlap with the detector to be considered to avoid auto-correlation, the azimuthal distributions of N_{ch} and E_T with respect to the reaction plane are measured. After performing Fourier analysis on the ϕ distribution, the first two Fourier moments v_1 and v_2 are determined as a function of η in different centrality bins [6,7]. In both N_{ch} and E_T, v_1, identified as the directed flow component, shows the typical S-shape across mid-pseudorapidity which is characteristic of the forward-backward correlation of directed flow. For semi-central events we observe directed flow signals of up to 10%. The second moment, v_2, has an almost constant positive value of 1-2% over the whole η range. The positive values of v_2 indicate an elliptical event shape with the major axis lying in the reaction plane, and this is different from the "squeeze out" (perpendicular to the reaction plane) observation at lower energy [9]. The magnitude of both v_1 and v_2 exhibit the same centrality dependence, which is of maximum magnitude in the mid-central collision and decreasing magnitude towards the peripheral and central ends.

II FLOW OF IDENTIFIED PARTICLES

The triple differential multiplicity $d^3N / dp_x\,dp_y\,dy$ (i.e. $d^3N / p_t dp_t\,d\phi\,dy$) is measured for protons and charged pions in the E877 spectrometer acceptance by defining the reaction plane orientation as the new x-z plane. The anisotropy could be observed by three kinds of signals: Fourier moments $v_n(p_t, y)$, inverse slope parameter $T_B(\phi, y)$, and mean transverse momentum into the reaction plane $\langle p_x \rangle(y)$ [7,8].

The first Fourier coefficients v_1 of protons have an almost linearly rising dependence on p_t as shown in Fig. 1. Charged pions on the other hand are characterized by a small negative v_1 values which rise towards positive values at $p_t \geq 0.4$ GeV/c. The azimuthal dependence of inverse slope parameters T_B of protons shown in Fig. 2 reach the maximum in the slice of $\phi=0$ and the minimum in the opposite ϕ direction because more protons are populated on the positive p_x side. The variation of T_B over ϕ is about 30%.

A common way of demonstrating directed flow effect is the evaluation of $\langle p_x \rangle$. To avoid bias from the limited acceptance, we extract $\langle p_x \rangle$ for the full phase space by fitting a boosted thermal source parameterization [7,8]. In

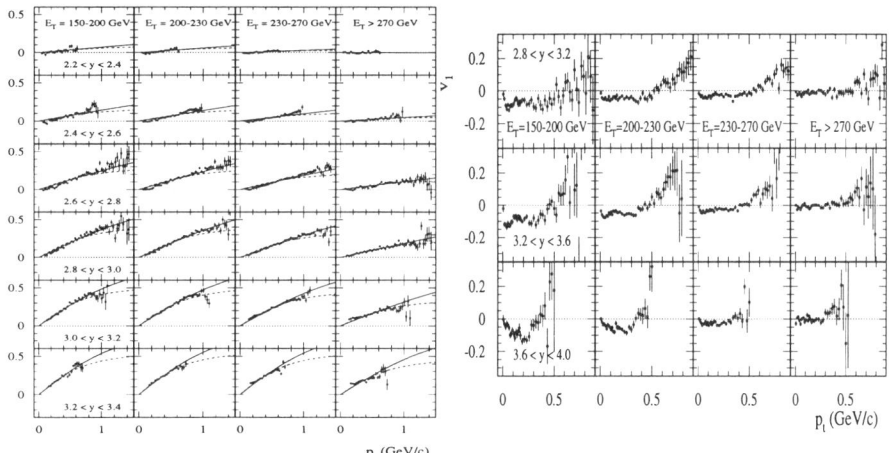

FIGURE 1. Left: $v_1(p_t)$ of protons in different rapidity bins (values inside the plot) and centralities (labeled by E_T at the top of each column). Right: $v_1(p_t)$ of π^+.

Fig. 2, $\langle p_x \rangle$ values of protons and charged pions are compared with the predictions from theoretical models RQMD [10] and ARC [11]. We would like to emphasize however that $\langle p_x \rangle(y)$ is one single integral value of the entire 2-d spectrum at each rapidity and thus $v_1(p_t, y)$ imposes more stringent tests on the models. One can notice that RQMD in the mean field mode reasonably well produces the integral values of $\langle p_x \rangle$ of protons at $2.8 < y < 3.0$, but the functional dependence $v_1(p_t)$ is completely at variance with the data as shown in Fig. 3.

III EXCITATION FUNCTIONS OF AZIMUTHALLY ANISOTROPIC FLOW

Preliminary data on the measured directed flow from E895 which runs at beam momenta at 2 and 4 GeV/c at the AGS are available [12]. Combining the BEVALAC [13] data and our measurement [6,8] on Au + Au system, it is observed in Fig. 4 that the magnitude of directed flow F, the slope of $\langle p_x \rangle$ at mid-rapidity normalized by projectile rapidity in the center of mass frame [7,13], rises with increasing beam energy for $E_k < 2$ GeV/c and then drops above this beam energy. Our measurement at 10A GeV/c is within errors of about the same magnitude as the preliminary value reported by E895 at 4 GeV/c.

As for the second moment of azimuthal asymmetry shown in Fig. 4, the values of v_2 at mid-rapidity are of the largest negative magnitude at $E = 0.5$

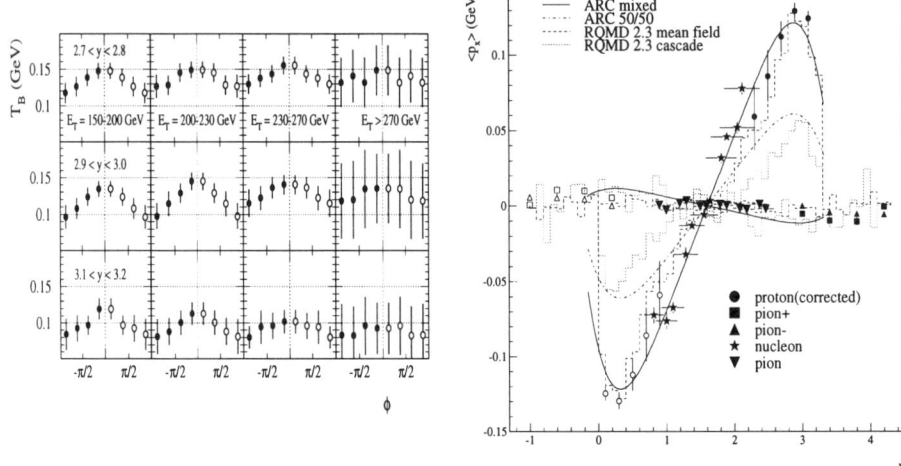

FIGURE 2. Left: $T_B(\phi)$ of protons in different rapidity bins (values inside the plot) and centralities (labeled by E_T at the top of each column). The azimuthal dependence of transverse spectra for protons at $2.9 < y < 3.0$. The four plots are at different centralities labeled by PCAL E_T. The open symbols are the reflected values. Right: $\langle p_x \rangle$ and cascade model comparison. The points around mid-rapidity(y=1.6) are the mapping of disentangled nucleon and pion flow.

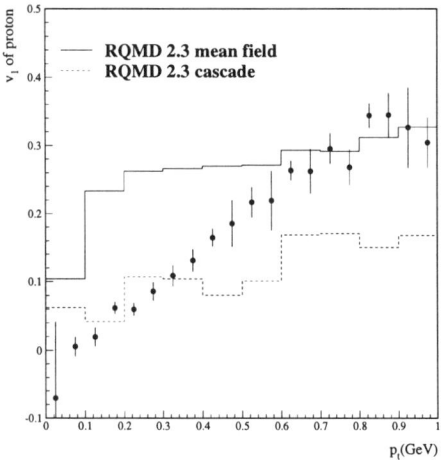

FIGURE 3. The comparison of $v_1(p_t, 2.8 < y < 3.0)$ for protons with the predictions from both modes of RQMD(version 2.3), cascade mode(dashed lines) and mean field mode(solid lines). The centrality bin is the top 13-9% central events.

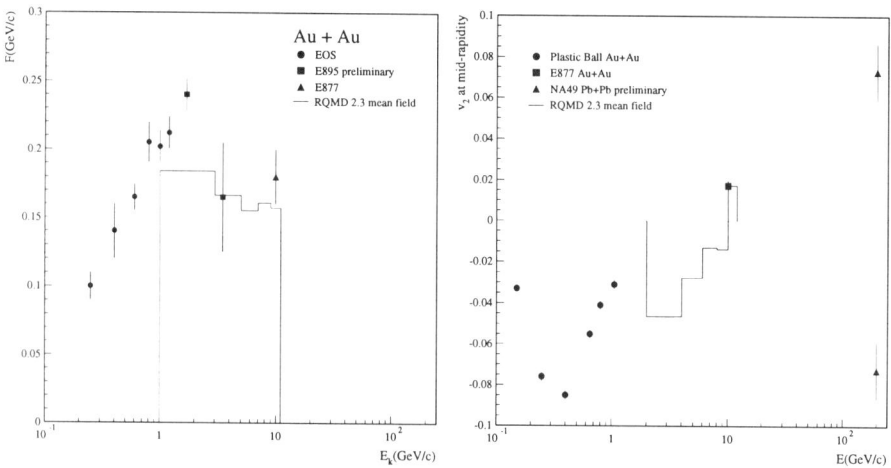

FIGURE 4. Left: The excitation function of directed flow as a function of beam kinetic energy per nucleon (E_k). Right: The excitation function of the second Fourier moment v_2 for protons at mid-rapidity as a function of beam energy per nucleon (E).

GeV/c per nucleon, where the phenomena of "squeeze-out" is maximum [9], and then decreases in magnitude. At 10A GeV/c, we measure a positive value. There has been a report from NA49 of a nonzero second moment at mid-pseudorapidity [14]. The orientation is not known yet. The RQMD predictions show qualitatively the same trend. This different kind of event shape may be understood as a consequence of geometrical configuration of participants without the re-scattering effects from fast moving spectators in the latter stage of collision at high beam energy [10].

SUMMARY

The two components, *directed flow* and *in plane "squeeze out"*, are identified in the E_T and N_{ch} azimuthal distribution with respect to the reconstructed reaction plane for Au+Au at AGS. The measured flow signals of identified particles are consistent with the observation in the measurement of global variables within the complementary acceptance. It is demonstrated that nucleons exhibit relatively large directed flow with a linearly rising dependence on p_t, whereas pions flow in the direction opposite to that of nucleons at low p_t.

The RQMD(2.3 cascade mode) fails to reproduce the amplitude of the directed flow signal $\langle p_x \rangle$ for protons and that necessitates the *mean field mechanism* to bring up the nucleon repulsion and quantitatively predict the observed positive second moment. Nevertheless, its predictions on p_t dependence of v_1 are inconsistent with the data. The other model ARC agrees with the measurement of $\langle p_x \rangle$ but fails in reproducing the "in-plane expansion" of the elliptic flow with an appropriate choice of repulsive and inelastic trajectories.

It is shown in the study of the excitation functions that the directed flow of protons decreases and the elliptic flow at mid-rapidity changes sign from BEVALAC to our measurements at AGS. The evolution of underlying physics with increasing beam energy needs to be understood. Theoretical treatments like the mean field potential [10], the diffractive trajectory [11] and the inclusion of QGP phase in hydrodynamics [15] should be scrutinized under the comparison with the more informative flow signals like $v_n(p_t, y)$.

REFERENCES

1. H.Stöcker, W. Greiner, Phys. Rep. **137**, 222 (1986)
2. J. Barrette *et al.*, E877 Collaboration, Phys. Rev. Lett. **73**, 2532 (1994).
3. S. Voloshin and Y. Zhang, Z. Physik **C70**, 665 (1996).
4. J. Barrette *et al.*, E877 Collaboration, Nucl. Phys. **A590**, 259c (1995).
5. W.-C. Chang, for the E877 collaboration, Proc. HIPAGS96,
   T. K. Hemmick, for the E877 collaboration, Nucl. Phys. **A610**, 63c (1996).
6. J. Barrette *et al.*, E877 Collaboration, Phys. Rev. **C55**, 1420 (1997).
7. W. Chang, Ph.D. Dissertation, SUNY at Stony Brook
   ("http://skipper.physics.sunysb.edu/~chang/physics.html").
8. J. Barrette *et al.*, E877 Collaboration, submitted to Phys. Rev. C.
9. H. A. Gustaffson *et. al.*, Phys. Rev. Lett. **52**, 1590 (1984),
   H. H. Gutbrod *et. al.*, Phys. Lett. **B216**, 267 (1989),
   H. H. Gutbrod *et. al.*, Phys. Rev. **C42**, 640 (1990).
10. H. Sorge, Phys. Rev. Lett. **78**, 2309 (1997)
11. D.E. Kahana, these proceedings
12. M. Gilkes, the E895 collaboration, private communication.
13. H. G. Ritter *et. al.*, the EOS collaboration, Nucl. Phys. **A583**, 491c (1995),
    M. D. Partlan *et. al.*, the EOS collaboration, Phys. Rev. Lett. **75**, 2100 (1995),
    M. A. Lisa *et. al.*, the EOS collaboration, Phys. Rev. Lett. **75**, 2662 (1995).
14. T. Wienold, for the NA49 collaboration, Nucl. Phys. **A610**, 76c (1996).
15. D.H. Rischke, Proc. HIPAGS96 and Nucl. Phys. **A610**, 88c (1996).

Multiplicities and Angular Distributions of Nucleus-Nucleus Interactions at SPS Energies: Protons to Lead

M.L. Cherry*, P. Deines-Jones*, A. Dabrowska†, J. Dugas*,
R. Holynski†, W.V. Jones*, D. Kudzia†, B.S. Nilsen*,
A. Olszewski†, M. Szarska†, A. Trzupek†, C.J. Waddington‡,
J.P. Wefel*, B. Wilczynska†, H. Wilczynski†, W. Wolter†,
B. Wosiek†, and K. Wozniak†

(KLM Collaboration)

*Louisiana State University, Baton Rouge, LA USA
†Institute for Nuclear Physics, Krakow, Poland
‡University of Minnesota, Minneapolis, MN USA

Abstract. Charged particle multiplicities from high multiplicity interactions of 158 GeV/n Pb ions on Pb targets were measured in nuclear emulsion chambers. These measurements are compared to measurements of central interactions of 200 GeV/n p, O, and S beams on silver or bromine and to simulations from the FRITIOF 7.02 and Venus 4.12 Monte Carlo event generators. Multiplicities in the central region are significantly lower than either simulation predicts. Venus, the only one of the two which attempts to incorporate reinteraction phenomena, predicts a significant narrowing of the pseudorapidity distribution for the highest multiplicity events, which is not observed in the data. However, we do find evidence for unexpectedly high spectator proton transverse momenta.

INTRODUCTION

The superposition model of nucleus-nucleus (AA) interactions has been highly successful. With the availability of Pb beams at the CERN SPS, superposition can now be tested over two orders of magnitude in mass from pp to Pb-Pb, and three orders of magnitude in the number of nucleon-nucleon (NN) collisions. The Pb-Pb system provides nearly the largest reaction volume achievable, and the highest energy densities attainable until RHIC begins colliding heavy ion beams. In this paper, we report Pb-Pb results from the

Krakow-Louisiana-Minnesota (KLM) emulsion chamber exposures in December 1994.

In order for a quark-gluon plasma or other collective behavior to occur in a high energy density collision, superposition must break down via some thermalization process such as reinteraction. To detect and understand such an event, 'ordinary' superposition and reinteraction physics must be well understood, especially if events exhibiting collective behavior are rare. Despite the fact that the SPS operates in an energy regime of high nuclear transparency, the Pb-Pb system is large enough that one might reasonably expect reinteraction to have an observable effect on the distribution of produced particles in the most central events.

For a sample of 60 high multiplicity central (maximum impact parameter $b_{max} = 4.2 \pm 0.6$ fm) Pb-Pb collisions, we present results showing the measured

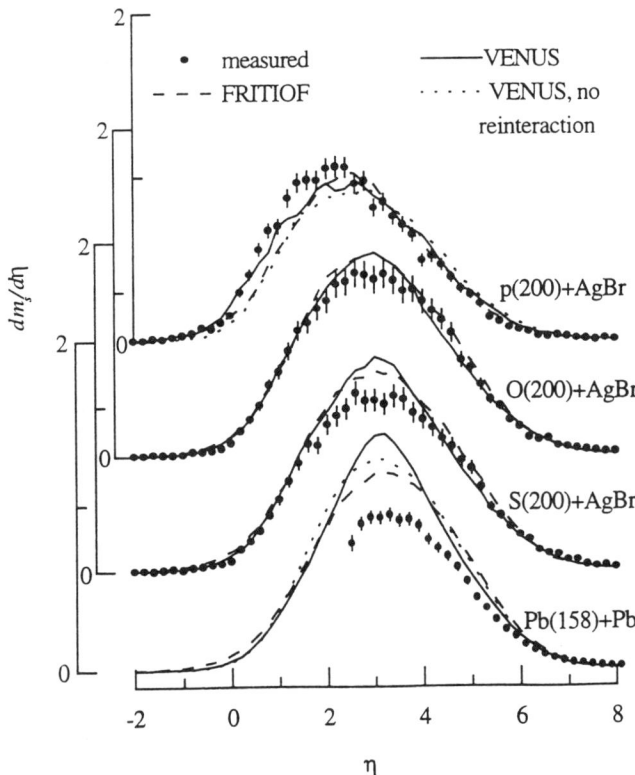

FIGURE 1. Shower particle pseudorapidity densities per wounded nucleon. For each projectile-target system, the same values of W are used to normalize the data, FRITIOF, and Venus.

multiplicities, produced particle and spectator pseudorapidity distributions, and the comparison with the Monte Carlo simulations. Descriptions of the experiment and the analysis procedures can be found in [1–3].

RESULTS

Central events are selected with relativistic charged particle multiplicities in excess of ~ 1000, two or fewer alphas in the forward direction, and no heavier projectile fragments. The average multiplicity of the selected sample of events is 1270 ± 50, corresponding to 170 ± 7 wounded projectile nucleons and an equal number of wounded target nucleons. Events are fully measured in the region forward of pseudorapidity $\eta = -\ln\tan\theta/2 = 2.9$, which includes all tracks forward of the pseudorapidity peak.

Figure 1 shows the shower particle multiplicity distributions as a function of pseudorapidity for 158 GeV/c per nucleon Pb-Pb interactions, and also for 200 GeV/n p-AgBr [4], O-AgBr, and S-AgBr [5]. In each case, the densities have been normalized by the calculated number of wounded nucleons for the data sample. The same normalizations have been used for the model calculations. The conclusion to be drawn from these results is that the Pb-Pb system has a significantly lower peak density than predicted by either model. A trend

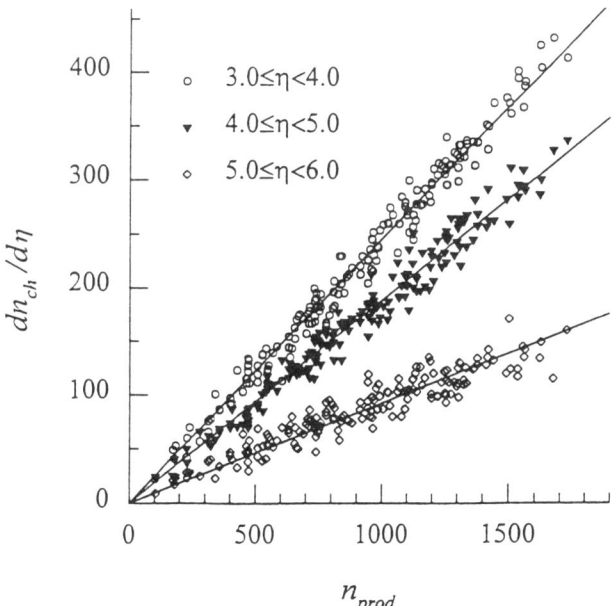

FIGURE 2. Dependence of Pb-Pb pseudorapidity density on multiplicity in three pseudorapidity intervals. The fitted lines are constrained to pass through the origin.

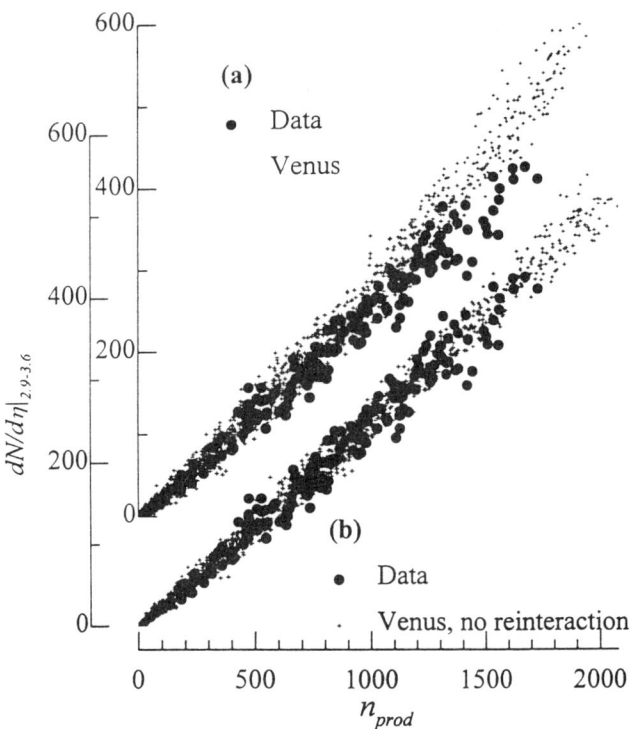

FIGURE 3. Comparison of measured Pb-Pb peak pseudorapidity densities with Venus (a) with reinteraction included, and (b) with reinteraction turned off but with all other parameters left unchanged.

is seen in going from lighter to heavier systems: The models slightly underestimate the p-AgBr central region, estimate the O-AgBr peak essentially correctly, slightly overestimate the S-AgBr, and significantly overestimate the Pb-Pb. The effect of reinteraction in Venus has been studied with separate runs in which reinteraction has been turned off, without adjusting any other parameters in the model. These runs are represented in Figure 1 as the lines with short dashes. As expected, reinteraction improves the p-AgBr fit in the target region [6]. However, turning reinteraction off moves the predictions closer to the Pb-Pb data, even without re-tuning the model's free parameters.

Based on superposition in the symmetric Pb-Pb system, one expects the produced particle multiplicity in the forward direction to be proportional to the total multiplicity. Figure 2 shows this behavior for Pb-Pb for three separate pseudorapidity windows. $dN/d\eta$ is expected to depend linearly on multiplicity (*i.e.*, the shape of the pseudorapidity distribution is expected to be independent of multiplicity and hence centrality) as long as second-order effects (*i.e.*, reinteraction) are unimportant. The peak Pb-Pb pseudorapidity density is plotted *vs.* the measured produced particle multiplicity in Figure 3, where the measured data are compared to the Venus predictions with and without reinteraction. With reinteraction (Figure 3a), the Venus simulated events deviate from linearity and differ in shape from the measured data at high multiplicity. Without reinteraction (Figure 3b), the Venus results are more consistent with the data.

Since the admixture of produced and spectator particles varies with centrality, one can derive separate produced and spectator distributions by comparing a central sample (almost all produced particles) with a semi-central one (with a significant spectator contribution). The resulting spectator distribution is shown in Figure 4 together with the best-fit Gaussian. Assuming these spectators have the same longitudinal momentum as the beam, the Gaussian half-width (1.46 ± 0.08 rad) corresponds to an r.m.s. transverse momentum of 230 ± 13 MeV/c. This is significantly larger than would be expected from an isotropic evaporation model (140 MeV/c). Thus, the spectator distribution suggests that there is a rescattered component.

DISCUSSION AND CONCLUSIONS

Among high energy heavy ion systems studied to date, the 158 GeV/n Pb-Pb system is unique in its combination of symmetry and large multiplicity, and consequently high track statistics in individual events. We have exploited these properties in studying centrality criteria, multiplicities, and the dependence of the shape of the pseudorapidity distribution on the multiplicity. Because of the Pb-Pb system's symmetry, any shape changes in the $dN/d\eta$ distributions must be due to reinteraction or the onset of non-superposition effects rather than changes in collision geometry. Multiplicities in these collisions are lower than predicted by the Monte Carlo event generators. In addition, we find no evidence of shape changes. Indeed, no direct evidence for reinteraction is observed in the central region. However, we do find evidence for reinteraction in the spectator distribution. This implies that in ultra-heavy systems at SPS energies, particle production in the center of mass is not significantly more isotropic in central events than it is in peripheral ones. This contrasts with the results from the AGS at 14.6 GeV/n, where pseudorapidity densities from heavy ion interactions become roughly isotropic at the highest multiplicities.

The fact that rescattering-induced narrowing is not observed is puzzling. In Venus, the narrowing of the distribution with reinteraction occurs largely because of a combination of greater proton stopping power and slower pions. The absence of narrowing in central events may indicate that the degree of slowing of protons and pions in Pb-Pb events is less than the model predicts. Without the additional energy available for particle production which comes from increased nucleon stopping power, increased heavy particle production

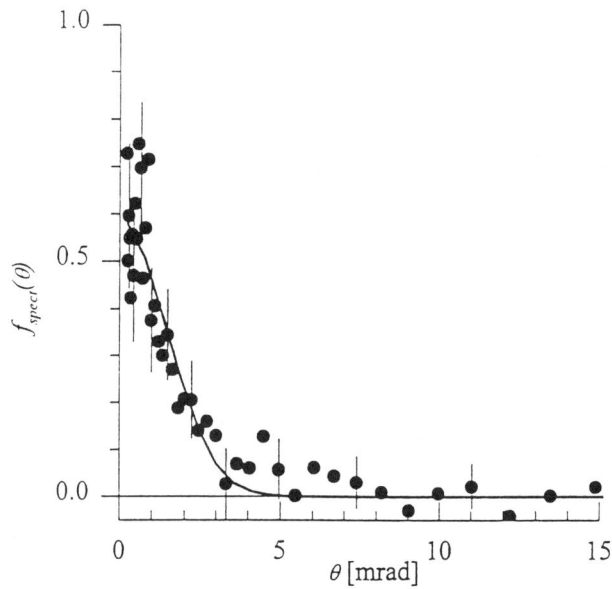

FIGURE 4. Derived Pb-Pb spectator distribution.

must come at the expense of smaller pion yields. Thus, heavy particle production accompanied by a smaller than expected increase in stopping power could explain both the observed shape independence and the low multiplicities.

This work has been supported by NSF PHY-921361, the Louisiana Space Grant Consortium, and LA Board of Regents grant NASA/LEQSF (94-97)IMP-02 at LSU, DOE-FG02-89ER40528 at Minnesota, and in Poland by the Comm. for Scientific Res. (2P03B18409) and Maria Sklodowska-Curie Fund (PAA/NSF-96-256). We thank Y. Takahashi and his EMU-16 colleagues for their generous assistance, the CERN staff, and A. Aranas and L. Wolf at LSU for their extensive microscope work.

REFERENCES

1. Deines-Jones, P. et al., *Phys. Rev.* **C53**, 3044 (1996).
2. Deines-Jones, P. et al., to be published (1997).
3. Wosiek, B. et al., Proc. Multiparticle Dynamics Symp., Stara Lesna (1996).
4. Anzon, E.V. et al., *Sov. J. Nucl. Phys.* **22**, 787 (1975).
5. Dabrowska, A. et al., *Phys. Rev.* **D47**, 1751 (1993).
6. Werner, K. et al., *Phys. Rep.* **232**, 87 (1993).

An Excitation Function of Particle Production at the AGS

James C. Dunlop

Massachusetts Institute of Technology
Laboratory for Nuclear Science
Cambridge, MA 02135, USA

for the E866 Collaboration:
BNL-Columbia-INS,Tokyo-Kyoto-LLNL-Maryland-MIT-
UC,Riverside-UC,Space Science Lab-Tokyo-Tsukuba-Yonsei
and the E917 Collaboration:
ANL-BNL-UC,Riverside-UIC-Maryland-MIT-Rochester-Yonsei

Abstract. Preliminary results on particle production from the recent energy scan at the AGS are presented. Results on π and K^+ yields at mid-rapidity as a function of both beam energy and centrality are examined. Indications are that the evolution of particle production with beam energy is smooth, with no evidence of an onset of anomalous production.

INTRODUCTION AND MOTIVATION

In collisions of heavy ions in the energy regime of the AGS we have the possibility of creating extended regions of high baryon density, through which it is hoped that a regime of new physics will be reached. One approach to the problem of separating out signals of new physics from the complicated hadronic background is to examine potential signatures over the fullest systematic range available. An excitation function fits well into this approach: by varying the beam energy, we vary the maximum baryon density achieved, and by selecting on events of differing centrality, we vary the size of the region at high baryon density. Hence, we propose to look at variables that may indicate new physics in a grid of these two quantities, beam energy and centrality. Within such a grid, we will search for signs of new physics, and provide a large data set with which to confront the various models of these collisions.

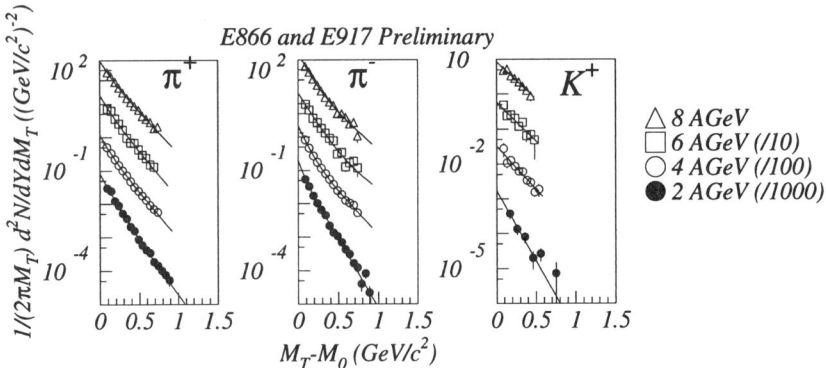

FIGURE 1. M_t Spectra at mid-rapidity vs beam energy. Plotted here are the double-differential yields in m_t within a window of $\pm 0.25 y_{cm}$ about y_{cm}, for the most central 525 mb ($\sim 8\%$ of σ_{int}) of the most central reactions. Error bars are statistical only.

PRELIMINARY RESULTS

The AGS ran lower beam energies in 1996, of which experiment E866 took the 2 and 4 A·GeV beam in January, while E917, the successor experiment to E866, took the 6 and 8 A·GeV beam in October. Both experiments are based on rotating spectrometers with time-of-flight walls for particle identification. E866 used the E802 spectrometer and a new forward spectrometer for the very forward angles, while E917 used the E802 spectrometer exclusively. The experiments also had equipment for global event characterization, including two devices for centrality selection. The first, the new multiplicity array, or NMA, is a large acceptance device made up of an array of lucite Cherenkov radiators, sensitive mostly to pions. The second is the zero-degree calorimeter, or ZCAL, which measures the forward-going energy in a collision and hence the number of nucleons not scattered. For a more detailed description of the experimental apparatus, see [1,2].

The data from this article is from the energy scans in 1996 for the lowest energies, and from the 1994 E866 data run at 10.7 A·GeV. We have analyzed only a small fraction of the data from the lower energies. The highest energy data were shown at Quark Matter last year [1]. For centrality cuts, the lower energy data used the NMA while the data at 10.7 GeV used the ZCAL.

Figure 1 shows the double differential yields in transverse mass and rapidity for our most central bin. The pions have a concave shape, and do not fit to a single exponential. Here we have fit the pion spectra to a sum of two exponentials. The kaon spectra are fit well by a single exponential.

With these fits we integrate over m_t to extract the dN/dy at mid-rapidity. The results are shown in figure 2. For each of the five beam energies, we have

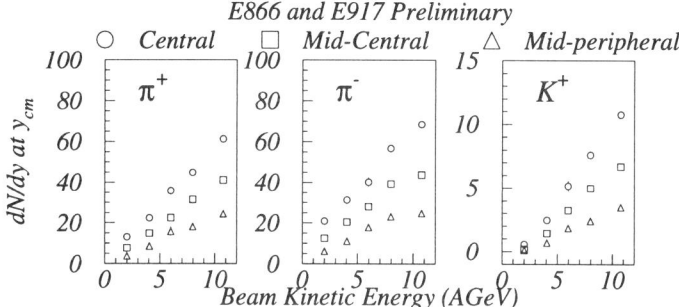

FIGURE 2. dN/dy at mid-rapidity vs beam energy and centrality. Plotted here is the dN/dy in a window of $\pm 0.25 y_{cm}$ about y_{cm}, for the three centrality bins and the five energies analyzed. The error bars are statistical only. The three centrality bins correspond to the uppermost 525 mb in the centrality selection device, the next 1050 mb, and the next 1050 mb, or $\sim$0-8%, 8-24%, and 24-40% of σ_{int}, respectively.

made three cuts in centrality. The error bars are statistical only; we estimate a systematic error of 20% at this stage of the analysis. Note the scales: there is an order of magnitude more kaons produced at the highest AGS energy than are produced at 2 GeV, with a smaller rise in the pions. However, the rise is very smooth, almost linear. We see no evidence of any sudden onset of increased particle production at any point in this grid.

Another thing to note is that the rise in particle production in the most

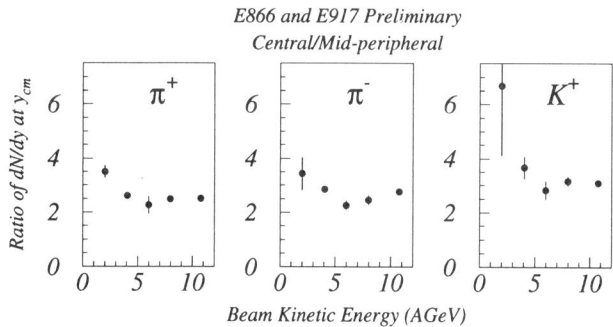

FIGURE 3. Ratio of dN/dy at mid-rapidity central collisions to mid-peripheral vs beam energy. Plotted here is the ratio of the dN/dy, within a window of $\pm 0.25 y_{cm}$ about y_{cm}, in the "central" to that in the "mid-peripheral" centrality bin, where "central" and "mid-peripheral" are as defined in the caption to figure 2. The error bars are statistical only.

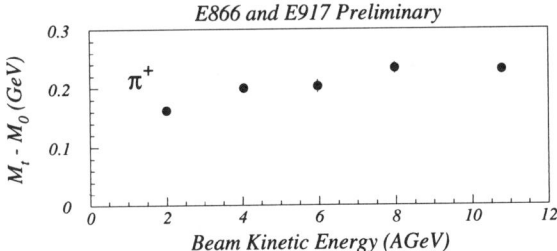

FIGURE 4. Mean transverse mass for π^+ at mid-rapidity vs beam energy. Plotted here is the mean transverse mass for π^+ in a window of $\pm 0.25 y_{cm}$ about y_{cm} vs beam kinetic energy, for the most central bin examined. Error bars are statistical only.

peripheral collisions tracks quite well with the rise in the most central collisions. This is made clearer by figure 3, in which we have divided the yields in the most central collisions by the yields in the most peripheral collisions analyzed. We note that, excepting a rise at 2 GeV, this ratio is constant with beam energy, indicating that the mechanisms that are driving the rise in yields as the beam energy increases are doing so equally well in the mid-peripheral as in the central collisions.

In contrast to the rise in yields, the mean transverse mass for π^+ rises less dramatically as a function of energy, as shown in figure 4. This clarifies the energetics of these collisions: the increasing energy available to the system with higher beam energy goes relatively more into particle production than into transverse motion.

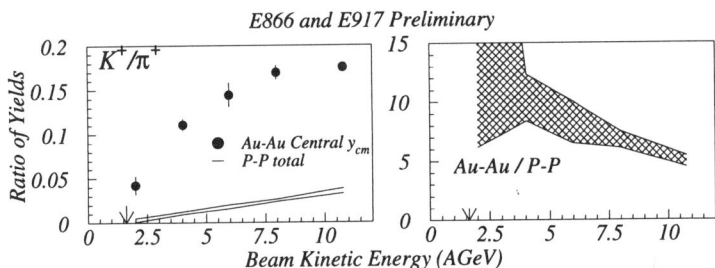

FIGURE 5. K^+/π^+ ratio of yields vs beam energy. Plotted on the left is the ratio of the K^+ and π^+ yields within $\pm 0.25 y_{cm}$ about y_{cm}, for the most central 525 mb in centrality. The error bars are statistical only. The p-p lines are ratios of total yields. The parametrizations of the K^+ yields are from a fit to the data in this energy range collected in [3,5], while those for the π^+ are from [3,4]. Plotted on the right is the ratio of the two quantities shown in left-hand plot.

The final figure we would like to show is of the K^+/π^+ ratio of yields at mid-rapidity, figure 5. This provides a relatively simple measure of strangeness enhancement in these collisions, in essence normalizing the production of strange mesons to that of the most abundantly produced particle. We see that the ratio rises, and that the rise is smooth. On the same plot we have plotted the K^+/π^+ ratio of total yields from parametrizations of data from p-p collisions. The K^+/π^+ ratio in central Au-Au collisions is relatively more enhanced compared to that in p-p collisions as the beam energy decreases. This indicates that the influence of secondary collisions is relatively larger at the lower energies. Note that the p-p parametrizations have not been corrected for Fermi motion or isospin.

CONCLUSIONS

We see no indications of any onset of sudden changes in particle production as a function of beam energy. Particle yields at mid-rapidity rise dramatically with beam energy, but quite smoothly and nearly linearly. The rise is similar in mid-peripheral and central collisions. In contrast, the mean transverse mass of the π^+ shows only a small increase. In terms of strangeness production, the K^+/π^+ ratio of yields at mid-rapidity rises smoothly as a function of beam energy, and is more enhanced relative to that in p-p collisions as the beam energy decreases. More detailed looks at the full rapidity and centrality range, and at the finer details of the transverse spectra, will greatly increase the systematic power of our observations, and allow us to make detailed confrontations of the models of these collisions.

ACKNOWLEDGEMENTS

Experiments E866 and E917 are supported in part by the DOE, NASA, Ministry of Education and KOSEF in Korea, and the Ministry of Education, Science and Culture of Japan.

REFERENCES

1. L. Ahle et al., *Nucl. Phys.* **A610**(1996) 139c, submitted to Phys. Rev. Lett.
2. T. Abbot et al., *Nucl. Instrum. Methods* **A290**(1990) 41.
3. M. Antinucci et al., *Nuovo Cim. Lett.* **6**(1973) 121.
4. A.M. Rossi et al., *Nucl. Phys.* **B84**(1975) 284.
5. A. Baldini et al., *Landolt-Börnstein New Series* **I\12b**, Springer-Verlag (1988).

LEXUS[1]

Sangyong Jeon

School of Physics and Astronomy
University of Minnesota
Minneapolis, Minnesota 55455

Abstract. We use a Glauber-like approach to describe very energetic nucleus-nucleus collisions as a sequence of binary nucleon-nucleon collisions. No free parameters are needed: all the information comes from simple parametrizations of nucleon-nucleon collision data. Produced mesons are assumed not to interact with each other or with the original baryons. Comparisons are made to published experimental measurements of baryon rapidity and transverse momentum distributions, negative hadron rapidity and transverse momentum distributions, average multiplicities of pions, kaons, hyperons, and antihyperons, and zero degree energy distributions for sulfur-sulfur collisions at 200 GeV/c per nucleon and for lead-lead collisions at 158 GeV/c per nucleon. Good agreement is found except that the number of strange particles produced, especially antihyperons, is too small compared with experiment. We call this model LEXUS: Linear EXtrapolation of Ultrarelativistic nucleon-nucleon Scattering to heavy ion collisions.

The goal of this talk is to construct a very simple model, LEXUS, of heavy ion collisions by linearly extrapolating nucleon-nucleon scatterings. Due to the length restriction, this talk is necessarily without details. See [1].

Let me emphasize two things before I present our model. First, LEXUS has no free parameters. All inputs are taken from available nucleon-nucleon scattering data. Hence, all the results presented here are absolutely normalized. Second, the purpose of LEXUS is *not* to describe everything in terms of sequences of nucleon-nucleon scatterings, but to have a baseline calculation that indicates what will happen if there is no new physics beyond scatterings of nucleons.

The basic idea behind LEXUS is fairly simple. It combines ideas of the Glauber model [2], Hüfner and Knoll's rows-on-rows collision model [3] and the evolution models of Hwa [4] and Kapusta and Csernai [5]. We consider a nucleus-nucleus collision as sequences of binary nucleon scatterings in free

---

[1] This is an abbreviated version of a paper to be published in Physical Review C **56** written by J.I.Kapusta and the present author.

space. Secondary particles are produced as in free space and escape without further scatterings.

Baryon rapidity distribution functions form the backbone of LEXUS. Projectile baryon rapidity distributions evolve according to the equation

$$W^P_{mn}(y_P) = \int_0^{y_0} dy'_P dy'_T W^P_{mn-1}(y'_P) W^T_{m-1n}(y'_T) Q(y_P - y'_T, y'_P - y'_T, y_P - y'_P) \,. \quad (1)$$

Here $W^P_{mn}(y_P)$ is the rapidity distribution of the m-th projectile nucleon after colliding with n target nucleons, y_0 is the beam rapidity, and Q is the evolution kernel. There is a similar equation for the target distributions $W^T_{mn}(y_T)$. However, due to the symmetry $W^T_{mn}(y_T) = W^P_{nm}(y_0 - y_T)$, only W^P_{mn}'s need to be calculated.

We use the following parametrization for our evolution kernel:

$$Q(a,b,c) = \lambda \frac{\cosh a}{\sinh b} + (1-\lambda)\delta(c) \,, \quad (2)$$

where λ ($= 0.6$) has the interpretation of hard inelastic scattering probability. As shown in Figure 1 this parametrization is good unless the rapidity is very close to the beam or the target rapidities.

Folding W^P_{mn} with the geometry of the two colliding nuclei and different probability factors yields various physical observables. Let us consider baryon distributions first. The simplest observable LEXUS can calculate is the proton rapidity distribution for nucleus-nucleus collisions. Figure 2 shows our calculations for NA35 and their published data [7]. Our calculation agrees fairly well with the data.

To calculate the transverse momentum distribution we regard each nucleon-nucleon collision as a step in a random walk. Experiments fix the width of the random walk to be $\langle p_T^2 \rangle_{NN} = 0.282$ (GeV/c)^2 [8]. Figure 3 shows that again our calculation and the data from NA35 agree well. Similar calculations for lead-lead at 158 (GeV/c) also show good agreements [1].

For the zero degree calorimeter (ZDC) energy distribution, Figure 4 shows that again LEXUS is in accordance with the NA49 data [9].

Now, consider hadron production. Let $\langle X(s) \rangle_{NN}$ represent the average number of mesons of type X produced in a nucleon-nucleon collision at center-of-mass energy $\sqrt{s}$. The average number of such mesons produced in a collision between the m-th projectile nucleon and the n-th target nucleon is obtained by replacing $Q(y_P - y'_T, y'_P - y'_T, y_P - y'_P)$ in Eq. (1) by $\lambda \langle X(s) \rangle_{NN}$. Folding with the geometrical factors, we get the results shown in table 1. LEXUS is good at reproducing negative hadron multiplicities, but it is not very good at reproducing the abundance of strange particles. This may be an indication where the new physics may lie. For the prediction for lead-lead collisions at 158 (GeV/c) see [1].

Let $\langle X(y_{cm}, y_{rel}, y) \rangle_{NN}$ represent the rapidity distribution of a meson of type X produced in a nucleon-nucleon collision. Experimentally, this distribution

	h^-	K^+	K^-	K^0_S	Λ	$\bar{\Lambda}$
NA35 2%	98 ± 3	12.5 ± 0.4	6.9 ± 0.4	10.5 ± 1.7	9.4 ± 1.0	2.2 ± 0.4
LEXUS 2%	102	$9.4^{+1.4}_{-2.3}$	$5.5^{+0.8}_{-1.3}$	$5.0^{+0.2}_{-0.8}$	$3.9^{+0.5}_{-2.0}$	$0.28^{+0.44}_{-0.16}$

TABLE 1. The average particle multiplicities in S+S collisions. The NA35 data [10] should be compared against the 2% most central collisions in LEXUS.

function is well represented by a Gaussian with a Landau width [11]. Folding this Gaussian with the nucleon distribution functions and the geometrical factors, one can then calculate the rapidity distribution functions for the hadrons produced in a nucleus-nucleus collision. As shown in Figure 5, LEXUS agrees well with the experimental negative hadron distribution.

The produced hadrons receive a transverse boost from the baryons. Taking this into account, the transverse momentum distribution of the negative hadron can also be calculated. As shown in Figure 6 LEXUS describes the experimental result quite well.

In conclusion, we have succeeded in constructing a linear extrapolation of nucleon-nucleon scatterings to nucleus-nucleus collisions. We know that this cannot be the whole story, nevertheless, it is important to have a baseline calculation to compare against with more detailed models and experimental data. Works in progress include dilepton productions, J/ψ production, and the description of pA collisions.

ACKNOWLEDGMENTS

We thank Marek Gaździcki, John Harris, Peter Jacobs, Spiros Margetis, Milton Toy, Flemming Videbaek and Nu Xu for helpful discussions of the experiments and data. We also thank Laszlo Csernai for helpful comments and discussions. This work was supported by the US Department of Energy under grant DE-FG02-87ER40328.

REFERENCES

1. Sangyong Jeon and J. I. Kapusta, to appear in Phys. Rev. C **56** (1997).
2. R. J. Glauber, in *Lectures in Theoretical Physics*, edited by W. E. Brittin and L. G. Dunham (Interscience, New York, 1959).
3. J. Hüfner and J. Knoll, Nucl. Phys. **A290**, 460 (1977).
4. R. C. Hwa, Phys. Rev. Lett. **52**, 492 (1984).
5. L. P. Csernai and J. I. Kapusta, Phys. Rev. D **29**, 2664 (1984); *ibid.* **31**, 2795 (1985).

6. Ole Hansen and F. Videbaek, Phys. Rev. C **52**, 2684 (1995). In this paper the authors look for a scaling behavior of the baryon rapidity distribution in terms of the variable y/y_0 which does not exist.
7. J. Bächler *et al.* (NA35 Collab.), Phys. Rev. Lett. **72**, 1419 (1994).
8. Much data has been assembled and parametrized in: E. E. Zabrodin *et al.*, Phys. Rev. D **52**, 1316 (1995).
9. T. Alber *et al.* (NA49 Collab.), Phys. Rev. Lett. **75**, 3814 (1995).
10. T. Alber *et al.* (NA35 Collab.), Z. Phys. C **64**, 195 (1994)
11. P. Carruthers and Minh Duong-van, Phys. Rev. D **8**, 859 (1973).

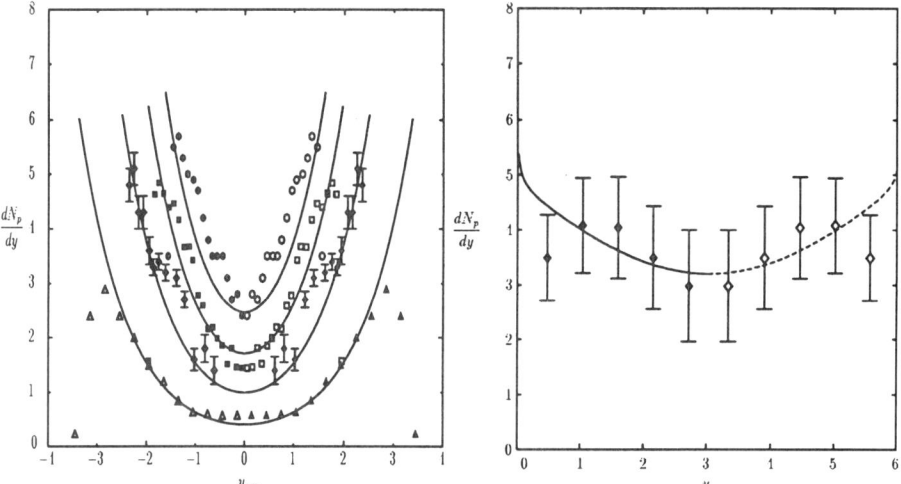

FIGURE 1. Proton rapidity distribution in $p + p \to p + X$ in the center of mass frame. From the top: $p_{lab} = 12, 24, 69$, and 400 GeV/c. The filled symbols indicate experimentally measured data points and the empty symbols are the reflected ones. The solid lines are $dN/dy = \lambda \cosh(y)/\sinh(y_0/2)$ with corresponding maximum rapidity y_0. The data were assembled in Ref. [6].

FIGURE 2. Proton rapidity distribution for a S+S collision at 200 GeV/c with a centrality of 2%. The solid line represents our calculation. Filled diamonds are data from NA35 [7]. Open diamonds and the dashed line are the reflection of the left half.

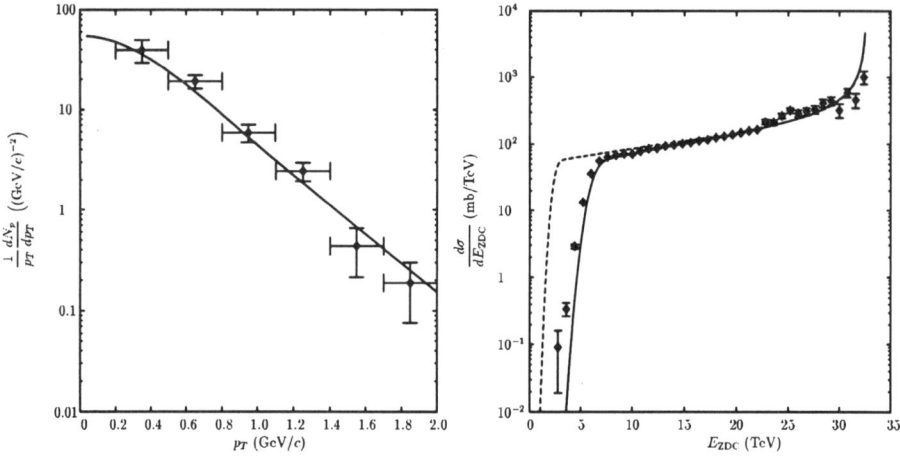

FIGURE 3. Proton transverse momentum distribution for a S+S collision at 200 GeV/c with a centrality of 2%. The solid line represents our calculation. Data are from NA35 [7]. Rapidity range is $0.2 < y < 3.0$.

FIGURE 4. The zero degree energy distribution for 158 GeV/c Pb+Pb collisions. The solid line represents LEXUS with an opening angle of 0.3 degree. The dashed line represents LEXUS with only the spectator nucleons. Filled circles are data from NA49 [9].

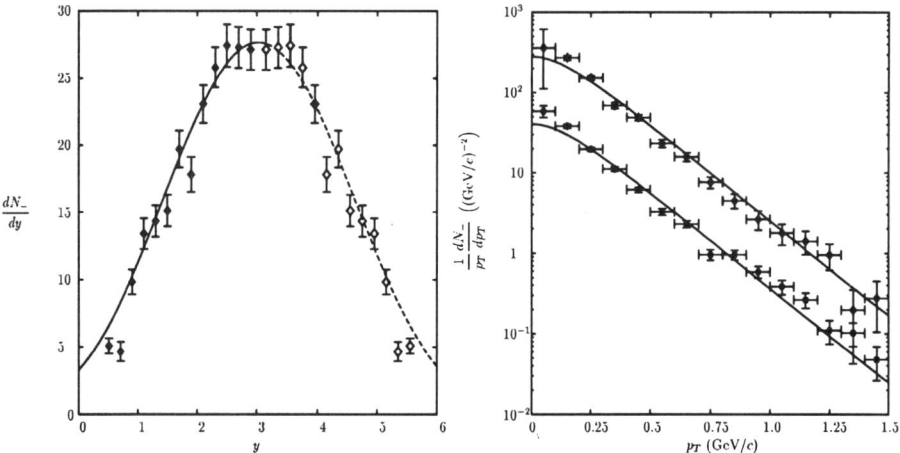

FIGURE 5. Rapidity distribution of h^- in 200 GeV/c S+S collisions with a centrality of 2%. The solid line represents our result. Filled diamonds are data from NA35 [7]. Open diamonds and the dashed line are the reflection of the left half.

FIGURE 6. Negative hadron transverse momentum distribution for a S+S collision at 200 GeV/c with a centrality of 2%. The solid line represents our result and the filled diamonds are NA35 data [7]. Rapidity range is $0.8 < y < 2.0$ for the upper curve and $2.0 < y < 3.0$ for the lower curve. The lower curve is also scaled by a factor of 1/10.

Peculiarities of Secondary Particle Generation Process in Pb-Pb Interactions at 158A GeV

N.M.Astafyeva[1], N.A.Dobrotin[2], I.M.Dremin[2],
E.L.Feinberg[2], L.A.Goncharova[2],
K.A.Kotelnikov[2], A.G.Martynov[2], N.G.Polukhina[2]

[1] *Space Research Institute, 117810 Moscow, Russia*
[2] *Lebedev Physical Institute, 117924 Moscow, Russia*[1]

Abstract. Experimental data on Pb-Pb central collisions were obtained by means of emulsion magnetic chambers containing the lead target irradiated with the Pb beam in the magnetic field of 1.8 Tesla at CERN in December, 1994 and November, 1996. 50 μm thick emulsion layers were placed perpendicular to the beam.

Results of data analysis show existence of certain peculiar patterns in several central collisions (e.g. ring-like events, jets, multiple narrow spikes in secondary particle pseudorapidity distributions, etc.) The methods of pattern recognition theory, in particular, wavelet technique (the method of localized spectral analysis), used in target diagram analysis, reveal the detailed features of such structures.

Our objective is studying properties of very dense and hot nuclear matter by event-by-event investigation of hadroproduction process peculiarities using emulsion magnetic chambers with a Pb target irradiated in the beam of high energy (158 GeV per nucleon) lead nuclei. The project is approved as EMU15 CERN experiment.

The emulsion chamber included a thin ($\sim 300\mu$m) lead target and 38 emulsion plates of 50μm thick deposited on the 25μm mylar base and placed perpendicularly to the beam of lead nuclei. For measuring charge signs and particle momenta, the chamber is placed in the transverse magnetic field of ~ 1.8 Tesla. The magnetic field, thin mylar base and small total thickness of the chamber help to reduce the background of tracks produced by secondary interactions.

---

[1] This work is supported partly by RFFI grants 96-02-19572 and 96-01-00340.

Fourteen chambers have been exposed to the Pb beam with a total number of about 10000 beam particles to each target (the diameter of the beam was 2-4 cm). Approximately 30 central Pb-Pb collisions were found in each chamber.

Excellent spatial resolution of nuclear emulsion made it possible to observe rather peculiar phenomena. Some of them were also found previously in both cosmic ray and accelerator experiments.

In [1], the ring-like structure was observed in target diagrams of secondary particle distributions for two interactions of superhigh energy cosmic-ray nuclei. The first one, with $\Sigma E_\gamma \sim 2*10^{15}$ eV, was registered by the stratospheric emulsion chamber [2], and the other one-in the installation of the Chacaltaya Japanese-Brazilian Collaboration [3]. Events of such kind had been found also by NA22 group [4] and excited a great interest. A possible interpretation has been proposed in [5]. Events with anomalously large number of narrow spikes in the rapidity distribution and events with the extremely high transverse momenta of secondaries were observed as well in [6] (similar to those found later by JACEE collaboration [7]); all such events are widely considered as essential for understanding features of high energy hadronic matter and possibly correspond to QGP production.

In Fig.1, we show the target diagram of secondary particle distribution for one event of Pb-Pb central interaction at 158A GeV with the charged-particle multiplicity equal to 1072. The diagram shows peculiarities in the track distribution such as the ring-like structure of the particle distribution and the jet-like bundles of particles with close polar and azimuth angles. Two circles correspond to pseudorapidities equal to 2.6 and 3.0. As an example of the traditional analysis, in Fig.2, we present the histogram of the pseudorapidity distribution. Two vertical broken lines show the region of the circles of Fig.1. The statistical analysis shows that these peculiarities cannot be caused by usual fluctuations.

The lego-plot is presented in Fig.3 for the above event. The equal binning in $\Delta\phi, \Delta\eta$ was made with $\Delta\phi = \Delta\eta = \pi/12$, where ϕ is azimuth angle and η - pseudorapidity. The numbers of particles belonging to each bin are plotted on the vertical axis. The pure phase space volume distribution would correspond to the flat plateau. However we can see that the graph reveals clearly the dynamical structure. One can guess that there is a ridge on the ring drawn in Fig.1 ($2.6 < \eta < 3.0$) which provides spikes in Fig.2 when integrated over azimuth angles.

We show that wavelet analysis yields more complete and quantitative results. The idea behind this method is to resolve any pattern at different locations with a variable resolution.

One of the examples of continuous wavelet is provided by the so-called Mexican hat (MHAT) wavelet which has the form:

$$\psi(\frac{x-b}{a}) = (1 - (\frac{x-b}{a})^2)\exp[-\frac{(x-b)^2}{2a^2}], \qquad (1)$$

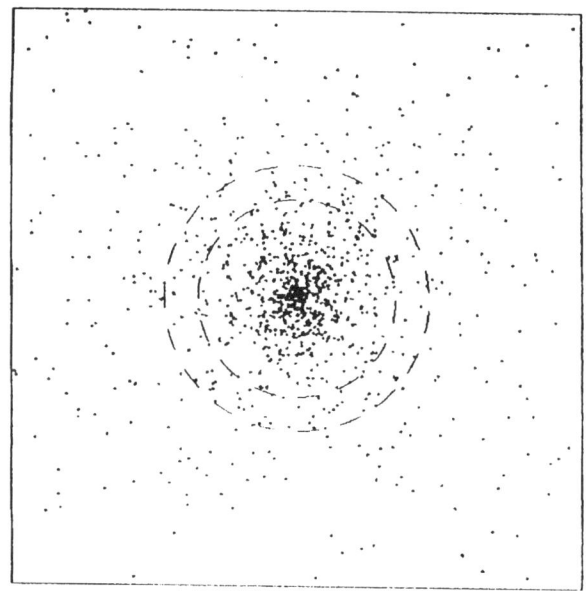

Fig. 1

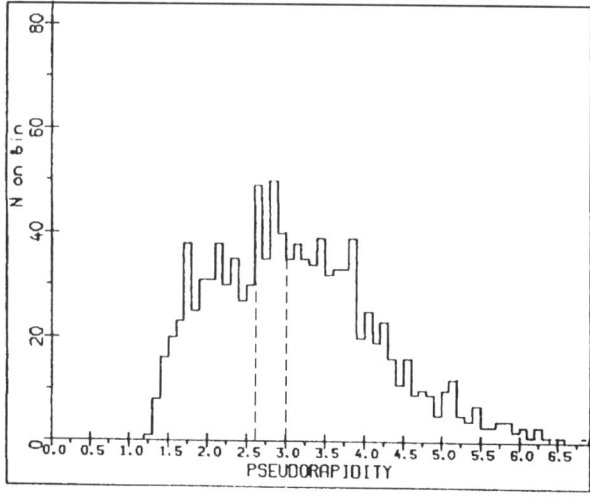

Fig. 2

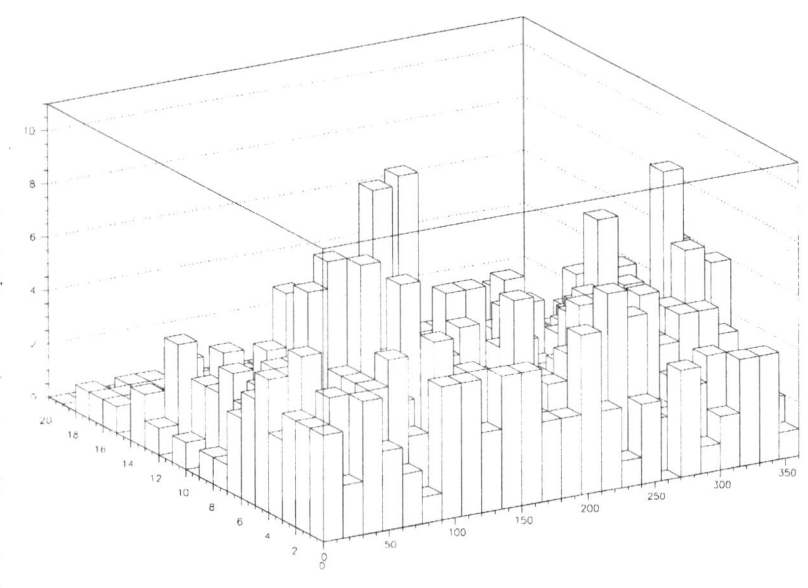

Fig. 3

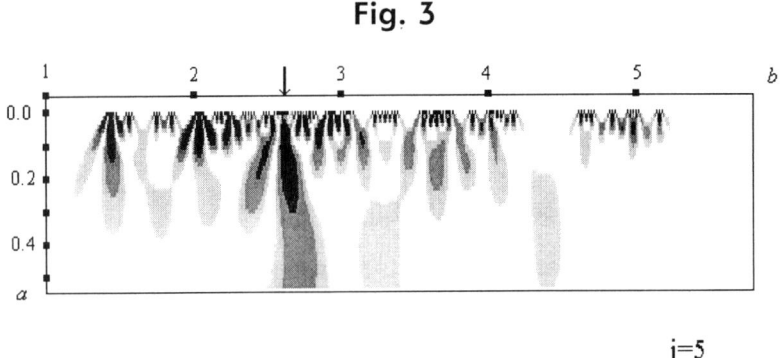

j=5

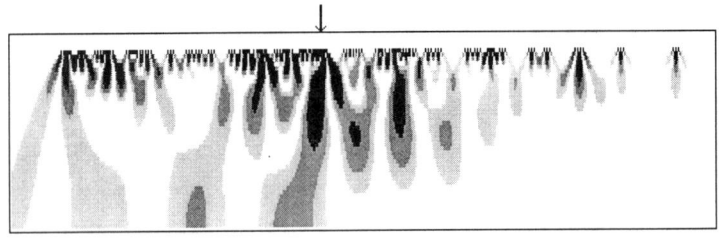

j=7

Fig. 4

i.e., at the location $x = b$ that efficiently takes into account the surrounding structure of any function $f(x)$ with the resolution given by the parameter a if the integral

$$W_{(a,b)} = |a|^{-1/2} \int f(x)\psi(\frac{x-b}{a})dx \qquad (2)$$

is calculated. The wavelet coefficients $W_{(a,b)}$ are the two-dimensional transform of the one-dimensional function $f(x)$. Their values squared $E_W = W^2$ are analogous to the power spectrum of Fourier transform. Namely, these wavelet power spectra were calculated [8] for the above event in different sectors of the target diagram. The 24 sectors with the azimuth width $\Delta\phi = \pi/12$ were drawn in the target diagram, and for each of them the functions $f_j(\eta)$ ($1 \leq j \leq 24$) were determined by experimental pseudorapidities of particles belonging to the sector chosen. The corresponding power spectra for two such sectors (with $j = 5$ and $j = 7$) are shown in Fig.4. The values of E_W as functions of the pseudorapidity location b and the resolution parameter a are shown by darker regions for larger E_W, i.e., Fig.4 presents equal height levels of E_W projected onto the (a, b) plane. The most dark streams indicated by arrows lie in the region of the ridge in Fig.3 (or within the ring in Fig.1). The comprehensive analysis of such spectra is in progress.

It has been shown [9] that in pattern recognition problems, the wavelet transform is about two orders of magnitude more efficient than the commonly used Fourier analysis.

At present time, we develop a techvision system for automation of the measuring complex which enables us to accelerate central Pb-Pb interaction treatment for event-by-event analysis.

REFERENCES

1. Apanasenko A.V. et al. JETP Lett. 30, 145(1979).
2. Apanasenko A.V. et al. Phys.Rep. C115 (1984) 153.
3. Japanese-Brazilian Collaboration,*14th ICRC, CKJ-Report*, 13 (1974) 1975.
4. Adamus M. et al.(NA22) Phys.Lett.B185 (1987), p.200.
   Dremin I.M.et al. Sov.J.Nucl.Phys. 52(1990) 840.
   Agababyan W.M. et al. Phys.Lett. B339(1996) 387.
5. Dremin I.M. JETP Lett.30 (1979) 140.
6. Dobrotin N.A. et al. Izv.A.S. USSR, ser.phys., v.44 (1980), p.494.
7. Burnet T.H. (JACEE) Phys.Rev.Lett. 50 (1983), p.2062.
8. Astafyeva N.M., Dremin I.M., Kotelnikov K.A., Pattern recognition in high multiplicity events, *Mod. Phys. Lett.* A (1997); hep-ex 9705003.
9. Abarbanel H.et al, Report on wavelets, *Cornell Mathematics*, 1994.

Coherent Photons and Pomerons in Heavy Ion Collisions

Spencer Klein[1] and Evan Scannapieco[1,2]

[1] *Lawrence Berkeley National Laboratory, Berkeley, CA, 94720, USA*
[2] *Department of Physics, University of California, Berkeley CA 94720-7304, USA.*

Abstract. Ultrarelativistic heavy ion beams carry large electromagnetic and strong absorptive fields, allowing exploration of a variety of physics. $\gamma\gamma$, γP, and PP interactions can probe a huge variety of couplings and final states. RHIC will be the first heavy ion accelerator energetic enough to produce hadronic final states via coherent couplings. Virtual photons from the nuclear EM fields can interact in $\gamma\gamma$ interactions, which can be exploited to study many particle spectroscopy and QCD topics. Because the photon flux scales as Z^2, $\gamma\gamma$ luminosities are large up to an energy of about $\gamma\hbar c/R \approx 3$ GeV/c. Photon-Pomeron interactions are sensitive to how different vector mesons, including the J/ψ, interact with nuclear matter. PP collisions rates are sensitive to the range of the Pomeron. Signals can be separated from backgrounds by using cuts on final state isolation (rapidity gaps) and $p_\perp$. We present Monte Carlo studies of different backgrounds, showing that representative signals can be extracted with good rates and signal to noise ratios.

I PHYSICS PROCESSES

When it is completed in 1999, the Relativistic Heavy Ion Collider [1] (RHIC) will be energetic enough to produce hadronic final states via $\gamma\gamma$, γP, and PP interactions that coherently couple to the nuclei as a whole. The electromagnetic field of the heavy nuclei can be considered as a flux, proportional to Z^2 of almost-real Weizsäcker-Williams virtual photons. The requirement that the photons couple to the entire nucleus limits the maximum photon energy to $\gamma\hbar c/R$, about 3 GeV/c for gold beams at RHIC. Thus, the maximum $\gamma\gamma$ energy is about 6 GeV; at energies of a few GeV, the $\gamma\gamma$ luminosity will be comparable to those of the next generation e^+e^- colliders.

$\gamma\gamma$ interactions can probe a wide variety of physics topics. Particle coupling to two photons is a measure of their internal charge; $q\bar{q}$ mesons couple strongly, but glueballs and mixed states ($q\bar{q}g$) should have much smaller couplings. Nonobservation in $\gamma\gamma$ collisions is therefore an important criteria for

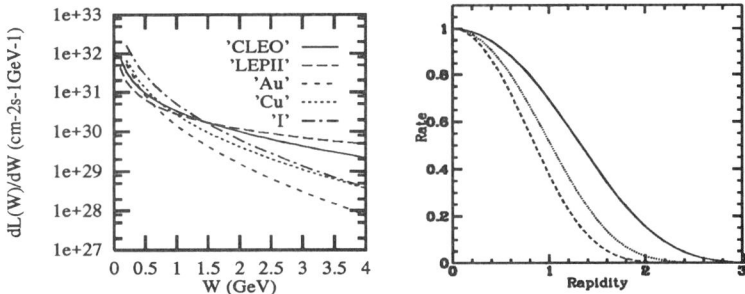

FIGURE 1. Left: Comparison of $\gamma\gamma$ luminosities at RHIC, for Au, I and Cu beams, with those of CESR (CLEO) and LEP II. Right: Rapidity distribution of produced particles for gold on gold collisions with W = 1 (solid), 2 (dotted) and 3 (dashed) GeV.

identification of glueball candidates [2]. $\gamma\gamma$ collisions can also probe spin 0 or 2 mesons with exotic quantum numbers. One interesting example is the near-threshold $\rho^0\rho^0$ resonance, produced in $\gamma\gamma$ collisions at e^+e^- colliders. According to the particle data book, "This process has not been explained by models in which only conventional resonances dominate" [3].

$\gamma\gamma$ collisions also produce large numbers of lepton and hadron pairs. Lepton pairs can be used to measure the $\gamma\gamma$ and relative hadronic luminosity, and to search for nonlinear QED effects due to the large coupling ($Z\alpha \approx 0.6$). Meson pairs can be used to measure form factors and for a variety of other QCD studies, and baryon pair production can test diquark based models [4].

Several authors have calculated the $\gamma\gamma$ luminosity produced by heavy ion colliders [5] [6] [7]. The $\gamma\gamma$ luminosity can be found by convoluting the photon fields of the two nuclei, and integrating over impact parameter b greater than twice the nuclear radius R_A. The latter criteria avoids events where hadronic collisions overshadow the $\gamma\gamma$ interaction, but cuts the usable luminosity by about 50%. Figure 1 compares the usable $\gamma\gamma$ luminosities for RHIC Au, Cu, and I beams to those of CLEO at CESR and LEP2. The lighter nuclei benefit from the higher AA luminosity and beam energy, and smaller nuclear radius, which more than compensate for the reduced Z. Because of the photon energy cutoff, final states are produced quite centrally, as shown in Fig. 1. Final states from $e^+e^- \to \gamma\gamma$ have a wider y distribution, and so, for a given setup, the experimental acceptance is lower.

Due to the nuclear form factor, the photons are almost real, with a maximum Q^2 given by the nuclear size, about $(30 \text{ MeV/c})^2$ for gold. This cutoff limits the perpendicular momentum of the photons, $p_\perp < \hbar c/R$; this helps separate coherent from incoherent interactions. This is illustrated in Fig 2. The Q^2 limit will slightly reduce the rate of pair production near threshold [8].

In addition to $\gamma\gamma$ collisions, the virtual photons can also couple to the Pomeron field of the other nucleus. The Pomeron can be thought of as representing the absorptive part of the nuclear potential. γP interactions using proton targets were studied extensively at HERA. RHIC can study these interactions in a nuclear environment. One reaction of interest is $\gamma P \to V$, where V is a vector meson. In the Vector Dominance Model, the photon can be considered to fluctuate into a spin 1 $q\bar{q}$ state, which then interacts with the nucleus. By studying production rates of various sized mesons, and varying the nuclear radius by changing the beams, the interactions between quark pairs and nuclei can be studied [9]. The kinematics are similar to $\gamma\gamma$ processes, so similar detection techniques can be used. RHIC will reach higher energies and luminosities than earlier NMC [10] and Fermilab E-665 [11] studies, producing 100,000's of exclusive ρ and ϕ mesons per year, large numbers of excited states, and the J/ψ. The latter is of special interest because the large quark mass may require perturbative treatment.

In addition to photon physics, double Pomeron interactions can be studied. Unobscured PP interactions can only occur in the impact parameter range $2R_A + 2R_P > b > 2R_A$, where R_P is the range of the Pomeron. The PP cross section is thus sensitive to the range of the Pomeron. As $\gamma\gamma$ and PP final states have similar kinematics, a statistical separation is required for this measurement. The relative rates will depend on A; PP couplings will dominate for small nuclei and $\gamma\gamma$ for large.

In many cases, the same final state can be produced through more than one intermediate state. Therefore, the possibility of interference exists. One place where it should occur is between the reactions $\gamma P \to V \to e^+e^-$ and $\gamma\gamma \to e^+e^-$ [12]. The $\gamma\gamma$ and γP interactions are out of phase, so the interference angle comes from the real part of the Pomeron and phase shift of the vector meson in the nuclear potential.

II EXPERIMENTAL FEASIBILITY

Measurements of coherent interactions are only possible if the signals can be separated from incoherent backgrounds [13]. This must be possible in both the final analysis and also at the trigger level; the latter appears to be the harder problem. The major backgrounds to be rejected are grazing nuclear collisions, photo-nuclear interactions, beam gas events, debris from upstream beam breakup, and cosmic ray muons; the latter two only affect triggering.

STAR (The Solenoidal Tracker at RHIC), is a general purpose large acceptance detector [14]. Time projection chambers track charged particles with pseudorapidity $|\eta| < 2$ and $2.5 < |\eta| < 4$. A silicon vertex tracker, time of flight system and TPC dE/dx help with particle identification. An electromagnetic calorimeter covers the range $-1 < \eta < 2$.

The coherent event trigger algorithms are based on requiring two or four tracks in the central TPC, with nothing else visible in the detector. Triggering uses a different set of detectors. The initial trigger selection uses scintillators

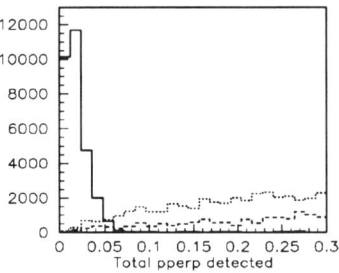

FIGURE 2. Comparison of $p_\perp$ between $\gamma\gamma$ and FRITIOF background events passing our cuts. The solid curve is for $\rho^0\rho^0$ production near threshold, the short dashes are beam-gas and the long dashes are peripheral nuclear backgrounds.

and wire chambers (the anode wires in the TPC endcaps, with a fast readout) surrounding the TPC to select events based on multiplicity and topology. Two to five possible tracks are required, with a reasonable topology. Timing may be used to help reject cosmic ray muons and beam gas. Higher level trigger algorithms use better hit location information to select events with two or four charged tracks, with tighter topological cuts. Then, the calorimeter and TPC tracking contribute, allowing cuts on neutral multiplicity, total charge, vertex position, and $p_\perp$. Monte Carlo studies have shown that these trigger algorithms have good acceptance for coherent events, and adequate background rejection.

We have simulated the $\gamma\gamma$ signals and backgrounds from grazing nuclear and beam gas [15]. We generated tables of $\gamma\gamma$ luminosity as a function of invariant mass and rapidity, and then created events based on these tables. $p_\perp$ spectra are included with a $1/R$ Gaussian form factor. Backgrounds were simulated with both the FRITIOF and Venus Monte Carlos; the $p_\perp$ spectra are compared in Fig. 2.

In order to study the effectiveness of different analysis techniques, we considered three sample analyses that are representative of a wide range of reactions; these analyses are listed in Table 1. The $f_2(1270)$ is a well understood $q\bar{q}$ meson, representative of a wide class of particles that decay to two charged particles. A spin 0 particle with the same mass and decay channel would have a slightly lower acceptance. $\rho^0\rho^0 \to \pi^+\pi^-\pi^+\pi^-$ near threshold (1.5 GeV$< M_{\rho\rho} <$ 1.6 GeV) is both a very interesting physics signal, and also a good benchmark for medium mass processes decaying to four charged particles. Finally, the 2960 MeV $J^{PC} = 0^{-+}$ $c\bar{c}$ resonance η_c was chosen as a real challenge; the luminosities are falling and the branching ratios are small.

For each analysis, we developed a set of cuts based on the required charged and neutral multiplicity and event topology. We required $p_\perp < 100$ MeV, and an appropriate invariant mass and found the rates and backgrounds given in Table 1. Although the FRITIOF and Venus background predictions are

TABLE 1. Rates and backgrounds for $\gamma\gamma$ events for gold on gold collisions at RHIC for 3 sample analyses. The $\rho^0\rho^0$ events were near threshold, with invariant masses between 1.5 and 1.6 GeV/c^2. The quantities in parenthesis assume particle identification by dE/dx and TOF.

Channel	Efficiency	Detected Events/Yr	FRITIOF Background	Venus Background
$f_2(1270) \to \pi^+\pi^-$	85%	920,000	53,000	100,000
$\rho^0\rho^0 \to \pi^+\pi^-\pi^+\pi^-$	38%	16,000	3,500	1,400
$\eta_c \to K^{*0}K^-\pi^+$ (w/ PID)	57%	70	210 (8)	510 (20)

different, this analysis shows that $f_2(1270) \to \pi^+\pi^-$ and threshold $\gamma\gamma \to \rho^0\rho^0 \to \pi^+\pi^-\pi^+\pi^-$ reactions should be clearly separable from backgrounds, while the $\eta_c \to K^{*0}K^-\pi^+$ may be accessible with particle identification.

We would like to thank our colleagues in the STAR collaboration. This work was supported by the DOE under contract DE-AC-03-76SF00098. E.S. has been partially supported by an NSF fellowship.

REFERENCES

1. *Conceptual Design of the Relativistic Heavy Ion Collider*, BNL-52195, May 1989, Brookhaven National Laboratory.
2. Paar, H., to appear in the Proceedings of *Photon '97*.
3. Particle Data Group, Barnett, R. M. *et al.*, Phys. Rev. **D54**, 1 (1996); minireview on pg. 557-9.
4. Miller, D. J., in *Lepton and Photon Interactions: XVI Intl. Symposium*, Ithaca, 1993, ed. P. Drell and D. Rubin.
5. Cahn, R. N. and Jackson, J. D., Phys. Rev. **D42**, 3690 (1990).
6. Baur, G. and Ferreira Filho, F., Nucl. Phys. **A518**, 786 (1990).
7. Hencken, K., Trautmann, D., and Baur G., Z. Phys. **C68**, 473 (1995).
8. Baur, G. and Baron, N., Nucl. Phys. **A561**, 628 (1993).
9. Brodsky, S., Kinoshinta, T., and Terazawa, H., Phys. Rev. **D4**, 1532 (1971).
10. Ashman, J. *et al.*, Z. Phys. **C39**, 169 (1988).
11. Baskin, B., these proceedings.
12. Leith, D. in *Electromagnetic Interactions of Hadrons VI*, ed. A. Donnachie and G. Shaw (Plenum Press, New York,1978).
13. Klein, S. R., in *Photon 95*, ed. D. J. Miller, S. L. Cartwright and V. Khoze (World Scientific, Singapore, 1996).
14. STAR Collaboration, *STAR Conceptual Design Report*, LBL-PUB-5347, 1992.
15. Klein, S. and Scannapieco, E., STAR Note 243, Feb. 1995, available at http://www.rsgi01.rhic.bnl.gov/star/starlib/doc/www/sno/ice/sn0243.html.

Future Perspectives at RHIC

Thomas S. Ullrich

Physics Department, Yale University
New Haven, CT 06520-8124

Abstract. In the light of the start of operations at the Relativistic Heavy Ion Collider (RHIC) at Brookhaven National Laboratory in 1999 and the many ongoing studies and simulation in this context I will discuss a few of the many physics perspectives at this facility from my personal point of view.

INTRODUCTION

The aim of high energy heavy-ion physics is the study of strongly interacting matter at extreme energy densities. Statistical QCD predicts that, at sufficiently high density, there will be a transition from hadronic matter to a plasma of deconfined quarks and gluons [1] – a transition which in the early universe took place in the inverse direction some 10^{-5} s after the Big Bang and which might still play a role today in the core of collapsing neutron stars [2].

The study of the phase diagram of nuclear matter, using methods and concepts from both nuclear and high-energy physics, constitutes a new and interdisciplinary approach to investigate matter and its interactions. It is of interest to explore and test QCD on its natural scale (Λ_{QCD}) and to address the fundamental questions of confinement and chiral-symmetry breaking. Moreover, it is of general relevance in understanding the dynamic nature of phase transitions involving elementary quantum fields, as the QCD phase transition is the only one accessible to laboratory experiments.

Already the early exploratory program at the Brookhaven National Laboratory Alternating Gradient Synchrotron (AGS) and the CERN Super Proton Synchrotron (SPS) has established the feasibility of high energy ion-ion experiments with their abundant particle production. It has shown that high energy densities of about 2.5 GeV/fm^3 can indeed be obtained in these reactions and has produced evidence for the onset of new collective phenomena. Some characteristic features of these reactions (e.g. momentum distributions and particle ratios) are already surprisingly close to the ones expected for a macroscopic system of hadronic matter heated to a temperature of close to 200 MeV in the initial phase of the collision. The hyperonic yield ratios right after hadronization appear to present a first glance at the position in

T, μ_B of the parton to hadron transition, and phase boundary [3]. Even more encouraging is the fact that recent data resulting from the CERN-SPS lead beams program show tentative evidence for the existence of a partonic phase from J/Ψ production data, apparently setting in with central Pb+Pb collisions that reach beyond the critical energy densities [4].

The Relativistic Heavy-Ion Collider (RHIC), presently under construction at Brookhaven National Laboratory (BNL) in New York, is a dedicated heavy-ion collider planned for experiments starting in 1999. RHIC will accelerate and collide a wide variety of ions up to Au at center-of-mass energies of $\sqrt{s} = 200$ GeV. It is thus the natural continuation of the present heavy-ion program exploring considerably higher energy densities than the one obtained with currently available accelerators.

One of the difficulties in interpreting existing data has been the lack of a well established and generally accepted theoretical framework for calculating the observables obtained in the experiments. A major difference between these earlier studies and the experiments to be carried out at RHIC is that at the the RHIC energy scale present theoretical estimates are that a large fraction of the energy transfer from the projectile frame to the midrapidity region may be calculated using Perturbative Quantum Chromodynamics (PQCD). This makes available a well established model for comparison with the data to be obtained at RHIC.

THE RHIC PROJECT

The scope of the RHIC project is to construct and operate a colliding beam facility which allows studies of phenomena in ultrarelativistic heavy-ion collisions [5]. The collider is located in the northwest section of the BNL site. Its construction began in 1991 and the completion of the complex, including detectors, is scheduled for the third quarter of 1999.

The RHIC detector program consists of two complementary major detector systems PHENIX and STAR and two smaller-scale experiments BRAHMS and PHOBOS. These detectors collectively cover most of the predicted signatures. At present, more than 700 physicists from 80 institutions internationally are involved in this project.

The Collider

The collider, which consists of two concentric rings of superconducting magnets, is constructed in an existing ring tunnel of ∼ 3.8 km circumference. It offers an extraordinary combination of energy, luminosity and polarization. The major performance parameters are summarized in Fig. 1.

RHIC is able to accelerate and store counter-rotating beams of ions ranging from those of gold to protons at the top energy of 100 GeV/nucleon for gold and 250 GeV for protons. The stored beam lifetime for gold in the energy range of 30 to 100

GeV/nucleon is expected to be approximately 10 hours. The layout of the tunnel

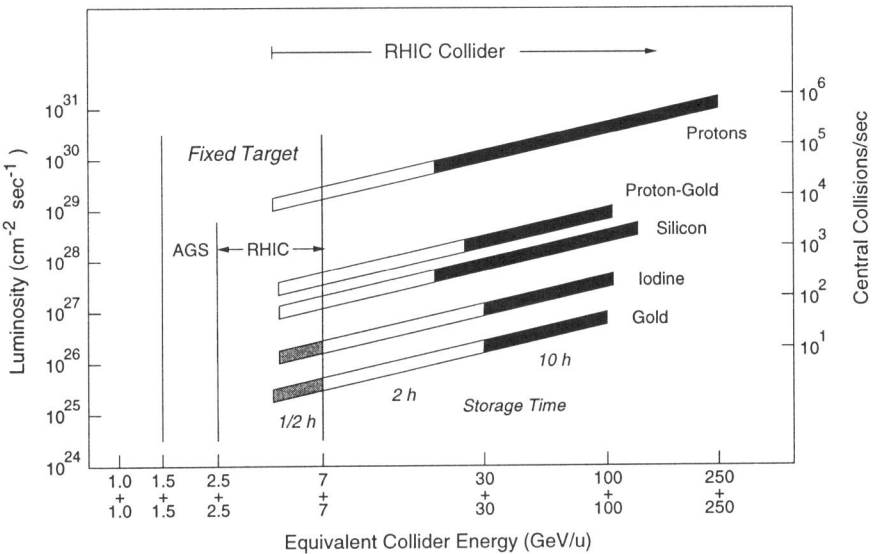

FIGURE 1. RHIC performance parameters.

and the magnet configuration allow the two rings to intersect at six locations along their circumference. The top kinetic energy will be 100+100 GeV/nucleon for gold ions. The operational momentum increase with the charge-to-mass ratio, resulting in kinetic energy of 125 GeV/nucleon for lighter ions and 250 GeV for protons. The collider will be able to operate a wide range from injection to top energies. The collider is designed for a Au-Au luminosity of about 2×10^{26} cm^{-2} s^{-1} at top energy. The luminosity is energy dependent and decreases approximately proportionally as the operating energy decreases. For lighter ions it will be significantly higher, with $\sim 1 \times 10^{31}$ cm^{-2} s^{-1} for pp collisions. The collider will allow collisions of beams of equal ion species all the way down to pp and of unequal species such as protons on gold ions.

Another unique aspect of RHIC is the ability to collide beams of polarized protons (70-80%) which allows the measurement of the spin structure functions for the sea quarks and gluons. In my talk, however, I will solely concentrate on the physics related with the relativistic heavy-ion program.

Experiments at RHIC

The collision environment at the RHIC facility will be complex and presents many new challenges not faced before in heavy-ion physics. Due to the large abundance of charged particles produced in these collisions ($dN_{ch}/d\eta \approx 1000$) the requirements on the detectors are strikingly different from those at existing colliders. Each of the RHIC experiments, BRAHMS, PHENIX, PHOBOS, and STAR have therefore developed unique capabilities to cope with these extreme conditions. In the following I will briefly introduce the four experiments under construction in alphabetical order.

BRAHMS

The **BR**oad **R**ange **H**adron **M**agnetic **S**pectrometer BRAHMS [6] experiment is designed to measure and identify charged hadrons ($\pi^\pm$, $K^\pm$, $(\bar{p}\atop p)$) over a wide range of rapidity and transverse momentum for all beams and energies available at RHIC. Because the conditions and thus the detector requirements at midrapidity and forward angles are different, the experiment uses two movable spectrometers for the two regions. As shown in Fig. 2, there is a midrapidity spectrometer to

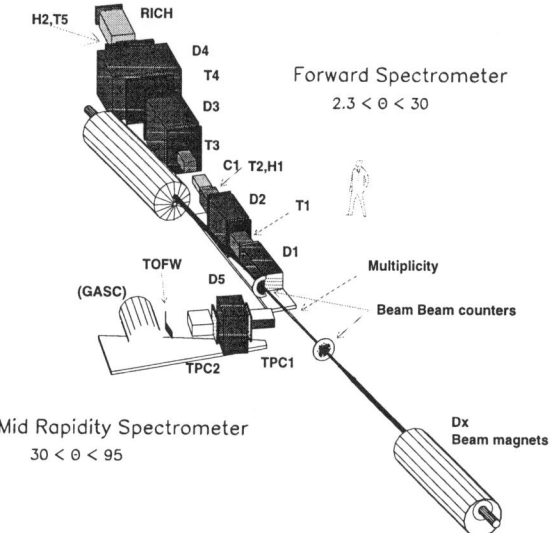

FIGURE 2. Layout of the BRAHMS spectrometers.

cover the the pseudorapidity range $0 \leq \eta \leq 1.3$ and a forward spectrometer to cover $1.3 \leq \eta \leq 4.0$. The latter employs four dipole magnets D1-D4, three time projection chambers (TPC) T1-T3, followed by drift chambers T4, T5. Particle identification is achieved with time-of-flight hodoscopes H1, H2, a threshold Cherenkov counter, and one ring-imaging Cherenkov counter (RICH). The solid angle acceptance of the forward arm is 0.8 mstr. The midrapidity spectrometer has been designed for charged particle measurements for $p \leq 5$ GeV/c. The spectrometer has two TPCs for tracking, a magnet D5 for momentum measurement, and a time-of-flight wall and segmented gas Cherenkov counter (GASC) for particle identification. It has a solid angle acceptance of 7 mstr. A set of beam counters and a silicon multiplicity array provide the experiment with trigger information and vertex determination.

PHENIX

The PHENIX detector [7] is designed for the measurement of leptons, photons and – to some extent – hadrons. It consists of four instrumented spectrometers: two central spectrometers and two forward muon arms of which one will be added as an future upgrade. The schematic layout of the detector is shown in Fig. 3.

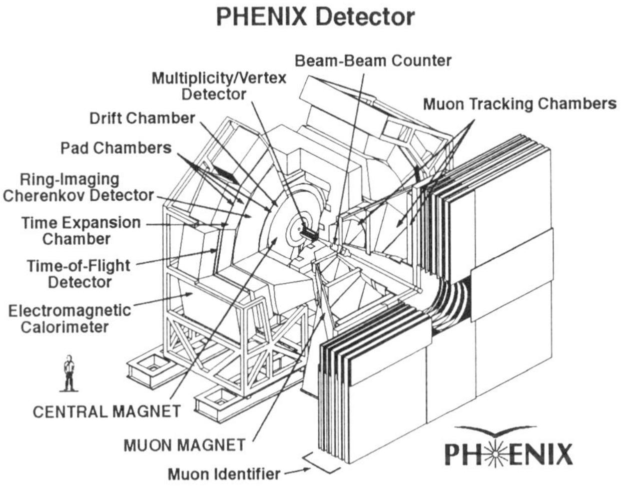

FIGURE 3. Schematic layout of the PHENIX detector.

The central detector is made of an axial field magnet and two almost identical arms placed left and right of the magnet. Each arm covers ± 0.35 units of pseudo-rapidity and $90°$ in ϕ. Viewing the detector radially from inside to out each central arm includes a silicon multiplicity vertex detector, an inner tracking system of two sets of drift chambers, and a pad chamber. It is followed by a RICH detector which provides electron/hadron separation up to 4 GeV/c. Outside of the RICH detector is an outer tracking system made of a time expansion chamber (TEC) sandwiched between two pad chambers. The last layer of the detector is an electromagnetic calorimeter (EMC) for photon measurement and for additional electron/hadron separation. A smaller part of one arm is equipped with high resolution TOF counters for hadron identification.

Each muon arm contains detector elements to perform charged particle tracking, momentum measurement, and muon identification. The steel yoke of the central magnet is used as the primary hadron absorber. Those particles which emerge from the pole tip enter the muon arm. The muon arm momentum measurement uses the radial field produced by the muon magnet. The trajectories are measured in multi-layer drift chambers that are arranged in three stations. Downstream of the muon magnet, muons are distinguished from pions and other shower products by means of a muon identifier, consisting of concrete absorber walls interleaved with planes of streamer tubes. The coverage of the muon detection extends from $\eta = 1.1 - 2.4$ ($10° - 35°$).

PHOBOS

The PHOBOS [8] detector is designed to detect as many of the produced particles as possible and to allow a momentum measurement down to very low $p_\perp$. The PHOBOS detector consists of two parts: a multiplicity detector covering almost the entire pseudorapidity range of the produced particles and a two arm spectrometer at mid-rapidity. Fig. 4 shows the detector, including the spectrometer arms, the multiplicity and vertex array, and the lower half of the magnet.

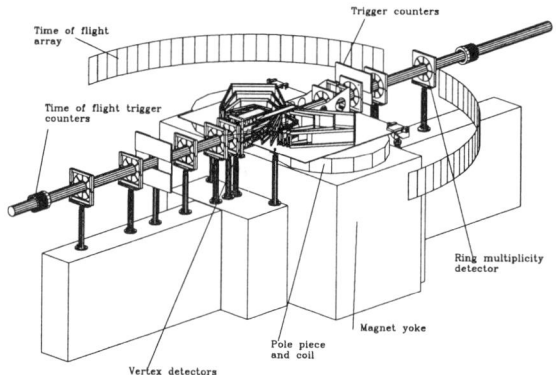

FIGURE 4. Schematic view of the PHOBOS detector.

One important aspect of the design is that all detectors are produced using a common technology, namely as silicon-pad or strip detectors. The multiplicity detector covers the range $-5.4 < \eta < 5.4$, measuring total charged multiplicity $dN_{ch}/d\eta$ over almost the entire phase space. For approximately 1% of the produced particles, information on momentum and particle identification will be provided by a two arm spectrometer located on either side of the interaction volume. Each arm covers about 0.4 rad in azimuth and one unit of pseudorapidity in the range $0 < \eta < 2$, depending on the interaction vertex, allowing the measurement of $p_\perp$ down to 40 MeV/c. Both detectors are capable of handling the 600 Hz minimum bias rate expected for all collisions at the nominal luminosity.

STAR

The **S**olenoidal **T**racker **A**t **R**HIC (STAR) [9] is a large acceptance detector capable of tracking charged particles and measuring their momenta in the expected high multiplicity environment. It is also designed for the measurement and correlations of global observables on an event-by-event basis and the study of hard parton scattering processes. A cutaway view of the detector is shown in Fig. 5.

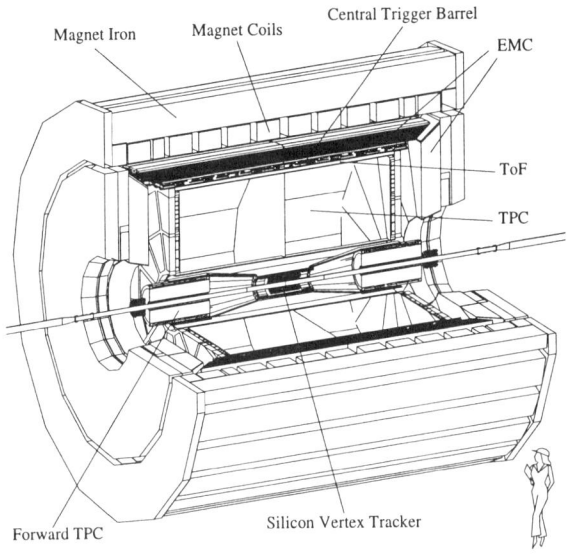

FIGURE 5. Schematic view of the STAR detector.

Viewing the detector from inside to out the first element is the silicon vertex tracker which consists of 3 layers of silicon drift detectors (SDD) arranged in cylindrical barrels. It allows the measurement of the primary vertex with high accuracy and the reconstruction of secondary vertices. The baseline detector is a cylindrical TPC located inside of a solenoidal magnet with a 0.5 T field covering 4 units of pseudorapidity ($|\eta| < 2$) with full azimuthal coverage. The central trigger barrel (CTB), consisting of 240 scintillator slats, lies just outside of the TPC outer field cage and provides fast charged-particle multiplicity information. The CTB is followed by a highly segmented TOF array to increase the π/K and p/K separation. The last layer of the detector is a Pb-scintillator sampling electromagnetic calorimeter which will be used to trigger on transverse energy and measure jets, photons, and electrons. It consists of a barrel ($|\eta| < 2, \Delta\phi = 2\pi$) and one endcap, both of which contain shower maximum detectors. Two proposed additional TPCs located in the forward regions will extend the STAR acceptance to near beam rapidities.

FUTURE PERSPECTIVES

The RHIC project offers an exciting discovery potential which will substantially influence and improve the understanding of the entire field. However, the collision environment at this new facility is extraordinarily complex, and presents many new challenges not faced before in nuclear or particle physics. This concerns not only the measurement of the proposed signatures but also their interpretation and theoretical understanding.

Extrapolating from present results, all parameters relevant to the formation of the QGP will be favourable: the energy density, the size and lifetime of the system, and relaxation times should all improve by a large factor compared to Pb-Pb collisions at the SPS. It should then be possible to obtain energy densities well above the confinement threshold, and to probe the QGP in its asymptotically free 'ideal gas' form. However, one must admit that a proper understanding of heavy-ion collisions involves more than the knowledge of a phase transition because in an actual collision, the hot phase is expected to be produced in a limited volume and last for a short time. Further complications at RHIC arise from the fact that many of the exciting physic processes studied at the AGS and SPS now turn into sources of background which aggravate the search for new signatures. Understanding RHIC physics therefore implies also a deep understanding of the results of the present heavy-ion programs.

In the following I will try to use the experience gained from the relativistic heavy ion experiments at the AGS and SPS to provide an overview of some of the techniques and strategies that will be employed in the search for the QGP at RHIC. Its is obvious that I cannot cover all aspects and details of every proposed signature and probe. This would certainly exceed the scope of this paper (and my knowledge). I rather decided to focus on a few topics which – to some extent – stand as examples for many other studies.

The First Months

It is evident that the four experiments will spend a large fraction of the first few months of RHIC running with the setup and the calibration of the individual detector components; software and algorithms will have to be adjusted in order to cope with the "real" experimental data. Many of the proposed signatures of a QGP, however, require complex and difficult analysis efforts including a profound understanding of the detector response. A valid question therfore is what physics results can one expect from the first months of RHIC running?

Considering the history of the heavy-ion programs at BNL and CERN it is very likely that the very first measurements reported from RHIC will be based on hadronic observables. All RHIC experiments already offer sufficient particle identification capabilities with their baseline detectors to study $p_\perp$ and rapidity distributions of the most abundant particles (kaons, pions and protons) in order to characterize the freeze-out properties of the final state and to determine the energy and baryon densities achieved in these collisions.

Regardless of whether or not a plasma can be detected, it is important that the basic conditions existing in RHIC collisions are well described. This is of fundamental interest in its own right. Hadronic measurements lead to complete information of the three-dimensional phase space of the state of nuclear matter (as a function of T, μ_B, and μ_s). Although it is generally agreed that no single hadronic parameter will serve as a "smoking gun" for the QGP, correlations between hadronic observables might yield conclusive evidence for a phase transition to a QGP. It is the input of these primary studies which will allow a much more accurate estimate of what can be expected in the following months and years of RHIC physics.

Proven Probes

As pointed out in several talks at this conference the suppression of J/Ψ production is one of the most promising signatures for a QGP created in nuclear collisions.

Quarkonium ground states (J/Ψ, Υ) are small and tightly bound resonances of heavy quarks. The J/Ψ has a radius much smaller then the normal hadronic scale and its binding energy of 0.6 GeV is much larger than $\Lambda_{QCD} \approx 0.2$ GeV. It therefore requires hard gluons to resolve and dissociate a J/Ψ. Confined matter for temperatures up to about 600 MeV does not contain sufficiently hard gluons to resolve the quark structure of the small J/Ψ, whereas deconfined matter for T > 200 MeV easily can [4]. J/Ψ suppression thus provides an unambiguous test for color deconfinement. Nuclear collisions up to central S-U interactions at 200 GeV/nucleon show only "normal" pre-resonance absorption in nuclear matter; in contrast, 160 GeV/nucleon Pb-Pb collisions lead to a further strong "anomalous" J/Ψ suppression, which can be interpreted as the onset of color deconfinement, though not necessary in an equilibrated medium. Up to now all data related to J/Ψ suppression in nucleus-nucleus collisions were essentially taken by the NA38/NA50 experiment at the CERN-SPS.

At RHIC the initial conditions are significantly more favourable to create a partonic state. The energy density is believed to considerably exceed 3 GeV/fm^3 and thus the mechanism to breakup J/Ψ should be much stronger.

The PHENIX experiment is capable of measuring the J/Ψ, Ψ' resonances through their leptonic decay channels, i.e. J/Ψ → e^+e^- in the central detectors and J/Ψ → $\mu^+\mu^-$ in the two muon arms. The particle identification (PID) in the detectors allow for electron-hadron separation at the level of $< 10^{-4}$ from a few hundred MeV to about 4 GeV. Assuming no suppression the estimated yield of J/Ψ's detected via the dielectron channel in minimum bias Au-Au interactions is ∼70k with an additional ∼1.4k in the Ψ' peak in one year of RHIC running. Similar estimates of the dimuon rate yield ∼1.1M J/Ψ's and ∼15k Ψ''s [10]. The signal-to-background ratio can be expected to be very large due to the decreasing combinatorial background at higher masses and the excellent mass resolution of < 1% for the J/Ψ. For the STAR detector the situation is somehow more difficult since its primary design aims for the measurement of hadronic probes and jets and it runs at a rather moderate data taking rate of 1 Hz. The detector, however, offers a large acceptance and sufficient PID from the TPC and the EMC for electrons and should thus be able to extract a statistically significant signal as well [11].

There is, however, one aspect which will complicate these measurements and their interpretation at RHIC which I want to discuss in briefly: Suppression of the J/Ψ and Ψ' resonance is usually expressed as the ratio of the integral yield in the resonance peak to the Drell-Yan continuum which, at lower center-of-mass energies, dominates the invariant mass region above $m_{\Psi'}$. It is this ratio which is then studied as a function of the transverse energy $E_\perp$ or of the average path length of the $c\bar{c}$ pair in nuclear matter L. The reason to use a ratio instead of the absolute cross-sections is many-fold: *(i)* the majority of systematic errors cancel out since both are hard processes with similar underlying kinematics and therefore suffer similar acceptance and efficiency losses and *(ii)* volume effects are minimized because both processes show – to some extent – the same nuclear dependence. A detailed study of absolute cross-sections would require a complicated and model dependent correction for mass number-, volume-, and/or impact parameter dependence such as to reflect the variation of energy density as well as a correction for efficiency and acceptance.

However, at RHIC energies the continuum above the J/Ψ and Ψ' resonances is not dominated by Drell-Yan pairs but by lepton-pairs originating from correlated charm decays; a process of minor importance at SPS energies. This is illustrated schematically in Fig. 6[1] for central Au-Au collisions at RHIC. As can be seen the Drell-Yan continuum dominates the spectrum at masses above 7 GeV/c^2; a region where statistical power is limited. For a proper interpretation of the J/Ψ production and suppression mechanism it is therefore of paramount importance to explicitly measure and study total charm production. The measurement of charm, however, is certainly one of the most challenging tasks at RHIC. It can be achieved in PHENIX by measuring the e-μ coincidences from the electron and muon arm respectively. Another, although

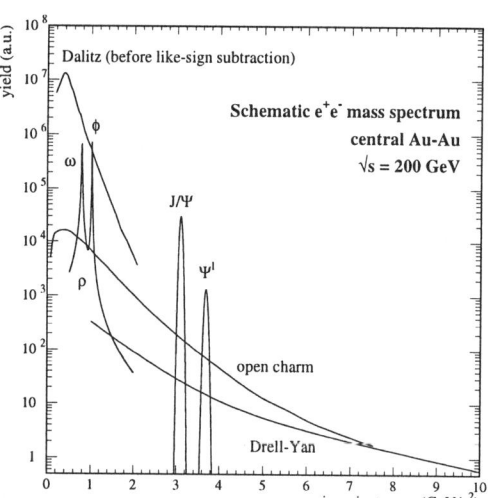

FIGURE 6. Schematic dielectron mass spectrum for central Au-Au collisions at $\sqrt{s} = 200$ GeV.

---

[1] This graph is derived from a plot in [10] but is strongly schematized to emphasize the importance of open charm.

even more challenging approach, is the reconstruction of D mesons in the STAR detector through the hadronic decay channels $K\pi$ and $K\pi\pi$. However, the large combinatorial background and the short decay length of the D mesons ($c\tau \simeq 300\mu$m) makes this measurement extremely difficult.

For RHIC this is one example of a measurement which, although obviously more favourable in terms of yields and accuracy, encounters new difficulties which arise in this new energy regime. *Todays signal is tomorrows background.*

New Probes

With the lower energy beams used in heavy ion studies at AGS and CERN the most basic input to the various models has been the flux and energy of the projectile nuclei. At RHIC energies ($\sqrt{s} = 200$) the basic inputs to the model are the flux and momentum distributions of the quarks and gluons in the colliding nuclei. This makes available a well established model for comparison with data to be obtained at RHIC: *perturbative QCD* (PQCD). There are predictions of two nuclear effects which might play a dominant role in Au-Au collisions at RHIC energies, namely gluon shadowing and jet quenching. From deep-inelastic scattering, it is well known that the quark structure functions with small x are depleted in a nucleus relative to a free nucleon. This phenomena is known as *shadowing* and applies for the quarks as well as the gluons which play such a vital role at RHIC. For the latter no direct information is available although it is plausible to expect that the gluon distributions are at least modified in a similar fashion. This, however, should be of interest at RHIC because it could significantly influence the initial conditions in such reactions. In addition, it is of fundamental interest in its own right as it pertains to the nuclear structure at the parton level not accessible via deep-inelastic reactions. Jet quenching, on the other hand, is of interest because it provides information on the final state interaction processes that may lead to partial thermal and chemical equilibrium in the produced dense partonic system. Jet quenching results from the energy loss of high-$p_\perp$ partons as they traverse the dense matter. PQCD estimates indicate that the induced gluon bremsstrahlung may dominate the energy loss mechanism which then depends sensitively on the Debye screening scale of the medium. Jet quenching therefore may vary significantly in the vicinity of the quark-gluon plasma transition.

At RHIC the study of jet quenching is a rather complex program since it involves systematic measurements in several colliding systems in order to deconvolute the two mechanism: quenching and shadowing. The first step is to measure the gluon structure function in pp collisions. This can be studied in the STAR experiment by measuring two-body kinematics in hard qg scattering, namely a γ and a jet, both with roughly equal energies and emitted back-to-back in the azimuthal angle. With the proposed installation of a single endcap EMC, x values down to 0.03 could be achieved. This is demonstrated in Fig. 7 taken from ref. [12]. The next step is to measure the gluon structure function for nucleons bound in a nucleus. One does this using p-Au running at RHIC. A comparison of pp with p-Au gives the measure

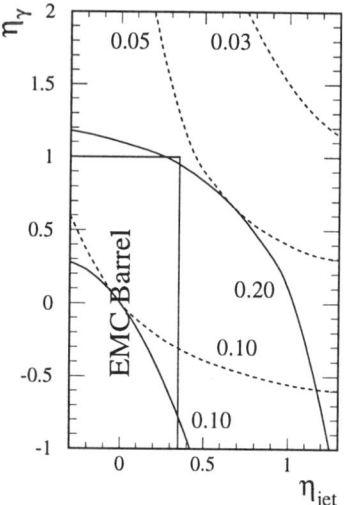

FIGURE 7. Acceptance of the STAR EMC. The dashed lines give the range for the gluon x values, the solid for the quarks. The small box shows the acceptance of the barrel only (no endcap).

for nuclear shadowing.

Given the gluon structure function in p-Au, one now takes advantage of the available calculational tool, perturbative QCD, to calculate how the inclusive high-$p_\perp$ spectrum should look like for Au-Au collisions and compares this with data. It is the difference of the two that might teach us about jet quenching as a possible signal of a QGP. It should be stressed that, due to two concurrent nuclear effects, a simple comparison of pp with Au-Au (e.g. $(d\sigma/dp_\perp)_{pp}/(d\sigma/dp_\perp)_{AA}$) does not provide a decisive answer.

This measurement is unique for RHIC and represents a new and interesting intersection with high-energy physics never accessible before. However, as already stressed above, this systematic study consists of a sequence of efforts both on the experimental and theoretical side. It certainly can be viewed as one of the challenging long term projects at RHIC.

Light Vector Mesons

As early as 1985 it has been suggested [13] to use the ϕ-meson as indicator of a quark-gluon plasma. In such a plasma one would expect a marked enhancement of $s\bar{s}$ pairs, which would increase the production of ϕ mesons. Due to the small ϕN cross section, the ϕ could escape the reaction volume without rescattering, carrying with it information about the initial conditions of the reaction.

At the CERN-SPS, two experiments have succeeded in measuring ϕ mesons in heavy ion collisions: HELIOS/NA34 and NA38/NA50. Both report excess yields over that expected from a simple extrapolation from pp to nucleus-nucleus collisions. None, however, had the mass resolution necessary to clearly resolve the ψ from ρ/ω resonances; typical values achieved so far are $\delta m/m = 8 - 10\%$ at the ϕ mass.

In addition to the question of the pure production cross sections there is an increasing interest in the study of the width and positions of the ρ, ω and ϕ peaks. They are sensitive to medium induced changes of the hadronic mass spectrum, especially to the possible drop of vector meson masses as a precursor of chiral symmetry restoration. In the absence of high baryon densities, modifications of the peak positions are predicted to be small except in the immediate vicinity of the phase transition, whereas the increase in the width of the ϕ-meson due to collision broadening could be substantial [14].

The interest in the subject has escalated as a results of low-mass dilepton measurements reported by the CERES/NA45 and HELIOS/NA34 collaboration [15,16]. Their data in the low-mass region have been quantitatively explained by introducing a decrease of meson masses – specifically that of the ρ meson [17]. However, the uncertainties in the data are too large to convincingly rule out other more conventional explanations. This makes it imperative to continue studies with the particular goal of achieving better mass-resolution and higher yields.

All four RHIC experiments will measure the ϕ resonance in the hadronic channel ($\phi \rightarrow K^+K^-$), PHENIX and possibly STAR in the leptonic ($\phi \rightarrow \ell^+\ell^-$) as well. The leptonic decay has the advantage that the produced leptons interact very little with the hadronic medium, providing the most direct measure of what happens to ϕs that decay early in the collision process. The hadronic channel has the advantage of *(i)* a higher branching ratio, *(ii)* a low Q value making the kaon momenta very sensitive to changes in the ϕ mass, and *(iii)* the fact that kaons are easy to detect. To give an example, Fig. 8 shows a simulated invariant mass distribution for the ϕ meson after background subtraction in PHOBOS [18]. The underlaying statistics can be achieved within 19 hours of data taking. Note the excellent resolution on the mass and width of 0.2 MeV and 0.5 MeV, respectively. For a full years run, this precision even allows a plot of the measured M and Γ as a function of other variables such as centrality or $p_\perp$.

However, the ϕ is not the best suited probe to study in-medium effects. The reason for this is its relatively long lifetime of $c\tau \simeq 46$ fm which has to be compared with the lifetime and size of the hot fireball of < 10 fm where these effects are strongest[2]. Hence most of the ϕs will decay outside the medium or – for low-$p_\perp$ ϕs – decay very late in the collision where the system already has cooled down and therefore only a small fraction will be sensitive to the effects in question. However, because of the good resolution in M and Γ one still might be sensitive to small deviations in the spectral shape.

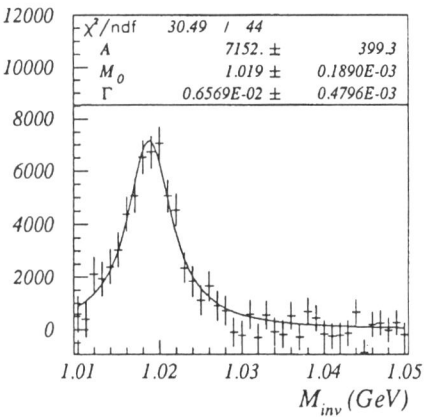

FIGURE 8. Simulated invariant mass distribution for the ϕ meson in PHOBOS measured in the K^+K^- channel after background subtraction (after ~ 1 day running).

---

[2] A simple estimate of the (HBT) freeze-out radius yields values of R $\approx$ 10 fm. The radius of the actual hot and dense zone will certainly be smaller.

Probably the most suitable meson to study in-medium effects is the ρ-meson with a lifetime of $c\tau \simeq 1.3$ fm. Unfortunately the study of the ρ via the $\ell^+\ell^-$ channel will be rather problematic because of the high combinatorial background. Even though the expected yield in a year of RHIC running is 110k, PHENIX – the experiment most suited for this measurement – estimates a signal-to-background ratio of 1:35 [10]. This is mainly because of the large width of the ρ ($\Gamma \simeq 150$ MeV) and the missing handles for rejecting π^0-Dalitz decays and photon conversions. Studies to improve the signal-to-background ratio are underway.

SUMMARY

RHIC is the continuation of an ongoing effort to study nuclear matter under extreme conditions. The collider is designed to cover a wide range of energies and ion masses and offers an excellent environment for new physics in the field of heavy-ion physics and beyond. The design of the four detectors is well matched to the physics goals of the project although some measurements might require future detector upgrades. Due to the increased center-of-mass energies new signatures can be uniquely studied at RHIC which cannot be observed at present accelerators. On the other hand, it must also be realized that the understanding of some, already successfully studied probes, will become much more difficult due to new background sources produced at these higher energies. Many fundamental questions can be answered in the first months of RHIC running, although it will certainly take some time to realize and take advantage of the full scientific potential of the RHIC program.

Acknowledgements

This work was supported by the Alexander von Humboldt Foundation (Lynen program) and the U.S. Department of Energy under contract DE-FG02-91ER40609. I would like to thank T.J. Hallman, W.B. Christie, M.D. Baker, and B.V. Jacak for the information provided. I am indebted to J.W. Harris for valuable comments and discussions.

REFERENCES

1. E. Laermann, Nucl. Phys. **A610** (1996) 1c-12c
2. G. Baym, Nucl. Phys. **A590** (1995) 233c-248c
3. R. Stock, *High Energy Nuclear Interactions and Heavy Ion Collisions*, see proceedings of ICHEP'96
4. D. Kharzeev, C. Lourenco, M. Nardi, H. Satz, Z. Phys. C **74** (1997) 307-318
5. See the RHIC Design Report, January 1996
6. BRAHMS Conceptual Design Report, October 1994
7. PHENIX Conceptual Design Report, January 1993

8. PHOBOS Conceptual Design Report, April 1994
9. STAR Conceptual Design Report, June 1992
10. Y. Akiba, *Physics with Collider Detectors at RHIC and the LHC*, Proceedings of the Pre-Conference Workshop Quark Matter '95 (1995) 131-141
11. T.J. LeCompte, STAR Note SN0287 (1997)
12. W.B. Christie, *Hard Scattering of Partons as Probe of Collisions at RHIC using the STAR Detector System*, Proceedings of the Pre-Conference Workshop Quark Matter '95 (1995) 63-71
13. A. Shor, Phys. Rev. Lett. 54 (1985) 1122
14. T. Hatsuda, Y. Koike and S.H. Lee, Nucl. Phys. **B394** (1993) 221
15. Th. Ullrich for the CERES collaboration, Nucl. Phys. **A610** (1996) 317c-330c
16. M. Masera for the HELIOS collaboration, Nucl. Phys. **A590** (1995) 93c
17. G.Q. Li, C.M. Ko and G.E. Brown, Phys. Rev. Lett. **75** (1995) 4007
18. M.D. Baker, *PHOBOS Physics Capabilities*, Proceedings of the Pre-Conference Workshop Quark Matter '95 (1995) 63-71

Facilities and Detectors

The BNL AGS Accelerator Complex Status And Future Plans

Mitsuyoshi Tanaka

*AGS Department, Brookhaven National Laboratory**
Upton NY 11973, USA

Abstract. This paper describes the present performance and capability of the BNL AGS accelerator complex and possible future intensity upgrade plans. In 1995, the AGS reached its design upgrade goal of $6.0 \cdot 10^{13}$ ppp with the Booster. The AGS with a new fast extracted beam (FEB) system is able to perform single bunch multiple extraction at 30 Hz per AGS cycle for the g-2 experiment and for RHIC injection.

INTRODUCTION

In July 1960, the BNL Alternating Gradient Synchrotron (AGS) accelerated protons to 30 GeV at an intensity of $\sim 10^{10}$ ppp--becoming the world's most powerful atom smasher. Since then, the AGS has been periodically upgraded, improved, and has evolved to satisfy its continuously changing user's wishes; e.g. the 50 MeV Linac was replaced with the 200 MeV Linac in 1970, the AGS was connected to the existing Tandem Van de Graaff Accelerator in 1983, the 1.5 GeV Booster was added in 1991, and the associated AGS subsystem upgrades were completed.

RECENT PERFORMANCE AND ACHIEVEMENTS

The AGS now accelerates heavy ions up to gold (Au^{77+}) at various energies for the heavy ion physics program and has recently increased the slow-extracted proton beam (SEB) intensity above $6.0 \cdot 10^{13}$ ppp at 25 GeV/c with a 1.6 sec uniform spill for various intensity hungry particle physics experiments such as rare K-decay experiments.

The accelerator also recently became equipped with a new FEB extraction system that allows the AGS to perform single bunch multiple extraction of a heavy ion or a (polarized) proton beam at a time interval of 33 ms up to 12 times

---

*Work performed under the auspices of the US Department of Energy.

**This paper published posthumously.

per AGS cycle for filling of the RHIC collider or for 3.0 GeV/c π/μ production for the 14 m diameter g-2 muon storage ring. In the fall of 1995, a single bunch of $0.25 \cdot 10^9$ gold ions was successfully extracted at 11.2 GeV/c/N from the AGS and transported through the 600 m long AGS-to-RHIC (AtR) transfer line to an internal beam dump. The RHIC injection test, which includes the full AtR line and 1/6 of the superconducting ring, was successfully completed in January 1997. The commissioning of V/V1 lines with protons was also completed for the g-2 ring, which is now storing muons for the muon anomalous magnetic moment (g-2) experiment (E821).

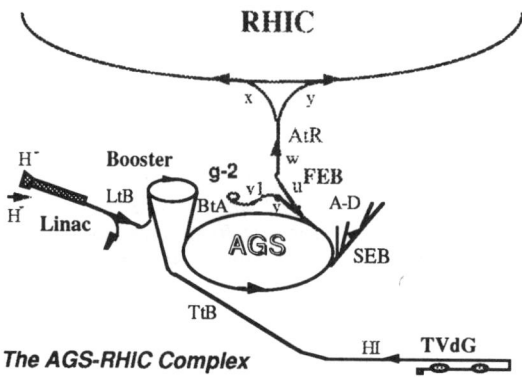

Figure 1. Schematic of the BNL AGS Accelerator Complex.

POSSIBLE FUTURE UPGRADE PLANS

So far we have focused mainly on increasing extracted beam intensity without much attention to the emittance blowup or beam loss. A better understanding of the critical high intensity effects can lead to better machine tuning, hence it could increase the intensity further. However, beyond 10^{14} ppp, various upgrades of the AGS are needed: 1) the AGS proton beam intensity could be further increased to 1.5×10^{14} ppp with the implementation of a barrier cavity rf system by avoiding bunched beam stacking during injection [1], 2) by adding a 1.5 GeV accumulator ring in the AGS tunnel to accept beam continuously from the Booster during 1.2 - 2.5 sec of the AGS cycle time, two to four times more beam pulses can be accumulated and be transferred into the AGS ring; 3) by increasing the AGS repetition rate to 2.5 Hz, the average beam current could be pushed to 40 µA [2]

SUMMARY

There were/are several proposals (LAMPF II, KAON, JHP etc.) for a high intensity hadron facility around the world, but only one that is already in operation and under construction. It is the BNL AGS complex. It is being done by a step-wise, physics driven approach in the most cost-effective way to minimize the downtime to the ongoing physics program and the cost without a major up front commitment in funding. The AGS average beam current was 1 µA just a few years ago, it is now 3 µA, and will become 5-7 µA with a barrier bucket rf system and tuning. With an accumulator, which is under design, it will reach 12 µA.

After 1999, the AGS becomes an injector for RHIC to fill its two rings with gold or polarized proton beams in a few minutes every 8-10 hours, the AGS could be available between fillings for forefront particle and nuclear physics program at an incremental cost.

ACKNOWLEDGMENTS

The work described in this paper has been performed collectively by the accelerator staff in the AGS Department and the RHIC Project.

REFERENCES

1. M. Blaskiewicz and J.M. Brennan, *A Barrier Bucket Experiment for Accumulating Debunched Beam in the AGS*, Proc. of the Fifth European Particle Accelerator Conf., Sitges, Spain, 1996, p. 2373.
2. J.M. Brennan and T. Roser, *High Intensity Performance of the Brookhaven AGS*, Proc. of the Fifth European Particle Accelerator Conf., Sitges, Spain, 1996, p. 530.

The Fermilab Long-Baseline Neutrino program

Maury Goodman
for the MINOS collaboration

HEP362, Argonne National Lab, Argonne Ill. 60439, USA

Abstract. Fermilab is embarking upon a neutrino oscillation program which includes a long-baseline neutrino experiment MINOS. MINOS will be a 10 kiloton detector located 730 km Northwest of Fermilab in the Soudan underground laboratory. It will be sensitive to neutrino oscillations with parameters above $\Delta m^2 \sim 3 \times 10^{-3} eV^2$ and $\sin^2(2\theta) \sim 0.02$.

I INTRODUCTION

The new neutrino beam from the Fermilab Main Injector is known as the NuMI program. There will be a short-baseline experiment, COSMOS based on an emulsion target, and a long-baseline experiment, MINOS which will include a new 10 kiloton detector and the existing 1 kiloton Soudan 2 detector. Here I will describe some of the considerations in the design of long-baseline neutrino oscillation experiments, and the physics goals and status of the design of the new MINOS detector.

II DESIGN OF A LONG-BASELINE NEUTRINO EXPERIMENT

The basic equation for a neutrino oscillation experiment is

$$P = \sin^2 2\theta \sin^2 1.27 \Delta m^2 L/E_\nu \tag{1}$$

The two physics parameters which the experimenter can measure are the mixing angle $\sin^2(2\theta)$ and the difference between the squares of the two neutrino masses Δm^2. In order to do this, one designs a test which is sensitive to the presence or absence of neutrino oscillations, yielding $<P>$ in the presence of oscillations, or the lowest possible P_{min} in the absence of evidence for oscillations. The distance between the neutrino source and the experiment

L, and the neutrino spectrum with an average energy $\langle E_\nu \rangle$ are basic design parameters of the experiment. Other important factors are the mass of the detector M and properties of the detector such as its energy resolution and spatial granularity.

It is useful to see how the sensitivity in parameter space changes as L and E are varied. This can be easily studied with the approximation shown in Figure 1. At high Δm^2 the effect of the oscillation washes out the effect of the energy spectrum and $P_{min} = \frac{1}{2}\sin^2(2\theta)$. At low Δm^2 we use the fact that

$$\left. \frac{\partial \Delta m^2}{\partial \sin^2(2\theta)} \right)_{\sin^2(2\theta)=1} = -\frac{1}{2} \qquad (2)$$

so the slope on a $\log[\sin^2(2\theta)] - \log[\Delta m^2]$ plot at low Δm^2 is -1/2. We can use these two-line limit plots to study scaling effects.

The L dependence has no clear optimum. As shown in Figure 2, as a fixed size detector is moved further from the ν source, the sensitivity in Δm^2 increases for a statistical test. (Note that this result does not apply to a background-free test.) However, due to the smaller number of events, the sensitivity to $\sin^2(2\theta)$ decreases. Thus the placement in L requires a value judgement concerning the region of neutrino oscillation parameter space to explore.

The energy dependence is more complicated. The parameter under primary control is not the neutrino energy, but the accelerator or proton energy used

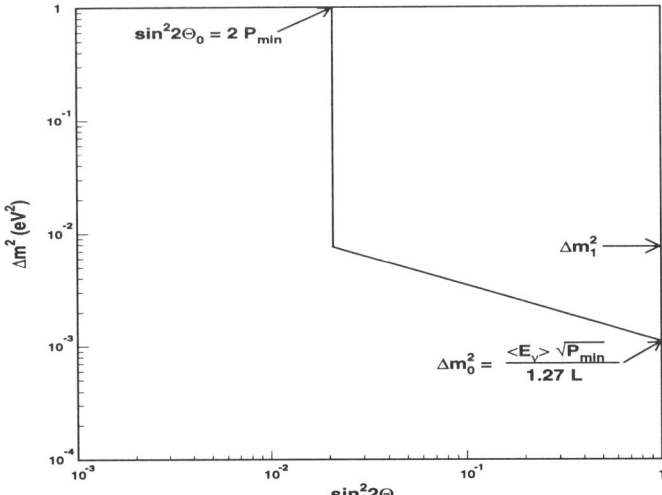

FIGURE 1. *Approximation for a neutrino oscillation limit curve showing the relationship between P_{min} and the limits at high and low Δm^2.*

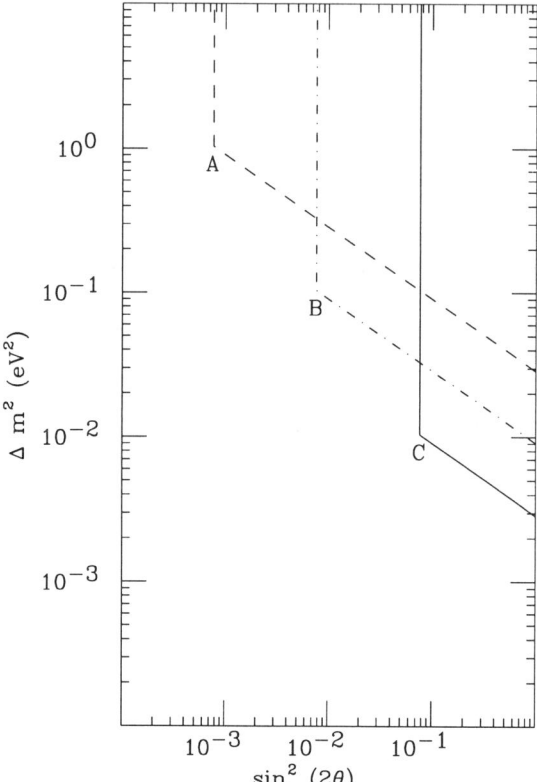

FIGURE 2. *Effect of moving a 1 kiloton detector from 7 km ("A") to 73 km ("B") to 730 km ("C") from the Fermilab NuMI beam. The pattern of these curves will be followed by any test in which $P_{min} \propto N^{-1/2}$.*

to create the neutrino beam. The average neutrino energy only increases logarithmically with E_p. There are four advantages to higher proton energy: 1) There are more π's and K's per proton, 2) The ν beam is more forward, 3) The ν cross section increases with E_ν, and 4) The ν_τ cross section relative to the ν_μ cross section rises with energy. There are also 3 advantages to lower proton energy: 1) There can be a larger proton current, 2) There are higher probability for π and K decays in a fixed length beam pipe and 3) The sensitivity in $\Delta m^2 \propto E_\nu$. To take advantage of ν_τ interactions, one wants $<E_\nu> \gg 4 GeV$ which is ν_τ CC threshold. Above 4 GeV, the ratio of the ν_τ to ν_μ cross section only increases slowly. A quantitative study for the NC/CC test of Δm^2_{min} and $\sin^2(2\theta)_{min}$ for $\nu_\mu \to \nu_\tau$ gives a fairly flat response from 100 to 400 GeV proton energy. [2]

A powerful test for $\nu_\mu \to \nu_\tau$ oscillations is the measure of the apparent R = NC/CC ratio in a ν_μ beam. In the presence of oscillations, the number of NC and CC events are:

$$n_{"cc"} = \frac{N(1 - P + \eta BP)}{1 + R} \text{ and } n_{"nc"} = \frac{N(R + \eta(1 - B)P)}{1 + R} \quad (3)$$

where B = 0.17 is the branching fraction for $\tau^- \to \mu^- X$, N is the number of events, P is the oscillation probability (suitably averaged) and η is the ratio of the ν_τ charged current cross section to the ν_μ charged current cross section. For the Main Injector energy spectrum, $\eta = 0.31$. The notation "cc" distinguishes events classified as charged current due to the presence of a muon in the final state from the actual charged current events, which for an incoming ν_e or ν_τ would be incorrectly classified as nc events. We can use equations 3 to calculate the modified R in the presence of oscillations.

$$R_{"nc"/"cc"} = \frac{n_{"nc"}}{n_{"cc"}} = \frac{R + \eta(1 - B)P}{1 - P + \eta BP} \quad (4)$$

This can be solved for P, and we use $\delta_R = R_{obs} - Rtrue$:

$$P = \frac{\delta_R}{\eta(1 - B) + R(1 - B\eta)} \quad (5)$$

This equation can be used to set a limit on P in the absence of oscillations by noting that $\sigma_R/R = R(1 + R)/\sqrt{NR}$ and that $\delta_R = 1.29\sigma_R$ for a 90% CL upper limit for a one-sided Gaussian. Plugging in the values of R, B and η, we obtain $P_{min} = 1.83/\sqrt{N}$, where N is the total number of nc + cc events expected.

As noted above, choosing the length of the baseline requires a physics value judgment. Such a value judgment was provided by the Fermilab PAC in June 1993 when they requested that a $\nu_\mu \to \nu_\tau$ experiment be sensitive to $\Delta m^2 \sim 10^{-2} eV^2$ and $\sin^2(2\theta) \sim 10^{-2}$. [3] Motivation for this choice of parameters comes from a desire to cover Cabbibo angle-like mixing for a wide range of

Δm^2, while covering the whole range of Δm^2 for the atmospheric anomaly at high $\sin^2(2\theta)$. These parameters lead to a requirement $P_{min} \sim 0.005$. For the Main Injector beam with $<E_\nu>= 15 GeV$ this leads to L = 830 km. The existing Soudan site at 730 km is an excellent match to this requirement. The requirement on N is 140,000 events. The latest design of the NuMI beam provides 3150 charged current events per kiloton per year at the far detector. We then solve for M (kT) in the equation

$$N = 140000 = 1.31(\frac{total\ events}{CC\ event}) \times 3150(\frac{CC}{kTyear}) * 4years * A * M \quad (6)$$

where $A \sim 0.75$ is the acceptance. This implies the need for a 11.3 kT detector, close to the 10 kT design of MINOS plus the one kiloton existing Soudan 2 detector.

III STATUS OF THE DESIGN OF THE MINOS EXPERIMENT

The MINOS experiment has been designed to take advantage of the high ν intensity available from the Fermilab Main Injector. The neutrino beam has been described elsewhere [4,5] The far detector will consist of a new 10 kiloton 8 meter diameter iron toroid with 600 or 1200 active detector planes between 2-4 cm iron plates Figure 3. There will also be a near MINOS detector of mass about one kiloton.

Three technologies for the active detector choice are currently under consideration: aluminum proportional tubes (APT) patterned after streamer chambers, liquid scintillator counters with fiber readout, and solid scintillator. In the latter two technologies, photodetector costs would be kept down by using multichannel 'pixel' photodetectors. APT's give good spatial resolution with strip readout and are a known technology. Issues include their calorimetric response, manufacturing and cost. Scintillators have good timing resolution and provide good EM calorimetry. Issues are the choice of photodetector, cost, and the light yield. MINOS plans to finalize its detector choice by the end of summer 1997.

Besides the NC/CC test, MINOS has been designed to detect $\nu_\mu \to \nu_\tau$ oscillations in a number of independent ways. A list of $\nu_\mu \to \nu_\tau$ tests to which the MINOS and Soudan 2 detectors will be sensitive in a wide band beam is given here:

1. Ratio of absolute rates in the near and far detectors.
2. $\frac{NC}{CC}$ ratio
3. $E_{tot}^{cc} = E_\mu + E_{had}$
4. E_{had}^{nc}

5. $R_{\frac{\mu}{\nu}}$ = Muons from Rock / ν interactions

6. Rate of stopping rock muons

7. Electron identification

8. Quasielastic ν_τ with $\tau \to \mu\nu\nu$ events

9. Missing p_t in $\tau \to hadrons$

10. $\tau \to \pi$ events

There are also a number of additional physics issues which the experiment plans to address. MINOS can search for $\nu_\mu \to \nu_e$ oscillations using several of the above tests. The experiment can also test the Harrison-Perkins-Scott scenario [6] in which both $\nu_\mu \to \nu_\tau$ and $\nu_\mu \to \nu_e$ oscillations would be detected. There are other signatures for $\nu_\mu \to \nu_\tau$ and $\nu_\mu \to \nu_e$ in a Narrow Band Beam. A decrease in the absolute rate of events and the CC energy distribution in the absence of a signal in the other tests would be evidence for $\nu_\mu \to \nu_{sterile}$ oscillations. Differences between ν_μ and $\bar{\nu}_\mu$ signals can be used to search for CP violation and matter enhanced $\nu_\mu \to \nu_e$ oscillations. MINOS can also search for gravitationally induced ν oscillations and look for anything in the beam which could lead to higher than expected event rates in the far detector. MINOS can also study the highest energy atmospheric neutrinos and will be the first large underground magnetic detector for cosmic ray studies.

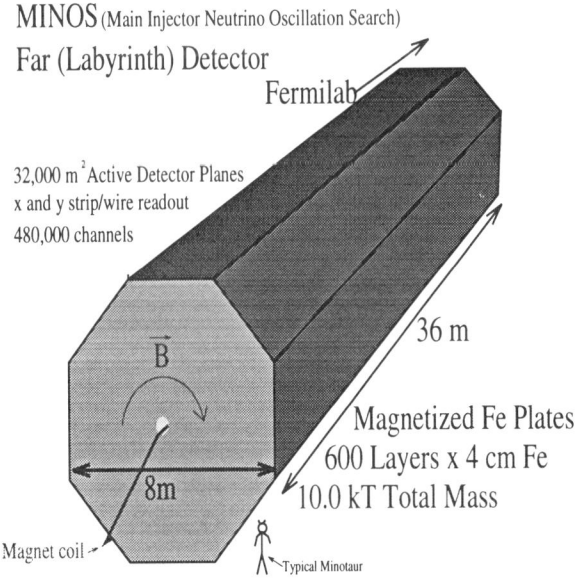

FIGURE 3. *The MINOS far detector schematic.*

IV OUTLOOK

The Fermilab NuMI beam, the Soudan site, the existing fine-grained Soudan 2 detector, and the planned new 10 kT MINOS detector are well matched to the needs of an ambitious new neutrino oscillation program from the Fermilab Main Injector. MINOS will start taking data whenever the NuMI beam comes on. The present best estimate is 2001.

Further information about the MINOS experiment can be found on the web at http://www.hep.anl.gov/ndk/hypertext/numi.html. A web page devoted to all experiments which study neutrino oscillations is called the "Neutrino Oscillation Industry" and can be found at http://www.hep.anl.gov/ndk/hypertext/nu_industry.html. A monthly email newsletter is also available by sending email to mcg@hep.anl.gov. To get the newsletter, send a message "subscribe". To get a one-line notice when the latest issue has been posted on the web, send "subscribe web".

REFERENCES

1. W.W.M. Allison et al., NIM **A376**, 36 (1996).
2. Crane and Goodman, p. 226 in "Particle and Nuclear Astrophysics and Cosmology in the Next Millenium", proceedings of the 1994 Snowmass Summer Study, World Scientific, Kolb and Peccei editors.
3. Fermilab Call for Proposals for a long-baseline neutrino experiment, June 1993.
4. Greg Bock, these proceedings.
5. D. Crane et al., Status Report: Technical Design of Neutrino Beams for the Main Injector, Fermilab-TM-1946, July 1995.
6. P.F. Harrison, D.H. Perkins and W.G. Scott, Phys.Lett. B 349 (1995) 137: B374 (1996) 111; also C. Giunti, C.W. Kim and J.D. Kim, Phys. Lett. B352(1995) 357.

First Commissioning Results from Hall A at TJNAF

Robert Michaels
for the Hall A Collaboration

Thomas Jefferson National Accelerator Facility, Newport News, VA 23606, USA

Abstract. The commissioning and resulting capabilities of Hall A at the Thomas Jefferson National Accelerator Facility for performing coincidence experiments with a continuous–wave electron beam at energies of 0.8–6.0 GeV will be described. The facility consists of a pair of high–resolution magnetic spectrometers to detect scattered particles, together with instrumentation to measure beam properties. The installation of the base equipment has been completed. Commissioning of the spectrometers, including their detector packages, is well advanced, both in single–arm mode and in coincidence mode.

Hall A at the Thomas Jefferson National Accelerator Facility is a new research facility for precise measurements of nucleon and nuclear structure at intermediate energies which will improve our understanding of the transition from low–energy phenomena to the regime where pQCD calculations are reliable. Jefferson Lab, with its continuous–wave beam of 0.8 to 6.0 GeV and currents up to 100 μA, is ideal for coincidence measurements, where one spectrometer detects the scattered electron and the other spectrometer detects a knocked–out proton or other hadron. The experimental program currently has 29 approved experiments involving physicists from 34 institutions. This contribution reports on the first results of the Hall A Collaboration from the past year of commissioning the newly constructed spectrometers and instrumentation in Hall A. The first two experiments, involving coincidence knockout of protons from oxygen using a water target, are underway. Hall A is one of three experimental halls. Hall B is currently in a commissioning phase, while Hall C has been operational and taking data for about two years.

The Hall A facility consists of a pair of identical spectrometers of QQDQ design, together with detectors for detecting the scattered particles, beamline equipment, and targets (fig. 1). In order to achieve the physics aims, the spectrometers need to have a large acceptance, with excellent resolution and absolute accuracy in the reconstructed four-vectors of the events and precise normalization of the cross section. These requirements imply the need for a high resolution in the momentum measurement, which is achieved mainly by the large size of the magnet system. Good knowledge of the transfer matrix for the spectrometer is necessary to reconstruct the event at the scattering point. Also required are good pointing accuracy for the location of the spectrometers and precise measurements of beam properties such as position, angle, current, and charge integrated during an experiment.

The detector package consists of scintillators for triggering, vertical drift chambers for reconstruction of particle trajectories, and aerogel and Cherenkov detectors and lead glass arrays for particle ID. In addition, one of the spectrometers is outfitted with a focal plane polarimeter for measuring proton polarization, an apparatus reported on separately in this session. The trigger is formed in programmable CAMAC electronics and is configurable to include various combinations of the scintillator and particle ID detectors at the trigger level. The targets used in the initial operation have included carbon, beryllium oxide, and water.

Calibration of the optical transfer matrix for the spectrometers is performed in the following way. A 0.5 cm thick tungsten plate with a rectangular array of holes is placed at the entrance of the spectrometer. The matrix which transfers the tracks from the focal plane, where they are reconstructed in the drift chambers, to the target is determined through a chi-square minimization procedure which reproduces the hole pattern. In addition, the elastic peak from ^{12}C(e,e') is moved in steps across the focal plane, for small changes in the magnetic fields. The properties which have been obtained from the commissioning process are listed in the table below. Improvement in some of these properties may be expected in the upcoming year of operation.

Figure 2a shows the momentum spectrum of electrons elastically scattered from carbon, and the first few excited states, showing a momentum resolution of 3×10^{-4} FWHM. Figure 2b shows the missing-energy distribution for ^{16}O(e,e'p) from a water target, showing a missing-energy resolution of about 1.3 MeV FWHM. The missing-energy is defined as $E_{\text{miss}} = E - E' - K_p - K_{A-1}$, where E and E' are respectively the energy of the incoming and outgoing electron, and K_p and K_{A-1} are respectively the kinetic energies of the knocked-out proton and the residual nucleus.

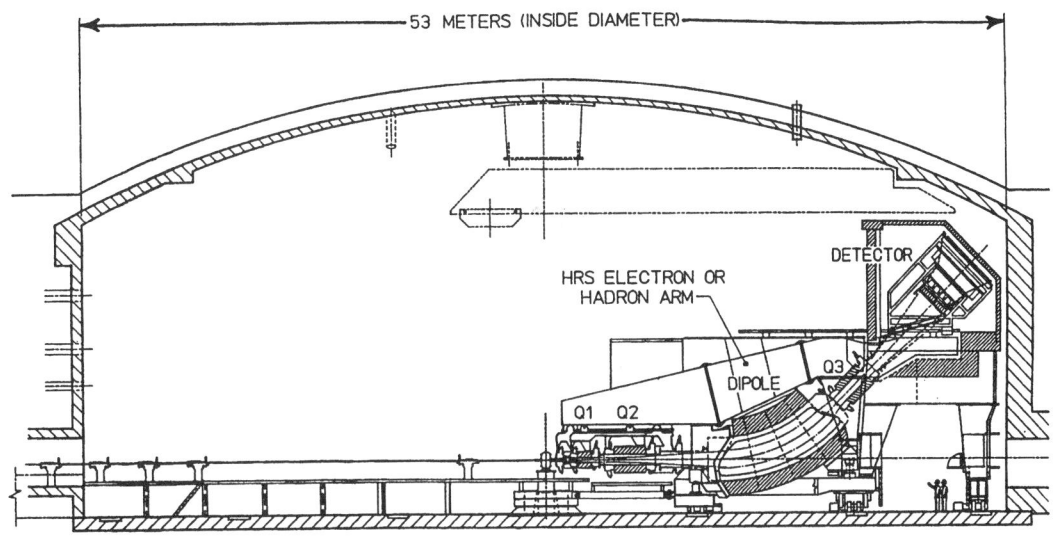

Figure 1. Cross Section of Hall A at TJNAF

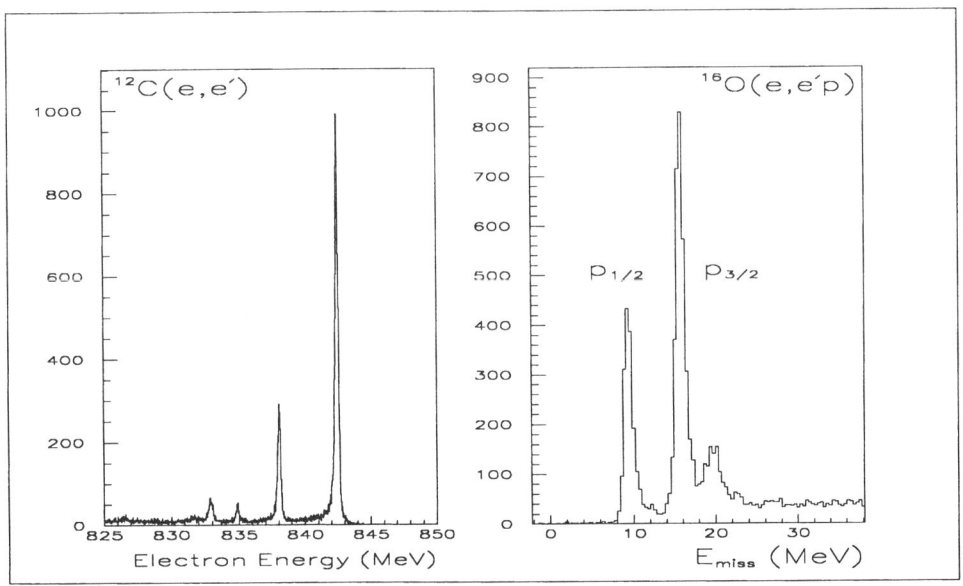

FIGURE 2. (a) Momentum distribution of electrons measured in elastic $^{12}C(e,e')$ scattering, showing a resolution of 3×10^{-4} FWHM. (b) Missing–energy distribution from $^{16}O(e,e'p)$ showing the p-shell states of oxygen with resolution 1.3 MeV FWHM.

TABLE 1. Properties of Hall A and Spectrometers

Luminosity	$\geq 10^{38}$ cm^{-2}sec^{-1}
Momentum Range	0.3 – 4.0 GeV/c
Momentum Acceptance	±4.5%
Momentum Resolution (FWHM)	3×10^{-4}
Solid Angle Acceptance	6.5 msr
Target Length Accepted at $\theta_{\text{scatt}}=16°$	7 cm
Target Resolution (FWHM)	4 mm
Coincidence Time Resolution (FWHM)	1 ns
Missing Energy Resolution (FWHM)	1.3 MeV

During the next year, Hall A will continue the development of its capabilities, interleaved with experiments that can make use of existing capabilities. The Møller polarimeter for measuring beam polarization is presently being commissioned. A cryogenic target for liguid hydrogen and deuterium targets will be installed in August 1997. Two apparatus for measuring the beam energy with accuracy of 10^{-4} will be installed in the next year. A Compton polarimeter, for measuring the beam polarization at 1% accuracy, is planned for next year.

K2K: KEK to Super-Kamiokande Long-Baseline Neutrino Oscillation Experiment

R. Jeffrey Wilkes*, for the K2K Collaboration [1]

*University of Washington
Seattle, Washington

Abstract.
K2K is a Japan-Korea-US collaboration which is constructing a long-baseline neutrino oscillation experiment employing a muon neutrino beam directed from KEK to Super-Kamiokande. The near detector complex at KEK is under construction and data taking will begin in early 1999. The experimental arrangement and anticipated data rates are described.

Recently, intense interest in long-baseline neutrino oscillation studies has been motivated by anomalies in measured neutrino fluxes from the Sun and from cosmic-ray interactions in the atmosphere. Measured solar ν_e fluxes are significantly lower than expectation from Standard Solar Models (SSM) [1]. Observed atmospheric-neutrino fluxes show a ratio ν_μ/ν_e which is substantially smaller than expectation based on the best theoretical modelling [2]. Flavor oscillations due to neutrino masses within current experimental limits could account for these anomalies without undue violence to theoretical fundamentals which are otherwise well established.

K2K is a Pacific-Rim collaboration (Japan-Korea-US) which will perform a long-baseline neutrino oscillation experiment in Japan (KEK E362) [3]. K2K was designed to decisively test neutrino oscillations in the region of mass-squared differences Δm^2 around $10^{-2}\mathrm{eV}^2$ and mixing angles $\sin^2 2\theta > 0.2$, which includes the area allowed by recent underground atmospheric-neutrino measurements [2]. It will be sensitive to $\nu_\mu \to \nu_e$ appearance and $\nu_\mu \to \nu_x$ disappearance effects (where ν_x is any neutrino *other* than ν_μ). Using an accelerator beam of high purity, it will be possible to largely eliminate uncertainties caused by the mixture of different neutrino flavors in the atmospheric

---

[1] See http://www.phys.washington.edu/~superk/k2k/members.html for collaboration membership.

neutrinos.

The K2K experimental arrangement will include a new neutrino beam line extended from the existing 12 GeV KEK Proton Synchrotron (KEK-PS) as a ν_μ source, a 1-kton water Čerenkov detector plus a fine-grained detector system within the KEK site as the "near" detector, and the existing Super-Kamiokande detector, located 250 km away, as the "far" detector. Many components of the near detector system are simply straightforward modifications of equipment available from previous projects at KEK. Civil construction on the beamline is underway, and data-taking is expected to begin in early 1999.

The layout of the experiment is shown in Fig.1. The KEK-PS can accelerate 6×10^{12} protons per spill to 12 GeV every 2.2 seconds. The new beam line, currently under construction using TRISTAN and SLAC dipole and quadrupole magnets, points about 1 degree downward, aimed underground towards Super-Kamiokande. Following a 200 m decay tunnel and a steel beam stop 6 m thick, an intense ν_μ beam having 97% purity with very small ν_e contamination ($< 1\%$) is produced. The energy of the ν_μ beam peaks at about 1 GeV and extends to 4 GeV, with a mean energy of 1.5 GeV. An upgrade of the KEK proton synchrotron to 50 GeV with higher intensity has been proposed for completion by ~ 2004.

The wide-band neutrino beam produced by two magnetic horns will be uni-

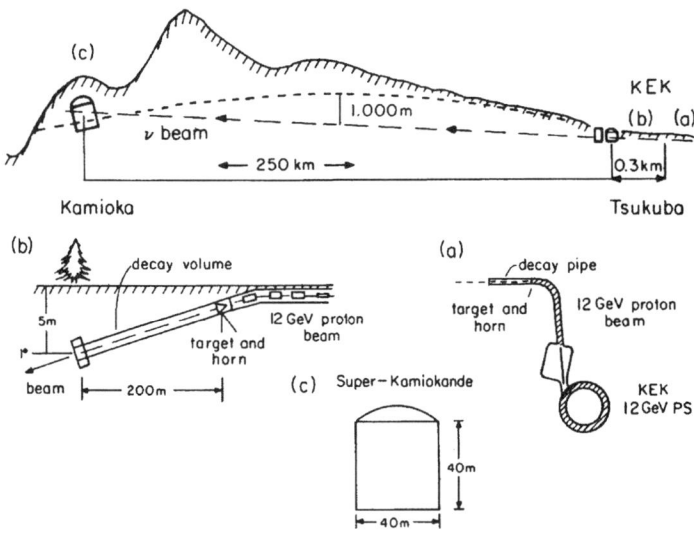

FIGURE 1. The overall configuration of the detectors. The front detector (b) and the Super-Kamiokande (c) are located at 300 m and 250 km from the proton target (a), respectively.

form to a few percent within the first few milliradians relative to the beam axis, so the spectra of neutrino beams at near and far detectors will be essentially identical. A target radius of 1 m at the near-detector covers a beam segment which does not significantly differ in profile and spectrum from the beam crossing Super-Kamiokande, where the corresponding area has radius over 800m. Beam pointing accuracy of 1 mrad is thus sufficient, and systematic corrections related to neutrino interactions in the near and far detectors will be minor.

Neutrinos are effectively produced from an extended source inside the decay tunnel, which is 200 m long and 3 m in diameter. Beam monitors along the decay tunnel will be used to reduce the uncertainty in the extrapolation of neutrino flux to large distances. A simulation shows that one can expect, for a spill of 6×10^{12} protons on target, an integral flux of 2×10^9 ν_μ's/m^2 at a distance of 300 m and 1.5×10^3 ν_μ's/m^2 at Super-Kamiokande.

The near detector, located 300 m from the proton beam target and 100 m from the end of the decay tunnel, consists of a 1-kton water Čerenkov detector plus a fine-grained detector with the ability to precisely observe the neutrino interaction vertex, provide tracking, and measure muon and electron energies (see Fig.2). The near detector is scheduled for completion by late 1998. The experimental area for the near detector is now under construction.

The 1-kton water-Čerenkov detector is a 10 m (height)×11 m (diameter) tank viewed by ∼200 PMTs. The fine-grained detector also uses water as target material for the neutrino beam, consisting of twenty 2.4m×2.4m×6cm segments totalling seven tons of water. The individual target segments are interleaved with scintillating-fiber tracking layers to provide good transverse vertex definition for neutrino interactions as well as track reconstruction capability. Sci-fi layers are presently being prepared in Japan and the water tanks will be made in the US.

A Pb-glass wall, originally built for TOPAZ and now available together with its electronics, will be located behind the scintillating-fiber detector to identify electrons. A muon ranger, consisting of iron plates and drift tubes that were used in VENUS, will be used to measure muon energy up to 3.5 GeV with good accuracy.

In order to cancel systematic uncertainties, K2K employs water targets for all detector systems and has water Čerenkov detectors at both near and far sites. The two near-detector components will be complementary. The water tank has its fiducial volume controlled less precisely due to coarser vertex resolution. However, unlike the fine-grained detector, the 1-kton detector can reconstruct π^0s, and due to the large size of the tank one can measure variations of the beam characteristics at larger distances from the beam axis. The fine-grained detector has a smaller target volume, but its fiducial volume is much more precisely defined due to accurate vertex measurements.

A 16-ton fiducial volume in the 1-kton detector will measure ν_e contamination and the fraction of weak neutral current (NC) events $\nu_\mu N \to \nu_\mu N' + \pi^0$, using the same methods as in the Super-Kamiokande detector. The fine-grained detector will also measure ν_e contamination. Quasi-elastic reactions ($\nu_\mu N \to \mu^- N'$) in the fine-grained detector will serve to measure beam uniformity and the neutrino energy spectrum.

The far detector for K2K is the Super-Kamiokande underground neutrino experiment, which has been taking data with more than 95% livetime since 1 April, 1996. Super-Kamiokande is a water Čerenkov detector consisting of 50 kton of ultra-pure water in a 41 m (height)×39 m (diameter) tank. Its inner volume is viewed by 11,150 20" diameter Hamamatsu photomultiplier tubes, providing photocathode coverage equivalent to 40% of the detector surface area. The inner detector is surrounded by a 2 m thick outer layer of water viewed by 1850 8" diameter Hamamatsu tubes. The outer detector is used to veto particles coming from outside as well as providing shielding against backgrounds emanating from the surrounding rock. A nominal 22-kton fiducial volume is used to study neutrino interactions. K2K will not interfere with normal operation of Super-Kamiokande. GPS clocks at both sites will be used to define appropriate DAQ time windows for KEK beam pulses arriving at Kamioka.

K2K will search for evidence of neutrino oscillations by comparing event

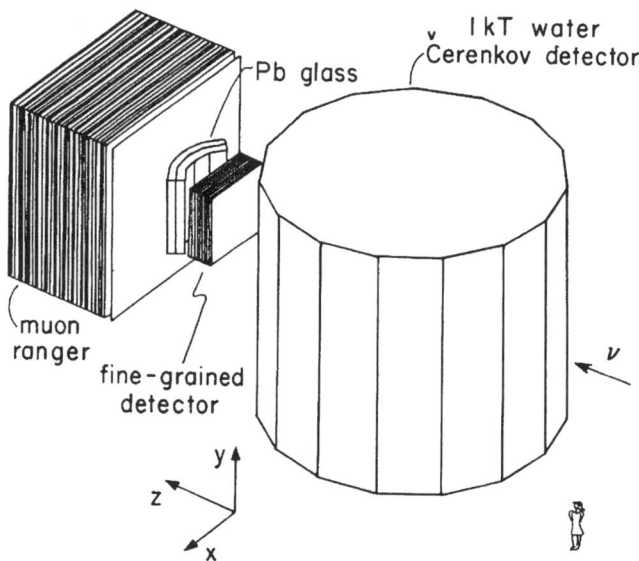

FIGURE 2. A schematic view of the near detector. The 1-kton cylindrical water Čerenkov detector is located in front of the fine-grained detector along the neutrino beam direction. The fine-grained detector consists of a scintillation fiber tracking device with water target, a lead-glass detector, and a muon ranger of iron plates with muon chambers along the neutrino beam direction.

rates in the near and far detectors. Signatures of oscillations will include ν_e appearance and ν_μ disappearance, signalled by decrease in the ν_μ flux, a change in the muon energy spectrum, and/or a difference in the NC/CC event rates. After collecting 10^{20} protons on target, the expected no-oscillation samples of various interactions per ton of water in the near detector and in the entire 22-kton fiducial volume of Super-Kamiokande are summarized in Table 1.

The statistical uncertainties of the event samples in Super-Kamiokande are thus 5% if all CC interactions are considered, and <16% for NC interactions (single π^0). The sensitivity to Δm^2 at the 95% confidence level is $\approx 1.5 \times 10^{-3} eV^2$ at the maximum mixing, i.e. $\sin^2(2\theta) = 1$, assuming a mean beam energy of 1.5 GeV.

In summary, the K2K long-baseline neutrino experiment will begin taking data in early 1999, with a newly constructed near detector at KEK, and the Super-Kamiokande experiment as the far detector. Construction of the neutrino beam line, experimental hall, and near detector components is well underway, and at time of writing (7/97) is on or ahead of the very tight schedule required.

REFERENCES

1. J.N. Bahcall, M. Pinsonneault, and G.J. Wasserburg, Rev. Mod. Phys. **67**, 781 (1995); J. N. Bahcall and M. Pinsonneault, Rev. Mod. Phys. **64**, 882 (1992); J.N. Bahcall and R.K. Ulrich, Rev. Mod. Phys. **60**, 297 (1988); S. Turck-Chieze et. al., Astrophys. J. **335**, 415 (1988).
2. Kamiokande: Y. Fukuda et. al., Phys. Lett. **B335**, 237 (1994). IMB: T. J. Haines, et al., Phys. Rev. Lett. **57** 1986 (1986); R. Becker-Szendy, et al., Phys. Rev. **D46** 9 (1992).
3. For further details on K2K, please see our Web page at http://www.phys.washington.edu/ superk/k2k/index.html

TABLE 1. Yields of neutrino interactions relevant to the oscillation search, per ton of water in the near detector, and in the 22-kton fiducial volume of Super-Kamiokande, for 10^{20} protons on target.

	Near (per ton)	Super-K (22-kton)
Total CC interactions	25,000	400
Quasi-elastic CC interactions	10,000	160
Single-π CC interactions	10,000	160
Multi-π CC interactions	5,000	80
Total NC interactions	5,000	80
Single-π NC interactions	2,500	40
Total ν_e CC interactions	250	4
Quasi-elastic ν_e CC interactions	100	2

The BaBar Detector at the SLAC B-Factory

Don H. Fujino[1]

Lawrence Livermore National Laboratory
Livermore, CA 94550
Representing the BaBar Collaboration

Abstract. The BaBar detector at the SLAC B-Factory will search for CP violation in B decays. I will describe the BaBar detector and its current status.

INTRODUCTION

One of the Holy Grails in particle physics is to understand the nature of CP violation. CP violation can arise naturally in the Standard Model from a complex phase in the Cabibbo-Kobayashi-Maskawa (CKM) matrix, which describes the weak couplings of the six quarks. The BaBar experiment[2] at the SLAC Asymmetric B-Factory is designed to observe CP violation in B meson decays that arise from $\Upsilon(4S) \to B^0 \bar{B}^0$.

When a neutral B meson decays into a CP eigenstate such as $B^0 \to \psi K_S^0$, ψK_L^0, or $\pi^+\pi^-$, it can decay directly into that final state, or it can mix into a $\bar{B}^0$ before decaying: $B^0 \to \bar{B}^0 \to \psi K_S^0$. Through interference effects from $B^0 - \bar{B}^0$ mixing, CP violation will introduce a time-dependent decay asymmetry between $B^0 \to f_{CP}$ and $\bar{B}^0 \to f_{CP}$, where f_{CP} is a CP eigenstate. The usual exponential decay behavior becomes modified by subtle oscillations in the decay lifetime differences.

$$\Gamma(B^0\bar{B}^0 \to B^0 f_{CP}) \propto e^{-\Gamma \Delta t}\left[1 + \sin\phi \sin(\Delta M \Delta t)\right]$$
$$\Gamma(B^0\bar{B}^0 \to \bar{B}^0 f_{CP}) \propto e^{-\Gamma \Delta t}\left[1 - \sin\phi \sin(\Delta M \Delta t)\right]$$

where $\sin\phi$ is the CP violating parameter, Δt is the decay time difference between the two B mesons from $\Upsilon(4S) \to B^0\bar{B}^0$ decays, and $\sin(\Delta M \Delta t)$ is the decay time oscillation from mixing.

---

[1] This work performed under the auspices of the US Department of Energy by the Lawrence Livermore National Laboratory under Contract W-7405ENG-48.
[2] For further details see the BaBar Technical Design Report, SLAC-R-95-457.

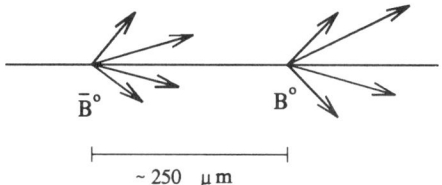

FIGURE 1. Vertex separation for $\Upsilon(4S) \to B^0\bar{B}^0$ decays in an asymmetric collider.

In the center-of-mass frame of the $\Upsilon(4S)$, the B's are nearly at rest, so there is no information on B decay times and the CP violating effects are lost. However, SLAC will boost the $\Upsilon(4S)$ frame by colliding a 9.0 GeV electron beam with a 3.1 GeV positron beam, so that the two B mesons will decay along the beam axis with an average separation of $\sim 250\mu$m (Figure 1.) This separation is easily resolved with a silicon vertex detector.

The BaBar detector must therefore satisfy three requirements:

1. Reconstruct B decays into CP eigenstates (ψK_S^0, ψK_L^0, $\pi^+\pi^-$, etc.)

2. Tag the flavor of the other B meson using charged kaons or leptons

3. Measure the relative decay time $\Delta t = t_2 - t_1$ of the two B mesons

THE BaBar DETECTOR

The SLAC B-Factory will achieve a peak luminosity of 3×10^{33} cm^{-2}s^{-1} by colliding 9.0 GeV e^- and 3.1 GeV e^+ beams. Such high luminosities and asymmetric beams require high currents (2 amp e^-, 1 amp e^+), 1700 bunches, 4 ns beam crossings, and separate high and low energy rings to transport the beams without introducing any parasitic crossing. A final bending magnet in the interaction region (IR) will ensure the e^+e^- beams collide head-on. The bending near the IR will generate intense synchrotron radiation that is not present in conventional e^+e^- machines.

The BaBar Collaboration is made up of 10 Countries, 78 Institutions and over 500 physicists. The BaBar detector has 6 major subsystems. Going out radially, BaBar has a silicon vertex tracker, a drift chamber, a Cerenkov detector for π/K separation, a CsI electromagnetic calorimeter, a 1.8 Tesla superconducting magnetic coil, and an instrumented flux return to detect muons and K_L^0. The detector is shown in Figure 2.

The Silicon Vertex Tracker (SVT) is a 5 layer double-sided detector which provides both ϕ and z information. It will locate each B decay vertex with a ΔZ resolution of 80 μm, and so will easily be able to resolve the 250μm average separation between the two B mesons. The five layers are at 3, 4, 5, 12, and 14 cm in radius; the outer two layers are arched to minimize the area of silicon required and to avoid large incident track angles. (See Figure 3).

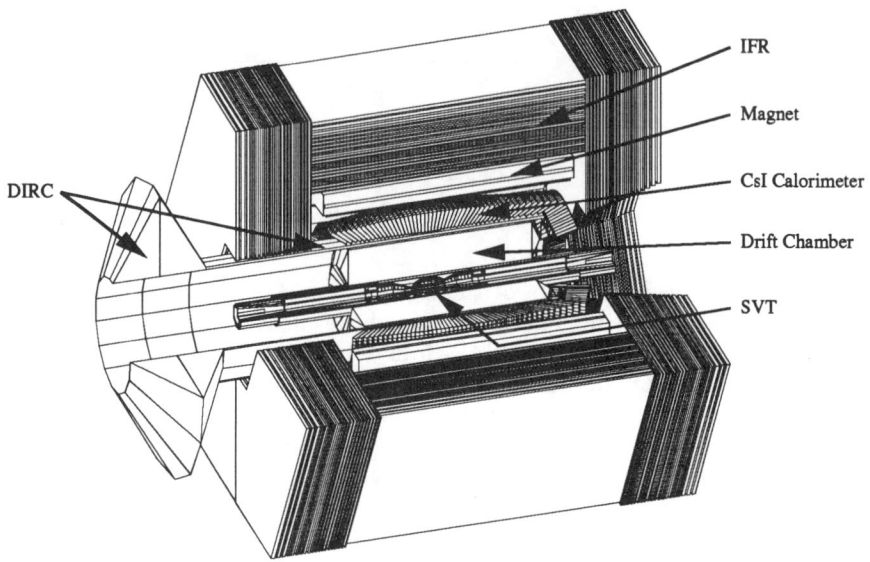

FIGURE 2. The BaBar detector.

The SVT will be able to find tracks independent of the drift chamber. It has 0.94 m^2 of silicon strips and 150,000 readout channels. Because the SVT will be in a high synchrotron radiation background, 33 KRad/year for the inner layer, the chips were designed rad-hard and AC coupled to reduce the noise. The SVT is constructed in two clam-shells which are mounted over the beampipe and the two conical bending magnets in the IR. A low mass space frame, constructed from carbon fiber rods spanning the length of the SVT, will provide the overall stiffness to the detector.

Just outside the SVT is the central drift chamber (DC). It must provide excellent momentum resolution for charged tracks in the range 100 MeV/c to 2.5 GeV/c to obtain good mass resolution for reconstructing exclusive B decays. The 100 MeV/c lower limit is defined by the inner radius of the drift chamber; lower momentum tracks must rely primarily on the SVT. The DC momentum resolution is estimated to be $\sigma_{P_T}/P_T = 0.21\% + 0.14\% \times P_T$. The constant term is from multiple Coulomb scattering and is kept minimal since

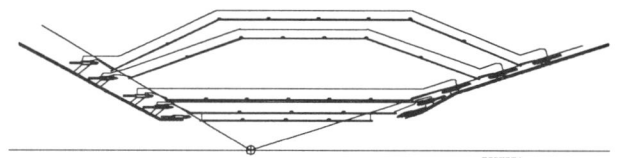

FIGURE 3. The 5-layer silicon vertex tracker.

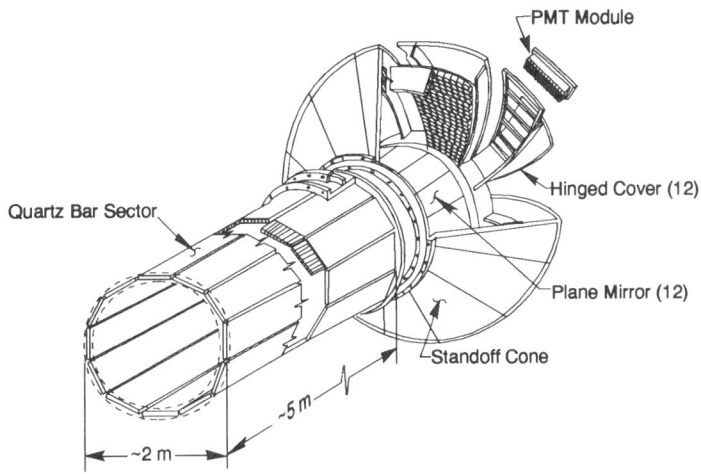

FIGURE 4. The DIRC detector.

most of the tracks from $B\bar{B}$ events have low momentum. This is achieved by using a helium-based gas in the chamber (80% helium, 20% isobutane), aluminum field wires, and an inner wall made of beryllium. The helium-based gas has a spatial resolution that is $\sim 30\%$ better than convential DC gases such as HRS gas (Ar:CO_2:CH_4 = 89:10:1). The drift chamber uses small cells arranged in 40 layers over a radius of 23 − 80 cm. They are grouped in 10 superlayers (AUVAUVAUVA), where A is an axial superlayer and UV are stereo superlayer pairs with a stereo angle of ±50 mrad.

Particle identification in the barrel region is provided by the Detector of Internally Reflected Cerenkov light (DIRC). The DIRC yields better than 4σ separation for charged kaons and pions over the entire momentum and $\cos\theta$ range for $\Upsilon(4S) \to B\bar{B}$ events. The drift chamber will provide limited π/K separation from specific ionization (dE/dx) information: over 3σ separation for $P < 0.7$ GeV/c and about 2σ separation for $P > 3$ GeV/c; however, it offers no discrimination for intermediate momenta. Kaon identification is crucial for tagging the flavor of the unreconstructed B meson and for distinguishing the CP eigenstate decay $B^0 \to \pi^+\pi^-$ from the penguin decay $B^0 \to K^+\pi^-$.

The DIRC is situated in the 10 cm gap between the drift chamber and the calorimeter and uses a novel method for Cerenkov ring imaging. The DIRC consists of 156 long quartz bars ($1.75 \times 3.5 \times 470$ cm^3) as Cerenkov radiators arranged in azimuth around the beamline (See Figure 4). Charged particles passing through the quartz bars radiate Cerenkov light which is internally reflected down the bar. The light exits in the backward region with its angular information still preserved into a large expansion region called the standoff cone. Six tons of water fill this region and the Cerenkov image expands ~ 120 cm to an array of 13400 photomultiplier tubes.

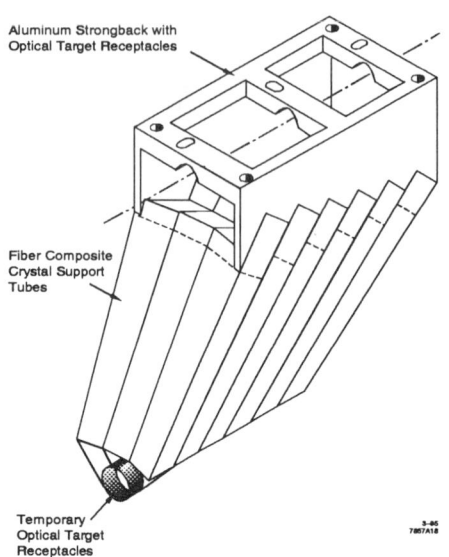

FIGURE 5. Calorimeter barrel module which houses an array of 3 × 7 CsI(Tl) crystals.

The CsI Calorimeter must reconstruct photons with high efficiency and energy resolution over a large dynamic range of 20 MeV to 5 GeV. Most photons from generic B decays are under 0.5 GeV, but those from $B^0 \to \pi^0\pi^0$ decays can have energies up to 4 GeV. The energy resolution is estimated to be $\sigma_E/E = 1\%/E(GeV)^{0.25} \oplus 1.2\%$, about a factor of two better than the CLEO CsI calorimeter. In addition, the calorimeter must reconstruct π^0's and identify electrons with momenta as low as $P > 0.5$ GeV/c. The calorimeter is composed of 5760 CsI(Tl) crystals arranged for the barrel and 900 crystals for the forward endcap. The barrel covers the range $-0.80 < \cos\theta_{lab} < 0.89$ and the forward endcap extends the coverage to $\cos\theta_{lab} = 0.97$. The barrel crystals are non-projective in $\cos\theta$ to reduce inefficiencies from gaps between crystals. The crystals are 16 − 17.5 radiation lengths, longer in the forward region. Two photodiodes on the back face of the crystal measure the light output. An array of 3 × 7 crystals are housed in a carbon fiber module. (See Figure 5). An aluminum support is bonded to the back of each module and is rigidly attached to an outer cylindrical strongback structure.

The Instrumented Flux Return (IFR) provides muon and K_L^0 identification. Muons will be identified for $P > 0.6$ GeV/c, significantly better than the limit of $P > 1.4$ GeV/c for unsegmented iron absorbers. K_L^0 mesons that interact in the IFR will have their direction measured from the energy deposition, but not their momentum. This is important for observing CP asymmetries in $B^0 \to \psi K_L^0$ decays. The IFR detector is composed of one barrel (in sextants) and two endcaps. There are a total of 20 layers of iron plates which increase in thickness

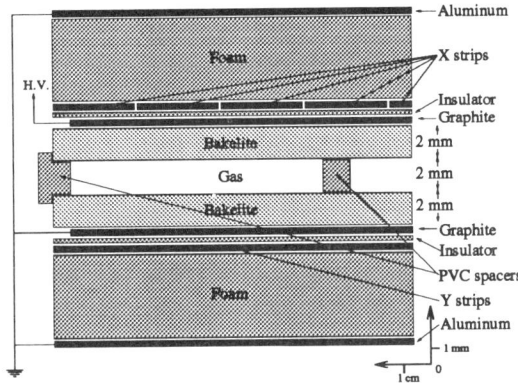

FIGURE 6. The RPC components in the IFR detector.

from 2 cm in the inner layers to 5 cm in the outer layers. This layout optimizes the muon and K_L^0 finding efficiency. Resistive Plate Chambers (RPC's) were chosen to instrument the 2500 m^2 of surface area since this technology is simple, reliable, and inexpensive (see Figure 6). RPC's have a 2 mm gas gap filled with Argon:Isobutane:Freon (48:48:4) at atmospheric pressure between two layers of Bakelite with bulk resistivity of $\sim 10^{11}\Omega$-cm. The two electrode plates are coated with a thin layer of graphite and connected to high voltage and to ground. Charged tracks generate a quenched spark that induces charge on X and Y pickup strips.

CONCLUSIONS

There is still a mountain of work to be accomplished before the B-Factory and BaBar detector are operational; however, the schedule has suffered no major slippages. The PEP II accelerator is ahead of schedule. The high energy ring has been built and is being commissioned. Magnets and vacuum systems are now being installed in the low energy ring. Progress continues towards completing the BaBar detector by the end of 1998. The steel for the magnetic flux returns has arrived at SLAC and the RPC's are being assembled into place. Many of the major items such as the low mass space frame for the SVT, the drift chamber endplates and robotic wire stringers, and the calorimeter structural support cylinder are either built or being fabricated. The silicon wafers, DIRC quartz ingots, and CsI(Tl) crystals are in full production. With the BaBar detector scheduled to move into the beamline by 1999, we will be in an excellent position to discover CP violation in B decays.

Acknowledgements: I would like to thank C. Wuest and B. Wisniewski for useful and stimulating discussions.

OPPIS DEVELOPMENT AT TRIUMF

A.N. Zelenski[1,2], G. Dutto[2], C.D.P. Levy[2], P.W. Schmor[2],
W.T.H. van Oers[3], G.W. Wight[2]

(1) Institute for Nuclear Research, Russian Academy of Sciences, 117312 Moscow, Russia
(2) TRIUMF, 4004 Wesbrook Mall, Vancouver, BC, Canada V6T 2A3
(3) Dept. of Physics, Univ. of Manitoba, Winnipeg, MB, Canada R3T 2N2

Abstract.
  The TRIUMF OPPIS (Optically Pumped Polarized Ion Source) provides a precision quality beam for the experiment on parity non-conservation in proton-proton scattering at 221 MeV beam energy. The pulsed OPPIS applications for the future RHIC and HERA polarization facilities are discussed. The polarization technique of the radioactive nuclide beam is proposed for β-NMR condensed matter studies with a new ISAC facility at TRIUMF.

INTRODUCTION

Collider experiments with polarized proton beams, approved at RHIC [1] and under consideration at HERA [2], will provide fundamental tests of QCD and the electroweak interaction. Polarized beam should allow better identification of new objects produced in proton-proton collisions and expand the limits of searches for possible manifestations of New Physics beyond the Standard Model. Such experiments will require the maximum available luminosity, and therefore polarization must be obtained as an extra beam quality without sacrificing intensity. Typical currents for unpolarized H$^-$ ion injectors are in the 20-50 mA range. With a lower current polarized source the use of multi-turn charge-exchange injection into a booster ring will partially compensate the loss, but only a 20-30 mA source will completely solve the problem. A 1.64 mA dc polarized H$^-$ ion current was obtained at the TRIUMF OPPIS, with a promise of further increase to the 2-3 mA range [3]. The ECR-type primary proton source used at the TRIUMF OPPIS has a comparatively low emission current density and high beam divergence, which limits further current increase and gives rise to inefficient use of the cw laser power for optical pumping. In pulsed operation, suitable for high energy accelerators, the ECR source shortcomings have been avoided by using an INR-type OPPIS with a high-brightness proton source situated outside the magnetic field [4]. Studies performed in collaboration with INR, Moscow and BINP, Novosibirsk have demonstrated the feasibility of producing 20 mA polarized H$^-$ ion currents using this scheme. Proposals on pulsed OPPIS developments for future polarization faciles at RHIC and HERA are considered below.

The TRIUMF OPPIS provides a precision quality beam for studies of parity non-conservation in proton-proton scattering at 221 MeV [5]. It operates very reliably and delivers the beam for about 40% of the cyclotron operational time.

β-NMR studies of surfaces with polarized radioactive nuclide beams were proposed for the new ISAC (Isotope Separator and Accelerator) facilities at TRIUMF [6]. The application of the optical pumping polarization technique for radioactive nuclides is described for ^{8}Li and ^{17}Ne beams.

POLARIZED BEAM FOR PARITY NON-CONSERVATION STUDIES

At present the TRIUMF OPPIS is heavily used for the parity violation experiment (E497). The goal of this experiment is the measurement of the parity violating analyzing power A_z for the scattering of a longitudinally polarized 221 MeV proton beam in a hydrogen target to an accuracy of $\pm 0.2 \times 10^{-7}$. This imposes severe constraints on the polarized beam quality. From the very beginning, OPPIS development at TRIUMF has been pushed by this demanding experiment. The initial expectation that spin-reversal-correlated modulations of the beam parameters should be smaller in the OPPIS than in the ABS has been demonstrated experimentally, although it took some time to understand its origins, develop the apparatus and find the optimal set of source parameters. The detailed results from the polarized beam development for the parity experiment were presented by A.Zelenski and D.Ramsey [7] at this conference.

The optimum beam current required at the target for the parity experiment is only 0.20 uA, but to minimize the helicity correlated modulations, most of the beam intensity must be sacrificed for beam quality. Therefore, high brightness source performance is required, and the ongoing high current OPPIS development at TRIUMF is of benefit to the parity experiment. At present, polarized beam quality meets the experimental requirements. The proposed extension of the parity experiment to 450 MeV has been approved at TRIUMF [8].

PROPOSAL FOR A POLARIZED 800 GEV PROTON BEAM AT THE HERA E-P COLLIDER

Studies of the hadron spin structure functions in collisions of polarized electrons with polarized helium-3, hydrogen and deuterium internal targets are in progress at DESY (HERMES experiment). The proposal to significantly expand the kinematic range of these studies and measure the gluon contribution to the proton spin in collisions of an 800 GeV polarized proton beam with a 30 GeV polarized electron beam was recently examined [2]. TRIUMF is a part of the SPIN Collaboration, which is working on a proposal for polarized proton acceleration in HERA to 800 GeV and experiments with the polarized beams.

The TRIUMF task is the development of the high intensity polarized H^- ion source. A polarized H^- ion current of 10-20 mA is required to provide sufficient luminosity of the polarized beam for the above experiments. The feasibility of 10 mA polarized H^- ion current production in the INR-type pulsed OPPIS was demonstrated in experiments with an atomic H injector at BINP, Novosibirsk and experiments at TRIUMF on optical pumping of high density Rb vapor in the presence of a high intensity proton beam [9]. A current of 20-30 mA may be feasible in the "combined" polarization scheme, where space charge compensation is easier to achieve [10]. The development of a pulsed OPPIS is in progress at TRIUMF. The atomic H injector is being constructed and tested on a test bench. A 16 mA pulsed H^- ion current with a beam energy of 6 keV was obtained on the test bench using a geometry which closely reproduced the pulsed OPPIS layout. After optimization it will be installed at the TRIUMF OPPIS setup, as shown in Fig. 1. The optical pumping of the high density, large diameter Rb vapor cell will be achieved using a pulsed Ti:sapphire laser. Nearly 100% Rb polarization has already

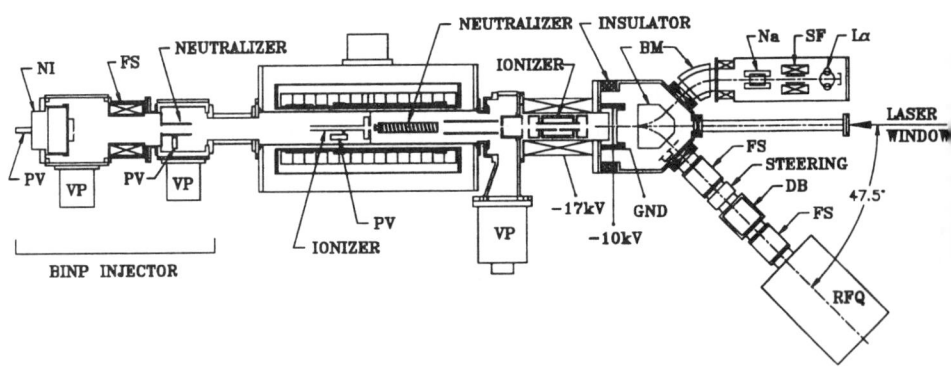

FIGURE 1. Pulsed OPPIS layout.

been obtained in an experiment with a pulsed laser under development. The H⁻ nuclear polarization will be measured and optimized using the powerful OPPIS diagnostic tools.

The high current, low energy polarized H⁻ ion beam must be accelerated immediately after the ionizer to 20-50 keV to prevent increase of the beam divergence due to space charge effects. This can be done by biasing the whole source to a potential of 20-50 kV. After acceleration in a two gap system, which will also provide the required focusing, the beam will be deflected by a bending magnet through 47.5 degrees to preserve longitudinal polarization. Alternatively, a 15 degree bend plus a solenoidal rotator can be used to align spin vertically. The beam will then be injected into an RFQ accelerator.

OPPIS INJECTOR FOR RHIC

The polarization facilities at RHIC will provide 70% polarized proton- proton collisions at energies up to $\sqrt{S} = 500$ GeV and a luminosity of 2×10^{32} cm^{-2} s^{-1} [1]. The polarized injector must produce in excess of 0.5 mA H⁻ ion current during the 300 μs pulse, or current $\times$ duration > 150 mA μs, within a normalized emittance of less than 2π mm mrad. This is an ideal application for the TRIUMF-type OPPIS, where a 1.64 mA current was obtained in dc mode. The required pulsed operation will greatly simplify and reduce the cost of the laser system, while providing the best source performance due to ample optical pumping laser power.

The OPPIS is not limited to H⁻ ion beams. A vector spin polarized D⁻ ion beam can be produced in the OPPIS using a dual optical pumping scheme, with an efficiency similar to that of the polarized H⁻ ion beam, as was demonstrated for the KEK OPPIS [11]. A large amount of nuclear spin polarized ³He atoms can be produced by direct optical pumping of helium atoms in metastable 2S states [12]. Hence, polarized ³He^{++} ion beams of 1 mA intensity can be produced via ionization in charge-exchange collisions, or in the ECR-ionizer.

There is discussion now between BNL, KEK and TRIUMF on upgrading the KEK OPPIS for the RHIC polarization facilities. This upgrade will provide a very high intensity polarized injector of H⁻ and D⁻ ion beams for high energy spin physics at RHIC.

POLARIZED RADIOACTIVE BEAMS FOR MATERIAL STUDIES AT ISAC

Implantation of several keV energy β-radioactive beams in a material surface (high temperature superconductors, or semiconductors being of greatest interest), and observation of the spin precession due to the local magnetic field

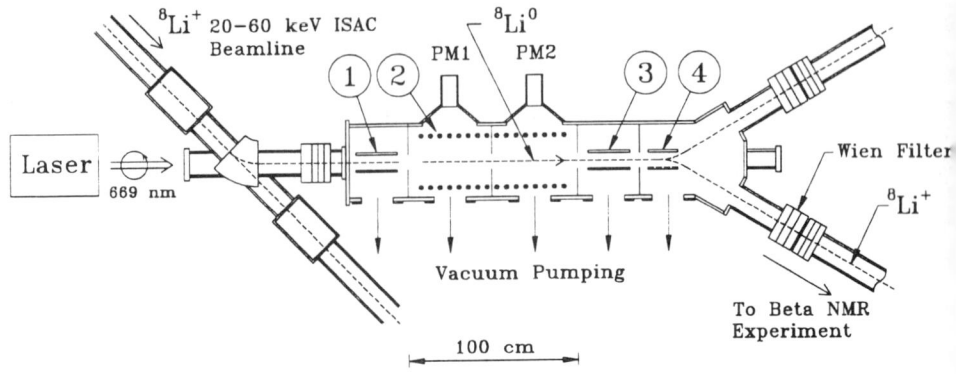

FIGURE 2. Proposed scheme for ^{8}Li beam polarization: 1) Na vapour neutralizer, 2) optical pumping region with magnetic shield and guiding magnetic field, 3) ionizer cell, 4) bender.

can be a useful tool for surface study, similar to the μSR technique for bulk materials [13]. The polarization precession can be detected by measuring the β decay asymmetry. For example, an ^{8}Li$^+$ ion beam intensity in excess of 10^8 s^{-1} will be available from the TRIUMF ISAC facility with a 10 μA proton beam at the production target. The polarization will be produced by direct optical pumping of ^{8}Li atoms in the setup shown in Fig. 2. The 20 keV Li$^+$ beam will be neutralized in a sodium vapor cell and then optically pumped by a colinear 671 nm wavelength dye laser beam (^{2}S$_{1/2}$ to ^{2}P$_{1/2}$ transition). The optical pumping region must be shielded from external magnetic fields and a homogeneous longitudinal field of a few G will be provided. The laser power density required for pumping both F=3/2, 5/2 states is about 0.5 W cm^{-2} for a multimode laser with a bandwidth of about 500 MHz, the latter being determined by the hyperfine splittings of 382 MHz in the ^{2}S$_{1/2}$ state and 44 MHz in the ^{2}P$_{1/2}$ state. After polarization, the Li beam is ionized to Li$^-$ in a second sodium cell with an efficiency of about 10%, or to Li$^+$ in a gaseous argon cell with about 30% efficiency. The ion beam is then bent and transported to the sample. This bending prevents direct deposition of sodium vapor on the sample surface and provides a convenient entrance for the laser beam. The ion beam can easily be transported a few meters, thus simplifying the obtaining of an ultrahigh vacuum in the analyzing chamber. It is estimated that about 10% of the primary beam can be optically pumped to 80-90% polarization.

Another choice of probe is a ^{17}Ne beam, which can be polarized by optical pumping in the metastable Ne* (^{3}P$_2$) state. In the sodium neutralization cell, approximately 10-20% of the initial 20 keV Ne$^+$ beam is produced in this metastable state, which can optically pumped using the ^{3}P$_2$-^{3}D$_3$ transition at

640 nm . Preferential selective ionization of the metastable atoms in charge-exchange collisions will permit the obtaining of nuclear polarizations as high as 50% in a Ne^+ ion beam suitable for implantation.

CONCLUSIONS

The powerful techniques of optical pumping and polarization transfer collisions are very successfully implemented in the high current OPPIS, which meets the demands of the new generation of high energy accelerators and colliders. The OPPIS also provides high quality beam for precision experiments at TRIUMF. We believe that development of new polarization facilities at RHIC and HERA will benefit from OPPIS technology and will, in turn, boost the further development of polarized sources.

ACKNOWLEDGEMENTS

We would like to thank J. Alessi, D. Barber, V. Davydenko, R. Kiefl, A. Krisch, Y. Mori, S. Page, T. Roser and T. Sakae for useful discussions. We acknowledge the support of SPIN Collaboration and INR-Moscow in this work.

REFERENCES

1. G.Bunce et al., *Particle World* **v3** (1992) 1-12.
2. *Prospects of the Spin Physics at HERA*, DESY-Zeuten, DESY Report (1995) 95-200.
3. A.N.Zelenski et al., *Proc. of the 1995 IEEE PAC*, Dallas (1995), 864.
4. A.N.Zelenski et al., *Nucl. Instr. Meth.* **A245** (1986) 223-229.
5. A.N.Zelenski et al., *Proc. 12th Int. Symp. on High Energy Spin Physics*, Amsterdam (1996) Ed. C.W. de Jager, World Scientific (1997) 634.
6. "A proposal for an intense radioactive beam facility", TRIUMF Report, TRI-95-1 (1995) (private communication).
7. A.N.Zelenski et al., "Polarized Beam for the TRIUMF Parity Violation Experiment", W.D.Ramsey et al., "The TRIUMF Parity Violation Experiment", this Proceeding.
8. TRIUMF experimental proposal E761, spokespersons J.Birchall, S.A. Page, W.T.H. van Oers.
9. A.N.Zelenski et al., *Rev. Sci. Instr.* **67** (1996) 1359.
10. A.N.Zelenski et al., *Proc. Conf. on Polarized Sources & Targets*, Cologne (1995) Ed. H.P. van Schieck, World Scientific (1996) 111.
11. M.Kinsho et al., *ibid.*, **10** p.126.
12. E.Otten, *ibid.*, p.3.
13. TRIUMF experimental proposals E815, E816 and E817, spokeperson R.Kiefl.

Polarized Beam for the TRIUMF Parity Violation Experiment

A.N. Zelenski[4,5], A.R. Berdoz[1], J. Birchall[1], J.D. Bowman[2],
J.R. Campbell[1], C.A. Davis[4], A.A. Green[1], P.W. Green[3], A.A.
Hamian[1], D.C. Healey[4], R. Helmer[4], S. Kadantsev[4], Y.
Kuznetsov[5], R. Laxdal[4], L. Lee[1], C.D.P. Levy[4], R.E. Mischke[2],
S.A. Page[1], W.D. Ramsay[1], S.D. Reitzner[1], G. Roy[3], P.W.
Schmor[4], J. Soukup[3], G.M. Stinson[3], V. Sum[1], T.S. Stocki[3],
N.A. Titov[5], W.T.H. van Oers[1], G.W. Wight[4]

(1) Dept. of Physics, Univ. of Manitoba, Winnipeg, MB, Canada R3T 2N2
(2) Los Alamos National Laboratory, Los Alamos, New Mexico, 87545, USA
(3) Dept. of Physics, Univ. of Alberta, Edmonton, AB, Canada T6G 2N5
(4) TRIUMF, 4004 Wesbrook Mall, Vancouver, BC, Canada V6T 2A3
(5) Institute for Nuclear Research, Russian Academy of Sciences, 117312 Moscow, Russia

Abstract.
Precision measurements of parity violation in proton-proton scattering impose very stringent requirements on the polarized beam quality. The contributions of random and spin-correlated beam modulations to the experimental accuracy are discussed. This analysis for the TRIUMF optically pumped polarized ion source and H$^-$ ion cyclotron with multiturn extraction is quite different from previous experiments at LAMPF, PSI and Bonn. The required beam quality is obtained by combining efforts at every stage: reduction of modulations at the source, minimization of transfer coefficients in the injection line and cyclotron, feedback systems for injected and accelerated beam stabilization, precision measurements of the beam energy, intensity and polarization profiles, and reduction of the sensitivities of the experimental apparatus to these modulations.

INTRODUCTION

Using spin to probe the proton structure is now generally recognized as a sensitive tool for testing QCD predictions of the Standard Model. Measurements of parity violation in proton-proton scattering can be considered in this context as complementary to high-energy studies of the proton structure, and may be sensitive to essential details which are otherwise inaccessible. The parity violation experiment underway at TRIUMF (E497) will measure the helicity dependence of p-p scattering, $A_z = \frac{1}{P_z}(\frac{\sigma^+ - \sigma^-}{\sigma^+ + \sigma^-})$ at 221 MeV. Direct quark model predictions[1] for A_z differ significantly from the meson exchange model[2]; to make a meaningful test, meeting the accuracy goal of $\pm 0.2 \times 10^{-7}$, as achieved previously at 13.6 (Bonn) and 45 MeV (SIN), is absolutely essential.

In contrast with the lower energy experiments, a transmission technique is used at TRIUMF, where an optically-pumped polarized H$^-$ ion source (OPPIS) and H$^-$ ion cyclotron with multiturn stripping extraction deliver a longitudinally polarized beam to a 40 cm liquid hydrogen target. The schematic layout of the TRIUMF experiment is presented in Fig.1. In this paper, the minimization of selected systematic error contributions due to the polarized beam properties is discussed. General descriptions of the experimental apparatus have been presented earlier[3].

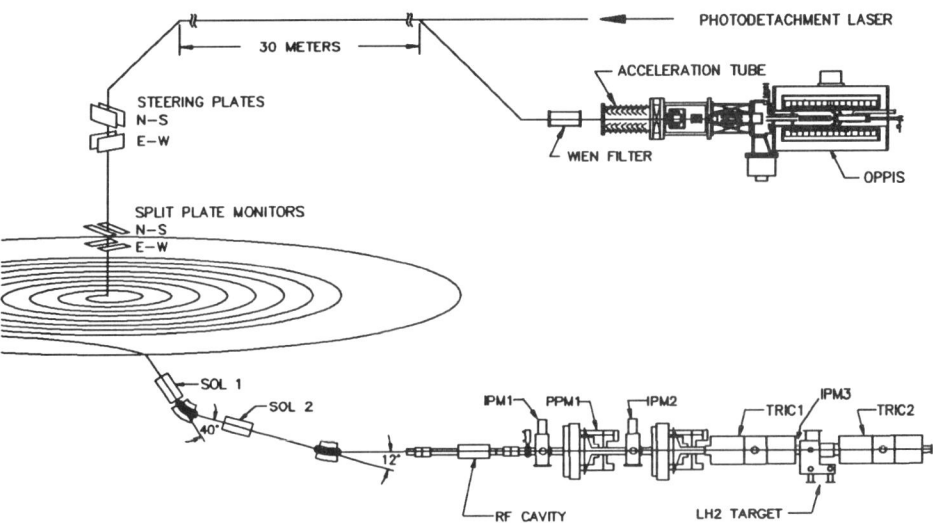

FIGURE 1. General layout of the TRIUMF parity experiment.

The optimum beam current at the parity apparatus is only 200 nA, but to achieve very small spin-correlated modulations, most of the source intensity must be sacrificed for beam quality. For example, the RF bunchers in the injection beamline which enhance the cyclotron transmission efficiency by a factor of 4 also increase the sensitivity to energy modulation by more than two orders of magnitude, and cannot be used for the parity experiment. Therefore, high brightness source performance is required, and the ongoing high-current OPPIS development[4] at TRIUMF is of benefit to the parity experiment.

There are two types of beam property fluctuation which affect the experimental accuracy. Modulations correlated with spin reversal are the most dangerous, since they contribute to systematic errors ΔA_z; in contrast, uncorrelated fluctuations contribute to the RMS noise in the measurements and increase the running time of the experiment. While spin-correlated modulations originate at the polarized ion source, beam transport through elements of the injection beamline, cyclotron, and extraction beamline can amplify and mix these modulations. Furthermore, the amplification and mixing factors can vary with time. To find tuning procedures that produce the best uncorrelated long and short term beam stability with the least possible amplification of the correlated modulations, various techniques for introducing relatively large artificial "spin-correlated" beam intensity, position, size and energy modulations have been developed. The sensitivities to these modulations are then measured at the parity experimental apparatus for various injector and cyclotron tunes.

Current Modulation

Spin-flip correlated current modulation $\frac{\Delta I}{I} = \frac{I^+ - I^-}{I^+ + I^-}$ introduces a systematic error ΔA_z via nonlinearity of the main detectors and associated electronics. The parity apparatus achieves minimal sensitivity to this type of modulation by precision analog subtraction of current signals from two identical parallel plate ionization chambers upstream and downstream of the liquid hydrogen target. A source of artificially enhanced "spin correlated" current modulation is required for tuning the precision subtractor circuitry for minimum current modulation sensitivity. This capability is provided by an auxilliary 16 W argon-ion laser beam which copropagates with the H^- beam along the 30 m long horizontal section of the injection beamline (as shown in Fig.1), neutralizing by photodetachment a fraction of the H^- ions in its path. A small mirror attached to an electromagnetic driver is used to interrupt the photodetachment laser beam synchronously with the parity spin sequence, so that the beam current in every second spin "off" data cycle is modulated at the 0.1% level according to the parity spin switching pattern. This technique provides online monitoring of the systematic error ΔA_z due to correlated current modulation.

Numerous improvements have been made in stabilization of the primary ECR proton source, the Rb vapor thickness, the injection beamline elements – the Wien filter was one of the most difficult – and cyclotron, in order to reduce the widths of the $\frac{\Delta I}{I}$ and A_z distributions. One of the most important improvements was active stabilization of the injected beam position (see Fig.1), which substantially reduces the beam noise due to low frequency mechanical vibrations of the injection beamline. The current drift and sparking rates have also been greatly reduced. As a result, spin-correlated current modulations and stability now meet the parity experiment tolerances of $|\frac{\Delta I}{I}| \sim 1 \times 10^{-5}$

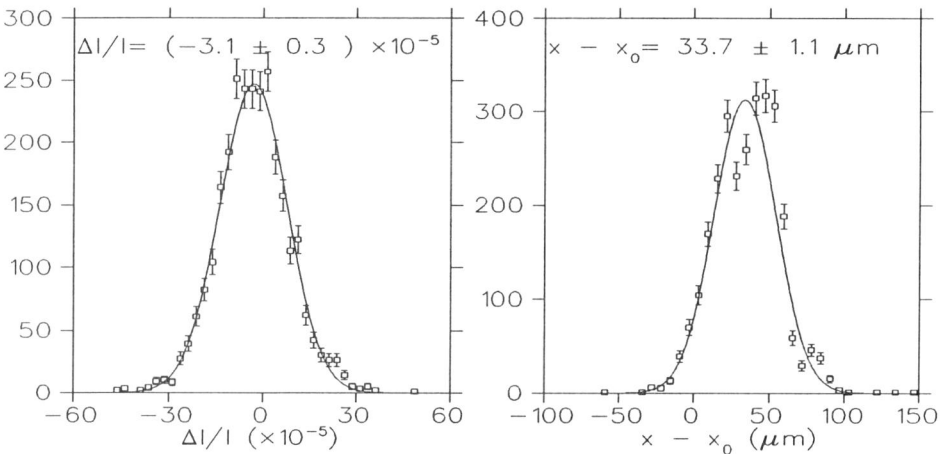

FIGURE 2. (a) Spin correlated intensity modulation measaured by the main parity detectors. (b) Beam position deviation from the symmtery axis (in x). These histograms represent 20 hours of integration time.

and $|(I - I_\circ)| \leq 2$ nA respectively. A typical $\frac{\Delta I}{I}$ histogram is shown in Fig. 2a.

Beam Positon Stabilization

Three intensity profile monitors (IPMs, see Fig. 1) are used to monitor the beam position and its spin correlated modulations. Each IPM is composed of two harps (one each for x and y); each harp is assembled from 31, 3 μm thick, 1.5 mm wide nickel foil strips which are soldered to a PC board using a precision mechano-optical setup, with positioning accuracy of ± 10 μm for each strip. The harp signals are individually amplified and digitized to provide the beam intensity distributions in x and y, from which the beam positions are extracted.

The amplified harp signals are also used in analog beam centroid evaluators (BCEs), which provide the feedback signals for the fast steering magnets which stabilize the beam position on the experimental symmetry axis. IPM3, which was recently installed just upstream of the LH_2 target, greatly improved the accuracy of the position stabilization loop by virtue of its extended lever arm. The advantage of the BCE-based position stabilization system over the split plate monitor-based system which was used initially is a reduced sensitivity to shape fluctuations in the beam intensity profile. As a result, the beam position deviations from the symmetry axis were reduced to within about 50 μm (see Fig. 2b), and all of the false asymmetries which are proportional to these displacements are now consistent with zero[5].

Correlated Energy Modulation (CEM)

The sensitivity $\frac{\Delta A_z}{\Delta E}$ arises from the fact that the beam energy is 27 MeV lower in the downstream ionization chamber than in the upstream one, due to energy loss in the liquid hydrogen target. This sensivitity was estimated to be 2.8×10^{-8} eV^{-1} from the energy dependence of $\frac{dE}{dx}$ in the detector gas, in excellent agreement with the value measured directly using an RF postaccelerator in the parity beamline[6]. Since the contributions to the false asymmetry, ΔA_z, caused by all of the measureable coherent modulations are known, the parity signal itself can be used to deduce CEM in the 221 MeV beam. CEM of the accelerated beam is caused by coherent modulation of the radial intensity distribution at the cyclotron extraction foil; this converts position modulation in the injected beam to energy modulation in the extracted beam. The conversion factor was estimated to be $\frac{dE}{dx,injected} \sim 100$ eV μm^{-1}, in agreement with direct measurements made by applying a large position modulation to the injected beam using electrostatic steering plates.

One source of spin correlated position modulation of the injected beam is correlated energy modulation of the beam produced in OPPIS, which is converted to position modulation as the beam passes through electrostatic steering plates, then back to energy modulation at the extraction foil as discussed above. This process amplifies the primary CEM by a factor of approximately 100, as measured using a magnetic spectrometer at 221 MeV[6]. The sensitivity of A_z to spin correlated energy modulation produced in the ion source was measured by applying a square wave voltage of 0.5 V amplitude to the isolated sodium ionizer cell in the OPPIS. The sensitivity $\Delta A_z \sim 0.4 \times 10^{-8}$ for 1 meV of primary CEM was measured, in agreement with the conversion

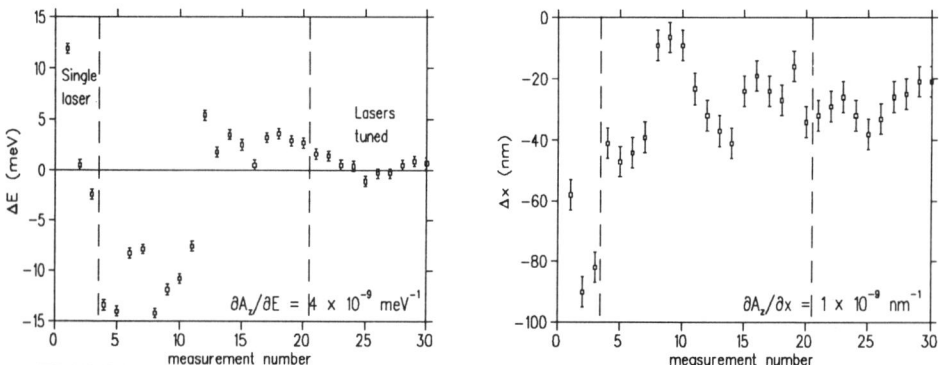

FIGURE 3. (a) Coherent energy modulation measured at the ion source. (b) Coherent position modulation measured at the ion source.

factor measurements. These studies indicate that a CEM of 10 meV in the OPPIS can produce a significant false asymmetry, ΔA_z. Pure spin correlated beam position modulation (CPM) produced at the OPPIS can also contribute directly to ΔA_z, with a measured sensitivity of approximately 0.1×10^{-8} for 1 nm of CPM.

CEM and CPM produced at the OPPIS were studied in a series of measurements using a pair of electrostatic steering plates as a beam energy analyzer, and an intensity profile monitor with 16 collector strips (similar to the parity IPMs) to measure the beam position downstream of the steering plates. This monitor was mounted on a swinging lever arm, and the coherent position modulation of the beam was measured for the left and right monitor positions. These two measurements allowed the separation of the energy modulation and position modulation components of the OPPIS' beam. The highly stable H$^-$ beam, together with synchronous detection techniques allowed an accuracy of 2 nm in 20 minutes of integration time for the spin correlated position measurements. Taking into account the CEM calibration factor of typically 10 nm meV^{-1}, an accuracy of 0.2 meV (in 20 minutes) was achieved for the coherent energy measurements. The results of the CEM and CPM measurements are presented in Fig. 3. The amplitudes of the modulations are quite sensitive to the pumping lasers' asymmetry between the two polarization states; after careful tuning of the lasers, the coherent energy modulation was reduced to 1-2 meV, and the coherent position modulation to the 20 nm level. The CPM and CEM produced by the OPPIS, and the corresponding sensitivities, will be regularly monitored during the parity data taking, so that corrections to A_z can be applied if necessary.

Transverse Polarization and Polarization Moments

The particle orbits are not separated in the TRIUMF cyclotron, and hence several turns are extracted simultaneously by the stripping foil. The proton spin precession during acceleration is different for different orbits, and averaging of several turns at extraction should reduce the transverse polarization components. Experimentally, a reduction factor $\frac{P_{x,extracted}}{P_{x,injected}}$ of about 0.05 was measured (see Fig. 4). A similar reduction should work for polarization moments which are produced at OPPIS and in the injection beamline. The upper limit for the magnitude of moments produced in the extraction beamline was estimated to be less than 10 μm, therefore large polarization moments that have been measured in previous tests (50-80 μm) can only have been produced in the cyclotron. Cyclotron tuning conditions were identified to keep the moments within the acceptable range for Parity.

The achieved polarized beam quality and performance of the parity experimental apparatus will allow a statistical accuracy of $\delta A_z = \pm(0.2 \times 10^{-7})$

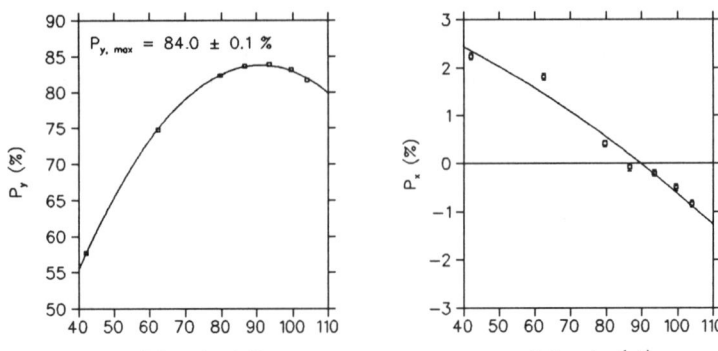

FIGURE 4. The transverse polarization components of the extracted beam as a function the Wien filter magnetic field.

in about 600 hours of datataking. Preliminary analysis of August 1996 and February 1997 runs gives the result $A_z = (1.1 \pm 0.6) \times 10^{-7}$[5], and demonstrates significant progress in tracking down systematic errors.

We thank the engineering and cyclotron operations staff for their support, patience, and helpful suggestions during this very challenging experiment.

REFERENCES

1. Grach I., and Shmatikov M., *Phys. Lett.* **B316**,467 (1993).
2. Driscoll D.E., and Miller G.A., *Phys. Rev.* **C40**, 2159 (1989).
3. J. Birchall et al., AIP Conf. Proc. **339** (1995) 136.
4. Zelenski A.N. et al., Proc. 1995 Particie Accelerator Conf., IEEE (1996) 864.
5. Ramsay W.D. et al., "The TRIUMF Parity Violation Experiment", these proceedings.
6. Zelenski A.N. et al., Proc. 12th Int. Symp. on High-Energy Spin Physics, Amsterdam (1996) Ed. C.W. de Jager, World Scientific (1997) 634.

Neutron-Antineutron Transition Search at HFIR Reactor

Yuri A. Kamyshkov

*Department of Physics, University of Tennessee, Knoxville, TN 37996
and Physics Division, Oak Ridge National Laboratory, Oak Ridge, TN 37831* [1]

Abstract. A new experiment to search for neutron-antineutron transitions was recently proposed for High Flux Isotope Reactor (HFIR) at Oak Ridge National Laboratory (ORNL). In this paper the physics motivation of a new search, the scheme and the discovery potential of the proposed HFIR-based experiment are discussed.

Physics Motivation

Baryon asymmetry of the universe [1] and the ideas of unification of particle and forces [2,3] are the two global concepts which motivated the experimental searches [4] of baryon instability for more than two decades. Neutron-antineutron transitions, first considered in [5–7] within the context of these concepts and violating the baryon number by two units ($\Delta B = 2$), may be a phenomenon preferred by nature, which is alternative or complementary to the proton decay ($\Delta B = 1$). The most recent reviews of theoretical and experimental situation related to baryon instability search can be found in [8]. The experimental status of proton decay search was also discussed by T. Haines [9] at this conference.

Possible nonconservation of the baryon number is closely related to the nonconservation of lepton number. Thus, for example, in the proton decay, the conservation of angular momentum (spin of proton) requires that at least one lepton should be present in the final state. This creates in general two possibilities corresponding to $\Delta B = \Delta L$ and $\Delta B = -\Delta L$; the first one conserving $(B - L)$ and the second one violating $(B - L)$ by two units. In neutron-antineutron transition, since leptons are not involved, the $(B - L)$ is violated by two units.

---

[1] managed by Lockheed Martin Energy Research Corp. for the U.S. Department of Energy under contract number DE-AC05-96OR22464.

The original $SU(5)$ unification model [3], where $(B-L)$ was conserved, favored the proton decay mode $p \to e^+ + \pi^0$ with predicted lifetime $\tau/B < 4 \cdot 10^{31}$ years. This model has been ruled out by the experiments where τ/B was measured to be $> 10^{33}$ years [8,9]. This situation raises a question whether the $(B-L)$ in general is conserved and motivates the experimental searches for $(B-L)$ nonconserving processes. There are also several other reasons to believe that $(B-L)$ might be not conserved in nature.

It was shown [10] that in baryogenesis the nonperturbative Standard Model effects at electroweak energy scale erase any baryon excess generated at the early moments of the universe $(T \gg 1 \text{ TeV})$ through $(B-L)$ conserving processes. At the same time, generating baryon excess through electroweak effects alone does not seem to be adequate to account for the observed baryon asymmetry and the dark matter [11,12]. Thus, a component with $\Delta(B-L) \neq 0$ might be required to explain the baryogenesis [13].

Standard Model weak interactions are not left-right symmetric like all other interactions: electromagnetic, strong, and gravitational. It is natural to think that the restoration of left-right symmetry should take place before the ultimate unification of all fundamental interaction can occur. In left-right symmetric unification models [2,14] the left-right symmetry is broken at the intermediate energy scale simultaneously with $(B-L)$ violation [15,7]. In such models the transitions $n \to \bar{n}$, as well as other $(B-L)$ nonconserving processes, might exist with the rates attainable by the modern experiments [16].

The smallness of neutrino masses in theory is explained by see-saw mechanism (see discussion in [13]) which implies the existence of heavy right-handed Majorana masses. Majorana neutrinos would violate both L and $(B-L)$ by two units. Atmospheric neutrino data [17] and LSND results [18], both calling for rather large Δm^2 values in neutrino oscillation scenario, would require the Majorana masses and the $(B-L)$ nonconservation energy scale to be in an intermediate energy range.

Observation of "atmospheric neutrino anomaly" [17] allows an interpretation [19] of measured excess of $e^\pm$ in the detectors as an observation of a proton decay into the mode $p \to e^+ + \nu + \nu$. $(B-L)$ is not conserved in such a decay. The "observed" rate of events corresponds to the proton lifetime of $\tau/B \approx 4 \cdot 10^{31}$ years and indicates the energy scale of $\sim 10^5$ GeV from where this process is originating.

An interesting possibility which might lead to the alternative mechanism of baryon (and $B-L$) number violating processes has been recently discussed by V. Kuzmin [20]. He assumed that interactions of quarks inside the baryons consisting of quarks from the different families (bus for example) might be mediated by the family-colored triplet scalar field coupled to the right components of the quarks. For neutral baryons such scalar field interaction will result in baryon-antibaryon ($bus \to \bar{b}\bar{u}\bar{s}$) oscillations with the characteristic time $\sim 10^{-12}$ s. Such transitions can be searched at B-factories. The $n \to \bar{n}$ transitions in this model would arise as radiative corrections to this new interaction with additional suppression by ~ 20 orders of magnitude in probability. This will result in n $\to$ n̄ characteristic transition time of $\tau_{n\bar{n}} \sim 10^8$ s, i.e, close to the existing experimental limits. As mentioned in [20], the 4-jet events, observed by ALEPH collaboration at LEP-II, which are peaked at ~ 105 GeV, produced with rather high cross section, and have no signature of b-quark jets in the final state, can be explained by this model as a 3-rd component family-colored scalar-antiscalar production.

The arguments, presented above, although allowing the alternative interpretations, let us think that the $(B-L)$ might not be conserved and the energy scale corresponding to $(B-L)$ nonconservation can be as low as $\sim 10^5 - 10^6$ GeV. Possible phenomena related to $\Delta(B-L) \neq 0$ would include: proton decay into modes $N \to l + mesons$ and $N \to ll\bar{l} + (mesons)$; Majorana masses for the neutrinos; neutrinoless double beta decay; transitions $bus \to \bar{b}\bar{u}\bar{s}$; intranuclear transitions of two nucleons into pions; and n $\to$ n̄ transitions. The question of whether such physics exists can be answered only experimentally. If it does exist, the experimental observation of n $\to$ n̄ transitions with free neutrons in a new reactor experiment will be its most clear and spectacular manifestation since (a) the detection signal for n $\to$ n̄ transition is clean and unambiguous and (b) the discovery potential for n $\to$ n̄ search can be experimentally advanced by a factor of $\sim 1,000$ relative to the present experimental limits [21].

Neutron-Antineutron Transition Search

The n $\to$ n̄ transitions can be searched (a) by utilizing free neutrons from reactors or neutron spallation sources and (b) with neutrons bound inside the nuclei.

The yield of antineutrons $N_{\bar{n}}$ in the beam of free neutrons in vacuum (in the absence of external fields) due to n $\to$ n̄ transitions depends on the observation time t as [7] $N_{\bar{n}} \propto N_n \cdot (t/\tau_{n\bar{n}})^2$, where N_n is the number of neutrons used in an experiment and $\tau_{n\bar{n}}$ is the characteristic n $\to$ n̄ transition time. It is assumed in this expression that neutrons and antineutrons have equal masses (as required by CPT theorem) and that the gravitational interaction with earth is the same for neutrons and antineutrons. In this way the *discovery potential* of an n $\to$ n̄ search experiment is proportional to the neutron flux and to the square of the neutron time-of-flight. High steady-flux reactors together with cold neutron moderators, which slow down the velocities of neutrons, would be, therefore, most appropriate for an n $\to$ n̄ search.

The general scheme of such an experiment is the following: neutrons emitted from the cold moderator are propagating in the vacuum volume (shielded against earth's magnetic field down to the level of a few nT) where the n $\to$ n̄ transition occurs. Produced antineutrons propagating along the initial neutron path would be detected as a few-meson star with total energy release of ~ 1.8 GeV resulting from the antineutron annihilation with a thin target film.

The recent most advanced experimental search for n $\to$ n̄ with free neutrons was performed [22] at the 58 MW research reactor at the Institute Laue-Langevin (ILL) in Grenoble. The experiment had a discovery potential of $N_n t^2 \sim 1.5 \cdot 10^9$ seconds and for one year of operation set a limit of $\tau_{n\bar{n}} \geq 8.6 \cdot 10^7$ s.

Intranuclear transition time τ_A is related to free neutron transition time $\tau_{n\bar{n}}$ as $\tau_A = T_R \cdot (\tau_{n\bar{n}})^2$, where T_R is the nuclear suppression factor. This factor has been evaluated by several authors during the last two decades. Most recent discussions and new reevaluations, as well as references to the previous works, can be found in [8]. According to [23], for oxygen, argon, and iron the suppression factor has a value $T_R \sim 2 \cdot 10^{23}$ s^{-1}.

Experimentally, the n $\to$ n̄ transitions have been searched in the nucleon stability experiments IMB, Kamiokande, and Fréjus [21]. For example, the limit for the intranuclear n $\to$ n̄ transition lifetime for iron nuclei set by Fréjus experiment is $\tau_{Iron} \geq 6.5 \cdot 10^{31}$ years which, according to the suppression factor from [23], corresponds to $\tau_{n\bar{n}} \geq (0.8 - 1.0) \cdot 10^8$ s.

During the next decade, the large new detectors SuperKamiokande and Icarus might improve the intranuclear n $\to$ n̄ transition limit. Thus, after a few years of operation the SuperKamiokande detector commissioned in April 1996 will be able to set an n $\to$ n̄ transition limit of $\tau_{Oxygen} \geq 10^{33}$ years [24].

The relative potentials of different methods of n $\to$ n̄ search were discussed in [25]. A new approach to intranuclear baryon instability search via the detection of traces of long-lived technetium isotopes in the deep-mined ores was presented by Yu. Efremenko at this conference [26].

The scheme and the discovery potential for a new proposed experiment with free neutrons at the HFIR reactor at ORNL are discussed below.

A New HFIR-Based Experiment

A new experiment for n $\to$ n̄ search at 100 MW HFIR reactor was proposed by a UT-ORNL group [27]. Proposed improvement in the discovery potential [28] is based on the properties of neutrons to be focused by means of reflection from the surfaces of certain materials. In this new approach an elliptical shape reflector intercepts the neutrons emitted from the cold source within a large solid angle and focuses them on the annihilation target situated at the distance 200-250 m from the source.

Schematic layout of the proposed HFIR-based experiment is illustrated in Figure 1. The gain in discovery potential (relative to the ILL-based experiment [22]) will result from the following factors: higher reactor power, larger area of the cold neutron emitting source, larger area of the annihilation detector, but, most essentially, from the use of a large-acceptance elliptical focusing

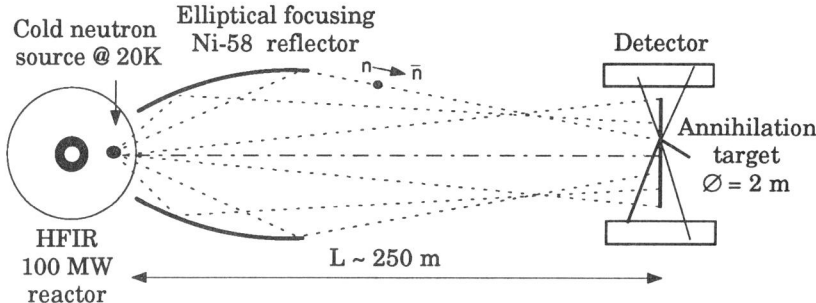

FIGURE 1. Conceptual layout of the experiment with a large elliptical focusing reflector for an n → n̄ transition search at the HFIR reactor (not to scale).

reflector. For three years of operation at HFIR the discovery potential can be increased by a factor of $\sim 1,000$ relative to the present experimental level.

Conclusions

The prospects of an n → n̄ transition search with free neutrons in reactor experiments and in intranuclear nucleon stability search experiments are compared in Figure 2. The HFIR reactor-based experiment should allow for three years of operation to increase the discovery potential of n → n̄ transition search by a factor of $\sim 1,000$ or to set the limit for free n → n̄ transition time $\geq 3 \cdot 10^9$ s. This will explore the stability of nuclear matter, although only in one particular mode of n → n̄ transitions, to the intranuclear transition lifetime limit of $\sim 10^{35}$ years which is not attainable by other baryon instability search experiments.

If n → n̄ transition would be found, it will establish a new phenomenon leading to a new physics at the energy scale of $\sim 10^5$ GeV, i.e., beyond the range of colliders. The new symmetry principles determining the history of the universe during the first moments of creation might be revealed; the left-right symmetry, broken in the Standard Model, may be restored. The discovery of n → n̄ transition might provide a major steering impact to the unification models and contribute to the understanding of baryon asymmetry in the universe. If and when such phenomenon is established, subsequent experiments with n → n̄ transition should allow, according to [29], to perform a most precision test of CPT invariance and the test of gravitational equivalence of baryonic matter and antimatter.

I would like to thank W. M. Bugg, Yu. V. Efremenko, V. A. Kuzmin, R. N. Mohapatra, and C. D. West for useful and stimulating discussions.

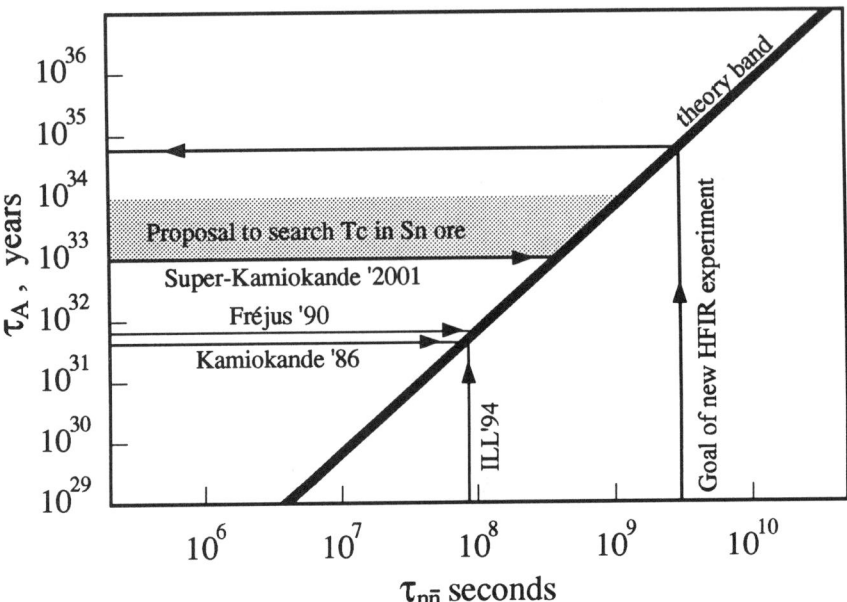

FIGURE 2. Comparison of n → n̄ search in intranuclear transitions (τ_A) to those in free neutron experiments ($\tau_{n\bar{n}}$). The slope and the width of the nuclear model band connecting these two processes corresponds to $\tau_A = T_R \cdot (\tau_{n\bar{n}})^2$, where T_R is the nuclear suppression factor taken from [23].

REFERENCES

1. Sakharov A., *JETP Lett.* **5** (1967) 24.
2. Pati J., and Salam A., *Phys. Rev.* **D10**, 275 (1973).
3. Georgi H., and Glashow S., *Phys. Rev. Lett.* **32**, 438 (1974).
4. Goldhaber M., "Search for Nucleon Instability (Origin and History)", *in Proceedings of International Workshop on Future Prospects of Baryon Instability Search in p-Decay and n → n̄ Oscillation Experiments*, Oak Ridge, 1 (1996).
5. Kuzmin V., *JETP Lett.* **12**, 228 (1970).
6. Glashow S., *preprint* HUTP-79/A059.
7. Mohapatra R., and Marshak R., *Phys. Lett.* **91B**, 222 (1980); *Phys. Rev. Lett.* **44**, 1316 (1980); *Phys. Lett.* **94B**, 183 (1980);
8. *Proceedings of International Workshop on Future Prospects of Baryon Instability Search in p-Decay and n → n̄ Oscillation Experiments*, Oak Ridge (1996).
9. Haines T., *Overview of Proton Decay*, these proceedings.
10. Kuzmin V., Rubakov V., and Shaposhnikov M., *Phys. Lett.* **155B**, 36 (1985).
11. Rubakov V., and Shaposhnikov M., *Phys. Usp.* **39**, 461 (1996).

12. Kuzmin V., *preprint* hep-ph/9701269 (1996).
13. Pati J., *in Proceedings of International Workshop on Future Prospects of Baryon Instability Search in p-Decay and $n \to \bar{n}$ Oscillation Experiments*, Oak Ridge, 7 (1996); *preprint* hep-ph/9611371 (1996).
14. Mohapatra R., and Pati J., *Phys. Rev* **D11**, 2558 (1975).
15. Davidson A., *Phys. Rev.* **D20**, 776 (1979).
16. Mohapatra R., *in Proceedings of International Workshop on Future Prospects of Baryon Instability Search in p-Decay and $n \to \bar{n}$ Oscillation Experiments*, Oak Ridge, 73 (1996); *preprint* hep-ph/9604414 (1996).
17. Haines T., *SuperKamiokande Results*, these proceedings; Clark R., *The Atmospheric Muon Neutrino Fraction Above 1 GeV*, these proceedings; M. Goodman *New Soudan ν_μ/ν_e Results*, these proceedings.
18. Mills G., *Results from the LSND Decay in Flight Oscillation Search*, these proceedings.
19. Mann W.A., *in Proceedings of International Workshop on Future Prospects of Baryon Instability Search in p-Decay and $n \to \bar{n}$ Oscillation Experiments*, Oak Ridge, 175 (1996).
20. Kuzmin V., *in Proceedings of International Workshop on Future Prospects of Baryon Instability Search in p-Decay and $n \to \bar{n}$ Oscillation Experiments*, Oak Ridge, 89 (1996); *preprint* hep-ph/9609253 (1996).
21. Particle Data Group, "Review of Particle Physics", *Phys. Rev* **D54**, 1 (1996).
22. Baldo-Ceolin M., et al., *Z. Phys* **C63**, 409 (1994).
23. Alberico W., *in Proceedings of International Workshop on Future Prospects of Baryon Instability Search in p-Decay and $n \to \bar{n}$ Oscillation Experiments*, Oak Ridge, 221 (1996).
24. Stone J., *in Proceedings of International Workshop on Future Prospects of Baryon Instability Search in p-Decay and $n \to \bar{n}$ Oscillation Experiments*, Oak Ridge, 357 (1996).
25. Kamyshkov Yu., *Nuclear Physics B (Proc. Suppl.)* **52A**, 263 (1997).;
    Kamyshkov Yu., *in Proceedings of International Workshop on Future Prospects of Baryon Instability Search in p-Decay and $n \to \bar{n}$ Oscillation Experiments*, Oak Ridge, 281 (1996).
26. Efremenko Yu., *Baryon Instability Search with Long-Lived Isotopes*, these proceedings;
    also Efremenko Yu., et al., *in Proceedings of International Workshop on Future Prospects of Baryon Instability Search in p-Decay and $n \to \bar{n}$ Oscillation Experiments*, Oak Ridge, 307 (1996).
27. Bugg W., et al., *Letter of Intent to the Oak Ridge National Laboratory to Search for the $n \to \bar{n}$ Transition Using a Detector to be Built at ORNL's High Flux Isotope Reactor*, UTK-PHYS-96-L1 (1996).
28. Kamyshkov Yu., *in the Proceedings of the ICANS-XIII meeting*, PSI, Villigen, 843 (1995).
29. Okun L., "Test of CPT", *preprint* ITEP-TH-55/96; hep-ph/9612247 (1996).

Status of the Focal Plane Polarimeter for Hall A at TJNAF

M. K. Jones*, F. T. Baker[||], L. Bimbot[‡], E. J. Brash[††],
R. Gilman[‡], C. Glashausser[‡], G. Kumbartzki[‡], J. McIntyre[‡],
C. F. Perdrisat*, V. Punjabi[†], G. Quéméner*, R. Ransome[‡],
P. M. Rutt[‡], K. Wijesooriya*, G. D. Zainea[††]
and the TJNAF Hall A Collaboration.

*College of William and Mary, Williamsburg, VA 23187, USA
[†]Norfolk State University, Norfolk, VA 23504, USA
[‡]Rutgers University, Piscataway, NJ 08855, USA
[||]Unviersity of Georgia, Athens, GA 30602, USA
[††]University of Regina, Regina, SK S4s 0A2, Canada

Abstract. The focal plane polarimeter (FPP) for Hall A at the Thomas Jefferson National Accelerator Facility (TJNAF) is ready for experiments. The ability to calibrate the FPP on-site using elastic scattering of polarized electrons from an unpolarized hydrogen target is demonstrated. The ratio of the proton electric form factor to its magnetic form factor has been measured at $Q^2 = 0.810 \ (GeV/c)^2$.

INTRODUCTION

Intermediate-energy proton polarimeters are generally based on nuclear scattering of the proton from a carbon analyzer. Due to the spin-orbit force an azimuthal asymmetry is measured in the differential cross section. The focal-plane polarimeter (FPP), installed in the Fall of 1996 in one of the Hall A high resolution spectrometers at the Thomas Jefferson National Accelerator Facility (TJNAF), is unique in several aspects, in that it is physically large, makes use of straw tube drift chambers and multiplexing electronics, and is the second FPP to be operated at an electron accelerator.

PHYSICAL DESCRIPTION OF THE FPP

The FPP consists of two straw tube drift chambers in front and two straw tube drift chambers behind the carbon analyzer. The straw chamber design is based on a design originally created for the EVA cylindrical straw chamber, experiment E850 at Brookhaven National Laboratory [1]. The straw tubes were made by wrapping together 10 micron thick aluminum foil and two 50 micron thick mylar layers plus heat setting glue. The inner diameter of the straws is 0.522 cm.

The front chambers were constructed identically and each has three U planes and three V planes adding up to a total of 2016 straws. The U and V planes are perpendicular to each other and rotated 45 degrees from the x-axis. The x-direction is the dispersive direction and z is in the proton momentum direction. For both the U and V planes, the three layers of straws are arranged so that the middle row is offset by half a straw diameter and the bottom and top layers line up. The active area of the front chambers is 60 cm in the y-direction by 209 cm in the x-direction. The distance between the front chambers is 117 cm.

To track the particle after scattering in the carbon analyzer two straw tube chambers are used. The rear chamber closest to the carbon analyzer has two U planes, two V planes and two X planes with the planes offset by a straw radius. This chamber has an active area of 124 cm in the y-direction by 272 cm in the x-direction. The X planes are used in reconstructing multiple-track events. The second rear chamber is 42 cm behind the first rear chamber and it has three U planes and three V planes with the middle plane offset by a straw radius. This chamber has an active area of 135 cm in the y-direction by 290 cm in the x-direction. The total number of straws for the rear chambers is 2274. The distance between the second front chamber and the first rear chamber is 97 cm.

The carbon analyzer is divided into five moveable sections. The sections have thicknesses of 1.9 cm, 3.8 cm, 7.6 cm, 15.2 cm and 22.9 cm with the thickest one closest to the front chambers. The ability to vary the carbon thickness allows optimization of the FPP performance. The distance between the second front chamber and the front of the 22.9 cm carbon analyzer is 26.5 cm. The large size of the rear chambers allows detection of all events scattered in the carbon with through angles less than 20 degrees.

Given the large number of straws, multiplexing of the readout of the straw signals was imperative to reduce costs. The readout of the straw signals is multiplexed in groups of eight which would allow about 100 kHz rate capability per straw. The Rutgers University electronics shop built readout boards which were small enough to be located near the straws (reducing possible noise pick-up) and could still contain the circuitry for amplifying, discriminating and multiplexing the straw signals. The multiplexing is done by using discriminator one-shots with fixed widths that vary from 25 to 105 ns for each group of eight. Multi-hit TDCs are used to measure the leading and trailing edge of the pulse. The leading edge determines the drift time and the difference between trailing and leading edges identifies the wire within the group of eight.

The gas used in the FPP chambers is a 62/38 mixture of argon/ethane. The threshold for the amplifier was set to correspond to chamber signals of about 2 μA. The efficiency of the chambers was measured to be 96%. Comparing FPP chamber tracks to those measured by the vertical drift chambers one can calculate the tracking efficiency of the FPP chambers to be 99%.

PRINCIPLES OF POLARIMETRY

The FPP will be used to measure the polarization of protons produced in electron and photon reactions. The components of outgoing proton polarization at the target are along the direction of the proton, P_z, in the dispersive direction, P_x, and in the transverse direction, P_y. While traveling through the spectrometer the proton polarization components at the target precess in the magnets. In the case of a simple dipole the proton polarization components at the focal plane (designated by a prime) are

$$P'_y = hP_y \quad P'_x = P_x \cdot \cos(\chi) - hP_z \cdot \sin(\chi) \tag{1}$$

In the equation χ is the spin precession angle and h is the beam helicity. The spectrometer's central bend angle is 45° for the Hall A dipole.

The FPP can only measure the transverse polarization components P'_x and P'_y but not the longitudinal component P'_z. The scattering of the proton in the carbon can be described in polar coordinates by scattering angles θ and ϕ. The ϕ-distribution can be made by summing over a range of θ. A ϕ-distribution which includes instrumental asymmetries would be

$$N(\phi) = N_o \cdot (1 + (a + a_i)\cos\phi + (b + b_i)\sin\phi + c_i \cos(2\phi) + d_i \sin(2\phi)) \tag{2}$$
$$a = -AP'_x \text{ and } b = AP'_y \tag{3}$$

where N_o is the average number of scattering events, A is the analyzing power of carbon and a_i, b_i, c_i and d_i are coefficients describing instrumental asymmetries.

For polarized electron scattering on a proton, $P_x \approx 0$. The other two components are

$$P_y = -2\sqrt{\tau(1+\tau)}G^p_E G^p_M \tan(\frac{\theta_e}{2})/I_o$$

$$P_z = \left(\frac{E_e + E'_e}{M}\right)(G^p_M)^2 \sqrt{\tau(1+\tau)} \tan^2(\frac{\theta_e}{2})/I_o \text{ with}$$

$$\tau = \frac{Q^2}{4M} \text{ and } I_o = (G^p_E)^2 + \tau(G^p_M)^2 \left[1 + 2(1+\tau)\tan^2(\frac{\theta_e}{2})\right] \tag{4}$$

In the equations $E_e, E'_e, \theta_e, M, Q^2$ denote the incident electron energy, the scattered electron energy, the scattered electron angle, the mass of the proton and the momentum transfer squared [2,3]. G^p_E and G^p_M are the Sachs electric and magnetic form factors of the proton.

For the reaction $^1H(\vec{e},e'\vec{p})$, Eq. 1 reduces to $P'_y = hP_y$ and $P'_x = -hP_z \sin(\chi)$. Therefore, the polarizations depend on the beam helicity which can be flipped between "plus" and "minus" states during the measurement. By taking the difference between ϕ-distributions, $N^+(\phi)$, gated on plus beam helicity, and $N^-(\phi)$, gated on minus beam helicity, one expects a ϕ-distribution

$$\frac{N^+(\phi)}{2N^+_o} - \frac{N^-(\phi)}{2N^-_o} = a\cos(\phi) + b\sin(\phi)$$
$$a = -AP'_x = hAP_z\sin(\chi) \text{ and } b = AP'_y = hAP_y \quad (5)$$

Thus the instrumental asymmetries have been eliminated and only asymmetries from the physics are left. Combining Eq. 5 and Eq. 4, one can compute

$$\frac{G^p_E}{G^p_M} = \frac{P_y}{P_z}\frac{E_e + E'_e}{M}\tan(\frac{\theta_e}{2}) = \frac{-b}{2a}\sin(\chi)\frac{E_e + E'_e}{M}\tan(\frac{\theta_e}{2}) \quad (6)$$

in which the ratio $\frac{G^p_E}{G^p_M}$ is independent of beam helicity and the analyzing power of carbon. Also using a and b a second independent quantity

$$hA = \frac{a\left[\frac{b(E_e+E'_e)\sin(\chi)}{2aM}\right]^2 + \tau\left[\cot^2(\theta_e) + 2(1+\tau)\right]}{\frac{(E_e+E'_e)}{M}\sqrt{\tau(1+\tau)}\sin(\chi)} \quad (7)$$

can be calculated. Since the beam helicity is measured, the analyzing power of the carbon can be determined and the FPP can be calibrated at Hall A.

With the reaction $^1H(\vec{e},e'\vec{p})$ the instrumental asymmetries of the FPP can be measured at the same time by taking the sum of the ϕ-distributions of plus and minus helicity beam (basically making the beam unpolarized). The ϕ-distribution can be described as

$$\frac{N^+(\phi)}{2N^+_o} + \frac{N^-(\phi)}{2N^-_o} = 1 + a_i\cos(\phi) + b_i\sin(\phi) + c_i\cos(2\phi) + d_i\sin(2\phi) \quad (8)$$

In this way the ratio of G^p_E to G^p_M, the analyzing power of carbon and the instrumental asymmetries of the FPP can be determined simultaneously.

TEST RESULTS

To test the FPP we measured the reaction $^1H(\vec{e},e'\vec{p})$ at Q^2 of 0.810 $(GeV/c)^2$ where the ratio of G^p_E to G^p_M and the analyzing power of carbon have been measured previously. The incident polarized electron beam had an energy of 2.4 GeV and a polarization of 37% which was flipped at a rate of 30 Hz. The electron scattering angle was 23.4 degrees. The scattered proton had a kinetic energy of 432 MeV. The average proton precession angle was 117.8 degrees. The target was a three-foil waterfall target. Scattering events from hydrogen were separated from oxygen events by a cut on the missing energy versus missing momentum histogram. Separate runs were taken with a carbon thickness of 15.2 and 22.9 cm. The ϕ-distributions were measured for θ between 5° and 20°.

In Figs. 1a and 1b the ϕ-distributions are shown for the plus helicity beam and the minus helicity beam. The ϕ-distributions are fitted by Eq. 2. The

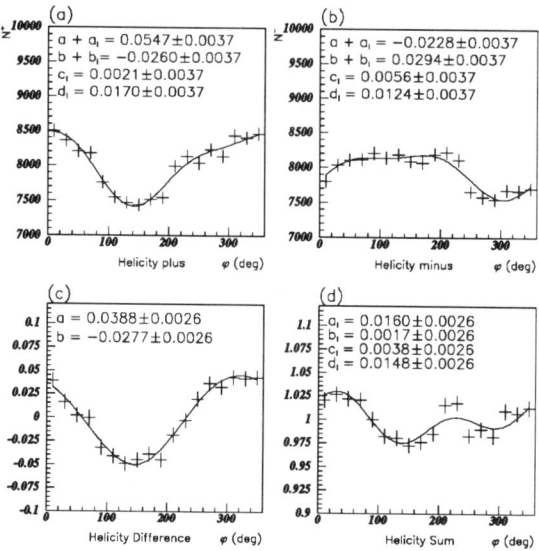

FIGURE 1. (a) The ϕ-distribution for plus helicity beam. (b) The ϕ-distribution for minus helicity beam. In (a) and (b) the line is a fit using Eq. 2. (c) The difference between the ϕ-distributions for the two beam helicity states. The line is a fit using Eq. 5. (d) The sum of the ϕ-distributions for the two beam helicity states. The line is a fit using Eq. 8.

ϕ-distribution for the sum of helicity states, as given by Eq. 8, is shown in Fig. 1d. The instrumental asymmetries are all less than 0.02. The design goal is to have instrumental asymmetries less than 0.005. An effort is underway to understand the origin of instrumental asymmetries and how they can be reduced.

The ϕ-distribution for the difference of helicity states, as given by Eq. 5, is shown in Fig. 1c. Using a and b determined by the fit with Eq. 5 one calculates $\mu G_E^p / G_M^p = 0.87 \pm 0.10$. When the data taken for the 22.9 cm carbon thickness and the 15.2 cm data are combined in a weighted average, $\mu G_E^p/G_M^p = 0.92 \pm 0.09$ is determined. The value is in good agreement with previous measurements as shown in Fig. 2. The run time was short for this commissioning, thus the error on G_E^p/G_M^p is dominated by statistics. In practice the error can be reduced to 0.02 with a reasonable amount of running time. Using Eq. 7 the analyzing power of the carbon is 0.37 ± 0.018 which is

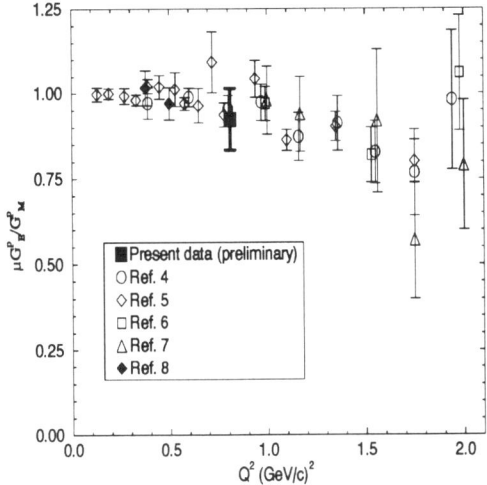

FIGURE 2. The ratio of $\mu G_E^p/G_M^p$ versus Q^2 All previous measurements were made using the Rosenbluth separation method [4–7], except for Ref. [8] which was the first to use a FPP to determine G_E^p/G_M^p.

in good agreement with the parametrization of Ref. [9] which gives a value of $0.36 \pm .007$. This result demonstrates that the FPP can be calibrated on-site.

CONCLUSIONS

Commissioning measurements using the FPP have shown that the detector can be calibrated on-site. Agreement with previous measurements of G_E^p/G_M^p demonstrate that the FPP is working properly and that corrections to the spin precession through the spectrometer from the simple dipole model are minimal. The FPP will be used in its first production experiment to measure the polarization components in the $^{16}O(\vec{e},e'\vec{p})$ reaction at electron beam energies of 2.4 GeV and recoil momenta from 85 MeV/c to 140 MeV/c. The next scheduled experiment at Hall A to use the FPP will measure G_E^p/G_M^p in a Q^2 range between 0.5 to 3.0 (GeV/c)^2 with an accuracy of between 0.02 and 0.035.

REFERENCES

1. M. Kmit, M. Montag, A. S. Carroll, F. J. Barbosa, and S. H. Baker, Brookhaven Informal Report EP&S 91-4.
2. R. G. Arnold, C. E. Carlson, F. Gross, *Phys. Rev. C* **23**, 363 (1981).
3. A. I. Akhiezer and M. P. Rekalo, *Sov. J. Part. Nucl.* **4**, 277 (1974)
4. Ch. Berger, V. Burkert, G. Knop, B. Langenbeck and K. Rith, *Phys. Lett.* **35B**, 87 (1971).
5. L. E. Price, J. R. Dunning, M. Goiten, K. Hanson, T. Kirk and R. Wilson, *Phys. Rev. D* **4**, 45 (1971).
6. J. Litt *et al.*, *Phys. Lett.* **31B**, 40 (1970).
7. W. Bartel *et al.*, *Nucl. Phys.* **B58**, 429 (1973).
8. B. Milbrath *et al.*, submitted to Phys. Rev. Lett.
9. M. W. McNaughton *et al.*, *Nucl. Instr. Meth.* **A241**, 435 (1985).

Tests of Fundamental Symmetries

The TRIUMF Parity Violation Experiment

A.R. Berdoz[b], J. Birchall[e], J.D. Bowman[d], J.R. Campbell[e],
C.A. Davis[e,f], A.A. Green[g], P.W. Green[a], A.A. Hamian[e],
D.C. Healey[f], R. Helmer[f], S. Kadantsev[c], Y. Kuznetsov[c],
R. Laxdal[f], L. Lee[e], C.D.P. Levy[f], R.E. Mischke[d], S.A. Page[e],
W.D. Ramsay[e], S.D. Reitzner[e], G. Roy[a], P. Schmor[f],
A.M. Sekulovich[e], J. Soukup[a], G.M. Stinson[a], T.J. Stocki[a],
V. Sum[e], N. Titov[c], W.T.H van Oers[e], R.J. Woo[e], A. Zelenski[c,f]

presented by W.D. Ramsay

(E497 Collaboration)
[a] *University of Alberta, Edmonton, Alberta T6G 2N5, Canada*
[b] *Carnegie Mellon University, Pittsburgh, PA 15213, USA*
[c] *Institute for Nuclear Research, 117312 Moscow, Russia*
[d] *Los Alamos National Laboratory, Los Alamos, NM 87545 USA*
[e] *University of Manitoba, Winnipeg, Manitoba, R3T 2N2, Canada*
[f] *TRIUMF, Vancouver, British Columbia, V6T 2A3, Canada*
[g] *University of the Western Cape, Bellville 7535, South Africa*

Abstract.
An experiment (E497) is underway at TRIUMF to measure the angle-integrated, parity violating longitudinal analyzing power, A_z, in proton-proton elastic scattering, to a precision of $\pm 0.2 \times 10^{-7}$. The experiment uses a 221 MeV longitudinally polarized proton beam incident on a 40 cm liquid hydrogen target. The beam energy is carefully chosen so that the contribution to A_z from the J=0 parity mixed partial wave ($^1S_0 - ^3P_0$) integrates to zero over the acceptance of the apparatus, leaving the experiment sensitive mainly (> 95%) to A_z arising from the $^3P_2 - ^1D_2$, J=2 wave. To minimize sources of systematic error, the TRIUMF ion source and cyclotron parameters have been refined to the extent that helicity correlated beam changes are at an extremely low level, and specialized instrumentation on the E497 beamline is able to measure residual helicity correlated modulations to a precision consistent with the goals of the experiment. A data taking run in February-March, 1997 logged approximately 12% of the desired data and produced a preliminary result, $A_z = (1.1 \pm 0.4 \pm 0.4) \times 10^{-7}$, where the error is statistical only.

INTRODUCTION

Figure 1 shows the overall layout of TRIUMF experiment E497. The polarized beam is prepared in an optically pumped polarized ion source, the spin is precessed into the vertical direction by a Wien filter, and the beam is accelerated to 221 MeV in the TRIUMF cyclotron. The extracted beam current is 200 nA and the polarization is typically 80%. The beamline precesses the spin into the longitudinal direction and transports the beam to the parity experimental area, where it passes through a series of beam diagnostic and control devices before it is scattered from a 40 cm thick liquid hydrogen target. Hydrogen filled transverse electric field ionization chambers measure the current before and after the target. Approximately 4% of the incident protons scatter due to the strong nuclear force between the incident and target protons. However, because of the simultaneous presence of the weak nuclear force, the scattering fraction is expected to be enhanced very slightly, by about one part in 10^7, if the incident proton spin is aligned with the beam direction, and reduced by the same fraction if the proton spin is opposite to the beam direction. This difference is expressed as the parity violating longitudinal analyzing power, $A_z = (\sigma^+ - \sigma^-)/(\sigma^+ + \sigma^-)$, where σ^+ and σ^- are the scattering cross sections for positive and negative helicity. The goal of E497 is to measure A_z with a precision of $\pm 0.2 \times 10^{-7}$.

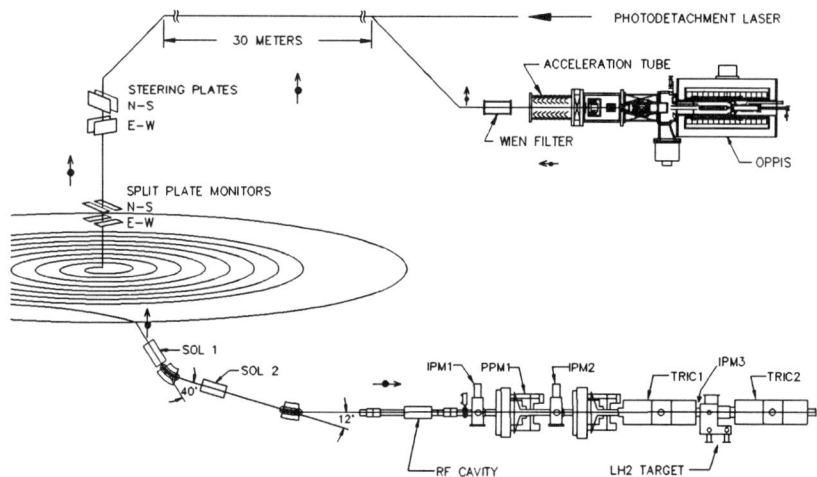

FIGURE 1. General layout of the TRIUMF parity experiment. (OPPIS: Optically Pumped Polarized Ion Source; SOL: Spin Precession Solenoid; IPM: Intensity Profile Monitor; PPM: Polarization Profile Monitor; TRIC: Transverse Field Ionization Chamber)

Choice of Energy

Figure 2 shows the results of meson exchange calculations by Driscoll and Miller [1] using the Bonn meson exchange potential [2,3] for the strong interaction and the DDH [4] predictions for the weak coupling parameters. The calculated A_z is shown broken down into contributions from the various parity mixed partial waves.

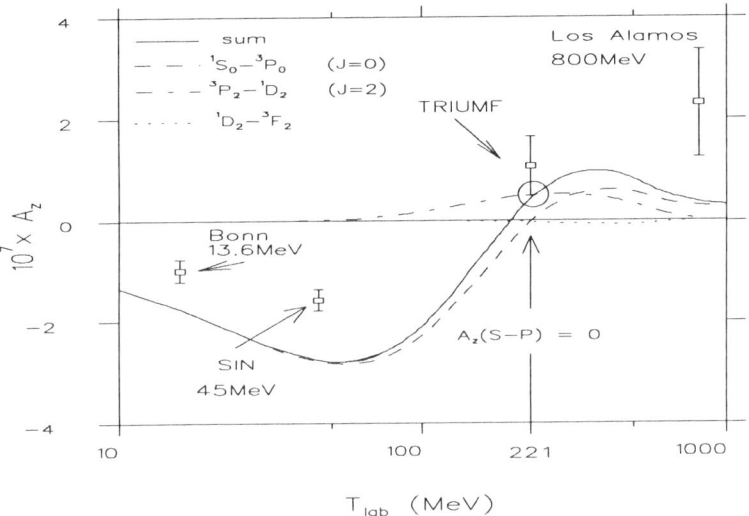

FIGURE 2. Partial wave contributions to A_z. The curves are from Driscoll and Miller[1]. Also shown are data from Bonn[5], SIN[6], Los Alamos[7], and the TRIUMF preliminary result.

Because the variation of A_z with angle is determined only by the well-known strong interaction, one can calculate the energy at which the lowest order, J=0, parity mixed partial wave ($^1S_0 - {^3P_0}$) will integrate to zero over the acceptance of the detectors. The 221 MeV energy of the TRIUMF experiment is chosen so that the measured A_z comes exclusively[1] from the J=2 parity mixed partial wave ($^3P_2 - {^1D_2}$) which, in the meson exchange model, comes from ρ-meson exchange. Theoretical predictions [1,8–11] for A_z at this energy, span a substantial range. It is hoped that the TRIUMF measurement, in selecting the effects of only one partial wave, will provide a definitive result.

---

[1]) At this energy the contribution to A_z from the $^1D_2 - {^3F_2}$ partial wave is only 5% of that from $^3P_2 - {^1D_2}$.

SYSTEMATIC ERRORS

If beam properties other than the spin direction change when the spin is flipped, it can affect the measured A_z. *Random* changes will appear as noise, and simply increase the time required to achieve a given statistical precision; *coherent* changes, that is changes which are synchronized with spin reversal, appear as a false A_z. Table 1 summarizes the corrections to the February-March 1997 data for all sources of systematic error that are measurable with the parity apparatus. Notice that the corrections are very small and, in most cases are zero to within statistics. This is the result of an exhaustive program of reducing all unwanted helicity correlated modulations to a bare minimum and of minimizing the sensitivity to these modulations.

Transverse Polarization Components

Ideally, the polarization is purely longitudinal. Unwanted transverse components, $P_t = P_x$ or P_y, are kept small (< 0.001) but couple to the large parity allowed analyzing power and, when combined with an off-center beam, generate a false A_z. The sensitivity to transverse components is found by setting the spin precession magnets for large P_t and measuring the false A_z as a function of beam position. This procedure also determines the "polarization neutral axis" – the beam position with minimum sensitivity to transverse polarization components. A beam position servo system then holds the average beam position on this axis to within about $50 \mu m$. During data taking, the polarization profile monitors (PPM1 and PPM2 in figure 1) continuously measure the transverse components and a correction is applied.

Even if the average transverse polarization is zero, the first moment of transverse polarization need not be. For example, the polarization may be up

TABLE 1. Corrections (ΔA_z) to the February-March 1997 Data.

Item	$10^7(\Delta A_z)$	Comment
P_x	0.01 ± 0.09	Correction very small
P_y	0.02 ± 0.20	for all data sets
yP_x	-0.001 ± 0.002	sensitivity extracted
xP_y	0.00 ± 0.02	from real data correlations
$\Delta\sigma$	-0.07 ± 0.20	sensitivity from separate measurement
Δx	-0.03 ± 0.15	correction ~ 0 for all sets
Δy	0.23 ± 0.16	some cancellation between sets
$\Delta I/I$	0.05 ± 0.05	using interleaved CIM data
Total	0.2 ± 0.4	

on the left of the beam and down on the right. Such unwanted polarization profiles also cause false A_z. For this reason, the first moments of transverse polarization, $<xP_y>$ and $<yP_x>$ are continuously monitored by the PPMs. They show a random variation from run to run, but typical values are a few μm for a one hour run, averaging to near zero over a 20 to 30 hour data set. In a drift space, first moments vary linearly with position along the beamline, making it possible to adjust the beam optics so that the first moments pass through zero at a point which minimizes their effect. The sensitivity to intrinsic first moments must be determined by looking at correlations between apparent A_z and the $<xP_y>$ and $<yP_x>$ measured by the PPMs. For the February, 1997 data, this sensitivity was consistent with zero.

Beam Size Modulation

Because the beam is different upstream and downstream of the liquid hydrogen target, a beam size change affects the current differently in the upstream and downstream detectors (TRIC1 and TRIC2 in figure 1). Actual coherent size modulation is typically only a few tenths of one μm on a $\sigma = 5mm$ beam, but it can cause a detectable shift in the measured A_z. To measure the sensitivity to beam size change, a relatively large coherent beam size modulation is introduced using a pair of fast ferrite-cored quadrupole magnets. The sensitivity measured in this way is then multiplied by the actual coherent size change measured during data taking to obtain the correction.

Beam Intensity Modulation

Coherent intensity modulation from the optically pumped polarized ion source is very small, usually only a few parts in 10^5. The coherent intensity modulation is measured constantly during data taking. In addition, the sensitivity is monitored by interleaving with the normal data a 10% subset with artificially enhanced ($\sim 0.1\%$) coherent intensity modulation (CIM).

Beam Position Modulation

If the beam position changes with spin reversal then a false A_z is introduced proportional to the magnitude of the coherent change and the mean position of the beam. The farther the beam is off-axis, the greater the sensitivity to coherent position change. Sensitivity to position modulation is measured by introducing relatively large position modulation using the same fast ferrite-cored steering magnets which are used for the beam position servo. This information is then combined with the actual beam position information recorded during the run to obtain the correction.

Overall Bias

Although corrections for all systematic errors measurable with the E497 apparatus appear to be well under control, the possibility remains that some additional effect could be present which cannot be measured, but which will cause a false A_z. One example is energy modulation. Such effects must be handled by reversing the sign of the spin in the beamline relative to that at the ion source or cyclotron. The real asymmetry from A_z will then reverse sign, but the false effect will not, but will simply appear as an overall bias on the measured A_z. For this reason, runs were made under four configurations representing all possible combinations of spin direction at the ion source, cyclotron, and parity apparatus (see figure 1). Only four different conditions are needed because changing the spin direction in the ion source is the same as reversing the definition of spin "up".

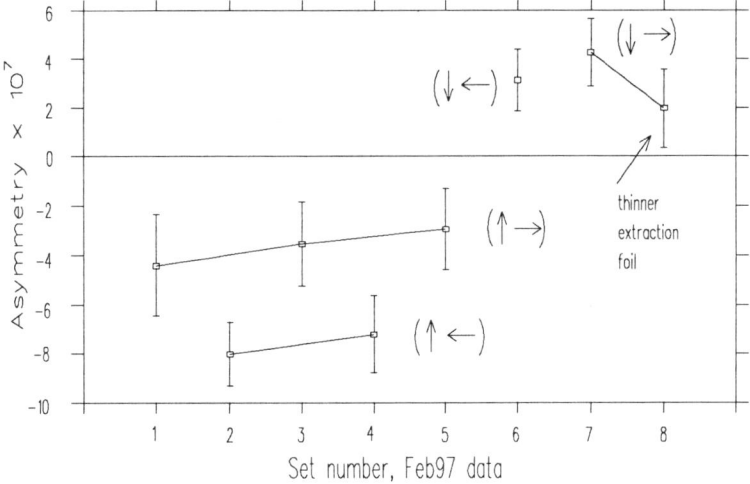

FIGURE 3. Data from the February-March, 1997 run after correction for all measurable sources of systematic error. In brackets, the left-hand arrow shows the orientation of the spin in the cyclotron and the right-hand arrow shows the direction at the parity apparatus. The sign of the plotted asymmetry is such that it is equal to $+A_z$ for positive helicity at the apparatus and equal to $-A_z$ for negative helicity at the apparatus. (A right arrow indicates positive helicity.)

RESULTS OF FEBRUARY-MARCH, 1997 RUN

A total of 231 runs of "real data" were taken during February and March. 80% was polarized, 10% was unpolarized and 10% was unpolarized with intentional coherent intensity modulation. 105 hours of polarized data passed all the cuts. Figure 3 shows the results from these data. Each point has already been corrected for the effects of coherent change in beam intensity, position, size, transverse polarization, and first moments of transverse polarization. The asymmetry apparently contains a part which is related to spin in the cyclotron and a part which is related to spin in the parity apparatus. To extract A_z, it is assumed that only that part of the asymmetry which reverses with spin at the parity apparatus is true A_z. In principle, it is enough to reverse the beamline helicity. If this is done fairly frequently, it will establish the constancy of the offset and the true A_z can be extracted. The reversal of the spin in the cyclotron was done in an attempt to locate the source of the offset.

The last point, set 8, on figure 3 requires some explanation. It was taken with the normal $6\,mg/cm^2$ stripping foil replaced with a $2.5\,mg/cm^2$ foil. This was done to test the theory that the bias might be arising from some interaction with the stripping foil. It appears the bias was reduced by the thin foil, but unfortunately the variation could be statistical. For the Parity data-taking run starting in August, 1997, The experiment will use a very thin, $200\mu g/cm^2$ stripping foil. This should make clear whether or not the large A_z offset arises from some interaction with the stripping foil.

The raw A_z from February-March, 1997 is $(1.3 \pm 0.4) \times 10^{-7}$ and, after all corrections are applied, the final A_z is $(1.1 \pm 0.4 \pm 0.4) \times 10^{-7}$, where the first error comes from statistical uncertainty in the raw A_z and the second error is dominated by statistical uncertainty in measurement of the helicity correlated quantities. Since the uncertainty in the correction is also statistical, the two errors can be combined, giving $A_z = (1.1 \pm 0.6) \times 10^{-7}$. We consider this number very preliminary because of the relatively large systematic shift which had to be removed by averaging over different beamline helicities.

REFERENCES

1. Driscoll D.E., and Miller G.A., *Phys. Rev.* **C39**, 1951 (1989).
2. Machleidt R., Holinde K., and Elster Ch., *Phys. Rep.* **149**, 1 (1987).
3. Machleidt R., *Adv. Nucl. Phys.* **19**, 189 (1989).
4. Desplanques B., Donoghue J.F., and Holstein B.R., *Ann. Phys.* (NY) **124** 449 (1980)
5. Eversheim P.D., *et al.*, *Phys. Lett.* **B256**, 11 (1991); Eversheim P.D., private communication (1994).
6. Kistryn S., *et al.*, *Phys. Rev. Lett.* **58**, 1616 (1987).
7. Yuan V., *et al.*, *Phys. Rev. Lett.* **57**, 1680 (1986).
8. Nessi-Tedaldi F., and Simonius M., *Phys. Lett.* **B215**, 159 (1988).
9. Driscoll D.E., and Miller G.A., *Phys. Rev.* **C40**, 2159 (1989).
10. Grach I., and Shmatikov M., *Phys. Lett.* **B316**, 467 (1993).
11. Iqbal M.J., and Niskanen J.A., *Phys. Rev.* **C49**, 355 (1994).

Tests of Chiral Perturbation Theory in Threshold Pion Photoproduction at SAL

Elie Korkmaz

Physics Department
University of Northern British Columbia
Prince George, BC V2N 4Z9, Canada

Abstract. The threshold pion photoproduction program at the Saskatchewan Accelerator Laboratoty is reviewed. Recently completed or currently in progress measurements are discussed for the reactions $\gamma p \to \pi^0 p$, $\gamma p \to \pi^+ n$, and $\gamma d \to \pi^0 d$, the latter being aimed at extracting information on the elementary process $\gamma n \to \pi^0 n$. Final or preliminary results are presented and compared to recent calculations based on chiral perturbation theory.

INTRODUCTION

The objective of my talk is to present an updated review of part of the threshold pion photoproduction program at the Saskatchewan Accelerator Laboratory (SAL). First, I will give a brief introduction to the topic of threshold pion photoproduction and its importance as a fundamental low energy process. The completed or ongoing measurements at SAL aimed at determining the threshold amplitudes of the reactions $\gamma p \to p\pi^0$, $\gamma p \to n\pi^+$, and $\gamma n \to n\pi^0$ will then be discussed. Final or preliminary results will be presented and compared to recent calculations generated within the framework of chiral perturbation theory (ChPT).

Recent progress in developing QCD-inspired effective field theories has made it possible to address the issue of low-energy hadron physics at a more fundamental level than generally provided by more conventional models. ChPT is one such effective field theory wherein the spontaneous and explicit breaking of the QCD chiral symmetry are explored to make predictions for many low-energy processes. The recently renewed interest in threshold pion photoproduction from the nucleon reflects the unique position that this process occupies as one of a small number of phenomena expected to provide crucial initial tests of ChPT applied to systems containing baryons [1].

TABLE 1. The multipole amplitudes, up to p-wave pions, that contribute to the reaction $\gamma N \to \pi N$.

$l_{\pi N}$	J^π	L	amplitude	e.m. multipole
0	$\frac{1}{2}^-$	1	E_{0+}	Electric Dipole
1	$\frac{3}{2}^+$	1	M_{1+}	Magnetic Dipole
1	$\frac{1}{2}^+$	1	M_{1-}	Magnetic Dipole
1	$\frac{3}{2}^+$	2	E_{1+}	Electric Quadrupole

At energies close to the reaction threshold, the s- and p-wave amplitudes that contribute to $\gamma N \to \pi N$ are shown in TABLE 1. The recently reported [2,3] ChPT threshold values of the s-wave amplitude E_{0+} are shown in TABLE 2 for the charge channels $\gamma p \to \pi^+ n$, $\gamma n \to \pi^- p$, $\gamma p \to \pi^0 p$, and $\gamma n \to \pi^0 n$. Also shown in TABLE 2 are the E_{0+} expressions and values based on the older Low Energy Theorems (LET) formulated using current algebra and PCAC. f is the πN coupling constant, $\mu = m_\pi/m_N$, and $\kappa_p(\kappa_n)$ is the proton(neutron) anomalous magnetic moment. The important thing to note here is that ChPT reproduces the LET leading order terms around the chiral limit ($\mu \to 0$) but allows for a systematic, model-independent evaluation of the higher-order terms (refer to [2,3] for the full ChPT expressions). These chiral corrections are small in the case of the charged pion channels because of the dominance of the zeroth-order Kroll-Ruderman term, $\frac{ef\sqrt{2}}{4\pi m_\pi}$. In neutral pion production, however, this term vanishes and the corrections become relatively very significant as seen in TABLE 2. Details on these ChPT calculations including uncertainties on the numerical values of the s-wave amplitudes are found in [2,3] and references therein.

TABLE 2. The ChPT and LET predictions (in units of $10^{-3}/m_\pi$) for the electric dipole amplitudes E_{0+} for $\gamma N \to \pi N$ at threshold.

reaction	amplitude	LET expression	LET	ChPT
$\gamma p \to \pi^+ n$	$E_{0+}(\pi^+)$	$\frac{ef\sqrt{2}}{4\pi m_\pi}[\,1 - \frac{3}{2}\mu + O(\mu^2)]$	27.5	28.2
$\gamma n \to \pi^- p$	$E_{0+}(\pi^-)$	$\frac{ef\sqrt{2}}{4\pi m_\pi}[-1 + \frac{1}{2}\mu + O(\mu^2)]$	-31.7	-32.7
$\gamma p \to \pi^0 p$	$E_{0+}(p\pi^0)$	$\frac{ef}{4\pi m_\pi}[\,-\mu + \frac{1}{2}(3+\kappa_p)\mu^2 + O(\mu^3)]$	-2.3	-1.16
$\gamma n \to \pi^0 n$	$E_{0+}(n\pi^0)$	$\frac{ef}{4\pi m_\pi}[\,\frac{\kappa_n}{2}\mu^2 + O(\mu^3)]$	0.4	2.13

THE $\gamma p \to \pi^0 p$ REACTION

The $\gamma p \to \pi^0 p$ reaction was studied at SAL in an experiment [4] which used a tagged photon beam, a LH2 target, and a π^0 spectrometer consisting of an array of 68 lead-glass detectors. Data were acquired for photon energies from threshold (E_γ=144.7 MeV) up to 25 MeV above threshold in 0.5 MeV steps. Both total and differential cross sections were measured independently.

Close to threshold, the cross section is dominated by the s- and p-wave amplitudes shown in TABLE 1 and can therefore be parametrized as:

$$\frac{d\sigma}{d\Omega} = \frac{q}{k}\left[A + B\cos\theta + C\cos^2\theta\right]$$

where k is the photon momentum, q the pion momentum and θ its angle, all in the c.m. system. In terms of the so-called natural p-wave amplitudes $P_1 = 3E_{1+} + M_{1+} - M_{1-}$, $P_2 = 3E_{1+} - M_{1+} + M_{1-}$, and $P_3 = 2M_{1+} + M_{1-}$, the coefficients A, B, C are: $A = |E_{0+}|^2 + |P_{23}|^2$, $B = 2.Re(E_{0+}.P_1^*)$, and $C = |P_1|^2 - |P_{23}|^2$, where $|P_{23}|^2 = \frac{1}{2}|P_2|^2 + \frac{1}{2}|P_3|^2$. In terms of these coeffcients, the total cross section for the process is simply:

$$\sigma_T = 4\pi\frac{q}{k}\left(A + \frac{1}{3}C\right)$$

Detailed multipole analyses of both the differential and total cross sections were completed and the results recently reported [5,6]. The extracted threshold value of the s-wave amplitude E_{0+} is in strong disagreement with the old value based on the incomplete LET. The observed energy dependence of this amplitude confirms the expected unitary cusp effect due to the opening of the $\gamma p \to \pi^+ n$ channel at 151.5 MeV and is in good agreement with the ChPT values [2] at both the π^0 and π^+ thresholds. This comparison is shown in TABLE 3. It should be emphasized here that despite the relatively slow convergence of the chiral series for this reaction [2], the SAL results provide strong support for the new ChPT calculations and the need for similar systematic treatments of other low-energy processes as well.

TABLE 3. The s- and p-wave amplitudes extracted from the SAL data along with their theoretical ChPT values. The s-wave amplitudes are in units of $10^{-3}/m_\pi$ while the reduced p-wave amplitudes are in units of $10^{-3}/m_\pi^3$. See text for more details.

	$E_{0+}(\pi^0_{thr})$	$E_{0+}(\pi^+_{thr})$	p_1	p_{23}	e_{1+}
SAL	-1.32 ± 0.08	-0.52 ± 0.16	10.26 ± 0.10	11.44 ± 0.09	-0.25 ± 0.17
ChPT	-1.16	-0.44	10.3	11.25	-0.1

Further support of ChPT is provided by the experimental values of the p-wave combinations P_1 and P_{23} defined above. Assuming all p-wave amplitudes to be real and proportional to (qk) near threshold, the SAL analysis [6], with minimum additional assumptions, allowed for determining the values of the reduced p-wave amplitudes $p_1 = P_1/(qk)$ and $p_{23} = P_{23}/(qk)$, as well as $e_{1+} = E_{1+}/(qk)$. These values are in good agreement with ChPT [2] as shown in TABLE 3. It should be noted that similar conclusions have also been reached based on the new $\gamma p \to \pi^0 p$ threshold data from MAMI [7]. Recently, a more rigourous analysis [8] of the MAMI data has been reported where the energy dependence of the p-wave amplitudes is investigated thouroughly in closer coordination with the ChPT calculations.

Finally, it should be mentioned that further studies of this reaction have recently been undertaken at MAMI using linearly polarized photon beams. These polarized data should make it possible to extract the individual p-wave amplitudes and help in constraining the imaginary part of E_{0+} (see discussion in [5–8]).

THE $\gamma p \to \pi^+ n$ REACTION

The goal of this experiment [9] was to determine the threshold value of the E_{0+} amplitude for $\gamma p \to \pi^+ n$ by measuring the cross section within 1-2 MeV of threshold. The measurement involved the use of a LH_2 target, a tagged photon beam, and a segmented neutron detector. Bremstrahlung photons were tagged in the range 142-158 MeV (the reaction threshold is 151.5 MeV) with a resolution of 0.25 MeV per tagger channel. The neutron detector, which consisted of an array of 84 rectangular cells of liquid scintillator and 12 charged-particle veto paddles, was positioned 3 m downstream of the target. Outgoing neutrons corresponding to 4π in the c.m. system were thus detected between 1^0 and 8^0 around the beam direction and their energies measured via time of flight, providing an independent measure of the incident photon energy. PSD techniques were employed to reject photon triggers at the electronic level. Further accidental background was measured concurrently with the aid of the sub-threshold tagger channels. Details of the measurement can be found in [11]. The neutron detection efficiency in the energy range of interest was measured at TRIUMF [10] using the neutron-tagging reaction (stopped $\pi^-)p \to n\gamma$.

Within 1-2 MeV of threshold, the contribution of the p-wave amplitudes is expected to be very small compared to the s-wave. We then have:

$$\frac{k}{q}\frac{d\sigma}{d\Omega} = |E_{0+}|^2$$

A flat angular distribution is expected and indeed observed as shown in FIGURE 1 for a beam energy of $E_\gamma = 152.5$ MeV. The value of E_{0+} extracted under this assumption is also shown. The other plot of FIGURE 1 shows the E_{0+} values extracted in a similar manner at different energies. Assuming a constant s-wave amplitude in this narrow energy range, a best value of $E_{0+} = (27.6 \pm 0.3) \times 10^{-3} \ m_\pi^{-1}$ is then implied, where the error is purely statistical. It is important to emphasize that this number is preliminary and may potentially change by a few per cent once all the normalization factors are finalized. The goal of stating this preliminary number here is primarily to illustrate the quality of the measurement and the level of accuracy achieved. The final results are expected within a few months.

It is important to note here that, because of the fast convergence of the chiral series for this reaction [3], a combination of the final experimental value of E_{0+} and its ChPT value provides an opportunity to impose strict constraints on the πN coupling constant, f, whose value remains a topic of debate. This is so because E_{0+} is, basically, directly proportional to f (see [3] and TABLE 2 where the E_{0+} values assume $f^2=0.079$). The same also applies to $E_{0+}(\gamma n \to \pi^- p)$ for which a new value of $(31.5 \pm 0.8) \times 10^{-3} \, m_\pi^{-1}$ has recently been reported [12] based on a study at TRIUMF of the inverse reaction $\pi^- p \to n\gamma$ at low pion energies.

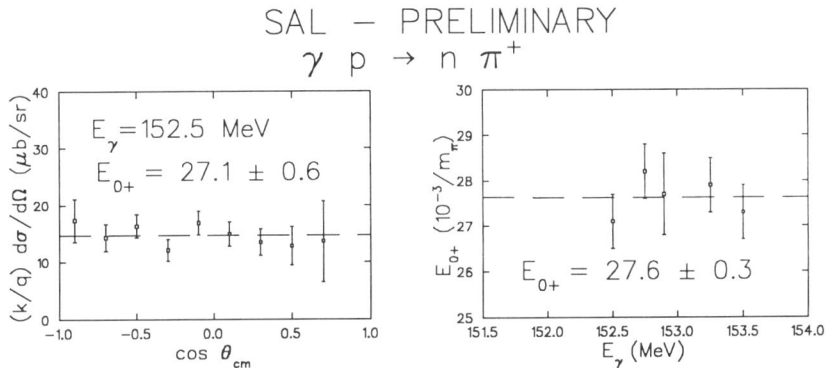

FIGURE 1. Left: The $\gamma p \to \pi^+ n$ differential cross section and the extracted E_{0+} amplitude at $E_\gamma=152.5$ MeV. Right: The $\gamma p \to \pi^+ n$ E_{0+} amplitudes for a number of energies close to threshold along with the best E_{0+} fit assuming no energy dependence.

THE $\gamma d \to \pi^0 d$ REACTION

The dramatic difference, shown in TABLE 2, between the ChPT and old LET predictions for $E_{0+}(\gamma n \to \pi^0 n)$ at threshold clearly states the need for an experimental determination of this amplitude. In an attempt to do just that, pion photoproduction measurements using a deuterium target are currently being carried out or planned at SAL. The first of these measurements involves the reaction $\gamma d \to \pi^0 d$ [13]. This study is now in progress with data taking almost completed. The experimental setup is basically the same as the one used for the $\gamma p \to \pi^0 p$ measurement with a LD$_2$ target replacing LH$_2$. Both differential and total cross sections were measured at photon energies from threshold up to 30 MeV above threshold.

A ChPT study of this reaction has very recently been carried out [1] and a value of $(-1.8 \pm 0.2) \times 10^{-3} \, m_\pi^{-1}$ for the $\pi^0 d$ s-wave amplitude E_d reported. This value results from relatively large two-body rescattering contributions in addition to the elementary $\gamma p \to \pi^0 p$ and $\gamma n \to \pi^0 n$ E_{0+} s- and p-wave amplitudes of [2]. At threshold, the $\gamma d \to \pi^0 d$ cross section is given by:

$$\frac{k}{q}\sigma_T = 4\pi \frac{8}{3}E_d^2$$

Extrapolation to threshold of the measured total cross section indicates that the value of E_d is noticeably smaller than the ChPT prediction. However, this is preliminary and one should await full analysis of the data including the measured angular distributions. Also, the effect of the deuteron break-up contribution to the measured pion yield is being investigated.

The second phase now being planned is a measurement of $d(\gamma, \pi^0 n)p$ in quasi-free kinematics (the proton acts as a spectator), where the outgoing neutron is detected in coincidence with the pion in the neutron detector array mentioned above. It is hoped that the quasi-free amplitude would prove less sensitive to rescattering effects and would, in any case, allow for an independent extraction of $E_{0+}(\gamma n \to \pi^0 n)$ at threshold.

SUMMARY

I have briefly outlined some elements of the threshold pion photoproduction program at SAL, showed final or preliminary results, and compared them to ChPT calculations. The final results support the ChPT approach to dealing with this important process. Studies at SAL and other laboratories are continuing with the aim of testing the extent of validity and usefulness of ChPT as a powerful tool providing deeper understanding of low energy hadronic phenomena in terms of the underlying QCD forces and symmetries.

REFERENCES

1. See U. van Kolck, 'Chiral Perturbation Theory in Few Nucleon Systems', these proceedings, and references therein.
2. V. Bernard, N.Kaiser, and Ulf-G. Meißner, Z. Phys. **C70**, 483 (1996).
3. V. Bernard, N.Kaiser, and Ulf-G. Meißner, Phys. Lett. **116** (1996).
4. SAL experiment No. 32, J.C. Bergstrom, spokesperson.
5. J.C. Bergstrom et al., Phys. Rev. C **53**, R1052 (1996).
6. J.C. Bergstrom, R. Igarashi, and J.M. Vogt, Phys. Rev. C **55**, 2016 (1997).
7. M. Fuchs et al., Phys. Lett. B **368** 20 (1996).
8. A.M. Bernstein et al., Phys. Rev. C **55**, 1509 (1997).
9. SAL experiment No. 27, E. Korkmaz, spokesperson.
10. TRIUMF experiment No 764, D.A. Hutcheon and E. Korkmaz, spokespersons.
11. SAL Annual Report, 1995.
12. M.A. Kovash et al., πN Newsletter No. 12 (1997).
13. SAL experiment No. 63, J.C. Bergstrom, spokesperson.

Isospin Splitting in the Pion-Nucleon Couplings from QCD Sum Rules

Thomas Meissner* and Ernest Henley†

*Department of Physics, Carnegie Mellon University, Pittsburgh, PA 15213[1]
†Department of Physics, University of Washington, Box 351560, Seattle, WA 98195[2]

Abstract. We use QCD sum rules for the three point function of a pseudoscalar and two nucleonic currents in order to estimate the charge dependence of the pion nucleon coupling constant $g_{NN\pi}$ coming from isospin violation in the strong interaction. The effect can be attributed primarily to the difference of the quark condensates $<\bar{u}u>$ and $<\bar{d}d>$. Assuming that the π^0 is a pure isostate we obtain for the splitting $(|g_{pp\pi_0}| - |g_{nn\pi_0}|)/g_{NN\pi}$ an interval of $0.8 * 10^{-2}$ to $2.3 * 10^{-2}$, the uncertainties coming mainly from the input parameters. In order to obtain the coupling to a physical π^0 we have to take $\pi - \eta$ mixing into account leading to an interval of $1.2*10^{-2}$ to $3.7*10^{-2}$. The charged pion nucleon coupling is found to be the average of $|g_{pp\pi_0}|$ and $|g_{nn\pi_0}|$. Electromagnetic effects are not included.

The effect of isospin violating meson nucleon couplings is of great importance in the investigation of charge symmetry breaking (CSB) phenomena [1]. On a microscopical level, isospin symmetry is broken by the electromagnetic interaction as well as the mass difference of up and down quarks $m_u \neq m_d$. We will examine the splitting between the pion nucleon coupling constants $g_{pp\pi^0}$, $g_{nn\pi^0}$ and $g_{pn\pi^+}$ using the QCD sum rule method, which has been established as a powerful and fruitful technique for describing hadronic phenomena at intermediate energies. We will only look at effects which arise from isospin breaking in the strong interaction. In the QCD sum rule method this is reflected by $m_u \neq m_d$ as well as by the isospin breaking of the vacuum condensates. Electromagnetic effects are not examined. Details of our calculation have been recently published [2].

We start from the three point function of two nucleonic (Ioffe) currents (η_N) and one pseudoscalar isovector ($P_a^{T=1}$) interpolating current with the appropriate isospin quantum numbers:

---

[1] email: meissner@yukawa.phys.cmu.edu
[2] email: henley@alpher.npl.washington.edu

$$A_{NN\pi^a}(p_1, p_2, q) = \int d^4x_1 d^4x_2 e^{ip_1x_1} e^{-ip_2x_2} \left\langle 0 | T\eta_N(x_1) P_a^{T=1}(0) \bar{\eta}_N(x_2) | 0 \right\rangle, \quad (1)$$

where a stands for + or 0 and N for proton or neutron, respectively. The momenta p_1 and p_2 are those of the nucleon, and $q = p_1 - p_2$ that of the pion. We are only keeping terms up to first order in isospin violation, i.e. $m_d - m_u$.

The phenomenological side of the QCD sum rules for the three point functions A are obtained by saturating the general expressions for the A's (1) with the corresponding nucleon and pion intermediate states. The overlap between the pion states and the pseudoscalar interpolating fields are given by current algebra and the axial Ward identity. For $a = 0$ and $N = p$ we have for example:

$$A_{pp\pi^0} = i\lambda_p^2 \frac{m_\pi^2 f_\pi}{m_0} \frac{g_{pp\pi^0}}{-q^2 + m_\pi^2} \frac{1}{p_1^2 - M_p^2} \frac{1}{p_2^2 - M_p^2} M_p \gamma_5 \slashed{q} + \ldots, \quad (2)$$

where the λ_p denotes the overlap between the proton Ioffe current and the corresponding single nucleon state. The difference between λ_p and λ_n is an input parameter for the sum rule analysis. It can be eliminated by using the isospin violating sum rule for the nucleon 2 point function which makes a prediction for the proton-neutron mass splitting [3]. The ... denote contributions from higher resonances and the continuum.

The three point function method works at large spacelike $q^2 \approx 1 \text{GeV}^2$. We neglect m_π^2 and identify the pion pole term $\frac{1}{q^2}$ on the r.h.s. of eq.(2) with the corresponding $\frac{1}{q^2}$ term in the operator product expansion (OPE) [4]. This treatment is justified in the isospin conserving case, where it has been shown that the non-$\frac{1}{q^2}$ terms give rise to a pion nucleon form factor $g_{\pi NN}(q^2)$ with a monopole cutoff mass of about 800MeV [5], which is almost identical to the result of lattice calculations.

In the OPE we include only the leading order term, which in this case is the the quark condensate $<\bar{q}q>$. The isospin splitting of the higher order condensates is practically unknown, the error due to their omission can be estimated to about 25% of the leading order contribution.

The Borel sum rule for the $\slashed{q}\gamma_5$ structure, which has a double nucleon pole, gives an interval of

$$8 * 10^{-3} < \left(\frac{(|g_{pp\pi^0}| - |g_{nn\pi^0}|)}{g_{NN\pi}} \right) < 23 * 10^{-3}. \quad (3)$$

where $g_{NN\pi} \equiv \frac{1}{2}(|g_{pp\pi^0}| + |g_{nn\pi^0}|)$.

A crucial input parameter is the isospin violation in the quark condensate $\gamma = \frac{<\bar{d}d>}{<\bar{u}u>} - 1$. The range for γ obtained from various other methods is rather large: $0.002 < -\gamma < 0.010$. This leads to a significant uncertainty in the final result for $(|g_{pp\pi^0}| - |g_{nn\pi^0}|)$.

Up to now we have treated the π^0 as pure isospin state. If we want to calculate the coupling to the physical pion we have to include $\pi - \eta$ mixing, which can be done at tree level chiral perturbation theory. The final result is:

$$12 * 10^{-3} < \left(\frac{(|g_{pp\pi^0}| - |g_{nn\pi^0}|)}{g_{NN\pi}} \right) < 37 * 10^{-3}. \tag{4}$$

It should be noted that in charge symmetry breaking NN potential calculations it is common to treat the $\pi - \eta$ mixing separately and to use the coupling to the pure isostate π^0.

Up to first order in isospin breaking the charged pion nucleon coupling is exactly the arithmetic average of the two neutral pion nucleon couplings:

$$\frac{|g_{pn\pi^+}|}{\sqrt{2}} = \frac{1}{2}[|g_{pp\pi^0}| + |g_{nn\pi^0}|]. \tag{5}$$

which is a simple consequence of the u and d quark contents of the three point functions and valid within the approximations considered.

Our result can be compared with those of other approaches [1] which analyze the isospin splitting of the pion nucleon couplings arising from the strong interaction, i.e., essentially the quark mass difference $m_d - m_u$ (Table 1).

TABLE 1.

	$(|g_{pp\pi_0}| - |g_{nn\pi_0}|)/g_{NN\pi}$
this work	$\approx 0.012\ldots 0.037$
Quark-Gluon Model	$\approx 0.010\ldots 0.014$
Quark-Pion Model	≈ 0.006
$\pi - \eta$ mixing	0.005 ± 0.0018
Chiral Bag Model	≈ 0.0067
Cloudy Bag Model	≈ -0.006

Direct experimental values are not available. The Nijmegen phase shift analysis for NN and $N\bar{N}$ scattering data, which includes electromagnetic effects, finds $\left(\frac{(|g_{pp\pi^0}| - |g_{nn\pi^0}|)}{g_{NN\pi}} \right) = 0.002$, but with an error of 0.008; thus, there is no evidence for a difference and it also find no evidence for a difference between $g_{pn\pi^+}$ and $g_{NN\pi^0}$ within the statistical errors [6].

REFERENCES

1. G.A. Miller, B.M.K. Nefkens and I. Slaus, Phys.Rep.**194**, 1 (1990).
2. T. Meissner and E. Henley, Phys.Rev. C **55**, 3093 (1997).
3. K-C. Yang, W-Y.P. Hwang, E.M. Henley and L.S. Kisslinger, Phys.Rev. D **47**, 3001 (1993).
4. L.J. Reinders, H. Rubinstein and S. Yazaki, Phys.Rep. **127**, 1 (1985).
5. T. Meissner, Phys.Rev. C **52**, 3386 (1995).
6. V. Stoks, R. Timmermans and J.de Swart, Phys.Rev. C **47**, 512 (1993).

Time Ordered Products, Vector-Meson Exchange, and Nuclear Charge Asymmetry

S. A. Coon*, B. H. J. McKellar†, and A. A. Rawlinson†

*Physics Department, New Mexico State University, Las Cruces, NM 88003, USA
†School of Physics, Research Center for High Energy Physics, University of Melbourne, Parkville, Victoria, Australia, 3052

Abstract. We investigate the "mixed propagator" definition of the $\rho\omega$ transition matrix element utilized in charge symmetry breaking vector-meson-exchange NN potentials. Different "mixed propagators" can be obtained, but some definitions can lead to unphysical results.

The study of fundamental symmetries and how they are broken has always been an important part of research in nuclear and particle physics. Theoretical estimates of the contribution of meson mixing to those parts of the NN potential which break charge symmetry rely on the transition matrix elements which describe the mixing of T=1 and T=0 mesons. Charge asymmetric NN potentials based upon the Coleman-Glashow tadpole picture extended to particle mixing [1-3] have been widely used to explain measures of charge asymmetry in low energy NN scattering and bound state data (the Nolen-Schiffer anomaly) in mirror nuclear systems [1,2,4], Coulomb displacement energies of isobaric analog states [5], isovector mass shifts of isospin multiplets in the 1s0d-shell, and isospin mixing matrix elements deduced from isospin-forbidden beta decays [6]. In addition, new tests of charge asymmetry in the neutron-proton system have been performed at TRIUMF [7] and planned at IUCF with high intensity, variable energy polarized neutron beams. The results of the first experiment at IUCF [8] are completely consistent with the mixing parameter obtained in Ref. [2] and, indeed, the experiment cannot be explained without $\rho\omega$ mixing. In all these studies, particle mixing and in particular $\rho\omega$ mixing, was found to be in reasonable agreement with experiment and/or to make the dominant contributions to explanations of the data.

Recently a plethora of papers have appeared which claim that this picture is false. They attribute a strong four-momentum dependence of the $\rho\omega$ transition matrix element such that the value attained from hadronic data (in the on-mass-shell timelike region) is suppressed and even changes sign for the virtual spacelike mixing needed in a NN potential [9]. A variety of models [10] involving fermion loops appear to support a picture of suppression of $\rho\omega$ mixing to such an extent that our present understanding of charge symmetry breaking is in question. One expects a q^2 variation of the $\rho\omega$ transition matrix element over a wide range of momentum transfer, either from general principles of analyticity and unitarity on the spectral function of the (mixed) vector meson propagator [11], some specific model [12] involving loops, or simply current mixing [1,13] rather than mass mixing (such as that of the tadpole picture). The question is "How large is this variation from the timelike region where the transition is observed to the spacelike region where it determines the charge asymmetric NN potential?"

A first attempt at obtaining a constraint on such behavior was the "node theorem" of Ref. [14]. The claim of Ref. [14] is that the $\rho\omega$ mixing amplitude vanishes at zero four-momentum transfer in any effective Lagrangian model, where there are no explicit mass mixing terms in the bare Lagrangian, and where the vector mesons have a local coupling to conserved currents which satisfy the usual vector current commutation relations.

We here discuss an example of $\rho\omega$ mixing in which where the vector mesons have a local coupling to conserved currents, but the mixing amplitude does not have a node. Our results are stated in the form of divergence conditions on the "mixed hadronic propagator" $M_{\mu\nu}(x,y) = \langle T[\rho_\mu(x)\omega_\nu(y)]\rangle_0$ which Refs. [11,14] use to *define* the $\rho\omega$ transition in NN force diagrams. Our findings are

1) The source terms (or currents) for the ρ and ω fields are divergenceless, so that the subsidiary condition on the ρ and ω fields in the Heisenberg picture remain $\partial_\mu \rho^\mu = \partial_\mu \omega^\mu = 0$.

2) A covariant T product, T^*, can be defined such that the usual covariant meson-nucleon interaction Hamiltonian, used in conjunction with this T^* product, gives a time evolution operator which satisfies the Tomonaga-Schwinger equation.

3) The mixed propagator $M^*_{\mu\nu}(x,y) = \langle T^*[\rho_\mu(x)\omega_\nu(y)]\rangle_0$ does not satisfy the condition $\partial^\mu_x M^*_{\mu\nu}(x,y) = 0$, nor does it satisfy the condition $\partial^\nu_y M^*_{\mu\nu}(x,y) = 0$ — although $\partial^\mu \rho_\mu = \partial^\mu \omega_\mu = 0$.

That is, 1) is the premise of Ref. [14], 2) is needed for a vector meson exchange NN amplitude to exist, and the nonzero divergence in 3) implies the lack of a node at zero four-momentum squared in the Fourier transform of $M^*_{\mu\nu}(x,y)$.

What is different in the earlier analysis [14]? That proof was based upon a definition of the "mixing tensor" in terms of the time ordered product of the source terms of the vector meson field; *i.e.* the vector currents themselves. It relied on the Feynman conjecture that the Schwinger terms in the divergence of the time-ordered product of vector currents are cancelled by the non-covariant terms in the T product of the currents. Specifically, the discussion of Ref. [14] was based on a procedure suggested by Itzykson and Zuber [15] to obtain this "desired" result. The latter exploited the fact that the time ordered product is not defined at equal times to define a new covariant product (of currents) $T^{\#}[A_\mu(x)A_\nu(y)]$ such that

$$M_{\mu\nu}^{\#}(x,y) = \langle T^{\#}[A_\mu(x)A_\nu(y)]\rangle_0 ,$$

has the property that

$$\partial_x^\mu M_{\mu\nu}^{\#} = 0$$

which leads to the node theorem of Ref. [14]. If $A_\mu(x)$ is the massive vector meson field, as it is in our model, then one can calculate the divergence of $M_{\mu\nu}^{\#}(x,y)$ explicitly and indeed one finds that it is zero. However, our analysis leading to items (1), (2) and (3) above shows that a scattering amplitude constructed from $M_{\mu\nu}^{\#}(x,y)$ as the vector propagator and an interaction term, $-\mathcal{L}_{int}$, in the Dyson-Wick procedure is Lorentz invariant, but it is *not* the matrix element of $U(\infty,-\infty)$, where $U(t,t_0)$ is the solution of the Tomonaga-Schwinger equation. To make it so, one can show that the correct NN scattering amplitude to all orders in the coupling constant, g, can be obtained using $M_{\mu\nu}^{\#}(x,y)$ as the propagator with the modified Lorentz invariant Hamiltonian density

$$\mathcal{H}_{int}^{\#} = -\tfrac{g^2}{2\mu^2}\bar{\psi}(x)\gamma^\mu\psi(x)\bar{\psi}(x)\gamma_\mu\psi(x) - \mathcal{L}_{int}$$

A contact, a four fermion, interaction in this equation is essential– without it the scattering matrix is not that of covariant perturbation theory. In addition, the failure of $M_{\mu\nu}^{*}(x,y)$ to satisfy the Feynman conjecture, illustrates the fact that if one chooses for the interpolating meson fields the currents themselves, via the current-field identity, as did the authors of Ref. [14], one finds that the the relationships between the time ordered products of fields and the time ordered products of currents may not be what is naively expected.

These results were shown for the case of a single vector meson interacting with a spin $\tfrac{1}{2}$ particle, but the model can be extended to include both ρ and ω mesons, and to include mixing between these mesons. A similarity transformation permits the meson sector of the model to be written in terms of non-interacting fields, and the results of the single meson case can be immediately taken over to the more interesting case. Then one comes to the final conclusion: these additional terms in the interaction Hamiltonian introduce a change in the charge symmetry breaking potential so that the potential is similar to that derived from a constant $\rho\omega$ mixing matrix element, in spite of the properties of the mixed propagator $M_{\mu\nu}^{\#}(x,y)$.

In conclusion, the "mixed propagator" is an off-shell Green function whose definition can be varied to obtain "desirable" properties. However, considerations of the off-mass-shell dependence of the $\rho\omega$ transition matrix element alone are not sufficient to determine the charge asymmetric NN potential. For this one must have an underlying theory which can be adjusted to preserve physical scattering amplitudes (see also Ref. [12]).

REFERENCES

1. P. C. McNamee, M. D. Scadron and S. A. Coon, Nucl. Phys. **A249**, 483 (1975); ibid., **A287**, 38 (1977); S. A. Coon and M. D. Scadron, Phys. Rev. **C26**, 562 (1982).
2. S. A. Coon and R. C. Barrett, Phys. Rev. **C36**, 2189 (1987).
3. S. A. Coon and M. D. Scadron, Phys. Rev. **C51**, 2923 (1995).
4. Y. Wu, S. Ishikawa, and T. Sasakawa, Phys. Rev. Lett. **64**, 1875 (1990); T. Suzuki, H. Sagawa, and A. Arima, Nucl. Phys. **A536**, 141 (1992).
5. T. Suzuki, H. Sagawa, and N. Van Gai, Phys. Rev. **C47**, R1360 (1993).
6. S. Nakamura, K. Muto, and T. Oda, Phys. Lett. B **311**, 15 (1993).
7. R. Abegg, *et al.*, Phys. Rev. Lett. **75**, 1711 (1995).
8. S. E. Vigdor, *et al.*, Phys. Rev. C **46**, 410 (1992).
9. T. Goldman, J. A. Henderson and A. W. Thomas, Few-Body Systems **12**, 123 (1992).
10. J. Piekarewicz and A. G. Williams, Phys. Rev. C **47**, 2462 (1993); G. Krein, A. W. Thomas and A. G. Williams, Phys. Lett. B **317**, 293 (1993); K. L. Mitchell, P. C. Tandy, C. D. Roberts, and R. T. Cahill, Phys. Lett. B **335**, 282 (1994); A. N. Mitra and K.C. Yang, Phys. Rev C **51**, 3404(1995); S. -F. Gao, C. M. Shakin, and W. -D. Sun, Phys. Rev. C **53**, 1374 (1996).
11. K. Maltman, Phys. Lett. B **362**, 11 (1995).
12. T. D. Cohen and G. A. Miller, Phys. Rev. C **52**, 3428 (1995).
13. R. G. Sachs and J. F. Willemsen, Phys. Rev. D **2**, 133 (1970).
14. H. B. O'Connell, B. C. Pearce, A. W. Thomas, and A. G. Williams, Phys. Lett. B **336**, 1 (1994).
15. C. Itzykson and J. -B. Zuber, Quantum Field Theory (McGraw-Hill, 1985) pp 217-224.

CPLEAR Experiment at CERN: Measurement of CP, T and CPT in the Neutral Kaon System

CPLEAR Collaboration

R. Adler[2], A. Angelopoulos[1], A. Apostolakis[1], E. Aslanides[11], G. Backenstoss[2],
P. Bargassa[13], C.P. Bee[9], O. Behnke [17], A. Benelli [9], V. Bertin[11], F. Blanc[7,13], P. Bloch[4],
P. Carlson[15], M. Carroll[9], J. Carvalho[5], E. Cawley[9], S. Charalambous[16], G. Chardin[14],
M.B. Chertok[3], A. Cody[9], M. Danielsson[15], M. Dejardin[14], J. Derre[14], A. Ealet[11],
B. Eckart[2], C. Eleftheriadis[16], I. Evangelou[8], L. Faravel [7], P. Fassnacht[11], C. Felder[2],
R. Ferreira-Marques[5], W. Fetscher[17], M. Fidecaro[4], A. Filipčič[10], D. Francis[3], J. Fry[9],
E. Gabathuler[9], R. Gamet[9], D. Garreta[14], H.- J. Gerber[17], A. Go[14], C. Guyot[14],
A. Haselden[9], P.J. Hayman[9], F. Henry-Couannier[11], R.W. Hollander[6], E. Hubert[11],
K. Jon-And[15], P.-R. Kettle[13], C. Kochowski[14], P. Kokkas[4], R. Kreuger[6,13], R. Le Gac[11],
F. Leimgruber[2], A. Liolios[16], E. Machado[5], I. Mandić[10], N. Manthos[8], G. Marel[14],
M. Mikuž[10], J. Miller[3], F. Montanet[11], A. Muller[14], T. Nakada[13], B. Pagels [17],
I. Papadopoulos[16], P. Pavlopoulos[2], J. Pinto da Cunha[5], A. Policarpo[5], G. Polivka[2],
R. Rickenbach[2], B.L. Roberts[3], T. Ruf[4], L. Sakeliou[1], P. Sanders[9], C. Santoni[2],
M. Schäfer[17], L.A. Schaller[7], T. Schietinger[2], A. Schopper[4], P. Schune[14], A. Soares[14],
L. Tauscher[2], C. Thibault[12], F. Touchard[11], C. Touramanis[4], F. Triantis[8], E. Van Beveren[5], C.W.E. Van Eijk[6], S. Vlachos[2], P. Weber[17], O. Wigger[13], M. Wolter[17],
C. Yeche[14], D. Zavrtanik[10] and D. Zimmerman[3]

[1]University of Athens, [2]University of Basle, [3]Boston University, [4]CERN, [5]LIP and University of Coimbra, [6]Delft University of Technology, [7]University of Fribourg, [8]University of Ioannina, [9]University of Liverpool, [10]J. Stefan Inst. and Phys. Dep., University of Ljubljana, [11]CPPM, IN2P3-CNRS et Université d'Aix-Marseille II, [12]CSNSM, IN2P3-CNRS, Orsay, [13]Paul-Scherrer-Institut(PSI), [14]CEA, DSM/DAPNIA CE-Saclay, [15]KTH-Stockholm, [16]University of Thessaloniki, [17]ETH-IPP Zürich

Presented by Marko Mikuž
University of Ljubljana and Jožef Stefan Institute, Ljubljana, Slovenia

Abstract. Using strangeness tagging at production time, CPLEAR measures $K^0/\bar{K}^0$ time-dependent asymmetries in pionic and semileptonic kaon decays. From those, a set of parameters describing CP, T and CPT violation in neutral kaon mixing and decay can be determined. Strangeness tagging at decay time with the lepton charge allows for time-reversal violation to be directly observed with a significance of more than three standard deviations. The precision on each of the CPT violation parameters is discussed. The mass equality of the K^0 and $\bar{K}^0$ is tested within $4. \times 10^{-19}$ GeV.

Introduction

Neutral kaons play a special role in tests of discrete symmetries. As the only system yet in nature where the violation of CP symmetry has been observed, it seems the obvious candidate for other surprises. In addition, experimental observation of interference effects allows measurements with a precision unachievable in other places.

The observed CP violation is consistent with T violation and CPT conservation, although no experiment was able to produce a direct measurement of T violation so far. The CPT theorem, regarded as an axiom of physics, leads to CPT conservation. Its derivation requires a minimum of assumptions: a local field theory, Lorentz invariance and the usual spin-statistics relation. Nevertheless, as any postulate in physics, it should be subject to a rigorous experimental test. And there are even some theoretical suggestions, especially in the framework of Quantum Gravity as a non-local theory, that CPT might indeed be violated at the level of the Planck scale [1].

Neutral Kaon System

The time evolution of the K^0 / $\bar{K}^0$ system can be described by a 2×2 Hamiltonian matrix $\hat{H}$. Because the weak interaction allows $K^0 \leftrightarrow \bar{K}^0$ transitions and kaon decays, the matrix is non-diagonal and complex [2]

$$\hat{H} = M - \frac{i}{2}\Gamma \qquad (1)$$

where M is the mass and Γ the decay matrix, both of them Hermitian if unitarity holds.

The smallness of the weak interaction allows for a perturbation expansion

$$M_{\alpha\beta} = <\alpha|H|\beta> + \sum_n \mathcal{P}\frac{<\alpha|H|n><n|H|\beta>}{m_K - m_n} \qquad (2)$$

$$\Gamma_{\alpha\beta} = 2\pi \sum_n <\alpha|H|n><n|H|\beta> \delta(m_K - m_n) \qquad (3)$$

where $\mathcal{P}$ denotes the principal value. The sum in M goes over all possible states while the delta function in Γ confines it to real states only.

Eigenstates of $\hat{H}$, K_L and K_S, have different lifetimes ($\tau_S = (89.27 \pm 0.09)$ ps; $\tau_L = (51.7 \pm 0.4)$ ns $\sim 580\,\tau_S$) and masses ($\Delta m = m_L - m_S = (530.4 \pm 1.4) \times 10^7\,\hbar/s = (3.491 \pm 0.009) \times 10^{-12}$ MeV) [3].

There are three phases resulting from CP and T operations

$$CP|K^0> = e^{+i\theta_{CP}}|\bar{K}^0> \quad CP|\bar{K}^0> = e^{-i\theta_{CP}}|K^0> \quad (4)$$
$$T|K^0> = e^{i\theta_T}|K^0> \quad T|\bar{K}^0> = e^{i\bar{\theta}_T}|\bar{K}^0> \quad (5)$$

which are, because of $CPT|K^0> = TCP|K^0>$, connected by $2\theta_{CP} = \bar{\theta}_T - \theta_T$. No observable is allowed to depend on these phases. As a proper phase choice simplifies the notation, I shall make use of the convention [4].

$$\theta_{CP} = 0 \Rightarrow \bar{\theta}_T = \theta_T. \quad (6)$$

The eigenstates of $\hat{H}$ are

$$|K_S> = \frac{1}{\sqrt{2}}((1+\epsilon_S)|K^0> + (1-\epsilon_S)|\bar{K}^0>) \quad (7)$$

$$|K_L> = \frac{1}{\sqrt{2}}((1+\epsilon_L)|K^0> - (1-\epsilon_L)|\bar{K}^0>)$$

where we introduced CP-violation parameters $\epsilon_S = \epsilon_K + \delta$ and $\epsilon_L = \epsilon_K - \delta$, with ϵ_K and δ the T- and CPT-violation parameters, expressed with the mixing and decay matrix elements as

$$\epsilon_K = \frac{H_{21} - H_{12}}{2(\Delta m + \frac{i}{2}\Delta\Gamma)} = (-\mathcal{I}mM_{12} + \frac{i}{2}\mathcal{I}m\Gamma_{12}) \cdot \frac{i\Delta m + \Delta\Gamma/2}{\Delta m^2 + (\Delta\Gamma/2)^2} \quad (8)$$

$$\delta = \frac{H_{22} - H_{11}}{2(\Delta m + \frac{i}{2}\Delta\Gamma)} = -\frac{i}{2}[(M_{22} - M_{11}) - \frac{i}{2}(\Gamma_{22} - \Gamma_{11})] \cdot \frac{i\Delta m + \Delta\Gamma/2}{\Delta m^2 + (\Delta\Gamma/2)^2}. \quad (9)$$

The argument of the common factor is the superweak angle $\tan\phi_{SW} = \frac{2\Delta m}{\Delta\Gamma} \sim \frac{\pi}{4}$, so each of the CP-violation parameters has a component parallel to ϕ_{SW} ($\mathcal{I}mM_{12}$ in ϵ_K, K^0 / $\bar{K}^0$ lifetime difference in δ) and one perpendicular to it ($\mathcal{I}m\Gamma_{12}$ in ϵ_K, K^0 / $\bar{K}^0$ mass difference in δ).

CPLEAR Experiment

The CPLEAR experiment makes use of the intense $\bar{p}$ beams available at LEAR to produce pure K^0 and $\bar{K}^0$ by tagging the accompanying charged kaon in $p\bar{p}$ annihilation

$$\bar{p}p \to K^0 K^- \pi^+ \quad \bar{p}p \to \bar{K}^0 K^+ \pi^- \quad (10)$$

both occurring at a branching ratio of $\approx 2 \times 10^{-3}$. The tagging of opposite strangeness states at production time maximizes the interference effects to be observed in K^0, $\bar{K}^0$ decays.

The experimental apparatus [5] is shown in Fig. 1. Ten chamber layers (2 of Proportional Chambers, 6 of Drift Chambers, 2 of Streamer Tubes) are used to trace charged particles resulting from annihilation and neutral

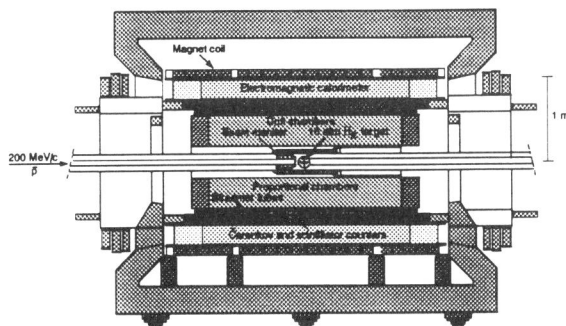

FIGURE 1. View of the CPLEAR detector

kaon decays. A 32-segment scintillator-Čerenkov-scintillator provides particle identification (kaons/pions/electrons). Photons are detected by a 18-layer fine-grain streamer tube / lead sampling calorimeter. Signals from all detectors are processed in a multilevel trigger, providing a rejection factor of over 1000 and allowing the detector to operate at a p̄ rate of 1 MHz. The material in the decay region up to the streamer tubes is minimized by using a gas target with mylar-kevlar walls and innovative low-mass chamber construction, thus reducing regeneration effects of neutral kaons. In 1995 a chamber was added at 1.7 cm radius to improve the trigger and tracking capabilities.

To isolate the interference term in K^0, $\bar{K}^0$ decays (see below) and to cancel systematic errors resulting from acceptance calculations, time-dependent asymmetries are formed from rates for each of the measured decay channels f,

$$A_f(\tau) = \frac{R(\bar{K}^0 \to f, \tau) - R(K^0 \to f, \tau)}{R(\bar{K}^0 \to f, \tau) + R(K^0 \to f, \tau)}. \quad (11)$$

In total, about 100 million K^0 and $\bar{K}^0$ decays were reconstructed. The results refer to the analysis of the complete data-set of 70 M $K^0, \bar{K}^0 \to \pi^+\pi^-$ decays with $\tau > 1\,\tau_S$ and 0.5 M $K^0, \bar{K}^0 \to \pi^+\pi^-\pi^0$ decays. Results on 1.8 M $K^0, \bar{K}^0 \to e\pi\nu$ decays are obtained from about 2/3 of the statistics available.

$\pi^+\pi^-$ Decays

Selection of $K^0, \bar{K}^0 \to \pi^+\pi^-$ decays [6] is based mainly on kinematical and geometrical constraints. As the events can be fully reconstructed, the background level is low throughout the interesting interference region.

The different acceptance for primary $K^+\pi^-$ versus $K^-\pi^+$ pairs can be parametrized by a normalization parameter α which depends only on the kinematics of the primary pair and is not correlated with the decay path. The high statistics of $K^0, \bar{K}^0 \to \pi^+\pi^-$ decays allows for a control of the overall normalization to a level of 6×10^{-4}. Geometrical effects are cancelled adequately by reversing the magnetic field several times a day.

The asymmetry, normalized to equal decay rates at production, reads

$$A_{+-}(\tau) = \frac{R(\bar{K}^0 \to \pi^+\pi^-) - \alpha R(K^0 \to \pi^+\pi^-)}{R(\bar{K}^0 \to \pi^+\pi^-) + \alpha R(K^0 \to \pi^+\pi^-)} = \frac{-2|\eta_{+-}|\cos(\Delta m \tau - \phi_{+-})e^{\frac{1}{2}(\Gamma_S - \Gamma_L)\tau}}{1 + |\eta_{+-}|^2 e^{(\Gamma_S - \Gamma_L)\tau}} \quad (12)$$

with the complex CP violation parameter $\eta_{+-} = |\eta_{+-}|e^{i\phi_{+-}}$ corresponding to the amplitude ratio

$$\eta_{+-} = \frac{A(K_L \to \pi^+\pi^-)}{A(K_S \to \pi^+\pi^-)}. \quad (13)$$

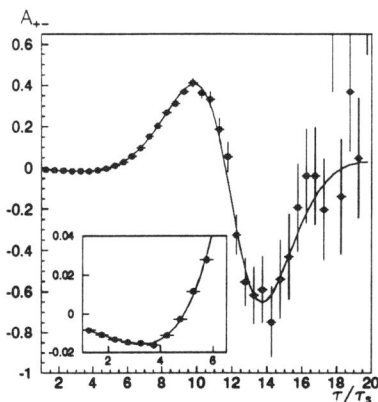

FIGURE 2. Two-pion time-dependent asymmetry. The region $0-6\,\tau_S$ is blown up in the insert.

Equation (12) is fitted to the measured asymmetry, shown in Fig. 2, leaving $|\eta_{+-}|$, ϕ_{+-} and α as free parameters while fixing Δm and Γ_S to the CPLEAR compilation value [13] and Γ_L to [3]. The preliminary results are

$$|\eta_{+-}| = (2.316 \pm 0.025_{stat} \pm 0.030_{syst}) \times 10^{-3}$$
$$\phi_{+-} = 43.5° \pm 0.5°_{stat} \pm 0.5°_{syst} \pm 0.4°_{\Delta m}$$

with ϕ_{+-} correlated to Δm by $\phi_{+-} = 43.5° + 0.293(\Delta m - 530.7)\frac{\text{deg}}{10^7 \hbar s^{-1}}$. The bulk of the systematic error originates from regeneration uncertainties. The

CPLEAR regeneration measurement [7] will render this contribution negligible, resulting in a systematic error of $0.2°$ only.

CP violation observed in $\pi^+\pi^-$ decays (η_{+-}) results from CP violation in mixing (ϵ_K, δ) or in the decay amplitudes

$$<\pi\pi_{I=0,2}|T|K^0> = (A_{0,2} + B_{0,2})e^{i\delta_{0,2}} \; ; \; <\pi\pi_{I=0,2}|T|\bar{K}^0> = (A^*_{0,2} - B^*_{0,2})e^{i\delta_{0,2}} \quad (14)$$

where $A_{0,2}$ and $B_{0,2}$ are weak decay amplitudes to isospin 0 and 2 states and $\delta_{0,2}$ the strong interaction phase-shifts.. The amplitude A is CPT conserving ($\bar{A} = A^*$) while B is introduced to explicitly violate CPT ($\bar{B} = -B^*$).

A straightforward calculation taking into account the isospin decomposition of $\pi^+\pi^-$ and the eqs. (7) for the $K_{S,L}$ eigenstates yields, keeping terms of first order in small parameters

$$\eta_{+-} = \epsilon_K - \delta + a + \epsilon' - a\omega, \quad (15)$$

where ϵ', a and ω denote the following combinations of the decay amplitudes

$$\epsilon' = \frac{1}{\sqrt{2}}[\frac{\mathcal{I}mA_2}{\mathcal{R}eA_0}e^{i(\delta_2-\delta_0+\frac{\pi}{2})} + \frac{\mathcal{R}eB_2}{\mathcal{R}eA_0}e^{i(\delta_2-\delta_0)}] \quad (16)$$

$$a = i\frac{\mathcal{I}mA_0}{\mathcal{R}eA_0} + \frac{\mathcal{R}eB_0}{\mathcal{R}eA_0} \qquad \omega = \frac{1}{\sqrt{2}}\frac{\mathcal{R}eA_2}{\mathcal{R}eA_0}e^{i(\delta_2-\delta_0)}. \quad (17)$$

Here, $\mathcal{I}mB$ was neglected, as a large value of $\mathcal{I}mB$ induces unitarity violation [9].

The parameter ϵ' describes direct CP violation in neutral kaon decays and ω quantifies the observed $\Delta I = 1/2$ rule ($|\omega| = 0.045$). Both ϵ' and a have a CPT conserving and a CPT violating part, which are perpendicular to each other. As by coincidence $|\delta_2 - \delta_0 + \frac{\pi}{2}| = 48° \pm 4° \approx \phi_{sw}$, the CPT conserving part of $(\epsilon' - a\omega)$ has a phase $\approx \phi_{sw}$ and $1/\sqrt{2}$ is taken as an adequate approximation of the corresponding sine and cosine. The CPT violating part has a phase $\approx \phi_{sw} + \pi/2$.

The dominant contributions of $\mathcal{I}m\Gamma_{12}$ to ϵ_K are given by (3) as

$$\epsilon_{K\perp} = \frac{i}{2}\mathcal{I}m\Gamma_{12} \cdot \frac{i\Delta m + \Delta\Gamma/2}{\Delta m^2 + (\Delta\Gamma/2)^2} \approx \frac{1}{\sqrt{2}\Gamma_S}\mathcal{I}m\Gamma_{12}e^{i(\phi_{sw}+\frac{\pi}{2})} \quad (18)$$

$$\approx (\frac{-\mathcal{I}mA_0}{\sqrt{2}\mathcal{R}eA_0} + \frac{\Delta\phi}{\sqrt{2}})e^{i(\phi_{sw}+\frac{\pi}{2})},$$

where

$$\Delta\phi = \frac{\Gamma_L}{\Gamma_S}[4BR(K_L \to l^+\pi^-\nu) \cdot \mathcal{I}m(x) - \quad (19)$$
$$BR(K_L \to \pi^+\pi^-\pi^0) \cdot \mathcal{I}m(\epsilon_S - \eta_{+-0}) -$$
$$BR(K_L \to \pi^0\pi^0\pi^0) \cdot \mathcal{I}m(\epsilon_S - \eta_{000})]$$

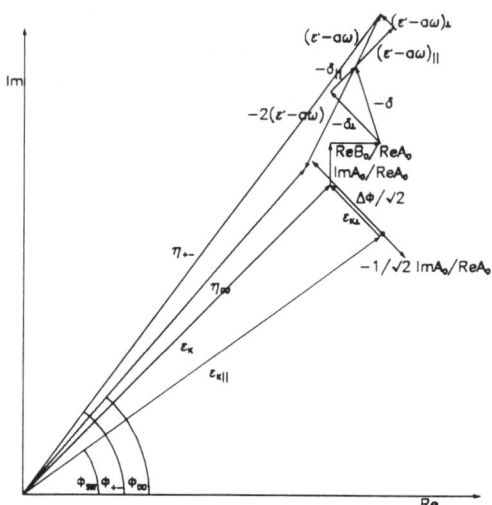

FIGURE 3. Relations between phenomenological parameters in the complex plane. The magnitude of all parameters is exaggerated with respect to η_{+-}, η_{00} and ϵ_K.

with η_{+-0}, η_{000} the CP violation parameters in three-pion decays and x the $\Delta Q = \Delta S$ violation parameter in semileptonic decays.

Equation (15) is depicted (not to scale) in Fig. 3 together with its equivalent for decays into two neutral pions ($\eta_{00} = \epsilon_L + a - 2(\epsilon' - a\omega)$). Concentrating on the deviation of the phase ϕ_{+-} from ϕ_{SW}, we observe that contributions from $\mathcal{I}m A_0$ cancel out, there is no contribution from the CPT conserving part $(\epsilon' - a\omega)_\parallel$ and no contribution from the K^0 / $\bar{K}^0$ time difference in δ. There remain three possible sources of the $\phi_{+-} - \phi_{SW}$ phase difference

- CPT violation in mixing – K^0 / $\bar{K}^0$ mass difference in δ
- CPT violation in decay – $\frac{\mathcal{R}e B_0}{\mathcal{R}e A_0}$ in a and $\frac{\mathcal{R}e B_2}{\mathcal{R}e A_0}$ in ϵ'
- $\Delta \phi$ in ϵ_K

Experimentally, CPT violation in $(\epsilon' - a\omega)_\perp$ can be constrained by the measurement of the $\phi_{+-} - \phi_{00}$ phase difference [3,10] to $(\epsilon' - a\omega)_\perp < 1.7 \times 10^{-5}$(90%C.L.), resulting in a constraint of $0.4°$(90%C.L.) to its contribution to $\phi_{+-} - \phi_{SW}$. $\Delta \phi$ is expected to be of $\mathcal{O}(10^{-7})$ in the Standard Model. However, relying on experimental limits [3] on x and η, expression (19) yields a limit on the contribution of $\Delta \phi$ to $\phi_{+-} - \phi_{SW}$ as high as $3°$. Hence it is obvious, that for a conclusion on limits of CPT violation from a comparison of ϕ_{+-} with ϕ_{SW}, an improvement on three-pion and semileptonic decays is essential.

$\pi^+\pi^-\pi^0$ Decays

The selection of $K^0, \bar{K}^0 \to \pi^+\pi^-\pi^0$ decays [11] is also based on kinematical and geometrical constraints. In addition, at least one of the photons from the π^0 is required to produce a shower in the calorimeter.

The $\pi^+\pi^-\pi^0$ state can have isospin 0,1,2 and 3. I=0 is suppressed by kinematics and the I=3 by the $\Delta I = 1/2$ rule. By symmetry considerations I=1 is CP=-1 and I=2 is CP=+1. Thus the CP violation parameter is defined as

$$\eta_{+-0} = \frac{\int d\Omega A_S^{CP-}(I=1) A_L^{*CP-}(I=1)}{\int d\Omega A_L^{CP-}(I=1) A_L^{*CP-}(I=1)} \qquad (20)$$

where the integral extends over phase space. As the CP conserving I=2 part of the K_S amplitude is odd in $p_{\pi^+} - p_{\pi^-}$ its contribution vanishes through integration. The CP violation asymmetry then reads

$$A_{+-0}(\tau) = -2[\mathcal{R}e(\eta_{+-0})\cos(\Delta m\tau) - \mathcal{I}m(\eta_{+-0})\sin(\Delta m\tau)]e^{-\Delta\Gamma\tau/2}. \qquad (21)$$

A fit to the measured asymmetry yields the values

$$\mathcal{R}e(\eta_{+-0}) = (-2 \pm 7(stat)\ ^{+4}_{-1}(syst)) \times 10^{-3}$$
$$\mathcal{I}m(\eta_{+-0}) = (-2 \pm 9(stat)\ ^{+2}_{-1}(syst)) \times 10^{-3}$$

which represent a good improvement over the previous best measurement [12] but are still a factor of 5 short of the expected level of CP violation.

Semileptonic Decays

The selection of semileptonic events [8] demands in addition to kinematical and geometrical constraints an electron identified by energy loss in the scintillators and number of photoelectrons in the Čerenkov detector. There are four measurable decay rates, labelled by the initial strangeness and final electron charge:

$$R^+ = R(K^0(\tau) \to e^+\pi^-\nu) \quad R^- = R(K^0(\tau) \to e^-\pi^+\bar{\nu})$$
$$\bar{R}^- = R(\bar{K}^0(\tau) \to e^-\pi^+\bar{\nu}) \quad \bar{R}^+ = R(\bar{K}^0(\tau) \to e^+\pi^-\nu).$$

By forming asymmetries from them, acceptances cancel and various relevant parameters can be isolated by a proper combination of the rates.

In the Standard Model, only $\Delta Q = \Delta S$ transitions are allowed to first order, resulting in a limit on the parameter

$$x = \frac{A(\bar{K}^0 \to l^+\pi^-\nu)}{A(K^0 \to l^+\pi^-\nu)} \qquad (22)$$

at $\mathcal{O}(10^{-6})$. We use this "$\Delta Q = \Delta S$ rule" as a strangeness tag at decay time - the current experimental errors [3] ($\sigma_{\mathcal{R}e(x)} = 18 \times 10^{-3}$, $\sigma_{\mathcal{I}m(x)} = 26 \times 10^{-3}$) are improved by our measurement.

Of special interest for the CPT test is the determination of Δm for its correlation with ϕ_{+-} and the determination of ϕ_{SW} as well as the measurement of $\mathcal{I}m(x)$ for its contribution to $\Delta \phi$ (eq. 19). The determination of Δm is best achieved by forming the asymmetry

$$A_{\Delta m}(\tau) = \frac{(R^+ + \bar{R}^-) - (R^- + \bar{R}^+)}{(R^+ + \bar{R}^-) + (R^- + \bar{R}^+)} = \frac{2\cos(\Delta m \tau) e^{-\frac{1}{2}(\Gamma_S + \Gamma_L)\tau}}{(1 + 2\mathcal{R}e(x))e^{-\Gamma_S \tau} + (1 - 2\mathcal{R}e(x))e^{-\Gamma_L \tau}} \quad (23)$$

assuming no CPT violation in $\Delta Q \neq \Delta S$ transitions. The preliminary results of a fit to the complete data-set are shown in Fig. 4 and yield

$$\Delta m = (529.2 \pm 1.8_{stat} \pm 0.5_{syst}) \times 10^7 \hbar s^{-1}.$$

This represents the most precise measurement of Δm with the same precision as the current world average [3].

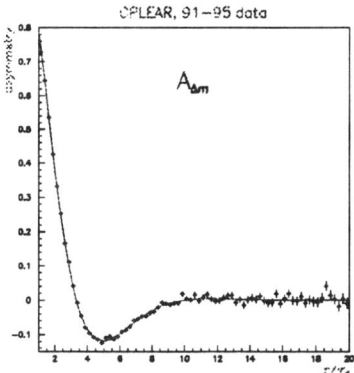

FIGURE 4. Time-dependent asymmetry $A_{\Delta m}$. Full line represents the fit to eq. (23)

The asymmetry between all K^0 and $\bar{K}^0$ semileptonic decays

$$A_2 = \frac{(\bar{R}^+ + \bar{R}^-) - (R^+ + R^-)}{(\bar{R}^+ + \bar{R}^-) + (R^+ + R^-)} \quad (24)$$

is sensitive to $\mathcal{I}m(x)$. The preliminary result from partial statistics

$$\mathcal{I}m(x + \delta) = (0.5 \pm 2.4_{stat} \pm 0.6_{syst}) \times 10^{-3}$$

represents a tenfold improvement over the previous world average [3].

Using the $\Delta Q = \Delta S$ rule for the strangeness tag at decay time, the asymmetry

$$A_T = \frac{\bar{R}^+ - R^-}{\bar{R}^+ + R^-} \stackrel{\tau \to \infty}{=} 4\mathcal{R}e(\epsilon_K) + 2\mathcal{R}e(y) \tag{25}$$

represents a direct test of T reversal [14] in absence of CPT violation in semileptonic decays, described by y. Two preliminary analyses, referring to different sets of data, each about 1/3 of the total, have been performed. The first normalized the relative $\frac{e^+\pi^-}{e^-\pi^+}$ efficiency to the measured value of semileptonic charge asymmetry δ_l, while the second used CPLEAR calibration data. Both yield compatible results ($A_T = (6.3 \pm 2.1_{stat} \pm 1.8_{syst}) \times 10^{-3}, (7.2 \pm 2.3_{stat} \pm 2.0_{syst}) \times 10^{-3}$). Systematic errors are predominantly uncorrelated, so the combined result

$$A_T = (6.7 \pm 1.6_{stat}) \times 10^{-3}$$

represents a direct experimental verification of T violation with a significance of more than three standard deviations.

The asymmetry

$$A_{CPT} = \frac{\bar{R}^- - R^+}{\bar{R}^- + R^+} \stackrel{\tau \to \infty}{=} 4\mathcal{R}e(\delta) - 2\mathcal{R}e(y) \tag{26}$$

in the same manner represents a direct test of CPT. The sum $A_T + A_{CPT} = 4\mathcal{R}e(\epsilon_K + \delta) = 4\mathcal{R}e(\epsilon_S)$ combined with the normalization factor in $\pi^+\pi^-$ decays $\alpha_{\pi^+\pi^-}$, which is proportional to $1 + 4\mathcal{R}e(\epsilon_L)$ yields

$$\frac{\bar{R}^- - \alpha_{\pi^+\pi^-} R^+}{\bar{R}^- + \alpha_{\pi^+\pi^-} R^+} + \frac{\bar{R}^+ - \alpha_{\pi^+\pi^-} R^-}{\bar{R}^+ + \alpha_{\pi^+\pi^-} R^-} = 8\mathcal{R}e(\delta) \tag{27}$$

independent of CPT violation in decays. In expression (27) R are the plain measured rates. Our preliminary result on 1/3 of data reads

$$\mathcal{R}e(\delta) = (0.1 \pm 3.7_{stat}) \times 10^{-4}. \tag{28}$$

Test of CPT

The improvements on $\mathcal{I}m(\eta_{+-0})$ and $\mathcal{I}m(x + \delta)$ made by the CPLEAR measurements greatly reduce the error on $\Delta\phi$. Assuming lepton universality ($x_e = x_\mu$) and the dominance of the I=1 amplitude in CP violation in $K^0 \to 3\pi^0$ decays ($\mathcal{I}m(\eta_{+-0}) = \mathcal{I}m(\eta_{000})$), the resulting uncertainty on the contribution to $\phi_{+-} - \phi_{SW}$ is diminished to $0.22°$. The assumption $\mathcal{I}m(x + \delta) \approx \mathcal{I}m(x)$ is justified by eq. (28) and the final result. Full use of these improvements is made by taking the values of ϕ_{+-} and ϕ_{SW} from the global fit of world data as performed by CPLEAR [13] and adopted by the PDG. The latest global fit values

$$\phi_{+-} = 43.56^\circ \pm 0.56^\circ \quad \phi_{SW} = 43.46^\circ \pm 0.08^\circ$$

show that the limitation to the test of CPT violation now comes from the error on ϕ_{+-} itself. The precision achieved allows for a bound

$$\left| \frac{M_{11} - M_{22}}{2\sqrt{2}\Delta m} + \frac{1}{\sqrt{2}} \frac{\mathcal{R}e B_0}{\mathcal{R}e A_0} \right| < 4.4 \times 10^{-5} (90\% \text{C.L.})$$

To interpret the phase difference in terms of a bound on the K^0 / $\bar{K}^0$ mass difference the following scenario is implied. Equation (2) allows a first order contribution to the mass matrix, while according to (3) contributions to the decay matrix start with the second order. Imagining a superweak type of CPT-violating interaction contributing differently in first order to $M_{11} = m_{K^0}$ and $M_{22} = m_{\bar{K}^0}$ implies an absence of CPT violating effects in decay. Hence, the mass difference in δ and $\Delta\phi$ remain the only sources of a phase difference between ϕ_{+-} and ϕ_{SW}. The comparison yields a limit

$$\left| \frac{M_{11} - M_{22}}{2\sqrt{2}\Delta m} \right| < 4. \times 10^{-5} (90\% \text{C.L.})$$

which can be expressed as a mass difference bound

$$|m_{\bar{K}^0} - m_{K^0}| < 4. \times 10^{-19} \text{ GeV } (90\% \text{C.L.}).$$

This represents the most precise test of CPT in the mass matrix.

REFERENCES

1. S. Hawking,Comm.Math.Phys.**87** (1982) 395.
2. V. Weisskopf, E. Wigner, Z.Phys.**63** (1930) 54.
3. Particle Data Group, Phys.Rev.**D54** (1996) 1.
4. C. D. Buchanan, Phys.Rev.**D45** (1992) 4088.
5. R. Adler et al., Nucl.Instr.Meth.**A379**(1996)76.
6. R. Adler et al., Phys.Lett.**B363** (1995) 243.
7. R. Adler et al.: Measurement of Neutral Kaon Regeneration Amplitude in Carbon at Momenta below 1 GeV/c, in preparation.
8. R. Adler et al., Phys.Lett.**B363** (1995) 237.
9. V. V. Barmin et al., Nucl.Phys.**B247**(1984)293.
10. B. Schwingenheuer et al., Phys.Rev.Lett.**74** (1995) 4376.
11. R. Adler et al., Phys.Lett.**B370** (1996) 167 and CERN-PPE/97-54(1997).
12. Y. Zou et al., Phys.Lett.**B329** (1994) 519.
13. R. Adler et al., Phys.Lett.**B369** (1996) 367, updated to include the whole CPLEAR statistics.
14. P. K. Kabir, Phys.Rev.**D2** (1970) 540.

Direct CP Violation in $B^\pm \to \rho^\pm \pi^+ \pi^-$ in the ρ^0-ω Interference Region

S. Gardner, H.B. O'Connell*, and A.W. Thomas[†]

*Department of Physics and Astronomy,
University of Kentucky, Lexington, KY 40506-0055 USA
[†]Department of Physics and Mathematical Physics,
and Special Research Centre for the Subatomic Structure of Matter,
University of Adelaide, Adelaide, S.A. 5005 AUSTRALIA

Abstract. We study direct CP violation in $B^\pm \to \rho^\pm \rho^0(\omega) \to \rho^\pm \pi^+ \pi^-$ and focus specifically on the rate asymmetry in the ρ^0-ω interference region. Here the strong phase is dominated by isospin violation, so that it can be essentially determined by $e^+e^- \to \rho^0(\omega) \to \pi^+\pi^-$ data. We find the CP-violating asymmetry to be of the order of 20% at the ω invariant mass. Moreover, it is robust with respect to the estimable strong-phase uncertainties, permitting the extraction of $\sin \alpha$ from this channel.

Experimental programs in the next years at HERA, KEK, and SLAC will study CP violation in the B-meson system in the hope of identifying physics beyond the Standard Model. In the Standard Model, the so-called unitarity triangle associated with the CKM parameters α, β, and γ requires that these angles sum to π [1]. Yet, the experiments in the neutral B sector which would measure α, e.g., determine merely $\sin 2\alpha$ [2], so that discrete ambiguities remain in α itself [3]. Here we consider CP violation in $B^\pm \to \rho^\pm \rho^0(\omega) \to \rho^\pm \pi^+ \pi^-$, with $\rho^0(\omega)$ denoting the ρ^0-ω interference region, proposed by Enomoto and Tanabashi [4]. The rate asymmetry, which is CP-violating, arises exclusively from a nonzero phase in the CKM matrix, so that the CP violation is termed "direct." The manner in which CP-violation is generated differs from that of the neutral meson case, so that the asymmetry depends on $\sin \alpha$. Its determination, then, removes the mod(π) ambiguity in α inherent in the $\sin 2\alpha$ measurement [3]. The rub, however, is that direct CP violation requires that both a strong and weak phase difference exist between two interfering amplitudes [5], so that information on the weak phase is generally obscured by strong interaction uncertainties. In the above case, however, data in $e^+e^- \to \pi^+\pi^-$ in the ρ^0-ω inteference region substantially constrains the strong phase [6]. Here we discuss the remaining uncertainties in the $\sin\alpha$ determination, and show how they may be obviated with further experimental data.

Resonances can play a strategic role in direct CP violation. Resonance information, such as the mass and width, can be used to constrain the strong phase, and their interference can significantly enhance the CP-violating asymmetry [7,8]. Both these effects are operative in hadronic B-decays in the ρ^0-ω interference region [4,9]. In $B^\pm \to \rho^\pm \rho^0(\omega)$, ρ-ω interference is the dominant source of strong phase, and it can be determined through fits to $e^+e^- \to \pi^+\pi^-$ data [6]. We also have computed the additional isospin violating effects arising from electroweak penguin contributions and within the ρ^- and ρ^0 hadronic form factors. Their impact is small, however, relative to that of the ρ-ω mixing contribution. The asymmetry we predict is of order of 20% at the ω invariant mass; the asymmetry is both large and robust with respect to the known strong phase uncertainties.

The CP-violating asymmetry in $B^\pm \to \rho^\pm \pi^+ \pi^-$ is significantly enhanced by ρ^0-ω mixing. To see why this is so, consider the amplitude A for $B^- \to \rho^- \pi^+ \pi^-$ decay:

$$A = \langle \pi^+\pi^-\rho^- | \mathcal{H}^T | B^- \rangle + \langle \pi^+\pi^-\rho^- | \mathcal{H}^P | B^- \rangle \,, \tag{1}$$

where A is given by the sum of the amplitudes corresponding to the tree and penguin diagrams, respectively. Defining the strong phase δ, the weak phase ϕ, and the magnitude r via

$$A = \langle \pi^+\pi^-\rho^- | \mathcal{H}^T | B^- \rangle \left[1 + re^{i\delta} e^{i\phi} \right] \,, \tag{2}$$

one has $\overline{A} = \langle \pi^+\pi^-\rho^+ | \mathcal{H}^T | B^+ \rangle [1 + re^{i\delta} e^{-i\phi}]$. Thus, the CP-violating asymmetry A_{CP} is

$$A_{CP} \equiv \frac{|A|^2 - |\overline{A}|^2}{|A|^2 + |\overline{A}|^2} = \frac{-2r\sin\delta\sin\phi}{1 + 2r\cos\delta\cos\phi + r^2} \,, \tag{3}$$

so that both δ and ϕ must be non-zero to yield a non-zero asymmetry. Here ϕ is $-\alpha$ [1].

To express δ in terms of the resonance parameters, let t_V be the tree amplitude and p_V be the penguin amplitude to produce a vector meson V. Thus, the tree and penguin amplitudes for $B^- \to \rho^- \pi^+ \pi^-$ can be written as

$$\langle \pi^+\pi^-\rho^- | \mathcal{H}^T | B^- \rangle = \frac{g_\rho}{s_\rho s_\omega} \tilde{\Pi}_{\rho\omega} t_\omega + \frac{g_\rho}{s_\rho} t_\rho \,, \tag{4}$$

$$\langle \pi^+\pi^-\rho^- | \mathcal{H}^P | B^- \rangle = \frac{g_\rho}{s_\rho s_\omega} \tilde{\Pi}_{\rho\omega} p_\omega + \frac{g_\rho}{s_\rho} p_\rho \,. \tag{5}$$

Note that $\tilde{\Pi}_{\rho\omega}$ is the effective ρ^0-ω mixing matrix element, g_ρ is the $\rho^0 \to \pi^+\pi^-$ coupling, $1/s_V$ is the vector meson propagator, $s_V = s - m_V^2 + im_V\Gamma_V$, with

m_V and Γ_V the vector meson mass and width, and s is the invariant mass of the $\pi^+\pi^-$ pair. $\tilde{\Pi}_{\rho\omega}$ is extracted from pion form-factor data, as measured in $e^+e^- \to \pi^+\pi^-$, and it is insensitive to the ambiguities in the ρ parametrization [10]. We have fit $\tilde{\Pi}_{\rho\omega}(s) = \tilde{\Pi}_{\rho\omega}(m_\omega^2) + (s - m_\omega^2)\tilde{\Pi}'_{\rho\omega}(m_\omega^2)$ to find $\tilde{\Pi}_{\rho\omega}(m_\omega^2) = -3500 \pm 300$ MeV^2 and $\tilde{\Pi}'_{\rho\omega}(m_\omega^2) = 0.03 \pm 0.04$ [10]. Using

$$re^{i\delta}e^{i\phi} = \frac{\langle \pi^+\pi^-\rho^- | \mathcal{H}^P | B^- \rangle}{\langle \pi^+\pi^-\rho^- | \mathcal{H}^T | B^- \rangle} = \frac{\tilde{\Pi}_{\rho\omega}p_\omega + s_\omega p_\rho}{\tilde{\Pi}_{\rho\omega}t_\omega + s_\omega t_\rho}, \tag{6}$$

and the definitions of Ref. [4]:

$$\frac{p_\omega}{t_\rho} \equiv r' e^{i(\delta_q + \phi)}, \quad \frac{t_\omega}{t_\rho} \equiv \alpha e^{i\delta_\alpha}, \quad \frac{p_\rho}{p_\omega} \equiv \beta e^{i\delta_\beta}, \tag{7}$$

one finds, to leading order in isospin violation,

$$re^{i\delta} = \frac{r'e^{i\delta_q}}{s_\omega}\left\{\tilde{\Pi}_{\rho\omega} + \beta e^{i\delta_\beta}\left(s_\omega - \tilde{\Pi}_{\rho\omega}\alpha e^{i\delta_\alpha}\right)\right\}. \tag{8}$$

Note that $\delta_\alpha, \delta_\beta$, and δ_q are "short-distance" phases, generated by putting the quarks in loops on their mass-shell [5]; this mechanism is the typical source of strong phase. A $J=0$, $I=0$ $\rho^\pm\rho^0$ final state is forbidden by Bose symmetry if isospin is perfect, so that β is non-zero only if electroweak penguin contributions and isospin violation in the $\rho^\pm$ and ρ^0 hadronic form factors are included. Numerically, $|\tilde{\Pi}_{\rho\omega}|/(m_\omega\Gamma_\omega) \gg \beta$. The resonant enhancement of the CP-violating asymmetry is thus driven by $\tilde{\Pi}_{\rho\omega}/s_\omega$. As $s \to m_\omega^2$, the asymmetry is maximized if $|\chi| \equiv |\tilde{\Pi}_{\rho\omega}|/m_\omega\Gamma_\omega \sim O(1)$ and $\delta_q + \eta \sim \pm\pi/2$, where $\eta = -\arg s_\omega$. Here $|\chi| \approx .53$ [10,1] and $\eta = -\pi/2$. Note that $\delta_q \lesssim -161°$ [4], so that $\delta_q + \eta \lesssim -251°$ at the ω mass.

The CP-violating asymmetry from Eqs. (3,8), then, is determined by the resonance parameters $\tilde{\Pi}_{\rho\omega}$, m_ω, Γ_ω, and the "short distance" parameters α, δ_α, β, δ_β, r', δ_q, as well as ϕ, the weak phase. The latter class of parameters are calculable within the context of the operator product expansion if the factorization approximation is applied, though a ratio of hadronic form factors enters as well. Here that ratio is modified from unity by isospin-violating effects only. Using the above method, we find asymmetries of the order of 20% at the ω invariant mass, and an asymmetry of this magnitude is retained even if the imaginary parts of the effective Wilson coefficients are set to zero [6].

From our earlier discussion, however, it is clear that the sign of $\sin\phi$ is of unique significance, so that it is useful to consider how it may be extracted. If $r < .5$ then the sign of the CP-violating asymmetry is determined by $\sin\delta$ and $\sin\phi$. As $s \to m_\omega^2$, the sign of $\sin\delta$ is determined by

$$\text{sgn}(\sin\delta) = \text{sgn}(\cos\delta_q \,\text{Im}\,\Omega + \sin\delta_q \,\text{Re}\,\Omega), \tag{9}$$

where $\Omega = \beta(\cos\delta_\beta - \chi\sin\delta_\beta) - i(\chi(1 - \beta\cos\delta_\beta) - \beta\sin\delta_\beta)$ in this limit, recalling $\chi = \tilde{\Pi}_{\rho\omega}/(m_\omega\Gamma_\omega)$. As $|\chi| > \beta$, the sign of $\sin\delta$ is determined by $-\chi\cos\delta_q$. The sign of χ is just that of $\tilde{\Pi}_{\rho\omega}$, but what of that of $\cos\delta_q$? To determine this, note that the "skew" of the asymmetry — the sign of the $(s - m_\omega^2)$ term multiplying $1/|s_\omega|$ in $\sin\delta$ from Eq. (8) — determines the sign of $\sin\delta_q$ as $|\chi| > \beta$. Note that the empirical s-dependence of $\tilde{\Pi}_{\rho\omega}(s)$ about $s = m_\omega^2$ does not cloud this interpretation [6]. In the factorization approximation, the sign of $\sin\delta_q$ is invariably that of $\cos\delta_q$ [4], yet one need not assume this. The sign of the asymmetry in $B^\pm \to \rho^\pm\omega \to \rho^\pm\pi^+\pi^-\pi^0$ is driven by $\sin\delta_q \sin\phi$, and it is also large [11]. The signs of the two asymmetries, whether they are the same or different, determines the sign of $\cos\delta_q$ once the sign of $\sin\delta_q$ is known. In this manner, the sign of $\sin\phi$, or specifically that of $-\sin\alpha$ [1], is determined without the need of the factorization approximation. The assumptions needed are that $r < .5$ and $|\chi| > \beta$ — both are borne out in our analysis [6]. The data needed in order to effect this extraction are the asymmetry and its shape in $B^\pm \to \rho^\pm\rho^0(\omega) \to \rho^\pm\pi^+\pi^-$ about $s = m_\omega^2$ and the asymmetry in $B^\pm \to \rho^\pm\omega \to \rho^\pm\pi^+\pi^-\pi^0$.

We thank H.J. Lipkin for helpful discussions, and A. Kagan, W. Korsch, and G. Valencia for useful comments and references.

REFERENCES

1. R.M. Barnett et al., *Phys. Rev. D* **54**, 1 (1996).
2. M. Gronau and D. London, *Phys. Rev. Lett.* **65**, 3381 (1990); A.E. Snyder and H.R. Quinn, *Phys. Rev. D* **48**, 2139 (1993).
3. Y. Grossman and H.R. Quinn, hep-ph/9705356.
4. R. Enomoto and M. Tanabashi, *Phys. Lett. B* **386**, 413 (1996); see also hep-ph/9706340.
5. M. Bander, D. Silverman, and A. Soni, *Phys. Rev. Lett.* **43**, 242 (1979).
6. S. Gardner, H.B. O'Connell, and A.W. Thomas, hep-ph/9705453.
7. D. Atwood and A. Soni, *Phys. Rev. Lett.* **74**, 220 (1995); *Z. Phys. C* **64**, 241 (1994).
8. D. Atwood, G. Eilam, M. Gronau, A. Soni, *Phys. Lett. B* **341**, 372 (1995); G. Eilam, M. Gronau, and R.R. Mendel, *Phys. Rev. Lett.* **74**, 4984 (1995).
9. H.J. Lipkin, hep-ph/9310318; *Phys. Lett. B* **357**, 404 (1995).
10. S. Gardner and H.B. O'Connell, hep-ph/9707385.
11. S. Gardner, H.B. O'Connell, and A.W. Thomas, in preparation.

A New Experiment to Measure the Electric Dipole Moment of the Neutron?

Steve Lamoreaux, Martin Cooper, Geoffrey Greene,
Seppo Penttilä, Michelle Espy, Larry Marek, Dale Tupa, and
Robert Krause

Los Alamos National Laboratory
Los Alamos New Mexico

John Doyle

Harvard University
Cambridge, Massachusetts

Robert Golub

Hahn-Meitner Institute
Berlin, Germany

Abstract. For nearly fifty years, the limits on the electric dipole moment of the neutron have provided information of great importance in our understanding of the fundamental symmetries of nature. Current experiments using bottled Ultra Cold Neutrons (UCN) provide the best experimental limits on the neutron EDM. While modest improvements may be expected by extension of current methods, major reductions in the experimental error appear unlikely due to statistical sensitivity and systematic effects. This situation is unfortunate as several theoretical notions (supersymmetry and the origin of the baryon asymmetry) suggest a magnitude for the neutron EDM which may be only one or two orders of magnitude below the current limit. Recently, Golub and Lamoreaux (1) have suggested a new method for the measurement of the neutron EDM that uses a novel feature of the interaction between low energy neutron and superfluid ^{4}He to provide a very high density of UCN in an experimental volume. The proposed method also promises a significant reduction in the dominant systematic effect using a polarized ^{3}He co-magnetometer in the same volume. Their careful analysis suggests that an improvement of two orders of magnitude in the uncertainty of the neutron EDM may be possible. A review of the current experimental situation is given and the prospects for the realization of such a new experiment are discussed.

INTRODUCTION

In 1950, Purcell and Ramsey (2) observed that: "It is generally assumed on the basis of some suggestive theoretical symmetry arguments that nuclei and elementary particles can have no electric dipole moments. It is the purpose of this note to point out that although these theoretical arguments are valid when applied to molecular and atomic moments whose electromagnetic origin is well understood, their extension to nuclei and elementary particles rests on assumptions not yet tested."

The above was the first explicit suggestion that fundamental symmetries such as parity and time reversal conservation are not inexorable but must be considered as experimental laws whose validity must be established by direct measurement. It is notable that this suggestion anticipated the "fall of parity" by seven years. In 1951, Smith, Purcell, and Ramsey (3) performed an experiment to search for a neutron electric dipole moment. This was the first experiment specifically intended as a search for parity violation. This pioneering measurement determined that the neutron electric dipole moment must be less than 5×10^{-20} ecm.

In the nearly a half-century since the seminal work of Purcell and Ramsey, a series of extremely sensitive and ingenious experiments has reduced the experimental limit on the neutron EDM by more than 5 orders of magnitude to 1.1×10^{-25} ecm (4). The most recent experiments have utilized Ultra Cold Neutrons (UCN) confined in material bottles.

Notwithstanding this impressive reduction in experimental effort, theoretical interest in the neutron electric dipole remains very high. Limits on the neutron EDM provide one of the most rigorous constraints on theories that seek to explain the CP violation observed in the neutral kaon decay. A thorough review of the implications of electric dipole moment limits on theories of CP violation is beyond the scope of this report. The reader is directed to He, McKellar, and Pakvassa (5), Ellis (6), Barr and Marciano (7), and Barr (8) for detailed discussions. Nonetheless, a few remarks on the current theoretical situation are appropriate.

It is important to note that in the standard model, a neutron EDM arises in second order in the weak coupling constant G_F, which implies a neutron EDM on the order of 10^{-32} ecm. This value is well beyond the sensitivity of any experiment under serious consideration at present. Thus, the pursuit of a lower experimental limit on the neutron EDM is driven by an interest in an extension of the standard model. Two particular theories should be noted as they suggest that the neutron electric dipole moment may be comparatively large and within the reach of new experimental search. It has been suggested, originally by Sakharov and later by Ellis, Gaillard, and Nanopolous (9) that the universal baryon-antibaryon asymmetry can be understood by invoking a CP-violating interaction in the early universe. The magnitude of the baryon-antibaryon asymmetry suggests that the neutron EDM might have a magnitude between 10^{-25} and 10^{-27}. Supersymmetric models suggest a neutron EDM of a comparable magnitude. As noted by Steven Weinberg (10), "endemic in supersymmetric theories are CP violations that go well beyond the standard model/, and for this reason it may be that the next exciting thing to come along will be the discovery of a neutron or electron-dipole moment. These dipole moments seem to me to offer one of the most exciting possibilities for progress in particle physics."

DETERMINATION OF THE NEUTRON EDM USING STORED NEUTRONS

The Hamiltonian for a neutron with magnetic moment u_n and electric dipole moment d_n in a magnetic field B and an electric field E is given by

$$H = \mu_n \frac{I \cdot B}{|I|} + d_e \frac{I \cdot E}{|I|}. \tag{1}$$

The neutron electric dipole moment is determined from the neutron's Larmor precession frequency for the cases where the applied electric and magnetic fields are parallel and anti-parallel. If $d_n \neq 0$, these two frequencies will differ by $\Delta w = (4dE/\hbar)$. The statistical sensitivity with which one can measure the neutron's Larmor frequency limits on the determination of the neutron EDM. In general this limit $\sigma(d_n)$ will be of the form

$$\sigma(d_n) \sim \alpha(PET\sqrt{N}), \tag{2}$$

where P is the degree of polarization of the neutrons, E is the magnitude of the applied magnetic field, T is the time in which the neutrons are allowed to freely precess and N_α is the total number of detected neutrons. The constant of proportionality (of order one) in Eq. (2) is determined by the details of the strategy for the determination of the frequency. In the classic Ramsey method of separated oscillatory field, this constant of proportionality is 1.

The design parameters of all neutron EDM experiments are intended to minimize $\sigma(d_n)$ by the maximization of P, E, T, and N. The most recent determinations of the neutron EDM, carried out at Grenoble (11), are approaching the limits of the state of the art in reaching high values of these parameters. Using magnetized transmission foils to polarize the UCN, values for P of about 70% have been obtained. Electric fields used in the current experiments, ~15 kV/cm, are at the practical limit for vacuum breakdown given the constraints imposed by the need for special materials. The coherence time, T (on the order of 100 s for the Grenoble experiment), is determined by the bottle lifetime. Finally the number of neutrons detected is given by the product of the UCN source intensity and duration of the experiment. Current UCN sources provide an initial bottle density of about $10/cm^3$. The duration of the data acquisition phase of an experiment, limited by the patience of the experimental team, the reliability the apparatus, and the generosity of the funding agencies, may be on the order of one year. An improved method for the determination of the neutron electric dipole moment must incorporate a significant increase in one or more of the parameters P, E, T, and N.

The statistical uncertainty in the determination of the neutron EDM is a major contribution to the total experimental error. Nonetheless, there are very significant systematic effects that contribute to the error in the current experiment and any experiment which would significantly reduce the experimental error must include a strategy to significantly reduce these systematic effects. The most troublesome one concerns the stability and uniformity of the static magnetic field which is applied to establish an axis of quantization and which determines the Larmor precession frequency. Spatial and temporal inhomogeneities in this magnetic field can lead to corresponding shifts in the observed resonance frequency. In particular, the extraction of neutron EDM from the difference in the Larmor frequencies for E and B parallel and antiparallel, implicitly assumes that the there is no change in the magnetic field upon reversal of the electric field. A systematic variation of the static magnetic field that is correlated with the reversal of the electric field is particularly dangerous as it would, if not properly accounted for, lead to a spurious non-zero EDM. This particular systematic effect is of great concern because there is a very straightforward mechanism by which the reversing electric field will create a small correlated magnetic field. Such a magnetic field arises from the small "leakage" currents that flow due the finite resistance in the HV system. Such currents are troublesome because, near the limits of HV breakdown, they are often irregular and highly asymmetric. Thus, simple monitoring of the magnitude of the leakage currents cannot be relied upon to account for the variation in the magnetic field. In the most recent determinations of the neutron EDM, systematic errors associated with magnetic field variations are comparable to, or larger than, the statistical error noted above.

The most straightforward approach to accounting for this systematic effect is to monitor the magnetic field in the neutron spin precession region with a sensitive magnetometer. A "sensitive" magnetometer, in this context, must have a greater sensitivity to magnetic field variations than does the neutron in order that the correction for magnetic field variations can be done with minimal increase in experimental error. To the extent that magnetometer performs the same spatial and temporal field average as do the neutrons, effects due to variations in the magnetic field may be canceled. In the most recent Grenoble experiment (12), the magnetic field was monitored by placing three spatially separated rubidium magnetometers around the neutron precession volume. While such a system reduces the effects of magnetic field variations, it is not an exact monitor as it measures the field in a slightly different region of space. Following the suggestions of Ramsey (13) and Lamoreaux (14), the current Grenoble experiment monitors the magnetic field by simultaneously filling the precession volume with a vapor of optically pumped Hg vapor along with the UCN, which will serve to reduce this systematic effect.

One may summarize by noting that, for a new experiment to significantly reduce the uncertainty in the measurement of the neutron EDM, it must provide a substantial improvement in statistical sensitivity and provide a robust strategy for the elimination of systematic effects related to variations in the static magnetic field.

A PROPOSED NEW EXPERIMENT

The proposed experiment is described by R. Golub and S. Lamoreaux (1). The basic idea is to produce and store UCN in a superfluid ^{4}He bath that contains a small concentration of polarized ^{3}He; the polarized ^{3}He serves as a UCN polarizer, magnetometer, and UCN spin precession analyzer. In principle, this new technique can achieve a factor of 1000 improvement in the experimental limit for the neutron EDM. This factor results from the possibility of an increased electric field (a factor of 5) due to the excellent dielectric properties of superfluid ^{4}He, an increase in total UCN stored (10,000 fold improvement), and an increased coherence time (up to 5 times); these changes give a direct increase in sensitivity by a factor of 500 in the shot noise limit, given in Eq. (2). However, the current limit is due to magnetic field systematics, and is almost a factor of 5 greater than the shot noise limit. With the proposed experiment, a net factor of 2,500 might be possible; the use of ^{3}He as a volume co-magnetometer is essential.

"Superthermal Process" of UCN Production

The so-called "superthermal processed" for producing UCN in superfluid ^{4}He was proposed by Golub and Pendlebury (15) and a number of aspects of the physical processes involved have been demonstrated (16). Briefly, the dispersion relation for the free neutron, relating kinetic energy $\hbar\omega$ to momentum $\hbar k$, is a parabola:

$$\omega = \hbar k^2/2m. \tag{3}$$

This parabola crosses the Landau-Feynman (L-F) dispersion curve for elementary excitations in superfluid ^{4}He at $2\pi/k^* = 8.9$ Å and $E^* = \hbar\omega = 11$K. The crossing point is in the linear region of the L-F curve. (The curves also intersect at $k = 0$). Neutrons in this range of wavelength are readily produced by a liquid deuterium or liquid hydrogen moderator.

Since both energy and momentum are conserved in the scattering process, neutrons at or near rest can only absorb phonons of energy E^*, where the dispersion curves cross. At low temperature, the 11K phonon density becomes very small; the upscattering process is strongly suppressed by the Boltzmann factor, $e^{-E^*/kT}$, when the superfluid temperature is less than 1 K. By the same argument, only neutrons with energy near E^* can scatter into the UCN energy region by emission of a single excitation.

A UCN source based on this process operates by the following principle. 8.9 Å neutrons can easily penetrate the walls of the storage container and enter the superfluid bath. These neutrons then downscatter, producing UCN which are trapped in the container. (The effective neutron potential U_F for liquid ^{4}He is about 20 neV, much less than the potential for most solid materials, so we can assume it is zero in the following discussion.) UCN produced in this way will remain trapped by the container walls in the superfluid He bath until they are lost through one of the possible loss mechanisms, including β-decay, absorption by ^{3}He, loss in the wall, etc. The UCN will reach a saturation density,

$$\rho_{ucn} = P\tau, \qquad (4)$$

where τ is the total loss rate,

$$\tau^{-1} = \tau^{-1}_{wall} + \tau^{-1}_\beta + \tau^{-1}_{3He} + \odot, \qquad (5)$$

and P is the UCN production rate (UCN/[cm^{3}sec)] due to the above mentioned downscattering process (17):

$$P = 7.2\frac{^2\Phi^*}{\lambda\Omega}\frac{1}{\lambda_u^3}\delta\Omega n \text{ cm}^{-3}\text{s}^{-1}, \qquad (6)$$

where the neutron spectral density is specified at 8.9 Å, λ_u is the shortest UCN wavelength that can be stored in the container, and $\delta\Omega$ is the source solid angle subtended at the superfluid bath.

The neutron-superfluid ^{4}He system is, in some sense, a two-level quantum system, and the production of UCN by the emission of a phonon can be compared to the spontaneous emission of radiation by an excited atom.

Attempts to make a UCN source based on the superthermal process in ^{4}He have so far been met with technical difficulties, primarily in regard to extraction of the UCN from the bath. Invariably, however, the thin material windows used to contain the liquid He, but allow the UCN to pass, become covered by condensable gases, increasing U_F and/or the UCN absorption. Typically, extracted densities have been a factor of 10 to 100 below that expected. Indirect measurement through the upscattering rate has confirmed that the expected density does indeed exist within the bath.

³He Polarization and Detection

The extraction problems encountered so far can be avoided by performing an EDM search directly in the liquid helium of the superthermal source. Such a system has a number of advantages; for example, because of the excellent dielectric properties of liquid helium, increasing the applied electric field by nearly an order of magnitude might be possible. Also, it appears feasible to use a dilute solution of polarized ³He as a UCN polarizer, spin analyzer, detector, and magnetometer.

³He only absorbs neutrons when the total spin is zero because the reaction occurs via the 0^+ excited state (18) as follows:

$$^3He + n \rightarrow p + T + 764 \, keV . \tag{7}$$

The polarization and cryogenic transport of polarized ³He have been studied (19). Furthermore, energetic charged particles produce ultraviolet scintillations in liquid helium with about 4 photons per keV of deposited energy. The reaction between ³He and neutrons in the liquid helium are easily detected, giving a detection of reactions with nearly 100% efficiency. See Doyle and Lamoreaux (20) for an application of these techniques to a measurement of the neutron β-decay lifetime, which may lead to a factor of 100 improvement in accuracy.

The ³He serves as a UCN polarizer by absorbing neutrons in the singlet state. To be effective, the rate of absorption should be slightly higher than other UCN loss mechanisms in the system. This condition implies a ³He concentration of 10^{-9}, or about 10^{13} atoms/cm³. At such low densities, a hexapole state selector can be used to produce essentially perfectly polarized ³He. Other techniques for producing higher densities have polarization limited to 70%; since the ³He serves three functions, it is expected that the experimental sensitivity will vary as the cube of the ³He polarization, in agreement with detailed calculations.

An EDM experiment based on these ideas, as originally proposed, could be sensitive to a neutron EDM by looking at the scintillation rate at the end of a double-pulse sequence, as a function of electric field polarity. It has been shown by solving the Schrödinger equation in the presence of a spin-dependent absorption probability that this technique is slightly less sensitive than the conventional bottle technique, however, this loss of sensitivity is more than made up for by elimination of the extraction losses and increase in electric field.

In the following discussion, let the subscript 3 refer to the ³He atoms and subscript n refer to the UCN. In the case where both species are polarized, the spin-dependent loss rate can be written,

$$\frac{1}{\tau_{abs}} = (1 - p_n p_3 \cos\theta_{n3})/\tau_{3He}, \tag{8}$$

where θ_{n3} is the angle between the spin polarization vectors and $|p_{n,3}| \leq 1$. Each loss (nuclear reaction) produces a scintillation pulse; the scintillation rate thus becomes a measure of the angle between the polarization vectors.

One could search for a neutron EDM by using the above UCN production/polarization technique. After the UCN are polarized (along a static field of magnitude B_0), the UCN and ^{3}He spins could be flipped by $\pi/2$; the spins then precess about the static field and there will be a modulation in the scintillation rate:

$$\phi(t) = 1 - p_3 p_n \cos[(\gamma_3 - \gamma_n) B_0 t + \Phi], \tag{9}$$

where $\phi(t)$ is the time-dependent scintillation rate, Φ is an arbitrary phase, and the gyromagnetic ratios are

$$\gamma_n/2\pi \approx -3 \text{ Hz/mG}$$

and

$$\gamma_3/2\pi \approx -3.33 \text{ Hz/mG}$$

The atomic EDM of ^{3}He is expected to be quite small due to the atomic screening; thus, if an electric field is applied along B_0 there will be a change in the frequency of the scintillation rate modulation. Unfortunately, the problem of measuring the magnetic field remains (although the effects are only 1/10 as large since the gyromagnetic ratios are nearly equal) and it has been demonstrated that experiments are presently limited by magnetic systematic effects. It might be possible to use SQUID magnetometers to detect the precessing ^{3}He magnetization, so that the ^{3}He could then serve as a direct magnetometer. However, the sensitivity is at best marginal, but recent technical advances have led us to reconsider this option.

Dressed Spin Magnetometry

In the above description, it is evident that a perfect experiment would be possible if the magnetic moments of the ^{3}He and neutron were equal; the fact that the magnetic moments are equal to within 10% reduces the sensitivity to background magnetic fields by an order of magnitude, and if the moments were exactly equal, there would be no effect at all. Unfortunately, we have no direct control over the physics responsible for the observed magnetic moments; however, these moments can be artificially modified by using "dressed atom" techniques (21), and it is possible make them equal (16).

In the presence of a strong oscillating magnetic field, the magnetic moment will be modified, or "dressed," yielding an effective gyromagnetic ratio

$$\gamma' = \gamma J_0(\gamma B_{RF}/\omega_{RF}) \equiv \gamma J_0(x), \qquad (10)$$

where γ is the unperturbed gyromagnetic ratio, B_{RF} and ω_{RF} are the amplitude and frequency of an applied oscillating magnetic (RF) field, and J_0 is the zeroth-order Bessel function. This effect can be qualitatively understood by taking the average of the spin in an oscillating magnetic field. Consider a spin pointing along $\hat{z}$ at $t = 0$. Now apply an oscillating field along $\hat{x}$; the spin precession frequency is time dependent,

$$\omega(t) = \dot{\theta}(t) = \gamma B_x(t) = \gamma B_{RF} \sin \omega_{RF} t \;,$$

so that the angle relative to $\hat{x}$ is

$$\theta = \gamma(B_{RF}/\omega_{RF})\cos\omega_{RF}t \;.$$

The average spin projection $\langle P_z \rangle$ along $\hat{z}$ is given by

$$\langle P_z \rangle = \frac{1}{T}\int_0^T \cos[\gamma B_{RF} t \cos\omega_{RF} t] = J_0(\gamma B_{RF}/\omega_{RF}) = J_0(x) \;.$$

A more sophisticated treatment shows that a spin will respond to a small (compared to the oscillating field amplitude) static field along $\hat{x}$, with an average magnetic moment $\gamma' = \gamma J_0(x)$; our simple estimate gives a picture of how the oscillating field dilutes the magnetic moment.

In practice, the oscillating field is at right angles to the static field B_0 around which the spins are precessing. In the absence of the oscillating field, one would see scintillation due to reactions occurring at a rate given by Eq. (9). Thus, there is an oscillation in the scintillation rate at the difference in the precession frequencies $[\delta\omega = (\gamma_n - \gamma_3)B_0]$. If the RF dressing field is now applied, the effective magnetic moments become modified, and

$$\delta\omega = (\gamma_n J_0(\gamma_n x) - \gamma_3 J_0(\gamma_3 x))B_0 \;. \qquad (11)$$

This formula has the property that $\delta\omega = 0$ when $\gamma_n x \approx 1.19$; this condition is referred to as "critical dressing." It can be achieved in practice with a dressing field frequency of order 1 kHz, an amplitude of 100 mG, and a spin precession frequency on the order of a few Hz.

If the neutron EDM is non-zero, the neutron precession frequency will be shifted by an amount $2d_n E J_0(\gamma_n x)$ (since the dressing dilutes the net spin projection). Thus, the value of $x = x_c$ to give $\delta\omega = 0$ is changed. By measuring the value of x_c vs. electric field direction, a neutron EDM would be evident. The important point is that the effect of static magnetic fields is canceled.

Experimentally, the neutron and helium spin vectors could be kept nearly parallel; the scintillation would increase or decrease as x is varied away from the value x_c such that $\delta\omega = 0$. Over the course of a storage period, x could be sinusoidally modulated at a low frequency ω_m and the value $x_c(\pm E)$ inferred from variations in the scintillation rate which occur at harmonics of ω_m. If the average value of $x \neq x_c$, there will be a first harmonic to the scintillation rate growing linearly in time. If $x = x_c$, there will be only a second harmonic component. In practice, a feedback system might be used to force the first harmonic signal to zero; the second harmonic then serves as a system calibration. (Note that the modulation in x and the subsequent modulation in the scintillation rate are 90° out of phase because the spin vectors must precess before the effects due to a change in x are manifest).

Current Experimental Efforts

The neutron lifetime experiment (20), nearing readiness to take data, will answer the rate uncertainties due to the superthermal production process. Many of the techniques employed to get the light out from the storage cell will be applicable to the EDM experiment.

We are presently working to determine whether the spatial distribution of ^{3}He, as a dilute solution in a superfluid bath, is adequate in regard to the magnetic field averaging as required in the proposed experiment. Our initial experiments will employ unpolarized ^{3}He, and we will use cold neutron tomography, as proposed in Ref. (1), to determine the spatial distribution. If this work shows that the distribution is random enough, we will begin to study the transport of polarized ^{3}He into the superthermal bath, and the extraction of the spent (depolarized) ^{3}He.

In addition, we are reinvestigating the use of SQUIDS as an alternative to the dressed spin technique. Current technology will allow the direct detection of the ^{3}He precession just at the required level of sensitivity. The issues to be addressed is noise generation from the application of high voltage to the measurement cell (which will likely be of high frequency character due to microdischarges in dielectrics), and vibrations from the cryogenic apparatus that must be coupled to the measurement apparatus.

There are other practical problems to be addressed such as engineering the high-voltage feed into the cryogenics and demonstrating the high polarization of ^{3}He that can be produced by a hexapole filter and transferred into the superthermal ^{4}He. Our group is forming an international collaboration whole goal is to produce a reviewable proposal within a two-year time frame. The proposal will be greatly strengthened by resolving experimentally as many of the outstanding issues as possible.

REFERENCES

1. R. Golub and S. Lamoreaux, *Phys. Lett.* **237**, 1 (1994).
2. E. M. Purcell and N. F. Ramsey, *Phys. Rev.* **78**, 807 (1950).
3. J. H. Smith, Ph.D. Thesis, Harvard University, 1951; J. H. Smith, E. M. Purcell, and N. F. Ramsey, *Phys. Rev.* **108**, 120, (1957).
4. I. B. Khriplovich and S. K. Lamoreaux, "CP Violation without Strangeness" (Heidelberg, Spring-Verlag, 1977).
5. He, McKellar, and Pakvassa, *J. Mod. Phys. A* **4**, 501 (1989).
6. Ellis, *Nucl. Instrum. Methods Phys. Res. A* **284**, 33 (1989).
7. Barr and Marciano, in *CP Violation*, ed. C. Jarlskog (World Scientific, 1989).
8. Barr, *Int. J. Mod. Phys. A* **8**, 2093 (1993).
9. Ellis, Gaillard, and Nanopolous, *Phys. Lett. B* **99**, 101 (1981).
10. Steven Weinberg, in *Proc. XXVI Int. Conf. on High Energy Physics*, Dallas, Texas, 1992.
11. Smith et al., *Phys. Lett. B* **234**, 191 (1990); Altarev et al., *Phys. Lett.* **276**, 242 (1992).
12. Pendlebury et al., *Phys. Lett. B* **136**, 327 (1984).
13. Ramsey, *Acta Phys. Hung* **55**, 117 (1984).
14. Lamoreaux, unpublished (1986).
15. R. Golub and J. M. Pendelbury, *Phys. Lett.* **62A**, 337 (1977).
16. R. Golub, D. J. Richarson, and S. K. Lamoreaux, "Ultracold Neutrons" (Bristol, Adam-Hilger, 1991).
17. S. K. Lamoreaux and R. Golub, *JETP Lett.* **58**, 844 (1993).
18. L. Passell and R. I. Schermer, *Phys. Rev.* **150**, 146 (1960).
19. C. G. Aminoff et al., *Rev. Phys. Appl.* **24**, 827 (1989).
20. J. M. Doyle and S. K. Lamoreaux, *Europhys. Lett.* **26**, 253 (1994).
21. C. Cohen-Tannoudji and S. Haroche, *J. Physique* **30**, 153 (1969).

Searching for Time Reversal Invariance Violation in Polarized Neutron Decay

L.J. Lising[*], J.M. Adams[*], J.M. Anaya[‡], T.J. Bowles[‡], T.E. Chupp[**], K.P. Coulter[**], M.S. Dewey[*], S.R. Elliott[†], S.J. Freedman[*], B.K. Fujikawa[*], A. Garcia[a], G.L. Greene[‡], S.-R. Hwang[**], G.L. Jones[*], J.S. Nico[*], H.G.R. Robertson[†], T.D. Steiger[†], W.A. Teasdale[‡], A.K. Thompson[*], E.G. Wasserman[*,b], F.E. Wietfeldt[*], and J.F. Wilkerson[†]

[*] *Lawrence Berkeley National Laboratory Berkeley, CA 94720*
[†] *University of Washington, Seattle, WA, 98195*
[⋆] *National Institute of Standards and Technology, Gaithersburg, MD 20899*
[**] *University of Michigan, Ann Arbor, MI 48109*
[‡] *Los Alamos National Laboratory, Los Alamos, NM, 87545*
[a] *University of Notre Dame, Notre Dame, IN 46556*
[b] *Present address: Abacus Concepts, Inc., Berkeley, CA, 94704*

Abstract. Time reversal invariance violation is tightly constrained in the Standard Model, and the existence of a T-violating effect above the predicted level would be an indication of new physics. A sensitive probe of this symmetry in the weak interaction is the measurement of the D-coefficient in neutron decay. This parameter characterizes the triple-correlation of neutron spin, electron momentum, and neutrino (or proton) momentum, which changes sign under time reversal. The emiT experiment, now on line, attempts to improve the measurement of D, whose current average is $0.3 \pm 1.5 \times 10^{-3}$.

INTRODUCTION

The origin of CP-violation has been a mystery since its observation in the kaon system in 1964. [1] The elusive nature of this phenomenon is compounded by the lack of any other evidence of CP-violation, or any direct observation of T-violation, which is implied under conservation of CPT. [2] An explanation within the Standard Model, via a phase in the Cabbibo-Kobayashi-Maskawa matrix of quark mixing, arises as a natural consequence of three generations of quarks. [3] While this formulation is adequate to describe the observations to date in kaons, it cannot address other suggestions of T-noninvariance such as Sakharov's mechanism for the evolution of the observed matter-antimatter asymmetry in the universe today. [4]

The neutron, a natural laboratory for precision studies for its neutrality and simplicity, presents several T-violating observables for which the Standard Model predicts values so small as to be experimentally inaccessible, yet could be measurable in alternative theories. Among the quantities are terms in the decay probability such as the "triple correlation",

$$D\hat{\sigma}_n \cdot \mathbf{p}_e \times \mathbf{p}_p. \tag{1}$$

Previous attempts to measure the D-coefficient have been consistent with zero with an average value of $0.3 \pm 1.5 \times 10^{-3}$. [5,6] The interpretation of correlations is usually complicated by the presence of final state effects, whereby T-invariant interactions mimic the T-violating signal at higher orders. However, for the neutron, the value of D from this effect is calculated to be near 5×10^{-5}, [7,8] leaving room for further exploration.

Several proposed extensions to the Standard Model allow a D-coefficient above the final state effects. [9] These include models with left-right symmetry which incorporate a heavy, right-handed W-boson, which can lead to T-violation if the boson mass eigenstates mix the two chiralities. Models with new, "exotic" fermions can similarly mix with their lighter counterparts. Models that allow quark to lepton transitions via "leptoquark" bosons can also show interference between the leptoquarks that couple to the electrons and neutrinos. Both the exotic fermions and L-R symmetry can lead to other T-violating effects, such as the kaon ϵ'/ϵ and the neutron electric dipole moment, both of which have been measured with great precision. Within the context of these models, one can estimate the expected contribution to the D-coefficient to be less than 3×10^{-5}, although this requires certain assumptions. However, the leptoquark prediction for D is not limited by these other two measurements as the leptoquark contribution to non-leptonic processes is not of lowest order.

EXPERIMENTAL METHOD

The emiT experiment is presently running at the National Institute of Standards and Technology's Cold Neutron Research Facility in Gaithersburg, Md. The facility utilizes a liquid hydrogen cold source to moderate thermal reactor neutrons, delivering them to users with minimal loss and a low gamma background by transport via reflecting guide tubes lined with Ni-58. A polarization of 96% is obtained by state-selective reflection inside a supermirror polarizer and is retained along a 5 gauss alignment field through the spin flipper and collimator regions into the detector. The detector consists of an octagonal array of four each proton and electron detectors as shown below.

The octagonal geometry maximizes the experiment's sensitivity to D by balancing the sine dependence of the cross product with the large angles favored by kinematics. The symmetry of the detector and flipping of the neutron spin every few seconds reduces our sensitivity to varying detector efficiencies and fluctuations in the beam, which cancel to first order in the extraction of D from the data. The electron segments are 50 cm long, 1/4-inch thick plastic scintillators with bialkali phototubes at each end. The recoil protons, whose maximum energy is only 750 eV, drift in a field free region until they near one of the four proton detectors, where they are accelerated through 36 kilovolts onto an array of windowless PIN diodes. The characteristic delay time of $0.5\mu s - 2.0\mu s$ between the recoil proton and electron pulses is used to distinguish signal from beam-related background. Careful study and modelling of systematics have allowed us to minimize and monitor the factors that could cause a false measurement of D. The two primary effects are due to deviations of the polarization from purely longitudinal, and to nonuniformities in detection efficiency over the face of the detectors. [10]

FIRST RUN

The first of our allotted reactor cycles was devoted to beamline development, including alignment and polarization measurements. The neutron flux was measured at the end of the neutron guide (before the polarizer) to be 1×10^9 neutrons/second, and 1×10^8 n/s at the end of our collimator series into the detector chamber. Following that, some time was devoted to testing

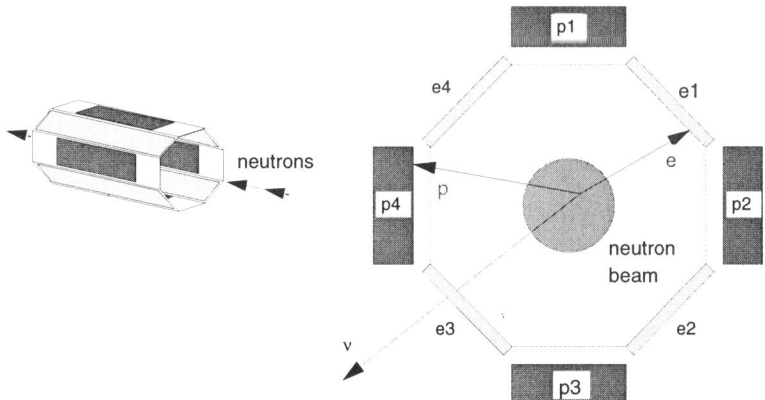

FIGURE 1. Detector geometry. Proton and electron detectors surround a 6 cm diameter longitudinally polarized beam.

and adaptaion of our data acquisition system, noise and background reduction, vacuum improvement, and most importantly, stabilization and control of the high voltage, including protection of electronics from spark damage and minimization of field emission background. The ratio of signal to background in the correct timing window varies widely depending on noise conditions, yet the background is reduced to less than 1/10 of the signal once the proper energy constraints are applied.

The expected rate for the full detector was 5-10 Hz in accepted neutron decays. We have already seen rates above 5 Hz and will recover more with repairs and improvements, which are ongoing. Analysis of the data is in progress, so far yielding a D-coefficient consistent with zero.

REFERENCES

1. Christenson, J. H. et al.,*Phys. Rev. Lett.* **13(4)**, 138-140 (1964).
2. Luders, G., *Ann. Phys.* **2**, 1-15(1957).
3. Kobayashi, M., and Maskawa, T., *Prog. Theo. Phys.* **49(2)**, 652-657 (1973).
4. Sakharov, A., *JETP Lett.* **5**, 24-27 (1967).
5. Steinberg, R.I. et al., *Phys. Rev. Lett.* **33(1)**, 41-44 (1974).
6. Erozolimskii, B.G. et al., *Sov. J. Nucl. Phys.* **28**, 48 (1978).
7. Jackson, J.D., Treiman, S.B., and Wyld, H.W.,*Nucl. Phys.* **4**, 206-212 (1957).
8. Callan, C.G. and Treiman, S.B., *Phys. Rev.* **162(5)**, 1494-1497 (1962).
9. Herczeg, P., in *Progress in Nuclear Physics*, Hwang, W.-Y.P., Ed., Elsevier Science Publishing Co., Inc., 1991, pp. 171-194.
10. Wasserman, E.G., Ph.D. Thesis (1994).

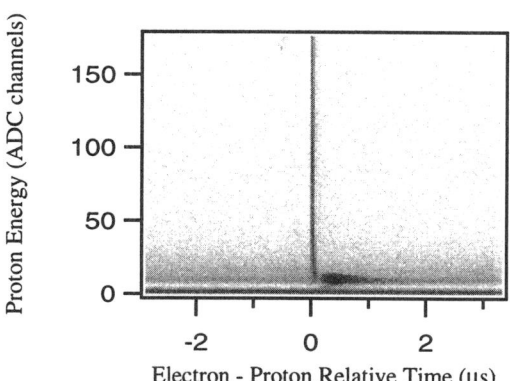

FIGURE 2. Proton detector energy versus relative electron-proton signal arrival time. The drift-delayed proton signal is well-separated from the narrow, high energy band of coincident events at zero time.

T-odd Correlations in Z^0 Decay into Three Jets and in Nuclear Beta Decay

H. E. Conzett

Nuclear Science Division, Lawrence Berkeley Laboratory
University of California, Berkeley, CA 94720

Abstract. The naive time-reversal (T_N) operation reverses momenta and spins *without* the interchange of initial and final states that is included in the time-reversal (T) operation. In a recent report of the measurement of the transverse analyzing-power A_y in the three-jet decay of polarized Z^0 bosons, A_y was considered to be a T_N-odd, but not T-odd, observable. This leads to an incorrect conclusion. In this paper it is shown that the T_N operation is an incomplete substitute for the combined conditions that T symmetry and hermiticity impose on the transition amplitudes in first-order electromagnetic or weak processes. Other misconceptions concerning relationships between T-odd operators and (so called) T-odd observables are discussed.

INTRODUCTION

A recent paper describes the production of polarized Z^0 bosons in $e^+ e^-$ annihilation with longitudinally polarized electrons and a measured correlation in the Z^0 decays into three jets [1]. The correlation, shown in Fig. 1, is that between the Z^0 polarization p^Z and the normal to the decay plane, $y = k_1 \times k_2$, defined by the momenta of the two highest-energy jets. With $p_y = p^Z \cdot y = p^Z \cos \omega$, the measured observable is the decay analyzing-power component A_y [2].

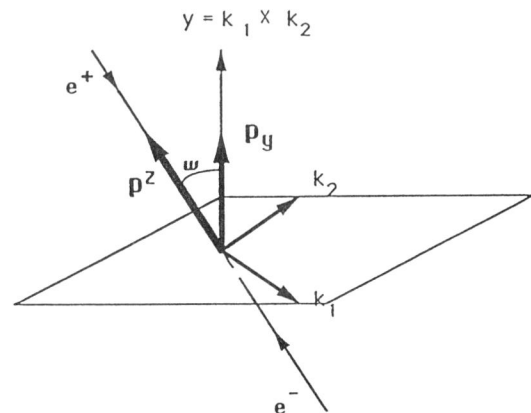

Figure 1

CP412, *Intersections Between Particle and Nuclear Physics:* 6th Conference
edited by T. W. Donnelly
© 1997 The American Institute of Physics 1-56396-712-X/97/$10.00

That is, the decay intensity for the Z^0 (spin-1) vector polarization component p_y is given by

$$I_y = I(1 + \frac{3}{2} p_y A_y), \qquad (1)$$

where I is the intensity for unpolarized Z^0 decays. In the notation of [1], $p^z = A_Z$, $p_y = A_Z \cos \omega$, and $A_y = \beta$.

It was claimed that the corresponding Z^0 spin-operator component,

$$S \cdot k_1 \times k_2 \equiv S_y, \qquad (2)$$

is odd under T_N, "naive time reversal" [4], which reverses momenta and spins without interchanging the initial and final states. As such, not being a true time-reversal operation, a nonzero value of the corresponding (T_N-odd) experimental observable A_y, in excess of the contribution from final-state interactions (FSI), would not signify any violation of T symmetry. However, an immediate inconsistency follows, because a long accepted test of T symmetry has been associated with exactly the same spin-operator component

$$S \cdot k_e \times k_v \equiv S_y \qquad (3)$$

in β-decay of polarized nuclei [5]. Since the spin operator S_y changes sign under T and is, thus, T-odd, the argument has been that the corresponding experimental observable A_y, is similarly T-odd and is required by T symmetry to vanish (ex FSI). However, that argument, without qualification with respect to the interaction dynamics, is itself in conflict with a theorem that states that there can be no null test of T symmetry [6,7]; i.e., T symmetry alone does not require any observable to vanish [8].

T-ODD OPERATORS AND OBSERVABLES

Consider a reaction $a + b \to c + d$ and examine the conditions imposed by T symmetry alone. Choosing the center of mass helicity frame, unit vectors along the coordinate axes are

$$z_i (z_f) = k_i (k_f) \qquad y = k_i \times k_f \qquad x_i (x_f) = y \times z_i (z_f) \qquad (4)$$

where $k_i (k_f)$ is the c.m. momentum of the projectile (ejectile). Then with the T transformation $k_i \leftrightarrow -k_f$, $\sigma \to -\sigma$, and noting that $\sigma_x \equiv \sigma \cdot x$ etc., one has the following transformations under the T operation:

$$T: \quad \sigma_x, \sigma_y, \sigma_z \to -\sigma_x, \sigma_y, \sigma_z. \qquad (5)$$

Here, then, by definition, σ_x, is a T-odd operator. However, it has been shown [7] that the conditions on the observables that follow from (5) are, for example

$$A_x = -P^t{}_x, \quad A_y = P^t{}_y, \quad A_z = P^t{}_z. \qquad (6)$$

That is, the analyzing-power components A_j are equal to the ($\pm$) polarizing-power components P^t_j in the inverse reaction. Thus, the even/odd character of the operators in (5) translates into the even/odd character of *pairs* of observables in (6), and it is interesting to note that the rather standard nuclear physics test of $A_y = P^t_y$ is a *T*-even test of *T* symmetry.

T SYMMETRY AND HERMITICITY

When, in combination with *T* symmetry, the dynamical restrictions of first-order electromagnetic or weak interactions are imposed on the amplitudes of the transition matrix, *M*, between initial and final helicity states, limited null-tests of *T* symmetry become available [7]. Specifically, from *S* matrix unitarity

$$SS^\dagger = (1 + 2ik_iM)(1 - 2ik_fM^\dagger) = 1 + 2i(k_iM - k_fM^\dagger) + 4k_ik_f MM^\dagger = 1 \quad (7)$$

so, to first order, the entire k_iM matrix is Hermitian. Then in a process $a(\alpha) + b(\beta) \to c(\gamma) + d(\delta)$, where $\alpha, \beta, \gamma, \delta$ are the particle helicities, from *T* symmetry and hermiticity *(H)* [9, 10],

$$T: \quad M_{\alpha\beta,\gamma\delta} = (-1)^{\alpha-\beta-\gamma+\delta} M_{\gamma\delta,\alpha\beta} \; (k_f/k_i), \quad (8a)$$
$$H: \quad M^*_{\alpha\beta,\gamma\delta} = (-1)^{\alpha-\beta-\gamma+\delta} M_{\gamma\delta,\alpha\beta} \; (k_f/k_i). \quad (8b)$$

In general, $M_{\alpha\beta,\gamma\delta}$ and $M_{\gamma\delta,\alpha\beta}$ are elements of the separate submatrices that corespond to the processes $ab \to cd$ and $cd \to ab$, respectively, and only for elastic scattering are they elements of the same *M* submatrix. The common phase factor in (8) comes from the interchange of initial and final states. Thus, from the *combination TH* the transition amplitudes

$$M_{\alpha\beta,\gamma\delta} = M^*_{\alpha\beta,\gamma\delta} \quad (9)$$

are real, whereas neither *T* nor *H*, *separately*, imposes any restriction on them. Since all observables are sums of bilinear combinations of these amplitudes, any observable that is given by the imaginary part of such a sum then vanishes from *TH*. The analyzing powers for polarized particles *a*, in reaction or decay processes, are given by [7]

$$A_j = Tr \, M \, \sigma_j M^\dagger / Tr \, M M^\dagger, \quad j = x, y, z. \quad (10)$$

Since the amplitudes $M_{\alpha\beta,\gamma\delta}$ and σ_x, σ_z are all real and σ_y is imaginary, only A_y is given by the imaginary part of such a sum and, thus, vanishes from *TH*. It is now seen that the T_N operation is an incomplete substitute for that of *TH*. That is, the *combination TH* twice interchanges initial and final states and yields the condition (9) on the amplitudes. T_N, however, which simply reverses the momenta and spins of a helicity amplitude $M_{\alpha\beta,\gamma\delta}$, leaves the helicities, thus the amplitude, unchanged. It is the (combined) *TH* operation that yields the essential condition (9).

CONCLUSIONS

Even though the T_N operation is a flawed substitute for the proper TH operation, its use has not invalidated any of the actual calculations of observables in various decay processes [1,4,11]. That is, the argument, that the T_N-odd operator S_y has a corresponding T_N-odd experimental observable which then vanishes in first order processes, is incorrect. It has, however, identified A_y, which is not the expectation value of S_y, as the vanishing observable; and, indeed, A_y does vanish properly from TH, as is explained after eq. (10). Thus, the calculations of FSI contributions to the transition amplitudes, and then to A_y via eq. (10), have not been affected by the incorrect argument

The T_N concept can, however, alter the basic interpretation of results. For example, reference [1] states that S_y (2) is CP-even, but that a non-zero value of the corresponding T_N-odd observable A_y would not signal CPT violation. Now, since A_y vanishes from TH, just as in the case of β decay, the opposite conclusion follows; i.e., it would signal T, and here, CPT violation, so the zero result is consistent with T and CPT symmetry.

Of course, in a more sensitive test, any non-zero value would have to exceed that contributed by the various final-state interaction processes before any T symmetry violation could be claimed. Since the calculated standard-model (SM) FSI contributions to A_y in Z^0 decay are $\approx 10^{-5}$ or less [12], reference [1] noted that A_y is a sensitive probe of such effects beyond the SM. As is argued here, it is also a limited null-test of T symmetry; i.e., the ultimate precision attainable in such a test is limited by that available in the calculation of the non-zero FSI contributions, and not by the experimental precision itself [7]. As an example, at the value of (D coefficient) $A_y = (-0.5 \pm 1.4) \times 10^{-3}$ achieved in neutron β-decay [13], the upper limit of these contributions, of order 10^{-5} [14], has not yet been approached.

Unfortunately, now, in Z^0 decay it appears that a value of A_y significantly larger than that from the FSI contributions in the SM would have an ambiguous interpretation. It could be due either to such non-SM contributions, or to T-symmetry violation, or to both.

ACKNOWLEDGMENT

This work was partially supported by the Director, Office of Energy Research, Office of High Energy and Nuclear Physics, Nuclear Physics Division of the U.S. Department of Energy under Contract DE-AC03-76SF00098.

REFERENCES

1. K. Abe et al., Phys. Rev. Lett. **75**, 4173 (1995).
2. I use the standard spin-polarization terminology and notation of the Madison Convention and its extensions; see, for example, [3].
3. H. E. Conzett, Rep. Prog. Phys. **57**, 1 (1994).
4. D. Atwood, G. Eilam, and A. Soni, Phys. Rev. Lett. **71**, 492 (1993), for example.
5. J. D. Jackson, S. B. Treiman, and H. W. Wyld, Phys. Rev. **106**, 517 (1957).
6. F. Arash, M. J. Moravcsik, and G. R. Goldstein, Phys. Rev. Lett. **54**, 2649 (1985).
7. H. E. Conzett, Phys. Rev. **C 52**, 1041 (1995).
8. The theorem does not include total cross-section observables, and it has been shown that a total cross-section T-odd spin-correlation observable does provide a null test of T symmetry; H. E. Conzett, Phys. Rev. **C 48**, 423 (1993), and references therein.
9. M. Simonius, *Lecture Notes in Physics* **30**, edited by J. Ehlers, K. Hepp and H. A. Weidenmüller (Springer-Verlag, Heidelberg, 1974), p. 38.
10. W. M. Gibson and B. R. Pollard, *Symmetry Principles in Elementary Particle Physics* (Cambridge University Press, London,1976), p. 196.
11. For example, A. de Rujula, R. Petronzio, and B. Lautrup, Nucl. Phys. **B146**, 50 (1978); G. Valencia and A. Soni, Phys. Lett. **B263**, 517 (1991); D. Atwood and A. Soni, Phys. Rev. Lett. **74**, 220 (1995); A. Brandenburg, L. Dixon, and Y. Shadmi, Phys. Rev. D **53**, 1264 (1996).
12. A. Brandenburg et al., Ref. [11].
13. Particle Data Group, Phys. Rev. D **54**, 60 (1996).
14. C. G. Callan and S. B. Treiman, Phys. Rev. **162**, 1494 (1967).

Search for Physics Beyond the Standard model via a Polarization-Asymmetry Correlation experiment on ^{107}In.

P. Schuurmans[1], J. Camps[1], N. Severijns[1], P. De Moor[1],
J. Deutsch[2], T. Otto[2], J. Govaerts[2], B.A. Brown[3], B. Holstein[4],
R. Kirchner[5], O. Naviliat-Cunic[6], R. Prieels[2], P.A. Quin[7],
E. Thomas[2], A. Van Geert[1], B. Vereecke[1] and L. Vanneste[1]

[1] *Inst. voor Kern- en Stralingsfysica, K.U. Leuven, B-3001 Leuven, Belgium*
[2] *Inst. de Physique Nucléaire, U.C. de Louvain, B-1348 Louvain-la-Neuve, Belgium*
[3] *Dep. of Physics and Astronomy, Michigan State U., East Lansing, MI 48824, USA*
[4] *Dep. of Physics and Astronomy, U. of Massachusetts, Amherst, MA 01002, USA*
[5] *G.S.I. Darmstadt, D-64220 Darmstadt 11, Germany*
[6] *Inst. für Teilchenphysik, E.T.H. Zürich, Hönggerberg, CH-8093 Zürich, Switzerland*
[7] *Dep. of Physics, U. of Wisconsin-Madison, WI 53706, USA*

Abstract. We report on a new precision measurement of the longitudinal polarization of positrons emitted by polarized ^{107}In nuclei. Preliminary results yield a lower limit of 303 GeV/c^2 for the mass of a possible, predominantly right-handed, W gauge boson, if interpreted in the framework of the manifest left-right symmetric model. In more general left-right symmetric models our result is complementary to those from muon decay experiments and to results from searches for a similar particle in collider experiments. Our data also provide new results on possible tensor contributions to the weak interaction.

INTRODUCTION

The standard electroweak model is very successful but has too many free parameters and ad-hoc assumptions to be accepted as the "ultimate" description of nature. Parity violation, in particular, is "explained" by assuming that only left-handed fermions participate in the charged current weak interactions. As a result, extensions of the Standard Model (SM) have been proposed in which the parity-symmetry of nature is recovered at some "fundamental" level by relating it to a spontaneous symmetry breaking mechanism similar to that at work elsewhere in the theory.

The simplest extension is offered by the so-called manifest left-right symmetric (MLRS) model [1,2]. Here, an additional charged gauge boson W_2 is introduced which, by spontaneous symmetry breaking, acquires a mass m_2 higher than the mass m_1 of the observed gauge boson W_1 ($m_2 >> m_1 = 80.2\ GeV/c^2$; $(m_1/m_2)^2 = \delta$). The two gauge-bosons couple with the same coupling constant g to the left- and right-handed fermions according to the mixing scheme :

$$W_L = W_1 cos(\zeta) + W_2 sin(\zeta),\quad W_R = -W_1 sin(\zeta) + W_2 cos(\zeta)$$

The strength of the charged current weak interaction processes implies the propagators $(q^2 + m_{1,2}^2)^{-1}$. It is thus easily seen that for $q^2 >> m_2^2$, i.e. at the temperature prevailing soon after the Big Bang, parity symmetry is restored while at our epoch with $q^2 << m_2^2$, parity violation is nearly maximal. In this minimal extension of the SM the interactions of all the basic fermion fields are influenced the same way, making experiments in all three sectors of the weak interaction (leptonic, semi-leptonic and non-leptonic) equivalent. In more general left-right symmetric (LRS) models [3,4] distinct left/right coupling constants g_L/g_R and Cabibbo-Kobayashi-Maskawa (CKM) matrix elements V^L/V^R are introduced, as well as other new features such as e.g. possible massive neutrinos.

STATUS OF W_R SEARCHES

We will restrict here to a brief summary of the large amount of experimental data. For the mixing angle, Aquino [5] recently obtained (in LRS-models and assuming a light ν_{e_R}) $-0.0006 < \zeta < 0.0028$ by combining the neutron lifetime and asymmetries with the unitarity of the CKM-matrix. Note that if $\zeta = 0$ as we will assume in the rest of this section, W_2 is identical to W_R.

Limits on the mass m_2 have been obtained in all three sectors of the weak interaction [6]. In nuclear β-decay a recent analysis [7] of the neutron lifetime and electron emission asymmetry data reported in the literature yields $m_2 > 222\ GeV/c^2$ (90% C.L.) while the recent results for the neutrino asymmetry [8] in neutron decay gives $m_2 > 282\ GeV/c^2$.

Measurements at the endpoint positron spectrum in the decay of highly polarized muons [9] yielded $m_2 > 482\ GeV/c^2$ (90% C.L.) assuming a light ν_R, although uncorrected systematic errors may weaken this limit.

Finally, in the past few years experiments at hadron colliders have been searching for a heavy charged vector boson W' other than W_1, but with couplings to quarks and leptons taken to be identical to those of W_1, in the $p\bar{p} \to W'X$ reaction. Investigation of $W' \to e\nu$ at the Tevatron collider has yielded $m(W') > 652\ GeV/c^2$ (CDF-collaboration [10]), resp. $> 720\ GeV/c^2$ (D_0-Collaboration [11]) (95% C.L.). These mass limits for W' are also valid for W_R if $m_{\nu_R} << m_{W_R}$. In the MLRS model W' is identical to W_2.

Whereas in the manifest left-right symmetric model the results of experiments in all three sectors of the weak interaction are equivalent, they probe different combinations of the parameters for the right-handed sector in general LRS models : e.g. experiments in nuclear β decay resp. μ decay probe the combinations (with $\zeta = 0$) [12] :

$$\frac{g_R^2 \, |V_{ud}^R| \sum_i' |U_{e\nu_i}^R| \, M_{W_L}^2}{g_L^2 \, |V_{ud}^L| \sum_i' |U_{e\nu_i}^L| \, M_{W_R}^2} \quad resp. \quad \frac{g_R^2 \sum_i' |U_{e\nu_i}^R| \sum_i' |U_{\mu\nu_i}^R| \, M_{W_L}^2}{g_L^2 \sum_i' |U_{e\nu_i}^L| \sum_i' |U_{\mu\nu_i}^L| \, M_{W_R}^2}$$

where the sum is over those neutrinos whose production is not suppressed through kinematics. If all neutrinos coupling to the W gauge bosons have negligible small masses (compared to $m_{W_{L,R}}$) the sum over the $U^{L,R}$ matrixelements vanishes.

METHOD AND FORMALISM

Here we report on a new measurement of the longitudinal polarization P of positrons emitted by polarized ^{107}In nuclei, in which the result of our previous measurement [13] was improved significantly. The new experiment was carried out with a better detection set-up and a much higher source strength. Also, this time the longitudinal polarization of positrons emitted by polarized and unpolarized nuclei $R = P^-/P^0$ were compared whereas earlier, positrons emitted in two opposite directions relative to the polarized nuclear spin were used. Thus, possible systematic effects related to the field reversals, needed to invert the nuclear polarization, were eliminated.

The so-called polarization-asymmetry correlation is potentially very sensitive to right-handed charged weak currents [14]. In the SM (i.e. $\delta = \zeta = 0$) the ratio of the two positron polarizations is given by :

$$R_0 \equiv \frac{P_0^-}{P_0^0} = \frac{\Theta^-}{\Theta^0} \frac{\beta^2 - (\bar{\beta} \cdot \bar{J} A)\Psi}{\beta^2(1 - \bar{\beta} \cdot \bar{J} A)}$$

with $\beta = v/c$ the relative velocity of the positrons, $\bar{J}$ the nuclear polarization and A the β asymmetry parameter. The functions Θ and Ψ are solid angle integrals that take into account the fact that the polarization of the positrons along the axis of the polarimeter magnetic field is measured thereby integrating over the finite acceptance of our apparatus, and the fact that small transverse components of the positron polarization are accepted by the polarimeter as well. The "−" superscript indicates the nuclear spin direction relative to the momentum of the observed positrons while the "0" superscript refers to the ^{107}In nuclei being unpolarized. The value R_0 can thus be obtained from a simultaneous measurement of the mean positron velocity and the emission asymmetry. Any deviation of the measured ratio from R_0 would either reflect contributions beyond the allowed approximation in nuclear β decay (recoil is opposite. The positron polarization can thus be obtained by observing the change in the population of these PT and PS states when the magnetic field direction in the polarimeter is reversed. During the experiment this is done every 4 minutes. The effective analyzing power is $\epsilon \approx 0.10$. In figure 1 an example of an experimental Ps decay spectrum, together with the response function, obtained by replacing the MgO powder by an Al pellet in which all positrons annihilate directly, is shown. Clearly, three regions are visible : a contribution from direct annihilation and the short lived PS state ($\tau \leq 2.5ns$), the PT state ($\tau \approx 7.5ns$) from which the polarization P is determined and the TR state ($\tau \approx 129ns$) which is insensitive to P and is therefore used for normalization.

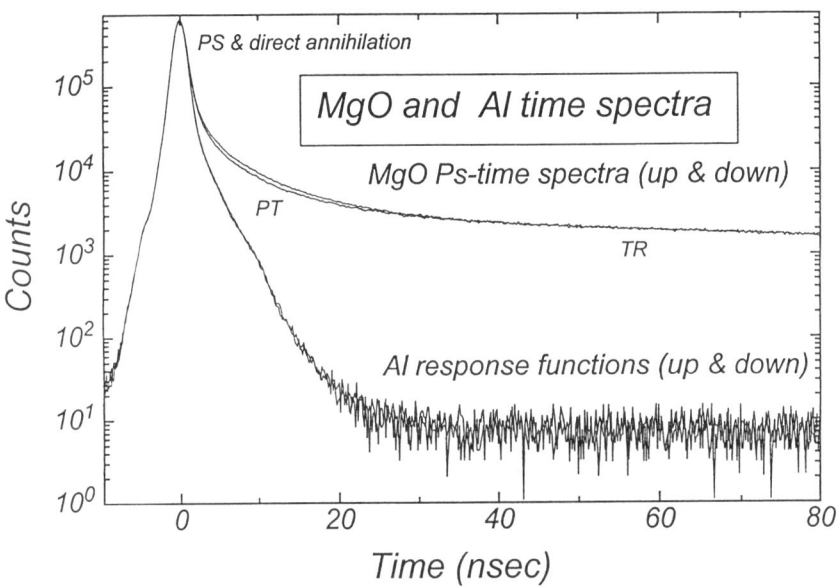

FIGURE 1. Experimental Ps decay spectrum and response function. The spectra obtained for two opposite field directions of the polarimeter magnetic field are superimposed.

ANALYSIS

From the above it is clear that for this experiment, 3 quantities need to be determined experimentally : the polarization ratio $R = P^-/P^0$, the beta asymmetry $\vec{\beta} \cdot \vec{J} A$ and the positron velocity β. The latter was calculated on the basis of the momentum acceptance and calibration of the spectrometer, the momentum distribution (phase space) of the positrons from the ^{107}In source effects), or indicate new physics beyond the Standard Model. Assuming right-handed currents and neglecting other types of new physics as well as recoil effects, the ratio R can be written (in the MLRS model and for $\zeta = 0$) as :

$$R = R_0[1 - \frac{4(\bar{\beta} \cdot \bar{J}A)\Psi}{(\beta^2 - \bar{\beta} \cdot \bar{J}A)\Psi}\delta^2] = R_0[1 - k\delta^2]$$

Clearly the enhancement factor k can become very large if both β and $\bar{\beta} \cdot \bar{J}A$ are close to unity ($\Theta \approx 1.027$).

EXPERIMENTAL SET-UP

The β decay of ^{107}In ($t_{1/2} = 32.4\ m$) is dominated by an allowed $9/2^+ \to 7/2^+$ pure GT transition with endpoint kinetic energy $E_0 = 2.26$ MeV and log ft = 5.65. The experimental set-up is shown schematically in figure 1 of [13]. The ^{107}In nuclei were produced with a heavy-ion fusion-evaporation reaction at the cyclotron CYCLONE in Louvain-la-Neuve. After mass separation with the LISOL on-line isotope separator, the 10^7 at/s ^{107}In beam was implanted at 50 keV into a polycrystalline Fe foil inside the KOOL on-line $^3He -^4 He$ dilution refrigerator where the nuclei were polarization to $\approx 80\%$ at 11 mK with the method of low temperature nuclear orientation. By raising the temperature to about 1 K, an unpolarized ^{107}In source was obtained.

The positrons emitted by the ^{107}In nuclei were energy selected with a magnetic spectrometer. To minimize effects from scattering in the iron foil and on the walls of the vacuum chamber of the spectrometer, the latter was coated with a rough-surfaced low-Z material (grey PVC) while the Fe foil was rotated towards the spectrometer axis. When entering the polarimeter (see insert of figure 1 in [13]) the positrons first passed through a C moderator, then crossed a 0.5mm plastic ΔE START scintillator and were finally guided by the strong polarimeter magnetic field of 9.6 kG into a MgO powder pellet where they were stopped and where about 45% of them formed positronium (Ps). The gamma rays of the Ps decay and of the positrons that annihilated immediately were observed with two BaF_2 STOP detectors. The signals of these were combined in "AND"-mode. Time-resolved Ps decay spectra were obtained with a time-to-digital converter of 156 ps channel width. A detailed description of the technique of time-resolved positronium hyperfine spectroscopy to determine positron polarizations was given previously [15]. It relies on the fact that in a magnetic field the $m = \pm 1$ triplet (TR) substates of the Ps are not affected, while the singlet state and the $m = 0$ triplet substate are perturbed by the field and mix, leading to a pseudo-singlet (PS) and a pseudo-triplet (PT) state. The population of the PS and PT states depends on the polarization P of the positrons, but also on the intensity of the magnetic field B and on the relative direction of $\bar{B}$ and $\bar{P}$: for the PT state it is proportional to $(1 \mp \epsilon P)$ for $\hat{B} \cdot \hat{P} = \pm 1$. For the population of the PS state the sign of P positrons is still being calculated. However, estimates show this to induce a minor effect only. Finally, also recoil terms, induced by matrix elements beyond the allowed approximation in nuclear β-decay, can cause a systematic effect. Using the formalism of Holstein [16] the contributions of recoil order terms up to the order $1/M$, with M being the mass of the daughter nuclear were computed. It was found that the terms b/Ac (weak magnetism) and d/Ac (induced tensor) are dominant. The effect of all terms together leads to $(R/R_0)_{recoil} = 0.9950(14)$. All measurements were corrected for this.

RESULTS AND DISCUSSION

The preliminary, weighted average result of all 13 runs carried out is $R/R_0 = 0.9898(82)$, or $(\delta + \zeta)^2 = 0.0021(17)$. This value improves the precision of the previous result [13] by more than a factor three. If interpreted in the manifest left-right symmetric model it corresponds to a lower limit of 303 GeV/c^2 (90%C.L.; $\zeta = 0$) for the mass of a charged W_R gauge boson with right-handed couplings. For comparison, the combined result of the data from

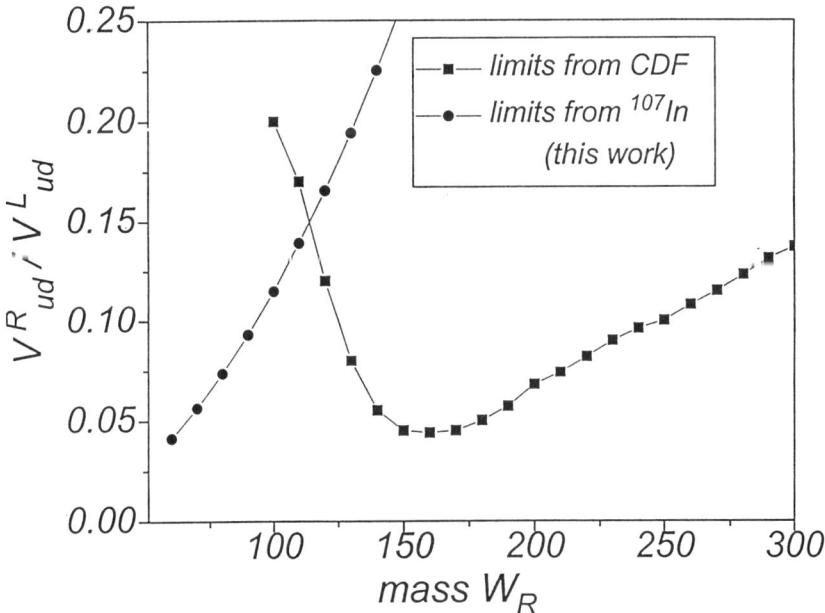

FIGURE 2. Limits on the up-down CKM matrix elements as a function of the mass of W_R. The area above the curves is excluded at 90% C.L.

and the contribution of positrons scattered from the iron implantation foil (about 6% of the total counts), yielding = 0.9517(15). The beta asymmetry was obtained from the intensity ratio of the triplet Ps state (being independent of the positron polarization) in the interval $80ns < t < 250ns$ for polarized and unpolarized nuclei : $N_{TR}^{pol}/N_{TR}^{unpol} = 1 - \bar{\beta} \cdot \bar{J}A$. Possible variations in the ^{107}In source intensity were taken into account by normalizing to the 205 keV γ ray in the decay of ^{107}In. The 4% anisotropy in the emission of this γ and the small residual nuclear polarization left over at 1 K, (i.e. 1.58(30) %) were also corrected for. The average $\bar{\beta} \cdot \bar{J}A$ in all 13 runs that were carried out was about 0.44. The ratio of the positron polarizations for polarized and unpolarized ^{107}In nuclei was obtained from:

$$\frac{[(N_{PT}/T)^p - (N_{PT}/T)^a]^{pol}}{[(N_{PT}/T)^p - (N_{PT}/T)^a]^{unpol}} = \frac{\epsilon P^-}{\epsilon P^0} = R$$

with $N_{PT}^{p,a}$ and T the integrals of the PT ($2.5ns < t < 30ns$) and TR ($80ns < t < 250ns$) region in the Ps decay spectrum. This ratio R should be a constant as a funcion of time in the PT-region as was verified for each Ps spectrum by dividing it into small bins of 1 ns wide and determining R(t) for each bin separately.

SYSTEMATIC EFFECTS AND CORRECTIONS

The most important potential source of systematic errors is the scattering of the positrons in the source foil and its backing. From measurements with a ^{68}Ga source, used with and without backing, as a function of the tilt angle of the source relative to the polarimeter axis is was found that about 6% of the count rate comes from positrons scattered in the source foil and its backing. Moreover, this scattering is mainly forward and consequently the polarisation-asymmetry correlation is preserved. This was verified by comparing the polarization of positrons emitted by the ^{68}Ga source with and without backing yielding 0.9969(95). Note also that due to the relative character of our measurements any effect of depolarization of the positrons, including the depolarization during the slowing down in the energy degrader and in the MgO powder, can only be of second order. A possible systematic error may also arise by small time shifts of the response functions corresponding to the two polarimeter magnetic field directions. These shifts were duly taken into account by shifting the corresponding time spectra. The optimization of the isotope separator beam between the different runs, could, in principle, change the location of the ^{107}In source, thus altering the acceptance of the spectrometer possibly giving rise to systematic errors. This was checked by comparing the polarizations of positrons emitted by unpolarized nuclei in the different runs. The average of the ratios thus constructed is 0.9993(43), in accord with the absence of any effect. A correction for the spin precession of the three previous relative positron polarization measurements on polarized ^{107}In and ^{12}N [13,17,7] is ($\delta^2 = 0.0004(26)$) which correspondings to a lower limit of 316 GeV/c^2. If interpreted in general LRS models our result is complementary to results from other sectors in the weak interaction, as was shown in [18]. This is also illustrated in fig.2 showing the limits on the "ud" CKM matrix elements as a function of the mass of W_R. Note that collider searches loose their sensitivity to a W' boson for masses below 100 GeV/c^2. Finally, if parity violation is assumed to be maximal, our result also provides stringent limits on possible tensor contributions to the nuclear beta decay Hamiltonian. Expressing the polarization ratio P^-/P^0 in terms of the tensor coupling constants C_T and C_T' yields :

$$Re[\frac{C_T + C_{T'}}{C_A}] = -0.015(12) \quad and \quad Im[\frac{C_T + C_{T'}}{C_A}] = -0.037(30)$$

REFERENCES

1. Pati J.C., and Salam A., *Phys. Rev. D* **10**, 275 (1974); Beg. M.A.B. et al., *Phys. Rev. Lett.* **38**, 1252 (1977).
2. Mohapatra R.N., *Unification and Supersymmetry*, Berlin : Springer Verlag, 1986, and references therein.
3. Herczeg P., *Phys. Rev. D* **34**, 3449 (1986).
4. Langacker P., and Sankar U., *Phys. Rev. D* **45**, 3449 (1986).
5. Aquino M., Fernandez A., and Garcia A., *Phys. Lett. B* **261**, 280 (1991).
6. Particle Data Group., *Phys. Rev. D* **50**, 1173 (1994).
7. Thomas E., Ph. D. Thesis, U.C. Louvain-la-Neuve, *unpublished* (1997);
8. Kuznetsov I.A., et al., *JETP Lett.* **60**, 315 (1994).
9. Jodidio A., et al., *Phys. Rev. D* **34**, 1967 (1986) & **37**, 237E (1988).
10. Abe F., et al., *Phys. Rev. Lett.* **74**, 2900 (1995).
11. Abacki S., et al., *Phys. Rev. Lett.* **76**, 3271 (1996).
12. Govaerts J., *Heavy charged gauge boson production at CDF and left-right symmetric models*, Univ. Cath. Louvain-la-Neuve (IPN) Internal report, July 11, (1995).
13. Severijns N., et al., *Phys. Rev. Lett.* **70**, 4047 (1993) & **73** 611E (1994).
14. Quin P.A., and Girard T.A., *Phys. Lett. A* **229**, 29 (1989).
15. Prieels R., *Proceedings of the 24th Rencontre de Moriond, January 1989*, Gif-sur-Yvette : Eds. Frontières, p. 287.
16. Holstein B.R., *Rev. Mod. Phys.* **46**, 789 (1974) & **48** 673E (1976).
17. Allet M., et al., *Phys. Lett. B* **383**, 139 (1996).
18. Quin P.A., et al., *Proceedings of the Conference on Particles and Nuclei (PANIC)*, Williamsburg (USA), May 1996, to be published.

The Mass-8 Experiment - Measuring the β-α angular correlations

J.F. Amsbaugh, M. Beck, L. De Braeckeleer[*], D.W. Storm, E. Swanson,
K.B. Swartz[§], J.P.S. van Schagen, D.C. Wright, Z. Zhao[+]

Physics Department, University of Washington, Seattle, WA 98195
[*] *Present address Physics Department, Duke University, Durham, NC 27706*
[§] *Present address Physics Department, Yale University, New Haven, CT 06520*
[+] *Present address McKinsey Co., Atlanta, GA 30303*

Abstract: The objective of the Mass-8 experiment is to perform a precision test of the conservation of the vector current hypothesis and a search for second class currents. We present preliminary data on the correlation coefficients of the β-α angular correlations of the β-delayed α-decays of ^{8}Li and ^{8}B.

Introduction

The standard model of electroweak interaction is being tested extensively by many experiments. The two questions investigated by the present experiment are the validity of the Conservation of the Vector Current hypothesis (CVC)[1] and the possible existence of Second Class Currents (SCC)[2]. CVC equates the vector part of the weak interaction with the vector part of the electromagnetic interaction. It is a fundamental property of the standard model. Present limits on the validity of CVC are at the level of 6 %[3]. SCC are defined by their transformation properties under G-parity; they only appear in the standard model at the level of isospin breaking:

First class currents:
$$GV_\mu G^{-1} = V_\mu$$
$$GA_\mu G^{-1} = -A_\mu$$

Second class currents:
$$GV_\mu G^{-1} = -V_\mu$$
$$GA_\mu G^{-1} = A_\mu$$

The investigation of CVC and SCC usually involves the comparison of the weak magnetism form factor from weak decays with the magnetic form factor from electromagnetic decays. The former can be obtained from e.g. precision measurements of β-spectral shapes, μ-capture, β-α and β-γ angular correlations. The latter can be retrieved from e.g. the isobaric analogue γ-transitions. Another process used for a test of CVC that does not involve nuclear physics is π/μ-decay (see ref. [3] for a summary). The experiment discussed here measures the $β^\pm$-α angular correlations in the A=8 system (Fig. 1):

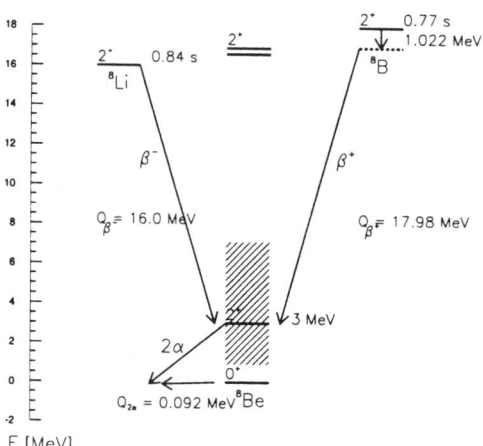

FIGURE 1. Decay scheme in A=8

$$\omega^{\pm}(E_\beta, \vartheta_{\beta\alpha}) = a_0 \cdot (1 + a_1 \cdot \cos\vartheta_{\beta\alpha} + a_2^{\pm} \cdot \cos^2\vartheta_{\beta\alpha}) \quad (1)$$

which together with the width of the isovector M1 γ-decay of the isobaric analogue state[4] in $^8Be^*$, $\Gamma_{M1}^{T=1}$, makes the CVC and SCC tests possible. In a simplified picture the a-coefficients can be written, with $M_n \cdot A$ the mass of ^8Be and v_α a mean α-velocity, as

$$a_1(E_\beta) = -\frac{2 \cdot p_\beta}{M_n \cdot A \cdot v_\alpha} \quad \text{(kinematical term)} \quad (2)$$

$$a_2^{\pm}(E_\beta) \approx \frac{p_\beta}{2 \cdot M_n \cdot A \cdot c} \cdot (\mp b + d_I \mp d_{II}) \quad (3)$$

and

$$b = M_n \cdot A \cdot \sqrt{\frac{6 \cdot \Gamma_{M1}^{T=1}}{\alpha \cdot E_\gamma^3}} \quad (4)$$

where b, d_I and d_{II} are induced nuclear matrix elements (table 1). Here d has been expanded into a first class and a second class part. The derivation of these coefficients can be found in ref. [5].

TABLE 1. Explanation of matrix elements.

b	weak magnetism (first forbidden vector)
c	allowed Gamov-Teller
d	induced Tensor

When deducing (3) for the A=8 system only the c, b and d matrix elements yield a significant contribution[4]. One tests CVC and SCC by comparing b extracted from (3) with b extracted from (4). The CVC and SCC tests are not independent: Only when assuming that there are no SCC can one test CVC and vice versa. The quantity extracted from the experimental data is

$$\delta^- = a_2^- - a_2^+ = \frac{E_\beta}{M_n \cdot A} \cdot \left(\frac{b}{c} + \frac{d_{II}}{c}\right)$$

The Experiment

The experimental setup to measure the β-α angular correlations is shown in fig. 2. The activity is produced in the target area and transported by a rotating arm into the counting area. This is in the center of two perpendicular planes of seven β-counters and four α-counters. The β-counters are scintillation counters consisting of the main β-energy counter with a veto counter around it and a ΔE-counter in front of it. The α-detectors are three multiwire gas-counters and one Silicon detector. In order to remove any β-efficiency dependence the angular correlation is extracted as

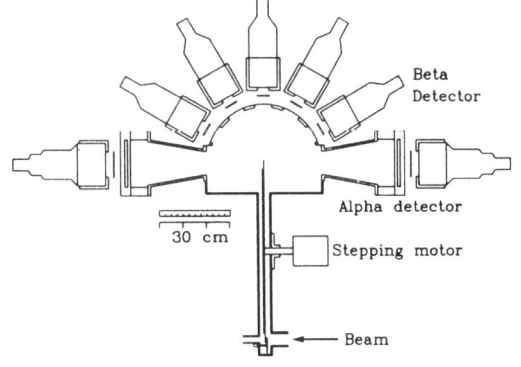

$$\omega(E_\beta, \vartheta_{ij}) = \frac{N_{ij}(E_\beta)}{N_{i3}(E_\beta)}$$

where i={1,...,7} is the index of for the β-counters, j={1,2} are the two gas-counters at $\vartheta_{\beta\alpha}=0°$

and 180° and α_3 is the gas-counter at 90°.

Preliminary Results

This is an ongoing experiment. The preliminary angular correlation coefficients are shown in fig. 3. The a_1-coefficients for ^{8}B deviate slightly more from the theoretical prediction than the ones for ^{8}Li, which follow the theory in most of the energy range. The deviations at the smallest and highest energies are most likely due to the response function of the β-detectors. The a_2-coefficients for ^{8}B as well as ^{8}Li are described by a straight line

$$a_2 = m \cdot \left(E_{\beta tot} - 0.511 keV\right)$$

except at the lowest and highest energies. The average slopes m of a_2 are

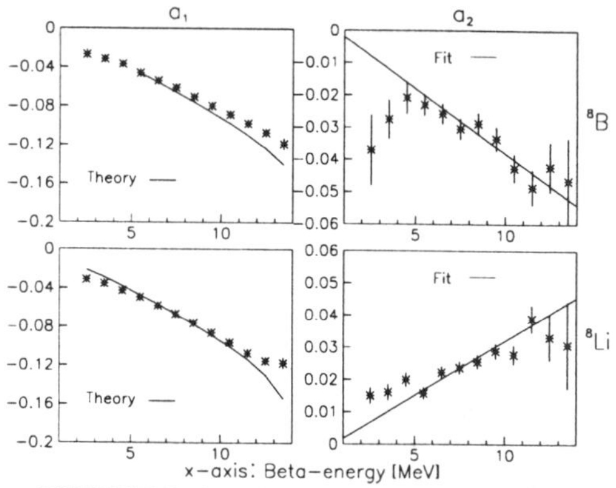

FIGURE 3. Angular correlation coefficients

^{8}B $m^+ = -4.02 \pm 0.17 GeV^{-1}$
^{8}Li $m^- = +3.37 \pm 0.09 GeV^{-1}$

which leads with $\delta^- = a_2^- - a_2^+$ to

$$\delta^- \cdot \frac{M_n}{E_\beta} = 6.9 \pm 0.2 GeV^{-1}$$

Estimates of the systematic uncertainties of a_2 are 3%, mostly due to the uncertainty of the response function. Table 2 shows the comparison of the δ^- of this experiment with results of previous measurements in the A=8 system.

TABLE 2. Comparison of experiments

$\delta^- \cdot \frac{M_n}{E_\beta} [1/GeV]$	Reference
$6.9 \pm 0.2_{stat} \pm 0.2_{sys}$	This work
6.5 ± 0.2	6
7.0 ± 0.5	7

We expect to improve the statistical uncertainties further by taking more ^{8}B data. We are presently studying the systematic uncertainties, especially the effect of the response function on the angular correlation coefficients, and expect to improve the systematic uncertainties considerably.

---

[1] R.P. Feynman and M. Gell-Mann, Phys. Rev. **109** (1958) 193
[2] S. Weinberg, Phys. Rev. **112** (1958) 1375
[3] L. Grenac, Ann. Rev. Nucl. Part. Sci. **35** (1985) 455; L. De Braeckeleer, Phys. Rev. C **45** (1992) 1935
[4] L. De Braeckeleer et al., Phys. Rev. C **51** (1995) 2778
[5] B.R. Holstein, Rev. Mod. Phys. **46** (1974) 789
[6] R.D. McKeown, G.T. Garvey, C.A. Gagliardi, Phys. Rev. C **22** (1980) 738
[7] R.E. Tribble, G.T. Garvey, Phys. Rev. C**12** (1975) 967

UCLA-HEP-97-003

A Search for Antiproton Decay at the Fermilab Antiproton Accumulator

Brent Corbin, for the APEX Collaboration

T. Armstrong[a], C. Buchanan[b], B. Corbin[b], S. Geer[c],
R. Gustafson[d], M. Hu[e], M. Lindgren[b], J. Marriner[c],
M. Martens[c], T. Müller[b], R. Ray[c], G. Snow[e], J. Streets[c],
W. Wester[c]

[a] *Pennsylvania State University, University Park, Pennsylvania 16802, USA*
[b] *University of California at Los Angeles, Los Angeles, California 90024, USA*
[c] *Fermi National Accelerator Laboratory, Batavia, Illinois 60510, USA*
[d] *University of Michigan, Ann Arbor, Michigan 48109, USA*
[e] *University of Nebraska-Lincoln, Lincoln, Nebraska 68506, USA*

Abstract. We report on the search for anti-proton decay at the Fermilab Antiproton Accumulator Ring. Experiment 868 (APEX) was designed to search for two-body $\bar{p}$ decay modes containing an electron in the final state ($\bar{p} \to e + X$) and to conduct an exploratory search for decays with a muon in the final state ($\bar{p} \to \mu + X$). Data were taken for three months in the Spring of 1995. Preliminary results yield lower limits on $\tau_{\bar{p}}/BR$ in the range of $10^5 - 10^6$ years for selected channels having an electron in the final state, improving on previous results by approximately 3 orders of magnitude. Additionally, we report the first preliminary results for the $\bar{p} \to \mu\gamma$ and $\bar{p} \to \mu\pi^0$ decay channels.

INTRODUCTION

CPT invariance implies that the lifetimes of a particle and its corresponding antiparticle should be equal. Measurements of the half-life of the proton have set (mode dependent) limits on the lifetime of the proton in the range of $\tau_p/BR > 10^{31-33}$ years. Laboratory measurements of the half-life of the antiproton have not yielded limits anywhere near this. Prior to the APEX run, the most sensitive laboratory measurements of the antiproton lifetime were made in the APEX test run (T861) [1], and ranged from $\tau_{\bar{p}}/BR(\bar{p} \to e^-K_L^0) > 9$ years to $\tau_{\bar{p}}/BR(\bar{p} \to e^-\gamma) > 1849$ years at 95% C.L..

The APEX experiment (E868) was designed to search for antiproton decay with a single-event sensitivity for the simplest mode ($\bar{p} \to e^-\gamma$) of approximately 10^6 years. It was installed on the 474 m circumference Antiproton Accumulator Ring at Fermilab, where it kept watch over an average of 10^{12} 8.9 GeV/c antiprotons during periods in which there was no stacking. Data was collected over a period of approximately three months (1 April - 30 June 1995) parasitic to CDF and D0 collider running. In all, APEX recorded some 10 Million triggers over 135 runs, totaling $\frac{\int N_{\bar{p}} dt}{\gamma_{\bar{p}}} = 3.31 \times 10^9 \bar{p}$-years, where $N_{\bar{p}}$ is the number of antiprotons circling in the accumulator, $\gamma_{\bar{p}}$ is the relativistic correction factor for the antiprotons (9.5), and the integral is taken over the lifetime of the experiment.

HARDWARE

To minimize the effects of our most likely source of background, beam-gas events, APEX installed an approximately 4 m long vacuum tank just upstream of the APEX detector. Titanium sublimation pumps were used to achieve a vacuum of 2.0×10^{-11} Torr within a conically shaped fiducial region. A removable wire target was mounted at the upstream end of the tank, which could be inserted into the beam line for the purpose of aligning and calibrating the detector components (see Figure 1).

Immediately downstream of the tank was the 1.5 m long tracking region of the APEX detector. Within this region there were three dE/dX counter stations used for triggering and crude identification of singly-charged relativistic particles. Each station was made up of four large (50 cm x 100 cm) scintillating planes, arranged such that there were horizontally-oriented planes above and below and vertically-oriented planes on each side of the beam line. The station located furthest downstream was situated behind a 1.27 cm thick lead preradiator, and was used primarily for identifying electromagnetic tracks.

The Scintillating Fiber Tracker was also located in this tracking region. The tracker consisted of 12 L-shaped panels, similar in size and orientation to the dEdX counters, arranged in groups of four to create three tracking stations. There were 384 2 mm diameter Bicron scintillating fibers on each L-shaped panel. The fibers were placed in two staggered layers. Each layer had an average center-to-center spacing of 2.6 mm, thus providing a pitch of 1.3mm. On each panel, the fibers were equally divided between two Hamamatsu R4135A multi-anode phototubes. The average track residual within a plane was observed to be about 620 μm. At the far end of the fiducial region (near the target) the vertex resolution along the beam line was measured to be about 13 cm for a track originating about 4 m upstream of the tracker, and the typical impact parameter for one of these tracks was somewhat less than 1 cm.

An electromagnetic calorimeter was placed just beyond the tracking region. It was made up of 144 cells (each measuring approximately 10 cm high by 10

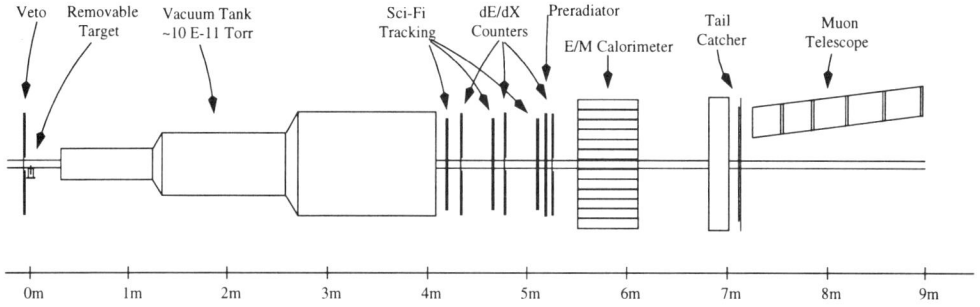

FIGURE 1. Side view of the APEX detector

cm wide by 60 cm deep) arranged so as to fill a 13 by 13 array with 6 cells missing from each corner and once cell missing around the beam line. The calorimeter was calibrated by reconstructing π^0's from beam-target events. It is also possible to see an η peak in the two-photon target data. The calorimeter was followed by a tail-catcher (scintillation counters behind a 20 cm thick lead wall), for identifying tracks that had punched-through the calorimeter (hadrons or muon candidates).

Finally, a muon telescope was located at the far downstream end of the detector. It was comprised of 5 alternating sections of iron (each 30.3 cm deep) and scintillator (each 1.91 cm deep). The telescope had a 30 cm by 30 cm cross section, and its central axis pointed towards the center of the vacuum tank. It was used for triggering and offline identification of muon candidates.

DIRECT SEARCHES

Using the tracker for kinematic reconstruction, the dE/dX counters and preradiator for particle identification and the calorimeter for energy reconstruction (and location of photons), the APEX collaboration has obtained limits on $\tau_{\bar{p}}$/BR for four channels by direct searches:

Summary of Direct Search Results

Mode	Surviving Events	Preliminary Result (90 % C.L.)
$\bar{p} \to e^- \gamma$	0	$> 1.3 \times 10^6$ years
$\bar{p} \to e^- \pi^0$	1	$> 3.3 \times 10^5$ years
$\bar{p} \to \mu^- \gamma$	0	$> 4.2 \times 10^4$ years
$\bar{p} \to \mu^- \pi^0$	0	$> 1.9 \times 10^4$ years
		NO SYSTEMATICS

It should be noted that the analysis is not yet fully optimized, and so these numbers should be considered preliminary. It is also worth noting that the one event that survives the $e^-\pi^0$ analysis cuts is not without its problems when viewed in the event display (it appears that the tracking algorithm may have been tricked by some extraneous hits in a tracking plane). This event may vanish in the final analysis, and the $e^-\pi^0$ limit would then go up accordingly.

INDIRECT SEARCHES

The $\bar{p} \to e^-\gamma$ analysis has been applied to Monte Carlo simulations of other decay channels containing an electron. One may infer the following limits on $\tau_{\bar{p}}/$ BR from the absence of surviving events in the $e^-\gamma$ analysis:

Summary of Indirect Search Results

Decay Channel	Lifetime Limit 90 % C.L.
$\bar{p} \to e^-\pi^0$	$> 6.4 \times 10^5$ years
$\bar{p} \to e^-\eta$	> 17000 years
$\bar{p} \to e^-\omega$	> 1700 years
$\bar{p} \to e^-K_L^0$	> 3600 years
$\bar{p} \to e^-K_S^0$	> 4700 years
$\bar{p} \to e^-\rho^0$	> 120 years
	NO SYSTEMATICS

Again, these numbers are to be considered preliminary. In particular, optimizing the analysis for each channel should improve the limits for the heavier decays.

REFERENCES

1. Geer S. *et al. Phys. Rev. Lett.* **72** (1994) 1596.

The SAMPLE Experiment

R. D. McKeown

W. K. Kellogg Radiation Laboratory[1]
California Institute of Technology, Pasadena, CA 91125

Abstract. The neutral weak magnetic form factor of the nucleon can be studied in parity-violating electron-nucleon scattering. The measurement of this form factor enables determination of the contribution of strange quark-antiquark to the proton's magnetic moment. We have recently obtained the first results on the neutral weak magnetic form factor using this method in the SAMPLE experiment at MIT/Bates.

INTRODUCTION

The electromagnetic form factors of the nucleon provide invaluable insight into nucleon structure. The weak form factors are equally significant, but our experimental knowledge is rather limited at this point. These form factors can be experimentally studied by measurement of parity violation in elastic electron proton scattering [1,2]. Kaplan and Manohar [3] have demonstrated that the neutral weak current can be used to determine the $\bar{s}s$ contributions to nucleon vector and axial vector form factors. Thus the study of neutral weak vector form factors of the proton can be used to determine the $\bar{s}s$ contribution to the charge and magnetization distributions of the nucleon.

In this paper, the first results on the neutral weak magnetic form factor of the proton from the SAMPLE experiment [4] are reported. This experimental result offers our very first glimpse at the role of $\bar{s}s$ pairs in the proton's magnetic moment.

THEORETICAL BACKGROUND

We begin with a discussion of the quark flavor structure of the nucleon electromagnetic and neutral weak currents. The electromagnetic current operator has the form

---

[1] Supported by the National Science Foundation.

$$\hat{V}_\gamma^\mu = \frac{2}{3}\bar{u}\gamma^\mu u - \frac{1}{3}\bar{d}\gamma^\mu d - \frac{1}{3}\bar{s}\gamma^\mu s. \tag{1}$$

The flavor stucture of this electromagnetic current operator implies that the electric and magnetic form factors can be written

$$G_{E,M}^\gamma = \frac{2}{3}G_{E,M}^u - \frac{1}{3}G_{E,M}^d - \frac{1}{3}G_{E,M}^s. \tag{2}$$

The neutral weak current operator is given by an expression analogous to to the electromagnetic current but with different coefficients:

$$\hat{V}_Z^\mu = (\frac{1}{4} - \frac{2}{3}\sin^2\theta_W)\bar{u}\gamma^\mu u + (-\frac{1}{4} + \frac{1}{3}\sin^2\theta_W)\bar{d}\gamma^\mu d + (-\frac{1}{4} + \frac{1}{3}\sin^2\theta_W)\bar{s}\gamma^\mu s. \tag{3}$$

Thus we also have expressions for the neutral weak form factors G_E^Z and G_M^Z in terms of the different quark flavor components. One can eliminate the up and down quark contributions to the neutral weak form factors by using the isovector and isoscalar (or proton and neutron) electromagnetic form factors, and obtain the expressions

$$G_{E,M}^{Z,p} = (\frac{1}{4} - \sin^2\theta_W)G_{E,M}^{\gamma,p} - \frac{1}{4}G_{E,M}^{\gamma,n} - \frac{1}{4}G_{E,M}^s. \tag{4}$$

This is a key result. It shows how the neutral weak form factors are related to the well known electromagnetic form factors plus a contribution from the strange (electric or magnetic) form factor. Thus measurement of the neutral weak form factor will allow (after combination with the electromagnetic form factors) determination of the strange form factor of interest. It should be mentioned that there are electroweak radiative corrections to the coefficients in this expression which have been computed. [5]

As noted above, the quantities $G_{E,M}^Z$ for the proton can be determined via elastic parity violating electron scattering. [1] The difference in cross sections for right and left handed incident electrons arises from interference of the electromagnetic and neutral weak amplitudes, and so contains products of electromagnetic and neutral weak form factors. The expression [6] for the left-right asymmetry in elastic scattering from the proton is given by

$$A = \left[\frac{-G_F Q^2}{\pi\alpha\sqrt{2}}\right] \frac{\varepsilon G_E^\gamma G_E^Z + \tau G_M^\gamma G_M^Z - \frac{1}{2}(1 - 4\sin^2\theta_W)\varepsilon' G_M^\gamma G_A^Z}{\varepsilon(G_E^\gamma)^2 + \tau(G_M^\gamma)^2} \tag{5}$$

where ε, ε', and τ are kinematic quantities, $Q^2 > 0$ is the four-momentum transfer, and θ is the laboratory electron scattering angle. There is one additional form factor in the numerator of the asymmetry expression: G_A^Z, the neutral weak axial form factor which is related to the isovector form factor

SAMPLE EXPERIMENT

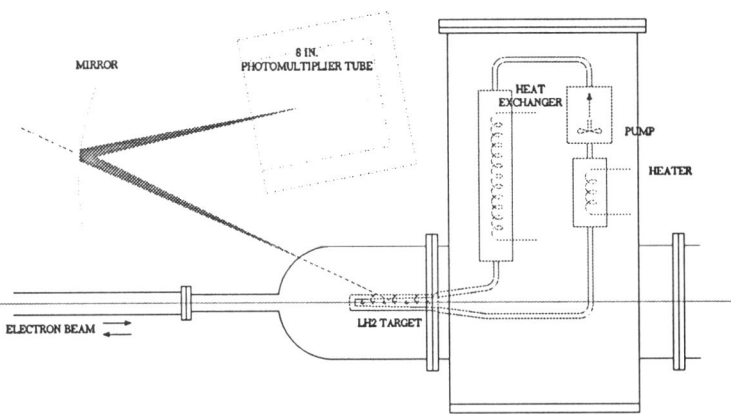

FIGURE 1. Schematic diagram of the layout of the SAMPLE target and detector system.

$G_A(Q^2 = 0) = 1.2601$ (from neutron beta decay) and $G_A^s = \Delta s \sim -0.1$ from polarized deep inelastic scattering

The goal of the SAMPLE experiment is to measure G_M^Z at low $Q^2 \simeq 0.1$ $(\text{GeV}/c)^2$ which enables determination of G_M^s. There have been a variety of theoretical predictions for $\mu_s \equiv G_M^s(Q^2 = 0)$ over the last few years, and a summary of them is presented in [7]. The typical magnitude is -0.3 nuclear magnetons (n.m.), and most predict $\mu_s < 0$.

SAMPLE EXPERIMENT

The SAMPLE experiment measures the left-right asymmetry at backward angles with 200 MeV incident electrons. The expected asymmetry for $G_M^s = 0$ is -7.2×10^{-6} or -7.2 ppm. At these kinematics the axial term is about 20% and the uncertainty due to the axial term is ±0.7 ppm.

Description of Apparatus

The experiment is performed using a 200 MeV polarized electron beam (40 μA) incident on a 40 cm long liquid hydrogen target. The scattered electrons are detected in a large solid angle (~ 1.5 sr) Cerenkov detector (similar to [8]) at backward angles $130° < \theta < 170°$. This results in an average $Q^2 \simeq 0.1$ $(\text{GeV}/c)^2$. A schematic of the apparatus is shown in Figure 1.

The detector consists of 10 large ellipsoidal mirrors that reflect the Cerenkov light into 8 inch diameter photomultiplier tubes. Each detector signal is integrated over the ~ 15μsec of the beam pulse and digitized. The beam intensity is similarly integrated and digitized. The ratio of integrated detector signal to integrated beam charge is the normalized yield which is proportional to cross section (plus background). The goal is to measure the beam helicity dependence of the cross section, or equivalently, the helicity dependence of the normalized yield. All 10 detectors are combined in software during the data analysis.

The polarized electron source is a bulk GaAs photoemission source, with polarization typically 35%. The beam helicity is randomly chosen for each of 10 consecutive beam pulses (the beam repitition rate is 600 Hz) and then the complement helicities are used for the next 10 pulses. The normalized detector asymmetry is computed for "pulse pairs" separated by 1/60 of a second to minimize systematic errors. The electron polarization is measured using a Moller apparatus on the beamline typically once per day.

Since one is attempting to measure a rather small asymmetry, it is essential that the experimental conditions (i.e., beam properties) remain as identical as possible when the helicity is reversed. There are many steps taken to insure this basic goal. An especially important one is to actively feedback the helicity correlated beam intensity exiting from the accelerator to a voltage adjustment on the Pockels cell. [9] This system reduces the helicity correlations in the beam intensity from typically 100-200 ppm to $\sim$ 1 ppm.

Residual differences in the beam parameters are monitored, continuously if possible. The position and angle at the target in both transverse dimensions (x and y), the beam energy, and the "halo" of the beam are continuously monitored for every beam pulse. The detector normalized yield asymmetry can be corrected for the helicity correlated effects observed in these beam monitors. Other properties, such as the size of the beam, can be monitored during special runs to search for such effects.

Finally, we also have the capability to manually reverse the beam helicity by rotating a $\lambda/2$ plate which reverses the helicity of the light (and the beam) relative to all electronic signals. A real parity violation signal will change sign under this "slow reversal". Electronic crosstalk and other effects will not change under "slow reversal", so this is an important test that our signal is real physics rather than a spurious systematic effect in the experiment.

We have studied the composition of the detected signal in great detail by using special runs at low beam current and computer simulations. The composition of the signal is 22.6 $\pm$ 0.5% photomultiplier (i.e., non-light related) background, 54.8$\pm$4.0% light from elastic scattering (primarily Cerenkov light) and $\sim$ 20.3 $\pm$ 3.7% scintillation light from soft EM radiation from the target. In addition, we estimate 2.3 $\pm$ 1.4% of the light is from decaying π^+ and π^0 generated in the target. The measured asymmetry is corrected for effects of these sources of backgound.

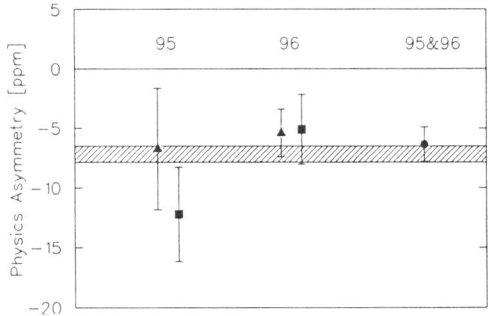

FIGURE 2. Results for the parity-violating asymmetry measured in the 1995 and 1996 running periods. For each running period we display the result for the "normal" (squares) and "reverse" (triangles) settings of slow helicity reversal, which are in good agreement. The combined result for both helicity states and both running periods is also shown (circle). The error bars include statistical errors only. The hatched region indicates the asymmetry band (due to the uncertain axial radiative correction) for $\mu_s = G_M^s = 0$.

Experimental Results

We have acquired a significant data sample from 2 runs. One in fall 1995 with about 6 Coulombs (beam on target) of good data, and a run in spring 1996 with about 30 Coulombs of good data. The complete experiment should require about 150 Coulombs of data so we have about 24% of the total at this point.

The resulting physics asymmetry for the normal and reverse slow helicity states in each of the two data runs are displayed in Figure 2. Combining the data from the 2 runs we obtain the result [15]:

$$A = -6.34 \pm 1.45 \pm 0.53 \text{ ppm}$$

where the first error is from statistics and the second is a systematic error that should be reduced in the future. The neutral weak magnetic form factor derived from this asymmetry is

$$G_M^Z(Q^2 = 0.1 \text{GeV}^2) = 0.34 \pm 0.09 \pm 0.04 \pm 0.05 \text{ n.m.}$$

where the last uncertainty is due to corrections to G_A^Z [5]. This is the first experimental measurement of this neutral weak magnetic form factor of the proton.

In the absence of a strange quark contribution we expect $G_M^Z = 0.40$ n.m.; our measurement of G_M^Z thus corresponds to a value of

$$G_M^s(Q^2 = 0.1 \text{ GeV}^2) = +0.23 \pm 0.37 \pm 0.15 \pm 0.19 \text{ n.m.}$$

SUMMARY AND OUTLOOK

The SAMPLE experiment is now working and ready to acquire much more data to further determine the strange magnetic moment of the proton. Efforts to reduce the systematic effects associated with beam helicity reversal and improve polarized beam reliability are in progress. In addition, the Bates laboratory will attempt to provide higher polarization beam from a strained GaAs crystal in the near future. After completing more of the full hydrogen data set, we plan to run the experiment with deuterium in the target. The deuterium asymmetry contains the isovector axial radiative correction, but is almost completely insensitive to the strange magnetic moment. Thus, we can reduce our uncertainty in μ_s by using deuterium and proton data together to almost eliminate the radiative correction uncertainty. [10]

Additional experiments to explore other features of neutral weak currents and strange form factors of the nucleon are planned at Mainz (MAMI-B) [11] and TJNAF (formerly known as CEBAF) [12–14] . These experiments provide a new and unique window on the quark structure of the nucleon, and should yield important information towards a more complete understanding of nucleon structure in the context of QCD.

REFERENCES

1. R. D. McKeown, Phys. Lett. **B219**, 140 (1989).
2. D. H Beck, Phys. Rev. **D39**, 3248 (1989).
3. D. Kaplan and A. Manohar, Nucl. Phys. B **310**, 527 (1988).
4. Bates proposal 89-06, R. D. McKeown and D. H. Beck, spokespersons.
5. M. J. Musolf and B. R. Holstein, Phys. Lett. **B242**, 461 (1990).
6. M. J. Musolf, et al.,, Phys. Rep. **239** 1 (1994).
7. E. J. Beise, et al., proceedings of SPIN96 symposium, preprint nucl-ex/9610011.
8. W. Heil, et al. Nucl. Phys. **B327**, 1 (1989).
9. P. A. Souder et al., Phys. Rev. Lett. **65**, 694 (1990).
10. Bates experiment 94-11 (M. Pitt and E. J. Beise, spokespersons).
11. Mainz proposal A4/1-93 94-11 (D. von Harrach, spokesperson).
12. TJNAF experiment E91-017 (D. Beck, spokesperson).
13. TJNAF experiment E91-010 (P. Souder and J. Finn, spokespersons).
14. TJNAF experiment E91-004 (E. J. Beise, spokesperson).
15. B. A. Mueller et al., Phys. Rev. Lett. **78**, 3824 (1997).

Increased Sensitivity to Possible Muonium to Antimuonium Conversion

V. Meyer[1], A. Grossmann[1], K. Jungmann[1], J. Merkel[1], G. zu Putlitz[1], I. Reinhard[1], K. Träger[1], P.V. Schmidt[1], L. Willmann[1], R. Engfer[2], H.P. Wirtz[2], R. Abela[3], W. Bertl[3], D. Renker[3], H.K. Walter[3], V. Karpuchin[4], I. Kisel[4], A. Korenchenko[4], S. Korenchenko[4], N. Kravchuk[4], N. Kuchinsky[4], A. Moiseenko[4], J. Bagaturia[5], D. Mzavia[5], T. Sakhelashvili[5], V.W. Hughes[6]

[1] *Physikalisches Institut, Universität Heidelberg, D-69120 Heidelberg, Germany* [2] *Physik Institut, Universität Zürich, CH-8057 Zürich, Switzerland* [3] *Paul Scherrer Institut, CH-5232 Villigen, Switzerland* [4] *Joint Institute of Nuclear Research, RU-141980 Dubna, Russia* [5] *Tblisi State University, GUS-380086 Tbilisi, Georgia* [6] *Physics Department, Yale University, New Haven CT 06520, USA*

Abstract. A new experimental search for muonium-antimuonium conversion was conducted at the Paul Scherrer Institute, Villigen, Switzerland. The preliminary analysis yielded one event fulfilling all required criteria at an expected background of 1.7(2) events due to accidental coincidences. An upper limit for the conversion probability in 0.1 T magnetic field is extracted as $8 \cdot 10^{-11}$ (00% CL).

The hydrogen like muonium atom ($M = \mu^+ e^-$) consists of two leptons from different generations. The close confinement of the bound state offers excellent opportunities to explore precisely fundamental electron-muon interaction. The dominant part of the binding in this system electromagnetic and can be calculated to very high accuracy in the framework of quantum electromagnetics (QED). Indeed, precision experiments on electromagnetic transitions in muonium have been employed both to verify bound state QED calculations and for determining most accurate values of fundamental constants [1].

Since the effects of all known fundamental forces in muonium are calculable very well, it renders the possibility to search sensitively for yet unknown interactions between both particles. A conversion of muonium into its antiatom

($\overline{M} = \mu^- e^+$) would violate additive lepton family number conservation and is not provided in standard theory. However, muonium-antimuonium conversion appears to be natural in many speculative theories, which try to extend the standard model in order to explain some of its yet not well understood features like parity violation in weak interaction and particle mass spectra. The interaction could be mediated by a doubly charged Higgs boson [2], heavy Majorana neutrinos [3], a neutral scalar [4], e.g. a supersymmetric τ-sneutrino [5] or a dileptonic gauge boson [6].

An experiment had been set up to search for spontaneous muonium-antimuonium conversion at the Paul Scherrer Institute (PSI) in Villigen, Switzerland [7]. It uses the powerful signature developed in an experiment at the Los Alamos Meson Physics Facility (LAMPF) USA, which requires the coincident identification of both constituents of the antiatom in its decay [8].

Muonium atoms were produced by stopping a beam of surface muons in a SiO_2 powder target, where a fraction of them forms muonium by electron capture, some of which diffuse through the target surface with thermal energies into vacuum. Energetic electrons from the decay of the μ^- in the antiatom can be observed in a magnetic spectrometer at 0.1 T magnetic field consisting of five concentric multiwire proportional chambers and a 64 fold segmented hodoscope. The positron in the atomic shell of the antiatom is left behind after the decay with 13.5 eV average kinetic energy. It can be electrostatically accelerated to 8 keV and guided in a magnetic transport system onto a position sensitive microchannel plate detector (MCP). Annihilation radiation can be observed in a 12 fold segmented pure CsI calorimeter surrounding the MCP.

The muonium production was monitored regularly by reversing all electric and magnetic fields of the instrument every five hours for a duration of 20 minutes. Targets had to be replaced twice a week because of observed deterioration of muonium production on a one week time scale. In the course of the experiment $5.7 \cdot 10^{10}$ muonium atoms were observed in the interaction volume for antimuonium decays. There was one event which passed all required criteria, i.e. fell into a 99% confidence interval of each relevant distribution. The expected background due to accidental coincidences is 1.7(2) events.

The preliminary combination of all data recorded in the experiment between 1993 and 1996 [7,9,10] results in an upper limit for the conversion probability in 0.1 T magnetic field of $P_{M\overline{M}} \leq 8 \cdot 10^{-11}$ (90 % CL). For an assumed effective (V-A)x(V-A) type four fermion interaction this corresponds to an upper limit for the coupling constant of $G_{M\overline{M}} \leq 3 \cdot 10^{-3} G_F$ (90 % CL), where G_F is the weak interaction Fermi coupling constant [9].

This new result allows to rule out definitively a certain Z_8 model with more than three particle generations [4] and to set a new lower limit of 2.6 TeV/c^2 on the mass of a dileptonic gauge boson in GUT models which is well beyond the value extracted from high energy Bhabha scattering [6]. It can be further shown in the framework of minimal left right symmetric and supersymmetric

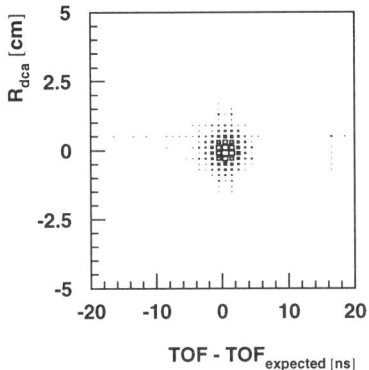 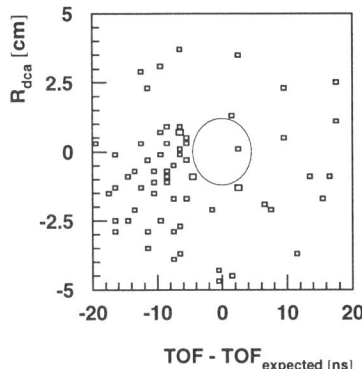

FIGURE 1. The distribution of the distance of closest approach (R_{dca}) between a track from an energetic particle in the magnetic spectrometer and the back projection of the position on the MCP detector versus the time of flight (TOF) of the atomic shell particle for a muonium measurement (left) and for all data recorded in 1996 while searching for antimuonium (right). One single event falls within 3 standard deviations region of the expected TOF and R_{dca} which is indicated by the ellipse. The events concentrated at early times and low R_{dca} correspond to a background signal from the allowed decay $\mu \to 3e + 2\nu$.

models [11] that lepton number violating muon decay ($\mu^+ \to e^+ + \nu_\mu + \overline{\nu}_e$) is not an option for explaining the excess neutrino counts in the LSND neutrino experiment at Los Alamos [12].

This work was supported by the Bundesminister für Bildung und Forschung (BMBF) of Germany, the Schweizer Nationalfond, the Russian Federation for Fundamental Research and a NATO research grant.

REFERENCES

1. Hughes V W and zu Putlitz G, in: *Quantum Electrodynamics*, Singapore, World Scientific Kinoshita T (ed.) p. 822 (1990); Jungmann K, in: *Atomic Physics 14*, New York, AIP Press, Wineland D et al. (ed.) p. 102 (1994)
2. Herczeg P and Mohapatra R N, Phys.Rev.Lett. **69** 2475 (1992)
3. Halprin A, Phys.Rev.Lett. **48** 1313 (1982)
4. Hou W S and Wong G G, Phys.Rev. D**53** 1537 (1996)
5. Mohapatra R N, Z.Phys. C**56** 117 (1992); Halprin A and Masiero A, Phys.Rev.D**48**, 2987 (1993)
6. Horrikawa K and Sasaki K, Phys.Rev. D**53** 560 (1994)
7. Abela R et al., Phys.Rev.Lett. **77** 1951 (1996)
8. Matthias B et al., Phys.Rev.Lett. **66** 2716 (1991)
9. Schmidt P, doctoral dissertation, University of Heidelberg, unpublished (1997)
10. Wirtz H P, doctoral dissertation, University of Zürich, unpublished (1997)
11. Herczeg P, conference "Beyond the Dessert", Castle Ringberg, Germany (1997)
12. Athanassopoulos C et al. Phys.Rev. C**54** 2685 (1996)

An Update on Cosmological Anisotropy in Electromagnetic Propagation

John P. Ralston* and Borge Nodland[†]

*Department of Physics and Astronomy
and Institute for Theoretical and Computational Science
University of Kansas, Lawrence, Kansas 66044
[†]Department of Physics and Astronomy
and Rochester Theory Center for Optical Science and Engineering
University of Rochester, Rochester, New York 14627

Abstract. We review evidence for a new phenomenon in the propagation of radio waves across the Universe, an anisotropic rotation of the plane of polarization not accounted for by conventional physics.

Radio waves propagating on cosmological distances provide an exceedingly sensitive laboratory to explore new phenomena. Recently [1], we reported an indication of anisotropy extracted from polarization measurements taken on distant radio galaxies. Here we review that work, subsequent reactions, and some new studies.

Galaxies monitored in the 100 MHz–GHz range are found to emit linearly polarized radio waves. The emission mechanism is believed to be synchrotron radiation. The observed plane of polarization does not usually align with the symmetry axis of the source (whose orientation angle is denoted ψ) [2]. For decades, this has been studied in terms of Faraday rotation in the intervening medium. Faraday rotation can be taken out in a model–independent way, because the Faraday angle of rotation $\theta_i(\lambda)$ for wavelength λ goes like RMλ^2, where RM is a constant depending on the integrated plasma parameters and magnetic field along the line of sight. Consistent linear dependence on λ^2 is indeed observed. However, the data fit requires more: for each source (i), the fits are given by $\theta_i(\lambda) = \text{RM}_i \lambda^2 + \chi_i$. The de–rotated polarization (whose orientation angle is χ) does not generally align with the galaxy major axis, nor with the axis at 90 degrees to the major axis; statistics on the offset of angles have puzzled astronomers for 30 years [2].

Analysis of the data is challenging. The data set is small, 160 sources. The angular distribution on the sky is highly non–uniform, as is the distribution in the distances $r(z)$ to the sources. Finally, for decades astronomers have used a particular arbitrary difference, $\chi - \psi$, as a reference variable. There is an underlying conceptual issue, that both the galaxy axis and the polarizations are projective variables that return to themselves after a rotation of π (not 2π). Straightforward use of $\chi - \psi$, binned between arbitrary intervals and defined mod π, will obscure correlations that may depend on the direction of travel of a wave; this is a crucial requirement if one is going to test for anisotropy. It also does not allow an analysis to keep the important distinction between an obtuse angle and the complementary acute one. To keep track of the difference between obtuse and acute angles, and to allow a test for directional dependency, we created [1] the angles β^+ and β^-, defined by

$$\beta^+ = \begin{cases} \chi - \psi & \text{if } \chi - \psi \geq 0 \\ \chi - \psi + \pi & \text{if } \chi - \psi < 0 \end{cases} \qquad \beta^- = \begin{cases} \chi - \psi - \pi & \text{if } \chi - \psi \geq 0 \\ \chi - \psi & \text{if } \chi - \psi < 0. \end{cases} \quad (1)$$

A test for directional dependency is allowed by defining the rotation angle β as $\beta = \beta^+$ when $\cos\gamma > 0$ and $\beta = \beta^-$ when $\cos\gamma < 0$, where γ is the angle between a spatial direction $\vec{s}$ and the propagating wavevector of the wave. We employed Monte Carlo methods to search for correlations [1]. We made thousands of fake galaxies with random χ and ψ, while keeping their positions on the dome of the sky the same as those of the real galaxies. As we varied $\vec{s}$ systematically across the dome of the sky, we calculated (for each $\vec{s}$) the probabilistic P-value of linear correlations in the observed data relative to the random sets (Procedure 1). In the full data set, we found a correlation described by $\beta = (r/\Lambda_s) \cos\gamma$. This indicates anisotropy. Making a cut on $z > 0.3$, which selects the most distant half of the data set, we found a striking correlation, with probabilistic P-values that the observed correlation would be produced by random angular fluctuations to be less than 10^{-3}; the effect is 3.7σ.

A separate study (Procedure 2) eliminated possible bias in Procedure 1's determination of $\vec{s}$. Procedure 2 determined the $\vec{s}$–direction that yielded the highest correlation for a specific random data set; the resulting "best" correlations were then accumulated for over a thousand different random sets. We then calculated the probabilistic P-value of finding the correlation of the observed data set relative to the best correlated random sets so constructed. This gave a P-value less than 0.006, corresponding to 2.7σ. The fits to the parameters Λ_s and $\vec{s}$ are $\Lambda_s = (1.1 \pm 0.08) 10^{25} \frac{h_0}{h}$m and $\vec{s} = (\text{Decl.}, \text{R.A.})^*_s = (0° \pm 20°, 21\text{hrs} \pm 2\text{hrs})$, where $h_0 = \frac{2}{3} 10^{-10} (\text{years})^{-1}$ and h is the Hubble constant. We do not find a significant correlation for $z < 0.3$; in our full data set we also do not find a significant correlation of $\beta = (\text{const}) r$.

Other searches have been conducted looking for systematic rotation depending only on distance $r(z)$. For example, Carroll, Field and Jackiw (CFJ) [3] looked at the same data set as we did. However, CFJ used $\chi - \psi$, which mixes up obtuse and acute angles, and doesn't allow for straightforward correlation analysis to be done. Moreover, one could easily question any systematic correlation with distance as possibly due to the evolution of a population: there was a "two–population" hypothesis 30 years ago [2], surmising that nearby sources emitted waves of polarization parallel to the sources' major axes, and that distant sources emitted waves of polarization perpendicular to the sources' major axes. We looked for an anisotropic correlation on the sky firstly because we had a theory that predicted it [1], and secondly because no population hypothesis could reasonably explain a signal if seen.

We have concentrated mainly on data analysis, because establishing that the effect is statistically significant is the first step. As for theoretical questions, in Ref. [1] we noted a gauge invariant modification of electrodynamics which has the same number of parameters as needed to explain the data, and is compatible with precision measurements of the Standard Model. Another possible explanation might be a domain wall from an axion–like particle condensate. A third explanation might be the twisting of polarization predicted by Brans [4], and Matzner and Tolman [5], from parallel transport in ordinary general relativity in an anisotropic cosmology. Several alternate cosmologies or theories of gravity [5] have also come to our attention, with claims that the anisotropic effect we saw would (or, in some cases, was) predicted.

Not unexpectedly, after publication of Ref. [1] there has been considerable controversy. Early charges that the data was "old and incomplete" were found to be spurious, as we have been vigorously reassured that the data set we used is the most complete one available [6]. The statistics have come under attack, but we have seen nothing valid to alter our conclusions. Eisenstein and Bunn (EB) [7] quickly criticized the work on the basis of a single scatter plot of β versus $r \cos \gamma$ included in Ref. [1]. EB observed that the data for $z > 0.3$ was not uniformly distributed along the β–axis. Without making any calculations, but "estimating by eye" as EB put it, the observed data set for $z > 0.3$ was claimed not to be better correlated than data with β-values randomly distributed about 90 degrees. EB conjectured that the anisotropic correlation would go away if one compared the observed data to data with random shufflings of the observed β-values, rather than comparing (as we did, and in fact, only in Procedure 1) the observed data to data with β-values obtained from uniform random distributions of χ and ψ. (The distributions of the observed χ and ψ are, in fact, uniform in the full data set).

There are several problems with EB's criticism, explained in more detail in our response to EB in Ref. [7]: (1) Procedure 2, the more demanding, was ignored; (2) The scatterplot showed data after a cut; this, combined with the fact that the data was linearly correlated, predicts the distribution in β to be as observed; (3) Since shuffled data from such a limited, cut region is pre-correlated, comparison of this data with the observed data may underestimate the statistical significance of a real correlation in the observed data; (4) Since eyeball estimates can be deceiving, we actually did the "EB calculation:" we found that our correlation survived, both in identifying the same anisotropy axis as found in Ref. [1], and in being statistically significant, with a P-value less than 2×10^{-2}. This was also independently confirmed numerically by P. Jain, who also reproduced our original numerical results in Ref. [1]. [We note that a more appropriate way to do shuffling is to shuffle the observed pairs (ψ, χ), not the β's; this resulted in an even stronger signal. Furthermore, we shuffled only the β's and (ψ, χ) pairs from the $z \geq 0.3$ set, although it is more reasonable to use the total set for this purpose].

Carroll and Field (CF) recently released a preprint [8], based on the "two–population hypothesis." Restricting themselves to $z > 0.3$, CF confirmed some of our calculations, and also restated their previous calculations (in Ref. [3]) using the variable $\chi - \psi$. With this variable, they found no indication of a correlation nor anisotropy. As explained above, the variable is unsuitable (as also pointed out in our response to Leahy in Ref. [9]), and the procedure used by CF will generally miss an anisotropic signal even in perfectly correlated data sets. Unfortunately, CF have not made a clear distinction between β and $\chi - \psi$ in their presentation, leading some readers to think that their calculation applies to our work, which used the direction-of-travel-dependent β. The same problem of not using β occurs in Leahy, Ref. [9].

Other reactions [9,10] have come from radio astronomers, who so far have not addressed the anisotropic correlations on the sky, which was the basic nature of our result in Ref. [1]. Instead, these studies use high resolution VLBI data examining special internal structures of selected objects, such as jets in the detailed radio maps of certain nice–looking quasars. There are some problems with the claims: (1) their data seems to be highly selected, with much smaller sets than we used, and ignoring parts of the quasars which are not "pretty;" (2) some of the logic appears to be circular, depending strongly on theoretical ideas engineered to understand the same data before birefringence was considered, (3) the methodology lacks statistical criteria If these problems are ignored, the claims state that no polarization rotation of the size we found is seen in the VLBI data. One actually goes beyond the scope of Ref. [1] when one compares the VLBI data and our data in this way: we reported an anisotropic correlation in our data, along with a thorough statistical characterization of its likelihood, and also pointed out the possibility that it was caused by systematic bias in our data.

Naturally, we believe that the VLBI data contains great potential for information, and should be studied systematically on its own basis. By this, we mean that the VLBI data should be subjected to the same kind of statistical Monte Carlo tests for anisotropy as the tests we applied to our data. One cannot deduce much from any particular polarization rotation found in one or a few galaxies. Much more relevant would be statistical significance of a possible anisotropic signal, as quantifed by the P–values obtained from analyses like ours.

It is also interesting to note that the frequency of the VLBI data is much higher than the frequency of the data we studied, so perhaps the studies in Refs. [9,10] really are looking at different phenomena. For example, we would think that the Brans mechanism [4] (if it exists) is independent of frequency. Regarding other hypotheses, there will be other predictions. For example, what is known generally about the frequency dependence of domain walls? It would seem to be model dependent to compare the different types of data in Refs. [9,10] and Ref. [1] prematurely.

Several recent articles report interesting progress. Obhukov et al. [11] claim that the correlation we observed could be caused by global rotation. Kühne [12], and Bracewell and Eshleman [13] have independently observed that the anisotropy axis extracted in Ref. [1] coincides tolerably well with the direction of the cosmic microwave background (CMB) dipole axis. This seems to be a strong clue, although our calculations on kinematic Doppler effects on Faraday rotation do not yield any mechanism to suggest a connection. It thus seems possible that some non–kinematic, grand medium effect may be at work, but we don't know what.

Finally, some critics seemed to be uncomfortable with the fact that our data for β versus $r \cos \gamma$ occupied both the first and third quadrants. We have taken this to heart by plotting $\beta \cos \gamma$ versus $r \cos^2 \gamma$ instead, which brings all the data into the first quadrant. The Jacobian of this transformation is a power of $\cos \gamma$, scaling both of the variables of interest by the same function. The plot is shown in Fig. 1, for the data of Fig. 1(d) in Ref. [1]. It must be noted that "eyeballing" scatter plots is dangerous; the plot in these variables happen to look nice, but one should rely on quantitative statistical measures. Yet the figure may make it more clear to the eye that there is a definite signal of anisotropy. The next question, yet unanswered, is: why?

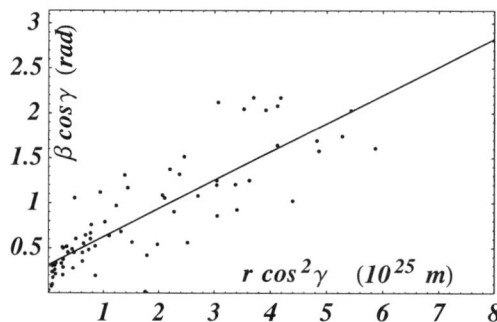

FIGURE 1. Plot of $\beta \cos \gamma$ versus $r \cos^2 \gamma$, for the data of Fig. 1(d) in Ref. [1]. This plot is equivalent to that of Ref. [1], but places all data points in the first quadrant. A definite correlation can be seen.

REFERENCES

1. B. Nodland and J. P. Ralston, *Phys. Rev. Lett.* **78**, 3043 (1997); preprint astro-ph/9704196.
2. J. N. Clarke *et al.*, *Mon. Not. R. Astron. Soc.* **190**, 205 (1980); F. F. Gardner and J. B. Whiteoak, *Nature* **197**, 1162 (1963).
3. S. M. Carroll, G. B. Field and R. Jackiw, *Phys. Rev.* D **41**, 1231 (1990).
4. C. H. Brans, *Astrophys. J.* **197**, 1 (1975).
5. R. A. Matzner and B. W. Tolman, *Phys. Rev. D* **26**, 2951 (1982); M. Sachs, *General Relativity and Matter* (Reidel, 1982); R. B. Mann and J. W. Moffat, *Can. J. Phys.* **59** 1730 (1981); C. M. Will, *Phys. Rev. Lett.* **62**, 369 (1989); R. B. Mann, J. W. Moffat, and J. H. Palmer, *ibid* **62**, 2765 (1989); J. W. Moffat, preprint astro-ph/9704300 (1997); D. V. Ahluwalia and T. Goldman, *Mod. Phys. Lett.* **A28**, 2623 (1993); Y. N. Obukhov, V. A. Korotky and F. W. Hehl, preprint astro-ph/9705243 (1997); A. Dobado and A. L. Maroto, preprint astro-ph/9706044 (1997).
6. P. Kronberg, private communication.
7. D. J. Eisenstein and E. F. Bunn, preprint astro-ph/9704247 (1997); B. Nodland and J. P. Ralston, preprint astro-ph/9705190 (1997).
8. S. M. Carroll and G. B. Field, preprint astro-ph/9704263 (1997).
9. J. P. Leahy, preprint astro-ph/9704285 (1997); B. Nodland and J. P. Ralston, preprint astro-ph/9706126 (1997).
10. J. F. C. Wardle, R. A. Perley, and M. H. Cohen, preprint astro-ph/9705142 (1997).
11. Y. N. Obukhov, V. A. Korotky and F. W. Hehl, preprint astro-ph/9705243 (1997).
12. R. W. Kühne, private communication and preprint submitted to *Phys. Rev. Lett.* (1997).
13. R. N. Bracewell and V. R. Eshleman, preprint aps1997jun13_006 (1997).

Tests of Fundamental Symmetries

S.A. Page[1] and B.R. Holstein[2]

(1) Dept. of Physics & Astronomy, Univ. of Manitoba, Winnipeg, MB, Canada R3T 2N2
(2) Dept. of Physics & Astronomy, Univ. of Massachusetts, Amherst, MA 01003-4525

Abstract. The sessions on Tests of Fundamental Symmetries at CIPANP97 spanned a wide range of topics. Experiments in this field are extremely challenging, often spanning a decade or more from proposal to completion; highlights included first reports of several new and preliminary results of high precision measurements probing symmetry violations in the strong and weak interactions.

PARITY VIOLATION

Parity violation is a unique feature of the weak interaction. First results of two high precision measurements of parity violation were reported in these sessions, yielding complementary information about the proton structure and the hadronic weak interaction at intermediate energy.

W.D. Ramsay[1] reported the first preliminary results of a proton-proton parity violation experiment at TRIUMF (E497). This experiment is designed to measure the helicity dependence of p-p scattering at 221 MeV. Also known as the longitudinal analyzing power, A_z, the parity violating effect is predicted to be of order 10^{-7} at this beam energy, which is carefully chosen to isolate a single parity-mixed partial wave contribution ($^3P_2$-$^1D_2$). This latter feature makes the interpretation of the experiment virtually model independent, since the selectivity arises from a strong interaction effect which is well known based on N-N scattering data. In the meson exchange model of the weak nucleon-nucleon interaction, this partial wave contribution is sensitive only to ρ meson exchange, and hence will determine for the first time an experimental value of the weak meson nucleon coupling constant h_ρ^{pp}. The TRIUMF experiment is performed in transmission mode, with a 200 nA beam at 80% polarization incident on a 40 cm liquid hydrogen target. Systematic errors can arise from helicity-correlated beam property modulations, such as position, size, current, transverse polarization, and energy. After many years of intensive effort involving the TRIUMF polarized ion source group and the parity collaboration, helicity correlated beam properties have been reduced to an extremely low level. A recent data taking run resulted in a preliminary measurement of $A_z = (1.1 \pm 0.4 \pm 0.4) \times 10^{-7}$, in agreement with the meson exchange predictions. All corrections to A_z that are directly measureable with the parity diagnostic apparatus are consistent with zero for this data set. Data taking will continue during 1997 and 1998 with the goal of reaching a combined statistical and systematic error of $\pm 0.2 \times 10^{-7}$.

R.D. McKeown[2] reported results of the SAMPLE experiment underway at MIT-Bates Linear Accelerator Center to measure parity violation in elastic electron scattering from the proton. This experiment makes use of the very successful Standard Model formalism for weak interactions to use parity violation as a probe of the proton structure. The parity violating asymmetry measured by comparing the elastic scattering cross section for opposite helicity states of the electron beam can be expressed as a product of electromagnetic and analogous weak form factors for the proton. At backward angles, and knowing the electromagnetic form factors independently, the weak magnetic form factor G_M^Z can be deduced from the measurements. G_M^Z in turn contains a contribution from strange quark-antiquark pairs; the SAMPLE experiment can therefore shed light on the strange quark contribution to the proton's magnetic moment (G_M^s). Using a 40 μA 200 MeV electron beam with typically 35% polarization incident on a 40 cm liquid hydrogen target, the SAMPLE apparatus detects backward scattered electrons by the Cerenkov light emitted in air and collected with an array of 10 large mirrors focussed onto large diameter photomultiplier tubes. Helicity correlated beam intensity modulations are controlled with a feedback system; corrections to the data amount to less than 1 ppm. The experimenters have reported a parity violating asymmetry: $A = -6.34 \pm 1.45 \pm 0.53$ ppm. This in turn implies a strange quark contribution to the proton magnetic moment of $G_M^S(Q^2 = 0.1 \ GeV^2) = +.23 \pm 0.37 \pm 0.15 \pm 0.19$ nm. The final goal of the experiment is to reduce the error on G_M^S to about ± 0.2 nm.

CP, T AND CPT TESTS

M. Mikuz[3] reported several new preliminary results on CP, T and CPT tests performed by the CPLEAR collaboration. The CPLEAR experiment has collected a vast amount of data on tagged K° and $\bar{K}^\circ$ decays produced in $p\bar{p}$ annihilation at LEAR; by tagging the opposite strangeness states at production time, the experiment achieves maximum sensitivity to symmetry violating interference effects in the subsequent decays. With a symmetric detection apparatus, time dependent asymmetries are measured by comparing decay rates of K° and $\bar{K}^\circ$ as a function of time and decay channel. Analysis is proceeding to extract in principle a complete set of parameters describing

CP, T and CPT violation in the neutral kaon system. At this conference, a number of experimental tests were reported, based on a large fraction of the total available data. These include a preliminary result for $\Delta m = m_L - m_S$ which is the most precise individual measurement of this quantity to date; a measurement of the T-violating asymmetry in semileptonic decays yielding the first experimental observation of direct T violation with greater than 3σ confidence level, as well as a new limit on the CPT violating mass difference $|m_{\bar{K}^0} - m_{K^0}| \leq 4 \times 10^{-19}$ GeV at 90% confidence level.

S. Gardner[4] reported results of a theoretical investigation of direct CP violation in hadronic B-decays. By focussing specifically on the decay modes $B^\pm \to \rho^\pm \pi^+ \pi^-$ in the region of phase space sensitive to $\rho - \omega$ interference, the strong phase is dominated by isospin violation and can be determined from e^+e^- data. A substantial CP-violating asymmetry of 20% is predicted, which, if verified experimentally, would permit the CKM parameter $\sin(\alpha)$ to be extracted from a study of this decay mode.

B. Corbin[5] reported preliminary results from the APEX experiment, setting new lower limits on the antiproton lifetime deduced in a search for two-body decay modes at the Fermilab Antiproton Accumulator. Analysis of a large data sample taken in 1995 is ongoing, with the most stringent preliminary limit coming from the direct search for $\bar{p} \to e^- \gamma$ of $> 1.3 \times 10^6$ years at 90% confidence level. While it will not be possible to obtain lifetime limits anywhere near as accurately as those for the proton, nonetheless the limit on CPT conservation from proton-antiproton decays will be improved significantly when the APEX results are finalized.

Two new experiments making use of polarized cold neutrons as a testing ground to search for T violation were also discussed in these sessions. The emiT experiment[6], currently underway at NIST, aims to improve the measurement of the time reversal violating D-parameter in neutron beta decay, sensitive to the P-even, T-odd triple correlation $\sigma_n \cdot (\vec{p}_e \times \vec{p}_p)$. The existing data for this observable average to $D = (0.3 \pm 1.5) \times 10^{-3}$, while final state interactions are expected to contribute only at the level of 5×10^{-5}. The protons and electrons from decays of a $10^8 s^{-1}$ beam of 96% longitudinally polarized neutrons from a liquid hydrogen cold source are detected in an array of detectors with fourfold symmetry. A first test run was devoted to beamline development; the experiment is running this summer.

An even more ambitious goal, to improve the current world limit on the P-odd, T-odd neutron electric dipole moment (d_n) by two orders of magnitude, was discussed by G.L. Greene[7]. The current upper limit is $d_n < 1.1 \times 10^{-25}$ e-cm; while the Standard Model prediction of $d_n < 10^{-31}$ e-cm is beyond experimental reach, extensions which account for the universal baryon asymmetry, as well as supersymmetry, predict $d_n > 10^{-27}$ e-cm, motivating the discussion of a new proposal to be considered at LANL. A novel method to produce a 'superthermal' source of ultracold neutrons is a key feature of the proposal, making use of the limited quantum states available for energy transfer between cold neutrons and superfluid ^{4}He to greatly enhance the density of ultracold neutrons compared to a thermal source at the same temperature. The concept for the experiment includes the creation of the high density superthermal cold neutron source and the measurement of d_n in the same volume of ^{4}He, in which a mixture of spin polarized ^{3}He would act as both a magnetic field and neutron polarization monitor; a new high intensity neutron spallation source at LANSCE also forms an essential part of the proposal.

CHIRAL SYMMETRY

Threshold pion photoproduction reactions are an ideal testing ground for predictions of chiral perturbation theory. E. Korkmaz[8] reviewed the threshold pion photoproduction program at the Saskatchewan Accelerator Laboratory (SAL). The program consists of measurements of the total cross section and angular distributions for the reactions: 1) $\gamma p \to p\pi°$, 2) $\gamma p \to n\pi^+$, and 3) $\gamma n \to n\pi°$, very close to threshold. Predictions of chiral perturbation theory (ChPT) as compared to the older Low Energy Theorems (LET) based on PCAC and current algebra are tested with experimental results. For reactions 1) and 3) the LET and ChPT predictions are markedly different for the threshold amplitudes, whereas they differ only by a few percent in case 2). SAL data for case 1) have been published, with results confirming ChPT predictions; preliminary data for case 2) were shown at the conference, with the energy dependence and overall scale of the cross section in agreement with ChPT. The $\gamma n \to n\pi°$ reaction is being studied with a deuterium target; in the first phase, the near threshold cross section for $\gamma d \to \pi° d$ has been measured, and preliminary analysis finds a result considerably smaller than the ChPT prediction. In the second phase, the reaction $d(\gamma, \pi° n)p$ will be studied in quasifree kinematics.

STANDARD MODEL TESTS

High precision measurements of nuclear beta decay and other weak interaction processes continue to challenge many fundamental assumptions of the Standard Model. At this conference, several new or preliminary experimental results were reported.

B. Fujikawa[9] reviewed the status of superallowed $0^+ \to 0^+$ beta decay tests and reported new results from a measurement of ^{10}C decay using the Gammasphere detector at LBNL. Precision Ft measurements now exist for daughter nuclei from Z=5 to Z=27. At issue has been the Z-dependence of the Ft values, which are sensitive to radiative and isospin breaking corrections; prior to the Gammasphere measurement on ^{10}C, an earlier measurement of the same decay hinted at a linear dependence of Ft that increased with Z. The new result reported here is approximately 2 σ larger and tends to favour a constant Ft for the superallowed beta decays. From the new world average value, a unitarity test of the first row of the CKM matrix can be performed, with the result that $|V_{ud}|^2 + |V_{us}|^2 + |V_{ub}|^2 = 0.9975 \pm 0.0019$, essentially consistent with the Standard Model.

P. Schuurmans[10] reported results of a new measurement of the longitudinal polarization of positrons from ^{107}In decay carried out at Louvain-la-Neuve. The experiment compared the polarization of positrons emitted by polarized and unpolarized ^{107}In nuclei, which provides a particularly sensitive test for right handed gauge bosons. The e^+ polarization is deduced from a novel technique which makes use of the dependence of the positronium lifetime in a magnetic field on the polarization of the positrons. The preliminary result presented here represents a factor of three improvement over an earlier measurement carried out by the collaboration, and together with all previously reported data places a lower limit of 316 GeV/c^2 on the mass of a possible charged W_R boson; the sensitivity of this measurement is significantly greater than collider searches for W_R boson below a mass limit of about 100 GeV/c^2.

M. Beck[11] reported preliminary results of a new measurement of $\beta - \alpha$ angular correlations in the A=8 isobaric triplet. By comparing the correlation coefficients for the decays ^{8}B $\to^8$Be + β^+ + ν_e $\to^4$He + ^{4}He and ^{8}Li $\to^8$Be + β^- + $\bar{\nu}_e$ $\to^4$He + ^{4}He as a function of E_β, a sensitive test of CVC can be performed. The measurements are ongoing at the University of Washington; previous limits on CVC conservation are at the level of 6%, and it is hoped that this experiment will be able to make a significant improvement at this level once completed. Preliminary data are consistent with previous measurements in the mass 8 system.

G.D. Cates[12] discussed preliminary results of a first measurement of the spin dependence of the reaction $\mu^- + ^3$He $\to^3$H + ν_μ, performed at TRIUMF. The capture rate for this reaction is dependent on the muon polarization and the direction of the triton recoil with respect to the μ^- polarization direction. Ultimately, a precision measurement of this type could be used to constrain the pseudoscalar form factor g_p, a fundamental quantity which is difficult to access experimentally. The experiment made used of highly polarized muonic ^{3}He, obtained by optical pumping in a high pressure cell containing ^{3}He gas and polarized Rb vapour. The triton asymmetry was determined by pulse shape analysis in a gridded ion chamber which was located inside the gas cell. This enabled the vector analyzing power for the reaction to be deduced for the first time from the ratio of the triton asymmetry to the muon polarization; a preliminary result at the ±20% level, consistent with PCAC, was deduced.

V. Meyer[13] discussed preliminary results of an ongoing muonium-antimuonium conversion search underway at PSI. The process $\mu^+e^- \to \mu^-e^+$, if observed, would violate lepton number conservation. A very sensitive apparatus has been developed, capable of identifying cleanly muonium (M) formation by tagging both the decay products in a spectrometer. By reversing the polarity of the spectrometer, searches for antimuonium ($\bar{M}$) decay can be made. Results are quoted in terms of a lepton flavour violating coupling constant $G_{M,\bar{M}}$ as a fraction of the Fermi coupling constant G_F. A large data sample has been collected in 1993, 1995 and 1996. Preliminary analysis of the combined data sets places a limit $G_{M,\bar{M}} < 10^{-3} G_F$ at 90% confidence level, improving previously reported limits by almost an order of magnitude.

COSMOLOGICAL PUZZLE?

The sessions concluded with a presentation by J. Ralston[14] summarizing results of a new statistical analysis of Faraday Rotation measurements on light emitted from distant radio galaxies. Ralston and his collaborator found a systematic anisotropy in the rotation of the plane of polarization of light about a preferred axis in space, on a cosmological distance scale, based on a sample of 160 galaxies. Monte Carlo simulations were used to test the analysis procedures, and confirmed the finding of a statistically significant effect. If confirmed by additional data, which could become available soon, this analysis could indicate an important and very interesting new cosmological effect.

REFERENCES

1. W.D. Ramsay et al., and A.N. Zelenski et al., these proceedings.
2. R.D. McKeown et al., these proceedings
3. R. Adler et al., these proceedings
4. S. Gardner et al., these proceedings
5. B. Corbin et al., these proceedings
6. L.J. Lising et al., these proceedings
7. G.L. Greene et al., these proceedings
8. E. Korkmaz, these proceedings
9. B. Fujikawa et al., these proceedings
10. P. Schuurmans et al., these proceedings
11. M. Beck et al., these proceedings
12. G.D. Cates, these proceedings
13. V. Meyer et al., these proceedings
14. J. Ralston et al., these proceedings

Hadron Spectroscopy

Interplay Between the $f_0(980)$ and $a_0(980)$ Mesons

J. Gunter

Dept. of Physics, Indiana University, Bloomington, IN 47404

Abstract. Many models attempt to classify the $a_0(980)$ and $f_0(980)$ as $qq\bar{q}\bar{q}$ or $K\bar{K}$ states. One of the predictions of these models is the mixing of these states. An interpretation of existing data from experiment E852 at Brookhaven National Lab as consistent with a_0/f_0 mixing at the few percent level is given.

The $f_0(980)$ and $a_0(980)$ mesons have a long and controversial history. Originally assigned to the $^3P_0$ $q\bar{q}$ scalar nonet, these states have masses too low and widths too narrow for this assignment in most models. Moreover, the near mass degeneracy of these states is not explained by $q\bar{q}$ models. The proximity of these states to $K\bar{K}$ threshold is a fact which only adds to the confusion by making precise experimental determinations of mass, width, and branching ratios difficult. In addition, the $f_0(980)$ appears differently in different reactions. This is chiefly due to the presence of other nearby f_0 resonances ($f_0(400-1200)$ and/or $f_0(1300)$). For example, in the reaction $\pi^-p \to \pi\pi n$ [1-3] the $f_0(980)$ appears as a dip in the cross-section due to destructive interference with the nearby state(s), while in $pp \to pp\pi\pi$ (central production) [4] the $f_0(980)$ appears as a shoulder, and in $\bar{p}p \to \eta\pi\pi$ [5] as well as J/ψ decay [6] it appears as a small peak.

Nearly as varied as the way the $f_0(980)$ appears experimentally are the theoretical interpretations as to the nature of the $f_0(980)$ and $a_0(980)$. The $a_0(980)$ has been interpreted as the ground state scalar $q\bar{q}$ system, a $qq\bar{q}\bar{q}$ system, a $K\bar{K}$-molecular system, and a dynamically generated enhancement related to $K\bar{K}$ threshold with no underlying fundamental state [7]. The $f_0(980)$ has been interpreted as as ground state $q\bar{q}$ system, a $qq\bar{q}\bar{q}$, a $K\bar{K}$-molecular system, and even as the scalar glueball. A large $K\bar{K}$ molecular component is expected to influence, for example, the branching ratios of ϕ radiative decay into $a_0(980)\gamma$ and $f_0(980)\gamma$ [8,9]. Another prediction of $K\bar{K}$ molecular models [10,11] is the existence of mixing between the states via their common $K\bar{K}$ modes. This mixing may be as large as $\sqrt{\alpha} \simeq 9\,percent$ [10]. As Barnes [11] noted, the observation of $a_0(980)$ production in $\pi^-p \to a_0(980)n$ with a momentum transfer

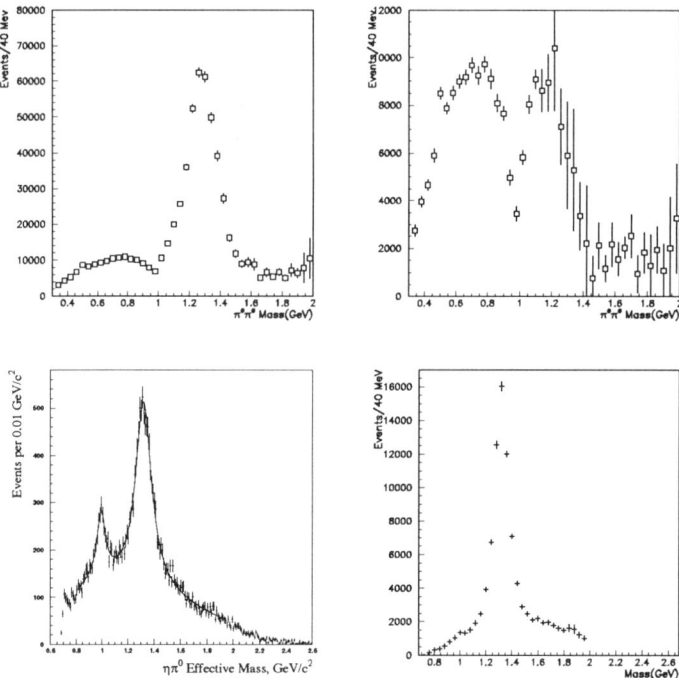

FIGURE 1. The $\pi^0\pi^0$ effective mass spectrum and the scalar component of the $\pi^0\pi^0$ system (as determined by partial wave analysis) are shown in the top left and right plots. The lower left (right) plot is the effective mass spectrum from $\eta\pi^0$ ($\eta\pi^-$) events.

(t) distribution characteristic of a one pion exchange production mechanism would be evidence of mixing.

Experiment E852 at Brookhaven National Lab has taken data using an 18.3 GeV/c negative pion beam. Three of the many reactions under analysis are $\pi^-p \to \pi^0\pi^0 n$ [3], $\pi^-p \to \eta\pi^0 n$ [12], and $\pi^-p \to \eta\pi^- p$ [13]. Figure 1 shows the acceptance-corrected mass spectra for these reactions. The $\pi^0\pi^0$ spectrum is dominated by the $f_2(1270)$. Partial wave analysis reveals a sharp dip in the $J = 0$ partial wave near $1.0\,GeV/c^2$ which is due to the $f_0(980)$ interfering with at least one other broad scalar state. The $\eta\pi^0$ system is dominated by the neutral $a_2(1320)$ and $a_0(980)$ mesons. In the $\eta\pi^-$ system the $a_2(1320)$ is again observed, but the $a_0(980)$ is practically absent.

Production of the $f_2(1270)$ meson is dominated by one pion exchange. In Fig. 2 the $|t|$-distributions for $f_2(1270)$ production and neutral $a_0(980)$ production are overlayed. Both vertical scales are arbitrary but the overall shapes are similar. This would be the case if the production mechanisms were the same. One pion exchange, however, does not have the correct quanumt num-

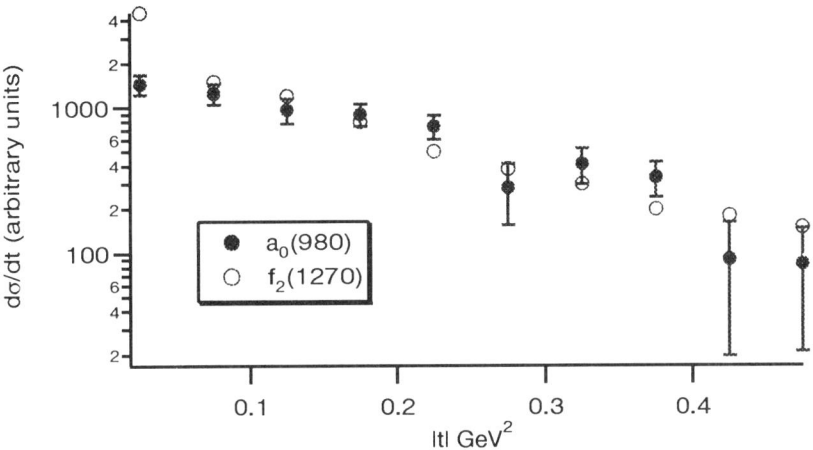

FIGURE 2. Shown are momentum transfer |t|-distributions for the production of $f_2(1270)$ and $a_0(980)$ mesons in pion induced reactions. The overall normalizations are arbitrary.

bers to produce the a_0 directly. The lightest exchange particle with quantum numbers capable of producing the $a_0(980)$ from in a pion induced reaction is the $b_1(1235)$ which leads to the expectation of a shallower t-distribution. It could be possible, however, to create the $a_0(980)$ by mixing $f_0(980)$ events produced by one pion exchange. Additionally, the production, via mixing, of $a_0(980)$ mesons from $f_0(980)$ mesons has no charged analog since there is no charged $f_0(980)$. This would explain the relatively small charged $a_0(980)$ signal as observed.

The $\eta\pi^0$ and $\pi^0\pi^0$ data have come from similar analyses of the same experimental data set. Thus, numerical comparison of the two spectra can be expected to be valid. Published data on cross-sections agree with the observed numbers of $a_2(1320)$ and $f_2(1270)$ mesons. With some assumptions, it is possible to make an estimate of the mixing probability for a produced $f_0(980)$ meson to mix into an $a_0(980)$ meson. Assumptions made are that the strength of the $f_0(980)$ resonance may be taken from the depth of the dip near $1.0\,GeV/c^2$ relative to the broad bump at lower mass in the $\pi^0\pi^0$ spectrum; that both states have the same widths; that the couplings for the $a_0(980)$ and the $f_0(980)$ to the unobserved $K\overline{K}$ channel are the identical; and that all of the observed neutral $a_0(980)$ mesons are created via to mixing. Given these assumptions the calculation of the mixing probability is essentially reduced to comparing plots in Fig 1. The depth of the dip in the scalar $\pi^0\pi^0$ spectrum multiplied by a factor of three (since only one-third of $f_0 \to \pi\pi$ decays are observed in the $\pi^0\pi^0$ final state) is compared to the height above background of the neutral a_0 peak scaled by a factor four to reflect the finer mass binning in the a_0 plot. This comparison leads to an estimated mixing probability of approximately 1 : 30.

REFERENCES

1. B. Hyams et al., Nucl. Phys. B64, 134 (1973).
2. D. Alde et al., Zeitz. Phys. C66 365 (1995).
3. B. Brabson et al., Proceedings of the VIth International Conference on Hadron Spectroscopy, Manchester, England, eds. M.C. Birse, G.D. Lafferty, and J.A. McGovern (World Scientific, Singapore, 1996) 494.
4. Armstrong et al., Zeitz. Phys. C51 351 (1991).
5. C. Amsler et al., Phys. Lett. B355 425 (1995).
6. J. Augustin et al., Nucl. Phys. B320 1 (1989).
7. O. Krehl, R. Rapp and J. Speth, Phys. Lett. B390 23 (1996).
8. F. Close, N. Isgur and S. Kumano, Nucl. Phys. B389, 513 (1993).
9. R. Gardner, "Radiative Phi Decays at Jefferson Lab", these Proceedings.
10. N. N. Achasov, S. A. Devyanin and G. N. Shestakov, Sov. Phys. Usp. 27(3), 161 (1984); Phys. Lett. B88 367 (1979); Modern Phys. Lett. A 8 (25), 2343 (1993); Yad. Fiz. 33, 1337 (1981); Sov. J. Nucl. Phys. 33(5), May 1981, 1982 American Institute of Physics.
11. T. Barnes, Phys. Lett. B165 434 (1985).
12. R. Lindenbusch, PhD Thesis, Indiana University Blooomington (1996).
13. N. M. Cason, "Exotic Meson Signal in the $\eta\pi^-$ System in π^-p Interactions at 18 GeV/c", these proceedings.

Search for the pentaquark[1] in Fermilab E791

Daniel Ashery[2]

School of Physics and Astronomy, Raymond and Beverly Sackler Faculty of Exact Sciences Tel Aviv University, Israel

Abstract. We report results of the first search for the pentaquark "$P_{\bar{c}s}$" which is predicted to be a doublet of states: $P^0_{\bar{c}s} = |\bar{c}suud\rangle$ and $P^-_{\bar{c}s} = |\bar{c}sddu\rangle$. A search was made for the decay $P^0_{\bar{c}s} \to \phi\pi p$ in data from Fermilab experiment E791, in which 500 GeV/c π^- beam interacted with nuclear targets. We present upper limits at 90% confidence level for the ratio of cross section times branching fraction of this decay to the decay $D^{\pm}_s \to \phi\pi^{\pm}$. The upper limits are 0.031 and 0.063 for $M(P^0_{\bar{c}s}) = 2.75$ and 2.86 GeV/c^2, respectively, assuming a $P^0_{\bar{c}s}$ lifetime of 0.4 ps. Preliminary results are presented for a $P^0_{\bar{c}s} \to K^*Kp$ decay.

The spectrum of observed hadrons fits into multiplets of two- and three-quark states. The mass differences within these multiplets can be explained by effective quark masses and the color-hyperfine (CH) interaction in the QCD Hamiltonian. Calculations done using the CH interaction predict the existence of particles made of more than three quarks. Jaffe [1] predicted the existence of the "H" dibaryon, $H = |uuddss\rangle$, and extensive efforts have been made to find it experimentally [2]. Lipkin [3] and Gignoux et al. [4] have proposed that a doublet of states, the $P^0_{\bar{c}s} = |\bar{c}suud\rangle$ and the $P^-_{\bar{c}s} = |\bar{c}sddu\rangle$, and their charge conjugate states, may exist and be stable against strong decays. These were named pentaquarks.

The threshold for strong decay of the pentaquark is 2907 MeV/c^2, above which it can decay to $D^{\pm}_s$ and a nucleon. Calculations done using the CH interaction predict pentaquark masses which vary from 150 MeV/c^2 to a few tens of MeV/c^2 below the D_s-nucleon threshold, depending on how SU(3)$_{flavor}$ symmetry breaking and the mass of the charm antiquark are taken into account [4]. Contributions to the binding energy from other components of the

---

[1] Supported in part by the Israel Science Foundation and the US-Israel Binational Science Foundation
[2] Representing Fermilab E791 Collaboration

Hamiltonian are more model dependent. Calculations done using an Instanton model [5], bag models [6,7] and a Skyrme model [8] conclude that, depending upon the choice of parameters, the pentaquark is bound or is a near-threshold resonance. If bound, the lifetime is expected to be similar to that of charm particles, with the exact value depending upon unknown internal structure. In a description of the pentaquark as an off-shell charm meson and a spectator baryon, it is assumed that the off-shell meson decays to the same decay products as the free meson. The pentaquark lifetime then should be similar to that of the $D_s^\pm$ charmed meson, 0.47 ps. A description of the pentaquark as a five-quark state allows more interactions among the quarks and consequently predicts a shorter lifetime. In the work described here, we have considered lifetimes ranging from 0.1 to 1.0 ps, and pentaquark masses between 2.75 and 2.91 GeV/c^2.

Only crude estimates of the pentaquark production cross-section exist in the literature. One mechanism considers a production of all five quarks in the interaction [9] and is based on an empirically motivated equation which predicts reasonably well the production cross-section of other charm particles. Another mechanism is the coalescence model, where pentaquark components such as the $D_s^\pm$ and a nucleon are produced in the reaction and fuse into one particle while in overlapping regions of phase-space [10]. Typically, the estimated pentaquark production cross-section is of the order of 1% of that of the $D_s^\pm$.

We report results from the first search for $P_{\bar{c}s}^0$ production, which was carried out in experiment E791 [11] at Fermilab. The experiment recorded 2×10^{10} events from interactions of a 500 GeV/c π^- beam in nuclear targets. The trigger included a loose requirement on transverse energy deposited in the calorimeters by particles coming from the interaction. Precision vertex and tracking information was provided by 23 silicon microstrip detectors (6 upstream and 17 downstream of the targets), ten proportional wire-chamber planes, and 35 drift-chamber planes. Momentum was measured using two dipole magnets. Two multicell, threshold Čerenkov counters were used for π, K and p identification [12].

We have searched for the pentaquark in its expected decay mode $P_{\bar{c}s}^0 \to \phi \pi p$, where the ϕ subsequently decays to $K^+ K^-$. We normalize the sensitivity of our search to $D_s^\pm \to \phi \pi^\pm$ decays which are similar enough that several systematic errors are common to both decay modes and cancel in the measured ratio of cross sections and branching fractions. We calculated the acceptance of our detector via Monte-Carlo simulation where the pentaquark was simulated with a lifetime of 0.4 ps and with two different masses, 2.75 and 2.86 GeV/c^2. In the pentaquark analysis, we searched for events with four tracks emerging from a decay vertex, consistent with being $K\,K\,\pi\,p$ according to information from the Čerenkov counters. It was required that the two kaons have opposite charge and that the total charge of the four tracks be zero. The invariant mass of the two kaons was required to be within ±5 MeV/c^2 of the ϕ mass. This process selects $P^0_{\bar{c}s}$ and $\bar{P}^0_{c\bar{s}}$ candidates equally. Topological, kinematical and other criteria were imposed to improve potential signal significance in the $\phi\pi p$ mass spectrum. A procedure for defining selection criteria was designed to minimize introducion of bias in this process. For each discrimination variable (track and vertex information, Čerenkov identification probabilities, kinematical data, etc.) a sensitivity function, $yield(signal)/\sqrt{background}$, was defined. This function was plotted versus the value of selection criteria for the discrimination variable of interest. Two sources of signal were used. One was $P^0_{\bar{c}s}$ generated with the Monte Carlo simulation. The other was the signal from the $D^0 \to K\pi\pi$ decay to optimize four-prong topological criteria and a sample of $\phi \to K^+K^-$ decays, selected independently from the $\phi\pi p$ sample, to optimize Čerenkov kaon identification criteria. The background was determined from the $\phi\pi p$ invariant mass spectrum, in a mass region outside the 2.75 to 2.91 GeV/c^2 range. The maximum in the sensitivity function determined the selection criterion for each discrimination variable. By comparing the results using the two sources of signal, we could select the criteria such that the sensitivity function was stable against small changes in criterion values.

The resulting criteria included topological requirements on the quality of the reconstructed production and decay vertices, their separation and isolation from other tracks. The momentum vector of the candidate pentaquark had to point back to the production vertex. Other selection criteria included kaon and proton identification and a minimal value for the the sum of the squared transverse momenta of the four tracks, relative to the direction of their summed momentum. Further background reduction criteria included elimination of $\phi\pi p$ vertices that contain $\Lambda \to \pi p$ candidates, a ϕ that points back to the production vertex or known particles decaying to four-prongs with the decay products (mis)identified as two kaons, a pion and a proton.

In Fig. 1(a), we show the final $\phi\pi p$ invariant mass spectrum for the optimized analysis cuts. Three events, above the appropriate threshold, which could be described as $(D_s^\pm, D^\pm \to \phi\pi^\pm) + p$ are shaded. In Fig. 1(b), we show the $\phi_{wings}\pi p$ invariant mass spectrum where ϕ_{wings} refers to K^+K^- candidates with invariant mass in a range *outside* the required ϕ mass window (between 5 and 10 MeV/c^2 below and above the ϕ mass). This spectrum contains mostly background events. In Fig. 1(c), we show the $\phi\pi p$ invariant mass spectrum with a tighter cut on the proton identification which gives essentially the same sensitivity for a pentaquark signal but with half the efficiency. In Fig. 1(d), we show the $\phi\pi$ invariant mass spectrum for the $D_s^\pm \to \phi\pi^\pm$ normalization sample. This sample was selected using the same selection criteria (where relevant) as were used to select pentaquark candidates. In this manner the systematic error on the ratio of efficiencies for the two decay modes was minimized.

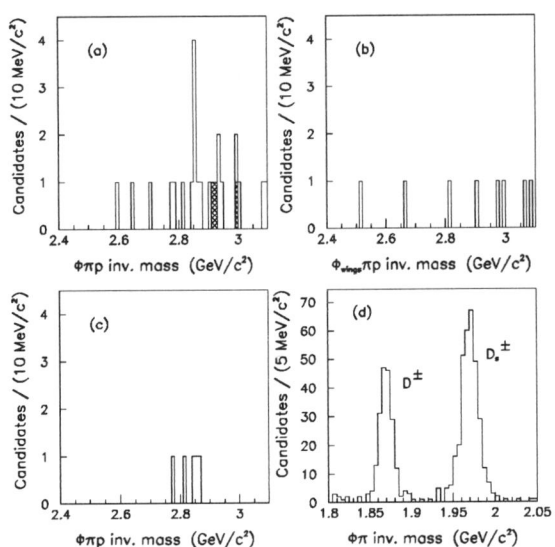

FIGURE 1. (a) The $\phi\pi p$ invariant mass spectrum obtained with the optimized selection criteria. Events in which the $\phi\pi$ invariant mass is consistent with the $D^\pm$ or $D_s^\pm$ masses are shaded. (b) Spectrum of $\phi_{wings}\pi p$ for the optimized selection criteria; see text for a full description. (c) The same spectrum as in (a), with a tighter proton identification criterion. (d) $\phi\pi$ invariant mass spectrum for the $D_s^\pm$ normalization sample.

The $\phi\pi p$ invariant mass spectrum (Fig. 1(a)) shows a concentration of seven events near 2.86 GeV/c^2 which is absent in the background spectrum of Fig. 1(b). Three of these seven events survive the tighter proton Čerenkov selection criterion described above, consistent with the expected efficiency of this criterion (Fig. 1(c)). Only two events outside the concentration survive this requirement. On the other hand, the proton candidate tracks project back to the production vertex with an impact parameter distribution different from that simulated by the Monte-Carlo. For that reason, and mainly because these results cannot be considered as statistically significant, we conclude that there is no convincing evidence for $P^0_{\bar{c}s} \to \phi\pi p$ decays in our data.

We use the spectrum of Fig. 1(a) to obtain 90% C.L. upper limits on (pentaquark production cross section) × (branching fraction to $\phi\pi p$). For a particular $\phi\pi p$ invariant mass, our limit is:

$$UL\left(\frac{\sigma \cdot B_{P \to \phi\pi p}}{\sigma \cdot B_{D_s \to \phi\pi}}\right) = \frac{UL(N_{\phi\pi p})/\varepsilon_{P \to \phi\pi p}}{N_{\phi\pi}/\varepsilon_{D_s \to \phi\pi}}, \quad (1)$$

where $UL(N_{\phi\pi p})$ is the 90% C.L. upper limit on the number of signal events in a mass window centered on the invariant mass of interest, given the number of events observed in the window and the expected number of background events [13]. The quantity $N_{\phi\pi}$ is the number of $D_s^\pm \to \phi\pi^\pm$ decays obtained from the normalization sample (Fig. 1d), and the quantities $\varepsilon_{P \to \phi\pi p}$ and $\varepsilon_{D_s \to \phi\pi}$ are the detection efficiencies for $P^0_{\bar{c}s} \to \phi\pi p$ and $D_s^\pm \to \phi\pi^\pm$, respectively. These efficiencies were calculated from Monte Carlo simulation; the $P \to \phi\pi p$ efficiency depends on the pentaquark mass. The background spectrum —for lack of more information— is assumed to be flat as indicated by Fig. 1(b).

We present limits for two different pentaquark masses: 2.75 GeV/c^2 and 2.86 GeV/c^2. Table 1 lists the numbers used in Eq. 1 and the resultant upper limits. These limits include a small correction factor to account for systematic uncertainties of 19% for $M(P^0_{\bar{c}s})$=2.75 GeV/c^2, and 16% for $M(P^0_{\bar{c}s})$=2.86 GeV/c^2. Assuming that the branching fractions of the $D_s^\pm \to \phi\pi^\pm$ and $P^0_{\bar{c}s} \to \phi\pi p$ decays are similar, the resulting upper-limits approach the range of the estimated ratio between the pentaquark and $D_s^\pm$ production cross sections.

	$P^0_{\bar{c}s} \to \phi\pi p$		$P^0_{\bar{c}s} \to K^*Kp$	
M(P^0) GeV/c^2	2.75	2.86	2.75	2.86
$UL(N_{P^0})$	3.4	11	5.5	3.3
$\frac{\varepsilon(P^0)}{\varepsilon(D_s)}$	0.38±0.07	0.62±0.09	0.21±0.02	0.29±0.02
$N(D_s)$	293±18		725±102	
90% CL Upper-Limit	0.031	0.063	0.04	0.02

TABLE 1. Upper-limit (90% C.L.) for the ratio of cross section times branching fraction for $\sigma \cdot$ BR of the $P^0_{\bar{c}s} \to \phi\pi p$ and $D_s^\pm \to \phi\pi^\pm$ decays and **preliminary** upper limits for the $P^0_{\bar{c}s} \to K^*Kp$ and $D_s^\pm \to K^*K^\pm$ decays. Results are given for two pentaquark masses and assuming a pentaquark lifetime of 0.4 ps.

A similar search is going on for another expected decay mode: $P^0_{\bar{c}s} \to K^*Kp$ with $K^* \to K\pi$. It is more difficult to study this mode because of the large width of the K* (50 MeV). This required application of tighter selection criteria which were optimized using the same procedure as described above. Preliminary results of this search are included in table 1.

The value of the upper-limit depends upon the pentaquark lifetime due to dependence of the acceptance on lifetime. The upper-limit, for a pentaquark with $M(P^0_{\bar{c}s})=2.86$ GeV/c^2, is a rapidly decreasing function of lifetime, from an upper-limit close to 1 for 0.1 ps, to the value listed in the table for 0.4 ps, and remaining about the same for larger lifetime values.

REFERENCES

1. R.L. Jaffe, Phys. Rev. Lett. **38**, 195 (1977).
2. J. Belz et al., Phys. Rev. Lett. **76**, 3277 (1996);
   R.W. Stotzer et al., Phys. Rev. Lett. **78**, 3646 (1997).
   and references therein.
3. H.J. Lipkin, New Possibilities for Exotic Hadrons, in Hadrons, Quarks and Gluons, Proceedings of the *XXIInd* Rencontre de Moriond, France 691 (1987); H.J. Lipkin, Phys. Lett. **B195**, 484 (1987).
4. C. Gignoux et al., Phys. Lett. **B193**, 323 (1987).
5. S. Takeuchi, S. Nussinov, and K. Kubodera, Phys. Lett. **B318**, 1 (1993).
6. S. Zouzou and J.M. Richard, *Few-Body Systems* **16**, 1 (1994).
7. S. Fleck et al., Phys. Lett. **B220**, 616 (1989).
8. D.O. Riska and N.N. Scoccola, Phys. Lett. **B299**, 338 (1993).
9. S. MayTal-Beck et al., Proc. of the 8th Meeting, Div. of Particles and Fields of the Amer. Phys. Soc., S. Seidel Ed., World Scientific, 1177 (1994) and references therein.
10. M. A. Moinester, D. Ashery, L. G. Landsberg, and H. J. Lipkin, Z. Phys. **A356**, 207 (1996).
11. J. A. Appel, Ann. Rev. Nucl. Part. Sci. **42**, 367 (1992), and references therein; D. J. Summers et al., Proceedings of the *XXVIIth* Recontre de Moriond, Les Arcs, France (1992) 417; S. Amato et al., Nucl. Inst. and Meth. **A324**, 535 (1993); E. M. Aitala et al., Phys. Rev. Lett. **76**, 364 (1996).
12. D. Bartlett et al., Nucl. Instr. and Meth. **A260**, 55 (1987).
13. Particle Data Group, Phys. Rev. **D54**, 166 (1996).

Searches for H Dibaryons at the AGS

Bernd Bassalleck

University of New Mexico
E-mail: Bassalleck@baryon.phys.unm.edu

Abstract. A summary is given of the present status (summer '97) of all recent and ongoing H dibaryon searches at the Brookhaven AGS. This includes direct searches either via H production or via detection of H decay products, as well as experiments with indirect implications on the existence of the H, like searches for H nuclei and $\Lambda\Lambda$ hypernuclei. Projectiles used in these experiments for H formation include protons, K^-, Ξ^-, and relativistic Si and Au ions. Recent results make the existence of a deeply-bound H appear very unlikely.

INTRODUCTION

Among non-standard ('exotic') hadrons containing more than the minimal number of quarks, the $S = -2$ H dibaryon has played a particularly prominent role. Originally predicted by Jaffe in the context of the MIT bag model already 20 years ago [1], this ($uuddss$) state with $J^P = 0^+$ and $I = 0$ easily classifies as the most talked-about and researched dibaryon. The reason for this is primarily that the importance of this particular 6 quark configuration appears to be inherent in QCD, as demonstrated by the fact that the H has appeared in many different models over the years. In addition, the unique (among dibaryons) potential of the H being stable against a strong decay (i.e. if $M_H < M_{\Lambda\Lambda}$) has attracted the experimentalists. Apart from the six quark sector, there have been a considerable number of theoretical speculations on the existence of strange quark matter or strange hadronic matter. The existence or non-existence of the most elementary of these systems, the H dibaryon, represents a key question for modeling the strong interactions of QCD. The motivation for H searches is thus fairly obvious. The long history of theoretical estimates of the H mass will not be recounted here. Let it suffice to say that even though the H appears in so many different models, the predicted H masses have varied so much that the theoretical guidance on its mass has been limited. A recent and particularly complete compilation of theoretical and experimental references on the H can be found in [2]. It is remarkable how much theoretical activity concerning the H is still going on.

Perhaps even more remarkable is the long list of recent and ongoing experiments (at BNL alone) attempting to find the H or set limits on its production. These recent experiments are the focus of this brief review.

H DIBARYON PRODUCTION MECHANISMS

As I will summarize below, a broad spectrum of projectiles and possible H production mechanisms is being employed in the many different searches at BNL. They range from (K^-, K^+) strangeness exchange reactions to stopping Ξ^-, as well as non-strange projectiles, i.e. protons and relativistic Si and Au ions. One motivation for this broad-based attack is the fact that theoretical model estimates for the mass of the H, its lifetime, and its production yield vary tremendously, making it impossible to cover the entire range of those parameters in one experiment. One important distinguishing feature among experiments is whether they are pure production experiments or whether they rely on detecting the H via its decay products. In the former case the experimental sensitivity and acceptance does not depend directly on the (unknown) lifetime nor on the decay modes of the H, whereas in the latter case such a sensitivity is built-in.

RECENT AND ONGOING BNL H SEARCHES

In this section I will summarize recent results and the present status of six direct H searches at the AGS. These experiments are:

• <u>E888</u>, a search in a 24 GeV/c proton-produced neutral beam: $p+A \to H+X$. The experiment consisted of two parts, a weak decay search and a dissociation search.

In the weak decay search the decays $H \to \Lambda n$ or $H \to \Sigma^0 n \to \Lambda \gamma n$ were looked for. The experimental signature consisted of $\Lambda \to \pi^- p$ decays with a large unbalanced $p_T(\Lambda) > 174$ MeV/c. A previous rare K_L setup (E791) was used, and details as well as results can be found in [3]. The two candidate events were not interpreted as a signal, but were instead used to set upper limit on H production cross sections. These 90% C.L. upper limits were in the range of $10^{-4} - 10^{-5}$ of those for Λ production, for H lifetimes longer than $\sim$ 6 ns. As the authors point out, the acceptance depended crucially on τ_H and on decay branching fractions. Since the publication [3], the two events have been thoroughly re-investigated and have been found to be unconvincing as H candidates. A more likely interpretation appears to be semi-leptonic K_L decays with mis-identified decay products [4].

In the dissociation search of E888 the decay sequence $H + A \to \Lambda \Lambda A \to \pi^- p \pi^- p A$ was looked for in a plastic scintillator dissociator. No evidence for H dissociation was observed, and a 90% C.L. upper limit on forward angle H

production of about ≤ 1 mb/sr was set for $\tau_H \geq 10$ ns. Results and details can be found in [5].

- E813, a search for $(\Xi^- d)_{atom} \to Hn$.

Several experiments (at BNL and KEK) have employed the (K^-, K^+) double strangeness exchange reaction in order to transfer two units of strangeness to a nuclear target and thus to end up with a $S = -2$ system. In E813 1.8 GeV/c K^- are used to produce Ξ^- via $K^- p \to \Xi^- K^+$. The K^- momentum is determined in the final stage of a high-intensity and high-purity kaon beamline, which is described in [6]. The K^+ is analyzed under small forward angles in a large-aperture dipole spectrometer.

The relatively low-energy Ξ^- from the above reaction are slowed down in W degraders, and a fraction of them will range out in a liquid deuterium target. The geometry of the combined LH_2/LD_2 target of E813 was optimized using a Monte Carlo simulation with the purpose of maximizing the stopping fraction of Ξ^-. Nevertheless, the majority of Ξ^- decay in flight. Si detectors between the hydrogen and deuterium vessels are used to tag the very slow Ξ^-, i.e. the ones with maximum stopping probability. Finally, H formation is searched for via $(\Xi^- d)_{atom} \to Hn$, where a monoenergetic neutron would carry the information on the mass of the H. The branching ratio for this last reaction is of course unknown. However, it has been calculated in [7]. Especially for a lightly-bound H this branching ratio was predicted to be quite large (tens of %), thus maximizing the sensitivity of E813 near $\Lambda\Lambda$-threshold. Momentum, time-of-flight, and aerogel Čerenkov detectors are used for particle identification.

Even though only a small fraction of the Ξ^- stop in the deuterium, the stopping probability per event is maximized in the analysis by carefully chosen cuts on the outgoing K^+ angle and on the Si detector pulse height. Data for E813 were taken in 1992, 1993, and 1995. Many more details and preliminary results from 1992 and 1993 can be found in [8–11]. The analysis of the largest (1995) data set is nearing completion. At this point there is no statistically significant and confirmed structure in the neutron time-of-flight spectrum. It is therefore expected that E813 will end up setting upper limits on the branching ratio for $(\Xi^- d)_{atom} \to H + n$ in the region of a lightly-bound H, i.e. binding energy less than about 100 MeV. The experiment has the advantage of being insensitive to the H lifetime and to its decay modes. It is also worth pointing out that E813 remains sensitive up to about 20 MeV *above* $\Lambda\Lambda$-threshold.

- E836, a search for the H in ^{3}He$(K^-, K^+)Hn$.

Here the same (K^-, K^+) reaction (and in fact the same K^- beamline and the same K^+ spectrometer) as in E813 was used on a single nuclear target, primarily ^{3}He, although some data also exist for ^{6}Li and ^{12}C. Effectively a pp-pair is converted into an H. The ^{3}He$(K^-, K^+)Hn$ reaction is particularly interesting since ^{3}He is a light nucleus with well-known wavefunction, and a detailed theoretical calculation exists for a direct comparison [12], which greatly helps in gauging the experimental sensitivity. As pointed out and calculated by Aerts and Dover [12], a relatively deeply-bound H will manifest itself as a well-separated, narrow peak in the K^+ momentum spectrum above the region of quasi-free Ξ^- production ($K^-+^3\text{He}\to K^+ + \Xi^- + p + n$). Therefore, the main thrust of E836 was a search for such a structure in the momentum spectrum above the endpoint of the quasi-free region. Just like E813, E836 was insensitive to the H lifetime and to its decay modes.

The experiment is completed and the results have been published [13]. No evidence for H production was observed in this high-sensitivity search. In a mass range extending from about 50 to 380 MeV/c^2 below the $\Lambda\Lambda$-threshold, the resulting 90% C.L. upper limits on the H production cross section are in the range of 0.058 to 0.021 μb/sr. These cross section limits are approximately one order of magnitude below the Aerts and Dover calculation, thus making an H bound by more than about 50 MeV appear very unlikely.

E836 results from a ^{6}Li target are also available [14], albeit without the virtue of a directly applicable theoretical model calculation, and with limited statistics. Again, no evidence for H production off ^{6}Li was found. It is also important to point out that the entire E836 results are consistent and compatible with the results from KEK E224, where the same (K^-, K^+) reaction was studied using a scintillating fiber target [15].

- <u>E864</u>, a search for rare composite objects in relativistic heavy ion collisions.

This experiment is primarily a high-sensitivity strangelet search using the relativistic Si and Au beams at the AGS [16]. In principle H dibaryons produced in reactions such as Au+Pb$\to H+X$ could be detected via time-of-flight and energy deposited in a calorimeter located almost 30 m downstream of the target. This long flight path limits the sensitivity of E864 to H particles which live for about 40 ns or longer (proper time). The H mass region lies on the high side tail of the (abundant) neutron mass spectrum. The analysis of such neutral events is progressing, but no limits have been worked out yet on the production of such long-lived H particles. If things work out as expected, the experiment might have a limit which is relevant to the large estimated production of the H in relativistic heavy ion collisions, where up to $\sim$10% of central collisions have been predicted to produce an H [17]. More data and more analysis will tell whether backgrounds and resolutions obtained will permit the estimated accuracy.

For more details on this experiment and first strangelet results see ref. [18]. Another (final?) data taking run is foreseen for 1998.

- <u>E810</u>, a search for the H produced in relativistic Si+Pb collisions.

This experiment was one of the early relativistic heavy ion experiments at the AGS. A byproduct is the search for H dibaryons through the decays $H \to \Sigma^-p$ and $H \to \Lambda p\pi^-$, with the H having been produced in collisions of

14.6 A GeV/c Si ions with a Pb target. Preliminary results based on 5000 central events were reported some time ago [19].

Charged tracks were measured in three TPC (Time Projection Chamber) modules in a magnetic field. The distance from the target to the first TPC module was ~ 40 cm. Centrally enriched events were selected using a cut on the highest multiplicity of negatively charged tracks. In order to search for potential $H \to \Sigma^- p$ decays, a pattern recognition was developed, which put together a positive and a negative track with a negative kink vertex sharing the same negative track. A fit is then made constraining the kink vertex to be consistent with a Σ^- decaying into a π^- plus an unobserved neutron.

20000 central Si+Pb events have now been analyzed and a surprising number of candidate H decays has been observed [20] in both of the decay channels. The events appear to reconstruct to an effective mass peak around 2210 MeV/c^2 (on top of some background), which would correspond to only about 20 MeV binding energy. From a global fit it is estimated that the H has a mass of $2.210\pm.015$ GeV/c^2 with 29 $\Sigma^- p$ events and 21 $\Lambda p\pi^-$ events. The signal in both channels is consistent with a $c\tau_H$ around 4 cm, i.e. $\sim c\tau_\Lambda/2$. For more details, incl. information on the Monte Carlo studies and the production model used therein, the reader is referred to [20]. One of the most intriguing aspects of this potential signal is the extraordinarily high yield observed, a factor of 25 to 50 times the $\Lambda\Lambda$ coalescence model of Baltz et al. [21], and of the same order of magnitude as Ξ^- production (measured at $\sim 1/2$ Ξ^- per central Si+Pb collision).

Clearly a confirmation of this result by the experiment discussed below (E896) is eagerly awaited.

- E896, a search for a short-lived H dibaryon and short-lived strange matter in 11.6 A GeV/c Au+Au collisions.

This is yet another relativistic heavy ion experiment at the AGS, designed to search for the unambiguous decay topology $H \to \Sigma^- p$ and to extend significant sensitivity into regions of short lifetimes ($\sim \tau_\Lambda/2$), thereby complementing E864. It attempts to set a more sensitive limit on H production than possible in pp or pA interactions due to the high yield of Λ hyperons in central Au+Au collisions and the corresponding high probability of $\Lambda\Lambda$ coalescence.

The experimental setup contains a 7 Tesla superconducting magnet with the target at its entrance. This sweeping magnet deflects charged particles. It is followed by an analyzing magnet containing a distributed drift chamber system for measuring the rigidity of the charged decay products in the sequence $H \to \Sigma^- p \to \pi^- np$. A downstream neutron detector and time-of-flight wall complete the system. For $c\tau_H$ in the range of 2-7 cm, a 1000 hour run, and an assumed branching ratio for $H \to \Sigma^- p$ of 0.3, the expected sensitivity is around $10^{-3} - 10^{-5}$ per central collision. For more details see ref. [22].

No data are available yet since the experiment had only an engineering run in early 1997, and expects a full data run in 1998. It is expected that E896 will easily be able to confirm the intriguing result of a potential $H(2210)$ seen in the E810 analysis [20].

The six above-mentioned H searches at the AGS are summarized in the following table.

TABLE 1. Recent and still ongoing H search experiments at the AGS. Listed are the experiment number, the main reactions involved, whether the experiment depends directly on lifetime or decay branching ratios of the H, and the present status including references.

Experiment	Reactions	Dependence on τ_H or H decay B.R.	Status
E888	$p + A \to H + X$ a) $H \to \Lambda n$ or $\Sigma^0 n$ b) $H + A \to \Lambda\Lambda A$	yes	completed/published [3,4] [5]
E813	$K^- + p \to K^+ + \Xi^-$ $(\Xi^- d)_{atom} \to Hn$	no	data taking completed analysis in progress [8–11]
E836	$K^- + {}^3He \to K^+ Hn$	no	completed/published [13]
E864	$Au + Pb \to H + X$ H via TOF and energy in calorimeter	yes	data taking (finish in 1998?) [16,18]
E810	$Si + Pb \to H + X$ $H \to \Sigma^- p$ or $\Lambda \pi^- p$ in TPC	yes	data taking completed analysis in progress [19,20]
E896	$Au + Au \to H + X$ $H \to \Sigma^- p \to \pi^- np$	yes	engineering run 1997 production run 1998 [22]

INDIRECT IMPLICATIONS ON THE H

Here I will briefly summarize a few recent and current experiments which are closely related to the existence or non-existence of the H, albeit somewhat less directly than the ones above.

- E886, a search for strangelets and H nuclei.

The possibility of binding an H to a light nuclear core was raised in ref. [23]. Using the AGS D6 beamline as a spectrometer, E886 searched for two possible bound states of the H, namely H-d and H-^{3}He. Both would be well separated in mass from known nuclear isotopes. The null result of this first search for such objects is reported in ref. [24]. Production cross section limits are given for long-lived objects in a limited region of p_t and rapidity phase space.

- E885 and E906, searches for $\Lambda\Lambda$ hypernuclei [25,26].

As has been pointed out numerous times, the existence and decay modes of $\Lambda\Lambda$ hypernuclei are intimately related to the existence and mass of the H. At issue is mainly whether the $\Lambda\Lambda$ hypernucleus decays by sequential weak decays of the two Λs or by the (presumably much faster) strangeness-conserving strong decay $_{\Lambda\Lambda}A \to H+X$. Observation of the weak decay would seem to exclude a bound H, except perhaps in a limited region of H binding energy (depending on the binding energy of the $\Lambda\Lambda$ hypernucleus). In three decades only three $\Lambda\Lambda$ hypernuclear candidates have been reported [27]. These three candidates were identified via their sequential weak decays in emulsions. In fact, the most recent KEK event [27], if taken at face value, would limit the H to a binding energy of less than ~ 30 MeV. Clearly such isolated single events require experimental confirmation.

E885 and E906 are counter (i.e. non-emulsion) experiments designed to settle this issue by detecting the decay products with significantly higher statistics. E885 relies on the well-proven (K^-, K^+) reaction on a C(diamond) target to produce $S = -2$ nuclei directly or by stopping Ξ^- produced in a quasifree (K^-, K^+) reaction. Decay product information is recorded in scintillating fiber stacks, which are viewed by gateable image intensifiers coupled to CCD cameras. The experiment took data in 1996, and the analysis is in progress. E906 also uses the (K^-, K^+) reaction, however, on a Be target. Charged decay products are recorded in a cylindrical detector system surrounding the target. An engineering run took place in 1997, and the full production run is expected in 1998. Both experiments use the K^- beamline and the K^+ spectrometer from E813/E836.

It is worth pointing out that another $\Lambda\Lambda$ hypernuclear emulsion experiment is being set up at KEK as well [28].

Unrelated to all the above, it is interesting to note that the first ever BNL H search via the reaction $p+p \to K^+K^+X$ [29] was recently repeated at KEK with much improved sensitivity [30]. Data taking finished during the summer of 1997.

CONCLUSIONS

A substantial amount of recent and still ongoing research activity surrounding the fascinating H dibaryon has been reviewed here. Many different experimental approaches have been taken, including a long list of projectiles and several dedicated, high-sensitivity searches. All indications from BNL as well as KEK are consistent with the notion that an H bound by more than about 30-50 MeV has become *very* unlikely. The theoretical prediction of rather copious H production in relativistic heavy ion collisions is also being tested at the AGS, and has produced a tentative, intriguing, but not yet confirmed lightly-bound H candidate. In addition, the important (for the H) issues surrounding ΛΛ hypernuclei are being tackled in a decisive fashion. We look forward to interesting conclusions from all these experiments over the next few years.

REFERENCES

1. Jaffe R.L., *Phys. Rev. Lett.* **38**, 195 (1977); **38**, 1617(E) (1977).
2. Stotzer R.W., Ph.D. dissertation, Univ. of New Mexico (1997).
3. Belz J. et al., *Phys. Rev. Lett.* **76**, 3277 (1996).
4. BNL E888 Collaboration, *Phys. Rev.* **C**, Addendum (1997).
5. Belz J. et al., *Phys. Rev.* **D53**, R3487 (1996).
6. Pile P.H. et al., *Nucl. Instr. and Meth.* **A321**, 48 (1992).
7. Aerts A.T.M. and Dover C.B., *Phys. Rev.* **D29**, 433 (1984).
8. Iijima T., Ph.D. dissertation, Kyoto Univ. (1995).
9. Merrill F., Ph.D. dissertation, Carnegie Mellon Univ. (1995).
10. Bassalleck B. et al., *Few-Body Systems Suppl.* 9, 51 (1995).
11. Ramsay W.D., in PANIC96, Proc. of the 14th Int. Conf. on Particles and Nuclei, ed. C.E. Carlson and J.J. Domingo, World Scientific, Singapore (1997), p. 602.
12. Aerts A.T.M. and Dover C.B., *Phys. Rev.* **D28**, 450 (1983).
13. Stotzer R.W. et al., *Phys. Rev. Lett.* **78**, 3646 (1997).
14. Bürger T., Ph.D. dissertation, Univ. of Freiburg/Germany (1997).
15. Ahn J.K. et al., *Phys. Lett.* **B378**, 53 (1996).
16. BNL E864 Proposal, Sandweiss J. and Majka R.D., spokesmen, 1990.
17. Dover C.B. et al., *Phys. Rev.* **C40**, 115 (1989); Dover C.B., *Nucl. Phys.* **A590**, 333c (1995).
18. Armstrong T.A. et al., nucl-ex/9706004, and talk by J. Hill at this conference.
19. Longacre R. et al., *Nucl. Phys.* **A590**, 477c (1995).
20. Longacre R., Observation of $H(2210)$ Dibaryon through Weak Decay Modes $\Sigma^- p$ and $\Lambda p \pi^-$, produced in Heavy Ion Collisions with 14.6 A GeV/c Si Beam on Pb Target, BNL Report (1997).
21. Baltz A.J. et al., *Phys. Lett.* **B325**, 7 (1994).
22. BNL E896 Proposal, Crawford H. and Hallman T., spokesmen, 1994.
23. Dover C.B., *Nuovo Cimento* **A102**, 521 (1989).
24. Rusek A. et al., *Phys. Rev.* **C52**, 1580 (1995).
25. BNL E885 Proposal, Davis C.A., Franklin G.B., and May M., spokesmen, 1992.
26. BNL E906 Proposal, Fukuda T. and Chrien R.E., spokesmen, 1994.
27. Aoki S. et al., *Prog. Theor. Phys.* **85**, 128 (1991), and references therein.
28. KEK E373 proposal, Imai K., spokesman.
29. Carroll A.S. et al., *Phys. Rev. Lett.* **41**, 777 (1978).
30. KEK E248 proposal, Kawai H., spokesman.

Search for Strange Quark Matter with AGS E864

John C. Hill* for the E864 Collaboration[†]

*Department of Physics and Astronomy
Iowa State University, Ames, IA 50011

[†]Ames Lab-Bari-BNL-Iowa State-Univ Mass-MIT-Penn State
Purdue-UCLA-USMA-Vanderbilt-Wayne State-Yale

Abstract. The E864 experiment at the AGS uses a large acceptance spectrometer in sensitive searches for strange quark matter (strangelets) produced in the collision of 11.6 A GeV/c Au beams with Pb or Pt targets. Such heavy ion collisions may provide conditions favorable to strangelet formation due to the high baryon density and strange quark content of the fireball formed in the collision. Strangelets could be formed by coalescence or formation of a quark-gluon plasma droplet that could evolve into a strangelet. The results of searches for strangelets with $Z=\pm 1$ and $Z=\pm 2$ using the 1994 and 1995 data sets are presented and plans are discussed for searches for positive, negative and neutral strangelets using the larger data set taken in 1996-7.

INTRODUCTION TO STRANGE QUARK MATTER

Ordinary hadronic matter consists of 'bags' of three quarks(baryons) or a quark-antiquark pair(mesons). Nuclei are conglomerates of B=1 baryons, not extended quark systems. Quark matter can be thought of as color singlet hadronic states with B>1. Such states are allowed by the standard model but are known by observation to be unstable if composed only of u and d quarks. Quark matter with only u and d quarks must be unstable, otherwise normal nuclei would have decayed to quark matter. Strange quark matter is defined as quark matter composed of u, d and s quarks.

There are stabilizing factors that indicate that strange quark matter might be stable or quasistable. As B>1 baryon systems are assembled the added s quarks would go into lower energy states than the corresponding u and d quarks thus lowering the energy of the system through the operation of the Pauli Principle. Also a system composed of equal numbers of u, d and s quarks would have zero charge. This implies a lowering of the energy of strange quark matter due to reduced Coulomb repulsion. A major destabilizing factor is the large mass of the s quark compared to that of the u and d quarks. The above considerations imply that the most stable varieties of strange quark matter will have a low value of Z/A(a useful experimental signature) and an increase in stability with mass number relative to ordinary nuclei due to the progressive lowering of the coulomb energy.

As early as 1979, based on QCD and the MIT bag models, Chin and Kerman [1] predicted that strange matter with A$\geq$10 might be metastable with lifetime$\geq 10^{-4}$s. Farhi and Jaffe postulated that low mass strange matter with B$\geq$6 may be metastable and coined the term strangelets [2]. Strangelets with Z/A$\leq$0.3 should be metastable for certain ranges of bag parameters.

In order to produce strangelets it is necessary to use processes that generate large numbers of s quarks such as relativistic heavy ion collisions. Small strangelets can be formed by coalescense of strange baryons, however it is expected that the production of objects with masses larger than 10 GeV/c^2 should be inhibited by the finite size of the colliding nuclei. Heavier strangelets could be formed by the formation of a quark-gluon plasma droplet followed by $\bar{s}$ distillation and cooling. The decay modes of strangelets have been studied by Berger and Jaffe [3]. These studies indicate that strangelets decaying by radiative weak or semi-leptonic decays might have half-lives long enough to be observed in E864 where the flight time is 4x10^{-8}s for central rapidity particles.

Previous searches for strangelets were made using Si beams by E814 and E858 on Cu and Au targets, respectively. E878 and E886 searched using AGS Au beams on Au and Pt targets, respectively. NA52 at CERN searched using S and Pb beams on Pb targets. E878, E886 and NA52 all used focusing spectrometers with limited acceptance range. E864 is a large acceptance spectrometer using Au beams from the AGS on Pb and Pt targets.

THE E864 SPECTROMETER

The E864 experiment uses an open geometry, large acceptance spectrometer, to search for strangelets expected to be produced with small cross sections. It measures particle production near CM rapidity to minimize production model dependence. The fused quartz Cerenkov beam counters define zero time and targets are Pb or Pt with interaction lengths of 10% or 30% and 60%, respectively. Multiplicity is determined by a segmented scintillation counter 10 cm downstream from the target. Central collisions are defined by gating the highest 10% of the pulse height distribution of the summed pulses. A view of the E864 spectrometer is shown in Fig. 1.

Two dipole analyzing magnets are used to measure the rigidity from the slope of the track in the bend plane. The three hodoscopes measure position, time and charge. Each unit has 206 vertical slats with PMT's at the top and bottom. The information is digitized with ADCs and TDCs. The time resolution ranges from 110 to 170 ps. The three straw tube units consist of a vertical plane and two planes tilted at ±20 deg. S2 and S3 allow more accurate tracking by giving a spatial resolution of ≃0.06 cm. The hadronic calorimeter consists of 13x58 identical 10x10x117 cm^3 towers of Pb/scintillating fiber construction, each read by a single PMT. [4] The PMT signals are digitized by ADCs and TDCs and the time resolution is better than 400 ps. The calorimeter provides a second mass measurement and the energy resolution σ_E/E is about $(34\%/\sqrt{E} + 3.4\%)$.

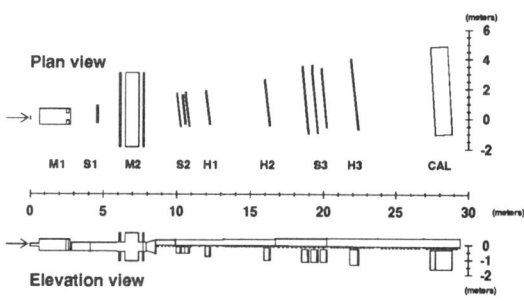

FIGURE 1. The E864 Spectrometer in the AGS A3 Beam Line.

A late energy trigger (LET) is used to select massive objects with TOF long relative to v≃c particles. It uses the energy and time signals from the 616 fiducial calorimeter towers and programmable lookup tables to define the energy and TOF accepted. Enhancement factors of 50 to 100 were obtained for strangelets. The data was stored in VME buffers and events constructed with VME event builders. The data was transferred to exabyte tapes at an event rate of ≃1300 per AGS spill. Typical mass resolutions were about 3% for particles with β <0.97. For a given B field a good variety of particles was obtained and low mass particles were swept out of the acceptance.

SEARCH FOR STRANGE QUARK MATTER

The first task in the strangelet search was to use the time of flight and reconstructed momenta associated with the tracks with appropriate cuts to establish a set of high mass candidates. Using an example from the -0.75T 1995 data set, a total of 85.5×10^6 LET triggers were used to establish a set of 8310, Z=-1 candidates, with tracking masses up to 10^3 GeV/c^2. This was due to charge exchange scattering of neutrons in the magnet exit window producing tracked protons with reconstructed momenta that were too large.

Most of these give m≃m(p⁺) in the calorimeter. A plot of calorimeter vs tracking mass for Z=-1 is shown in Fig. 2. The requirement of tracking and calorimeter mass agreement leaves no candidates with M>5 GeV/c².

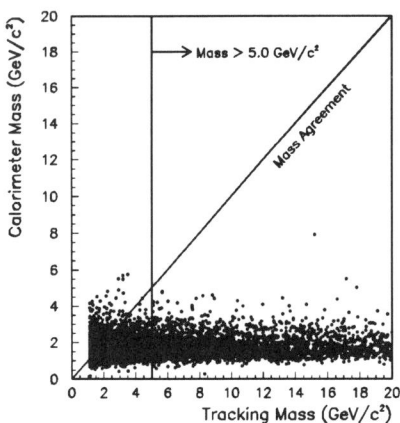

FIGURE 2. Z=-1 Strangelet Candidate Tracking vs Calorimeter Mass

Limits for strangelet production were determined by calculating 90% confidence level (CL) upper limits in the 10% most central Au+Pb interactions at an Au momentum of 11.5 GeV/c. We assume a strangelet production model separable in y and $p_\perp$:

$$\frac{d^2\sigma}{dp_\perp \, dy} \propto \left[p_\perp \exp\left(\frac{-2p_\perp}{<p_\perp>} \right) \right] \left[\exp\left(\frac{-(y - y_{cm})^2}{2\sigma_y^2} \right) \right]$$

where σ_y is the RMS width of the rapidity distribution and $<p_\perp>$ is the mean transverse momentum of the strangelet. In order to calculate the total acceptance and efficiency, two models of the rapidity width were used which were $\sigma_y = 0.5$ or $0.5/\sqrt{A}$. The results from the 1994 and 1995 strangelet searches using the +1.5T (solid lines) and -0.75T (dashed lines) magnetic field settings are shown in Fig. 3. The thick and thin lines are for Z=±1 and Z=±2, respectively. Shown are 90% C.L. limits in 10% central collisions for strangelets with lifetimes greater than 40 ns. The number of 10% central collisions was 26.5×10^6 in 1994 and in 1995 14.9×10^8 for +1.5T and 47.0×10^8 for -0.75T. The 1995 data was taken using the LET.

It is necessary to convert our limits for 10% central events to minimum bias in order to compare with model predictions. A. Baltz et al. [5] predict that

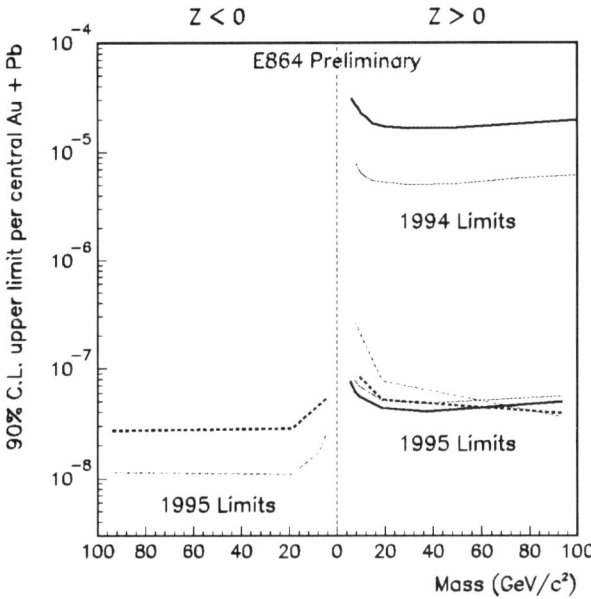

FIGURE 3. Limits for Z=±1 and Z=±2 Strangelet Production.

roughly 50% of the cross section for an A=6 hypernucleus with S=2 resides in the 10% most central events. Using this assumption our central limits are divided by a factor of 5 to get minimum bias limits. Comparisons between the E864 limits and predictions from various models are shown in Fig. 4. Predictions for the formation of strangelets by coalescence from Ref. 5 are shown as filled squares. It should be noted that this model over predicts the experimentally observed yields for ^{3}He and ^{4}He [6] so further work is needed to interpret our sensitivity in the context of coalescence models.

Another mechanism for strangelet formation in heavy ion collisions involves the formation of QGP followed by evolution to a strangelet via the process of strangeness distillation. Liu and Shaw [7] made predictions for strangelet formation from the QGP which are shown by filled circles joined by a line. Their predictions are for Si beams but one might expect higher production rates with Pb targets. The predictions by Crawford et al., [8] are shown by filled triangles. The predictions were for Si+Au collisions but they provided a formula which we applied to derive their Au+Au predictions.

E864 data was used to set 90% CL upper limits on the production of QGP followed by condensation to a strangelet. The limit for Z=±1,±2 strangelets with masses between 10 and 100 GeV/c^2 and halflives above 40 ns is 2.5×10^{-8} for a 10% central collision of a 11.6 GeV/c Au projectile on a Pb target.

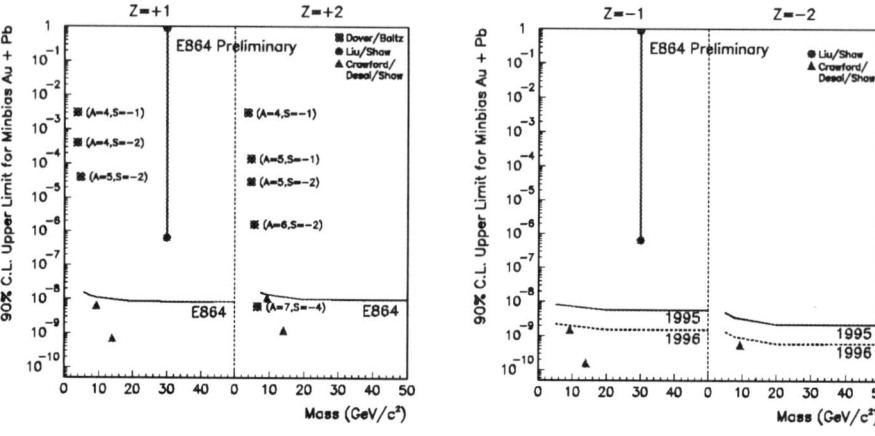

FIGURE 4. Comparison of E864's Limits with Strangelet Models.

CONCLUSIONS AND FUTURE PROSPECTS

In runs of E864 in 1995 with the completed calorimeter and LET, no strangelets were observed. The increased search sensitivity resulted in 90% CL upper limits for Z=±1,±2 strangelets of $\simeq(1\text{-}6)\times 10^{-8}$ per 10% most central Au+Pb interaction at an Au momentum of 11.6 GeV/c. In the 1996-7 data run 224×10^6 and 268×10^6 LET triggers were obtained for the +1.5T and -0.75T field settings respectively. This should decrease the limits for positive and negative strangelets by factors of about 10 and 3 respectively.

Using the calorimeter, E864 can set limits on the production of neutral strangelets. No limits have been reported in the literature but a realistic goal for E864 is $<2.5\times 10^{-7}$ for A>40 and $<10^{-5}$ for A=6. Analysis is in progress.

This work was supported by U.S. Department of Energy, U.S. National Science Foundation and the Istituto Nazionale di Fisica Nucleare of Italy.

REFERENCES

1. S.A. Chin and A. K. Kerman, *Phys. Rev. Lett.* **43**, 1292 (1979).
2. E. Farhi and R.L. Jaffe, *Phys. Rev. D* **30**, 2379 (1984).
3. M.S. Berger and R.L. Jaffe, *Phys. Rev. C* **35**, 213 (1987).
4. K.N. Barish et al., *Nucl. Instr. Meth.*, (1997), submitted.
5. A. Baltz et al., *Phys. Lett. B* **325**, 7 (1994).
6. J.K. Pope for the E864 Collaboration, in *Proceedings of Heavy Ion Physics at the AGS (HIPAGS)*, C.A. Pruneau et al., editors (Wayne State Univ. publ., WSU-NP-96-16, 1996), p. 119.
7. H.C. Liu and G.L. Shaw, *Phys. Rev. D* **30**, 1137 (1984).
8. H. Crawford et al., *Phys. Rev. D* **45**, 857 (1992).

Exotic Meson Signal in the $\eta\pi^-$ System in π^-p Interactions at 18 GeV/c

Neal M. Cason for the E852 Collaboration*

University of Notre Dame, Notre Dame, IN 46556

Abstract. The $\eta\pi^-$ system has been studied in the reaction $\pi^-p \to \eta\pi^-p$ at 18 GeV/c. A mass-dependent analysis of the results of a partial-wave analysis shows that the asymmetry observed in the angular distribution can be understood in terms of interference between the $a_2(1320)$ and an exotic meson with $J^{PC} = 1^{-+}$.

The $\eta\pi$ system is particularly interesting to study in the search for non-$q\bar{q}$ exotic mesons since, if a resonance were found in a relative P wave (L=1), it would have $J^{PC} = 1^{-+}$, quantum numbers not allowed for a normal $q\bar{q}$ meson. Having isospin I=1, such a state could not be a glueball (2g, 3g, ...), but it could be a hybrid ($q\bar{q}g$) or a multiquark ($q\bar{q}q\bar{q}$) meson.

The $\eta\pi^-$ system has been studied in several recent experiments [1,2] following the GAMS experiment [3] where indications of an exotic meson were observed in $\eta\pi^0$ production. Aoyagi et al. [1], in a π^-p experiment at 6.3 GeV/c at KEK, observed a rather narrow P-wave enhancement in the $\eta\pi^-$ system at 1.3 GeV/c^2. In their analysis, the P-wave amplitude had a phase variation which closely followed that of the $a_2(1320)$ meson (a well-known D-wave resonance). Beladidze et al. [2], in the VES experiment at IHEP, (π^-N interactions at 37 GeV/c) also reported a P-wave signal in the $\eta\pi^-$ state, but their signal was broader and had a significantly different phase variation from that of the KEK experiment.

Here we report results on the analysis of the $\eta\pi^-$ system produced in the reaction $\pi^-p \to \eta\pi^-p$ at 18 GeV/c. The data sample was collected at the AGS at Brookhaven using the MPS [4,5] facility augmented with additional detectors [6–9]. The trigger required a recoil charged particle; one forward charged particle; and a lead-glass calorimeter trigger-processor signal which enhanced the fraction of η's relative to π^0's in the sample. A total of 47 million triggers were recorded. Of these, 47,200 events of this type were selected for this analysis.

Shown in Fig. 1a is the 2γ effective mass distribution for events in the $a_2(1320)$ mass region showing the presence of the η with very little background. The $a_2(1320)$ is the dominant feature of the $\eta\pi^-$ mass spectrum shown in Fig. 1b.

The distributions of $\cos\theta$ and ϕ, the $\eta\pi$ decay angles in the Gottfried-Jackson frame after correction for acceptance, are shown in Figs. 2a and 2b respectively for events in the $a_2(1320)$ mass region. There is a forward-backward asymmetry in $\cos\theta$. This asymmetry for $|\cos\theta| < 0.8$ is plotted as a function of $\eta\pi^-$ mass in Fig. 2c. The

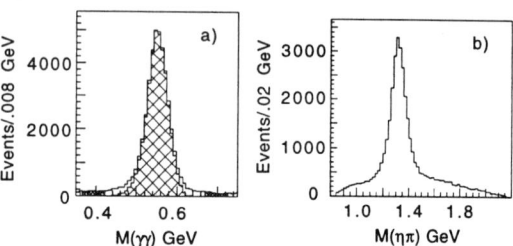

FIGURE 1. a.) The two-photon effective mass distribution. The cental cross-hatched region shows the final data sample and the shaded side bands show the region used to estimate background. b.) The $\eta\pi^-$ effective mass distribution.

asymmetry is large, statistically significant and mass dependent. Since the presence of only even values of L would yield a symmetric distribution in $\cos\theta$, the observed asymmetry requires that odd-L partial waves be present to describe the data.

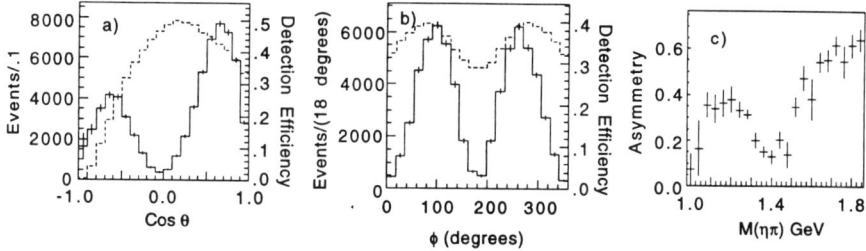

FIGURE 2. Distributions of a.) the cosine of the decay angle and b.) the azimuthal angle for the $\eta\pi^-$ system in the Gottfried-Jackson frame for $1.22 \leq M(\eta\pi^-) \leq 1.42$; shown in c.) is the forward-backward asymmetry as a function of $M(\eta\pi^-)$. The dashed lines and the right hand scales in a.) and b.) give the acceptance in $\cos\theta$ and ϕ.

A partial-wave analysis (PWA) [10,11] based on the extended maximum likelihood method has been used to study the spin-parity structure of the $\eta\pi^-$ system. The partial waves are parameterized in terms of the quantum numbers J^{PC} as well as m, the *absolute value* of the angular momentum projection, and the reflectivity ϵ [12]. In our naming convention, a letter indicates the angular momentum of the partial wave in standard spectroscopic notation, while a subscript of 0 means $m = 0$, $\epsilon = -1$, and a subscript of $+(-)$ means $m = 1$, $\epsilon = +1(-1)$. Thus, S_0 denotes the partial wave having $J^{PC}m^\epsilon = 0^{++}0^-$, while P_- signifies $1^{-+}1^-$, D_+ means $2^{++}1^+$, and so on. We consider partial waves with $m \leq 1$, and we assume that the production spin-density matrix has rank one. Thus, PWA fits shown or referred to in this letter include all partial waves with $J \leq 2$ and $m \leq 1$ (i.e. S_0, P_0, P_-, D_0, D_-, P_+, and D_+). A non-interfering, isotropic background term of fixed magnitude is used.

Results of the PWA fit of 38,200 events in the range $0.98 < M(\eta\pi^-) < 1.82$ GeV/c^2 and $0.10 < |t| < 0.95$ GeV^2 are shown in Fig. 3 where the acceptance-corrected numbers of events predicted by the PWA fit for the D_+ and P_+ amplitudes and their phase difference $\Delta\Phi$ are shown as a function of $M(\eta\pi^-)$. There are eight ambiguous solutions in the fit [11,13,14]. We show the range of fitted values for these ambiguous solutions in the vertical rectangular bar at each mass bin, and the maximum extent of their errors is shown as the error bar.

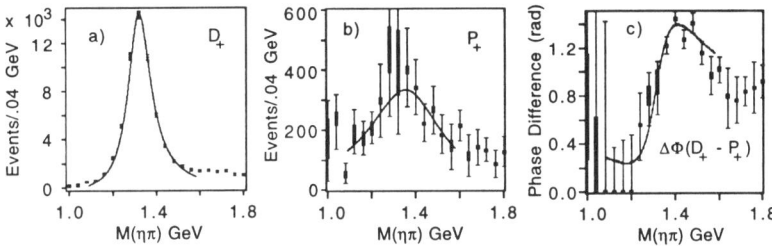

FIGURE 3. Results of the partial wave amplitude analysis. Shown are a) the fitted intensity distributions for the D_+ and b) the P_+ partial waves, and c) their phase difference $\Delta\Phi$. The range of values for the eight ambiguous solutions is shown by the central bar and the extent of the maximum error is shown by the error bars. Also shown as curves in a), b), and c) are the results of the mass dependent analysis described in the text.

The $a_2(1320)$ is clearly observed in the D_+ partial wave (Fig. 3a). A broad peak is seen in the P_+ wave at about 1.4 GeV/c^2 (Fig. 3b). The phase difference $\Delta\Phi$ increases through the $a_2(1320)$ region, and then decreases above about 1.5 GeV/c^2 (Fig. 3c). This phase behavior will allow us to study the nature of the P_+ wave.

These results for the P_+ and D_+ intensities and their phase difference are quite consistent with the VES results [2] as can be seen in Fig. 4a. In particular, the shape of the phase difference as a function of mass is virtually identical to that reported by VES. Our results are compared with those of the KEK experiment [1] in Fig. 4b. Although error bars on some of the points in the KEK analysis are quite large, it is clear that their analysis does not result in the significant phase difference variation observed by both VES and in our analysis. It is also true that our P-wave intensity distribution is quite compatible with the broad structure observed by VES (not shown) and not consistent with the narrow structure observed by KEK (not shown).

In an attempt to understand the nature of the P_+ wave observed in our experiment, we have carried out a *mass-dependent* fit to the results of the mass-independent amplitude analysis over the $\eta\pi$ mass range from 1.1 to 1.6 GeV/c^2. In this fit, we have assumed that the D_+-wave and the P_+-wave decay amplitudes are resonant and have used relativistic Breit-Wigner forms for these amplitudes. One can view this fit as a test of the hypothesis that the correlation between the fitted P-wave intensity and its phase (as a function of mass) can be fit with a resonant Breit-Wigner amplitude.

Results of the fit are shown as the smooth curves in Fig. 3a, 3b, and 3c. The mass and width of the $J^{PC} = 2^{++}$ state shown in Fig. 3a are (1317 ±1 ±2) MeV/c^2 and (127 ±2 ±2) MeV/c^2 respectively [16]. (The first error given is statistical and the second is systematic.) The mass and width of the fitted $J^{PC} = 1^{-+}$ state (Fig. 3b) are found to be

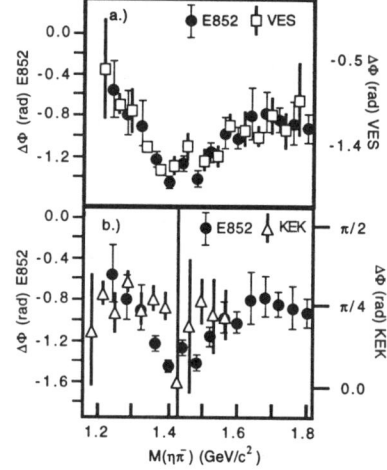

FIGURE 4. Comparison of the results of this amplitude analysis with those of a.) Ref. 2 and b.) Ref. 1.

(1370 ±16 $^{+50}_{-30}$) MeV/c^2 and (385 ±40 $^{+65}_{-105}$) MeV/c^2 respectively. The systematic errors have been determined from consideration of the range of solutions possible because of the ambiguous solutions in the PWA.

The fit to the resonance hypothesis has a χ^2/dof of 1.49. The production phase difference can be fit by a mass-independent constant (of 0.6 rad). If one fits the data to a non-resonant (constant phase) P_+ wave, and also assumes a Gaussian intensity distribution for the P_+ wave, one obtains a fit with a χ^2/dof of 1.55. In this case, the observed phase dependence on mass is attributed to a rapidly varying production phase[1]. Such a phase variation cannot be excluded, but is not expected for any known model. Note that for this non-resonant hypothesis one must have a separate hypothesis for the observed structure in the P_+ intensity — a structure which is explained naturally by the resonance hypothesis. We thus conclude that there is credible evidence for the production of a $J^{PC} = 1^{-+}$ exotic meson.

We would like to express our deep appreciation to the members of the MPS group. Without their outstanding efforts, the results presented here could not have been obtained. We would also like to acknowledge the invaluable assistance of the staffs of the AGS and BNL, and of the various collaborating institutions. This research was supported in part by the National Science

---

[1] The fit requires a linear production phase difference with a slope of -4.3 rad/GeV.

Foundation, the US Department of Energy, and the Russian State Committee for Science and Technology.

REFERENCES

* E852 Institutions for this analysis are: University of Notre Dame, Notre Dame, IN 46556, USA; Brookhaven National Laboratory, Upton, Long Island, NY 11973, USA; Institute for High Energy Physics, Protvino, Russian Federation; University of Massachusetts Dartmouth, North Dartmouth, MA 02747, USA; Moscow State University, Moscow, Russian Federation; Northwestern University, Evanston, IL 60208, USA; and Rensselaer Polytechnic Institute, Troy, NY 12180, USA.

1. H. Aoyagi et al., Phys. Lett. B **314**, 246 (1993).
2. G.M. Beladidze et al., Phys. Lett. B **313**, 276 (1993).
3. D. Alde et al., Phys. Lett. B **205**, 397 (1988).
4. S. Ozaki, "Abbreviated Description of the MPS", Brookhaven MPS note 40, unpublished (1978).
5. S.E. Eiseman et al., Nucl. Instr. & Meth. **217**, 140 (1983).
6. Z. Bar-Yam et al., Nucl. Instr. & Meth. A **386**, 253 (1997).
7. T. Adams et al., Nucl. Instr. & Meth. A **368**, 617 (1996).
8. R.R. Crittenden et al., Nucl. Instr. & Meth. A **387**, 377 (1997).
9. S. Teige et al., Proceedings of the Fifth International Conference on Calorimetry in High Energy Physics, eds. Howard A. Gordon and Doris Rueger (World Scientific, Singapore, 1995) 161.
10. S.U. Chung, "Formulas for Partial-Wave Analysis", Brookhaven BNL-QGS-93-05, unpublished (1993).
11. S.U. Chung, "Amplitude Analysis for Two-pseudoscalar Systems", Brookhaven BNL-QGS-97-041 (1997); submitted to Phys. Rev.
12. The naturality of the exchanged particle is given by the reflectivity of the wave. See S.U. Chung and T.L. Trueman, Phys. Rev. D **11**, 633 (1975).
13. S.A. Sadovsky, "On the Ambiguities in the Partial-Wave Analysis of $\pi^- p \to \eta \pi^0 n$ Reaction", Inst. for High Energy Physics IHEP-91-75, unpublished (1991).
14. E. Barrelet, Nuovo Cimento A **8**, 331 (1972).
15. F. v. Hippel and C. Quigg, Phys. Rev.**5**, 624 (1972).
16. R.M. Barnett et al., Phys. Rev. D **54**, 1 (1996).
17. S. U. Chung, 'Spin Formalisms,' CERN Yellow Report CERN 71-8 (1971).

PWA Analysis of BES Data on $J/\psi \to \gamma\pi^+\pi^-\pi^+\pi^-$

Y.C. Zhu

*Institute of High Energy Physics,
Beijing 100039, China*

Abstract. The BES data on $J/\psi \to \gamma\pi^+\pi^-\pi^+\pi^-$ has been analyzed by using the partial wave amplitude analysis method. We fit with resonances having $J^{PC} = 2^{++}$ at 1275 MeV, 0^{-+} at 1385 MeV, 0^{++} at 1505 MeV, 2^{++} at 1565 and 1697 MeV, 0^{++} at 1775 MeV, 2^{++} at 2010 MeV, 0^{++} at 2100 MeV and 2^{++} at 2220 MeV. The 0^{++} resonances decay dominantly to $\sigma\sigma$, while 2^{++} resonances at high mass region can decay to $f_2(1270)\sigma$, $\sigma\sigma$ and $\rho\rho$, and 2^{++} at low mass region decay dominantly to $\rho\rho$.

INTRODUCTION

MARK III and DM2 measured their data on $J/\psi \to \gamma\pi^+\pi^-\pi^+\pi^-$ in 1986 and 1989 respectively [1,2]. MARK III claimed a pseudoscalar resonance at 1.55 GeV/c². This was supported by DM2, and in addition they claimed that the structures at 1.80 and 2.10 GeV/c² were also $J^{PC} = 0^{-+}$.

Both MARK III and DM2 analyzed their data in term of $J/\psi \to \gamma\rho\rho \to \gamma\pi^+\pi^-\pi^+\pi^-$, but they didn't consider $J/\psi \to \gamma\sigma\sigma \to \gamma\pi^+\pi^-\pi^+\pi^-$. In 1995, D. Bugg *et al.* reanalyzed MARK III data on $J/\psi \to \gamma\pi^+\pi^-\pi^+\pi^-$ [3]. They proposed that the $\sigma\sigma$ could be a decay mode in addition to $\rho\rho$, where σ is $\pi - \pi$ S-wave. Inclusion of this decay mode results in the observation of two $I = 0$ scalar resonances decaying exclusively via $\sigma\sigma$. Those states have mass in the region 1.5 and 2.1 GeV/c². An additional scalar state is required at 1.75 GeV/c² and decays dominantly to $\sigma\sigma$, but also shows significant decays via $\rho\rho$.

We have analyzed the BES data on $J/\psi \to \gamma\pi^+\pi^-\pi^+\pi^-$ by using the partial wave amplitude analysis method provided by D. Bugg. The full Monte Carlo simulation, which was not used by D. Bugg in his reanalysis of Mark III data, is employed in BES analysis, for the proper Monte Carlo simulation for Mark III data is no longer available at that time. The main background channel $J/\psi \to \pi^0\pi^+\pi^-\pi^+\pi^-$ has been much carefully studied in order to determine the background in $J/\psi \to \gamma\pi^+\pi^-\pi^+\pi^-$.

EVENTS SELECTION

Candidates for the decay $J/\psi \to \gamma\pi^+\pi^-\pi^+\pi^-$ are selected by requiring exactly four charged tracks in the drift chamber with zero total charge and exactly one photon. The minimum energy of 50 MeV is required for photon. Then the four-constraint kinematic fits are performed by assuming the final states as $\gamma\gamma 4\pi$, $\gamma\gamma 4K$, $\gamma\gamma 2K2\pi$, $\gamma\gamma 2p2\pi$, $\gamma 4\pi$, $\gamma 4K$, $\gamma 2K2\pi$, $\gamma 2p2\pi$, 4π, $4K$, $2K2\pi$, and $2p2\pi$ respectively for each event. The value of $\chi^2(\gamma 4\pi)$ is required to be the smallest one and probability of $\chi^2(\gamma 4\pi)$ is larger than 5% to improve the resolution and to select the good photon.

ANALYSIS METHOD

The maximum likelihood method is employed to fit the data. The probability density function $f(x_i;\alpha)$ with x_i as a set of independently measured quantities and α as a set of unknown parameters is given as below,

$$f(x_i;\alpha) = \frac{d\sigma}{d\Phi}/\sigma,$$

where the total cross section $\sigma = \int |A_{total}|^2 \, d\Phi$ is calculated by using the Monte Carlo integration. The differential cross section is given by

$$\frac{d\sigma}{d\Phi} = |A^{0-} + A^{0+} + A^{2+} + \cdots|^2 + BG,$$

here, BG is the background term, A^{0-}, A^{0+}, A^{2+} are the amplitudes for pseudoscalar, scalar and tensor components respectively.

Fit is made to the data in order to maximize the value of

$$\mathcal{L} = \prod_{i=1}^{N_{event}} f(x_i;\alpha).$$

Actually, $S = -\ln\mathcal{L}$ is minimized in the fit.

The background function

$$BG = 1.30(e^{-0.0856(3.499-m)^2})^2 \times (e^{-0.1020(3.462-m)^2})^2$$

was determined from Monte Carlo simulation.

ANALYSIS AND RESULTS

Possible resonances and possible decay modes for each resonance have been included in the fit. The resonances and decay modes which make real improvements in log likelihood function have been listed in Table 1

The 4π invariant mass distribution of $J/\psi \to \gamma(\pi^0)\pi^+\pi^-\pi^+\pi^-$ is shown in Figure 1, which presents peaks at approximately 1.50, 1.75 and 2.10 GeV/c². The final full fit is performed, the 4π mass spectrum for the final full fit is shown in Figure 2. The contribution of every component in the final full fit are shown in Figure 3.

The obtained branching ratios are given in Table 2, and the branching fraction for $M_{4\pi}$ less than 3 GeV/c² is

$$B(J/\psi \to \gamma\pi^+\pi^-\pi^+\pi^-) = (6.7 \pm 0.3 \pm 0.8) \times 10^{-3}$$

In order to check the results of final full fit, a slice fit method is employed, by which the data is fitted in each mass bin. 10 bins are used with 100 MeV width for each bin and extended from 1350 to 2350 MeV. The comparisons of the full fit and the slice fit are plotted in Figure 4.

The solid line presents the full fit results, while the dot stands for the slice analysis results. They are in agreement with each other. From the slice analysis, the tensor and scalar contribution of both bin fit and full fit confirms that around 2000 MeV region, tensor contribution is prominent.

TABLE 1. Resonances fitted to the BES data and decay modes used in the final fit. L is the orbital angular momentum between the two final state particles following the decay of the resonance.

J^P	Resonance	Measured		Used in final fit		decays	L or $^{2S+1}L_2$	
		Mass (MeV)	Γ (MeV)	Mass (MeV)	Γ (MeV)			
2^+	$f_2(1270)$	1310^{+150}_{-40}		1275	185	$\rho\rho$	$^1D_2$	$^3D_2$
0^-	$\eta(1420)$	1385^{+20}_{-20}	190^{+15}_{-15}	1385	190	$\rho\rho$	$L=1$	
0^+	$f_0(1500)$	1505^{+15}_{-20}	105^{+20}_{-20}	1505	120	$\sigma\sigma$	$L=0$	
2^+	$f_2(1560)$	1470^{+25}_{-25}	180^{+70}_{-60}	1565	150	$\rho\rho$	$^5S_2$ $^1D_2$	$^3D_2$
2^+	$f_2(1710)$	1690^{+25}_{-25}	360^{+250}_{-220}	1697	175	$\rho\rho$	$^5S_2$	
0^+	$f_0(1750)$	1775^{+25}_{-25}	190^{+80}_{-60}	1775	190	$\sigma\sigma$	$L=0$	
2^+	$f_2(2010)$	2010^{+50}_{-70}	350^{+100}_{-70}	2010	350	$f_2(1270)\sigma$	$^5S_2$	
						$\sigma\sigma$	$^1D_2$	$^3D_2$
						$\rho\rho$	$^1D_2$	$^3D_2$
0^+	$f_0(2100)$	2055^{+35}_{-30}	205^{+220}_{-80}	2100	205	$\sigma\sigma$	$L=0$	
2^+	$f_2(2220)$	2220^{+20}_{-20}	105^{+70}_{-55}	2220	80	$f_2(1270)\sigma$	$^5S_2$	
						$\sigma\sigma$	$^5S_2$	$^3D_2$
						$\rho\rho$	$^5S_2$	$^3D_2$

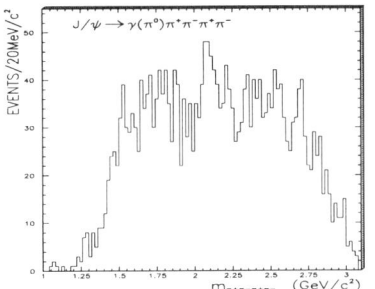
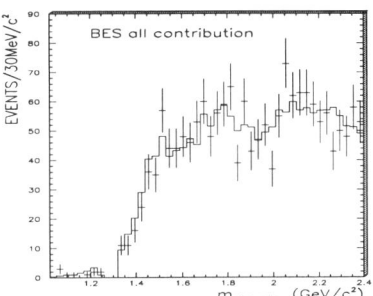

FIGURE 1. The 4π mass spectrum from BES data on $J/\psi \to \gamma(\pi^0)\pi^+\pi^-\pi^+\pi^-$

FIGURE 2. The 4π mass spectrum from the final full fit

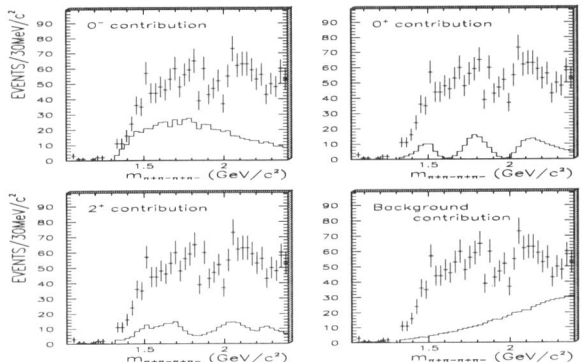

FIGURE 3. component contribution

TABLE 2. Branching ratios for various final states in J/ψ decays, integrated up to $M(4\pi) = 2400$ MeV. Errors are statistical. There is in addition a $\pm 15\%$ normalization error common to all channels.

Process	Branching ratios
$B(J/\psi \to \gamma f_2(1270)) \times B(f_2(1270) \to \pi^+\pi^-\pi^+\pi^-)$	$(2.1 \pm 0.3) \times 10^{-4}$
$B(J/\psi \to \gamma \eta(1420)) \times B(\eta(1420) \to \pi^+\pi^-\pi^+\pi^-)$	$(2.5 \pm 0.4) \times 10^{-3}$
$B(J/\psi \to \gamma f_0(1500)) \times B(f_0(1500) \to \pi^+\pi^-\pi^+\pi^-)$	$(4.0 \pm 0.6) \times 10^{-4}$
$B(J/\psi \to \gamma f_2(1560)) \times B(f_2(1560) \to \pi^+\pi^-\pi^+\pi^-)$	$(4.3 \pm 0.7) \times 10^{-4}$
$B(J/\psi \to \gamma f_2(1710)) \times B(f_2(1710) \to \pi^+\pi^-\pi^+\pi^-)$	$(5.5 \pm 0.8) \times 10^{-4}$
$B(J/\psi \to \gamma f_0(1750)) \times B(f_0(1750) \to \pi^+\pi^-\pi^+\pi^-)$	$(5.5 \pm 0.8) \times 10^{-4}$
$B(J/\psi \to \gamma f_2(2010)) \times B(f_2(2010) \to \pi^+\pi^-\pi^+\pi^-)$	$(1.0 \pm 0.2) \times 10^{-3}$
$B(J/\psi \to \gamma f_0(2100)) \times B(f_0(2100) \to \pi^+\pi^-\pi^+\pi^-)$	$(6.1 \pm 0.9) \times 10^{-4}$
$B(J/\psi \to \gamma f_2(2220)) \times B(f_2(2220) \to \pi^+\pi^-\pi^+\pi^-)$	$(2.6 \pm 0.4) \times 10^{-4}$

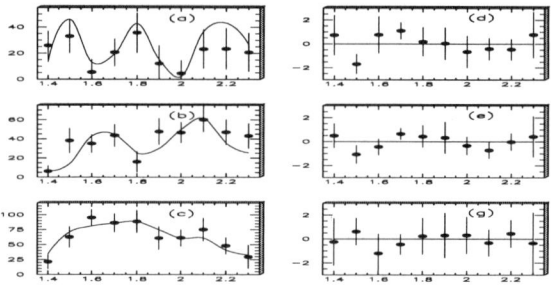

FIGURE 4. Numbers of events in 100 MeV slices of $M(4\pi)$ (data point) from (a) 0^+, (b) 2^+ and (c) 0^- and the full fit (curves); the discrepancy in phase difference (d) $\phi(0^+)-\phi(0^-)$, (e) $\phi(2^+)-\phi(0^-)$ and (f) $\phi(2^+)-\phi(0^+)$ between the data and fit(radians); the value plotted is that for each slice minus the value from the fit.

DISCUSSION AND CONCLUSION

The BES results are consistent with the reanalysis results of MARK III data. But the BES data seems more definite in favor of 2^+ at high mass region. A very important conclusion is that a 2^+ contribution is definitely needed over a broad mass region around 2 GeV, for it is just located at the mass region where the 2^+ glueball is expected, and J/ψ radiative decays are supposed to be one of the best places to search for the glueballs. Recently, the WA102 group found a broad 2^+ resonance at 1930 MeV in their 4π mass spectrum [6], this agrees with our observation.

The BES data on $J/\psi \rightarrow \gamma\pi^+\pi^-\pi^+\pi^-$ shows two resonances in 1.7 GeV region which were identified as $f_J(1710)$: a scalar with mass 1775 MeV and width 190 MeV, and a tensor with mass 1697 MeV and width 175 MeV. This agree with the results of BES data on $J/\psi \rightarrow \gamma K^+K^-$ channel [7].

There are some tentative evidences for the existence of a fairly narrow $f_2(2220)$, but they are not conclusive. Whether this $f_2(2220)$ is really $\xi(2230)$ needs further studies.

REFERENCES

1. R.M. Baltrusaitis *et al.*, Phys. Rev. **D 33** (1986) 1222.
2. D. Bisello *et al.*, Phys. Rev. **D 39** (1989) 701.
3. D.V. Bugg *et al.*, Phys. Lett. **B 353** (1995) 378.
4. Particle Data Group, Phys. Rev. **D 54** (1996) 346.
5. Particle Data Group, Phys. Rev. **D 54** (1996) 368.
6. D. Barberis *et al.*, Phys. Lett. **B 397** (1997) 339.
7. J.Z. Bai *et al.*, Phys. Rev. Lett., **77** (1996) 3959.

In-Flight Hadron Spectroscopy at LEAR

D. W. Hertzog

Department of Physics
University of Illinois at Urbana-Champaign, Urbana, Illinois, USA

Abstract.
In this paper, I present the most recent hadron spectroscopy results stemming from in-flight $\bar{p}p$ reactions at LEAR. In the mass range from approximately 2.0 to 2.43 GeV, scans for both broad and narrow resonances have been performed. The PS185, Crystal Barrel and Jetset experiments figure prominently in the results presented. Particular attention is given to the latest searches for the $\xi(2220)$.

INTRODUCTION

The Low-Energy Antiproton Ring (LEAR) at CERN operated for more than a decade before its closure at the end of 1996. During that time, a considerable effort was directed toward elucidating the hadronic spectrum, with both exotic and ordinary meson searches in place. At this conference, C. Meyer presented an excellent overview of the status of exotic spectroscopy, much of the excitement of which comes from LEAR experiments. The focus of this presentation is to highlight the latest results from LEAR experiments which feature $in - flight$ antiproton-proton annihilations. Most of these results are new, some preliminary, and generally they are less well known compared to the annihilation $at - rest$ studies.

The main aim of the spectroscopy effort at LEAR is the search for exotic hadrons. Glueball states are expected to be created in antiproton-proton annihilations and thus seen in the debris of such interactions. The discovery [1] of the $f_0(1500)$ is an example of the success of this program. This state is the leading candidate for the scalar glueball and its mass can be used to set the scale for the tensor glueball search. Using extrapolations from lattice QCD work, the tensor glueball is expected to be found at a mass of approximately 1.5 times the scalar [2]. This puts it in the 2.0 to 2.5 GeV range which is appropriate for the in-flight annihilation studies reported on here.

In prior experimental work, resonances have been reported in this range including the three broad tensor "g_T states" at masses of 2.01, 2.30, and 2.34 GeV [3] which were seen at Brookhaven in $\pi^- p \to \phi\phi n$ and interpreted by the authors as glueball candidates. Of special interest at this session is the narrow $f_J(2220)$, also known as the $\xi(2220)$ which was first observed in radiative J/ψ decays [4,5] and has the peculiarity of a narrow width and a flavor-neutral decay pattern, including a strong decay to $\bar{p}p$. In baryon-antibaryon systems, a structure in the cross section just above threshold in $\bar{p}p \to \bar{\Lambda}\Lambda$ was reported by PS185 and has been reinvestigated with a larger event sample. In other $\overline{N}N$ systems, a strong resonance of the $\bar{p}p$ system was observed in a central production experiment and has been investigated recently in a complementary $\bar{p}p$-induced reaction at LEAR.

BARYON-ANTIBARYON

The PS185 Collaboration has studied the near-threshold production of hyperon-antihyperon pairs since the beginning of LEAR. J. Franz reported on the latest results in another session. One of the lingering issues raised in both the first [6] and second [7] examinations of the production cross section of $\bar{\Lambda}\Lambda$ very near to threshold (excess energy $\epsilon < 6$ MeV) is a structure in the total cross section appearing at approximately $\epsilon = 1$ MeV. Carbonell and Protosov [8] associated this with a subthreshold quasi-nuclear bound state. The PS185/2 Collaboration has re-examined the region with a high-precision study and a special emphasis on reducing potential systematic errors in the cross section extraction. A final result is given in the thesis of Jones [9]. No structure appears and the cross section rises very smoothly with a greater ratio of S−wave to P−wave production compared to that which was reported previously. A parallel and independent analysis is ongoing at a second institution and a final combined result is expected toward the end of the year.

Narrow baryonium resonance searches were part of the motivation to build LEAR in the first place. After an early round of experiments with negative results, this subject has nearly died. Recently however, the WA56 Collaboration working at CERN's Omega spectrometer, has reported strong evidence for a narrow $\bar{p}p$ state at a mass of 2.02 GeV, steming from the central production reactions of the type $\pi p \to p_{fast}(\bar{p}p)\pi_{slow}$. This result comes from a reanalysis [10] of past work, not from new data. The striking result shows a significant Breit-Wigner lineshape with a width between 10 and 20 MeV. The Jetset collaboration at LEAR accumulates the reaction $\bar{p}p \to \bar{p}p\pi^+\pi^-$ in its sample of four charged kaon triggers. Analyzing this reaction for evidence of any resonance in the $(\bar{p}p)$ intermediate state, even with special cuts to imitate "centrality", yields no resonance. Jetset quotes [11] a peak cross section less than 200 nb at the 95% C.L.

BROAD SCANS

Several experiments at LEAR made cross section and spin-observable measurements over a broad range of momenta. For example, in the earliest years of LEAR operations experimental results refuted states like the $S(1936)$ meson. Later, in more detailed experiments, a complete mapping of the differential cross section and analyzing power versus incident antiproton momentum was made by PS172 [12] for the reactions $\bar{p}p$ to $\pi^+\pi^-$ and K^+K^-. Although very rich structure was found, the findings have been mainly useful in the determination of the parameters for nucleon-antinucleon potentials. No significant spectroscopy results have been extracted from the data apart from setting limits on what is excluded.

Second-generation LEAR experiments were designed for spectroscopy searches and used channels which were selected to enhance the appearance of a possible structure. In Jetset, a strangeness filter like the $\phi\phi$ final state and to a lesser extent $K_s K_s$ and $\phi\omega$ were used. The final results of the broad scan in $\bar{p}p \to K_s K_s$ are completed [13] and indicate no structures apart from an enhancement in the cross section which may be attributed to the $f_2(2150)$ state seen in numerous other nucluon-antinucleon measurements, but never before in one involving strange quarks.

The $\phi\phi$ results show an enhancement above threshold followed by a smooth fall off [14]. To hunt for structures, a partial-wave analysis is underway making use of the power of the vector nature of the ϕ to establish three meaningful angles in the analysis. Preliminary results indicate a strong 2^{++} wave with resonance-like behavior near 2200 MeV. However, it is unlikely that all three g_T states can be supported.

$\xi(2220)$ SEARCH

After more than a decade following its discovery, the $\xi(2220)$ remains an intriguing mystery in hadron spectroscopy. Its width is narrow, it appears to have a flavor-neutral decay pattern and its mass is where one expects to find the tensor glueball. What is this state?

The ξ was first seen in the radiative J/ψ decays to $K\overline{K}$ by Mark III [4] but not by DM2 [15]. In other channels, limits were established. Following its discovery numerous hadron-induced searches have been made yielding mixed results and no clearer picture. GAMS [16] found a signal in $\eta\eta'$ at about the right mass, but the width and significance of their finding is difficult to deduce making an unambiguous association with the ξ state uncertain. Later LASS found the hint of a signal in K^- induced reactions [17] which they interpreted to have a prefered value of $J = 2$ or 4. Again, the significance of the LASS data and the association with the ξ are uncertain. Also in K^- central production, a group at Serpukhov found a broader ξ in the $K_s K_s$ final state which, if associated with the ξ significantly increases the width value to 80 ± 30 MeV. In $\bar{p}p$-induced reactions, several searches [18,19] failed to find the $\xi(2220)$. At least two of these looked only at the ratio $K^+K^-/\pi^+\pi^-$, which made them less meaningful if the ξ decays to pions. In direct $K_s K_s$ production, PS185 found a null result [20] and set strong limits. A straight-forward interpretation of these searches is that the ξ does not couple strongly to $\bar{p}p$.

The recent interest has been sparked by last year's BES report [5]. In addition to confirming the $K\overline{K}$ final states at about the same branching fractions, they also observed significant signals in $\pi^+\pi^-$ and $\bar{p}p$. The latter channel is the entrance channel for the LEAR work and indicates that some coupling must be present. Together with the $\pi\pi$ result, a flavor-neutral decay pattern is established. This gives credence to the glueball interpretation of the state. At this conference, CLEO [21] updated a previous finding by Argus [22] that the ξ does not couple to $\gamma\gamma$, again supporting the glueball interpretation.

A dedicated $\xi(2220)$ search was one of the main aims of the Jetset program. Scans were performed in both 1991 and 1993 in the center-of-mass region around 2230 MeV and included final state channels such as $\phi\phi$, $K_s K_s$, $\pi^0\pi^0$ and $\eta\eta$. Before any of the analysis was published, the BES result appeared which stimulated the excitement that indeed, in this $\bar{p}p$ search, a signal might be found. However, the $K_s K_s$ results from the 1991 Jetset scan, like those from PS185, indicated another null result. The combined Jetset/PS185 $K_s K_s$ data are shown in Figure 1. In the final evaluation of the cross section, a fit is made to the data in order to extract the sensitivity (or non-sensitivity) to a possible resonance state. The procedure for establishing these limits and merging the data sets from PS185 and Jetset was developed by P. Reimer and is described in Ref. [13]. From such an analysis, coupled with the radiative J/ψ results, the limits of allowable branching ratios are set. A plot of the BR for $\xi \to \bar{p}p$ versus the BR for $\xi \to K_s K_s$ is shown in Figure 2. The solid diagonal line represents the ratio established by BES and the dashed lines indicate the 1σ limits. The $\bar{p}p \to K_s K_s$ limits yield the hyperbolas which restrict the product of the individual branching ratios to values which are below approximately 1% each. Such limits are severe and surprising since the $K\overline{K}$ decay modes were thought to be strong.

Continuing, Jetset has completed the analysis of its $\phi\phi$ cross section data in the region around $\sqrt{s} = 2230$ MeV [14]. The steps are very detailed featuring 13 measurements from both the 1991 and 1993 scans. No structure in the cross section is evident, see Figure 1. Limits on the double branching ratio $BR(\xi \to \bar{p}p)BR(\xi \to \phi\phi)$ are found to be typically less than 10×10^{-5} which is about as sensitive as those from $K_s K_s$. The importance of this result should not be underestimated. If the $\xi(2220)$ is a glueball, then one might expect an appreciable coupling to V-V states like $\phi\phi$. If it is of a more conventional nature, for example an $s\bar{s}$ meson, then the coupling to $\phi\phi$ would likely be non-negligible, especially in comparison with the non-strange observed channels such as $\pi\pi$ and $\bar{p}p$. The overall strength of the cross section [23] for $\bar{p}p \to \phi\phi$ is in the $2-3\mu b$ range which is approximately 100 times greater than expectations drawn from strict enforcement of the OZI rule. The excess could be associated with gluonic intermediate states.

At this conference, BES showed [27] preliminary results of their $J/\psi \to \gamma 4\pi$ analysis. Although numerous peaks were identified, one of them may be able to be associated with the $\xi(2220)$. Its strength would not account for the missing large fraction of decays and its width at 105 MeV is wider than the other channels. We will return to this point in the conclusion.

An obvious next question is whether the $\xi(2220)$ couples to $\eta\eta$. Last summer two exciting preliminary results appeared. The Jetset detector featured a forward electromagnetic calorimeter and a barrel gamma "veto" detector which was upgraded to permit a low-resolution determination of gamma position and energy. Using a special trigger, a sample of $\bar{p}p \to 4\gamma$ events was accumulated and the analysis produced results for the two-body final states $\pi^0\pi^0$, $\pi^0\eta$ and $\eta\eta$ when two gammas went to the forward detector and two went to the barrel. For all three reactions many events were lost which did not feature this topology and much of the angular acceptance was missing. Nevertheless, "yields" could be extracted versus momentum which is adequate for the $\xi(2220)$ search. In the first pass through the data, a possible structure appeared in the $\eta\eta$ cross section at exactly the right mass for the ξ. This fueled the beginning of a reanalysis effort, a new proposal at Brookhaven [24],

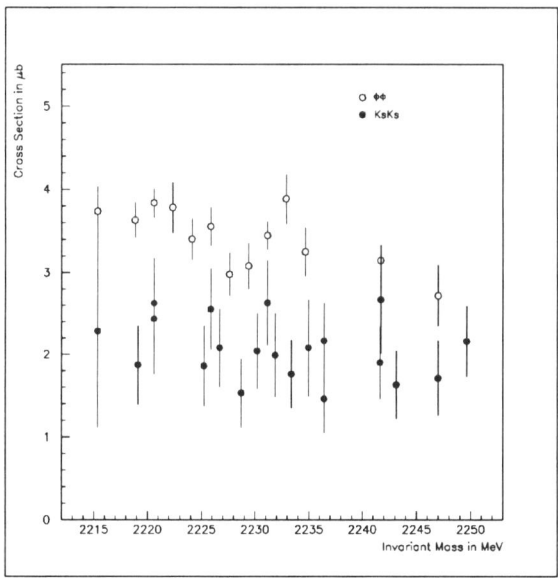

FIGURE 1. Recent $\xi(2220)$ scans from Jetset and PS185 data on $\phi\phi$ and $K_s K_s$ final states.

and a redirection of a part of Crystal Barrel's in-flight program for Fall 1996. At about the same time, BES announced preliminary results featuring a very strong coupling of the $\xi(2220)$ to $\eta\eta$ (and to $\pi^0\pi^0$ and $\eta\eta'$). For an instant, the ξ story appeared to be resolved. The missing strength was in this channel and also in $\eta\eta'$ and all other branching ratios looked compatible. With their high-resolution CsI EM calorimeter, Crystal Barrel was well poised to make a definitive measurement and in November 1996, they performed a 9-point scan centered at 2230 MeV. With a fast preliminary analysis practically "on-line" they determined that no strong signal was present of the type Jetset had seen [25]. Meanwhile, the reanalysis of the Jetset data was in progress and several months later the new results confirmed that no strong and narrow signal was present in the $\eta\eta$ data [26]. Since neither of these results are finalized, they are not included in Figure 1. Barring the truly unexpected, the cross section data of both groups will accomodate no narrow structures. To date, the BES findings have not been published either, but if there remains a strong signal in their neutral channels, it will be quite difficult to reconcile all of these findings.

What could be wrong? Let us start by assuming that the experimental results are all correct within the quoted errors. Jetset and Crystal Barrel

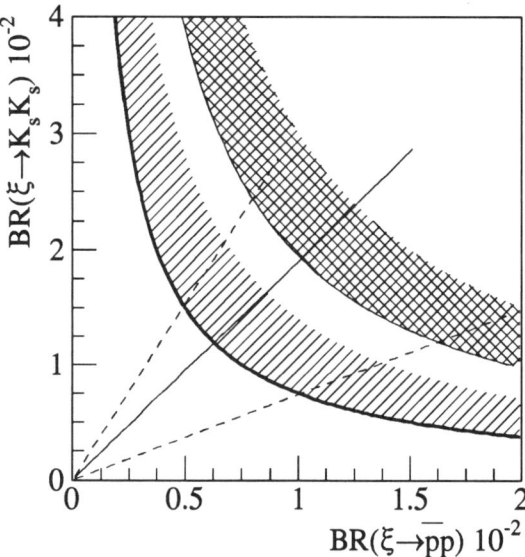

FIGURE 2. Region of allowable branching ratios is enclosed by the two dashed diagonal lines and is limited by the hyperbolas. The most-restrictive hyperbola limit is deduced from an analysis of the combined Jetset/PS185 data. The least-restrictive boundary is from the Jetset data alone.

examined only a very narrow region around 2230 MeV. Their sensitivity to the $\xi(2220)$ is greatest for small widths, such as those below 30 MeV. If the central production $f(2220)$ states are the same as the ξ, then they push the average width to higher values, more likely above 50 MeV. Also, the preliminary 4π BES results shown at this conference featured a 105 MeV wide peak around 2220 MeV. Higher width values would be missed by the $\bar{p}p$ searches. If the mass is significantly different from 2230 MeV, even by as little as 10 MeV, the $\bar{p}p$ searches are less sensitive. These points can be overcome in the new BNL experiment E924 which will look over a broader range with all data taken simultaneously and highlighting the neutral final state channels.

The BES collaboration has upgraded their detector. The upgrade will permit them to make better all-neutral measurements and to once-again measure the $\bar{p}p$ channel. If the $\bar{p}p$ strength is found to be lower by several factors, then this represents another possibility which would put the experiments in better agreement. At present, the error on the branching ratio to $\bar{p}p$ is about 50% so there is room for it to move. In conclusion, the $\xi(2220)$ is not resolved, it remains interesting, and over the next few years the story will continue to unfold. At stake is the possible identification of the tensor glueball.

ACKNOWLEDGMENTS

I would like to thank my colleagues in the PS185 and Jetset (PS202) Collaborations at LEAR. In particular, a large part of the ξ analysis has been carried out by P. Reimer, J. Ritter and R. Jones. This work has been supported in part by the National Science Foundation, under contract NSF PHY 94-20787.

REFERENCES

1. E. Aker et al., Phys. Lett. **B260**, 249 (1991).
2. G. S. Bali et al., Phys. Lett. **B309**, 378 (1993); and J. Sexton, A. Vacarino and D. Weingarten, Phys. Rev. Lett. **75**, 4563 (1995).
3. A. Etkin et al., Phys. Lett. **B165**, 217 (1985), and A. Etkin et al., Phys. Lett. **B201**, 568 (1988).
4. R. M. Baltrusaitis et al., Phys. Rev. Lett. **56**, 107 (1896).
5. J. Z. Bai et al., Phys. Rev. Lett. **76**, 3502 (1996).
6. P. Barnes et al., Phys. Lett. **B229**, 432 (1989).
7. P. D. Barnes et al., Phys. Lett. **B331**, 203 (1994).
8. J. Carbonell, K. Protasov and O. Dalkarov, Nuc. Phys. **A558**, 353c (1993).
9. T.D. Jones, *A measurement of $\bar{p}p \to \bar{\Lambda}\Lambda$ near threshold* Ph. D. Thesis, University of Illinois at Urbana-Champaign, 1996 (unpublished).
10. A. Ferrer et al., Nucl. Phys. **A558**, 191c (1993) and A. Ferrer et al., Proceedings of the Hadron95 Conference, Manchester, July 1995.

11. C. Evangelista *et al.*, Zeitschrift für Physik C, Particles and Fields, 1997, in press.
12. A. Hasan *et al.*, Nucl. Phys. **B378**, 3 (1992).
13. C. Evangelista *et al.*, submitted to Phys. Rev. D, May 1997.
14. M. Lo Vetere *et al.*, Proceedings of the 4th Biennial Low-Energy Antiproton Conference, Dinkelsbuhl, Germany, Aug. 1996.
15. J.E. Augustin *et al.*, Phys. Rev. Lett. **60** (1988) 2238.
16. D. Alde *et al.*, Phys. Lett. **B177**, 120 (1986).
17. D. Aston *et al.*, Nucl. Phys. **B301**, 525 (1988).
18. J. Sculli *et al.*, Phys. Rev. Lett. **58**, 1715 (1987).
19. G. Bardin *et al.*, Phys. Lett. **B195**, 292 (1987).
20. P. D. Barnes *et al.*, Phys. Lett. **B309**, 469 (1993).
21. R. Galik, this conference.
22. H. Albrecht *et al.*, Z. Phys. **C48**, 183 (1990).
23. L. Bertolotto *et al.*, Phys. Lett. **B345**, 325 (1995).
24. AGS E924, *Study of the $\xi(2220)$ in $\bar{p}p$ to neutral final states using the Crystal Ball in the AGS D6 line.*
25. K.K. Seth, Proceedings of the Recontres de Moriond, QCD and Hadronic Interactions, Les Arcs, France, March 1997.
26. J. Ritter, private communication.
27. Y. Zhu, this conference.

Search for Two-Photon Production of $f_J(2220)/\xi(2230)$ at CLEO

Richard S. Galik
CLEO Collaboration

*Newman Laboratory of Nuclear Studies
Cornell University - Ithaca, NY 14853*

Abstract.
We use the CLEO detector at the Cornell e^+e^- storage ring, CESR, to search for the two-photon production of the glueball candidate $f_J(2220)$ in its decay to $K_s K_s$. We present a restrictive upper limit on the product of the two-photon partial width and the $K_s K_s$ branching fraction, $\Gamma_{\gamma\gamma} \cdot \mathcal{B}_{K_s K_s}$ for this narrow resonance. We use this limit to calculate a lower limit on the stickiness, which is a measure of the two-gluon coupling relative to the two-photon coupling. This limit on stickiness indicates that the $f_J(2220)$ has substantial glueball content.

INTRODUCTION

Resonances that are purely gluonic in content (*i.e.*, no valence quarks) have been long predicted but experimentally elusive, as evidenced by the number of talks on this subject at this conference. One might expect to find these "glueballs" in processes such as radiative J/ψ decay ($J/\psi \to \gamma X$) in which one of the three gluons in the dominant amplitude is replaced by a photon, leaving a "glue-rich" environment from which to form the particle X.

The processes $gg \to X$ and $\gamma\gamma \to X$ are very similar, both starting with two massless pointlike vector bosons. However the gluon couples to QCD color whereas the photon couples to EM charge. If X is a meson both processes will have approximately equal amplitudes; it will be roughly as easy to produce via $\gamma\gamma$ as by gg; The relative amplitudes for these two production modes is often expressed as the "stickiness", S, given by [1]

$$S \sim \frac{<X|gg>}{<X|\gamma\gamma>}, \qquad (1)$$

with S being large for a glueball and roughly unity for a conventional meson.

OUTLINE OF OUR STUDY

The $f_J(2220)$, often called $\xi(2230)$ in the literature, is a narrow resonance observed in six experiments in radiative J/ψ decay [2,3], in hadro-production [4–6], and in Z^0 decays [7]; five of these report seeing the f_J decay to $K_s^0 K_s^0$. In the chain $J/\psi \to \gamma f_J \to K_s^0 K_s^0$, the vector nature of the J/ψ and γ and the decay to identical spinless bosons lead to the assignment of even spin and positive parity and C-parity. Spin analyses allow $J = 2^{++}$ or 4^{++}.

CLEO at CESR might appear an odd place to study the f_J. CESR is an e^+e^- storage ring operating at $E_{bm} \sim 5.2$ GeV. CLEO is a "4π" solenoidal detector/spectrometer operated by some 200 physicists from 24 institutions with the main areas of study being B mesons, the Υ resonances, and $\tau^+\tau^-$ and $c\bar{c}$ pairs. However, $\gamma\gamma$ production processes are also studied, using one photon from each of the incident leptons. In the bulk of such interactions the angles of the photons and scattered leptons are quite small. The photons are approximately real (meaning $J_z = \pm 1$) and have a $1/E_\gamma$ energy spectrum. Two such "real" photons produce resonances with $C = +1$ and $J = 0, 2$. So, the most easily produced states have $J^{PC} = 0^{++}, 2^{++}$.

The plan then is as follows: (*i*) look for $f_J(2220) \to K_s^0 K_s^0$ events that come from $e^+e^- \to e^+e^-\gamma\gamma$; (*ii*) assume $J = 2$, use a Monte Carlo simulation to calculate the photon flux, model the production and obtain the CLEO detector acceptance; (*iii*) estimate systematic uncertainties to get a result for the product of the two-photon width and branching fraction, $\Gamma_{\gamma\gamma} \cdot \mathcal{B}_{K_s K_s}$; (*iv*) combine this result with those from radiative J/ψ decay to evaluate the gluon/quark content of the $f_J(2220)$.

SELECTION CRITERIA

We use 3.0 fb^{-1} of e^+e^- luminosity in our study [8]. To select $e^+e^- \to e^+e^-\gamma\gamma$ events we use the fact that the scattered leptons carry most of the incident energy down the beampipe and that the two photons are at small angles with respect to the beam axis. We thus require that both the total visible energy in the detector and the net momentum transverse to the beam axis be small.

To select the $K_s^0 K_s^0$ final state we first need two pairs of charged pions, so we require there be four well-fit charged tracks with $\Sigma Q = 0$. We loosely require the pairs to form vertices and then demand that the $r - \phi$ distance between the vertices be greater than 5 mm. Finally we calculate the invariant masses of the two pairs and demand that the point $(m_{\pi\pi}, m_{\pi\pi})$ be within 10 MeV/c^2 of (m_{K^0}, m_{K^0}) in the two-dimensional plane; this is about 3σ in terms of our mass resolution. The data before imposition of this last criterion is shown in Fig. 1. There is a clear enhancement and little non-KK background (estimated at less than 5%).

All of these selection criteria are rather loose so as to minimize possible systematic biases. Based on our simulations and on calibration channels in the data we estimate contributions of 8%, 7%, and 7% to the systematic uncertainty from effects of triggering, track finding, and event selection.

We now combine the two kaon candidates to look for resonances produced by the $\gamma\gamma$ collisions that decay to $K_s^0 K_s^0$. The region near 2200 MeV/c^2 is shown in Fig. 2, showing no evidence of the $f_J(2220)$. In this plot the bin width is roughly our experimental resolution. The paucity of events is *not* due a fall-off in our efficiency, which is roughly constant this region, but rather due to the steep W dependence of the $e^+e^- \to e^+e^- KK$ cross section. The spectrum at lower invariant mass *does* show a strong enhancement at the $f_2'(1525)$ and we compute a $\Gamma_{\gamma\gamma} \cdot \mathcal{B}_{K_s K_s}$ for that resonance that is consistent with that given by the Particle Data Group [9].

LIMITS ON $\gamma\gamma$ PRODUCTION AND STICKINESS

From the simulation we obtain the expected line shape for the $f_J(2220)$ under the assumption that $J^{PC} = 2^{++}$ and using the resonance parameters obtained by BES [3] and Mark III [2] of $m_f = 2234 \pm 4$ MeV/c^2 and $\Gamma_{tot} = 19 \pm 11$ MeV. Excluding the region $(m_f \pm 40)$ MeV/c^2 we linearly fit the background near the $f_J(2220)$. Then we chose a window centered on m_f with width $\pm \delta m$ so as to maximize the ratio of the square of the expected signal

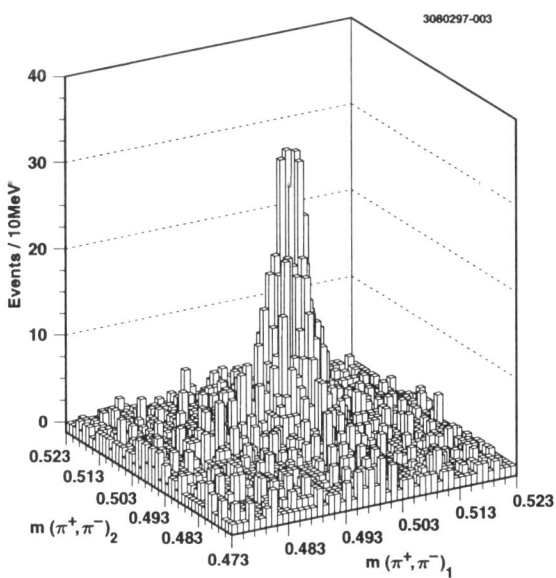

FIGURE 1. Invariant masses of the two pion pairs from the data.

fraction (s) in that window divided by the amount of fitted background (b) in that window; *i.e.*, we chose a window size to maximize s^2/b. This mass window is indicated by the two vertical arrows in Fig. 2.

We count the number of events in the window (in this case the number is four) and use the PDG [9] prescription for obtaining a limit in the presence of background (here that's 4.9 events at 95% C.L.). The solid curve in Fig. 2 is the linear background plus this limit.

The aforementioned systematic uncertainties and the 16% uncertainty in the level of the background are incorporated by a Monte Carlo technique. Then, comparing the limit from the data to the yield from the simulation we obtain the following for the $f_J(2220)$:

$$[0.52 \cdot \Gamma_{\gamma\gamma}^{2,0} + 1.08 \cdot \Gamma_{\gamma\gamma}^{2,2}] \cdot \mathcal{B}_{K_s K_s} < 1.3 \text{eV} \quad (95\% CL) \qquad (2)$$

$$\Gamma_{\gamma\gamma} \cdot \mathcal{B}_{K_s K_s} < 1.3 \text{eV} \quad (95\% CL). \qquad (3)$$

Note that our efficiency is dependent on which of the two allowed helicities is used and that our second limit takes the 1:6 ratio that comes from Clebsch-Gordon coefficients [9]. This product is extremely small in that most mesons have $\Gamma_{\gamma\gamma} \cdot \mathcal{B}_{K_s K_s}$ on the order of 50 eV. The analysis was redone for (m_f, Γ_{tot}) parameters corresponding to one standard deviation shifts from the central values, showing only a small change in the limit.

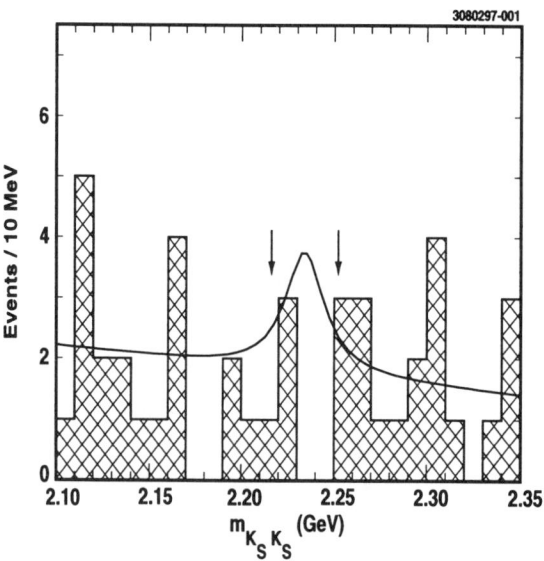

FIGURE 2. Mass spectrum from the data. The curve is the background fit with our 95% C.L. of any possible signal at the $f_J(2220)$.

Finally we investigate the stickiness given by [1]:

$$S_X = N_L \left(\frac{m_X}{k_\gamma}\right)^{2L+1} \cdot \frac{\Gamma(J/\psi \to \gamma X)\mathcal{B}_{K_s K_s}}{\Gamma(X \to \gamma\gamma)\mathcal{B}_{K_s K_s}} \qquad (4)$$

Here k_γ is the photon energy in radiative J/ψ decay and L is the angular momentum between the two vectors (taken as 0 in our case). N_L is a normalization factor used to make $S=1$ for $q\bar{q}$ mesons; we use the $f_2(1270)$ to obtain this normalization [9]. In the numerator we use the combined Mark III [2] and BES [3] result of $\mathcal{B}(J/\psi \to \gamma f_J)\mathcal{B}(f_J \to K_s K_s) = (2.2 \pm 0.6)10^{-5}$. Incorporating the uncertainties in the $f_2(1270)$ and J/ψ measurements via a Monte Carlo technique we then obtain:

$$S_{f_J(2220)} > 82 \qquad (95\% CL) \qquad (5)$$

significantly larger than the stickiness of any of the known mesons.

Some suggested explanations have been put forward for our results.

- Perhaps the $f_J(2220)$ does not exist. While this is possible, the fact that experiments using three different production mechanisms have observed the $f_J(2220)$ seems rather convincing.

- Perhaps the $f_J(2220)$ has $J^{PC} = 4^{++}$. This would change the stickiness normalization and our acceptance by small factors while increasing the stickiness phase space factor by a factor of about 80. This $J^{PC} = 4^{++}$ would have an even larger S limit.

- Perhaps the $f_J(2220)$ is a pure $s\bar{s}$ state. Such mesons *do* have larger S values, but typically 10-20, not >80!

This leaves us with the conclusion that the $f_J(2220)$ (or $\xi(2230)$) is an *excellent* candidate for a tensor glueball.

REFERENCES

1. M. Chanowitz "Resonances in Photon-Photon Scattering" Proceedings of the VIth International Workshop on Photon- Photon Collisions (1984).
2. Mark III Collaboration, R. Baltrusaitis et al., *Phys. Rev. Lett.* **56**, 107 (1986).
3. BES Collaboration, J.Z. Bai et al., *Phys. Rev. Lett.* **76**, 3502 (1996).
4. GAMS Collaboration, Alde et al., *Phys. Lett.* B **177**, 120 (1986).
5. LASS Collaboration, D. Aston et al., *Phys. Lett.* B **215**, 199 (1988).
6. MSS Collaboration, B.V. Bolonkin et al., *Nuc. Phys.* B **309**, 426 (1988).
7. L3 Collaboration, G. Fotconi et al., Contributed to DPF '96.
8. This work is available in preprint form as CLNS97/1947 or hep-ex/9703009; it has been submitted to Physical Review Letters.
9. Particle Data Group, *Phys. Rev.* **D54** (1996).

Our Present Understanding of the N and Δ Baryons

D. Mark Manley

*Department of Physics and Center for Nuclear Research,
Kent State University, Kent, Ohio 44242*

Abstract. Recent experiments and theoretical calculations relevant to the spectroscopy of N and Δ baryons are discussed, with an emphasis on baryons in the first and second resonance regions.

INTRODUCTION

Several experimental and theoretical initiatives are underway to explore the underlying quark structure of the N and Δ baryons. Such studies will also help elucidate their production and decay mechanisms. New experimental efforts have focussed on baryons at c.m. energies $W < 1600$ MeV. The established resonances in this mass range include the $P_{33}(1232)$ (first resonance region), and the $P_{11}(1440)$, $S_{11}(1535)$, and $D_{13}(1520)$ (second resonance region).

THE $P_{33}(1232)$ AND $P_{11}(1440)$ RESONANCES

The $\Delta(1232) \to \gamma N$ transition continues to be of great theoretical and experimental interest. In the first resonance region, γN scattering is dominated by an M_{1+} (or $M1$) transition. Precise new data are beginning to provide details on the small E_{1+} (or $E2$) amplitude, which is generally thought to be crucial for understanding deformation in the $N \to \Delta$ transition and for understanding the tensor forces between quarks.

One method to determine the small $E2$ amplitude is to perform a multipole analysis of $\gamma N \to \pi N$ data [1,2]. This method, like others, is complicated not only by the relative smallness of the $E2$ amplitude compared with the dominant $M1$ amplitude, but also by the need to use a model to separate the *resonant* $E2$ amplitude from the *total* $E2$ amplitude, which contains a nonresonant background contribution. Because of these complications, it is desirable also to use other methods for determining the small $E2$ amplitude.

One such method that recently has been exploited is the study of Compton scattering by the proton, $\gamma p \to \gamma p$. Measurements of Compton scattering at $\theta_{cm} = 75°$ and 90° have been performed at Mainz [3,4], and additional new measurements have been performed with the LEGS facility at BNL [5]. When results of the two Mainz experiments are combined, it is found [3] that the *resonant M1* strength of the $\Delta(1232)$ given by the recent partial-wave analysis (PWA) of $\gamma N \to \pi N$ by Arndt et al. [1] must be reduced by $(-2.8 \pm 0.9)\%$. The combined $\gamma p \to \gamma p$ data in the first resonance region at $\theta_{cm} = 75°$ and 90° leads to the following precise value for the $M1$ combination of $A_{1/2}$ and $A_{3/2}$ helicity amplitudes: $A_{3/2} + \frac{1}{\sqrt{3}} A_{1/2} = (-333 \pm 3) \times 10^{-3}$ GeV$^{-1/2}$. At present, the corresponding $E2$ combination, $A_{3/2} - \sqrt{3} A_{1/2}$, is not so well constrained by the present data, because they were taken only at central angles.

R. A. Arndt et al. recently published results from the VPI group's newest PWA of $\gamma N \to \pi N$ data (the SM95 solution) [1]. Most of the post-1993 data are for the $\gamma p \to \pi^+ n$ reaction at energies in the first and second resonance regions ($W < 1600$ MeV). New data for the other two charge channels are primarily in the first resonance region. Of particular interest are the M_{1-} magnetic multipole amplitudes for the $P_{11}(1440)$ resonance. The γn amplitude in particular displays major changes compared with results from the previous SP93 solution [2]. While the precise energy dependence of the M_{1-} amplitudes remains uncertain, both corresponding $A_{1/2}$ helicity amplitudes appear to have the *wrong signs* compared with state-of-the-art quark-model calculations [6]. Since the lowest hybrid ($qqqG$) baryon is predicted to have the same quantum numbers as the $P_{11}(1440)$, and to lie nearby in mass, this disagreement is a problem of ongoing interest.

PROPERTIES OF THE $S_{11}(1535)$

Since 1995, there has been much experimental [7-10] and theoretical effort [11-20] to investigate properties of the $S_{11}(1535)$ resonance. Of note are the beautiful low-energy data from Mainz for the $\gamma p \to \eta p$ [9] and $\gamma n \to \eta n$ reactions [10]. The measurements with a deuterium target are consistent with $\sigma_T(\gamma n \to \eta n)/\sigma_T(\gamma p \to \eta p) = |A^n_{1/2}|^2/|A^p_{1/2}|^2 \approx \frac{2}{3}$, where $A^p_{1/2}$ is the $S_{11} \to \gamma p$ helicity-$\frac{1}{2}$ amplitude. Several papers [12,15,16,18-20] have discussed the surprising result that the $A_{1/2}$ helicity amplitudes as determined by the precise new Mainz data are about twice as large as the values obtained from PWAs [1,2] of $\gamma N \to \pi N$ data. As well known, the $S_{11}(1535)$ dominates $\sigma_T(\gamma p \to \eta p)$ near threshold, whereas it contributes only weakly to $\gamma N \to \pi N$. In the most recently published PWA of pion photoproduction data by the VPI group [1], the $A^n_{1/2}$ coupling for the $S_{11}(1535)$ proved difficult to fit, and the value obtained, $(-20 \pm 35) \times 10^{-3}$ GeV$^{-1/2}$, was quite different from the value, $(-2 \pm 2) \times 10^{-3}$ GeV$^{-1/2}$ obtained in their earlier analysis [2]. The new uncertainty is also *much larger* than previously estimated. In Zhenping Li's recent chiral quark-model fit [15] of the Mainz data [9] for $\gamma p \to \eta p$, he obtained $A^p_{1/2} = 111 \times 10^{-3}$ GeV$^{-1/2}$ and a total width of $\Gamma = 198$ MeV for the $S_{11}(1535)$. This value of $A^p_{1/2}$ agrees with the value, $(125 \pm 25) \times 10^{-3}$ GeV$^{-1/2}$, published earlier by Krusche et al. [9]. Very recently, Krushe et al. [20] have shown that the precise new differential and total cross-section measurements from Mainz [9,10], combined with the total cross section for η electroproduction close to the threshold point recently measured at Bonn [21], allow the value of the total width of the $S_{11}(1535)$ *to be narrowed significantly*, when compared with the large range, 100–250 MeV, given in the most recent *Review of Particle Properties* (RPP) [22]. Taking the mass to be $M = 1544$ MeV, and the ηN branching ratio to be $b_{\eta N} = 0.45$, Krusche et al. found the values $A^p_{1/2} = (120 \pm 11 \pm 15) \times 10^{-3}$ GeV$^{-1/2}$ and $\Gamma = 212 \pm 20$ MeV, which agree with the values obtained earlier by Li's quark-model fit. The analysis of Krusche et al. made use of an approximately model-independent parameter ξ, which was introduced by Mukhopadhyay et al. to describe both η photoproduction [16] and electroproduction [19]. Krusche et al. [20] concluded that the $S_{11}(1535)$ parameters proposed in the RPP are *inconsistent* with the new η photoproduction data.

Tom Vrana recently completed a successful unitary fit of S_{11} amplitudes for $\pi N \to \pi N$, $\pi N \to \pi\pi N$, $\pi N \to \gamma N$, $\pi N \to \eta N$, and $\gamma N \to \eta N$ [23]. He obtained a good description for all of the above reactions with the $S_{11}(1535)$ parameters, $M = 1545$ MeV, $\Gamma = 126$ MeV, $b_{\eta N} = 55\%$, $b_{\pi N} = 32\%$, $b_{\pi\pi N} = 13\%$, and $A^p_{1/2} = 87 \times 10^{-3}$ GeV$^{-1/2}$. He concluded that the discrepancy between prior analyses of the π and η photoproduction data was likely due to ignoring the influence of either the $S_{11}(1650)$ and/or coupled-channel effects.

NEUTRAL BARYON SPECTROSCOPY WITH THE CRYSTAL BALL

The Crystal Ball Spectrometer, developed at SLAC, was recently delivered to Brookhaven National Laboratory and installed in the AGS C6 line [24]. This detector consists of 672 NaI(Tl) crystals covering most of a 4π geometry. Energy calibration data were obtained during the first week of April, 1997 using the stopped π^- reaction, $\pi^- p \to \gamma n$. This reaction produces a monoenergetic γ ray ($E_\gamma = 129.4$ MeV), which deposits its energy in only a few crystals. After the April run, a LH$_2$ target was installed in the Crystal Ball. During the final two weeks of May, 1997, data for AGS Expt. E913 were measured at 12 π^- energies up to $W \approx 1530$ MeV. The measured $\pi^- p \to \eta n$ data will help determine the properties of the $S_{11}(1535)$ resonance, which is the only known resonance having a large ($\sim 50\%$) ηN branching fraction. The $\pi^- p \to \pi^\circ \pi^\circ n$ data will help determine the properties of the $P_{11}(1440)$ resonance, which strongly enhances the cross section near threshold. The $\pi^- p \to \pi^\circ n$ data will be used to help improve partial-wave analyses of $\pi N \to \pi N$ reactions, and the $\pi^- p \to \gamma n$ data will provide a check on precise new measurements of the inverse reaction, $\gamma n \to \pi^- p$, which will be obtained using a deuterium target in the CLAS detector in Hall B at JLab. Plans are underway to measure additional data with the Crystal Ball next year.

KAON PHOTOPRODUCTION

Zhenping Li and collaborators have successfully used the chiral quark model to describe available data for the photoproduction of pseudoscalar mesons [15,25–27]. Here I briefly describe their work as it pertains to kaon photoproduction [26,27]. The six reactions that occur are $\gamma p \to K^+\Lambda$, $\gamma n \to K^\circ\Lambda$, $\gamma p \to K^+\Sigma^\circ$, $\gamma n \to K^\circ\Sigma^\circ$, $\gamma p \to K^\circ\Sigma^+$, and $\gamma n \to K^+\Sigma^-$. Most of the available data are for the $\gamma p \to K^+\Lambda$ and $\gamma p \to K^+\Sigma^\circ$ reactions. Contributions from *all* s-channel resonances were included. The relative strengths and phases of each term in the s, t, and u channels were determined by the quark-model wave function with the exact $SU(6) \otimes O(3)$ limit. A general feature found is that *charged* pseudoscalar meson photoproductions should be *forward peaked* at energies just above threshold because of dominance by the *contact term* in the low-energy region [27]. (Contributions from this term vanish for neutral-meson final states.) Differential cross-section data for $\gamma p \to K^+\Lambda$ and $\gamma p \to K^+\Sigma^\circ$ support this conclusion.

Li *et al.* also found several features that highlight the roles of s-channel resonances in the quark model for $\gamma N \to K\Sigma$ [27]. In the threshold region, a crucial role is played by s-wave resonances, which are determined by the E_{0+} transition. The $S_{11}(1650)$ and $D_{15}(1675)$ resonances are important for kaon photoproduction on a *neutron* target, but give no contribution on a *proton* target, due to the Moorhouse selection rule. A peak near threshold in $\sigma_T(\gamma n \to K^+\Sigma^-)$ is predicted, due to the $S_{11}(1650)$. As the energy increases, resonances with larger quantum number N become important, and those with $L = N$ become dominant. Finally, all four isospin channels for $\gamma N \to K\Sigma$ should be dominated by $I = \frac{3}{2}$ resonances, particularly, $F_{37}(1950)$, $F_{35}(1905)$, $P_{33}(1920)$, and $P_{31}(1910)$. This leads to the observed maximum in $\sigma_T(\gamma p \to K^+\Sigma^\circ)$ and to a predicted minimum in $\sigma_T(\gamma p \to K^\circ\Sigma^+)$ at $W \approx 1900$ MeV. [Only $I = \frac{1}{2}$ resonances contribute to $\gamma N \to K\Lambda$ reactions.]

SUMMARY AND CONCLUSIONS

At present, there are essentially no new data on baryon resonances above the second resonance region ($W > 1700$ MeV). In addition, very few new experiments to study baryon resonances are planned using hadronic reactions. (One exception is the new program to study neutral resonances at the Brookhaven AGS using the Crystal Ball Spectrometer.) A major program to study baryon resonances using electromagnetic probes will soon begin in Hall B at Jefferson Lab (commissioning began in June, 1997). Some resonance experiments have already been completed in Hall C at JLab, as discussed by others at this conference. There have also been new experiments to study nucleon resonances at Mainz, Bonn, and BNL. This new era of experiments has already inspired many theoretical and phenomenological developments. Future experiments at JLab will explore the entire resonance region and will investigate a variety of resonance decay modes, including γN, πN, $\pi \Delta$, ρN, ωN, and ηN. Some of these experiments may provide evidence for the "missing resonances", which, although predicted by quark models, have not been observed, presumably due to their small πN branching fractions. The next several years should be very exciting for the field of baryon resonances.

REFERENCES

1. R. A. Arndt et al., Phys. Rev. C **53**, 430 (1996).
2. Zhujun Li et al., Phys. Rev. C **47**, 2759 (1993).
3. J. Peise et al., Phys. Lett. B **384**, 37 (1996).
4. C. Molinari et al., Phys. Lett. B **371**, 181 (1996).
5. G. Blanpied et al., Phys. Rev. Lett. **76**, 1023 (1996).
6. S. Capstick, Phys. Rev. D **46**, 2864 (1992).
7. S. A. Dytman et al., Phys. Rev. C **51**, 2710 (1995).
8. J. W. Price et al., Phys. Rev. C **51**, R2283 (1995).
9. B. Krusche et al., Phys. Rev. Lett. **74**, 3736 (1995).
10. B. Krusche et al., Phys. Lett. B **358**, 40 (1995); *Erratum*: Phys. Lett. B **376**, 331 (1996).
11. Ch. Sauermann et al., Phys. Lett. B **341**, 261 (1995).
12. G. Knöchlein et al., Z. Phys. A **352**, 327 (1995).
13. M. Batinić et al., Phys. Rev. C **51**, 2310 (1995).
14. R. A. Arndt et al., Phys. Rev. C **52**, 2120 (1995).
15. Zhenping Li, Phys. Rev. C **52**, 4961 (1995).
16. N. C. Mukhopadhyay et al., Phys. Lett. B **364**, 1 (1995).
17. V. V. Abaev and B. M. K. Nefkens, Phys. Rev. C **53**, 385 (1996).
18. Zhenping Li and R. Workman, Phys. Rev. C **53**, R549 (1996).
19. M. Benmerrouche et al., Phys. Rev. Lett. **77**, 4716 (1996).
20. B. Krusche et al., Phys. Lett. B **397**, 171 (1997).
21. M. Wilhelm, Ph.D. thesis, University of Bonn; B. Schoch et al., Prog. Part. Nucl. Phys. **34**, 43 (1995).
22. R. M. Barnett et al. Phys. Rev. D **54**, Part I, 1 (1996).
23. T. Vrana, Proceedings of this conference.
24. CERN Courier, April 1997, p. 9.
25. Zhenping Li, Phys. Rev. D **50**, 5639 (1994).
26. Zhenping Li, Phys. Rev. C **52**, 1648 (1995).
27. Zhenping Li, Ma Wei-Hsing, and Zhang Lin, Phys. Rev. C **54**, R2171 (1996).

Spectroscopy of Nonstrange Baryons: Recent Results from PNPI

Igor V. Lopatin

Petersburg Nuclear Physics Institute,
Gatchina, Leningrad district, 188350 Russia

Abstract. A summary of experiments carried out at PNPI in the framework of the general program for stydying πp elastic scattering is given. Results of the phase shift analysis PNPI-94 performed using new experimental data are discussed. The new setup built at PNPI for measuring DCS of $\pi^- p$ charge exchange scattering is described briefly and the first preliminary results are presented. New spin rotation data obtained recently for $\pi^\pm p$ elastic scattering by the PNPI–ITEP collaboration are presented; these new results lead to the conclusion that Tables of PDG need to be revised in part concerning characteristics of πN resonances.

Strong-interaction physics at intermediate energies has a need for better determinations of the masses, widths and decay rates of baryons. There exist various theoretical models which predict rather differing spectra and characteristics of these baryons, and unambiguous determination of such spectra from experimental data is required to choose the most adequate theoretical approach. As to nonstrange excited baryons (i.e. pion-nucleon resonances), the only way to extract this information from experimental data-base is to perform a phase shift analysis of πp scattering data. Three such attempts were made in the phase shift analyses (PSAs) KH [1], CMU [2] and VPI [3]. Unfortunately, spectra and characteristics of πN resonances predicted by these analyses differ rather essentially. The reason of such discrepancy may be twofold: 1) uncompleteness and unsufficient quality of the experimental database; 2) invalidity of some suppositions and approximations used in performing PSAs.

The main aim of PNPI pion-nucleon research program is to solve these problems by means of obtaining new precise experimental data and performing a new phase shift analysis.

During last decade, the differential cross sections and the polarization parameters P, A, R for $\pi^\pm p$ elastic scattering were measured at PNPI at many energies. The total list of results obtained in the experiments on the pion channel of the PNPI synchrocyclotron is presented in Table.

DCS⁻		DCS⁺		P⁻		P⁺		A⁻, R⁻	
T_π, MeV	Num. exp. pnts	T_π, MeV	Num. exp. pnts	T_π, MeV	Num. exp. pnts	T_π, MeV	Num. exp. pnts	T_π, MeV	Num. exp. pnts
288	10	277	9	450	12	335	10	450	10
344	10	290	12	490	14	370	9	530	12
401	14	293	17	530	12	410	9	560	12
425	10	308	9	560	14	510	10	600	12
442	14	334	16	600	14	530	9		
465	11	400	10			560	8		
490	11	442	10			580	8		
530	11	466	10						
560	11	490	11						
600	11	530	11						
640	11	560	13						
		600	12						

Experimental data obtained at PNPI [4] were used together with the results of LAMPF [5] for performing a new phase shift analysis PNPI-94 [6]. A principally new feature of this analysis was employing the generator of discrete ambiguities which has made the search for a unique solution more effective. Another peculiarity was that the analysis did not imply the assumption of charge symmetry of πN partial amplitudes. The PSA PNPI-94 resulted in obtaining the most precise amplitudes for energies of the incident pions from 160 to 600 MeV (corresponding values of c.m.s. energies $\sqrt{s} = 1210 \div 1510$ MeV). These amplitudes differ in many essential features from the amplitudes obtained in the "old" PSAs KH and CMU.

One of the most interesting results obtained in the PSA PNPI-94 is observation of charge splitting in the P_{33} phase shifts; quantitatively this effect can be characterized by the difference $\delta_{33}^{++} - \delta_{33}^0$. This difference depends on energy and varies from + 2 degrees at $T_\pi = 200$ MeV to – 2 degrees at $T_\pi = 450$ MeV changing a sign at $T_\pi = 350$ MeV – see Figure 1.

Shown by curve in Figure 1 are results of fitting [6] made taking into account the Breit-Wigner resonance term and nonresonance background. Following values for the masses (M) and widths (Γ) of the P_{33}-resonances were obtained after such parametrization:

$$M^0 = (1233.1 \pm 0.3) \text{ MeV}, \quad M^{++} = (1230.5 \pm 0.2) \text{ MeV},$$
$$M^0 - M^{++} = (2.6 \pm 0.4) \text{ MeV}, \quad \Gamma^0 - \Gamma^{++} = (5.1 \pm 1.0) \text{ MeV}.$$

These values are very close to those obtained earlier on the base of analysis of total cross section data.

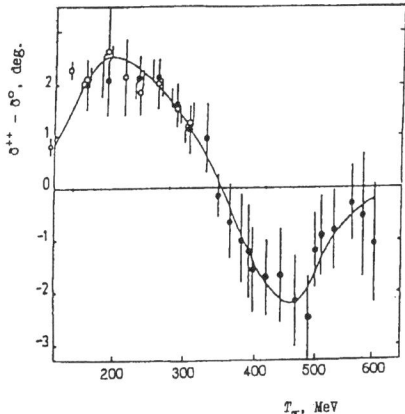

FIGURE 1. Energy dependence of the phase shift difference $\delta_{33}^{++} - \delta_{33}^{0}$ obtained as a result of PSA PNPI–94 (•) and by D.Bugg [7] (o).

At present, an accuracy in determining the characteristics of πN resonances is limited mostly by a lack of high quality experimental data on $\pi^- p$ charge exchange scattering. To improve the situation and fill the gap in the data-base, measurements of DCS for the reaction $\pi^- p \to \pi^0 n$ are now under way at PNPI [8] in the energy range from 300 to 600 MeV (corresponding values of momenta are 417 to 725 MeV/c). Being performed with a small momentum step, these measurements will allow to study also cusp effects in the region of the η production threshold ($P_\pi = 685$ MeV/c) arising because of opening the new inelastic channel $\pi^- p \to \eta n$.

The experiment is carried out by detecting the recoil neutrons in coincidence with gammas from the decay $\pi^0 \to 2\gamma$. The energy of neutrons is measured using the time-of-flight technique. To detect gammas and to measure their energies two different types of total absorption electromagnetic calorimeters were used in the experiment: one is made of sixteen CsI(Na) crystals, another consists of eight Cherenkov lead glass blocks. A special hodoscope of beam counters was placed in the dispersive part of the pion channel with the aim to divide the total momentum acceptance of the channel into several narrow momentum bins.

In the first stage of the experiment, DCS of $\pi^- p$ charge exchange scattering to backward angles were measured. Preliminary results of the first run are presented in Figure 2. Some irregularities in the momentum dependence of DCS in the region around $P_\pi = 685$ MeV/c may be treated as a manifestation of the cusp effects. More careful investigation of this effect is being planned.

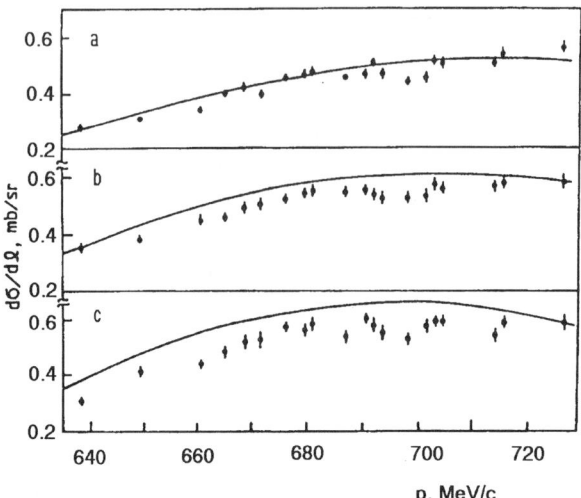

FIGURE 2. Preliminary results of measuring differential cross sections of $\pi^- p$ charge exchange scattering for following c.m.s. angles: a) 157°; b) 166°; c) 175.5°. Shown by curves are the predictions of PSA PNPI–94.

The most of πN resonances are located at c.m.s. energies $\sqrt{s} > 1750$ MeV that corresponds to $T_\pi > 1000$ MeV. Characteristics of these resonances determined in the three global PSAs KH, CMU and VPI differ rather essentially. The only way to remove such ambiguity is to measure the spin rotation parameters A and R. Such measurements were performed recently by the PNPI–ITEP collaboration [9]. The experiment was carried out on a pion channel of the ITEP accelerator at the momentum of positive pions $P_\pi = 1430$ MeV/c. Main parts of the experimental setup were a polarized proton target with a polarization vector lying in the horizontal plane and a multiplate proton polarimeter consisting of optical spark chambers with carbon electrodes. The special automatic television system was designed and created at PNPI for performing filmless read-out from these optical chambers.

Obtained results are shown in Figure 3 in comparison with the predictions of different PSAs. Experimental data confirm the predictions of the analysis VPI and contradict obviously to the predictions of the analyses KH and CMU. This result is very important since all characteristics of πN resonances presented in Tables of the Particle Data Group are based just on PSAs KH and CMU. The fact that the experimental data do not confirm the predictions of these analyses leads to the conclusion that these Tables need to be revised.

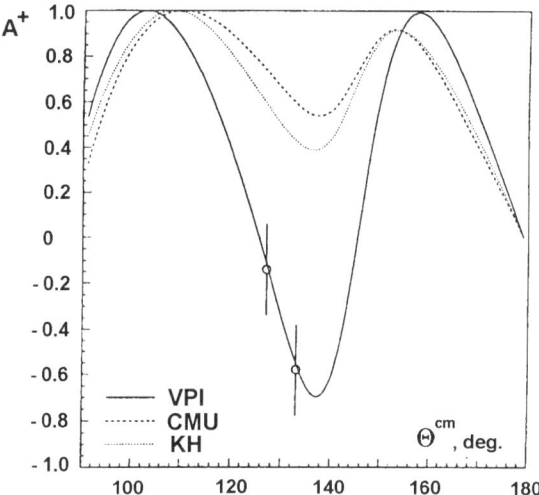

FIGURE 3. Results of measurements of the spin rotation parameter A in π^+p elastic scattering at 1430 MeV/c (open circles) in comparison with the predictions of different PSAs.

It means that a new global phase shift analysis has to be performed in the energy range up to 2000 MeV with the aim to determine unambiguously the characteristics of all πN resonances. For performing such analysis new spin rotation data covering the energy range up to 2000 MeV are needed.

REFERENCES

1. Höhler G., *Handbook of Pion-Nucleon Scattering, Physics Data* **No 12–1**, Fachinformationzentrum, Karlsruhe, 1979.
2. Cutkosky R.E. et al., *Phys. Rev.* **D 20**, 2839 (1979).
3. Arndt R.A. et al., *Phys. Rev.* **D 52**, 2120 (1995).
4. Kruglov S.P., Proc. the 4th Int. Symp. on Pion-Nucleon Physics (Bad Honnef, Germany, 1991), in πN *Newsletter* **No 4**, 14 (1991).
5. Sadler M.E., Proc. the Int. Conference on Mesons and Nuclei at Intermediate Energies (Dubna, Russia, 1994), p.3, 1994.
6. Abaev V.V. and Kruglov S.P., *Z. Phys.* **A 352**, 85 (1995).
7. Bugg D.V., πN *Newsletter* **No 6**, 7 (1992).
8. Lopatin I.V. et al., *Few Body Systems Suppl.* **9**, 241 (1995).
9. Alekseev I.G. et al., *Phys. Lett.* **B 351**, 585 (1995).

Pentaquark Phenomenology

Harry J. Lipkin [a,b,c1]

[a] *Department of Particle Physics Weizmann Institute of Science, Rehovot 76100, Israel*

[b] *School of Physics and Astronomy, Raymond and Beverly Sackler Faculty of Exact Sciences, Tel Aviv University, Tel Aviv, Israel*

[c] *High Energy Physics Division, Argonne National Laboratory, Argonne, IL 60439-4815, USA*

Abstract. Heavy flavor hadron scattering and bound states give experimental information otherwise unobtainable on effective $(qq)_6$ and $(\bar{q}q)_8$ interactions. All constituent quark model successes in (uds) hadron spectroscopy depend only on $(qq)_{3*}$ and $(\bar{q}q)_1$ interactions. Reliable vertex detectors open new pentaquark search directions. Any event in with a proton emitted from a secondary vertex indicates a particle decaying weakly by proton emission and the discovery of a new particle if its mass is higher than that of known charmed baryons. There is no combinatorial background and striking decay signatures like $p\phi\pi^-$ are no longer needed.

INTRODUCTION - THE BLIND MEN AND THE ELEPHANT

Why look for an anticharmed strange baryon? Who cares whether it is bound? Answer: It can help understanding how QCD makes hadrons [1] QCD does not yet explain everything. Present attempts to describe hadrons recall the story of the blind men and the elephant.

- A pion is a Goldstone Boson - A proton is a Skyrmion

- A pion is two-thirds of a proton. The simple quark model prediction $\sigma_{tot}(\pi^- p) \approx (2/3) \cdot \sigma_{tot}(pp)$ [2,3] still fits experimental data better than 7% up to 310 Gev/c [4].

---

[1] Supported in part by grant No. I-0304-120-.07/93 from The German-Israeli Foundation for Scientific Research and Development and by the U.S. Department of Energy, Division of High Energy Physics, Contract W-31-109-ENG-38.

- Mesons and Baryons are made of the same quarks. Describing both as simple composites of asymptotically free quasiparticles with a unique effective mass value predicts hadron masses, magnetic moments and hyperfine splittings [5-7].
- Lattice QCD can give all the answers.
- Lattice calculations disagree on whether the H dibaryon is bound. No hope of settling this question until much bigger lattices are available.
- Light (uds) SU(3) symmetry and Heavy Quark symmetry (cbt) are good
- Light (uds) SU(3) symmetry is bad. All nontrivial hadron states violate SU(3). All light V, A and T mesons have good isospin symmetry with flavor mixing in (u.d) space and no $s\bar{s}$ component; e.g. ρ, ω.
- The s-quark is a heavy quark. Flavor mixing in mass eigenstates predicted by SU(3) is not there. Most nontrivial strange hadron states satisfy (scb) heavy quark symmetry with no flavor mixing.; e.g. ϕ, ψ, Υ.

An underlying dynamics describes many meson and baryon properties primarily in terms constituent quarks which are the same in mesons and baryons.

$$\langle m_s - m_u \rangle_{Bar} = M_\Lambda - M_N = 177\,\text{MeV} =$$
$$= \frac{M_N + M_\Delta}{6} \cdot \left(\frac{M_\Delta - M_N}{M_{\Sigma^*} - M_\Sigma} - 1 \right) = 190\,\text{MeV}. \tag{1a}$$

$$\langle m_s - m_u \rangle_{mes} = \frac{3(M_{K^*} - M_\rho) + M_K - M_\pi}{4} = 180\,\text{MeV} =$$
$$= \frac{3M_\rho + M_\pi}{8} \cdot \left(\frac{M_\rho - M_\pi}{M_{K^*} - M_K} - 1 \right) = 178\,\text{MeV}. \tag{1b}$$

The mass difference $m_s - m_u$ has the same value ±3% when calculated in two independent ways from baryon masses and meson masses [5-7]. The same approach applied to $m_b - m_c$ gives

$$\langle m_b - m_c \rangle_{Bar} = M(\Lambda_b) - M(\Lambda_c) = 3356\,\text{MeV}, \tag{2a}$$

$$\langle m_b - m_c \rangle_{mes} = \frac{3(M_{B^*} - M_{D^*}) + M_B - M_D}{4} = 3338\,\text{MeV}. \tag{2b}$$

The ratio $\frac{m_s}{m_u}$ has the same value ±2.5% for mesons and baryons.

$$\left(\frac{m_s}{m_u} \right)_{Bar} = \frac{M_\Delta - M_N}{M_{\Sigma^*} - M_\Sigma} = 1.53 = \left(\frac{m_s}{m_u} \right)_{Mes} = \frac{M_\rho - M_\pi}{M_{K^*} - M_K} = 1.61 \tag{3}$$

Three magnetic moment predictions with no free parameters [8,9]

$$\mu_\Lambda = -0.61\,\text{n.m.} = -\frac{\mu_p}{3} \cdot \frac{m_u}{m_s} = -\frac{\mu_p}{3} \frac{M_{\Sigma^*} - M_\Sigma}{M_\Delta - M_N} = -0.61\,\text{n.m.} \tag{4}$$

$$-1.46 = \frac{\mu_p}{\mu_n} = -\frac{3}{2} \qquad (5a)$$

$$\mu_p + \mu_n = 0.88 \, \text{n.m.} = \frac{M_p}{3m_u} = \frac{2M_p}{M_N + M_\Delta} = 0.865 \, \text{n.m.} \qquad (5b)$$

QCD calculations have not yet explained such remarkably successful simple constituent quark model results. A search for new experimental input to guide us is therefore of interest.

EXOTIC HADRONS AND HEAVY FLAVORS - A WINDOW INTO QCD

Additional input comes from two striking features of the hadron spectrum:

- Absence of strongly bound multiquark exotic states like a dipion with a mass less than two pion masses or a dibaryon bound by 100 MeV.

- Nuclear structure described by three-quark clusters called nucleons.

The constituent quark model gives a very simple answer [10]. The one gluon exchange ansatz for $\frac{V(q\bar{q})_8}{V(q\bar{q})_1}$ and $\frac{V(qq)_6}{V(qq)_{3*}}$ gives:

- Color-exchange color-electric interaction saturates [11] - no forces between color singlet hadrons.

- Color electric energy unchanged by color recoupling.

- Color magnetic qq forces repulsive for a single flavor - attractive between s quark in D_s or B_s and u and d quarks in proton.

- Energy gain by color-spin recoupling can bind H (hexaquark) [12], charmed and beauty pentaquarks [1,13–18] $P_c = \bar{c}suud$; $P_b = \bar{b}suud$.

The validity of this simple picture still remains to be confirmed by experiment since no experimental information is yet available about short-range color-sextet or color-octet two-body interactions. All constituent quark model successes with a two-body color-exchange interaction [6–8,11] and all hadron spectroscopy without exotics including scattering depend only upon $(\bar{q}q)_1$ and $(qq)_{3*}$ interactions Baryon-nucleon scattering in the (u,d,s) sector is dominated by a short-range color-magnetic repulsion (well-known repulsive core in the nucleon-nucleon interaction). Meson-hadron scattering in the (uds) sector must have either a quark or an antiquark in the meson with the same flavor as a quarks in the other hadron. A qq pair of the same flavor has a repulsive color-magnetic interaction keeping the two hadrons apart. A $\bar{q}q$ pair of the same flavor can annihilate and produce a hadron resonance. Thus hadron-hadron scattering in the (u,d,s) sector is dominated either by qq repulsion or by resonances produced by $\bar{q}q$ annihilation.

Only with more than three flavors can the $(qq)_6$ or $(\bar{q}q)_8$ interactions be observed in realistic scattering experiments (unrealistic cases like $K^-\Delta^-$ and ϕN are excluded) with no common flavor between beam and target. Thus the possible existence of exotic hadrons remains crucial to understanding how QCD makes hadrons from quarks and gluons [19].

The H dibaryon [12] was shown to have a gain in color-magnetic energy over the $\Lambda\Lambda$ system [12,20]. But a lattice calculation [21] showed a repulsive Λ-Λ interaction generated by quark exchange [22,23] not included in simple model calculations which could well prevent the six quarks from coming close enough together to feel the additional binding of the short range color-magnetic interaction. Pentaquarks, shown [13-18] to have a color-magnetic binding roughly equal to the H, have no possible quark exchange force in the lowest decay channel $D_s N$ [22]. The simplest lattice calculation can easily be done in parallel with the more complicated H calculation both in the symmetry limit where all light quarks have the same mass and with $SU(3)$ symmetry breaking. Comparing results may provide considerable insight into the physics of QCD in multiquark systems even if the pentaquark is not bound. However, no such lattice calculation has been done or is planned.

GOOD VERTEX DETECTORS CREATE NEW BALL GAME

Without vertex detectors a pentaquark decay signal appears against a large combinatorial background. Peaks in a mass spectrum can arise from statistical fluctuations in the background. Standard statistical considerations and good signatures like $p\phi\pi^-$ are needed to analyze data.

Good vertex detectors eliminate all combinatorial background. An event with a decay proton from a secondary vertex cannot be a statistical fluctuation of known physics. If its mass differs from that of known weakly decaying baryons, it indicates a new as yet unknown particle. Searches for such secondary-vertex protons are open searches for new weakly decaying baryons and might even find new physics beyond the standard model. This point has not been noted in previous articles on pentaquark searches [24,28]

A weakly-decaying baryon should produce not only a peak in the mass spectrum, but also a tail below corresponding to decays where neutral particles have escaped detection. Events containing a muon or electron are expected and particularly significant since weak decays via a W always have semileptonic modes.

The original suggestion [13] directing the search to striking signatures like $p\phi\pi^-$ no longer holds with good vertex detectors. One might better begin with an extremely stringent cut on the proton to select *all* events with protons really from a secondary vertex. Even and odd prong events correspond respectively to decays of neutral and charged particles. A charged pentaquark with the structure of a D_s bound to a neutron rather than a proton would be more apt to decay to a final state containing a neutron rather than a proton, unless the final state baryon is a Δ° or $N^{*\circ}$ decaying to $p\pi^-$. This immediately suggests looking for $p\pi^-$ resonances in all odd prong events.

Quasi-two-body events with protons definitely coming from a secondary vertex have unique energies providing a striking signal. Selection of a proton with an energy near the appropriate value would enhance signal/noise ratios by a considerable factor in modes like $p\pi^-$ decay which would be impossible signatures without vertex detectors.

The particle-ID and vertex detector in the Fermilab experiment [27] are not sufficiently reliable to identify events as definitely arising from secondary protons. Their small number of events do not provide sufficient evidence for a new particle, but are not easily dismissed as due to known systematics. Better experiments with reliable vertex detectors are needed to resolve this question.

ACKNOWLEDGEMENT

Many stimulating discussions with the Tel Aviv experimental group [25,28] are gratefully acknowledged and in particular Danny Ashery, Sharon May-Tal Beck, Gilad Hurvits, Jechiel Lichtenstadt and Murray Moinester.

REFERENCES

1. Harry J. Lipkin, Nucl. Phys. A, in press and references therein
2. E. M. Levin and L. L. Frankfurt, Zh. Eksperim. i. Theor. Fiz.-Pis'ma Redakt (1965) 105; JETP Letters (1965) 65
3. H.J. Lipkin and F. Scheck, Phys. Rev. Lett. 16 (1966) 71
4. Harry J. Lipkin, Physics Letters B335 (1994) 500
5. Ya. B. Zeldovich and A.D. Sakharov, Yad. Fiz 4 (1966)395; Sov. J. Nucl. Phys. 4 (1967) 283
6. I. Cohen and H. J. Lipkin, Phys. Lett. 93B, (1980) 56
7. Harry J. Lipkin, Phys. Lett. B233 (1989) 446; Nuc. Phys. A507 (1990) 205c
8. A. De Rujula, H. Georgi and S.L. Glashow, Phys. Rev. D12 (1975) 147
9. Harry J. Lipkin, Nucl. Phys. A478, (1988) 307c
10. H.J. Lipkin, Phys. Lett. 198B (1987) 131
11. H.J. Lipkin, Phys. Lett. 45B (1973) 267
12. R. L. Jaffe, Phys. Rev. Lett. 38, (1977) 195
13. Harry J. Lipkin, in Hadrons, Quarks and Gluons, Proceedings of the Hadronic Session of the XXIInd Rencontre de Moriond, Edited by J. Tran Thanh Van, Editions Frontieres, Gif Sur Yvette - France (1987), p.691
14. Harry J. Lipkin, In The Elementary Structure of Matter, Proceedings of the Workshop, Les Houches, France, 1987 Edited by J.-M. Richard et al, Springer-Verlag (1987) p.24

15. Harry J. Lipkin, in Hadron '87, Proceedings of the Second International Conference on Hadron Spectroscopy, KEK Tsukuba, Japan, edited by Y. Oyanagi, K. Takamatsu and T.Tsuru, KEK Report 87-7 (1987), p.363.
16. Harry J. Lipkin, In Proceedings of PANIC '87, XI International Conference on Particles and Nuclei, Nucl. Phys. A478, 307c (1988)
17. Harry J. Lipkin, Phys. Lett. 195B, (1987) 484
18. C. Gignoux, B. Silvestre-Brac and J. M. Richard, In The Elementary Structure of Matter, Proceedings of the Workshop, Les Houches, France, 1987 Edited by J.-M. Richard et al, Springer-Verlag (1987) p.42; Phys. Lett. **B193** (1987) 323
19. Harry J. Lipkin, In Intersections Between Particle and Nuclear Physics, Proc. Conf. on The Intersections Between Particle and Nuclear Physics, Lake Louise, Canada, 1986 Edited by Donald F. Geesaman AIP Conference Proceedings No. 150, p. 657
20. J. L. Rosner, Phys. Rev. D 33 (1986) 2043
21. P. MacKenzie and H. Thacker, Phys. Rev. Letters 65, 2539 (1985)
22. Harry J. Lipkin, In Proceedings of the International Symposium on The Production and Decay of Heavy Flavors, Stanford (1987) Edited by Elliott D. Bloom and Alfred Fridman, Annals of the New York Academy of Sciences, Vol. 535 (1988) p.438
23. H. Thacker, private communication
24. Harry J. Lipkin, in Proceedings of the Rheinfels Workshop 1990 on Hadron Mass Spectrum, St.Goar at the Rhine, Germany, Sept. 3-6, 1990, Nucl. Phys. B(Proc. Suppl.) 21 (1991) 258
25. S. May-Tal Beck, (for FNAL E791 Collab.) In Proceedings of the 1994 Annual Meeting of the Division of Particles and Fields, Albuquerque, N.M., S. Seidel, Ed., World Scientific, (1995) 1177
26. E.M. Aitala et al., FERMILAB-Pub-97/118-E and to be published
27. Daniel Ashery, These Proceedings
28. M. A. Moinester, D. Ashery, L. G. Landsberg and H. J. Lipkin, Zeitschrift fur Physik A 356 (1996) 207

The Signatures Of Glueballs In J/ψ Radiative Decays

Zhenping Li

Physics Department, Peking University
Beijing, 100871, P.R. China

Abstract. In this talk, I shall discuss the signatures of glueballs in the J/ψ radiative decays. Further experimental and theoretical investigations are suggested.

The existence of glueballs and hybrids in nature has been one of the important predictions of the quantum chromodynamics(QCD). Considerable progress has been made recently in identifying the glueball candidates. In the scalar meson sector, the $f_0(1300)$, $f_0(1500)$ [1] and $f_0(1780)$ [2] have been established in the recent experiments and this has raised the possibility that these three scalars are the mixed states between the ground state glueball and two nearby $q\bar{q}$ nonet. The study has shown [3] that the observed properties of these states are incompatible with them being $q\bar{q}$ states, and one of them, in particular $f_0(1500)$, might be a ground state glueball. Moreover, the discovery [4] of non-strange decay modes of the state $\xi(2230)$ in addition to the strange decay channel observed in earlier experiments have also fueled the speculation of it being a tensor glueball state. The observed relative strength of each decay mode of the $\xi(2230)$ shows a remarkable flavor symmetry, which is one of the important characteristics of a glueball state. In this talk, I shall concentrate on the theoretical and experimental aspects of the J/ψ radiative decays, which has become increasingly important in identifying the glueball candidates.

The advantage of studying the glueball productions in J/ψ radiative decays is that the properties of the glueballs can be investigated not only via their decays but also through their productions. According to the perturbative QCD, the production of a light meson state R in the J/ψ radiative decay proceeds by the sequence $J/\psi \to \gamma + gg \to \gamma + R$. In leading order pQCD, its amplitude A is given by

$$A = \frac{1}{2}\sum \int \frac{d^4k}{(2\pi)^4} \frac{1}{k_1^2} \frac{1}{k_2^2} <(Q\bar{Q})_V|\gamma g^a g^b><g^a g^b|R>. \quad (1)$$

The summation is over the polarization vectors $\epsilon_{1,2}$ and color indices a,b of the intermediate gluons, whose momenta are denoted as $k_{1,2}$. Thus, there are three major components in evaluating the J/ψ radiative decays; the inclusive process $J/\psi \to \gamma + gg$ whose amplitude $<(Q\bar{Q})_V|\gamma g^a g^b>$ has been given reliably in pQCD, the process $gg \to R$ and the loop integral. The process $gg \to R$ for a glueball state has not been investigated before. We find [5] that it is reasonable to assume the amplitude $<g^a g^b|R>$ for both $q\bar{q}$ and glueball states having the form

$$\psi(R) = \begin{cases} \frac{1}{\sqrt{3}} P_{\mu\nu} G_{\mu\rho}^{1a} G_{\nu\rho}^{2a} F_0(k_1^2, k_2^2) & \text{for } 0^{++} \\ \epsilon_{\mu\nu} G_{\mu\rho}^{1a} G_{\nu\rho}^{2a} F_2(k_1^2, k_2^2) & \text{for } 2^{++} \end{cases} \quad (2)$$

where $P_{\rho\sigma} \equiv g_{\rho\sigma} - \frac{P_\rho P_\sigma}{m^2}$ for a resonance with mass m and momentum P_μ, and $\epsilon_{\rho\sigma}$ are the tensor for a tensor resonance, and satisfy the relations

$$\sum_\epsilon \epsilon_{\rho\sigma} \epsilon_{\rho'\sigma'} = \frac{1}{2}(P_{\rho\rho'} P_{\sigma\sigma'} + P_{\rho\sigma'} P_{\sigma\rho'}) - \frac{1}{3} P_{\rho\sigma} P_{\rho'\sigma'}. \quad (3)$$

A direct consequence from Eq. 2 is that ratio of the two gluon width between the scalar and the tensor states is

$$\frac{\Gamma(0^{++})}{\Gamma(2^{++})} = \frac{15}{4} \quad (4)$$

for both $q\bar{q}$ and glueball states assuming equal masses and form factors. Qualitatively one would expect that the total width for a tensor glueballs should be of order $O(25MeV)$ if the width for the scalar is at $O(100MeV)$ suggested by the states $f_0(1500)$ and $f_0(1780)$. Of course, these are circumstantial arguments for the $\xi(2230)$ being a tensor glueball as its total width is around $20 \sim 30$ MeV, and the experimental determination of the spin of $\xi(2230)$ is calling for.

The form factor $F(k_1^2, k_2^2)$ in Eq. 2 is well established for the $q\bar{q}$ states, while there is little information on this form factor for glueball states. Its determination for glueball states depends how much we understand the structure of their wavefunctions. Assuming that the $q\bar{q}$ and glueball states have the same form factor, the branching ratio for the J/ψ radiative decaying into a resonance R with the mass m has a general form [7];

$$B(J/\psi \to \gamma + R_J) = B(J/\psi \to \gamma + gg) C_{R_J} \Gamma(R_J \to gg) \frac{x|H_J|^2}{8\pi(\pi^2-9)} \frac{m}{M^2}, \quad (5)$$

where M is the mass of the state J/ψ and $x = 1 - \left(\frac{m}{M}\right)^2$. The coefficient C_R in Eq. 5 depends on the spin parity of the final resonance R, and it is

$$C_{R_J} \equiv 1 \ (0^{-+}); \frac{2}{3} \ (0^{++}); \frac{5}{2} \ (2^{++}). \tag{6}$$

The quantity $B(J/\psi \to \gamma + gg)$ is a branching ratio for the inclusive process. It is determined by the vertex $J/\psi \to \gamma gg$, and its numerical value has been well determined, which gives $B(J/\psi \to \gamma + gg) \approx 0.06 \sim 0.08$. The $H_J(x)$ in Eq. 5 is a loop integral, and it has been evaluated in the case of $R = q\bar{q}$ [6]. The quantity $\Gamma(R_J \to gg)$ represents the width of the resonance R decaying into the two gluon state gg, which determines the vertex $R \to gg$. Generally the decay of a resonance R into the two gluon state gg is not the same as its total decay width, since the gluon hadronization is not the major decay mode for a light $q\bar{q}$ meson. Thus, one can define a branching ratio $b(R_J \to gg)$ so that $\Gamma(R_J \to gg) = b(R_J \to gg)\Gamma_T$, which measures the gluonic content of a resonance R. Cakir and Farrar [7] argued that

$$b(R(q\bar{q}) \to gg) = O(\alpha_s^2) \simeq 0.1 \sim 0.2 \tag{7}$$

for a normal $q\bar{q}$ meson, while

$$b(R(G) \to gg) \simeq 0.5 \sim 1 \tag{8}$$

for a glueball state.

It is the branching ratio $b(R \to gg)$ for a resonance R that can be extracted from the data for $B(J/\psi \to \gamma + R)$ and Γ_T with the theoretical input of the loop integral $x|H(x)|^2$ in Eq. 5. The numerical results from Ref. [6] show that the loop integral $x|H(x)|^2 \approx 35 \sim 40$ for the scalar and tensor states with masses around 1.5 GeV in J/ψ radiative decays. Thus, one can rewrite Eq. 5 as [5]

$$10^3 B(\psi \to \gamma 2^{++}) = \left(\frac{m}{1.5 \ GeV}\right)\left(\frac{\Gamma_{R \to gg}}{26 \ MeV}\right)\frac{x|H_T|^2}{34}$$
$$10^3 B(\psi \to \gamma + 0^{++}) = \left(\frac{m}{1.5 \ GeV}\right)\left(\frac{\Gamma_{R \to gg}}{96 \ MeV}\right)\frac{x|H_S|^2}{35}, \tag{9}$$

A straightforward evaluation shows that the branching ratios $b(R \to gg)$ extracted from the J/ψ radiative decay data for the established $q\bar{q}$ mesons, such as $f_2(1270)$ and $f_2(1525)$, clearly satisfy Eq. 7, while the existing data for $f_0(1500)$ and $f_J(1710)$ may well be the examples of Eq. 8. The particle data group [8] gives

$$B(J/\psi \to \gamma + f_0(1500)) = (0.82 \pm 0.15) \times 10^{-3}, \tag{10}$$

which is in good agreement with the recent BES results [11] in $J/\psi \to \gamma f_0(1500) \to \gamma \pi^0 \pi^0$, which translates into

$$b(f_0(1500) \to gg) = 0.64 \pm 0.11, \tag{11}$$

assuming that $\Gamma_T \approx 120 \pm 20$ MeV. This suggests that the resonance $f_0(1500)$ should have a large glueball component in its wavefunction. The recent results from the BES group suggested that the $f_J(1710)$ be separated into $f_2(1690)$ and $f_0(1780)$ states, and the scalar $f_0(1780)$ is consistent with the analysis in Ref [10] in which a scalar with mass 1.75 GeV and 160 MeV width is reported. Its decay into the 4π channel is very large and dominated by the $\sigma\sigma$ contributions. The branching ratio $B(J/\psi \to \gamma + f_0(1750))$ is found to be

$$B(J/\psi \to \gamma f_0(1750))B(f_0(1750) \to 4\pi) \approx 1.0 \times 10^{-3} \tag{12}$$

with the total width 160 MeV. This corresponds to

$$b(f_0(1710) \to gg) \geq 0.5. \tag{13}$$

Thus, Eqs. 11 and 13 suggest that the ground state glueball should be mixed with the nearby $SU(3)$ $q\bar{q}$ nonet. The experimental consequences of the configuration mixings between the $q\bar{q}$ nonet and glueballs in the J/ψ radiative decays and the $\gamma\gamma$ collisions have been discussed extensively in Ref. [5].

Now, we examine the $\xi(2230)$ and its implications. The data from BES collaboration [4] for $J/\psi \to \gamma + \xi(2230)$ in the $K\bar{K}$ and $p\bar{p}$ channels are [4]

$$B(J/\psi \to \gamma + \xi(2230))B(\xi(2230) \to K^+K^-) = (3.3 \pm 2.5) \times 10^{-5}$$
$$B(J/\psi \to \gamma + \xi(2230))B(\xi(2230) \to K^0_s K^0_s) = (2.7 \pm 2.0) \times 10^{-5}$$
$$B(J/\psi \to \gamma + \xi(2230))B(\xi(2230) \to p\bar{p}) = (1.5 \pm 1.0) \times 10^{-5}, \tag{14}$$

while the recent results from JETSET [9] has set a very strict upper limit on $B(\xi \to p\bar{p})B(\xi \to K\bar{K})$

$$B(\xi \to p\bar{p})B(\xi \to K\bar{K}) < 7.5 \times 10^{-5}. \tag{15}$$

This gives a lower limit for the branching ratio $B(J/\psi \to \gamma + \xi(2230))$

$$b(J/\psi \to \gamma\xi(2230)) \geq 2.0 \times 10^{-3}. \tag{16}$$

Because $\xi(2230)$ has a narrow width, $\Gamma_T = 20$ MeV, the resulting branching ratio $b(R \to gg)$ from Eq. 9 would be

$$b(\xi(2230)|J = 0^{++} \to gg) \geq 10 \pm 5$$
$$b(\xi(2230)|J = 2^{++} \to gg) \geq 2 \pm 1. \tag{17}$$

This suggests that a scalar $\xi(2230)$ may have already been excluded by the data, while a tensor $\xi(2230)$ is still possible considering the uncertainties in our approach. To clarify these questions requires the measurements with better statistics in both $p\bar{p} \to K\bar{K}$ and $j/\psi \to \gamma\xi(2230) \to \gamma p\bar{p}$. Eq. 15 also suggests that the two body final states are not the major decay modes for the $\xi(2230)$, and the recent analysis in $J/\psi \to \gamma 4\pi$ shows [13] a $f_2(2220)$

state, whose decay is dominantly via $f_2(1270)\sigma$. Further analysis with better statistics are needed to confirm that $f_2(2220)$ is indeed $\xi(2230)$. The analysis $J/\psi \to \gamma K\bar{K}\pi\pi$ channel would be important, as the flavor symmetry would also suggest that $\xi(2230) \to f_2(1525)\sigma$ would be another important channel.

Theoretically, the remaining question is the theoretical uncertainties of the loop integral $x|H(x)|^2$ in Eq. 5. The form factor $F(k_1^2, k_2^2)$ for glueballs is still unknown, and to obtain it requires better knowledge of glueball wavefunctions. Another source of such uncertainty is the relativistic effects in $R(q\bar{q}) \to gg$, which was shown [12] to be very important for light quark mesons. Thus, a lot of more theoretical and experimental works remains to be done to understand the nature of glueballs and their differences with the normal $q\bar{q}$ states, which in turn will help us to identify the glueball states with more confidence.

Acknowledgment

The extensive discussions with F. Close, G. Farrar, C. Meyer, Zhu Yucan and Shen Xiaoyan are gratefully acknowledged. This work was supported in part by Peking University.

REFERENCES

1. V.V. Anisovich et al., Phys. Lett. **323B**, 233(1994); C. Amsler et al., ibid. **342B**, 433(1994); **291B**, 347(1992); **340B**, 259(1994).
2. BES Collaboration, J.Z. Bai et al., Phys. Rev. Lett. **77**, 3959(1996).
3. C. Amsler and F.E.Close, Phys Lett **B353** (1995) 385; Phys Rev **D53**, 295(1996).
4. J. Bai et al., BES collaboration, Phys. Rev. Lett. **76**, 3502(1996).
5. F. E. Close, G. Farrar and Zhenping Li, Phys. Rev. **D55**, 5749(1997).
6. J. G. Körner, J.H.Kühn, M. Krammer, and H. Schneider, Nucl. Phys. **B229**, 115(1983), J. G. Köner, J. H. Kühn, and H. Schneider, Phys. Lett. **120B**, 444(1983).
7. M.B. Cakir and G. Farrar, Phys. Rev. **D50**, 3268(1994).
8. Particle Data Group, Phys. Rev. **D54**, 1(1996).
9. A. Buzzo, et al. JETSET Collaboration, to be published.
10. D. V. Bugg, et al., Phys. Lett. **353B**, 378(1995).
11. BES Collaboration, to be published.
12. Zhenping Li, F. E. Close and T. Barnes, Phys. Rev. **D43**, 2621(1991).
13. Zhu Yucan, private communication.

Glueball Spectroscopy in a Many Body QCD Hamiltonian Approach[1]

S. R. Cotanch, A. Szczepaniak, E. S. Swanson and C.- R. Ji

North Carolina State University, Raleigh, NC 27695-8202

Abstract. The low-lying glueball mass spectrum is calculated utilizing a new, comprehensive, relativistic many-body approach based upon the Coulomb gauge QCD Hamiltonian. Central to the formalism is the renormalization procedure for regularizing divergences and the Nambu-Goldstone realization of chiral symmetry. Standard many- body techniques are adopted to compute the constituent quark, gluon masses and condensates and glueball spectrum. Results are in good agreement with QCD sum-rule and quenched lattice gauge theory.

INTRODUCTION

Establishing that glueballs, hadrons with no explicit (valence) quark degrees of freedom, can be produced in the laboratory would be a crowning achievement for QCD and the Standard Model. Analyses of $p\bar{p}$ and J/ψ decay data suggest the existence of gluon rich states, especially the scalar $J^{\pi} = 0^{++}$ with mass about 1500 MeV, but clear confirmation is hampered by possible mixing complications with conventional meson states in this mass region. Lattice gauge measurements also provide evidence for glueballs but offer limited structure information which might clarify the degree of meson mixing.

The purpose of this talk is to present theoretical predictions for gluonia which confirm lattice measurements. Further, and most significantly, because our approach is based upon a parton, QCD Hamiltonian formulation, it provides direct dynamical insight into the nature of glueballs and the prospect for understanding their decays and mixing with the quark sector. In previous publications [1-3] we have reported elements of our formalism which we submit is an attractive framework for investigating hadron structure. The next two sections briefly detail and highlight our findings.

---

[1] Supported by DOE grants DE-FG05-88ER40461, DE-FG05-90ER40589 and DE-FG-96ER40944.

THE EFFECTIVE HAMILTONIAN

The development of the effective Hamiltonian used in our calculations is a multi-step procedure for which, due to space limitations, only a synopsis is given. Beginning with the Coulomb gauge QCD Hamiltonian [4] H, we define $V(g)$ by $H = H_0 + V(g)$, with $H_0 = H$ for coupling constant $g = 0$, and introduce a perturbative Fock space spanned by eigenfunctions Φ_n of H_0 having eigenvalues E_n^0. Next V is expanded in powers of g, $V(g) = gV_1 + g^2V_2 + ...$, which is valid in the high energy regime. Then the divergent matrix elements of V in this basis are regularized by a cut-off procedure with energy scale Λ_0 such that $\langle\Phi_n|V|\Phi_m\rangle = 0$ for $|E_n^0 - E_m^0| > \Lambda_0$. The attending, but unacceptable, Λ_0 dependence can be removed by a renormalization scheme developed by Głazek and Wilson [5] in which a similarity transformation with unitary matrix S generates an equivalent Hamiltonian, $H^\Lambda = SH^{\Lambda_0}S^{-1}$, that is also band diagonal but at a much lower energy scale Λ, of order the hadron masses. The procedure is actually implemented by a series of infinitesimal similarity transformations from scale Λ_0 to Λ enabling all intermediate quantities to remain small, consistent with the perturbative expansion above, and to map the cut-off sensitivity to Λ which determines the appropriate counter-terms which must be added to H^{Λ_0}. These counter-terms eliminate the cut-off sensitivity and divergences when Λ_0 goes to infinity. The counter-term remnants in this limit are further adjusted by fitting known observables.

In our application we have carried out this procedure to order g^N with $N = 2$ which is insufficient to generate confinement and this is the major deficiency in our approach. Although it is doubtful that any practical value of N will generate confinement, we are investigating if larger values produce confinement precursors and for now introduce a linear confining potential with slope specified by the string tension value .18 GeV^2 leading to a semi-phenomenological Hamiltonian H_{phen}.

Finally, motivated by the legacy of many-body nuclear physics, we have determined an improved Fock state truncation approximation can be obtained by performing a BCS (Bogoliubov) similarity transformation [6]. This generates our effective Hamiltonian $H_{eff} = S(\phi)H_{phen}S(\phi)^{-1}$ with BCS rotation angle ϕ from the the bare parton Fock state creation/annihilation operator basis to the dressed, quasi-particle basis. It is significant to note that our approach explicitly preserves chiral symmetry since H_{eff} essentially commutes with the chiral charge operator and, as shown elsewhere [1,2], the Gell-Mann-Oakes-Renner relation, $(f_\pi m_\pi)^2 = -2m_q\langle\bar{q}q\rangle$, directly follows. Here f_π, m_π, m_q and $\langle\bar{q}q\rangle$ are the pion decay constant, pion mass, current quark mass and quark condensate, respectively. We therefore obtain the Nambu-Goldstone realization of chiral symmetry since in the chiral limit, $m_q \to 0 \Rightarrow m_\pi \to 0$, consistent with Goldstone's theorem.

APPROXIMATE MANY-BODY SOLUTIONS

Results are now presented from approximately solving $H_{eff}\Psi = E\Psi$ for the vacuum, using the BCS approximation, and for the gluon sector using the Tamm-Dancoff (TDA) and the random phase approximations (RPA).

For the BCS vacuum a variational calculation, $\delta\langle 0|H_{eff} - E|0\rangle = 0$, is performed. This leads to the gap or Schwinger-Dyson self-energy equation from which the quasi-particle energies and BCS angle can be determined: $\sin\phi(k) = M_q/\epsilon_q$ and $\sinh\phi(k) = M_g^2/2k\epsilon_g$ with energy dispersion $\epsilon_c = (M_c(k)^2 + k^2)^{1/2}$. The constituent quark and gluon masses are then obtained by extrapolating to zero momentum yielding $M_q(0) \simeq 180\ MeV$, $M_g(0) \simeq 950\ MeV$, respectively. The corresponding quark and gluon condensates are then computed to be $\langle \bar{q}q \rangle \simeq -(100\ MeV)^3$ and $\langle \alpha G^2 \rangle \simeq .035\ GeV^4$. Note that while the constituent quark mass is roughly consistent with quark/bag model studies the quark condensate is lower than the lattice value of about $-(250\ MeV)^3$. The gluon condensate is in reasonable agreement with sum rule results for both a running and fixed, $\alpha_s \simeq .4$, coupling constant.

In both the TDA and RPA applications the calculation is limited to two valence gluons which generates glueball states having the following quantum numbers: $J^{\pi c}(^{2S+1}L_J) = 0^{++}(^1S_0), 0^{-+}(^3P_0), 2^{++}(^1D_2 + {}^5S_2), 2^{-+}(^3P_2 + {}^3F_2)$. The TDA equations of motion are truncated at the 1 particle-1 hole level and lead to the glueball mass spectrum displayed in Fig. 1. Also displayed are three different lattice gauge calculations [7–9] that use the quenched approximation which is the consistent comparison since our treatment excludes the quark sector. We believe this is the first QCD based model to reproduce the lattice measurements. Finally, our RPA calculation, which is a coupled channels treatment involving an additional wavefunction component, produces to within about 1% the same spectrum. This is because the off-diagonal coupling interactions, which entail differences in gluon energies, are suppressed for this low momentum calculation due to the large gluon constituent mass. Since the quark constituent mass is much smaller we expect significant TDA-RPA differences in our quark sector calculation for meson and baryon masses which is in progress.

SUMMARY AND CONCLUSIONS

The important results emerging from our approach are: 1) a renormalized effective Hamiltonian that preserves chiral symmetry and phenomenologically provides confinement; 2) the BCS vacuum solution exhibits dynamical chiral symmetry breaking and produces reasonable constituent quark and gluon masses; 3) the BCS gluon condensate agrees with QCD sum-rule results but the quark condensate is too low; 4) the TDA and RPA glueball solutions are essentially identical and reproduce quenched lattice gauge results.

We conclude that our results support the existence of a scalar glueball state in the mass range 1.5 to 1.7 GeV which has been strongly conjectured by previous experimental and lattice gauge studies.

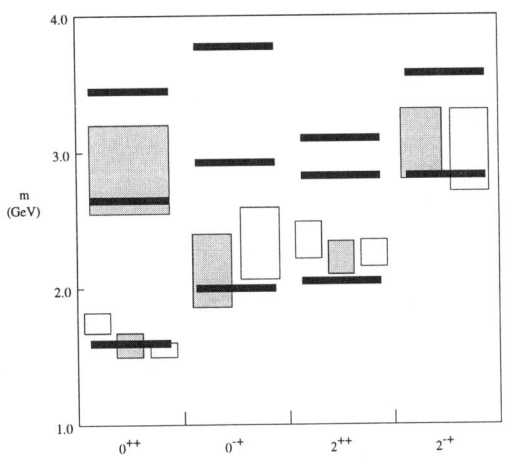

Fig. 1. Lattice gauge (boxes) and model glueball spectra (black bars). Light, open and dark boxes represent lattice measurements from Refs. 7,8 and 9, respectively.

REFERENCES

1. Szczepaniak, A., Swanson, E. S., Ji, C.-R., and Cotanch, S. R., *Phys. Rev. Lett.* **76**, 2011 (1996).
2. Cotanch, S. R., Szczepaniak, A., Swanson, E. S., and Ji, C.-R., *Proceedings of the International Workshop on Quark Confinement and the Hadron Spectrum II*, Como, Italy (World Scientific, 1997).
3. Szczepaniak, A., and Swanson, E. S., *Phys. Rev.* D **55**, 3987 (1997).
4. Lee, T. D., *Particle Physics and Introduction to Field Theory* (Harwood Academic Publishers, N. Y., 1981).
5. Głazek, St., and Wilson, K. G., *Phys. Rev.* D **48**, 5863 (1993); *ibid.* **49**, 4214 (1994).
6. Le Yaouanc, A., *et al*, *Phys. Rev.* D **29**, 1233 (1993); *ibid.* **31**, 137 (1994).
7. UKQCD Collaboration, Bali, G. S., *et al*, *Phys. Lett.* B **309**, 378 (1993); Report No. hep-lat/9304012 (unpublished).
8. Chen, H., Sexton, J., Vaccarino, A., and Weingarten, D., *Nucl. Phys.* (Proc. Suppl.) **22**, 357 (1994); Report No. hep-lat/9401020 (unpublished).
9. Teper, M., Report No. OUTP-95-06P (to be published).

Electromagnetic Form Factors of the Nucleon

R. Bijker[1] and A. Leviatan[2]

[1] *Instituto de Ciencias Nucleares, Universidad Nacional Autónoma de México, A.P. 70-543, 04510 México D.F., México*
[2] *Racah Institute of Physics, The Hebrew University, Jerusalem 91904, Israel*

Abstract. We reanalyze the world data on the electromagnetic form factors of the nucleon. The calculations are performed in the framework of an algebraic model of the nucleon combined with vector meson dominance.

I INTRODUCTION

The electromagnetic form factors of the nucleon and its excitations (baryon resonances) provide a powerful tool to investigate the structure of the nucleon [1]. These form factors can be measured in electroproduction as a function of the four-momentum squared $q^2 = -Q^2$ of the virtual photon. Especially the transition to the region of high momentum transfer, for which the methods of perturbative QCD apply, is currently of much interest [2].

In this contribution we present an analysis of the world data for the elastic form factors $G_{E/M}^{p/n}$. Our method is a combination of a recently introduced algebraic model of the nucleon [3] and vector meson dominance.

II ALGEBRAIC MODEL

The algebraic approach provides a unified treatment of various constituent quark models [3], such as harmonic oscillator quark models and collective models. In this contribution we employ a collective model of the nucleon in which baryon resonances are interpreted as vibrational and rotational excitations of an oblate top. There are two fundamental vibrations: a breathing mode and a two-dimensional vibrational mode, which are associated with the $N(1440)P_{11}$ Roper resonance and the $N(1710)P_{11}$ resonance, respectively. The negative parity resonances of the second resonance region are interpreted as rotational excitations. Since each vibrational mode has its own characteristic frequency, this collective model has no problem with the relative energy of the Roper resonance with respect to the negative parity resonances.

In [4] we studied the elastic electromagnetic form factors of the nucleon. These calculations include anomalous magnetic moments for the proton and the neutron, as well as a flavor dependent distribution functions of the charge and magnetization. Supposedly, the anomalous magnetic moments and the flavor dependence arise as effective parameters, since the coupling to the meson cloud surrounding the nucleon was not included explicitly. According to [4] the electric and magnetic form factors of the nucleon, when folded with a distribution of the charge and magnetization, can be expressed in terms of a common intrinsic dipole form factor

$$G_E^p(Q^2) = G_M^p(Q^2) = g(Q^2) = 1/(1+\gamma Q^2)^2 ,$$
$$G_M^n(Q^2)/G_M^p(Q^2) = -2/3 , \qquad G_E^n(Q^2) = 0 . \qquad (1)$$

Note that the Sachs form factors of Eq. (1) do not contain anomalous magnetic moments nor involve flavor dependent distribution functions. In order to study the coupling to the meson cloud we express the Sachs form factors in terms of the isoscalar and isovector Dirac ($F_1^{S/V}$) and Pauli ($F_2^{S/V}$) form factors

$$F_1^S(Q^2) = g(Q^2)\frac{1+\frac{1}{3}\tau}{1+\tau} , \qquad F_1^V(Q^2) = g(Q^2)\frac{1+\frac{5}{3}\tau}{1+\tau} ,$$
$$F_2^S(Q^2) = -F_2^V(Q^2) = -\frac{2}{3}g(Q^2)\frac{1}{1+\tau} , \qquad (2)$$

with $\tau = Q^2/4M^2$.

III MESON CLOUD COUPLINGS

The effects of the meson cloud surrounding the nucleon are taken into account in a similar way as in [5], *i.e.* by including the coupling to the isoscalar vector mesons ω and ϕ and the isovector vector meson ρ. These contributions are studied phenomenologically by parametrizing the Dirac and Pauli form factors as

$$F_1^S(Q^2) = g(Q^2)\frac{1+\frac{1}{3}\tau}{1+\tau}\left[\beta^S + \beta_\omega \frac{m_\omega^2}{m_\omega^2+Q^2} + \beta_\phi \frac{m_\phi^2}{m_\phi^2+Q^2}\right] ,$$
$$F_1^V(Q^2) = g(Q^2)\frac{1+\frac{5}{3}\tau}{1+\tau}\left[\beta^V + \beta_\rho \frac{m_\rho^2}{m_\rho^2+Q^2}\right] ,$$
$$F_2^S(Q^2) = -\frac{2}{3}g(Q^2)\frac{1}{1+\tau}\left[\alpha^S + \alpha_\omega \frac{m_\omega^2}{m_\omega^2+Q^2} + \alpha_\phi \frac{m_\phi^2}{m_\phi^2+Q^2}\right] ,$$
$$F_2^V(Q^2) = \frac{2}{3}g(Q^2)\frac{1}{1+\tau}\left[\alpha^V + \alpha_\rho \frac{m_\rho^2}{m_\rho^2+Q^2}\right] . \qquad (3)$$

The large width of the ρ meson ($\Gamma_\rho = 151$ MeV) is taken into account by making the replacement [5]

$$\frac{m_\rho^2}{m_\rho^2 + Q^2} \to \frac{m_\rho^2 + 8\Gamma_\rho m_\pi/\pi}{m_\rho^2 + Q^2 + (4m_\pi^2 + Q^2)\Gamma_\rho \alpha(Q^2)/m_\pi} \tag{4}$$

with

$$\alpha(Q^2) = \frac{2}{\pi} \left[\frac{4m_\pi^2 + Q^2}{Q^2}\right]^{1/2} \ln\left(\frac{\sqrt{4m_\pi^2 + Q^2} + \sqrt{Q^2}}{2m_\pi}\right) . \tag{5}$$

The coefficients $\beta^{S/V}$ and $\alpha^{S/V}$ are determined by the electric charges and the magnetic moments of the nucleon, respectively.

For small values of the momentum transfer the Dirac and Pauli form factors are dominated by the meson dynamics and reduce to a monopole form, whereas for high values they show the Q^2 dependence as predicted by perturbative QCD

$$F_1^{S/V} \sim 1/Q^4 , \qquad F_2^{S/V} \sim 1/Q^6 . \tag{6}$$

IV RESULTS

Recently, the electromagnetic form factors of the nucleon have been remeasured (or reanalyzed) [6,7]. In Figs. 1–4 we show a compilation of the world data on the electromagnetic form factors of the nucleon. In the present calculation the coefficients β_M and α_M (with $M = \rho, \omega, \phi$) and the scale parameter γ in the dipole form factor $g(Q^2)$ are determined in a simultaneous fit to all four electromagnetic form factors of the nucleon with $Q^2 \leq 10$ (GeV/c)^2 (solid lines, Fit 1). As usual, the form factors are scaled by the standard dipole fit $F_D = 1/(1 + Q^2/0.71)^2$. The oscillations around the dipole values are due to the meson cloud couplings. The magnetic form factors show an interesting behavior. Whereas G_M^p first decreases with respect to the dipole, the new measurements of G_M^n show an increase with respect to F_D [7,8]. This behavior is reproduced by the present calculation.

The electric form factor of the neutron is the least known. Unlike for the proton, the Rosenbluth separation of G_E^n from G_M^n for a neutron target is difficult for all values of Q^2: for small Q^2 because of the small size of G_E^n compared to G_M^n, and for large Q^2 because the magnetic component dominates both the angular dependent and angular independent term in the cross section. For this reason we have carried out a second fit, in which G_E^n is excluded from the fitting procedure and replaced by the proton and neutron charge radii (dashed lines, Fit 2). In this calculation the neutron charge radius (the

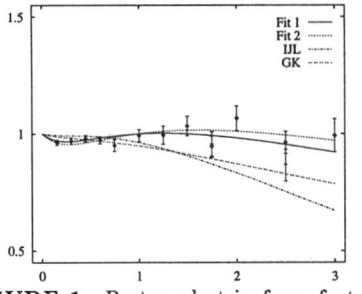

FIGURE 1. Proton electric form factor G^p_E/F_D as a function of Q^2 in (GeV/c)^2.

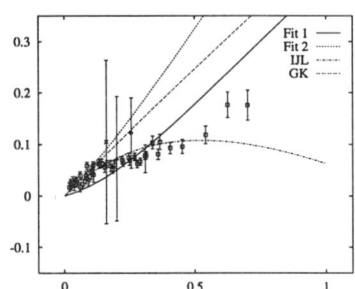

FIGURE 2. Neutron electric form factor G^n_E/F_D.

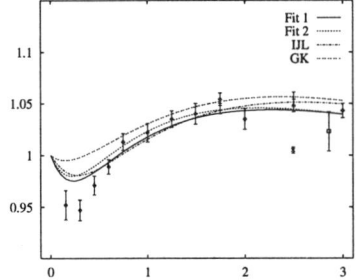

FIGURE 3. Proton magnetic form factor $G^p_M/\mu_p F_D$.

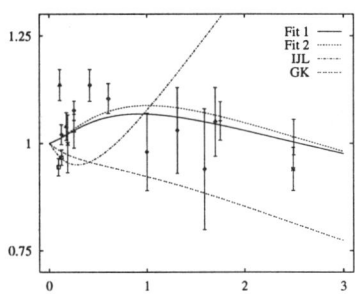

FIGURE 4. Neutron magnetic form factor $G^n_M/\mu_n F_D$.

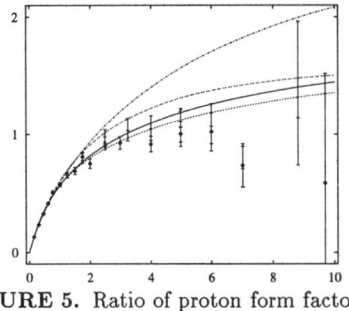

FIGURE 5. Ratio of proton form factors $Q^2 F^p_2/F^p_1$.

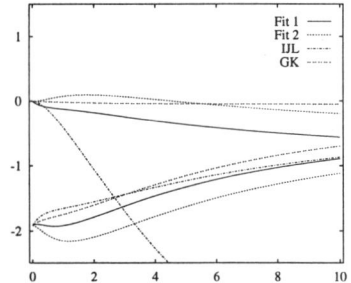

FIGURE 6. Neutron form factors F^n_1/F_D and F^n_2/F_D.

slope of G^n_E in the origin) is reproduced, but the existing data for G^n_E are overpredicted. The changes for the other form factors are minor.

In Fig. 5 we show the scaling property of the proton form factors: $Q^2 F^p_2/F^p_1 \sim 1$. The Dirac and Pauli form factors of the neutron are presented in Fig. 6. Since in our calculations the Dirac form factor is small $F^n_1 \approx 0$, we find $G^n_E \approx -\tau G^n_M$. For $\tau \approx 1$ ($Q^2 \approx 4M^2$) the electric and magnetic form factor become comparable in size. The same effect was pointed out in [9].

For comparison we also show the results of two other calculations: the vector meson dominance model of Iachello, Jackson and Lande [5] (dash-dotted lines, IJL), and a hybrid model (interpolation between vector meson dominance and pQCD) by Gari and Krümpelmann [9] (dash-dashed lines, GK).

V CONCLUSIONS

We have presented a simultaneous analysis of the elastic electromagnetic form factors of the nucleon in the context of an algebraic model of the nucleon combined with vector meson dominance. For a phenomenological approach (as the present one) a good data set is a prerequisite. Whereas the proton form factors are relatively well known, there is still some controversy about the neutron form factors. New measurements of the polarization asymmetry in which the ratio of the electric and magnetic form factor of the neutron is extracted [7,10] may help to clarify the experimental situation.

In addition to the elastic form factors discussed in this contribution, there is currently much interest in the inelastic transition form factors [2]. We plan to extend the present approach to include the resonance form factors as well, in order to analyze all electromagnetic form factors within the same framework.

ACKNOWLEDGEMENTS

It is a pleasure to thank P.E. Bosted, E.E.W. Bruins and P. Stoler for sharing their respective compilations of the world data on the nucleon form factors. The work is supported in part by grant No. 94-00059 from the United States-Israel Binational Science Foundation (BSF), Jerusalem, Israel (A.L.) and by DGAPA-UNAM under project IN105194 (R.B.).

REFERENCES

1. See *e.g.* Baryons '95, Proceedings of the 7th International Conference on the Structure of Baryons, Eds. B.F. Gibson, P.D. Barnes, J.B. McClelland and W. Weise, World Scientific, Singapore, 1996.
2. P. Stoler, these proceedings, and in [1].
3. R. Bijker, F. Iachello and A. Leviatan, Ann. Phys. (N.Y.) **236**, 69 (1994).
4. R. Bijker, F. Iachello and A. Leviatan, Phys. Rev. C **54**, 1935 (1996).
5. F. Iachello, A.D. Jackson and A. Lande, Phys. Lett. B **43**, 191 (1973).
6. For a recent compilation see *e.g.* P.E. Bosted, Phys. Rev. C **51**, 409 (1995).
7. M. Ostrick, these proceedings.
8. B.H. Schoch, in [1].
9. M. Gari and W. Krümpelmann, Z. Phys. A **322**, 689 (1985); Phys. Lett. B **173**, 10 (1986).
10. V.D. Burkert, in [1].

Strange Skyrmion Molecules

Vladimir B.Kopeliovich, Boris E.Stern

*Institute for Nuclear Research of the Russian Academy of Sciences,
60th October Anniversary Prospect 7A, Moscow 117312*

Abstract. Composed skyrmions with B=2, strangeness content close to 0.5 and the binding energy of several tens of Mev are described. These skyrmions are obtained starting from the system of two B=1 hedgehogs located in different $SU(2)$ subgroups of $SU(3)$ and have the mass and baryon number distribution of molecular (dipole) type. The quantization of zero modes of skyrmion molecules and physics consequences of their existence are discussed.

1. The effective theory of chiral (meson) fields proposed at first by Skyrme provides an attractive possibility to describe mesons and baryons starting from the lagrangian written in terms of the chiral fields only. The existence of bound states of skyrmions opened new prospects for the applications of this theory also in nuclear physics [1]-[4]. The attempts have been made to describe the nuclei with $B = 2$ [1, 2], 3 [3] and 4 [4] as quantized bound skyrmions. For $B = 2$ most serious attempt has been done recently [2] where the nonzero modes were quantized for two $B = 1$ skyrmions in most attractive orientation. The binding energy of the deuteron was found to be $\sim 6 Mev$, close to the observed value $2.23 Mev$.

Baryonic systems with strangeness appeared as $SO(3)$ solitons [5] or as $SU(3)$ quantized bound $SU(2)$ skyrmions [6, 7].

An important question is that about the lowest in energy configuration for each value of the baryon number B in the $SU(3)$ configuration space. Such configuration should be used as a starting one for the quantization of zero as well as nonzero modes to get the observable spectrum of physical states. We investigated this problem in the $SU(3)$ extension of the Skyrme model in the sector with baryon number $B = 2$. The method has been developed allowing for minimization of the energy functional depending on 8 functions of 3 variables (instead of 3 functions in $SU(2)$ case) [8].

2. The expressions for the energy of the soliton can be obtained from the well known lagrangian of the Skyrme model extended to $SU(3)$. The static energy of skyrmions can be written as

$$E_{stat} = E_2 + E_4 + E_M \tag{1}$$

with

$$E_2 = \frac{F_\pi^2}{8}(\vec{L}_1^2 + \vec{L}_2^2 + ... + \vec{L}_8^2) \tag{2}$$

$$E_4 = \frac{1}{4e^2}\Big\{(\vec{s}_{12}+\vec{s}_{45})^2+(\vec{s}_{45}+\vec{s}_{67})^2+(\vec{s}_{67}-\vec{s}_{12})^2+\frac{1}{2}\Big((2\vec{s}_{13}-\vec{s}_{46}-\vec{s}_{57})^2+(2\vec{s}_{23}+\vec{s}_{47}-\vec{s}_{56})^2+$$

---

[1] The work supported by Russian Fund for Fundamental Research, grant 95-02-03868a

$$+(2\vec{s}_{34}+\vec{s}_{16}-\vec{s}_{27})^2+(2\vec{s}_{35}+\vec{s}_{17}+\vec{s}_{26})^2+(2\vec{s}_{36}+\vec{s}_{14}+\vec{s}_{25})^2+(2\vec{s}_{37}+\vec{s}_{15}-\vec{s}_{24})^2 \Big) \Big\} \quad (3)$$

$\vec{s}_{ik} = [\vec{L}_i \vec{L}_k]$, $U^+ \vec{dU} = i\lambda_k \vec{L}_k$, $k = 1,...8$. In the phenomenological approach we accept here $e = 4.12$, $F_\pi = 186 Mev$ [9].

$$E_M = F_\pi^2 m_\pi^2 (3 - v_1 - v_2 - v_3)/8 + (F_K^2 m_K^2 - F_\pi^2 m_\pi^2)(1 - v_3)/4, \quad (4)$$

v_1, v_2, v_3 are real parts of the diagonal matrix elements of the matrix U.

The baryon number of $SU(3)$ skyrmions can be written also in terms of $\vec{L}_i$ in a form where its symmetry in different $SU(2)$ subgroups of $SU(3)$ is clear [10]:

$$B = -\frac{1}{2\pi^2} \int \Big((\vec{L}_1 \vec{L}_2 \vec{L}_3) + (\vec{L}_4 \vec{L}_5 \tilde{\vec{L}}_3) + (\vec{L}_6 \vec{L}_7 \tilde{\tilde{\vec{L}}}_3) +$$

$$+ \frac{1}{2}[(\vec{L}_1, \vec{L}_4 \vec{L}_7 - \vec{L}_5 \vec{L}_6) + (\vec{L}_2, \vec{L}_4 \vec{L}_6 + \vec{L}_5 \vec{L}_7)] \Big) d^3r \quad (5)$$

The contributions of 3 $SU(2)$ subgroups enter the baryon number on equal footing. $\tilde{L}_3 = (-L_3 + \sqrt{3})/2$, $\tilde{\tilde{L}}_3 = (L_3 + \sqrt{3})/2$.

The inequality can be derived [10]:

$$E_{stat} - E_M \geq 3\pi^2 F_\pi B/e \quad (6)$$

This inequality was obtained at first by Skyrme for the $SU(2)$ model and is the particular case of the Bogomol'ny-type bound.

After the introduction of the time-dependent collective coordinates according to the relation $U(r,t) = A(t)U(r)A^+(t)$ the integration by parts is possible in the Wess-Zumino term in the action [11], and we obtain for the WZ-term contribution to the lagrangian describing skyrmions:

$$L^{WZ} = \frac{-iN_c}{48\pi^2} \epsilon_{\alpha\beta\gamma} \int TrA^+ \dot{A}(R_\alpha R_\beta R_\gamma + L_\alpha L_\beta L_\gamma) d^3x = \frac{N_c}{24\pi^2} \omega_k \int WZ_k d^3x = \omega_k L_k^{WZ} \quad (7)$$

Functions WZ_k can be expressed through the chiral derivatives $\vec{L}_k$:

$$WZ_i = WZ_i^R + WZ_i^L = (R_{ik}(U_0) + \delta_{ik})WZ_k^L, \quad (8)$$

$i, k = 1, ...8$, angular velocities ω_i are defined as usually, and

$$WZ_8^L = -\sqrt{3}(\vec{L}_1 \vec{L}_2 \vec{L}_3) + (\vec{L}_8 \vec{L}_4 \vec{L}_5) + (\vec{L}_8 \vec{L}_6 \vec{L}_7) \quad (9)$$

The real orthogonal matrix $R_{ik}(U_0) = \frac{1}{2} Tr \lambda_i U_0 \lambda_k U_0^\dagger$. WZ_i^R are defined by expressions (9) with substitution $\vec{L}_k \to \vec{R}_k$. The result of calculation of WZ-term depends on the orientation of the soliton in the $SU(3)$ configuration space. It was argued in [10] that approximate relation takes place:

$$Y_R^{min} = 2dL^{WZ}/(\sqrt{3} d\omega_8) \simeq N_c B(1 - 3C_S)/3 \quad (10)$$

where the scalar strangeness content C_S is defined in terms of v_i:

$$C_S = <1 - v_3>/<3 - v_1 - v_2 - v_3>, \quad (11)$$

<> means averaging or integration over the 3-dimensional space. When solitons are located in the (u,d) $SU(2)$ subgroup of $SU(3)$ only L_1, L_2 and L_3 are different from zero, $C_S = 0$ and (10) goes over into the well known quantization condition by Guadagnini [12].

3. As the first step we studied the ansatz similar to one used often for the description of $SU(3)$ rotations in the procedure of zero-modes quantization of skyrmions:

$$U = U_L U_4(\nu) U_8(\rho) U_R \qquad (12)$$

where U_L and U_R belong to the same (u,d) $SU(2)$ subgroup of $SU(3)$ and depend on 3 functions each, left and right baryon numbers B_L and B_R connected with U_L and U_R can be arbitrary integer numbers. $U_4 = exp(-i\nu\lambda_4)$, $U_8 = exp(-i\rho\lambda_8/\sqrt{3})$. ν and ρ are the functions of 3 variables.

The following step after the choice of the ansatz is the choice of the boundary conditions on the functions which enter it. The choice $\nu_0 = 0$ corresponds to the incident $C_S = 0$. For the baryon number located only in U_L or U_R it was found [8] that $SU(2)$ $B = 1$ hedgehog and $B = 2$ torus are the local minima in $SU(3)$, i.e. $\nu = 0$ everywhere identically. For $B_L = B_R = 1$ and for topological centers of both skyrmions located at different points direct calculation shows that the point $\nu = 0$ also is a local minimum in $SU(3)$ relative to the rotations into "strange" direction.

The other parametrization we investigated is [8]:

$$U = U_L(u,s) U(u,d) U_R(d,s) \qquad (13)$$

with $U(u,d) = \exp(ia\lambda_2)\exp(ib\lambda_3)$ and $U_L(u,s)$ and $U_R(d,s)$ being deformed interacting $SU(2)$ skyrmions.

The maximal interference between L and R skyrmions in (13) takes place for $a = \pi/2$, but this corresponds to the case when both skyrmions are located in the same $SU(2)$ subgroup of $SU(3)$, and some new results are not expected for this case. New local minimum was found for $a = b = 0$.

The WZ-term can be written in terms of $SU(2)$ currents $\vec{l}_k$ and $\vec{r}_k$ and functions a, b. It follows from (9) that at large relative distances, for arbitrary but not overlapping solitons, and for $a = b = 0$

$$Y_R^{min} = \frac{1}{2\sqrt{3}\pi^2}\int W Z_8^L d^3x = \frac{1}{4\pi^2}\int [(\vec{l}_1 \vec{l}_2 \vec{l}_3) + (\vec{r}_1 \vec{r}_2 \vec{r}_3)] d^3x = -(B_L + B_R)/2 \qquad (14)$$

where B_L and B_R are the baryon numbers located in left (u,s) and right (d,s) $SU(2)$ subgroups of $SU(3)$. Relation (10) holds since $C_S = 1/2$ for both (u,s) and (d,s) skyrmions. (14) does not hold in general case, for overlapping solitons, since there is no conservation law for the components of the WZ term. Acorrding to calculation [10] the contribution $-(B_L+B_R)/2$ also appears with some additional terms which turned out to be small numerically.

To find the configurations corresponding to the local minimum we used the stochastic variational method extended to the case when the energy functional depends on 8 functions of 3 variables [8]. The binding energy of the molecule (13) turned out to be about half of the torus, i.e. $\sim 75 Mev$ for our choice of parameters and $m_K = m_\pi$, $(M_1 = 1702 Mev)$, the distance between centers of skyrmions is about $\sim 1\ Fm$.

4. The expression for the rotation energy density of the system depending on the angular velocities of rotations in $SU(3)$ collective coordinates space can be obtained by means of substitutions $\vec{L}_i \to \vec{\omega}_i/2$ in (2) and $s_{ik} \to (\tilde{\omega}_i \vec{L}_k - \tilde{\omega}_k \vec{L}_i)/2$ in (3), [10] with $\tilde{\omega}$ being some linear combination of ω_i. 8 diagonal moments of inertia and 28 off-diagonal define the rotation energy - quadratic form in $\omega_i \omega_k$.

In the case of strange skyrmion molecule we obtained 4 different diagonal moments of inertia [10]: $\Theta_1 = \Theta_2 = \Theta_N$; Θ_3; $\Theta_4 = \Theta_5 = \Theta_6 = \Theta_7 = \Theta_S$ and Θ_8. Numerically the difference between Θ_N and Θ_3 is small, Θ_8 is a bit greater than Θ_S. In view of the symmetry properties of the configuration many off-diagonal moments of inertia are equal to zero. Few of them are different from zero, but at least one order of magnitude smaller than diagonal inertia: Θ_{46}, Θ_{57}. By this reason we neglected them for the estimates.

The lagrangian of the system can be written in terms of angular velocities of rotation and moments of inertia in such form (in body-fixed system):

$$L_{rot} = \frac{\Theta_N}{2}(\omega_1^2+\omega_2^2)+\frac{\Theta_3}{2}\omega_3^2+\frac{\Theta_S}{2}(\omega_4^2+\omega_5^2+\omega_6^2+\omega_7^2)+\frac{\Theta_8}{2}\omega_8^2+\Theta_{45}(\omega_4\omega_5-\omega_6\omega_7)+... \quad (15)$$

When two $B = 1$ hedgehogs in different subgroups, (d,s) and (u,s), are located at large distances, the relations for the $B = 2$ moments of inertia Θ in terms of $B = 1$ inertia θ appear:

$$\Theta_N = 2\theta_S; \Theta_S = \theta_N + \theta_S; \Theta_3 = \theta_N/2; \Theta_8 = 3\theta_N/2 = 3\Theta_3. \quad (16)$$

For interacting hedgehogs in the molecule these relations are fulfilled only approximately.

The hamiltonian of the system can be obtained by canonical quantization procedure [13] and is the bilinear function of the generators J_i^R. For the states belonging to definite $SU(3)$ irrep the rotation energy can be written in such simplified form:

$$E_{rot} \simeq \frac{C_2(SU_3) - 3Y_R^2/4}{2\Theta_S} + \frac{N(N+1)}{2}\left(\frac{1}{\Theta_N} - \frac{1}{\Theta_S}\right) + \frac{3(Y_R - Y_R^{min})^2}{8\Theta_8} \quad (17)$$

$C_2(SU_3)$ is the second order Casimir operator of the $SU(3)$ group, N is the "right" isospin. The terms linear in angular velocities present in the lagrangian due to the WZW- term are canceled in the hamiltonian, but they lead to the quantization condition discussed in previous section. Note that Y_R^{min} can take arbitrary noninteger values because it is the quantity similar to the strangeness content C_S, not the quantum number. Y_R is the quantum number and can take only integer values. The usual space angular momentum $J = 0$ here.

It is clear from expression (17) that for $\Theta_8 \to 0$ the right hypercharge $Y_R = Y_R^{min} = \frac{2}{\sqrt{3}}L_8^{WZ}$, otherwise the quantum correction due to ω_8 will be infinite. For solitons located in (u,d) subgroup $\Theta_8 = 0$ and $Y_R = \frac{2}{\sqrt{3}}L_8^{WZ} = B$ - the quantization condition [12] with $N_c = 3$.

For the skyrmion molecule [8] $L_8^{WZ} \approx -\sqrt{3}/2$, or $Y_R^{min} \approx -1$, as it was explained above. The last term in (17) is absent for $Y_R = -1$, and because of the evident constraints

$$(p+2q)/3 \geq Y_R \geq -(q+2p)/3 \quad (18)$$

the following lowest $SU(3)$ multiplets are possible: octet, $(p,q) = (1,1)$, decuplet $(3,0)$ and antidecuplet $(0,3)$. The right isospin $N = 1/2$ for the octet and decuplet, $N = 3/2$ for $\bar{10}$.

5. The mass splittings inside $SU(3)$ multiplets are defined as usually by flavor symmetry breaking (FSB) part of the mass terms in the lagrangian density L_M, see (4). When (u,d) $SU(2)$ solitons are rotated in "strange" direction by means of $U_4 = exp(-i\nu\lambda_4)$ matrix, (4) leads to the substitution $F_\pi^2 m_\pi^2 \to F_\pi^2 m_\pi^2 + sin^2\nu(F_K^2 m_K^2 - F_\pi^2 m_\pi^2)$. For the ansatz (13) after averaging over all phases in the matrix $A(t)$ except ν the mass term in the energy density can be rewritten according to (4) for $F_K = F_\pi$

$$E_M = \frac{F_\pi^2 m_\pi^2}{4}[(3 - v_1 - v_2 - v_3)(\frac{1}{2} + (\frac{m_K^2}{m_\pi^2} - 1)C_S) + (\frac{m_K^2}{m_\pi^2} - 1)(2v_3 - v_1 - v_2)\frac{sin^2\nu}{2}] \quad (19)$$

In the flavor symmetric, FS case the first part of the mass term is included into the classical mass M_{cl} which is minimized. In the FSB case the second part also is included into minimized M_{cl}, and the mass term is squeezed about ~ 3 times in comparison with the FS case [8, 10].

The mass splittings inside $SU(3)$ multiplets are defined by the term δM proprtional to $-(v_1 + v_2 - 2v_3) < \frac{1}{2}sin^2\nu >$ which is not included into M_{cl} and considered as a perturbation in both cases (it is negative!). ν is the angle of rotation into "nonstrange" direction. For two undeformed hedgehogs at large relative distances $v_1 + v_2 - 2v_3 \to 2(1 - cosF)$ where F is the profile function of the $B = 1$ hedgehog.

For the octet the allowed strangeness of states is $-1, -2, -3$, for decuplet it ranges from -1 to -4, the nonstrange dibaryons appear in $\bar{10}$, 27-plet, etc. The masses of dibaryons calculated according to FS and FSB schemes differ, but not very much since the increase of the total mass term in FS case is compensated by the decrease of E_{rot} in comparison with FSB case.

The relative binding $\epsilon = (M_1 + M_2 - M)/(M_1 + M_2)$ is less sensitive to the way of calculation. M_1 and M_2 are masses of final baryons available due to strong interactions, calculated within the same approach (theory-to-theory comparison). It was found that for the octet $\epsilon \simeq 0.14$, for decuplet $\simeq 0.11$ and for $\bar{10}$ it ranges from 0.04 for the $S = 0$ state to 0.09 for the state with $S = -3$. Inclusion of the configuration mixing leads usually to the increase of the mass splittings by $\sim 0.3 - 0.4$ [14]. Since the results for the mass splittings depend on the starting configuration one should use some interpolating procedure, similar, e.g. to the slow rotator approximation used successfully in [9] for the description of the hyperons mass splittings.

6. Our results can be formulated as follows. First, the known local minima in $SU(2)$ configuration space, $B = 1$ hedgehog and $B = 2$ torus are also the local minima in $SU(3)$ configuration space. New local minimum with large strangeness content, close to 0.5 is found with the binding energy about half of that of the torus. The quantization of zero modes of this configuration - strange skyrmion molecule - leads to another branch of predictions of families of strange dibaryons with the smallest uncertainty due to poor known Casimir energy (CE) which controls the absolute values of masses of both $B = 1$ and $B = 2$ states. CE makes contribution of the order of N_c^0 into the masses of the configurations [15]. Since the dipole-type configuration

does not differ much from the $B = 2$ configuration within the product ansatz which we used as a starting one in our calculations [9], the CE of the dipole can be close to twice of that for $B = 1$ soliton and cancel in the binding energies of dibaryons. The values of masses and bindings we obtained cannot be, however, taken too seriously not only because of poor known Casimir energy but also because the non-zero modes contributions closely connected with CE (breathing and vibrational modes at first) have not been taken into account. These effects not only decrease the binding energies [2], but can make unbound many of the states.

The prediction by chiral soliton models of the rich spectrum of baryonic states with different values of strangeness remains one of the intriguing properties of such models. The comparison with predictions of the quark or quark-bag models [16, 17] is of special interest, as well as further experimental searches for strange dibaryons.

The problem of the H-dibaryon discussed in [5] is that of parity doubling: the $SO(3)$ soliton has no definite parity, special procedure of symmetrization should be done, therefore. Similar problem exists for the strange molecules also, and the procedure of (anti)symmetrization should be made, similar to the H-particle case, providing the state of definite parity and removing the e.d.m. of the molecule.

We are thankful to Bernd Schwesinger for valuable discussions and suggestions on the initial stages of the work.

REFERENCES

1. V.B.Kopeliovich, Yad.Fiz. 47(1988)1495; ibid. 56(1993)160;
   E.Braaten, L.Carson, Phys.Rev. D38(1988)3525
2. R.A.Leese, N.S.Manton, B.J.Schroers, Nucl.Phys. B442(1995)228
3. L.Carson, Nucl.Phys. A535(1991)479;
   N.R.Walet, Nucl.Phys. A586(1995)649; hep-ph/9603273
4. T.Walhout, Nucl.Phys. A547(1992)423
5. A.P.Balachandran et al, Nucl.Phys. B256(1985)525;
   R.L.Jaffe, C.L.Korpa, Nucl.Phys. B258(1985)468
6. V.B.Kopeliovich, B.E.Stern, JETP Lett. 45(1987)203
7. V.B.Kopeliovich, Yad.Fiz. 51(1990)241; Phys.Lett. 259B(1991)234; V.Kopeliovich,
   B.Schwesinger, B.Stern, Phys.Lett.242B(1990)145; Nucl.Phys.A549(1992)485
8. V.B.Kopeliovich, B.Schwesinger, B.E.Stern, JETP Lett. 62(1995)185
9. B.Schwesinger, H.Weigel, Phys.Lett. B267(1991)438
10. V.B.Kopeliovich, JETP, in print; JETP Lett. 64(1996)426
    V.B.Kopeliovich, B.E.Stern, $SU(3)$ skyrmions, hep-th/9612211.
11. E.Witten, Nucl.Phys. B223(1983)422,433
12. E.Guadagnini, Nucl.Phys. B236(1984)35
13. G.Adkins, C.Nappi, E.Witten, Nucl.Phys. B228(1983)552
14. H.Yabu, K.Ando, Nucl.Phys. B301(1988)264
15. B.Moussalam, Ann. of Phys. (N.Y.) 225(1993)264
16. R.L.Jaffe, Phys.Rev.Lett. 38(1977)195
17. T.Goldman et al, Phys.Rev.Lett. 59(1987)627

Hadron Spectroscopy Summary

Jim Napolitano* and Alan Schwartz[†]

*Department of Physics, Rensselaer Polytechnic Institute, Troy, NY 12180-3590
[†]Department of Physics, Princeton University, Princeton, NJ 08544

Abstract. We summarize the four parallel sessions on "Hadron Spectroscopy" at the Sixth International Conference on the Intersections of Particle and Nuclear Physics, held at Big Sky, MT, May 27-June 2, 1997.

INTRODUCTION

Hadron spectroscopy helps us to understand the fundamental degrees of freedom for QCD in the confinement regime. In fact, spectroscopy has always pointed physicists towards the underlying dynamics by illuminating the relevant degrees of freedom and their group structure. Examples in atomic physics, nuclear physics, and even condensed matter physics abound. Historically, hadron spectroscopy led directly to the naive quark model, based on the observed spectrum of mesons and baryons.

Today, hadron spectroscopy aims to test detailed models of QCD at low energy. There are many such models, and it is important to look for experimental signatures that are unique to a subset of these models. Possibly the most fruitful class of these signatures would be the discovery of states which cannot be accomodated by the naive quark model. Our sessions concentrated on these kinds of problems.

This summary does not precisely follow the order of the sessions, but instead divides the physics up according to the following topics:

- Mesons, including manifestly exotic mesons, glueballs, and "oddballs"
- Baryons
- H-dibaryons and strangelets
- Pentaquarks

For more details on meson spectroscopy, the reader is referred to Curtis Meyer's plenary talk. Many recent results on the spectroscopy of charmed hadrons were discussed by Guy Blaylock, also in the plenary session.

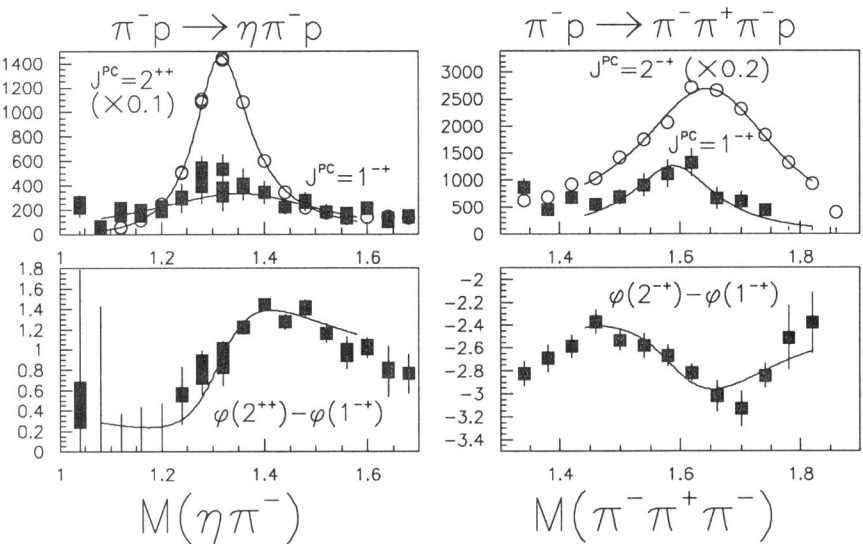

FIGURE 1. Evidence for exotic mesons from BNL E852. The left plots show the $\eta\pi^-$ final state, and the right plots show $\pi^-\pi^+\pi^-$. The top plots show the intensities for the exotic $J^{PC} = 1^{-+}$ wave and a nearby non-exotic wave, while the bottom plots show the phase difference between the two. The lines are simultaneous fits using Breit-Wigner forms.

MESONS

A meson is a hadron with integer spin. Our sessions discussed three types of "exotic" mesons which we review here in turn.

Manifestly exotic mesons

Some quantum numbers are impossible for a fermion-antifermion pair such as $q\bar{q}$. These include $J^{PC} = 0^{--}, 0^{+-}, 1^{-+}, 2^{+-}$, and so on. If mesons with any of these sets of quantum numbers are experimentally established, then it is unequivocal evidence for non-quark degrees of freedom in hadrons.

Neal Cason and the BNL E852 collaboration presented evidence for $J^{PC} = 1^{-+}$ exotics produced in 18 GeV/c π^-p interactions. Two exotic states were discussed, one decaying to $\eta\pi^-$ [1] with mass near 1370 MeV/c^2, and another decaying to $\pi^-\pi^+\pi^-$ with mass around 1593 MeV/c^2. Their results are summarized in Figure 1. The partial wave analysis demonstrates the existence of $J^{PC} = 1^{-+}$ strength, and the phase motion is shown to be consistent with a Breit-Wigner hypothesis. The collaboration is analyzing more data on these channels as well as data on other final states.

Glueballs

Meson states should also exist which have *no* quark degrees of freedom and instead are "pure glue." These are neutral isoscalar hadrons, and the lightest of them are expected to have $J^{PC} = 0^{++}, 0^{-+}$, and 2^{++}. Lattice gauge theories predict a mass near 1500 MeV/c^2 for the 0^{++} glueball, and around 2 GeV/c^2 for the 0^{-+} and 2^{++} states. In his presentation, Steve Cotanch described an analytic calculation based on a relativistic many-body Hamiltonian approach, which reproduces the results of lattice QCD [2]. This strongly suggests that glueballs have masses near those of the ordinary (non-exotic) isoscalar mesons, and would therefore mix with them. This leads to complicated dynamical signatures as well as overpopulation of the meson nonets.

The Crystal Barrel collaboration at CERN has clearly established the existence of an "extra" isoscalar $J^{PC} = 0^{++}$ meson, the $f_0(1500)$. The decays of this and related states suggest a strong mixing with a glueball; details of such mixing can be found in Curtis Meyer's plenary talk.

Our sessions concentrated on evidence for the $\xi(2230)$ observed in radiative J/ψ decay. The ξ was first seen by the Mark III collaboration at SPEAR decaying to K^+K^- and $K_S^0 K_S^0$, and has been observed more recently by the BES collaboration who reported these decay channels as well as $\xi \to \pi^+\pi^-$ and $\xi \to p\bar{p}$. The quantum numbers $J^{PC} = 2^{++}$ are suggested and the state is frequently referred to as $f_2(2230)$.

Zhenping Li discussed the phenomenology connected with the $f_2(2230)$ (and other states) and explained why it might be the lightest tensor glueball [3]. Yucan Zhu reviewed recent measurements by BES, including tacit evidence for $f_2(2230) \to \rho^0 \rho^0 \to \pi^+\pi^-\pi^+\pi^-$. David Hertzog showed us results of annihilation-in-flight spectroscopy using $\bar{p}$ beams at LEAR, including fine-binned energy scans for the reaction $\bar{p}p \to K_S K_S$ with JETSET and Crystal Barrel. These scans show no evidence for production of the $f_2(2230)$ and put limits on the product of its branching ratios to $p\bar{p}$ and $K\bar{K}$ which are barely consistent with the BES results.

Rich Galik presented another $f_2(2230)$ search result, this time in the reaction $\gamma\gamma \to K_S K_S$ at CLEO [4]. As shown in Figure 2, there is no evidence for its production. However, both Galik and Li point out that a glueball should have a small coupling to $\gamma\gamma$ (as measured by its "stickiness"), and thus the CLEO result can be interpreted as evidence for the glueball nature of the $f_2(2230)$.

"Oddballs"

Some mesons clearly do not fit into the $q\bar{q}$ scheme of the quark model even though they have conventional quantum numbers, yet we are hard-pressed to identify them as glueballs. The best example of these are the $a_0(980)$ and $f_0(980)$, an apparent isovector/isoscalar set which has anomalously small mass

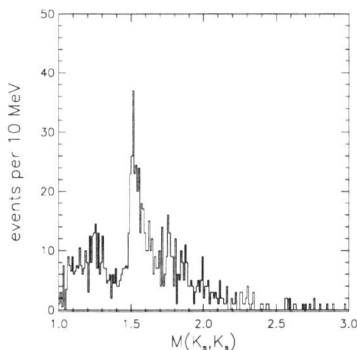

FIGURE 2. CLEO result of a search for the $f_2(2230)$ in the reaction $\gamma\gamma \to K_S K_S$. The $f'_2(1525)$, a well known $s\bar{s}$ state, is clearly visible but there is no evidence of a peak near 2230 MeV/c^2.

and width for the $S=1$ and $L=1$ $q\bar{q}$ quantum numbers implied by the quark model. These mesons have a strikingly large branching ratio to $K\bar{K}$ despite being barely above threshold, and some interpret them as $K\bar{K}$ molecules.

Jeff Gunter from BNL E852 showed us several peculiar features of these mesons as observed in 18 GeV/c π^-p interactions. One observation, shown in Figure 3, suggests that the $a_0(980)$ is produced via π exchange, even though $a_0 \to \pi\pi$ is forbidden. Based on the $K\bar{K}$ molecule model, this may proceed via $f_0(980)$ production followed by isospin mixing to $a_0(980)$. This suggestion is supported by the collaboration's observation of $f_0(980)$ production in this reaction and observation of a strong suppression of $a_0^\pm(980)$ production relative to $a_0^0(980)$.

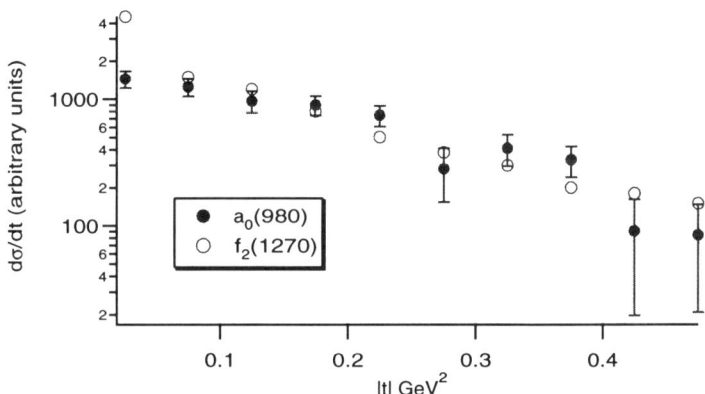

FIGURE 3. BNL E852 measurement of the t-dependence of $a_0(980)$ and $f_2(1270)$ production in 18 GeV/c π^-p interactions. The dependence is quite similar for both reactions, although the $f_2(1270)$ can be produced by π exchange while the $a_0(980)$ cannot.

BARYONS

If one's objective is to identify something peculiar in baryon spectroscopy, then it is important to pick one's problem carefully and look closely. This is especially true for baryons with no strangeness or charm. Excited states are numerous and broad, and there are no exotic quantum numbers. Consequently, one needs to rely on detailed analyses and conclusions about dynamics which can be drawn from them.

Igor Lopatin described to us a series of experiments at the Petersburg Nuclear Physics Institute aimed at carrying out a complete partial wave analysis (PWA) of the excited N^* (isospin $\frac{1}{2}$) and Δ (isospin $\frac{3}{2}$) systems. A number of results were discussed, including a measurement of the isospin splitting of the $\Delta(1232)$. They find $M_{\Delta^0} - M_{\Delta^{++}} = 2.6 \pm 0.4$ MeV/c^2.

Tom Vrana presented results from a new Pittsburgh/Argonne analysis which fits the results of an existing partial wave decomposition with a coupled channel prescription incorporating unitarity. Impressive agreement is achieved across many waves, at least to within the gross uncertainties of the PWA. In particular, they obtain a well determined value for the $\gamma N \to \eta N$ photocoupling through the $S_{11}(1535)$ baryon (i.e. isospin $\frac{1}{2}$ and $J^P = \frac{1}{2}^-$).

The peculiar features of the "Roper Resonance", also known as the $P_{11}(1440)$ baryon (i.e. isospin $\frac{1}{2}$ and $J^P = \frac{1}{2}^+$), were touched on by Zhenping Li, Roelof Bijker, and Mark Manley. This state has a very low mass in terms of the harmonic oscillator shell model, as well as anomalous behavior in photo- and electroproduction. Li put forth the suggestion that the Roper is in fact a hybrid state, with explicit glue degrees of freedom. Bijker discussed the electromagnetic form factors for the Roper and other baryon excitations in the framework of a collective quark model which more naturally explains the low mass. In his review talk, Manley also pointed out that partial wave analysis of the P_{11} wave has not been entirely consistent with itself as new results are added to the data set.

New experiments at Jefferson Laboratory and with the Crystal Ball at Brookhaven will help settle these issues involving the light quark baryons.

Charmed baryons provide a different look at the three-quark system, and may in fact help us learn about the light quark baryons by invoking the heavy quark limit. Mike Zoeller provided us with some of the latest charmed baryon spectroscopy results using CLEO. These included Σ_c^+ production and decay in $B \to \Sigma_c^+ X$, as well as the first evidence for $\Xi_c'^+(\frac{1}{2}^+) \to \Xi_c^+(\frac{1}{2}^+)\gamma$. This subject was also covered in Guy Blaylock's plenary talk as well as in several talks in the parallel sessions on charm physics.

Implications for light quark spectroscopy based on the heavy quark physics leads us to finally consider exotic multiquark states, where s and c quarks might lead to exceptional binding when there are more than three quarks present. We discuss these now.

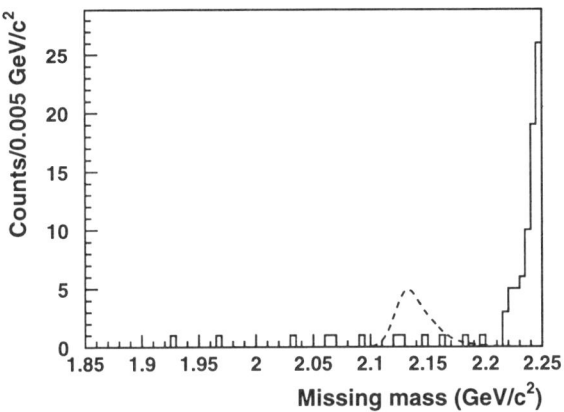

FIGURE 4. Results of an H-dibaryon search from BNL E836 showing the missing mass in the ^{3}He$(K^-, K^+)Xn$ reaction, treating the neutron as a spectator. The dashed line shows the estimated signal for $M_H = 2.13$ GeV/c^2.

H-DIBARYONS AND STRANGELETS

As originally suggested by Jaffe, a *uuddss* six-quark state which is a spin, isospin, and color singlet may be bound against strong decay. That is, it would have a mass less than twice the mass of the Λ^0 and would therefore decay weakly, most likely to $\Lambda^0 n$, $\Sigma^0 n$, or $\Sigma^- p$. Vladimir Kopeliovich discussed such systems in terms of $SU(3)$ skyrmions. However, it seems clear that for the time being, it is the experimentalists' responsibility to try to establish whether or not such objects exist.

This field is truly an "intersection" between Particle and Nuclear Physics, with experiments being performed with approaches typical of each of these disciplines. Bernd Bassalleck gave us a complete summary of the experimental status, mainly based on experiments at the Alternating Gradient Synchrotron (AGS) at Brookhaven. These included "nuclear" physics spectroscopy experiments based on the (K^-, K^+) reaction (BNL E813 and E836), a high energy experiment using a 24 GeV/c proton beam incident on a nuclear target (E888), and also a potentially positive signal from a relativistic heavy ion experiment (E810).

Results from E836 [5] are shown in Figure 4. This experiment studied missing mass in the reaction ^{3}He$(K^-, K^+)Xn$ where the neutron is treated as a spectator and the "target" is therefore two protons. Consequently, this search is independent of the decay mode or lifetime of the H. The estimated signal for $M_H = 2.13$ GeV/c^2 based on recent calculations of the ^{3}He wave function is shown in the figure. Analysis of data from E813, which used the same apparatus as E836 but which employed a dual $H2/D2$ target to search for $p(K^-, K^+)\Xi^-$ followed by $\Xi^- d \to Hn$, is in progress.

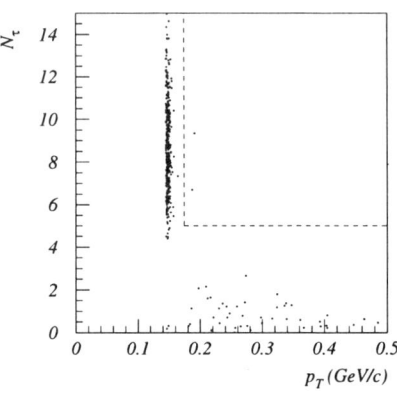

FIGURE 5. Results of an H-dibaryon search in BNL E888. Candidate events are plotted as the number of Λ^0 lifetimes versus the Λ^0 transverse momentum. There are two events in the signal region (the dashed box) which do not appear to be true dibaryon decays, but which are nevertheless included in the limit calculations.

A different approach was taken by E888 [6], which looked inclusively in $p-$Cu and $p-$Pt reactions using an existing neutral beamline established to study K_L decays. The beam energy was 24 GeV. H-particles produced at the target would decay to $\Lambda^0 n$ or $\Sigma^0 n$ (followed by $\Sigma^0 \to \Lambda^0 \gamma$) upstream of the detector elements. The signal was therefore a Λ^0 (observed via its $p\pi^-$ decay mode) having high transverse momentum with respect to the line of flight connecting the production and decay vertices, with the decay vertex located a large number of Λ lifetimes away from the nearest beamline element to which the momentum vector projects. Figure 5 shows the results of this search. Backgrounds with high p_T but typical Λ^0 lifetimes are most likely from Λ^0 production in downstream beam elements, while events from $K_L \to \pi\mu\nu$ decays, which leak into the Λ^0 mass window, appear at low p_T. Two events appear in the signal region and are used to calculate a conservative upper limit on the H production cross section. This limit is significantly below theoretical estimates. The two events have recently been thoroughly studied and found to be unlikely to have arisen from H decay.

We also heard of a potentially positive signal from an H-dibaryon search in BNL E810 which measured multiparticle production with a TPC inside the MPS magnet, in high energy collisions of Si nuclei on a Pb target. Reconstruction of the $\Sigma^- p$ invariant mass shows a peak which is not present in the opposite charge topology. A follow-up experiment at BNL is underway.

The H-dibaryon would be the lightest form of "strange matter", i.e. anomalously massive ($A \gg 1$) particles with nonzero strangeness. John Hill informed us about searches for such particles at BNL, primarily from E864 which searched inclusively in high energy Au+Pb collisions. Preliminary limits of singly or doubly charged particles with atomic mass $A \geq 5$ and various amounts of strangeness seem to be well below most "coalescence" predictions, but at about the same order of magnitude as more recent calculations assuming Quark-Gluon Plasma production.

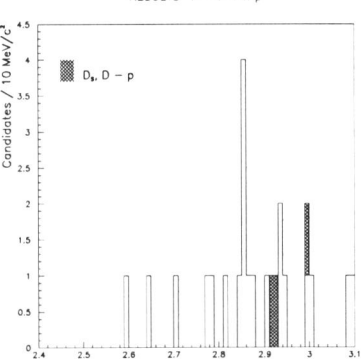

FIGURE 6. Results of a search for the $P^0 = \bar{c}suud$ pentaquark from FNAL E791. The invariant mass of $\phi\pi^-p$ combinations are plotted for events with a detached vertex, indicative of weak charm decay.

PENTAQUARKS

Finally, we learned about the pentaquark, which has flavor quantum numbers $\bar{c}suud$ (P^0) or $\bar{c}sudd$ (P^-). This particle was originally predicted by Harry Lipkin, who at Big Sky reminded us that the presence of a heavy *anti*quark can lead to tight binding by virtue of its color hyperfine interaction. Such an object would again be stable against strong decay, and so decay weakly.

Danny Ashery discussed a search for the P^0 as part of FNAL E791, a charm hadroproduction experiment using a 500 GeV/c π^- beam incident on Pt and C targets. The signature was $P^0 \to \phi\pi^-p$ where all tracks originate at a secondary vertex. Results appear in Figure 6. The total mass of a D_S ($\bar{c}s$) and proton (uud) is 2.91 GeV/c^2. The authors discount the small excess near 2.85 GeV/c^2 and quote an upper limit on the production cross section.

ACKNOWLEDGEMENTS

We would like to express our sincere appreciation to all the contributors and chairpersons in the sessions on Hadron Spectroscopy. We would especially like to thank Dick Mischke and Susan Ramsay and all the other organizers for putting together an outstanding conference.

REFERENCES

1. Thompson D.R., et al., hep-ex/9705011 submitted to *Phys. Rev. Lett.* (1997).
2. Szczepaniak A., et al., *Phys. Rev. Lett.* **76**, 2011 (1996).
3. Close F., Farrar G., and Li Z., *Phys. Rev.* **D55**, 5749 (1997).
4. CLEO collaboration, hep-ex/9703009, submitted to *Phys. Rev. Lett.* (1997).
5. Stotzer R.W., et al., *Phys. Rev. Lett.* **78**, 3646 (1997).
6. Belz J., et al., *Phys. Rev. Lett.* **76**, 3277 (1996).

Lepton Probes
of Hadron Structure

Measurement of the Neutron's electric Formfactor $G_{E,n}$ in $D(\vec{e},e'\vec{n})p$ and $^3\vec{He}(\vec{e},e'n)pp$ [1]

Michael Ostrick*
for the A3–Collaboration at the Mainz Microtron MAMI

*Institut für Kernphysik, Universität Mainz, D-55099 Mainz
Department of Physics & Astronomy, University of Glasgow, Glasgow, U.K.
Institut für Physik, Universität Mainz, D-55099 Mainz
Physikalisches Institut, Universität Tübingen, D-72076 Tübingen
Physikalisches Institut, Universität Bonn, D-53115 Bonn
Ecole Normale Supérieure, Paris, France

Abstract. Form factors for elastic electron scattering parametrise the coherent electromagnetic response of a composite system. They provide information vital for our understanding of the dynamical properties of the nucleon ground state. The proton form factors have been measured over the last 40 years through Rosenbluth–separation over a large range of momentum transfer. Neutron form factors now benefit from coincidence experiments at modern cw accelerators. Least known is the electric formfactor of the neutron $G_{E,n}$. Here double polarisation experiments start to exhibit their full potential and deliver first results.

NUCLEON FORM FACTORS IN ELECTRON SCATTERING

One of the fundamental problems addressed by nuclear and particle physics over the past half of the century is the underlying structure of protons and neutrons. Elastic electron scattering leaves the constituents bound after the collision and the dynamics which keeps the nucleon together should manifest itself in these processes. Therefore accurate data on the elastic response of the nucleon are of vital importance. It is especially worth to look at the electrostatic response of the neutron. As not beeing dominated by an overall electric charge it is particulary suited for an investigation of internal degrees of freedom.
Electron scattering can be described to lowest order in the electromagnetic

---

[1] Supported by the Deutsche Forschungsgemeinschaft (SFB 201), BMBF(06 TU 669), DAAD and the U.K. Science and Engineering Research Council

coupling constant by the exchange of a virtual photon that is characterised by momentum transfer ($Q^2 = \vec{q}^2 - \omega^2 > 0$) and degree of linear transverse polarisation $\epsilon = (1 + 2\vec{q}^2/Q^2 \tan^2(\vartheta_e/2))^{-1}$. The longitudinal polarisation vanishes in the case of real photons ($Q^2 = 0$).

The cross section for electron scattering can be expressed in terms of two structure functions $R_L(Q^2, \omega)$ and $R_T(Q^2, \omega)$ describing the response of the target.

$$\frac{d\sigma}{d\Omega_e} = \left(\frac{d\sigma}{d\Omega}\right)_{Mott} \cdot \frac{1}{1+\tau} \cdot \left(R_L + \frac{\tau}{\epsilon} R_T \right); \quad \tau = \frac{Q^2}{4M_p^2} \quad (1)$$

In the case of elastic scattering off a spin $\frac{1}{2}$ target this formula reduces to the Rosenbluth form

$$\frac{d\sigma}{d\Omega_e} = \left(\frac{d\sigma}{d\Omega}\right)_{Mott} \cdot \left(\frac{G_E^2 + \tau G_M^2}{1 + \tau} + 2\tau G_M^2 \tan^2 \frac{\Theta}{2} \right).$$

The longitudinal response is given by the charge form factor $G_E^2(Q^2)$ and the transverse response by the magnetic form factor $G_M^2(Q^2)$. They can be interpreted as the distributions of charge and magnetism in the nuclear Breit frame. Due to the different angular weight G_E^2 and G_M^2 can, in principle, be separated (Rosenbluth–separation). Most of our knowledge about the form factors of the proton comes from the application of this technique. A recent compilation of the data is given in [2].

Neutron form factors arise more difficulties. As a target of sufficient density it is only available in nuclei where it underlies the Fermi motion and is necessarily accompanied by protons. The deuteron is usually chosen to extract the e-n-cross section in single arm experiments by subtracting the proton contribution [2]. In the case of the electric neutron form factor the smallness of $G_{E,n}^2$ as compared to $\tau G_{M,n}^2$ makes a Rosenbluth separation difficult. Nevertheless, it was successfully applied in the range $Q^2 \simeq 1.75 - 4\ GeV^2/c^2$ [3].

In the intermediate momentum transfer range the most precise data on $G_{E,n}$

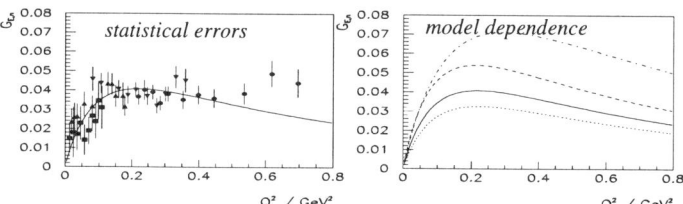

FIGURE 1. Left: electric form factor of the neutron extracted from elastic electron deuteron scattering using the Paris potential. Right: compilation of best fits through the data extracted using different potentials [4]

came for a long time from the study of the elastic deuteron structure functions

A and B which are closely related to $R_L^D(Q^2)$ and $R_T^D(Q^2)$. $R_L^D(Q^2)$ contains $(G_{E,n} + G_{E,p})^2$ and the occurence of the interference term $G_{E,n} \cdot G_{E,p}$ leads to a higher sensitivity to $G_{E,n}$. Unfortunately the necessary unfolding of the deuteron wavefunction introduces a substantial model dependence in $G_{E,n}$ ([4] and fig. 1).

POLARISATION EXPERIMENTS

In the search for a more sensitive observable with less model dependence Arnold, Carlson and Gross [5] suggested to exploit the spin transfer from electrons with helicity P_e to neutrons.

$$P_n^x = -P_e * \frac{2\sqrt{\tau(1+\tau)}\tan(\vartheta_e/2) \cdot G_E G_M}{G_E^2 + \tau/\epsilon G_M^2}; \quad P_n^y = 0 \quad (2)$$

$$P_n^z = P_e * \frac{2\tau\sqrt{1+\tau+(1+\tau)^2\tan^2(\vartheta_e/2)}\tan(\vartheta_e/2) \cdot G_M^2}{G_E^2 + \tau/\epsilon G_M^2} = \vec{P}_n \cdot \hat{q}$$

An interference term $\sim G_{E,n} \cdot G_{M,n}$ shows up in the transverse neutron polarisation P_n^x, which is perpendicular to $\vec{q}$ in the electron scattering plane.
Similarly the asymmetry of the cross section for $\vec{e}-\vec{n}$ scattering with polarised neutrons in the initial state (P_T) is given by

$$A = P_T^x \cdot P_n^x - P_T^z \cdot P_n^z. \quad (3)$$

Several ways to approximate the ideal case of the $\vec{n}(\vec{e}, e'n)$– or the $n(\vec{e}, e'\vec{n})$– reaction with real targets have been investigated.
Polarised $^3\vec{H}e$ is assumed to carry a high n–polarisation in quasifree scattering [6]. The absolute magnitude will even cancel in the asymmetry ratio A_x/A_z. In addition the exclusive reaction avoids proton contributions provided that final state interactions can be neglected.
For the spin transfer in $D(\vec{e}, e'\vec{n})p$ Arenhövel [7] found a large sensitivity to $G_{E,n}$ and little dependence on the deuteron wave function for quasifree kinematics. In this case the recoil neutron polarisation must be analysed. Elastic n-p-scattering offers a reasonable analysing power $\mathcal{A}(\Theta_{n'}, T_n)$ for polarisation components $P_n^\perp$ transverse to the direction of neutron momentum leading to a ϕ'-asymmetry of the scattered neutrons:

$$I(\Theta_{n'}, \phi') = I_0 \cdot (1 + \mathcal{A}(\Theta_{n'}, T_n) \cdot P_n^\perp \cdot sin\phi') \quad (4)$$

It has been demonstrated [8] that the $n-p$-neutron scattering in the detection process inside a plastic scintillator itself can be used for the spin analysis with great advantage.

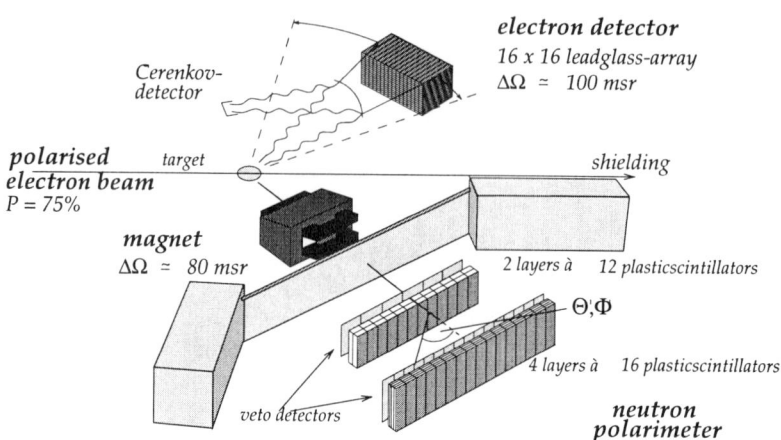

FIGURE 2. set up for the $G_{E,n}$ experiments at MAMI. The magnet is only used for the $D(\vec{e},e'\vec{n})p$ experiment

$G_{E,N}$ EXPERIMENTS AT MAMI

At MAMI $^3\vec{H}e(\vec{e},e'n)pp$ and $D(\vec{e},e'\vec{n})p$ experiments were performed with one common large solid angle detector system (fig. 2). It consisted of a Pb-glass array with a Čerenkov detector for the electrons and two walls of plastic scintillators for neutron detection. The source of polarised electrons is based on photoelectron emission of GaAsP [9,10]. Meanwhile with a strained layer crystal a beam polarisation of about 75% at currents up to $8\mu A$ have been achieved.

In the $^3\vec{H}e(\vec{e},e'n)pp$ -*experiment* the neutron detector was operated as a time-of-flight device. According to eq.3 a cross section asymmetry with respect to $\vec{e}$-helicity reversal was measured. ^{3}He gas was spinpolarised by metastability exchange scattering with optically pumped metastable ^{3}He atoms in their $1s2s^3S_1$ state [11] and compressed to 1 bar in the target cell by a Toepler compressor. The results of a first exploratory experiment [12] together with the preliminary result of the full run averaged over $Q^2 = 0.3 - 0.5 GeV^2$ are shown in fig. 3.

A new experiment at MAMI to access higher momentum transfers ($Q^2 \simeq 0.6\,GeV^2/c^2$) based on a magnetic sprectrometer and a higher target density has just started running [13]. In the $D(\vec{e},e'\vec{n})p$ -*experiment* the neutron detector operated in polarimeter mode. Through the position of the hit in the scintillator walls the scattering angles $\Theta_{n'}$ and $\phi_{n'}$ can be reconstructed (fig. 2).

The determination of the desired recoil polarisation P_n^x requires the knowledge of the effective analysing power of the polarimeter which depends crucially on

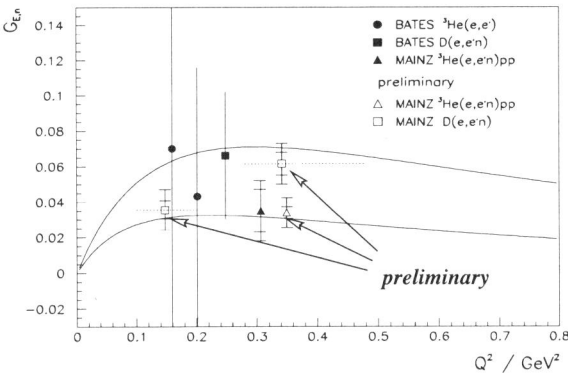

FIGURE 3. Compilation of $G_{E,n}$ data from double polarisation experiments [12,14,15,16]. Also shown is the span of model dependence for data extracted from elastic Deuteron structure functions [4].

the kinematical cuts applied to select the n-p–scattering events in the first detector plane. An experimental calibration of the polarimeter at a polarised neutron beam can be circumvented by using the spin precession in a magnetic field. On their way L through a magnet with a field $\vec{B}$ perpendicular to the scattering plane the neutrons undergo a spin precession by the angle $\chi = \mu_n/(2\cdot\beta_n)\cdot\int_L B(l)dl$. Behind the magnet the transverse polarisation $P_n^\perp$

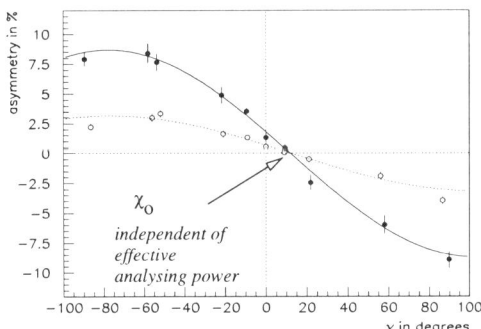

FIGURE 4. Amplitudes of reconstructed ϕ'-asymmetries for various precession angles χ, averaged over detector acceptance, for two different cuts on the analysing reaction.

receives contributions from both, P_n^x and P_n^z, leading to an amplitude in the azimutal asymmetry

$$A(\chi) = \mathcal{A}\cdot(P_n^x\cdot\cos\chi - P_n^z\cdot\sin\chi) \tag{5}$$

The angle of zero crossing χ_0 ($A(\chi_0) = 0$) is directly related to the ratio of the electric and magnetic form factors

$$\tan \chi_0 = \frac{1}{\sqrt{\tau + \tau \cdot (1+\tau) \cdot \tan^2 \vartheta_e/2}} * \frac{G_E}{G_M}. \qquad (6)$$

and depends neither on the absolute values of the polarimeter's analysing power nor on the electron beam polarisation. Asymmetry data for various settings of the magnetic field are shown in figure 4. Kinematical cuts change the amplitude, i.e. the effective analysing power, but not the zero crossing χ_0. By changing the electron energy we measured at two two different mean values of Q^2. The preliminary values for $G_{E,n}$ averaged over the full acceptance are shown in figure 3 along with results from other double polarisation experiments.

CONCLUSION

Among electromagnetic form factors $G_{E,n}$ is least known. New double polarisation experiments have started to improve on the sensitivity and model independence. Data have been taken for $D(\vec{e}, e'\vec{n})p$ and $^3\vec{He}(\vec{e}, e'n)pp$ at MAMI with statistical errors of typically 10%. The first preliminary result for $^3\vec{He}(\vec{e}, e'n)pp$ is in accordance with [4] using the Paris potential and with the result of the exploratory experiment [12]. The value obtained in $D(\vec{e}, e'\vec{n})p$ is somewhat higher. Final results will be published soon.

REFERENCES

1. S.Kopecki et al., Phys. Rev. Lett. 74, 2427 (1995)
2. P.Bosted, Phys.Rev. C51, 409 (1995)
3. A.Lung et al., Phys.Rev.Lett. 70, 718 (1993)
4. S.Platchkov et al., Nucl.Phys. A510, 740 (1990)
5. R.G. Arnold et al., Phys.Rev.C23, 363 (1981)
6. J.Friar, Phys.Rev. C42, 2310 (1990)
7. H. Arenhövel, Ph.Lett. B199, 13 (1987); Z.f.Phys. A331, 509 (1988)
8. T.N.Taddeucci et al., Nucl.Inst.Meth. A241, 448 (1985)
9. K.Aulenbacher, dissertation, Mainz 1994
10. W. Hartmann et al., Nucl.Inst.Meth. A286, 1 (1990)
11. G.Eckert et al., Nucl.Instr.Meth. A320, 53 (1992)
12. M.Meyerhoff et al., Phys.Lett. B327, 201 (1994)
13. Measurement of the Electric Form Factor of the Neutron via $^3\vec{He}(\vec{e}, e'n)$, proposal to the joint MAMI/ELSA PAC, Mainz 1995
14. A.K.Thompson et al., Phys.Rev.Lett. 68, 2901 (1992)
15. C.E.Jones et al., Phys.Rev. C41, 110 (1993)
16. T.Eden et al., Phys. Rev. C50, R1749, (1994)

The E2/M1 Ratio in Δ Photoproduction

S. Hoblit[1,5], G. Blanpied[4], M. Blecher[6], A. Caracappa[1],
C. Djalali[4], G. Giordano[2], K. Hicks[7], M. Khandaker[6,1],
O.C. Kistner[1], A. Kuczewski[1], M. Lowry[1], M. Lucas[4],
G. Matone[2], L. Miceli[1,4], B. Preedom[4], D. Rebreyend[4],
A.M. Sandorfi[1], C. Schaerf[3], R.M. Sealock[5], H. Ströher[8],
C.E. Thorn[1], S.T. Thornton[5], J. Tonnison[6,1], C.S. Whisnant[4],
H. Zhang[7], X. Zhao[6]
(The LEGS Collaboration)

[1] *Physics Department, Brookhaven National Laboratory, Upton, NY 11973*
[2] *INFN-Laboratori Natzionali di Frascati, Frascati, Italy*
[3] *Università di Roma "Tor Vergata" and INFN-Sezione di Roma2, Rome, Italy*
[4] *Department of Physics, University of South Carolina, Columbia, SC 29208*
[5] *Department of Physics, University of Virginia, Charlottesville, VA 22901*
[6] *Physics Department, Virginia Polytechnic Inst. & SU, Blacksburg, VA 24061*
[7] *Department of Physics, Ohio University, Athens, OH 45701*
[8] *II Physikalisches Institut, Universität Giessen, Giessen, Germany*

Abstract. New high-precision measurements of $p(\vec{\gamma}, \pi)$ and $p(\vec{\gamma}, \gamma)$ cross sections and beam asymmetries have been combined with other polarization ratios in a simultaneous analysis of both reactions. The E2/M1 mixing ratio for the N → Δ transition extracted from this analysis is EMR = -3.0% ± 0.3 (stat+sys) ± 0.2 (model).

The well-isolated N → Δ resonance serves as a sensitive test for models of nucleon structure [1-4]. To lowest order, N → Δ is a simple M1 quark spin-flip transition. Small L=2 components in the N and Δ wavefunctions allow this excitation to proceed via an electric quadrupole transition. The most sensitive observable to E2 strength is the beam asymmetry in $p(\vec{\gamma}, \pi^\circ)$ [5]. In a recent Mainz measurement of $p(\vec{\gamma}, \pi)$ an EMR of -2.5% was extracted using the π° channel alone [6]. As will be shown, this value is artificially inflated by a factor of 2 due to multipole ambiguities. We report an improved value for the EMR that is constrained by new measurements and two new observables.

At any energy, a minimum of 8 independent observables are required to specify the photo-pion amplitude [7]. Such complete information has never been available and previous analyses have relied on at most four observables, usually measured separately with independent systematic errors. Although the $\tau = 3/2$ M1 and E2 components can extracted from a multipole fit, many observables are needed to avoid multipole ambiguities [8]. In the present work, p($\vec{\gamma}, \pi^\circ$), p($\vec{\gamma}, \pi^+$) and p($\vec{\gamma}, \gamma$) cross sections and beam asymmetries were all measured simultaneously to provide new constraints on the photo-pion multipoles.

At LEGS, polarized tagged γ-ray beams between 209 and 333 MeV were produced by backscattering laser light from 2.6 GeV electrons at the National Synchrotron Light Source. Beams, with linear polarizations greater than 80% and known to $\pm 1\%$, were flipped between orthogonal states at random intervals between 150 and 450 seconds.

One goal of this experiment was the first complete separation of Compton scattering and π°-production. The two reactions were distinguished by comparing their γ-ray and proton-recoil energies. High energy γ-rays were detected in a large NaI(Tl) crystal, while recoil protons were tracked through wire chambers and stopped in an array of plastic scintillators. A schematic of this arrangement and a spectrum showing the separation of the two channels is given in [9]. All detector efficiencies were determined directly from the data itself, an important advantage. Charged pions were detected in 6 NaI detectors, including the large crystal used for the Compton and π° channels. The high resolution of the NaI detectors was essential in determining π^+ efficiencies, which were simulated with GEANT [10] using GCALOR to model hadronic interactions [11]. Systematic effects were combined in quadrature with statistical errors ($\sim 1\%$) for a net measurement error.

In the vicinity of the Δ peak, the spin-averaged π°, π^+, and Compton cross sections determined in this experiment are all consistently higher than earlier measurements from Bonn [12-15] while for energies lower than $\sim$270 MeV substantial agreement is observed. Of the previous π^+ cross section measurements, those from Tokyo [16] are in closest agreement to the present work. The present work is also in very good agreement with two recent Compton measurements from Mainz at 90° and 75° [17,18]. All LEGS cross sections are locked together with a common systematic scale uncertainty, due to possible flux and target thickness variations, of 2%.

To obtain a consistent description of these results we have performed an energy-dependent analysis, expanding the π-production amplitude into electric and magnetic partial waves, $E_{\ell\pm}^\tau$ and $M_{\ell\pm}^\tau$, with relative πN angular momentum ℓ, and intermediate-state spin $j = \ell \pm \frac{1}{2}$ and isospin $\tau = \frac{1}{2}$ or $\frac{3}{2}$. In order to reproduce our angular distributions in the region of the Δ, we must vary the D wave contributions. To reduce ambiguities [8], we truncate our fit at F waves, while keeping the Born terms up to order $\ell = 19$.

The (γ, π) multipoles were parameterized with a K-matrix-like unitarization,

$$A^\tau_{\ell\pm} = \left(A^\tau_B(E_\gamma) + \alpha_1\epsilon_\pi + \alpha_2\epsilon_\pi^2 + \alpha_3\Theta_{2\pi}(E_\gamma - E_\gamma^{2\pi})^2\right)$$
$$\times \left(1 + iT^\ell_{\pi N}\right) + \beta \cdot T^\ell_{\pi N}. \qquad (1)$$

Here, E_γ and ϵ_π are the beam and corresponding π^+ kinetic energies, and A^τ_B is the full pseudo-vector Born multipole, including ρ and ω t-channel exchange [19]. The VPI[SM95] values are used for the πN scattering T-matrix elements [20]. Below 2π threshold, $E_\gamma^{2\pi} = 309$ MeV, $T^\ell_{\pi N}$ reduces to $\sin(\delta_\ell)e^{i\delta_\ell}$, $\delta_\ell(E_\gamma)$ being the elastic πN phase shift. Thus, eqn. 1 explicitly satisfies Watson's theorem [21] below $E_\gamma^{2\pi}$ and provides a consistent, albeit model-dependent, procedure for maintaining unitarity at higher energies. The β term was fixed at zero for all multipoles except $M^{3/2}_{1+}$, $E^{3/2}_{1+}$, and $M^{1/2}_{1-}$, the first two describing M1 and E2 N $\rightarrow$ P$_{33}$ excitation and the latter allowing for a possible tail from the P$_{11}$ resonance. The other terms describe the non-resonant background, with the α_i included to account for non-Born contributions. Each fitted multipole contains a term in α_1, while the additional α_2 term is used only in $E^{1/2}_{0+}$, $M^{3/2}_{1+}$ and $E^{3/2}_{1+}$. The α_3 term containing the unit Heavyside step function $\Theta_{2\pi}$ (=1 for $E_\gamma > 309$ MeV) is used only in the E_{0+} amplitudes to accommodate possible effects from S-wave 2π production.

Once the (γ, π) multipoles are specified, the imaginary parts of the six Compton helicity amplitudes are completely determined by unitarity, and dispersion integrals can be used to calculate their real parts, where we have implemented the computation of L'vov and co-workers [22]. This requires the evaluation of the dispersion integrals at energies outside the range of the present work. For this we have used the VPI-SM95 solution up to 1.5 GeV [20], and estimates from Regge theory for higher energies. The polarizabilities can also be extracted from this analysis, but they have only small effects on the N $\rightarrow$ Δ amplitudes.

We report here a summary of the results of a fit to the parameters of the (γ, π) multipoles, minimizing χ^2 for both predicted (γ, π) and (γ, γ) observables. In this fit we have used p($\vec{\gamma}, \pi^\circ$), p($\vec{\gamma}, \pi^+$) and p($\vec{\gamma}, \gamma$) cross sections only from the present experiment, since these are locked together with a small common scale uncertainty, and augmented our beam asymmetry data with other published polarization *ratios* (in which systematic errors tend to cancel). These include our earlier $\Sigma(\pi^\circ)$ data [5],{T(π°), T(π^+)} data from Bonn [23], {T(π°), P(π°), T(π^+), P(π^+)} data from Khar'kov [24,25], and the few beam-target asymmetry points {G(π^+), H(π^+)} from Khar'kov [26]. Systematic scale corrections were fitted following the procedure of ref. [27]. To minimize the effect of 2π-production we have limited the fitting interval from 200 MeV to 350 MeV. The reduced χ^2 for this analysis is $\chi^2_{df} = 997/(644 - 34) = 1.63$.

The EMR for N $\rightarrow$ Δ is just the ratio of fitted β coefficients in eqn. 1 for the $E^{3/2}_{1+}$ and $M^{3/2}_{1+}$ multipoles, -0.0296 $\pm$ 0.0021. The fitting errors reflect

TABLE 1. Dependence of the EMR on p(γ,π) cross sections. Rows 1 and 3 summarize our multipole fit to p(γ,π) and p(γ,γ) using unpolarized p(γ,π) results from this work in row 1, and substituting only the Bonn cross sections from [12,13] in row 3.

Source	$\frac{d\sigma}{d\Omega}(\gamma,\pi)$	EMR(%)	χ^2_{df}
$(\gamma,\pi) + (\gamma,\gamma)$ fit	LEGS	-3.0 ± 0.3	1.63
fit to DMW	LEGS	$-3.0 + 0.2 / -0.3$	
$(\gamma,\pi) + (\gamma,\gamma)$ fit	Bonn	-1.3 ± 0.2	1.89
Sato-Lee [3]	Bonn	-1.8 ± 0.9	

all statistical and systematic uncertainties. The full *unbiased estimate* of the uncertainty is $\sqrt{\chi^2}$ larger [28]. We have studied the variations that result from truncating the multipoles at D waves, using a different πN phase shift solution [29], allowing for differences in energy calibration between photoproduction and πN scattering, and varying the assumptions used to compute the Compton dispersion integrals [22]. The EMR is most sensitive to the multipole order and to the energy scale. Combining these *model* uncertainties in quadrature leads to our final result:

EMR = -3.0% ± 0.3 (stat+sys) ± 0.2 (model) .

To investigate the effect of the difference in the p(γ,π) cross sections between our results and the Bonn data, we have repeated this analysis substituting the values from [12,24] for our own. This reduces the EMR substantially (Table 1, row 3).

In ref. [6], a fit to the recent Mainz π° cross section and $\Sigma(\pi^\circ)$ data, neglecting non-Born contributions beyond S and P waves, was used to extract an EMR of -2.5% ± 0.2 (stat) ± 0.2 (sys). The Mainz data agrees with Bonn cross sections [12] and LEGS $\Sigma(\pi^\circ)$ data, and thus should correspond to row 3 of table I, and the factor of 2 difference between this value and their reported results reflect the ambiguities in the multipoles constrained by only 2 observables.

Various theoretical techniques have been used to separate the N → Δ component. Our result can be directly compared with models, such as DMW [30] and Sato & Lee [3], that report ratios of γNΔ couplings deduced with a K-matrix type unitarization equivalent to eqn. 1. We have refit the DMW parameters to our multipoles, with the result EMR = $-3.0\% + 0.2/-0.3$. This, and the result of Sato & Lee who fitted their parameters to the Bonn cross sections and our $\{\Sigma(\pi^\circ), \Sigma(\pi^+)\}$ data, are listed in table I and are consistent with the set of (γ,π) cross sections that were used to fix their parameters.

To summarize recent data and analyses, there are two new sets of measurements of p(γ,π) and p(γ,γ), the Mainz experiments reported in [6,17,18] and the LEGS experiment reported here and in [9]. While Compton cross sections measured in the two labs agree, p(γ,π) cross sections do not. A consistent analysis applied to both groups of data yields EMR values different by more than a factor of 2. The source of this difference is the p(γ,π) cross section scale, and the advantage of the LEGS data lies in the fact that both p(γ,π) and p(γ,γ) channels are locked together with a small common systematic scale uncertainty.

LEGS is supported by the U.S. Dept. of Energy under Contract No. DE-AC02-76-CH00016, by the Istituto Nazionale di Fisica Nucleare, Italy, and by the U.S. Nat. Science Foundation.

REFERENCES

1. A. Wirzba and W. Weise, Phys. Lett. **B188**, 6 (1987).
2. R. Bijker, F. Iachello and A. Leviatan, Ann. Phys. **236**, 69 (1994).
3. T. Sato and T.-S.H. Lee, Phys. Rev. **C54**, 2660 (1996).
4. A.J. Buchmann, et al., Phys. Rev. **C55**, 448 (1997).
5. LEGS Collaboration, G. Blanpied et al., Phys. Rev. Lett. **69**, 1880 (1992).
6. R. Beck et al., Phys. Rev. Lett. **78**, 606, (1997); H.-P. Krahn, thesis, U. Mainz (1996).
7. W. Chiang and F. Tabakin, Phys. Rev. **C55**, 2054 (1997).
8. A. Donnachie, Rep. Prog. Phys. **36**, 695 (1973).
9. LEGS Collaboration, G. Blanpied et al., Phys. Rev. Lett. **76**, 1023 (1996).
10. GEANT 3.2.1, CERN Library W5013, CERN (1993).
11. *GCALOR*, C. Zeitnitz and T.A. Gabriel, Nucl. Instr. Meth. **A349**, 106 (1994).
12. H. Genzel et al., Z. Physik **268**, 43 (1974).
13. G. Fischer et al., Z. Physik **253**, 38 (1972).
14. K. Büchler et al., Nucl. Phys. **A570**, 580, (1994).
15. H. Genzel et al., Z. Physik **A279**, 399 (1976).
16. T. Fujii et al., Nucl. Phys. **B120**, 395 (1977).
17. C. Molinari et al., Phys. Lett. **B371**, 181 (1996).
18. J. Peise et al., Phys. Lett. **B384**, 37 (1996); J. Ahrens, priv. comm.
19. Shin-nan Yang, J. Phys. G11, L205 (1985); priv. comm.
20. *SAID*, telnet vtinte.phys.vt.edu; R. Arndt, I. Strakovsky and R. Workman, Phys. Rev. **C53**, 430 (1996).
21. K. Watson, Phys. Rev. **95**, 228, (1954).
22. A. L'vov et al., Phys. Rev. **C55**, 359 (1997).
23. H. Dutz et al., Nucl. Phys. **A601**, 319, (1996); Gisela Anton, priv. comm.
24. A. Belyaev et al., Nucl. Phys. **B213**, 201 (1983).
25. V.A. Get'man et al., Nucl. Phys. **B188**, 397 (1981).
26. A.A. Belyaev et al., Sov. J. Nucl. Phys. **40**, 83 (1984); *ibid*, **43**, 947 (1986).
27. G. D'Agostini, Nucl. Inst. Meth. **A346**, 306 (1994).
28. J.R. Wolberg, *"Prediction Analysis"*, Van Nostrand Co., NY, p. 54-66 (1967).
29. G. Höhler et al., *Handbook Pi-N Scattering*, Phys. Data **12-1**, Karlsruhe (1979).
30. R. Davidson, N. Mukhopadhyay and R. Wittman, Phys. Rev. **D43**, 71 (1991).

Excited Baryon Form Factors at High Q^2

Paul Stoler,
Rensselaer Polytechnic Institute, Troy NY 12180

for Jefferson Lab Collaboration E94-014

G. Adams[1], A. Ahmidouch[3], C. Armstrong[9], K. Assamagan[3], S. Avery[3], K.Baker[3], P. Bosted[2], V. Burkert[4], J. Dunne[4], T. Eden[3], R. Ent[4], V. Frolov[1], D. Gaskell[4], P. Guèye[3], W. Hinton[3], C. Keppel[3],W.Kim[5], M.Klusman[1], D.Koltenuk[6], D.Mack[4], R.Madey[3], D.Meekins[4], R.Minehart[7], J.Mitchell[4], H.Mkrtchyan[9], J.Napolitano[1], G.Niculescu[3], I.Niculescu[3], M.Nozar[1], J.Price[1], P. Stoler[1], V. Tadevosyan[9], L. Tang[3], M. Witkowski[1], S. Wood[4]

[1]*Rensselaer Polytechnic Institute,* [2]*American University,* [3]*Hampton University,* [4]*Thomas Jefferson National Accelerator Facility,* [5]*Kyungpook National University,* [6]*University of Pennsylvania,* [7]*University of Virginia,* [8]*College of William and Mary,* [9]*Yerevan Physics Institute*

Abstract: The role of resonance electroproduction at high Q^2 is discussed in the context of exclusive reactions.as well as the alternative theoretical models which are proposed to treat exclusive reactions in the few GeV^2/c^2 region of momentum transfer. Jefferson Lab experiment 94-014, which measured the excitation of the $\Delta(1232)$ and $S_{11}(1535)$ via the reactions $p(e, e'p)\pi^0$ and $p(e, e'p)\eta$ respectively at $Q^2 \sim 2.8$ and 4 GeV^2/c^2 is described. and the state of analysis reported.

Hadronic resonance electroproduction at high Q^2 is an example of coherent exclusive reactions, which are distinguised from inclusive deep inelastic scattering (DIS) reactions in that DIS interacts incoherently with, and maps out the electromagnetic properties of the partons from which the hadron is composed, while coherent reactions measure electromagnetic properties of the entire hadron, selecting a subset of the hadron's Fock state expansion having size dependent upon Q^2 $\sim (1/\lambda^2)$, as indicated in Figure 1. below. At high Q^2 the selected hadronic state is small,

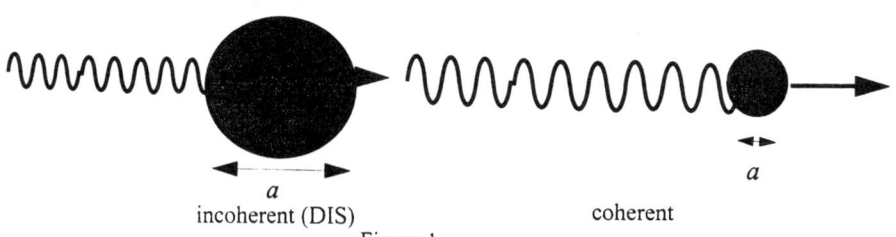

Figure 1.

and if the reaction were imbedded in a nucleus would exhibit the phenomena of *color transparency*. Examples of coherent exclusive reactions include the measurement of nucleon and pion elastic scattering form factors, $\langle f|T|i\rangle$, in which the initial and final states are identical, and hadron resonance electroproduction, in which the final state $|f\rangle$ is a resonant excitation of the initial hadron.

The total cross section for virtual photon absorption on a proton in the resonance region is shown in Figure 2 below.. Note that in addition to the large number of overlapping resonances

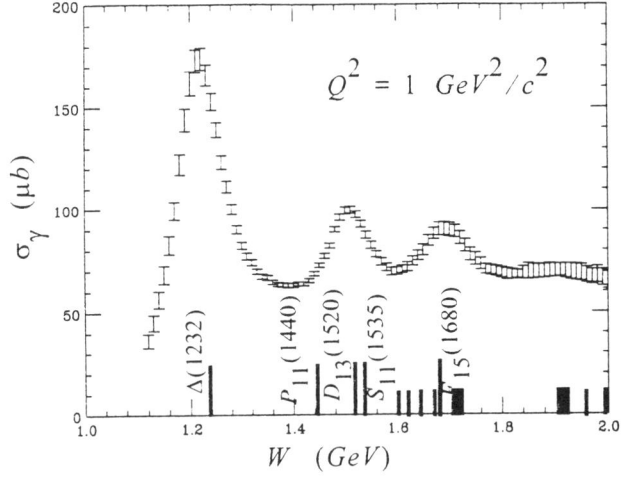

Figure 2.. The virtual photon cross section in the resonance region at $Q^2=1$ GeV^2/c^2. Of the approximately 20 resonances in this region only the most prominent are explicitly indicated.

there is a very significant non-resonant background.

Among the models for describing exclusive reactions, and in particular resonance production, the earliest have been based on effective Lagrangians, schematically illustrated in Figure 3., for which the degrees of freedom are hadronic currents. Although these models are not fundamental in terms of the quark and gluon structure of the hadron, they are still very useful for fitting and extracting the basic resonance and underlying non-resonant background parameters which are then used as experimental tests of quark and gluon based theories.

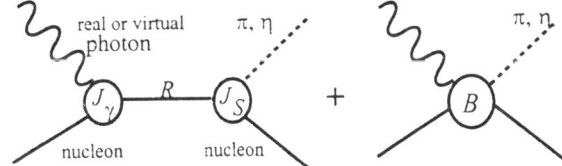

Figure 3. Schematic representation of the effective Lagrangian approach to resonance plus non-resonant backgrond. The total amplitude is schematically expressed as

In the language of quarks and gluons the best description of the reactions depends upon the range of Q^2 considered.

At low Q^2 (~0) the most commonl approaches have been elaborations of the constituent quark model (CQM) based on the formulation of Isgur and Karl (Is-79). An example of the remarkable success and at the same time failure of the model is the $\Delta(1232)$ resonance. On the one hand the most naive version of the model, in which the quarks move independently in a comon potential, predicts that the Δ(1232) is excited via a simple spin-flip of one of the quarks, as shown in Figure4.

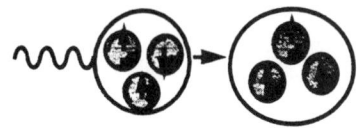

Figure 4.

This occurs via a pure M_{1+} multipole. For a spherical basis $E_{1+}=0$ since Y_2 cannot connect two Y_0 states. The inclusion of a residual one gluon exchange interaction between quarks gives the delta a small quadrupole moment and thus there is a small non-zero E_{1+} component which contributes. This was recently measured at $Q^2 = 0$ at Mainz (Be-97) for which it is found that $E_{1+}/M_{1+} \sim .03$. This near vanishing of E_{1+}/M_{1+} is one of the most striking confirmations of the CQM. On the other hand, one of the glaring deficiencies of the CQM is the inability to correctly give the absolute values of the dominant M_{1+} amplitude. For example Ca-92 obtain M_{1+} = -215 compared with $M_{1+}(exp)$ = 282 ± 1.3 obtained by Da-97.

At very high Q^2 it is widely believed that the reaction can be described by hard perturbative QCD (pQCD) processes involving only the minimal valence *current* quarks as shown in Figure 5. However, a controversial issue is how high a Q^2 is necessary for these hard processes to dominate (see St-97 and references within). Among the well known signatures for pQCD are constituent scaling rules, which for baryons predict the form factor $\sim 1/Q^4$. In the case of the Δ(1232) there is another important signature. Helicity conservation tells us that $E_{1+}/M_{1+} = 1$, which is in remarkable contrast to the CQM prediction.

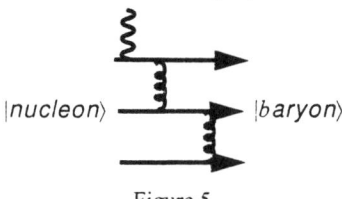

Figure 5.

Many believe that at currently attainable momentum transfers one is in a transition region in which more complicated soft processes play a very important role, and the reaction is dominated by Feynman like processes, schematically shown in Figure 6. A QCD sum rule calculation by Ba-96, which models these soft processes predict an $E_{1+}/M_{1+} \sim - 0.15$, which is very different from either the pQCD or CQM expectations. The only existing moderately high Q^2 data (Ha-79) have been fit by Bu-95 and Da-97, who find it is small, but with large statistical error.

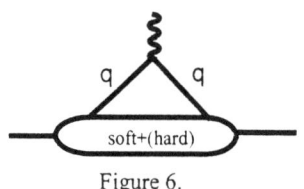

Figure 6.

Jefferson Lab experiment 94-014, which ran in Hall C in Nov. - Dec. 1996, has as Iobjectives to assess the limits of the CQM, and investigate the interplay of both soft and hard perturbative processes with increasing Q^2. The reactions measured were $p(e, e'p)\pi_0$, η and the resonances studied were the Δ(1232) and the $S_{11}(1535)$. The Reasons for choosing these resonances are as follows. The $\Delta(1232)$ is isolated, and has been extensively studied at low Q^2. It also has an anomolously large rate of decrease as a function of Q^2, and most importantly as discussed above, different models make very different predictions about the amplitudes. The $S_{11}(1535)$ has also been extensively studied at lower Q^2 since it is the only strongly excited resonance which has a large branching ratio (~50%) to the η, for reasons which are not understood, thus effectively isolating it, and its $J^\pi = 1/2^-$ structure results in a simple $l = 0$ decay angular distribution of the η. Also, its form factor appears to fall less with Q^2 than the nearby resonances so that it remains strongly excited at high Q^2.

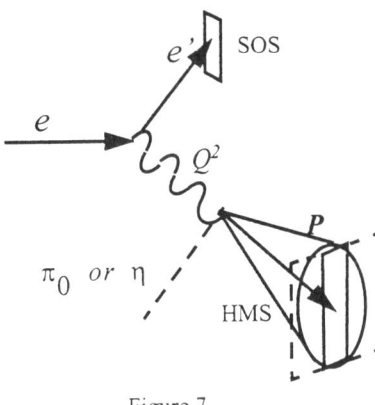

Figure 7.

Figure 7. schematically illustrates the experiment. $p(e,e'p)\pi^0$ from the $\Delta(1232)$ at $Q^2 = 2.8$ and 4 GeV^2/c^2, and $p(e,e'p)\eta$ from the $S_{11}(1535)$ at $Q^2 = $ 2.4 and 3.6 GeV^2/c^2, were measured with electron energies 3.2 and 4.0 GeV respectively. The experiment utilized about 200 hrs of beam at a current of nearly 100 μA, producing about 50,000 events each for the $\Delta(1232)$ and $S_{11}(1535)$ at each Q^2 setting. For each beam energy the electrons were detected by the SOS spectrometer, which was fixed in angle and momentum to cover the entire W range from elastic through about 1.6 GeV. Protons were detected by the HMS spectrometer. At high momentum transfer the protons emerge in a rather narrow cone around the q vector corresponding to 4π in the cm. as shown in Figure 7. At 4 Gev about 5 angular and 5 momentum settings of the HMS were sufficient to cover a large part of 4π with 50% overlap between adjacent angular settings. Since the experiment was kinematically complete the identification of π^0's and η's was accomplished by missing mass reconstruction on an event by event basis, as were the kinematic variables Q^2, and W, and the resonance center of mass decay angles θ_{cm}. This is shown in Figure 8 for one run corresponding to about 1.5% of the total data. The and reconstructions are clearly visible in the projection on the m^2 axis on the right, as is the multipion continuum. The projection on W with a cut on the missing masses of the and shows the clean separation of the and the S11(1535) by means of π^0 and η production respectively. The kinematic acceptance of this run relative to the widths of the two resonances is illustrated by the solid curves, which are arbitrarily normalized. The reconstructed cm. decay angles are typically about $\delta(\varphi \approx 3^0$ and $\delta\cos\theta \approx 0.04$. For the $\Delta(1232)$ the 2 pion background is totally eliminated, whereas for the $S_{11}(1535)$ only a small multipion background remains. The remaining radiative tail can be further reduced by angle cuts around the incident and scattered beam directions. An example of the cm. angular distribution is shown in Figure 8

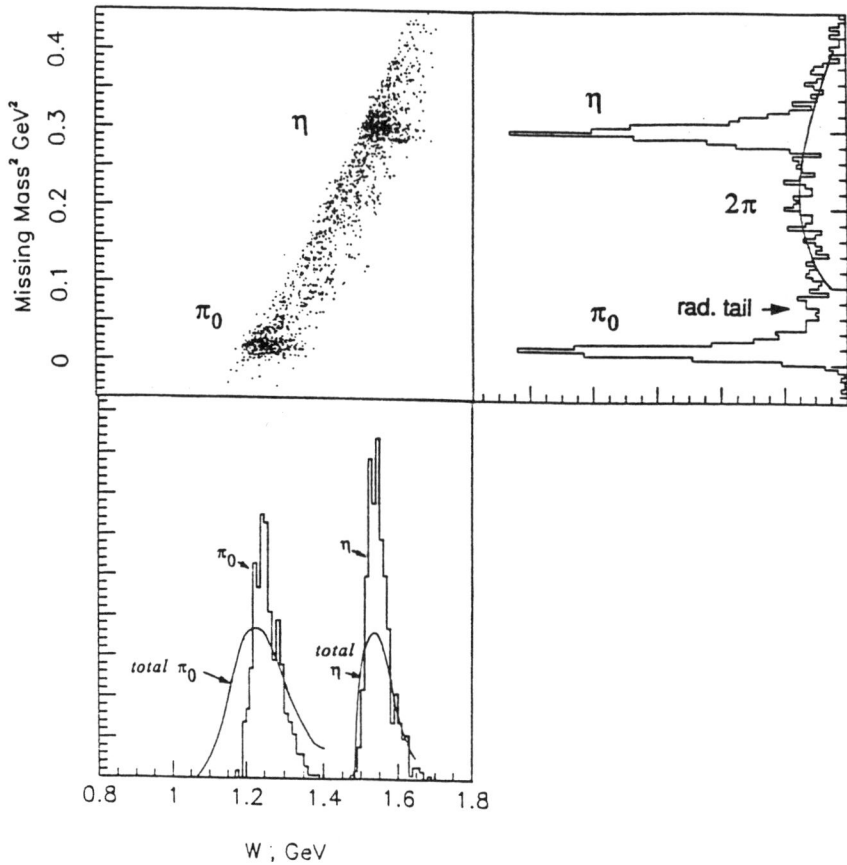

Figure 7. Missing mass squared m^2 vs. W for the reaction obtained in experiment 94-014. This corresponds to one kinematic setting in $\theta_0(HMS)$ and $p_0(HMS)$.

Figure 8. Preliminary center of mass angular distribution of π^0 production at the $\Delta(1232)$, at $W=1.235$ GeV and $\varphi \sim 0^0$ at $Q^2 = 2.8$ GeV^2/c^2, obtained in experiment 94-014. This represents a few percent of the available data at this Q^2. The different symbols correspond to data taken at different HMS kinematic settings. Although the overlap between points from different kinematic settings is reasonable, radiative corrections have not yet been carried out, and improvements in HMS and SOS acceptance and optics corections are in progress.

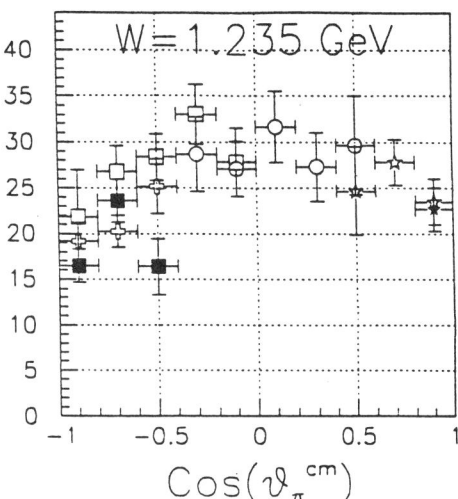

References:

Ba-96 Belyaev VM, Radyushkin AV. *Phys. Rev.* D53, 6509 (1996)
Be-97 R. Beck et al., *Phys. Rev. Lett.*, 78, 606 (1995)
Bu-95 V. Burkert and L. Elouadrhiri, *Phys. Rev. Lett.*, 75, 3614 (1995)
Ca-92 S. Capstick, *Phys. Rev.* D46, 2864 (1992).
Da-97 R. Davidson and N. C. Mukhopadhyay, *Phys Rev C*, to be published.
Ha-79 R. Haiden, DESY Report No. F21-79-03, unpublished, (1979)
Is-79 N. Isgur and G. Karl, *Phys. Rev.* D19, 2653 (1979)
St-97 G. Sterman and P. Stoler, *Ann. Rev. Nucl. Part. Sci.*, to be published.

Polarized Parity Violating Electron Scattering in ^3He and ^3H

S.L. Mintz, G. M. Gerstner, M.A. Barnett, and M. Pourkaviani

Physics Department
Florida International University
Miami, Florida 33199

Abstract. We calculate the asymmetry parameter, A, for the electron scattering reaction, $e+^3He \longrightarrow e+^3He$, run with right and left handed polarized electrons. We have calculated this reaction at energies from .1 GeV to 6 GeV. We present here the results for this reaction run at 1 GeV as an example. We obtain both A and a figure of merit. We find a sharp and unexpected variation in A due to cancellation between the weak and electromagnetic form factors. This is caused by the different q^2 dependences of the weak and electromagnetic form factors. Unfortunately in this region the figure of merit is small.

INTRODUCTION

It has long been suggested that parity violating polarized electron scattering might be useful for studying the weak interaction in nucleons and nuclei[1] and more recently for studying the role of strange quarks[2] in the nuclear medium. The quantity which is usually calculated in studying these processes is called the asymmetry and is defined as:

$$A = \frac{\frac{d\sigma(L)}{d\Omega} - \frac{d\sigma(R)}{d\Omega}}{\frac{d\sigma(L)}{d\Omega} + \frac{d\sigma(R)}{d\Omega}} \quad (1)$$

where L and R indicate the electron polarization.

The numerator of this quantity contains only parity violating terms which come from the interference between the one photon exchange and one Z-boson exchange diagrams for electron nucleus scattering. These terms are contained in the square of the following matrix element:

$$M = \frac{e^2}{q^2}\bar{u}\gamma_\mu u <f|J^\mu_{em}(0)|i> + \frac{G}{\sqrt{2}}\bar{u}\gamma_\mu[g_V + g_A\gamma_5]u <f|J^\mu_{weak(3)}(0)|i> \quad (2)$$

where for our case i and f are both either ^3He or ^3H. Our ability to calculate the asymmetry is for the most part tied to our ability to obtain the nuclear matrix elements of J^μ_{em} and J^μ_{weak}. We discuss this in the next section.

MATRIX ELEMENTS

It is well known[3] that the current matrix elements for spin 1/2 - spin 1/2 transitions may be written as:

$$<k|J^\mu_m(0)|k> = \bar{u}(F_a(q^2)\gamma^\mu + \frac{iF_b(q^2)\sigma^{\mu\nu}q_\nu}{2m_p})u \qquad (3)$$

where we use a and b generically for vector current form factors for both weak and electromagnetic currents. We can write a similar equation for the weak axial current:

$$<k|A^\mu(0)|k> = \bar{u}(\gamma^\mu\gamma_5 F_A(q^2) + \frac{q^\mu\gamma_5 F_P(q^2)}{m_\pi})u. \qquad (4)$$

Thus we need to know the form factors which appear in the current matrix elements given above and which contain the nuclear structure. Fortunately these form factors have been found[3] and are all of the form:

$$F(q^2) = F(0)\cos^2(\frac{-q^2}{\alpha m_\pi^2})(1 - \frac{q^2}{\beta m_\pi^2})^{-2} \qquad (5)$$

with various values of α and β. This is a crucial point because the q^2 dependences are not the same for the weak form factor, F_V^3 and the electromagnetic form factor F_1. The neutral weak current which enters into this calculation may be written as:

$$<f|J^{weak(3)}_\nu(0)|i> = <f|V^{(3)}_\nu - A^{(3)}_\nu - 2\sin^2\theta_W J^{em}_\nu|i>. \qquad (6)$$

This current matrix element contains a term of the form:

$$F'_V = F^{(3)}_V - 2\sin^2\theta_W F_1 \qquad (7)$$

which would normally play a dominant role in the asymmetry. However because of the different q^2 dependence and similar sizes of $F^{(3)}_V$ and F_1, F'_V can rapidly change its value and go from positive to negative at large q^2. In the asymmetry which consists of interference terms, the large term is normally $F'_V F_1$. However because of the above affect it will happen that for some q^2 values other terms will dominate. For this reaction, because $F^{(3)'}_M$ and $F^{(3)}_A$ are large, the normally suppressed axial terms become dominant at these q^2 values. This leads to very rapid fluctuations in the asymmetry.

RESULTS AND DISCUSSION

Making use of the previous equations we calculate the asymmetry and a figure of merit (F-O-M) for incident electrons of 1 GeV. The results are shown in figure 1. It can be seen that the asymmetry changes sign at about 32 degrees. At this value of θ the vector current part of the numerator has shrunk and is dominated by the axial part. These values for the asymmetry are reasonably large. Unfortunately the figure of merit is already quite small by this point. We note similar effects in the ^3H case.

It would clearly be very interesting to try to determine if this apparent structure is indeed present. If found ,it it would help confirm whole nucleus CVC at high q^2 and might make measurements of the the form factors possible over a wider range of q^2 than can be done at present. Thus these results are both interesting and unexpected.

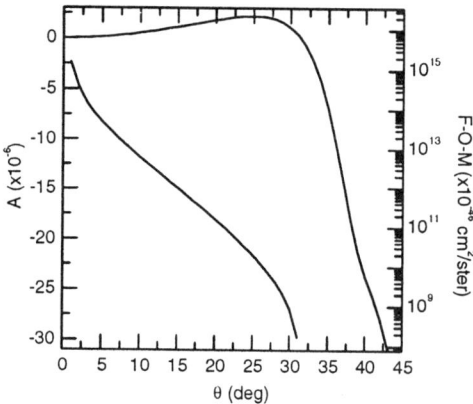

FIGURE 1. Plot of the Asymmetry and Figure of Merit as a function of laboratory scattering angle for incident 1 GeV electrons.The asymmetry is the upper curve (left scale, $A \times 10^{-6}$). The F-O-M is the lower curve (right scale,F-O-M $\times 10^{-46}$).

REFERENCES

1. Feinberg G., *Phys. Rev.* **D 12**,3575(1975).
2. Musolf M.J. and Donnelly T. W., *Nucl. Phys.* **A546**,509(1992).
3. Mintz S.L. et al.,*Nucl. Phys.***A598**,367(1996).

Proton Propagation Through Nuclei and The Quasi-Free Reaction Mechanism Studied with (e, e'p) Reactions.

D. Dutta[15], D. Abbott[3], Ts. A. Amatuni[24], A. Ahmidouch[8], C. Armstrong[23],
J. Arrington[2], K. A. Assamagan[7], O. K. Baker[7], S. Barrow[18], K. Beard[7],
D. Beatty[18], S. Beedoe[14], E. Beise[12], E. Belz[5], C. Bochna[9], H. Breuer[12],
E. Bruins[11], R. Carlini[3], J. Cha[7], N. Chant[12], C. Cothran[22],
W. J. Cummings[1], S. Danagoulian[14], D. Day[22], D. DeSchepper[11],
J.-E. Ducret[21], F. Duncan[12], J. Dunne[3], T. Eden[7], R. Ent[3], J. Fedchak[1],
H T. Fortune[18], V. Frolov[19], D. F. Geesaman[1], H. Gao[10], R. Gilman[20],
P. Gueye[7], J. O. Hansen[1], W. Hinton[7], R. Holt[10], C. Jackson[14],
H. E. Jackson[1], C. Jones[1], S. Kaufman[1], J. J. Kelly[12], C. Keppel[7],
M. Khandaker[13], W. Kim[9], E. Kinney[5], A. Klein[17], D. Koltenuk[18],
L. Kramer[11], W. Lorenzon[18], A. Lung[2], K. McFarlane[13], D. Mack[3],
R. Madey[7], P. Markowitz[6], J. Martin[11], A. Mateos[11], D. Meekins[3],
M. Miller[10], R. Milner[11], J. Mitchell[3], H. Mkrtchyan[24], R. Mohring[12],
G. Niculescu[7], I. Niculescu[7], T. G. O'Neill[1], D. Potterveld[1], J. W. Price[19],
J. Reinhold[1] C. Salgado[13], J. P. Schiffer[1], R. E. Segel[15], P. Stoler[19],
R. Suleiman[8], V. Tadevosyan[24], L. Tang[7], B. Terburg[10], W. Turchinetz[11],
D. van Westrum[5], Pat. Welch[16], C. Williamson[11], S. Wood[3], C. Yan[3],
J-C. Yang[4], J. Yu[18], B. Zeidman[1], W. Zhao[11], and B. Zihlmann[22].

[1] *Argonne National Laboratory, Argonne IL 60439* [2] *California Institute of Technology, Pasadena CA 91125* [3] *TJNAF, Newport News VA 23606* [4] *Chungnam National University, Taejon Korea* [5] *University of Colorado, Boulder CO 80309* [6] *Florida International University, University Park, FL 33199* [7] *Hampton University, Hampton VA 23668* [8] *Kent State University, Kent OH 44242* [9] *Kyungpook National University, Taegu, South Korea* [10] *University of Illinois, Champaign-Urbana IL 61801* [11] *Massachusetts Institute of Technology, Cambridge MA 02139* [12] *University of Maryland, College Park MD 20742* [13] *Norfolk State University, Norfolk VA 23504* [14] *North Carolina A & T, Greensboro NC 27411* [15] *Northwestern University, Evanston IL 60201* [16] *Oregon State University, Corvallis OR 97331* [17] *Old Dominion University, Norfolk, VA 23529* [18] *University of Pennsylvania, Philladelphia PA 19104* [19] *Rensselaer Polytechnic Institute, Troy NY 12180* [20] *Rutgers University, New Brunswick NJ 08903* [21] *CE Saclay, Gif-sur-Yvette France* [22] *University of Virginia, Charlottesville VA 22901* [23] *William and Mary, Williamsburg, VA 23187* [24] *Yerevan Physics Institute, Yeravan, Armenia*

QUASI-FREE $(e,e'p)$ SCATTERING

Quasi-free $(e,e'p)$ scattering is the process in which the incident electron directly knocks out a proton from the target nucleus, with the scattered electron and the knocked out proton detected in coincidence. Important kinematic quantities which can be measured in these processes are missing energy, E_m and missing momentum, P_m. These can be interpreted as measures of the separation energy and initial momentum of the proton in the nucleus. The definitions of these quantities are given below;

$$E_m = \omega - E_{p'} + M_p - T_{A-1} \tag{1}$$

$$\vec{P}_m = \vec{q} - \vec{p}\,' \tag{2}$$

Here ω is the energy transfer, $\vec{q}$ is the 3 momentum transfer from the electron to the nuclear system, $\vec{p}\,'$ is the momentum of the outgoing proton, M_p is the proton mass and T_{A-1} is the kinetic energy of the recoiling A-1 nucleons. The coincidence scattering cross-section for such a process in the plane wave impulse approximation (PWIA) can be written as,

$$\frac{d^6\sigma}{d\omega d\Omega_{e'} dE_{p'} d\Omega_{p'}} = P_{p'} E_{p'} \sigma_{ep} S(E_m, P_m) \tag{3}$$

where $E_{p'}$ and $P_{p'}$ are the energy and momentum of the scattered proton, σ_{ep} is the cross-section of the electron scattering off an off-shell proton [3] and $S(E_m, P_m)$ is the spectral function which gives the probability of finding a proton with separation energy and momentum (E_m, P_m) in the nucleus. If we compare the experimental cross-section with the cross-section calculated using the expression above, we get a measure of the proton attenuation. Under the PWIA it is assumed that there are no final state interactions and that the kinematics of the knocked out proton remains unchanged. Thus the ratio of the measured to the calculated cross-section tells us the fraction of the knocked out protons which emerge from the nucleus without any interaction with the residual nucleons which we define as the transparency T.

$$T = \frac{\int_V d^3 P_m dE_m N_{expt}(E_m, P_m)}{\int_V d^3 P_m dE_m N_{PWIA}(E_m, P_m)} \tag{4}$$

Here V is the finite experimental phase space volume of integration.

RESULTS

The PWIA calculations are done by using a realistic Monte Carlo [2,4,5] simulation of the apparatus, which uses model spectral functions based on the

Abstract. Jefferson Lab experiment E91-013 measured the energy dependence of proton propagation in nuclei, using the quasi-free $(e,e'p)$ reaction. The ratios of the experimental $(e,e'p)$ cross-sections integrated over the quasi-free region to PWIA calculations are presented as a function of momentum transfer, $(0.6 < Q^2 < 3.3 \text{ GeV}^2)$ and target nucleus (C, Fe and Au). As a first step towards a longitudinal and transverse separation of the quasi-free cross-section, a super ratio of the measured to the calculated cross-sections at forward and backward angles is presented.

INTRODUCTION

The propagation of nucleons through nuclei is one of the basic nuclear many body problems. Since electrons can probe the entire nuclear volume and knock protons from deep in the nuclear interior, quasi-free electron-proton scattering is an excellent tool to study proton propagation, if single nucleon currents dominate the reaction mechanism. This assumption can be directly tested by comparing the longitudinal and transverse nuclear response. The longitudinal response is expected to be predominantly single particle in nature while the transverse response can have significant contributions from two-body and n-body currents. Previous experiments at Bates at lower momentum transfer, Q, observed significantly more transverse quasi-free strength than longitudinal strength above the two-body threshold and concluded that multi-nucleon absorption mechanisms were important [1]. Quasi-free experiments at higher Q, seemed to support the single nucleon picture, but did not perform a longitudinal-transverse separation. Experiment NE18 at SLAC found that the measured spectral functions agreed very well with PWIA calculations [2] at Q^2 ranging from 1 to 6.8 GeV^2. Our experiment was designed to study the reaction mechanism with a longitudinal-transverse separation at two momentum transfers and use this understanding of the reaction mechanism to interpret the $(e,e'p)$ results in terms of nucleon propagation through the nucleus.

With these goals in mind, the experiment was carried out at the Thomas Jefferson National Accelerator Facility (Jefferson Lab), in experimental Hall C. The accelerator at Jefferson Lab provided a continuous electron beam with currents of about 20μA and energies ranging from 0.8 to 4.0 GeV. The experiment was performed using the two magnetic spectrometers (The High Momentum Spectrometer, HMS and the Short Orbit Spectrometer, SOS) in the hall. For each momentum transfer, data were taken at several proton angles over the range of quasi-free scattering. The two spectrometers had similar detector packages which consisted of two drift chambers for tracking, two planes of hodoscopes for trigger and time-of-flight, a threshold Čerenkov counter and a Lead-Glass calorimeter for particle identification.

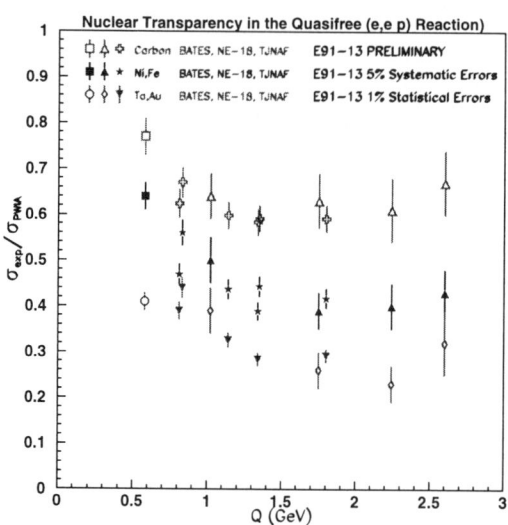

FIGURE 1. Transparency vs momentum transfer for C, Fe and Au. The present experiment has systematic uncertainty of ±5%. NE18 results are from reference [6] and Bates results on C, Ni and Ta targets are from reference [7]

independent particle shell model. Radiative corrections are included in the Monte Carlo and the de Forest prescription cc1 is used for the off-shell e-p cross-section. Both the data and the simulation were integrated over a finite volume defined by cuts $E_m < 80$ MeV and $|P_m| < 300$ MeV. Fig.1 shows the transparency as a function of momentum transfer Q and target nucleus. The transparency at a particular Q is determined by taking a weighted average of the transparency obtained at different proton angles around the momentum transfer direction. In addition, the transparency was multiplied by a factor (C 1.11, Fe 1.22 and Au 1.28) to correct for correlations, following the procedure of reference [6]. In Fig.1 our results are shown along with the results from the two previous experiments, the Bates [7] and the NE18 [6] experiments. The results from our experiment have less than 1% statistical uncertainty and systematic uncertainty of ±5%, which we expect to reduce significantly. Our results are in agreement with the previous experiments and have better accuracy. One can also notice that the transparency is relatively flat beyond Q of 1 GeV. The pair of points at Q of both 0.8 GeV and 1.4 GeV correspond to the forward and backward angle data, with the backward angle data being higher in all cases.

These data, taken with a range in virtual photon polarization, $\Delta\epsilon$, of around 0.5, will enable us to perform a Rosenbluth separation at these kinematics.

At this stage we do not have separated cross-sections, but as a first step we have calculated a super-ratio defined as:

$$R = \frac{\text{data(backward)}/\text{MonteCarlo(backward)}}{\text{data(forward)}/\text{MonteCarlo(forward)}} \quad (5)$$

The top panels of Fig.2 show the super-ratio as a function of missing energy for carbon at the two different Q^2. Each plot is for proton angles along the momentum transfer directions. The lower panels are plots of missing energy distributions which show the location of the p and s-shell peaks. The super-ratio is unity for the p-shell in both cases. At the lower Q^2, we see some excess transverse strength beyond 50 MeV in missing energy, while at the higher Q^2, we do not see any excess transverse strength over the full missing energy region. We are currently in the process of reducing our systematic uncertainties and doing a L/T separation of the data.

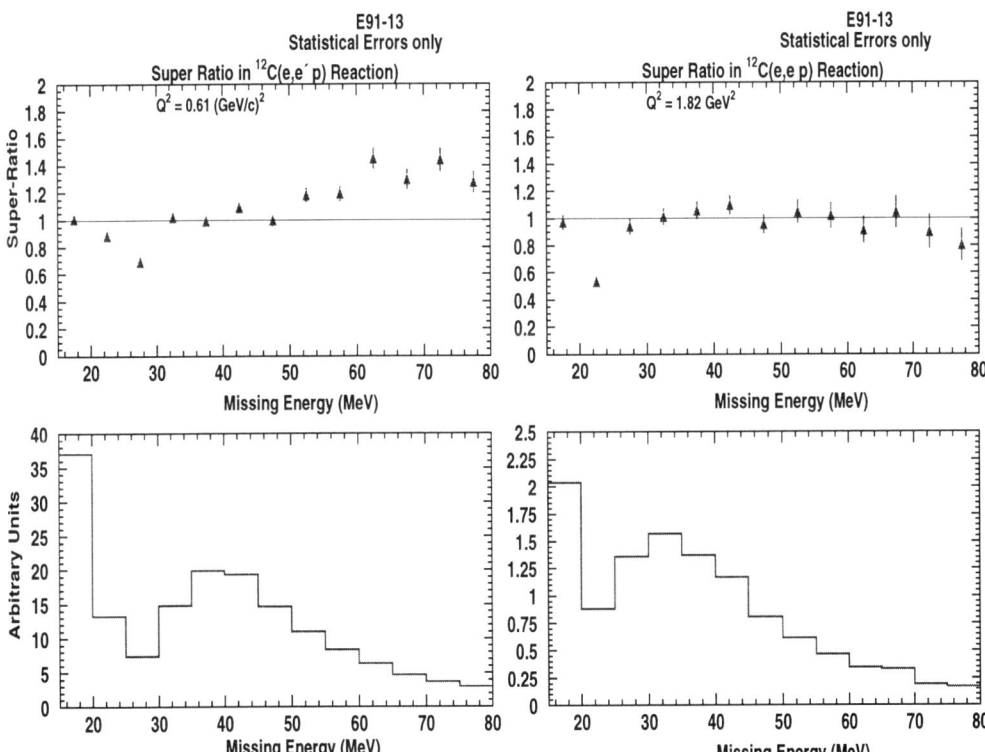

FIGURE 2. Top: super-ratio (ratio of the backward angle data to Monte Carlo over forward angle data to Monte Carlo) as a function of missing energy at Q^2 of 0.6 and 1.8 GeV^2. Bottom: missing energy distribution for the two Q^2 values.

ACKNOWLEDGMENTS

This work was supported in part by the U.S. Department of Energy and by the National Science Foundation.

REFERENCES

1. P. Ulmer et al., *Phys. Rev. Lett.* **59**, 2259 (1987).
2. N.C.R.Makins, Massachussetts Institute of Technology Ph.D. thesis (1994).
3. T. de Forest, *Nucl. Phys.* **A 392**, 232 (1983).
4. N.C.R. Makins et al., *Phys. Rev. Lett.* **72**, 1986 (1994).
5. T.G. O'Neill, California Institute of Technology Ph.D. thesis (1994).
6. T.G. O'Neill et al., *Phys.Lett.* **B 351**, 87 (1995).
7. G. Garino et al., *Phys. Rev.* **C 45**, 780 (1992).

The submitted manuscript has been authored by a contractor of the U. S. Government under contract No. W-31-109-ENG-38. Accordingly, the U. S. Government retains a nonexclusive, royalty-free license to publish or reproduce the published form of this contribution, or allow others to do so, for U. S. Government purposes.

η-Meson Photoproduction Dynamics and Missing Resonances

B. Saghai[1], F. Tabakin[2], J. Ajaka[3], and P. Hoffmann-Rothe[3]

[1]*Service de Physique Nucléaire, DAPNIA, CEA-Saclay, 91191 Gif-sur-Yvette, France*
[2]*Department of Physics & Astronomy, University of Pittsburgh, Pittsburgh, PA 15260*
[3]*Institut de Physique Nucléaire, 91406 Orsay, France*

Abstract. The general nodal structure approach is applied to the recent $\gamma\vec{p} \to \eta p$ T-asymmetry data from ELSA. The reaction mechanism is found to require, in addition to the dominant S_{11} and D_{13} resonances, contributions from P_{13} and D_{15} resonances. This finding is confirmed within a simple dynamical approach. An indication on the presence of a predicted P_{13} nucleonic resonance is observed.

INTRODUCTION

Using a density matrix approach [1] in a multipole truncated framework, we have examined the energy dependent evolution of the nodes that can occur in meson photoproduction spin observables [1–3] and have obtained general *model independent constraints* on the cross section and on all of the other 15 spin observables asymmetries for pseudoscalar meson photoproduction processes: $\gamma p \to \pi^+ n$, $K^+ \Lambda$ and ηp.

The angular structure of selected spin observables were then proven [?,3] to provide powerful means for deepening understanding of the underlying reaction mechanisms, and especially [3] for studying a possible role played by the Roper resonance and for revealing some of the low-mass missing nucleonic resonances. A rather large number of missing baryonic resonances have been predicted by quark-based studies [4] of the baryon spectrum. These undiscovered resonances are typically weakly coupled to the πN channel, but should appear in other meson-nucleon systems, such as ηN. These well identified observables can be measured at CEBAF, ELSA, ESRF, and MAMI.

RESULTS AND DISCUSSION

In previous publications, we had anticipated the interest in the target asymmetry T, and produced predictions [3]. It was shown that the profile function $T(\theta)$ is of Legendre class $\mathcal{L}_{1b}$ and hence has the general form: $T(\theta) = \sin\theta \sum_{L=0}^{n} a_L \cos^L \theta$. The polynomial coefficients can be expressed as functions of imaginary parts of bilinear products of the electric, $E_\ell^\pm$, and magnetic, $M_\ell^\pm$, multipole amplitudes. The conventions and expressions in Ref. [3], involve a simple notation in which $S \equiv E_0^+$, while P denotes the P−wave $J = 1/2$ (E_1^-, M_1^-) multipoles. Similarly, $P' \equiv [P$−wave $J = 3/2$ (E_1^+, M_1^+)], $D \equiv [D$−wave $J = 3/2$ (E_2^-, M_2^-)], $D' \equiv [$ D−wave $J = 5/2$ (E_2^+, M_2^+)]. Using that abbreviated notation, the structures of a_0 to a_3 are described by:

$$a_0 \to \boxed{S}P' \oplus P\boxed{D} \oplus PD' \oplus P'\boxed{D} \oplus P'D',$$

$$a_1 \to P' \oplus \boxed{D} \oplus D' \oplus \boxed{S}\boxed{D} \oplus \boxed{S}D' \oplus PP' \oplus \boxed{D}D',$$

$$a_2 \to PD' \oplus P'\boxed{D} \oplus P'D', \quad a_3 \to D' \oplus \boxed{D}D'.$$

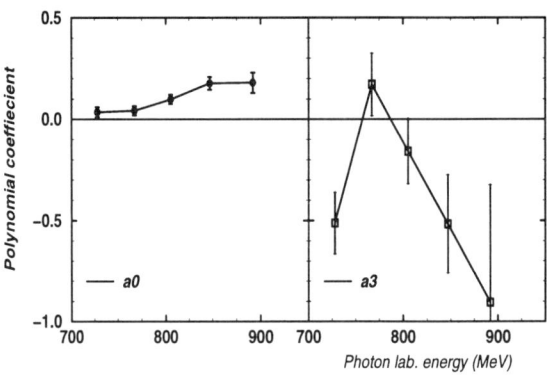

FIGURE 1. Polynomial coefficients a_0 and a_3 for $T(\theta) = \sin\theta \sum_{L=0}^{3} a_L \cos^L \theta$ as functions of energy as obtained by fitting the data from Ref. [5]. The curves are eye guides.

Here we apply [6] our method to the recent $\gamma\vec{p} \to \eta p$ data from Bonn [5], which provides angular distributions of the polarized target asymmetry T. Notice that if the intervening resonances were limited to S_{11} and D_{13}, only a_1 would be nonzero. As shown in Fig. 1, a_0 and a_3 assume finite values at all five measured energies. From the above expressions for a_0 to a_3 coefficients, our analysis (Fig. 1) shows clearly that, in addition to the dominant $S_{11}(1535)$ and $D_{13}(1520)$ resonances [7], these data require contributions from P_{13} and D_{15} resonances. Moreover, contributions from P_{11} resonances can not be excluded by the present data base.

Finally, in Fig. 2, we show the results of a simple dynamical approach [6], where electric and magnetic multipole amplitudes are expressed in terms of various nucleonic resonances (described by "relativized" energy-dependent Breit-Wigner forms), plus a smooth background including S- and P- waves.

This analysis, fitting the Bonn T-asymmetry data [6], confirms the presence of the P_{13} and D_{15} resonances in the dynamics of the η photoproduction. Here, the best agreement with the data is obtained by introducing a P_{13} missing resonance with M=1880 MeV (and Γ=150 MeV). Investigations using more realistic dynamical models [7,8] are anticipated.

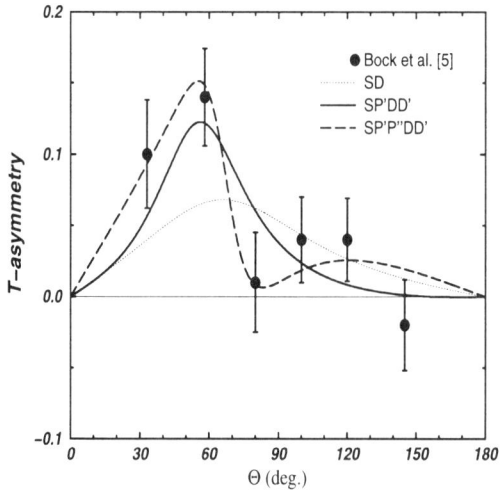

FIGURE 2. T-asymmetry angular distribution for the reaction $\gamma\vec{p} \to \eta p$ at E_γ^{lab} =767 MeV. Curves result from a simple dynamical model including the dominant $S_{11}(1535)$ and $D_{13}(1520)$ resonances (SD), an additional P_{13} and D_{15} resonances ($SP'DD'$). The effect of a predicted P_{13} resonance is also shown ($SP'P''DD'$). Data are from Ref. [5]

REFERENCES

1. C.G. Fasano, F. Tabakin and B. Saghai, *Phys. Rev. C* **46**, 2430 (1992).
2. B. Saghai and F. Tabakin, *Phys. Rev. C* **53**, 66 (1996).
3. B. Saghai and F. Tabakin, *Phys. Rev. C* **55**, 917 (1997); F. Tabakin and B. Saghai, in *Proceedings of the N* Workshop*, Seattle, Sept. 1996, Editors T.-S. Harry Lee and W. Roberts, *underpress*.
4. R. Koniuk and N. Isgur, *Phys. Rev. D* **21**, 1868 (1980); S. Capstick and W. Roberts, *ibid D* **49**, 4570 (1994).
5. A. Bock, PhD Thesis, Bonn University (1996); G. Anton, A. Bock, and J. P. Didelez, *private communication* (1996).
6. B. Saghai, F. Tabakin, J. Ajaka, and P. Hoffmann-Rothe, *in preparation*.
7. M. Benmerrouche, Nimai C. Mukhopadhyay, and J.F. Zhang, *Phys. Rev. D* **51**, 3237 (1995).
8. Zhenping Li, *Phys. Rev. C* **52**, 1648 (1995).

Measurements of the reactions $^{12}C(\nu_\mu, \mu^-)^{12}N_{g.s.}$ and $^{12}C(\nu_\mu, \mu^-) X$

Eric D. Church

University of California, Riverside, CA 92521,
representing the LSND collaboration

Abstract. Charged current scattering of ν_μ on ^{12}C has been studied [1] using a π^+ decay-in-flight ν_μ beam at the Los Alamos Meson Physics Facility. The observed flux-averaged cross section for the exclusive reaction $^{12}C(\nu_\mu, \mu^-)^{12}N_{g.s.}$ of $(6.6 \pm 1.0 \pm 1.0) \times 10^{-41} cm^2$ agrees well with reliable theoretical expectations. A measurement was also obtained for the inclusive cross section to all accessible ^{12}N states, $^{12}C(\nu_\mu, \mu^-)X$. This flux-averaged cross section is $(11.2 \pm 0.3 \pm 1.8) \times 10^{-40}$ cm^2, which is approximately half of that given by a recent Continuum Random Phase Approximation (CRPA) calculation [2], but is in agreement with other calculations [3].

I INTRODUCTION

LSND identifies $\nu_\mu\, ^{12}C \to \mu^- + X$ reactions by detecting the μ^- and the subsequent electron from the decay $\mu^- \to e^- + \bar{\nu}_e + \nu_\mu$. As a result of this coincidence requirement a clean beam excess sample of events can be obtained with relatively loose selection criteria. For analysis of the exclusive process $\nu_\mu\, ^{12}C \to \mu^- + ^{12}N_{g.s.}$ we further require detection of the e^+ from the beta decay of $^{12}N_{g.s.}$. Additional information in the inclusive reaction may be taken from the rate of neutron production; the presence of a neutron can be established from the neutron capture reaction $n + p \to d + \gamma$.

The data reported here were obtained in 1994 and 1995 at the Los Alamos Meson Physics Facility (LAMPF) primarily using neutrinos produced at the A6 proton beam stop, though some neutrinos are also produced at upstream targets A1 and A2. The neutrino source and detector are described in detail elsewhere [4].

II ANALYSIS

Events are required to have very few hits in the veto at both the muon time and the electron time and there must not be too many hits in the tank at the muon time – the purpose of both of which is to suppress cosmic ray muons and their decay (Michel) electrons. Additionally, the electron energy must conform to the well-known Michel energy distribution, and the time and space differences between the muon and electron must satisfy $\delta t_{\mu e} \lesssim 4$ muon lifetimes (in oil) and $\delta r_{\mu e} \leq 200$ cm, respectively. Finally, the electron vertex must reconstruct inside the very well-understood portion of the detector volume. The distribution of $\delta t_{\mu e}$, shown on the left in Fig. 1, agrees well with the 2.03 μs μ^- lifetime in mineral oil. The best fit, which is also shown, corresponds to a lifetime of $1.98 \pm 0.06 \mu$s.

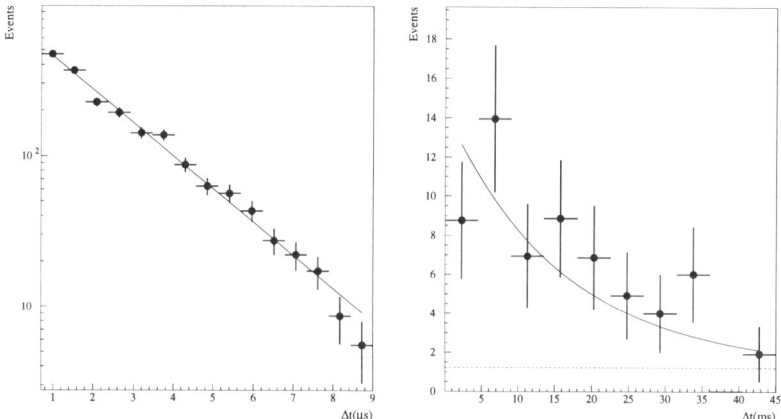

FIGURE 1. The distribution of time between the μ^- and the decay e^- in the inclusive sample, $^{12}C(\nu_\mu,\mu^-)X$. Adjacent is the distribution of time between the e^- and the decay e^+ in the exclusive sample. The expected background (dashed) and background plus signal (solid) with a 15.9 ms $^{12}N_{g.s.}$ lifetime is also shown.

For the exclusive reaction, for which three particles are detected and the mean beta decay lifetime is 15.9 ms, the positron must have fewer than 4 in-time veto hits, the e^-–e^+ vertex separation must be small, and the time of e^+ emission must be less than 45 ms. Shown on the right in Fig. 1 is the distribution of the time for positron emission, $\delta t_{e^-e^+}$ from the $^{12}N_{g.s.}$, as well as the expected distribution.

The cosmic ray background which remains after all selection criteria have been applied is well measured with the beam-off data and subtracted using the duty ratio, the ratio of beam-on time to beam-off time.

The presence of a neutron is established by detection of the gamma ray from the neutron's capture on a proton in the detector via the reaction n + p

→ d + γ. The distribution of the likelihood ratio R, described in Ref. [5], for correlated γ's from neutron capture is very different from that for uncorrelated (accidental) γ's.

III RESULTS

Table 1 shows the number of beam excess events and the number of background events for the exclusive reaction, $^{12}C(\nu_e,e^-)^{12}N_{g.s.}$. The flux-averaged cross section is measured to be $<\sigma> = (6.6 \pm 1.0 \pm 1.0) \times 10^{-41} cm^2$, where the first error is statistical and the second systematic.

Table 1 also shows the number of beam excess events and the number of background events for the inclusive reaction, $^{12}C(\nu_\mu, \mu^-)X$. The measured flux-averaged cross section for the inclusive reaction is $<\sigma> = (11.2 \pm 0.3 \pm 1.8) \times 10^{-40} cm^2$, where the first error is statistical and the second systematic.

TABLE 1. Beam-excess and background events for both the exclusive and inclusive reactions.

Corrected beam excess events – exclusive	72.4 ± 9.5
$\bar{\nu}_\mu + {}^{12}C \to \mu^+ + {}^{12}B_{g.s.}$ background	2.0 ± 0.4
accidental e^+ background	13.6 ± 1.4
$\nu_e + {}^{12}C \to \mu^- + {}^{12}N_{g.s.}$	56.8 ± 9.6
Corrected Beam-excess events – inclusive	1942 ± 46
$\bar{\nu}_\mu + p \to \mu^+ + n$ background	140 ± 22
$\bar{\nu}_\mu + {}^{12}C \to \mu^+ + X$ background	46 ± 23
$\nu_\mu + {}^{13}C \to \mu^- + X$ background	18 ± 9
$\nu_\mu + {}^{12}C \to \mu^- + X$	1738 ± 56

Last, after subtraction of background neutron-producing reactions we conclude that 1.9±2.6% of the events from the reaction $^{12}C(\bar{\nu}_\mu,\mu^+)X$ have an associated neutron compared to the CRPA calculation [2] of 5.9%.

REFERENCES

1. C. Athanassopoulos, *et. al.*, LA-UR-97-1848 (available at http://nu1.lampf.lanl.gov/~lsnd) submitted to Phys. Rev. C.
2. E. Kolbe *et al.*, Phys. Rev. C **52**, 3437 (1995).
3. N. Auerbach, N. Van Giai, O.K. Vorov, nucl-th/970503.
   A. C. Hayes, paper presented at the Joint APS/AAPT meeting, April, 18-21, 1997.
4. C. Athanassopoulos, *et. al.*, Nucl. Inst. and Meth. A **388** 149 (1997).
5. C. Athanassopoulos *et al.*, Phys. Rev. C **54**, 2685 (1996).
   C. Athanassopoulos *et al.*, Phys.Rev.Lett. **77** 3082 (1996).
6. C. Athanassopoulos *et al.*, Phys. Rev. C **55**, 2078 (1997).

New Results from the SMC

Andrzej Sandacz[*][1]

*Sołtan Institute for Nuclear Studies
ul. Hoża 69, 00681 Warsaw, Poland*

Abstract. We present a preliminary new measurement of the spin-dependent structure function g_1^p. The data cover the range $0.003 < x < 0.7$ extending to small x values not accessible in other experiments. They do not support an earlier suggestion of rising g_1^p as $x \to 0$. Further, from the analysis of combined world data on spin structure functions g_1 we obtain the estimates of moments Γ_1 for proton, deuteron and neutron. We conclude that the data agree with the Bjorken sum rule, and that they are at variance with the Ellis-Jaffe sum rules. We also determine the flavour-singlet axial coupling, a_0, which is 0.28 ± 0.05, and the strange quark axial coupling, a_s, which is negative and significantly different from zero. Finally, for the total SMC statistics we present the preliminary measurements of the semi-inclusive spin asymmetries and of the polarisation distributions of u and d valence quarks, and of non-strange sea quarks.

NEW RESULTS ON g_1^p

In this presentation we report on a preliminary measurement of the spin-dependent structure function g_1^p obtained by scattering longitudinally polarised muons of 190 GeV energy on longitudinally polarised protons in the kinematic range $0.003 < x < 0.7$ and $1 \text{ GeV}^2 < Q^2 < 100 \text{ GeV}^2$. The data were collected in 1996 with the high-energy muon beam M2 of the CERN SPS and using ammonia [1] as the target material. They complement our earlier measurements taken at the same beam energy with the butanol target [2].

The experimental set-up, the data taking procedure, the evaluation of the cross-section asymmetry A_1^p and of the spin structure function g_1^p are similar to those described in detail in Ref. [2]. The beam polarisation was determined by measuring the cross-section asymmetry for the scattering of polarised muons on polarised atomic electrons. The average polarisation is $P_\mu = -0.77 \pm 0.03$. The ammonia target had an average proton polarisation of approximately 0.89.

In 1996 we obtained approximately 9 million inclusive events after cuts. This is approximately twice the number of events for 1993 [2]. Together with the

---

[1] On behalf of the Spin Muon Collaboration

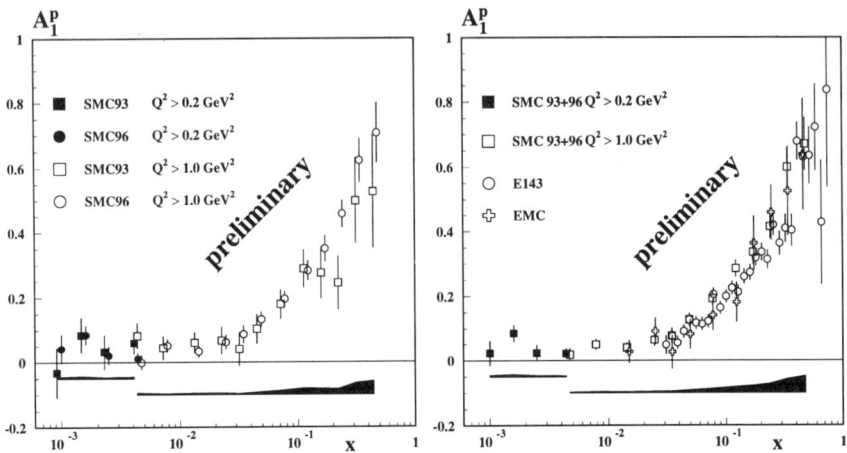

FIGURE 1. Left: A_1^p at measured Q^2 for SMC 1993 and 1996 data separately. The bars represent the statistical errors and the shaded bands show the systematic errors. The solid circles and squares show SMC data of $Q^2 < 1.0$ GeV^2. Right: A_1^p at measured Q^2 for all SMC proton data, as well as for E143 and EMC.

higher proton polarisation and a greater proportion of polarizable protons in ammonia compared to butanol it results in about 1.7 times smaller statistical errors for 1996 data.

In Fig. 1 (left) the SMC results on A_1^p are compared for 1996 and 1993 data. The statistical errors of 1996 data are significantly smaller and the two data sets are compatible within errors. We then combine both data sets and compare them with other experimental data in the right figure. The measurements of A_1^p from different experiments are consistent and complementary; the E143 data [4] are more precise in the common x region, but the SMC data extends into lower x, which is important for the evaluation of the first moment Γ_1^p. It should be noted that the average Q^2 for the SMC and EMC [3] data is approximately 10 GeV^2, whereas it is about 3 GeV^2 for the E143 data. Thus the agreement between the CERN and SLAC experiments indicates no significant Q^2 dependence of A_1^p within the errors. Although the SMC measurements of A_1 cover the range $0.0008 < x < 0.7$ and 0.2 GeV$^2 < Q^2 < 100$ GeV^2, for the determination of g_1 and its first moment Γ_1 we use only data at $Q^2 > 1$ GeV^2, which correspond to the range $0.003 < x < 0.7$.

In order to estimate the moments Γ_1, the structure functions g_1 have to be evaluated at a fixed value Q_0^2. It is done using a perturbative QCD evolution in next-to-leading order [2]. To determine parametrisations of the spin dependent parton distributions we use the published data from the SLAC experiments E142 [5] and E143 [4,6], from the EMC [3] as well as the SMC data for deuteron [7] and proton, including preliminary 1996 data.

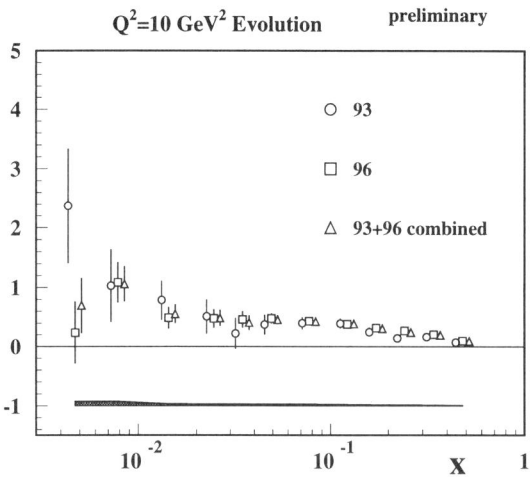

FIGURE 2. The structure function g_1^p evolved to $Q_0^2 = 10$ GeV^2 using NLO QCD evolution. The 1993 and 1996 data sets are shown separately as well as the combined data. The statistical errors are shown as bars, the shaded band represent the systematic errors.

The SMC measurements of g_1^p evolved to $Q_0^2 = 10$ GeV^2 are presented in Fig. 2. Our data from 1993 and 1996 are in agreement within the statistical errors. The more accurate 1996 data do not support an earlier suggestion of rising g_1^p as $x \to 0$.

MOMENTS Γ_1 AND AXIAL COUPLINGS

For the combined analysis of the first moments $\Gamma_1 = \int_0^1 g_1(x)\,dx$ and of axial couplings we used the data from the SLAC experiments E143, E142, from the EMC and all deuteron and proton SMC data, including preliminary 1996 data.

TABLE 1. Moments Γ_1 and axial couplings obtained from the combined analysis of world data on g_1 at $Q_0^2 = 5$ GeV^2. Γ_1^p, Γ_1^n and Γ_1^d are compared with Ellis-Jaffe sum rules predictions with NLO QCD corrections, $\Gamma_1^p - \Gamma_1^n$ is compared with the Bjorken sum rule prediction with NNLO QCD corrections.

	Experiment	Theor. prediction
Γ_1^p	0.136 ± 0.009	0.1668 ± 0.0046
Γ_1^d	0.038 ± 0.006	0.0704 ± 0.0041
Γ_1^n	-0.059 ± 0.016	-0.0146 ± 0.0044
$\Gamma_1^p - \Gamma_1^n$	0.192 ± 0.021	0.1814 ± 0.0028
a_0	0.281 ± 0.053	
a_s	-0.100 ± 0.022	

The analysis is similar to that described in Ref. [2]. The combined results are evaluated at an intermediate Q^2 of 5 GeV^2 to avoid large Q^2 evolutions.

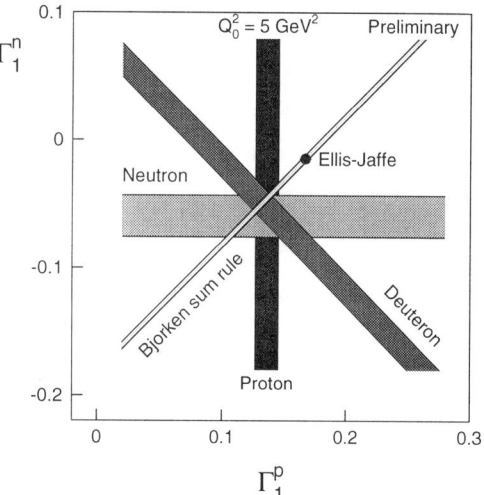

FIGURE 3. Comparison of the combined experimental results for Γ_1^p, Γ_1^n and Γ_1^d with the predictions for the Bjorken and the Ellis-Jaffe sum rules. The Ellis-Jaffe prediction is shown by the black ellipse inside the Bjorken sum rule band.

The results are given in Table 1 and in Fig. 3. The data agree with the Bjorken sum rule, which at present is verified experimentally with about 10% accuracy. The data significantly disagree with the Ellis-Jaffe sum rules for proton, deuteron and neutron. The values of the axial couplings, $a_0 = 0.28 \pm 0.05$ and $a_s = -0.10 \pm 0.02$, are at variance with the Ellis-Jaffe assumption of $a_s = 0$ and the prediction $a_0 = 0.579 \pm 0.025$ [8].

The flavour-singlet axial coupling a_0 and the corresponding couplings of the individual quark flavours can be understood as quark spin contributions to the proton spin, up to a gluonic contribution which is due to $U(1)$ anomaly of the singlet axial vector current [9]. In the Adler-Bardeen scheme, the axial couplings are decomposed into quark and gluon spin contributions Δq and Δg as

$$a_q = \Delta q - \frac{\alpha_s}{2\pi}\Delta g \qquad (q = u, d, s). \tag{1}$$

The axial coupling a_q depends on Q^2, whereas Δq is Q^2-independent in the AB scheme. When we make the assumption that $\Delta s = 0$, our measurement of a_0 corresponds to a gluon polarisation $\Delta g = 2.2 \pm 0.5$ at $Q^2 = 5$ GeV^2 in the AB scheme. From our QCD fits of the spin dependent parton distributions mentioned in the previous section, the first moment of the polarised gluon distribution at $Q^2 = 5$ GeV^2 is $\Delta g \approx 1.3$.

SEMI-INCLUSIVE DATA

We present preliminary results from the analysis of the total SMC statistics of the semi-inclusive data. The analysis is similar to that described in Ref. [10], where the semi-inclusive measurements for a part of the SMC data were published.

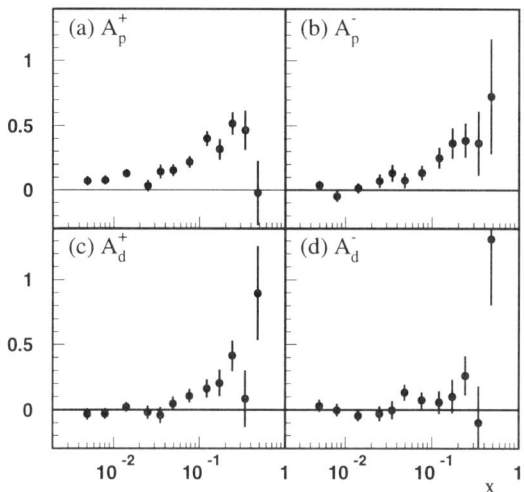

FIGURE 4. Semi-inclusive asymmetries of spin-dependent cross sections of (a) positive hadrons on proton, (b) negative hadrons on proton, (c) positive hadrons on deuteron and (d) negative hadrons on deuteron. The error bars are statistical.

In Fig. 4 we present the semi-inclusive spin symmetries $A_p^{+(-)}$ and $A_d^{+(-)}$ for positively and negatively charged hadrons from deep inelastic scattering of polarised muons on polarised protons and deuterons in the range $0.003 < x < 0.7$.

In the quark parton model the semi-inclusive spin asymmetries depend on the quark distibutions, polarised as well as unpolarised, and on the quark fragmentation functions, whereas the inclusive asymmetries $A_1^{p(d)}$ depend only on the quark distributions [10]. Using the semi-inclusive asymmetries presented in Fig. 4 together with the SMC inclusive asymmetries A_1^p (shown earlier) and A_1^d [7] we obtained the polarisation distributions of u and d valence quarks, and of non-strange sea quarks[2].

---

[2] Our data have little sensitivity to the strange quarks polarisation distribution, which was then parametrised as in Ref. [10].

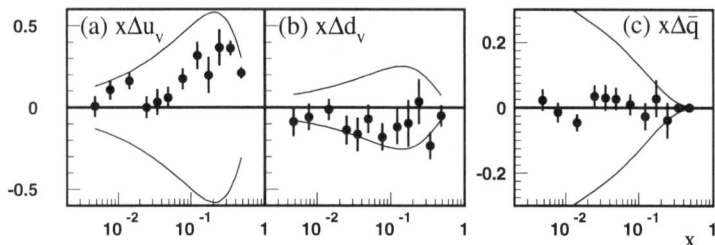

FIGURE 5. Quark spin distribution functions (a) $x\Delta u_v(x)$, (b) $x\Delta d_v(x)$, (c) $x\Delta \bar{q}(x)$ obtained with the assumption $\Delta \bar{u}(x) = \Delta \bar{d}(x) \equiv \Delta \bar{q}(x)$. The points at the two highest x bins were obtained when the sea polarisation was set to zero. The curves correspond to the upper and lower limits $\pm q(x)$ from the unpolarised quark distributions (GRV 94 LO [11]) evaluated at $Q^2 = 10$ Gev^2.

Our results, shown in Fig. 5, indicate a sizable polarisations of the valence quarks, in paricular at small x, which is consistent with the observation of $g_1^p \neq g_1^n$ at the smallest x. In the same region we observe that the spin contribution of the non-strange sea is small and consistent with zero, although the unpolarised sea is large.

REFERENCES

1. SMC, B. Adeva et al., submitted to Nucl. Instr. and Meth., CERN-PPE/97-66.
2. SMC, D. Adams et al., submitted to Phys. Rev. D, CERN-PPE/97-22 (hep-ex/9702005).
3. EMC, J. Ashman et al., Phys. Lett. **B206** (1988) 364; Nucl. Phys. **B328** (1989) 1.
4. SLAC E143, K. Abe et al., Phys. Rev. Lett. **74** (1995) 346.
5. SLAC E142, P.L. Anthony et al., Phys. Rev. **D54** (1996) 6620.
6. SLAC E143, K. Abe et al., Phys. Rev. Lett. **75** (1995) 25; Phys. Lett. **B364** (1995) 61; Phys. Rev. Lett. **76** (1996)587.
7. SMC, D. Adams et al., Phys. Lett. **B396** (1997) 338.
8. J. Ellis and R.L. Jaffe, Phys. Rev. **D9** (1974) 1444; **D10** (1974) 1669.
9. G. Altarelli and G.G. Ross, Phys. Lett. **B212** (1988) 391; R.D. Carlitz, J.C. Collins and A.H. Mueller, Phys. Lett. **B214** (1988) 229; S. Forte, CERN-TH.7453/94(1994) (hep-ph/9409416) and references therein.
10. SMC, B. Adeva et al.,, Phys. Lett. **B369** (1996) 93.
11. H. Plothow-Besch, Int. J. Mod. Phys. **A10** (1995) 2901.

MEASUREMENT OF THE NEUTRON SPIN STRUCTURE FUNCTION g_1^n AND A pQCD NLO ANALYSIS.

Stephen Rock for the E154 Collaboration.

American University
Washington DC 20016

Abstract. We have measured the spin structure function of the neutron g_1^n in the kinematic region $0.014 \leq x \leq 0.7$ and $1 \leq Q^2 \leq 15$. The experiment was done at SLAC by scattering longitudinal polarized 48.3 GeV electrons from a He^3 target polarized in the longitudinal and transverse directions. g_1^n becomes more negative with decreasing x in a manner inconsistent with naive single pole Regge theory. In the kinematic region $0.08 \leq x \leq 0.4$ we can determine the Q^2 dependence of g_1^n using data from both our spectrometers. A Next to Leading Order Perturbative QCD analysis has been made with this and previous data to extract the parton distributions and make an estimate of the very low x behavior.

INTRODUCTION

Deep inelastic scattering (DIS) of polarized leptons by polarized nucleons has been the cornerstone for studying the internal spin structure of the proton and neutron. Although the first experiments [1,2] found large asymmetries in the spin-dependent scattering of electrons by protons which wereconsistent with the early quark-parton model (QPM) predictions [3], subsequent experiments [4–6] performed at higher energies found that the proton asymmetries at low values of x disagreed with the early QPM predictions. In fact, higher energy proton measurements were inconsistent with one of the QPM sum rules derived by Ellis and Jaffe [7] based upon an unpolarized strange sea. First measurements of spin-dependent scattering of polarized leptons off polarized neutrons found small negative asymmetries, and, along with the proton results, provided the first tests of the fundamental Bjorken Sum Rule [8]. However, the neutron results suffered either from large statistical uncertainties at low x [9,10], or from a limited beam energy [11,12]. This talk presents results on a precision measurement of the neutron spin structure function g_1^n and on a

NLO pQCD analysis of the current world results on g_1^p, g_1^d, and g_1^d. The experiment (E154) collected 10^8 events at the Stanford Linear Accelerator Center (SLAC) in October and November of 1995 using 48.3 GeV polarized electrons scattered from polarized ^{3}He to achieve x values as low as 0.014.

The asymmetries $A_\parallel (A_\perp)$ measured in DIS of longitudinally polarized electrons by longitudinally (transversely) polarized nucleons can be used to find the nucleon spin structure function g_1 using:

$$g_1(x,Q^2) = F_2(x,Q^2) \frac{1+\gamma^2}{2xD'(1+R(x,Q^2))}[A_\parallel + \tan(\theta/2) A_\perp],$$

where Q^2 is the squared four-momentum transfer of the virtual photon; x is the fraction of nucleon momentum carried by the struck quark; γ and D' are factors depending on the scattered electron's initial and final energies and the electron scattering angle θ; $F_2(x,Q^2)$ is the unpolarized nucleon spin structure function and $R(x,Q^2) = \sigma_L/\sigma_T$ is the longitudinal to transverse virtual photoabsorption cross section ratio. The asymmetries $A_\parallel (A_\perp)$ may also be used to find the virtual photon-nucleon asymmetries $A_1(x,Q^2)$.

EXPERIMENTAL APPARATUS

Polarized electrons were obtained using a strained GaAs cathode illuminated by circularly polarized light from a flashlamp-pumped Ti:sapphire laser. The electron spin direction was reversed randomly on a pulse-to-pulse basis by reversing the helicity of the laser light. The beam polarization was measured to be 0.82 ± 0.02 over the duration of the experiment using a single arm Møller polarimeter located upstream of the target.

The polarized ^{3}He target consisted of double-chamber glass cells filled with $\sim$9.5 atm of ^{3}He (as measured at 20°C). NMR techniques calibrated by proton NMR and by frequency shift techniques, were used to measure the polarization of the ^{3}He nuclei. The average polarization was 0.38 ± 0.02 over the duration of the experiment.

Two new single-arm spectrometers, at central scattering angles of 2.75° and 5.5°, were used to analyze scattered electrons. Each spectrometer utilized a pair of threshold Čerenkov counters operating with nitrogen at a pressure corresponding to a pion momentum threshold of approximately 19 (16) GeV in the 2.75° (5.5°) arm. Each Čerenkov counter was read out by a Flash ADC that recorded the pulse shape in 1 ns time slices covering the full beam pulse of $\approx 250ns$. Ten (eight) planes of hodoscopes, with each finger connected to mulit-hit TDCs, were used for tracking in the 2.75° (5.5°) spectrometer. The resulting momentum resolution ranged from $\pm$ 2% at low momentum to $\pm$ 4% at high momentum. At the rear of each spectrometer a 200 block lead glass calorimeter was arranged in a fly's eye configuration with each element recorded in an ADC and from 1 to 3 multi-hit TDCs with different thresholds. This gave an energy resolution of 3% + $(8/\sqrt{E(\text{GeV})})$%.

ANALYSIS

Events were classified as electrons if they passed a threshold cut in both Čerenkov counters in coincidence with a cluster in the lead glass with energy within 20% of the momentum determined from the track.

After corrections for hadronic and pair-symmetric backgrounds, dilutions and polarizations, the asymmetries $A_\parallel$ and $A_\perp$ were formed. The asymmetries were corrected for radiative processes to find the single-photon exchange Born results. Uncertainties in the radiative corrections were estimated by varying the input models over a range consistent with the measured data.

Corrections due to the nuclear wave function of the polarized ^{3}He nucleus were applied [13-16] using the recent proton data [5,6] to evaluate the proton contributions; however these contributions had only a small impact on the results. No other corrections were made for the fact that the polarized neutron is embedded in the ^{3}He nucleus.

RESULTS

Results for A_1^n vs Q^2 for several different values of x for this experiment and others [9-11] are shown in Figure 1 along with our pQCD NLO analysis of world data [17]. The data are consistent with both a no dependence on Q^2 (solid line) and with our pQCD NLO analysis (dashed line).

Fig. 2a shows g_1^n from this experiment along with those of the SLAC E142 experiment [11]. The results from both experiments are evolved to $Q^2 = 5$ (GeV/c)^2 under the assumption that g_1/F_1 is independent of Q^2. Using the NLO fit for the evolution makes little difference in this kinematic range. Good agreement with the E142 results is seen in the overlapping x range. Over the range of this experiment, we find a neutron spin structure function integral of $\int_{0.014}^{0.7} g_1^n(x)dx = -0.036 \pm 0.004$ (stat.) ± 0.005 (syst.).

A notable feature of Fig. 2a is the strong x-dependence observed at low x. Figure 2b shows g_1^n in the low x region from this experiment and from SMC. Our data is clearly inconsistent with an often used Regge inspired assumption that that g_1^n is constant with x for $x \leq 0.1$, although it is still possible that g_1^n is constant for $x \leq 0.03$ (dashed line). Also shown are a power law fit, as well as our pQCD NLO fit. Unfortunately, the new data do not adequately constrain the low-x region such that the integral of g_1^n can be reliably extracted.

PERTURBATIVE QCD NEXT TO LEADING ORDER ANALYSIS

We did a Next to Leading Order pQCD analysis [17] in both the $\overline{MS}$ and Adler-Barden [18] schemes. Unlike most NLO analyses, we do not assume $SU(3)_{flavor}$ symmetry and do not fix the normalization of the non-singlet distributions by the axial charges $\Delta q_3 = F + D$ and $\Delta q_8 = 3F - D$, where F and D are the antisymmetric and symmetric SU(3) coupling constants of hyperon beta decay. We followed the perscription of Gluck, Reya and Voss (GRV) [19], starting the evolution at $Q_o^2 = .34$ and using

$$\Delta f(x, Q_0^2) = A_f x^{\alpha_f} f(x, Q_0^2) , \qquad (1)$$

where Δf are the parton asymmetries and f are the unpolarized parton distributions of GRV. The published world data was used as input. Figure 3 shows a sample of $xg1_n$ and xg_1^p data as well as our NLO fit with error bands, all evaluated at $Q^2 = 5$. xg_1^n appears to approache zero slowly with decreasing x while xg_1^p is very small and may cross zero.

We have calculated the Bjorken integral $\Gamma_1^{p-n} = \int_0^1 dx \, (g_1^p - g_1^n)$ at $Q_o^2 = 5$ using SLAC data for the proton and neutron. Our NLO fit was used in the measured region to evolve the data to Q_o, and also to determine the low and high-x regions. Fortunately, the behavior of the purely non-singlet

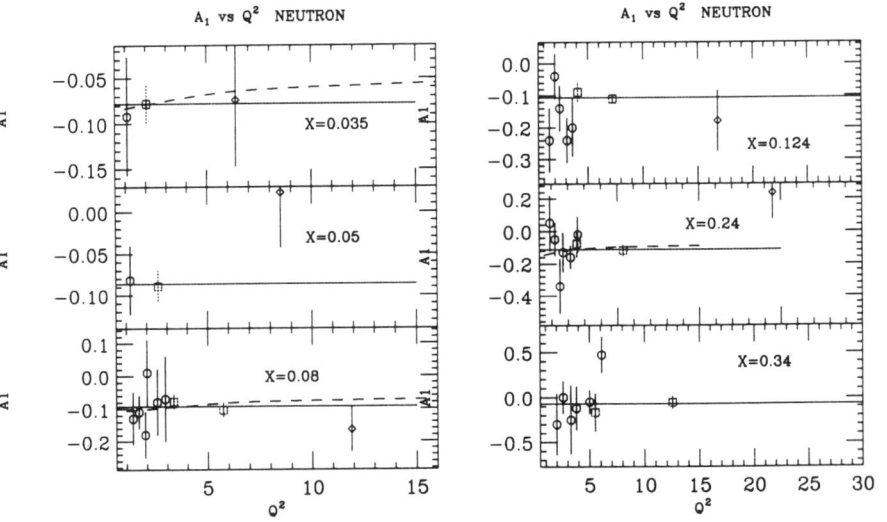

FIGURE 1. A_1^n for this experiment and E142 and SMC at different values of x as a function of Q^2. The solid line is at a constant value, while the dashed line corresponds to the pQCD NLO fit described in the text.

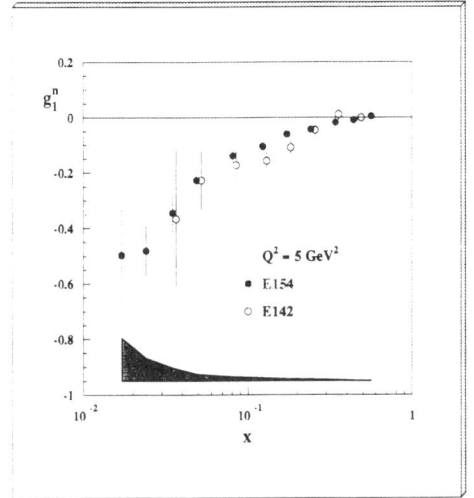 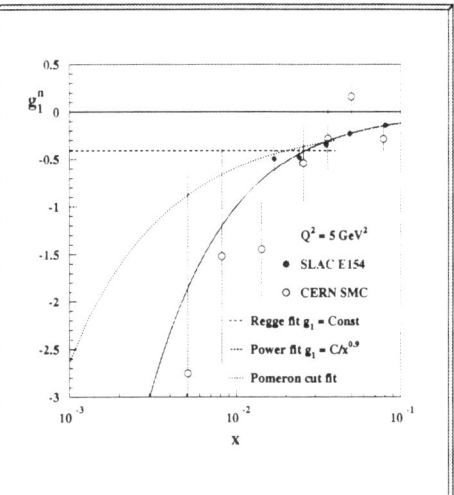

FIGURE 2. xg_1^n for this experiment and SLAC E142. The bottom band represents the E155 systematic errors. The plot on the right shows the low-x region with

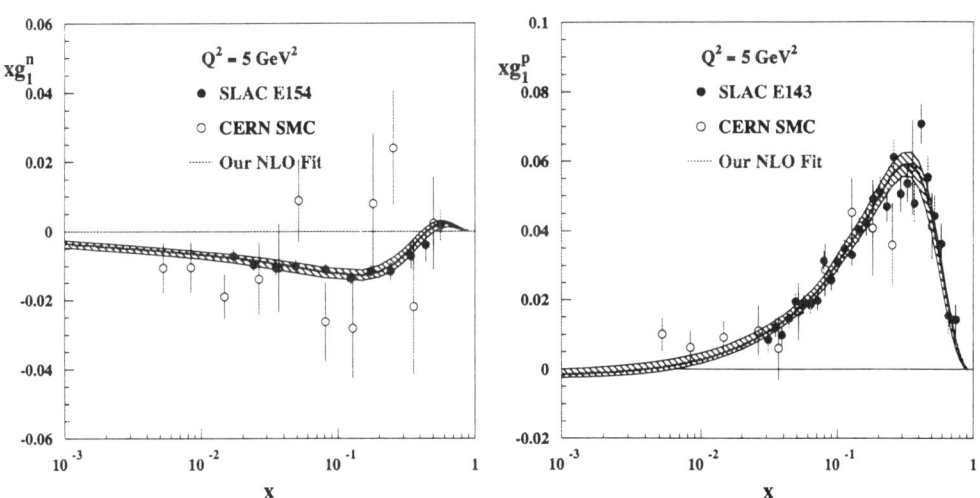

FIGURE 3. xg_1^n and xg_1^n data compared to our NLO pQCD analysis. Neutron results are on the left, proton on the right.

combination $(g_1^p - g_1^n)(x)$ is softer at low x than its singlet counterpart (g_1^n) making Γ_1^{p-n} less sensitive to low-x extrapolations. We obtain: $\Gamma_1^{p-n} = 0.171 \pm 0.005 \pm 0.010 \pm 0.006$ where the errors are statistical, experimental systematic and NLO uncertainties respectively. This is in excellent agreement with the predicted value of 0.188 evaluated with $\alpha_S(M_Z^2) = 0.109$ (used in this analysis following GRV).

CONCLUSION

We have found relatively large negative values of g_1^n at low x. One possible explanation for this behavior can be associated with sea and gluon spin contributions [18-20]. A breakdown in the simple Regge theory description at low x is also a possible consequence. Further precision data using proton and deuteron targets over the same kinematic range are expected to be of great use in unraveling the behavior of the nucleon spin structure functions at moderately low x (down to $x \approx 0.01$). High precision low x measurements of the nucleon spin structure functions are still needed to understand how g_1^n converges at low x and to extract the neutron integral $\int_0^1 g_1^n(x) dx$.

REFERENCES

1. M. J. Alguard et al., Phys. Rev. Lett **37**, 1258 (1976); **37**, 1261 (1976).
2. G. Baum et al., Phys. Rev. Lett. **51**, 1135 (1983).
3. R. D. Carlitz and J. Kaur, Phys. Rev. Lett. **37**, 673 (1977).
4. J. Ashman et al., Phys. Rev. Lett. **B206**, 364 (1988); Nucl. Phys. **B328**, 1 (1989).
5. D. Adams et al., Phys. Lett. **B329**, 399 (1994).
6. K. Abe et al., Phys. Rev. Lett. **74**, 346 (1995).
7. J. Ellis and R. L. Jaffe, Phys. Rev. **D9**, 1444 (1974); **D10**, 1669 (1974).
8. J. D. Bjorken, Phys. Rev. **148**, 1467 (1966); Phys. Rev. **D1**, 1376 (1970).
9. D. Adams et al., Phys. Lett. **B357**, 248 (1995).
10. D. Adams et al., CERN-PPE-97-008, submitted to Phys. Lett. (1997).
11. P. L. Anthony et al., Phys. Rev. Lett. **71**, 959 (1993); Phys. Rev. **D54**, 6620 (1996).
12. K. Abe et al., Phys. Rev. Lett. **75**, 25 (1995).
13. C. Ciofi degli Atti et al., Phys. Rev. **C48**, 968 (1993).
14. B. Blankleider and R. M. Woloshyn, Phys. Rev. **C29**, 538 (1984).
15. J. L. Friar et al., Phys. Rev. **C42**, 2310 (1990).
16. R.-W. Shulze and P. U. Sauer, Phys. Rev. **C48**, 38 (1993).
17. K. Abe et al., SLAC-PUB-7461, hep-ph/9705344, submitted to Phys. Lett. B.
18. R. D. Ball, S. Forte and G. Ridolfi, Phys. Lett. **B378**, 255 (1996).
19. M. Gluck et al., Phys. Rev. **D53**, 4775 (1996).
20. T. Gehrmann and W. J. Stirling, Phys. Rev. **D53**, 6100 (1996).

Inclusive Nucleon Resonance Electroproduction: Recent Results From Jefferson Lab

Cynthia Keppel

Hampton University, Hampton, Virginia 23668
Thomas Jefferson National Accelerator Facility,
Newport News, Virginia 23606

Abstract. High statistics measurements of inclusive nucleon resonance electroproduction cross sections have been performed in Hall C at Jefferson Lab. The invariant mass range $1 < W^2 < 4$ GeV^2 was probed for four-momentum transfer values between 0.5 and 4.0 (GeV/c)^2 for both liquid hydrogen and deuterium targets. The cross sections will be used in conjunction with existing deep inelastic and elastic data for precision experimental tests of parton-hadron (Bloom-Gilman) duality in the nucleon structure functions. Preliminary results of such testing is discussed. The data are additionally being analyzed in order to study resonance transition form factors, in particular to compare proton and neutron resonance cross sections.

PARTON-HADRON DUALITY

Over 20 years ago Bloom and Gilman observed the behavior of elastic scattering and of the electroproduction of nucleon resonances to be closely related to the behavior of deep inelastic electron-nucleon scattering [1,2]. Precisely, the prominent resonances in inclusive electron-proton scattering do not disappear with increasing four-momentum transfer squared (Q^2) relative to the background under them, but instead fall at roughly the same rate. Also, the smooth scaling limit seen at high Q^2 and large missing mass squared (W^2) for the structure function $\nu W_2(\omega')$ is an average of the resonance enhancements at the same ω', but lower Q^2 and W^2. Here, ω' is an "improved" scaling variable and is equal to $1+ W^2 / Q^2 = (2M\nu + M^2)/Q^2$. These observations are termed Bloom-Gilman duality, or local duality. Bloom and Gilman quantified the latter observation with the following equation:

$$\frac{2M}{Q^2}\int_0^{\nu_m} \nu W_2(\nu, Q^2)d\nu = \int_1^{(2M\nu_m+m^2)/Q^2} \nu W_2(\omega')d\omega'. \quad (1)$$

Here, ν is the energy transfer. This observed duality relationship between resonance electroproduction and scaling behavior as observed in deep inelastic scattering suggests a common origin for both phenomena.

The description of hadrons and their excitations in terms of elementary quark and gluon constituents is one of the fundamental challenges in physics today. Quantum chromodynamics (QCD) is the theory of strong interactions that describes particles in terms of these elementary quantities. A QCD-based explanation of why the resonance structure functions average to the F_2 scaling curve was offered by De Rujula, Georgi, and Politzer in 1977 [3,4]. While original studies were somewhat qualitative, enormous progress has been made in understanding QCD in the past two decades and recent work has focused once again on Bloom-Gilman duality [5-8].

There exists a large body of precision deep inelastic lepton-nucleon scattering data. Combined with the new precision resonance data, it is possible to rigorously study the observations and predictions of duality. Duality will be tested for the neutron by subtraction of the kinematically-matched proton data, using smearing and deuteron wave function modelling. Parton-hadron duality will also be investigated on the deuteron itself as a hadron.

Figure 1 displays preliminary Jefferson Lab inclusive nucleon resonance electroproduction spectra from hydrogen, plotted as a function of the Bloom-Gilman scaling variable ω'. The deep inelastic scaling limit curve is from a SLAC global fit [9]. Note that ω' is plotted on a logarithmic scale. Obtained at differing central Q^2, the resonance data all appear to follow the smooth scaling limit curve,

Figure 2 depicts the ratio of the integrals in Equation 1, resonance over deep inelastic, integrated from the elastic peak to $W^2 = 4$ GeV^2. The stars represent preliminary Jefferson Lab resonance data. The circles represent integrals obtained from existing SLAC data. Both are in ratio to the integral over F_2 from reference [9]. Evidence for parton-hadron duality is observed and may be displaying a Q^2 dependence.

REFERENCES

1. E.D. Bloom and F.J. Gilman, Phys. Rev. D4, 2901 (1970)
2. E.D. Bloom and F.J. Gilman, Phys. Rev. Lett. 25, 1140 (1970)
3. A. DeRujula, H. Georgi, and H.D. Politzer, Phys. Lett. B64, 428 (1977)
4. A. DeRujula, H. Georgi, and H.D. Politzer, Annals Phys. 103, 315 (1977)
5. C.E Carlson and N.C. Mukhopadhyay, Phys. Rev. D47, R1737 (1993)
6. X. Ji and P. Unrau, Phys. Rev. D52, 72 (1995)
7. V.M. Belyaev and A.V. Radyushkin, Phys. Lett. B359, 194 (1995)
8. G. West, preprint, hep-ph/9612403 (1996)
9. L.W. Whitlow et al., Phys. Lett. B250, 193 (1990)

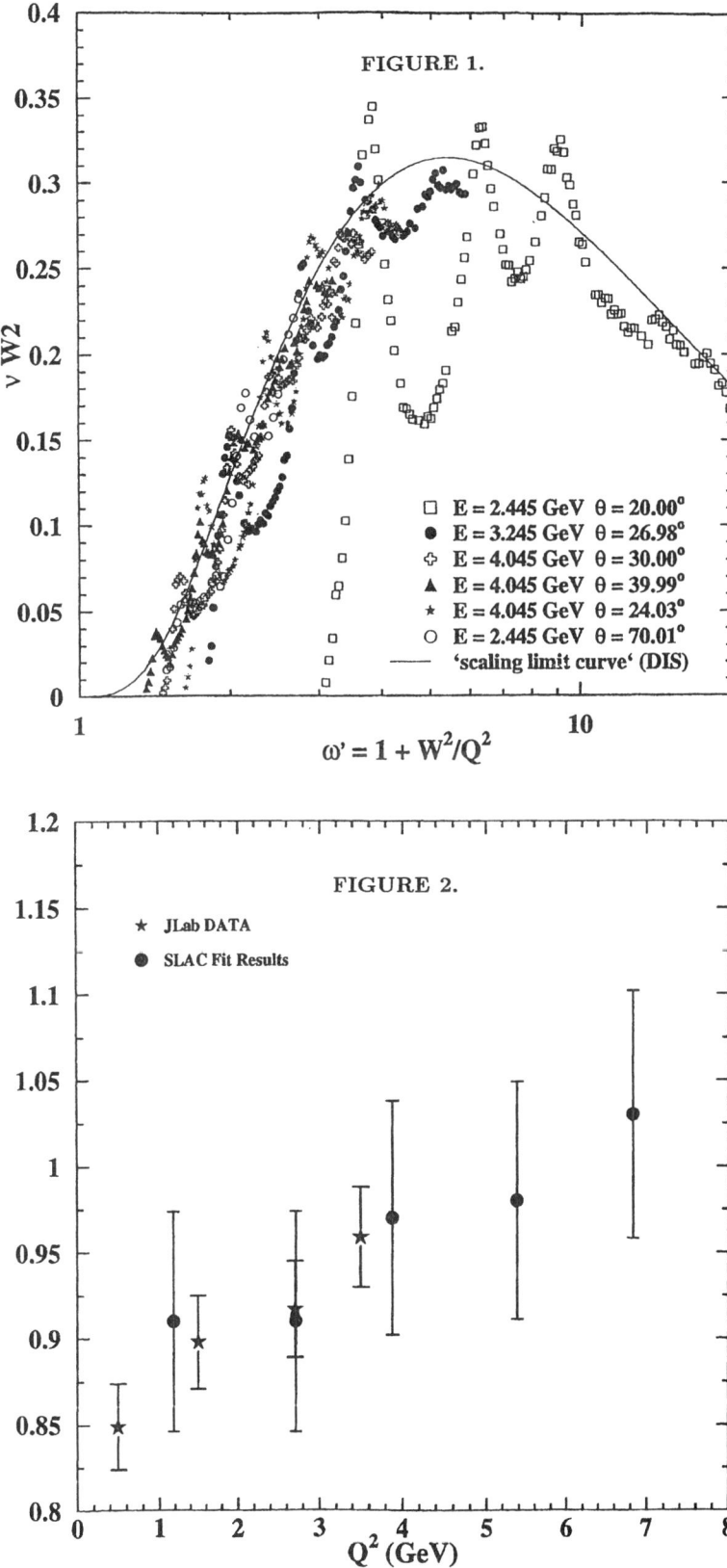

Measurements of $R_d = \sigma_L/\sigma_T$ for $0.03 < x < 0.1$

P. Bosted and J. Fellbaum
(for The SLAC E143 Collaboration[1])

American University, Washington DC 20016

Abstract. Measurements were made at SLAC of the cross section for scattering 29 GeV electrons from deuterium at a laboratory angle of 4.5 degrees, corresponding to $0.03 < x < 0.1$ and $1.3 < Q^2 < 2.7$ GeV^2. Values of $R_d = \sigma_L/\sigma_T$ were extracted in this kinematic range by comparing to cross sections measured at a higher beam energy by the NMC collaboration. The results are in reasonable agreement with QCD calculations and with extrapolations of the R1990 fit to world data.

INTRODUCTION

The spin-averaged cross section for virtual photon absorption on a nucleon can be expressed as the sum of two components: the longitudinal cross section $\sigma_L(x, Q^2)$, and the transverse cross section $\sigma_T(x, Q^2)$, where Q^2 is the four-momentum squared of the photon, and x is the light-cone momentum fraction of the struck parton. The ratio $R = \sigma_L/\sigma_T$ is sensitive to the spin of the struck partons: at large Q^2, R is zero for spin 1/2 quarks (neglecting transverse momentum), while R is very large if the struck partons are spin 1 gluons. Thus R can be used as a measure of the gluon content of nucleons. It is also of practical interest: for example, in experiments measuring the spin structure function g_1, knowledge of R is needed to extract g_1 from the asymmetry in scattering longitudinally polarized leptons from polarized nucleons.

The spin-averaged cross section for lepton scattering from a nucleon can be written as

$$\frac{d\sigma(x, Q^2, \epsilon)}{dx dQ^2} = \Gamma(x, Q^2, \epsilon)[\sigma_T(x, Q^2) + \epsilon \sigma_L(x, Q^2)]$$

---

[1] Work supported by the National Science Foundation and the Department of Energy.

where Γ is the virtual photon flux, $\epsilon^{-1} = 1 + 2(1 + Q^2/4M^2x^2)\tan^2(\theta/2)$, and θ is the lepton scattering angle. Measurements of R are obtained by making cross section measurements at fixed (x, Q^2) as a function of ϵ.

A good parameterization was made of the world data on R (primarily from SLAC) in 1990 [1] (R1990), but the fit is limited in validity to $x > 0.07$, where input data existed. To test the accuracy of the extrapolation of this fit to lower x, we made cross section measurements in the range $0.03 < x < 0.1$ as part of SLAC experiment E143 [2], whose primarily goal was the measurement of g_1 for the proton and deuteron.

E143 CROSS SECTIONS

We measured the cross section for scattering of 29.1 GeV electrons from a 1.7 gm/cm^2 carbon target at angles near 4.5 degrees. The acceptance of this large-momentum-range spectrometer was calibrated using data with $x > 0.1$ by comparing to a fit [3] to world cross sections. The central momentum of the spectrometer was then lowered in several steps to put scattered electrons with momenta between 7 and 14 GeV into the well-measured acceptance region. Small adjustments were made to the spectrometer model until the overlaps between spectra with different central momentum settings were in good agreement. In the end, the corrections to the acceptance compared to the original model were in the range of a few percent.

Absolute cross sections for a pure deuteron target were then obtained taking into account the experimental efficiency for detecting electrons, the trigger dead time, and applying radiative corrections [4] and a fit to the A-dependence of lepton-nucleon scattering [5]. The results are shown in Fig 1a. The errors are dominated by a systematic normalization error due the combined uncertainties in target thickness, beam charge, spectrometer acceptance and detection efficiency. The uncertainty due to radiative corrections is largest at low x (about 3%), decreasing to about 1% at the highest x. The Q^2 for the points in Fig. 1a vary linearly with x and range from 1.3 to 2.7 GeV^2, and ϵ ranges from 0.5 to 0.8. The curve on Fig. 1a is the predicted cross section using the NMC fit to F_2 [3] and the R1990 fit [1] to R. The good agreement between data and model for $x < 0.1$ is an indication that the extrapolation of R1990 to $x = 0.03$ works reasonably well.

RESULTS FOR R_D

To determine values for R, the E143 cross sections were compared with cross sections measured by the NMC [3] at the same (x, Q^2) points, but at much higher beam energies and hence higher values of ϵ. The results are shown in Fig. 1b, and are significantly higher than the data recently published by NMC [3] at Q^2 values approximately five times larger than for E143, in accord with the expectation that R should decrease rapidly with increasing Q^2 at fixed x (power law falloff expected at low Q^2, followed by logarithmic falloff at high Q^2). The new data are in reasonable agreement with the error band of the R1990 fit, plotted at the Q^2 values of E143, although there is a tendency for the data to be slightly higher than the fit at low x. The data are also in agreement with the Bodek, Rock, and Yang QCD-based fit [6] with three representative parton distributions. The new data are not precise enough to distinguish between these distributions, which differ principally in the relative gluon content of the nucleon. This situation will improve as measurements of R from neutrino scattering gain in precision (see [7] for preliminary results).

REFERENCES

1. L. W. Whitlow *et al.*, Phys. Lett. B250 (1990), 193.
2. SLAC E143, K. Abe *et al.*, Phys. Rev. Lett. 74 (1995), 346; Phys. Rev. Lett. 75 (1995), 25.
3. NMC, M. Arneodo *et al.*, Nucl. Phys. B482, 3 (1997).
4. Y. S. Tsai, Report No. SLAC–PUB–848, 1971; Rev. Mod. Phys. 46 (1974), 815.
5. J. Gomez *et al.*, Phys. Rev. D 49, 4348 (1994).
6. A. Bodek, S. Rock, and U. Yang, 1996 UR-1355 (Z. Phys. C to be published).
7. U. K. Yang *et al.*,, J. Phys. G 22, 775 (1996).

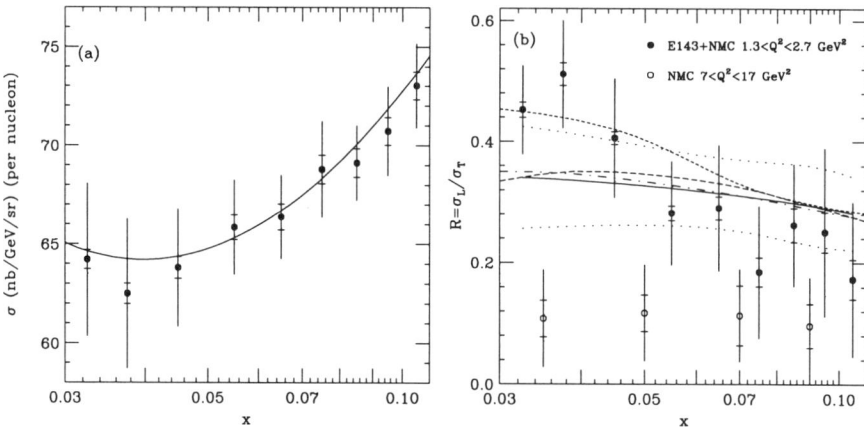

FIGURE 1. (a) Cross sections from this experiment (E143) for 29.1 GeV electron scattering from deuterium at 4.5 degrees. Inner (outer) error bars are statistical (systematic). The curve is calculated using the NMC fit to F_2 and the R1990 fit to R. (b) R_d from this experiment (E143) combined with low Q^2 NMC data (solid circles) and from NMC alone (open circles). Inner (outer) error bars are statistical (systematic). The solid curve is the R1990 fit, with the dotted curves showing the error band. The long-dashed, short-dashed, and dot-dashed curves are from the QCD-based fits of Bodek, Rock and Yang with three representative parton distributions. All curves are plotted at the Q^2 values of E143.

Diffractive Deep Inelastic Scattering at HERA

Ralf Todenhagen*

on behalf of the H1 and ZEUS collaborations

*Max-Planck-Institut für Kernphysik, Saupfercheckweg 1, 69117 Heidelberg

Abstract. At HERA, about 10% of the deep inelastic events present a large rapidity gap in the hadronic final state, attributed to the exchange of a colour singlet system. A pomeron and a secondary reggeon exchange are needed in order to explain the data in terms of Regge phenomenology. A QCD analysis of the extracted structure function for the pomeron exchange indicates parton distributions dominated by a gluon carrying a large fraction of the exchange momentum at low Q^2. This behaviour is confirmed in measurements of the hadronic final states. The usage of leading proton spectrometers allows the measurement of additional kinematic variables.

INTRODUCTION

Soon after the initial startup of the HERA machine a special class of deep inelastic scattering events attracted a lot of attention. Unlike the usual event topology this type of events exhibited a large angular region around the proton beam pipe without any energy deposit in the detectors - a large rapidity gap [1,2]. These events can be interpreted as deep inelastic scattering off a colourless component of the proton, mainly the pomeron, the exchange of which makes the dominant contribution in high energy diffractive interactions of hadrons [3,4]. Diffractive deep inelastic scattering offers thus the possibility to probe the partonic structure of the pomeron and shed new light on the nature of diffractive scattering [5].

DIFFRACTIVE STRUCTURE FUNCTION

Figure 1 displays the sketch of a diffractive deep inleastic scattering event. In the following we will mainly consider single photon dissociation, i.e. events in which the sytem Y going into the forward direction consists only of a single proton carrying most of the momentum of the incoming proton.

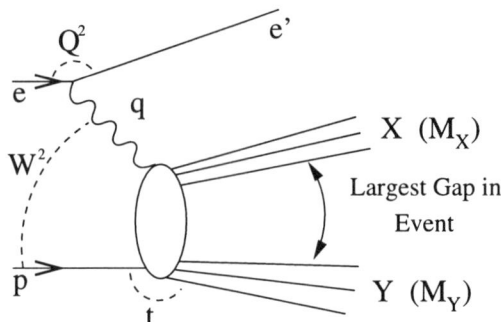

FIGURE 1. Sketch of a diffractive event explaining the kinematic variables.

The variables $Q^2 = -q^2$ and $W^2 = (p+q)^2$ are used in ordinary deep inelastic scattering to parameterise the data. q and p are the four momenta of the virtual photon and the proton respectively. For diffractive events two additional variables are necessary to describe the more complex hadronic final state. A possible choice is the four momentum transfer squared at the proton vertex $t = (Y - p)^2$ and the momentum fraction of the proton carried by the diffractive final system X, $x_{I\!\!P} \simeq \frac{Q^2 + M_X^2}{Q^2 + W^2}$. $\beta = x/x_{I\!\!P}$ defines the Bjorken-x variable for the 'diffractive system' X.

Following Ingelman and Schlein [5] the cross section for rapidity gap events can be parameterised in terms of a diffractive structure function. This function is defined in complete analogy with the inclusive structure function:[1]

$$\frac{d^4\sigma_{ep\to e'XY}}{dt\, dx_{I\!\!P}\, d\beta\, dQ^2} = \frac{4\pi\alpha^2}{\beta Q^4}(1 - y + \frac{y^2}{2})F_2^{D(4)}(\beta, Q^2, x_{I\!\!P}, t).$$

Usually the scattered proton leaves the interaction region undetected and it is not possible to measure t. Experimentally the structure function $F_2^{D(3)}(x_{I\!\!P}, \beta, Q^2)$ is defined by integrating over the unmeasured variable t for $|t| < 1\,\text{GeV}^2$. The data points in figure 2 show results for the quantity $x_{I\!\!P} \cdot F_2^{D(3)}$ extracted by the H1-collaboration [6].

In the Ingelman Schlein model the diffractive structure function factorizes into a pomeron flux factor and the structure function of the pomeron:

$$F_2^{D(3)} = f_{I\!\!P/p}(x_{I\!\!P}) \cdot F_2^{I\!\!P}(\beta, Q^2).$$

Regge theory based models suggest that the flux factor has a power law behaviour for each exchange, and in particular: $f_{I\!\!P/p}(x_{I\!\!P}, t) \sim 1/x_{I\!\!P}^{2\alpha(t)-1}$, where $\alpha(t) = \alpha(0) + \alpha' \cdot t$ denotes the Pomeron trajectory. The curves in figure 2 correspond to a fit based on a model of this type allowing in addition the exchange

---

[1] The longitudinal part of the cross section is neglected here.

H1 Preliminary 1994

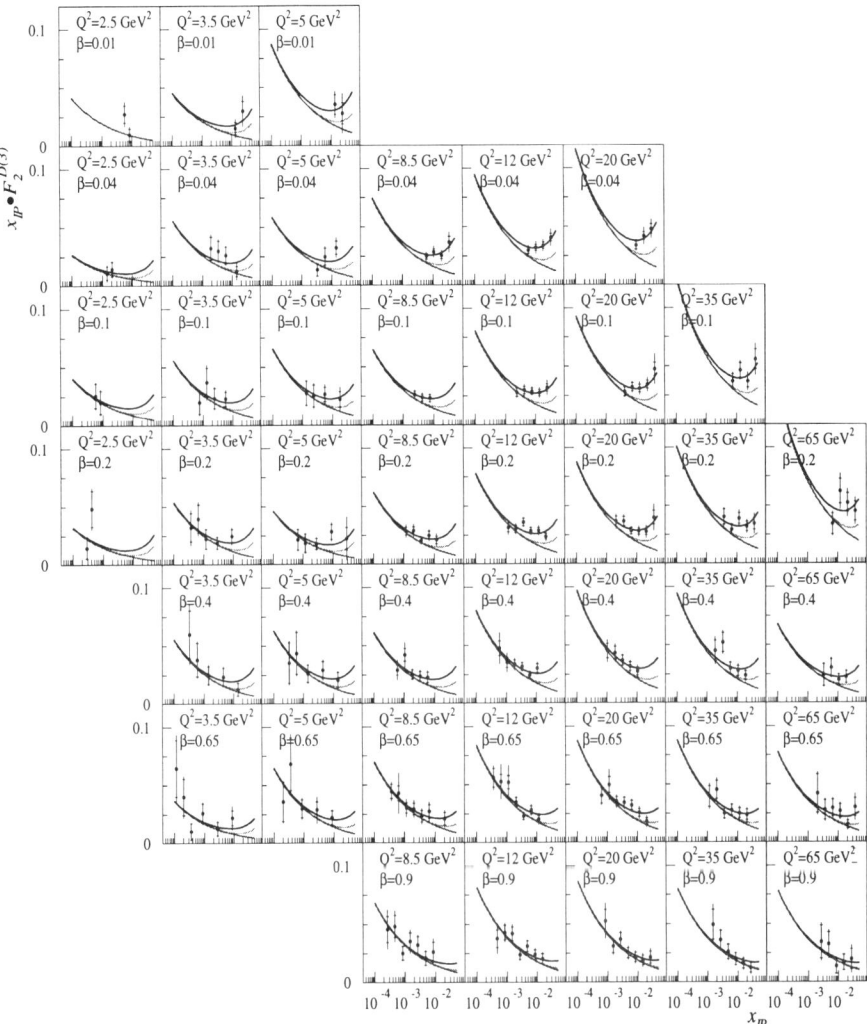

FIGURE 2. Measured values for $x_{I\!P} F_2^D(3)$. The lowest curve represents a fit with pomeron exchange only. The central line corresponds to a fit with pomeron and subleading reggeon exchange and the uppermost curve shows the full fit including the interference term.

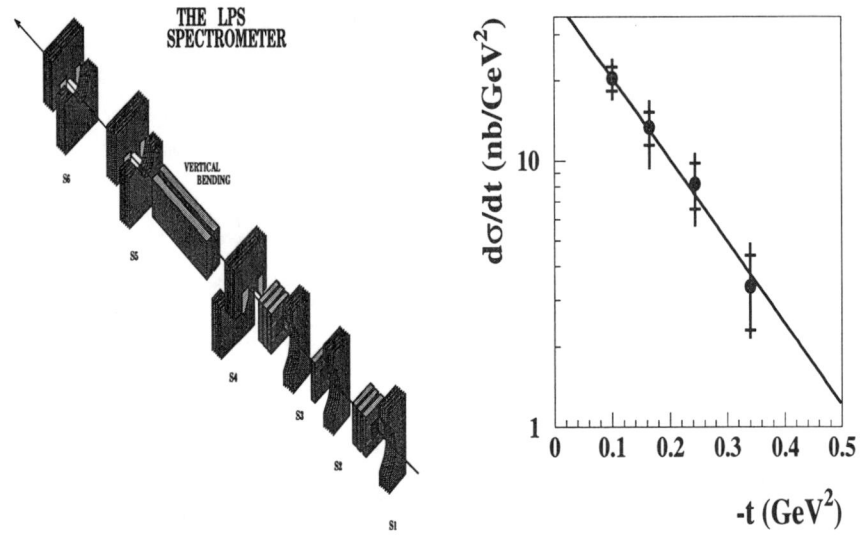

FIGURE 3. Schematic view of the ZEUS leading proton spectrometer (left picture) and the preliminary t-distribution for the 1994 data set (right picture).

of a subleading trajectory with the f_2 meson quantum numbers, and for an interference term. The resulting fit describes the data well and the measured value of the pomeron Regge intercept $\alpha_{I\!P}(0) = 1.18 \pm 0.02(stat.) \pm 0.04(syst.)$ is slightly higher than for soft hadronic interactions.

In a subsequent step the H1 collaboration has extracted the structure function of the colour singlet exchange integrated over the $x_{I\!P}$ and t ranges measured. A QCD analysis of this structure function using standard DGLAP [9] evolution equations suggests a leading gluon behaviour for the exchanged object. Within the measured Q^2 range 80-90% of the momentum is found to be carried by gluons [6].

The ZEUS collaboration applies a different method to select diffractive events by defining diffractive events as those which form the excess over the exponential fall-off in the $\ln M_X^2$ distribution observed in the main detector [7]. The extracted diffractive structure function agrees well with the H1 data and the measured values for $\alpha_{I\!P}$ are also slightly above the expectation for soft hadronic interactions [8].

LEADING PARTICLE DETECTORS

Instead of selecting events with a large rapidity gap it is possible to tag diffractive events by measuring the leading proton. In addition this method allows the determination of the four momentum transfer at the proton vertex. ZEUS and H1 have installed leading proton spectrometers for this purpose.

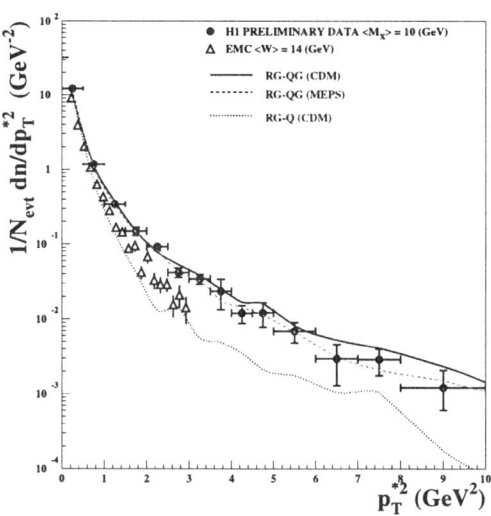

FIGURE 4. Distribution of p_T^{*2} for charged tracks in the $\gamma^* I\!P$ system

The ZEUS device is shown in figure 3. It consists of six silicon strip detectors placed in Roman pots in the proton beam pipe between 20 and 90 m from the interaction region. Horizontal bending magnets of the HERA proton beam line are used for momentum analysis of the secondary protons [10].

The right hand side of figure 3 shows the preliminary t-distribution of protons from diffractive events [11,12]. These events are defined by requiring a longitudinal momentum fraction of at least 97% carried by the final state proton. The extracted slope is $b = 7.1 \pm 1.1^{+0.7}_{-1.0} GeV^{-2}$ for a fit of the type $d\sigma/d|t| \sim \exp(-b|t|)$ indicated by the line in the figure. The ZEUS collaboration has also extracted the diffractive structure function for these events. The measured F_2^D values are in good agreement with the measurements of F_2^D from both experiments based on the presence of a rapidity gap or the $\ln M_X^2$ method [8].

HADRONIC FINAL STATES

The QCD analysis of the diffractive structure function suggests a leading gluon behaviour for the exchanged colour singlet object. Further investigations of the hadronic final state (the system X in figure 1) can be used to check this conclusion. In the following two examples are presented. Further information as well as other final state properties are given in [6,13–16] and the references therein.

Figure 4 shows the corrected distribution of p_T^{*2} for charged tracks measured in the central trackers of H1. The momentum is defined in the rest frame of the system X with respect to the photon-pomeron axis. The data are compared with EMC data collected for inclusive deep inelastic muon proton scattering in a similar Q^2 range. These data were measured in an x region where the valence quarks are dominating the parton distributions and the p_T^{*2} spectra are therefore characteristic for an hadronic object which is predominantly made of quarks. The H1 data were measured at β values which are comparable to the EMC x values. The p_T^{*2} distribution extends to much higher values disfavouring the hypothesis of a quark dominated pomeron. The measured values are also compared with the expectation for different Monte Carlo models based on the RAPGAP Monte Carlo program [17]. The MC models labelled RG-QG(CDM) and RG-QG(MEPS) in figure 4 present the predictions based on the pomeron structure function obtained from the DGLAP fit to $F_2^{D(3)}$ mentioned above, the difference of the two curves being in the hadronization scheme. These predictions are observed to reproduce well the data. This is not the case for the prediction labelled RG-Q(CDM), for which only quarks are allowed at the starting scale of the QCD evolution. Further details of this analysis can be found in [6] and [15].

For their deep inelastic scattering data collected the H1 collaboration could extract a clear signal for charm production in the channel $ep \to e'(D^{*\pm}X)Y$ where the $D^{*\pm}$ meson decays in a system $(D^0\pi^\pm)$, the D^0 meson decaying into $K^-\pi^+$ in turn. The cross section for diffractive production of $D^{*\pm}$ is measured to be $(380^{+150}_{-120}(stat.)^{+140}_{-110}(syst.))$pb [6]. This crossection is compared with different Monte Carlo models based on RAPGAP. Again the data favour the gluon dominated pomeron model yielding a cross section of ~ 200pb compared to a cross section of ~ 10pb for the quark dominated pomeron.

The ZEUS collaboration reaches the same conclusion in [14] where the cross section for diffractive production of $D^{*\pm}$ is $875 \pm 248(stat.)^{+295}_{-199}(syst.)$ pb and the cross section for the complete deep inelastic production of $D^{*\pm}$ is measured to be 3.9 ± 0.4 nb[2]. Therefore the diffractive subsample contains approximatly 20% of all $D^{*\pm}$ events. Comparing this number with the fraction of 10% of events with a large rapidity gap this supports the evidence for the large gluon component of the pomeron, since in this case one would expect an enhanced charm production via photon gluon fusion type processes.

SUMMARY

Since the first observation of large rapidity gap events at HERA an extended program of measurement has yielded an increasing understanding of this phenomenon. A Regge analysis of the cross section indicates the exchange of the pomeron and of subleading trajectories. Using the DGLAP evolution equations, the partonic content of the pomeron is described as dominated by a

---

[2]) Note that the kinematic range for the ZEUS and the H1 results are slightly different.

leading gluon component at the starting scale. The structure of the hadronic final state is found to be consistent with this result. In future measurements there will be improved precision especially for charm and high p_T final states and new ways of tagging diffractive events using leading particle detectors. This will increase the precision and diversity of existing data and allow an even more detailed look into the reaction mechanism.

REFERENCES

1. Derrick M. et al., *Phys. Lett.* **B315**, 481 (1993).
2. Ahmed T. et al., *Nucl. Phys.* **B429**, 377 (1994).
3. Ahmed T. et al., *Phys. Lett.* **B348**, 681 (1995)
4. Derrick M. et al., *Z. Phys.* **C 68**, 569 (1995)
5. Ingelman G. and Schlein P. E., *Phys. Lett.* **B152**, 256 (1985).
6. J.P. Phillips,"Diffractive Hard Interactions in Electron Proton Collisions", in Proc. of the 28th International Conference on High Energy Physics, Warsaw Poland, July 1996, ed. Z. Ajduk and A.K. Wroblewski, ISBN 981-02-2874-0
7. M. Derrick et al., *Z. PhysC* **70**, 391 (1996)
8. Grothe M., in *5th International Workshop on DIS and QCD*, Chicago(1997)
9. Dokshitzer Yu., *JETP* **46**, 641 (1977)
    Gribov V. and Lipatov L., *Sov. Journ. Nucl. Phys.* **15** 78 (1972)
    Altarelli G. and Parisi G., *Nucl. Phys.* **B126** 298 (1977)
10. M. Derrick et al., *Z. Phys* **C 73** 253 (1997)
11. G. Barbagli,"Diffractive Cross Sections and Final States at large Scales",in Proc. of the 28th International Conference on High Energy Physics, Warsaw Poland, July 1996, ed. Z. Ajduk and A.K. Wroblewski, ISBN 981-02-2874-0
12. N. Cartiglia, in *5th International Workshop on DIS and QCD*, Chicago(1997)
13. C.M. Cormack, in *5th International Workshop on DIS and QCD*, Chicago(1997)
14. J. Terrón, in *5th International Workshop on DIS and QCD*, Chicago(1997)
15. H1 Collaboration, ICHEP 28, Warsaw Poland 25-31 July 1996, **pa02-062**
16. H1 Collaboration, ICHEP 28 Warsaw Poland 25-31 July 1996, **pa02-068**
17. H. Jung, *Comp. Phys. Comm.* **86** (1995) 147.

Vector Meson and Heavy Flavor Production at HERA

Song Ming Wang

University of Iowa, Department of Physics and Astronomy,
Iowa City, Iowa 52242, U.S.A.
on behalf of the H1 and ZEUS Collaborations

Abstract. Results based on recent investigation of vector meson production using H1 and ZEUS detectors at HERA are presented. Measurements of elastic production of ρ^0, ω, ϕ and J/ψ mesons at $Q^2 \approx 0$ and at large Q^2 are discussed, along with results on inelastic photoproduction of J/ψ meson via photon-gluon fusion.

INTRODUCTION

Elastic and inelastic photo- and leptoproduction of vector mesons has been studied in fixed target experiments and has provided information about the hadronic features of the photon, the structure of proton, and the nature of diffraction. At HERA the high flux of almost real and virtual photons, emitted by the lepton beam, make it possible to extend the above studies to a higher photon-proton center of mass energy, $W_{\gamma^* p}$. By measuring the production of different vector mesons at various values of Q^2 it may be possible to study the transition from a "soft" non-perturbative regime to a "hard" regime where perturbative QCD calculations may become applicable. Such studies can also provide direct or indirect measurements of the parton distributions of the proton.

In elastic production of vector mesons, the incoming e^+/e^- beam emits a virtual photon, which scatters off the proton. The hadronic final state consists of a vector meson and the proton remains intact. This process is described either in terms of "soft" non-perturbative interactions or "hard" interactions. In the "soft" approach [1], based on the Vector Dominance Model (VDM) and Regge theory, the virtual photon fluctuates into a vector meson and interacts hadronically with the proton, through the exchange of a soft pomeron Regge trajectory $\alpha(t) = \alpha(0) + \alpha' t$ with an intercept $\alpha(0) = 1.08$ and slope $\alpha' = 0.25$ GeV^{-2} (t is the four-momentum transfer squared at the proton vertex). This approach predicts a weak dependence of the cross section on energy, of the type $\sigma_{\gamma^*p \to Vp}(W_{\gamma^*p}) \approx W_{\gamma^*p}^{0.22}$. The differential cross section as a function of t can be parameterized by an expoential function, $\sim e^{-b|t|}$. The model predicts a logarithmic dependence of slope b on W_{γ^*p}, $b = b_o + 4\alpha'_{pom} \ln \frac{W_{\gamma^*p}}{W_0}$. This dependence of b on W_{γ^*p} is known as shrinkage.

Conversely in the "hard" models, the $q\bar{q}$ pair into which the photon fluctuates into, interacts with the proton through the exchange of a gluon ladder. In these models [2] the cross section is proportional to the square of the gluon momentum density in the proton. These models are only valid in the presence of a hard scale which can either come from t, Q^2 or Mass$_V$ (Mass$_V$ is the mass of the vector meson).

Inelastic J/ψ photoproduction includes photon-gluon fusion, proton-diffractive dissociation, and resolved-photon processes. This paper will only cover the measurements of inelastic J/ψ photoproduction in the photon-gluon fusion process. In the color singlet (CS) model [4] the exchanged photon interacts with a gluon from the proton through the subprocess $\gamma g \to c\bar{c}$ (c is the charm quark). A color singlet state is then achieved by the emission of a hard gluon from the $c\bar{c}$ system. The predicted cross section is proportional to the gluon momentum density in the proton. Recently a NLO calculation [5] has become available which is reliable in the kinematic range $z < 0.8$ and $P_t > 1$ GeV (z is the ratio between J/ψ and the incoming photon energies in the proton rest frame, P_t is the transverse momentum of J/ψ with respect to the beam axis). In the color octet (CO) model [6], the contribution from $c\bar{c}$ pairs produced in a color-octet state is also included. The color-octet $c\bar{c}$ pair is assumed to emit a soft gluon in order to conserve color. The non-perturbative parameters that are involved in the calculation of this model are extracted from fits to the CDF data on J/ψ production [7].

HERA collides 820 GeV protons with 27.5 GeV of electrons or positrons (e^- in '93, and e^+ in '94 and '95). H1 and ZEUS are both general purpose detectors instrumented with high resolution calorimeters and tracking chambers. In photoproduction events the scattered e^+/e^- is not detected in the central detector, thus restricting Q^2 below 2 GeV^2. The median Q^2 for the accepted events is approximately 10^{-4} GeV^2. Events with scattered e^+/e^- detected in the central detector are classified as large Q^2 events, with $Q^2 > 2$ GeV^2. The results presented here are based on the data collected in '93, '94, and '95 with corresponding luminosities per experiment of approximately 0.5 pb^{-1}, 3 pb^{-1} and 6 pb^{-1} respectively.

ELASTIC VECTOR MESON PRODUCTION

In this paper the results of the reaction of $ep \to eVp(V = \rho^0, \omega, \phi, J/\psi)$ at $Q^2 \approx 0$ GeV^2, intermediate Q^2 (0.1 − 1.0 GeV^2), and large Q^2 ($Q^2 > 2$ GeV^2) from H1 and ZEUS are presented [8-16]. The vector mesons are measured in the following decay channels : $\rho^0 \to \pi^+\pi^-$, $\omega \to \pi^+\pi^-\pi^0$, $\phi \to K^+K^-$

and $J/\psi \to e^+e^-/\mu^+\mu^-$. The reaction is taken to be elastic when there is little or no energy deposited in the main calorimeter other than that from the decay particles of the vector meson and the scattered e^+/e^- (for the case of production at large Q^2).

The cross section $\sigma_{\gamma^*p \to Vp}$ as a function of W_{γ^*p} was studied at various Q^2. In the case of photoproduction of light vector meson (ρ^0, ω, ϕ) $\sigma_{\gamma^*p \to Vp}$, shown in figure 1(a), shows a weak dependence on W_{γ^*p} which can be described by $W_{\gamma^*p}^{0.22}$. This is consistent with the prediction from models based on the exchange of a "soft" pomeron. However for the photoproduction of J/ψ meson the cross section $\sigma_{\gamma^*p \to J/\psi p}$ shows a stronger dependence on W_{γ^*p} which can be described by models based on pQCD using an appropriate gluon distribution of the proton. In these models [2], $\sigma_{\gamma^*p} \propto (xg(x))^2$, and $x \sim W_{\gamma^*p}^{-2}$ (x is the fraction of the proton's momentum carried by the gluon). Thus the steep rise in $\sigma_{\gamma^*p \to J/\psi p}$ with W_{γ^*p} is due to steep rise in $xg(x)$ as x becomes small. The mass of J/ψ sets the scale of the interaction, which is large enough for the application of pQCD. In figure 1(a), the three solid curves are from ref. [3] and correspond to three different gluon distributions which are consistent with that extracted from the analysis of scaling violations in the proton structure function F_2 measured at HERA.

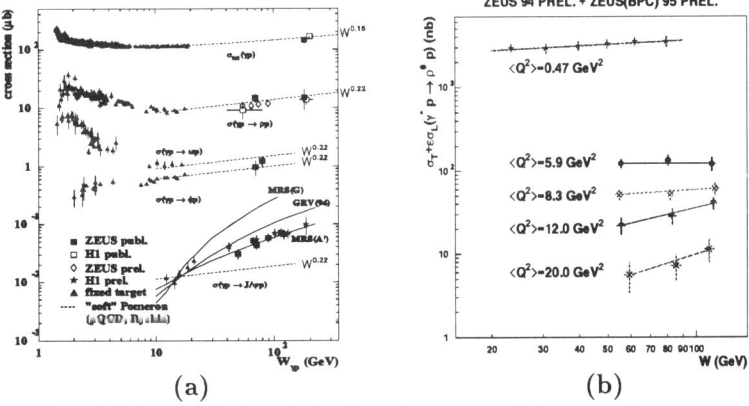

FIGURE 1. (a) A compilation of elastic vector meson photoproduction cross section results vs W_{γ^*p} from fixed target experiments and from H1 and ZEUS; (b) ZEUS 94 prel. and 95 prel. result on $\sigma_T + \epsilon\sigma_L(\gamma^*p \to \rho^0 p)$ vs W_{γ^*p} at different Q^2 values. σ_T and σ_L are the transverse and longitudinal cross sections respectively. ϵ is the ratio of the flux of longitudinally polarized photon to the flux of transversely polarized photon.

For the case of vector meson production at large Q^2, the comparison of $\sigma_{\gamma^*p \to Vp}$ for ρ^0 and ϕ mesons measured by both ZEUS and H1 (at $Q^2 \sim 6$, 10 and 20 GeV^2 for ρ^0, and at $Q^2 \sim 8$ and 15 GeV^2 for ϕ) with the results from NMC [17] (scaled to the same Q^2) show a steeper rise of $\sigma_{\gamma^*p \to Vp}$ with W_{γ^*p} than in the case of photoproduction. But if the result of $\sigma_{\gamma^*p \to \rho^0 p}$ at $Q^2 \sim 6$ GeV^2 from ZEUS are compared with E665's result [18], the rise of

TABLE 1. ZEUS preliminary result on the values of k obtained from fits of $\sigma_{\gamma^*p} \propto W_{\gamma^*p}^k$ to the elastic ρ^0 production at different Q^2.

$<Q^2>$ (GeV^2)	k	$<Q^2>$ (GeV^2)	k
0.47	$0.18 \pm 0.05 \pm 0.13$	12.0	$0.86 \pm 0.46 \pm 0.40$
5.9	$0.00 \pm 0.33 \pm 0.27$	20.0	$1.15 \pm 0.92 \pm 0.56$
8.3	$0.22 \pm 0.38 \pm 0.26$		

TABLE 2. Slope b for elastic J/ψ production at different Q^2.

Experiment	H1	ZEUS
b (GeV^{-2}) ($Q^2 \approx 0$ GeV^2)	$4.0 \pm 0.2 \pm 0.2$	$4.6 \pm 0.4^{+0.4}_{-0.6}$
b (GeV^{-2}) (large Q^2)	$3.8 \pm 1.2^{+2.0}_{-1.6}$	$4.5 \pm 0.8 \pm 1.0$ (prel.)

$\sigma_{\gamma^*p \to \rho^0 p}$ with W_{γ^*p} is not as steep. Since the experimental conditions may be quite different for different experiments, ZEUS has attempted to study this dependence of $\sigma_{\gamma^*p \to \rho^0 p}$ on W_{γ^*p} at different Q^2 values within one experiment. In figure 1(b) the points are fitted with the function $\sigma_{\gamma^*p} \propto W_{\gamma^*p}^k$. The results for k are shown in table 1. There is an indication of an increase of k with Q^2 but more data are needed to settle the question unambiguously. For J/ψ production at large Q^2, preliminary results from H1 and ZEUS show a W_{γ^*p} dependence consistent with that measured in photoproduction.

The differential cross section $d\sigma/d|t|$ was studied in the range $0 < |t| < 0.6$ GeV^2 for ρ^0, ω, ϕ, and $0 < |t| < 1.0$ GeV^2 for J/ψ. All these distributions can be described by exponential functions $d\sigma/d|t| \propto e^{-b|t|}$, with different values of the slopes b. The study of the slope b as a function of Mass$_V^2$ for photoproduction, shown in figure 2(a), indicates that b decreases with Mass$_V$. The result suggests that the size of the vector meson decreases with Mass$_V$.

The slope b was also studied as a function of Q^2 for ρ^0, ϕ, and J/ψ mesons. Figure 2(b), a plot of b as a function of Q^2 for ρ^0 production, shows that b decreases with Q^2. A similar behavior was also observed for elastic ϕ production. This trend can be attributed to the decrease of the $q\bar{q}$ pair transverse separation with increasing Q^2. For the case of J/ψ production, the slope b seems to be similar both in photoproduction and production at large Q^2, a consequence of the small size of J/ψ meson: in both regimes b measures the size of the proton only, the contribution from the J/ψ being negligible (cf. table 2).

INELASTIC J/ψ PHOTOPRODUCTION

In inelastic J/ψ photoproduction, H1 [14] and ZEUS [8] tagged J/ψ using the leptonic decay channels ($J/\psi \to e^+e^-/\mu^+\mu^-$). To select inelastic events, it

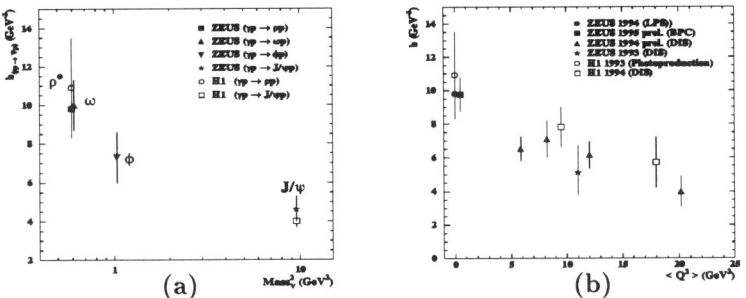

FIGURE 2. (a) The slope b as a function of Mass$_V^2$ for elastic photoproduction of vector meson; (b) the slope b as a function of Q^2 in the elastic production of ρ^0.

was required that there should be some activity in the main calorimeter or the central tracking detector other than that related to the J/ψ decay products. A cut in $0.4 < z < 0.9$ was applied to enhance the signal from inelastic J/ψ production via the photon-gluon fusion mechanism. The contributions from other inelastic processes were estimated from Monte Carlo studies and were subtracted from the selected sample. Figure 3(a) shows the differential cross section $d\sigma_{\gamma^*p \to J/\psi X}/dz$ measured by H1 and ZEUS. The solid line, which is the prediction from the NLO calculation of the color-singlet (CS) model [4], agrees well with the data. The dashed line, which is the prediction from a specific LO calculation of the color-octet (CO) model [6], overestimates the data at high z.

The cross section $\sigma_{\gamma^*p \to J/\psi X}$ was also measured as a function of W_{γ^*p}, cf. figure 3(b). The NLO calculation [4] agrees with data in both the shape and normalization.

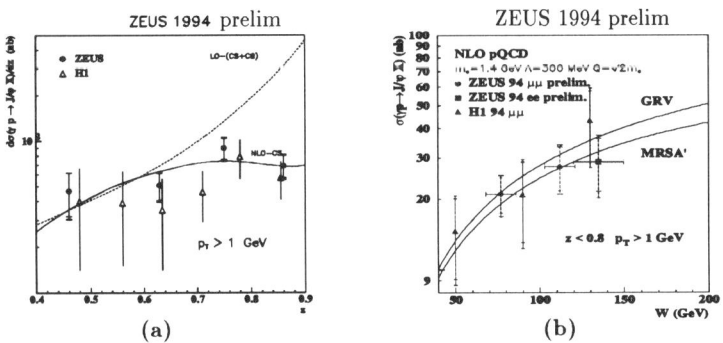

FIGURE 3. (a) Differential cross section $\sigma_{\gamma^*p \to J/\psi X}$ as a function of z for inelastic J/ψ photoproduction; (b) cross section $\sigma_{\gamma^*p \to J/\psi X}$ as a function of W_{γ^*p} for inelastic J/ψ photoproduction.

SUMMARY

The measurements of elastic photoproduction of ρ^o, ω and ϕ mesons are in agreement with the expectations based on "soft" pomeron model, whereas the elastic photoproduction of J/ψ and elastic production of ρ^0, ϕ and J/ψ at higher Q^2 favor the "hard" perturbative QCD model. This indicates that the variables Q^2 and Mass$_V$ can be used to define a transition region between "soft" and "hard" diffraction.

Inelastic J/ψ photoproduction, via photon-gluon-fusion, can be described by color-singlet (CS) model in NLO within a restricted kinematic range.

REFERENCES

1. A. Donnachie and P. V. Landshoff, *Phys. Lett.* **B296**, 227 (1992).
2. M.G. Ryskin, *Z. Phys.* **C57**, 89 (1993); S.J. Brodsky et al., *Phys. Rev.* **D50**, 3134 (1994); L. Frankfurt et al., *Phys. Rev.* **D54**, 3194 (1996); A. Martin et al., Preprint HEP–PH/96-09-448 (1996); J. Nemchik et al., *Phys. Lett.* **B374**, 199 (1996); D.Yu. Ivanov, *Phys. Rev.* **D53**, 3564 (1996); I.F. Ginzburg and D.Yu. Ivanov, *Phys. Rev.* **D54**, 5523 (1996); J.R. Forshaw et al., *Z. Phys.* **C68**, 137 (1995).
3. M. G. Ryskin et al., Preprint HEP–PH/95-11-228 (1995).
4. E.L. Berger and D. Jones, *Phys. Rev.* **D23**, 1521 (1981);
   R. Baier and R. Rückl, *Nucl. Phys.* **B201**, 1 (1982).
5. M. Krämer et al. , *Phys. Lett.* **B348**, 657 (1995);
   M. Krämer, *Nucl. Phys.* **B459**, 3 (1996).
6. M. Cacciari and M. Krämer, DESY preprint DESY 96-005 (1995).
7. CDF Coll, FERMILAB-CONF-94-136-E, June 1994, contribution to the XXVII International Conference on High Energy Physics, Editors P.J. Bussey and I.G. Knowles, Glasgow, Scotland, July 1994.
   CDF Coll, FERMILAB-CONF-95-226-E, July 1995, submitted to the Int. Symp. on Lepton Photon Interactions, Beijing 1995.
8. Talks by L. Adamczyk, L. Bellagamba, F. Gaede, I. Korzhavina, T. Monteiro, and A. Wegner at the 5th International Workshop on Deep Inelastic Scattering and QCD, April, 1997, http://www.hep.anl.gov/dis97
9. ZEUS Collab., M. Derrick et al., *Z. Phys.* **C69**, 39 (1995).
10. H1 Collab., S. Aid et al., *Nucl. Phys.* **B463**, 3 (1996).
11. ZEUS Collab., M. Derrick et al., *Phys. Lett.* **B377**, 259 (1996).
12. ZEUS Collab., M. Derrick et al., *Z. Phys.* **C73**, 73 (1996).
13. ZEUS Collab., M. Derrick et al., *Phys. Lett.* **B380**, 220 (1996).
14. H1 Collab., S. Aid et al., *Nucl.Phys.* **B472**, 3 (1996).
15. ZEUS Collab., J.Breitweg et al., DESY report DESY 97-060 (1997) , to be publ. in *Z. Phys.*
16. H1 Collab., S. Aid et al., *Nucl.Phys.* **B468**, 3 (1996).
17. NMC Collab., M. Arneodo et al., *Nucl. Phys.* **B429**, 503 (1994).
18. E665 Collab., M.R. Adams et al., *Z. Phys.* **C72**, 237 (1997).

Jets in DIS and Photoproduction at HERA

Leif Jönsson

Physics Department
Lund University
Box 118, 221 00 Lund
Sweden

Abstract. Results on jet shapes, inclusive jet production, di-jet cross sections and jet rates are presented in order to illustrate the composition of 'direct' and 'resolved' photon processes in ep scattering at HERA.

INTRODUCTION

Production of jets with large transverse energies from first order α_S processes in ep scattering offers a possibility of probing the partonic structures of both the exchanged photon and the proton. Two distinct processes are contributing. One is the 'direct' process in which the photon couples directly, in a pointlike fashion, to the partons in the proton and thereby will be sensitive to the parton content of the proton. This process dominates in the kinematic region where Q^2 is much larger than p_T^2 of the produced jets. In the 'resolved' process, on the other hand, the photon interacts via its own parton content with the proton. The 'resolved' process becomes important as p_T^2 of the produced jets is sufficiently large compared to Q^2 i.e. the transverse scale of the probe is small compared with the transverse dimension of the target photon. In this case the parton of the proton will resolve the structure of the photon.

At HERA the 'direct' and 'resolved' processes are distinguishable by measuring the fraction of the initial photon energy entering into the hard scattering, $x_\gamma = (\sum_i' E_{T_i} e^{-\eta_i})/(2yE_e)$, where E_{T_i} is the transverse energy of jet i, η_i is the pseudorapidity and $y = E_\gamma/E_e$ is the fraction of the electron energy, E_e, taken by the photon. A cut at $x_\gamma = 0.75$ gives a fairly clean separation between the two processes.

JET SHAPES

At sufficiently high transverse energies, E_T, the profiles of jets are expected to be dominated by the parton emission rather than by the long distance fragmentation process or soft interaction with the remnant jet. Therefore it should be possible to calculate the jet shapes in perturbative QCD and comparisons with experimental data on the hadron level can be performed.

The ZEUS experiment has studied jet shapes in photoproduction using the cone algorithm with $R = \sqrt{\Delta\eta^2 + \Delta\phi^2} = 1$. Experimentally two jets are combined if they have an overlapping transverse energy exceeding 75% of the least energetic jet. For each jet found with the cone algorithm, the procedure is to define another concentric cone with a smaller radius, r, and determine the fraction of the jet energy falling inside this cone.

Fig. 1a shows the fractional jet energy, $\psi(r)$, of inclusive jets as a function of the inner cone radius, r, for four bins in rapidity, requiring the transverse energy to be greater than 14 GeV. Data are compared to the predictions of the LO Monte Carlo program PYTHIA [1] which includes contributions from both 'direct' and 'resolved' processes with a $p_{T_{min}}$ cut off at 8 GeV/c. The MRSA [2] and GRV-HO [3] parametrizations are used for the proton and photon parton densities, respectively, and multiple interactions with the remnant jet as simulated by PYTHIA is taken into account. As can be seen from fig. 1a jets from 'direct' processes are expected to be narrower than those from 'resolved' processes in the full rapidity range investigated. The data are consistent with a dominance of 'resolved' processes but a clear discrepancy between PYTHIA and data is observed in the most forward region. Since 'direct' photon processes are dominated by quark initiated jets through the BGF process ($\gamma g \to q\bar{q}$) and the 'resolved' process has a larger fraction of gluon jets produced in the processes $q_\gamma g_p \to qg$ and $g_\gamma g_q \to gg$, the discrepancy might be due to a possible underestimation by PYTHIA of the fraction of gluon jets in this region.

In fig. 1b the same data is compared to NLO calculations [4] in which the photon emission is given by Weizsäcker-Williams approximation and the structure of the photon is described by the GRV model. The parton density in the proton is taken from CTEQ4M [5]. The jet profiles are calculated using the cone algorithm. Two partons are combined into one jet if they fulfil the condition $R \leq min[(E_{T_i} + E_{T_j})/max(E_{T_i}, E_{T_j}), R_{sep}]$, where R_{sep} has been introduced in order to account for jet broadening due to multiple interactions with the remnant jet and can be assigned any value between $(1-2) \cdot R_{cone}$. The data are well described by the NLO calculations provided the R_{sep} value is properly chosen in each rapidity bin.

INCLUSIVE JET CROSS SECTIONS

The H1 experiment has measured the single inclusive jet cross section in a range of photon virtualities from $Q^2 = 0$ to 50 GeV^2 using the k_T cluster algorithm in the $\gamma^* p$ rest frame. This algorithm assigns particles to either the remnant jet(s) or to the jets of the hard subsystem. The jets were required to have transverse momenta of 4 or 5 GeV/c depending on whether they were produced in deep inelastic scattering or in photoproduction and they had to fall inside the rapidity range $-2.5 < \eta^* < -0.5$. Fig. 3 shows the inclusive jet

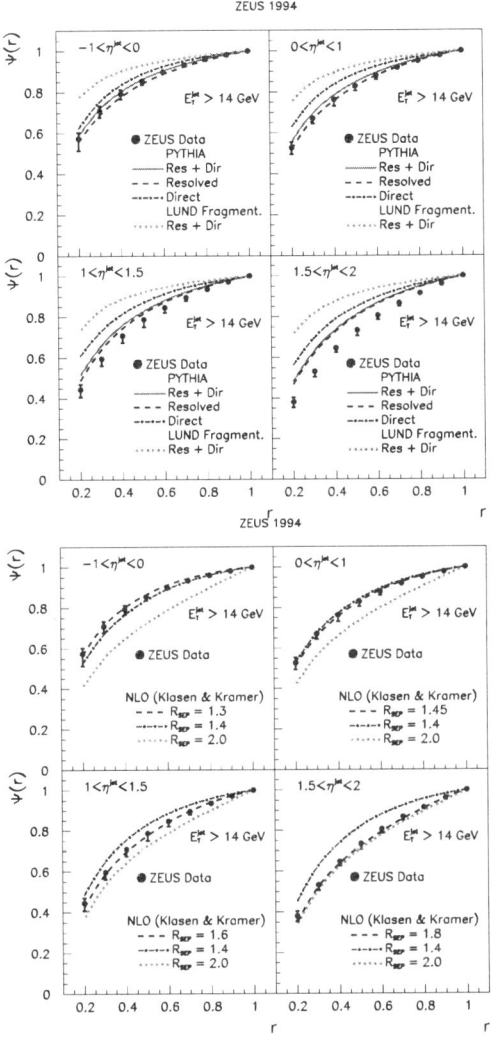

Figure 1: The fractional jet energy in four η bins compared to the predictions of a) PYTHIA and b) NLO calculations

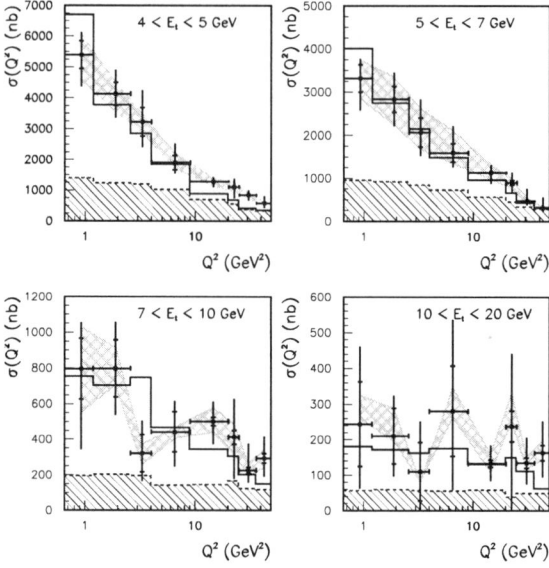

Figure 2: The inclusive jet cross section as a function of Q^2 for different E_T bins compared to the 'direct' and 'resolved' contributions according to the HERWIG model.

cross section as a function of Q^2 in bins of E_T together with the predictions of the HERWIG Monte Carlo program [6]. HERWIG uses first order α_S QCD matrix elements (ME) to calculate hard scattering processes and adds QCD parton showers (PS) to account for higher order contributions. This description corresponds to the 'direct' photon process. Contributions from 'resolved' processes based on the equivalent photon approximation can also be switched on. In order for HERWIG to give a good description of the measured jet profiles a 15% probability for soft underlying events had to be included at $Q^2 = 0$ GeV^2.

The hatched histograms in fig. 2 correspond to 'direct' contributions which is fairly independent of Q^2 in the region $Q^2 < p_T^2$ and can not at all reproduce data. The inclusion of 'resolved' processes results in the full histograms which give a good description of data in the full E_T range.

Another way of investigating the influence of 'resolved' photon processes is to determine what fraction of the initial photon energy, $f = \sum E_i/(yE_e)$, goes into the photon remnant. Here $\sum E_i$ is the energy of the particles allocated to the photon remnant. For 'direct' processes this fraction should be zero whereas for 'resolved' processes a significant fraction of the photon energy is expected to be taken by the photon remnant. This effect is also observed experimentally from the f-distributions which in the low Q^2 range, dominated by 'resolved' processes, are peaked at values greater than zero and at high Q^2 values, which is dominated by 'direct' processes, are peaked at zero. Consequently, the inclusion of 'resolved' processes is needed for HERWIG to give agreement with data at low Q^2 while at large Q^2 the MEPS model is doing well.

DI-JET CROSS SECTIONS

Cross sections for photoproduction of di-jets have been measured by the ZEUS collaboration and compared to NLO calculations [7]. At small Q^2 the ep cross section can be calculated according to the Weizsäcker-Williams approximation as a convolution of the photon flux and the γp cross section. The respective parton distributions for the proton and the photon have been represented by the GRV and CTEQ3M [8] parametrizations. In this investigation the k_T clustering algorithm has been used with a distance parameter defined in the $\eta - \phi$ space. The inclusive di-jet cross section $d\sigma/d\bar{\eta}$ is presented separately for the event samples with $x_\gamma > 0.75$ and $0.3 < x_\gamma < 0.75$, where $\bar{\eta}$ is the average pseudorapidity of the two reconstructed jets with the requirement $|\Delta\eta| < 0.5$. This requirement ensures the two jets to lie well inside the detector acceptance and in addition it increases the sensitivity to the parton distribution in the incoming particles. For each of the two event samples four different cuts on $E_T^{min} = 6, 8, 11$ and 15 GeV have been applied.

An overall agreement between data and the NLO predictions for the 'direct' cross sections is obtained. On the other hand the 'resolved' two-jet cross sections are only reasonably described for the higher E_T^{min} cuts. The discrepancy at lower E_T^{min} cuts is assumed to be due to effects from multiple interactions with the remnant jet which are not taken into account in the NLO calculations.

JET RATES

The rate of jets from first order α_s processes has been studied by the H1 collaboration. The cone algorithm was used to define the jets which were required to have transverse momenta greater than 5 GeV/c. Data were corrected for detector effects and compared to model predictions. In addition to the MEPS model, which has been described earlier, the colour dipole model (CDM) [9] and NLO order calculations as given by the DISENT program [10] were used for comparison with data. In the CDM model the gluon emission is treated as radiation from colour dipoles stretched between quark-antiquark pairs. These colour dipoles radiate independently and therefore the gluons are not ordered in transverse momenta in contradiction to what is the case for the DGLAP evolution used in the MEPS model which leads to a strong ordering in transverse momenta.

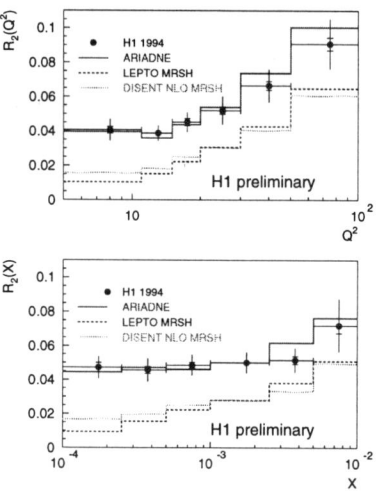

Figure 3: The fractional 2-jet rate as a function of a) Q^2 and b) x, compared to predictions of the MEPS and CDM models, and NLO calculations.

Fig. 3a and b show the jet rate as a function of Q^2 and x. The MEPS model clearly fails to describe the data as does the NLO order calculation which is performed on the parton level. However, since the hadronisation effects turn out to be small, the comparison with data is still relevant. The CDM model on the other hand gives an almost perfect description of the jet rates. The interpretation is, however, not straight forward since the CDM model does not use parton densities in the calculation of cross sections but rather suppression factors due to the extended photon and proton remnants [11]. One possible interpretation is that the CDM model simulates the effect of a 'resolved' photon contribution. If this hypothesis is correct, the x_γ spectrum should exhibit a dependence on Q^2 since the contribution from 'resolved' photon processes should depend on this variable. The CDM model produces such a dependence and it is in fact also observed in data. Another possible interpretation is that the deviation from the MEPS model is due to new small x dynamics à la BFKL [12] which gives no ordering in transverse momenta as is also true for the CDM model.

SUMMARY

Photoproduction of jets with high transverse energies are dominated by processes where the photon interacts with the proton via its partonic structure.

Such processes have a larger fraction of gluon jets than those where the photon couples directly to the proton. As the jets go more to the forward region, the effect of multiple interactions with the remnant jet gets increasingly important. Inclusive jet cross sections measured in a Q^2 range up to 50 GeV^2 can only be described by 'direct' interaction of the photon if $Q^2 > p_T^2$. As Q^2 becomes smaller than p_T^2, the contribution from 'resolved' photon processes has to be included to get agreement with data. Jet rates of first order α_S processes from deep inelasic scattering up to 100 GeV^2 are not reproduced by models based on 'direct' processes or by NLO calculations. Considering the previous results discussed, it seems reasonable that deviations at low Q^2 could be due to 'resolved' photon processes but the deviations at high Q^2 need a more careful study to be understood.

Acknowledgements. I would like to thank J. Hartmann and C. Foudas for providing me with recent Zeus results. Among my colleagues in H1 especially T. Ebert, H. Jung, H. Küster and J. Spiekermann have contributed to the material presented.

References

[1] H.U. Bengtsson and T. Sjöstrand, *Comp. Phys. Comm.* **46** (1987) 43; T. Sjöstrand, *Comp. Phys. Comm.* **82** (1994) 74.

[2] A.D. Martin, W.J. Stirling and R.G. Roberts, *Phys. Rev.* **D50** (1994) 6734.

[3] M. Glück, E. Reya and A. Vogt, *Phys. Rev.* **D46** (1992) 1973.

[4] M. Klassen and G. Kramer, DESY 97-002, 1997.

[5] H.L. Lai, J. Huston, S. Kuhlmann, F. Olness, J. Owens, D. Soper, W.K. Tung, H. Weerts, MSUHEP-60426, June 1996, hep-ph/9606399.

[6] G. Marchesini, B.R. Webber et al., *Comp. Phys. Comm.* **67** (1992) 465.

[7] M. Klassen and G. Kramer, DESY 96-246, 1996.

[8] H.L. Lai, J. Botts, J. Huston, J.G. Morfin, J.F. Owens, J.W. Qiu, W.K. Tung, H. Weerts, *Phys. Rev.* **D51** (1995) 4763.

[9] G. Gustafson, U. Petterson, *Nucl. Phys.* **B306** (1988); G. Gustafson, *Phys. Lett.* **B175** (1986) 453 B. Andersson, G. Gustafson, L. Lönnblad, U. Petterson, *Z. Phys.* **C43** (1989) 625.

[10] S. Catani and M. Seymour, CERN-TH-96-342, 1996, hep-ph/9612236.

[11] J. Rathsman, *Phys. Lett.* **B393** (1997) 181.

[12] E.A. Kuraev, L.N. Lipatov and V.S. Fadin, *Sov. Phys.* **JETP 45** (1972) 199; Y.Y. Balitsky and L.N. Lipatov, *Sov. J. Nucl. Phys.* **28** (1978) 282.

Exotic Searches at HERA

Masahiro Kuze

Institute of Particle and Nuclear Studies, KEK
3-2-1 Midoricho, Tanashi, Tokyo 188 Japan
(On behalf of the H1 and ZEUS collaborations)

Abstract. The search for new physics beyond Standard Model is one of the most important physics programs at the e-p collider HERA. Searches for various exotic phenomena using up to 1995 data are summarized as an introduction, and the new results from the H1 and ZEUS collaborations including 1996 data, on the excess of high-Q^2 high-x neutral current DIS events are described.

INTRODUCTION

At HERA, electrons (positrons) and quarks (or partons in general) in the proton interact with each other at a very short distance. At $Q^2 = 10^4$ GeV^2, the probed distance is of the order of 10^{-16} cm which is one over a thousand of the proton radius. This is one order of magnitude smaller than the region explored by fixed-target DIS experiments. Also, the available total energy $\sqrt{s}$ is 300GeV at maximum, which is larger than LEP2 total energy. Various searches for exotic physics manifestations have been done by both H1 and ZEUS experiments, which are summarized in the following.

If electrons or quarks are not elementary particles, excited states of these particles can be produced in e-q interaction, which subsequently decay into various final states (for example, $e^* \to e\gamma$). Searches for such excited fermions have been done and are reported in [1,2].

If there are right-handed electron neutrinos (N_R) and they are heavier than right-handed $W(W_R)$, N_R can be created in right-handed charged current process and can decay in the chain $N_R \to eW_R, W_R \to q\bar{q}'$. The final states will look like 2-jet neutral current events in which dijet mass and e-jet-jet mass make resonant peaks. ZEUS has reported about the search results in [3].

The effect of new physics with a very high mass scale at a lower energy can be approximated by an effective four-fermion point-like interaction called contact interaction. The mass scale appears in the Lagrangian as the denominator ($mass^{-2}$) in the effective coupling. The existence of such new physics will alter the neutral current DIS cross sections at high Q^2. These cross sections were measured and compared with Standard Model [4,5].

Existence of leptoquark is suggested by many models beyond Standard Model. It is a new boson with both lepton and quark quantum numbers, and HERA is an ideal place to produce it. If it decays into eq final state, it is identical to neutral current DIS events and makes a resonant peak in the cross section at $x = M^2/s$, where M is the leptoquark mass. It could also decay into charged current mode ($\nu q'$), or to leptons with other flavors ($\mu q, \tau q$) which have a distinct characteristic of lepton flavor violation. Searches for these final states, using data up to 1994, are reported in [4,6-8].

Supersymmetry (SUSY) is widely believed to be the most promising model which explains many of the theoretical problems in Standard Model. If R-parity [$R = (-1)^{3B+L+2S}$, where B, L, S being baryon, lepton number and spin respectively] is violated, a squark can be singly produced in e-q interaction, e.g. $e^+ d \to \tilde{u}, \tilde{c}, \tilde{t}$. They can decay into ed final state, which shows the same signature as leptoquarks, but depending on SUSY parameters also R-parity conserving decay can happen where the squark decays into quark and chargino or neutralino. If R-parity is not violated, SUSY particles can only be produced in pair. At HERA, $eq \to \tilde{e}\tilde{q}$ can occur through t-channel neutralino exchange. For both cases H1 has made extensive searches and reported in [9,10]

Above searches are based on e^-p collision data in 1993 and 1994 (about $1 pb^{-1}$ available for analysis) and e^+p data in 1994 and 1995 (about $10 pb^{-1}$). All searches gave negative results and new limits on the mass or coupling of the new physics were obtained. This also means that cross sections of neutral and charged current DIS were confirmed to be consistent with Standard Model up to $Q^2 \approx 10^4 \text{GeV}^2$ where the statistics become too few.

In 1996, HERA continued to provide e^+p collisions and we got another $\approx 10 pb^{-1}$ of data usable for analysis. Therefore, both experiments combined e^+p data of 1994 to 1996 and compared the number of high-Q^2 events with Standard Model. The results [11,12] are described in the following.

NEUTRAL CURRENT DIS EVENTS AT HIGH Q^2

The cross section of neutral current (t-channel exchange of γ and Z) e^+p scattering can be written as:

$$\frac{d^2\sigma}{dxdQ^2} = \frac{2\pi\alpha^2}{xQ^4}[\{1 + (1-y)^2\}F_2(x,Q^2) - \{1 - (1-y)^2\}xF_3(x,Q^2)],$$

where α is the electromagnetic coupling. Contribution from longitudinal structure function F_L can safely be neglected in the high-Q^2, high-x kinematic region considered in the analysis. The structure functions F_2 and F_3 contain the information of quark distribution functions, together with the coefficients consisting with the electric charge and weak coupling constants of positron and quark, as well as the propagation factor coming from the Z exchange and $\gamma - Z$ interference. As $Q^2 = sxy$, the $1/Q^4$ dependence of the cross section means $1/y^2$ at fixed x, thus the cross section falls down as y increases. On the other hand, in the case of scalar leptoquarks for example, y shows a flat distribution since it is related to the decay angle in leptoquark rest frame. Therefore, to separate the effect of new physics, one should look at the events at high-y, namely high-Q^2 region.

Both experiments make similar event selections. A primary event vertex obtained from tracking information is required to lie in the interaction region. The longitudinal momentum $E - P_z$ is required to be close to twice the positron beam energy to make sure that no significant energy has escaped to the beam pipe in the incident positron direction. Then the scattered positron was searched for as an isolated electromagnetic cluster in the calorimeter with high transverse energy, with a matching track where applicable. There are additional cuts to remove background events such as QED Compton, where one finds positron and photon back to back in the transverse plane. The selection efficiency in the kinematic region studied is about 80% for both experiments.

As we observe the scattered positron and the current jet from the scattered quark in the detector, one can use four variables, the energy and polar angle from each, to calculate the kinematics. As there are only two independent kinematic variables, there is a redundancy and one can use different methods to confirm consistency of the event. H1 uses the energy and angle of the positron, which is called Electron Method. It does not depend on the hadronic measurement. ZEUS uses Double Angle Method which uses the positron angle and jet angle. It is insensitive to the energy scale uncertainty to the first order. Both experiments use the other method as a cross-check.

H1 analyzed $14.2pb^{-1}$ of data and obtained 443 events in the region $0.1 < y < 0.9, Q^2 > 2500 \text{GeV}^2$, of which 122 events have $Q^2 > 5000 \text{GeV}^2$. (With Double Angle Method, 460 events in $Q^2 > 2500 \text{GeV}^2$.) Figure 1 shows the events in (M, y) plane, where M is the invariant mass of the eq system calculated from x variable as $M = \sqrt{sx}$. (For example, $M = 200 \text{GeV}$ corresponds to $x = 0.443$). The total number of events well matches the expectation from Standard Model which is 427 ± 38, but there are 7 events in

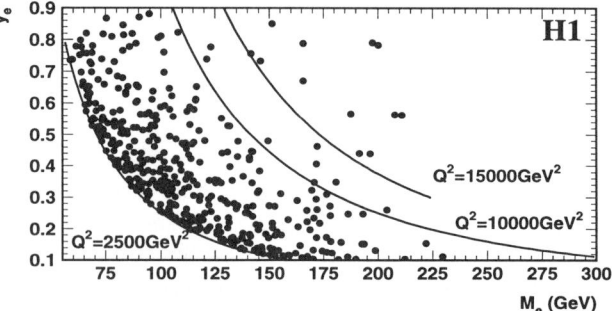

FIGURE 1. (M, y) distribution of H1 events.

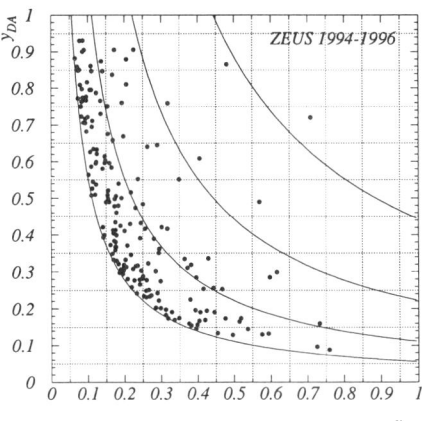

FIGURE 2. (x,y) distribution of ZEUS events.

$M > 180$GeV, $y > 0.4$ and their mass values are clustered around 200GeV. ZEUS has 191 events with $Q^2 > 5000$GeV^2 from $20.1 pb^{-1}$ of data. Figure 2 is the (x,y) scatter plot of the events. Again, total number of events is consistent with 197 ± 9.9 expected from Standard Model, but 4 events are observed in the region $x > 0.55, y > 0.25$.

H1 quotes 7GeV for the resolution of mass reconstruction. At $M = 200$GeV, this corresponds to 7% resolution in x. In ZEUS, the x resolution (in the region $x > 0.45$) is 9% for $y > 0.25$ and improves to 6% for $y > 0.5$. The Q^2 resolution by ZEUS is typically 5% at large x and y. The kinematic variables for the events at high x, high y were cross-checked with different methods and found to be consistent within the errors.

Then the statistical significance of these events was evaluated. ZEUS made a cut $y > 0.25$ and looked at x distribution (Figure 3). The number of events above certain x^* was integrated. From the number of expected events from Standard Model, one can calculate the Poisson probability that at least the number of events in the data (or more) is observed by statistical fluctuation. One gets high probability at low x^* which means the total number of events agree with expectation, and gets the minimum probability of 0.6% at $x^* = 0.57$, where 0.7 events are expected and 4 events are observed. Then the probability of observing such fluctuation (0.6% or lower) in Standard Model at *any* x^* was calculated, by making many hypothetical experiments using the Monte Carlo events. This probability turned out to be 7.2%.

H1 made a window in mass axis and counted the number of events within the window, after making a cut in y. Various combination of window width and y cut were tried, and in all cases the lowest probability is found near $M = 200$GeV. For example, with the width 25GeV and $y > 0.4$, the probability is 2.6×10^{-4} at $M = 200$GeV. The number of observed events is 7 and the expectation is 0.95 ± 0.18. Similarly, the probability to find such fluctuation in Standard Model *anywhere* in the mass range is 0.9%.

The excess was evaluated also in Q^2 projection. ZEUS observed two events with $Q^2 > Q^{2*} = 35000 \text{GeV}^2$ where 0.145 ± 0.013 are expected. The probability of such an excess at *any* Q^{2*} is 6%. H1 observed 12 events in $Q^2 > 15000 \text{GeV}^2$ where one expects 4.71 ± 0.76 events (Figure 4), giving Poisson probability of 0.6%. For this region, total number of events in H1 and ZEUS is 24, where the sum of the expected number is 13.4 ± 1.0.

Extensive studies were made about the possibility of having such events from other processes in Standard Model than neutral current DIS. Processes studied include prompt photon production, dijet photoproduction, QED Compton, two-photon processes and production of W, Z. Monte Carlo events were generated and processed through the selection criteria to see how many of these survive. The result was that negligible number of events is expected in the region studied, at most 0.1 events for both experiments.

Finally, the systematic errors in the Standard Model expectation were evaluated. Since HERA experiments are the first to measure the proton structure functions at this Q^2 and x, one needs to make extrapolation of the measurements at the same high x but much lower Q^2, using NLO QCD evolution based on DGLAP equations. It is also sensitive to the error in α_s. From the errors of fixed target experiments and the errors in α_s, the uncertainty in the cross sections coming from these sources was estimated to be around 6.5%. Also the effect of higher order QED radiative corrections was checked and estimated to be 2%. The errors from electroweak parameters are negligibly small.

It should be noted that CTEQ4HJ parameterization, which is inspired from the anomalous high-E_T jet production in CDF and has a large gluon contribution at high x, gives only a small change in our process, less than 5%. This is because the cross section is dominated by u-quark contribution at high x.

From the experimental side, one has 2.3% error in luminosity measurement. The error coming from detector simulation is 4 to 5% in both experiments. Largest contribution to this error is the energy scale calibration. As a result, the overall uncertainty in the expected number of events in the kinematic region studied is at a few % level; ZEUS quotes a value of 8.4% as the total systematic error.

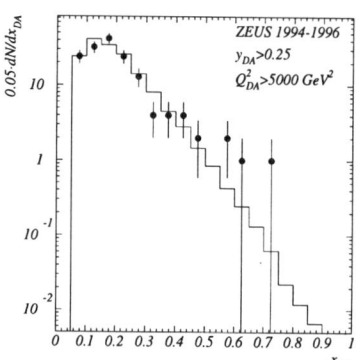

FIGURE 3. x distribution of ZEUS.

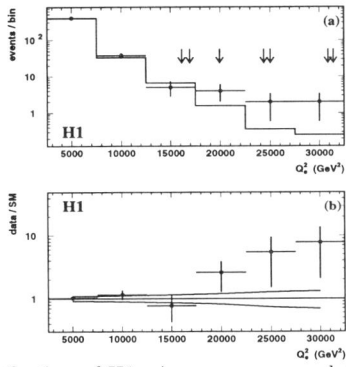

FIGURE 4. (a) Q^2 distribution of H1. Arrows correspond to events $M > 180$GeV and $Q^2 > 15,000$GeV^2. (b) Ratio to Standard Model with the band of systematic errors.

CONCLUSION

The excess of events at high Q^2 and high $x(M)$ observed by H1 and ZEUS cannot be accounted for by systematic effects, nor by the background from other processes. The only explanation is statistical fluctuation, and the probability is the order of 10^{-2} in each of the experiments. The distribution of the events agrees with Standard Model up to $Q^2 \approx 15,000$GeV^2, but the excess is observed at higher Q^2. The H1 events are clustered around $M = 200$GeV$(x = 0.44)$, but the ZEUS events are more spread at higher $x, x > 0.55$. However, the statistics are not enough to say anything about the resonance structure.

Definitely we need more data, and this year HERA has been running since March in e^+p mode. We expect to double the statistics with 1997 data. As this DIS kinematic region was not explored by previous experiments, it is very important to pursue the cause of this excess.

REFERENCES

1. H1 collab., S. Aid et al., *Nucl. Phys. B* **483**, 44 (1997).
2. ZEUS collab., J. Breitweg et al., DESY 97-112, submitted to *Z. Phys. C*.
3. ZEUS collab., Contributed Paper pa12-011 for ICHEP96 Warsaw.
4. H1 collab., S. Aid et al., *Phys. Lett. B* **353**, 578 (1995).
5. ZEUS collab., Contributed Paper pa04-036 for ICHEP96 Warsaw.
6. H1 collab., S. Aid et al., *Phys. Lett. B* **369**, 173 (1996).
7. ZEUS collab., M. Derrick et al., *Phys. Lett. B* **306**, 173 (1993).
8. ZEUS collab., M. Derrick et al., *Z. Phys. C* **73**, 613 (1997).
9. H1 collab., S. Aid et al., *Z. Phys. C* **71**, 211 (1996).
10. H1 collab., S. Aid et al., *Phys. Lett. B* **380**, 461 (1996).
11. H1 collab., C. Adloff et al., *Z. Phys. C* **74**, 191 (1997).
12. ZEUS collab., J. Breitweg et al., *Z. Phys. C* **74**, 207 (1997).

Hadron Dynamics

Color Transparency - Color Coherent Effects in Nuclear Physics

Gerald A. Miller

*Department of Physics,
University of Washington,
Seattle, WA 98195-1560*

Abstract. Efforts to observe color transparency in (e,e'p), (p,pp) and coherent diffractive $\pi \to$ two jets reactions are reviewed.

INTRODUCTION

Color transparency is the vanishing of initial and final state interactions, predicted by QCD to occur in high momentum transfer quasielastic nuclear reactions. The nuclear processes (e,e'p) and (p,pp) are examples. The experiments need to be done with good enough resolution so that the basic high momentum transfer reaction (electron-proton or proton-proton) is elastic. The energy transfer to the recoiling nucleus must also be small. If this is the case, the quark-gluon calculation of the process is coherent: one computes the scattering amplitude by adding all of the terms that lead to the same final state. The cross section is the absolute square of that amplitude. Thus color transparency effects are also color coherent effects.

Color transparency occurs [1,2] under the following three conditions:

- To obtain an appreciable amplitude for a high momentum transfer reaction on a nucleon leading to a nucleon, the colored constituents must be close together. Small objects, or point-like-configurations PLCs are produced in high momentum transfer exclusive processes.

- If the quark-gluon constituents of a color singlet object are close together, their color electric dipole moment is small and the soft interactions with the medium are suppressed.

- If the small color singlet can remain small as it moves through the nucleus, it can escape without further interaction.

Understanding each of the above points represents a serious research process; see the review [3]. I make only brief remarks about each item here.

Verifying or disproving the existence of point-like-configurations is the goal of color transparency studies. Originally, studies of perturbative QCD calculations of hadronic elastic form factors showed that the dominant contributions came from configurations which consisted of the fewest possible quarks (and no gluons) each at the same impact parameter. These calculations have been challenged. Frankfurt, Miller and Strikman [4] studied the existence of point-like-configurations within the framework of a wide variety of non-perturbative models. The most realistic models, those which feature correlations between quarks, do admit the existence of a point-like-configuration. But the existence or non-existence of point-like-configurations can not be decided by theory- experiments are needed.

That point-like-configurations have small forward scattering amplitudes is the least controversial of the three key points. Such an effect occurs in theories such as QED and QCD for which there is a concept of neutrality. The significant experimental evidence for reduced interactions comes from the existence of scaling in low-x deep inelastic scattering, hadron-proton total cross sections (see the review [3]) and in diffractive production of ρ mesons [5].

The point-like-configuration is not a physical state; it is therefore a wave packet which undergoes time evolution. The point-like-configuration initially has no size, so any evolution necessarily causes expansion. If the point-like-configuration expands while it is in the nucleus, it undergoes typical baryonic interactions and color transparency does not occur even if the point-like-configuration is made. One can estimate [3] the laboratory expansion time in terms of a product of a rest frame expansion time of order 1 fm and a time dilation factor which is expected to be only about 2 for the (e,e'p) experiments done at SLAC and planned at TJNAF and is about 5 or 6 for the (p,pp) experiments at BNL. Expansion effects must therefore be included to interpret these experiments.

CURRENT (E,E'P) AND (P,PP) DATA

Color transparency (CT) and color coherent effects have been recently under intense experimental and theoretical investigation. The (p,pp) experiment of Carroll et al. [6] found evidence for color transparency, see Fig. 2 of Ref. [7] while the NE18 (e,e'p) experiment [8] did not. The key feature in the (p,pp) data is the magnitude of the cross section, which is too large to be explained by standard treatments fo final state interactions. This experiment was done for p_L =6, 10, and 12 GeV/c, at scattering angles such that $Q^2 = 4.8, 8.6$ and 10.4 GeV^2/c^2. The expansion times are roughly 3.4, 5.7 and 6.8 fm. No such enhancement was found in the electron scattering data for $Q^2 = 1, 3, 5$, and 7 GeV^2/c^2. These Q^2 are reasonably large, but the outgoing proton momenta are 1.1, 2.3, 3.5 and 4.6 GeV/c, so the expansion times take on the small values of about 0.65, 1.3 2.0 and 2.6 fm.

The Q^2 of the NE18 experiment seem to be large enough to form a small color singlet object, but the expansion times are small. Thus this failure to observe significant color transparency effects is caused by the rapid expansion of the point like configuration to nearly normal size (and nearly normal absorption) at the relatively low momenta of the ejected protons [9], [10]. In particular, models of color transparency which reproduce the (p,2p) data and include expansion effects predict small CT effects for the NE18 kinematics, consistent with their findings, see the discussion and Fig. 11 of Ref. [3].

If this intepretation is correct, extending the electron scattering experiments to a $Q^2 \approx 10$ Gev^2/c^2 such that the outgoing proton momentum is about 6 GeV/c should allow the observation of enhanced cross sections.

EVA

The current (p,pp) experiment running at BNL has a new detector EVA [11] which should allow much improved measurements. DR. S. Durant, explains in this session how this detector works and presents new preliminary data for p_L=6 and 7.5 GeV/c, taken for quasielastic kinematics such that the initial bound proton is at rest ($\alpha = 1$). They find that the ratio of the measured to Born approximation cross sections $d\sigma/d\sigma_B$ (which is unity if color transparency is fully manifest) rises by about 30% between those two incident momenta.

I ran our [7] program for this situation (before the meeting) and the results are displayed in Fig. 1. The effects of color transparency lead to an enhancement over the usual treatment of initial and final state interactions (labelled initial, final state interactions). The (p,pp) reaction is more complicated than the (e,e'p) reaction because the PLC involves all six quarks at the same point. One expects other configurations to contribute. The simplest way to account for such is to assume that the PLC is complemented by another configuration of average size, which we call the blob-like configuration. Two models of such configurations are in the literature [12,13]. Here we use the version of Ref. [13]. In our treatment such effects are not extremely important [7] for the kinematics of the earlier experiment [7], but seem to stand out at $\alpha = 1$. Including them leads to a more rapid rise with p_L as indicated by the preliminary data. Future data taken at higher energies will be very interesting.

VICTORY OVER THE EXPANSION ENEMY?

The practical problem in looking for color transparency effects in experiments at Q^2 from about one to a few GeV^2 is that the assumed PLC expands rapidly while propagating through the nucleus. To observe color trans-

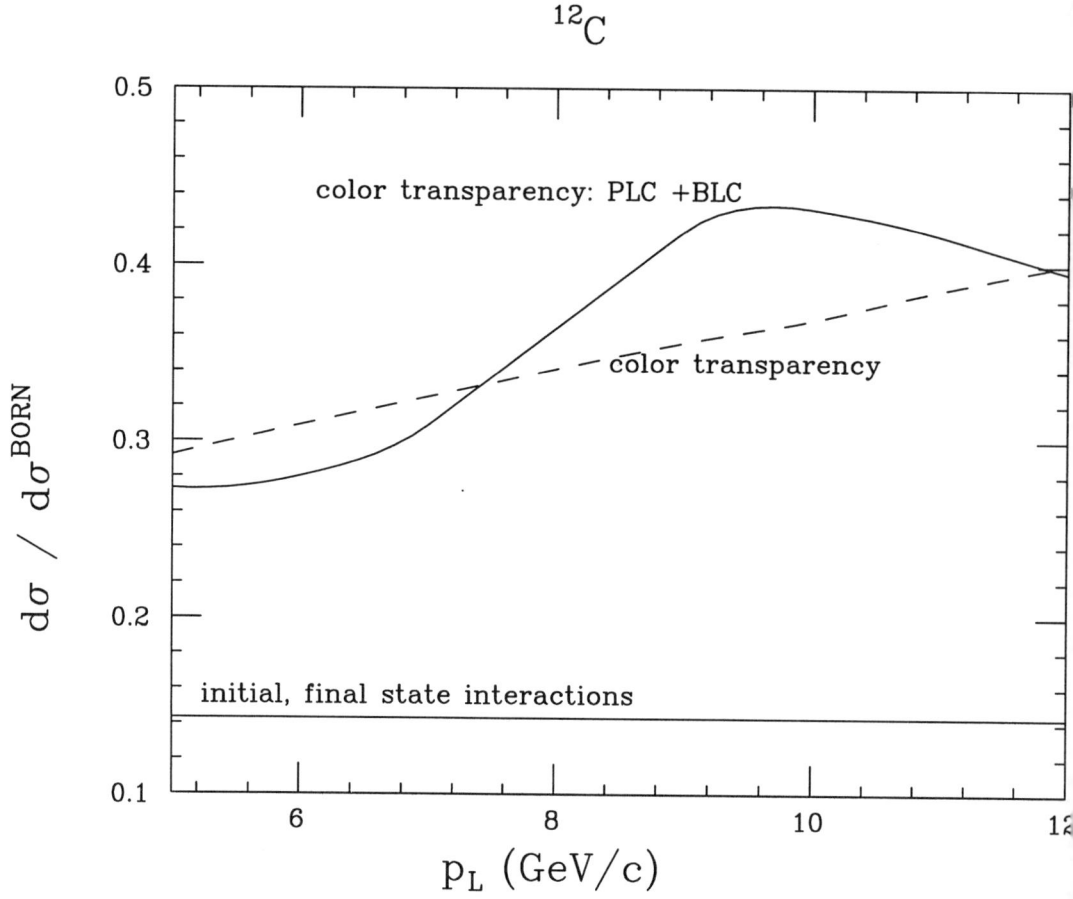

FIGURE 1. Color transparency at quasielastic kinematics

parency at intermediate values of Q^2 it is necessary to suppress the effects of wavepacket expansion. The propagation distances are small for the the lightest nuclear targets, but then, the transparency is close to unity and the effects of CT in $(e, e'p)$ reactions tend to be small. However, if one studies a process where the produced system can **only** be produced by an interaction in the final state, the color coherent effects would be manifest as a decrease of the probability for final-state interactions with increasing Q^2. Thus, the measured cross section would be compared with a vanishing quantity, the relevant ratio of cross sections would vary between unity and infinity and, a big signal could be possible. The first calculations [14] showed that substantial CT effects are observable in the (e,e'pp) reactions on ^{4,3}He targets. Later calculations [15] showed that substantial effects of color transparency are possible to observe using the deuteron as a target in the process $e\,d \to e'$pn. Such experiments are planned to run at the Jefferson Lab [16].

Another idea involves pionic degrees of freedom. For some reactions involving nucleons, the initial and final state interactions are expected to be dominated by exchanges of pions. These interactions are also hindered in high momentum transfer nuclear quasielastic reactions because the probability for a PLC to emit a quark-anti-quark pair is suppressed by color neutrality. The vanishing of pion exchange interactions has been called [4] "chiral transparency".

One example is the quasielastic production of the Δ^{++} in electron scattering - the $(e, e'\Delta^{++})$ reaction. The initial singly charged object is knocked out of the nucleus by the virtual photon and converts to a Δ^{++} by emitting or absorbing a charged pion. But pionic coupling to small-sized systems is suppressed, so this cross section for quasielastic production of Δ^{++}'s should fall faster with increasing Q^2 than the predictions of conventional theories. The first calculations [17] indicate that large effects are possible.

COHERENT NUCLEAR DIFFRACTION OF HIGH ENERGY PIONS INTO TWO JETS

Consider a coherent nuclear process in which a high-energy pion undergoes diffractive dissociation into a $q\bar{q}$ pair of high relative transverse momentum $\vec{k}_\perp$ but the nucleus remains intact. This is a very simple way to select a PLC in a hadronic projectile [18].

If the final $q\bar{q}$ pair carries of all of pion's energy, only the $q\bar{q}$ component of the light-cone wave function is needed for calculations. In this case, only the small sized pionic configurations pass through the nucleus without absorption. Observing high $k_\perp$ jets insures that only small $q\bar{q}$ separations are involved [18]. The jets we consider involve high (greater than about 1 GeV/c) but not very high values of $k_\perp$ less than about 3 or which is a huge enhancement factor for heavy nuclei. 5 GeV/c (for currently accessible energies).

The small sizes and the large beam momentum greatly simplify any computations of the matrix elements. The large beam momentum insures that the momentum transfer to the high momentum $q\bar{q}$ system is transverse, so the $q\bar{q}$-nucleon interaction is essentially independent of the longitudinal momentum of $q\bar{q}$ pair. Furthermore, the effects of expansion should be negligible.

The dominance of configurations of small size is a key feature of PQCD predictions for coherent diffraction of pions into two jets. Thus, even in a nuclear target, color screening implies that the coherent $q\bar{q}$ system can only weakly interact. In leading-logarithmic approximation and in the light-cone gauge only two gluons connect the pion - two jet system with the nucleus. Thus the hadronic system propagating through the nucleus suffers no initial-state or final-state absorption, and the nuclear dependence of the $\pi + A \to 2$ jets + A forward amplitude will be approximately additive in the nucleon number A, so the cross section would be proportional to A^2, a huge enhancement for heavy nuclei. The Fermilab experiment E791 [19], using a 500 GeV pion beam, will measure an integral over t so that the expected A dependence is $A^{4/3}$. Preliminary indications [19] are that this coherent process could be measurable.

Although the vector meson (two jets) suffers no final state interactions, the forward amplitude is not strictly additive in nuclear number since the gluon distribution itself is shadowed. This effect must be taken into account at very high energies; see the discussion in [3].

QUANTUM INVISIBILITY

This manuscript summarizes some of the methods that researchers are using to observe color transparency or color coherent effects. A clear and convincing signal of color transparency would represent the discovery of a new physical phenomenon. Initial and final state interactions are the bane of a nuclear physicist's everyday existence- their disappearance would be a surprise indeed.

Initial and final state interactions, which usually cause the nucleus to act similarly to a black disk, can be said to cast a shadow behind the nucleus. Only visible objects can cast a shadow, so that color transparency is a quantum invisibility.

REFERENCES

1. A.H. Mueller in Proceedings of Seventeenth rencontre de Moriond, Moriond, 1982 ed. J Tran Thanh Van (Editions Frontieres, Gif-sur-Yvette, France, 1982)p13.
2. S.J. Brodsky in Proceedings of the Thirteenth intl Symposium on Multiparticle Dynamics, ed. W. Kittel, W. Metzger and A. Stergiou (World Scientific, Singapore 1982,) p963.

3. L.L. Frankfurt, G.A. Miller and M. Strikman, Ann. Rev. Nucl. Part. Sci. **45** (1994) 501.
4. L. Frankfurt, G.A. Miller and M. Strikman, Comm. Nucl. Part Phys. **21** (1992) 1; Nucl. Phys. **A555** (1993) 752.
5. L. Frankfurt, W. Koepf, M. Strikman, Phys.Rev. **D54** (1996) 3194.
6. A.S. Carroll *et al.*, Phys. Rev. Lett. **61** (1988) 1698
7. B.K. Jennings and G.A. Miller, Phys. Lett. **B318** (1993) 7
8. N.C.R. Makins *et al.*, Phys. Rev. Lett. **72** (1994) 1986; T.G. O'Neill *et al.*, Phys. Lett. **B351** (1995) 87; Nucl. Phys. **A555** (1993) 752.
9. G.R. Farrar, H. Liu, L.L. Frankfurt & M.I. Strikman, Phys. Rev. Lett. **61** (1988) 686.
10. B.K. Jennings and G.A. Miller, Phys. Lett. **B236** (1990) 209; Phys. Rev. **D44** (1990) 692; B.K. Jennings and G.A. Miller, Phys. Rev. Lett. **70** (1992) 3619.
11. A.S. Carroll et al. BNL expt 850.
12. S. J. Brodsky & G.F. De Teramond, Phys. Rev. Lett. **60** (1988) 1924.
13. J. P. Ralston and B. Pire, Phys. Rev. Lett. **61** (1988) 1823; J.P. Ralston and B. Pire, Phys. Rev. Lett. **65** (1990) 2343.
14. K. Egiyan *et al.*, Nucl. Phys. **A580**(1994) 365.
15. L. Frankfurt, W.R. Greenberg, G.A. Miller, M.M. Sargsyan and M.I. Strikman, Z. Phys. **A352** (1995) 97; Phys.Lett. **B369** (1996) 201.
16. TJNAF Expt. K. Griffieon, spokesman
17. L. Frankfurt, T.S.H. Lee, G.A. Miller, M. Strikman Phys. Rev. **C55** (1997) 909.
18. L. Frankfurt, G.A. Miller, and M. Strikman Phys. Lett. **B304** (1993) 1.
19. FNAL E791, R. Weiss-Babai, Tel-Aviv. Univ., Israel
629;

A-Dependent Effects in Jet Production

M. D. Corcoran
Rice University
Houston, Texas 77005

Abstract

Strong A-dependent effects have been seen in jet production in several different experiments. The observed effects are consistent with final-state rescattering effects, leading to both enhanced cross sections as well as smearing of jet-jet coplanarity. A-dependent effects are also seen in forward energy flow in jet production from nuclei. Results are presented from Fermilab experiments E609 and E683.

INTRODUCTION

A-dependent effects in hard processes have been known for more than 20 years. The earliest example is the "Cronin effect" [1] in which the cross section for the production of single high p_t particles from a heavy nucleus increases faster than the atomic mass A. Many subsequent experiments found similar effects under a variety of conditions.

The generally-accepted picture of hard scattering inside a nucleus that has emerged is as follows. After a hard process, one or more energetic partons propogates through the nucleus. The uncertainty principle leads to the expectation that hadronization of the parton into a jet occurs well outside the nucleus, an expectation that is consistent with data. As the parton propagates through nuclear matter, it can rescatter, thereby changing its direction and possibly also losing energy. Studying A-dependent effects in jet production therefore gives us information about the propagation of partons through nuclear matter.

Traditionally A-dependent effects have been characterized by the parameter α in the expression $\sigma(pA) = \sigma(pp)A^\alpha$. If all of the nucleons act independently α would be 1. If the cross section falls steeply with respect to the variable being considered (such as p_t), smearing of the angle of the outgoing parton will lead to $\alpha > 1$. A value of α less than one indicates shadowing of some sort.

Another way to study nuclear rescattering is through its contribution to the unbalance in p_t of the outgoing partons in dijet final states. We can define $\vec{k_t}$ as the vector unbalance in p_t of the dijet system. $\vec{k_t}$ has contributions from several sources, such as primordial Fermi motion, gluon radiation, and hadronization, as well as experimental effects. When the target is a heavy nucleus, nuclear rescattering can make a contribution to $\vec{k_t}$. The variable $k_{t\phi}$ can be used to quantify nuclear rescattering effects. $k_{t\phi}$ measures the ϕ component of $\vec{k_t}$ in the transverse plane, as shown in figure 1.

Two Fermilab experiments, E609 and E683, have measured $k_{t\phi}$ in dijet production under somewhat different conditions. E609 studied jet production in pA interactions at $\sqrt{s}=27$ GeV, and E683 studied jet production in γA and πA interactions over a $\sqrt{s}$ range from 20 to 25 GeV. Both experiments [2], [3] report a substantial nuclear contribution to

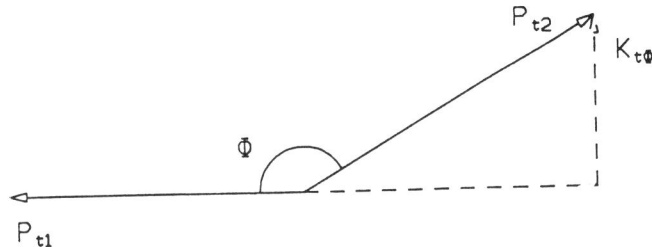

Figure 1: Definition of $k_{t\phi}$.

$k_{t\phi}$ as shown in Figure 2, indicating a smearing of the azimuthal opening angle between the jets ($\Delta\phi$) due to the presence of the nucleus. Neither set of data has been corrected for instrumental contributions to $k_{t\phi}$ due to jet-finding errors or calorimeter energy resolution. The E609 data shows dijet events with average p_t ¿ 4 GeV, while the E683 data has a p_t threshold of 3 GeV. A direct comparison between the two experiments is difficult due to the different jetfinders used, the different acceptances, and the different values of $\sqrt{s}$. Nevertheless it appears that the nuclear enhancement of $k_{t\phi}$ may be somewhat larger for the pA data compared to either γA or πA.

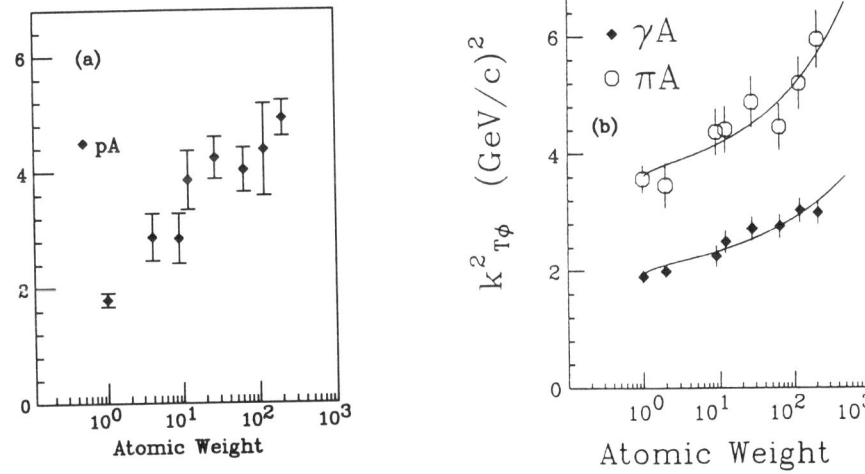

Figure 2: $k_{t\phi}$ for (a) E609 data and (b) E683 data. The fits shown in (b) are of the form $\langle k_{t\phi}^2 \rangle = C_0 + C_1(A-1)^\alpha$. For the γA data $\alpha=0.32\pm0.08$, and for the πA data α is 0.39±0.15. Although no fit is shown, the pA data are also consistent with $\alpha=0.3$.

In addition to an enhancement of $k_{t\phi}$, nuclear effects are also seen in forward energy flow. Both E609 and E683 had forward calorimeters to measure the energy from CM polar angles between 0° and 20-30°. In pA interactions[4], the forward energy is greatly decreased by the presence of the nucleus. Some of this energy appears in the main calorimeter at intermediate angles (CM polar angles from 20° to 120°). But substantial energy is lost to the target region, as seen in figure 3.

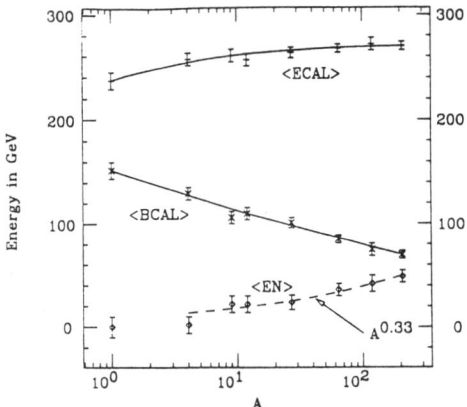

Figure 3: Energy flow for the E609 pA dijet events. Shown are the average energy in the forward direction $\langle\text{BCAL}\rangle$, average energy in the central calorimeter $\langle\text{ECAL}\rangle$, and, by energy conservation, the average energy in the nuclear target region $\langle\text{EN}\rangle$.

The E683 data for both γA and πA interactions show a similar effect. Figure 4 shows E683 dijet events with the beam energy in the range of 200-300 GeV and the average p_t of the two jets greater than 3 GeV. It is interesting to divide the E683 data into bins of x_{beam}, where x_{beam} is the momentum fraction carried by the interacting beam parton. As required by energy conservation, the magnitude of the forward energy

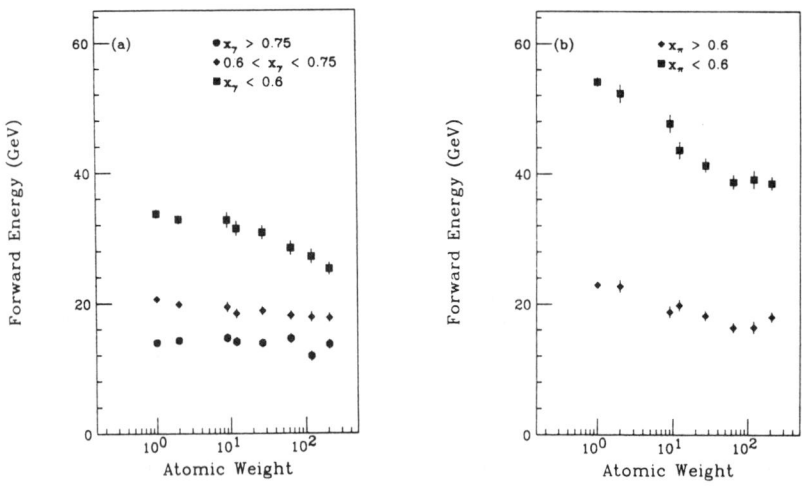

Figure 4: Forward energy flow for E683 (a) γA and (b) πA data. The data are divided into bins of x_{beam}, as indicated.

flow increases with decreasing x_{beam}. However, the fraction of forward energy lost due the the presence of the nucleus is greater for smaller values of x_{beam}, as can be seen in figure 4. In fact, for the largest bin of x_{beam} reached in the γA data, there seems to be no A-dependent change in the forward energy flow at all. The x_{beam} distribution for the photon data is harder than that for the pion data, so larger values of x_{beam} are reached by the γp data compared to the πp data.

Theoretical work in this area is scarce. Sterman and Qiu[5] have attempted to predict A-dependence effects using non-leading power contributions in perturbative QCD. More theoretical guidance in this area would be welcomed.

Results in pA, πA and γA interactions are in contrast to lack of A-dependent effects in Drell-Yan dilepton production, where it has been known for some time that A-dependent effects are small[6]. It is a long-standing puzzle in this field as to why there seems to be substantial multiple scattering in the nucleus in the final state but not the initial state.

REFERENCES

1. J. Cronin et al., Phys. Rev. Lett. **31** 1426 (1973).

2. M. D. Corcoran et al., Physics Letters **259**, 209 (1991).

3. D. Naples et al., Phys. Rev. Lett. **72**, 2341 (1994).

4. R. C. Moore et al., Physics Letters **244**, 347 (1990).

5. M. Luo, J. Qiu, and G. Sterman, Physics Letters, **B279** 377 (1992).

6. D. M. Alde et al, Phys. Rev. Lett., **66**, 2285 (1991); P. Bordalo et al., Physics Letters, **B193**, 373 (1987).

High P_T Production of Neutral Mesons and Direct Photons

George Ginther [1]

University of Rochester, Rochester NY 14627

Abstract. Perturbative QCD (pQCD) calculations of the inclusive production of mesons and direct photons at high p_T have been tested using high-statistics measurements from Fermilab experiment E706. Inclusive π^0 and direct-γ cross sections in the kinematic range $3.5 < p_T < 12$ GeV/c with central rapidities have been measured for 530 and 800 GeV/c proton beams and a 515 GeV/c π^- beam incident on nuclear targets. Current Next-to-Leading Log (NLL) pQCD calculations fail to describe the data for usual choices of scales. Other aspects of our data indicate the presence of a substantial initial state parton transverse momentum (k_T) in the hard scattering. Incorporating a kinematic model of k_T effects improves the agreement between the calculations and the measured cross sections. Gluon distributions are extracted from Deep Inelastic Scattering (DIS), Drell-Yan (DY), and our direct-γ data using fitting procedures similar to those used by the CTEQ collaboration; jet cross section calculations using this gluon distribution function are consistent with D0 and CDF measurements.

E706 is designed to measure large p_T production of direct-γ's, neutral mesons, and associated particles. The apparatus features a large lead and liquid argon electromagnetic calorimeter, and a charged particle spectrometer [1]. The experiment accumulated $\approx$10 events/pb of π^- beam data, $\approx$9 events/pb of proton beam data at 0.5 TeV/c, and $\approx$11 events/pb of proton beam data at 0.8 TeV/c, on Be, Cu, and H targets.

These high statistics precision samples of hard scattering data provide an opportunity to probe various elements of pQCD calculations. The direct-γ process has long been expected to provide an accurate determination of the distributions of gluons in nucleons and mesons, especially for large values of x. Inclusive meson production at large p_T tests the QCD description of a different mix of hard scattering interactions and provides insights into the parton fragmentation mechanism. In general, our data are not described satisfactorily by NLL pQCD calculations for the standard choices of parameters.

Kinematic distributions observed in the data indicate that the interacting partons have significant initial-state transverse momentum. When the impact of this k_T is incorporated in the pQCD calculations, the resulting cross sections are much more compatible with our measurements, and the gluon

---

[1] For the Fermilab E706 Collaboration

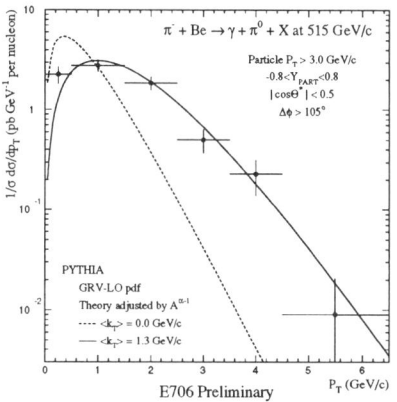 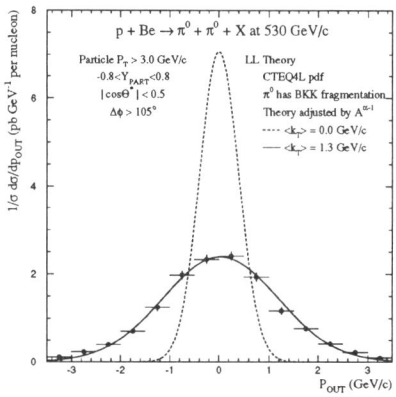

FIGURE 1. Left: The p_T distribution of direct-γ π^0 pairs compared to PYTHIA results for two choices of $\langle k_T \rangle$. Right: The p_{out} distribution for $\pi^0\pi^0$ pairs. The dashed curve represents a LL calculation with $\langle k_T \rangle=0$, while the solid curve is for $\langle k_T \rangle=1.3$ GeV/c.

distribution, extracted using CTEQ-style fits to DIS, DY, and our direct-γ data, is consistent with jet cross sections obtained at CDF and D0.

PARTON TRANSVERSE MOMENTUM

The presence of higher-order QCD contributions can be investigated experimentally through studies of processes sensitive to the transverse motion of the partons prior to the hard scatter. This motion is presumably due to initial-state gluon radiation, and a variety of data indicate that the resultant k_T values are larger than expected from NLL pQCD calculations [2,3].

The kinematic distributions of $\pi^0\pi^0$, $\gamma\pi^0$, and $\gamma\gamma$ pairs (selected via a two arm trigger) provide information on the k_T of the interacting partons. The p_T distribution for high-mass direct-γ π^0 pairs produced in 515 GeV/c π^--nucleon collisions is shown in Fig. 1 compared to PYTHIA calculations with $\langle k_T \rangle$ values of 0 and 1.3 GeV/c. Fragmentation generates significant pair p_T even when the incident partons have no tranverse momentum, but an additional $\langle k_T \rangle \approx 1.3$ GeV/c provides a much better representation of the data.

Another kinematic variable that is sensitive to k_T is the out-of-plane momentum, p_{out} (the component of the momentum of one high-p_T particle perpendicular to the plane defined by the beam and the other member of the pair). Figure 1 shows the p_{out} distribution for high-mass pairs of π^0's produced in proton-nucleon interactions. The solid curve is for a LL calculation assuming that the incident partons have Gaussian transverse momentum distributions [4], with an average of $\langle k_T \rangle=1.3$ GeV/c; the dashed curve is for $k_T=0$. The effective values of $\langle k_T \rangle$ depend only weakly on the choice of nuclear target. Similar results hold for other samples of our data.

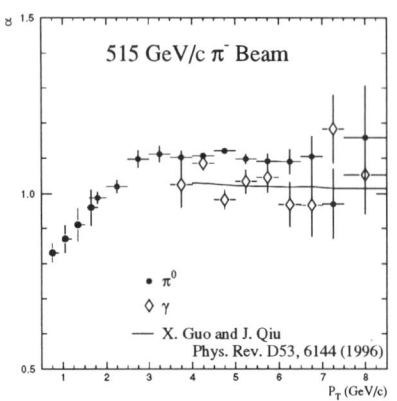 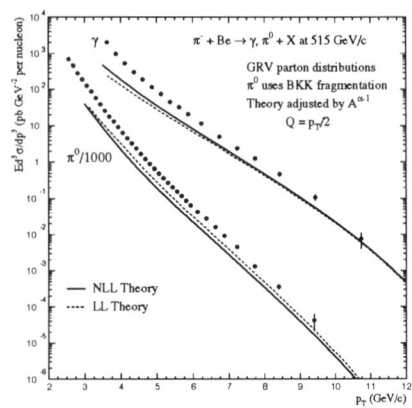

FIGURE 2. Left: α for π^0's and direct-γ's as functions of p_T. Right: π^0 and direct-γ inclusive cross sections as functions of p_T compared to LL and NLL pQCD calculations.

INCLUSIVE RESULTS

The nuclear dependence of the inclusive cross sections can be simply parameterized as A^α. Figure 2 shows α as a function of p_T for π^0 and direct-γ production as determined using the Be and Cu target data. The production of high p_T neutral mesons exhibits a clear nuclear enhancement, in contrast to the direct photons which show significantly smaller nuclear effects. Cross sections obtained using π^- beam incident on Be are shown in Fig. 2 compared to results of LL and NLL calculations. Constant correction factors for nuclear dependence have been applied to the pQCD calculations.

The π^0 and direct-γ cross sections as functions of p_T for 530 GeV/c proton-nucleon collisions are shown in Fig. 3. The results of NLL pQCD calculations [5–7] using CTEQ4M parton distribution functions [8] (and BKK fragmentation functions for the π^0 [9]) are compared with data for several choices of the renormalization and factorization scales (all QCD scales are set equal). The NLL calculations are quite sensitive to the choice of these scales (an indication that the NLL calculations are not fully adequate), but even for rather small scale choices, the NLL calculations fail to accommodate our data.

Since inclusive spectra fall rapidly with increasing p_T, the introduction of k_T has a significant effect on the predicted cross sections. Discrepancies between data from various direct-γ experiments and results from pQCD calculations may be related to such effects [10]. The LL calculations with $\langle k_T \rangle$ larger than 1 GeV/c yield a significantly better description of our data than those with no k_T (not shown). To introduce supplemental k_T smearing into the inclusive NLL pQCD calculations for direct-γ (and π^0) production, we have estimated k_T correction factors (as functions of p_T) via ratios of results from LL pQCD calculations with various $\langle k_T \rangle$ values compared to results without k_T. These k_T factors were applied to the results of NLL pQCD calculations. Since this

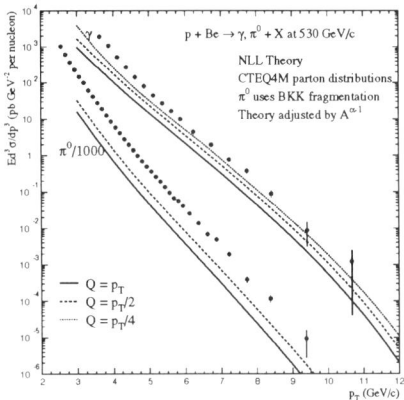
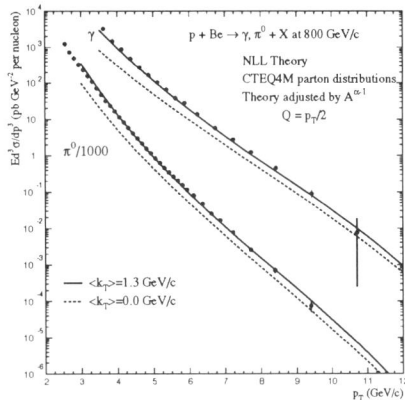

FIGURE 3. Left: π^0 and direct-γ inclusive cross sections as functions of p_T compared to NLL pQCD calculations for several choices of scales (the π^0 pQCD calculation for $Q = p_T/4$ breaks down below 6 GeV/c). Right: Cross sections compared to the results of NLL pQCD calculations without supplemental k_T and with $\langle k_T \rangle$=1.3 GeV/c.

procedure is intended to estimate k_T effects beyond those already in the NLL calculations, we considered $\langle k_T \rangle$ to be a free parameter in these comparisons. As shown in Fig. 3, reasonable representations of both the direct-γ and π^0 data at 800 GeV/c are obtained for $\langle k_T \rangle \approx$ 1.3 GeV/c. Analogous results hold at 530 GeV/c, and for π^--nucleon interactions at 515 GeV/c.

More theoretical work is needed to sort out alternative explanations and to achieve a rigorous QCD treatment (see, e.g., [11]). Until such treatment is achieved, we believe it is reasonable to use the simple implementation of k_T smearing (described above) in studies of the gluon distribution.

GLUON DISTRIBUTION FUNCTION

NLL calculations supplemented with k_T factors were used in CTEQ-style fits to DIS, DY, and our incident proton direct-γ data to evaluate the gluon distribution function [12]. Shape constraints were emphasized; fits with reasonable normalizations relative to our data correspond to $\langle k_T \rangle$'s of 1.15 and 1.3 GeV/c at 530 and 800 GeV/c, respectively. (A difference of $\approx$0.2 GeV/c between $\langle k_T \rangle$'s is consistent with the $\sqrt{s}$ dependence of the $\langle p_T \rangle$ of DY pairs.) The gluon distribution function from this NLL fit is compatible with the CTEQ4M result for $x \lesssim 0.7$. Ratios of the E706 inclusive direct-γ cross sections for incident proton beam to the results of the NLL pQCD calculations, supplemented by the appropriate k_T factors and using this gluon distribution function are shown in Fig. 4. The jet cross sections calculated using this gluon distribution are consistent with CDF and D0 measurements (Fig. 4). (Similar conclusions were obtained using the LL formulation.)

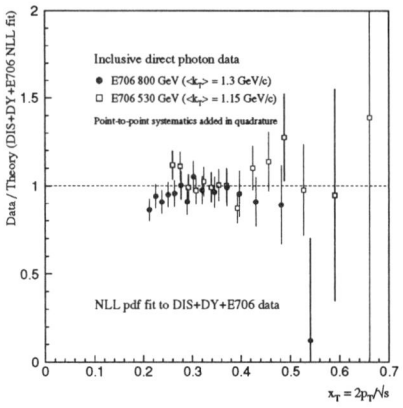
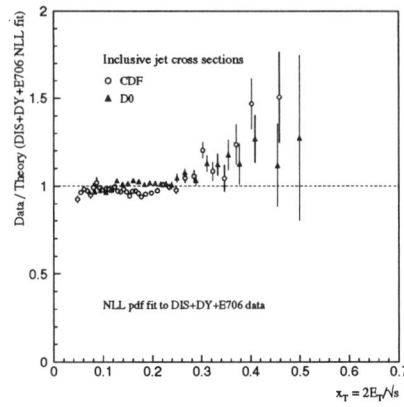

FIGURE 4. Left: Ratios of our direct-γ cross sections as functions of $x_T \equiv 2p_T/\sqrt{s}$ to the NLL calculations using the gluon determined by fitting data from DIS, DY, and the E706 direct-γ data. Right: Ratios of jet cross sections as functions of the scaled E_T measured by D0 and CDF to the NLL calculations using the same gluon distribution.

CONCLUSIONS

Our high statistics direct-γ data are sensitive to the gluon distribution at medium and large x values. An improved theoretical understanding of soft-gluon effects in inclusive direct photon production will facilitate the global determination of the gluon distribution function. Our data can also be used to improve the understanding of pQCD in this kinematic domain.

We thank the CTEQ collaboration for help with the studies of the gluon distribution function, and the DOE, NSF, and UGC of India for support.

REFERENCES

1. Apanasevich, L., et al., FERMILAB-Pub-97/030-E, hep-ex/9702014 (1997).
2. Cox, B., and Malhotra, P. K., *Phys. Rev. D* **29**, 63–66 (1984).
3. Bonvin, E., et al., *Phys. Lett.* B **236**, 523–527 (1990).
4. Owens, J. F., *Rev. Mod. Phys.* **59**, 465–503 (1987).
5. Aurenche, P., et al., *Phys. Lett.* **140B**, 87–92 (1984).
6. Berger, E. L., and Qiu, J., *Phys. Rev. D* **44**, 2002–2024 (1991).
7. Aversa, F., et al., *Nucl. Phys.* **B327**, 105–143 (1989).
8. Lai, H. L., et al., *Phys. Rev. D* **55**, 1280–1296 (1997).
9. Binnewies, J., et al., *Phys. Rev. D* **52**, 4947–4960 (1995).
10. Huston, J., et al., *Phys. Rev. D* **51**, 6139–6145 (1995).
11. Sterman, G., *Proceedings of ICHEP '96*, Warsaw, Poland, July 1996 (World Scientific, Singapore, 1997, Vol. I, pp.103–120) and private communication.
12. The DIS and DY data samples which contributed to the CTEQ4 fits [8] were also used in the fits described in this paper.

Quasi-Elastic Hadronic Scattering at High Momentum Transfer

Yael Mardor[1]

School of Physics and Astronomy, Raymond and Beverly Sackler Faculty of Exact Sciences, Tel Aviv University, Tel Aviv 69978, Israel.

Abstract.
We measured the high-momentum quasi-elastic $^{12}C(p,2p)$ reaction (at $\theta_{cm} \simeq 90°$) for 6 and 7.5 GeV/c incident protons. The validity of the QE picture and the impulse approximation was verified up to large Fermi momenta were it might be questionable. Transverse and longitudinal Fermi momentum distributions of the target proton were measured and compared to independent particle models which do not reproduce the large momentum tails. We also observed that the transverse Fermi distribution gets wider for smaller longitudinal components, in contrast to expectations from a simple Fermi gas model.

Hard quasi-elastic scattering can be used [1] to explore the momentum distribution of nucleons in nuclei as well as the nuclear transparency of hadrons. In both cases the interpretation of the measurements can be extracted from quasi-elastic scattering only by using the impulse approximation. The practical meaning of this approximation is that the quasi-elastic cross section can be factorized in the following way:

$$\left. \frac{d^4\sigma}{dt d\vec{P}_f}(s,t) \right|_{q.e.} = \left. \frac{d\sigma}{dt}(s,t) \right|_{free} \cdot n(\vec{P}_f) \cdot T(s,t) \tag{1}$$

where s and t are the Mandelstam variables. We define E_f and $\vec{P}_f$ to be the target nucleon Fermi energy and momentum, where $\vec{P}_f = (\vec{P}_{ft}, P_{fz}) = (P_{fx}, P_{fy}, P_{fz})$, z is the beam direction and y is perpendicular to the scattering

---

[1] Representing the E850 collaboration: J. Aclander[1], J. Alster[1], D. Barton[2], G. Bunce[2], A. Carroll[2], N. Christensen[3], H. Courant[4], S. Durrant[2], S. Gushue[2], S. Heppelmann[5], E. Kosonovsky[1], Y. Mardor[1], M. Marshak[4], Y. Makdisi[2], E. Minor[5], I. Navon[1], H. Nicholson[6], E. Piasetzky[1] T. Roser[2], J. Russell[7], S. Sutton[6], M. Tanaka[2] (deceased), J-Y Wu[5]. [1] *Tel Aviv University,* [2] *Brookhaven National Laboratory,* [3] *University of Auckland,* [4] *University of Minnesota,* [5] *Pennsylvania State University,* [6] *Mount Holyoke College,* [7] *University of Massachusetts Dartmouth.*

plane. The momentum distribution of the nuclear g.s. is $n(\vec{P_f})$. For scattering at $\theta_{cm} \simeq 90°$ the free cross section is almost constant as a function of t and the nuclear transparency, T, depends very weakly on t. Therefore, the shape of the quasi-elastic $d^2\sigma/d\vec{P}_{ft}$ for fixed s should depend only on n.

We measured quasi-elastic $^{12}C(p,2p)$ scattering at the AGS using the EVA detector (exp E850) [1], at incident beam momenta of 5.9 GeV/c and 7.5 GeV/c and large momentum transfers (near $\theta_{cm} = 90°$).

The data were analyzed in terms of P_{fy} and the light cone variable $\alpha = \frac{E_f - P_{fz}}{m}$. To a good approximation ($E_0 = P_0$ where E_0 and P_0 are the energy and momentum of the incident beam) s is proportional to α: $s \sim 2m^2 + 2mP_0\alpha$.

We test the validity of the factorization, by comparing the transverse Fermi distributions of two data sets, with different beam momenta P_0 (5.9 GeV/c and 7.5 GeV/c) for a few fixed values of α (or s). They must have the same shape if they depend only on nuclear properties.

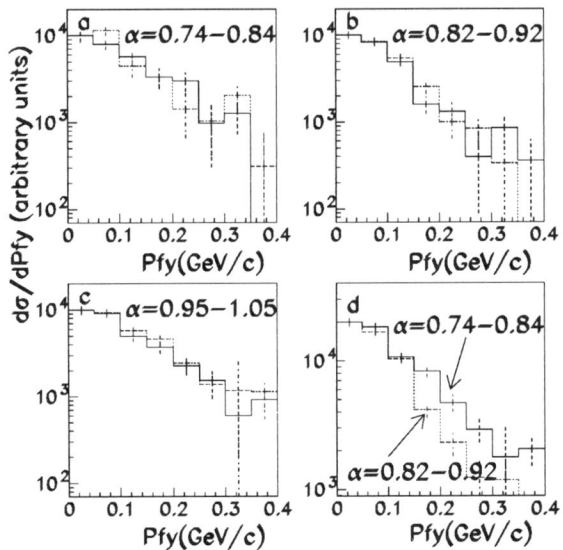

FIGURE 1. a-c: $|P_{fy}|$ distributions for different α slices. The 6 GeV/c (solid line) and the 7.5GeV/c (dashed line) distributions were normalized at $P_{fy} = 0$. d: $|P_{fy}|$ distributions for $\alpha = 0.79$ (solid line) compared to that for $\alpha = 0.87$ (dashed line).

Figs 1(a)- 1(c) show the distributions at 6 and 7.5 GeV/c for different slices of α. They were normalized at $P_{fy} = 0$ and one can see that, within experimental errors, the shapes are independent of the incident energy. This is what one expects from the impulse approximation.

In Fig. 1(d) we compare the transverse Fermi distributions for two different α ranges. It can be seen clearly that the transverse Fermi distribution gets wider for smaller α. This is in contradiction to what one expects from a simple Fermi gas model. We do not have a clear answer as to the source of this observed phenomenon.

The variable s depends strongly on P_{fz}: $s = (P_0 + P_f)^2 \simeq 2E_0 E_f + 2m^2 - 2P_0 P_{fz}$. Since the cross section varies as $1/s^{10}$, the cross section for high P_{fz} (low α) nucleons is strongly enhanced [2], providing a high statistics measurement for large Fermi momenta in the beam direction.

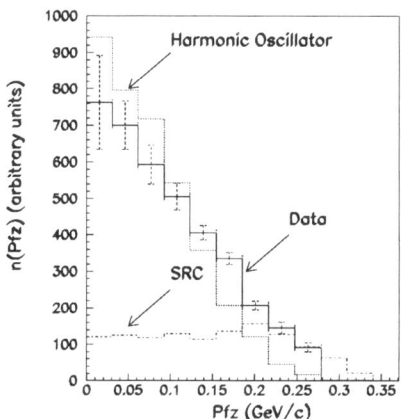

FIGURE 2. The P_{fz} distribution (solid line). The dotted line is a HO model calculation normalized to the same number of events as the data. The dashed dotted curve represent a model for SRC contribution with arbitrary normalization.

Figure 2 shows the longitudinal Fermi momentum distribution. The histogram consists of the summed events for both incident momenta, corrected with the free cross section ($\simeq 1/s^{10}$). Minor corrections still need to be applied. The distribution is cut by acceptance at $P_{fz} = 0.28 GeV/c$. The data are compared to predictions of a harmonic oscillator model [3] and to a model for SRC [4].

This work was supported by the Department of Energy, the National Science Foundation, the Israel-US Binational Science Foundation and the Israel Academy of Sciences.

REFERENCES

1. A.S. Carroll et al., Phys. Rev. Lett. 61, 1698 (1988); S. Heppelmann et al., Phys. Let. B232, 167 (1989); Exp. E850 Proposal to BNL, 1988 (unpublished).
2. G.R. Farrar et al., Phys. Rev. Lett. 62, 1095 (1989)
3. M. Sargsyan, Private Communication.
4. L.L. Frankfurt and M. Strikman, Phys. Rep. 76, 215 (1981) ; Phys. Rep. 160, 235 (1988).

High p_t quasi-exclusive scattering with resonance production

Israel Mardor[1]

School of Physics and Astronomy, Raymond and Beverly Sackler Faculty of Exact Sciences, Tel Aviv University, Tel Aviv 69978, Israel.

Abstract.
We measured exclusive resonance production at large p_t in nuclei. We bombarded C and CD_2 targets with 6 and 7.5 GeV/c protons and detected three charged particle events, two of which with high p_t. We present missing energy distributions which show a clear quasi-exclusive peak (a reaction on a neutron in the nucleus) and invariant mass distributions for those quasi-exclusive events. Results for Carbon and Deuterium targets are compared and the relation to Color Transparency is discussed.

We discuss an on-going analysis of high p_t resonances, produced exclusively inside nuclei. We measured the reaction $pA \to pp\pi^-(A-1)$ at the AGS using the detector EVA [1] (exp E850). The momenta of the two protons and the π^- are measured and $A - 1$ is a spectator. The targets were Carbon and Deuterium. The kinematic region is $p_{beam} = 5.9, 7.5\ GeV/c$, $\theta_{CM} \approx 90°$, $Q^2 \approx 4.8, 6.2\ (GeV/c)^2$. This reaction is interesting because the π^- with one of the protons may be the decay products of a high p_t baryonic resonance, produced inside the nucleus.

The goals of the project are: 1. The measurement of the nuclear transparency to resonances, especially its incoming energy dependence. 2. The measurement of the ratio of resonance transparency to $(p, 2p)$ transparency. If this ratio is about unity, it means that two hadrons are emerging from the nucleus which in turn suggests that mostly resonances are produced. A very low ratio may be the result of three particles produced directly at the main

---

[1] Representing the E850 collaboration: J. Aclander[1], J. Alster[1], D. Barton[2], G. Bunce[2], A. Carroll[2], N. Christensen[3], H. Courant[4], S. Durrant[2], S. Gushue[2], S. Heppelmann[5], E. Kosonovsky[1], Y. Mardor[1], M. Marshak[4], Y. Makdisi[2], E. Minor[5], I. Navon[1], H. Nicholson[6], E. Piasetzky[1] T. Roser[2], J. Russell[7], S. Sutton[6], M. Tanaka[2] (deceased), J-Y Wu[5]. [1]*Tel Aviv University,* [2]*Brookhaven National Laboratory,* [3]*University of Auckland,* [4]*University of Minnesota,* [5]*Pennsylvania State University,* [6]*Mount Holyoke College,* [7]*University of Massachusetts Dartmouth.*

vertex. 3. The measurement of the ratio of quasi-exclusive resonance production to quasi-elastic proton scattering. This ratio may shed light on models which treat point like configurations as a superposition of resonances [2], and thus give another handle on understanding Color Transparency [3].

At this stage of we are able to show that the extraction of the above ratios is feasible with our apparatus. The main difficulty is the identification of the quasi-exclusive signal of the basic reaction $pn \to px^0 \to pp\pi^-$. The important variable is the missing energy, $E_{miss} = (E_{beam} + M_n) - (E_{p_1} + E_{p_2} + E_{\pi^-})$. For quasi-exclusive events it is zero, up to the nuclear binding energy. Figure 1 shows missing energy plots for the C and CD_2 targets at 5.9 GeV/c incoming momentum.

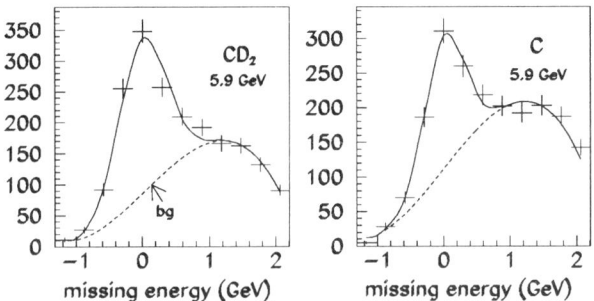

FIGURE 1. Missing energy plots. The solid line is a fit to a Gaussian plus a polynomial (dashed line).

The events in these plots passed the following kinematical cuts: $|P_{fy}|, |P_{fx}| < 0.3 \ GeV/c$ and $0.8 < |\alpha| < 1.1$, where $\alpha = \frac{E_f - P_{fz}}{M_n}$. E_f is the Fermi energy, P_{fy}, P_{fx} are the transverse Fermi momentum components and P_{fz} is the longitudinal Fermi momentum.

The plots clearly show a peak at the right place. We assume that the missing energy plots are composed of the quasi-exclusive signal, which is modeled as a Gaussian around zero with a width comparable with our experimental resolution ($\sigma \approx 290 MeV/c$), while the background is modeled by a third order polynomial. The background parameters are determined by events that have an extra charged particle in them and are thus known to be inclusive.

We now turn to the invariant mass spectrum for the quasi-exclusive resonances. Unfortunately, it is impossible in this kinematical region to determine which of the two protons is the decay product and which is the recoil particle, so we use both choices. This means that the plots in Figure 2 have two entries for each event, one is correct and one is wrong. The result is the true signal on top of a significant background. The plots for D are the result of subtracting the C signal from the CD_2 signal.

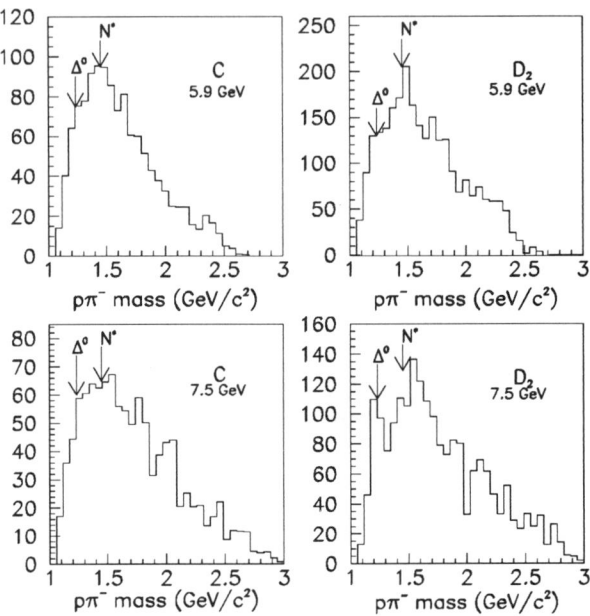

FIGURE 2. Invariant mass plots. Each plot includes 2 entries per event (see text).

The plots in Figure 2 show interesting structure and we can draw from them two conclusions: 1. The resonances are more pronounced in D than in C. This is surprising since in this kinematic region the resonances are supposed to decay either at the edge or outside the C nucleus. 2. The production ratio of Δ^0 to N^* seems to increase with incoming momentum and with Q^2 for both C and D. Given the different quantum numbers of the Δ^0 and N^*, which may naively suggest different geometrical sizes, their production ratio may provide new input to the ongoing investigation of Color Transparency.

This work was supported by the Department of Energy, the National Science Foundation, the Israel-US Binational Science Foundation and the Israel Academy of Sciences.

REFERENCES

1. Experiment E850 Proposal to Brookhaven National Laboratory, 1988 (unpublished).
2. B.K. Jennings and G.A. Miller, Phys. Rev. **D44**, 692 (1991).
3. P. Jain *et. al.*, Phys. Rep. **271**, 67 (1996) and references therein.

A Preliminary Measurement of the $\bar{u}/\bar{d}$ Asymmetry in the Proton Sea

P. E. Reimer[f], T. C. Awes[i], M. E. Beddo[h], C. N. Brown[c], J. D. Bush[a],
T. A. Carey[f], T. H. Chang[h], W. E. Cooper[c], C. A. Gagliardi[j],
G. T. Garvey[f], D. F. Geesaman[b], E. A. Hawker[j], X. C. He[d],
L. D. Isenhower[a], S. B. Kaufman[b], D. M. Kaplan[e], P. N. Kirk[g],
D. D. Koetke[k], G. Kyle[h], D. M. Lee[f], W. Lee[d], M. J. Leitch[f],
N. Makins[b], P. L. McGaughey[f], J. M. Moss[f], P. M. Nord[k], B. K. Park[f],
V. Papavassiliou[h], J. C. Peng[f], G. Petitt[d], M. E. Sadler[a], J. Selden[h],
P. W. Stankus[i], W. E. Sondheim[f], T. N. Thompson[f], R. S. Towell[a],
R. E. Tribble[j], M. A. Vasiliev[j], Y. C. Wang[g], Z. F. Wang[g], J. C. Webb[h],
J. L. Willis[a], D. Wise[a], G. R. Young[i], B. Zeidman[b]

[a] *Abilene Christian University, ACU Station, Box 7963, Abilene, TX 79699*
[b] *Argonne National Laboratory, 9700 S Cass Ave., Argonne, IL 60439*
[c] *Fermi National Accelerator Laboratory, P.O. Box 500, Batavia, IL 60510*
[d] *Georgia State University, Atlanta, GA 30303*
[e] *Illinois Institute of Technology, Physics Department, Chicago, IL 60616*
[f] *Los Alamos National Laboratory, P-25, MS-H846, Los Alamos, NM 87545*
[g] *Louisiana State University, Baton Rouge, LA 70803*
[h] *New Mexico State University, Las Cruces, NM, 88003*
[i] *Oak Ridge National Laboratory, Physics Division, P.O. Box 2008, Oak Ridge, TN 37831*
[j] *Texas A & M University, Cyclotron Institute, College Station, TX 77843*
[k] *Valparaiso University, Niels Sci. Ctr., Valparaiso, IN 46383*

Abstract. The NuSea (E866) experiment at Fermilab has been using Drell-Yan scattering to study the $\bar{u}/\bar{d}$ quark content of the proton by comparing of the yield between liquid hydrogen and liquid deuterium targets. A preliminary ratio of Drell-Yan yields, $\sigma^{pd}/2\sigma^{pp}$ as a function of x is shown. This data confirms the previous indications from the NMC and NA51 experiments that $\bar{u} \neq \bar{d}$.

Until recently it was assumed that the up-down quark content in the nucleon sea was flavor symmetric. The first evidence that this may not be the case was found using the Gottfried Sum Rule [1]. The Gottfried Sum is given by

$$I_{GS} = \int_0^1 [F_2^p(x) - F_2^n(x)] \frac{dx}{x}.$$

Here, $F_2^p(x)$ and $F_2^n(x)$ represent the proton and neutron structure functions and x represents the fraction of the proton's momentum carried by the struck parton. This expression can be expanded by substituting the parton distribution functions for the structure functions, $F_2(x) = \sum_i e_i^2 x \left(q_i(x) + \bar{q}_i(x) \right)$, where the sum is over all the partons in the proton or neutron. With this substitution and the assumption of charge symmetry ($u_p(x) = d_n(x)$, $\bar{u}_n(x) = \bar{d}_p(x)$, etc.), the Gottfried Sum simplifies to

$$I_{GS} = \frac{1}{3} - \frac{2}{3} \int_0^1 \left[\bar{d}_p(x) - \bar{u}_p(x) \right] dx.$$

If the proton sea is flavor symmetric, the integral vanishes, and the Gottfried Sum reduces to 1/3.

Using deep inelastic scattering on hydrogen and deuterium targets, the NMC collaboration [2] was able to determine the F_2^p and F_2^n structure functions in the range of $0.004 \leq x \leq 0.8$. After extrapolating through the unmeasured regions, NMC found that

$$I_{GS} = 0.235 \pm 0.026 \neq \frac{1}{3}.$$

Several explanations have been proposed for the difference between the NMC result and 1/3. The extrapolation over the unmeasured region in x is subject to any anomalous behavior at small x. FNAL E665 extended NMC's measurements to smaller x and uncovered evidence for nuclear shadowing [3]. Such an effect would produce an even smaller value for I_{GS}. A violation of charge symmetry in the nucleon could also explain these results, but a violation at the level necessary to account for the difference has not been seen elsewhere. This leaves an up-down flavor asymmetry in the nucleon sea as the most likely explanation.

Since deep inelastic scattering measures the quark weighted sum of *all* of the quark distributions, it was suggested by Ellis and Stirling [4] that the Drell-Yan process be used to measure the up-down asymmetry of the sea. In the Drell-Yan process [5], a quark from the beam (or target) annihilates with an antiquark from the target (or beam) to produce a virtual photon, which decays into a lepton pair. By comparing the production rates of the lepton pairs from various nuclear targets, the antiquark sea is probed. This approach has been by the E772 collaboration at Fermilab [6] and by the NA51 collaboration [7]. E772 reanalyzed data which had been collected to study the nuclear dependence of the Drell-Yan process. Their results were consistent with a symmetric sea. NA51 measured Drell-Yan yields from protons incident on hydrogen and deuterium targets. The NA51 spectrometer had a limited range of kinematic acceptance, and so was only able to measure the $\bar{u}/\bar{d}$ ratio near $x = 0.18$ and $x_f = 0$. NA51 observed approximately 6000 events and found

$$\left. \frac{\bar{u}}{\bar{d}} \right|_{x=0.18} = 0.51 \pm 0.04 \pm 0.05.$$

This showed that the nucleon sea was not flavor symmetric. However, it did not determine the x dependence of this ratio.

The Fermilab E866 experiment used an 800 GeV extracted proton beam to measure Drell-Yan muon pair production from 50 cm long flasks of liquid hydrogen and liquid deuterium. An empty flask was also used for background determination. The experimental apparatus was the forward x_F spectrometer [8] shown in Fig. 1. The acceptance of the spectrometer was defined by the first two magnets (SM0 and SM12). Within the second magnet, the beam was intercepted by a copper beam dump. A large absorber wall, also located within the second magnet, removed scattered hadrons and allowed only muons to pass through the detection system. The muons passed through four tracking stations and a momentum analyzing magnet. The trigger was formed by a pair of triple hodoscope coincidences matching the topology of a muon pair produced at the target.

Data were collected over a six month period ending in March of 1997. More than 1.3×10^{17} protons were incident on the targets, and over 3.5×10^5 Drell-Yan events were recorded, as well as more than 2×10^4 Υ and 10^6 J/ψ events. Three spectrometer magnet settings were used in order to shift the peak acceptance of the spectrometer to low, intermediate and high mass muon pairs.

A preliminary analysis has been completed for each of the three data sets. The results presented here are based on only the high mass data set. This data set was chosen because it does not suffer from several systematic effects which are present in the other two data sets. The combined mass spectrum, as well as the mass spectrum for only the high mass data set is shown in Fig. 2. Also shown in Fig. 2 is the kinematic coverage of our data in terms of x_2, the fractional momentum carried by the target parton.

Based on only the high mass data set, a preliminary ratio of Drell-Yan yields from hydrogen and deuterium as a function of x_2 is shown in Fig. 3, along

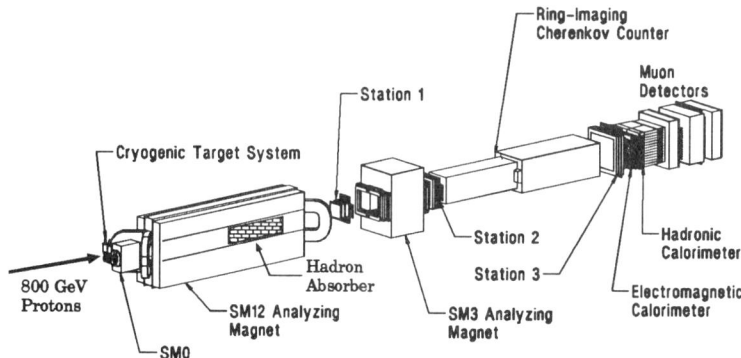

FIGURE 1. The Meson East spectrometer at Fermilab used by the E866 experiment. The detector was approximately 60 m long. The RICH and Calorimeters were not instrumented.

with the predictions from several common parameterizations of the proton, which have been weighted by the detector's acceptance. The "CTEQ4M ($\bar{u} = \bar{d}$)" curve needs special comment. For this curve, the CTEQ4M distribution functions [9] were modified to force $\bar{u} = \bar{d} = (\bar{u} + \bar{d})/2$ to simulate a totally flavor symmetric sea. Inspection of this plot reveals that for values of x_2 below approximately 0.2, the parameterizations which have $\bar{u} \neq \bar{d}$ match or are slightly above the data, but reflect the general trend of the data. For x_2 above 0.2, the data clearly tend toward a more flavor symmetric sea. Unfortunately, the statistical uncertainties are large in this region, but should improve after the analysis of the complete data set. While interesting, this region is perhaps better studied with a higher intensity, lower energy proton beam, such as the one soon to be available from the new Main Ring Injector nearing completion at Fermilab.

The extraction of $\bar{d}(x)/\bar{u}(x)$ from the ratio of cross sections is not a simple exercise. Simplifications may be made for $x_1 \gg x_2$, in which case

$$\left. \frac{\sigma^{pd}}{2\sigma^{pp}} \right|_{x_1 \gg x_2} \approx \frac{1}{2}\left[1 + \frac{\bar{d}(x)}{\bar{u}(x)}\right].$$

This simplification is not applicable to all of our data, however. Thus, the full parton distribution functions, which themselves are established based on the large body of existing data, must be used. At present, a procedure, based

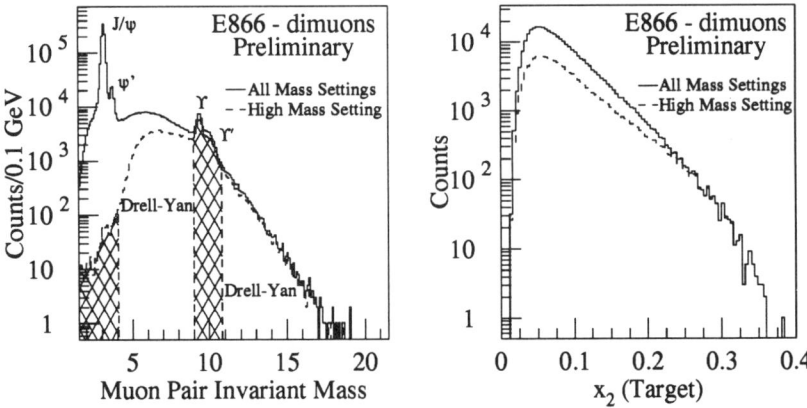

FIGURE 2. Reconstructed mass spectrum for muon pairs is shown on the left. The solid line represents the sum of all three spectrometer settings. The dashed line shows only the data obtained with the high mass settings. The cross hatched region was excluded from the analysis because of the ψ and Υ resonances. On the right is shown the distribution in x_2 of the data. Again, the solid line represents the sum of all data sets and the dashed line is only the high mass data set. The events in the hashed region of the mass plot have been excluded in producing the x_2 plot.

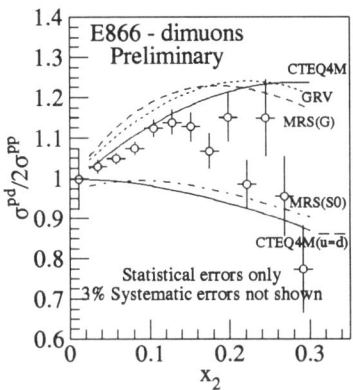

FIGURE 3. The Drell-Yan cross section ratio for $\sigma_{p+d}/2\sigma_{p+p}$ as a function of x_2. Also shown is this ratio for several common parameterizations of the proton. For the CTEQ4M ($\bar{u} = \bar{d}$) curve, we have set $\bar{u} = \bar{d} = \left[(\bar{u} + \bar{d})/2\right]_{CTEQ4M}$.

on the CTEQ4M parameterization, which modifies the $\bar{d}(x)/\bar{u}(x)$ ratio while requiring the sum $\bar{d}(x) + \bar{u}(x)$ to remain constant, is being developed. Using this procedure, work is currently underway to extract the $\bar{d}(x)/\bar{u}(x)$ ratio.

In conclusion, the preliminary results from FNAL E866 are consistent with a violation of the Gottfried Sum rule as first observed by NMC, and confirm the NA51 result that $\bar{u} \neq \bar{d}$. For $x < 0.2$ the data are consistent with common parameterizations of the proton. At larger x, the data tend toward a more flavor symmetric sea. After full analysis of all of the data collected, we expect to achieve our goal of 1% systematic error on $\sigma^{pd}/2\sigma^{pp}$ for $0.03 < x_2 < 0.15$ and will extract the $\int \left(\bar{d}(x) - \bar{u}(x)\right) dx$ portion of the Gottfried Sum over the measured x_2 region. This work was supported in part by the U. S. Department of Energy and the National Science Foundation.

REFERENCES

1. K. Gottfried, Phys Rev. Lett. **18**, 1174 (1967).
2. M. Arneodo *et al.*, Phys. Rev. **D50**, R1 (1994).
3. M. R. Adams *et al.*, Phys. Rev. Lett. **74**, 1525 (1995).
4. S. D. Ellis and W. J. Sterling, Phys. Lett. **B256** 258 (1991).
5. S. D. Drell and T. M. Yan, Ann. Phys. **66**, 578 (1971).
6. P. L. McGaughey *et al.*, Phys. Rev. Lett. **69**, 1726 (1992).
7. A. Baldit *et al.*, Phys. Lett. **B332**, 224, (1994).
8. G. Moreno *et al.*, Phys. Rev. **D43** 2815 (1991).
9. H. L. Lai *et al.*, Phys. Rev. **D55**, 1280 (1997).

Hadroproduction of Charm in FNAL E769 and E791 [1]

Jean Slaughter

Physics Department
Yale University
New Haven, Connecticut 06520-8121

Abstract. I present some recent results from Fermilab charm hadroproduction experiments E769 and E791 that illustrate elements of the QCD parton model of charm hadroproduction. The results from E769 on the beam flavor dependence of the x_F and p_t^2 distributions for D's produced by incident K, p and π beams are used to demonstrate the effect of the parton distribution functions in the incident beam particle. E791 data on charm pairs are used to show features of the underlying $gg \to c\bar{c}$ interaction.

INTRODUCTION AND THEORETICAL FRAMEWORK

Our conceptual framework for charm production assumes that the process can be factorized into perturbative and non-perturbative elements. There are three components to this model: (1) Parton distribution functions describing the fraction of the incident momentum carried by the gluons(g) of the incoming beam and target. An intrinsic transverse momentum, k_t, is also usually given to the interacting partons. (2) A gg interaction, calculated [1] in perturbative QCD in next-to-leading order (NLO). (The $q-\bar{q}$ contribution can be neglected at fixed target energies.) (3) Hadronization, the process whereby the c quark fragments into a hadron and subsequently decays.

This conceptual framework has been implemented in several models that make predictions which can be compared to data. Two implementations of these models, (but not the only ones), frequently used in experimental papers are HVQMNR [2] and PYTHIA [3] [4]. HVQMNR is a Fortran program that implements the full NLO calculation for heavy quark pair production of Mangano, Nason, and Ridolfi(MNR). PYTHIA is a Monte Carlo implementation of the model which uses the LUND string model of hadronization. The

---

[1] Presented at the Intersections Conference, Bigsky, Montana, May 27-June 2, 1997

basic hard scatter is done in LO QCD. Parton showers are used to include higher order effects. The experimental quantities that are typically compared to the model predictions are: (1) The total cross section as a function of incident beam energy, beam species(i.e., π, K, p), and target material (atomic number A). (2) The relative inclusive cross sections for the different species (e.g., the ratio of D^0 to D^+ production), again as a function of beam type and target material. (3) The single particle inclusive distributions in terms of Feynman-x ($x_F \equiv p_z/p_{z_{max}}$) and transverse momentum squared (p_t^2), and correlations between x_F and p_t^2. (4) Production asymmetries, i.e. the ratios of x_F and p_t^2 distributions for two species such as D^- and D^+, either integrated over x_F or p_t^2 or as a function of x_F or p_t^2. (5) Correlations between a charm particle and a π with approximately the same velocity as the heavy particle. (6) Correlations between one charm particle and the other charm particle in the event. The usual correlation variables are the angle $\triangle\phi$ between the two particles in the plane transverse to the beam direction, the mass and P_t^2 of the pair, the difference in rapidity of the two particles and the sum of the rapidities of the two particles. Also, the combinations of species can be measured, i.e. how often the produced pair consists of a $\Lambda_c D$, a $D^o \overline{D^o}$, etc.

DESCRIPTION OF FNAL E769 AND E791 EXPERIMENTS

Both E769 and E791 used the same basic TPL spectrometer [5] [6] at Fermilab. It is an open geometry, two magnet spectrometer with silicon microstrip detectors, 35 planes of drift chambers, electromagnetic and hadronic calorimeters, and a scintillation counter muon wall. Charm particles are identified by reconstructing the production and the decay vertices using the precision tracking information provided by the silicon microstrip detectors in the beam and downstream of the target. E791 upgraded the spectrometer by adding more planes of silicon microstrip detectors and greatly increasing the capacity of the data acquisition system. The main features of the experiments are listed in Table 1.

TABLE 1. Characteristics of FNAL experiments E769 and E791.

	E769	E791
Focus	Production	Decay and Production
Beams	$\pi^\pm, K^\pm, p$	π^-
Targets	Be, Al, Cu, W	Pt, C
P of beam	250 GeV/c	500 GeV/c
Triggers	400 M	20,000 M
Charm	4 K	200 K

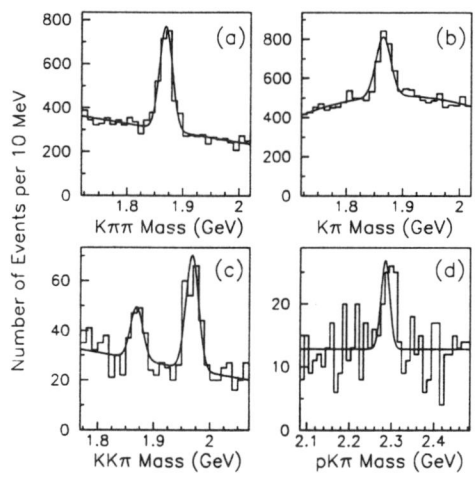

FIGURE 1. From Reference [7]. Example of combined E769 Signals, π^+, π^-, K^+, K^-, p incident, all targets. a) $D^\pm \to K^\mp \pi^\pm \pi^\pm$, b) $D^o \to K^\mp \pi^p m$ (and C.C.), c) $D^\pm$, $D_s^\pm \to K^\mp K^\pm \pi^\pm$ ($\phi\pi$ and K^*K modes), d) $\Lambda_c^\pm \to p^\pm K^\mp \pi^\pm$.

E769 - BEAM FLAVOR DEPENDENCE OF X_F AND P_T^2 DISTRIBUTIONS

E769 is unique among charm hadroproduction experiments because it has a variety of targets and incident beam species. Figure 1 shows a number of their integrated data samples. Published results include the following list of topics: (1) total cross sections [7], $x_F > 0.0$, for $D^\pm$, $D^0/\overline{D^0}$, D_s, $D^{*\pm}$ with π^+, π^-, K^+, K^-, p incident, (2) differential cross sections [8] [9] [10] $\frac{d\sigma}{dx_F}$ and $\frac{d\sigma}{dp_t^2}$ for $D^\pm$, $D^0/\overline{D^0}$, D_s, $D^{*\pm}$ with $\pi^\pm$, $K^\pm$, p incident, (3) production asymmetries [11] [8], for $D^\pm$, $D^0/\overline{D^0}$, D_s, $D^{*\pm}$, Λ_c with $\pi^\pm$, $K^\pm$, p incident, (4) cross section vs. target A [8] [12], for $D^\pm/D^0/\overline{D^0}$, $D^{*\pm}$ with $\pi^\pm$ incident.

One of the most interesting results from E769 is the beam flavor dependence of the $\frac{d\sigma}{dx_F}$ and $\frac{d\sigma}{dp_t^2}$ distributions, shown in Figure 2. Since the targets and the underlying $gg \to c\bar{c}$ process are the same, the differences in the distributions reflect the differences in the g distributions in the projectiles. We see that the π and K distributions are the same but differ from the p distribution.

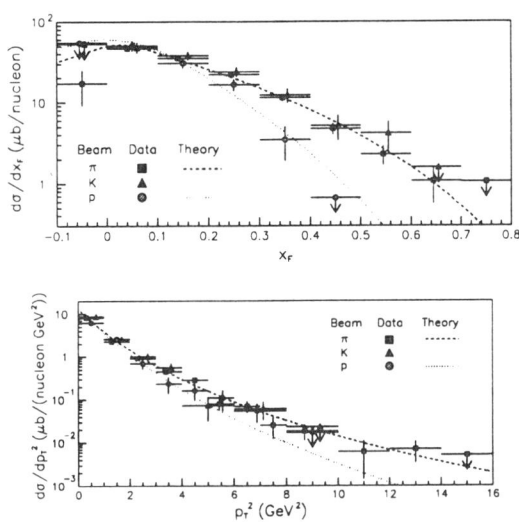

FIGURE 2. From Reference [8]. Top, $\frac{d\sigma}{dx_F}$ and bottom, $\frac{d\sigma}{dp_t^2}$, both for $\pi^\pm$, $K^\pm$ and p beam particles.

E791 - CHARM PAIR CORRELATIONS

The strength of E791 is in its very large samples of reconstructed charm particles. As an example, Figure 3 shows the mass distribution for $D^\pm$, $D_s^\pm \to \phi\pi^\pm$. Besides a number of papers on D decay topics, E791 has reported results on $D^\pm$ production asymmetries [6] as a function of x_F and p_t^2, on $D-\pi$ production correlations [13], and on Σ_c^0 and Σ_c^{++} production [14].

E791 has the largest sample, ≈ 800, of fully reconstructed charm pairs [15] to date. Figure 4 gives the $\triangle\phi$ distribution. In LO the two charm particles are produced back to back: $\triangle\phi = 180°$ and P_t, the vector sum of the individual p_t's of the pair, should be 0. NLO effects and intrinsic k_t smear out the $\triangle\phi$ distribution but hadronization effects should be negligible. The data are compared to three models: the MNR prediction for quarks, the PYTHIA prediction for quarks, and the PYTHIA prediction for mesons. In all cases the default parameters, as suggested by the authors of the models, are used. The theoretical and experimental distributions are normalized to unit area. The $\triangle\phi$ distribution is less peaked than the predictions. Figure 5 gives $\triangle\phi$ for three ranges in P_t^2, the scaler sum of the p_t's of the D and the $\overline{D}$. Figure 5 shows that, as expected and as already shown with less statistics by E653, the two particles become more back to back as the scaler P_t^2 increases.

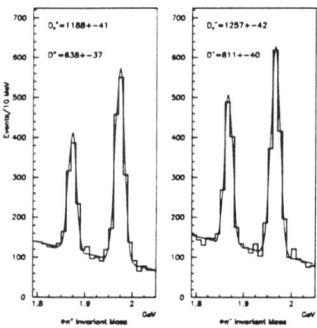

FIGURE 3. From E791. $D^{\pm}$, $D_s^{\pm} \to \phi\pi^{\pm}$.

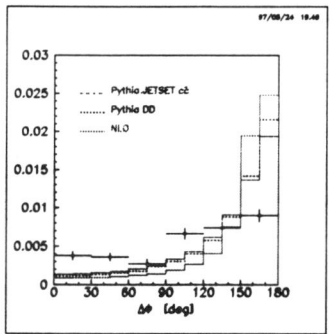

FIGURE 4. Preliminary $\Delta\phi$ distribution

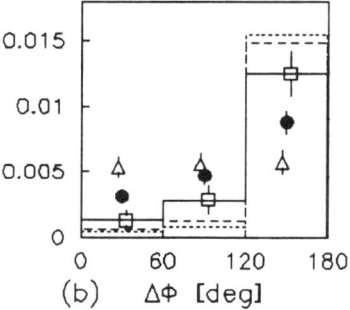

FIGURE 5. From E791. Preliminary distribution for $\Delta\phi$ for three different bins in the scaler sum of the p_t^2's of the two D's. The histograms are from HVQMNR. The three bins are: $0.0 - 1.5$ GeV^2/c^2, (solid, $\triangle$), $1.5 - 4.0$ GeV^2/c^2, (dashed, •) and $4.0 - 12.0$ GeV^2/c^2, (dotted, □).

CONCLUSIONS AND SUMMARY

Despite the progress in recent years, there is still not a complete set of precise measurements in either hadroproduction or photoproduction of charm. E791 will be reporting more results. Two new FNAL experiments, SELEX and FOCUS, are underway at the present time. Besides more data, future progress in understanding charm hadroproduction and photoproduction hinges on a simultaneous "fit" of all the available data to see if a common set of physical parameters can be obtained.

I would like to thank the organizers of the conference for the opportunity to participate. This work was supported by DOE Grant DE-FG-02-92ER40704.

REFERENCES

1. M. Mangano, P. Nason, G. Ridolfi,*Nucl. Phys.* B **405**, 507 (1993).
2. S. Fixone, M. Mangano, P. Nason, hep-ph/9702287,CERN-TH/97-16 and S. Fixone, M. Mangano, P. Nason, G. Ridolfi, *Nucl. Phys.* B **412**, 225 (1994).
3. T. Sjostrand, Comput. Phys. Comm. **82**, 74 (1994).
4. B. Andersson, H. Bergtsson, G. Gustafson, LU TP 83-4 (1983).
5. J.A. Appel, *Ann. Rev. Nucl. Part. Sci.* **42**, 367 (1992) and references therein.
6. E.M. Aitala et al. (E791), **Phys. Lett. B 371**, 157 (1996).
7. G.A. Alves *et al.* (E769), *Phys. Rev. Lett.* **77**, 2388 (1996).
8. G.A. Alves *et al.* (E769), *Phys. Rev.* D **49**, 4317 (1994).
9. G.A. Alves *et al.* (E769), *Phys. Rev. Lett.* **77**, 2392 (1996).
10. G.A. Alves *et al.* (E769), *Phys. Rev. Lett.* **69**, 3147 (1992).
11. G.A. Alves *et al.* (E769), *Phys. Rev. Lett.* **72**, 812 (1994).
12. G.A. Alves *et al.* (E769), *Phys. Rev. Lett.* **70**, 722 (1993).
13. E.M. Aitala et al. (E791),UMS/HEP/96–001,FERMILAB Pub–96/206–E, (accepted by Physics Letters B).
14. E.M. Aitala et al. (E791), *Phys. Lett.* B **379**, 292 (1996).
15. J. Leslie, (E791) Ph.D dissertation, University of California at Santa Cruz, unpublished, 1996.

Dynamics of Open Charm Production

Robert W. Gardner

Department of Physics, Indiana University
Bloomington, Indiana 47405

Abstract. We discuss results from recent fixed-target experiments which shed light on production dynamics of open charm.

INTRODUCTION

The study of production dynamics of charmed hadrons reveals several interesting problems in applying perturbative QCD to systems at the scale of the charm quark mass. A picture is emerging where observables such as the cross section as a function of p_t^2 (transverse momentum-squared) or x_f (feynman x) are described by a mix of perturbative and non-perturbative processes. Measurements of correlations between charm pairs and of charm-anticharm production asymmetries provide additional probes of the effects induced by nonperturbative processes. In this talk I focus mostly on photoproduction; the dynamics of hadroproduced charm is covered in detail by Prof. Jean Slaughter elsewhere in these proceedings.

OPEN CHARM PRODUCTION

In the photoproduction of charm quarks, at lowest order and at energies significantly above charm threshold, the production is expected to be dominated by a photon-gluon fusion process while for charm hadroproduction a very similar process occurs where a gluon from the projectile fuses with a gluon from target to form a $c\bar{c}$ pair. The next-to-leading order (NLO) production mechanisms are similar but include contributions from processes with extra quarks or gluons in the final state. These cross sections have been computed by Frixione, Mangano, Nason, and Ridolfi [1] (FMNR) and a recent compilation of comparisons to a large number of experiments is available [2]. These calculations require a choice of charmed quark mass, renormalization scale, and gluon momentum fraction distribution. The agreement between the predictions and the data is, for the most part, good when uncertainties in these quantities are taken into account. In many cases good agreement can only be obtained after supplementing the NLO prediction with a prescription which accounts for the nonperturbative effects. An exception is the $c\bar{c}$ total cross section energy dependence in which the theoretical cross section does not depend explicitly on a hadronization model. It is left to the experimenter to unfold the observed cross section for charmed hadrons back into that for quarks which necessarily introduces some model dependence.

The nonperturbative, soft physics effects that seem to play a role are numerous. Fragmentation effects can change the direction of the charmed mesons relative to the charmed quarks as well as significantly degrade their energy and thus their p_t. Conversely the intrinsic k_t of the target gluon (and projectile gluon in the hadroproduction case) can induce an stiffening of the inclusive p_t^2 distribution of the observed charmed hadrons. For example, the photoproduction data clearly do not agree with the FMNR predictions for *bare* charmed quarks where hard gluon emission causes a high end tail to persist in the p_t^2 distribution. However it is possible to match the p_t^2 spectrum by supplementing the NLO calculations with nonperturbative effects such as intrinsic gluon k_t and momentum loss described by Peterson fragmentation [2,3]. Interestingly the photoproduction data can be described fairly well by the PYTHIA-JETSET Monte Carlo which augments the lowest order photon-gluon fusion model with the Lund string fragmentation model [4].

Correlations between charm pairs offer more stringent tests of the NLO calculations. The story is similar: the NLO calculations which have been supplemented with nonperturbative effects do a reasonable job describing the data [2].

CHARM-ANTICHARM ASYMMETRIES

Another interesting probe of the nonperturbative aspects of open charm production are measures of charm-anticharm production asymmetries or "leading particle" effects. In both photoproduction and hadroproduction, the charm and anticharm quark are produced symmetrically at leading order QCD. At next-to-leading order, small ($\approx 1\%$) asymmetries in quark momenta are induced by the interference between the contributing amplitudes. A significant degree of $c\bar{c}$ asymmetry can arise through the fragmentation process. Figure 1 illustrates the mechanism responsible for the photoproduced $c\bar{c}$ asymmetry in the PYTHIA-JETSET model. In this model the struck gluon leaves the target nucleon in a color octet state which can be divided into a color antitriplet pole ("diquark") and color triplet pole ("bachelor quark"). The color field between the target diquark and the charm quark, and the field between the target bachelor quark and antichark quark, are treated as strings having uniform energy per unit length corresponding to a linear confinement potential. The two strings are broken into $q\bar{q}$ pairs (or diquark-antidiquark pairs) resulting

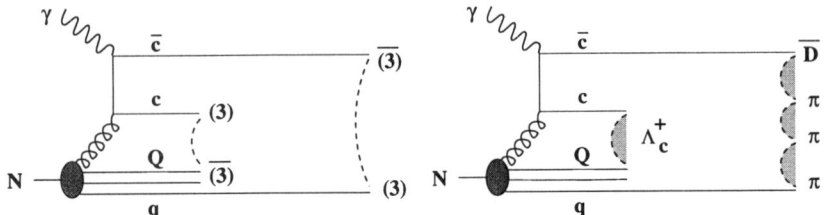

FIGURE 1. The arrangement of color poles and an example fragmentation process for charm photoproduction in the PYTHIA-JETSET model.

Reprinted from *Physics Letters B,* **370** 222 P. L. Frabetti. "Charm-anticharm Asymmetrics in High Energy Photoproduction," pg. 224, 229 ©1996 with kind permission of Elsevier Science - NL, Sara Burgerhartstreet 25, 1055 KV Amsterdam, The Netherlands.

in a final state configuration of colorless hadrons. Because the c-quark must dress with the q-remnant and the $\bar{c}$-quark must dress with the Q-remnant a difference between the bachelor quark and diquark momenta fraction distributions will lead to an asymmetry between the momenta spectra of the charmed and anti-charmed hadrons. Since the diquark and quark remnants are actually *effective* color poles, rather than actual quarks or diquarks, there is no reason to assume that their momentum fractions are the same as the quark momentum distributions measured in deep inelastic leptoproduction. In the default PYTHIA-JETSET model, the effective quark is drawn with a momentum fraction which has a soft $1/x$ dependence expected for a sea quark distribution. At fixed target energies the model puts nearly all of the nucleon momentum fraction into the effective diquark. The resulting asymmetries predicted by the model are much stronger than those observed in the E687 photoproduction experiment [5] as shown in Figure 2. An alternative effective quark distribution called the "counting rule" option draws the effective quark momentum fraction according to a much harder spectrum which results in 1/3 of the nucleon remnant momentum on average being carried by the effective quark and 2/3 by the diquark. This model predicts significantly less charm-anticharm production asymmetry and is in good agreement with the experimental data.

The charm-anticharm hadroproduction asymmetry has been studied by several fixed-target experiments [6]. The highest statistics result comes from the E791 Collaboration [7] which used an ≈74,000 event sample of $D^{\pm}$ mesons produced with a 500 GeV/c π^- beam incident on diamond and platinum foils. Their result, expressed as a "leading particle" asymmetry with the D^- being the "leading particle" (it shares a valence quark with the beam particle), has the trend of an asymmetry which favors more D^- production at larger values of x_f. The trend is the same trend observed by the photoproduction data of Figure 2 (the convention is opposite between the two experiments) but with much larger magnitude. The experimenters found their data to be well matched by the PYTHIA-JETSET model after tunes were made for the

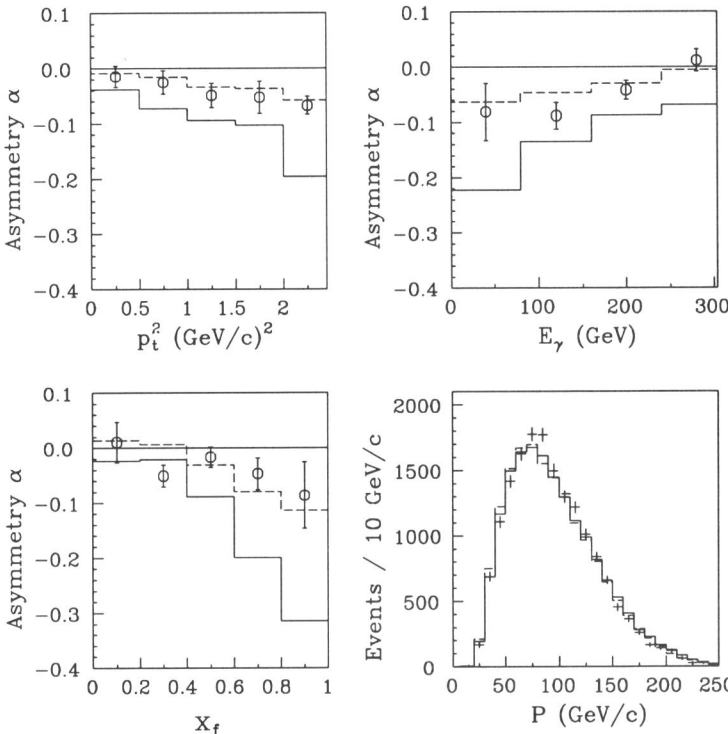

FIGURE 2. The asymmetry $\alpha = (N(D^+) - N(D^-))/(N(D^+) + N(D^-))$ in bins of $p_t^2(D^\pm)$, incident photon energy, and $x_f(D^\pm)$ for accepted $D^\pm$ photoproduced from E687. The data is indicated with error bars, the sea-like effective quark predictions are solid and the counting rule effective quark predictions are dashed. The lower right corner compares the total lab momentum spectrum for $D^\pm$ candidates to the two models. In contrast to the asymmetry, the lab momentum spectrum is insensitive to the assumed effective quark momentum distribution.

Reprinted from *Physics Letters B,* **370** 222 P. L. Frabetti. "Charm-anticharm Asymmetrics in High Energy Photoproduction," pg. 224, 229 ©1996 with kind permission of Elsevier Science - NL, Sara Burgerhartstreet 25, 1055 KV Amsterdam, The Netherlands.

intrinsic k_t and charm quark mass.

SUMMARY

A variety of soft, nonperturbative processes play an important role in describing open charm production. QCD-inspired models of hadronization plus leading order photon-gluon fusion go a long way in describing much of the data. Next-to-leading order calculations augmented with simple nonperturbative prescriptions adequately describe many of the observed distributions. Measurements such as charm-anticharm production asymmetries and leading particle effects provide useful tools in studying nonperturbative processes in photo- and hadroproduced open charm.

ACKNOWLEGEMENTS

I wish to thank my colleagues from the Fermilab E687 experiment and especially Prof. Jim Wiss from the University of Illinois with whom I collaborated closely on the charm photoproduction analysis. I also wish to thank Tom Carter and Jean Slaughter from Fermilab E791 for useful and interesting discussions of the hadroproduction data.

REFERENCES

1. S. Frixione, M.Mangano, P.Nason, G. Ridolfi, Nucl. Phys B405 (1993) 507.
2. S. Frixione, M.Mangano, P.Nason, G. Ridolfi, CERN-TH/97-16.
3. C. Peterson, D. Schlatter, I. Schmitt, P. Zerwas, Phys. Rev. D 27 (1983) 105.
4. T.Sjöstrand, Computer Phys. Comm. 39 (1986) 347;
   T.Sjöstrand, M.Bengtsson, Computer Phys. Comm. 43 (1987) 367;
   H.-U.Bengtsson, T.Sjöstrand, Computer Phys. Comm. 46 (1987) 43.
5. E687 Collab., P.L. Frabetti et al., Phys. Lett. B370 (1996) 222.
6. Recent experiments include:
   ACCMOR Collab. S. Barlag et al., Z. Phys. C49 (1991) 555;
   WA82 Collab. M. Adamovich et al., Phys. Lett. B305 (1993) 402;
   E769 Collab. G. A. Alves et al., Phys. Rev. Lett. 72 (1994) 812.
7. E791 Collab., E.M. Aitala et al., Phys. Lett. B371 (1996) 157.

Beauty hadroproduction at fixed target in WA92 experiment

Claudia Gemme

University of Genova and INFN, Genova, Italy

Abstract. Using a sample of 10^8 triggered events, produced in π-Cu interactions at 350 GeV/c, we have identified 26 events due to beauty production. From these data we measure the total cross-section and some preliminary distributions of beauty production characteristics.

Heavy quark hadroproduction can in principle be described by perturbative QCD; complete calculations are nowadays available at Next To Leading Order [1]. These predictions are nevertheless affected by theoretical uncertainties but, due to the size of the b quark mass, the measurement of the beauty hadroproduction cross-section is a good test of perturbative QCD. At fixed target energies the beauty cross-section is of the order of nanobarns, or $\sim 10^{-6}$ of the inelastic cross-section. Identifying the decay vertices of beauty mesons is a very difficult task: the richest sample of events fulfilling this request is so far of 9 events reported by experiment E653 [2].

Experiment WA92 collected data using a 350 GeV/c π^- beam incident on a 2 mm copper target in the CERN Ω Spectrometer. The main features of the experimental apparatus were a high-precision silicon microstrip detector array which provided excellent track visualization and a fast online secondary vertex trigger. The silicon tracking array consisted of a Decay Detector (DkD), made of 17 planes of 10 μm pitch silicon microstrips covering the first 3.2 cm downstream of the target, followed by a Vertex Detector (VxD) consisting of 12 planes of 25 μm pitch and 5 planes of 50 μm pitch. WA92 adopted a combination of several independent trigger components, in order to keep the acceptance of beauty events as high as possible. An interaction trigger was in 'AND' with two over three of the following triggers: the high p_T trigger, obtained with 2 bow-tie shaped hodoscopes, accepting only tracks with p_T larger than 0.6 GeV/c, the RPC muon trigger, selecting muons coming from the target region, and the Beauty Contiguity Processor (BCP) trigger, which used the information given by planes of the beam telescope and of the VxD to select events with secondary vertices. More details on the description and performance of the experimental apparatus and of the trigger scheme can be found elsewhere [3].

The experiment collected 10^8 events in 1992 and 1993 with a total luminosity of $8.1\,\mathrm{nb}^{-1}$. To evaluate the acceptances we fully simulated minimum bias, $b\bar{b}$ and $c\bar{c}$ events. Minimum bias events were generated using Fluka, while events with heavy quarks were generated using a combination of Pythia 5.4, providing the description of hard processes, Jetset 7.3 and Fluka. The tracking of the particles in the experimental apparatus was simulated in detail with Geant 3.21. The beauty acceptance at the trigger level resulted $\sim 30\%$ (*i.e.* 155 beauty events on tape per nanobarn of cross–section); $\sim 2\%$ of real data were kept.

In order to search for beauty decays, events were selected with secondary activity in the DkD, rejecting events with large energy releases close to the secondary vertices to discard hadronic interactions. This was possible thanks to the DkD analogue read-out. We then defined three classes of events:

1. *μ events*, characterized by a high p_T muon from a secondary vertex and at least one other secondary vertex in the fiducial volume (this choice is justified by the large semileptonic branching ratio of B decays and exploits the RPC muon trigger).

2. *multivertex events*, *i.e.* events with at least three secondary vertices in the fiducial volume and satisfying a high-p_T track request (this category has been suggested by the chained topology of beauty meson decays and by the availability of the secondary vertex trigger (BCP)).

3. *non-pointing D events*, defined as events with a fully-reconstructed Cabibbo-favoured D decay and not pointing to the primary vertex, and at least one another secondary vertex (this sample, which also exploits the BCP trigger, is justified by the large fraction of B mesons decaying into a D and by the rich sample of fully-reconstructed charm events [4]).

The selected events went through class-dependent cuts. *μ events* were selected by imposing some very strict topological and kinematical cuts, while for *multivertex* and *non-pointing D* events a neural network procedure [5] was applied. In this way a total of a few thousand events were selected to go through a graphical scanning analysis, reducing the triggered beauty sample to about 10%.

As we are aware of the possible subjectivity of scanning results, we defined rules as rigorously as possible. The most relevant scanning rules were defined in common for the three classes of events and mainly consisted of a visual quality check of tracks and vertices: the vertices had to be well reconstructed and well separated from each other, the confirmed secondary vertices had to be in the fiducial volume (at least one of them in the DkD), the absence of a large energy release around secondary vertices had to be confirmed, the tracks had to have the majority of their hits correctly associated and had to be linked correctly to their vertex in both projections. Both for the *multivertex events* and the *non-pointing D events* a possible kinematical association of a beauty–charm decay chain was checked. Furthermore for the *multivertex events* at least one track linked to a B decay candidate vertex had to have $p_{Tf} \geq 1$ GeV/c with respect to the decaying particle line of flight.

The total number of beauty candidates is so divided in the three categories: 12 beauty events in the μ *events* sample, 12 beauty events in the *multivertex events* sample, 5 beauty events in the *non-pointing D events* sample. One event is common to the three classes and one to the *multivertex* and μ *events* classes.

We identify three possible background sources to our beauty event candidates: charm events, nuclear interactions without nuclear breakdown and in-flight decays of pions and kaons. To study the background from primary charm, we simulated $c\bar{c}$ events which went through the analysis and selection chain as the real data; after renormalization the number of $c\bar{c}$ background events surviving the selection is 1.2±0.7. White interactions, *i.e.* events in which a secondary coherent nuclear interaction is produced, without nuclear breakdown and without consequent large energy release, were studied "whitening" secondary black interactions, *i.e.* erasing the big clusters of energy release in the DkD. No events survive the selection procedure and the scanning. A possible background source for μ *events* could be due to events in which the triggering muon is produced by the in-flight decay of a pion or a kaon. We generated "fake-μ" events starting from the real reconstructed events, forcing pions to decay in flight. Again no events survive the selection.

Assuming an A^1 dependence, the total beauty hadroproduction cross–section is measured to be:

$$\sigma_{b\bar{b}} = 5.6 \pm 1.2_{stat} \pm^{0.7}_{0.8}{}_{syst} \; nb/N$$

where sources of systematics are the error on luminosity ($\simeq 6\%$), the error on beauty acceptance from Montecarlo statistics ($\simeq 10\%$), the error on background contamination from montecarlo statistics ($\simeq 8\%$).

As a check, we can evaluate the beauty hadroproduction cross–section independently from each of the three categories of events:

$$\sigma_{mvtx} = 5.2 \pm 1.5_{stat} \pm^{0.8}_{1.0}{}_{syst} \; nb/N$$

$$\sigma_{npD} = 7.2 \pm 3.2_{stat} \pm^{1.9}_{2.4}{}_{syst} \; nb/N$$

$$\sigma_{\mu} = 6.2 \pm 2.0_{stat} \pm^{1.1}_{1.3}{}_{syst} \; nb/N$$

Our measurement is in good agreement with next-to-leading order perturbative QCD calculations, which anyway present large uncertainty bands, as can

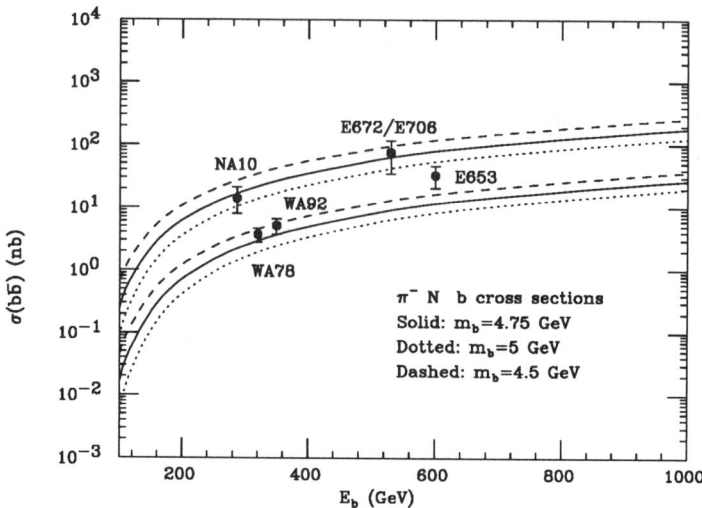

FIGURE 1. *Beauty cross-section measurements in $\pi^- N$ interactions compared with theoretical predictions at different b quark mass hypothesis. Theoretical uncertainty bands are obtained varying the factorization and renormalization scales.*

be seen in fig. 1, where this measurement and the ones previously published by different experiments are compared to the NLO QCD predictions.

Out of our 26 events in 15 events both beauty decays can be identified (requiring one track with $p_{Tf} \geq 1$ GeV/c or a minimum mass ≥ 2 GeV/c^2 or the invariant mass ≥ 2 GeV/c^2). For these events the distribution of the correlation variable $\Delta\phi$, defined as the azimuthal angle between the two vertices in the plane perpendicular to the beam direction, is shown in fig. 2a. This distribution is still preliminary, being uncorrected for acceptance, even if a first study indicates that it is rather flat, thus not affecting the distribution much. The sharp peaking at 180^0 for beauty pairs is well predicted by NLO QCD calculations, as shown is fig. 2a.

To obtain kinematical information from the 41 identified beauty decays, it is necessary to estimate the initial momenta taking into account the unseen decay products. We use an estimator very similar to that one already used by this collaboration for the charm analysis [6]. It exploits a combination of two different approaches: one closes the decay by adding one missing neutral pion, the other one uses an approximate Lorentz boost. Figs. 2b and 2c show the preliminary single–differential distributions for x_F and p_T^2, uncorrected for acceptance, and compared with NLO QCD calculations. As acceptance for the p_T^2 distribution is rather flat, it could be noticed that there is good agreement between our data and next-to-leading order QCD without any corrections due to fragmentation or higher order corrections, as expected thanks to the large

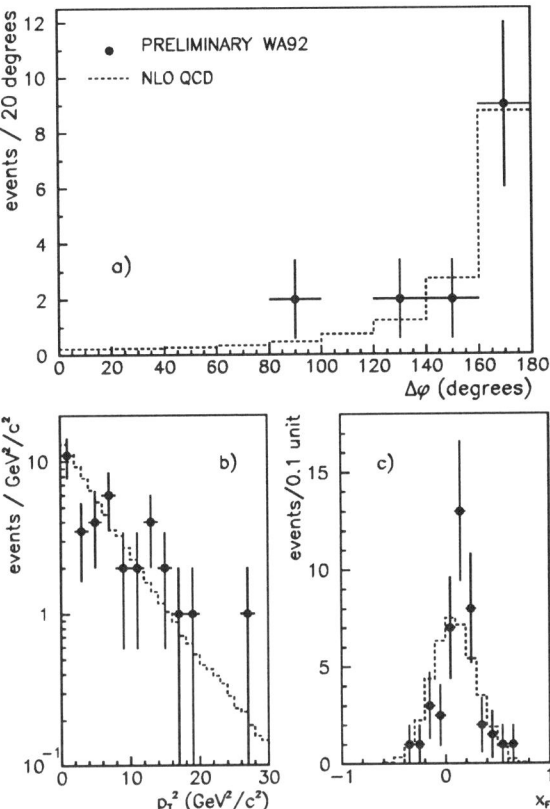

FIGURE 2. Distributions of a) $\Delta\phi$, b) p_T^2 and c) x_F. The dotted line shows for comparison the results of NLO QCD calculation.

beauty mass. The x_F distribution is shifted towards positive values, therefore a significant comparison with QCD predictions is still not possible without acceptance correction which has to take into account detector effects.

REFERENCES

1. Frixione S., Mangano M.L., Nason P., Ridolfi G., *Nucl. Phys.* B **431**, 453 (1994).
2. Kodama K., *et al, Phys. Lett.* B **303**, 359 (1993).
3. Adamovich M. *et al, Nucl. Instr. and Meth.* A **379**, 252 (1996).
4. Adamovich M. *et al, Nucl. Phys.* B **459**, 3 (1997).
5. Malferrari L., *Nucl. Instr. and Meth.* A **368**, 185 (1995).
6. Adamovich M. *et al, Phys. Lett.* B **385**, 487 (1996).

Structure Functions from Chiral Soliton Models[1]

H. Weigel*, L. Gamberg†, and H. Reinhardt*

*Institute for Theoretical Physics, Tübingen University
Auf der Morgenstelle 14, D-72076 Tübingen, Germany
†Department of Physics and Astronomy, University of Oklahoma
440 W. Brooks Ave, Norman, Oklahoma 73019-0225, USA

Abstract. We study nucleon structure functions within the bosonized Nambu–Jona–Lasinio (NJL) model where the nucleon emerges as a chiral soliton. We discuss the model predictions on the Gottfried sum rule for electron–nucleon scattering. A comparison with a low–scale parametrization shows that the model reproduces the gross features of the empirical structure functions. We also compute the leading twist contributions of the polarized structure functions g_1 and g_2 in this model. We compare the model predictions on these structure functions with data from the E143 experiment by GLAP evolving them from the scale characteristic for the NJL-model to the scale of the data.

The purpose of this investigation is to provide a link between two successful although seemingly unrelated pictures of baryons. On one side we have the quark parton model which successfully describes the scaling behavior of the structure functions in deep inelastic scattering (DIS) processes. The deviations from these scaling laws are computable in the framework of perturbative QCD. On the other side we have the chiral soliton approach which is motivated by the large N_C expansion of QCD, N_C being the number of color degrees of freedom. For $N_C \to \infty$, QCD is known to be equivalent to an effective theory of weakly interacting mesons. Although this theory is not explicitly known it can be modeled by assuming that at low energies only the light mesons (pions, kaons, ρ, ω) are relevant. When modeling the meson theory one requires the symmetry structure of QCD. In particular besides Pioncarè invariance we require chiral symmetry and its spontaneous breaking. Baryons emerge as non–perturbative (topological) configurations of the meson fields, the so–called solitons. The link between these two pictures can be established by computing structure functions within a chiral soliton model for the nucleon

---

[1] Supported in part by the Deutsche Forschungsgemeinschaft (DFG) under contract Re 856/2–3 and the US–DOE grant DE–FE–02–95ER40923.

from the hadronic tensor

$$W^{ab}_{\mu\nu}(q) = \frac{1}{4\pi} \int d^4\xi \, e^{iq\cdot\xi} \langle N(P)| \left[J^a_\mu(\xi), J^{b\dagger}_\nu(0) \right] |N(P)\rangle , \qquad (1)$$

which describes the strong interaction part of the DIS cross–section. In eq (1) $|N(P)\rangle$ refers to the nucleon state with momentum P and $J^a_\mu(\xi)$ to the hadronic current suitable for the process under consideration. In most soliton models – due to the non–perturbative nature of the soliton configuration – the current commutator (1) remains intractable. However, the Nambu and Jona–Lasinio (NJL) model [1] of quark flavor dynamics, which can be bosonized by functional integral techniques [2], contains simple current operators. Most importantly, the bosonized version of the NJL–model contains soliton solutions [3]. This paves the way to compute structure functions in the soliton approach.

In order to extract the leading twist contributions to the structure function one computes the hadronic tensor in the Bjorken limit

$$q_0 = |\mathbf{q}| - M_N x \quad \text{with} \quad |\mathbf{q}| \to \infty \quad \text{and} \quad x = -q^2/2P\cdot q \quad \text{fixed} . \qquad (2)$$

Here we confine ourselves to presenting the key issues of the calculation, details may be traced from refs. [4–6].

THE NUCLEON FROM THE CHIRAL SOLITON IN THE NJL MODEL

In this section we briefly summarize the basic features of the chiral soliton in the NJL–model and discuss how states with nucleon quantum numbers are generated. For more details see refs. [3,7] and quotations therein.

The NJL–model Lagrangian contains a quartic quark interaction which is chirally symmetric. Derivatives of the quarks fields only appear in form of a free Dirac Lagrangian, hence the current operator is formally free. Upon bosonization the action may be expressed as [2]

$$\mathcal{A} = \text{Trln}_\Lambda \left(i\slashed{\partial} - mU^{\gamma_5} \right) + \frac{m_0 m}{4G} \text{tr} \left(U + U^\dagger - 2 \right) \qquad (3)$$

where we have confined ourselves to the interaction in the pseudoscalar channel. The associated pion fields $\boldsymbol{\pi}$ are contained in the non–linear realization $U = \exp(i\boldsymbol{\tau}\cdot\boldsymbol{\pi}/f_\pi)$. In eq (3) tr denotes discrete flavor trace while Tr also includes the functional trace. The parameters of the model are the coupling constant G, the current quark mass m_0 and the UV cut–off Λ. The constituent quark mass m arises as the solution to the Schwinger–Dyson (gap) equation and characterizes the spontaneous breaking of chiral symmetry. A Bethe–Salpeter equation of the pion field can be derived from eq (3) which allows one to express the pion mass $m_\pi = 135\text{MeV}$ and decay constant $f_\pi = 93\text{MeV}$ in terms of the model parameters. Fixing these quantities leaves one parameter undetermined which maybe expressed in terms of the constituent quark mass m. Subsequently an energy functional for non–perturbative but static field configurations $U(\boldsymbol{r})$ can be extracted from (3). It can be expressed as a regularized sum of single quark energies ϵ_μ. For the hedgehog *ansatz*, $U_H = \exp(i\boldsymbol{\tau}\cdot\hat{\boldsymbol{r}}\Theta(r))$ the assoicated one–particle Dirac Hamiltonian becomes

$$h = \boldsymbol{\alpha}\cdot\boldsymbol{p} - m\,\exp\left(i\gamma_5\boldsymbol{\tau}\cdot\hat{\boldsymbol{r}}\Theta(r)\right)\,,\quad h\Psi_\mu = \epsilon_\mu\Psi_\mu\,. \tag{4}$$

The distinct level (v), which is bound in the background of U_H, is referred to as the valence quark state. Its explicit occupation guarantees unit baryon number. The chiral angle $\Theta(r)$ of the soliton is determined by self–consistently minimizing the energy functional. This soliton configuration does not yet carry nucleon quantum numbers. To generate them the (unknown) time dependent field configuration is approximated by elevating the zero modes to time dependent collective coordinates $U(\boldsymbol{r},t) = A(t)U_H(\boldsymbol{r})A^\dagger(t)$, $A(t)\in \mathrm{SU}(2)$. Upon canonical quantization the angular velocities, $\boldsymbol{\Omega} = -2i\mathrm{tr}(\boldsymbol{\tau}A^\dagger\dot{A})$, are replaced by the spin operator $\boldsymbol{J}$ via $\boldsymbol{\Omega} = \boldsymbol{J}/\alpha^2$ with α^2 being the moment of inertia[2] while the nucleon states $|N\rangle$ emerge as Wigner D–functions. To compute nucleon properties the action (3) is expanded in powers of $\boldsymbol{\Omega}$ corresponding to an expansion in $1/N_C$. In particular the valence quark wave–function $\Psi_\mathrm{v}(\boldsymbol{x})$ acquires a linear correction

$$\Psi_\mathrm{v}(\boldsymbol{x},t) = \mathrm{e}^{-i\epsilon_\mathrm{v} t}A(t)\left\{\Psi_\mathrm{v}(\boldsymbol{x}) + \sum_{\mu\neq\mathrm{v}}\Psi_\mu(\boldsymbol{x})\frac{\langle\mu|\boldsymbol{\tau}\cdot\boldsymbol{\Omega}|\mathrm{v}\rangle}{2(\epsilon_\mathrm{v}-\epsilon_\mu)}\right\} = \mathrm{e}^{-i\epsilon_\mathrm{v} t}A(t)\psi_\mathrm{v}(\boldsymbol{x}). \tag{5}$$

Here $\psi_\mathrm{v}(\boldsymbol{x})$ refers to the spatial part of the body–fixed valence quark wave–function with the rotational corrections included.

STRUCTURE FUNCTIONS IN THE VALENCE QUARK APPROXIMATION

The starting point for computing the unpolarized structure functions is the symmetric part of hadronic tensor in a form suitable for localized fields [9],

$$W^{lm}_{\{\mu\nu\}}(q) = \zeta\int\frac{d^4k}{(2\pi)^4}\,S_{\mu\rho\nu\sigma}\,k^\rho\,\mathrm{sgn}\,(k_0)\,\delta\left(k^2\right)\int_{-\infty}^{+\infty}dt\,\mathrm{e}^{i(k_0+q_0)t}$$

$$\times\int d^3x_1\int d^3x_2\,\exp\left[-i(\boldsymbol{k}+\boldsymbol{q})\cdot(\boldsymbol{x}_1-\boldsymbol{x}_2)\right]$$

$$\times\langle N|\left\{\bar{\hat{\Psi}}(\boldsymbol{x}_1,t)t_lt_m\gamma^\sigma\hat{\Psi}(\boldsymbol{x}_2,0)-\bar{\hat{\Psi}}(\boldsymbol{x}_2,0)t_mt_l\gamma^\sigma\hat{\Psi}(\boldsymbol{x}_1,t)\right\}|N\rangle. \tag{6}$$

Note that the quark spinors are functionals of the soliton. Here $S_{\mu\rho\nu\sigma} = g_{\mu\rho}g_{\nu\sigma} + g_{\mu\sigma}g_{\nu\rho} - g_{\mu\nu}g_{\rho\sigma}$ and $\zeta = 1(2)$ for the structure functions associated

---

[2)] Generalizing this treatment to flavor SU(3) indeed shows that the baryons have to be quantized as half–integer objects. For a review on solitons in SU(3) see *e.g.* [8].

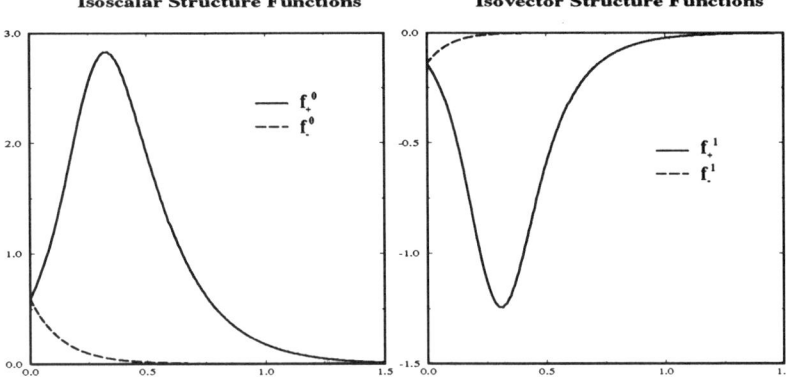

FIGURE 1. The unpolarized structure functions obtained after extracting the collective part of the nucleon matrix elements. Here we used $m = 350\mathrm{MeV}$.

with the vector (weak) current and t_m is a suitable isospin matrix. The matrix element between the nucleon states ($|N\rangle$) is taken in the space of the collective coordinates. In deriving eq. (6) the *free* correlation function for the intermediate quark fields has been assumed. In the Bjorken limit (2) the momentum, k, of the intermediate quark is highly off-shell and hence not sensitive to momenta typical for the soliton configuration. Thus the use of the free correlation function is a valid treatment in this kinematical regime.

The valence quark approximation ignores the vacuum polarization in (6), *e.g.* the quark field operator $\hat{\Psi}$ is substituted by the valence quark contribution (5). For small constituent quark masses $m \sim 400\mathrm{MeV}$ this is well justified since this level provides the dominant share to static observables [3,7]. The structure function $F_2(x)$ can be obtained from (6) by an appropriate projection[3]. After computing the collective coordinate matrix elements all physical relevant processes are described in terms of four structure functions $f_{\pm}^{0,1}$. The superscript denotes the isopsin combination of $t_l t_m$ while the subscript refers to forward and backward moving intermediate quarks in (6). In figure 1 the predictions for these four structure functions are displayed. Although the problem is not formulated Lorentz–covariantly these structure functions are reasonably well localized in the interval $x \in [0, 1]$. Furthermore the contributions of the backward moving quarks are quite small, however, they increase with m. Note that for consistency with the Adler sum rule also the moment of inertia must be restricted to the valence quark contribution [4,5]. For $m = 350\mathrm{MeV}$ this, however, is almost 90%.

We continue by presenting the numerical results for the structure functions for physical processes. In figure 2 we display the linear combination relevant for the Gottfried sum rule

$$x\left(F_2^{ep} - F_2^{en}\right) = -\left(f_+^1 - f_-^1\right)/3 \qquad (7)$$

---

[3] In the Bjorken limit the Callan–Gross relation $F_2(x) = 2xF_1(x)$ is satisfied.

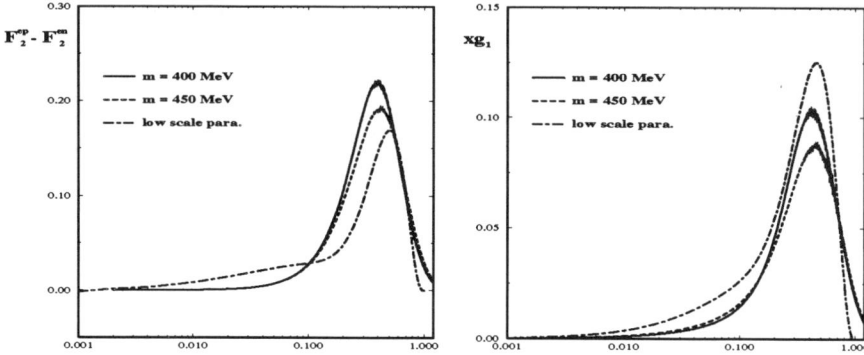

FIGURE 2. Comparison of the model structure functions with the low–scale parametrization of ref [10]. Left panel: The structure function $F_2(x)$ for electron–nucleon scattering. Right panel: The polarized structure function xg_1 for the nucleon.

and compare it to the low–scale parametrization of the empirical data [10]. This is obtained from a next–to–leading order QCD evolution of the experimental to a low–energy regime, where soliton models are valid. The agreement improves with increasing constituent quark masses. Apparently the model reproduce the gross features of the low–scale parametrization. Moreover the integral of the Gottfried sum rule

$$S_G = \int_0^\infty \frac{dx}{x} \left(F_2^{ep} - F_2^{en}\right) = \begin{cases} 0.29, & m = 400\text{MeV} \\ 0.27, & m = 450\text{MeV} \end{cases} \qquad (8)$$

agrees reasonably well with the empirical value $S_G = 0.235 \pm 0.026$ [11]. In particular the deviation from the naïve value (1/3) [12] is in the direction demanded by experiment.

Figure 2 also shows the comparison of the model prediction for the polarized structure function $g_1(x)$ with the corresponding low–scale parametrization [10]. In this case the agreement improves with decreasing m.

No low–scale approximation is available for the polarized structure function $g_2(x)$. We have therefore projected the predicted structure function onto the interval $x \in [0,1]$ [13] and subsequently performed a leading order QCD evolution to the scale of the experiment, see ref. [6] for details. The resulting polarized structure functions are displayed in figure 3. Apparently the model reproduces the empirical data quite well, although the associated error bars are sizable.

CONCLUSIONS

In this talk we have presented a calculation of nucleon structure functions within a chiral soliton model. We have argued that the soliton approach to the bosonized version of the NJL–model is most suitable since (formally) the

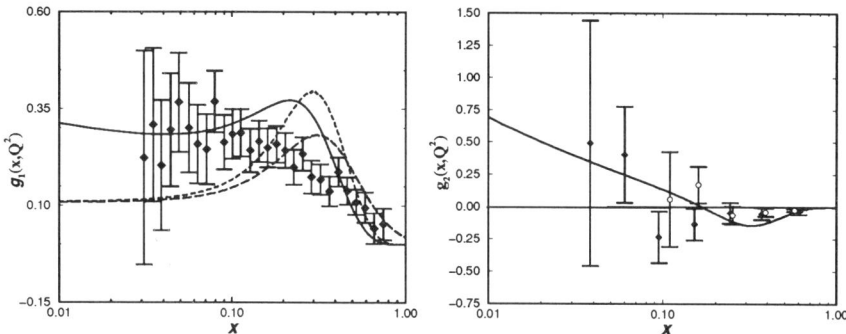

FIGURE 3. The polarized structure functions g_1 and g_2 after projection and QCD evolution. Left panel: The dashed (dotted) line denotes the (projected) low scale model prediction.

required current operator is identical to the one in a free Dirac theory. Hence there is no need to approximate the current operator by *e.g.* performing a gradient expansion. Although the calculation contains a few (well–motivated) approximations it reproduces the gross features of the empirical structure functions at low energy scales. This happens to be the case for both the polarized as well as the unpolarized structure functions.

Future projects will include to extend the valence quark approximation, improvements on the projection issue and the extension to flavor SU(3).

REFERENCES

1. Nambu, Y., and Jona–Lasinio, G., *Phys. Rev.* **122** (1961) 345; **124** (1961) 246.
2. Ebert, D., and Reinhardt, H., *Nucl. Phys.* **B271** (1986) 188.
3. Alkofer, R., Reinhardt, H., and Weigel, H., *Phys. Rep.* **265** (1996) 139.
4. Weigel, H., Gamberg, L., and Reinhardt, H., *Mod. Phys. Lett.* **A11** (1996) 3021.
5. Weigel, H., Gamberg, L., and Reinhardt, H., *Phys. Lett.* **399** (1997) 287.
6. Weigel, H., Gamberg, L., and Reinhardt, H., *Phys. Rev.* **D55** (1997) 6910.
7. Christov, C., *et al.*, *Prog. Part. Nucl. Phys.* **37** (1996) 91.
8. Weigel, H., *Int. J. Mod. Phys.* **A11** (1996) 2419.
9. Jaffe, R. L., *Phys. Rev.* **D11** (1975) 1953;
   Jaffe, R. L., and Patrascioiu, A., *Phys. Rev.* **D12** (1975) 1314.
10. Glück, M., Reya, E., and Vogt, A., *Z. Phys.* **C67** (1995) 433;
    Glück, M., Reya, E., Stratmann, M., and Vogelsang, W., *Phys. Rev.* **D53** (1996) 4775.
11. Arneodo, M., *et al.* (NMC), *Phys. Rev.* **D50** (1994) R1.
12. Gottfried, K., *Phys. Rev. Lett.* **18** (1967) 1174.
13. Jaffe, R., L., *Ann. Phys.* (NY) **132** (1981) 32.

Quark and Proton Spin Structure in the Instanton Liquid Model of QCD

Andree Blotz * and Edward Shuryak[†]

Los Alamos National Laboratory, Los Alamos, NM 87545
[†]*State University of Stony Brook, Stony Brook, NY 11794*

Abstract. Within an instanton based model of QCD we address the important question of how much of the proton spin is carried by the spins of the quarks and how much is due to orbital angular momentum and the spins of the gluons. Since this question arises already on the level of a single quark inside the proton, we study axial vector correlation functions for a quark in the so called Random Instanton Liquid Model (RILM) as well as for the Interacting Instanton Liquid Model (IILM).

INTRODUCTION

Recent lattice calculations support the decade old picture of a ground state of QCD which is filled with a liquid of instantons and anti-instantons. These classical solutions of Euclidean field equations do not only describe the qualitative features of low energy QCD but also quantitatively a large variety of hadronic and gluonic two-point correlation functions [1]. As a first step to investigate the structure of the hadrons we have calculated the three point correlation function of the pion with an external electro-magnetic field [2]. The result was not only the reproduction of the monopole shape formfactor but showed also very nicely that the formfactor of the pion is basically the formfactor of the instanton. Hereby the experimentally measured formfactor confirmed an average instanton size $\bar{\rho} \simeq 0.35$fm, which is the value used for more than a decade now [3].

Based on this successful model we are asking now where 'the proton really gets its spin' [4]. For this aim we calculate three point correlation function for the axial vector current in coordinate space and project onto the axial vector coupling as well as the induced pseudoscalar coupling constant [5]. To confirm the finding we also evaluate the divergence of the axial current, which in the chiral limit is given by the anomaly, within a quark state.

THE QUARK PROPAGATOR

The propagator of a single quark in the Instanton Liquid Model is shown in Fig. 1 in dependence of the quark separation x and for the spin flip and spin non-flip amplitude. Both curves are normalized to the free spin non-flip amplitude, so that the short distance behaviour of the spin non-flip amplitude reflects asymptotic freedom The figure shows a fit with a constant constituent quark mass (cqm), where the quark mass turns out to be $\simeq$ 300MeV, but it neither describes the shorter distances < 1fm nor the larger ones. A reliable fit is obtained by assuming a momentum dependent quark mass. For small momenta then, the mass corresponds roughly to the constituent quark mass $\simeq 360 - 370$MeV, whereas for larger momenta it approaches zero.

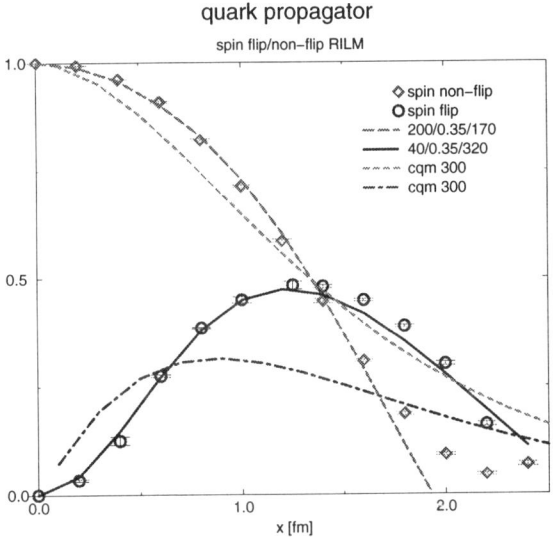

Fig. 1: The two-point correlation function of a quark, normalized to the free correlator and in the non interacting configuration (RILM). The instanton size is $\rho = 0.35$fm and the instanton density is $n = 1\text{fm}^{-4}$ [5].

I THE DIVERGENCE OF THE AXIAL CORRELATOR

Using this momentum dependent quark propagator one can determine the axial vector coupling g_A either from axial vector current correlator [5] or from the anomaly. This is because the divergence of the flavor singlet axial current in QCD is non-vanishing in the chiral limit due to

$$\partial_\mu j^{GI}_{\mu,5}(x) = 2m\bar{q}i\gamma_5 q + \frac{N_f}{16\pi^2}G_{\mu\nu}\tilde{G}_{\mu\nu} \qquad (1)$$

In Fig. 2 we have evaluated the anomaly term $G_{\mu\nu}\tilde{G}_{\mu\nu}$ as well as the connected and disconnected contribution to $\bar{q}i\gamma_5 q$ for a constituent quark. As can be seen the anomaly term (GG disc) almost cancels with the disconnected quark term (mP disc) and moreover the total contribution of all 3 terms has a different sign than the anomaly term.

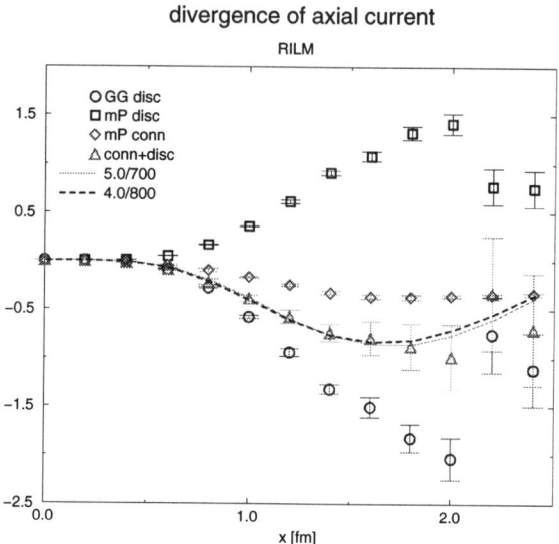

Fig. 2: The three-point correlation function $\Pi_3^-(x,y)$ of a quark, normalized to the free correlator and for the random configuration (RILM).

The final value of the quark spin expectation value from this calculation depends on the interactions of the instantons. Whereas random configurations of instantons and anti-instantons suggest that g_A is indeed close to one, the interacting ensemble $IILM$ shows a reduction of almost 50%. This is an clear indication that instanton effects might explain the proton spin puzzle.

REFERENCES

1. T. Schaefer and E. Shuryak, Instantons in QCD, HEP-PH 9610451, 1996.
2. A. Blotz and E. Shuryak, Phys. Rev. **D55**, 4055 (1997).
3. E. V. Shuryak, the QCD Vacuum, Hadrons and the Superdense Matter, World Scientific, Singapore, 1988.
4. R. L. Jaffe, Phys. Today **48**, NO.924 (1996).
5. A. Blotz and E. Shuryak, Structure of a constituent quark in the instanton liquid, 1997.

Nucleon strangeness and spin crisis?

M. D. Scadron

Physics Department, University of Arizona, Tucson, AZ, USA

Abstract

We summarize three alternative extensions of the quark-valence picture of nucleon strangeness and spin, including (i) gluon spin in QCD, (ii) Bjorken sum rule and QCD, (iii) D–E meson mixing and tadpole leakage. All three approaches suggest that $\sim 40\%$ of the nucleon's spin (as well as its momentum) resides with quarks, but that less than 6% is due to strange quarks.

I. Quark Valence Picture

Given that pions are composed of $\bar{q}q$ nonstrange (NS) quarks and (static) nucleons are composed of qqq NS quarks uud and ddu, the strangeness-antistrangeness quantum numbers appear when e.g. $\pi^- p \to \Lambda K^0$. The static nonrelativistic quark model then requires the SU(6) valence (V) values for the axial-vector quark spin component ($\Delta q \sim \langle N | \bar{q} \gamma_\mu \gamma_5 q | N \rangle S^\mu$) matrix elements of the nucleon to be

$$\Delta u_V = \tfrac{4}{3}, \qquad \Delta d_V = -\tfrac{1}{3}, \qquad \Delta s_V = 0. \tag{1}$$

This quark valence picture in Eq. (1) was first tested using the Sachs form factors for on-shell nucleons $\Gamma_\mu = G_E(q^2) P_\mu / m_N + G_M(q^2) r_\mu / 4m_N^2$, where $P = (p' + p)/2$, $q = p' - p$, $r_\mu = 2\varepsilon_{\mu\alpha\beta\gamma} q^\alpha P^\beta \gamma^\gamma \gamma_5$ and $G_M(q^2 = 0) = \mu_N$ is the nucleon total magnetic moment. The corresponding proton and neutron magnetic moments expressed in terms of the valence quark spins of Eqs. (1) are [1]

$$\mu_p = \mu_u \Delta u_V + \mu_d \Delta d_V + \mu_s \Delta s_V, \tag{2a}$$

$$\mu_n = \mu_d \Delta u_V + \mu_u \Delta d_V + \mu_s \Delta s_V, \tag{2b}$$

where $\mu_p, \mu_n = 2.793, -1.913$ are scaled to nucleon magnetons $e/2m_N$ and the corresponding quark magnetons are in the ratio of quark charges and constituent quark masses as

$$\mu_u / \mu_d \approx e_u / e_d = -2, \qquad \mu_s / \mu_d = e_s m_d / e_d m_s \approx \tfrac{2}{3}. \tag{2c}$$

Then substituting the quark valence spin values in Eq. (1) and the first quark magneton ratio of Eq. (2c) into the quark spin versions of the nucleon magnetic moment valence expressions Eqs. (2a,b), one obtains the predicted ratio [1]

$$\frac{\mu_p}{\mu_n} = \frac{\left(\frac{\mu_u}{\mu_d} \frac{\Delta u_V}{\Delta d_V}\right) + 1 + 0}{\frac{\Delta u_V}{\Delta d_V} + \frac{\mu_u}{\mu_d} + 0} = \frac{9}{-6} = -1.50, \tag{2d}$$

very close to the observed magnetic moment ratio [2] $\mu_p/\mu_n \approx -1.460$.

It is clear that the above quark valence picture cannot be the whole story, however, because the nonstrange λ_3 isovector difference of quark spins determined by neutron β decay [2],

$$\Delta u - \Delta d = g_A = 1.2601 \pm 0.0025 \approx 1.26 , \tag{3}$$

is 25% less than the valence value in Eq. (1); i.e. $\Delta u_V - \Delta d_V = \frac{5}{3}$. Moreover, the λ_8 component of the nucleon axial-vector current found from the various hyperon semileptonic weak decays

$$\Delta u + \Delta d - 2\Delta s = g_A \left[\frac{3f-d}{f+d}\right]_A \approx 0.58 \tag{4}$$

for the empirically determined [4] ratio $(d/f)_A \approx 1.74$ (or equivalently $(f/d)_A \approx 0.58$), is 40% less than the valence λ_8 prediction $\Delta u_V + \Delta d_V - 2\Delta s_V = 1$. Moreover, these quark-spin difference SU(3) relations, (3) and (4) have more dynamical significance. Adding and subtracting these quark spin relations (3) and (4) lead to two more (equivalent) quark spin SU(3) relations [5]

$$\Delta u - \Delta s \approx 0.92 , \qquad \Delta d - \Delta s \approx -0.34 . \tag{5}$$

For example, the ratio of the first difference in (5) to the λ_3 difference in (3) gives

$$\frac{\Delta u - \Delta s}{\Delta u - \Delta d} = \left[\frac{2f}{f+d}\right]_A \approx 0.73 , \tag{6}$$

which in turn predicts the sigma to nucleon magnetic moment difference to be 0.73, also close to the experimental ratio of 0.77 . However, these more dynamical quark spin differences (3-5) do a poor job when combined with the (static) nucleon magnetic moment difference from (2a,b), predicting instead

$$\mu_p - \mu_n = (\mu_u - \mu_d)(\Delta u - \Delta d) \approx (e/2m_d)g_A \approx 4.71(e/2m_N) , \tag{7}$$

or a smaller-than-expected constituent quark mass of $m_d \approx 250$ MeV.

II. Dynamical QCD Models

To apply these four quark spin difference relations in (3, 4) and (5) to more dynamical problems, one must introduce a (chiral) model to set the overall quark spin scale. In QCD the gluon spin Δg must also be accounted for via the helicity sum rule resulting in the spin $\frac{1}{2}$ nucleon:

$$s_N = \tfrac{1}{2} = \tfrac{1}{2}\Delta\Sigma + \Delta g + \Delta L_z , \tag{8a}$$

where ΔL_z is the orbital angular momentum of the constituent quarks and $\Delta\Sigma = \Delta u + \Delta d + \Delta s$ is the total quark spin of the nucleon ($\Delta\Sigma_V = 1$ in the valence picture of Eq. (1)). The SLAC scaling experiments of 1970 suggest a 1 GeV scale of [6,5] $-\Delta L_z = k_\perp r_c \approx 1.29$ for $k_\perp \approx 300$ MeV and a proton transverse radius of $r_c \approx 0.85$ fm (the proton charge radius). Then the helicity sum rule (8a) at a 1 GeV scale reduces to

$$\tfrac{1}{2}\Delta\Sigma + \Delta g \approx 1.79 \ . \tag{8b}$$

On the other hand [7], $\Delta q \to \Delta q - (\alpha_s \Delta g)/2\pi$ in QCD, but the gluon spin does not affect the SU(3) quark spin difference relations (3–5) because Δg is flavor-independent. However, Δg does enter spin singlet-dependent relations such as the EMC measurement [8] of the polarized proton structure function integral at $q^2 \approx 10$ GeV^2

$$\int_0^1 dx g_1^p(x) = \tfrac{1}{2}\left[\tfrac{4}{9}\Delta u + \tfrac{1}{9}\Delta d + \tfrac{1}{9}\Delta s\right] - \tfrac{1}{3}\left(\tfrac{\alpha_s \Delta g}{2\pi}\right) \approx 0.126 \ . \tag{9}$$

The solution of the quark spin difference relations (3–5) together with the helicity sum rule and EMC relations (8) and (9) leads to the unique solution [5]

$$\Delta u = 0.87 \ , \quad \Delta d = -0.39 \ , \quad \Delta s = -0.05 \ , \quad \Delta\Sigma = 0.43 \ . \tag{10}$$

An alternative QCD-dependent determination of the various quark spins follows from exploiting the Bjorken sum rule [10] as it pertains to QCD beyond leading order [11]. Using the recent EMC [8] and SMC data [12] for proton structure functions and SLAC E142 data [13] for neutron structure functions, the Bjorken sum rule (BSR)

$$\int_0^1 [g_1^p(x) - g_1^n(x)]dx = \tfrac{1}{6}g_A[1 - \alpha_s(Q^2)/\pi] \tag{11}$$

is shown to be valid to within 12% [14]. Given this BSR (11), perturbative QCD corrections calculated up to $\mathcal{O}\left[(\alpha_s/\pi)^3\right]$ and estimated through $\mathcal{O}\left[(\alpha_s/\pi)^4\right]$ lead to the quark spin relations [11] (including higher twist effects)

$$\begin{array}{ll} \Delta u = 0.85 \pm 0.03 \ , & \Delta d = -0.41 \pm 0.03 \\ \Delta s = -0.08 \pm 0.03 \ , & \Delta\Sigma = 0.37 \pm 0.07 \end{array} \tag{12}$$

Also to order $\mathcal{O}\left[(\alpha_s/\pi)^4\right]$, ref. [11] finds $\alpha_s(2.5 \text{ GeV}^2) = 0.375 \, ^{+0.062}_{-0.081}$ (compatible with $\alpha_s(1 \text{ GeV}^2) \approx 0.50$ in ref. [9]), corresponding to $\alpha_s(M_Z^2) = 0.122 \pm 0.007$ as found in the PDG tables [2] due to averaging data of PEP/PETRA, TRISTAN, LEP, SLC, CLEO.

III. Dynamical Tadpole Leakage

A third scheme to estimate the quark axial-vector spins Δu, Δd, Δs which naturally links up to the quark-valence picture of Sec. I is due to axial-vector tadpole graphs. For simplicity we shall label the axial-vector hadronic states $f_1(1285)$ and $f_1(1420)$ by their early names, D and E respectively. Then we use the nonstrange (NS) and strange (S) hadronic axial-vector basis to write [15]

$$|D\rangle = \cos\phi_A \, |A_{NS}\rangle - \sin\phi_A \, |A_S\rangle \ , \quad |E\rangle = \sin\phi_A \, |A_{NS}\rangle + \cos\phi_A \, |A_S\rangle \ , \tag{13}$$

where $|A_{NS}\rangle = |\overline{u}u + \overline{d}d\rangle/\sqrt{2}$ and $|A_S\rangle = |\overline{s}s\rangle$. The axial-vector mixing angle in (13) is determined by the present decay rate ratio [2]

$$\frac{\Gamma_{DK\overline{K}\pi}}{\Gamma_{EK\overline{K}\pi}} \approx \tan^2 \phi_A \approx \frac{2.4 \text{ MeV}}{53 \text{ MeV}} \approx 0.045, \qquad \phi_A \approx 12°. \tag{14}$$

Given eqs. (13) and (14), the quark spin Δs tadpole graph gives [16,5]

$$\Delta s = -\left[\frac{1}{m_D^2} - \frac{1}{m_E^2}\right] \cos\phi_A \sin\phi_A \langle 0|\overline{s}\gamma_\mu\gamma_5 s|A_S\rangle \langle A_{NS}N|N\rangle^\mu. \tag{15a}$$

To link up with Sec. I, the nucleon vertex in (15a) is of the valence form, with $\langle A_{NS}N|N\rangle^\mu = 1 \cdot S^\mu$, where S^μ is the spin 1 polarization four vector, along with $\langle 0|\overline{s}\gamma_\mu\gamma_5 s|A_S\rangle = S_\mu m_{A_S}^2$ (with $m_{A_S}^2 = m_D^2 \sin^2\phi_A^2 + m_E^2 \cos^2\phi_A \approx 2$ GeV^2). Then the tadpole equation (15a) requires

$$\Delta s = -\left[\frac{1}{m_D^2} - \frac{1}{m_E^2}\right] \cos 12° \sin 12° m_{A_S}^2 \approx -0.05, \tag{15b}$$

since [2] $m_D \approx 1282$ MeV and $m_E \approx 1427$ MeV.

Encouraged by the proximity of (15b) to our earlier determinations in (10) and in (12) for Δs, we extend this tadpole analysis in order to estimate the quark spin Δu in nucleons, leading to

$$\Delta u = \left[\frac{\cos^2 \phi_A}{m_D^2} + \frac{\sin^2 \phi_A}{m_E^2}\right] \langle 0|\overline{u}\gamma_\mu\gamma_5 u|A_{NS}\rangle \langle A_{NS}N|N\rangle^\mu. \tag{16a}$$

The first two terms in brackets in (16a) sum to 0.603 GeV^{-2}, while the second factor is $\langle 0|\overline{u}\gamma_\mu\gamma_5 u|A_{NS}\rangle = \frac{1}{\sqrt{2}}\langle 0|\overline{u}\gamma_\mu\gamma_5 u|\overline{u}u\rangle$ and the latter up flavor matrix element is $\langle 0|\overline{u}\gamma_\mu\gamma_5 u|\overline{u}u\rangle = S_\mu \cdot 2$ GeV^2 in the SU(3) conserving limit as above (15b). Then the tadpole version of Δu in (16a) becomes

$$\Delta u \approx 0.603 \text{ GeV}^{-2} \cdot 2 \text{ GeV}^2/\sqrt{2} \approx 0.85, \tag{16b}$$

again close to Δu in (10) and in (12).

Finally we compute the quark spin Δd in this tadpole picture driven again by D and E spin 1^+ poles. Since they too simulate the axial-vector current, we can invoke the λ_3 axial current quark spin difference relation found from neutron β decay, $\Delta u - \Delta d = g_A$. This latter relation was used in part to fix the quark spins in (10) and it was also needed to derive the Bjorken sum rule (11) which in turn led to the quark spins in (12). So assuming the tadpole-driven $\Delta u \approx 0.85$ in (16b), the tadpole-driven Δd is -0.41. Collecting the tadpole-driven quark spins of (15b) and (16b), we have

$$\Delta u \approx 0.85, \qquad \Delta d \approx -0.41, \qquad \Delta s \approx -0.05, \qquad \Delta\Sigma \approx 0.39. \tag{17}$$

IV. Conclusion

The consistent pattern of the above three dynamical schemes giving (10), (12) and (17) makes us question whether such a compatible picture is really a quark spin "crisis". While it is true that the quark spin Δs in (10, 12, 13) is a nonvanishingly small (and negative) 5%–8% of the valence values in Eq. (1), $\Sigma_V = 1$, $\Delta s_V = 0$, the momentum distribution of strange quarks in nucleons was observed by the Columbia-Chicago-Fermi-Rochester collaboration [17] to be a similar small 6% fraction. Moreover the SLAC scaling experiments of the 1970s found that the overall momentum distribution was about 50% due to quarks and presumably 50% due to QCD gluons. In like manner, the total quark spin $\Delta \Sigma$ in (10, 12, 17) is about 40% and implied Δg glue is 60% of the nucleon spin. Even as a crisis, the above quark spins always form an orderly pattern.

References

[1] M. Beg, B. W. Lee and A. Pais, *Phys. Rev. Lett.* **13**, 514 (1964).

[2] Particle Data Group, R. M. Barnett et al, *Phys. Rev.* D **54**, Part I, 1 (1996).

[3] N. Isgur and G. Karl, *Phys. Rev.* D **20**, 1190 (1979); V. Elias, M. Tong and M. D. Scadron, ibid., D **40**, 3670 (1989).

[4] Z. Djumbowski and J. Franklin, Temple University preprint TUHE-89-11(1989); *J. Phys.* G **17**, 213 (1991); F. E. Close, *Phys. Rev. Lett.* **64**, 361 (1990); M. Roos, XXV International Conference on High Energy Physics, Singapore (1990); R. L. Jaffe and A. V. Manohar, *Nucl. Phys.* B **337**, 509 (1990).

[5] M. D. Scadron, *Z. Phys.* C **54**, 595 (1992).

[6] See e.g. M. Anselmino and M. D. Scadron, *Phys. Lett.* B **229**, 117 (1989).

[7] A. V. Efremov and O. V. Teryaev, JINR preprint E2-88-287 (1988); G. Altarelli and G. C. Ross, *Phys. Lett.* B **212**, 391 (1988); R. D. Carlitz, J. C. Collins and A. H. Mueller, *Phys. Lett.* B **214**, 229 (1988); E. Leader and M. Anselmino, Santa Barbara preprint NSF-ITP-88-142 (1988).

[8] J. Ashman et al, *Phys. Lett.* B **206**, 364 (1988); *Nucl. Phys.* B **328**, 1 (1989).

[9] See e.g. A. De Rújula, H. Georgi and S. Glashow, *Phys. Rev.* D **12**, 147 (1975). For an infrared cutoff $\Lambda \approx 250$ MeV with 3 quark flavors, QCD requires $\alpha_s(1 \text{ GeV}^2) = \pi d / \ln\left[1 \text{ GeV}^2/\Lambda^2\right] \approx 0.50$ for $d = 12/(33-2N_f) = \frac{4}{9}$. At $q^2 = 10$ GeV^2, $\alpha_s \sim 0.25$ but then $\Delta g \sim 3$ to 4; see e.g. M. Glück and E. Reya, *Z. Phys.* C **43**, 679 (1989).

[10] J. D. Bjorken, *Phys. Rev.* **148**, 1467 (1966); D **1**, 1376 (1970); J. Kodaira et al, ibid D **20**, 627 (1979); *Nucl. Phys.* B **159**, 99 (1979).

[11] J. Ellis and M. Karliner, *Phys. Lett.* B **341**, 397 (1995).

[12] B. Adeva et al, *Phys. Lett.* B **302**, 533 (1993).

[13] P. L. Anthony et al, *Phys. Rev. Lett.* **71**, 959 (1993).

[14] J. Ellis and M. Karliner, *Phys. Lett. B* **313**, 131 (1993). The BSR error was later shown in ref. [11] to be less than 10%.

[15] See e.g. H. F. Jones and M. D. Scadron, *Nucl. Phys. B* **155**, 409 (1979); M. D. Scadron, *Phys. Rev. D* **29**, 2076 (1984).

[16] M. Anselmino and M. D. Scadron, *Nuovo Cim. A* **104**, 1091 (1991).

[17] C. Foudas et al, *Phys. Rev. Lett.* **64**, 1207 (1990); S. A. Rabinowitz et al, ibid **70**, 134 (1993).

Hunting the $d'(2065)$ with various probes

Gerhard J. Wagner

Physikalisches Institut der Universität Tübingen[1]
D-72076 Tübingen, Germany

- for the LEPS Collaboration: *PSI-Karlsruhe-Moscow-Tübingen*
- for the CHAOS Collaboration: *TRIUMF-Boulder-Jerusalem-Karlsruhe-Melbourne-Moscow-Regina-Sacramento-Trieste-Tübingen-Vancouver-Victoria*
- for the WASA/PROMICE Collaboration: *Uppsala-Dubna-Jülich-Lodz-Moscow-Novosibirsk-Osaka-Tübingen-Warsaw*

Abstract. A status report is given of the various experiments performed in search of existence and properties of the recently postulated narrow πNN-resonance with mass 2.06 GeV/c^2.

INTRODUCTION

Pionic double charge exchange reactions (DCX) on nuclei have been found to exhibit a resonance-like cross section at incident pion energies of about 50 MeV on all nuclei studied. Despite several theoretical attempts (e.g. [1]) no satisfactory explanation in terms of a conventional reaction model was found to date. This led us to postulate a narrow πNN resonance, called d', with a mass of about 2.06 GeV as the origin of this feature [2]. The conjecture is based on QCD-inspired constituent quark models [3,4] predicting dibaryon states with the required properties.

Assuming $J^P = 0^-$ and isospin even, and adjusting the partial width $\Gamma_{\pi NN} \approx$ 0.5 and the spreading width (due to collision damping in nuclear matter) $\Gamma_{spr} \approx 5$ MeV, and accounting through the use of shell-model wave function

---

[1] Supported by BMBF (06 TÜ 669), DFG (Mu 705/3, Graduiertenkolleg), DAAD (313/S-PPP-2194)

for Fermi motion in nuclei, a satisfactory description of all available DCX data was given [2]. In order to prove the existence and study the properties of the suspected d' a series of experiments has been initiated.

DCX REACTIONS ON HEAVY NUCLEI

In continuation of our early DCX experiments [2] with the LEPS spectrometer at PSI we have recently measured [5] the ground state transitions on ^{16}O and ^{40}Ca. These proceed dominantly via cross-shell transitions which are expected to be particularly weak in the conventional DCX mechanism which assumes sequential single charge exchange processes. Instead, the measured excitation functions show very pronounced low-energy resonances exceeding in magnitude even the signals of the Δ-resonance at about 160 MeV (Fig. 1).

The solid curves in Fig. 1 are based on the d' hypothesis and on simple shell model wave functions. These detailed measurements allow an improved determination of the spreading widths. The derived values of $\Gamma_{spr} \approx 10$ and 20 MeV for ^{16}O and ^{40}Ca, respectively, agree with estimates of the collision damping.

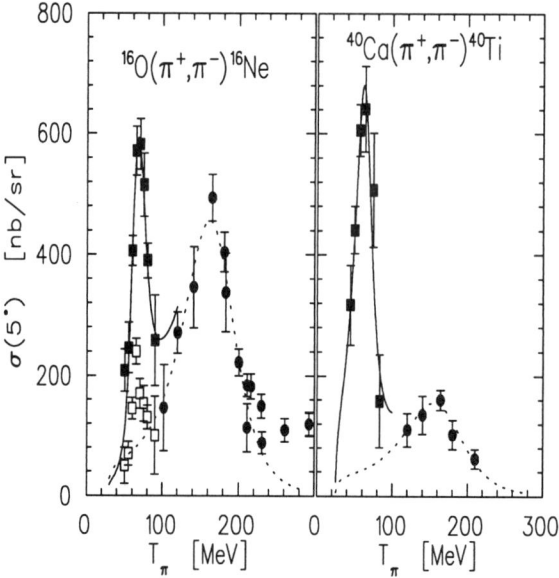

FIGURE 1. Energy dependences of the forward angle cross sections for the GSTs in ^{16}O and ^{40}Ca. The open symbols show our results for the 0_2^+ state in ^{16}Ne. Data for $T_\pi \geq 100$ MeV are from LAMPF [6]. Dotted lines give the parametrization of the $\Delta\Delta$-process, solid lines are d' calculations with $\Gamma = 10$ and 20 MeV for ^{16}O and ^{40}Ca, respectively. From [5].

These measurements clearly corroborate our earlier findings on DCX cross sections. Moreover, with their higher accuracy they provide valuable information about the collision damping of the possible d' in the nuclear medium. They do not, of course, eliminate the suspicion that possibly medium effects alone might be responsible for the observed signal. Therefore our further efforts are directed towards d' production outside the nuclear medium.

DCX REACTIONS ON FEW NUCLEON SYSTEMS

The lightest target nuclei on which DCX reactions are possible are ^{3}He and ^{4}He. The final states are in the continuum with three and four, respectively, identical nucleons. In Ref. [7] we showed that as a result of Pauli blocking the cross section for the conventional process rises slowly above the DCX threshold (about 30 MeV for ^{4}He). In the case of additional d' production the cross section now being much less subject to Pauli blocking was predicted to rise steeply at the d' threshold of about 80 MeV (depending on the precise d' mass). Based on that prediction the CHAOS collaboration at TRIUMF has performed an inclusive ^{4}He(π^+, π^-) experiment (E 725) using a liquid ^{4}He target in the CHAOS detector [8].

The resulting data points (Fig. 2) show the expected [7] fast rise of the cross section at the d' threshold and match well with previous measurements at higher energies. The conventional model used is a semi-classical Monte-Carlo simulation of two single charge exchange processes using a Fermi gas ansatz for the ^{4}He nucleus. This model also yields a satisfactory fit to the measured π^- momentum distributions at higher energies. The observed excess of the measured cross sections over the conventional model prediction at about 100 MeV supports the d' hypothesis. Subtle effects through final state interactions, however, can not easily be ruled out as an origin of the excess.

Therefore the CHAOS collaboration has in addition performed an exclusive DCX experiment (E 719) on ^{4}He, this time using a gas target with consequently tiny yields, to determine the invariant mass of the d' from its observed $pp\pi^-$ decays. First results will be presented at MENU 97.

Presently a ^{3}He$_{LQ}(\pi^-, \pi^+)3n$ experiment (E 785) is under way at TRIUMF. In the case of a $d'n$ intermediate state, the neutron time-of-flight spectra in coincidence with π^+ should exhibit a narrow signal corresponding to the d' mass.

PRODUCTION OF THE FREE d'

Based on estimates of the d' production by photoabsorption on deuterium [9] an attempt was made at MAMI to study the $\gamma + d \rightarrow d' \rightarrow np\pi^0$ reaction using the TAPS detector for π^0 and neutron detection. As only a third of the requested run time was made available I consider the chances as slim that the ongoing analysis will reveal the expected small d' signal above the background of the conventional π^0 production cross section. Naturally, as soon as the existence of the d' will have been established a careful measurement of its electromagnetic properties will be mandatory.

In contrast, our attempts to produce the free d' hadronically look very promising. Based on estimates [10] of the d'-production in pp-collisions the WASA/PROMICE collaboration at the CELSIUS ring in Uppsala has investigated the $pp \rightarrow pp\pi^-\pi^+$ reaction at 750 MeV [11]. Using a scintillator hodoscope without magnetic field the π^+ were identified by their afterpulses from weak decays. For identified 4-prong events the π^+ momenta alone determine the $pp\pi^-$ invariant mass spectrum. Fig. 3 clearly shows a narrow, statistically weak (4σ) signal at the expected energy above the continuum of expected 2π-production.

In order to improve the statistical accuracy and to exclude an artifact produced by the segmented hodoscope, additional data were taken in 1996 at proton energies of 750 and 775 MeV. These data are being analyzed presently.

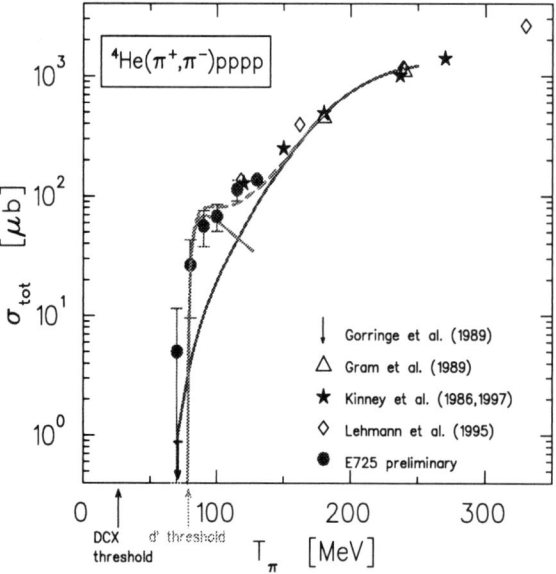

FIGURE 2. Angle and energy integrated DCX cross sections on ^{4}He as measured by E725 at TRIUMF (full points) together with existing recent data at higher energies. The full curves show the predicted conventional cross section (slow rise) and d' cross sections (rapid rise). The data follow closely the incoherent sum of both (dashed curve).

CONCLUSIONS

I have presented accumulating evidence for the existence of the d'. I have also discussed ongoing efforts which potentially may lead to an unambiguous proof of the existence of a narrow $\pi\mathcal{N}\mathcal{N}$ resonance. Once this will be established the task will shift towards closer determination of its properties. Given its weak signals it will be very challenging to determine quantum numbers and form factors. The information presented here on collision damping may be a first step towards its size and hence its nature as a $\pi\mathcal{N}\mathcal{N}$-molecule or as a proper dibaryon state. Should, however, the d' turn out to be non-existent, then the signals observed so far will have to be explained in a conventional way. And the hunt for the d' will have served as an incentive for precision measurements which are of interest in other contexts.

Acknowledgement: While the work of many collaborators within three collaborations was essential to obtain these results I wish to acknowledge in particular the tireless efforts of the "hunting party" at Tübingen, most notably R. Bilger, W. Brodowski, H. Clement, K. Föhl, J. Gräter, J. Kreß and R. Meier.

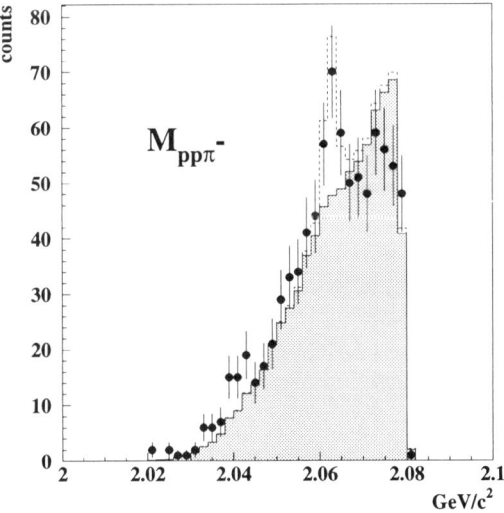

FIGURE 3. Invariant-mass spectrum $M_{pp\pi^-}$ of the reaction $pp \to pp\pi^-\pi^+$ at $T_p = 750$ MeV. The solid histograms show the MC simulations of the conventional 2π production process, the dashed ones show the result with inclusion of the d' production process. From [11].

REFERENCES

1. Kagarlis, M.A. and Johnson, M.B., *Phys. Rev. Lett.* **73**, 73 (1994)
2. Bilger, R. et. al. *Z. Phys.* **A343**, 491 (1992) and *Phys. Rev. Lett.* **71**, 42 (1993)
3. Mulders, P.G. Aerts, A.T., and de Swart, J.J. *Phys. Rev.* **D21**, 2653 (1980)
4. Kondratyuk, L.A., Martemyanov, B.U., and Schepkin, M.G., *Sov. J. Nucl. Phys.* **45**, 776 (1987)
5. Föhl, K., Ph.D. thesis, Tübingen (1996) and Föhl, K., et al., submitted for publication
6. Gilmann, R. et al., *Phys. Rev.* **C34**, 1895 (1986) and **C35**, 1334 (1997)
7. Clement, H, Schepkin, M., Wagner, G.J., and Zaboronsky, O., *Phys. Lett.* **B337**, 43 (1994)
8. Smith, G.R. et al., *NIM* **A362**, 349 (1995)
9. Bilger, R. et al., *Nucl. Phys.* **A596**, 586 (1996)
10. Schepkin, M., Zaboronsky, O., and Clement, H., *Z. Phys.* **A345**, 407 (1993)
11. Brodowski, W. et al., *Z. Phys.* **A355**, 5 (1996)

The d'-dibaryon in a colored cluster model

A. J. Buchmann, Georg Wagner and Amand Faessler

Institut für Theoretische Physik, Universität Tübingen, Auf der Morgenstelle 14, D-72076 Tübingen, Germany

Abstract. We calculate the mass and structure of a $J^P=0^-$, T=0 six-quark system using a colored diquark-tetraquark cluster wave function and a nonrelativistic quark model Hamiltonian. The calculated mass is some 350 MeV above the empirical value if the same confinement strength as in the nucleon is used. If the effective two-body confinement strength is weaker in a compound six-quark system than in a single baryon, as expected from a simple harmonic oscillator model, one obtains $M_{d'}$=2092 MeV close to experiment.

INTRODUCTION

In pionic double charge exchange reactions on nuclei at 50 MeV and forward angles, there is considerable experimental evidence for a narrow resonance in the πNN-system with quantum numbers $J^P=0^-$, T=0. This resonance has been named d'-dibaryon. Experimentally, it has a mass of $M_{d'}$=2065 MeV, and a free hadronic decay width of approximately $\Gamma_{d'} \approx 0.5$ MeV [1].

In the present work, we investigate the mass and hadronic structure of a $J^P=0^-$, T=0 six-quark system in a colored diquark-tetraquark cluster model using the Resonating Group Method (RGM) [2]. This method determines the orbital configuration of the six-quark system dynamically, i. e. according to a given model Hamiltonian. Our microscopic approach allows to test the underlying assumption of the bag-string model [3], that the d' is a stretched diquark-tetraquark system. The bag-string model prediction for the d' mass $M_{d'} \simeq 2100$ MeV employs a single, non-antisymmetrized $q^2 - q^4$ dumbbell configuration. The present potential model description of the d' improves previous bag-string model calculations in the following respects: (i) the center of mass energy is exactly removed, (ii) the Pauli principle for the whole six-quark system is respected, (iii) by virtue of the antisymmetrizer, the $q^3 - q^3$, $q^2 - q^4$, and $q^1 - q^5$ partition into colored clusters, as well as the q^6 compound state are automatically included. The model accurately reproduces the mass of the deuteron, which is the only established dibaryon.

A major purpose of this work is to study the effect of quark exchange interactions (Pauli principle) between the colored clusters on the mass and wave function of the d'. By comparing the RGM solutions with previous quark shell model results [4] employing a six-quark "bag" basis, we obtain additional information on the amount of clusterization in the system. We employ different confinement parametrizations and study how our results depend on the model of confinement. The central question is whether the present model supports a $J^P=0^-$, T=0 state with a mass compatible with experiment.

MODEL DESCRIPTION

The spontaneous breaking of chiral symmetry of low-energy QCD by the physical vacuum is responsible for the constituent quark mass generation, as well as for the appearence of pseudoscalar and scalar collective excitations of the vacuum (π and σ fields), that couple to the constituent quarks. The nonrelativistic quark model Hamiltonian for n-quarks with equal masses $m_q = 313$ MeV$=m_N/3$ (in SU$_F(2)$) contains therefore besides the residual one-gluon-exchange interaction, modelling asymptotic freedom at short distances, and besides the long-range effective two-body confinement potential, regularized one-pion- and one-sigma-exchange between constituent quarks. Several two-body confinement potentials, that differ in their radial dependence, have been considered [2]. As usual, the few parameters of the Hamiltonian are fitted to the nucleon and Δ ground state masses.

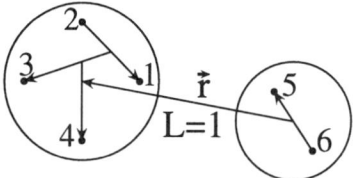

FIGURE 1. The colored diquark-tetraquark cluster model for the d'.

The six-quark wave function $|\Psi_{d'}\rangle = \mathcal{A}(\chi_{L=1}(\vec{r}) \otimes |D\rangle \otimes |T\rangle)$ sketched in figure 1 is expanded in the cluster basis into the internal wave functions of the tetraquark $|T\rangle$ and diquark $|D\rangle$ clusters, respectively, and the relative wave function $\chi_{L=1}(\vec{r})$ between the two clusters, projected onto angular momentum L=1. The antisymmetrizer $\mathcal{A}$ of Eq. (1), neglected in previous calculations [3], contains the quark permutation operators P_{ij}^{XSTC} in orbital (X), spin (S), isospin (T) and color-space (C)

$$\mathcal{A} = 1 - 8P_{46}^{XSTC} + 6P_{35}^{XSTC} P_{46}^{XSTC} , \qquad (1)$$

which ensures that the Pauli principle is respected for the whole six-quark system, and allows for a continuous transition from the compound q^6 six-quark state to the $q^2 - q^4$ clusterized state and back.

RESULTS AND DISCUSSION

Table 1 shows our main results for two different treatments for the effective two-body confinement. If we assume that the Hamiltonian for baryons and dibaryons is the same, in particular that there exists a universal effective two-body confinement strength for both, baryons and dibaryons (set I), the mass of the $J^P=0^-$, $T=0$ state is with $M_{d'}=2440$ MeV nearly 400 MeV higher than suggested by experiment. The use of different radial functions for the confinement potential (e. g. linear or error-function) leads to some reduction of the predicted mass, but $M_{d'}$ is still 300-200 MeV higher than the experimental resonance position. A comparison of the results for the d' mass with and without quark exchange (M_{noQEX}) show that the quark exchange interactions required by the Pauli principle contribute an additional energy of 80-120 MeV.

TABLE 1. We show the diquark M_D and tetraquark masses M_T, the dibaryon mass $M_{d'}$, the harmonic oscillator parameter b_6 minimizing the d' mass, the d' mass neglecting the Pauli principle M_{noQEX} and the rms radius for the intercluster coordinate $r_{d'}^{\text{RGM}}$ for the two parameter sets discussed in the text.

Set	M_D [MeV]	M_T [MeV]	$M_{d'}$ [MeV]	b_6 [fm]	M_{noQEX} [MeV]	$r_{d'}^{\text{RGM}}$ [fm]
I	643	1456	2440	0.75	2316	1.10
II	621	1309	2092	0.95	2013	1.39

The comparison of the cluster model results [2] with previous shell model results [4] reveals an overall quantitative agreement for the d' mass, the orbital structure, and the size parameter b_6 of the d'. We show in figure 2, that the d' wave function does not display a pronounced clusterization but resembles rather closely the pure six-quark harmonic oscillator (H.O.) wave function given by the dot-dashed curve. The relative wave function has already died out at distances of about 2.5 fm between the clusters. The quark exchange diagrams enhance the radial color "attraction" between the two colored clusters.

The rms radius in the relative cluster coordinate $r_{d'}^{\text{RGM}}$ is in general smaller than the sum of the corresponding diquark and tetraquark radii. In other words, there is considerable overlap between the clusters. Therefore, the bag-string model assumption of inert colored clusters at the ends of a stretched bag [3] is not satisfied in the present calculation. The Pauli principle, i. e. the fact that the quantum numbers of the d' are incompatible with two colorless ground state nucleons, together with the confinement forces prevent large interquark distances and no distinct clusterization is observed.

At present, there is no theory of confinement. Although we do not know how to calculate the effective confinement strength for three- and compound six-quarks systems from first principles, we can gain some qualitative insight within the harmonic oscillator model of confinement. If we assume a universal

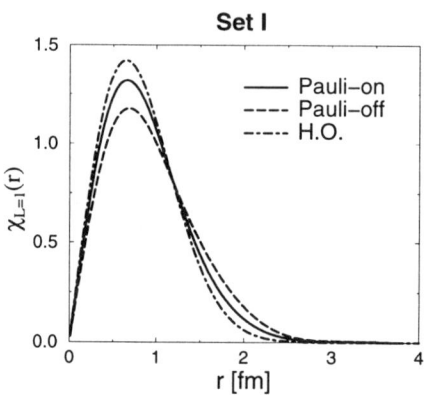

FIGURE 2. The relative RGM wave function between the tetraquark and diquark clusters with (Pauli-on) and without (Pauli-off) inclusion of the quark exchange diagrams for set I. The RGM wave functions are compared to a single six-quark shell model state (H.O.).

confining mean field for any quark in a hadron (set II), one derives for the quadratic confinement, that the effective two-body confinement strength in the six-quark system $a_c^{(6)}$ is considerably smaller than in the baryon: $a_c^{(6)} \approx a_c^{(3)}/3$. Besides the dependence of the effective two-body confinement strength on the number of quarks and the color representation of the system, the larger characteristic hadronic size of a six-quark bag as compared to a baryon is mostly responsible for the weakening of the six-quark confinement. This assumption leads to a larger d' and a mass $M_{d'}=2092$ MeV close to the experimental d' mass. Conversely, the empirical d' mass may be interpreted as evidence for a weaker effective two-body confinement strength in a compound six-quark system.

Recently, we have calculated the decay width of the d' [5], showing that for a d'-mass of $M_{d'} \approx 2100$ MeV, as predicted by the weaker confinement hypothesis, the calculated pionic decay width is compatible with experiment.

REFERENCES

1. Wagner G. J., contribution to these Proceedings;
   Bilger R., Clement H. A., and Schepkin M. G., *Phys. Rev. Lett.* **71**, 42 (1993).
2. Buchmann A. J., Wagner Georg, and Faessler Amand, submitted for publication;
   Buchmann A. J., Wagner Georg, Tsushima K., Glozman L. Ya., and Faessler Amand, *Prog. Part. Nucl. Phys.* **36**, 383 (1996).
3. Mulders P. J. G., Aerts A. T. M., and de Swart J. J.,
   Phys. Rev. **D21**, 2653 (1980); *Phys. Rev. Lett.* **40**, 1543 (1978).
4. Wagner Georg, Glozman L. Ya., Buchmann A. J., and Faessler Amand,
   Nucl. Phys. **A594**, 263 (1995).
5. Itonaga K. et al., *Nucl. Phys.* **A609**, 422 (1996);
   Obukhovsky I. T. et al., accepted for publication in *Phys. Rev. C*, (1997).

Dibaryons from Hadron Supersymmetry and a Diquark Model

D. B. Lichtenberg*

*Physics Department, Indiana University [1]
Bloomington, Indiana 47405

Abstract. We use the ideas of hadron supersymmetry and diquarks to obtain predictions concerning the masses of certain dibaryons from the known masses of mesons and baryons without using any free parameters. We find that the H dibaryon, consisting of $uuddss$ quarks, is unbound against strong decay. Likewise, we find that the H_c dibaryon, in which one s quark of the H is replaced by a c quark, is unstable against strong decay. However, in our model the H_b dibaryon may be stable against strong decay.

INTRODUCTION

This talk is based for the most part on work done with Enrico Predazzi and Renato Roncaglia [1], but I shall say several things here that they are not responsible for.

The deuteron is the only known stable dibaryon. There is some evidence for excited dibaryons, including the so-called d' dibaryon, that decay strongly into two baryons. Wagner [2] and Buchmann [3] have discussed the d' and other unstable dibaryons at this conference.

Here I want to address the question of whether there exist any dibaryons that decay weakly because they are bound against strong decay. I am particularly interested in dibaryons that, in the constituent quark model, have an internal structure unlike the deuteron, which is a loosely bound state of two quark clusters, each containing three quarks in a color-singlet state.

I shall concentrate on three dibaryons, the H [4], composed of the quarks $uuddss$, and the H_c and H_b, in which one of the s quarks is replaced by a c or b quark respectively. In our model, each of these dibaryons is composed of three diquark clusters, and each diquark is in a color-antitriplet state. In

---

[1] Work supported in part by the Department of Energy.

this model the H is made of $(ud)(us)(ds)$, the H_c is made of $(ud)(us)(dc)$ or $(ud)(ds)(uc)$, and the H_b is made of $(ud)(us)(db)$ or $(ud)(ds)(ub)$. Two quarks in parentheses denote a color-antitriplet diquark. We take these diquarks to have spin zero, because the chromomagentic energy of a spin-zero diquark is lower than that of a spin-one diquark with the same quark content. Please note that the H dibaryon of Jaffe's original model [4] has the same quark content as ours but a different internal structure.

The H dibaryon has the quantum numbers of two Λ baryons and will decay strongly into two Λ's if it has sufficiently high mass. In Jaffe's original model the H is bound with respect to two Λ's, and can decay weakly to a Λ plus neutron. However, Jaffe was well aware that his model was oversimplified and held open the possibility that the H is unbound.

Since Jaffe's work, many other theoretical papers have been written on the H, some finding that it is bound and others that it is not. Even two lattice calculations led to different conclusions, one group [5] finding that the H is not bound, and another group [6] finding that it is.

We use the idea of broken hadron supersymmetry [7] to obtain lower limits on the masses of the H, H_c, and H_b dibaryons from the known masses of mesons and baryons without using any free parameters. In our model the H and H_c are unbound, and the H_b may be bound. Although we believe that our model is reasonable, we cannot prove that it is better than all other models. Physics is an experimental science, and it is up to experimentalists to tell us whether these dibaryons are bound or not. If the H and H_c turn out not to be bound, it will give us more confidence in other predictions of broken hadron supersymmetry.

INGREDIENTS OF THE MODEL

Our model makes use of the following four ingredients: (1) constituent quarks, (2) a diquark approximation, (3) hadron supersymmetry, and (4) supersymmetry breaking. I shall discuss each of these ingredients in turn.

A constituent quark is complex particle, containing a current quark surrounded by a cloud of quark-antiquark pairs and gluons. The size and mass of a constituent quark are larger than those of a current quark. I shall not attempt to estimate the size of constituent quarks, but I do need an estimate of their masses. To be specific, I take the masses from a paper concerned only with mesons and baryons [8]. The quark masses are (in MeV)

$$m_u = m_d = 300, \quad m_s = 475, \quad m_c = 1640, \quad m_b = 4985. \tag{1}$$

An essential feature of our model is the idea of a diquark, or two-quark system considered as a single particle with a size bigger than zero [9]. A diquark can be either in a color-sextet or color-antitriplet state. The Coulomb-like contribution from one-gluon exchange between the two quarks in a diquark is repulsive in a sextet state and attractive in an antitriplet state, leading us to believe that the color-sextet diquark has a significantly higher mass. From here on, I confine myself to color-antitriplet diquarks.

The third ingredient is hadron supersymmetry, first postulated by Miyazawa [7] and discussed in terms of QCD by Catto and Gürsey [10]. The idea is that a diquark and an antiquark are both color-antitriplets of QCD. Because to first approximation, the forces of QCD depend only on color, it should be possible to make a transformation by replacing an antiquark (a fermion) in a hadron by a diquark (a boson) without changing the forces. (This supersymmetric transformation changes the baryon number and parity of the state.)

However, the supersymmetry must be broken, or the pion and nucleon would have the same mass. There are at least three sources of supersymmetry breaking: An antiquark and diquark do not have the same mass or size, and there exist spin-dependent forces which are different for the two states. We break the symmetry by taking into account mass differences and spin-dependent forces, but we neglect size differences.

BREAKING THE SYMMETRY

Our aim is to break the supersymmetry without introducing any adjustable parameters into the model. We first consider the breaking arising from diquark-antiquark mass differences and then the breaking arising from spin differences. I shall only sketch the procedure, omitting details.

In considering mass breaking, it is useful to eliminate the effect of spin as much as possible. We do this by averaging over the different possible spin states of hadrons having the same quark content. If the spin-dependent forces arise from the chromomagnetic interaction of one-gluon exchange, then a prescription exists for carrying out the spin-averaging process for mesons and baryons [11,8].

Once we have removed the effects of spin, then, replacing an antiquark by a diquark in a hadron should not change the forces. In other words, a diquark should act like a fictitious antiquark with the same mass as the diquark. This result enables us to obtain the masses of dibaryons from the masses of mesons and baryons as follows:

The mass M_{12} of a hadron containing two clusters of consituents can be written as

$$M_{12} = m_1 + m_2 + E_{12}, \qquad (2)$$

where m_1 and m_2 are the cluster masses and E_{12} is the interaction energy, which turns out to be a smooth function of the reduced mass of the constituents [8]. We obtain this function from the observed meson masses with the aid of the quark masses of Eq. (1). We next assume that E_{12} is independent of whether the two constituents are quarks, diquarks, or even more complicated clusters, so long as they are a color-triplet and a color-antitriplet combined to form a color-singlet hadron. Then, for any baryon, we can find the mass of a diquark cluster that, combined with the third quark, will give the baryon mass. This method gives us the diquark masses averaged over spin.

Next we estimate the chromomagnetic interaction energy for spin-zero and spin-one diquarks. We obtain this information from the observed mass differences of baryons containing a given quark content; for example, for uud baryons, we use the observed difference between the mass of the Δ and proton. The procedure has already been worked out [11,8].

Once we have the masses of the spin-zero diquarks, we can combine two of them to form a quadriquark cluster, and then combine the quadriquark with the remaining diquark in order to obtain the mass of the dibaryon.

There is the unfortunate complication that our procedure neglects the Pauli principle for identical quarks in different diquarks. The effect of the Pauli principle is to raise the dibaryon mass, as has been explicitly calculated for the d' dibaryon [12]. This means that our estimates of dibaryon masses are only lower limits for dibaryons having the configuration of our model.

RESULTS AND DISCUSSION

Using the procedures described in the previous sections, we obtain a lower limit on the mass of the H dibaryon, namely

$$M(H) \geq 2320 \quad \text{MeV}, \tag{3}$$

a value that is about 90 MeV above the mass of two Λ baryons. Therefore, according to our model the H is able to decay strongly into two Λ's. Likewise, we find

$$M(H_c) \geq 3460 \quad \text{MeV}, \tag{4}$$

a value about 60 MeV above the sum of the masses of the Λ and Λ_c. Therefore, the H_c also decays strongly in our model. On the other hand, we find

$$M(H_b) \geq 6730 \quad \text{MeV}, \tag{5}$$

a value 20 ± 50 MeV *below* the sum of the masses of the Λ and Λ_b. In this case, we cannot tell in our model whether or not the H_b is bound against strong decay, and we urge that a search be made for the H_b dibaryon.

Our model has several advantages and several disadvantages. Among the advantages are (1) No specific Hamiltonian is needed. (2) No hard numerical calculations are required. (3) The model has no free parameters. Among the disadvantages are (1) Because we do not have a Hamiltonian, we have no dibaryon wave functions and so cannot calculate decay rates. (2) We must exclude color-sextet diquarks because such objects are not related to antiquarks by supersymmetry. (3) There is no configuration mixing in our model. (4) We do not take take into account the Pauli principle for identical quarks in different diquarks.

In a talk at this meeting, Negele [13] stated that lattice gauge calculations have shown that instantons are mainly responsible for binding in light quark systems. Because we do not use a Hamiltonian, Negele's result does not seem to alter our main conclusions. Even if instantons, rather than chromomagnetic interactions, cause the spin splitting in hadrons, I think that our treatment of the spin splitting need not be altered because the phenomenological consquences of the two types of spin-dependent interactions appear to be qualitatively the same. On the other hand, if the spin splittings were caused by pion exchange [14], which depends on *flavor*, then our treatment would have to be changed.

In conclusion, our model allows us to obtain predictions of lower limits on the masses of dibaryons from a broken supersymmetry using as input the observed masses of mesons and baryons. Although the ideas of our model are motivated by QCD, we cannot be sure that a good treatment of QCD will yield qualitatively similar results.

REFERENCES

1. Lichtenberg, D., Roncaglia, R., and Predazzi, E., *J. Phys. G* to be published (1997).
2. Wagner, G., talk at this conference (1997),
3. Buchmann, A., talk at this conference (1997).
4. Jaffe, R., *Phys. Rev. Lett.* **38**, 195 (1977).
5. Mackenzie, P., and Thacker, H., *Phys. Rev. Lett.* **55**, 2539 (1985).
6. Iwasaki Y., Yoshie, T., and Tsuboi, Y., *Phys. Rev. Lett.* **60**, 1371 (1988).
7. Miyazawa, H., *Prog. Theor. Phys.* **36**, 1266 (1966).
8. Roncaglia R., Lichtenberg, D,, and Predazzi, E., *Phys. Rev. D* **52**, 1722 (1995).
9. Anselmino, M., Predazzi, E., Ekelin, S., Fredriksson, S., and Lichtenberg, D., *Rev. Mod. Phys.* **65**, 1199 (1993).
10. Catto, S., and Gürsey, F., *Nuovo Cimento* **86** 201, (1985).
11. Anselmino, M., Lichtenberg, D., and Predazzi, E., *Z. Phys. C* **48**, 605 (1990).
12. Buchmann, A., Wagner, G., Tsushima, K., Glozman, L, and Faessler, A., *Prog. Part. Nucl. Phys.* **36**, 383 (1996).
13. Negele, J., talk at this conference (1997).
14. Genovese, M., Richard, J-M., Pepin, S., and Stancu, Fl., *Intern. Workshop on Diquarks III*, Torino, (Oct. 28–30,1996), proceedings to be published.

Search for Dibaryons in Reactions of K^- Mesons with Deuterium and ^{3}He Nuclei at 0.87 GeV/c

Henryk Piekarz

Florida State University, Tallahassee, Florida 32306

Abstract. The reactions $K^- d \to \pi^- X^+$, $K^- {}^3He \to \pi^+ n X^0$, and $K^- {}^3He \to \pi^+ p X^-$ at 0.87 GeV/c were studied using missing mass magnetic spectrometer. The X stands for a two-baryon, strangeness -1 resonant system. In order to enhance p-wave production of a dibaryon state all reactions were studied at large scattering angles; 10° to 26° for deuterium target, and 20° for ^3He one. After known backgrounds were reproduced an excess of events is observed in all three reactions. The mass distribution and the final state particles associated with the excess events suggest compatibility of these events with expectations for two-baryon resonant state decays.

Motivation

With recent discovery of the *top* quark all fundamental blocks of matter of the Quantum Chromodynamics (QCD) model have been found. This success of QCD at high energies proves that the color force is a fundamental one for mediating hadronic interactions. Consequently, the strong force used to describe the hadronic interactions at low energies is a residue of the color force. This in turn leads to a suggestion that the multiquark hadrons other than $3q$, or $q\bar{q}$ structure may exist. In particular, strangeness -1 dibaryons of $(q^4)_3 \otimes (q^2)_3$. structure (1) received great deal of attantion as their asymmetric quark arrangement requires p-wave production mechanism. This is contrary to the potential models of hyperon-nucleon interaction mediated with meson exchanges where the p-wave production is expected to be much suppressed.

Experimental Procedure

The reactions of K^- mesons with deuterium and ^3He target nuclei are expected to produce dibaryons with strangeness -1 of any allowed quark substructure. The kaon beam momentum of 0.87 GeV/c combined with large scattering angles (10° to 26°) allows one to examine these reactions in a wide range of momentum transfers (0.2-0.5) GeV/c, thus creating conditions for enhancement of reactions produced in p-wave relative to a dominantly s-wave produced background.

The principle of the experimental procedure is based on the measurement of the missing mass X in the (K^-, π^-) reaction with the deuterium target, or missing mass NX in the (K^-, π^+) reaction with the ^3He target. In the case of the deuterium target the missing mass is a two-baryon system, while in the case of the ^3He target it is a three-baryon system. The knowledge of the kaon and pion momenta from the measurement in the spectrometer, and the target mass allow one to calculate the mass of the produced resonant system. The missing mass resolution was measured using the $H_1(K^-, \pi^-)\Sigma^+$ reaction. The Σ particle mass resolution was found to be about 6 MeV/c^2 (fwhm) with very little dependence on the scattering angle. Consequently, if the mass width of the candidate dibaryon state is narrower than spectator nucleon energy spread (about 24 MeV/c^2) such a resonance will be distinguished from the quasi-elastic background.

The most crucial factor in these experiments is the ability to distinguish the final state particles produced in the resonant state decay from those emitted in the quasi-elastic background processes with one nucleon (deuterium target), or two nucleons (^3He target) as spectators. The momenta of the spectator nucleons are typically well below 150 MeV/c with a maximum at values close to zero. Contrary to that the momenta of particles from the decay of the two-baryon resonant state produced in reactions induced by 0.87 GeV/c kaons have wide and for most part flat distribution ranging up to 500 MeV/c. The range hodoscope detector consisting of 12 Φ segments surrounding closely experimental target was used to detect these final state charged particles. The minimum detectable proton momentum was set to about 200 MeV/c. This allowed to reduce the quasi-elastic backgrounds by about two orders of magnitude while the resonance detection with Λp system in the final state was reduced only by about a factor of 4. The details of the experimental procedure are described in (2).

Reproduction of Backgrounds and Data Analysis

The acceptance of background reactions was studied using a simulation program. In this program beam particles were generated according to the beam phase space observed in the experiment. The interaction vertex was generated taking into account the shape and size of the space where the beam and target overlap (which changes with the spectrometer angle), as well as the beam flux decrease as the beam traversed the target. For the generated quasi-elastic processes the kaon was interacting with a nucleon inside the deuterium or ^3He nucleus. The momentum of a spectator nucleon(s) was generated according to the Hulthén wave function (3) and with a random direction. The interaction energy was determined, and the event was assigned a weight given by the cross-section for that channel at the appropriate *cms* energy. The resonant states were generated with a flat mass spectrum so that acceptances could be calculated as a function of mass. The particles from the resonance decay (Λp or $\Sigma^0 p$ for deuterium target, and Λn or $\Sigma^- n$ for ^3He target), as well as from the decay of Λ and Σ particles produced directly in the quasi-elastic processes were tracked in the target hodoscope until they ranged out. The effects of accidental hits, δ rays, photon conversion and secondary pion scattering in the target hodoscope were included in the simulation.

For the $K^- d \to \pi^- X^+$ reaction data were taken at five scattering angles: 10°, 14°, 18°, 22° and 26°, and with two target hodoscope triggers: charged particle multiplicity 2 and 3 at each angle. The projected quasi-elastic backgrounds were simultaneously fit to all data, which consisted of 10 independent sets. In the fitting procedure the background reaction cross-sections were allowed to vary within their errors. The fitting program produced remarkably consistent result for data at all scattering angles, and it revealed existence of two narrow maxima in addition to the known backgrounds. The observed and simulated missing mass distributions with details of the analysis of deuterium data are presented in (2). The first maximum at about 2130 MeV/c^2 has the highest cross-section at low scattering angles (so it is s-wave produced), while the second maximum which is at about 2140 MeV/c^2 peaks between 18° and 22° scattering angle suggesting its p-wave origin. The analysis of charged particle multiplicity associated with these maxima indicates that both strongly couple to the Λp system, and as both have the same spin and charge, the mass difference is due to a different orbital momentum of these states. Assuming that these maxima are due to dibaryon states, we denote them D_s^+ and D_p^+, where + is for charge of the two-baryon state, and s and p stand for the s-wave and p-wave production, respectively.

The data for $K^- \, ^3He \to \pi^+ n X^0$, and $K^- \, ^3He \to \pi^+ p X^-$ reactions were taken at 20° only where the deuterium experiment suggested highest production rate of a p-wave resonance. The first reaction proceeds on a di-proton with a neutron as a spectator. As the di-proton in ^3He is in a singlet state of spin 0, the production of a spin 0 dibaryon state is expected. The second reaction proceeds on a neutron in ^3He with one proton being attached to form a new two-baryon system while a second proton is a spectator. This allows for a production of a spin 1 dibaryon state. The dominant backgrounds are the quasi-elastic production of the Σ^-, and the K^- decays to K_s^0. Both these backgrounds are strongly suppressed by requesting the charged particle multiplicity 2 in the target hodoscope. The K_s^0 background is additionally suppressed by using a cut on the interaction vertex. The simulated backgrounds were fit to data by allowing both the quasi-elastic and K_s^0 backgrounds to vary only within the errors of the production cross-sections. The observed and simulated missing mass distributions with details of the analysis of ^3He data are presented in (2). It turned out that the best fit to data required an assumption of two gaussian shaped maxima with masses 3015.6 MeV/c^2 and 3049.5 MeV/c^2, respectively. The width of these maxima (26 MeV/c^2

and 29 MeV/c^2) is much smaller than that of the quasi-elastic processes with two spectator nucleons (about 40 MeV/c^2) thus indicating that these maxima couple to two-baryon systems. In addition, the analysis of the charged particle multiplicity associated with these maxima suggests that the two-baryon system of the 3049.5 MeV/c^2 maximum couples to the Λn particles, while for the 3015.6 MeV/c^2 maximum the final state $\Sigma^- n$ two-baryon system is the most probable one. The mass of the two-baryon system is obtained by subtracting the mass of the appropriate spectator nucleon. We denote these projected dibaryon states as D_p^0 and D_p^-, respectively with 0 and - representing charge of the two-baryon state, while p stands for the p-wave production.

Results and Discussion

The properties of the observed maxima in both deuterium and ^3He experiments, and their possible assignment to dibaryon states are summarized in a Table below. The properties of all p-wave produced candidate dibaryon states

Reaction Final State Scattering Angle	Mass MeV/c^2	Width MeV/c^2	Cross-section $\mu b/sr$	L^P S I	Dibaryon Configuration Predicted Mass (GeV/c^2)
$K^-d \to \pi^- X^+$ $X^+ \to \Lambda p$ 22°	2128.5 ±.2 (stat) ±2.3 (syst)	12.8 ±.8 (stat) ±1.2 (syst)	29.3 ±3.3 (stat) ±1.0 (syst)	0^+ 1 1/2	D_s^+ $(q^6)_1$ 2.17
$K^-d \to \pi^- X^+$ $X^+ \to \Lambda p$ 22°	2144.3 ±1.0 (stat) ±2.3 (syst)	13.6 ±2.7 (stat) ±2.0 (syst)	25.6 ±4.5 (stat) ±2.2 (syst)	1^- 1 1/2	D_p^+ $(q^4)_3 \otimes (q^2)_{3^*}$ 2.15
$K^{-3}He \to \pi^+ X^0 n_s$ $X^0 \to \Lambda n$ 20°	2109.9 ±0.4 (stat) ±3.0 (syst)	16.3 ±2.9 (stat) ±1.3 (syst)	0.98 ±0.04 (stat) ±0.15 (syst)	1^- 0 1/2	D_p^0 $(q^4)_3 \otimes (q^2)_{3^*}$ 2.11
$K^{-3}He \to \pi^+ X^- p_s$ $X^- \to \Sigma^- n$ 20°	2077.3 ±1.1 (stat) ±3.0 (syst)	9.6 ±5.6 (stat) ±12.0 (syst)	1.46 ±0.16 (stat) ±0.22 (syst)	1^- 1 3/2	D_p^- String Model 2.06-2.13

agree remarkably well with theoretical predictions. The mass difference between spin 1, D_p^+ state and spin 0, D_p^0 state is about 34 MeV/c^2, to be compared to the predicted value of 40 MeV/c^2 (1). The D_p^0 dibaryon formation requires exchange of 4 quarks while only 2 quarks have to be exchanged to form the D_p^+ state. Consequently, the production of the D_p^0 state should be suppressed relative to that of the D_p^+ one, as observed in the experiment.

Probably the most intrigueing result is the observation of a maximum at 2077 MeV/c^2 with suggested identification as D_p^- dibaryon state coupling to a $\Sigma^- n$ of isospin 3/2 . As the mass of this maximum is well below the ΣN threshold only a weak decay of such a dibaryon state is possible. In Ref. (1) the isospin 3/2 dibaryon states were assumed to be less bound than those of isospin 1/2, but in a study of a dibaryon string model (4) a weakly bound $\Sigma^- n$ state is predicted to exist. Although only 2 quarks have to be exchanged to produce the D_p^- state in reaction of K^- mesons with ^3He nuclei, its production is much suppressed due to necessity of breaking-up the di-proton system in ^3He nucleus. Earlier search (5) for a $\Sigma^- n$ dibaryon in the reaction of K^- mesons with a deuterium target produced a negative result, but the experiment was not sensitive to dibaryon mass below 2110 MeV/c^2.

The maximum at 2128.5 MeV/c^2 is strongly produced in the s-wave (2), and its mass is slightly below the ΣN mass. Both the large width (13 MeV/c^2), and the strong coupling to the Λp instead of to the ΣN system do not support an earlier suggestion (6) of a threshold cusp effect being responsible for this maximum.

The masses of all p-wave produced dibaryon candidates are only within a few MeV/c^2 from the predicted values in (1). The predicted mass of the lowest $(q^6)_1$ singlet dibaryon state is at 2.17 GeV/c^2, thus being some 40 MeV/c^2 above the mass of the D_s^+ dibaryon candidate at 2128.5 MeV/c^2. The assumed physical parameters of dibaryon structure have a very strong impact on the color-magnetic energy of the system, and thus on the dibaryon mass. As these quantities are not known, and difficult to project, the mass prediction can not be accurate. The remarkable agreement of theory and experiment for the orbitally excited dibaryon candidates may be simply accidental. However, the mass difference of the D_p^+ and D_p^0 dibaryon candidate states is significant as it relates to the spin-dependent term of the color-magnetic interaction.

References

1. P.J.Mulders. A.T.Aerts and J.J.deSwart, Phys.Rev.D21, 1980, p.2653
2. H.Piekarz, FSU-HEP-931201 Report, 1993.
3. L.Heulthen, M.Sugawara, Handbuch der Physik, 1967, Vol 39, Ch.1.
4. L.A.Kondratyuk,YU.V.Ralchenko and A.V.Vasilets, 1988, ITEP 17.
5. M.May et al., Phys.Rev.C25, 1982, p.1079
6. R.H.Dalitz and A.Deloff, Czech.J.Phys.,1982, B32, p.1021.

Diffractive Phenomena at Tevatron [1]

A. Santoro[2]

For the DØ Collaboration
LAFEX/CBPF
Rio de Janeiro, RJ, Brazil
Fermilab
P.O.Box 500, Batavia, Il
60510, USA

Abstract.
Preliminary results from the DØ experiment on jet production with rapidity gaps in $p\bar{p}$ collisions are presented. A class of dijet events with a forward rapidity gap is observed at center-of-mass energies $\sqrt{s} = 1800$ GeV and 630 GeV. The number of events with rapidity gaps at both center-of-mass energies is significantly greater than the expectation from multiplicity fluctuations and is consistent with a hard single diffractive process. A class of events with two forward gaps and central dijets are also observed at 1800 GeV. This topology is consistent with hard double pomeron exchange. We also present proposed plans for extending these analysis into Run II through the use of a forward proton detector.

I INTRODUCTION

The results coming from Tevatron (CDF and DØ) and Hera (H1 and ZEUS) [1,2] and the progress in the phenomenological and theoretical side of diffractive physics are very important to understand the complexity of the pomeron in high energy physics. Soft and hard diffraction are an important part of strong interactions. Soft diffraction is well described phenomenologically by the Regge Model (RM) with a hadronic pomeron [3]. For hard diffraction [4] we have many interesting models proposing different configurations for the pomeron as a composite object. Among the leading models for a hard pomeron we have the "economical" two-gluon model [5] and the "hot spot of gluons" or the BFKL pomeron [6].

Hard diffraction can be best studied at the Tevatron due to the large diffractive mass accessible. For hard single diffraction we have aproximately $M_x=400$ GeV and for double pomeron exchange $M_x=100$ GeV, where M_x is the diffractive mass produced by interactions with pomeron. We believe that these subjects are certainly a good bridge between the soft (RM) and hard (QCD) [4] interactions. We will present here some preliminary results from DØ for hard single diffraction and for double pomeron exchange [2].

---

[1] Published Proceedings from VI Conference on the Intersections of Particle and Nuclear physics, Big Sky, Montana, USA, May 27–June 2 1997.

II HARD SINGLE DIFFRACTION

An experimental signature of hard diffractive events is the presence of a rapidity gap (lack of particle production in a rapidity or pseudorapidity[3] region), along with a hard scattering (jet production, W production, etc.). Since the pomeron is a color singlet, radiation is suppressed in events with pomeron exchange, typically resulting in a large rapidity gap [9]. In hard single diffraction a pomeron is emitted from one of the incident particles (proton or antiproton) and undergoes a hard scattering with the second proton, often leaving a rapidity gap in the direction of its parent particle (antiproton or proton). We examine the process $p + \bar{p} \rightarrow j + j + X$ and look for the presence of a forward rapidity gap along the direction of one of the initial beam particles.

The existence of a diffractive signal in the experimental data may be observed as a larger number of rapidity gap events in the forward multiplicity distribution than expected from the non-diffractive background. Given sufficient detector resolution, sensitivity, and statistics, two components in the multiplicity distribution can be resolved and the relative fraction of rapidity gap events in excess of expectations from a smoothly falling multiplicity distribution can be estimated.

The DØ detector [7] is used to provide experimental information on the fraction of jet events with forward rapidity gaps. This analysis primarily utilizes the uranium-liquid argon calorimeters which have full coverage for a pseudorapidity range of $|\eta| < 4.1$. The transverse segmentation of the projective calorimeter towers is typically $\Delta\eta \times \Delta\phi = 0.1 \times 0.1$. The electromagnetic (EM) section of the calorimeters is used to search for rapidity gaps. The EM section is particularly useful for identifying low energy particles due to its low level of noise and ability to detect neutral pions. A particle is tagged by the deposition of more than 200 MeV of energy in a single EM calorimeter tower.

The data used in this study were obtained using a forward trigger requiring at least two jets above 12 GeV in the same hemisphere with both jets having $\eta > 1.6$ or $\eta < -1.6$. Since the pomeron only carries a few per cent of the initial proton momentum, the jet system is expected to be boosted in diffractive jet production, thus a forward jet trigger can be utilized to provide an enhanced sample of diffractive events. Offline, two jets above trigger threshold are required and events with multiple $p\bar{p}$ interactions or spurious jets are removed. Jets are reconstructed using a cone algorithm with radius, $R = \sqrt{\Delta\eta^2 + \Delta\phi^2} = 0.7$. The number of EM towers (n_{EM}) above a 200 MeV energy threshold is measured in the hemisphere opposite the leading two jets in the region $2 < |\eta| < 4.1$. The (n_{EM}) distribution for the forward trigger is shown in Fig. 1 for $\sqrt{s}$ of (a) 1800 GeV and (b) 630 GeV.

---

[3] Pseudorapidity or $\eta = -ln[tan(\frac{\theta}{2})]$, where θ is the polar angle defined relative to the proton beam direction.

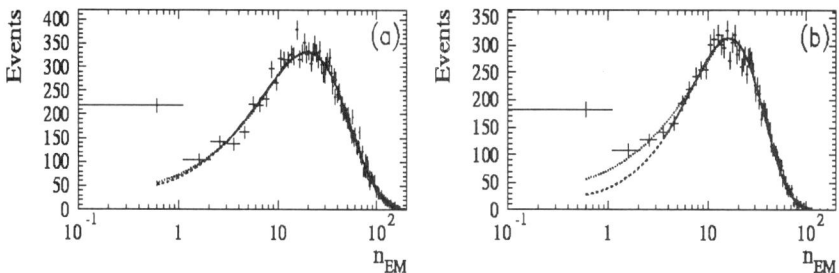

FIGURE 1. Number of electromagnetic calorimeter towers (n_{EM}) above a 200 MeV energy threshold for the region $2 < \eta < 4.1$ opposite the forward jets for center-of-mass energies of (a) 1800 GeV and (b) 630 GeV. The curves are negative binomial fits to the data excluding low multiplicity bins.

The distributions at both center-of-mass energies show a peak at zero multiplicity in qualitative agreement with expectations for a diffractive signal component. Negative binomial fits to the leading edge and the whole distribution (excluding $n_{EM} = 0$) have been used to estimate the non-diffractive background. A fractional excess of rapidity gap events is defined to be the number of zero multiplicity events in excess of those predicted by the fit divided by the total number of events in the sample. The fractional excess observed in the forward region for the $\sqrt{s} = 1800\,\text{GeV}$ sample is $0.67 \pm 0.05\%$, where the error includes only statistical uncertainties and a systematic uncertainty based on the choice of range for the fit. An excess of rapidity gap events is also clearly observed at 630 GeV with a magnitude of $1 - 2\%$. Systematic studies have not been completed, but effects such as gap detection efficiency are expected to reduce the number of observed rapidity gaps, and correcting for these effects is expected to give a modest increase in the magnitude of the signal measurement. The observed fractional excess is relatively insensitive to the calorimeter energy threshold, and rapidity gap events ($n_{EM} = 0$) typically have zero multiplicity in other available detectors, such as hadronic calorimeters, forward tracking, beam hodoscopes, and forward muon chambers.

The forward gap fraction measurement for the $\sqrt{s} = 1800\,\text{GeV}$ sample has been extended to unrestricted jet topologies using an inclusive jet trigger and we observe that the gap fraction increases with the boost of the jets, consistent with the expected behavior of diffractive events discussed earlier.

III DOUBLE POMERON EXCHANGE

The same experimental methods may be applied to a search for hard double pomeron exchange. In this process both incoming protons emit a pomeron and the two pomerons interact to produce a jet system. Rapidity gaps are expected to be produced along each forward beam direction, since there is no color connection between the jet system and the beam particles. In this analysis we have selected an enhanced sample of forward rapidity gap events with a dedicated single gap trigger, which, in addition to the jet requirements, vetoes on forward particles in either beam direction using the scintillator beam hodoscopes which bracket the DØ collision region. Events were selected to have a rapidity gap ($n_{EM} = 0$) in the direction of the online veto. These data consist of about 40,000 single gap events at $\sqrt{s} = 1800$ GeV, compared to the approximately 200 events observed in the forward trigger sample after background subtraction. This enhanced diffractive sample is used to search for double forward gap events, in which we require no towers above threshold in both forward calorimeter regions along with two jets with $E_T > 15$ GeV and $|\eta| < 1.0$. This is an expected topology for events produced in hard double pomeron exchange. The n_{EM} distribution for the veto-trigger is plotted in Fig. 2 for the forward region ($2 < |\eta| < 4.1$) opposite the tagged rapidity gap. We clearly observe a sample of double gap events, although an interpretation of them in terms of hard double pomeron exchange requires further study.

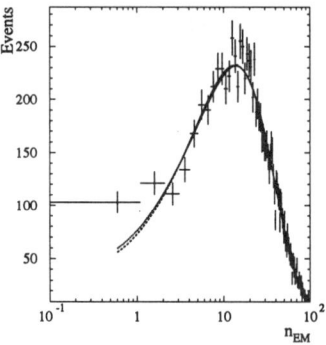

FIGURE 2. The n_{EM} distribution opposite the tagged gap for single gap trigger data. The zero multiplicity events are double gap events in this sample. The curves are negative binomial fits to the data excluding low multiplicity bins as described in the previous section.

IV CONCLUSION AND PROSPECTS FOR THE NEXT FUTURE - RUN II AT THE TEVATRON

We have observed the presence of forward rapidity gaps in events with high E_T jet production with the DØ detector at Fermilab. The fraction of forward rapidity gap events observed is in excess of those expected to be produced via multiplicity fluctuations at center-of-mass energies of 1800 GeV and 630 GeV. This is consistent with expectations from hard single diffractive jet production and provides the first experimental evidence for this process at $\sqrt{s} = 1800$ GeV. We also observe a class of events containing high E_T central jets and two forward rapidity gaps, consistent with a hard double pomeron exchange event topology.

All these results motivated us to propose a new set of sub-detectors (the so called Roman Pots [8]) to be introduced in the two arms of the DØ Detector. This forward proton detector (FPD) is composed of four quadrupole spectrometers and one dipole spectrometer. They will tag the scattered proton and/or anti-proton, allowing us to observe directly a number of processes like double pomeron exchange and improve our studies of all diffractive topics in Run II at the Tevatron.

We acknowledge the support of the US Department of Energy and the collaborating institutions and their funding agencies in this work.

REFERENCES

1. A. Brandt et al. (UA8 Collaboration), Phys. Lett. B **297**, 417 (1992); S. Abachi et al.(DØ Collaboration) *Phys. Rev. Lett.76,734 (1996)*; P. Melese (CDF Collaboration), *Proceedings of the 11th Topical Workshop on Proton-Antiproton Collider Physics*, Abano Terme, Italy, 1996; A. Doyle, *Workshop on HERA Physics, "Proton, Photon, and Pomeron Structure"*, GLAS-PPE/96-01; F. Abe et al. (CDF Collaboration), Phys. Rev. Lett. 78 (1997) 2698.
2. A. Brandt (DØ Collaboration) *Proceedings of the 11th Topical Workshop on Proton-Antiproton Collider Physics*, Abano Terme, Italy, 1996.
3. E. L. Berger and P. Pirila, *Phys.Rev. D12,3448 (1975)*; *Phys.Lett. 59B,361 (1975)*; E. L. Berger and R. Cutler, *Phys.Rev. D15, 1903 (1977)*; G. Cohen-Tannoudji, A. Santoro and M. Souza, *Nucl.Phys. B125, 445 (1977)*; G. Alberi and G. Goggi, *Phys.Report 74,1 (1981)*; K. Goulianos *Phys.Report 3,169 (1983)*; A. Donnachie and P. V. Landshoff, *Phys.Lett. B296,227 (1992)*; S. V. Levonian *28th.ICHEP Proceedings vol.I,17 Warsaw (1996)*
4. G. I. Ingelman and P. E. Schlein, *Phys. Lett. 152B,256 (1985)*; E. L. Berger, J. C. Collins, D. E. Soper, G. Sterman, *Nucl.Phys. B286,704 (1987)*; G. Sterman, *28th. ICHEP Proceedings, vol.I,103 (1996)*
5. S. Nussivov, *P.R.L. 34,1286 (1975)*;*P.R. D14,246 (1976)*; F. E. Low, *P.R. D12,163 (1975)*; J. D. Bjorken, *P.R. D47,101 (1993)*
6. Ia. Ia. Balitski, L. N. Lipatov, *Sov. J. Nucl. Phys. 28,822 (1978)*; E. A. Kuraev, L. N. Lipatov, V. S. Fadin, *Sov. Phys. JETP 45,199 (1977)*;
7. S. Abachi et al. (DØ Collaboration), *Nucl. Instrum. and Meth. in Phys. Res. A* **338**, *185 (1994)*.
8. G. Matthiae, *Rep. Prog. Phys. 57,743 (1994)*.
9. H. Chehime and D. Zeppenfeld, *preprint MAD/PH/814 (1994)*.

Exclusive Near Threshold Two-Pion Production with the MOMO Experiment at COSY

S. Bavink*, F. Bellemann*, A. Berg*, J. Bisplinghoff*,
G. Bohlscheid*, J. Ernst*, C. Henrich*, F. Hinterberger*,
R. Ibald*, R. Jahn*, L. Jarczyk**, R. Joosten*, A. Kozela***,
H. Machner[†], A. Magiera**, R. Maschuw*, T. Mayer-Kuckuk*,
G. Mertler*, J. Munkel*, P. v. Neumann-Cosel[††],
D. Rosendaal*, P. v. Rossen[†], H. Schnitker*, K. Scho*,
J. Smyrski**, A. Strzalkowski**, R. Tölle[†], R. Wurzinger[+]

* *Institut für Strahlen- und Kernphysik, Universität Bonn*
** *Institute for Physics, Jagellonian University Krakow, Poland*
*** *Institute for Nuclear Physics, Krakow, Poland*
[†] *Institut für Kernphysik, Forschungszentrum Jülich*
[††] *Institut für Kernphysik, Technische Hochschule Darmstadt*
[+] *IPN Orsay, France*

Abstract. Near threshold two pion production via the reaction $pd \to {}^3He\pi^+\pi^-$ was measured kinematically complete with the MOMO experiment at COSY. A remarkable deviation of the obtained two pion invariant mass spectra from phase space as well as a predominat sidewise and back to back emission of the two mesons was observed.

The MOMO experiment focusses on near threshold two meson production via the reactions $pd \to {}^3He\pi^+\pi^-$ and $pd \to {}^3HeK^+K^-$. It takes advantage of the high quality of the cooled external COSY beam and the high resolution spectrometer BIG KARL. The setup consists of a high granularity scintillating fibers meson detector near the target with a ± 45 deg. opening angle, and the spectrometer, which is used for ${}^3He$-identification. The large solid angle and high resolution of this detection method will yield precision data on the low energy (T<90MeV) meson-meson interaction and probe into questions like meson-nucleon resonances and KK-molecule.

ππ Relative Energy

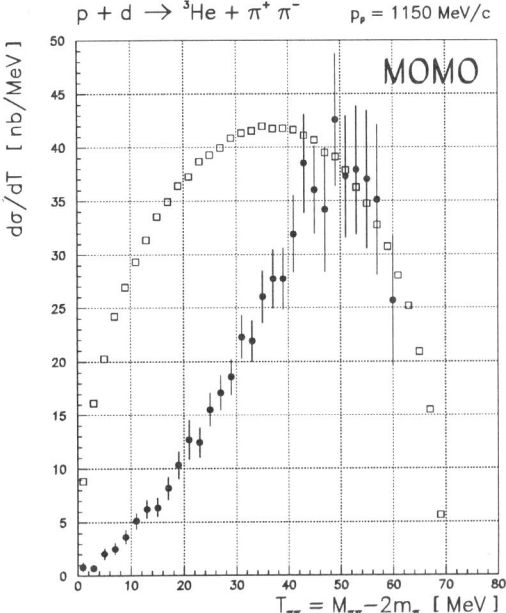

FIGURE 1. Two-pion invariant mass spectrum at 70 MeV above threshold plotted in units of the $\pi^+\pi^-$ relative energy

The MOMO vertex detector consists of 672 scintillating fibers (round, 2.5mm diameter) arranged in three planes tilted 60 deg. versus each other. The fibers are read out by 16-fold photomultipliers. The MOMO scattering chamber houses a 4mm LD_2 target with extremely thin windows. In recent COSY beam times the reaction $pd \rightarrow {}^3He\pi^+\pi^-$ was measured at three different proton beam momenta, corresponding to 35 MeV, 70 MeV and 90 MeV center of mass energy above the reaction threshold. Beam intensities ranged up to some 10^9 protons per beam burst. The ^3He particles could be unambigiously identified by time of flight and energy loss measurements. The two-pion hits on the vertex detector could be uniquely identified by their hit patterns. Good events must be coplanar in respect to the total meson momentum axis, which is defined by the beam and the ^3He momenta.

In total some 25 000 kinematically complete $\pi^+\pi^-$- events were observed. The obtained two - pion invariant mass spectra show a strong deviation from phase space at all three energies. Fig.1 depicts the 1150 MeV/c data. Since there is no absolute normalization of the phase space, the shape of our experimental spectra can be interpreted either as missing strength at low relative

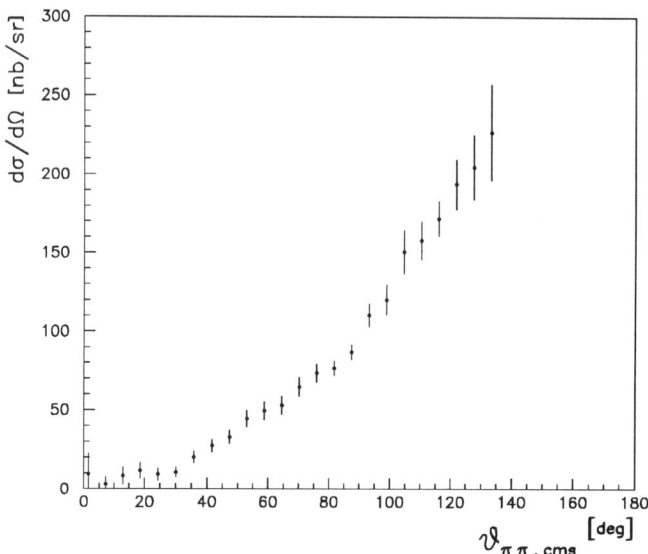

FIGURE 2. Relative angle between the two emitted pions at 70 MeV above threshold in the total center of mass system

energies or as an extra bump at large relative energies. It should be noted, that almost the same shape of the two pion invariant mass spectrum was observed in a recent study [1] of the reaction $\pi^+ + d \to pp\pi^+\pi^-$. In addition, inclusive single arm data of the $np \to d(\pi\pi)^0$ reaction at a comparable c.m.s. energy indicate a similar deviation from phase space [2]. There are several speculative explanations for the observed effect, e.g. pion - pion interaction, charge exchange ($\pi^+\pi^- \to \pi^0\pi^0$), reaction dynamics, ABC effect (bump at about 50 MeV).

The pion angular distributions measured by the MOMO experiment show a remarkable sidewise peaking (in the c.m.s.) and a preferential back to back emission of the two pions (see Fig.2), probably indicating a two delta production mechanism. It should be further noted, that the also obtained $\pi^3 He$ invariant mass spectra follow phase space, which excludes any significant detector acceptance problems. Analysis of the 35 MeV and 90 MeV data is in progress.

REFERENCES

1. TRIUMF CHAOS collaboration, *Phys. Rev. Lett.* **77**, 603 (1996).
2. Hollas, C., L. et al., *Phys. Rev.* **C 25**, 2614 (1982).

Microscopic In-Medium NN Cross-Sections up to 2 GeV

R. Machleidt and F. Sammarruca

Department of Physics, University of Idaho, Moscow, Idaho 83844

The cross sections for nucleon-nucleon (NN) scattering in the nuclear medium (nuclear matter) are an important ingredient for microscopic calculations of heavy ion reactions. Presently, in most such calculations, either the free NN cross sections or phenomenological parametrizations are used.

We have embarked on a program to derive NN in-medium cross sections—including pion production—at intermediate energies (1–2 GeV), in a microscopic way. We start from a relativistic, fieldtheoretic model for the NN interaction which includes Δ-isobar excitation and describes the free-space elastic and inelastic NN data up to about 1.5 GeV well [1]. In the nuclear medium, we take the following effects into account:

- Pauli blocking,

- "dispersive" effects (i.e., effects due to the change of the baryon energies in the medium),

- relativistic "Dirac" effects (i.e., medium effects on the Dirac spinors representing baryons),

- "screening" (i.e., effects due to the pion self-energy in the medium).

The most important in-medium cross sections at intermediate energies are the $NN \to N\Delta$, the $N\Delta \to NN$, and the $N\Delta \to N\Delta$ cross sections. In recent years, several calculations of these cross sections have been published in the literature. Some of these calculations include relativistic effects, some include screening, some solve explicitly a Lippmann-Schwinger type of equation in the medium, others use just Born terms with energy-dependent coupling constants instead of the T-matrix. Our preliminary calculations show that all effects listed above are important. Thus, for reliable predictions, a comprehensive calculation taking all effects carefully into account is needed.

REFERENCES

1. R. Machleidt, Adv. Nucl. Phys. **19**, 189 (1989).

The N-N Interaction inside Nuclei: Evidence for Partial Chiral Restoration?

E.J. Stephenson* and F. Sammarruca[†]

Indiana University Cyclotron Facility, Bloomington, IN 47408
[†] *University of Idaho, Moscow, ID 83843*

Abstract. Using a G-matrix to generate a density-dependent effective NN interaction, we examined the effects of Pauli blocking, nuclear binding, strong relativistic potentials, and modified meson masses on the polarization observables for natural parity states in ^{28}Si and the 4^- and 6^- unnatural parity transitions in ^{16}O and ^{28}Si. Changing meson masses in accordance with the expectation from partial chiral restoration produced no systematic improvement in the ability to describe the data.

INTRODUCTION

Changes to the effective meson-exchange nucleon-nucleon (NN) interaction with density (ρ) and temperature must be included in models where a quark-gluon plasma is formed in relativistic heavy ion collisions. We now understand the effects from Pauli blocking and binding in the nuclear mean field [1–3], as well as strong relativistic potentials [4], that are required to explain the binding energy and density of nuclear matter [5,6]. It has been suggested that partial chiral restoration [7] would reduce all meson masses (except for the pion) about 20% at full nuclear density, an idea that may describe excess dilepton production in heavy ion collisions at CERN [8,9].

Nucleon-induced reactions at intermediate energy should be sensitive to such mass reductions, with effects most clearly seen in measurements of polarization transfer observables [10,11]. Starting from differences between data and the predictions of standard calculations [12], initial studies of isovector, spin-flip transitions reported evidence for ρ-meson mass reductions in ^{16}O [13], ^{10}B [14], and ^{28}Si [15]. But expected spin-orbit effects appeared only for ^{28}Si [15], and contributions to isoscalar transitions through exchange were ignored. Thus a proper evaluation should include both $\Delta T=0$ and 1 transitions, as well as a broader range of meson mass modifications. Such an investigation

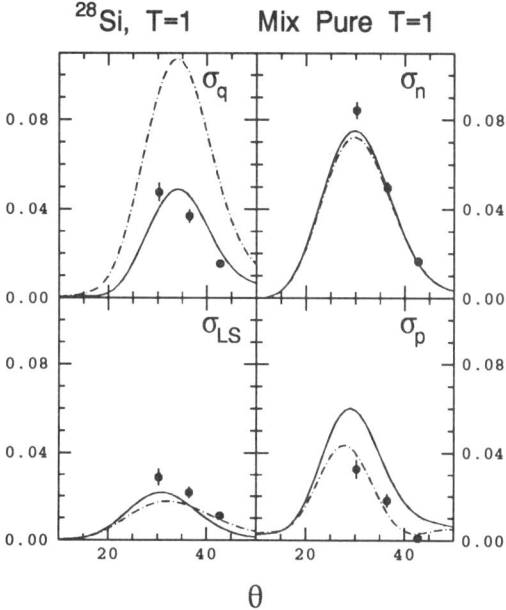

FIGURE 1. Data for the polarized cross sections, σ_q, σ_n, σ_{LS}, and σ_p, for the ^{28}Si ΔT=1 6^- transition. The dashed (solid) curves include Pauli, binding, and relativistic medium effects with an isospin mixing angle of $\alpha = 135°$ (122°).

must simultaneously account for the traditional medium effects as well as test these new suggestions. We report here the first investigation using effective NN interactions generated by a G-matrix calculation that includes all of the aforementioned medium effects.

DESCRIPTION OF THE CALCULATION

The G-matrix calculation was based on an earlier treatment of Pauli blocking, nuclear binding, and relativistic potentials in nuclear matter [5] using an interaction that describes free space NN scattering well [16]. The complex matrix elements so generated were parametrized by a series of density-dependent Yukawa terms matched to the interaction on-shell. That connection [17] also produced a model separation of the direct and exchange components of the t-matrix in preparation for an r-space DWBA calculation of the (p,p') reaction. Observables for the individual transitions were calculated using the DWBA program LEA [18]. Optical model distortions were generated microscopically from the effective t-matrix based on charge densities from electron scattering

[19] and an N=Z symmetry assumed for the nuclei considered here.

In this preliminary investigation, we considered the natural parity transition to the 2^+ first excited state of ^{28}Si [20], the $\Delta T=0$ and 1 6^- transitions in ^{28}Si [15,21], and the three 4^- transitions in ^{16}O [12,22]. The proton bombarding energies were between 180 and 200 MeV.

The structure of the 2^+ transition was based on an expansion of its charge density [20]. The 4^- and 6^- transitions were constrained by the transverse formfactors reported from (e,e') inelastic scattering [23] for the $\Delta T=1$ states. Again using N=Z symmetry, the same formfactors were assumed for the $\Delta T=0$ transitions.

The isospin of these transitions is constrained by the cross section ratios between (π^+,π^+) and (π^-,π^-) scattering for ^{16}O and ^{28}Si [24], which still allows considerable latitude. For example, the nominally $\Delta T=1$ 6^- transition in ^{28}Si may be parametrized using neutron and proton terms and a mixing angle α as $(\cos\alpha)|p\rangle + (\sin\alpha)|n\rangle$. Large-basis shell model calculations [25] predict some $\Delta T=0$ strength in the region of the $\Delta T=1$ state which, in addition to the strong $\Delta T=0$ state, could provide a basis for mixing. When we varied α from the $\Delta T=1$ value of 135° (dashed curves in Fig. 1), the best agreement (solid curves) was found for 122°. For σ_q, this result differs from the earlier chiral restoration explanation (see Fig. 2 in [15]). For the 4^- transitions in ^{16}O, we changed the mixing angle for the nominally $\Delta T=1$ transition from 135° to 129°, and for the upper $\Delta T=0$ transition from 57° to 82°.

RESULTS AND DISCUSSION

A wide variation among the experimental errors for different polarization observables led us to choose the rms deviation rather than χ^2 to achieve balance in the fit. We ignored cross section data to avoid large systematic errors. An example of the rms deviation contour is shown in Fig. 2. The most important constraint was the balance between the attractive σ and repulsive ω masses required by the analyzing power for the natural parity transition, as shown in the top panel. The good agreement obtained in more conventional treatments of medium effects is reflected here by the closeness of this curve to the free mass values (denoted by the cross in Fig. 2). Varying the ρ-meson mass in proportion to the ω [7] produces the rms deviation curve in the lower panel. While the minimum in this curve appears for meson masses larger than the free values, this change does not significantly reduce the rms deviation below what it is for free masses, and thus does not represent a systematic improvement in the ability to describe the polarization data.

No significant improvement is achieved if the ρ- and ω-meson masses are allowed to vary independently, or if the π- and η-meson masses are also varied. The same result is found when the isospin mixing angles α are reset to their original values. It must be noticed that significant differences between

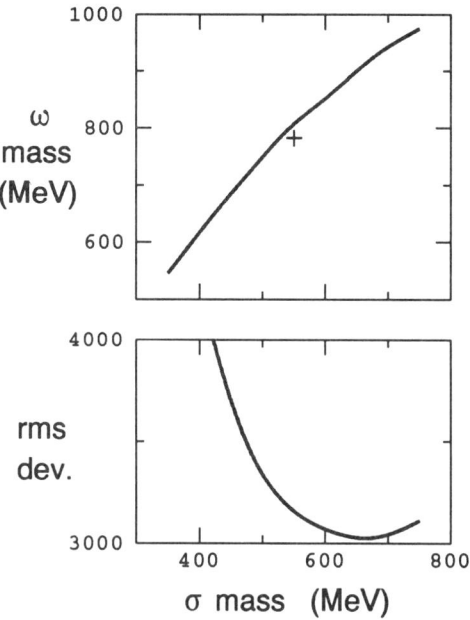

FIGURE 2. The top panel shows the constraint between the ω and σ meson masses at full nuclear density provided by the natural parity transition. The cross shows the free mass values. The lower panel shows the variation of the rms deviation in arbitrary units when the ρ-meson mass varies in proportion to the ω-meson mass.

calculations and data still remain even after adjustment to the isospin mixing angle (see $sigma_p$ in Fig. 1).

The result here differs from earlier work [13–15] because the G-matrix alters the effects of meson mass changes, most notably by inverting the sign of changes to the imaginary t-matrix. Beyond that, improvements are not systematic. A variation in a particular mass may cause improvements one place while making agreement worse in another. Such balancing occurs at all scales, within angular distributions, between different observables, between $\Delta=0$ and 1, and for natural versus unnatural parity transitions. This emphasizes the importance of considering a wide range of data when investigating medium effects, and in part explains why different conclusions were reached with less inclusive sets [13–15].

Since our description of the (p,p') polarization measurements is still less than satisfactory, it is important to continue to examine our calculations for what may be more appropriate degrees of freedom. We will investigate the G-matrix calculations of Tjon [26] which use the Dirac equation to incorporate relativity and consider virtual excitation of the Δ resonance. Medium effects are altered in the presence of coupling to the Δ resonance, which, in the present context, may provide an additional degree of freedom.

REFERENCES

1. von Geramb H.V., *The Interaction between Medium Energy Nucleons in Nuclei - 1982*, New York: AIP Press, 1983, pp. 44-77.
2. Nakayama K., and Love W.G., *Phys. Rev. C* **38**, 51 (1988).
3. Ray L., *Phys. Rev. C* **41**, 2816 (1990).
4. Furnstahl R.J., and Wallace S.J., *Phys. Rev. C* **47**, 2812 (1993).
5. Machleidt R., *Advances in Nuclear Physics* **19**, 189 (1989).
6. Horowitz C.J., and Serot B.D., *Nucl. Phys.* **A368**, 503 (1981).
7. Brown G.E., Buballa M., Li Z.B., and Wambach J., *Nucl. Phys.* **A593**, 295 (1995).
8. Li G.Q., Ko C.M., and Brown G.E., *Phys. Rev. Lett.* **75**, 4007 (1995).
9. Drees A., proceedings of this conference.
10. Bleszynski E., Bleszynski M., and Whitten C.A. Jr., *Phys. Rev. C* **26**, 2063 (1982).
11. Moss J.M., *Phys. Rev. C* **26**, 727 (1982).
12. Wissink S.W., *Spin and Isospin in Nuclear Interactions*, New York: Plenum, 1991, pp. 253-279.
13. Stephenson E.J., and Tostevin J.A., *Spin and Isospin in Nuclear Interactions*, New York: Plenum, 1991, pp. 281-286.
14. Baghaei H., et al., *Phys. Rev. Lett.* **69**, 2054 (1992).
15. Stephenson E.J., et al. *Phys. Rev. Lett.* **78**, 1636 (1997).
16. Machleidt R., Sammarruca F., and Song Y., *Phys. Rev. C* **53**, R1483 (1996); and Machleidt R., Sammarruca F., and Song Y., *Few-Body Systems*, Suppl. **9**, 410 (1995).
17. Macfarlane M.H., and Redish E.F., *Phys. Rev. C* **37**, 2245 (1988).
18. Kelly J.J., program LEA, private communication.
19. de Vries H., de Jager C.W., de Vries C., *At. Data and Nucl. Data Tables* **36**, 495 (1987).
20. Chen Q., et al. *Phys. Rev. C* **41**, 2514 (1990).
21. Liu J., Indiana University Ph.D. thesis (1995).
22. Olmer C., *Antinucleon- and Nucleon-Nucleus Interactions*, New York: Plenum, 1985, pp. 261-275.
23. Clausen B.L., Peterson R.J., and Lindgren R.A., *Phys. Rev. C* **38**, 589 (1988).
24. Carr J.A., Halderson D., Holtkamp D.B., and Cottingame W.B., *Phys. Rev. C* **27**, 1636 (1983).
25. Carr J.A., Bloom S.D., Petrovich F., and Philpott R.J., *Phys. Rev. Lett.* **62**, 2249 (1989).
26. von Faassen E.E., and Tjon J.A., *Phys. Rev. C* **30**, 285 (1984).

Elastic pp Scattering Excitation Functions at Intermediate Energies

Frank Hinterberger for the EDDA Collaboration[1] at COSY

Inst. f. Strahlen- und Kernphysik, Univ. Bonn, Nussallee 14-16, D-53115 Bonn

Abstract. Excitation functions of proton–proton elastic scattering cross sections have been measured in narrow momentum steps in the momentum range 1.1 to 3.3 GeV/c (kinetic energy range 0.5 to 2.5 GeV) and the angular range $35° \leq \Theta_{cm} \leq 90°$ with a detector providing $\Delta\Theta_{cm} \approx 1.4°$ resolution and 82% solid angle coverage. Measurements have been performed continuously during projectile acceleration in the Cooler Synchrotron COSY with an internal polypropylene (CH_2) fiber target, taking particular care to monitor luminosity as a function of beam energy. The results provide excitation functions and angular distributions of unprecedented precision and internal consistency. No evidence for narrow structures was found. The measured cross sections are compared to a recent phase shift analysis. In phase 2 of the experiment excitation functions of the analyzing power A and the polarization correlation parameters A_{NN}, A_{SS} and A_{SL} will be measured using a polarized proton beam and a polarized atomic beam target.

The proton-proton elastic scattering is one of the basic reactions involving strong interaction. At 1-2 GeV kinetic energies it provides a focus on the short-range part of the nucleon-nucleon interaction, i.e. in the framework of mesonexchange models, on the role of the ω-meson [2]. Measuring excitation functions of spin-averaged and spin-dependent cross sections over a wide momentum range allows to disentangle the various contributions to central, spin-orbit, spin-spin and spin-tensor components. Looking at the world data set [3,4], there is a lack of data at higher energies and there are inconsistencies in the normalization of angular distributions measured at different energies. In addition, energy dependent structures have been observed in certain spin observables [5]. In this context, there is still the open question of the existence or nonexistence of dibaryons [3]. Another motivation is to provide precise data for the improvement and extension of phase shift analyses and potential models at higher energies.

The idea of the EDDA experiment [6] is to measure excitation functions over a wide momentum range using the internal proton beam of the cooler synchrotron COSY [7] during beam acceleration. In phase 1, unpolarized differential cross sections were measured using 4 μm × 5 μm CH_2 fiber targets and 5 μm diameter C fiber targets for background subtraction. The targets were covered by a 20 μg/cm^2 Aluminum film to make them electrically conducting. In phase 2, the analyzing power A and the polarization correlation parameters A_{NN}, A_{SS} and A_{SL} will be measured as a function of momentum using a polarized atomic beam target and a polarized proton beam.

The detector consists of two concentric cylindrical hodoscopes with a vertex resolution of 1.3 mm covering 82 % of 4π. The phase 1 measurements were done using only the outer detector shell since the scattering vertex was well defined by the beam spot on the fiber target. A crucial point of the experiment is the luminosity monitor. Relative normalization as a function of momentum was accomplished by measuring (i) the secondary electron current scaling as the restricted energy loss rate and (ii) the δ-electrons scaling as the Rosenbluth cross section. The two methods give the same relative normalization within 2.5 % for all beam energies. Absolute normalization is established at 1455 MeV/c with reference to the angular distribution measured by Simon et al. [8] with 1% total uncertainty. Beam parameters like position, angle, width and momentum were continuously measured during the acceleration ramp. Data were taken with a constant momentum growth of 1.15 (GeV/c)/s and an average luminosity of 5×10^{29} cm^{-2}s^{-1}. Two examples of excitation functions are shown in Fig. 1 and compared to a recent phase shift analysis [9]. The exci-

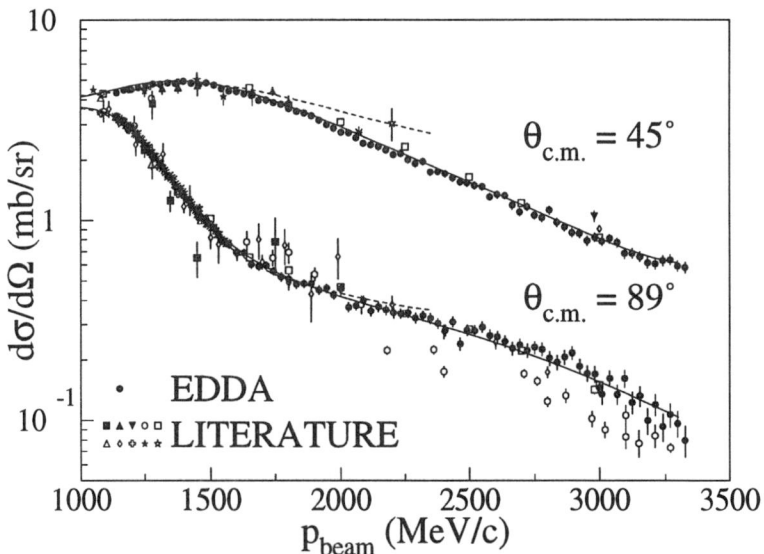

FIGURE 1. Excitation functions for elastic pp scattering at $\Theta_{cm} = 45° \pm 1°$ and $89° \pm 1°$. The solid (dashed) curves are from the SM97 (SM94) phase shift analysis [9].

tation functions show a smooth dependence on beam momentum. Especially near 2.9 GeV/c, where a narrow anomaly in the polarization correlation A_{NN} has been observed [5] we do not see any anomaly in our data. We deduce an upper limit of 0.1 for the elasticity of a possible $^1S_0$ resonance in that region.

The EDDA experiment provides consistent proton-proton elastic scattering differential cross sections covering $\Theta_{cm} = 35° - 90°$ and $p_{lab} = 1.1 - 3.3$ GeV/c to a precision of typically 5 %. Refined data analysis, including recent additional data, is striving to achieve 2 % relative accuracy in momentum dependence, and a more stringent limit to the existence of resonant excursions. On the basis of the new EDDA data, Arndt and coworkers were able to extend their partial-wave analysis from 1.6 to 2.5 GeV kinetic energy [9]. The method of using the internal synchrotron beam works well and can now be used to measure spin observables with a polarized proton beam and a polarized atomic beam as target. The support by the COSY group, the FZ Jülich and the BMBF is gratefully acknowledged.

REFERENCES

1. The EDDA Collaboration, Spokesmen J. Bisplinghoff[1], F. Hinterberger[1], W. Scobel[2]: D. Albers[2], F. Bauer[2], J. Bisplinghoff[1], R. Bollmann[2], M. Busch[1], K. Büßer[2], P. Cloth[3], R. Daniel[1], O. Diehl[1], F. Dohrmann[2], H.P. Engelhardt[1], J. Ernst[1], P.D. Eversheim[1], O. Felden[1], J. Flammer[2], R. Gebel[1], J. Greiff[2], A. Groß[2], R. Groß-Hardt[1], K. Hebbel[2], A. Heine [1], F. Hinterberger[1], T. Hüskes[1], M. Igelbrink[2], R. Jahn[1], I. Koch[2], M. Jeske[1], U. Lahr[1], R. Langkau[2], T. Lindemann[2], J. Lindlein[2], R. Maier[3], R. Maschuw[1], T. Mayer-Kuckuk[1], F. Mosel[1], M. Pfuff[2], D. Prasuhn[3], H. Rohdjeß[1], D. Rosendaal[1], U. Roß[1], P. von Rossen[3], H. Scheid[1], N. Schirm[2], M. Schulz-Rojahn[1], F. Schwandt[1], V. Schwarz[1], W. Scobel[2], B. Steeg[2], S. Steinbeck[2], G. Sterzenbach[3], S. Thomas[1], H.J. Trelle[1], E. Weise[1], A. Wellinghausen[2], W. Wiedmann[1], K. Woller[2], R. Ziegler[1]
[1]Inst. f. Strahlen- und Kernphysik, Univ. Bonn, [2]I. Inst. f. Experimentalphysik, Univ. Hamburg, [3]Inst. f. Kernphysik, FZ Jülich
2. Machleidt, R., *Adv. in Nucl. Phys.* **19**, 189 (1989).
3. Lechanoine-Leluc, C., and Lehar, F., *Rev. Mod. Phys.* **65**, 47 (1993).
4. Arndt, R.A., et al., *Phys. Rev.* **C 50**, 2731 (1994), SAID solution SM94.
5. Ball, J., et al., *Phys. Lett.* **B 320**, 206 (1994).
6. Bisplinghoff, J., and Hinterberger, F. *Particle Production Near Threshold, AIP Conf. Proc.* **221**, 312 (1991); Scobel, W. *Phys. Scr.* **48**, 92 (1993); Rohdjeß, H., *Proc. Int. Conf. on Physics with GeV-Particle Beams, Jülich, 1994* Singapore: World Scientific, 1995, p. 334; Albers, D., et al. *Phys. Rev. Lett.* **78**, 1652 (1997)
7. Maier, R., et al., *Proc. Int. Conf. on Physics with GeV-Particle Beams, Jülich, 1994* Singapore: World Scientific, 1995, p. 504.
8. Simon, A.J., et al., *Phys. Rev.* **C 48**, 662 (1993); **53**, 3 (1996).
9. Arndt, R.A., et al., SAID solution SM97, Preprint nucl-th/9706003.

New Perspectives in Multi-Nucleon Pion Absorption on Light Nuclei

A. Lehmann[1,8], D. Androić[9], G. Backenstoss[1], D. Bosnar[9], H. Breuer[4], H. Döbbeling[8], T. Dooling[7], M. Furić[9], P.A.M. Gram[3], N.K. Gregory[5], A. Hoffart[2,8], C.H.Q. Ingram[8], A. Klein[7], K. Koch[8], J. Köhler[1], B. Kotliński[8], M. Kroedel[1], G. Kyle[6], A.O. Mateos[5], K. Michaelian[8], T. Petković[9], M. Planinić[9], R.P. Redwine[5], D. Rowntree[5], U. Sennhauser[8], N. Šimičević[5], R. Trezeciak[2], H. Ullrich[2], M. Wang[6], M.H. Wang[6], H.J. Weyer[1,8], M. Wildi[1], K.E. Wilson[5]

(LADS collaboration)

[1] *University of Basel, CH-4056 Basel, Switzerland*
[2] *University of Karlsruhe, D-76128 Karlsruhe, Germany*
[3] *LAMPF, Los Alamos, New Mexico 87545*
[4] *University of Maryland, College Park, Maryland 20742*
[5] *Massachusetts Institute of Technology, Cambridge, Massachusetts 02139*
[6] *New Mexico State University, Las Cruces, New Mexico 88003*
[7] *Old Dominion University, Norfolk, Virginia 23529*
[8] *Paul Scherrer Institute, CH-5232 Villigen PSI, Switzerland*
[9] *University of Zagreb, HR-10000 Zagreb, Croatia*

Although very important in the energy region of the Δ-resonance, the absorption of a pion by nucleons is one of the least understood reactions of the pion-nuclear interaction. We know that the basic mode involves an isoscalar nucleon pair in the absorption process (2NA), but a significant fraction of the total absorption cross section leads to at least three energetic, unbound nucleons in the final state (3NA). This could be understood by a scattering of the incident pion prior to (initial state interaction, ISI) or by an interaction of one of the emerging nucleons with the nuclear environment after (final state interaction, FSI) the actual absorption of the pion. However, since previous investigations on the light system ^{3}He revealed no explicit evidence of these two-step processes (ISI+2NA, 2NA+FSI) being the origin of the 3NA mode, it was concluded that new reaction dynamics might exist in pion absorption [1]. In recent times, there are also discussions whether this $3N$ process may be an effect of the three-nucleon force.

The 4π solid angle Large Acceptance Detector System (LADS) [2] was built at PSI to study the multi-nucleon absorption mode in detail. Due to the kinematic completeness of this detector an accurate measurement of in particular the 3NA integrated and differential cross sections on the nuclei ^{3}He and ^{4}He for five incident pion energies across the $\Delta(1232)$-resonance was done, and a comprehensive analysis of the reactions ^{3}He(π^+,ppp), ^{4}He$(\pi^+,ppp)n$, and ^{4}He$(\pi^+,ppn)p$ in terms of ISI+2NA, 2NA+FSI, and coherent 3NA was performed.

With this measurement the existence of a significant 3NA component on ^{3}He and ^{4}He was conclusively confirmed [3-5]. The partial cross sections of the reactions ^{3}He(π^+,ppp), ^{4}He$(\pi^+,ppp)n$, and ^{4}He$(\pi^+,ppn)p$ are summarized in Table 1. These yields correspond to 3NA fractions in the total absorption cross section which increase with the incident pion energy and the nuclear mass (e.g., about 15% at 70 MeV on ^{3}He, and about 40% at 239 MeV on ^{4}He). The ratios $\sigma_{ppp}^{^4He}/\sigma_{ppp}^{^3He}$ and $\sigma_{ppn}^{^4He}/\sigma_{ppp}^{^4He}$ of these 3NA partial cross sections were compared to simple estimates for ISI+2NA and 2NA+FSI, and it was found that those are difficult to understand in terms of two-step processes. On the other hand, a simple consideration of the isospins in the initial and final state assuming a one-step 3NA process [6] can roughly explain these ratios. This is discussed in more detail in Ref. [5].

TABLE 1. 3NA cross sections on ^{3}He and ^{4}He with three unbound nucleons in the final state. The uncertainties include errors due to the normalization and due to different models added in quadrature.

	T_π (MeV)	70	118	162	239	330
^{3}He(π^+,ppp)	$\sigma_{ppp}^{^3He}$ (mb)	2.8±0.6	6.3±0.7	7.6±0.5	3.8±0.3	1.2±0.2
^{4}He$(\pi^+,ppp)n$	$\sigma_{ppp}^{^4He}$ (mb)	2.0±0.7	3.8±0.5	5.9±0.7	4.3±0.4	2.6±0.3
^{4}He$(\pi^+,ppn)p$	$\sigma_{ppn}^{^4He}$ (mb)	7.2±1.3	9.8±1.3	10.9±1.4	6.0±0.7	4.9±1.4

A complete set of variables was defined for the 3NA distributions and simultaneously fitted with simple semi-classical models for ISI+2NA and 2NA+FSI, and with a $3N$ phase space model (3N-PS) as an approximation of a coherent $3N$ process. A signal of the ISI+2NA process was identified [7,8], and we find that the contribution of this process increases with the incident pion energy. The signatures of 2NA+FSI are less well defined, but the fits indicate that the importance of this process, especially in the reaction ^{4}He$(\pi^+,ppn)p$, decreases with the pion energy (see Table 2). These trends are roughly consistent with expectations from the πN and NN elastic cross sections. This is also the case for the stronger yield of 2NA+FSI in the reaction ^{4}He$(\pi^+,ppn)p$ compared to ^{4}He$(\pi^+,ppp)n$. However, a large fraction (60% or more) of the 3NA yield cannot be explained by these two-step models, neither on ^{3}He nor on ^{4}He.

The distributions of this remaining 3NA cross section are described fairly well by the 3N-PS model, if the incident pion is allowed to be in an orbital angular momentum state of at least p-wave relative to the absorbing $3N$ system. There are also some significant structures in the 3NA differential cross sections, in particular in the Dalitz plots of the reaction ^{4}He$(\pi^+,ppp)n$ [4], that are not explainable by our simple models. However, these models are composed of an incoherent superposition of elementary scattering and absorption processes, which allows no interference.

TABLE 2. Fractions of the yields of the possible absorption mechanisms in the 3NA cross sections on ^{3}He and ^{4}He. The uncertainties reflect the variations of the results by fitting different sets of distributions.

	T_π (MeV)	70	118	162	239	330
^{3}He(π^+,ppp)	(ISI+2NA)	26±7 %	17±3 %	21±2 %	26±2 %	28±5 %
	[(2NA+FSI) + 3N-PS]	74±7 %	83±3 %	79±2 %	74±2 %	72±5 %
^{4}He$(\pi^+,ppp)n$	(ISI+2NA)	1±1 %	1±1 %	4±2 %	11±7 %	11±5 %
	(2NA+FSI)	6±7 %	4±5 %	1±3 %	0±1 %	1±2 %
	3N-PS	15±5 %	23±3 %	30±3 %	35±6 %	23±5 %
^{4}He$(\pi^+,ppn)p$	(ISI+2NA)	10±13 %	4±4 %	5±5 %	20±4 %	16±7 %
	(2NA+FSI)	23±11 %	20±9 %	7±9 %	1±1 %	0±1 %
	3N-PS	45±10 %	48±8 %	53±8 %	43±3 %	49±5 %

In summary, we have discussed the multi-nucleon absorption cross sections on ^{3}He and ^{4}He with three unbound, energetic nucleons in the final state. The ratios of the different 3NA cross sections are difficult to understand in terms of only two-step processes. Similar difficulties show up in the decomposition of the 3NA yield into mechanisms. Although a part of this yield is identified to be due to two-step processes, a large fraction is well described by a simple $3N$ phase space model. The question whether these problems are caused by a new coherent $3N$ process must await its final answer for realisitic quantum mechanical calculations, which take into account interferences between the initial state and the final states.

REFERENCES

1. H.J. Weyer et al., Phys. Rep. **195**, 295 (1990).
2. T. Alteholz, et al., Nucl. Instrum. Methods **A373**, 374 (1996).
3. T. Alteholz, et al., Phys. Rev. Lett. **73**, 1336 (1994).
4. A. Lehmann, et al., Phys. Rev. **C55**, 2931 (1997).
5. A. Lehmann, et al., Preprint PSI-PR-97-16 (1997), submitted to Phys. Rev. C.
6. A. Mateos and N. Šimičević, Phys. Rev. **C47**, R1842 (1993).
7. G. Backenstoss, et al., Phys. Lett. **B379**, 60 (1996).
8. D. Androić, et al., Phys. Rev. **C53**, R2591 (1996).

Pion Double Charge Exchange on Nucleus and the Two-Pion Intermediate States

A.B.Kaidalov and A.P.Krutenkova

State Research Center, Institute of Theoretical and Experimental Physics, Moscow, 117259 Russia

Abstract. The inclusive pion double charge exchange process (DCX) on nuclei is considered in the framework of the Gribov–Glauber approach. It is shown that inelastic rescatterings related to production of two pions in the intermediate state give an important contribution to DCX at kinetic energies above ~ 0.5 GeV. This mechanism dominates at energies $\gtrsim 1$ GeV and allows to explain the weak energy dependence of the DCX cross section observed experimentally.

The standard mechanism of DCX corresponds to two sequential single charge exchanges (SSCX) of pion on nucleons of a nucleus (with π^0 in the intermediate state, i.e. elastic rescattering in the Glauber picture). The mechanism gives a reasonable description of experimental data on DCX at kinetic energies of incident pions $T_0 \lesssim 0.5$ GeV and predicts a strong decrease of a small angle DCX at higher energies. Recently the inclusive DCX cross section has been measured for the reactions $A(\pi^-, \pi^+)X$ on ^6Li and ^{16}O at energies $T_0 = 0.6$, 0.75 and 1.1 GeV ($\langle\theta\rangle \approx 5^0$) [1]. To detect just DCX, the kinematics was chosen in such a way as to forbid production of an extra pion, $\Delta T = T_0 - T \leq m_\pi$, where T is the kinetic energy of the produced pion. It appeared that the measured differential cross section decreases with energy rather weakly and exceeds the theoretical prediction at the highest energy by an order of magnitude. Thus other mechanisms of DCX are needed to explain the observed energy dependence.

As it was shown in [2] the inelastic rescatterings (two pions in the intermediate state) give an important contribution to the cross section of DCX at energies $T_0 > 0.5$ GeV and allow to understand the experimentally observed pattern. We used there the Gribov formalism for a description of inelastic Glauber rescatterings (IR) on nuclei [3].

The results are given in Fig. 1 (IR1 and IR2 take into account the inelastic rescatterings, including, for IR2, also the experimental limitation on ΔT).

An overall agreement between experimental data and predictions of the model with inelastic rescatterings is quite satisfactory taking into account that several simplifying assumptions were made.

So we can conclude: a) for pion DCX the inelastic rescatterings are dominant already at $T_0 \gtrsim 1$ GeV, while for ordinary hadron-nucleus interactions it is not very large (20-30% of elastic rescattering); b) the theoretical interpretation of the DCX experiments (inclusive and exclusive) at $T_0 \gtrsim 1$ GeV is incorrect without IR. The available data on the total cross section of the reaction $\pi^- p \to \pi^+ \pi^- n$ and $\pi^- p \to \pi^0 n$ at $T_0 \sim 2 - 5$ GeV show that the value of inelastic rescatterings at these energies is still rather large. Thus we expect large deviations from the standard DCX mechanism also in this energy region.

REFERENCES

1. Abramov B.M. et al., *Yad. Fiz.* **59**, 399 (1996) (English translation: *Phys. Atomic Nucl.* **59**, 376 (1996)).
   Abramov B.M. et al., *Few-Body Systems Suppl.*, **9**, 237 (1995).
2. Kaidalov A.B., and Krutenkova A.P. Nucl-th 9704042.
3. Gribov V.N., *ZhETF*, **56**, 892 (1969) (English translation: *Sov. Phys. JETP*, **29**, 483 (1969)).

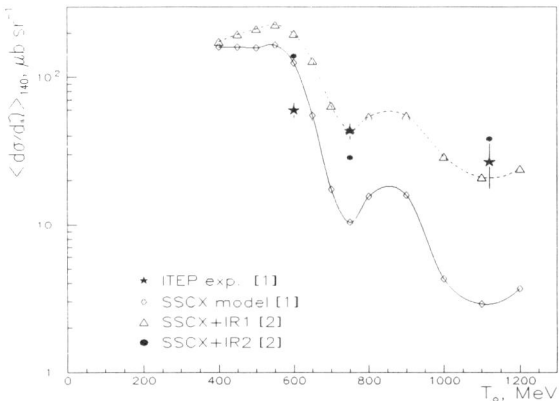

FIGURE 1. The cross section of the inclusive pion double charge exchange on ^{16}O integrated over the region $\Delta T = 0 \div 140\ MeV$

Study of Hyperonic Atoms at the FNAL Main Injector

Yuri M. Ivanov and Anatoli A. Petrunin

Petersburg Nuclear Physics Institute
Gatchina 188350, Russia
e-mail: yumi@hep486.pnpi.spb.ru

Abstract. Possible experiments with Σ, $\overline{\Sigma}$, Ξ, and Ω atoms at the FNAL Main Injector are discussed which aim to improve hyperon masses and moments, to measure, at first, Σ-nuclear spin-orbit strong interaction, and to search for Σ^-n hypernucleus.

INTRODUCTION

In experiment [1] at IHEP there was developed a new method to produce and study hyperonic atoms using crystal spectrometer with target irradiated by high energy proton beam (70 GeV, $4 \cdot 10^{12}$ protons per 9 s). The measured 5g-4f transition in Σ-C atom provided the most precise determination of Σ^- mass [2].

A more intensive proton beam of the FNAL Main Injector (120 GeV, $3 \cdot 10^{13}$ protons per 3 s) makes possible to advance previous research. To estimate hyperonic atom production at the FNAL Main Injector we used GEANT 3.21 with FLUKA [3] as hadronic shower generator. The applicability of the simulation tools was tested by comparison with measurements at IHEP. The Monte-Carlo results for carbon targets are shown in Table. 1.

TABLE 1. Stops per incident proton at the MI.

Particle	C, ϕ 0.4 cm × 20 cm	C, ϕ 3 cm × 20 cm
Σ^-	$3.0 \cdot 10^{-5}$	$2.2 \cdot 10^{-4}$
Ξ^-	$< 5.0 \cdot 10^{-7}$	$1.6 \cdot 10^{-6}$
$\overline{\Sigma}^-$	$< 5.0 \cdot 10^{-7}$	$5.0 \cdot 10^{-7}$
Ω^-	$< 5.0 \cdot 10^{-7}$	$< 2.0 \cdot 10^{-8}$

Σ^- MASS AND MAGNETIC MOMENT

The most suitable transition for measurement of Σ^- mass and magnetic moment is the 5g-4f of Σ-C atom. MC simulation shows that experiment at the FNAL MI can improve by an order of magnitude a precision of Σ^- mass and simultaneously can bring a new determination of Σ^- magnetic moment with accuracy better than 1%. The improved Σ^- mass and magnetic moment provide possibility to study ΣN strong interaction and to measure the binding energy of Σ^-n system, if it exists.

Σ^- SPIN-ORBIT STRONG INTERACTION WITH NUCLEI

Theoretical calculations for Σ-nucleus spin-orbit force have large uncertainties and are changed from about zero up to 4/3 of nucleon force [4,5]. Attempts to measure this interaction were undertaken at CERN, KEK, BNL, but no clear experimental data on this problem were obtained.

We estimated the most suitable transitions for studying ΣN spin-orbit strong interaction with crystal spectrometer. These are the 4f-3d transition in Σ-C atom and the 5g-4f and 6g-5f transitions in Σ-O atom with energies 50.7, 42.7 and 23,2 keV, respectively. Strong interaction broadens and shifts atomic energy levels versus electromagnetic values, and these effects depend on magnitude of spin-orbit force. In case of the 4f-3d transition in Σ-C atom the spin-orbit contributions are so considerable that two main components of the fine structure are displaced each other and their widths are differed by about 4 times. From MC simulations the estimated accuracy of spin-orbit strength measurements is expected to be about several percents.

SEARCH FOR Σ^-N HYPERNUCLEUS

A bound Σ^-n system was unsuccessfully searched for in K^- capture by D and ^{4}He at rest (for references see [6]). In case of ^{4}He the upper limit (1 event) was obtained [7] which is equal to about 0.3% per Σ^- produced in K^- ^{4}He capture (for other nuclei this limit should to be varied due to nuclear structure effects).

If bound Σ^-n state exists, the low energy Σ^- hyperons may pickup neutrons from nuclei, where they were produced, and the estimated cross-section is equal to about several mb. The MC simulations show that measurements of X-rays from Σ^-n-atoms are possible in this case.

From measured Σ^-n-atomic transitions the binding energy and spin of Σ^-n can be found. The estimated accuracy of binding energy determination (after improvement of Σ^- mass) is about several percents.

MEASUREMENTS WITH Ξ, $\overline{\Sigma}$ AND Ω-ATOMS

An applying a focusing spectrometer of reflecting type can provide possibility to improve Ξ^- and $\overline{\Sigma}^-$ masses by an order of magnitude and to study strong interaction of Ξ^- with nuclei. The observation of Ξ-atoms could be a first step towards a study of Ω-atoms which aims to measure Ω^- spin and electric quadrupole moment providing information on tensor quark-quark interaction [8].

ACKNOWLEDGEMENTS

We are grateful to Prof.A.A.Vorobyev for initiating this work and our colleagues for help and useful discussions.

The work was supported in partly by Russian Foundation for Basic Researches (grant 97-02-17087)

REFERENCES

1. Gur'ev, M.P., et al., *JETP Lett.*, **57**, 400 (1993).
2. Particle Data Group, *Phys. Rev.*, **D54**, 1 (1996).
3. Fasso, A., et al., "FLUKA: present status and future developments", in *Proceedings of the IV Int. Conf. on Calorimetry in High Energy Physics*, La Biodola (Elba), September 19-25, 1993.
4. Bouyssy, A., *Nucl. Phys.*, **A381**, 445 (1982).
5. Dover, C.B., et al., *Phys. Rep.*, **184**, 1 (1989).
6. Budick, B., *Nucl. Phys.*, **A329**, 331 (1979).
7. Burnstein, R.A., et al., *Phys. Rev.*, **177**, 1945 (1969).
8. Gershtein, S.S., and Zinoviev, Yu.M., "On quadrupole moment of Ω^- hyperon", preprint IHEP 80-177, Serpukhov, 1980.

On the Structure of the first Excited Nucleon States

Christian Schütz, Johann Haidenbauer and Josef Speth

Institut für Kernphysik, Forschungszentrum Jülich.
D-52425 Jülich, Germany

Abstract. Effective meson Lagrangians are the appropriate theoretical tool for the non–perturbative regime of QCD which in the present work is applied to pion–nucleon scattering. We focus our interest on the first excited states of the nucleon. We discuss the interplay between genuine poles and resonance behavior due to the meson–baryon dynamics. In the case of the Δ–isobar we find that the non–pole background plays a crucial role for the quantitative understanding of the Δ width. The P_{11} partial wave of πN scattering is described without inclusion of a genuine N^* (Roper) resonance. In the pion production region the effect of coupling to the reaction channels $\pi\Delta$ and σN is investigated.

INTRODUCTION

The understanding of the structure of hadrons and the interaction between them is the most important unsolved problem in medium–energy physics. Nowadays, quantum chromodynamics (QCD) is considered to be the underlying theory of the strong interaction, with quarks and gluons as fundamental degrees of freedom. In the perturbative regime, i.e. at short distances which are probed with high momentum transfer, QCD results obtained within perturbation theory agree quantitatively with the experimental data. However, QCD does not yet provide much help in producing a description of the structure of the strongly interacting particles at large distances since solving the QCD equations for a many–body system in the non-perturbative regime is at the present stage far beyond our abilities. Therefore it is still important to construct models and constrain them by symmetries and experimental informations. For us it seems important that such models should be able to describe aspects of the non–perturbative as well as the perturbative regime. In the former mesons and baryons have definitely retained their importance as efficient, collective degrees of freedom for a wide range of nuclear phenomena. Meson–exchange models for the strong interaction and the vector–dominance for the electromagnetic form-factors of the proton and pion are well-known examples.

Our modern picture of hadrons is that of a core of valence quarks surrounded by a sea of $q\bar{q}$ pairs and gluons. For low energy processes (up to a few GeV) it seems that much of the detailed dynamics of the quarks can be neglected and the cloud can be approximated by the correlated, color-neutral $q\bar{q}$ states that we know as mesons. This is the basis of the success of the meson-exchange models. It is obvious that such a description must eventually break down since, at some energy scale, the quark and gluonic degrees of freedom must become important. It should be kept in mind, however, that the radii of hadrons are given by the confining region as well as the meson cloud so that the range of applicability of the meson-cloud model may be much larger than thought from naive estimates based on, e.g., electric or magnetic radii.

A MESON–EXCHANGE MODEL OF PION–NUCLEON INTERACTION

Recently, we recently developed a dynamical model for correlated 2π exchange in the πN interaction [1]. If this model for correlated two-pion ex-

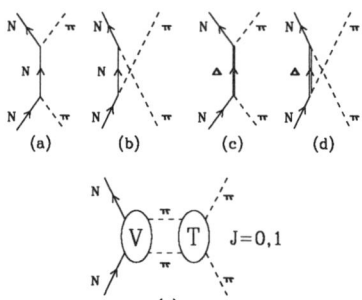

FIGURE 1. Contributions to our πN interaction model.

change in the $J = 0$ (σ) and $J = 1$ (ρ) channels of Fig. 1 (e) is supplemented by direct and exchange pole diagrams involving the nucleon and the Δ-isobar [2] (cf. Fig. 1 (a)...(d)), we are able to obtain a quantitative description of all relevant S and P partial waves of elastic πN scattering. In Fig. 2, we summarize the results for elastic πN scattering, which we obtain after iteration of this Born term.

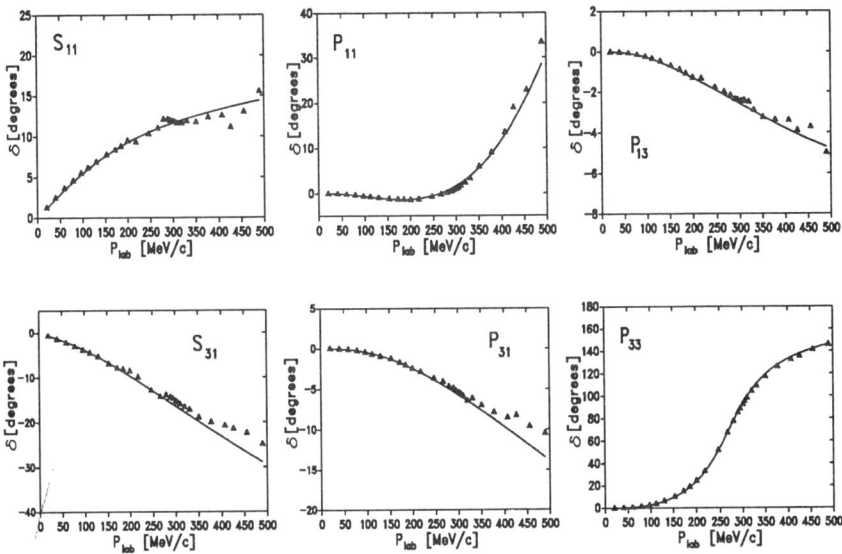

FIGURE 2. πN scattering phase shifts in S and P waves, as function of the pion laboratory momentum. The empirical information is taken from Ref. [3].

THE NATURE OF THE Δ–ISOBAR AND THE ROPER RESONANCE

In this work we focus our interest on the first resonant partial waves of πN interaction. Within our πN interaction model we are able to disentangle meson–cloud effects arising from the underlying meson–baryon dynamics (like e.g. the crossed pole diagrams and the correlated 2π exchange in Fig. 1) from genuine three–quark resonance contributions which correspond to the direct pole diagrams in our model. Therefore we can investigate whether there is any need for a three–quark core to describe the resonant partial waves and if so, we are able to calculate within our model the contributions from the non–resonant background.

In Chew–Low theory it was shown that it is possible to generate the Δ resonance from the crossed nucleon pole of Fig. 1 (b). Therefore it is not surprising that we also find within the model described above that this process plays an important role for the understanding of the P_{33} partial wave of πN scattering. It dominates the non–pole part of the interaction in this partial wave. However, to generate a resonant behavior without inclusion of a direct Δ pole (cf. Fig 1 (c)), one needs an extremely hard πNN form factor which is not consistent with the description of the other partial waves and therefore unrealistic.

How large is the effect of the non–resonant background ? To answer this

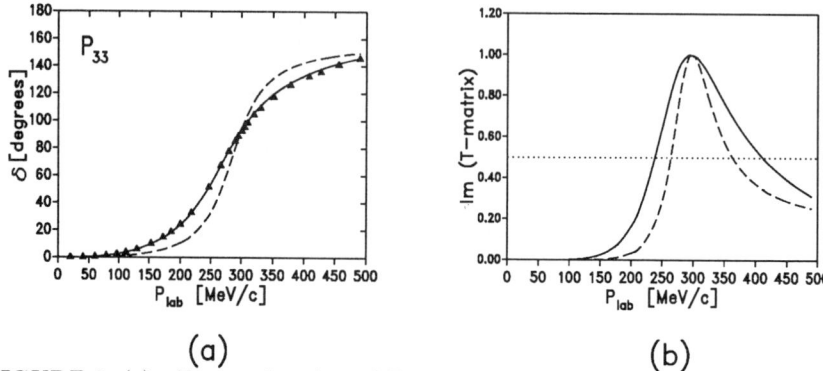

FIGURE 3. (a) πN scattering phase shift in the P_{33} partial wave, (b) imaginary part of the T matrix in this partial wave. The solid line denotes the result of our full model whereas the dashed line is the prediction if we do not include the non–pole part of the interaction.

question, we have plotted in Fig. 3 (a) a simplified model which does not include any non–resonant background and which is based on the same bare $N\Delta\pi$ coupling as our full model. As one can already see from the phase shift, the background plays a crucial role for the quantitative understanding of the Δ width. From the imaginary part of the T matrix plotted in part (b) of this figure one can conclude that the non–pole interaction accounts for about 40% of the Δ width.

One of the most interesting partial waves in πN scattering is the P_{11} partial wave. It is a challenge for models of πN interaction to describe quantitatively the repulsion at low energies followed by the attraction leading to the Roper resonance at higher energies. Within the model presented here, the rise of the P_{11} phase shift in the elastic region of the interaction can be reproduced without any contribution from a genuine N^* particle, cf. Fig. 2. The non–pole part of the interaction in this partial wave is dominated by correlated 2π–exchange in the ρ channel. If this strong attraction is counterbalanced by the direct nucleon pole contribution we are able to obtain the quantitative description of the data shown above.

However, to really learn about the nature of the $N^*(1440)$ resonance, we have to extend the energy range of our model. For energies above 1.3 GeV, pion–nucleon interaction gets inelastic due to coupling to the $\pi\pi N$ channel. Here the dominant contributions are coupling to the $\pi\Delta$ and the σN channel, where σ denotes an isoscalar $\pi\pi$ S–wave state.

Therefore we have extended our model of elastic πN interaction to a coupled–channel model involving the reaction channels πN, $\pi\Delta$ and σN. Self–energy contributions of the Δ as well as the σ are taken into account. The coupling constants occuring in the transition $\pi N \to \pi\Delta$ and the direct $\pi\Delta$ interaction have been obtained from the $SU(2) \times SU(2)$ quark model [5]. The (effective) σNN coupling is taken from Ref. [6].

As one can see from the dash–dotted lines in Fig. 4, already the coupled channel model involving the πN and the $\pi\Delta$ channels can generate a resonant behavior in the P_{11} partial wave at the correct energy to explain the $N^*(1440)$ dynamically. However, coupling exclusively to the $\pi\Delta$ channel accounts only for a small part of the inelasticity. The description of the inelasticity can be improved considerably if additional coupling to the reaction channel σN is taken into accout. The strong inelasticity indicates that there is a considerable direct σN interaction which is modelled by the effective σ exchange in our calculation. In our full $\pi N/\pi\Delta/\sigma N$ interaction model the $N^*(1440)$ resonance thus can be understood purely in terms of the meson–baryon dynamics. There is no need to include a direct pole diagram corresponding to a three–quark contribution to the $N^*(1440)$.

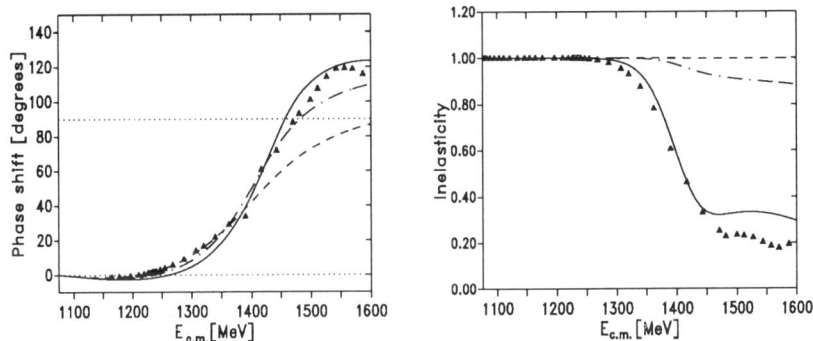

FIGURE 4. Phase shift and inelasticity in the P_{11} partial wave of πN scattering compared to the KA84 phase shift solution [7]. The dashed line denotes the extrapolation of our elastic model, the dash–dotted line takes into account coupling to the reaction channel $\pi\Delta$, and the solid line is the result of our full $\pi N/\pi\Delta/\sigma N$ coupled–channel model.

REFERENCES

1. Schütz C., Durso J.W., Holinde K., Pearce B.C., and Speth J., *Phys. Rev. C* **51**, 1374 (1995).
2. Schütz C., Durso J.W., Holinde K., and Speth J., *Phys. Rev. C* **49**, 2671 (1994).
3. Koch R., and Pietarinen E., *Nucl. Phys.* **A 336**, 331 (1980).
4. Chew G.F., and Low F.E., *Phys. Rev.* **101**, 1570 (1956).
5. Brown G.E., and Weise W., *Phys. Rep.* **C22**, 281 (1975).
6. Durso J.W., Jackson A.D., and Verwest B.J., *Nucl. Phys.* **A 345**, 471 (1980).
7. Höhler G., private communication.

ϕ Meson Couplings to the Nucleon and Strange Vector Currents

Ulf-G. Meißner*, V. Mull, J. Speth* and J.W. Van Orden[†]

FZ Jülich, IKP (Th), D-52425 Jülich, Germany
[†] *TJNAF, Newport News, VA 23606, USA*

Abstract. We investigate within a meson–exchange model the OZI allowed coupling of the ϕ meson to the nucleon with the inclusion of kaon loops and hyperon excitations. All parameters of the model have previously been determined from a variety of hadronic reactions. A strong cancellation of the various contributions is observed which results in a small ϕNN coupling. We also show that a realistic isoscalar spectral function including the correlated $\pi\rho$ exchange leads to sizeably reduced strange vector form factors based on the dispersion–theoretical analysis of the nucleons' electromagnetic form factors explaining our previous result.

A MODEL FOR THE ϕNN COUPLINGS

The ϕ–meson is believed to be an almost pure $\bar{s}s$ state whereas the nucleon consists, at least in the naive constituent quark model, of valence u and d quarks only. Therefore, the ϕ cannot couple directly to the nucleon, at least in this scenario. This is in line with the phenomenological OZI–rule. On the other hand, it has been known for quite some time that ϕ production is enhanced beyond expectation from the OZI–rule in various hadronic reactions [1]. Such an enhancement could point towards the existence of resonant gluonic intermediate states (glueballs) or a sizeable $\bar{s}s$ component in the nucleon [1]. However, one should be aware that OZI–forbidden transitions can go via two-step processes in which each individual transition is OZI–allowed. In this way, the ϕ–meson can couple via non–glueball, i.e. hadronic intermediate states, to the nucleon. In previous work it has been shown that the cross sections for the reactions $\bar{p}p \to \bar{\Lambda}\Lambda$ and $\bar{\Sigma}\Sigma$ as well as into $\bar{K}K$ and $\bar{K}^*K$ can be well understood in the framework of meson–exchange models [2]. In a more recent work it has also been shown that the sizeable cross sections for the annihilation reaction $\bar{p}p \to \phi\phi$ can be explained in a purely hadronic framework by two–step processes with $\bar{\Lambda}\Lambda$ and $\bar{\Sigma}\Sigma$ intermediate states [3]. Within the same meson–exchange model with the parameters used in the previous calculations and thus these parameters being completely *fixed*, we now want to address the question of what is the effective coupling constant of the ϕ–meson to the nucleon. We have performed a calculation of the ϕNN vertex function in a hadronic picture, considering $\bar{K}K$, $\bar{K}^*K - \bar{K}K^*$, $\bar{\Lambda}\Lambda$ and $\bar{\Sigma}\Sigma$ intermediate states. There are sizeable cancellations between the various contributions from graphs with intermediate K's, K^*'s and diagrams with the direct hyperon interactions leading to a very small ϕ coupling,

$$\frac{g_{\phi NN}^2}{4\pi} \simeq 0.005 \,, \quad \kappa_\phi \simeq \pm 0.2 \,. \tag{1}$$

Here, κ denotes the tensor–to–vector coupling ratio. The various contributions are tabulated in table 1 of ref. [4]. The sign of the tensor coupling is very sensitive to the details of the calculation. The smallness of these couplings amounts to a "resurrection" of the OZI rule. Also given in that table are the coupling constants g^{Born} and f^{Born}. These are calculated within the Born approximation. In the final calculation, we consider the rescattering of the K–mesons within a T–matrix approach and the interaction between the hyperons by an optical potential. Both interactions have been used previously and are *not* adjusted to the present process. The Born results differ considerably from the full calculation, in the case of g even in sign. This clearly shows the importance of the final state interactions (or higher loops and unitarization in another language) in addition to the cancellation between the K and K^* contributions.

STRANGE VECTOR FORM FACTORS

The spectral functions of the isoscalar form factors $F_{1,2}^{(0)}$ encode information about the strange vector current since the photon couples to a certain extent via mesons with strangeness (here the ϕ and the S) to the nucleon. As explained in some detail in ref. [4], correlated $\pi\rho$ exchange dominates the isoscalar spectral function in the 1 GeV mass region, letting one expect that its inclusion in the dispersion–theoretical treatment of the nucleons em form factors together with a weakly coupled ϕ will lead to a consistent picture. Assuming that the strange form factors have the same large momentum fall–off as the isoscalar electromagnetic ones [5] [6] and neglecting the small $\omega - \phi$ mixing, it is straightforward to extract the strange Dirac and Pauli form factors following the formalism outlined in [5] [6]

$$F_1^s(t) = t\, L(t)\, a_1^\phi L_\phi^{-1} \frac{M_\phi^2 - M_S^2}{(t - M_\phi^2)(t - M_S^2)} \,, \quad F_2^s(t) = F_1^s(t) \frac{a_2^\phi}{a_1^\phi} \frac{1}{t} \,, \tag{2}$$

with $L_\phi^{-1} = 1/L(M_\phi^2)$ (see [7] for definitions). Clearly, the size of these strange form factors is given by the strength of the ϕ–nucleon couplings (as encoded in the residua $a^\phi_{1,2}$). In particular, we notice that the sign of the strange radius $r^2_{1,s}$ is determined from the sign of a^ϕ_1 whereas the sign of the strange magnetic moment, $\mu_s = F^{(s)}_2(0)$, is fixed by the sign of the tensor coupling $\sim a^\phi_2$. A best fit to the available data as compiled in [8] is obtained with $M_{S'} = 1.63\,\text{GeV}$, $M_{\rho'''} = 1.72\,\text{GeV}$ and $a^\omega_1 = 0.677$ (for $g_{\phi NN} = -0.24$ and $\kappa_\phi = 0.2$). The χ^2/datum of the fit is 1.02. All contraints are fulfilled to high numerical accuracy. The corresponding strange form factors can easily be calculated from Eqs.(2). Notice that $F^{(s)}_1(t)$ varies very weakly between $t = -1\ldots-10\,\text{GeV}^2$. Furthermore, the strange magnetic moment and radius are

$$\mu_s = 0.003\,\text{n.m.} \; , \quad r^2_s = 0.002\,\text{fm}^2 \; , \qquad (3)$$

respectively. These are orders of magnitude smaller than in previous analysis [5] [6] where the ϕ–pole subsumed the non–strange physics of the isoscalar spectral function in the mass region of about 1 GeV, i.e. the sizeable effect of the $\pi\rho$ correlations. We note that the inclusion of $\omega - \phi$ mixing, which is at the heart of Jaffe's analysis, would increase the size of the strange matrix elements, but not dramatically. The main reason for the suppression observed here are the small ϕNN couplings which are independent of the precise parameters entering the meson–exchange model. As already stated, the ϕNN tensor coupling could change sign, leading to a small and negative strange magnetic moment. The value for μ_s found here is sizeably smaller than the experimental value reported in ref. [9] but within the uncertainty, $G^s_M(Q^2 = 0.1\,\text{GeV}^2) = +0.23 \pm 0.37 \pm 0.15 \pm 0.19\,\text{n.m.}$ The theoretical uncertainty on the numbers given in Eq.(3) is certainly of the same size as the given numbers. A more precise estimate of these error bars can only be given when the meson–exchange potential has been fine–tuned to include effects like e.g. $\omega - \phi$–mixing.

REFERENCES

1. J. Ellis, E. Gabathuler and M. Karliner, *Phys. Lett.* **B217**, 173 (1989).
2. J. Haidenbauer, K. Holinde and J. Speth, *Nucl. Phys.* **A562**, 317 (1993).
3. V. Mull, K. Holinde and J. Speth, *Phys. Lett.* **B334**, 295 (1994).
4. Ulf-G. Meißner, V. Mull, J. Speth and J. W. Van Orden, [hep-ph/9701296], to appear in *Phys. Lett.* **B** (1997).
5. R.L. Jaffe, *Phys. Lett.* **B229**, 275 (1989).
6. H.-W. Hammer, Ulf-G. Meißner and D. Drechsel, *Phys. Lett.* **B367**, 323 (1996).
7. P. Mergell, Ulf-G. Meißner and D. Drechsel, *Nucl. Phys.* **A596**, 367 (1996).
8. H.-W. Hammer, Ulf-G. Meißner and D. Drechsel, *Phys. Lett.* **B385**, 343 (1996).
9. B. Mueller et al. (SAMPLE collaboration), *Phys. Rev. Lett.* **78**, 3824 (1997).

Nucleon Strangeness Content through Vector Meson Dominance[1]

Stephen R. Cotanch[†] and Robert A. Williams[*]

[†] *North Carolina State University, Raleigh, NC 27695-8202*
[*] *Thomas Jefferson National Accelerator Facility, Newport News, Virginia 23606 and Department of Physics, Hampton University, Hampton, Virginia 23668*

Abstract. We document that $N(\pi, e^+e^-)N$ measurements will provide new information about the nucleon time-like form factor in the ϕ meson region including the value of the ϕN coupling constant, $g_{\phi NN}$. Using vector meson dominance and a field theoretical model which describes pion photoproduction data, we report crossing predictions for the $N(\pi, e^+e^-)N$ cross section which exhibit dramatic, several order of magnitude enhancements from ϕN coupling in the nucleon form factors. These narrow, dual peaked resonances provide a novel experimental signature for OZI violation and the related strangeness content of the nucleon.

INTRODUCTION

Although the simple quark model has been quite successful in describing hadron structure, it is at odds with experiments related to the spin, momentum distribution and flavor content of the proton [1]. This paper addresses the last issue, the strangeness content of the nucleon, through the assessment of $g_{\phi NN}$ which can be inferred from the nucleon time-like form factors. Such information is currently incomplete since momentum conservation requires $q_\gamma^2 \geq 4M_N^2$ using e^+e^- annihilation reactions. However, $N(\pi, e^+e^-)N$ entails a three-body final state with an effectively unrestricted virtual photon mass $q_\gamma^2 \geq 4M_e^2 \sim 0$ which permits kinematically accessing the important ϕ region.

To motivate this experiment, we have performed calculations [2] for $N(\pi, e^+e^-)N$ and document a dramatic, novel effect. Assuming the validity of vector meson dominance (VMD), we predict several order of magnitude cross section resonances for $\sqrt{q_\gamma^2}$ near the (ρ, ω, ϕ) masses from coupling to the nucleon. As discussed in the next section, measuring these resonances will directly determine $g_{\phi NN}$ which characterizes the nucleon strangeness content.

---

[1]) Supported by DOE grant DE-FG05-88ER40461 and NSF grant HRD-9154080

THEORETICAL DETAILS AND RESULTS

Our calculations are based upon a consistent combination of VMD with a specific covariant quantum hadrodynamic model that is gauge invariant and incorporates crossing and duality constraints. This formalism (see ref. [2,3] for full details) has been used to comprehensively describe pseudoscalar meson electromagnetic production and radiative capture. For the baryon form factors we utilized a hybrid VMD model [4], implementing Sakurai's universality hypothesis, to analyze the baryon octet EM form factors obtaining a good description of the nucleon data. In Fig. 1 we display the electric proton form factor, $G_E^p(q^2)$, with and without the ϕ contribution. Our ϕN coupling constant relative to ωN is $g_{\phi NN}^2 / g_{\omega NN}^2 = 0.14$. There is considerable uncertainty in $g_{\phi NN}$ whose magnitude is governed by the currently unknown nucleon strangeness content as well as the small, but much better known, u and d quark content of the ϕ. To establish the validity of our reaction model we have reproduced photoproduction data using Born amplitudes (i.e. N, π and ρ graphs) with a phenomenological treatment of the hadronic vertex factors and final state absorption effects. With all model parameters fixed by photoproduction and form factor data, crossing symmetry directly predicts the $p(\pi^-, e^+e^-)n$ cross section shown in Fig. 2. The modest ϕ contribution to the nucleon form factors leads to a three order of magnitude enhancement in the cross section. The dashed curve is for $g_{\phi NN} = 0$ and documents the effect of the $\rho\pi\phi$ coupling due to ρ-exchange between π and N. This smaller peak represents the expected, minimal ϕN production and provides a realistic ϕN background for determining excess production from a nonnegligible $g_{\phi NN}$.

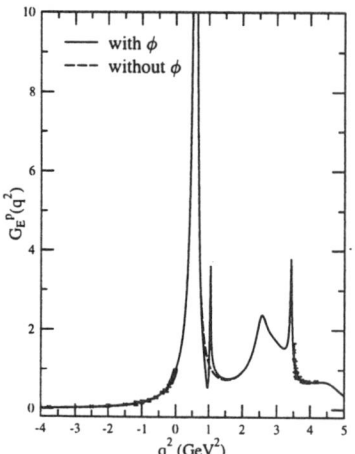 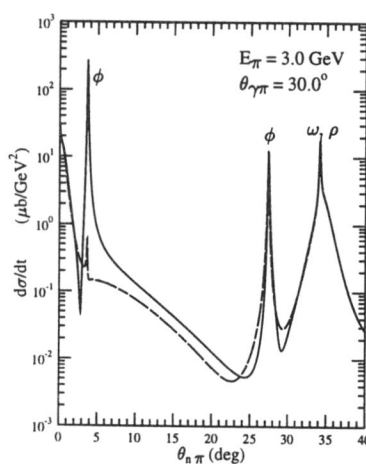

Figure 1. Proton electric form factor.   Figure 2. $p(\pi^-, e^+e^-)n$ cross section.

The dual peak structure has a sensitive energy and angular dependence which experimentally permits separately studying ϕ production from the proton (u channel), ρ-exchange (t channel $\rho\pi\phi$ contribution) or neutron (s channel). For the kinematics in Fig. 2 s is much larger than u so that u channel diagrams dominate and therefore such measurements will mainly provide proton form factor information. Similarly we have verified that the $n(\pi^+, e^+e^-)p$ reaction essentially provides information about the neutron form factors which can be obtained from $d(\pi^+, e^+e^-)pp$ experiments. We have also derived a high t theorem for the ratio of the ϕ to ω peak cross sections, $R = d^3\sigma(q^2 = M_\phi^2)/d^3\sigma(q^2 = M_\omega^2) = g_{\phi NN}^2/g_{\omega NN}^2 f$, for the same s and t (note the triple cross section has a single peak structure versus q^2) where f is a known kinematic factor. We have confirmed numerically that $R \simeq .14f$, roughly 30 times larger than the OZI prediction $R = tan^2(\delta)f' = 4.2x10^{-3}f'$. Here δ is the small deviation from ideal mixing of the u, d quark components in the ϕ and f' is another known factor of order unity.

Finally, we note possible contributions to $g_{\phi NN}$ from OZI evading kaon loops. However, two independent analyses [5,6] report the loop diagrams effectively cancel and ref. [6] calculates a remnant value $g_{\phi NN}^2/4\pi \simeq .005$. This is substantially smaller than our value of .13 and if used in this analysis would result in a reduction of the 4° resonance peak in Fig. 2 by over an order of magnitude. If loop suppression proves valid, an experimental determination of $g_{\phi NN}^2/4\pi \gg .005$ would constitute direct evidence for either substantial OZI violation (with no nucleon strangeness) or, and more probable, a $s\bar{s}$ component in the nucleon. Interestingly, a most recent ϕ low energy photoproduction analysis [7] yields $g_{\phi NN}^2/4\pi = .034$, intermediate between the above two values. Photoproduction involves space-like photons with no resonant ϕ signature to dominate competing mechanisms, such as ϕ diffractive scattering. It is therefore essential that complimentary (π, e^+e^-) experiments be performed.

SUMMARY AND CONCLUSIONS

We conclude that the (π, e^+e^-) process is ideal for investigating the nucleon time-like form factors and strangeness content. Measurements should clearly distinguish between alternative values of $g_{\phi NN}$ with attending ramifications.

REFERENCES

1. Ellis J., Gabathuler E., and Karliner M., *Phys. Lett.* B **217**, 173 (1989).
2. Williams, R. A., and Cotanch, S. R., *Phys. Rev. Lett.* **77**, 1008 (1996).
3. Williams, R. A., Ji, C.- R., and Cotanch, S. R., *Phys. Rev.* D **41**, 1449 (1990); *Phys. Rev.* C **43**, 452 (1991); *ibid.* **46**, 1617 (1992); *ibid.* **48**, 1318 (1993).
4. Williams, R. A., and Truman, C. P., *Phys. Rev.* C **53**, 1580 (1996).
5. Isgur, N., and Geiger, P., *Phys. Rev.* D **55**, 299 (1997).
6. Meißner, Ulf-G., *et al*, Report No. JLAB-THY-97-02; hep-ph/9701296.
7. Williams, R. A., preprint.

Sum Rule Analyses for the Light Quark Masses Revisited

K. Maltman*,**, R. Gupta† and T. Bhattacharya†

*Department of Math and Stats, York University, North York, ON CANADA,** Special Research Center for the Subatomic Structure of Matter, University of Adelaide, Adelaide 5005 Australia, †T8, Los Alamos National Laboratory, Los Alamos, NM 87545

Abstract. We explore existing sum rule analyses of the u, d and s quark masses. We argue that certain assumptions about the behavior of unmeasured continuum scalar and pseudoscalar spectral functions can lead to significant overestimates of these masses, and illustrate the source of the problem in the analogous isovector vector channel.

The commonly quoted $\overline{MS}$ u, d and s masses are obtained via QCD sum rules analyses of correlators of the products of either the divergences of two vector, or two axial vector currents. A recent analysis of the world's lattice data, however, produces light quark masses lower by as much as a factor of 2 [1], and hence suggests that these analyses be revisited. For $m_u + m_d$, the analysis [2] starts with the light quark isovector axial current correlator

$$\Psi_5(q^2) \equiv i \int d^4x e^{iq\cdot x} \langle 0|T\{\partial^\mu A_\mu^{(-)}(x), \partial^\nu A_\nu^{(+)}(0)\}|0\rangle$$

$$= (m_d + m_u)^2 i \int d^4x e^{iq\cdot x} \langle 0|T\{P^{(-)}(x), P^{(+)}(0)\}|0\rangle \qquad (1)$$

and forms finite energy sum rules (FESR) by integrating products of the form $t^n \Psi_5(t)$ over a contour enclosing the positive real t axis from 0 to s_0 closed over the circle of radius s_0 in the complex t plane. For $n > 0$ the contour integral is zero. When s_0 is sufficiently large, one may use the operator product expansion (OPE) and perturbative QCD (pQCD) to evaluate the integral over the circular portion of the contour. The remaining contributions along the positive t axis are determined from a model for the continuum part of the corresponding hadronic spectral function (the pion pole contribution is known in terms of f_π, m_π). The model assumes the shape is that of a sum of Breit-Wigners, the overall normalization of which produces 100% dominance of the threshold region by resonance contributions, with the threshold value obtained from tree-level chiral perturbation theory (ChPT). The relative contributions of the $\pi(1300)$ and $\pi(1800)$ are adjusted to obtain a match between the ratios of the OPE and model sides of the $n = 0$ and $n = 1$ sum rules and the resulting spectral function then used to extract $m_u + m_d$ from the $n = 0$ sum rule. We concentrate here on the question of the ansatz for the overall normalization [3] since more than half of the quoted $m_u + m_d$ value is associated with the continuum portion of the spectral integral. Note also that the inverse $n = -1$ weighting produces a sum rule relating the t^{-1} weighted integral over the *continuum portion only* of the hadronic spectral function to the scale-independent fourth order ChPT low-energy constant (LEC) combination $2L_8^r - H_2^r$. The failure of this combination to be constant as a function of s_0 (see Figure 4 of Ref. [2]) indicates deficiencies in the model spectral function. The extraction of the strange quark mass [4,5] similarly starts from the correlator

$$\Psi(q^2) = i \int d^4x \, e^{i q \cdot x} \langle 0 | T \{ \partial^\mu V_\mu(x) \partial^\nu V^\dagger_\nu(0) \} | 0 \rangle$$
$$= (m_s - m_u)^2 \, i \int d^4x \, e^{i q \cdot x} \langle 0 | T \{ S(x) S^\dagger(0) \} | 0 \rangle \qquad (2)$$

where V_μ is the strangeness-changing vector current, and S the corresponding scalar current. A conventional QCD sum rule analysis is performed, relating the exponentially weighted integral over the hadronic spectral function to the Borel transformed value of the correlator at large spacelike values of q^2, for which an OPE/pQCD evaluation is valid. The threshold $t = (m_K + m_\pi)^2$ spectral function can be determined from extrapolation of K_{e3} data using the Omnes representation, and the unknown continuum spectral function is then modelled, as in the case of the light quark analysis, by assuming the form to be a sum of Breit-Wigners, and the normalization to be such as to produce 100% resonance saturation at threshold.

There are phenomenological reasons to question the "resonance saturation at threshold" ansatz. If one first imagines constructing an effective hadronic theory including both the Goldstone bosons and the higher resonances, then in the Goldstone-boson sector, the leading (second) order LEC's are the same as they would be in the effective theory in which the higher resonance degrees of freedom are integrated out: the contributions of the resonance degrees of freedom to the LEC's of the Goldstone-bosons-only chiral Lagrangian first appear in the fourth order LEC's (L_k^5 and H_k^r in the notation of Gasser and Leutwyler [6]) [7]. Contributions to the hadronic spectral function computed using the Goldstone-bosons-only effective Lagrangian and associated with the second order LEC's should thus not be associated with the resonance degrees of freedom. Since it is observed, phenomenologically, that the resonance contributions to the fourth order LEC's essentially saturate these LEC's at a renormalization scale of order the vector meson mass [7], we propose a phenomenological alternative for normalizing the continuum spectral function, namely that the overall scale should be set so the tail of the resonance contributions at threshold reproduces the relevant LEC contributions, rather than the whole threshold spectral function. Since this alternate ansatz is purely phenomenological, and hence subject to potential objections, we test both the original ("100% resonance saturation") and alternative ansatze in a case where the behavior of the continuum spectral function in the region of a resonance peak is known. Consider, then, the isovector vector correlator

$$\Pi_{33}^{\mu\nu}(q^2) \equiv \left(q_\mu q_\nu - q^2 g_{\mu\nu}\right) \Pi_{33}(q^2) = i \int d^4x e^{iq\cdot x} \langle 0|T\{V_3^\mu(x)V_3^\nu(0)\}|0\rangle \quad (3)$$

where V_3^μ is the $I = 1$ vector current. The ρ contribution to the spectral function of Π_{33}, ρ_{33}, at the ρ peak, is known in terms of the ρ decay constant, and the behavior near threshold through the two-loop ChPT calculation of Ref. [8]. We assume, in exact analogy with the strange quark mass analysis [4,5], that the continuum spectral function is given by a ρ Breit-Wigner having the usual p-wave s-dependent width, the overall normalization of which is set so the tail of the resonance produces the entire threshold spectral function. Doing so, we find that the spectral function at the ρ peak, is overestimated by a factor of 5.3. In contrast, normalizing the Breit-Wigner form to reproduce only the contribution proportional to the fourth order LEC L_9^r, one finds that the correct value at the ρ peak is reproduced to within a few %. Employing this same ansatz in the case of the scalar correlator used to extract m_s suggests that the continuum contributions to the spectral integral, and hence to the resulting strange quark mass squared, have been overestimated by a factor of 4. We are not able, however, to provide a modified estimate of m_s because the sum rule analysis turns out to have no stability plateau for continuum thresholds compatible with keeping only the two known strange scalar resonances in the model hadronic spectral function [3].

REFERENCES

1. Gupta, R. and Bhattacharya, T., it hep-lat/9605039.
2. Bijnens, J., Prades, J. and de Rafael, E., *Phys. Lett.* **B348**, 226 (1995).
3. Bhattacharya, T., Gupta, R. and Maltman, K., *hep-ph/9703455*.
4. Jamin, M. and Münz, M., *Z. Phys.* **C66**, 633 (1995).
5. Chetyrkin, K.G., Pirjol, D. and Schilcher, K., *hep-ph/9612394*.
6. Gasser, J. and Leutwyler, H., *Nucl. Phys.* **B250**, 465 (1985).
7. Ecker, G., Gasser, J., Pich, A. and de Rafael, E., *Nucl. Phys.* **B321**, 311 (1989).
8. Golowich, E. and Kambor, J., *Phys. Rev.* **D53**, 2651 (1996).

The three Nucleon System in the Skyrme Model

Niels R. Walet

Dept. of Physics, UMIST, P.O. Box 88, Manchester M60 1QD, U.K.

Abstract. We construct part of a collective Hamiltonian for the three baryon system. Large-amplitude quantum fluctuations play an important rôle in the intrinsic wave function of the ground-state, changing its symmetry from tetrahedral to cubic. Apart from the tetrahedron describing the minimum of the potential, we identify a "doughnut" and a "pretzel" as the most important saddle points in the potential energy surface. We show that it is likely that inclusion of fluctuations through these saddle points lead to an energy close to the triton's value.

I INTRODUCTION

A while ago, Carson [1] has studied the application of the Skyrme model to the system of three baryons. In these studies one starts from the minimum energy solution where the baryon density has tetrahedral symmetry. One then makes the approximation that this minimum describes the triton. This approximation leads to a tremendous over-binding, which was attributed to the neglect of simple quantum effects, especially vibrational zero-point motion. It has also been mentioned that anharmonic modes might play a role as well.

In this contribution I shall show that the low-energy potential "landscape" has a lot of structure, which has both important qualitative and quantitative effects. Beyond the tetrahedral solutions, the most salient features in this landscape are the $B = 3$ doughnut and the "pretzel" which has a planar symmetry similar to the doughnut but has two instead of one hole.

The approach taken is this work, described in detail in Ref. [3], is to use the techniques discussed in in Ref. [2]. These are based on a mode-following approach, where we start from the harmonic fluctuations around a stationary solution, and follow those into the anharmonic regime.

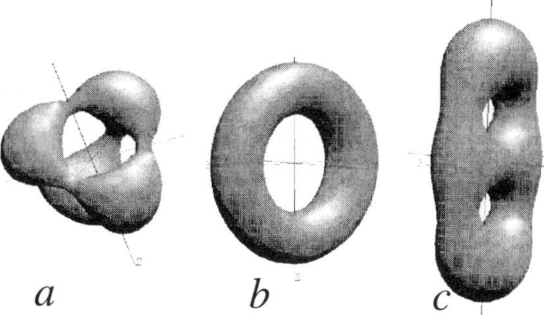

FIGURE 1. A plot of a surface of constant baryon density in (a) the tetrahedron, (b) the $B = 3$ doughnut and (c) the pretzel.

II QUANTISATION OF THE COLLECTIVE COORDINATES

The minimal number of collective coordinates to describe the internal motion of the $B = 3$ system is nine non-zero modes and nine zero modes (i.e., 6 per baryon).

A Harmonic approximation

The most straightforward way to get a "quantum" result for $B = 3$ is to quantise in harmonic approximation. We find the following energy balance:

$$\begin{array}{rcccccc}
 & & E_0 & & \frac{1}{2}\sum \hbar\omega & & E_{\text{rot}} \\
E_{B=3} & = & 206.6 f_\pi/e & + & 1.76 f_\pi e & + & .00353 f_\pi e^3 \\
3E_{B+1} & = & 220.8 f_\pi/e & + & 0 & + & \frac{9}{4}\frac{1}{140.1} f_\pi e^3 \\
\hline
\Delta E & = & -14.2 f_\pi/e & + & 1.76 f_\pi e & - & 0.0125 f_\pi e^3
\end{array} \quad (1)$$

For three reasonable choice of parameters this takes on the values 319.2, 386.5 and 242.1 MeV. Here we see the effect of the over-estimate of quantum corrections in harmonic approximation. Notice that without these corrections the value would be -168, -132 and -320 MeV, respectively. Since we expect anharmonicities to play a large rôle, we shall now study the effect of large amplitude fluctuations.

B Approximate large-amplitude dynamics

We have used our mode following approach to gain information about the structure of the potential landscape. We find a doublet of modes at the tetrahedron, both of which describe motion towards a doughnut, and a triplet of modes describing motion towards the pretzels. The dougnut is a saddle-point in energy. It has two sets of unstable modes, the lowest of which corresponds to the direction towards the tetrahedron. The other unstable mode leads to yet another saddle point. This solution has the double-hole structure of a pretzel. The potential energy of all these states, *i.e.*, ignoring contributions from the rotational kinetic energy, is not very different

$$E_{\text{tetrahedron}} = 206.56 f_\pi/e, \quad E_{\text{pretzel}} = 211.54 f_\pi/e,$$
$$E_{\text{doughnut}} = 214.06 f_\pi/e, \quad 3E_{B=1} = 220.8 f_\pi/e.$$

Each of the modes through tetrahedron and pretzel describes a circuit through two pretzels and another tetrahedron, as denoted schematically in Fig. 2.

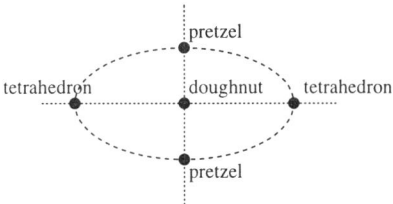

FIGURE 2. A schematic sketch of a few of the connections through a given doughnut.

Note that the lowest mode for the tetrahedron is the one connecting to the doughnut, which is the higher of the two saddle points. Obviously this figure only sketches one out of three connections, the whole structure of the collective manifold being highly complex.

As a fast and relatively simple way to study the modes one can assume that one can independently quantise the motion along the paths represented schematically by the straight line and the ellipse in Fig. 2, since these represent orthogonal modes at the tetrahedron.

Understanding the paths through the doughnut(s) requires a bit of thought, since there are only two modes at each tetrahedron for these modes, and not three. Inspired by the chemical VSEPR model we found a simple analogy that helps describe this. It is based on placing particles on a sphere with pairwise repulsive interaction, and ends up being described by one set of spherical cooridnates. Studying the potential in this model we find that at the minima in potential we have three paths escaping; since this is a two-dimensional surface there are only two corresponding modes.

For the pretzel we find that there are three – periodic – paths that describe tunnelling possibilities from one tetrahedron to one that is related by it through an element of the octahedral and not in the tetrahedral group. Since these have a saddle point at lower energy than the paths through the doughnuts, one expects that they are even more important than those solutions. In essence these paths can be visualised as the edges of a cube, where each closed path is equivalent to one edge. We can estimate the spectrum using a simple Hückel model. We find that the harmonic oscillator length parameter for each of the wells is comparaable to the length of the circuit for any reasonable value of Skyrme-model parameters. In this case the wave function in the circuit becomes approximately constant, and the energy is the average of the potential along the circuit. Since the contribution of the rotational kinetic energy tends to flatten the potential even more. If we just naively solve the one-dimensional problem on the circle, we find an energy of $209.5 f_\pi/e + .00319 f_\pi e^3$, a considerable rise in energy. Furthermore, since the pretzel is a lot bigger than the tetrahedron, we find an enhanced value for the r.m.s. radius (by about 10%).

For the paths through the doughnut we cannot readily make such an estimate. Even though the harmonic oscillator length parameter is again close to the length of the path, which would imply a similar estimate as above, we cannot ignore the two-dimensional nature of the problem here. The simplest symmetric wave function carries non-zero kinetic energy, and a straightforward calculation shows that $\langle K \rangle = \frac{16}{5}\hbar^2$. The best possible estimate for the contribution to the ground state energy from the two lowest modes is

$$\Delta E = \frac{1}{R^2}\langle K \rangle + \frac{1}{L}\int (V - E_0)dl + \frac{1}{L}\int E_{\rm rot} dl. \qquad (2)$$

Here R is the radius of the sphere associated with a tunnelling path of length L, $L = 2\arctan(\sqrt{\frac{1}{2}})R$. The integral is along one of the tunnelling paths. We estimate that $\langle L^2 \rangle \approx 3.5\hbar^2$, and calculate the remaining quantities from our results.

Our final model is an independent quantisation of the five lowest non-zero modes, treating the remaining four in harmonic approximation. Taking all this together, and subtracting three times the $B = 1$ result, we find

$$E_{\rm gs} = -1.3 f_\pi/e + 0.781 f_\pi e - 0.007 f_\pi e^3. \qquad (3)$$

If we use our three parameter sets, we find values of 185.5 MeV for ANW, 186.5 MeV for ANW' and 211.6 MeV for R. These values are still high above the threshold, but note that we have overestimated the effect of the lowest five modes by treating them as independent. Furthermore, some of the higher modes lead to a separated single hedgehog and a doughnut. Our experience from the $B = 2$ case suggests that in such cases the harmonic approximation may be relatively poor. We have found a reduction of the energy, however, which is at least promising for a more mature approach to the problem.

REFERENCES

1. L. Carson, Phys. Rev. Lett, **66** (1991) 406; Nucl. Phys. **A535** (1991) 479.
2. N. R. Walet, Nucl. Phys. **A586** (1995) 649.
3. N. R. Walet, Nucl. Phys. **A606** (1996) 429.

Baryon Magnetic Moments and Axial Coupling Constants with Relativistic and Exchange Current Effects

C. Helminen*, K. Dannbom*, L.Ya. Glozman[†] and D.O. Riska*

*Department of Physics, P.O. Box 9, 00014 University of Helsinki, Finland
[†] Institute for Theorethical Physics, University of Graz, 8010 Graz, Austria

Abstract. The predictions for the axial coupling constants of the baryons are improved by the large relativistic corrections to the constituent quark current operators, while the predictions for the magnetic moments are worsened. The relativistic corrections to the baryon magnetic moments can be compensated for by the exchange current corrections that are associated with spin and flavor dependent quark-quark interactions with a form suggested by pseudoscalar meson exchange. This is shown by a calculation of the magnetic moments of the light and strange baryons within a phenomenological spin and flavor dependent interquark interaction model, which in combination with a linear confining interaction yields a spectrum that is close to the empirical one. We also consider the possibility that a part of the spin and flavor dependent interaction could be due to axial vector and vector exchange.

The electromagnetic and axial current operators will have significant relativistic corrections if the constituent quarks are treated as Dirac particles due to the small mass of the constituent quarks. The effect of these corrections is to simultaneously reduce the magnitude of the axial coupling constants and the magnetic moments of the baryons in the static quark model. The standard overprediction of the axial coupling constants of the nucleons and most of the strange hyperons is thus reduced, while the mostly satisfactory predictions for the magnetic moments of the baryons in the impulse approximation of the static quark model are worsened by the relativistic corrections. The magnitude of the corrections depends on the average velocity of the confined quarks and thus on the model for the hyperfine interaction between the quarks [1].

Flavor dependent interquark interactions, by the requirement of current conservation, implies the presence or two-body exchange magnetic moment operators. The exchange current corrections that are associated with spin and flavor dependent interquark interactions, with the same operator structure as that of the pseudoscalar meson octet exchange interaction, increase the magnitudes of the magnetic moments of the baryons that are predicted by the static quark model [2]. Thus they in principle should counteract the reduction caused by the relativistic corrections. Since there is no corresponding pure pseudoscalar octet exchange contribution to the axial current of the baryons the relativistic corrections to the axial coupling constants remain uncompensated. We have constructed the exchange current so as to be consistent with and to satisfy the continuity equation with a pseudoscalar boson exchange interaction even when this is modified phenomenologically at short range, using the explicit phenomenological model for the hyperfine interaction between the quarks given in Ref. [3], which yields a satisfactory description of the nucleon and Δ-spectra in combination with a static linear confining interaction. The results for the axial coupling constants of the baryon octet in this model are given in Table 1.

The hyperfine interaction is of the form

$$V(\vec{r}) = \frac{1}{3} f(r) \vec{\sigma}^1 \cdot \vec{\sigma}^2 \vec{\lambda}^1 \cdot \vec{\lambda}^2 , \qquad (1)$$

where the form for $f(r)$ is taken to be that of Ref. [3]. The associated exchange magnetic moment will then be

$$\vec{M} = -\frac{1}{2} g(r) \{ (\vec{\tau}^1 \times \vec{\tau}^2)_3 + \lambda_4^1 \lambda_5^2 - \lambda_5^1 \lambda_4^2 \} \vec{\sigma}^1 \times \vec{\sigma}^2 , \qquad (2)$$

where the function $g(r)$ is determined by $f(r)$ as

$$g(r) = -\frac{1}{3} \{ 2 \int_r^\infty dr' r' f(r) - \frac{1}{r} \int_r^\infty dr' \int_{r'}^\infty dr'' r'' f(r'') \} . \qquad (3)$$

The volume integral of this model for the hyperfine interaction does not vanish, and therefore only part of it can be intepreted as being due to pseudoscalar exchange mechanisms (or, alternatively, vector exchange mechanisms, in which case the exchange current contribution would be somewhat reduced). The remaining part was treated phenomenologically, using an appropriate exchange current operator for an interaction with a nonvanishing volume integral - in this case an exchange current operator that corresponds to the axial vector (A) Fermi invariant. Another possibility would have been the tensor (T) invariant exchange current operator with a form very similar to the axial vector exchange current operator. The construction of the flavor octet exchange current operator was non-relativistic, and therefore we also considered the lowest order relativistic corrections to the exchange magnetic moment operators in a way suggested by the relativistic corrections that appear in the single quark current operator.

The results remain qualitative, but even so we find that the exchange current contributions are more than sufficient to compensate the reduction of the impulse approximation values for the magnetic moments caused by the relativistic corrections. When including exchange current contributions associated with the confining interaction, assuming that this can be effectively viewed as a (relativistic) scalar exchange interaction [2], possible overpredictions are substantially reduced, giving results that for most of the magnetic moments are close to the corresponding empirical values (Table 2). The full result of the calculations can be found in Ref. [4].

TABLE 1. Axial coupling constants of the baryon octet

	Prediction 1[a]	Prediction 2[b]	Experiment
n → p	1.35	1.17	1.26
$\Sigma^\pm \to \Lambda$	0.66	0.57	0.62
$\Sigma^- \to \Sigma^0$	0.76	0.66	0.67
$\Lambda \to p$	1.01	0.88	0.88
$\Sigma^- \to n$	0.28	0.24	0.34
$\Xi^- \to \Lambda$	0.34	0.30	0.31
$\Xi^- \to \Sigma^0$	0.97	0.84	1.36
$\Xi^0 \to \Sigma^+$	1.38	1.20	?
$\Xi^- \to \Xi^0$	-0.27	-0.23	-0.28

[a] Results in the large color limit
[b] Results when including lowest $1/N_C$ corrections

TABLE 2. Magnetic moments of the ground state baryon octet and the Δ^{++} and the Ω^- (in nuclear magnetons)

	Prediction	Experiment
p	2.81	2.79
n	-2.06	-1.91
Λ	-0.76	-0.61
Σ^+	2.46	2.46
Σ^0	0.86	?
$\Sigma^0 \to \Lambda$	-1.69	$|1.61|$
Σ^-	-0.76	-1.16
Ξ^0	-1.62	-1.25
Ξ^-	-0.56	-0.65
Δ^{++}	4.52	4.52
$\Delta^+ \to p$	2.92	3.1
Ω^-	-1.65	-2.019

REFERENCES

1. C. Hayne and N. Isgur, *Phys. Rev.* **D 25** (1982) 1944.
2. A. Buchmann et. al., *Nucl. Phys.* **A 569** (1994) 661.
3. L.Ya. Glozman et. al., *Phys. Lett.* **B 381** (1996) 311.
4. K. Dannbom et. al., *Nucl. Phys.* **A 616** (1997) 555.

A Dynamical η'–Mass from an Infrared Enhanced Gluon Exchange

Lorenz von Smekal[*], Almut Mecke[†] and Reinhard Alkofer[†]

[*]*Physics Division, Argonne National Laboratory, Argonne, IL, 60439*
[†]*Institut für Theoretische Physik, Universität Tübingen, 72076 Tübingen, Germany*

ANL-PHY-8768-TH-97   UNITU-THEP-14/1997   hep-ph/9707210

Abstract. The pseudo–scalar flavor–singlet meson mixes with two gluons. A dimensional argument by Kogut and Susskind shows that this can screen the Goldstone pole of the chiral limit in this channel, if the gluon correlations are infrared enhanced. Using a gluon propagator as singular as σ/k^4 for $k^2 \to 0$ we relate the screening mass to the string tension σ. In the Witten–Veneziano action to describe the η–η' mixing this relation yields masses of about 810MeV for the η', 430MeV for the η and a mixing angle of about $-30°$ from the phenomenological value $\sigma \approx 0.18 \text{GeV}^2$. The very weak temperature dependence of the string tension should make this mechanism experimentally distinguishable from exponentially temperature dependent instanton model predictions.

More than twenty years ago Kogut and Susskind pointed out that for dimensional reasons a non–vanishing contribution to the mass of the pseudo–scalar flavor–singlet meson in the chiral limit can result from its mixing with two non–perturbatively infrared enhanced gluons corresponding to a momentum space propagator $D(k) \sim \sigma/k^4$ for $k^2 \to 0$ [1]. Such infrared enhanced gluon correlations are known to lead to an area law in analogy to the Schwinger model in two dimensions. The identification of the string tension σ shows that effects due to infrared enhanced gluons can be expected to be complementary to instanton models. In particular, a description of the η–η' mixing driven by the string tension [2], provides an interesting alternative to the standard solution of the $U_A(1)$ problem by instantons.

Phenomenologically, this mixing is described by the $\eta_8 - \eta_0$ mass matrix [3],

$$\frac{1}{2} \begin{pmatrix} \eta_8 & \eta_0 \end{pmatrix} \begin{pmatrix} \frac{4}{3}m_K^2 - \frac{1}{3}m_\pi^2 & \frac{2}{3}\sqrt{2}(m_\pi^2 - m_K^2) \\ \frac{2}{3}\sqrt{2}(m_\pi^2 - m_K^2) & \frac{2}{3}m_K^2 + \frac{1}{3}m_\pi^2 + \frac{2N_f}{f_0^2}\chi^2 \end{pmatrix} \begin{pmatrix} \eta_8 \\ \eta_0 \end{pmatrix} \quad (1)$$

where the screening mass in the flavor–singlet component, $m_0^2 := 2N_f\chi^2/f_0^2$, is given by a non–vanishing topological susceptibility,

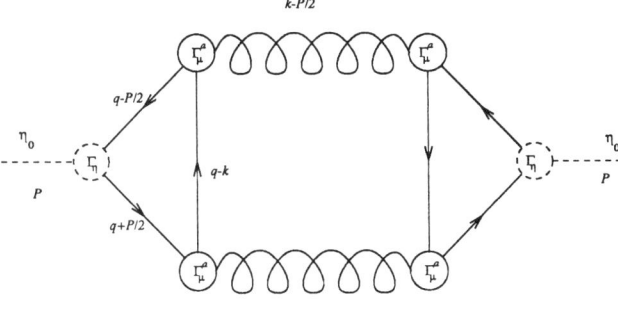

FIGURE 1. The diamond diagram $\Pi(P^2)$, a factor 2 arises from crossed gluon exchange.

$$\chi^2 := \frac{g^2}{(32\pi^2)^2} \int d^4x \, \langle \tilde{G}G(x) \, \tilde{G}G(0) \rangle \quad \text{with} \tag{2}$$

$$\tilde{G}G = \epsilon^{\mu\nu\rho\sigma} 2\partial_\mu \, \text{tr}(A_\nu \partial_\rho A_\sigma - ig\frac{2}{3} A_\nu A_\rho A_\sigma) \, .$$

In the Instanton Liquid Model the topological susceptibility, given by the density of instantons, is $\chi^2 \approx 1\text{fm}^{-4}$, and the mass eigenvalues are $m_\eta \approx 530\text{MeV}$, $m_{\eta'} \approx 1170\text{MeV}$ together with a mixing angle of $\theta \approx -11.5°$ [4].

Here, we concentrate on the mixing of the flavor–singlet pseudo–scalar with two uncorrelated gluons. According to the Kogut–Susskind argument, for infrared enhanced gluons $\sim \sigma/k^4$, the corresponding diagram, see fig. 1, can contribute to the topological susceptibility for the meson momentum $P \to 0$. To explore this conjecture and its quantitative consequences, we use the following model interaction for quarks in the Landau gauge,

$$g^2 D_{\mu\nu}(k) = P_{\mu\nu}(k) \begin{pmatrix} \frac{8\pi\sigma}{k^4} & + & \frac{16\pi^2/9}{k^2 \ln(e+k^2/\Lambda^2)} \end{pmatrix} \, . \tag{3}$$

The second term, subdominant in the infrared, was added to simulate the effect of the leading logarithmic contribution of perturbative QCD for $N_f = 3$. Strictly speaking, a quark interaction of the form (3) cannot arise from gluons alone in Landau gauge, since the product $g^2 D_{\mu\nu}$ is not renormalization group invariant for any finite number of flavors or colors. Even though this is assumed in the *Abelian* approximation, ghost contributions do implicitly enter in the RG invariant interaction (by the dressing of the quark–gluon vertex function). In fact, three quite different approaches to the pure gauge theory are available at present to suggest that the strong infrared enhancement of the interaction might be generated by ghost contributions in Landau gauge [5].

From the axial anomaly, the quark triangle $\Gamma^{ab}_{\mu\nu}$ in fig. 1 has the limit,

$$P \to 0, k^2 = 0: \quad \Gamma^{ab}_{\mu\nu} \to \delta^{ab} \epsilon_{\mu\nu\rho\sigma} k^\rho P^\sigma \sqrt{N_f} f_0^{-1} g^2/(8\pi^2)$$

This model independent form, determining the coupling of two gluons to the pseudo–scalar flavor–singlet bound state in the infrared, is particularly suited for the present calculation, since the contribution to χ^2 is obtained from $P \to 0$, and since the gluon interaction (3) weights the integrand so strongly in the infrared ($\sim \sigma/(k \pm P/2)^4$). With this, all contributions containing ultraviolet dominant terms of the interaction (3) vanish for $P \to 0$, and we obtain [2],

$$m_0^2 = \lim_{P^2 \to 0} \Pi(P^2) = \frac{2N_f}{f_0^2} \chi^2 = \frac{3N_f}{f_0^2} \frac{\sigma}{\pi^4}. \tag{4}$$

The phenomenological string tension $\sigma = 0.18\text{GeV}^2$ and $f_0 \approx f_\pi = 93\text{MeV}$ thus yield $m_0^2 \approx 0.346\text{GeV}^2$, and the physical mass eigenstates are, $m_{\eta'} \approx 810\text{MeV}$ and $m_\eta \approx 430\text{MeV}$, with a corresponding mixing angle $\theta \approx -30°$.

Using free constituent quarks of a mass of about 300MeV in the triangle to suppress spurious ultraviolet contributions, from $f_0^2 \simeq f_\pi^2(1 + \Pi'(P^2)|_{P^2 \to 0})$ with $\Lambda \approx 500\text{MeV}$ in (3), we obtain an additional contribution to the decay constant of the flavor–singlet of about 30% as compared to the pion [2].

As these values are reasonably close to experiment, we conclude that the $U_A(1)$–anomaly might be encoded in the infrared behavior of QCD Green's functions. Whether the Kogut–Susskind mechanism or the instanton based solution to the $U_A(1)$ problem is realized in nature, can be assessed from their respective temperature dependences. If the origin of the η' mass is predominantly due to instantons, the $\eta - \eta'$ mixing angle is expected to vary exponentially with temperature, leading to a significant change of η and η' production rates in relativistic heavy ion collisions [6]. On the other hand, lattice calculations indicate that the string tension is almost temperature independent up to the deconfinement transition. This offers the possibility to study the physics of the $U_A(1)$ anomaly experimentally.

We thank T.-S. H. Lee, H. Reinhardt and C. D. Roberts for helpful discussions. RA gratefully acknowledges the hospitality of the Physics Division at ANL. This work was supported by the DFG under contract Al 279/3-1 and the US-DOE, Nuclear Physics Division, contract # W-31-109-ENG-38.

REFERENCES

1. J. Kogut and L. Susskind, Phys. Rev. **D10** 3468, (1974).
2. A. Mecke, L. v. Smekal and R. Alkofer, in preparation; A. Mecke, Diploma Thesis, Tübingen University, April 1997; R. Alkofer, talk presented at the *Argonne Theory Institute*, Argonne National Laboratory, July 22 - 27 (1996).
3. G. Veneziano, Nucl. Phys. **B 159** 461 (1979).
4. R. Alkofer, M. Nowak, J. Verbaarschot, I. Zahed, Phys. Lett. **233B** 205 (1989).
5. D. Zwanziger, Nucl. Phys. **B 412** 657 (1994); H. Suman and K. Schilling, Phys. Lett **373B** 314 (1996); L. v. Smekal, A. Hauck and R. Alkofer, hep-ph/9705242.
6. R. Alkofer, P. A. Amundsen and H. Reinhardt, Phys. Lett. **218B** 75 (1989).

The submitted manuscript has been authored by a contractor of the U. S. Government under contract No. W-31-109-ENG-38. Accordingly, the U. S. Government retains a nonexclusive royalty-free license to publish or reproduce the published form of this contribution, or allow others to do so, for U. S. Government purposes.

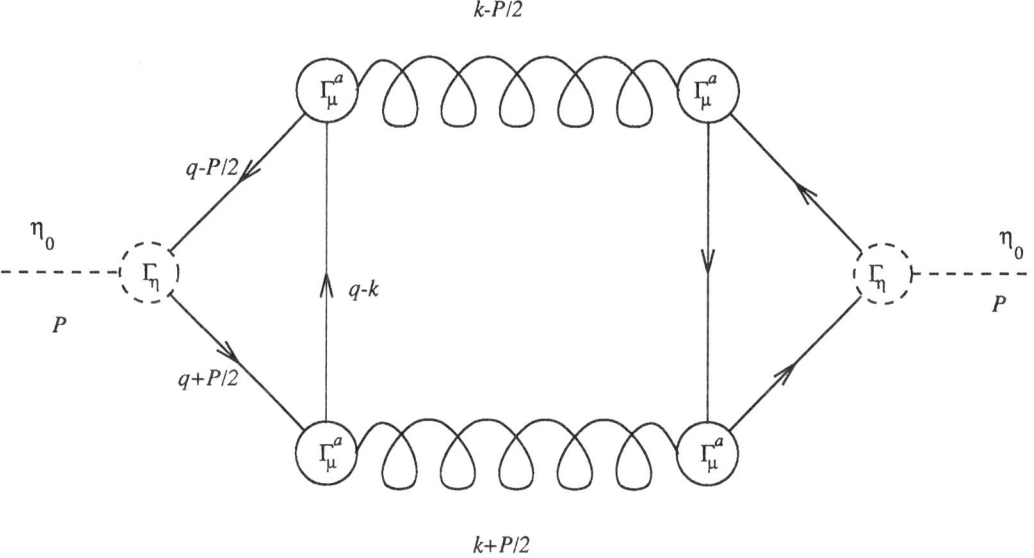

Figure 1: The diamond diagram $\Pi(P^2)$, a factor 2 arises from crossed gluon exchange.

Strong $U_A(1)$ breaking in radiative η decays

Makoto Takizawa[*], Yukio Nemoto[†] and Makoto Oka[†]

[*]*Showa College of Pharmaceutical Sciences, Machida, Tokyo 194 Japan*
[†]*Department of Physics, Tokyo Institute of Technology, Meguro, Tokyo 152 Japan*

Abstract. We study the $\eta \to \gamma\gamma$, $\eta \to \gamma\mu^-\mu^+$ and $\eta \to \pi^0\gamma\gamma$ decays using an extended three-flavor Nambu-Jona-Lasinio model that includes the instanton induced $U_A(1)$ breaking interaction. We find that the η-meson mass and η decay widths are in good agreement with the experimental values when the $U_A(1)$ breaking is rather strong.

In order to understand the role of the $U_A(1)$ anomaly in the low-energy QCD, it is important to study the η-meson decays as well as its mass and decay constant. Among the η-meson decays, $\eta \to \gamma\gamma$, $\eta \to \gamma\mu^-\mu^+$ and $\eta \to \pi^0\gamma\gamma$ decays are interesting. They have no final state interactions and involve only neutral mesons so that the electromagnetic transitions are induced only by the internal structure of the mesons.

We have studied [1–3] the $\eta \to \gamma\gamma$, $\eta \to \gamma\mu^-\mu^+$ and $\eta \to \pi^0\gamma\gamma$ decays in the framework of the three-flavor Nambu-Jona-Lasinio (NJL) model so that the quark structure of the η meson is explicitly taken into account. In this approach the effects of the explicit breaking of the chiral symmetry by the current quark mass term and the $U_A(1)$ anomaly on the η decay amplitudes can be calculated consistently with those on the η-meson mass, η decay constant and mixing angle within the model applicability.

We work with the following NJL model lagrangian density:

$$\mathcal{L} = \mathcal{L}_0 + \mathcal{L}_4 + \mathcal{L}_6, \tag{1}$$

$$\mathcal{L}_0 = \bar{\psi}\left(i\partial_\mu\gamma^\mu - \hat{m}\right)\psi, \tag{2}$$

$$\mathcal{L}_4 = \frac{G_S}{2} \sum_{a=0}^{8} \left[\left(\bar{\psi}\lambda^a\psi\right)^2 + \left(\bar{\psi}\lambda^a i\gamma_5\psi\right)^2 \right], \tag{3}$$

$$\mathcal{L}_6 = G_D \left\{ \det\left[\bar{\psi}_i(1-\gamma_5)\psi_j\right] + \det\left[\bar{\psi}_i(1+\gamma_5)\psi_j\right] \right\}. \tag{4}$$

Here the quark field ψ is a column vector in color, flavor and Dirac spaces and $\lambda^a (a = 0\ldots 8)$ is the $U(3)$ generator in flavor space. The free Dirac lagrangian

$\mathcal{L}_0$ incorporates the current quark mass matrix $\hat{m} = \text{diag}(m_u, m_d, m_s)$ which breaks the chiral $U_L(3) \times U_R(3)$ invariance explicitly. $\mathcal{L}_4$ is a QCD motivated four-fermion interaction, which is chiral $U_L(3) \times U_R(3)$ invariant. The 't Hooft determinant $\mathcal{L}_6$ represents the $U_A(1)$ anomaly. It is a 3×3 determinant with respect to flavor with $i, j = $ u, d, s.

Quark condensates and constituent quark masses are self-consistently determined by the gap equations in the mean field approximation. The pseudoscalar channel quark-antiquark scattering amplitudes are then calculated in the ladder approximation. We assume the isospin symmetry, i.e., $m_u = m_d$. From the pole positions of the scattering amplitudes, the pseudoscalar meson masses are determined. Because of the $SU(3)$ symmetry breaking, the flavor $\lambda^8 - \lambda^0$ components mix with each other. Thus we solve the coupled-channel $q\bar{q}$-scattering problem for the η meson. The mixing angle θ is obtained by diagonalization of the $q\bar{q}$-scattering amplitude. The η decay constant f_η is determined by calculating the quark-antiquark one-loop graph.

For the η' meson, since the NJL model does not confine quarks, the η'-meson state has the unphysical imaginary part which corresponds to the $\eta' \to q\bar{q}$ decays. Therefore we do not apply our model to the η' meson in this article.

We evaluate the $\eta \to \gamma\gamma$ decay amplitude by calculating the quark triangle diagrams. It should be noted that the low-energy theorem for the $\pi^0 \to \gamma\gamma$ decay amplitude is reproduced in this approach if one takes chiral limit. The $\eta \to \gamma\mu^-\mu^+$ decay amplitude is evaluated in the similar manner. On the other hand, the $\eta \to \pi^0\gamma\gamma$ decay amplitude is evaluated by calculating the quark box diagrams. Note that the strange quark loops do not contribute.

Let us now move to the discussions of our numerical results. The recent experimental results of the $\eta \to \gamma\gamma$, $\eta \to \gamma\mu^-\mu^+$ and $\eta \to \pi^0\gamma\gamma$ decay widths are $\Gamma_{\eta \to \gamma\gamma} = 0.510 \pm 0.026\,\text{keV}$, $\Gamma_{\eta \to \gamma\mu^-\mu^+} = 0.41 \pm 0.06\,\text{eV}$ and $\Gamma_{\eta \to \pi^0\gamma\gamma} = 0.93 \pm 0.19\,\text{eV}$ respectively.

In our theoretical calculations, the parameters of the NJL model are the current quark masses $m_u = m_d, m_s$, the four quark coupling constant G_S, the six-quark determinant coupling constant G_D and the covariant cutoff Λ. We take G_D as a free parameter and study η meson properties as functions of G_D. We use the light current quark masses $m_u = m_d = 8.0$ MeV to reproduce $M_u = M_d \simeq 330$ MeV ($\simeq 1/3 M_N$) which is the value usually used in the nonrelativistic quark model. Other parameters, m_s, G_S, and Λ, are determined so as to reproduce the isospin averaged observed masses, m_π, m_K, and the pion decay constant f_π.

We obtain $m_s = 193$ MeV, $\Lambda = 783$ MeV and $M_{u,d} = 325$ MeV which are almost independent of G_D. The ratio of the current s-quark mass to the current u,d-quark mass is $m_s/m_u = 24.1$, which agrees well with $m_s/\hat{m} = 25 \pm 2.5$ ($\hat{m} = \frac{1}{2}(m_u + m_d)$) derived from Chiral Perturbation Theory (ChPT). We have obtained $f_K = 97$ MeV which is about 14% smaller than the observed value.

Table 1 summarizes the fitted results of the model parameters and the cal-

TABLE 1. The parameters of the model, the η meson mass, the η decay constant, the mixing angle and the $\eta \to \gamma\gamma$, $\eta \to \gamma\mu^-\mu^+$ and $\eta \to \pi^0\gamma\gamma$ decay widths for each G_D^{eff}.

G_D^{eff}	G_S^{eff}	m_η [MeV]	f_η [MeV]	θ [deg]	$\Gamma_{\eta\to\gamma\gamma}$ [keV]	$\Gamma_{\eta\to\gamma\mu^-\mu^+}$ [eV]	$\Gamma_{\eta\to\pi^0\gamma\gamma}$ [eV]
0.00	0.73	138.1	92.4	-54.74	1.42	1.84	2.88
0.10	0.70	285.3	92.3	-44.61	1.27	1.51	2.46
0.20	0.66	366.1	91.9	-33.52	1.11	1.19	2.06
0.30	0.63	419.1	91.4	-23.24	0.94	0.94	1.71
0.40	0.60	455.0	91.0	-14.98	0.79	0.73	1.42
0.50	0.57	479.7	90.9	-8.86	0.67	0.59	1.20
0.60	0.54	497.3	91.0	-4.44	0.57	0.48	1.04
0.70	0.51	510.0	91.2	-1.25	0.50	0.41	0.92
0.80	0.47	519.6	91.4	1.09	0.45	0.35	0.84
0.90	0.44	527.0	91.7	2.84	0.41	0.31	0.77
1.00	0.41	532.8	91.9	4.17	0.37	0.28	0.71

culated results of the η meson properties. We define dimensionless parameters $G_D^{\text{eff}} \equiv -G_D(\Lambda/2\pi)^4 \Lambda N_c^2$ and $G_S^{\text{eff}} \equiv G_S(\Lambda/2\pi)^2 N_c$.

The experimental values of the $\eta \to \gamma\gamma$, $\eta \to \gamma\mu^-\mu^+$ and $\eta \to \pi^0\gamma\gamma$ decay widths are reproduced at about $G_D^{\text{eff}} = 0.7$. The calculated η-meson mass at $G_D^{\text{eff}} = 0.7$ is 7% smaller than the observed mass. $G_D^{\text{eff}} = 0.7$ corresponds to $G_D\langle \bar{s}s \rangle/G_S = 0.44$, suggesting that the contribution from $\mathcal{L}_6$ to the dynamical mass of the up and down quarks is 44% of that from $\mathcal{L}_4$. It reminds us of the instanton liquid picture of the QCD vacuum [4] although the $U_A(1)$-breaking interaction we have used is derived in the dilute instanton gas approximation.

The mixing angle at $G_D^{\text{eff}} = 0.7$ is $\theta = -1.25°$. This disagrees with the "standard" value $\theta \simeq -20°$. We have reanalyzed the η-η' mixing angle in the context of (1) the η-η' masses in the $1/N_C$ expansion approach, (2) the $\eta(\eta') \to \gamma\gamma$ decays in the PCAC with the chiral anomaly approach and (3) the $J/\psi \to \gamma\eta(\eta')$ decays in the simple quark model with the flavor singlet dominance of the OZI violation. We have found that the theoretical uncertainty is not small enough to give a definite value of the η-η' mixing angle [5].

REFERENCES

1. Takizawa M., and Oka M., *Phys. Lett.* B **359**, 210 (1995); **364**, 249 (1995) (E).
2. Nemoto Y., Oka M., and Takizawa M., *Phys. Rev.* D **54**, 6777 (1996).
3. Takizawa M., Nemoto Y., and Oka M., *Phys. Rev.* D **55**, 4083 (1997).
4. Schäfer T., and Shuryak E.V., *Phys. Rev.* D **53**, 6522 (1996); **54**, 1099 (1996).
5. Takizawa M., and Oka M., in preparation.

Low Energy QCD From an Effective Quark-Quark Interaction

Thomas Meissner* and Michael Frank[†]

*Department of Physics, Carnegie Mellon University, Pittsburgh, PA 15213[1]
[†]Institute for Nuclear Theory, Seattle, WA 98195[2]

Abstract. We consider a model truncation of QCD which is based on an effective quark-quark interaction. The truncation allows for a phenomenological description in a framework which maintains the global symmetries of QCD and permits a $1/N_c$ expansion. The applied truncation leads to the Schwinger-Dyson equation for the quark self energy in the rainbow approximation, which is solved numerically for a given model form of the gluon 2 point function $D(q^2)$. Meson bound states appear as solutions of the homogeneous ladder Bethe-Salpeter equation. This approach allows for a detailed and systematic investigation of nonperturbative phenomena at low and intermediate energies. A systematic chiral low energy expansion is performed leading to a model prediction of all the chiral coefficients (Gasser Leutwyler coefficients). We demonstrate how the $U_A(1)$ anomaly and the splitting between η and η' can arise in this approach. It turns out that a necessary condition is a $\frac{1}{q^4}$ infrared singularity for the gluon 2 point function. Within the truncation a general technique for calculating nonperturbative quark and gluonic vacuum condensates can be developed. We demonstrate this in case of the mixed condensate $<\bar{q}G\sigma q>$. Final results for this condensate as well as $<\bar{q}q>$ are presented.

I A PATH FROM QCD TO LOW ENERGY CHIRAL PHYSICS

A global color symmetry model (GCM) that is based upon an effective quark-quark interaction arises from the QCD partition function by formally integrating over the gluon fields and truncating the expansion in gluon n point functions after $n = 2$ (for a review c.f. [1]). In the chiral limit $m_0 = 0$ the generating functional reads:

$$\mathcal{Z}_{\rm GCM} = \int \mathcal{D}q\mathcal{D}\bar{q}\, e^{-\left\{\int \bar{q}\slashed{\partial}q + \frac{g_s^2}{2}\int dxdy \left[\bar{q}(x)\gamma_\mu \frac{\lambda^a}{2}q(x)\right] D^{ab}_{\mu\nu}(x-y)\left[\bar{q}(y)\gamma_\nu \frac{\lambda^b}{2}q(y)\right]\right\}}$$

---

[1]) email: meissner@yukawa.phys.cmu.edu
[2]) email: frank@phys.washington.edu

$$= \int \mathcal{D}q\mathcal{D}\bar{q}\mathcal{D}Ae^{-\{\int \bar{q}(\partial - ig_s A)q + \int \int \frac{1}{2}AD^{-1}A\}}. \qquad (1)$$

Here $j^a_\nu(x) \equiv \bar{q}(x)\frac{\lambda^a}{2}\gamma_\nu q(x)$ is the quark color current, and for convenience a gauge for the gluon propagator $D^{ab}_{\mu\nu}(x-y) = \delta_{ab}\delta_{\mu\nu}D(x-y)$ is employed. This model truncation respects all global symmetries of QCD, in particular chiral symmetry. What is lost is invariance under local color gauge transformations. Fixing a model form for the gluon propagator $D(q^2)$ specifies a quark-quark interaction. Furthermore the model allows a $\frac{1}{N_c}$ expansion. The lowest order ($\mathcal{O}(N_c)$) consists in solving the Dyson-Schwinger equation for the quark self energy including rainbow gluon dressings and the ladder Bethe-Salpeter equation for mesonic bound states. One then is able to change the degrees of freedom from quarks to hadrons by performing an appropriate variable transformation in the generating functional path integral. For the moment only the Goldstone bosons are taken into account. We obtain a non-local hadronic interaction between the Goldstone fields, which can be systematically expanded in the momenta of the Goldstone fields and the current quark mass leading exactly to the form given by Gasser and Leutwyler. The chiral low energy coefficients L_i are now determined by the dynamics of the quark-quark interaction and can be numerically calculated from the quark self energy functions. They turn out to be compatible with the phenomenological results [2].

II TRIANGLE DIAGRAMS, $U_A(1)$ BREAKING AND THE η' MASS

Including higher mass mesonic states other than the Goldstone fields and integrating them out generates triangle diagrams such as shown in Fig.1.

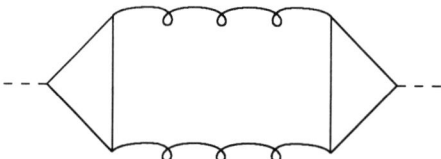

FIGURE 1.

Those diagrams are suppressed by one order of $\frac{1}{N_c}$ compared with ones mentioned in the last section. They break the $U_A(1)$ symmetry anomalously (triangle anomaly). This itself is however not enough to generate a finite mass for the the iso-singlet pseudoscalar Goldstone boson η'. This fact is commonly known as the "$U_A(1)$ Problem". In order make the the η' massive it is necessary and sufficient that the model gluon 2 point function $D(q^2)$ has

TABLE 1.

	$-\langle \bar{q}q \rangle^{\frac{1}{3}}$ [MeV]	$-\langle g_s \bar{q}\sigma G q \rangle^{\frac{1}{5}}$ [MeV]
this paper	150 − 180	400 − 460
QCD sum rules	210 − 230	375 − 395
quenched lattice	225	402 − 429
instanton liquid	272	490

a $\frac{1}{q^4}$ singularity in the infrared $q^2 \to 0$ [3]. This provides a mechanism for an η' mass without employing explicitly topological gauge field configurations (instantons).

III QUARK AND GLUON CONDENSATES

The quark condensate $<\bar{q}q>$ is simply given as the trace over the scalar component of the quark propagator S_0

$$<\bar{q}q> = (-)\text{Tr}_{\gamma C}[S_0]. \quad (2)$$

As it can be seen from eq.(1) in our model truncation the integration over the gluon field A is quadratic. Therefore the integration over any any gluonic vacuum expectation value can be performed exactly and renders effectively the quark color current j together with the gluon 2 point function D, e.g.

$$\int \mathcal{D}AA e^{-\frac{1}{2}AD^{-1}A+jA} = (jD)e^{\frac{1}{2}jDj}. \quad (3)$$

This allows the calculation of the vacuum expectation value of any gluon and combined quark-gluon operator [4]. As the simplest example we have studied the mixed condensate $g_s \langle \bar{q} G_{\mu\nu} \sigma^{\mu\nu} q \rangle$. The results for $<\bar{q}q>$ and $<\bar{q}G \cdot \sigma q>$, both evaluated at a renormalization point of $\mu = 1\text{GeV}$ are shown in Table 1 and compared with other nonperturbative approaches.

Very recently the method has been extended to study the space time structure of nonlocal vacuum condensates $<\bar{q}(x)q(0)>$ [5].

REFERENCES

1. P. Tandy, nucl-th/9705018, Prog.Part.Nucl.Phys. **39** (1997) (in press).
2. M. Frank and T. Meissner, Phys.Rev. C **53**, 2410 (1996).
3. M. Frank and T. Meissner, hep-ph/9703270, subm. to Phys.Rev. C.
4. T. Meissner, hep-ph/9702293, Phys.Lett. B (in press).
5. L.S. Kisslinger and T. Meissner, hep-ph/9706423, subm. to Phys.Rev. C.

Instanton-Monopole Correlations in Lattice QCD [1]

M. Feurstein, H. Markum and S. Thurner

Institut für Kernphysik, TU Wien, Wiedner Hauptstraße 8-10, A-1040 Vienna, Austria

Abstract. We analyze the interplay of topological objects in four dimensional QCD. The distributions of color magnetic monopoles obtained in the maximum abelian gauge are computed around instantons in four-dimensional full QCD. We find an enhanced probability of encountering monopoles inside the core of an instanton. Moreover we present evidence that nontrivial values of the chiral condensate are predominantly found at the locations of instantons and monopoles.

Classical gauge field configurations with non-trivial topology are believed to play the essential role in the quark-binding mechanism. In the scenario of the dual superconductor, which is in essence confirmed [1], abelian monopoles condense leading to confinement. Large and interacting instantons could bind quarks if they form an instanton liquid. These two pictures are profoundly distinct in describing the mechanism: In the monopole scenario color electric flux tubes form [2], whereas in the instanton paradigm instantons are seen as *traps* for quarks. Fermions hop from one instanton to another. If more than one quark is travelling through the instanton liquid, this trapping leads to an effective attractive interaction, which could explain quark-binding [3]. Hence the question arises, whether and if, how instantons and monopoles are related to each other. Several groups investigated the relation between monopoles and instantons for semi-classical configurations [4]. In a series of papers we presented evidence that those correlations exist also in realistic equilibrium configurations [5]. In this contribution we present results on the local correlation functions between the chiral condensate $\bar\psi\psi(x)$, the topological charge density $q(x)$, and the monopole density $\rho(x)$. They might be interpreted as a first direct confirmation of the fermion hopping mechanism. These findings indicate that both confinement mechanisms have the same topological origin and that both approaches should be united.

---

[1] Supported by Fonds zur Förderung der wissenschaftlichen Forschung under Project P11456-PHY.

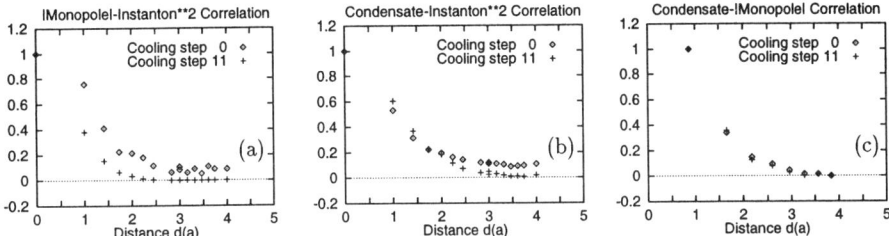

FIGURE 1. Correlation functions in the presence of dynamical quarks in the confinement ($\beta = 5.2$). The monopole-instanton correlation (a) is similar as in pure $SU(3)$. The correlation of the quark condensate and the topological charge density (b) is cooling dependent, whereas the correlation between the condensate and the monopole density (c) is not. All correlations extend over two lattice spacings and indicate local correlations of the chiral condensate and topological objects.

To investigate monopole currents one has to project $SU(N)$ onto its abelian degrees of freedom, such that an abelian $U(1)^{N-1}$ theory remains. We employ the maximum abelian gauge. For the definition of the monopole currents $m_i(x, \mu)$ we use the standard method [6]. From the monopole currents we define the local monopole density as $\rho(x) = \frac{1}{3\cdot 4 V_4} \sum_{\mu,i} |m_i(x,\mu)|$. There exist several definitions of the topological charge on the lattice. We use the field theoretic charge definition which is a straightforward discretization of the continuum charge expression $q(x) = \frac{g^2}{32\pi^2} \epsilon^{\mu\nu\rho\sigma} \text{Tr}(F_{\mu\nu}(x) F_{\rho\sigma}(x))$. To get rid of the renormalization constants we apply the "Cabbibo-Marinari cooling method" [7]. To measure correlations between topological quantities we calculate functions of the type

$$\langle q(0)q(d)\rangle, \langle \rho(0)\rho(d)\rangle, \langle \rho(0)q^2(d)\rangle, \langle q^2(0)\bar{\psi}\psi(d)\rangle, \langle \rho(0)\bar{\psi}\psi(d)\rangle, \qquad (1)$$

which are normalized after subtracting the corresponding cluster values.

Figure 1 shows correlation functions of full $SU(3)$ QCD with 3 flavors of Kogut-Susskind quarks of equal mass $ma = 0.1$ in the confinement region. The ρq^2-correlation (a) looks similar to the corresponding function in pure QCD [5]. For pure $SU(2)$ gauge theory we have reported the corresponding screening masses [8] which are in the order of the expected glueball masses. In the case of the $\bar{\psi}\psi q^2$-correlation (b) exponential fits show that an increasing number of cooling steps results in a narrower correlation function. The $\bar{\psi}\psi\rho$-correlation (c) on the other hand is not sensitive to cooling and has the same exponential decay as the $\bar{\psi}\psi q^2$-correlation after some cooling steps.

In Fig. 2 (a) a time slice of a typical configuration from $SU(3)$ theory with dynamical quarks on the $8^3 \times 4$ lattice in the confinement phase is shown. This particular configuration possesses a cluster with a positive topological charge

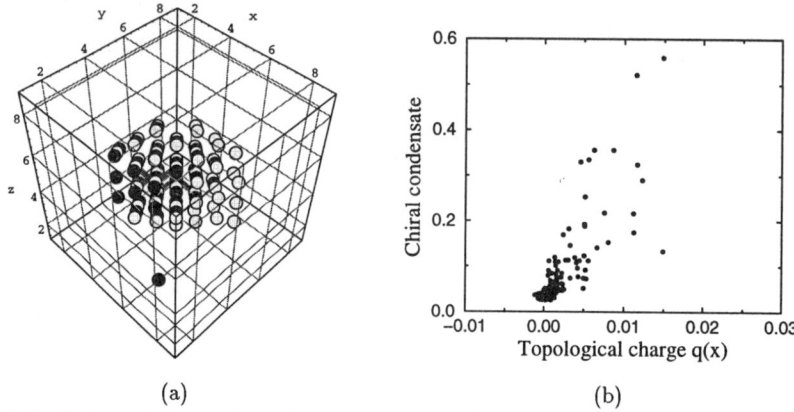

FIGURE 2. (a) Time slice of a single gauge field configuration of $SU(3)$ theory with dynamical quarks after 10 cooling steps. The dark dots represent the topological charge distribution, associated with an instanton, the light dots the distribution of the chiral condensate. Monopole loops are represented by lines. It turns out that chiral symmetry breaking occurs at the positions of the instantons. (b) Scatter plot of $\bar{\psi}\psi(x)$ against $q(x)$ in the volume of the same time slice of the same gauge field configuration. A linear correlation of the topological charge density and the local values of the quark condensate is suggested.

corresponding to an instanton. We display the topological charge density by dark dots if $|q(x)| > 0.003$. The quark-antiquark density is indicated by light dots if a threshold for $\bar{\psi}\psi(x) > 0.066$ is exceeded. At 10 cooling steps clusters of nonzero topological charge density and quark condensate are clearly resolved. Combining the above finding of Fig. 1 showing that the correlation functions between $\bar{\psi}\psi(x)$ and $q^2(y)$ (or $\rho(y)$) are rather insensitive under cooling together with 3D images like Fig. 2 (a), we conclude that instantons go hand in hand with clusters of $\bar{\psi}\psi(x) \neq 0$ also in the uncooled QCD vacuum. Figure 2 (a) demonstrates that the local chiral condensate attains its maximum values at the same positions as the extreme values of the topological charge density. This behavior is further substantiated in Fig. 2 (b) where the $\bar{\psi}\psi(x)$-values are plotted against $q(x)$ for all points x in the same configuration at 10 cooling steps. At first sight a linear relationship between the topological charge density and the virtual quark density is suggested.

We comment on the behavior of the topological structure and the quark condensate when crossing the phase transition. The normalized correlation functions do not change qualitatively. The strength of the correlations becomes, however, 2-3 orders of magnitude smaller. This means that the topological activity becomes much weaker as a whole in the deconfinement phase, as expected. Most of the configurations have trivial net topological charge. This does not exclude the existence of instanton-antiinstanton pairs which become more difficult to be resolved at the smaller physical volume at $\beta = 5.4$.

In summary, our calculations of correlation functions between topological objects and the quark condensate suggest that the local chiral condensate takes a non-vanishing value predominantly in the regions of instantons and monopole loops. We found that the clusters of non-vanishing quark condensate have a size of about 0.4 fm, which corresponds to the instanton sizes observed in the same configurations. It was well known before that the chiral condensate is related to the topological charge and topological susceptibility. We demonstrated that exactly at those places in euclidean space-time, where tunneling between the vacua occurs, amplified production of quark condensate takes place. This supports the fermion hopping picture in the instanton liquid approach. On the other hand monopole condensation has been confirmed quite relyably, so that from a lattice point of view *both* of the two distinct mechanisms seem to be realized. This necessitates a more mature model for the quark-binding mechanism and future work should focus on additional low energy features for further evidence of the instanton liquid model [9].

REFERENCES

1. Di Giacomo, A., *Nucl. Phys.* B (Proc. Suppl.) **47**, 136 (1996).
2. 't Hooft, G., in *High Energy Physics*, Proceedings of the EPS International Conference, Palermo, Italy, 1975, edited by Zichichi, A. (Editrice Compositori, Bologna, 1976); Mandelstam, S., *Phys. Rep.* **23C**, 245 (1976).
3. Shuryak, E.V., *Nucl. Phys.* **B302**, 559 (1988); Schäfer, T., and Shuryak, E.V., hep-ph/9610451.
4. Chernodub, M.N., and Gubarev, F.V., *JETP Lett.* **62**, 100 (1995); Hart, A., and Teper, M., *Phys. Lett. B* **371**, 261 (1996); Bornyakov, V., and Schierholz, G., *Phys. Lett. B* **384**, 190 (1996); Fukushima, M., Sasaki, S., Suganuma, H., Tanaka, A., Toki, H., and Diakonov, D., *Phys. Lett. B* **399**, 141 (1997); Brower, R.C., Orginos, K.N., and Tan, C.-I, *Phys. Rev. D* **55**, 6313 (1997); Reinhardt, H., hep-th/9702049.
5. Thurner, S., Markum, H., and Sakuler, W., in *Confinement 95*, Proceedings of the International Workshop, Osaka, Japan, 1995, edited by Toki, H., *et al.* (World Scientific, 1996) p. 77; hep-th/9506123; Thurner, S., Feurstein, M., Markum, H., and Sakuler, W., *Phys. Rev. D* **54**, 3457 (1996); Feurstein, M., Markum, H., and Thurner, S., *Phys. Lett. B* **396**, 203 (1997).
6. Kronfeld, A.S., Schierholz, G., and Wiese, U.-J., *Nucl. Phys.* **B293**, 461 (1987).
7. Di Vecchia, P., Fabricius, K., Rossi, G.C., and Veneziano, G., *Nucl. Phys.* **B192**, 392 (1981); *Phys. Lett. B* **108**, 323 (1982); *Phys. Lett. B* **249**, 490 (1990).
8. Feurstein, M., Markum, H., and Thurner, S., *Nucl. Phys.* B (Proc. Suppl.) **53**, 550 (1997).
9. Feurstein, M., Ilgenfritz, E.-M., Müller-Preußker, M., and Thurner, S., hep-lat/9611024.

Meson and Lepton Decays

First Physics from KTeV

John Belz

Department of Physics and Astronomy
Rutgers University, Piscataway, NJ 08855

Abstract. The KTeV experiment at Fermilab is nearing the end of a year–long run to collect data on CP violation and rare kaon decays. I report on the status of this run and present preliminary results from several of the decay modes under study. A total of 5.7 million (before offline cuts) decays of the $\Re e(\epsilon'/\epsilon)$ statistics–limiting $K_L^0 \longrightarrow \pi^0\pi^0$ mode were collected. A search for light supersymmetric particles yielded a null result, excluding the existence of the R^0 ($g\tilde{g}$) over a mass range of $1.2 \rightarrow 4.6$ GeV/c^2 and a lifetime range of $2 \times 10^{-10} \rightarrow 7 \times 10^{-4}$ seconds. I report on the first observation of the decay $K_L^0 \longrightarrow \pi^+\pi^- e^+ e^-$, along with a preliminary branching ratio for this decay of $(2.6 \pm 0.5) \times 10^{-7}$. Finally, I report on an improved (preliminary) upper limit on the branching ratio for $K_L^0 \longrightarrow \pi^0 \nu \bar{\nu}$ of 1.8×10^{-6} (90% c.l.).

KTeV consists of two experiments: E832, a search for the "direct" component of CP violation in $K \longrightarrow \pi\pi$ decays; and E799–II, a study of rare kaon decays. First physics results from the two experiments are presented here.

All CP violation phenomena observed to date can be understood as arising from a small admixture of the CP–even eigenstate $|K_1>$ into the primarily CP–odd long–lived neutral kaon:

$$|K_L^0> \sim |K_2> + \epsilon|K_1>, \tag{1}$$

the only particle known to break the CP symmetry. More than 30 years of experimental scrutiny have not shed any additional light on this picture.

A second mechanism exists within the standard model, wherein the decay amplitude of the CP–odd $|K_2>$ into a CP–even final state is nonzero. This "direct" CP violation is experimentally accessible by virtue of the expected inequality of the amplitudes

$$<\pi^+\pi^-|H_w|K_2> \neq <\pi^0\pi^0|H_w|K_2> \tag{2}$$

(where H_w is the weak Hamiltonian) due to the different isospin decompositions of the final states. We can define a new parameter ϵ' such that

$$\eta_{+-} = \frac{<\pi^+\pi^-|H_w|K_L>}{<\pi^+\pi^-|H_w|K_S>} = \epsilon + \epsilon' \qquad (3)$$

$$\eta_{00} = \frac{<\pi^0\pi^0|H_w|K_L>}{<\pi^0\pi^0|H_w|K_S>} = \epsilon - 2\epsilon' \qquad (4)$$

which leads to the experimentally observable

$$\left|\frac{\eta_{+-}}{\eta_{00}}\right|^2 \approx 1 + 6\Re(\epsilon'/\epsilon). \qquad (5)$$

The current experimental situation with regard to $\Re(\epsilon'/\epsilon)$ is unresolved. The two best numbers come from Fermilab E731 [1] and CERN NA31 [2]:

$$\text{E731}: \qquad \Re(\epsilon'/\epsilon) = (7.4 \pm 5.2(\text{stat.}) \pm 2.9(\text{syst.})) \times 10^{-4} \qquad (6)$$

$$\text{NA31}: \qquad \Re(\epsilon'/\epsilon) = (23.0 \pm 3.6(\text{stat.}) \pm 5.4(\text{syst.})) \times 10^{-4} \qquad (7)$$

Theoretically, recent predictions within the standard model place $\Re(\epsilon'/\epsilon)$ in the range of $0 \rightarrow 10 \times 10^{-4}$ [3-5]. The detector for KTeV, the successor to E731, was designed to improve the uncertainty in $\Re(\epsilon'/\epsilon)$ to the 1×10^{-4} level, resolve the experimental discrepancy and perhaps find unambiguous evidence for direct CP violation for the first time.

E832 extracts $\Re(\epsilon'/\epsilon)$ by the simultaneous collection of K_L^0 and K_S^0 decays into $\pi^+\pi^-$ and $\pi^0\pi^0$. The technique utilizes dual beams, as illustrated in Figure 1. An active (scintillator) regenerator is used as a K_S^0 source. The regenerator moves between the two beams once per Tevatron pulse. The simultaneous detection of both long- and short-lived decays into charged and neutral two pion modes, along with trigger, data acquisition and event reconstruction which is blind to the particular beam, results in many systematic uncertainties cancelling in the double ratio.

Decays with charged products are collected in a spectrometer consisting of four planar drift chambers, two on either side of a dipole analyzing magnet. Each chamber measures positions in two orthogonal views. Each view consists of two planes of wires, with cells arranged in a hexagonal geometry. Each chamber has approximately 100 μm single-hit position resolution per plane. The spectrometer magnet imparts a transverse momentum impulse of 411 (205) MeV/c to charged particles in E832 (E799) configuration. Helium bags integrated into the spectrometer minimize multiple scattering. The invariant mass resolution for the decay $K_L^0 \rightarrow \pi^+\pi^-$ is better than 2 MeV/c^2.

A 1.9m by 1.9m pure CsI electromagnetic calorimeter is used to reconstruct the energy of photons and electrons to a precision of better than a percent. The calorimeter has two 15 cm by 15 cm holes in the middle to allow passage of the beams. The calorimeter is also used to match tracks in the two views and reject background from semileptonic kaon decays. A set of 12 photon vetos located throughout the detector provides a hermetic photon coverage up to angles of 100 mr.

A counter bank (muon veto) located at the downstream end of the detector is used to reject $K_{\mu 3}$ decays, and for detection of muonic decay modes in the rare–decay search (E799).

Data for the direct CP-violation search E832 was collected in two periods: A six–week run in the fall of 1996 and a spring–summer run in 1997. A total of approximately 5.7 million (before serious offline cuts) events of the $\Re(\epsilon'/\epsilon)$ statistics–limiting $K_L^0 \longrightarrow \pi^0\pi^0$ decay were collected, simultaneous with regenerator beam $K \longrightarrow \pi^0\pi^0$ and vacuum and regenerator $K \longrightarrow \pi^+\pi^-$ decays. Figure 2 shows the distribution of $K \longrightarrow \pi^+\pi^-$ decays downstream of the active regenerator, for a fraction of the 1996 data set.

In addition to the $\Re(\epsilon'/\epsilon)$ measurement, E832 will probe other aspects of CP violation and kaon decays with unprecedented sensitivity. These additional topics include (1) the form factor and search for direct CP violation in $K_{L,S}^0 \longrightarrow \pi^+\pi^-\gamma$, (2) charge asymmetry in $K \longrightarrow \pi^{\pm}l^{\mp}\nu$, (3) CPT studies ($\Delta\Phi$), (4) $K_L^0 \longrightarrow \pi^+\pi^-\pi^0$, $K_L^0 \longrightarrow \pi^0\pi^0\pi^0$ Dalitz plot parameters, (5) measurement of $K^*(892)$ radiative width and first direct observation of $K^*(1410)$, and (6) Search for supersymmetric particles (R^0).

The decay $K_L^0 \longrightarrow \pi^+\pi^-\gamma$ is potentially an additional source of information on direct CP violation. It arises from two processes; the internal bremsstrahlung (IB) of a CP-even $K \longrightarrow \pi^+\pi^-$ decay, and the primarily CP-

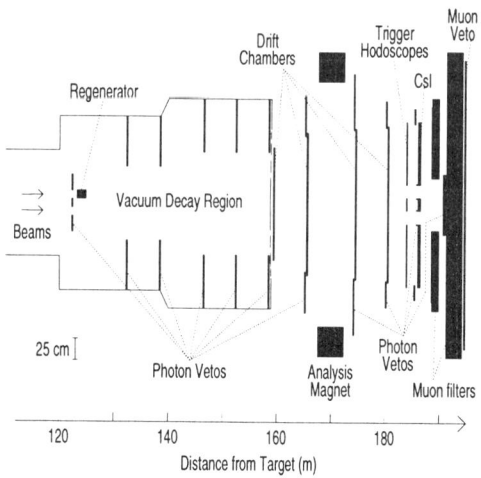

FIGURE 1. Schematic of the KTeV Detector (E832 Configuration).

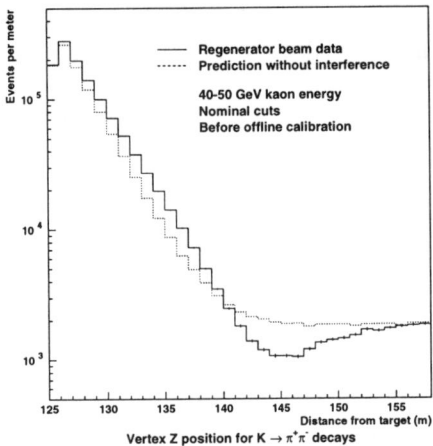

FIGURE 2. $K \longrightarrow \pi^+\pi^-$ interference downstream of regenerator.

odd direct emission (DE) process in which the photon is emitted directly from the primary decay vertex before the hadronization of the pions. Interference between these terms, possible only if the DE term has a CP violating electric dipole component, could result in a difference between the parameter

$$\eta_{+-\gamma} = \frac{<\pi^+\pi^-\gamma|H_w|K_L>}{<\pi^+\pi^-\gamma|H_w|K_S>} = \epsilon + \epsilon'_{+-\gamma} \qquad (8)$$

and its two-body counterpart η_{+-} (Equation 3) and is thus another potential window on direct CP violation.

In Figure 3 the statistics for this mode are shown for the regenerator and vacuum beams, for one day of KTeV data. Based on these statistics, we can conclude that KTeV will substantially improve the world samples for both K_L^0 and K_S^0 decays, collecting approximately 150k (140k) K_L^0 (K_S^0) $\longrightarrow \pi^+\pi^-\gamma$ decays compared to a world sample of 4k (13k) decays. With these numbers, KTeV should reduce the statistical uncertainty in $\eta_{+-\gamma}$ to better than a percent of itself.

Another E832 physics result, which has been submitted to *Physical Review Letters*, is the results of a search for the R^0, the lightest bound state of gluon and a light gluino in certain supersymmetric models [6]. The R^0 is predicted to decay via the sequence

$$R^0 \longrightarrow \tilde{\gamma}\rho, \rho \longrightarrow \pi^+\pi^-, \qquad (9)$$

have a mass in the $\lesssim 2$ GeV/c^2 range, and a lifetime of 10^{-10} to 10^{-7} seconds. This makes the KTeV spectrometer an ideal place to search for these exotics.

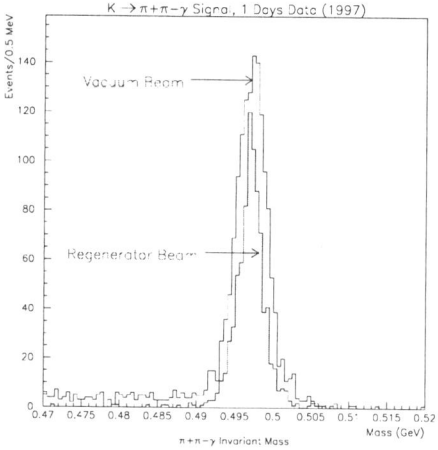

FIGURE 3. $K^0_{L,S} \longrightarrow \pi^+\pi^-\gamma$ Signal from one days E832 data.

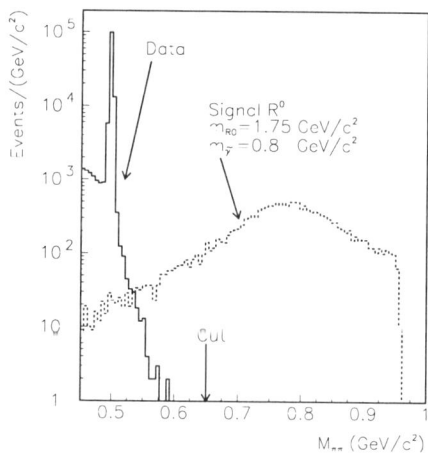

FIGURE 4. Search for R^0 (g$\tilde{g}$) in 1 day of KTeV data.

Figure 4 illustrates the results of the E832 R^0 search. No high–mass $\pi^+\pi^-$ candidates are observed. Based on this plot, from 1 day of data, KTeV severely restricts the lightest stable SUSY state in this theory (R^0) well beyond 10^{-4} in the R^0/K_L ratio, over a mass range of $1.2 \to 4.6$ GeV/c^2 and a lifetime range of $2 \times 10^{-10} \to 7 \times 10^{-4}$ seconds.

Data was collected in E799–II (rare decay) mode for six weeks in the winter of 1997. For this run, the active regenerator was removed from the detector (Figure 2) and eight transition radiation detectors (TRD's) were inserted between the most downstream drift chamber and the trigger hodoscope. The TRD's were designed to provide a additional factor of 200 in π/e rejection.

A significant accomplishment achieved in the winter run was the first observation of the rare decay $K_L^0 \longrightarrow \pi^+\pi^-e^+e^-$. This decay, arising from a combination of intermediate $K_L^0 \longrightarrow \pi^+\pi^-\gamma^*$ IB and DE states, is interesting as the sixth K_L^0 decay mode with the potential of manifesting CP violation, and the first new such mode observed in 30 years. In this case, CP violation is observed in the asymmetry of the angle between decay planes of the e^+e^- and $\pi^+\pi^-$ pairs. The asymmetry analysis is still underway. Figure 5 shows the signal as a peak in the $e^+e^-\pi^+\pi^-$ invariant mass, with statistics from one

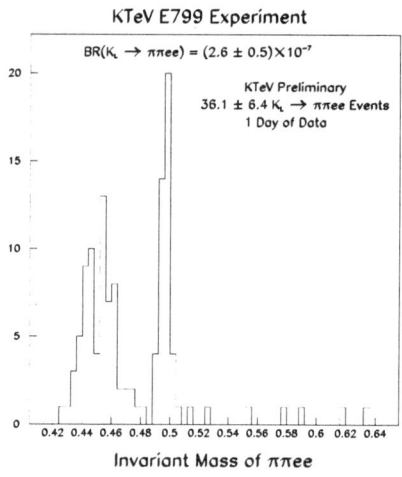

FIGURE 5. $\pi^+\pi^-e^+e^-$ Invariant mass distribution based on one day of data from winter 1997 E799 run. The peak at the kaon mass represents the first observation of the decay mode $K_L^0 \longrightarrow \pi^+\pi^-e^+e^-$.

day of data. Based on this data, KTeV reports a preliminary branching ratio for this decay of $(2.6 \pm 0.5) \times 10^{-7}$.

The decay $K_L^0 \longrightarrow \pi^0 \nu \bar{\nu}$ has gained considerable interest lately as an essentially pure direct CP violating rare decay. Here, KTeV reports a new upper limit in the search for this decay. The data from which this limit is derived was collected in a special 1-day run with a modified E832 detector. For this run, the regenerator was removed and one of the two beams was plugged, the other further collimated. The search looked for high transverse momentum (P_T) $\gamma\gamma$ pairs, where the transverse momentum is measured assuming the decay occurred within the neutral beam.

The results of this search, along with our current understanding of the

Status of BNL E787: Search for the Decay $K^+ \to \pi^+ \nu \bar{\nu}$

Toshio Numao

TRIUMF

4004 Wesbrook Mall, Vancouver, B.C., Canada

For the BNL E787 collaboration [1]

Abstract. The present status of the AGS experiment E787, search for the decay $K^+ \to \pi^+ \nu \bar{\nu}$ and other rare K^+ decay modes, is presented. The beamline and the detector were upgraded recently to observe several events if it occurs at a level of standard model predictions $\sim 10^{-10}$. Data taking and the analysis of new data are in progress.

INTRODUCTION

E787 experiment at Brookhaven National Laboratory (BNL) searches the rare decay $K^+ \to \pi^+ \nu \bar{\nu}$ at a sensitivity below 10^{-10}. It is a theoretically clean mode that proceeds through second order weak interactions, and the branching ratio is predicted at a level of 10^{-10} [1]. Long distance contributions are expected to be very small and the uncertainties in charm quark contributions are to be less than a few % in amplitude. Since the top-quark contribution is proportional to the CKM matrix element $|V_{td}|^2$, this mode provides a clean measurement of the parameter V_{td}. Together with measurements of the neutral mode $K^0 \to \pi^0 \nu \bar{\nu}$, the measurement of the decay $K^+ \to \pi^+ \nu \bar{\nu}$ will determine the unitarity triangle. The experiment is also sensitive to decays $K^+ \to \pi^+ X^0$ and $K^+ \to \pi^+ X^0 X^0$, where X^0 represents any weakly interacting neutral particle, e.g., axion, majoron and photino.

The signature of the decay $K^+ \to \pi^+ \nu \bar{\nu}$ is a single pion track without any other activity in the detector system. The π^+ momentum spans up to 227 MeV but the region below the $K^+ \to \pi^+ \pi^0$ decay peak is omitted from the search because of a possible low momentum tail due to pion-nucleus interaction while the π^+ slows down.

---

[1] E787 collaboration: BNL-KEK-Osaka-Princeton-TRIUMF.

RESULTS FROM PHASE I

An 800-MeV/c K^+ beam from the AGS at BNL was stopped in a finely segmented target of scintillation fibers. A decay product was momentum-analysed by a cylindrical drift chamber and stopped in a stack of plastic scintillators for range and energy measurements. The tracking detector was surrounded by nearly 4π sr photon counters of lead/scintillator sandwiches. The pion was positively identified by the kinematics (range, energy and momentum) as well as by observation of the full pion decay sequence, $\pi \to \mu\nu$ followed by $\mu \to e\nu\bar{\nu}$.

Data taking of the Phase-I experiment was completed in 1991. No $K^+ \to \pi^+\nu\bar{\nu}$ decay candidates were observed in the search region, π^+ range between 34 and 40 cm and kinetic energy between 115 and 135 MeV. This resulted in an upper limit on the branching ratio, 2.4×10^{-9} (90 % c.l.) [2]. Several new K^+-decay modes were also measured for the first time. Table I summarizes the Phase-I results. For subsequent data, the beam line and detector were upgraded.

Table I. Results from the pre-upgrade data.

Modes	90 % c.l. upper limits or measurements*	
	E787	before E787 [5]
$K^+ \to \pi^+\nu\bar{\nu}$	2.4×10^{-9} [2]	1.4×10^{-7}
$K^+ \to \pi^+\mu^\pm\mu^\mp$	* $5.0 \pm 0.4 \pm 0.7 \pm 0.6 \times 10^{-8}$	2.4×10^{-6}
$K^+ \to \pi^+\gamma\gamma$	* $6.0 \pm 1.4 \pm 0.7 \times 10^{-7}$ ($100 < P_\pi < 180$ MeV/c)	8.4×10^{-6}
$K^+ \to \mu^+\mu^+\mu^-\nu$	4.1×10^{-7} [3]	
$\pi^0 \to \nu\bar{\nu}$	8.3×10^{-7} [4]	6.5×10^{-6}

* also indicates the result is preliminary.

UPGRADE

The new beam line LESB-III commissioned in 1992 extracts kaons at $0°$ and has two DC separators, which improved the K/π ratio to 4/1. Together with the upgrade of the AGS, the K^+ beam flux was increased by nearly an order of magnitude. Taking advantage of the high kaon flux, the channel momentum was eventually reduced by 10 % to 710 MeV/c to reduce the thickness of the beam degrader and improve the fraction of stopping kaons in the target; this optimized the amount of useful data because kaon interactions in the beam degrader produce γ-rays and neutrons causing accidental hits in the detector.

The improvements in the tracking devices include a new target consisting of ~3-m long 5mm×5mm square scintillating fibers which run all the way to the phototube. The photon yield increased by a factor of four and the fraction of dead material was reduced from 30 % to 10 %. The mass in the active volume of the new central drift chamber [6] was reduced by using Al cathode wires and by separating active and inactive regions of the chamber with self-supporting foils with cathode strips for induced charge readout; inactive volumes were filled with light gas (nitrogen). The active volume of the chamber was extended to a smaller radius region by 3 cm. With these improvements, the momentum resolution was improved from 2.4 % to 1.2 % for the $K^+ \to \pi^+\pi^0$ peak. Also, kinematic π/μ separation was improved significantly, reducing the muon related background. Nine inner layers of range stack scintillators, which had been grouped into three, were now demultiplexed and read out individually. This resulted in a factor of 1.5 more light output. The tracking chambers embedded in the range stack were also replaced with newly constructed 2-layer straw chambers. The amount of inactive mass in the supporting wall of the chamber was reduced by a factor of six.

The capability of photon detection was also improved significantly. The beam hole was an escape route of γ-rays that caused photon detection inefficiencies. The downstream end of the hole was covered by the ~5 radiation-length "light guide" part of the target and the upstream end was improved by replacing the target-end section of the beam degrader with a lead-glass Čerenkov counter. The radial gaps between the target (and the degrader) and the endcaps were covered each by a calorimeter of lead/scintillator sandwiches. The new endcap detectors [7] consisted of pure CsI crystals which reduced the fraction of invisible energy. With the newly installed CCD readout system, the timing resolution of the endcap was improved from 2 ns to 0.7 ns.

All target fibers and endcap crystals were read by 8-bit 500 MHz CCD's [8]; this almost doubled the size of data (to nearly 100 kbytes/event). The trigger logic and the data acquisition system were improved to keep up with the increase in the kaon stopping rate. The total trigger rate was kept around 100/spill and the dead time always around 20-30 %.

PRESENT STATUS

The kaon stopping rate in the target is 1–1.5 M/spill and more than 10^{12} stopped kaons were accumulated each year in 1995 and 1996. In the off-line analysis, the data went thought the PASS-1 analysis that sorted out different triggers and very modestly removed unreconstructed events. Then, at the stage of the PASS-2 analysis many streams of data were generated for various types of background studies in addition to the one for the $K^+ \to \pi^+\nu\bar{\nu}$ search.

The acceptance was estimated to be 0.2–0.3 %. This put the sensitivity of the 1995 data alone at the upper end of the standard-model prediction.

Because of the proximity of the expected sensitivity of the present data to the "signal" region, background estimations have been carefully performed. Major sources of background are: $K^+ \to \pi^+\pi^0$ when two photons are missed and kinematic parameters (range, momentum and energy) of the pion are misreconstructed higher, $K^+ \to \mu^+\nu$ when the muon is misidentified as a pion with larger kinematic parameters and a second pulse is accidentally produced in the stopping counter, an extra pion that is scattered into the detector without leaving multi-hit information in the beam counters, and a pion from a K_L^0 decay that is produced by a charge-exchange reaction of a beam K^+. For the estimate of particular backgrounds, two groups of cuts were selected. In the case of the $K^+ \to \pi^+\pi^0$ background estimation, they were a group of photon-veto cuts and a group of cuts on kinematic parameters. The rejection factor of one group of cuts was obtained by inverting the cuts for the other group thus enhancing the background. Effects of run-time hardware and software problems are taken into account in the estimate when the full sample of data is used. This method assumes the two groups of cuts are uncorrelated; e.g., in the study of the $K^+ \to \pi^+\pi^0$ background, photon vetoing and kinematic parameters are assumed to be independent. However, if one of γ-rays overlaps with a track, the photon-veto inefficiency and the observed kinetic energy are not independent any more. These kinds of effects were carefully examined for all background types by enhancing particular effects and Monte Carlo calculations. Also, clean-up cuts in PASS-2 were chosen to minimize such effects. The background from each type was consistent with $<< 10^{-10}$ in terms of the branching ratio. Presently, the analysis of the 1995 data is in the final stage, and PASS-1 analysis of the 1996 data has been completed.

In 1995, a few days were dedicated for the study of the radiative decay $K^+ \to \mu^+\nu\gamma$ in the kinematical region, with high γ-ray energy and high muon energy, where it is sensitive to the structure dependent contribution SD_+. The branching ratio was measured to be $BR(SD_+) = (1.331 \pm 0.120 \pm 0.183) \times 10^{-5}$ (preliminary) which corresponds to $|F_V + F_A| = 0.165 \pm 0.007 \pm 0.011$.

CONCLUSION

The estimated sensitivity of the data taken after the upgrade of the beamline and the detector is at a level of standard model predictions. In the near future, the beam intensity at the production target is expected to increase. A longer beam spill length is also expected to double the total kaon flux without changing the instantaneous rate. The kaon momentum may be lowered further as the beam intensity increases; this improves the kaon stopping fraction and reduces the chance coincidence rates. Also, the search will extend to the pion momentum region below the $K^+ \to \pi^+\pi^0$ peak, where pion-nucleus interaction may cause a higher background level; after the upgrade, the photon vetoing capability has been enhanced at least by a factor of three, and the detection efficiency of extra energy or a kink in the target fiber (indicating nuclear interaction) has been significantly improved. Together with all these additional improvements, an order-of-magnitude improvement of sensitivity over the new data may be possible in the near future.

REFERENCES

1. G. Buchalla, these proceedings; A.J. Buras and R. Fleischer, TUM-HEP-275/97 and TTP97-15, 1997.
2. S. Adler et al., Phys. Rev. Lett. **76** (1996) 1421.
3. M.S. Atiya et al., Phys. Rev. Lett. **63** (1989) 2177.
4. M.S. Atiya et al., Phys. Rev. Lett. **66** (1991) 2189.
5. See references in Review of Particle Properties, Phys. Rev. **D50** (1994) 1173.
6. E.W. Blackmore et al., to be published in Nucl. Instr. Method and TRI-96-39, 1996.
7. I-H. Chiang et al., IEEE trans. NS **42** (1995) 394.
8. D.A. Bryman et al., to be published in Nucl. Instr. Method and TRI-96-51, 1996.

Experimental Studies of Rare K^+ and π^0 Decays

S. Eilerts for the BNL E865 collaboration

University of New Mexico
Email: seilerts@unm.edu

Abstract. Experiment E865 at the BNL AGS is a search for the lepton flavor violating decay $K^+ \to \pi^+\mu^+e^-$ with an expected sensitivity of 3×10^{-12}. The experimental apparatus involves a magnetic spectrometer and detectors capable of identifying the species of the final state particles with a low probability of error. In addition to the search for $K^+ \to \pi^+\mu^+e^-$, several other rare K^+ decay modes are observed, including the decays $K^+ \to \pi^+\mu^+\mu^-$ and $K^+ \to \pi^+e^+e^-$. Also, via $K^+ \to \pi^+\pi^0$, the rare decays $\pi^0 \to e^+e^-e^+e^-$ and $\pi^0 \to e^+e^-$ are studied. The experiment is described and preliminary results from the 1995, 1996, and 1997 runs are presented.

I INTRODUCTION

Lepton number violation, forbidden by the Standard Model of particle interactions, is allowed under such Standard Model extensions as technicolor and supersymmetry. Experiment E865 at the Brookhaven National Lab Alternating Gradient Synchrotron (BNL AGS) is a search for the lepton flavor violating decay $K^+ \to \pi^+\mu^+e^-$. This experiment is sensitive to new physics beyond the Standard Model at a mass scale of 100 TeV. The existing upper limit of 2.1×10^{-10} was established by the experiment's predecessor, E777 at the BNL AGS [1].

With this apparatus we can also measure rare K^+ and π^0 decays and make high statistics measurements of more common decays. Of the rare decays, $K^+ \to \pi^+e^+e^-$ and $K^+ \to \pi^+\mu^+\mu^-$ are highly suppressed due to absence of flavor changing neutral currents. Measurement of the branching ratios and form factors for these decays provide a direct test of Chiral Perturbation Theory (CHPT). Via $K^+ \to \pi^+\pi^0$, the rare processes $\pi^0 \to e^+e^-$ and $\pi^0 \to e^+e^-e^+e^-$ can be studied. A precise determination of the branching ratio for the decay $K^+ \to \pi^0 e^+\nu_e$ leads to a more accurate determination of the CKM matrix element $|V_{us}|$. $\pi\pi$ scattering can be studied via the decay $K^+ \to \pi^+\pi^-e^+\nu_e$. Finally, the structure dependence in $K^+ \to \mu^+\nu_e e^+e^-$ can be

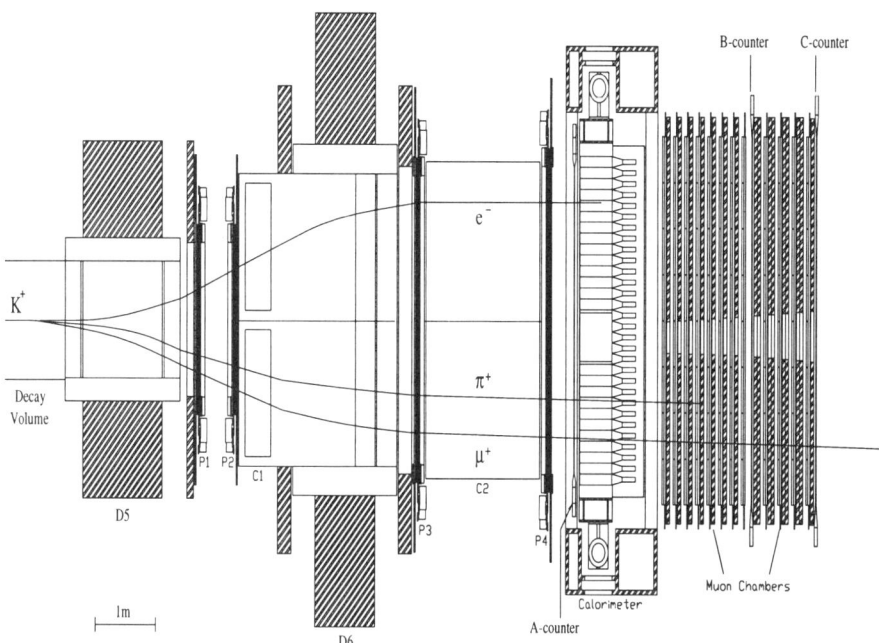

FIGURE 1. E865 Detector Plan View.

investigated with a high statistics measurement. The decays $K^+ \to \pi^+ e^+ e^-$ and $\pi^0 \to e^+ e^-$ were also measured by this experiment's predecessor [2,3].

The experiment occupies the A2 6 GeV/c unseparated beamline at the AGS. The experimental layout is shown in Fig. 1. The detector is preceded by a 5 meter long decay tank. A 48D48 dipole magnet spreads decay products out, and a 120D36 dipole magnet provides momentum determination. Particle trajectories are such that positive particles are accepted on one side and negatives on other side. The apparatus consists of proportional wire chambers (PWC's) P1 through P4, two gas Cerenkov counters C1 and C2 to identify electrons, trigger hodoscope A, an electromagnetic shower calorimeter, and muon detectors consisting of 12 proportional wire chambers and two hodoscopes (B and C) interspersed between 12 steel plates. The design intensity of 1.2×10^{13} protons per pulse on target produces a beam of 2×10^9 protons and pions along with 7×10^7 kaons.

II CURRENT STATUS

Initial engineering runs were conducted in 1993 and 1994 in which the beamline was built and tested and the detector elements were installed. After final detector tests in the beginning of 1995, the experiment took $K^+ \to \pi^+ \mu^+ e^-$

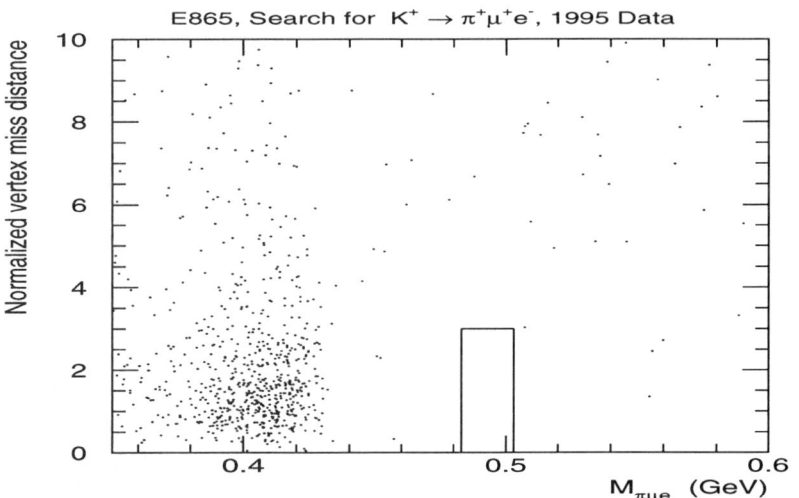

FIGURE 2. Vertex fit vs. reconstructed mass for $K^+ \to \pi^+\mu^+e^-$. There are no events in the expected signal area indicated by the box

data for 11 weeks in 1995 and 16 weeks in 1996. Data for other processes was also collected in parallel.

Data collected in 1995 and 1996 has undergone an initial reduction and preliminary analysis for some event channels has been done. Figure 2 shows the vertex fit vs. $\pi^+\mu^+e^-$ mass for events with a $\pi^+\mu^+e^-$ signature from 1995 data. There are no events in the expected signal region. Based on 1995 data alone, the upper limit on the branching ratio for $K^+ \to \pi^+\mu^+e^-$ is 2.0×10^{-10} at the 90% C.L. This result can be combined with the upper limit from E777 for a new upper limit on the $K^+ \to \pi^+\mu^+e^-$ branching ratio of 1.0×10^{-10} at the 90% C.L. The expected sensitivity from the 1996 data alone is at the 4.0×10^{-11} level.

The E865 apparatus has the capability to make precision measurements of the decays $K^+ \to \pi^+e^+e^-$ and $K^+ \to \pi^+\mu^+\mu^-$. Figure 3a shows the reconstructed Kaon invariant mass from the combined 1995 and 1996 data sets for $K^+ \to \pi^+e^+e^-$ with the mass of the e^+e^- pair over 150 MeV. There is an excess of 8000 events in the signal area, over one order of magnitude more statistics compared to the previous measurement of this decay. Approximately one half of the 1997 run was dedicated to a measurement of the decay $K^+ \to \pi^+\mu^+\mu^-$. Figure 3b shows the reconstructed $\pi\mu\mu$ mass for events with a $\pi^+\mu^+\mu^-$ signature. There is a clear signal of more than 200 events for this decay. Background events with a low $\pi\mu\mu$ mass are attributed to the $K^+ \to \pi^+\pi^+\pi^-$, where two of the pions have decayed in flight to muons or have been misidentified.

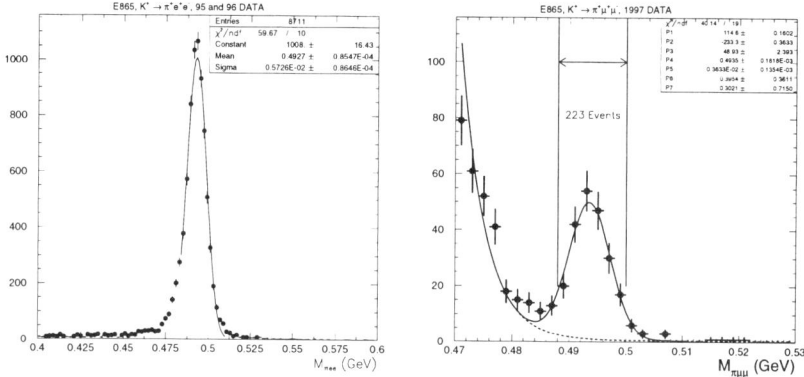

FIGURE 3. a) Reconstructed mass for $K^+ \to \pi^+ e^+ e^-$ events. b) Reconstructed mass for $\pi^+ \mu^+ \mu^-$ events.

We also have the opportunity to measure rare π^0 decays having at least one electron-positron pair in the final state via $K^+ \to \pi^+ \pi^0$. Figure 4a shows the reconstructed $e^+ e^- e^+ e^-$ mass for events with a $\pi^+ e^+ e^- e^+ e^-$ signature from the 1995 and 1996 data sets. The distribution peaks at the π^0 mass, so these events are attributed to $\pi^0 \to e^+ e^- e^+ e^-$. Figure 4b shows the reconstructed $e^+ e^-$ mass for events with a $\pi^+ e^+ e^-$ signature from 1995 data. There is a peak in the vicinity of the π^0 which is attributed to the decay $\pi^0 \to e^+ e^-$. Background to this decay includes $\pi^0 \to e^+ e^- \gamma$ and $\pi^0 \to e^+ e^- e^+ e^-$, where the photon or an electron-positron pair goes undetected. These events can reconstruct an $e^+ e^-$ mass at nearly the π^0, and can only be reduced by requiring

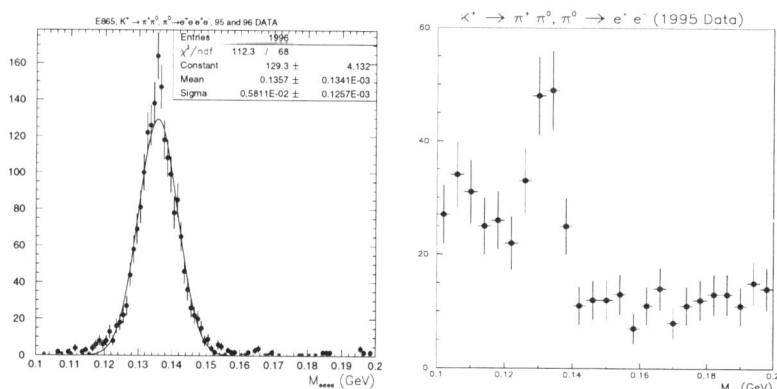

FIGURE 4. a) Reconstructed mass for $\pi^0 \to e^+ e^- e^+ e^-$ events. b) Reconstructed $e^+ e^-$ mass for events with a $\pi^+ e^+ e^-$ signature.

the $\pi^+e^+e^-$ mass to be at the Kaon mass. Background from $K^+ \to \pi^+e^+e^-$ is continuous in the e^+e^- mass spectrum shown.

Two detector upgrades were completed before the beginning of the 1997 run. A change to the data acquisition system now allows us to take beam at the design intensity of 1.2×10^{13} protons per pulse. A pixel hodoscope was installed in the beamline upstream of the decay volume for the purpose of measuring the Kaon position before it decays. This was necessary to improve reconstruction of decays that involve neutrinos. After spending the first half of the run testing the pixel hodoscope and collecting $K^+ \to \pi^+\mu^+\mu^-$ data, $K^+ \to \pi^0 e^+ \nu_e$ data was collected for a few days. We hope to make a new measurement of the CKM matrix element $|V_{us}|$ to better than 1% accuracy. The remainder of the run was spent collecting $K^+ \to \pi^+\pi^-e^+\nu_e$ data.

III CONCLUSION

Data analysis is in progress for a wide variety of rare K^+ and π^0 decays from 1995-1997 data. Based on 1995 data, the upper limit on the branching ratio for $K^+ \to \pi^+\mu^+e^-$ is 2.0×10^{-10} at the 90% C.L. This result can be combined with the upper limit from E777 for a new upper limit on the $K^+ \to \pi^+\mu^+e^-$ branching ratio of 1.0×10^{-10} at the 90% C.L. The expected sensitivity from the 1996 data alone is at the 4.0×10^{-11} level. A 24 week run in 1998-1999 will be necessary to complete the experiment.

REFERENCES

1. A. M. Lee et al, Phys. Rev. Lett. **64**, 165 (1990).
2. C. Alliegro et al, Phys. Rev. Lett. **65**, 278 (1992).
3. A. Deshpande et al, Phys. Rev. Lett. **71**, 27 (1993).

Status report of the NA48 experiment at the CERN SPS

Cinzia Talamonti*

University of Edinburgh

For the NA48 collaboration

Abstract. The aim of the NA48 experiment at the CERN SPS is to measure direct CP violation in neutral kaon decays thus determining the parameter ε'/ε with an accuracy of 2×10^{-4}. The advantages of NA48 with respect to previous experiments are high statistics and reduced systematic effects. The principle of the experiment and the performance of the detector components are presented.

INTRODUCTION

In 1964, CP violation was discovered in the decay of the long-lived kaon into two pions [1]. Since then, considerable experimental and theoretical effort has been devoted to understanding its origin. After thirty years, no evidence of CP-violation has been observed in particle physics outside the neutral kaon system. The cause of violation of CP symmetry is still an open issue in physics. The study of direct CP-violation has turned out to be difficult, both experimentally and theoretically, and the understanding of its magnitude and origin is still far from satisfactory. The latest theoretical predictions for the value of $\Re(\varepsilon'/\varepsilon)$ are in the range $1 \times 10^{-4} \leq \Re(\varepsilon'/\varepsilon) \leq 15 \times 10^{-4}$ [2-5]. The experimental measurement of ε'/ε should allow discrimination between different theoretical hypotheses. E731 at FNAL and NA31 at CERN have measured $\Re(\varepsilon'/\varepsilon)$ with a precision better than 10^{-3}. NA31 has obtained a value of $(2.0 \pm 0.7)10^{-3}$ [6] while E731 has found a value of $(0.74 \pm 0.59)10^{-3}$ [7]. The first one indicates evidence of direct CP violation and the second is compatible with a null result. To clarify the situation requires new measurements, at CERN, NA48 and at FNAL, KTeV, aim for a more precise determination of $(\varepsilon'/\varepsilon)$ with a accuracy on $\Re(\varepsilon'/\varepsilon)$ of $\sim 2 \times 10^{-4}$.

PRINCIPLE OF THE EXPERIMENT

The experimental technique

Experimentally the measurement of $\varepsilon\prime/\varepsilon$ is obtained by evaluating the double ratio of the decay rates of K_L and K_S into two neutral and two charged pions:

$$\Re(\varepsilon\prime/\varepsilon) \cong \frac{1}{6}\left(1 - \frac{\Gamma(K_L \to \pi^0\pi^0)}{\Gamma(K_S \to \pi^0\pi^0)} \cdot \frac{\Gamma(K_S \to \pi^+\pi^-)}{\Gamma(K_L \to \pi^+\pi^-)}\right) \qquad (1)$$

with simultaneous observation of all four decay modes in the same experimental setup. The basic NA48 scheme [8] features two nearly collinear beams of K_L and K_S, produced by protons hitting two different targets, and distinguished by tagging of the protons producing the K_S component. A magnetic spectrometer with four drift chambers measures the momentum of the charged particles, and a fast liquid krypton calorimeter measures the energy, the position and the time of a neutral decay. Provided that the position of the two beams coincide in the detector, the advantage of this method is that at any given energy and vertex position, the detection efficiencies cancel. Differences and variations in detection efficiencies for K_L and K_S decays become unimportant, as do rate dependent effects introduced by accidental activity in the detector elements.

The statistical and systematic errors

The statistical error on the double ratio R is dominated by the statistical precision of the rarest decay. In order to determine $\Re(\varepsilon\prime/\varepsilon)$ with a precision of 10^{-4}, it is necessary to collect about 5×10^6 $K_L \to \pi^0\pi^0$ decays, one order of magnitude more than the previous experiments, NA31 and E731.

It is mostly the systematic error which limits the precision attainable on the measurement of R. In fact, due to different beam divergences, energy spectra and decay vertex positions, the acceptances of K_L and K_S decays are not exactly the same. In order to have a negligible difference in momentum spectra, the production angles of the K_L and K_S beams have been carefully chosen (2.4 and 4.2 mrad respectively). This way, the decay spectra are similar when the kaon momentum is between 70 and 170 GeV.

Since the distribution of decay vertex are different (one is flat (K_L) and the other is exponential (K_S)), a weight W, depending on kaon momentum p and on the longitudinal coordinate z of kaon decay vertex, will be applied to K_L events. Using this technique, the spatial distribution of K_L decay vertices reproduces that of the K_S beam.

Another effect which can contribute to the systematic error is due to the imperfect knowledge of the fiducial volume. For the neutral decays, an accurate knowledge of the energy scale is required and this is achieved by adjusting the reconstructed longitudinal decay vertex position to a counter edge placed at the beginning of the K_S beam.

Background events, mainly three-body decays, are another important source of systematic errors. In case of neutral events, most of the decays $K_L \to \pi^0\pi^0\pi^0$ are rejected by using anti-counter rings which veto photons outside the fiducial region, and an electromagnetic calorimeter with excellent resolution in energy and space. The background of charged events is given by $K_L \to \pi^\pm e^\mp \nu(\bar{\nu})$, $K_L \to \pi^\pm \mu^\mp \nu(\bar{\nu})$ and by $K_L \to \pi^+\pi^-\pi^0$. The information of the muon veto counters, located at the rear end of the detector, is used at the trigger level to reject $K_L \to \pi^\pm \mu^\mp \nu(\bar{\nu})$ events. The invariant mass is reconstructed at the trigger level as well, in order to discard $K_L \to \pi^+\pi^-\pi^0$ and $K_L \to \pi^\pm e^\mp \nu(\bar{\nu})$ decays (at 90% level).

THE K_L AND K_S BEAMS

The NA48 experiment compares the $\pi^0\pi^0$ and $\pi^+\pi^-$ decay rates from distinct K_L and K_S beams which enter a common decay region along a path entirely contained in vacuum. The four decay modes are thus recorded at the same time and from the same fiducial length. The two beams are nearly collinear, converging at an angle comparable with the beam divergence.

A 450 GeV/c proton beam with a nominal flux of 1.5×10^{12} ppp strikes a beryllium target, producing the K_L beam at an angle of 2.4 mrad. After a first collimator which limits acceptance, charged particles (including the remaining primary protons) are deviated from the K_L line by a sweeping magnet. A fraction of the primary protons (3×10^7 ppp) is channelled back towards the K_L beam by a bent silicon crystal [9]. It then passes through a tagging counter, and is eventually deflected and focused on the K_L line. It then hits a second target, placed 72 mm above the first one and at a distance of 120 m from it, to produce a K_S beam at an angle of 4.2 mrad. This beam is defined by a collimator and converges with the K_L one at an angle of 0.6 mrad. The exit of the K_S collimator coincides with the last of three collimators designed to define the K_L beam, so that background from the collimators themselves cannot reach the detector. After the K_S collimator, a detector, vetoes K_S decays which occurred before the collimator exit and defines the fiducial region. A schematic view of the the kaon beam lines and of the detector layout is shown in Fig.1.

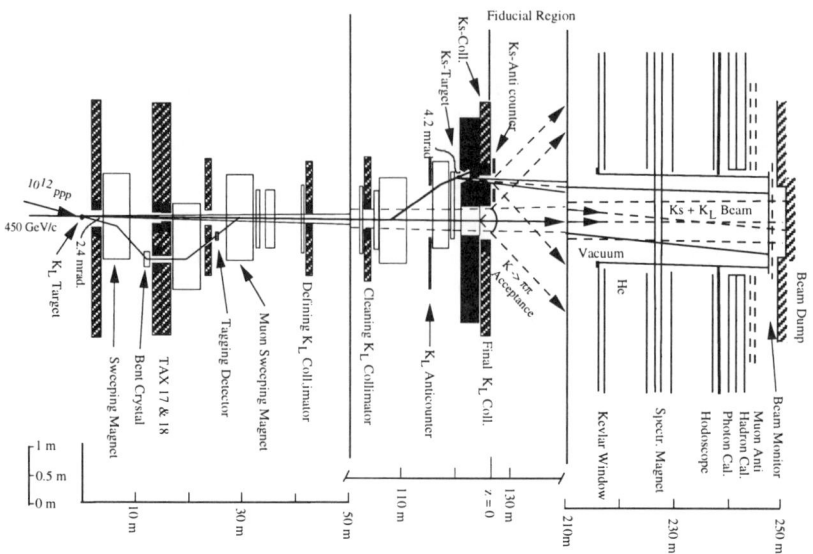

FIGURE 1. The Na48 kaon beams and detector layout.

THE NA48 DETECTOR

The tagging counter

The K_L and K_S assignment for each decay is done by measuring the time difference between the passage of a proton in the tagging counter [10] upstream of the K_S target and the event time in the detector. Events with a time difference inside of a given interval Δt will be called "K_S", any other events will be called "K_L". It has been designed to cope with proton rates above 10MHz and to provide a detection efficiency close to 100% combined with a time resolution better than 500 ps. To ensure that the rate in each counter is less than 1MHz the tagger is designed with two sets of staggered scintillation foils arranged alternately in the horizontal and vertical planes. The depth of the foils in the beam direction is 4 mm and the width varies from 200 μm on the beam-axis to 3000 μm at the beam edges. Thus the entire beam profile is covered. The time resolution is less than 200 ps, with a double pulse resolution below 4 ns. Inefficiency and accidental activity of the tagging counter result in a correction to R, but being decay mode independent they cannot generate a fake non-zero value for $\Re(\varepsilon'/\varepsilon)$, nevertheless the efficiency will be monitored at the 10^{-4} level.

The magnetic spectrometer

A magnetic spectrometer [11] is used to analyse the $\pi^+\pi^-$ decay: two set of drift chambers on each side of a central dipole magnet provide a measurement of charged particle momenta. Each chamber has four views X, Y, U and V to avoid ambiguities in the position measurement. Each view consist of two planes of 256 wires, at a distance of 1 cm from each other. The spatial resolution of each view is $\simeq 100\mu$m, which corresponds to $\Delta p/p = 0.6\%$ accuracy on mean momentum measurement. The field integral of the main field component (B_y) is equivalent to a transverse momentum kick of 250 MeV/c.

During the 1995 and 1996 test runs, very good performance was obtained with the magnetic spectrometer for the $K_{S,L} \to \pi^+\pi^-$ decays. The four drift chambers worked reliably and provided high detection efficiency and position resolution at rates close to 1MHz. The kaon invariant mass resolution obtained off-line from fully reconstructed $\pi^+\pi^-$ events is 3 MeV/c^2 in both K_L and K_S beams (Fig. 2). The good resolution on the decay vertex position, in the transverse plane, provides a clear separation of the K_L and K_S sources, allowing a measurement of the tagging efficiency in the charged decay mode.

The charged hodoscope

The timining information provided by this detector is used in conjunction with the tagger to identify the K_L or K_S origin of charged decays, and also in the trigger to select two-body decays. The detector consist of two planes of scintillators counters, at a distance of 50 cm, located after the magnetic spectrometer along the beam line. Each plane consists of 64 scintillators, arranged vertically and horizontally. The time resolution for a two pion event is better than 250 ps.

The electromagnetic calorimeter

The characteristic of the decays to be detected, and the high rate of the experiment, impose same stringent requirements on the electromagnetic calorimeter. In addition, a large sensitive area ($\sim 6m^2$) has to be covered. Good energy resolution is needed: $\sigma_E/E \sim 1\%$ at 10 GeV with a constant term lower than 0.5%. A spatial resolution $\sigma_{x,y} \sim 1.5\ mm$ is also needed. The tranverse scale has to be accurate to 0.1/1000 mm to keep the systematic error induced by a possible difference in energy scale for charged and neutral decays at the desired level. For K_S identification a timing accuracy $\sigma_t \leq 0.5\ ns$ has to be achieved. High single rate capability($\sim 1MHz$) is needed to collect the required statitiscs, high granularity and short sensitive time would reduce the effect of accidentals and background. The NA48 collaboration has built a fully sensitive liquid krypton ionization chamber with electrodes parallel to the

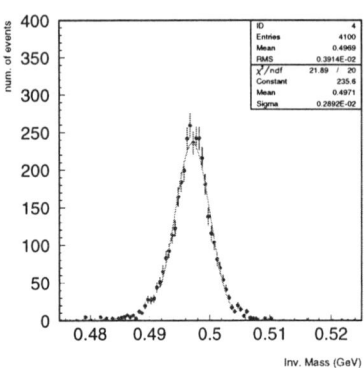
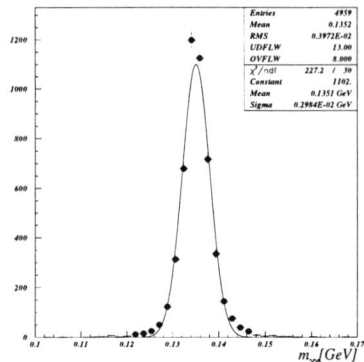

FIGURE 2. Invariant mass distributions of the $\pi^+\pi^-$ measured in the magnetic spectrometer and of two photons measured in the calorimeter.

shower direction and a tower structure segmentation into $2 \times 2 cm^2$ cells. Its transverse dimension is about 2.6m and its thickness is 125cm, corresponding to about 27 X_0 of liquid krypton. The total krypton volume in the cryostat is about $10 m^3$. To suppliment the time measurement of the detected photons, scintillating fibers have been inserted vertically in the liquid krypton at a depth of 9.5 X_0. The required performance has been achieved after several test beam runs with a prototype [12]. The energy resolution achieved with an electron test beam is

$$\sigma_E/E = 3.5\%/\sqrt{E} \oplus 40 MeV/E \oplus .42\% \qquad (2)$$

During the 1996 test run period only the 8% of the LKr electronics was available; the entire calorimeter was read-out by instrumenting "supercells" of 2x8 single cells. In this way physics data could be taken with both beams. The performance of the calorimeter was tested during the calibration period with an electron beam and the resolution obtained is in good agreement with prototype results. The neutral decays in two π^0 with four clusters in the Lkr calorimeter were identified and their rates were found to agree with the predicted ones. The π^0 mass resolution is about 3.5 MeV/c^2 (Fig.2) due to the poor granularity of the supercells and their associated noise, but it is compatible with the proposal (1 MeV/c^2).

The hadronic calorimeter

The most important task of this subdetector is to provide an energy threshold signal which is part of trigger, and is used to reject background coming from the three body decays. The front and the back modules consist of alternate iron and scintillator planes oriented either horizontally or vertically. The thickness corresponds to 7.2 interaction lengths. The energy resolution of the hadron calorimeter is $65\%/\sqrt{E}$.

The muon veto system

To reject the background due to the $K_L \to \pi^{\pm}\mu^{\mp}\nu(\overline{\nu})$ it is used the information provided by the μ-counters, located at the end of the beam line. A total of 28 scintillator strips are arranged in three planes, preceded by an iron layer 80 cm thick. The strips are alternatively vertical and horizontal, partially overlapping. The efficiency of this system is better than 99% for muons above 5 GeV.

PERSPECTIVES AND CONCLUSION

The beam and detector for the NA48 experiment are ready to take data for the study of direct CP violation in the K^0 system. Data taking will start in August. The beams have achieved their design goals, including the novel use of a bent silicon crystal for deflecting protons at 450 Gev/c and tagging them with time resolution of better than 500 ps at 10 MHz beam rates. The detector is also complete, the required mass resolutions have been achieved in the magnetic spectrometer, and will be achieved in the liquid krypton calorimenter when the full electronics is commissioned in July 1997.

REFERENCES

1. J.H. Christenson, J.W. Cronin, V.L. Fitch, R. Turlay, *Phys. Rev. Lett.* 13, 214 (1964).
2. G. Buchalla *et all*, *Nucl. Phys.* B337, 313 (1990);
3. A.J. Buras *et all*, *Nucl. Phys.* B408, 209 (1993);
4. M. Ciuchini *et all*, *Phys. Lett.* B301, 263 (1993);
5. A.J. Buras *et all*, Preprint hep-ph/9608365, August 1996.
6. G.D. Barr *et all*, *Phys. Lett.* B317, 233 (1993).
7. L.K. Gibbson *et all*, *Phys. Rev. Lett.* 70, 1203 (1993).
8. G.D. Barr *et all*, *CERN/SPSC/90-22/P253*.
9. N. Doble *et all*, *Nucl. Instr. Meth.* B119, 181 (1996).
10. P. Grafstrom *et all*, *Nucl. Instr. Meth.* A344, 487 (1994).
11. D. Bederede *et all*, *Nucl. Instr. Meth.* A367, 88 (1995).
12. G.D. Barr *et all*, *Nucl. Instr. Meth.* A370, 413 (1996).

Radiative ϕ Decays at Jefferson Lab

Robert W. Gardner

*Department of Physics, Indiana University
Bloomington, Indiana 47405*

Abstract. We discuss the status and future prospects of a program to study radiative ϕ decays in at Jefferson Lab.

INTRODUCTION

At Jefferson Laboratory the energies of photons in the range 3-6 GeV are well above threshold for ϕ production in the reaction: $\gamma p \to \phi p$. These reactions are described by the vector meson dominance model in which the $s\bar{s}$ component of the photon's wavefunction is picked out by a diffractive exchange of a soft pomeron with the proton target. The cross section is approximately constant with center of mass energy and is strongly peaked in the forward direction with a steeply falling $|t|$ distribution. Thus the ϕ's materialize in the lab along the beam direction and have angular distributions consistent with s-channel helicity conservation. These properties allow the recoiling target proton to be be used as a tag of the diffractively photoproduced ϕ.

Alternatively, at low $|t|$ the production of hadronic final states can take place for example by exchange of a π or ω meson which drops off rapidly as a function of center of mass energy (i.e. photon energy). Knowledge of the production mechanism may provide discriminating power in separating exotic events from conventional states.

With a large sample of tagged ϕ's one can study its rare decay modes. As an example, rare radiative decays of the ϕ offer clues into the structure of two low mass scalar mesons, the isovector $a_0(980)$ and the isoscalar $f_0(975)$. These states have been the focus of much attention in the last few years, both theoretically and experimentally. Although they have been identified with the $^3P_0$ $q\bar{q}$ scalar nonet, the low values for their masses and widths have called these assignments into question. Experimentally there is evidence which argues for these states being non-$q\bar{q}$ objects. For example the recently observed isoscalar $f_0(1365)$ and isovector $a_0(1450)$ [1] states are more plausible candidates to fill out the $^3P_0$ $q\bar{q}$ scalar nonet. There are several ideas for what the true identity of these states are, among them weakly bound $K\overline{K}$ "molecules" or four quark states $qq\overline{q}\overline{q}$ of various configurations [2].

Close, Isgur, and Kumano [3] have shown how the measured ratio of the ϕ patial widths into these two decay modes can resolve the structure of these scalar states. In the $^3P_0$ $q\bar{q}$ picture the ratio of partial widths, $\Gamma(\phi \to a_0\gamma)/\Gamma(\phi \to f_0\gamma)$ is expected to be $\approx$ 0 by OZI suppression. The $K\overline{K}$ and $qq\overline{q}\overline{q}$ possibilities can be distinguished by the electric dipole nature of these decays through the relative phases in the I=0 and I=1 wave functions and the spatial distributions of quarks and antiquarks. Their predictions are summarized in Table 1.

TABLE 1. Predictions of ϕ partial widths into the two predictive radiative decay modes: $R = \Gamma(\phi \to a_0(980)\gamma)/\Gamma(\phi \to f_0(975)\gamma)$

Hypothesis	R (theory)	Comment
Quarkonia ($q\bar{q}$)	0	OZI suppression
Naive molecule	1	dipole transition
Diquark-antidiquark	9	dipole transition

Measuring these decay modes is one of the principal goals of the Radiative-ϕ experiment E-94-016. (See Refs. [4] for previous talks on this experiment.) As a by-product of these studies, a determination of the branching ratio for $\phi \to a_0\gamma$ or $f_0\gamma$ will allow a determination of the branching ratio $\phi \to K_s K_s \gamma$ or $\phi \to K_L K_L \gamma$ [3]. If these modes were found to be present they would pose serious backgrounds to CP violation searches at ϕ factories.

The fundamental constituents of QCD include quarks and gluons, and there is no reason to expect that in addition to mesons made from quarks and antiquarks there should not also be glueballs (pure glue) and hybrids (quarks and glue). Some of these states will have exotic J^{PC} quantum numbers, that is, they will lie outside the quark model. Thus one of the important goals of present spectroscopy experiments is to search for such states. Of importance here is that these states must be observed in the presence of a very large conventional quarkonia background. Most searches for these states have been undertaken using hadron beams. Data from photoproduction experiments, which would provide an important cross-check due to the differring production mechanism, is sorely lacking.

Lattice calculations in the quenched approximation have made predictions for masses and widths of mesons with exotic quantum numbers [5]. Observation of such states would provide information on the dynamical properties in the low energy regime, as well as provide calculational guidance.

There are currently efforts to test CP and CPT violation in ϕ decay at e^+e^- machines in Frascati (DAPHNE) and Novisibirsk. The experiments produce the ϕ's at rest and measure the asymmetry between the decay modes $K^0 \to \pi^+\pi^-$ and $K^0 \to \pi^0\pi^0$ as a function of the separation of the K_S^0 and K_L^0 from the ϕ decay. This not only provides a measure of ϵ'/ϵ, but also a new test of CPT violation, which can be non-negligible in string theory [6].

The ϕ's produced at TJNAF have a boost in the laboratory and thus would provide a significantly different collection of detection, resolution, and background systematics than those faced by the e^+e^- collider facilities. Exploiting these ϕ samples for CP and CPT tests deserves careful consideration.

OVERVIEW OF THE RADIATIVE-ϕ EXPERIMENT

The Radiative-ϕ Collaboration consists of physicists, engineers, students and technicians from DoE and NSF supported universities and Jefferson Lab[1]. The experiment is located in the alcove of Hall B downstream of the CLAS detector. Installation of a Stage-I detector began in December 1996 and was essentially complete by mid-April 1997. A schematic of the setup is shown in Figure 1. Briefly, photons incident from the left impinge on a one inch long beryllium target which is surrounded by a projective proton recoil detector (RPD). Forward charged and neutral particles emerging from the interaction are detected by a downstream lead glass electromagnetic calorimeter (LGD) and a segmented charged particle veto wall (CPV), which is not shown in the figure. Also not shown is a monitoring system which consists of a plexiglass sheet located just upstream of the front face of the LGD and supplied by a laser-scintillator light source. The RPD and LGD/CPV were mounted on separate transporter carriages which were remotely controllable (vertically and horizontally). Additionally the longitudinal position of the beryllium target was remotely controllable. The relatively simple design of this spectrometer is sufficient to detect and measure all neutral final states which is what is required here.

The recoil detector for the Stage-I Radiative-ϕ experiment consists of 18 scintillator counters surrounding the one inch beryllium target. The counters are arranged so that recoil protons emerging from the beryllium target between the angles 40-60 degrees form trigger coincidences. This simple projective configuration is adequate since the short beryllium target can be thought of as approximately a point source of recoil protons. Additional advantages are the low cost, simplicity of operation and ease of trigger formation. The disadvantages of this design are that recoil protons suffer the effects of Fermi motion within the beryllium nucleus, nuclear rescattering, and significant multiple

---

[1] The collaborating institutions are the Catholic University of America, University of Connecticut-Storrs, Indiana University-Bloomington, Jefferson Laboratory, University of Notre Dame, Rensselaer Polytechnic Institute, University of Richmond, University of Virginia, and the College of William and Mary.

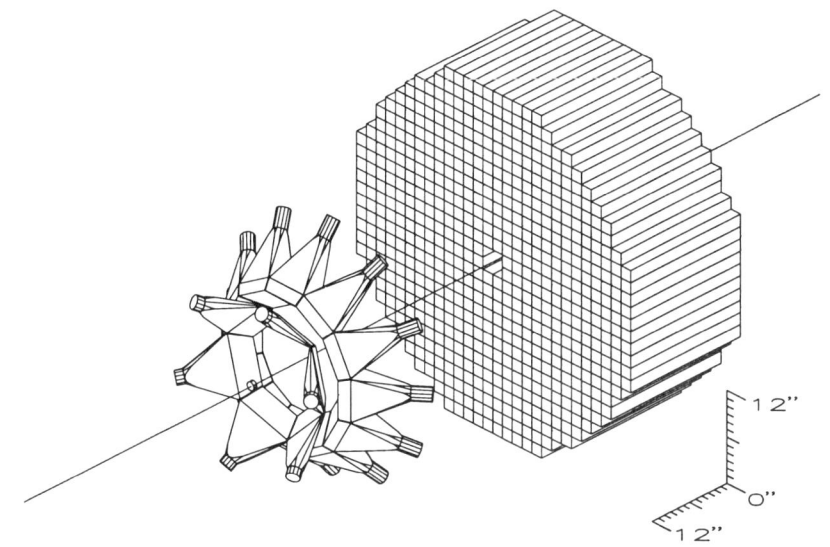

FIGURE 1. Schematic drawing of the Stage-I apparatus for the Radiative-ϕ experiment.

coulomb scattering (the target diameter is approximately 2.5 cm) on exiting the target material.

The lead glass calorimeter for the Radiative-ϕ experiment is comprised of 640 4cm×4cm×45cm and 3.8cm×3.8cm×45cm lead glass blocks. The differring block sizes is not a deliberate design choice but are a consequence of making due with available materials. Shims were used in the stacking process to accomodate the two block sizes. Approximately 320 of these blocks were instrumented for the Stage-I detector using Cockroft-Walton bases designed and contructed by Indiana University [7]. The detector will be used to reconstruct all neutral decay modes of the ϕ, for example, $\phi \to \eta\gamma$, $\eta \to \gamma\gamma$.

The first level trigger consists of a coincidence between the projective recoil proton detector elements and a tagged photon energy in the upper 20% of the energy spectrum (endpoint energy of 4 GeV). Events with forward charged tracks (either from the primary interaction or from secondary decays) were vetoed by the charged particle veto wall. The second level trigger consists of a threshold requirement on an electromagnetic energy sum sampled from the lead glass blocks.

TEST RUN RESULTS

In early May 1997 we commissioned the experiment utilizing approximately two days of live low energy photon beam (1.6 GeV endpoint) and three hours of high energy 4 GeV beam. From this brief period we were able to gain

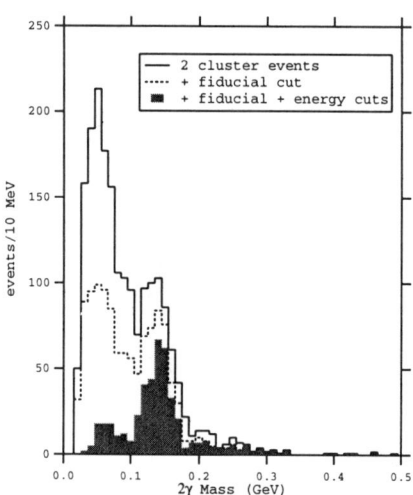

FIGURE 2. From the 3 hour test run: 2γ invariant mass and the effects of various cleanup cuts.

balance, equitime, calibrate, and perform a number of useful studies with the stage-I detector.

We utilized the 4.0 GeV beam to establish an initial calibration of the lead glass detector. An algorithm to find clusters of struck blocks consistent with isolated photons in the LGD was exercised. Starting with an initial set of calibration constants from the laser calibration system, the 2γ invariant mass was used to iteratively set the energy scale. Figure 2 shows the result of an online calibration performed the same afternoon as the data were recorded. The outer histogram is the 2γ invariant mass from 2-cluster events with no cuts applied. The dashed histogram results after removing events with hits in the blocks surrounding the beam; this serves as a shower containment cut. The filled histogram is the distribution after an additional minimum energy is required for each of the showers. A fit to the resulting π^0 peak which consists of ≈ 150 events and a width of 16 MeV. After scaling for the differences in event energies, this resolution is consistent with the performance achieved by the E852 LGD [8] which utilized 4cm×4cm×45cm blocks.

OUTLOOK

A second commissioning run is scheduled for the last weekend in July 1997. Our goal is to complete the calibration of the stage-I LGD and record approximately one million hadronic triggers. From these data we expect to be able to study ω, η, and possibly η' decays (though our acceptance for the latter is quite low with this stage-I detector). We additionally will investigate properites of hadronic events from photon interactions with a polyethelene (nucleon-like) target.

Our next major goal is to build a fully instrumented LGD with acceptance suitable for ϕ reconstruction. To improve the trigger efficiency and to better exploit the kinematic correlation for ϕ meson production, we anticipate upgrading the target region to employ a liquid hydrogen target surrounded by a recoil detector having cylindrical geometry. The advantages of the hydrogen target are the absence of Fermi motion and nuclear rescatterring effects, and less multiple coulomb scattering for the recoil proton. To maintain a good interaction rate it will be about 30 cm long, and this extended length (as opposed to the essentially point-like beryllium target in use for the the stage-I experiment) implies measuring a recoil proton track segment (slope and interecept). We are exploring options, such as designs based on straw tubes or scintillating fibers, for this more sophisticated recoil detector and associated trigger.

I would like to thank Professor Alex Dzierba of Indiana University who was originally scheduled for this talk and who supplied massive help putting it together. I would also like to thank the members of the Radiative-ϕ collaboration and the enthusiastic and helpful staff of Jefferson Laboratory.

REFERENCES

1. Crystal Barrel Collab., V.V. Anisovish *et al.*, Phys. Lett. B323 (1994) 233; C. Amsler *et al.*, Phys. Lett. B333 (1994) 277.
2. J. Weinstein, N. Isgur, Phys. Rev. Lett. 48 (1982) 659; Phys. Rev. D 41 (1990) 2236.
3. F. Close, N. Isgur and S. Kumano, Nucl. Phys. B389 (1993) 513.
4. A. Dzierba "Measuring Rare Radiative Decays of the ϕ Meson at CEBAF", Workshop on Physics and Detectors for DAPHNE '95, R. Baldini et al editors, INFN Frascati Physics Series, 1995; A. Dzierba "Scalar Mesons and Radiative Decays of the Phi", DAFCE Workshop, Frascati, Italy, Nov. 1996, to be published by Elvesier; A. Dzierba, "Photoproduction Opportunities at CEBAF: Meson Spectroscopy and the Physics of Flying Phi's", invited talk at Workshop on CEBAF at Higher Energies, (1994) 13.
5. UKQCD Collab., P. Lacock, C. Michael, P. Boyle, P. Rowland Phys.Rev. D54 (1996) 6997; MILC Collab., C. Bernard *et al.*, hep-lat/97077008; D. Weingarten, hep-ph/9607212.
6. V.A. Kostelecky, R. Potting, Nucl. Phys. B359 (1991) 545.
7. P. Smith *et. al*, Indiana High Energy Physics Internal Note (May, 1997), submitted to Nucl. Instrum. Meth.
8. R.R. Crittenden *et al.*, Nucl. Instrum. Meth. A 387 (1997) 377.

STATUS OF THE BNL MUON $(g-2)$ EXPERIMENT

J.P. Miller[a], L.M. Barkov[i], J. Benante[b], D.H. Brown[a], H.N. Brown[b], G. Bunce[b], R.M. Carey[a], A. Chertovskikh[i], J. Cullen[b], P. Cushman[h], G.T. Danby[b], P.T. Debevec[g], H. Deng[m], W. Deninger[g], S.K. Dhawan[m], A. Disco[m], V.P. Druzhinin[i], L. Duong[h], W. Earle[a], K. Endo[j], E. Efstathiadis[a], F.J.M. Farley[m], G.V. Fedotovich[i], X. Fei[m], J. Geller[b], J. Gerhaeuser[e], S. Giron[h], D.N. Grigorev[i], V.B. Golubev[i], M. Grosse Perdekamp[m], A. Grossmann[e], U. Haeberlen[f], E.S. Hazen[a], D.W. Hertzog[g], H. Hirabayashi[j], H. Hseuh[b], B.J. Hughes[a], V.W. Hughes[m], S. Ichii[j], K. Ishida[k], J.W. Jackson[b], L. Jia[b], K. Jungmann[e], D. Kawall[m], B.I. Khazin[i], J. Kindem[h], T. Kinoshita[c], F. Krienen[a], S. Kurokawa[j], R. Larsen[b], Y.Y. Lee[b], I. Logashenko[i], M. Mapes[b], R. McNabb[h], W. Meng[b], Yu. Merzliakov[i], D. Miller[h], Y. Mizumachi[n], V. Monich[i], W.M. Morse[b], Y. Orlov[c], J. Ouyang[a], C. Pai[b], C. Pearson[b], I. Polk[b], C. Polly[g], R. Prigl[b], G. zu Putlitz[e], S. Rankowitz[b], S.I. Redin[m], O. Rind[a], B.L. Roberts[a], N. Ryskulov[i], J. Sandberg[b], T. Sato[j], S. Sedykh[g], Y.K. Semertzidis[b], S. Serednyakov[i], Yu.M. Shatunov[i], R. Shutt[b], L. Snydstrup[b], E. Solodov[i], A. Soukas[b], A. Stillman[b], L.R. Sulak[a], T. Tallerico[b], M. Tanaka[b], F. Toldo[b], C. Timmermans[h], A. Trofimov[a], D. Urner[g], P. von Walter[e], D. Winn[d], K. Woodle[b], W.A. Worstell[a], A. Yamamoto[j], D. Zimmerman[h].

[a]*Department of Physics, Boston University, Boston, MA 02215, USA,* [b]*Brookhaven National Laboratory, Upton, NY 11973, USA,* [c]*Newman Laboratory, Cornell University, Ithaca NY 14853, USA,* [d]*Fairfield University, Fairfield, CT 06430, USA,* [e]*Physikalisches Institut der Universität Heidelberg, 69120 Heidelberg, Germany,* [f]*MPI für Med. Forschung, D69120 Heidelberg, Germany,* [g]*Department of Physics, University of Illinois, Urbana, IL 61820, USA,* [h]*Department of Physics, University of Minnesota, Minneapolis,*

MN 55455, USA, ^{i}*Budker Institute of Nuclear Physics, Novosibirsk, Russia,* ^{j}*KEK, Japan,* ^{l}*Riken, Japan,* ^{n} *Science University of Tokyo, Tokyo, Japan,* ^{m}*Department of Physics, Yale University, New Haven, CT 06511, USA.*

Abstract.
The muon ($g-2$) experiment at Brookhaven has just completed a 3-month run for checkout and initial data-taking. In the first two months beam was taken in a parasitic mode where one out of ten AGS pulses was delivered for commissioning of the beam line, quadrupoles, detectors, and data acquisition system. This was followed by four weeks of dedicated data collection. The main components of the experiment, which include the pion/muon beam line, the superconducting inflector, the superferric storage ring with its pulsed electric quadrupoles and magnetic field measurement system, and the detector system based on lead-scintillating fiber electron calorimeters, have been satisfactorily commissioned. The muon ($g-2$) precession frequency is clearly seen as a large signal. It is estimated that over 25×10^6 decay positrons with energies greater than 1.5 GeV have been detected.

INTRODUCTION

To the current level of experimental precision (4 ppb), the anomalous g factor for the electron,

$$a_e = \frac{(g-2)_e}{2} \qquad (1)$$

arises entirely from virtual photons and electron-positron pairs. [1] The agreement between experimental and theoretical values of a_e, which is limited by the accuracy of about 25 ppb with which the fine structure constant α is known, provides one of the most decisive tests of QED. [2]

For the muon, the relative contributions to a_μ of heavier particles or new contact interactions are much larger than for the electron because they usually scale as $(m_\mu/m_e)^2 \simeq 4 \times 10^4$. In particular, the contribution from diagrams including virtual hadrons is about 60 ppm, and was observed for the first time in the most recent CERN experiment, [3] which measured a_μ with a precision of 7 ppm (with approximately 1 ppm systematic error).

The goal of the new BNL muon $g-2$ experiment is to measure a_μ to 0.35 ppm, a factor 20 improvement over the CERN experiment. With this precision the electroweak contribution [4–7] of 1.3 ppm, which arises from virtual processes involving the Z and W vector bosons, should be observed, thus offering an important test of electroweak renormalization. In addition, a sensitive and powerful test for contributions beyond the standard model will be provided, especially for μ and W substructure or anomalous couplings, but also for supersymmetry and leptoquarks. [4,8–11]

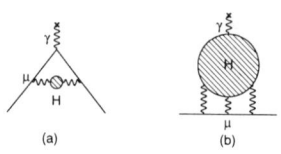

FIGURE 1. (a) The lowest order hadronic contribution to a_μ. (b) The hadronic light-by-light contribution.

HADRONIC AND WEAK CONTRIBUTIONS

The hadronic contribution to a_μ cannot presently be calculated from first principles of QCD. However, using dispersion theory the lowest order term in a_μ^{had}, which is shown in Fig. 1a, can be expressed in terms of $R(s) = (\sigma_{tot}(e^+e^- \to hadrons))/(\sigma_{tot}(e^+e^- \to \mu^+\mu^-))$. Present experimental knowledge of R, [12,13] including recent data from τ decay, [14] determines that $a_\mu^{had}(tot) = 6\ 828(98) \times 10^{-11}$ or (59.0 ± 0.84) ppm of a_μ. The higher order hadronic contribution [15] (excluding the contribution from Fig. 1b) is $-101(6) \times 10^{-11}$. The contribution of the light-by-light scattering diagram of Fig. 1b has recently been estimated to be $a_\mu^{had} = -92(32) \times 10^{-11}$ or (0.79 ± 0.27) ppm of a_μ. [16,17] The uncertainty on this contribution should be reduced when additional calculations have been completed. [17] Further reductions in the total error of a_μ^{had} are anticipated with the planned additional measurements of R from the CMD2 detector in the e^+e^- collider at Novosibirsk with $E_{cm}(max) = 1.4$ GeV, from the ϕ factory at Frascati, and from the Beijing e^+e^- collider in the energy region above and below the J/ψ energy.

The electroweak contribution to a_μ arises in lowest order from the diagrams shown in Figs. 2a and 2b, and was calculated in the early 1970's. Higher order contributions involving electroweak radiative corrections are shown in Figs. 2c, 2d and 2e. The first order weak contribution of $195(4) \times 10^{-11}$ is reduced to $151(4) \times 10^{-11}$ (1.3 ppm of a_μ) when the second order terms are included. [5–7]

The full theoretical value from the standard model is :

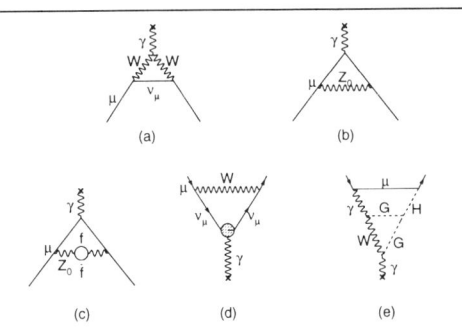

FIGURE 2. Weak contributions to the muon anomalous g factor. Single-loop contributions from (a) virtual W and (b) virtual Z gauge bosons. These two contributions enter with opposite sign, and there is a partial cancellation. The two-loop contributions fall into three categories. An example of each category is given. (c) Fermionic loops which involve the coupling of the gauge bosons to quarks; (d) Bosonic loops which appear as corrections to the one-loop diagrams; and (e) A new class of diagrams where G is the longitudinal component of the gauge bosons.

$$\begin{aligned} a_\mu &= a_\mu^{QED} + a_\mu^{had} + a_\mu^{weak} \\ &= 116\ 584\ 705.7(1.9) \times 10^{-11} (\pm 0.016\ ppm) \\ &\quad + 6\ 828\ \ (98) \times 10^{-11} (\pm 0.84\ ppm) \\ &\quad + 151\ \ (\ 4) \times 10^{-11} (\pm 0.03\ ppm) \\ &= 116\ 591\ 685(98) \times 10^{-11} (\pm 0.84\ ppm) \end{aligned} \quad (2)$$

The theoretical error of ± 0.84 ppm is dominated by the uncertainty of a_μ^{had}.

THE BROOKHAVEN EXPERIMENT

A new experiment to measure the anomalous g factor of the muon more precisely has been constructed at Brookhaven National Laboratory. First data were taken in June and July 1997.

The principle of the measurement is similar to the third CERN experiment. [3] Polarized muons are stored in a uniform dipole magnetic field with electrostatic quadrupoles providing weak vertical focusing. The muon spin precesses relative to the momentum vector with the frequency

$$\vec{\omega}_a = -\frac{e}{m_\mu c}\left[a_\mu \vec{B} - \left(a_\mu - \frac{1}{\gamma_\mu^2 - 1} \right) \vec{\beta} \times \vec{E} \right], \quad (3)$$

assuming that the electric dipole moment of the muon is zero. The dependence of ω_a on the electric field can be eliminated by storing muons with the "magic"

$\gamma_\mu=29.3$, corresponding to a muon momentum $p_\mu = 3.094$ GeV/c. In this ideal case $a_\mu - 1/(\gamma_\mu^2 - 1) = 0$, and the focussing electric field does not affect the spin precession frequency which then depends only on the average magnetic field seen by the muons. In practice there is a small electric field correction to ω_a at the ≤ 1 ppm level since not all muons are precisely at the magic momentum, and a pitch correction at a similar level is introduced because of oscillations of the muon orbits induced by the vertical focussing. a_μ is extracted from $f_a = \omega_a/2\pi \approx 230 kHz$ through

$$a_\mu = \frac{f_a/f_p}{\mu_\mu/\mu_p - f_a/f_p} \quad (4)$$

where $f_p \approx 63 MHz$ is the free proton precession frequency in the same magnetic field seen by the muons. The ratio of muon to proton magnetic moments, μ_μ/μ_p, is currently known to 0.30 ppm from the Zeeman effect in muonium and μSR, and to 0.15 ppm from the hyperfine structure interval in muonium, [18,19] and will be improved by a recent experiment. [20]

The source of the stored muons is the AGS proton beam, which delivers eight bunches with a total of $\approx 40 \times 10^{12}$ protons at 24 GeV/c onto a nickel production target. The cycle of eight bunches is repeated every 2.6 s, with the individual bunches having a $\sigma = 27$ ns and spaced apart by 33 ms.

From each bunch roughly 4×10^7 pions at ≈ 3.1 GeV/c are transported from the production target along a 116 m beam line through a hole in the back of the storage ring magnet yoke. After passing through a field-free region supplied by a superconducting inflector magnet, [21] the pion beam enters the toroidal storage region which has a radius of 7.112 m and a 9 cm diameter cross section.

From the pion decay, $\pi^+ \to \mu^+ \nu_\mu$, a small fraction of muons are launched onto stable orbits and are stored. Pulsed electric quadrupoles and a set of collimators are used to scrape the muon beam and thus reduce losses at later times. The bunched structure of the muon beam just after the pion injection time at the cyclotron period of 149 ns has been observed. The rate at which the bunches spread out with time gives a measure of the momentum distribution of stored muons.

The storage ring is a continuous superferric 'C' magnet with the open side facing inward, see Fig. 3. The 1.4513 T storage ring B-field is excited by three superconducting coils which carry a current of 5177 A. The field strength and changes in homogeneity are monitored by 366 NMR probes [22] spaced around the ring outside of the storage volume. In addition, the field inside the storage region is mapped several times every week using a trolley containing a matrix of 17 NMR probes which operates in vacuum inside the storage region. When integrated over azimuth the field in the storage region is currently uniform to about 25 ppm. The dipole field is stable to a few ppm/hour, with the field changes caused primarily by the thermal expansion and contraction of the iron yoke, which affects the magnet gap. In next year's run, we expect to collect most of our data by injecting muons directly into the ring, with a pulsed full-aperture magnetic kicker providing the impulse needed to put the muons onto stable orbits. The result will be a substantially higher data rate and reduced detector background. As we approach the statistics required to achieve an 0.35 ppm measurement in subsequent runs, the requirement on the magnetic field homogeneity will be more stringent. Magnet shimming will resume after this run to achieve the design goal of 1 ppm homogeneity in $\int \mathbf{B} \cdot d\mathbf{l}$, and an enclosure for the magnet will be constructed in order to stabilize the temperature.

The decay positrons from $\mu^+ \to e^+ \nu_e \bar{\nu}_\mu$ range in energy from 0 to 3.1 GeV, Those positrons above about 1GeV are detected with Pb-scintillating fiber calorimeters placed symmetrically at 24 positions around the inside of the storage ring. Because of the parity violating nature of the weak decay, the highest energy positrons are preferentially emitted along the muon spin direction. The muon spin precession is reflected in the decay positron spectrum, N(t), where we expect :

$$N(t) = N_0 e^{-t/\tau_\mu} \left[1 + A \cos\left(\omega_a t + \phi\right)\right]. \tag{5}$$

The asymmetry parameter, A, is energy dependent, and for a positron energy lower threshold of 1.8 GeV is ~ 0.4.

The arrival times of the positrons are recorded in multi-hit TDCs, and the calorimeter pulses are also sampled by a custom 400 MHz waveform digitizer which delivers both time and pulse height information. An important experimental consideration is the necessity of reducing systematic early-to-late shifts of the event times to the level of 20 ps over a period of 200 μs (muon decays are recorded for 640 μs). A laser and LED system are used to monitor such potential time shifts, as well as gain shifts. The principal difficulty is overcoming the effects of the 'flash' seen in the detectors during and after pion injection. The calorimeter photomultiplier tubes must be gated off during this intense burst of light, and the recovery of the tubes involves possible changes in gain and pulse times, since the pulses sit on top of a residual, decaying DC light level.

In addition to the calorimeters, several detector stations are outfitted with a finely segmented hodoscope array of 20×32 small scintillating elements connected to a multianode phototube, which will provide position sensitive information on the muon decay positron. Additional spatial and temporal event information is derived from four stations each equipped with five scintillator paddles oriented horizontally. Finally, a traceback chamber is in operation which allows the trajectories of the decay positrons to be traced back to their origin in the storage ring, with the objective of providing information on the stored muon phase space. The latter will be convoluted with the magnetic field map to determine the average field seen by the muons.

When the experiment is fully commissioned next year, it will have several significant advantages over the third CERN experiment:

FIGURE 3. A view of the $(g-2)$ storage ring and its detectors.

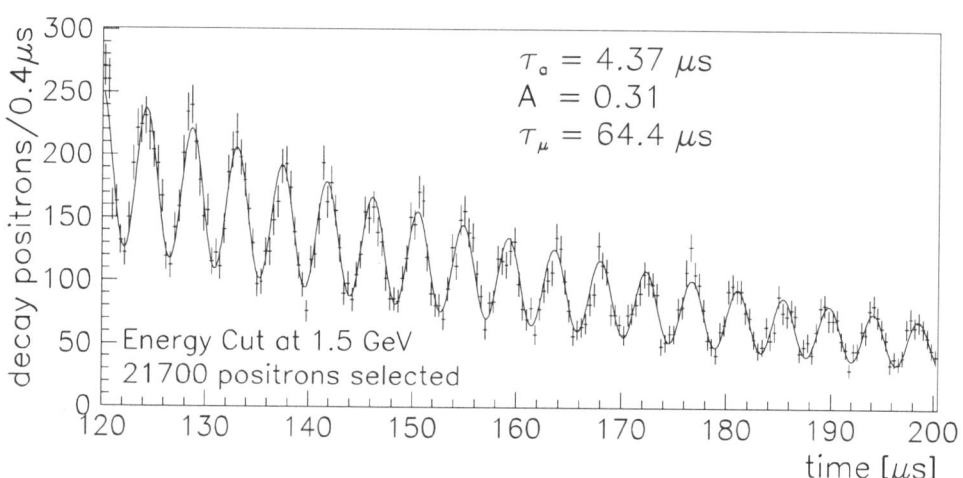

FIGURE 4. Preliminary analysis of the time spectrum of the decay positrons from a short run. The exponential muon decay is clearly modulated at the $(g-2)$ precession period, τ_a. The solid line represents a fit to the data.

- The AGS supplies ten times larger pion flux.

- Once the full-aperture muon kicker is installed, we will have the option of muon injection, which will provide ten times more stored muons with fifty times less 'flash' in the detector system compared to pion injection.

- The super-ferric magnet is more stable and, once completely shimmed, will have a much more uniform magnetic field, than the conventional ferric magnet employed at CERN. The plan is to produce a magnet which will give an average B-field which is the same for all muons to the 1 ppm level averaged in azimuth, regardless of path followed by the muon in the storage region, then use accurate determinations of the B-field and muon storage distribution to reduce the uncertainty in average field seen by the ensemble of muons to 0.1 ppm. The contribution of the field to the systematic error in a_μ would then also be held to 0.1 ppm.

- The NMR trolley permits precision B-field measurements throughout the storage region without having to remove the surrounding vacuum chamber.

- To determine the average stored muon phase space in the storage region, we will have the previously described electron traceback in addition to the conventional analysis (used by CERN) of the debunching with time of the stored muons.

- The asymmetry in the decay electron time spectrum is improved by 40% compared to CERN, and this is attributed to having improved energy resolution in the electromagnetic calorimeters and improved vacuum chamber design which results in less degradation of the positron energy from showering in the chamber walls.

After more than a decade in preparation, we have seen for the first time at BNL the exponential muon decay, modulated by the $(g-2)$ precession which is shown in Figure 4. At present more than 25×10^6 positrons with energies greater than 1.5 GeV have been detected, corresponding to a statistical precision of better than 10^{-5} on a_μ.

REFERENCES

1. R. van Dyke et al., *Phys. Rev. Lett.* **59**, 26 (1987); Robert S. Van Dyck Jr. in *Quantum Electrodynamics*, ed. T. Kinoshita, (World Scientific, Singapore, 1990), p. 322.
2. T. Kinoshita, *Rep. Prog. Phys.* **59**, 1459 (1996).
3. J. Bailey et al., *Nucl. Phys. B* **B150**, 1 (1979); F. J. M. Farley and E. Picasso in *Quantum Electrodynamics*, ed. T. Kinoshita, (World Scientific, Singapore, 1990), p. 479.

4. T. Kinoshita and W.J. Marciano in *Quantum Electrodynamics*, ed. T. Kinoshita, (World Scientific, Singapore, 1990), p. 419.
5. S. Peris, M. Perrottet, and E. de Rafael, *Phys. Lett.* B **355**, 523 (1995).
6. A. Czarnecki, B. Krause and W. Marciano, *Phys. Rev. Lett.* **76**, 3267 (1996).
7. A. Czarnecki, B. Krause and W.J. Marciano, *Phys. Rev.* D **52**, R2619 (1995).
8. P. Méry, S.E. Moubarik, M. Perrottet, F.M Renard, *Z. Phys.* C **46**, 229 (1990).
9. J. Lopez, D. Nanopoulos, and X. Wang, *Phys. Rev.* D **49**, 366 (1994); U. Chattopadhyay and P. Nath, *Phys. Rev.* D **53**, 1648 (1996); T. Moroi PRD **53**, 6565 (1996).
10. G. Couture and H. Konig, *Phys. Rev.* D **53**, 555 (1996).
11. F.M. Renard, S. Spagnolo and C. Verzegnassi, *Constraints on anomalous gauge couplings from present LEP1 and future LEP2, BNL data*, Preprint hep-ph/9705274 (1997).
12. S. Eidelman and F. Jegerlehner, *Z. Phys.* C **67**, 585 (1995).
13. D.H. Brown and W.A. Worstell, *Phys. Rev.* D **54**, 3237 (1996).
14. R. Alemany, M. Davier and A. Hocker, submitted to *Z. Phys. C* (1997).
15. B. Krause, *Phys. Lett.* B **390**, 392 (1997).
16. J. Bijnens, E. Pallante and J. Prades, *Phys. Rev. Lett.* **75**, 1447 (1995) and *Phys. Rev. Lett.* **75**, 3781 (1995).
17. M. Hayakawa, T. Kinoshita and A. I. Sanda, *Phys. Rev. Lett.* **75**, 790 (1995); M. Hayakawa, T. Kinoshita and A. Sanda, *Phys. Rev.* D **54**, 3137 (1996) and work in progress.
18. F. G. Mariam et al., *Phys. Rev. Lett.* **49**, 993 (1982).
19. E. R. Cohen and B. N. Taylor, *Rev. Mod. Phys.* **59**, 1121 (1987).
20. M. G. Boshier et al., *Phys. Rev.* A **52**, 1948 (1995).
21. F. Krienen, D. Loomba and W. Meng, *Nucl. Instrum. Methods* A **283**, 5 (1989).
22. R. Prigl et al., *Nucl. Instrum. Methods* A **374**, 118 (1996).

Charmless B Decays at CLEO

Paula A. Pomianowski

Virginia Tech / CLEO Collaboration

Abstract. Many rare charmless B decays have been measured or searched for by the CLEO collaboration. These decays are useful for understanding CP violation in the B sector, as well as for overdetermining the Unitarity Triangle. The most recent preliminary results are presented here.

CP VIOLATION AND CHARMLESS B DECAYS

In the Standard Model, CP violation arises due to a single phase entering the CKM matrix:

$$V = \begin{bmatrix} V_{ud} & V_{us} & V_{ub} \\ V_{cd} & V_{cs} & V_{cb} \\ V_{td} & V_{ts} & V_{tb} \end{bmatrix}.$$

Written as vectors in a complex plane, $V_{ud}V_{ub}^* + V_{cd}V_{cb}^* + V_{td}V_{tb}^* = 0$ describes the 'Unitarity Triangle' shown in Figure 1, where η controls the amount of CP violation present and ρ controls the distribution among various B decays. In general, each CP violating process in B physics is sensitive to α, β, or γ.

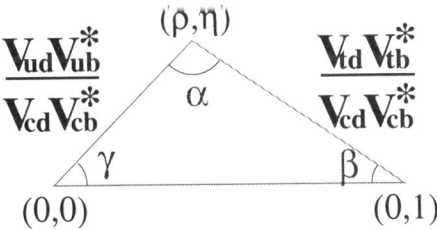

FIGURE 1. Unitarity Triangle

Two types of CP violation are of interest: direct CP violation and indirect CP violation. In direct CP violation, interference occurs between two amplitudes, when two competing decay mechanisms with different weak phases decay to the same final state. The strong interaction contribution to the competing channels must also have different phases, which complicates the

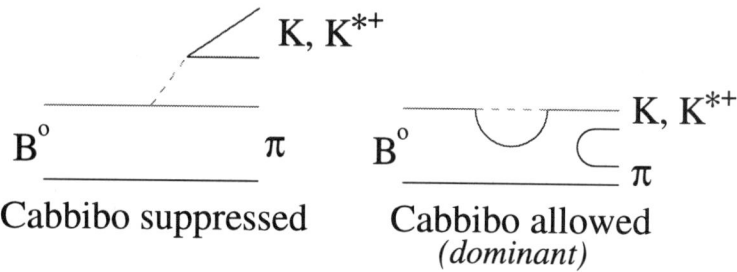

FIGURE 2. Spectator and Penguin Diagrams for $B^0 \to K^{(*)}\pi$

theoretical interpretation of these decays. The decay mode $B \to K\pi$ is an excellent candidate to study direct CP violation, where at most a few percent asymmetry can arise due to the interference between the spectator and penguin diagrams shown in Figure 2.

In indirect CP violation, interference occurs through B^0 mixing, where the decay $B^0 \to f$ interferes with $B^0 \to \bar{B}^0 \to f$. Measuring this requires tagged, time dependent rate asymmetry measurements, as can be done at an asymmetric B factory or hadron collider. $B^0 \to \pi^+\pi^-$, a $b \to u$ spectator transition, is a CP eigenstate and is the 'Golden Mode' for measuring $\sin 2\alpha$. However, this measurement is contaminated by "penguin pollution". To disentangle penguin and spectator effects, measuring modes such as $B^0 \to \pi^0\pi^0$, $B^+ \to \pi^\pm\pi^0$, etc., is crucial.

It is of great interest to measure as many two-body charmless B decay modes as possible, as this will help to overdetermine sides and angles of the Unitarity Triangle, an important check of the validity of the Standard Model.

A brief description of the basic analysis techniques used by the CLEO collaboration is given in the following section. The results for different modes follow, and are summarized at the end with theoretical expectations.

CLEO II EXPERIMENT AND EVENT SELECTION

The CLEO II detector [1] operates at the Cornell Electron Storage Ring. Electron-positron collisions at and just below the 10.6 GeV $\Upsilon(4S)$ resonance have provided a data sample of 3.3 M $B\bar{B}$, 4.8 M $c\bar{c}$, and 3.9 M $\tau^+\tau^-$ events. Charged particle tracking information is obtained from three concentric cylindrical wire drift chambers. Specific ionization measurements (dE/dx) from tracking provides particle identification, and electromagnetic showers are reconstructed with excellent resolution in the cesium iodide crystal calorimeter.

Because the $\Upsilon(4S)$ is just above the threshold of $B\bar{B}$ production, B's are produced essentially at rest and therefore decay isotropically. Hence, $e^+e^- \to B\bar{B}$ events at CLEO have a spherical distribution of particles, as opposed to continuum background events which have a 2-jet structure. To suppress the continuum, a cut is made on the angle between the thrust axis of a B candidate and the thrust axis of the remainder of the event. This $\cos\theta_{\text{thrust}}$ is flat for $B\bar{B}$ events and strongly peaked to 1 for continuum events. Another constraint is the beam energy. Because each B has energy equal to the beam energy, $\Delta E \equiv E_B - E_{\text{beam}}$ should be ≈ 0. The next handle is beam-constrained mass. If E_{beam} is used to calculate mass ($M_B = \sqrt{E_{\text{beam}}^2 - p^2}$), true B's from $B\bar{B}$ events will peak at the B mass (5.28 GeV/c^2). Finally, a Fisher Discriminant ($\mathcal{F}$) is formed, a linear combination of 11 variables which characterize the distribution of particles around the candidate axis and which were chosen to maximize the difference between signal and background distributions. Two of the 11 variables are the angle between the candidate thrust axis and the beam axis, and the angle between the B flight direction and the beam axis; the remaining 9 are the sum of all momenta in 10° cones around the candidate thrust axis.

To distiguish between modes like $B^0 \to \pi^+\pi^-$ and $B^0 \to K^+\pi^-$, dE/dx information is used to identify hadron species. For momenta below 0.8 GeV/c, good separation is obtained whereas for momenta above 2 GeV/c, the relativistic rise of charged pions again creates a usable K/π separation of $\approx 1.7\sigma$.

$B \to \pi\pi$, $K\pi$, AND KK DECAYS

Previously, the CLEO collaboration reported [2] a combined measurement of $\mathcal{B}(B^0 \to \pi^+\pi^- + K^+\pi^-) = (1.8^{+0.6+0.2}_{-0.5-0.3} \pm 0.2) \times 10^{-5}$. Since then, 30% more data has become available. A reoptimization of the analysis has yielded a 20% higher efficiency. The event shape variables were also adjusted, and a maximum likelihood fit to M_B, ΔE, $\mathcal{F}, dE/dx_1, dE/dx_2$,

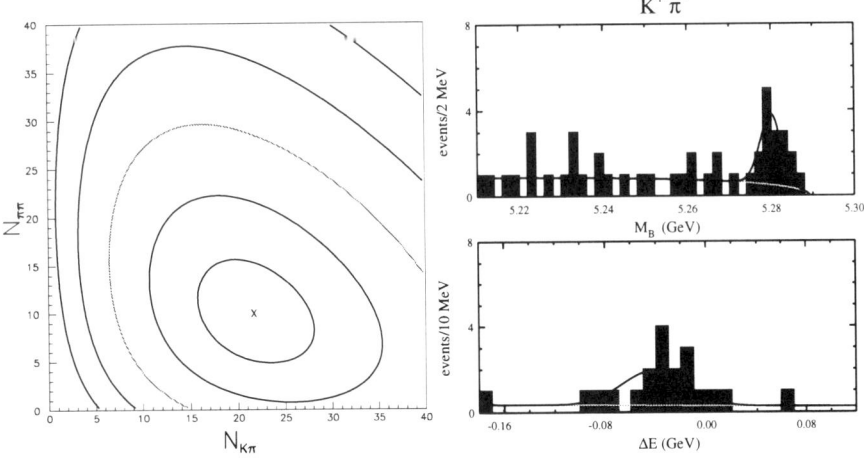

FIGURE 3. Maximum Likelihood Fit for $B^0 \to K\pi$ vs. $B^0 \to \pi\pi$

and $\cos\theta_B$ performed, as shown in Figure 3. This has allowed the separation of $K\pi$ and $\pi\pi$ yields: $N_{K\pi} = 21.7^{+6.8}_{-6.0}$, a 5.6$\sigma$ effect, and $N_{\pi\pi} = 10.0^{+6.0}_{-5.1}$, a 2.2$\sigma$ effect, and therefore new preliminary branching fractions: $\mathcal{B}(B^0 \to K^+\pi^-) = (1.5^{+0.5+0.1}_{-0.4-0.1} \pm 0.1) \times 10^{-5}$, $\mathcal{B}(B^0 \to \pi^+\pi^-) < 1.5 \times 10^{-5}$, and $\mathcal{B}(B^0 \to K^+K^-) < 0.43 \times 10^{-5}$. Figure 3 also shows the M_B and ΔE plots for the $B^0 \to K^+\pi^-$ mode.

A maximum likelihood fit for the decays $B^+ \to \pi^+\pi^0$ and $B^+ \to K^+\pi^0$ have also been performed. The combined yield of $20.0^{+6.8}_{-5.9}(5.5\sigma)$ yielding $\mathcal{B}(B^+ \to h^+\pi^0) = (1.6^{+0.5+0.2}_{-0.5-0.2} \pm 0.1) \times 10^{-5}$ has been separated into $N(K^+\pi^0) = 8.7^{+5.3}_{-4.2}$, a 2.7$\sigma$ effect, and $N(\pi^+\pi^0) = 11.3^{+6.3}_{-5.2}$, a 2.8$\sigma$ effect. As these are not 3σ measurements, only (preliminary) upper limits for the separate modes can be given: $\mathcal{B}(B^+ \to K^+\pi^0) < 1.6 \times 10^{-5}$ and $\mathcal{B}(B^+ \to \pi^+\pi^0) < 2.0 \times 10^{-5}$.

The decay mode $B^+ \to h^+K^0$ has a signal yield of $N(h^+K^0_S) = 9.8^{+4.5}_{-4.0}(4.4\sigma)$, and $\mathcal{B}(B^+ \to h^+K^0) = (2.4^{+1.1+0.2}_{-1.0-0.2} \pm 0.2) \times 10^{-5}$. The maximum likelihood fit separates $B^+ \to h^+K^0$ into π^+K^0 and K^+K^0 as follows: $N(\pi^+K^0_S) = 9.2^{+4.3}_{-3.8}(3.2\sigma)$ with $\mathcal{B}(\pi^+K^0) = (2.3^{+1.1+0.2}_{-1.0-0.2} \pm 0.2) \times 10^{-5}$, and $N(K^+K^0_S) = 0.5^{+3.8}_{-0.5}(0.2\sigma)$ with $\mathcal{B}(K^+K^0) < 1.6 \times 10^{-5}$.

Finally, a search for $B^0 \to K^0\bar{K}^0$ is made. This mode is expected to be pure penguin which makes it useful for separating out 'penguin pollution' effects. Zero events were found, with an expected background of 0.6 ± 0.2, giving an upper limit of $\mathcal{B}(B^0 \to K^0\bar{K}^0) < 1.7 \times 10^{-5}$.

For a summary of the $B \to K\pi, \pi\pi$, and KK results, see Table 1.

TABLE 1. Preliminary $B \to K\pi, \pi\pi$, and KK Results

Mode	Yield	Signif	BR (10^{-5})	UL (10^{-5})
$K^\pm\pi^\mp$	$21.7^{+6.8}_{-6.0}$	5.6σ	$1.5^{+0.5+0.1}_{-0.4-0.1} \pm 0.1$	
$K^\pm\pi^0$	$8.7^{+5.3}_{-4.2}$	2.7σ		1.6
$K^0\pi^\pm$	$9.2^{+4.3}_{-3.8}$	3.2σ	$2.3^{+1.1+0.2}_{-1.0-0.2} \pm 0.2$	
$K^0\pi^0$ *	$2.3^{+2.2}_{-1.5}$			4.0
$\pi^\pm\pi^\mp$	$10.0^{+6.0}_{-5.1}$	2.2σ		1.5
$\pi^\pm\pi^0$	$11.3^{+6.3}_{-5.2}$	2.8σ		2.0
$\pi^0\pi^0$ *	$1.2^{+1.7}_{-0.9}$			0.9
$K^\pm K^\mp$	$0.0^{+1.3}_{-0.0}$	0.0σ		0.4
$K^\pm K^0$	$0.5^{+3.8}_{-0.5}$	0.2σ		1.6
$K^0\bar{K}^0$	0			1.7
$h^\pm\pi^\mp$	$31.7^{+8.4}_{-7.3}$	7.8σ	$2.2^{+0.6}_{-0.5}$	
$h^\pm\pi^0$	$20.0^{+6.8}_{-5.9}$	5.5σ	$1.6^{+0.5+0.2}_{-0.5-0.2} \pm 0.1$	
$h^\pm K^0$	$9.8^{+4.5}_{-4.0}$	4.4σ	$2.4^{+1.1+0.2}_{-1.0-0.2} \pm 0.2$	

* Previously published CLEO Result [2]

OTHER CHARMLESS TWO-BODY DECAY MODES

Other charmless decay modes are useful to measure γ and to study 'penguin pollution'. In addition, these modes have extra analysis handles, such as resonance mass. The resonance decays used in the following results are $\rho \to \pi\pi$, $K^* \to K^+\pi$ or $K_S^0\pi$, $\phi \to K^+K^-$, $\omega \to \pi^+\pi^-\pi^0$, $\eta \to \pi^+\pi^-\pi^0$ or $\gamma\gamma$, and $\eta' \to \eta\pi^+\pi^-$ with $\eta \to \gamma\gamma$. Because these modes are cleaner, the event shape cuts can be loosened.

The modes $B^+ \to \omega h^+$ and $B^+ \to \eta' h^+$ are seen to be large. Our maximum likelihood fit yields $9.5^{+5.3}_{-4.2}$ $\omega\pi^+$ events, a 3.3σ effect and $8.6^{+4.9}_{-3.9}$ ωK^+ events, a 2.9σ effect. This gives the following branching fractions: $\mathcal{B}(B^+ \to \omega\pi^+) = (1.2^{+0.7}_{-0.5} \pm 0.2) \times 10^{-5}$ and $\mathcal{B}(B^+ \to \omega K^+) = (1.2^{+0.7}_{-0.5} \pm 0.2) \times 10^{-5}$.

Similarly, the fit for $B^+ \to \eta'h^+$ yields $N(\eta'K^+) = 12^{+4.1}_{-3.4}$ (5.5σ) and $N(\eta'\pi^+) = 1.4^{+2.0}_{-1.1}$ (2.0σ). Hence, $\mathcal{B}(B^+ \to \eta'K^+) = (7.8^{+2.7}_{-2.2} \pm 1.0) \times 10^{-5}$ and $\mathcal{B}(B^+ \to \eta'\pi^+) < 4.4 \times 10^{-5}$. It is currently being debated if the large size of $\mathcal{B}(B^+ \to \eta'K^+)$ is surprising or not. Lipkin [3] believes that a large $\eta'K/\eta K$ ratio arises due to interference between two penguin diagrams. Others are proposing additional mechanisms which will enhance the $\eta'K^+$ rate, such as $b \to c\bar{c}s$ coupling to the $c\bar{c}$ component of η' [4] [5], $c\bar{c} \to g \to \eta'$ [6], or anomalous $gg\eta'$ coupling [7].

For brevity, other modes are summarized in Table 2.

TABLE 2. Preliminary Results for Other Modes

Mode	Yield	Expected BKG	UL (90% CL)
$B^0 \to \omega\rho^0$	0	1.2 ± 0.3	$< 3.4 \times 10^{-5}$
$B^0 \to \omega K^{*0}$	1	1.7 ± 0.4	$< 3.8 \times 10^{-5}$
$B^+ \to \omega K^{*+}$	0	0.9 ± 0.3	$< 11 \times 10^{-5}$
$B^+ \to \eta h^+$	0	2.2 ± 0.4	$< 0.8 \times 10^{-5}$
$B^0 \to \eta\rho^0$	1	0.8 ± 0.2	$< 8.4 \times 10^{-5}$
$B^0 \to \eta K^{*0}$	1	1.3 ± 0.3	$< 3.3 \times 10^{-5}$
$B^+ \to \eta K^{*+}$	0	0.5 ± 0.2	$< 24 \times 10^{-5}$
$B^0 \to \eta'K^{*0}$	0	0.3 ± 0.1	$< 9.9 \times 10^{-5}$
$B^+ \to \eta'K^{*+}$	0	0.3 ± 0.1	$< 29 \times 10^{-5}$

SUMMARY OF CLEO CHARMLESS B DECAY MEASUREMENTS

All modes are summarized with theoretical expectations in Figure 4. With the addition of more statistics and improved technique, it is now possible to separate $B^0 \to h^+\pi^-$ into $K^+\pi^-$ and $\pi^+\pi^-$. Modes ωh^+ and $\eta' K^+$ are measured to be larger than predicted, which provides evidence that additional mechanisms are needed to describe these rates. Finally, many of the upper limits are now approaching theory.

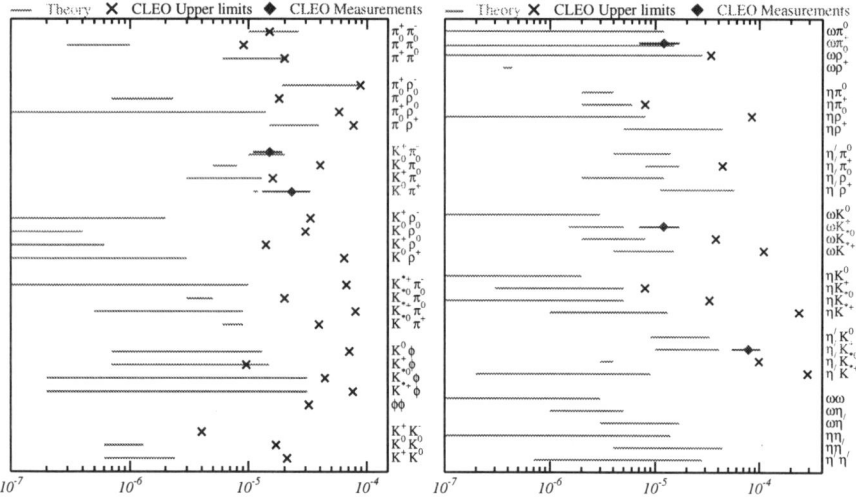

FIGURE 4. Summary of PRELIMINARY Results

REFERENCES

1. CLEO Collab., Y. Kubota et al., Nucl. Inst. Meth. Phys. Res. A **320**, 66 (1992).
2. CLEO Collab., D. Asner et al., Phys. Rev. D **53**, 1039 (1996).
3. H. Lipkin, Phys. Lett. B **254**, 247 (1991).
4. K. Berkelman, CLEO Internal Note, CBX96-79 and CBX96-79a (1996).
5. I. Halperin and A. Zhitnitsky, HEPPH-9704412 (1997); I. Halperin and A. Zhitnitsky, HEPPH-9705251 (1997); E. Shuryak and A. Zhitnitsky, HEPPH-9706316 (1997).
6. I. Dunietz et al., HEPPH-9612421 (1996).
7. Atwood and Soni, HEPPH-9704357 (1997).

Search for the CP-violating decay $K_L^0 \to \pi^0 \nu \bar{\nu}$ at BNL E926

Toshio Numao

TRIUMF
4004 Wesbrook Mall, Vancouver, B.C., Canada

For the BNL E926 collaboration [1]

Abstract. The present status of the BNL experiment E926 for a measurement of the CP-violating rare decay $K_L^0 \to \pi^0 \nu \bar{\nu}$ is described. The experiment employs kaon time-of-flight and full reconstruction of the π^0 to suppress backgrounds to a level well below the expected signal around $(3 \pm 2) \times 10^{-11}$.

INTRODUCTION

The decay $K_L^0 \to \pi^0 \nu \bar{\nu}$ violates the CP conservation law. The final state ($\pi^0 \nu \bar{\nu}$) is a CP even state while the K_L^0 is primarily a CP odd state with a small (ϵ) admixture of a CP even state. The contribution from the mixing alone was estimated to be of the order of $\epsilon^2 \cdot BR(K^+ \to \pi^+ \nu \bar{\nu}) \sim 10^{-15}$ [1]. The dominant contribution comes from the amplitude between the CP-odd part of K_L^0 and the CP-even final state via second-order weak-interaction diagrams that have a top quark as an intermediate state. Because of the cancellation of the charm quark contributions and "known" hadronic matrix elements of the weak current from isospin symmetry to the decay $K^+ \to \pi^0 e^+ \nu$, the theoretical uncertainty for this mode is only of the order of 1 %. The branching ratio is proportional to $Im^2(V_{td})$ or η^2 in the Wolfenstein parametrization and expected at $(3 \pm 2) \times 10^{-11}$, where the uncertainty comes from the poor constraints in CKM parameters. The observation of the decay $K_L^0 \to \pi^0 \nu \bar{\nu}$ will confirm the standard-model origin of CP violation and determines the height of the unitarity triangle, thus tightly constraining the triangle when combined with a measurement of the decay $K^+ \to \pi^+ \nu \bar{\nu}$. A phenomenological upper limit for this mode 1.1×10^{-8} comes from the isospin relationship to the charged mode [2]. The present experimental limit is 1.8×10^{-6} (90 % c.l.) [3].

---

[1] E926 collaboration: BNL-INR-Kyoto-New Mexico-TJNAF-TRIUMF-VPI-Yale

CONCEPTS OF EXPERIMENT

The signature of the decay $K_L^0 \to \pi^0 \nu \bar{\nu}$ is two γ-rays from the π^0 decay and nothing else. The only possible measurable quantities of the decay products are energies, locations, directions and timing of the γ-rays. With them, the π^0 mass, its direction and the vertex can be reconstructed. The direction and the location of a K_L^0 can be constrained from the vertex and the production target position as well as by beam collimation. However, without the information related to the energy of the kaon no constraints are available for the kaon mass or the missing mass of the particle X in the decay $K_L^0 \to \pi^0 X$, which is essential to eliminate many types of background. The time of flight (TOF) of the K_L^0 can be measured between the production target and the vertex, if the primary proton beam is pulsed and the kaon momentum is low.

Most decay modes of K_L^0 include two charged particles or more than two γ-rays as the final products— the only exception is the decay $K_L^0 \to \gamma\gamma$ that occurs at a level of 10^{-4} but this can be suppressed by kinematics to a negligible level. Therefore, complete coverage of the decay volume by efficient charged-particle detectors and photon detectors is also essential. Experiment E926 [4] at BNL AGS is designed to measure kaon TOF and all possible γ-ray kinematics, and has a nearly 4π coverage for γ-ray and charged-particle detection to suppress backgrounds to a level, an order of magnitude below the expected signal.

E926 EXPERIMENT

In order to achieve reasonable TOF resolution, the width of the beam bunch needs to be at most 300 ps. A preliminary test at the AGS indicates that a further improvement over the expected width is possible. The neutral beam extraction angle is set at 45° to soften the K_L^0 spectrum (peaks at 0.65 GeV/c) and facilitate the TOF measurement. This large angle extraction also helps to suppress background arising from $\Lambda \to n\pi^0$ decays to a negligible level. The vacuum of the K_L^0 beam line (especially, the 3-m long decay volume) is designed to be better than 10^{-7} Torr to reduce n- and K_L^0-nucleus interactions that produce π^0's.

The detector system shown in Fig. 1 consists of a preradiator, a forward endcap calorimeter and photon veto detectors. The preradiator provides the measurement of γ-ray directions, positions and a part of the energies. The requirements include an angular resolution of 25 mrad and conversion efficiency of ~ 0.7. The present design is based on 42 modules of sandwiches (each consisting of a layer of 2 mm thick scintillator, a 1-cm thick wire chamber with cathode strips, and a sheet of 0.035 radiation-length metallic radiator as a mechanical support).

The endcap calorimeter is located downstream of the preradiator and provides energy measurements of the γ-rays. The expected energy resolution for the calorimeter is 2–$4\%/\sqrt{E(GeV)}$ with a thickness of 18 radiation length. The KLOE group at DAΦNE has constructed a large lead/scintillator calorimeter consisting of thin (0.5 mm) corrugated lead sheets alternating with planes of 1-mm diameter scintillating fibers [5]. They achieved energy resolution of $4.4\%/\sqrt{E(GeV)}$ that was limited by sampling fluctuation. In the present experiment, reduction of the thickness of lead by a factor of three (0.17 mm) is expected to increase the visible fraction of light and improves resolution to $2.8\%/\sqrt{E(GeV)}$.

In order to veto on extra γ-rays, the decay volume is surrounded by the endcaps and a cylinder of a 18 radiation-length thick sampling calorimeter. The beam hole is covered by a "catcher" detector located 15 m downstream from the decay volume. The catcher, layers of lead/lucite sandwiches, is designed to detect only γ-rays but to be insensitive to the copious neutrons in the beam line.

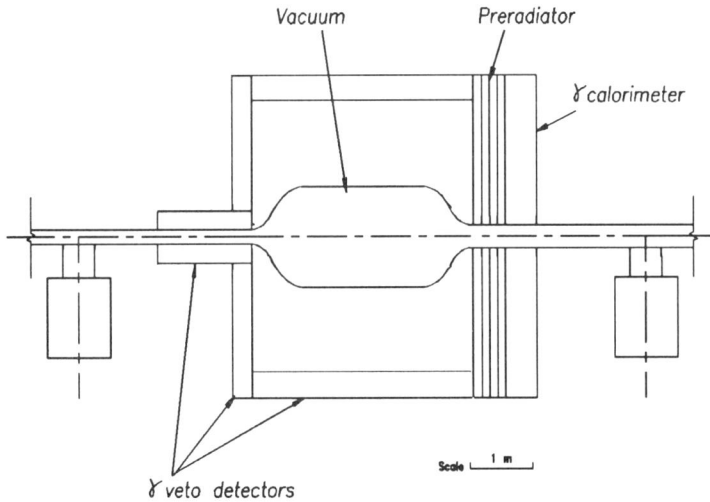

FIGURE 1. Side view of the detector.

SENSITIVITY AND BACKGROUND

Assuming 8000 hours of running time at 5×10^{13} protons/spill, 1.5×10^{14} K_L^0 decays in the fiducial volume are expected. The acceptance is estimated to be approximately 1.6 %, which includes the solid angle, photon conversion and reconstruction factors, phase space and other cuts. This results in 70 events, or a single event sensitivity of $\sim 4 \times 10^{-13}$.

A summary of possible backgrounds is shown in Table I. The largest potential background comes from the decay $K_L^0 \to \pi^0\pi^0$ ($BR \sim 10^{-3}$) when two γ-rays are undetected. The detection inefficiency for missing extra two γ-rays from this process is around 10^{-8} based on measurements at comparable energies [7]. In the case when the two detected γ-rays come from the same pion (even pairing), a significant difference between two- and three-body decays in the pion momentum in the rest mass frame gives tight kinematic constraints. In the case when the two detected γ-rays come from different pions (odd pairing), they are not reconstructed to a pion. Additional kinematic handle comes from missing energy and mass that reduce $K_L^0 \to \pi^0\pi^0$ background to a manageable level.

Table I. A summary of background estimations.

Modes	Background level (Normalized to $K_L^0 \to \pi^0\nu\bar{\nu}$)
$K_L^0 \to \pi^0\pi^0$	0.1
$K_L^0 \to \pi^\pm e^\mp \nu$	0.0001
$K_L^0 \to \gamma\gamma$	0.0006
$nA \to \pi^0 A'$	0.005
$\Lambda \to \pi^0 n$	0.0005

Since $K_L^0 \to \gamma\gamma$ is a two-body decay, there are strong kinematical constraints; e.g., the total K_L^0 energy is equal to the sum of two γ-ray energies, and the sum of the transverse momenta is zero.

The toughest charged mode to suppress is the decay $K_L^0 \to \pi^- e^+ \nu$ where both π^- and e^+ become invisible in the first counter by the reaction $\pi^- + p \to \pi^0 + n$ and annihilation in flight, respectively; the charged-particle vetoing inefficiency for this mode was estimated to be 2×10^{-7} [6]. In the present design, this background is well suppressed, together with charged-particle vetoing, by two kinematic constrains (the pion mass and the center-of-mass energy of two "photons").

When a high energy neutron interacts with the residual gas in the decay volume, it can produce a π^0 without leaving any other trace in the detector. For this source, kinematics again plays an important role. Further reduction comes from the difference in the acceptable momentum region for the same time window; the TOF of the acceptable K_L^0 momentum region $0.5 \leq P_{K_L} \leq 1.3$ GeV/c corresponds to neutron momentum region $1 \leq P_n \leq 2.4$ GeV/c, where the n/K_L^0 ratio drops by a factor of 5.

STATUS

The experiment was approved at BNL in 1996, and there are many tests and prototype detectors in progress. One crucial element for this experiment is the K_L^0 yield at $45°$. Cross section measurements are being done at Brookhaven at the time of this conference. More tests for beam bunching are planned in order to improve the timing resolution. Several prototype detectors, including the "catcher" veto counter, have been constructed and are tested in the beam.

REFERENCES

1. L. Littenberg, Phys. Rev. **D39** (1989) 3322.
2. Y. Grossman and Y. Nir, Phys. Lett. **B398** (1997) 163; G. Buchalla, these proceedings.
3. J. Belz (FNAL E799 collaboration), these proceedings.
4. I-H. Chiang et al., BNL AGS proposal E926, 1996.
5. A. Antonelli et al., Nucl. Instr. Method **A354** (1995) 352.
6. T. Inagaki (KEK E391 collaboration), KEK 96-181, 1997.
7. M.S. Atiya et al. (BNL E787 collaboration), Nucl. Instr. Method **A321**, 129 (1992).

CP and CPT Violation Searches with K_S Mesons

T. Alexopoulos§, C. Bhat*, D. Bergman†, T.J. Devlin†,
J. Doornbos‡, A. Eichenbaum†, A. Erwin§, P. Martin*,
S.R. Schnetzer†, S.V. Somalwar†1, R. Stone†, M. Thompson§,
G.B. Thomson†2, H. White*

Fermi National Accelerator Laboratory, Batavia, Illinois 60510
†*Rutgers University, Piscataway, New Jersey 08855*
‡*TRIUMF, Vancouver, B.C., Canada V6T 2A3*
§*University of Wisconsin, Madison, Wisconsin 53706*

(CPT Collaboration)

Abstract. We describe a proposal for a Fermilab Main Injector experiment designed to maximize the interference between K_L and K_S mesons near their production target. Of the many accessible physics topics, one of the most interesting is a test of CPT symmetry conservation (a measurement of the difference in phase between η_{+-} and ϵ) which is sensitive to the Planck scale. The measurement of several CP violation parameters, and of the branching ratios of rare K_S decays will also be possible. The experiment will use an RF-separated K^+ beam to make the K^0 beam by charge exchange.

The CPT experiment (P894) will exploit the unique features of Fermilab's Main Injector for an ambitious program of kaon physics. This experiment has a number of new features to it.

- Sensitivity at the level of the Planck scale: 1.2×10^{19} GeV. The main CPT test in the experiment comes through an accurate measurement of the phase difference between η_{+-} and ϵ. This probes the $K^0 - \overline{K}^0$ mass difference at the level $M_{K^0} - M_{\overline{K}^0} \sim M_K^2/M_{Planck}$.

- Sensitivity to CP violation parameters never before (or poorly) measured. We will be able to measure $\eta_{+-0}, \eta_{000}, \eta_{+-\gamma}$, and η_{+-} to much better accuracy than they are presently known.

---
[1] Conference Speaker
[2] Scientific Spokesperson

- Creation of a pure K^+ beam. We will be able to use the new Main Injector at Fermilab to make an RF-separated K^+ beam of unprecedented intensity and purity. This beam will also be used by the "CKM" experiment, and will be an extremely valuable facility for the laboratory.

- Creation of a pure K^0 Beam. It will be made by charge exchange from the K^+ beam. In addition it will be a short beam in which decays will be dominated by those of the K_S.

- We will be able to reduce the uncertainty in the Bell-Steinberger relation by several orders of magnitude.

- Sensitivity to x, the $\Delta S = \Delta Q$ rule violation parameter, 50 times better than it's presently known, by studying the time dependence of semileptonic decays.

- We will be able to search for rare decays of the K_S, such as $K_S \to \pi^0 e^+ e^-$.

The K_L/K_S system forms a finely balanced interferometer that can be affected by small perturbations like CP violation or CPT violation (if it exists). The experiment is designed to maximize this interference to search for these effects. It consists of an RF-separated K^+ beam that strikes a target at the entrance of a magnetized collimator which defines a short neutral beam coming from that target. The K^+'s make K^0's copiously by charge exchange, with only a very small component of $\overline{K}^0$'s. This is the situation that maximizes the interference between the K_S and K_L decays of the K^0 mesons in the beam. The detector consists of a standard Vee spectrometer, a lead glass electromagnetic calorimeter, and a muon detector.

I CPT IN K^0 DECAYS

The CPT theorem is based on the assumptions of locality, causality, Lorentz invariance, the spin-statistics theorem, and the assumption of asymptotically free wave functions. All quantum field theories (including the standard model of the elementary particles) assume CPT symmetry invariance. But there is a theoretical hint of the level at which CPT symmetry might be violated. This comes from the fact that gravity cannot be included in a quantum field theory. Many physicists think that there must be a more general theory that has quantum field theory embedded in it. In this more general theory CPT symmetry may be violated.

One expects to see effects of quantum gravity at what is called the Planck scale: at energies of $M_{Planck}c^2 = \sqrt{\hbar c^5/G} = 1.2 \times 10^{19}$ GeV, or at distances of the order of 10^{-33} cm. Since it is hard to see such effects in ordinary processes, one would look in a place where quantum field theories predict a null effect, then if something is observed it could be ascribed to quantum gravity.

Therefore, it would be very interesting to test CPT symmetry conservation at the Planck scale.

In K^0 physics, one can observe CPT violating effects through mixing or decays (called indirect or direct CPT violation). In mixing, one introduces a parameter Δ which is both CP and CPT violating:

$$\begin{cases} K_S = K_1 + (\epsilon + \Delta)K_2 \\ K_L = K_2 + (\epsilon - \Delta)K_1 \end{cases} \qquad (1)$$

There are several measurements that would signify CPT violation:

- A difference between the phase of ϵ and the phase of η_{+-}.
- Certain interference terms between K_L and K_S in semileptonic decays.
- Evidence for a non-zero Δ in the Bell-Steinberger relation.
- A difference between the phases of η_{+-} and η_{00}.

In this report we will concentrate on the first method, measuring the phase of η_{+-} and comparing it to the calculated value of the phase of ϵ.

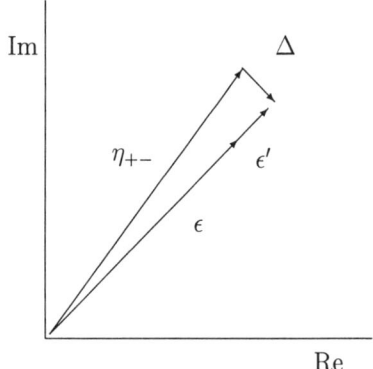

The figure above shows the relationships between $\epsilon, \epsilon', \Delta$, and η_{+-}. As shown, ϵ' and Δ are greatly enlarged for clarity. The size of $|\epsilon'/\epsilon|$ is of order 10^{-4}, and the phase of ϵ' is very close to that of ϵ, so the phase of the vector $\epsilon + \epsilon'$ is the same, to good accuracy, to the phase of ϵ. We can see from the figure that the component of Δ perpendicular to ϵ, $\Delta_\perp$, is

$$\Delta_\perp = |\eta_{+-}|(\phi_{+-} - \phi_\epsilon) \qquad (2)$$

where ϕ_{+-} (ϕ_ϵ) is the phase of $\eta_{+-}(\epsilon)$. In general, in terms of the elements of the kaon decay matrix Γ and mass matrix M, Δ is given by:

$$\Delta = \frac{(\Gamma_{11} - \Gamma_{22}) + i(M_{11} - M_{22})}{(\Gamma_S - \Gamma_L) + 2i(M_L - M_S)} \quad (3)$$

The mass term has a phase perpendicular to ϵ and the decay term is parallel to ϵ. We can solve Eqns. 2 and 3 for $M_{11} - M_{22}$, which is the mass difference between the K^0 and $\overline{K^0}$ mesons:

$$\frac{|M_{K^0} - M_{\overline{K^0}}|}{M_{K^0}} = \frac{2(M_L - M_S)}{M_{K^0}} \frac{|\eta_{+-}|}{\sin \phi_{SW}} |\phi_{+-} - \phi_\epsilon| \quad (4)$$

where $\tan \phi_{SW} = 2(M_L - M_S)/(\Gamma_S - \Gamma_L)$.

In Eqn. 4, $(M_L - M_S)$ is 10^{-6} eV, and when one divides by M_{K^0} the ratio is 7×10^{-15}. $|\eta_{+-}|$ is 2×10^{-3}. These factors, at current accelerator energies, let us approach the Planck scale.

$$\frac{|M_{K^0} - M_{\overline{K^0}}|}{M_{K^0}} = \frac{M_{K^0}}{M_{Planck}} = 4.1 \times 10^{-20} \quad (5)$$

II EXPERIMENTAL METHOD

The best experimental limit on CPT violation came from Fermilab experiment E773. This limit is (at 90% confidence level),

$$\frac{|M_{K^0} - M_{\overline{K^0}}|}{M_{K^0}} < 1.3 \times 10^{-18} \quad (6)$$

The E773 result stands at 31 times the Planck scale.

In the KTeV experiment we expect to make an improvement of a factor of 3 to 5. The size of the interference term, $2|\eta_{+-}||\rho|\cos(\Delta mt + \phi_\rho - \phi_{+-})\exp(-t/2\tau_s)$, is limited by the regeneration amplitude $|\rho| \simeq 0.03$. Further, ϕ_1 and ϕ_ρ are hard to disentangle. KTeV is probably pushing this method as far as it can go.

After KTeV we will remain a factor of 6 to 10 away from the Planck scale. To close that gap we propose an experiment near the kaon production target. The interference term is then $2D|\eta_{+-}|\cos(\Delta mt - \phi_{+-})\exp(-t/2\tau_s)$. Here ϕ_{+-} appears alone, and $|\rho|$ is replaced with the dilution factor, $D = (K^0 - \overline{K}^0)/(K^0 + \overline{K}^0)$ at the target. To maximize D, we will make our K^0 beam from a K^+ beam by charge exchange. Then at medium to high Feynman x, $D \simeq 1$. The charge exchange cross section is large, about 20% of the total cross section. To maximize the flux of K^+ made from the 120 GeV/c protons from the Fermilab Main Injector we choose a K^+ momentum of 25 GeV/c. We would use a magnetic collimator to define the K^0 beam, similar to the one in the Proton Center beam line at Fermilab. In the calculations described below we will assume the use of a vee spectrometer, a lead glass electromagnetic calorimeter, and a muon detector.

III PRECISION

E773 measured ϕ_{+-} to 1 degree accuracy. To reach the Planck scale we must achieve 0.03 degree accuracy. We have calculated the statistical sensitivity of the experiment assuming a beam of $2 \times 10^8 K^+$ per spill, striking a 10 cm tungsten target at 9 mrad, for 1×10^7 seconds, with the spill structure of the Fermilab Main Injector, a solid angle of 36 μster for the K^0 beam, a magnetic collimator 1.5 meters thick, a 13.5 meter long decay region, followed by a vee spectrometer, a lead glass electromagnetic calorimeter, and a muon detector.

We calculated the distribution of events in momentum and proper time for the resulting 15-billion $K^0 \to \pi^+\pi^-$ events. The fitting parameters were $|\eta_{+-}|, \phi_{+-}, D$, and three parameters describing the normalization and shape of the kaon momentum spectrum. The uncertainty in ϕ_{+-} was 0.02 degrees, better than is needed to place a 90% confidence limit on CPT symmetry violation at the Planck scale.

IV CP-VIOLATION MEASUREMENTS

In this experiment we will be able to measure several CP-violation parameters much more accurately than they are now known. η_{+-0} and η_{000} are the amplitude ratios for $(K_S \to 3\pi)/(K_L \to 3\pi)$. To date every decay where CP violation has been seen is of the K_L meson, and all CP violation parameters are measurements of the $\epsilon - \Delta$ part of the K_L as shown in Eqn. 1. Our measurements of η_{+-0} and η_{000} would be the first observation of CP violation outside the K_L system. If only indirect CP violation existed η_{+-0} and η_{000} would be equal to $\epsilon + \Delta$. Because Δ enters with opposite sign this would be an interesting test of CPT conservation. The contribution of direct CP violation in these decays is relatively large: up to 10% is predicted in the standard model, with a known phase. These CP violation parameters are also the place where the largest contribution is predicted from beyond-the-standard-model effects. For example, the right-left symmetric model, where there are right-handed intermediate bosons as well as left-handed ones, predicts a 50% difference between η_{+-0} and η_{+-}.

$\eta_{+-\gamma}$ is the ratio of the CP-violating part of the $K_L \to \pi^+\pi^-\gamma$ decay amplitude to that of the K_S. This decay proceeds by either an inner bremsstrahlung photon being emitted from one of the final state pion lines, or by a direct emission photon that comes from the decay vertex. The K_S decay is CP conserving and is dominated by the inner bremsstrahlung amplitude. The part of the K_L decay coming from inner bremsstrahlung is supressed because it is CP violating, allowing a CP-conserving direct emission M1 contribution to be seen clearly. The parameterization used for this decay is: $\eta_{+-\gamma} = \epsilon + \epsilon'_{+-\gamma}$. The standard model predicts that the direct CP-violating contribution to this decay ($\epsilon'_{+-\gamma}/\epsilon$) be of order 10^{-2}. This is considerably larger than ϵ'/ϵ for the

2π decays, which is about 3×10^{-4}, because there is no $\Delta I = 1/2$ rule suppression in $\epsilon'_{+-\gamma}$. The product of branching ratio times direct CP violation size is about the same for the $\pi^+\pi^-\gamma$ decay as for the 2π decays that are used to measure ϵ'/ϵ, so one might think that the difficulty of measuring direct CP violation would be the same using the two methods. However, the 2π method is much more difficult because of the nothing-in-nothing-out character of the $K_L \to \pi^0\pi^0$ decay.

η_{+-} was the first CP violation parameter to be measured, but it is only known to 1% accuracy. Measuring the phase of η_{+-} very accurately is part of one CPT symmetry conservation test in this experiment. Measuring the magnitude of η_{+-} to about 0.1% accuracy would allow us to perform another, equally powerful, CPT conservation test using the Bell-Steinberger relation. This relation is derived from the conservation of probability, and, through Eqn. 1, it includes the CPT violation parameter Δ. To evaluate the Bell-Steinberger relation one also has to know η_{00} to about the same accuracy as η_{+-}, but we do not have to measure η_{00} since we will be able to constrain it using the KTeV measurement of ϵ'.

V RARE DECAYS OF THE K_S

A great deal of attention has been paid to measurements of rare decays of the K_L meson, but little work has been done on the K_S. Some of the most interesting searches (for K_L and K_S both) are for decays that test strangeness-changing neutral currents: $K_S \to \pi^0 e^+ e^-, \mu^+\mu^-$, and e^+e^-. Of these the first is the most interesting because it has another use as well. The decay $K_L \to \pi^0 e^+ e^-$ is an important decay because direct CP violation is expected to contribute about 1/3 of the rate. The other 2/3 come from indirect CP violation and from a CP conserving contribution. Each of these contributions should be approximately equal. The only drawback to using this decay to study direct CP violation is that the branching ratio is small, perhaps 1×10^{-11}. To measure the direct CP violating rate the other two contributions must be subtracted. The CP conserving rate can be calculated accurately, but theoretical calculations of the rate for the indirect CP violating part of the K_L decay have very large uncertainties. The only way to learn the size of this contribution is to measure the branching ratio of the K_S decay. Then a simple calculation results in the K_L rate.

VI THE RF-SEPARATED K^+ BEAM

One of the novel aspects of the CPT experiment is its use of an RF-separated K^+ beam. In building such a beam one must solve two main problems: designing the optics of the beam to accomplish the separation, and building the necessary RF cavities. In addition there are two experiments proposed for the Main Injector that require K^+ beams, the "CKM" experiment and the present one. A great saving could occur if the same beam could be used for both experiments.

One of us (J.D.) designed the optics of such a beam for the proposed KAON accelerator at TRIUMF, and has adapted the design for this application. The goals of the beam design are to have a a flux of 2×10^8 25 GeV/c K^+/spill, with 5×5 mm^2 spot size (for the CPT experiment), a flux of 3×10^7 22.8 GeV/c K^+/spill, with 50 - 100 μrad divergence in x and y (for the CKM experiment), an impurity $\leq 10\%$, and to have a simple change-over between the CPT and CKM experiments.

We have a beam design that accomplishes all these goals. With 5×10^{12} Main Injector protons incident per spill, we can achieve the necessary intensity, spot size, and divergence for each experiment. The resulting beam purity is about 99%. Of the K^+ mesons that enter the acceptance of the beam, about 40% strike the beam stopper that eliminates the pions and protons, and the beam is about 1 K^+ lifetime long. To switch from one experiment to the other, the polarity of one bending magnet is reversed, slit openings are changed, some magnet currents adjusted, the stopper changed (from 1.5 cm thickness for CPT to 0.5cm for CKM), and two magnets are moved by about a foot. The whole changeover should take about a shift.

VII CONCLUSION

The Main Injector at Fermilab is a very powerful accelerator. It operates in a new energy regime, and its large flux makes possible a new range of experiments in many areas. The Main Injector era will be the first time that it will be possibile to build a high intensity secondary beam of pure K^+ mesons (and a tertiary beam of pure K^0 mesons, and even a short-lived K^0 beam). The CPT experiment seeks to exploit this possibility to do an experiment with the largest possible interference between K_S and K_L mesons. We will test CPT symmetry conservation at the Planck scale, measure CP violation parameters, test the $\Delta S = \Delta Q$ rule, and search for rare K_S^0 decays.

Detecting K Mesons Leptonic Decays with KLOE

Patrizia de Simone*
Representing the KLOE Collaboration

*Laboratori Nazionali di Frascati dell'INFN, Frascati, Italy

Abstract. The KLOE detector is presently under construction at DAΦNE, the Frascati $\phi-factory$, with the aim of studying $\mathcal{CP}$ violation in the $K^\circ \bar{K}^\circ$ system. We discuss the measurements of the slope of the scalar form factor λ_o for the K's, which will be done at the DAΦNE start up time when the expected luminosity will be $\mathcal{L}_o = 10^{32}$ cm^{-2}s^{-1}. At the design luminosity $\mathcal{L}_o = 10^{33}$ cm^{-2}s^{-1} about 4.2×10^{10} kaons will be produced per year: this will be a unique opportunity to collect enough K_{l4} decays which provides the only means for the determination of the $\pi\pi$ phase shifts.

ϕ PRODUCTION AT DAΦNE

The Frascati $\phi - factory$ DAΦNE (Double Annular $\phi - factory$ for Nice Experiments) [1] is an e^+e^- collider optimized for operation at a center of mass energy equal to the ϕ mass.

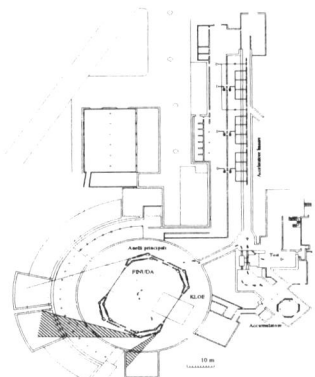

FIGURE 1. The DAΦNE machine complex.

TABLE 1. The DAΦNE machine parameters.

E_{beam}(MeV)	510
energy spread (MeV)	0.4
$\mathcal{L}_o$(cm^{-2}s^{-1})	$10^{32} \rightarrow 10^{33}$
σ_x(cm)	2.1×10^{-1}
σ_y(cm)	2.1×10^{-3}
σ_z(cm)	3.
particles/bunches	8.9×10^{10}
N bunches	$30 \rightarrow 120$
time between collisions	$10.8 \rightarrow 2.7$ ns
crossing half angle	10 to 15 mrad
commissioning date	end of 1997

TABLE 2. ϕ decays (M_ϕ = 1019.41 MeV, Γ_ϕ = 4.41 MeV, Γ_{ee} = 1.37 KeV, $J^{PC} = 1^{--}$)

Mode	Br(%)	β_K	$\beta\gamma c\tau_K$ (cm)	p_{lab}(MeV/c)	# per year
K^+K^-	49.5	0.249	95.4	127	2.5×10^{10}
$K_S K_L$	34.4	0.216	343.8 (K_L)	110	1.7×10^{10}
			0.59 (K_S)		
$\rho\pi$	12.9	-	-	182	6×10^9
$\pi^+\pi^-\pi^o$	1.9	-	-	462	1×10^9
$\eta\gamma$	1.28	-	-	362	6×10^8
other	$\sim 1.$	-	-	-	5×10^8

The machine complex of DAΦNE is shown in Fig. 1. The initial luminosity is expected to be 10^{32}cm^{-2}s^{-1}. In one or two years the luminosity will reach 10^{33}cm^{-2}s^{-1}. The DAΦNE machine parameters are listed in Table 1. At the target luminosity of DAΦNE, given that the cross section for $e^+e^- \to \phi$ at the ϕ resonance peak is $\simeq 5\mu$b, about 5000 ϕ's will be produced per second. The vector meson ϕ decays into other mesons with the Br's given in Table 2 together with other relevant kinematic quantities, as well as the number of particles produced per year. One note therefore that DAΦNE is indeed a kaon-factory, and because the ϕ decays at rest the kaons are produced in pairs collinear and monocromatic, the observation of one kaon guarantees the existence of the other with determined direction and identity, i.e. K's can be tagged, and an absolute normalization of the K_S, K_L fluxes is available.

THE KLOE DETECTOR

The main physics motivation of the KLOE experiment is the measurement of the $\mathcal{CP}$ violation parameters to an accuracy of $\mathcal{O}(10^{-4})$. This will be done with the classical method of the double ratio and also by kaon interferometry experiments [2].
Concerning the $\mathcal{CP}$ studies, the main tasks of the KLOE detector are the control of the efficiencies for the decays of interest, and the rejection of the background from the copious K_L decays to states other than $\pi\pi$. The main components of the KLOE detector are the drift chamber [3] and the electromagnetic calorimeter (EmC) [4], both embedded in a

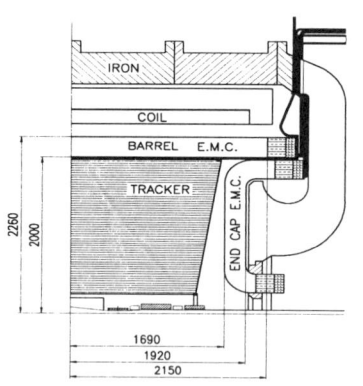

FIGURE 2. Cross section along the beam axis of 1/4 of the KLOE detector.

magnetic field of 0.6 T, generated by a superconducting solenoid (see Fig. 2). Because of the decay length of K_L's, to maximize the fiducial volume for the charged decay detection, the chosen radius of the drift chamber is 2 m. The K_L decay vertices are uniformly distributed in the tracking volume, and to obtain high and uniform track and vertex reconstruction efficiency the wires are in an all-stereo configuration in order to have an uniform filling of the chamber with square cells (2×2 cm^2 for the first 12 layers, and 3×3 cm^2 for the 46 outer layers). The expected resolution of the single position measurements is $\sigma_{r,\phi} \leq 200$ μm; this will give us the possibility to measure the vertices with a resolution at least of 500 μm in(r, ϕ), and of 2 mm in z. A very good momentum resolution is required, $\sigma_{p_\perp}/p_\perp \simeq 0.5\%$, and because of the low momenta of the particles ($50 \geq p \geq 300$ MeV/c), a helium based gas mixture has been chosen to minimize the multiple scattering ($90\% He - 10\% iC_4 H_{10}$). Also, the KLOE chamber needs all its walls to be very thin in order for them to be transparent to γ's from π^o decays to be detected in the calorimeter; the whole mechanical structure is made of carbon fiber.

The main tasks of the KLOE EmC are the measurements of the neutral vertices with a resolution of $\simeq 1$ cm, and the rejection of the background due to the $3\pi^o$ mode. To reach this purposes the EmC has to provide good energy resolution, excellent timing performance, resolution on the γ apices measurement of $\simeq 1$ cm, high efficiency for the detection of low energy γ (down to 20 MeV) and hermetic coverage. The KLOE EmC is a very fine sampling lead (0.5 mm thick) scintillating fiber(1 mm) calorimeter with photomultipliers read-out, covering $\sim 98\%$ of the solid angle. The barrel is a cylinder of 4 m inner diameter and 4.3 m length, and is made of 24 modules 23 cm thick, with fibers running parallel to the beam line. The end-caps are made of 32 C shaped modules of variable length with fibers running perpendicular to the beam line. Test done with prototypes and with the final barrel modules gave the following results for the relevant parameters: $\sigma_E/E \simeq 4.7\%/\sqrt{E(GeV)}$, and $\sigma_t \simeq 55$ ps$/\sqrt{E(GeV)}$.

K_{l3} DECAYS

A convenient parametrization of the K_{l3} matrix element of the hadronic current is in term of the form factors $f_+(t)$ and $f_o(t)$ [5], where t is the square of the 4-momentum transfer to the leptons. Analyses of K_{l3} data frequently assume a linear dependence $f_{+,o}(t) = f_{+,o}(0)(1 + \lambda_{+,o}\frac{t}{m_\pi^2})$. The measurement of the Dalitz plot distribution, $\rho(E_\pi, E_l)$, of K_{l3} data allows to determine the form factors in the range $m_l^2 \leq t \leq (M_K - m_\pi)^2$.

Tables 3 and 4 show respectively the measured average values [1] and the

---

[1] Note that the form factor f_o cannot be determined from the K_{e3} mode because its contribution to the decay probability is suppressed by the factor $(m_e/M_K)^2 \simeq 10^{-6}$.

TABLE 3. $\lambda_{+,o}$ determined with the Dalitz plot analysis of the $K_{\mu 3}$ and K_{e3} data.

	$K^o_{\mu 3}$	$K^\pm_{\mu 3}$	K^o_{e3}	$K^\pm_{e3}$
λ_+ [6]	0.034 ± 0.005	0.032 ± 0.008	0.0300 ± 0.0016	0.0285 ± 0.0043
λ_0 [7]	the $K_{\mu 3}$ data analysis done by Chounet et al. favors a negative slope for $f(t)$		-	-
λ_0 [6]	0.025 ± 0.006	0.004 ± 0.007		

TABLE 4. Theoretical predictions for $\lambda_{+,o}$.

	Fubini-Furlan	Callan-Treiman	χ_{PT} theory
λ_+	-	-	0.031
λ_0	-0.025 ± 0.013	0.017 ± 0.013	0.017

theoretical predictions of the parameters λ_+ and λ_o.
The average values of λ_+ measured with the neutral and the charged $K_{\mu 3}$ decays agree, are fairly consistent with the results from the analysis of the K_{e3} data, and agree with the chiral theory (χ_{PT}) prediction. The situation is completely different for the λ_o parameter; the experimental results gives different indications that favor different theoretical predictions. Nowadays, the χ_{PT} predictions for λ_o are generally accepted as being correct, however new, precise measurements will help to determine this still unknown parameter.

The number of collected $K_L \to \pi^\pm \mu^\mp \nu_\mu$ events used to calculate the average value of the λ_o parameter shown in Table 3 is $\simeq 4. \times 10^6$ [6]. The increase in statistic obtainable in the first year of running at DAΦNE is $\simeq \times 14$:

$$N(K^o_{\mu 3}) = \underbrace{5 \times 10^9}_{N\phi} \times \underbrace{.334}_{Br(\phi \to K_S K_L)} \times \underbrace{.27}_{Br(K^o_{\mu 3})} \times \underbrace{(e^{-40/350} - e^{-170/350})}_{a_{FV}} \times \underbrace{.44}_{\epsilon_V} \simeq 5.6 \times 10^7$$

where a_{FV} is the K_L acceptance, and ϵ_V is the 0th-order vertex reconstruction efficiency.

The first step to reconstruct one $K^o_{\mu 3}$ event is the measurement of the $\pi\mu$ vertex coordinates from which the K_L direction can be determined (the expected resolution on the K_L momentum is $\sigma(P_K) \simeq 3$ MeV). Then, the momentum and the energy of the $\pi\mu$ pair in the kaon rest frame can be evaluated. Without any particle identification applied there are two possible pairs of energy values ($E_{\pi 1} = \sqrt{p_1^2 + m_\pi^2}$; $E_{\mu 1} = \sqrt{p_2^2 + m_\mu^2}$; $E_{\pi 2} = \sqrt{p_2^2 + m_\pi^2}$; $E_{\mu 2} = \sqrt{p_1^2 + m_\mu^2}$) and the probability density function f, that is, the Dalitz function $\rho(E_\pi, E_l)$ normalised to the whole space phase, becomes

$$f_{sim}(E_\pi, E_\mu; \lambda_+, \lambda_0) = \frac{1}{2}[f(E_{\pi 1}, E_{\mu 1}; \lambda_+, \lambda_0) + f(E_{\pi 2}, E_{\mu 2}; \lambda_+, \lambda_0)]$$

To have a chance that f_{sim} will describe correctly the Dalitz plot distribution, the measured number of $K^o_{\mu 3}$ decays in each bin of the (E_π, E_μ) plane has to be weighted with ϵ_V. This will be controlled at KLOE at

TABLE 5. Predicted $\sigma(\lambda_o)$ from $K^o_{\mu 3}$.

	$\sigma(\lambda_o)$
KLOE	$0.96/\sqrt{N1} = 1.2 \times 10^{-4}$
	$0.96/\sqrt{N2} = 6.4 \times 10^{-5}$
Donaldson [8]	$4. \times 10^{-3}$

TABLE 6. Predicted $\sigma(\lambda_o)$ from $K^\pm_{\mu 3}$.

	$\sigma(\lambda_o)$
KLOE	$0.73/\sqrt{N1} = 1.5 \times 10^{-4}$
	$0.73/\sqrt{N2} = 7.3 \times 10^{-5}$
Merlan [9]	1.5×10^{-2}

the level of $\Delta\epsilon/\epsilon \simeq 10^{-3} \div 10^{-4}$ using, for example, the very copious K_L (K_S) $\to \pi^+\pi^-\pi_o$ ($\pi^+\pi^-$) events ($\simeq 9.5 \times 10^7$ in 1 year of running). Then, there will be some contamination due to the $K_L \to \pi^+\pi^-$ events ($Br(K_{\mu 3})/Br(K_L \to \pi^+\pi^-) \simeq 133$) that will be easily removed through kinematic cuts.

The "a priori" error estimates on λ_o have been calculated and are shown in Table 5 for two samples of $K^o_{\mu 3}$ events; the first sample (N_1) corresponds to the DAΦNE initial luminosity, while the second one (N_2) corresponds to a later luminosity of 5×10^{32}cm^{-2}s^{-1}.

The number of collected $K^\pm \to \pi^o\mu^\pm\nu_\mu$ events used to evaluate the λ_o avarage value shown in Table 3 is $\simeq 10^5$ [6]. The increase in statistic obtainable in the first year of running at DAΦNE is a factor of $\simeq 250$:

$$N(K^+_{\mu 3}) = \underbrace{5 \times 10^9}_{N\phi} \times \underbrace{.495}_{Br(\phi \to K^+K^-)} \times \underbrace{.032}_{Br(K^+_{\mu 3})} \times \underbrace{.71}_{\epsilon_{\pi^o}} \times \underbrace{.44}_{\epsilon_V} \simeq 2.5 \times 10^7$$

The steps to reconstruct one charged $K_{\mu 3}$ event are the measurement of the $K\mu$ vertex coordinates, and of the energies and apices of the photons from the π^o. The main background is due to $K^\pm \to \pi^\pm\pi^o$: $Br(K^\pm \to \pi^\pm\pi^o)/Br(K^\pm \to \pi^o\mu^\pm\nu) \simeq 6.7$. To identify and to reject this background, some constrained kinematic fit techniques will be applied, and taking advantage of the good timing performance of the KLOE EmC, the $\pi\mu$ discrimination can be accomplished using TOF measurements. The Table 6 shows the evaluated accuracies on the λ_o parameter for two samples of $K^\pm_{\mu 3}$ events corresponding to the initial (N_1) and to a later (N_2) DAΦNE luminosity. The "a priori" evaluation of $\sigma(\lambda_o)$ in the $K^\pm_{\mu 3}$ case considers a perfect subtraction of the $K^\pm \to \pi^\pm\pi^o$ background.

K_{l4} DECAYS

The K_{l4} decays have a partial rate decay of the form

$$d^5\Gamma = G_F^2|V_{us}|^2 N(s_\pi, s_l) J_5(s_\pi, s_l, \theta_\pi, \theta_l, \phi) ds_\pi ds_l d(\cos\theta_\pi) d(\cos\theta_l) d\phi$$

where s_π, s_l, θ_π, θ_l and ϕ are a set of kinematic variables necessary to describe the K_{l4}, defined in ref. [5]. The quantity J_5 can be written as an expansion of simple functions of ϕ and θ_l multiplying nine intensities

TABLE 7. K_{l4} statistic in 1 year of running at $\mathcal{L}_o = 5 \times 10^{32}$ cm^{-2}s^{-1}.

channel	Br	# events(PDG)	DAΦNE	impr.
1) $K^\pm \to \pi^+\pi^- e^\pm \nu_e$	3.9×10^{-5}	3×10^4	3×10^5	10
2) $K^\pm \to \pi^+\pi^- \mu^\pm \nu_\mu$	10^{-6}	7	8.4×10^3	1.2×10^3
3) $K^\pm \to \pi^0\pi^0 e^\pm \nu_e$	2.1×10^{-5}	< 50	2×10^5	$> 4 \times 10^3$
4) $K_L \to \pi^0\pi^\pm e^\mp \nu_e$	5.2×10^{-5}	729	5×10^4	68

$I_i(s_\pi, s_l, \theta_\pi, F, G, H, R)$. The explicit dependences may be found in ref [5]. The form factors F, G, H, and R can be written in a partial wave expansion in the variable θ_π. The partial wave amplitudes f_l, g_l, r_l, and h_l depend on s_π and s_l, and their phases coincide with the phase shifts δ_l^I in elastic $\pi^+\pi^-$ scattering. Introducing the partial wave expansions of the form factors in the $I_i (i=1,...,9)$, one can get the expression of the intensity distribution J_5 in terms of the $\pi^+\pi^-$ phase shifts, which allows their determination as a function of s_π in a global analysis, together with the form factors f_s, f_p, etc. The decay $K^+ \to \pi^+\pi^- e^+ \nu_e$ has been already used by L. Rosselet et al. to determine the δ_l^I and the related isoscalar S-wave scattering length a_o^o [10]: $a_o^o = 0.26 \pm 0.05$. This result must be compared with the $SU(2)_R \times SU(2)_L$ prediction: $a_o^o = 0.20 \pm 0.01$. Low energy $\pi^+\pi^-$ scattering is one of the few places where chiral symmetry allows a precise prediction within the framework of QCD. The standard picture of the vacuum structure in QCD [11] would have to be revised, should the central value $a_o^o = 0.26$ be confirmed with a substantially smaller error.

The Table 7 shows the number of K_{l4} events collected until now for each observed channel [6], and the number of K_{l4} events that will be produced at DAΦNE in 1 year of running at the luminosity of 5×10^{32} cm^{-2}s^{-1}.

The reconstruction of the channels 1 and 2 requires the measurement of the 4-prong $K^\pm \pi^+ \pi^- l^\pm$ vertex coordinates. The main background for these channels is due to the $K^\pm \to \pi^+\pi^-\pi^\pm$ events (Br $\simeq 5.59\%$). The identification of the channel 3 requires the reconstruction of the $K^\pm e^\pm$ vertex, and of 2 + 2 e.m. clusters in the EmC not associated at any charged tracks, and then the γ's pairing into π^o's. The main background for this channel is due to the $K^\pm \to \pi^o\pi^o\pi^\pm$ events (Br $\simeq 1.73\%$). Finally, the identification of the channel 4 requires the reconstruction of the $\pi^\pm e^\mp$ vertex, from which the K_L flight direction can be determined, and of 2 e.m. clusters in the EmC not associated at any charged tracks, and then the γ's pairing into π^o's. The background related to this channel is due to $K_L \to \pi^+\pi^-\pi^o$ events (Br $\simeq 12.33\%$). A part from channel 2, the backgrounds will be easily removed thanks to the big difference on mass between pions and electrons.

The experimental accuracies of the parameters describing the $K^\pm \to \pi^+\pi^- e^\pm \nu_e$ decay has been evaluated using the maximum likelihood method [12]. In particular, concerning the $\pi^+\pi^-$ phase shifts, the error on the isoscalar S-wave scattering length has been estimated using the parametrization introduced by J.L.Basdevant et al. [13]. For 30000 K_{e4} events the estimated error is $\delta a_o^o = 0.029$ to be compared with 0.05 in ref. [10]. This estimate at the "Rosselet statistics" is purely statistical, but apply to a "perfect" detector, i.e., one which covers the whole phase space with unity efficiency everywhere. This is close to being true for KLOE which is a hermetic detector, operating at DAΦNE producing self-tagging $K^\pm$ pairs, with high reconstruction efficiency of neutral and charged low energy particles that will be controlled at the level of $\simeq 10^{-3} \div 10^{-4}$. So, in one year of running at the luminosity of 5×10^{32} cm^{-2}s^{-1}, a factor of 5 improvement in the error on the $\pi^+\pi^-$ scattering length is expected: $\delta a_o^o = 0.01$.

CONCLUSIONS

The capabilities of the KLOE detector and the over constrained kinematic of the ϕ's decays at DAΦNE, will make possible in the first year of running (10^{32}cm^{-2}s^{-1}) to measure the scalar form factor λ_o in the $K_{\mu 3}^o$ decays with an error $\simeq 30$ times smaller than the error reached in previous experiments. When $\mathcal{L}_o = 5 \times 10^{32}$cm^{-2}s^{-1} will be reached we expect, that the $\pi^+\pi^-$ scattering length a_o^o should be measurable to an accuracy of about 0.01, sufficient to determine if the existing discrepancy between present measurements and predictions is statistically significant.

REFERENCES

1. G.Vignola, Proc. Workshop on Physics and detectors for DAΦNE, ed. G.Pancheri (Frascati,1991) p.11.
2. J.Lee-Franzini, **The second DAΦNE Physics Handbook**, ed. L.Maiani (Frascati,1995) p.761.
3. A.Calcaterra et al., Nucl. Instr. and Meth. A 367 (1995) 104-107.
4. J.Lee-Franzini et al., Nucl. Instr. and Meth. A 360 (1995) 201-205.
5. J.Bijnens et al., **The second DAΦNE Physics Handbook**, ed. L.Maiani (Frascati,1995) p.315.
6. M.Aguilar-Benitez et al.,*Review of Particle Properties*, Phys. Rev. D45, (1994).
7. L.M.Chounet et al., Phys. Lett. 32B (1972) 201.
8. Donaldson et al., Phys. Rev. D9 (1974) 2960.
9. Merlan et al., Phys. Rev. D9 (1974) 107.
10. L.Rosselet et al., Phys. Rev. D15 (1977) 574.
11. M.Gell-Mann et al., Phys. Rev. 175 (1968) 2195.
12. M.Baillargeon and P.J. Franzini, **The Second DAΦNE Physics Handbook** ed. L.Maiani (Frascati,1995) p.413.
13. J.L. Basdevant et al., Nucl. Phys. B72 (1974) 413.

Measurement of the Michel Rho Parameter in Direct Muon Decay

Leo Piilonen,[a] J.F. Amann,[b] R.D. Bolton,[b] Y. Chen,[c]
M.D. Cooper,[b] P.S. Cooper,[d] M. Dzemidzic,[c] W. Foreman,[b]
C.A. Gagliardi,[e] D. Haim,[a] R. Harrison,[b] G. Hart,[b]
G.E. Hogan[b] E.V. Hungerford III,[c] C.C.H. Jui,[f] J.E. Knott,[g]
D.D. Koetke,[h] T. Kozlowski,[b] M.A. Kroupa,[b] K. Lan,[c]
F.S. Lee,[a] F. Liu,[e] R. Manweiler,[h] B.W. Mayes II,[c]
R.E. Mischke,[b] C. Pillai,[b] L. Pinsky,[c] S. Schilling,[b]
T.D.S. Stanislaus,[h] K.M. Stantz,[g] J.J. Szymanski,[g]
R.E. Tribble,[e] X.L. Tu,[e] L. A. Van Ausdeln,[e] W. von Witsch,[c]
D. Whitehouse,[b], B.K. Wright,[i] S.C. Wright,[j] Y. Zhang,[a]
and K.O.H. Ziock[i]

[a] *Virginia Polytechnic Institute and State University, Blacksburg, VA 24061*
[b] *Los Alamos National Laboratory, Los Alamos, NM 87545*
[c] *University of Houston, Houston, TX 77004*
[d] *Fermi National Acceleratory Laboratory, Batavia, IL 60510*
[e] *Texas A&M University, College Station, TX 77843*
[f] *University of Utah, Salt Lake City, UT 84112*
[g] *Indiana University, Bloomington, IN 47405*
[h] *Valparaiso University, Valparaiso, IN 46383*
[i] *University of Virginia, Charlottesville, VA 22901*
[j] *University of Chicago, Chicago, IL 60637*

Abstract. We report on the status of LAMPF experiment E-1240 to measure the Michel ρ parameter in direct muon decay. This experiment ran in 1993, and the data are currently being analyzed. The expected precision on the ρ parameter is ± 0.0008. This result will provide better constraints on new physics, particularly on the charged vector bosons' mixing angle ζ in the manifestly left-right symmetric extension of the Standard Model.

INTRODUCTION

The RHO experiment at the Clinton P. Anderson Meson Physics Facility (LAMPF) is a precise measurement of the Michel ρ parameter [1] in the direct decay of the muon, $\mu^+ \to e^+ \nu_e \bar{\nu}_\mu$. This experiment was proposed in 1991, [2] and was approved for running in 1992 and 1993, using a subset of the apparatus of the MEGA experiment. [3]. Deviation of the positron energy spectrum from its expected shape—parameterized in the four-fermion approximation of the Standard Model [4] by the quantity $\rho \equiv 3/4$—would be an indication of the inadequacy of this theory to account for *all* known basic interactions and/or members of the subatomic zoo. The most precise determination of the ρ parameter was carried out in 1966 [5] and dominates in the overall best-fit determination [6] of 0.7518 ± 0.0026 for this parameter. The present experiment's goal is to reduce the combined statistical and systematic uncertainty to ± 0.0008 and thereby tighten the constraints on possible new physics beyond the Standard Model.

MUON DECAY

Direct muon decay, $\mu^+ \to e^+ \nu_e \bar{\nu}_\mu$, is a particularly useful tool in exploring the weak interaction since it involves only the light point-like leptons. The small momentum transfer permits us to treat the left-handed weak boson—and, *afortiori*, any additional heavier gauge bosons that might mediate this decay—as infinitely massive. In this four-fermion contact interaction limit, the theoretical expression for the differential decay rate of polarized muons *without radiation* is given by

$$\frac{d^2\Gamma}{dx\,d(\cos\theta)} \propto \sqrt{x^2 - x_\circ^2} \left\{ 3x(1-x) + \frac{2}{3}\rho\left(4x^2 - 3x - x_\circ^2\right) + 3\eta\, x_\circ(1-x) \right. \\ \left. + P_\mu \xi \sqrt{x^2 - x_\circ^2} \cos\theta \left[1 - x + \frac{2}{3}\delta\left(4x - 2 - \sqrt{1 - x_\circ^2}\right) \right] \right\} \quad (1)$$

where neither the neutrinos nor the polarization of the outgoing positron are measured. In this expression, $x = E_e/E_e(\max) \simeq 2E_e/m_\mu$ refers to the positron's fractional energy, $x_\circ = m_e/E_e(\max)$ is the minimum value of x, θ is the angle between the muon polarization vector and the positron momentum vector, P_μ is the polarization of the muon, and ρ, δ, η, and ξ are four of the Michel parameters. Integrating over equal ranges of the polar angle θ in the forward and backward hemispheres results in the cancellation of the parity-violating term proportional to ξ, leaving only the isotropic term. For positron energies near the endpoint, this isotropic term is maximally (minimally) sensitive to the parameter ρ (η).

The decay rate can be written instead as a contact interaction in terms of the lepton currents; assuming the most general derivative-free, lepton-number-conserving interaction, this formulation uses a set of ten complex coupling constants that can be related to the Michel parameters. In particular, the ρ parameter can be expressed in terms of these coupling constants via

$$1 - \frac{4}{3}\rho \simeq |g_{LR}^V|^2 + |g_{RL}^V|^2 + 2|g_{LR}^T|^2 + 2|g_{RL}^T|^2 + \text{Re}\left(g_{LR}^S g_{LR}^{T\star} + g_{RL}^S g_{RL}^{T\star}\right) \quad (2)$$

The superscript (V, S, or T) on each coupling constant refers to the vector, scalar or tensor nature of the contact interaction, while the subscript pair (LR, for example) refers to the handedness of the positron and muon. In the Standard Model, only one coupling constant—g_{LL}^V—is nonzero, so ρ is predicted to have the value 3/4.

In the manifestly left-right symmetric extension of the Standard Model, additional right-handed vector bosons are introduced to eliminate the apparent asymmetry in the weak interaction at low energies due to spontaneous symmetry breaking. If the left- and right-handed W bosons are admixtures of the mass eigenstates (with mixing angle ζ), then the coupling constants g_{LR}^V and g_{RL}^V would be nonzero. For small mixing values of ζ, the ρ parameter takes on the value

$$\rho = \frac{3}{4}\left(1 - 2\zeta^2\right) \quad (3)$$

The current best constraint on the mixing angle ζ comes from the existing best-fit determination of ρ stated in the Introduction. This experiment would provide a better constraint on this mixing angle.

EXPERIMENTAL METHOD

The experiment was carried out in the Stopped Muon Channel (SMC) at LAMPF, using the positron spectrometer of the existing MEGA detector. This being an opportunistic measurement, the apparatus configuration was not optimized for the measurement of the Michel parameters; in particular, the geometry of the positron spectrometer restricted the polar angle to roughly $\pi/4 < \theta 3\pi/4$ and the positron energy to $0.75 < x \leq 1$. However, within this acceptance, the design of the apparatus—ultra-low-mass wire chambers, helium along most of the positron trajectory, and many independent detection elements—permitted a very precise determination of the positron momentum with very little distortion due to multiple scattering or energy loss. For tracks spiraling between one and two times between the target and the scintillator barrel in the 1.5 T solenoidal magnetic field, the spectrometer's acceptance was essentially independent of the positron momentum.

Figure 1 shows a typical "zero-loop" positron track viewed along the muon beam axis, as the positron spirals through the eight cylindrical multiwire proportional chambers (MWPCs) before striking one of the two scintillator barrels

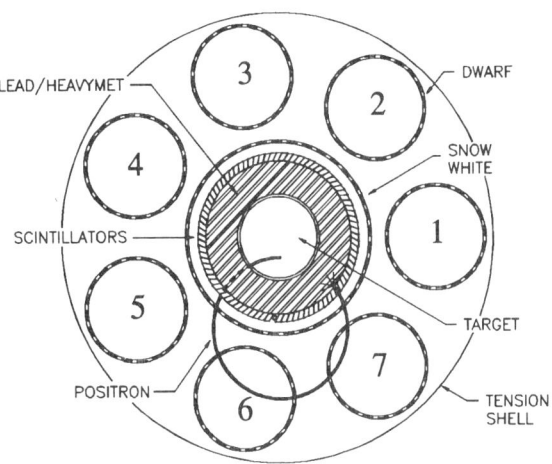

FIGURE 1. End view of a zero-loop positron track in the MEGA positron spectrometer, looking upstream.

(upstream and downstream) and showering in the underlying lead annulus. The apparatus is described in more detail in Ref. [7].

In this experiment, the SMC was tuned to deliver either surface muons (produced by pions decaying essentially at rest at the surface of the primary proton target in the LAMPF beamline) or decay muons (produced by energetic pions decaying within the first half of the SMC), in order to alternate the muon polarization in the target. For the surface muon tune, the flux was reduced using beamline jaws and an upstream beam collimator to a rate of about 5 kHz. The flux of decay muons, without collimation, was typically much lower than this. The muons stopped in a vertical 250 μm Mylar target and decayed. The outgoing positron was tracked by the MWPCs as it spiraled in the 1.5 T magnetic field. Events were triggered by a hit in either of the scintillator barrels at the upstream and downstream ends of the spectrometer.

The detection efficiency of the scintillators was measured in several special runs using an independent trigger based on pulses in the unstriped central section of the Snow White MWPC's cathodes.

During the Summer 1993 run, over 500 million triggers were recorded in a five day period. Table 1 summarizes the distribution of these triggers for the various running conditions. The statistical precision for the three primary data sets—surface tune with 1.5 T field, surface tune with 14.25 T field, and decay tune with 1.5 T field—is approximately ±0.0006.

TABLE 1. Summary of the running conditions and data set sizes for the Summer 1993 run

Trigger	Muon Beam	Magnetic Field	# of Triggers	$\Delta\rho$ (statistical)
Scint	Surface	15.00 kG	135M	0.00076
Scint	Surface	14.25 kG	104M	0.00087
Scint	Decay	15.00 kG	162M	0.00068
MWPC	Surface	15.00 kG	34M	—
MWPC	Surface	14.25 kG	26M	—
MWPC	Decay	15.00 kG	16M	—

DATA ANALYSIS

The value of the ρ parameter is extracted from the data by fitting the measured positron energy spectrum to the best combination of two simulated spectra—one generated using a direct muon decay spectrum (with radiation!) with a value of $\rho = 3/4$, and another generated using the derivative of the muon decay spectrum with respect to the ρ parameter, evaluated at $\rho = 3/4$. This technique demands a thorough understanding of the detector geometry and performance, as well as the inclusion of all of the relevant physical processes in the Monte Carlo simulation. The fidelity of the simulation is tested by making detailed low-level comparisons with the data of distributions such as the beam spot, wire chamber hit frequencies, and scintillator hit frequencies.

The detector geometry is determined from alignment surveys and from reconstruction of energetic cosmic ray muons. The chamber and scintillator efficiencies are extracted from the data. Similarly, the average beam polarization can be extracted from the data (assuming a purely left-handed interaction, of course).

The EGS4-based [8] simulation includes first-order radiative corrections in the generation of the decay positrons. The positron tracking has been enhanced over what is provided in EGS4 by the inclusion of single- and plural-scattering and of fluctuations in ionization energy loss below the cutoff for delta ray production. The chamber response includes finite-size image charges on the cathodes, realistic pulse development with pileup on both anodes and cathodes, additional UV photon hits in the vicinity of the track-chamber crossing, and wire-by-wire discriminator threshold settings for latching the pulses.

STATUS

The analysis of the experiment is dominated by our understanding of the systematic uncertainties. The relevant ones are listed in Table 2, along with

TABLE 2. Evaluation of the dominant systematic uncertainties in the determination of the Michel ρ parameter. "Expected Uncertainties" refer to the numbers in the proposal[2] for this experiment.

Quantity	Expected Uncertainty	Required Uncertainty	Achieved to date
Energy scale	0.2 keV	10 keV	1 keV
Alignment	5 μm	40 μm	50 μm
Energy resolution	2 keV	30 keV	30 keV
Beam stop depth in tgt	1 μm	150 μm	25 μm
Scintillator efficiency	0.001	0.003	0.01
MWPC efficiency	0.001	0.01	0.001–0.05

an indication of where we stand at present in determining and minimizing these uncertainties. The dominant task is the matching of the simulation's efficiencies and cluster widths for the MWPC cathodes. Additional work on the beam spot distribution and the scintillator response is also in progress. We expect to reach the desired level of uncertainty for all identified sources of systematic error.

REFERENCES

1. L. Michel, Proc. Phys. Soc. **A63**, 514 (1950).
2. J. F. Amann et al., *A Proposal to Measure the Michel Parameter Rho with the MEGA Positron Spectrometer*, LAMPF Research Proposal 1240, 1991.
3. M. D. Cooper et al., *MEGA: Search for the Rare Decay $\mu^+ \to e^+ \gamma$*, LAMPF Research Proposal 969, 1985.
4. S. Glashow, Nucl. Phys. **22**, 529 (1961); S. Weinberg, Phys. Rev. Lett. **19**, 1264 (1967); A. Salam, in *Elementary Particle Theory: Relativistic Groups and Analyticity (Nobel Symposium No. 8)*, edited by N. Svartholm (Almqvist & Wiksell, Stockholm, 1968).
5. J. Peoples, Columbia University, Ph.D. thesis, Nevis 147, 1966 (unpublished).
6. S. Derenzo, Phys. Rev. **181**, 1854 (1969).
7. V. Armijo et al., Nucl. Instr. and Methods **A303**, 298 (1991); S. Stanislaus et al., Nucl. Instr. and Methods **A323**, 198 (1992); V. Armijo et al., *Construction and Performance of MEGA's Low Mass, High Rate Cylindrical MWPCs*, to be submitted to Nucl. Instrum. and Methods.
8. W.R. Nelson, H. Hirayama and D.W.O. Rogers, SLAC Report 265 (1985).

Tau Decays at LEP

Randall J. Sobie

Institute of Particle Physics of Canada
and University of Victoria, Department of Physics and Astronomy
P.O. Box 3055, Victoria, British Columbia, V8W 3P6 Canada

Abstract. The measurements of the properties of the tau lepton are becoming increasingly more precise. We will show that these results can be used to test the Standard Model. In addition, we will illustrate how the hadronic decays of the tau can be used to study the strong interaction.

INTRODUCTION

The tau lepton is the third member of the lepton family. Since the tau is the heaviest of the three charged leptons, one might expect that the tau is more sensitive to new physics than the electron or muon.

To use the tau as a laboratory for studying the Standard Model, it is important that the properties of the tau be well measured. The tau mass has been measured by the BES Collaboration using a novel technique [1]. Many groups have measured the tau lifetime using a variety of techniques [2]. The branching ratios of the leptonic decay modes are well measured with a precision of 0.4-0.5% [2], while the hadronic decays of the tau have been resolved into many final states [3]. A complete list of published results are summarized in the Particle Data Group compilation [4].

In this review we will focus on two aspects of tau physics. The first part will examine whether the tau is a heavier version of the muon or electron (i.e. lepton universality). In the second part we will show how the hadronic decays of the tau can be used to study the strong interaction.

CHARGED CURRENT COUPLINGS

In the Standard Model it is assumed that both the charged current and neutral current couplings to the gauge bosons are identical for all three lepton species. The ratio of the electron and muon charged current couplings can be obtained by comparing the $\tau^- \to e^- \bar{\nu}_e \nu_\tau$ to the $\tau^- \to \mu^- \bar{\nu}_\mu \nu_\tau$ decay. It can also be found by comparing $\pi^- \to e^-\nu_e$ and $\pi^- \to \mu^-\nu_\mu$ decays. The two comparisons probe separately the couplings to transverse and longitudinal W bosons. The g_μ/g_e ratio is obtained from tau decays by

$$\left(\frac{g_\mu}{g_e}\right)_T^2 = \left(\frac{1}{0.9726}\right) \frac{B(\tau^- \to \mu^-\bar{\nu}_\mu\nu_\tau)}{B(\tau^- \to e^-\bar{\nu}_e\nu_\tau)}$$

where the numerical factor accounts for the small difference in phase space for the two modes. Using the world average values, one obtains $(g_\mu/g_e)_T = 1.0008 \pm 0.0028$ and $(g_\mu/g_e)_L = 1.0012 \pm 0.0015$ for the tau and pion decays respectively [2].

Similarly the ratio of the tau and muon couplings can be obtained by comparing $\mu^- \to e^-\bar{\nu}_e\nu_\mu$ and $\tau^- \to e^-\bar{\nu}_e\nu_\tau$ decays or $\tau^- \to h^-\nu_\tau$ and $h^- \to \mu^-\nu_\mu$ decays (where h^- is a π^- or K^-). The g_τ/g_μ ratio is obtained by

$$\left(\frac{g_\tau}{g_\mu}\right)_T^2 = 0.9996 \frac{\tau_\mu}{\tau_\tau} \left(\frac{m_\mu}{m_\tau}\right)^5 B(\tau^- \to e^-\bar{\nu}_e\nu_\tau)$$

and

$$\left(\frac{g_\tau}{g_\mu}\right)_L^2 = \frac{2m_\mu^2}{m_\tau^2} \frac{B(\tau^- \to h^-\nu_\tau)}{H_\pi + H_K}$$

where $B(\tau^- \to h^-\nu_\tau) = B(\tau^- \to \pi^-\nu_\tau) + B(\tau^- \to K^-\nu_\tau)$. In addition,

$$H_h = (1+\delta_h) \frac{\tau_\tau m_\tau}{\tau_h m_h} \left(\frac{1-(m_h/m_\tau)^2}{1-(m_\mu/m_h)^2}\right)^2 B(h^- \to \mu^-\nu_\mu)$$

where δ_h are radiative corrections [5]. Using the average of all experimental measurements gives $(g_\tau/g_\mu)_T = 1.0003 \pm 0.0029$ and $(g_\tau/g_\mu)_L = 1.0067 \pm 0.0064$ [2].

All the ratios of the couplings are consistent with unity suggesting that the assumption of lepton universality in the charged current interaction is good to a level between 0.2-0.7%.

NEUTRAL CURRENT COUPLINGS

The Standard Model also assumes that the vector and axial-vector couplings of the leptons to the Z^0 boson are identical. One way to test this assumption is to measure the polarization of the tau leptons in the $e^+e^- \to \tau^+\tau^-$ reaction at LEP.

All fermions are in fact polarized at LEP but it can only be measured for the tau lepton. The tau polarization at the Z^0 pole can be written in terms of the electron and tau asymmetries as

$$P_\tau(\cos\theta) = -\frac{A_\tau(1+\cos^2\theta) + 2A_e\cos\theta}{(1+\cos^2\theta) + 2A_eA_\tau\cos\theta}$$

where $\cos\theta$ is the scattering angle of the τ^- relative to the electron direction. The lepton asymmetries are defined by $A_l \equiv 2g_{v_l} g_{a_l} / (g_{v_l}^2 + g_{a_l}^2)$ with g_{v_l} and g_{a_l} being the vector and axial-vector coupling constants.

The tau polarization is determined by measuring the kinematical quantities of the tau decay products. For example, the pion energy (x_π) spectrum in the $\tau \to \pi\nu_\tau$ decay is proportional to $[1 + (2x_\pi - 1)P_\tau]$.

The average results from the LEP Collaborations are $A_e = 0.1382 \pm 0.0076$ and $A_\tau = 0.1401 \pm 0.0067$, which is consistent with the hypothesis of lepton universality [6]. Combining these results with other measurements, one can determine the vector and axial-vector coupling constants, g_V and g_a. Fig. 1 gives the 68% contours are shown for g_V versus g_a for the three lepton species as well as the combined result. The surface shown in fig. 1 is the Standard Model prediction. The extent of the surface reflects the dependence on the top quark mass $(175 \pm 9$ GeV$)$ and Higgs mass $(300^{+700}_{-240}$ GeV$)$.

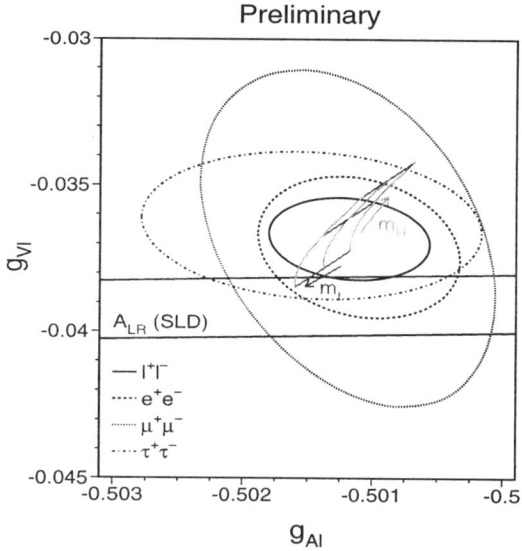

FIGURE 1. Contours of 68% probability in the $g_V - g_a$ from LEP measurements. The solid contour results from a fit assuming lepton universality. Also shown is the one standard-deviation band resulting from the A_{LR} measurement of SLD.

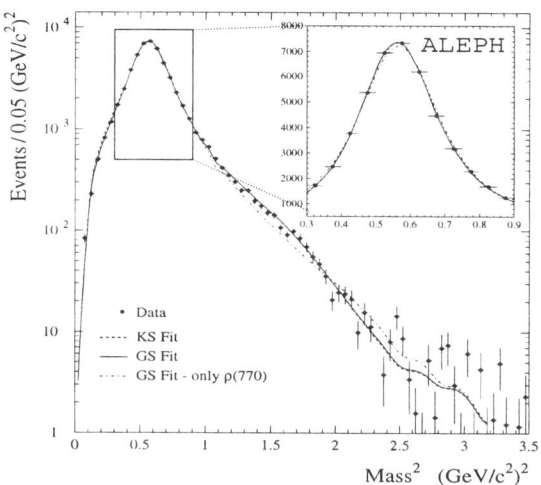

FIGURE 2. Pion form factor measured by ALEPH in $\tau^- \to \pi^-\pi^0\nu_\tau$.

MICHEL PARAMETERS

The $V - A$ structure of the tau leptonic decays can be tested by measuring the lepton energy spectrum. The distribution of a polarized tau is given by

$$\frac{d\Gamma}{d\Omega dx} = x^2 \left(3(1-x) + 2\rho(\frac{4}{3}x - 1) + \eta \frac{m_l}{m_\tau} \frac{6(1-x)}{x} \right.$$
$$\left. - P_\tau \xi \cos\theta \left[(1-x) + 2\delta(\frac{4}{3}x - 1) \right] \right)$$

where $x = E_l/E_\tau$. The Michel parameters obtained from tau decay are shown in Table 1 and are in good agreement with the predicted values. We also list the Michel parameters measured in muon decay for comparison.

TABLE 1. Michel Parameters

	Prediction	Average	Muon
ρ	0.75	0.741 ± 0.014	0.7518 ± 0.0026
η	0	0.046 ± 0.075	-0.007 ± 0.013
ξ	1	1.04 ± 0.09	1.0027 ± 0.0085
$\xi\delta$	0.75	0.73 ± 0.07	0.7506 ± 0.0074

DECAY DYNAMICS

The hadronic decays of the tau lepton are a good laboratory for studying the strong interaction. Both ALEPH [7] and CLEO [8] have measured the spectral functions in tau decays. In figure 2 we show the pion form factor obtained from $\tau^- \to \pi^-\pi^0\nu_\tau$ decays as measured by ALEPH [7] where there is clear evidence for the $\rho(770)$ and $\rho'(1270)$ resonances. Fits for the parameters of the ρ resonances depend on the model used in the fits and the details can be found in ref. [7,8]. The spectral functions obtained from tau decay can compared with those from low energy $e^+e^- \to \pi^+\pi^-$ data assuming CVC. The tau data are in good agreement (and of comparable precision) with the e^+e^- data.

A number of experiments have studied dynamics of the $\tau^- \to \pi^-\pi^+\pi^-\nu_\tau$ decay. For example, the OPAL Collaboration have tested two models [9]. Both models assume that the decay of the tau proceeds through the a_1 resonance ($a_1^- \to \rho^0\pi^-$ and $\rho^0 \to \pi^+\pi^-$). The first model by Kühn and Santamaria (KS) [10] has the mass and width of the a_1 as the only free parameters. The second model by Isgur, Morningstar and Reader (IMR) [11] has the mass and three coefficients of a non-resonant polynominal background as the free parameters. The results of fits to the OPAL data are shown in fig. 3. Fig. 3(a) shows the Q^2 distribution while figs. 3(b-d) show projections of the Dalitz plot for various regions of the Q^2 distribution. The KS model (solid line) gives a good description of the Q^2 distribution while overestimating the ρ resonance region in the Dalitz plot projections. The IMR model (dashed line) also describes the Q^2 distribution but gives a better description of the ρ resonance region.

There have been many new measurements of tau decays involving charged and neutral kaons [3]. In general there is good agreement between the different experiments and the theoretical predictions for the K^-, $(K\pi)^-$ and $(KK\pi)^-$ decay modes [12]. One noticeable exception are the $(K\pi\pi)^-$ decay modes. The $(K\pi\pi)^-$ is assumed to proceed through the axial-vector K_1 resonances. The prediction for the branching ratio is dependent on the width used for the K_1 resonances. An improved description of the experimental data is obtained when the width of both the $K_1(1270)$ and $K_1(1400)$ are increased from 90 and 175 MeV to 250 MeV [12].

SUMMARY

The tau lepton is rapidly becoming an important probe for studying the Standard Model. We have shown that the assumption of lepton universality appears to valid in both the charged current and neutral current sector to better than 1%. The measurement of the Michel parameters in the tau leptonic decays are consistent with a $V - A$ form for the interaction. In addition we have shown that tau decays are a clean laboratory for studying the strong interaction.

REFERENCES

1. BES Collaboration, J.Z. Bai et al., Phys. Rev. D53 (1996) 20.
2. P. Weber, "Review of tau lifetime and leptonic branching ratio measurements", Fourth Int. Workshop on Tau Lepton Physics, St. Estes, Col., Sept. 1996.
3. H. Evans, "Charged Current Measurements: a Tau96 Overview", Fourth Int. Workshop on Tau Lepton Physics, St. Estes, Col., Sept. 1996.
4. Particle Data Group, R.M.Barret et al., Phys. Rev. D54 (1996) 1.
5. R. Decker and M. Finkemeier, Phys. Lett. B334 (1994) 199.
6. The LEP Collaborations, "A Combination of Preliminary LEP and SLD Electroweak Measurements and Contraints on the Standard Model", LEPEWWE-97-01, Jan 1997.
7. ALEPH Collaboration, R. Barate et al., CERN Preprint CERN-PPE-97-013.
8. J. Urheim, "The hadronic current in tau leptonic decays to two pseudscalar mesons", Fourth Int. Workshop on Tau Lepton Physics, St. Estes, Col., Sept. 1996.
9. OPAL Collaboration, K. Ackerstaff et al., CERN Preprint CERN-PPE-97-020.
10. J.H. Kühn and A. Santamaria, Z. Phys. C48 (1990) 445.
11. N. Isgur, C. Morningstar and C. Reader, Phys. Rev. D39 (1989) 1357.
12. M.Finkemeier, J.H.Kühn and E.Mirkes, "Theoretical Aspects of $\tau \to Kh(h)\nu_\tau$ Decays and Experimental Comparisons", Fourth Int. Workshop on Tau Lepton Physics, St. Estes, Col., Sept. 1996.

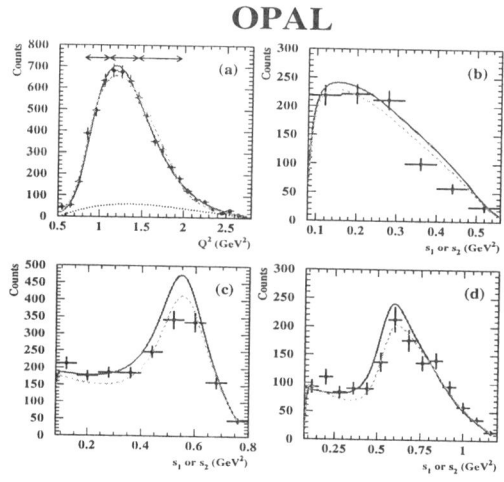

FIGURE 3. Model fits to the OPAL $\tau^- \to \pi^-\pi^+\pi^-\nu_\tau$ data. The sold line is the fit using the KS model while the dashed line uses the IMR model. The dotted line is the non-resonant contribution used in the IMR model.

A Precision Measurement of Muon Decay

Dennis H. Wright* representing the E614 collaboration:
*Kurchatov Institute, Texas A & M University, TRIUMF,
University of Alberta, University of British Columbia,
Universitè de Montrèal, University of Regina,
University of Saskatchewan, Valparaiso State University,
Virginia Polytechnic Institute & State University*

*TRIUMF, Vancouver, B.C., Canada V6T 2A3

Abstract. TRIUMF Experiment 614 will perform the first measurement of the double differential cross section of muon decay. A fit of the decay positron spectrum in both energy and angle will provide a simultaneous measurement of the Michel parameters ρ, ξ and δ to a precision of a few parts in 10^4 and the parameter η to a few parts in 10^3. These values will place constraints upon the non-Standard Model weak couplings 3-10 times tighter than those now existing. Within the context of left-right symmetric models E614 will be sensitive to right-handed Ws with masses up to 820 GeV.

A precise measurement of the Michel parameters of ordinary muon decay provides the most sensitive details of the structure of the weak interaction in a purely leptonic system and a means of testing various extensions of the Standard Model. The goal of the E614 collaboration, the measurement of the parameters ρ, ξ and δ to better than 3 parts in 10^4 and η to better than 3 parts in 10^3, will allow significant improvement over current limits on the non-V-A components of the weak charged current.

The experiment will be performed by stopping surface muons in a 75 μm thick aluminum target and detecting the angle and energy of the decay positrons. Positron tracks will be detected in 26 pairs of 4mm thick, low-mass drift chambers and analyzed in a 2.2T solenoidal magnetic field. All but the largest angle positron tracks will be contained in the spectrometer, allowing the double-differential positron cross section, $d^2\sigma/dE/d(cos\theta)$ to be measured for the first time over a large portion of phase space. The positron will be tracked over several loops of its helix, ensuring a typical energy resolution of

<1 MeV (FWHM).

The positron cross section consists of two parts. The angle-independent part is governed by the Michel parameters ρ and η, while the angle-dependent part is governed by the Michel parameters ξ and δ. Deviations of any of these parameters from their Standard Model values produces a change in the shape of the cross section. By collecting 10^9 muon decay events and by performing various calibration experiments to reduce systematic errors, the following sensitivities can be reached (at 90% CL):

$$0.74977 < \rho < 0.75023 \tag{1}$$

$$0.99970 < \xi < 1.00030 \tag{2}$$

$$0.74964 < \delta < 0.75036 \tag{3}$$

$$|\eta| < 0.003 . \tag{4}$$

The limits on the weak coupling constants implied by the above precision can be obtained by first writing the Michel parameters in a model-independent way:

$$\rho = \frac{3}{4} - \frac{3}{4}[\ |g^V_{LR}|^2 + |g^V_{RL}|^2 + 2|g^T_{LR}|^2 + 2|g^T_{RL}|^2 + Re(g^S_{RL}g^{T*}_{RL} + g^S_{LR}g^{T*}_{LR})\ ] \tag{5}$$

$$\xi\delta = \frac{3}{4} - \frac{3}{4}[|g^V_{LR}|^2 + |g^V_{RL}|^2 + 4|g^T_{LR}|^2 + 2|g^T_{RL}|^2 + 2|g^V_{RR}|^2 + \frac{1}{2}|g^S_{RR}|^2 \tag{6}$$

$$+ \frac{1}{2}|g^S_{LR}|^2 + Re(g^S_{RL}g^{T*}_{RL} - g^S_{LR}g^{T*}_{LR})\ ]$$

$$\xi = 1 - [\frac{1}{2}|g^S_{RR}|^2 + \frac{1}{2}|g^S_{LR}|^2 + 2|g^V_{RR}|^2 + 4|g^V_{RL}|^2 - 2|g^V_{LR}|^2 - 2|g^T_{LR}|^2 \tag{7}$$

$$+ 8|g^T_{RL}|^2 + 4Re(g^S_{RL}g^{T*}_{RL} - g^S_{LR}g^{T*}_{LR})\ ]$$

$$\eta = \frac{1}{2}Re[g^V_{LL}g^{S*}_{RR} + g^V_{RL}(g^{S*}_{LR} + 6g^{T*}_{LR}) + g^V_{LR}(g^{S*}_{RL} + 6g^{T*}_{RL}) + g^V_{RR}g^{S*}_{LL}] . \tag{8}$$

The above relations were derived from Fetscher and Gerber [1] and assume the most general, derivative-free weak interaction. The superscripts indicate the scalar, vector or tensor nature of the couplings while the subscripts refer to the chirality of the electron and muon currents. To obtain new upper limits on the couplings implied by the above constraints, the magnitudes of the coupling constants were varied independently from zero up to the current limit for each coupling, and the relative phases were varied from 0 to π. For each combination of couplings and phases which satisfied constraints 5-8, maximum values of the couplings were saved. These values are shown in column 5 of Table 1. Column 1 shows the current limits as quoted in reference [2].

	Current Limits	E614(A)	E614(B)	E614(C)	E614(D)
g_{RR}^S	0.066	—	—	0.034	0.045
g_{RR}^V	0.033	0.012	0.017	0.017	0.022
g_{LR}^S	0.125	—	—	0.034	0.046
g_{LR}^V	0.060	0.012	0.015	0.015	0.018
g_{LR}^T	0.036	—	0.010	—	0.013
g_{RL}^S	0.424	—	—	—	—
g_{RL}^V	0.110	0.012	0.012	0.012	—
g_{RL}^T	0.122	—	0.008	—	—
g_{LL}^S	0.55	—	—	—	—
g_{LL}^V	>0.96	>0.99977	>0.99942	—	—

Table 1. Upper limits (90% CL) for weak coupling constants with current limits taken from reference 2. Limits set by E614 assume (A) V, A couplings only, (B) V,A and T couplings, (C) V,A and S couplings or (D) most general (V,A,S and T) derivative-free couplings.

In the most general case D), limits on all of the RR and LR couplings are improved over current values while none of the RL and LL limits are. However, since very few extensions of the standard model include both scalar and tensor couplings, three additional cases were studied: A) only vector and axial vector couplings are allowed, B) all but scalar couplings are allowed, and C) all but tensor couplings are allowed.

Case A is a minimal extension of the standard model, in which only three independent vector coupling constants are needed. In this case the limit on the total deviation from V-A is quite stringent, as seen by the deduced lower limit on g_{LL}^V. Adding the assumption that $g_L = g_R = g$, non-zero g_{RL}^V and g_{LR}^V indicate the admixture of right-handed currents while non-zero g_{RR}^V is dependent upon the mass of the right-handed vector boson W_R. A measurement of the parameter ξ will therefore set limits on both the left-right mixing ζ and the right-handed boson mass M_R. The allowed region set by the product $P_\mu \xi$ at the 95% CL is given by [3]

$$\frac{1 - P_\mu \xi}{4} = \zeta^2 + \frac{M_L^4}{M_R^4} + \zeta \frac{M_L^2}{M_R^2} \tag{9}$$

and is shown in Fig. 1. The lower limit on the the W_R mass is therefore

$$M_R > 820 \text{ GeV}/c^2$$

and the limit on the mixing angle ζ is

$$-0.0110 < \zeta < 0.0095 \ .$$

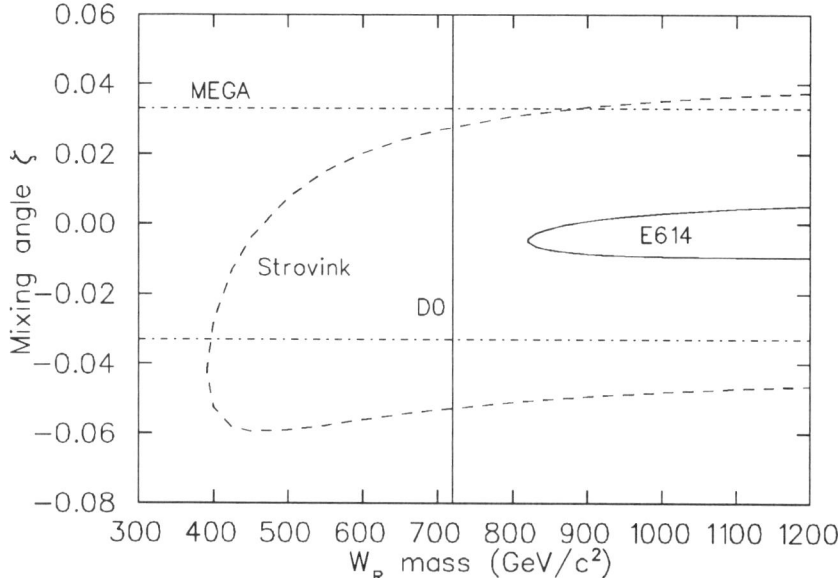

FIGURE 1. 95 % CL allowed regions: solid line - proposed E614 limits; solid vertical line - limit from D0; dot-dashed line - MEGA (ρ) limits; dashed line - Strovink muon decay experiment.

Also shown in Fig. 1 is the allowed region at 95% CL as defined by previous measurements. It is bounded from below in mass by the recent D0 result [4] and in ζ by the preliminary ρ measurement result from MEGA [5]. The previous muon decay result of Strovink [6] is also shown. The D0 result assumes that $g_R = g_L$, that the left- and right-handed CKM matrices are equal and that right-handed neutrino masses are not bigger than certain limits. Because it is a purely leptonic probe, E614 is much less sensitive to assumptions about the CKM matrices. Even so, the mass limit for E614 exceeds the D0 limit and the total area in $M_R - \zeta$ space excluded by E614 is much larger.

REFERENCES

1. W. Fetscher and H.-J. Gerber, ETH Zurich reprint ETHZ-IMP PR/93-1 February 1993.
2. W. Fetscher and H.-J. Gerber, Phys. Rev. D45, VI.16, 1 June 1992.
3. P. Herczeg, Phys. Rev. D34, 3449 (1986).
4. D0 collaboration, Phys. Rev. Lett. 76, 3271 (1996).
5. MEGA collaboration, private communication (1997).
6. A. Jodidio et al., Phys. Rev. D37, 237 (1988); Phys. Rev. D34, 1967 (1986).

Search for T-Violation in $K^+ \to \mu^+ \pi^o \nu_\mu$ Decay

Michael D. Hasinoff
for the KEK-246 Collaboration[1]

*Department of Physics and Astronomy
The University of British Columbia
Vancouver, B.C. Canada V6T 1Z1*

Abstract. An experiment to search for the T-Violating transverse muon polarization in $K^+ \to \mu^+ \pi^o \nu_\mu$ decay ($K_{\mu3}$) is now underway at the 12 GeV proton synchroton at KEK. The experiment uses a stopped K^+ beam and a large Superconducting Toroidal Sepectrometer. The expected limit on P_T will set constraints on non-standard models of CP-violation such as the multi-Higgs doublet model. The experiment should finish data-taking in late 1998.

INTRODUCTION

The Standard Model of the Electro–Weak interaction has been extremely successful in describing all of the existing experimental data. Nevertheless, it does contain several ad-hoc parameters and it is not complete (e.g., in the quark sector it is necessary to add imaginary phases in the standard three generation Cabbibo-Kobayashi-Maskawa (CKM) matrix in order to accommodate CP-violation). Many extensions to the Standard Model have been proposed; most of these extended models contain additional imaginary phases since it has been suggested [1] that the Standard Model CP-violation might not be large enough to explain the large Baryonic Asymmetry of the Universe (BAU). Hence it is very useful to search for other possible mechanisms of CP-violation. Since CPT invariance is normally assumed to be true, T-violation is equivalent to CP-violation.

---

[1] Representing the KEK–246 collaboration—Inst. of Particle & Nuclear Studies/KEK, Tokyo Inst. of Technology, Univ. of Tokyo, Univ. of Tsukuba, Osaka Univ., Inst. for Nucl. Research (Moscow), Univ. of British Columbia, Univ. of Montréal, Univ. of Saskatchewan, TRIUMF, Korea Univ., Yonsei Univ., Princeton Univ., Virginia Polytechnic Institute & State Univ., National Taiwan Univ.

The measurement of a non-zero triple-vector correlation, such as a component of the muon polarization normal to the decay plane in $K^+ \to \mu^+\pi^0\nu_\mu$ decay ($K_{\mu 3}$), would be an indication of the violation of time-reversal invariance. This transverse muon polarization (P_T), is defined as a T-odd triple product

$$P_T \equiv \frac{\vec{s}_{\mu^+} \cdot (\vec{p}_{\mu^+} \times \vec{p}_{\pi^0})}{|\vec{p}_{\mu^+} \times \vec{p}_{\pi^0}|} \quad (1)$$

where $\vec{s}_{\mu^+}$ and $\vec{p}_{\mu^+(\pi^0)}$ are the muon spin vector and the muon(pion) momentum vectors, respectively. The possibility of using this correlation to look for T-violation was first suggested by Sakuri [2], and several experiments have been performed using both neutral [3,4] and charged kaons [5]. The unique feature of P_T in $K_{\mu 3}$ decay is that it does not have contributions from the Standard Model at the tree level and higher order effects are $\sim 10^{-6}$. Furthermore, since there is only one charged particle in the final state, the final state interaction, which can mimic a T-violation effect, has been calculated [6] to be $\sim 10^{-6}$ as well. Consequently, this measurement has the potential to reveal new physics beyond the Standard Model.

Left-right symmetric extensions to the Standard Model do not produce a non-zero P_T since they also involve vector or axial-vector interactions. However, both scalar(S) and pseudoscalar(P) interactions could generate relatively large values of P_T. The three-Higgs-doublet model [8] and the Leptoquark model [9] can both give P_T values as large as 10^{-3}. Recent SUSY model calculations [10] also produce P_T as large as 2×10^{-4}.

The matrix element for $K_{\mu 3}$ decay can be described [7] by two form factors, $f_+(q^2)$ and $f_-(q^2)$. P_T is then equal to the imaginary part of the parameter $\xi = f_-(q^2)/f_+(q^2)$ multiplied by a kinematic factor. The previous measurement [5] used in-flight decays and obtained $P_T = (-3.1 \pm 5.3) \times 10^{-3}$ with a corresponding value of the T-violating parameter, Im $\xi = -0.016 \pm 0.025$, for the maximal case when Re $\xi = 0$.

KEK-246 EXPERIMENT

The present KEK experiment uses a low momentum (660 MeV/c) separated K^+ beam from the K5 channel along with the high acceptance Superconducting Toroidal Spectrometer(STS). A Fitch-type Cherenkov counter is used to select K's which are then degraded in BeO and stopped in an active target (consisting of 256-5 mm square fibres) located at the centre of the magnet (see Fig.1). The charged decay products are also tracked inside the target before being momentum analyzed (by wire chambers C1-C4) in any one of the 12 identical magnet gaps. Upon exiting the magnet gap the μ's are degraded by a wedge-shaped Cu degrader and then stopped in a segmented high-purity Al target. The muon polarization is measured by detecting an asymmetry in the e$^+$ angular distribution. A transverse magnetic field (~ 150 G) is applied at the position of the muon stoppers so that P$_T$ can be preserved while, at the same time, the in-plane muon polarization can be precessed. This serves to reduce any spurious effect due to the large in-plane μ^+ polarization. The π^o's from $K_{\mu 3}$ decay are detected in a highly segmented CsI(Tl) barrel detector (768 crystals) which completely surrounds the target except for the beam holes and 12 openings at the entrance to the STS magnet gaps.

In the present experiment the stopped beam method is used, in contrast to the previous experiment, which used in-flight K^+ decays. Our detector has nearly full $K_{\mu 3}$ kinematic coverage so that we can simultaneously observe $K_{\mu 3}$ events with the π^o going forward or backward along the detector axis. Such a forward/backward comparison increases our sensitivity to P$_T$ and reduces our systematic errors. Moreover our kinematic coverage also includes a region where P$_T$ should vanish so that we can perform a null check of our apparatus. Secondly, the isotropic decay from stopped K^+'s significantly reduces any spurious asymmetry from an asymmetric K^+ stopping distribution in the target.

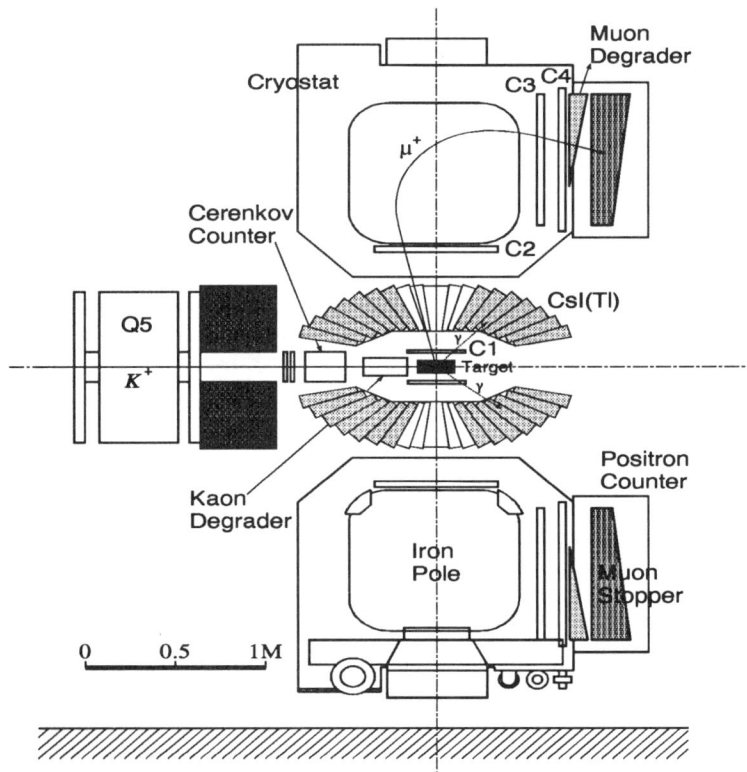

FIGURE 1. Side view of one magnet gap for the E246 apparatus.

Finally, the e^+ polarimeters can be located far from the beam axis where the normal beam associated backgrounds are lower.

SENSITIVITY TO P_T

For a $K_{\mu 3}$ event having the π^o moving along the detector axis (forward or backward), the decay plane is almost radial to the target and parallel to the beam axis. P_T is then directed azimuthally in a screw sense around the detector axis and it would show up as a difference in the e^+ counting rate between the clockwise(cw) and counter-clockwise(ccw) side counters surrounding that muon stopper in the polarimeter. Since the STS magnet has perfect $30°$ rotational symmetry, we can sum all the cw- and ccw-counts and form the ratio

$$\frac{\sum_{i=1}^{12} N_i(cw)}{\sum_{i=1}^{12} N_i(ccw)} \simeq 1 \pm 2A\alpha P_T \qquad (2)$$

which is proportional to P_T. Here $N_i(cw)$ and $N_i(ccw)$ are the e^+ counts in the cw- and ccw-side counters for the ith polarimeter sector, A is the positron asymmetry coefficient and α is a kinematical attenuation factor caused by the finite acceptance of the spectrometer. The $\pm$ signs in eqn(2) correspond to the forward and backward π^o events as shown schematically in Fig. 2.

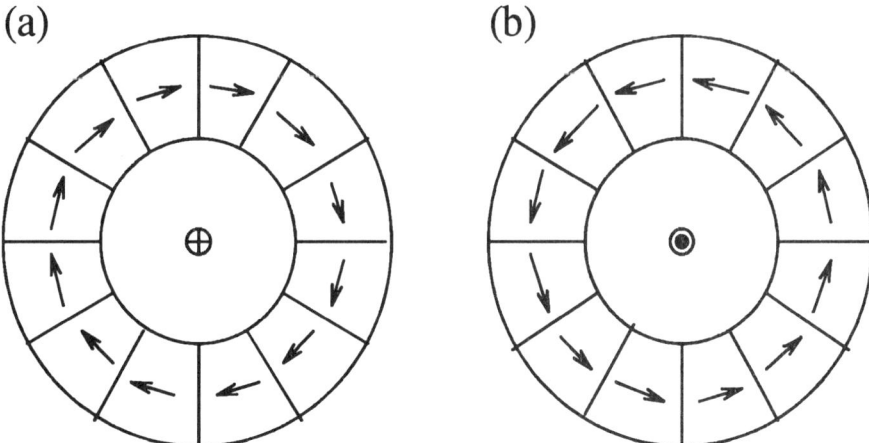

FIGURE 2. Direction of P_T as viewed from downstream of the detector; (a) forward π^o events and (b) backward π^o events.

Since the sign of P_T is opposite in these two special geometries, the double ratio

$$\frac{\left[\sum_{i=1}^{12} N_i(cw)/\sum_{i=1}^{12} N_i(ccw)\right]_{fwd}}{\left[\sum_{i=1}^{12} N_i(cw)/\sum_{i=1}^{12} N_i(ccw)\right]_{bwd}} \simeq 1 + 4A\alpha P_T \qquad (3)$$

will provide another factor of 2 increase in sensitivity to P_T and also a substantial reduction in nearly all potential systematic errors.

The major source of background in this experiment is the $K^+ \to \pi^+\pi^o$ decay ($K^+_{\pi 2}$) which has a much larger branching ratio (21 % vs 3.2 % for $K_{\mu 3}$). Kinematic cuts on the $\pi^+ - \pi^o$ opening angle and the π^o energy as well as a limit on the μ^+ momentum (100 MeV/c $\leq p_{\mu^+} \leq$ 180 MeV/c) are used to reject such events. This momentum region includes about 65 % of the total $K_{\mu 3}$ spectrum. Only a small fraction of in-flight π^+ decays should contaminate the final $K_{\mu 3}$ muon spectrum.

The total acceptance for the $K_{\mu 3}$ events is estimated to be $\sim 1.2 \times 10^{-3}$ for the forward and backward π^o events with $|\cos\theta_{\pi^o}| > 0.5$. The analyzing power, $A\alpha$, is expected to be ~ 0.18. The kinematic factor, $P_T/\text{Im }\xi$, for our acceptance has been estimated to be ~ 0.26. To date (Feb'96–Apr'97) we have accumulated about $\sim 1/3$ of our statistics and have almost exceeded the published limit for P_T. We anticipate further improvement of the slow-extracted proton beam intensity during the second period of data-collection (Nov'97–Oct'98). Assuming the present KEK beam intensity of 1×10^{12} protons/sec we anticipate a final statistical error, $\Delta P_T \sim 1.5 \times 10^{-3}$, corresponding to Im $\xi \sim 6 \times 10^{-3}$. Monte Carlo simulations indicate that the various systematic errors should be even smaller.

REFERENCES

1. Cohen, A.G., Kaplan, D.B., and Nelson, A.E., *Ann. Rev. Nucl. Sci.* **43**, 27 (1993).
2. Sakuri, J.J., *Phys. Rev.* **109**, 980, (1958).
3. Schmidt, M. et al., *Phys. Rev. Lett.* **43**, 556, (1979).
4. Morse, W.M. et al., *Phys. Rev.* **D21**, 1750, (1980).
5. Blatt, S.R., et al., *Phys. Rev.* **D27**, 1056, (1983).
6. Zhitnitskii, A.R., *Sov. J. Nucl. Phys.* **31**, 529, (1980).
7. Cabbibo, N., and Maksymowicz, A., *Phys. Lett.* **9**, 352, (1964).
8. Garisto, R., and Kane, G., *Phys. Rev.* **D44**, 2038, (1991).
9. Bélanger, G. and Geng, C.Q., *Phys. Rev.* **D44**, 2789, (1991).
10. Wu, Guo-Hong, and Ng, John N., *Phys. Lett.* **B392**, 93, (1997).

Physics with Strangeness and Charm

STRANGENESS PHOTOPRODUCTION WITH THE SAPHIR DETECTOR[1]

J. BARTH, M. BOCKHORST, W. BRAUN, R. BURGWINKEL, K.H. GLANDER, S. GOERS,
J. HANNAPPEL, N. JÖPEN, U. KIRCH, F. KLEIN, F.J. KLEIN, **D. MENZE**, W.
NEUERBURG, E. PAUL, W.J. SCHWILLE, M.-Q. TRAN, R. WEDEMEYER, F. WEHNES,
B. WIEGERS, F.W. WIELAND, J. WISSKIRCHEN

Physikalisches Institut, Universität Bonn, 53115 Bonn, Germany

J. ERNST, H.G. JÜNGST, H. KALINOWSKY, E. KLEMPT, J. LINK, H.v. PEE, R. PLÖTZKE

ISKP, Universität Bonn, 53115 Bonn, Germany

M. SCHUMACHER, F. SMEND

II. Physikalisches Institut, Universität Göttingen, 37073 Göttingen, Germany

T. MART

Jurusan Fisika, FMIPA, Universitas Indonesia, Depok 16424, Indonesia

C. BENNHOLD[2]

Department of Physics, The George Washington University, Washington DC, USA

Statistically improved data of total cross sections and of angular distributions for differential cross sections and hyperon recoil polarizations of the reactions $\gamma p \to K^+ \Lambda$ and $\gamma p \to K^+ \Sigma^0$ have been collected with the SAPHIR detector at photon energies between threshold and 2.0 GeV. Here total cross section data up to 1.5 GeV are presented. The opposite sign of Λ and Σ polarization and the change of sign between forward and backward direction could be confirmed by higher statistics. A steep threshold behaviour of the $K^+ \Lambda$ total cross section is observed.

1 Experimental Data

Using the SAPHIR detector [1] data of $\gamma p \to K^+ \Lambda$ and $\gamma p \to K^+ \Sigma^0$ were taken and analyzed. Starting from 3 reconstructed tracks a $p\pi^-$ sub sample was preselected by requesting that the invariant mass of a track pair with one positive and one negative charge was within the range of the Λ mass. From the p and π^- tracks the secondary vertex of the Λ decay has been determined while the remaining (K^+) track was used in addition to identify the primary vertex. The $K^+\Lambda$ channel was separated from the $K^+\Sigma$ one by kinematical fits. The separation of background contributions was achieved by cuts in the missing mass, in the invariant mass distribution of the $p\pi^-$ system and in the probabilities of the vertex fits. Total cross sections, a complete set of angular distributions of the Λ and Σ polarization and differential cross sections for both reactions have been determined in a photon energy range between threshold and 2.0 GeV .

2 Results and Discussions

The data presented here have been analysed in the course of a thesis [2]. They have been taken with a trigger on 2 charged particles in the final state [1] and include the reanalysed data of [3] together with additional new data to improve the statistics. Fig. 1 shows the total cross section data in the energy range between threshold and 1.5 GeV. The analysis of the

---

[1]This work is supported by the Bundesminister für Forschung und Technologie (BMFT) and by the Deutsche Forschungsgemeinschaft (DFG).
[2]Supported by DOE grant DE-FG02-95-ER40907

data at higher energies is still in progress. The total cross section of the reaction $\gamma p \to K^+\Lambda$ rises rapidly from threshold up to a pronounced maximum. A comparison with a chiral perturbation theory [4] and with combined channel calculations in chiral SU(3) dynamics [5] shows qualitative agreement in the near threshold region. In the latter reference the maximum is predicted as a cusp structure near the $K^+\Sigma^0$ threshold [5]. While conceptually appealing until now these calculations are limited to s-wave amplitudes.
In the case of $K^+\Sigma^0$ the rise of the total cross section near threshold is smooth and overestimated by both models.

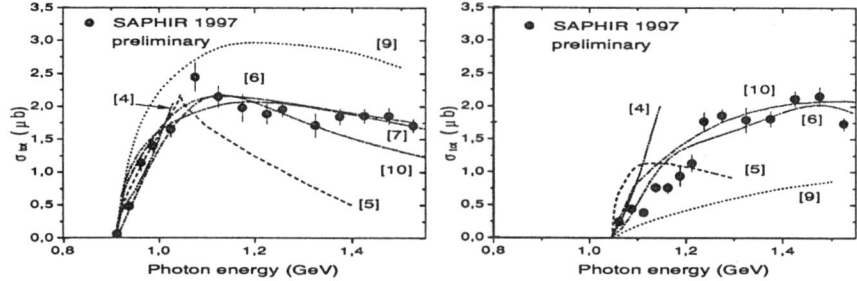

Figure 1: Total cross sections for $\gamma p \to K^+\Lambda$ (left) and $\gamma p \to K^+\Sigma^0$ (right) in comparison with theoretical calculations. Black circles = SAPHIR data [2]. The errors are statistical ones combined quadratically with the sytematic errors from the normalization by the primary photon flux. The numbers at the curves correspond to the references: [4],[5] = chiral calculations, [6],[7] = isobaric models, [9] = Regge calculations, [10] = quark model.

Between threshold and about 1.5 GeV the isobar model constitutes the most widely used method of analysis [6, 7, 8]. In this phenomenological approach a number of tree-level diagrams with s-, t- and u-channel resonances with couplings fitted to the data are included. The quality of the data until now does not allow to determine the resonance contributions uniquely; different models include different sets of resonances. The problem of overestimating the total cross sections at energies above 1.5 GeV has been partially solved either by making use of hadronic form factors [6] or the inclusion of additional t-channel resonances [7].

A Regge-based model [9] has been developed that describes high energy (E_γ = 6 - 12 GeV) photoproduction of πN and KY data. While this description leads to good agreement with data at very high energies, it severely overestimates the $K\Lambda$ and underestimates the $K\Sigma$ total cross sections.

In addition to the above mentioned models, different quark models [10, 11] describe the kaon photoproduction cross section data with few parameters.

Fig. 2 shows the hyperon polarisations in two different energy bins. The data show opposite signs of the Λ and Σ^0 polarizations. Furthermore, there is a sign change between forward and backward direction. Up to now there is no theoretical model which is able to describe these features in detail. The statistics of our data in the near threshold region is not sufficient to compare them with chiral calculations [4, 5]. In fig. 2 the predictions of [4] have been extrapolated to 1.25 GeV. Although this is not the threshold region the prediction for the Λ polarization shows a negative sign in forward direction, for the Σ polarization a change in sign between forward and backward direction. The extrapolation of the Regge calculations down to 2 GeV also show the gross features of the angular distributions [9].

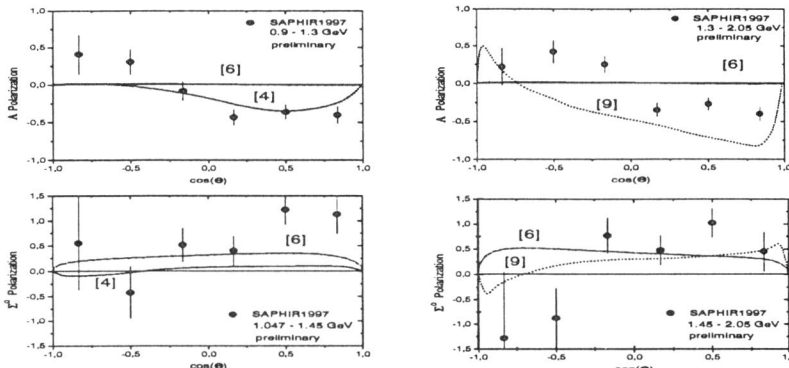

Figure 2: Angular distributions of the Λ (above) and Σ^0 (below) polarizations, in dependence on E_γ (left: lower energies, right: higher energies) and in comparison with theoretical calculations. Black circles = SAPHIR data [2]. The numbers at the curves are explained in fig. 1. The prediction of [6] of the Λ polarization is so small that it cannot be distinguished from zero in this plot.

I would like to thank M. Guidal, J.M. Laget, Z. Li and B. Saghai for helpful discussions.

References

[1] W.J. Schwille et al. , Nucl. Instr. Meth. A **344**(1994)470.

[2] M.Q. Tran, Ph.D. Thesis, Bonn preprint, BONN-IR-97-11(1997).

[3] SAPHIR collaboration, M. Bockhorst et al., Z. Phys. C **63** (1994)37, L. Lindemann, Ph.D. Thesis, Bonn preprint, BONN-IR-93-26 (1993), H. Jüngst, Ph.D. Thesis, Bonn preprint, BONN-IR-95-19 (1995).

[4] S. Steininger et al., Physics Letters **B301**(1997)446.

[5] N. Kaiser et al., Nucl. Phys. **A612**(1997)297.

[6] T. Mart and C. Bennhold, Nucl. Phys. **A585**(1995)369c, T. Mart, C. Bennhold, and C.E. Hyde-Wright, Phys. Rev. **C51**(1995)R1075, T. Mart and C. Bennhold, Few-Body Systems Suppl. **99**(1995)1, T. Mart, Ph.D. Thesis, Mainz preprint, KPH12/96(1996), C. Bennhold et al., Washington preprint, Nucl-th/9703004(1997).

[7] R.A. Adelseck and B. Saghai, Phys. Rev. **C42**(1990)108, J.C. David et al., Phys. Rev. **C53**(1996)2613.

[8] R.A. Williams, C.-R. Ji, S.R. Cotanch, Phys. Rev. **C46**(1992)1617.

[9] M. Guidal et al., preprint Saclay, DAPNIA 97-26 (1997).

[10] Z. Li, Phys. Rev. C **52**, 1648 (1995).

[11] V. Keiner, Z. Phys. **A352**(1995)215.

Kaon Productions Off Nucleons And The Structure Of Baryon Resonances

Zhenping Li

Physics Department, Peking University
Beijing, 100871, P.R. China

Abstract. The recent investigations in the chiral quark model show that kaon productions of nucleons play an important role in understanding the structure of baryons. The evidences of a third S_{11} resonance in the second resonance region and two narrow states around 2 GeV suggest a set of the molecular type baryons with the hidden strangeness. Confirming these states requires further theoretical and experimental studies of the strangeness production.

There have been considerable recent progress in the investigation of baryon resonances in the meson photoproductions. New experimental data [1] for the η photoproduction in the threshold region have been published. These data, in particular the data from the Mainz accelerator MAMI, played a very important role in studying the structure of the resonance $S_{11}(1535)$. On the theoretical side, a new approach based on the chiral quark model has been developed [2] for the meson photoproductions. Comparing to other models at the hadronic levels, the quark model approach to the meson photoproductions introduces the quark and gluon degrees of freedom into the reaction mechanism, and it relates the photoproduction data directly to the spin flavor structure of baryon resonances. Here, we would like to highlight some important features that one could learn from the forthcoming kaon production data at TJNAF and the Bonn accelerator ELSA.

An important feature that we have learnt from the η photoproduction off nucleons is the enhancement of the resonance $S_{11}(1535)$ and the suppression of the resonance $S_{11}(1650)$ in the ηN channel. In Ref. [3], we showed that this phenomenon can not be explained in the framework of the constituent quark model. The solution of this problem might come from the existence of a third S_{11} resonance in the second resonance region. We indicated that there are considerable circumstantial evidences suggesting the presence of a third S_{11} resonance with mass $1.7 \sim 1.8$ GeV and width around 0.2 GeV, which certainly can not be accommodated by the quark model. Moreover, the data from SPHINX collaboration in $p+C$ coherent diffractive productions [4] show two new states,

$$X(2000) \to \Sigma^0 K^+, \quad M_{X(2000)} = 1996 \pm 7, \quad \Gamma_{X(2000)} = 99 \pm 21$$
$$X(2050) \to \Sigma^*(1385)K^+, \quad M_{X(2050)} = 2052 \pm 6, \quad \Gamma_{X(2050)} = 35 \pm 29,$$

in which the masses and widths are in the unit of MeV. The resonances with such small width at 2 GeV are unlikely to be normal qqq states. Notice that the threshold energies for $K^*\Lambda$ and $K^*\Sigma$ productions are 2007 and 2084 MeV respectively. A common feature for the S_{11} resonances in the second resonance region and the states $X(2000)$ and $X(2050)$ is that the masses of these states are just below the threshold energies of the kaon and K^* productions. This suggests that a set of molecular type baryons with hidden strangeness, $K\Lambda$ or $K\Sigma$ state for the S_{11} resonances in the second resonance region, $K^*\Lambda$ for the $X(2000)$ and $K^*\Sigma$ for the $X(2050)$, may indeed exists bellow the threshold energies of Kaon and K^* productions in addition to the well known $\bar{K}N$ candidate $\Lambda(1405)$. The $K\Lambda$ or $K\Sigma$ quasi bound state was first proposed for the resonance $S_{11}(1535)$ in Ref. [5], while the interpretation of the $K^*\Lambda$ and $K^*\Sigma$ states for the $X(2000)$ and $X(2050)$ was suggested in Ref. [6]. In Ref. [3] we pointed out that a pure $K\Lambda$ or $K\Sigma$ configuration for the resonance $S_{11}(1535)$ is inconsistent with the data for the electromagnetic form factor of this resonance, and it should be strongly mixed with the normal qqq S_{11} states.

How the $K\Lambda$ or $K\Sigma$ bound state is mixed with normal qqq S_{11} states has not been investigated theoretically. However, the kaon productions would be very important channels to study the structure of these resonances experimentally. Notice that the masses of the S_{11} resonances are sandwiched between the threshold energies of the kaon and the η productions, the enhancement of the contributions from the S_{11} resonances are expected in the η and kaon productions. This is particularly true for the η photoproduction in the threshold region, where the dominance by the contributions from the resonance $S_{11}(1535)$ is well established. A similar behavior should be expected for the kaon photoproduction; our investigations [7] in the $\gamma N \to K\Sigma$ reactions found that the resonance $S_{11}(1650)$ is indeed enhanced in the threshold region of the reaction $\gamma n \to K^- p$. Thus, the presence of the third S_{11} resonances could be tested in the kaon production experiments. Moreover, as the new states $X(2000)$ and $X(2050)$ are just below the $K^*\Lambda$ and $K^*\Sigma$ threshold energies, these states are expected to be enhanced in the reactions $\gamma N \to K\pi\Lambda$ and $\gamma N \to K\pi\Sigma$, as there is no pomeron exchange to contaminate the cross sections. Unfortunately, there are few data available for these reactions to either confirm or refute the presence of these two states. Therefore, the experimental and theoretical studies of the $K\pi$ productions should become one of the top priorities in the strangeness productions. It is very interesting to note that the preliminary data from ELSA [8] indeed have some hints of resonance structures in the

1.7 GeV region of the reaction $\gamma p \to K^+\Lambda$ and around 2.0 GeV region in both reactions $\gamma p \to K^+\Lambda$ and $\gamma p \to K^+\Sigma^0$, which might correspond to the third S_{11} resonance and the resonances $X(2000)$ and $X(2050)$. Of course further experiments with better precision, in particular the data for the polarization, are needed. Furthermore, the data in $\pi N \to K\Lambda$ or $\pi N \to K^+\Sigma$ are also required so that these new states could be fully established.

Thus, understanding the reaction mechanism of kaon productions has become increasingly important. There has been considerable recent progress in understanding the kaon photoproductions in the traditional isobaric models [9] and in the newly developed chiral quark model [2]. An important feature from the quark model approach in the kaon production via $\gamma N \to K\Sigma$ is that the resonances $F_{37}(1950)$, $F_{35}(1905)$, $P_{33}(1920)$ and $P_{31}(1910)$ belonging to **56** multiplet in the quark model play very important role in the reaction $\gamma N \to K\Sigma$. Better data in these reactions could provide us important information on the structure of these resonances as well.

In conclusion, the high precision data for kaon photoproductions will provide us very important insights into the structure of baryon resonances that could not be possible from the pion photoproductions, and establish the molecular baryons with the hidden strangeness. It may also help us to resolve the puzzle with the S_{11} resonances in the second resonance region. Without doubt, the strangeness productions will be an important and exciting field in the near future.

The author acknowledges the collaborations with R. Workman, Wei-Hsing Ma and Zhang Lin on the works presented here. Discussions with B. Saghai and R. Schumacher are gratefully acknowledged. This work was supported in part by Peking University.

REFERENCES

1. S. Dytman *et al.*, Phys. Rev. **C51**, 2170(1995); J. Price *et al.*, Phys. Rev. **C51**, R2283(1995); B. Krushe, *et al.*, Phys. Rev. Lett. **74**, 3736(1995).
2. Zhenping Li, Ye Hongxing and Lu Minghui, "An Unified Approach To The Meson Photoproductions Off Nucleons In The Quark Model" To Appear on Phys. Rev. C. (1997).
3. Zhenping LI, and R. Workman, Phys. Rev. **C53**, R549(1996).
4. SPHINX Collaboration, S.V. Golovkin *et al.*, Z. Phys. **C68**, 585(1995).
5. N. Kaiser, P.B. Siegel, and W. Weise, Phys. Lett. **B362**, 23(1995).
6. R. Schumacher "Evidence For And Against The States Hidden Strangeness States Near 2 GeV", Submitted to Phys. ReV.
7. Zhenping Li, Ma Wei-Hsing and Zhang Lin, Phys. Rev. **C54**, R2171(1996).
8. D. Menze, in this proceedings.
9. J. C. David, C. Fayard, G.H. Lamot, and B. Saghai, Phys. Rev. **C53**, 2613(1996), and references therein.

Off-shell effects in electromagnetic production of strangeness

C. Fayard[1], G. H. Lamot[1], T. Mizutani[2], and B. Saghai[3]

[1] *IPN-Lyon, IN2P3/CNRS, Université Claude Bernard, 69622 Villeurbanne, France*
[2] *Department of Physics, VPI and State University, Blacksburg, VA 24061, USA*
[3] *Service de Physique Nucléaire, DAPNIA, CEA-Saclay, 91191 Gif-sur-Yvette, France*

Abstract. Previous approaches to the photo- and electro-production of strangeness on the proton, based upon effective Lagrangian, is extended to incorporate the so called *off-shell effects* (*OSE*) required while dealing with spin $\geq 3/2$ baryonic resonances. Results for $K^+ \Lambda$ channels are presented.

INTRODUCTION

An effective Lagrangian-based formalism [1], including the nucleonic (spin $\leq 5/2$), hyperonic (spin $1/2$) and two kaonic resonances ($K^*(892)$, $K1(1270)$), has recently been proven to describe well enough all the available data for the electromagnetic strangeness production and $K^- p$ radiative capture processes; namely,

$$\gamma\, p \to K^+ \Lambda,\ K^+ \Sigma^0,\ K^0 \Sigma^+;\ E_\gamma^{lab} \leq 2.1\ \text{GeV},$$

$$e\, p \to e'\, K^+ \Lambda,\ e'\, K^+ \Sigma^0,$$

$$K^- p \to \gamma \Lambda,\ \gamma \Sigma^o\ \text{(branching ratios with stopped kaons)}.$$

However, the importance of OSE for spin $3/2$ nucleonic resonances in the photoproduction of π and η mesons has recently been demonstrated [2].

In the past, two methods have been used to introduce the spin $3/2$ (and eventually $5/2$) nucleonic resonances in the strangeness sector: *i)* the invariant amplitudes are expressed as sums of resonant and non-resonant parts [3], with the latter contributions bringing in an undesirable behavior of the observables as energy increases; *ii)* an *ad-hoc* prescription is used [1,4] to preserve gauge invariance: the mass of the resonance appearing in the numerator of the spin $3/2$ propagator and in the expression of the spin $3/2$ vertex is replaced by the total invariant energy $\sqrt{s}$. The correct treatment of an interacting baryon, with spin higher than $1/2$, in the effective Lagrangian approaches [5] has to take into account the effects related to the off-shell behavior of the exchanged particles (or resonances) at the relevant vertices and propagators.

RESULTS AND DISCUSSION

Here, we present results of such a treatment and illustrate the sensitivity of different observables *via* a dynamical model quite similar to the Saclay-Lyon model [1]. Namely, a model containing, besides extended Born term and the above mentioned t-channel resonances, the following u- and s-channel resonances: $\Lambda(1405)$, $\Lambda(1670)$, $\Lambda(1810)$, $\Sigma(1660)$, $N(1720)$, with the latter

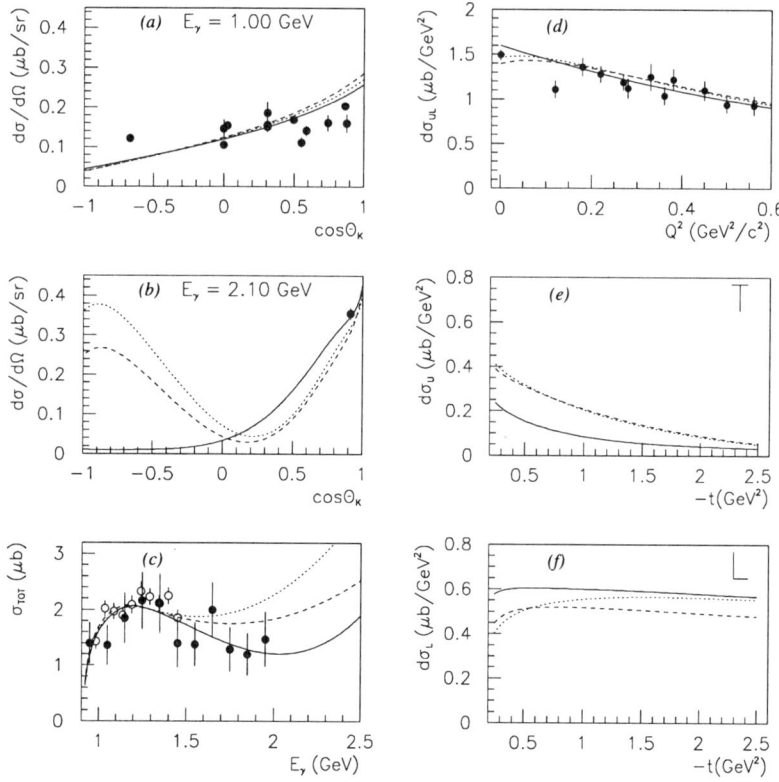

FIGURE 1. Observables for $K^+\Lambda$ channels. Results for $\gamma p \to K^+\Lambda$ reaction are: a) and b) angular distributions at $E_\gamma^{lab} = 1.0$ and 2.1 GeV, respectively, and c) total cross section. Results for $ep \to e'K^+\Lambda$, are: d) differential cross section $d\sigma_{UL}$ as a function of momentum transfer (Q^2) at s=5.02 GeV^2, t=-0.15 GeV^2, and ϵ=.72; e) and f) transverse (T) $d\sigma_U(t)$, and longitudinal (L) $d\sigma_L(t)$ components at Q^2=-1.0 (GeV/c)^2 and ϵ=.72. The curves are explained in the text. References to the data are given in Ref. [1].

one being the only spin 3/2 resonance of the model. The choice of $N(1720)$ was dictated by the present data after we examined possible contributions from all known spin 3/2 nucleonic and hyperonic resonances according to the procedure explained in Ref. [1].

In Fig. 1, the dotted curves correspond to this model *without* any OSE included [1]. The full curves differ from the latter by a proper OSE treatment [5] of the $N(1720)$. To *illustrate* the manifestation of off-shell effects, we have also added an hyperonic spin 3/2 resonance $\Lambda(1890)$ at the top of this model (dashed curves).

The photoproduction channel at low energy (Fig. 1a) does not show a significant sensitivity to OSE, while at higher energies (Fig. 1b), the backward hemisphere is drastically affected by the OSE. This behavior pulls down the total cross section at higher energies (Fig. 1c, full curve) as required by the existing data. Moreover, the preliminary results from SAPHIR collaboration [6], support strongly the need for taking into account the OSE as reported in Figs 1b and 1c (full curves).

For the electroproduction process, the unpolarized component of the differential cross section $d\sigma_{UL} = d\sigma_U + \varepsilon_L \, d\sigma_L$ depicted in Fig. 1d, shows no significant sensitivity to the OSE. However, its transverse (Fig. 1e) and longitudinal (Fig. 1f) components show sizeable differences according to the treatments investigated here.

The forthcoming electroproduction measurements at TJNAF/CEBAF [7] and photoproduction data from ELSA [6] are awaited for to clear up the importance of off-shell effects in the strangeness sector.

ACKNOWLEDGMENTS

We would like to thank Zhenping Li and Nimai Mukhopadhyay for fruitful discussions and Dietmar Menze and Reinhard Schumacher for helpful exchanges on the experimental results and projects.

REFERENCES

1. J.C. David, C. Fayard, G.H. Lamot, and B. Saghai, *Phys. Rev. C* **53**, 2613 (1996).
2. M. Benmerrouche, Nimai C. Mukhopadhyay, and J.F. Zhang, *Phys. Rev. D* **51**, 3237 (1995), and references therein.
3. F. M. Renard and Y. Renard, *Nucl. Phys.* **B25**, 490 (1971); Y. Renard, *ibid* **B40**, 499 (1972).
4. R. A. Adelseck, C. Bennhold, and L. E. Wright, *Phys. Rev. C* **32**, 1681 (1985).
5. C. Fayard, G.H. Lamot, T. Mizutani, and B. Saghai, *in preparation*.
6. D. Menze *et al.*, SAPHIR Collaboration, *in these Proceedings*.
7. O. K. Baker *et al.*, CEBAF-Hall C Collaboration, *in these Proceedings*.

Hyperon-Production with Anti-protons at LEAR

Jürgen Franz

Fakultät für Physik, Universität Freiburg, D-79104 Freiburg, Germany
Representing the PS185 Collaboration[1]

Abstract. Recent results from the experiment PS185 at LEAR/CERN on the production of antihyperon-hyperon ($\overline{Y}Y$) pairs are reported. An overview is given for the observables σ, $d\sigma/dt$, P, C_{ij} and S_F in the channel $\overline{p}p \to \overline{\Lambda}\Lambda$. The results are compared with other measured antihyperon-hyperon pairs: $\overline{\Sigma}^0\Lambda$+c.c., $\overline{\Sigma}^+\Sigma^+$ and $\overline{\Sigma}^-\Sigma^-$.

INTRODUCTION

The exclusive production of antihyperon-hyperon ($\overline{Y}Y$) pairs in $\overline{p}p$ collisions is an especially attractive reaction channel in investigating $\overline{N}N$ annihilation mechanisms. The strangeness produced in the final state provides information on the dynamics of flavor creation. Numerous calculations have been done in different approaches like meson-exchange, quark-inspired reaction mechanisms or amplitude analyses in the near-threshold region.

A discussion of various theoretical aspects appear in the contribution to this conference from M. Alberg. Here the experimental results are given and systematic features and apparent trends are pointed out. More details can be found in recent papers of the PS185 collaboration [1–5].

---

[1] The PS185 Collaboration at LEAR/CERN: B.Bassalleck, S.Eilerts, D.Fields, P.Kingsberry, J.Lowe, R.Stotzer (University of New Mexico, Albuquerque); H.Dutz, W.Meyer, G.Reicherz, B.Schoch (Universität Bonn); P.D.Barnes, G.Franklin, C.A.Meyer, B.Quinn, K.Paschke, R.Schumacher, V.Zeps (Carnegie-Mellon University); N.Hamann* (CERN); H.Dennert, W.Eyrich, J.Hauffe, A.Hofmann*, M.Moosburger, F.Stinzing, S.Wirth (Universität Erlangen-Nürnberg); W.Dutty, H.Fischer, J.Franz, E.Rössle, M.Ruh, H.Schledermann, H.Schmitt, R.Todenhagen (Universität Freiburg); R.Bröders, R.v.Frankenberg, R.Geyer, K.Kilian, W.Oelert, K.Röhrich, K.Sachs, T.Sefzick, G.Sehl M.Ziolkowski (Institut für Kernphysik der KFA Jülich); R.A.Eisenstein, P.G.Harris, D.W.Hertzog, S.A.Hughes, T.D.Jones, P.E.Reimer, R.L.Tayloe (University of Illinois, Urbana); T.Johansson, S.Ohlsson, S.Pomp (Uppsala University); W.Breunlich, N.Nägele (Institut für Mittelenergiephysik der ÖAW, Vienna).   *Deceased

EXPERIMENT

The PS185 experiment, performed at the Low-Energy Antiproton Ring (LEAR/CERN), was designed for high acceptance measurements for $\bar{p}p \to \bar{Y}Y$ from threshold to 2 GeV/c. The setup consists of a segmented target (C and CH_2), a tracking arrangement of 10 multiwire proportinal chamber and 13 driftchamber planes, a scintillator hodoscope and a baryon number identifier. The target segments with their surrounding scintillators were adapted to the requirements of the $\bar{Y}Y$ channel to be measured [5]. In combination with the scintillator hodoscope a neutral or double charge trigger could be formed in a flexible way. The track imaging detector had a 4π–acceptance for the decay (anti)baryons and in the three drift chamber planes inside the baryon identifier solenoid at least one of the decay (anti)baryons could be registered. For more details see [1–5].

CROSS SECTIONS

The majority of data has been taken for the reaction $\bar{p}p \to \bar{\Lambda}\Lambda$ and the precisely measured observables in this channel serve as a reference when comparing to the other hyperon channels. The integrated $\bar{\Lambda}\Lambda$ cross section rises rapidly from threshold and reaches its maximum of about 90 μb towards the highest $\bar{p}$ momenta at LEAR, i.e. $p_{\bar{p}} = 2\text{GeV/c}$ which corresponds to an excess energy $\epsilon_{\bar{\Lambda}\Lambda} = 200\text{MeV/c}$. The excess energy for $\bar{p}p \to \bar{Y}Y$ is defined as $\epsilon_{\bar{Y}Y} = \sqrt{s} - m_{\bar{Y}} - m_Y$. The energy dependence of the $\bar{\Lambda}\Lambda$ cross section is smooth. A preliminary analysis of a fine scan in the very threshold region, $\epsilon_{\bar{\Lambda}\Lambda} \lesssim 6$ MeV, shows no evidence of a resonance like structure at $\epsilon \approx 0.8$ MeV which has been suggested on the basis of previous data sets with less statistics.

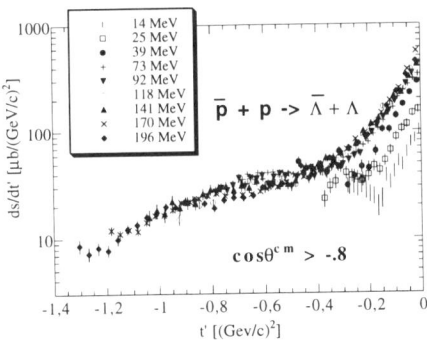

FIGURE 1. Differential cross sections $d\sigma/dt'$, for $\bar{p}p \to \bar{\Lambda}\Lambda$ in the LEAR momentum range.

The steep rise of the cross section at threshold indicates that already at very low excess energies a sizeable p-wave contribution is required. The presence of p-waves and contributions from higher partial waves is obvious from the shape of the differential cross section for $\overline{\Lambda}\Lambda$ production which is anisotropic and shows a strong peak in forward $\overline{\Lambda}$ direction. In Figure 1 the differential cross sections, $d\sigma/dt'$, for $\overline{p}p \to \overline{\Lambda}\Lambda$ are plotted for the full range of LEAR momenta versus the shifted 4-momentum transfer squared, t'. This variable $t' = t(\cos\theta^{cm}) - t(\cos\theta^{cm} = 1)$ is linearly related to $\cos\theta^{cm}$, where $t = m_p{}^2 + m_\Lambda{}^2 - \frac{s}{2} + \frac{1}{2}\sqrt{(s - 4m_p{}^2)((s - 4m_\Lambda{}^2)}\cos\theta^{cm}$. As can be seen from the semilogarithmic plot $d\sigma/dt'$ has a common feature at all energies, namely the same exponential slope in forward direction and a changeover to a more flat angular distribution at $|t'| \approx 0.15 - 0.2 \text{GeV}^2$. This behaviour is known for a strongly absorptive reaction mechanism. A fit of the forward data to an ansatz of diffractive scattering from a black disc, i.e. $\sigma \propto \exp(-bt')$, shows b-parameter values $\approx 8 - 10(\text{GeV/c})^2$ (Fig.2) which corresponds to a black disc radius of $\approx 1.1 - 1.3\text{fm}$.

For the $\overline{p}p \to \overline{\Sigma}{}^0\Lambda + \overline{\Lambda}\Sigma^0$ reaction we find a similar behaviour. The $\sigma(\overline{\Sigma}{}^0\Lambda)/\sigma(\overline{\Lambda}\Lambda)$ ratio stays the same within errors when compared at the same excess energies ϵ. The ratios (≈ 0.27) are in agreement with predictions from quark line models. The $\overline{\Sigma}{}^0\Lambda$ differential cross section rises steeper in forward

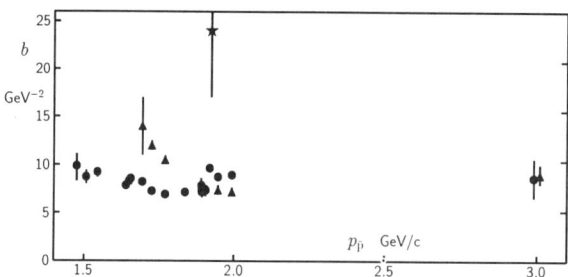

FIGURE 2. Exponential slope parameter b for $\overline{p}p \to \overline{\Lambda}\Lambda$ ($\bullet$), $\overline{\Sigma}{}^0\Lambda$ ($\triangle$) and $\overline{\Sigma}{}^+\Sigma^+$ ($\star$).

direction and hence the slope parameters b are larger (Fig. 2). At higher $\overline{p}$-momenta, however, the b values for the two processes approach each other and agree within their error bars.

The $\overline{\Sigma}{}^+\Sigma^+$ reaction shows the asymmetric angular distribution with an even stronger forward peak for the antihyperon. Its integrated cross section is not too far from estimations in simplified quark line models. The $\overline{\Sigma}{}^-\Sigma^-$ channel, however, has an unexpectedly high cross section close to threshold ($\epsilon_{\overline{\Sigma}\Sigma} = 7\text{MeV}$). This reaction has to proceed via higher order processes because two antiquark-quark pairs have to change flavor or in the meson exchange picture at least two mesons have to be involved in order to exchange the double charge. The $\overline{\Sigma^-}\Sigma^-$ differential cross section at $\epsilon = 32\text{MeV}$ has a more flat angular distribution in forward direction and we have an idication that this reaction has a strong rise in backward direction. This new feature, however, has to be confirmed by more elaborate checks.

SPIN OBSERVABLES

The big advantage of measurements of Λ and Σ (anti)hyperons is their asymmetric weak decay to nonstrange mesons and baryons which offers access to all spin observables. The polarization data for $\overline{\Lambda}\Lambda$ production ($P_{\overline{\Lambda}} = P_\Lambda$) show a cross over from positive to negative values at an approximately constant value of $|t'| \approx 0.15 - 0.2 \text{GeV}^2$ irrespective of the incident beam momentum. This occurs roughly at the same $|t'|$ value where $d\sigma/dt'$ changes from the steep forward to the flat backward distribution and this is probably an indication of a characteristic size associated with the strong absorption. Another zero crossing of $P_{\overline{\Lambda}}$ occurs at larger $|t'|$ values for the measurements at highest LEAR momenta.

The non-zero spin correlation parameters C_{ij} which include the combined decay information show approximately stable patterns when plotted versus the momentum transfer variable t'. The singlet fraction S_F can be derived from the C_{ij}-matrix as $S_F = \frac{1}{4}(1 - \vec{\sigma}_{\overline{\Lambda}} \cdot \vec{\sigma}_\Lambda) = \frac{1}{4}(1 + C_{\overline{x}x} - C_{\overline{y}y} + C_{\overline{z}z})$. We found at all measured $\overline{p}$-momenta $S_F = 0$ which means that the $\overline{\Lambda}\Lambda$ pair is always produced in a triplet state, i.e. with the spins of $\overline{\Lambda}$ and Λ aligned parallel. This is not the case for the reaction $\overline{p}p \to \overline{\Sigma^0}\Lambda$ where S_F differs from zero which is indicative of combined singlet and triplet production [3].

Our last measurement at LEAR – which was shut down in January 1997 – was performed after a major upgrade of the experiment. A "frozen spin" target was added to our hyperon decay spectrometer to allow a measurement of the depolarization D_{nn} and the spin transfer K_{nn} in the reaction $\overline{p}p \to \overline{\Lambda}\Lambda$. As pointed out in [6,7] the measurement of these spin observables has the potential of distinguishing meson exchange calculations from predictions of quark inspired models and may give insight to the question whether the proton wave function contains an admixture of polarized $\overline{s}s$ quark pairs.

REFERENCES

1. Eisenstein, R.A., *Physics of Atomic Nuclei* **57**, 1680 (1994).
2. Barnes P.D., et al., *Phys. Rev. C* **54**, 1877 (1996).
3. Barnes P.D., et al., *Phys. Rev. C* **54**, 2831 (1996).
4. Johansson T., *Invited Contribution at LEAP'96*
5. Barnes P.D., et al., *Phys. Lett. B* **402**, 227 (1997).
6. Bassalleck R., et al. *Proposal*, CERN/SPSLC 95-13, SPSLC/P287, (1995).
7. Alberg M., et al., *Phys. Lett. B* **356**, 113 (1995).

Antihyperon-hyperon Production in a Quark Model

M.A. Alberg[*,†] E.M. Henley[†] P.D. Kunz[‡] and L. Wilets[†]

[*] Department of Physics, Seattle University, Seattle, WA 98122
[†] Department of Physics, University of Washington, Seattle, WA 98195
[‡] Nuclear Physics Laboratory, University of Colorado, Boulder, CO 80309

Abstract. A quark model which includes both scalar and vector contributions to the reaction mechanism is used in a DWBA calculation of total and differential cross-sections, polarizations, and spin correlation coefficients for the reactions $\bar{p}p \to \bar{\Lambda}\Lambda, \bar{\Lambda}\Sigma, \bar{\Sigma}\Lambda$, and $\bar{\Sigma}\Sigma$ at laboratory momenta from threshold to 1.92 GeV/c. The free parameters of the calculation include the scalar and vector strengths, a quark cluster size parameter, and parameters of the unknown antihyperon-hyperon interactions. For $\bar{\Lambda}\Lambda$ production, we also calculate the target depolarization D_{nn}.

INTRODUCTION

Exclusive antihyperon-hyperon production provides a test for models of strangeness production and final state interactions. Good fits to data can be obtained for either meson exchange or quark models of the reaction. Haidenbauer et al. [1] have proposed that the measurement of target depolarization D_{nn} in $\bar{p}p \to \bar{\Lambda}\Lambda$ might help to determine which model is a better description of the reaction mechanism. It has also been suggested [2] that the measurement of this spin observable in the recent PS185 experiment [3] may test dynamical mechanisms invoked to explain the proton spin puzzle.

THEORY

The reaction mechanism in our quark model for $\bar{p}p \to \bar{Y}Y$ includes both scalar "$^3P_0$" and vector "$^3S_1$" contributions to $\bar{q}q$ annihilation and creation. We use a realistic $\bar{N}N$ potential in a DWBA calculation of total and differential cross-sections, polarizations, and spin correlation coefficients. We find excellent agreement with experiment for $\bar{Y}Y$ potentials which differ from those predicted on the basis of SU(3) arguments. Our results are presented in detail

TABLE 1. Born approximation for D_{nn}

model	reaction mechanism	D_{nn}
meson exchange	"tensor" $\vec{\sigma}_1 \cdot r\, \vec{\sigma}_2 \cdot r$	-1
	S_{12}	$-2/3$
quark	vector "$^3S_1$"	$+2/3$
	scalar "$^3P_0$"	≥ 0

elsewhere [4]. Meson-exchange calculations are also in reasonable agreement with experiment.

RESULTS

For the reaction $\bar{p}p \to \bar{\Lambda}\Lambda$, we have examined the suggestion of Haidenbauer et al. [1] that a measurement of the depolarization, D_{nn}, of a polarized proton target may discriminate between the two models for the reaction mechanism. As shown in Table 1, in Born approximation, meson exchange (dominated by tensor terms) and quark model predictions for D_{nn} differ even in sign. We find that these sign differences persist when initial and final state interactions are taken into account [5]. The sign of D_{nn} also distinguishes between two alternative dynamical mechanisms that have been invoked to explain the proton spin puzzle. Models with negatively polarized $\bar{s}s$ pairs in the proton wave fuction predict $D_{nn} < 0$, whereas models with positively polarized gluons predict $D_{nn} > 0$.

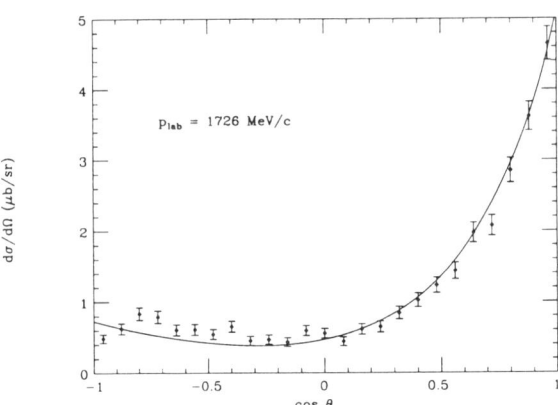

FIGURE 1. Differential cross section for $\bar{p}p \to \bar{\Lambda}\Sigma + c.c.$

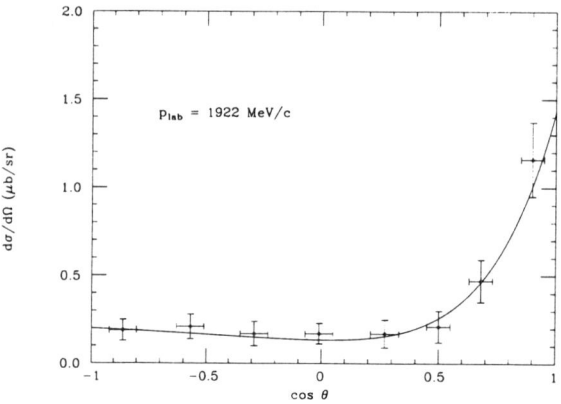

FIGURE 2. Differential cross section for $\bar{p}p \rightarrow \bar{\Sigma}^+\Sigma^+$

For $\bar{p}p \rightarrow \bar{\Sigma}\Lambda + c.c.$ we find a non-zero singlet fraction, in agreement with experiment. Our calculation of the differential cross section is compared to experiment [6] in Fig. 1.

Our preliminary results for $\bar{p}p \rightarrow \bar{\Sigma}^+\Sigma^+$ are shown in Fig. 2, also in excellent agreement with experiment [7].

ACKNOWLEDGMENTS

This work has been supported in part by the US Department of Energy, Contract # DE-FG03-97ER41014 and by the National Science Foundation.

REFERENCES

1. J. Haidenbauer, K. Holinde, V. Mull and J. Speth, *Phys. Lett.* B **291**, 223 (1992).
2. M. Alberg, J. Ellis and D. Kharzeev, *Phys. Lett.* B **356**, 113 (1995).
3. B. Bassalleck et al., Proposal P287, CERN/SPSLC 95-13 (accepted as PS185-3, completed Fall 1996).
4. M. Alberg, E.M. Henley, L. Wilets and P.D. Kunz, *Nucl. Phys.* A **560**, 365 (1993).
5. M. Alberg, E.M. Henley, L. Wilets and P.D. Kunz, *Phys. Atom. Nucl.* **57**, 1678 (1994).
6. P.D. Barnes et al., *Phys. Rev.* C **54**, 2831 (1996).
7. P.D. Barnes et al., *Phys. Lett.* B **402**, 227 (1997).

Recent Charm Physics Results from CLEO

Don H. Fujino[1]

*Lawrence Livermore National Laboratory
Livermore, CA 94550
Representing the CLEO Collaboration*

Abstract. In this talk I present recent charm physics results from the CLEO experiment. Final state interactions and W-annihilation effects in charmed mesons decays are discussed. These include an isospin analysis of $D \to K\bar{K}$, observation of the candidate W-annihilation decay $D_s^+ \to \omega\pi^+$, and evidence of non-factorizable effects in $D_s^+ \to \eta\pi^+$, $\eta'\pi^+$, $\eta\rho^+$, and $\eta'\rho^+$ decays. Presented next are CLEO's observations of the spin $\frac{3}{2}^+$ excited charmed baryons Σ_c^{*++} and Σ_c^{*0}, and the excited charmed-strange baryons Ξ_c^{*+} and Ξ_c^{*0}. I conclude with future prospects in charm physics with CLEO's new silicon detector.

INTRODUCTION

The external and internal spectator diagrams do an excellent job of explaining most of the features in charmed meson decays. For example, the lifetime hierarchy, *i.e.* the fact that the D^+ meson lifetime is a factor of $2 - 2.5$ times longer than the D^0 or D_s^+ lifetime can be attributed to the destructive interference between the external and internal diagrams for D^+ decays. The D^+ decays ($c\bar{d} \to su\bar{d}\bar{d}$) have two $\bar{d}$ quarks in the final state and so decays such as $D^+ \to \bar{K}^0\pi^+$ can occur through both the external and internal spectator diagrams. There is no interference in D^0 or D_s^+ decays since the anti-quarks in the final state are not identical for D^0 decays ($c\bar{u} \to su\bar{d}\bar{u}$) and D_s^+ decays ($c\bar{s} \to su\bar{d}\bar{s}$).

However, the charm quark is still light enough that non-factorizable effects can have non-negligible contributions when compared with the simple spectator decays in many exclusive decays of D mesons. Two effects I will explore with the recent CLEO results are final state interactions (FSI) and W-exchange and W-annihilation decays.

---

[1] This work performed under the auspices of the US Department of Energy by the Lawrence Livermore National Laboratory under Contract W-7405ENG-48.

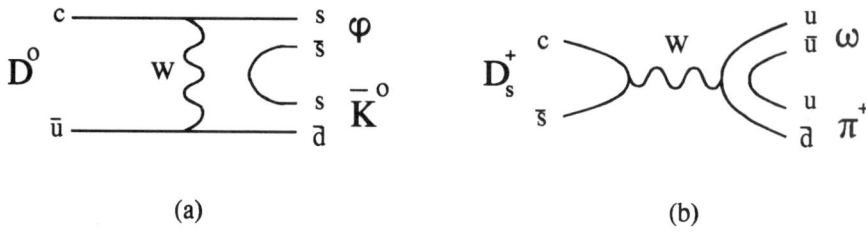

FIGURE 1. (a) W-exchange decay for $D^0 \to \phi \bar{K}^0$ and (b) W-annihilation decay for $D_s^+ \to \omega \pi^+$.

Final state interactions occur when the outgoing mesons get rescattered through intermediate resonances that lie near the D mass. For example, the exiting kaon and pion in the $D^0 \to K^-\pi^+$ decay can get rescattered through a highly excited $\bar{K}^0$ state and emerge as the final state $\bar{K}^0 \pi^0$. This can enhance color-suppressed decays, since FSI feeds into these channels from the color-allowed decays. For the $D \to K\pi$ system, $\mathcal{B}(D^0 \to \bar{K}^0\pi^0)/\mathcal{B}(D^0 \to K^-\pi^+) = 0.62 \pm 0.13$, which is much larger than the naive expectation of 1/9 from color matching for quarks.

The above example is considered elastic FSI because the isospin and spin content of the initial and rescattered mesons do not change. On the other hand, inelastic FSI change the isospin and spin structure of the outgoing mesons. Some examples of inelastic FSI include $K\rho \leftrightarrow K^*\pi$, $KK \leftrightarrow \pi\pi$, and $K^*K^* \leftrightarrow KK$ rescattering.

W-exchange and W-annihilations diagrams for D decays such as $D^0 \to \phi\bar{K}^0$ and $D_s^+ \to \omega\pi^+$ are shown in Figure 1. These diagrams are helicity suppressed compared to the spectator diagrams, and the ratio of decay rates is $\Gamma_{annihilation}/\Gamma_{spectator} \sim M_q^2/M_D^2$, where M_q is the light quark mass. The only clear evidence of W-annihilation decays is in the purely leptonic D_s^+ decays [1]. The $D_s^+ \to \mu\nu$ decay has been observed by WA75, CLEO, BES, and most recently by E653 with a branching fraction of $\sim 0.5\%$. The L3 Collaboration [2] has recently observed $D_s^+ \to \tau\nu$ with a branching fraction of $(7.4 \pm 3.6)\%$.

CHARM PHYSICS WITH CLEO

The CLEO Collaboration consists of over 200 physicists from 24 universities. The data were selected from hadronic events collected by the CLEO II detector at the Cornell Electron Storage Ring (CESR). The CLEO II detector [3] is a large solenoidal detector with 67 tracking layers and a CsI electromagnetic calorimeter that provides efficient π^0 reconstruction. Charged kaon and pions are identified using specific ionization (dE/dx) and, when available, time-of-flight (TOF) information. The dE/dx and TOF information provide good

K/π separation up to 0.7 and 1.1 GeV/c, respectively. For high momentum particles the dE/dx relativistic rise provides greater than 2σ K/π separation, which is essential for separating rare B processes such as $B^0 \to \pi^+\pi^-$ from $B^0 \to K^+\pi^-$.

The data used in most of the analyses presented in this talk consist of an integrated luminosity of 4.8 fb^{-1} taken at and just below the $\Upsilon(4S)$ resonance, corresponding to $\sim 3 \times 10^6$ $B\bar{B}$ events and $\sim 5 \times 10^6$ $e^+e^- \to c\bar{c}$ events. In general, the charm physics analyses at CLEO use a common set of techniques to enhance the charmed hadron signal of interest. These include cuts on the D momentum, event shape, and D^* mass difference. The momentum spectrum for charmed hadrons from $e^+e^- \to c\bar{c}$ events is quite hard, so requiring the scaled D momentum to satisfy $x_p = p_D/\sqrt{E_{beam}^2 - M_D^2} > 0.5$ removes a lot of the combinatoric background from low momentum tracks. Charmed hadrons from B decays will have a momenta $x_p < 0.5$, and as a side-effect get eliminated. The event shape can be used to enhance charm events since $e^+e^- \to c\bar{c}$ events are jetty, whereas $B\bar{B}$ events decay isotropically because the B mesons are produced nearly at rest. Finally, the D^* mass trick is a powerful tool to enhance the D signal. The D^0, D^+, and D_s^+ mesons are required to come from $D^{*+} \to D^0\pi^+$, $D^{*+} \to D^+\pi^0$, and $D_s^{*+} \to D_s^+\gamma$ decays, respectively. The $D^* \to D\pi$ decay has limited phase space and the resolution of the mass difference, $\Delta M = M_{D\pi} - M_D$, is on average 10 times better than the invariant mass of the reconstructed D meson. So for example, a $D^{*+} \to D^0\pi^+$ tag reduces the D^0 signal by 4× but reduces the background by 20 – 40×.

ISOSPIN ANALYSIS OF $D \to KK$

CLEO [4] has recently measured the branching fraction to the Cabibbo-allowed decay $D^+ \to K_s^0\pi^+$ and the Cabibbo-suppressed decay $D^+ \to K_s^0 K^+$. To extract the signals, we require a $D^{*+} \to D^+\pi^0$ tag and $x_p > 0.55$. We observe 70 ± 12 events for $D^+ \to K_s^0 K^+$ and 473 ± 26 events for $D^+ \to K_s^0\pi^+$ (shown in Figure 2.) The broad peak on the right of the $D^+ \to K_s^0 K^+$ signal are $D^+ \to K_s^0\pi^+$ events where the pion is mis-identified as a kaon, and the excess on the left is due to feed-down from $D \to K_s^0\pi\pi$ events.

The branching fractions are computed to be $\mathcal{B}(D^+ \to \bar{K}^0 K^+) = (0.70 \pm 0.12 \pm 0.07 \pm 0.05)\%$ and $\mathcal{B}(D^+ \to \bar{K}^0\pi^+) = (3.17 \pm 0.21 \pm 0.19 \pm 0.21 \pm 0.32)\%$, where the second error is systematic, the third is due to uncertainty in the normalization branching fraction, and the fourth error is due to possible interference from doubly Cabibbo suppressed decays (DCSD). The ratio of Cabibbo suppressed to allowed modes is 3.6σ higher than $\tan^2\theta_C \approx 0.05$. This is due to the destructive interference occurring in the $D^+ \to \bar{K}^0\pi^+$ decay mode.

Using CLEO's [1] previous results on $D^0 \to K^+K^-$, $K^0\bar{K}^0$, and $\bar{K}^0\pi^0$, we have measured the isospin amplitudes and relative phase shift for $D \to K\bar{K}$

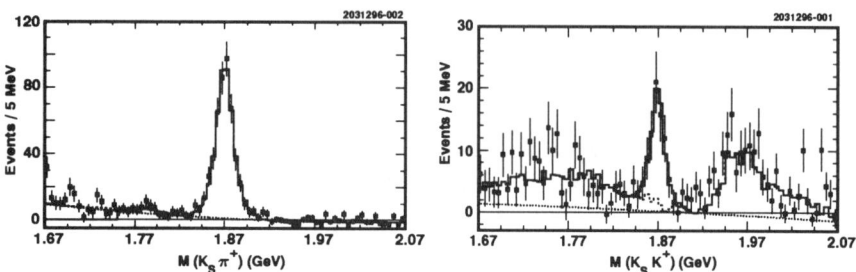

FIGURE 2. The invariant mass distribution for $K_s^0 \pi^+$ and $K_s^0 K^+$ combinations.

TABLE 1. Isospin analysis of $D \to KK$, $K\pi$, and $\pi\pi$.

Decay Mode	Amplitude Ratio	$\cos\delta$
$D \to KK$	$\left|\frac{A_1}{A_0}\right| = 0.61 \pm 0.11$	0.88 ± 0.09
$D \to K\pi$	$\left|\frac{A_{3/2}}{A_{1/2}}\right| = 0.27 \pm 0.03$	-0.12 ± 0.22
$D \to \pi\pi$	$\left|\frac{A_2}{A_0}\right| = 0.72 \pm 0.17$	0.14 ± 0.16

for the first time, and improved the isospin analysis for the $D \to K\pi$ system. The three decay modes $D^+ \to K^+K^-$, $D^+ \to K^0\bar{K}^0$, and $D^+ \to \bar{K}^0 K^+$ can be described by two isospin amplitudes: A_0 and A_1, which correspond to the isospin 0 and 1 states for $K\bar{K}$. The decay rate amplitudes are linear combinations of the two isospin amplitudes.

$$Amp(D^+ \to K^+K^-) = \sqrt{1/2}(A_1 + A_0)$$
$$Amp(D^+ \to K^0\bar{K}^0) = \sqrt{1/2}(A_1 - A_0)$$
$$Amp(D^+ \to \bar{K}^0 K^+) = \sqrt{2}A_1$$

Similar isospin relations can be expressed for $D \to K\pi$ and $D \to \pi\pi$.

In the absence of FSI both A_0 and A_1 are real. When elastic FSI is turned on via resonant rescattering, the magnitude of the two isospin amplitudes do not change, but phase shifts can be introduced. This causes the K^+K^- and $K^0\bar{K}^0$ amplitudes to mix while the summed rate $\Gamma(D^+ \to K^+K^-)+\Gamma(D^+ \to K^0\bar{K}^0)$ stays unchanged. The $D^+ \to \bar{K}^0 K^+$ amplitude remains invariant. Table 1 shows the isospin amplitude ratio and the relative phase shift (δ). The $\Delta I = 1/2$ rule in $K \to \pi\pi$ decays, which suppresses the $K^+ \to \pi^+\pi^0$ decay and implies $|A_2/A_0| \approx 0.05$, does not hold for $D \to \pi\pi$ decays which has a large A_2 amplitude. The striking difference in behavior is not well understood.

In the $D \to K\pi$ and $\pi\pi$ systems, the color suppressed modes are apprecia-

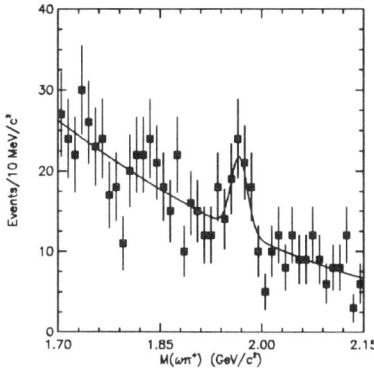

FIGURE 3. The invariant mass distribution for $\omega\pi^+$ combinations from tagged D_s^{*+} decays.

ble because $\delta \approx 90°$. The ratio of color suppressed to color allowed branching fractions are large: $\mathcal{B}(D^0 \to \bar{K}^0\pi^0)/\mathcal{B}(D^0 \to K^-\pi^+) = 0.62 \pm 0.13$ and $\mathcal{B}(D^0 \to \pi^0\pi^0)/\mathcal{B}(D^0 \to \pi^+\pi^-) = 0.63 \pm 0.18$. Turning off elastic FSI by setting $\delta \to 0$ would decrease the ratios significantly to 0.14 and 0.00, respectively.

On the other hand, the large ratio $\mathcal{B}(D^0 \to K^0\bar{K}^0)/\mathcal{B}(D^0 \to K^+K^-) = 0.12 \pm 0.03$ is unexpected. Because the $D^0 \to K^0\bar{K}^0$ decay can only occur at tree-level via W-exchange, assuming this rate is negligible we would expect $|A_1/A_0| = 1$ and the $D^0 \to K^0\bar{K}^0$ rate to be non-zero solely through FSI. But δ is consistent with being zero, so elastic FSI alone cannot be the explanation. Moreover, $|A_1/A_0|$ is 3.5σ from unity. So either W-exchange contributions are large or "inelastic" FSI, which can change both the magnitude and phase of A_0 and A_1, are appreciable.

OBSERVATION OF $D_S^+ \to \omega\pi^+$

D_s^+ decays into final states with no strangeness are strong candidates for W-annihilation decays, since D_s^+ spectator decays will always have an $\bar{s}$ quark in the final state. The $D_s^+ \to \omega\pi^+$ decay was expected to be a smoking gun for W-annihilation (see Figure 1) since the decay has no FSI contributions; the quantum numbers of the $\omega\pi^+$ system ($J^P = 0^-$ and $I^G = 1^+$) do not correspond to any known resonances. This is not true for $D_s^+ \to \pi\pi$ or $\rho\pi$ which are expected to have significant FSI.

However, this simple model is flawed. The $u\bar{d}$ quark pair from W-annihilation cannot hadronize into $\omega\pi^+$ due to G-parity conservation, which would require a second-class axial current. If, however, three gluons are connected to the initial state $c\bar{s}$ quark line, the modified W-annihilation process can occur with $W^+ \to u\bar{d} \to \pi^+$ and $ggg \to \omega$. Since the gluons also can carry spin, this process is no longer helicity suppressed. To further complicate the picture, although resonant FSI are not possible, Kamal et al. [5] and Buccella et al. [6] have suggested that non-resonant FSI can feed into $D_s^+ \to \omega\pi^+$ in the range 0.3 – 3%.

CLEO [7] has searched for the decay $D_s^+ \to \omega\pi^+$ with $\omega \to \pi^+\pi^-\pi^0$. Only $\omega \to \pi^+\pi^-\pi^0$ combinations in the central Dalitz region were selected, and $\omega\pi^+$ candidates were required to come from D_s^{*+} decays. We observed 36 ± 10 $D_s^+ \to \omega\pi^+$ events, shown in Figure 3. To verify there is no contribution from other $D_s^+ \to 4\pi$ decays, the ω mass cut was loosened and the 4π invariant mass was required to be in the D_s^+ signal region. A fit to the 3π invariant mass after a D_s^+ sideband subtraction yielded a comparable ω signal of 32 ± 12 events. The $\omega\pi^+$ combinations that failed the D_s^{*+} tag still contain an excess of 133 ± 57 events at the D_s^+ mass. The tagged and untagged results were combined and the branching fraction was normalized to the $D_s^+ \to \eta\pi^+$ decay to obtain $\mathcal{B}(D_s^+ \to \omega\pi^+)/\mathcal{B}(D_s^+ \to \eta\pi^+) = 0.16 \pm 0.04 \pm 0.03$. This is the first measurement of $D_s^+ \to \omega\pi^+$. The decay cannot occur from a simple spectator diagram, but must be due to FSI and/or W-annihilation.

MEASUREMENT OF $D_S^+ \to \eta\pi^+, \eta'\pi^+, \eta\rho^+,$ AND $\eta'\rho^+$

The decays $D_s^+ \to \eta\pi^+, \eta'\pi^+, \eta\rho^+,$ and $\eta'\rho^+$ comprise 25 – 30% of the total D_s^+ branching fraction. An earlier CLEO analysis showed a 2σ departure from the factorization hypothesis that relates the two-body hadronic decay $D_s^+ \to \eta'\rho^+$ to the semileptonic decay $D_s^+ \to \eta'\ell\nu$:

$$\Gamma(D_s^+ \to \eta'\rho^+) = 6\pi^2 a_1^2 f_\rho^2 |V_{ud}|^2 \frac{d\Gamma}{dq^2}(D_s^+ \to \eta'\ell\nu)\Big|_{q^2=m_\rho^2}$$

CLEO measured $\Gamma(D_s^+ \to \eta'\rho^+)/\Gamma(D_s^+ \to \eta'\ell\nu) = 14.8 \pm 5.8$ and $\Gamma(D_s^+ \to \eta\rho^+)/\Gamma(D_s^+ \to \eta\ell\nu) = 4.3 \pm 1.1$, whereas theory predicts both ratios to be 2.9 [8].

With the full CLEO II dataset we have reexamined these branching fractions and the factorization hypothesis. For the $D_s^+ \to \eta\pi^+$ and $\eta'\pi^+$ modes, the η is reconstructed in the $\gamma\gamma$ and $\pi^+\pi^-\pi^0$ channels. To construct the $\eta\rho^+$ and $\eta'\rho^+$ invariant mass distributions, only the $\eta \to \gamma\gamma$ mode was used. Non-resonant $\eta^{(\prime)}\pi^+\pi^0$ contributions were reduced by requiring the $\pi^+\pi^0$ mass to be within 170 MeV/c^2 of the ρ^+ mass and the helicity angle to be $|\cos\theta_\pi| > 0.45$ since the ρ^+ must have zero helicity. There can still be non-negligible non-resonant feedthrough, so the $D_s^+ \to \eta\rho^+$ and $\eta'\rho^+$ branching fractions are extracted by relaxing the ρ cuts and fitting the Dalitz plot. Shown in Table 2 are the measured branching fractions along with the theoretical predictions. The variety of theoretical predictions assume form factors for the two-body decays from the pole model and from semileptonic decays. The predictions are consistent with the $\eta\pi^+$, $\eta'\pi^+$, and $\eta\rho^+$ branching fractions but cannot explain

TABLE 2. CLEO results and theoretical predictions for the $D_s^+ \to \eta\pi^+$, $\eta'\pi^+$, $\eta\rho^+$, and $\eta'\rho^+$ branching fractions.

D_s^+ Mode	$\Gamma/\Gamma(D_s^+ \to \phi\pi^+)$				
	CLEO	VKK [9]	BSW [10]	BLP [6]	HK [11]
$\eta\pi^+$	$0.48 \pm 0.03 \pm 0.04$	0.9	1.0	0.3	0.6
$\eta'\pi^+$	$1.03 \pm 0.06 \pm 0.07$	0.8	0.6	1.3	1.6
$\eta\rho^+$	$2.98 \pm 0.20 \pm 0.39$	2.0	2.0	1.8	2.9
$\eta'\rho^+$	$2.78 \pm 0.28 \pm 0.30$	0.7	0.6	0.6	0.4

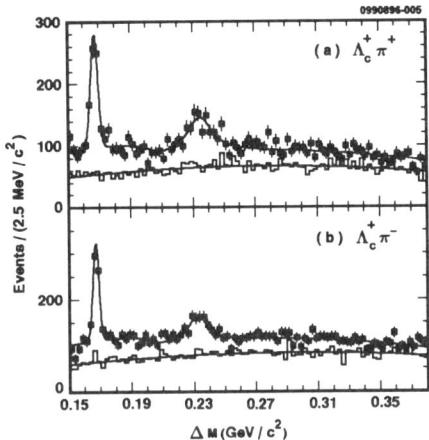

FIGURE 4. Mass difference spectra for (a) $\Lambda_c^+\pi^+$ candidates and (b) $\Lambda_c^+\pi^-$ candidates. The histogram shows the spectra for normalized sidebands of the Λ_c^+.

the large rate measured for $D_s^+ \to \eta'\rho^+$. Nor can FSI both resonant and non-resonant explain the discrepancy. In terms of the original factorization hypothesis, we now measure $\Gamma(D_s^+ \to \eta'\rho^+)/\Gamma(D_s^+ \to \eta'\ell\nu) = 12.0 \pm 4.3$ and $\Gamma(D_s^+ \to \eta\rho^+)/\Gamma(D_s^+ \to \eta\ell\nu) = 4.4 \pm 1.2$, compared with the theoretical prediction of 2.9. The uncertainty is primarily due to the uncertainty in the measured semileptonic decay rates. Ball et al. [12] has suggested that a modified W-annihilation diagram may be enhancing this mode, where two gluons connected to the initial $c\bar{s}$ quark line hadronize to the η'. This is analogous to the W-annihilation diagram for $D_s^+ \to \omega\pi^+$.

CHARMED BARYON SPECTROSCOPY

The family of charmed baryons is quite diverse. There is the isosinglet Λ_c^+ and the isotriplet Σ_c, the charmed-strange baryons Ξ_c and Ξ_c', and finally the

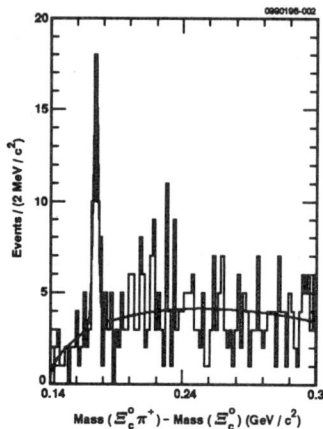 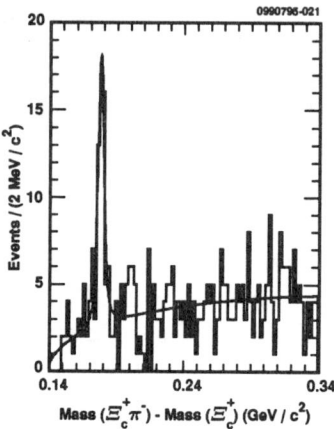

FIGURE 5. Mass difference spectra for (a) $\Xi_c^0 \pi^+$ candidates and (b) $\Xi_c^+ \pi^-$ candidates.

doubly-strange charmed baryon Ω_c. Most of the ground state charmed baryons with $J^P = \frac{1}{2}^+$ have been well measured. CLEO [13–15] has now observed many of the spin $\frac{3}{2}^+$ excitations of charmed baryons. These include the Σ_c^{*++} and Σ_c^{*0} which decay into $\Lambda_c^+ \pi^+$ and $\Lambda_c^+ \pi^-$, respectively; and the charmed-strange baryons Ξ_c^{*+} and Ξ_c^{*0} which are observed via the decay into $\Xi_c^0 \pi^+$ and $\Xi_c^+ \pi^-$, respectively. The Σ_c^{*++} and Σ_c^{*0} are part of an isospin triplet, but the Σ_c^{*+} has yet to be seen, primarily because of the difficulty of reconstructing its decay to $\Lambda_c^+ \pi^0$.

The search for these spin excitations is similar to reconstructing D^* mesons. We accumulate large samples of the ground state charmed baryons, add a charged pion, apply an $x_p > 0.5$ cut, and examine the mass difference spectrum. For the Σ_c^{*++} and Σ_c^{*0} search, we start with ~ 15000 Λ_c^+ candidates from 13 decay modes. These include $\Lambda_c^+ \to pK^-\pi^+$ (~ 8400 evts), pK_s^0 (~ 1000 evts), $\Lambda\pi^+$ (~ 1100 evts), and $\Lambda\pi^+\pi^0$ (~ 900 evts). The $\Lambda_c^+\pi^+$ like-sign and $\Lambda_c^+\pi^-$ unlike-sign mass difference distributions are shown in Figure 4. Results of the fits to the Σ_c^* signals with a Breit-Wigner convoluted with the detector resolution are shown in Table 3. The broad peaks at $\Delta M \approx 233$ MeV/c^2 are the signals for Σ_c^{*++} and Σ_c^{*0}, in very good agreement with theoretical predictions. Both the $\Lambda_c^+\pi^+$ and $\Lambda_c^+\pi^-$ have nearly identical mass splittings, as is expected for isospin partners. The narrow peak at $\Delta M \approx 167$ MeV/c^2 is the ground state Σ_c. The Σ_c^* natural width is large because of the available phase space.

To search for the excited Ξ_c^{*+} and Ξ_c^{*0} baryons, we collect a large sample of the ground state Ξ_c^+ and Ξ_c^0 baryons. We use about 300 Ξ_c^+ candidates from the decays $\Xi_c^+ \to \Xi^-\pi^+\pi^+$, $\Xi^0\pi^+\pi^0$, and $\Sigma^+ K^{*0}$; and about 300 Ξ_c^0 candidates from the decays $\Xi_c^0 \to \Xi^-\pi^+$, $\Omega^- K^+$, $\Xi^-\pi^+\pi^0$, and $\Xi^0\pi^+\pi^-$. Combining

TABLE 3. Spin $\frac{3}{2}^+$ excited charmed baryons: Σ_c^* and Ξ_c^*.

Decay Mode	Signal (evts)	ΔM (MeV/c^2)	Γ (MeV/c^2)
$\Sigma_c^{*++} \to \Lambda_c^+ \pi^+$	677^{+101}_{-93}	$234.5 \pm 1.1 \pm 0.8$	$17.9^{+3.8}_{-3.2} \pm 4.0$
$\Sigma_c^{*0} \to \Lambda_c^+ \pi^-$	504^{+93}_{-83}	$232.6 \pm 1.0 \pm 0.8$	$13.0^{+3.7}_{-3.0} \pm 4.0$
$\Xi_c^{*+} \to \Xi_c^0 \pi^+$	35 ± 9	$174.3 \pm 0.5 \pm 1.0$	< 3.9
$\Xi_c^{*0} \to \Xi_c^+ \pi^-$	55 ± 12	$178.2 \pm 0.5 \pm 1.0$	< 5.5

the Ξ_c candidate with a charged pion and calculating the mass difference $\Delta M = M(\Xi_c \pi) - M(\Xi_c)$, we observe a narrow peak at $\Delta M = 174.3 \pm 0.5 \pm 1.0$ for $\Xi_c^0 \pi^+$ and a narrow peak at $\Delta M = 178.2 \pm 0.5 \pm 1.0$ for $\Xi_c^+ \pi^-$ combinations (see Figure 5 and Table 3). We identify these states as the Ξ_c^{*+} and Ξ_c^{*0}, respectively, because the theoretical mass predictions are consistent with our measurements and the two mass differences are nearly identical as is expected for isospin partners.

FUTURE PROSPECTS IN CHARM PHYSICS

In the Fall of 1995 the Silicon Vertex detector (SVX) was installed in the CLEO detector, and we have to date collected ~ 3 fb^{-1} of data, roughly 60% of the CLEO II dataset. The SVX is a three-layer, double-sided silicon strip detector that provides both $r\phi$ and z measurements. This will give a big boost to the charm physics program at CLEO. For the first time we will have decay length information for charmed hadrons. Vertex constrained fits to three-prong D^+ or two-prong D^0 decays will greatly reduce the combinatoric backgrounds. The $D^{*+} \to D^0 \pi^+$ mass difference resolution has improved three-fold due to the SVX performance, Kalman track fitting, and vertex constrained fitting. Finally, the CLEO drift chamber is now using a helium based gas in place of Argon-Ethane, which reduces the multiple scattering and reduces the Lorentz angle to provide improved hit efficiency over the entire drift cell.

Below is a wish list of future charm physics analyses that will benefit strongly from the SVX and improved tracking capabilities.

- Measure the D^0 and D^+ lifetimes with $1 - 2\%$ precision

- Untangle the DCSD and $D^0 \bar{D}^0$ mixing contributions to $D^0 \to K^+ \pi^-$ decays from the decay time information.

- Measure the natural width of D^{*+} to a precision of ~ 50 KeV

- Measure Cabibbo suppressed decays such as $D^+ \to \omega \ell^+ \nu$

- Analyze $D^+ \to \bar{K}^{*0} \ell^+ \nu$ form factors

The CLEO detector continues to evolve and will undergo a major upgrade in two years time when basically everything inside the CsI calorimeter will be replaced. The CESR peak luminosity is currently 4×10^{32} cm^{-2}s^{-1}; there is steady progress to increase the luminosity by a factor of 10. This will necessitate new rare earth and superconducting quadrupole magnets in the final focus. CLEO will build a new four-layer silicon detector and drift chamber which uses a helium-based gas. Finally, to identify pions and kaons with $> 4\sigma$ separation over the entire momentum and $\cos\theta$ range a ring imaging Cerenkov detector will be constructed.

In summary, the recent CLEO results in charm physics have explored final state interactions and W-annihilation processes in charmed meson decays. Isospin analyses of $D \to K\bar{K}$, $K\pi$, and $\pi\pi$ show significant FSI. The candidate W-annihilation decay $D_s^+ \to \omega\pi^+$ has been observed. The $D_s^+ \to \eta'\rho^+$ decay rate cannot be explained by factorization and may have a decay mechanism similar to that of $D_s^+ \to \omega\pi^+$. We have now observed many of the spin $\frac{3}{2}^+$ excited charmed baryons: Σ_c^{*++}, Σ_c^{*0}, Ξ_c^{*+}, and Ξ_c^{*0}. Discovery of the missing spectral states including the Ξ_c' and new orbitally excited charmed baryons are just around the corner. With luminosity and detector advancements, CLEO will continue to have a rich charm physics program in the years to come.

Acknowledgements: I would like to thank H. Yamamoto, J. Urheim, L. Gibbons, J. Bartelt, D. Kim, and M. Bishai for useful and stimulating discussions.

REFERENCES

1. Particle Data Group, R.M. Barnett *et al.*, Phys. Rev. D **54**, 1 (1996).
2. L3 Collaboration, preprint CERN-PPE/96-198.
3. Y. Kubota *et al.*, Nucl. Inst. and Meth. **A320**, 66 (1992).
4. M. Bishai *et al.*, Phys. Rev. Lett. **78**, 3261 (1997).
5. A.N. Kamal, N. Sinha, and R. Sinha, Phys. Rev. D **39**, 3503 (1989).
6. F. Buccella, M. Lusignoli, and A. Pugliese, Phys. Lett. B **379**, 249 (1996).
7. R. Balest *et al.*, preprint CLNS 97/1479, CLEO 97-8.
8. A.N. Kamal, Q.P. Xu, and A. Czarnecki, Phys. Rev. D **49**, 1330 (1994).
9. R.C. Verma, A.N. Kamal, and M.P. Khanna, Z. Phys. C **65**, 255 (1995).
10. M. Bauer, B. Stech, and M. Wirbel, Z. Phys. C **34**, 102 (1987).
11. I. Hinchliffe and T.A. Kaeding, Phys. Rev. D **54**, 914 (1996).
12. P. Ball, J.M. Frere, and M. Tytgat, Phys. Lett. B **365**, 367 (1996).
13. G. Brandenburg *et al.*, Phys. Rev. Lett. **78**, 2304 (1997).
14. L. Gibbons *et al.*, Phys. Rev. Lett. **77**, 810 (1996).
15. P. Avery *et al.*, Phys. Rev. Lett. **75**, 4364 (1995).

Charm- & Strangeness-Production in Σ^--Nucleus-Interactions

Eva B. Wittmann
WA89-Collaboration

*Max-Planck-Instut für Kernphysik;
Postfach 103980, D-69029 Heidelberg, Germany.*

Abstract. We report on a study of the leading particle effect both in strange and charm final states. In the strange sector the inclusive cross sections for Ξ^- hyperon production in high-energy Σ^-, π^- and neutron induced interactions were measured by the WA89 experiment at CERN. A strong leading effect for Ξ^- produced by Σ^- is observed. The influence of the target mass on the Ξ^- cross section is explored by comparing reactions on copper and carbon nuclei. This constitutes the first measurement using three different projectiles and two different targets within the same experiment thus ensuring small systematic uncertainties. In the charm sector we studied the leading particle effect in D^-/D^+, D_s^-/D_s^+ and $\Lambda_c^+/\bar{\Lambda}_c^-$ production in high energy Σ^- interactions. The study of D_s^-/D_s^+ production represents the first measurement of charm production with a leading strange quark.

INTRODUCTION

The predictive power of Quantum Chromodynamics is currently limited to interactions with high momenta transferred where the perturbative approach (pQCD) is applicable. The process of hadronisation, i.e. the recombination of quarks into observable hadrons does not fall into this category and needs different approaches to be understood. Hadroproduction of charmed particles includes both processes since the production of the relatively heavy charm quark can be described within pQCD while the kinematic distributions of charmed particles reflect the process of hadronisation.

The WA89-Experiment at CERN-SPS measured in 1993/1994 the hadroproduction of charm and strange hadrons. Data were obtained using a 345 GeV/c Σ^--beam incident upon copper and carbon-targets. A large acceptance spectrometer containing a silicon vertex detector and a ring imaging Cherenkov counter provided high vertex precision and good particle identification. Both criteria are vital ingredients for charm reconstruction. A detailed description of the detector can be found elsewhere [1] and references there in.

THE ROLE OF THE BEAM VALENCE QUARKS IN PRODUCTION

The role of the beam valence quarks in the hadronisation process can be explored by studying the so called 'leading particle effect'. This is an enhancement in the differential cross section in the forward direction when the beam and the final state particle have at least one valence quark in common. To compare leading and nonleading production one can (1) either look at the same final state with different beams of leading or nonleading quark content or (2) look at leading and nonleading final states with the same beam. In the study of strange production the first method is used, while in the study of charm production the second method is applied.

Leading Particle Effect in Strange Final States

We report on a study of the leading particle effect in inclusive Ξ^- production in high energy Σ^-, π^- and neutron induced interactions [1]. Secondary Σ^- and π^- beams with average momenta of 345 GeV/c and a neutron beam with average momentum of 260 GeV/c were produced by primary 450 GeV/c protons from the CERN SPS. The studied reaction modes together with the

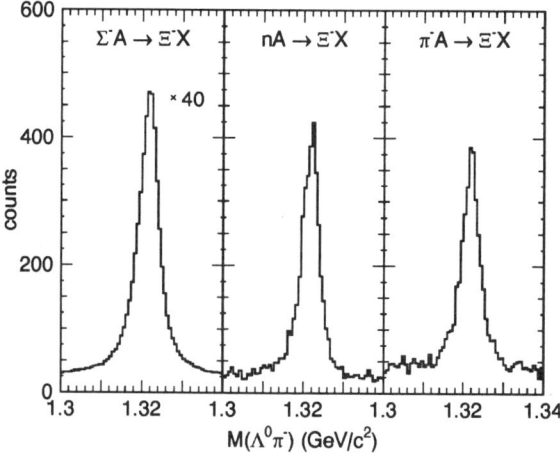

FIGURE 1. Invariant mass distributions of $\Lambda\pi^-$ pairs produced in Σ^- (left), neutron (centre) and π^- (right) reactions. The data of both targets have been added.

collected statistics are summarized in the following table and the reconstructed candidates are shown in Fig. 1.

$\Sigma^-(sdd)$ $\approx$ 178000 candidates   leading sd-quarks
$\pi^-(d\bar{u}) \rightarrow \Xi^-(sds)$ $\approx$ 3000 candidates   leading d-quark
$n\ (ddu)$ $\approx$ 3500 candidates   leading d-quark

Our measurement of the differential cross-section of the Ξ^- produced by the three different beam particles (Fig. 2) show comparable Ξ^- cross sections of the different beam particles in the central region thus indicating that the initial (strangeness) content in the projectile is not relevant and all quarks are produced in the fragmentation process. In turn, in the projectile fragmentation region the quark overlap between projectile and produced particle is reflected in the production yield thus resulting in a strong leading effect for the Σ^- projectile at large x_F.

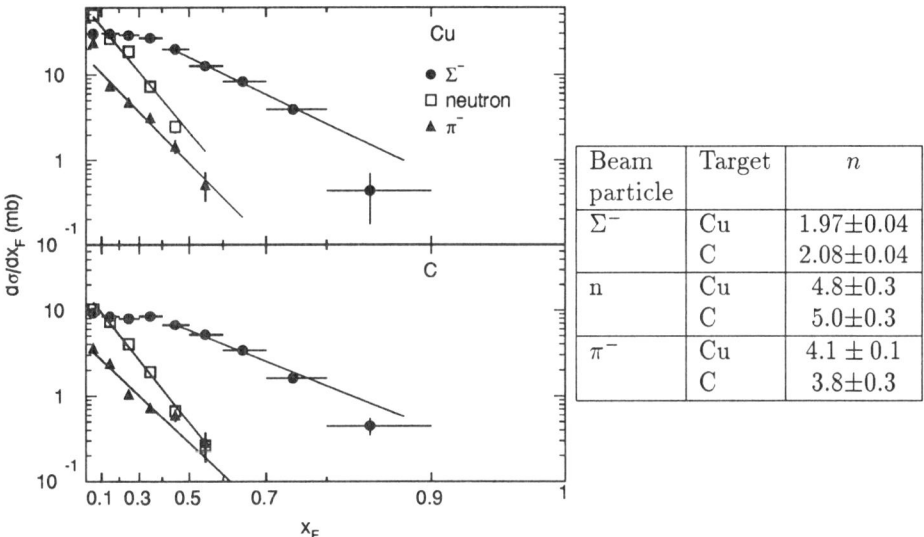

FIGURE 2. The differential cross section of inclusive Ξ^--production $d\sigma/dx_F$ as a function of x_F integrated over p_t in case of neutron, π^- and Σ^- interactions with copper (top part) and carbon (bottom part). The lines represent fits proportional to $(1-x_F)^n$ with the parameters n as given in table on the right.

Leading Particle Effect in Charm Final States

We also report on a study of the leading particle effect in D^-/D^+, D_s^-/D_s^+ and $\Lambda_c^+/\bar{\Lambda}_c^-$ production by Σ^- of 345 GeV/c. The reaction modes are summarized in the following table:

leading d-quark: non-leading:
$\Sigma^-(sdd) \longrightarrow D^-(\bar{c}d)$ $\Sigma^-(sdd) \longrightarrow D^+(c\bar{d})$
$\Sigma^-(sdd) \longrightarrow \Lambda_c^+(cud)$ $\Sigma^-(sdd) \longrightarrow \bar{\Lambda}_c^-(\bar{c}\bar{u}d)$
leading s-quark: non-leading:
$\Sigma^-(dds) \longrightarrow D_s^-(\bar{c}s)$ $\Sigma^-(dds) \longrightarrow D_s^+(c\bar{s})$

The reconstruction efficiency for the particle- and the referring antiparticle final state are equal because one final state is the charge conjugate of the other and our detector is symmetric for both signs of charge.

The D^- (D^+) are reconstructed in the dominant hadronic mode $K^+\pi^-\pi^-$ ($K^-\pi^+\pi^+$) of which the branching fraction is $9.1 \pm 0.6\%$ [2]. The invariant mass distributions are shown in Fig. 3.

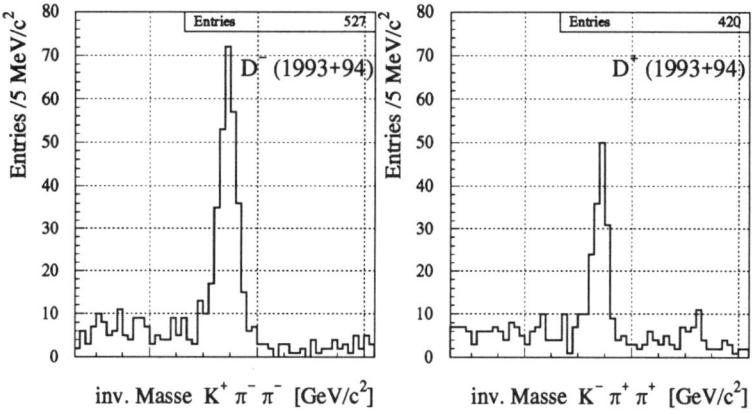

FIGURE 3. Invariant Mass distributions of leading $K^+\pi^-\pi^-$ (left) and non-leading $K^-\pi^+\pi^+$ (right) candidates produced in high energy Σ^- interactions.

We determined the cross section ratio R of the nonleading $D^+(c\bar{d})$ and the leading $D^-(\bar{c}d)$ production mode to be $R = \sigma(D^+)/\sigma(D^-) = 0.6 \pm 0.2$ for $x_F > 0.2$ (WA89 preliminary). This signals a clear leading particle production. The value is consistent with the value of $R = 0.69 \pm 0.15$ for $x_F > 0$ measured by E769 in proton induced D^+/D^--production [4].

Additionally the production asymmetry $A(D^-, D^+)$

$$A(D^-, D^+) = \frac{\sigma(D^-) - \sigma(D^+)}{\sigma(D^-) + \sigma(D^+)}$$

(which is related to the ratio R by $R = \frac{1-A}{1+A}$) was determined for two bins of x_F (see Fig. 4):

$x_F[0.1 - 0.3]$: $A(D^-, D^+) = 0.24 \pm 0.13$
$x_F[0.3 - 0.8]$: $A(D^-, D^+) = 0.51 \pm 0.1$

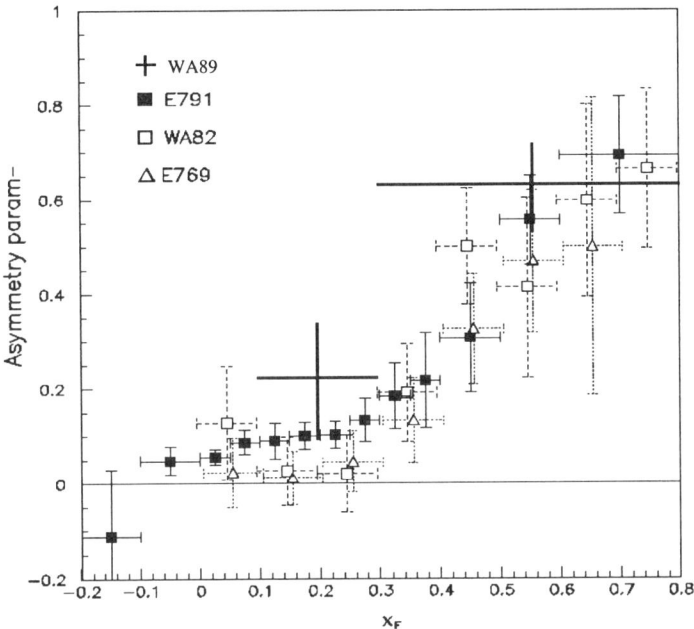

FIGURE 4. WA89 preliminary measurement of the asymmetry $A(D^-, D^+)$ as defined in the text compared to the results of other experiments.

Considering the large statistical uncertainties, the agreement between our results and data from other experiments is good.

The phenomenon of leading production is also observed in D_s^-/D_s^+ and $\Lambda_c^+/\bar{\Lambda}_c^-$ production by Σ^-. The D_s^- (D_s^+) are reconstructed in the hadronic decay mode $K^+K^-\pi^-$ ($K^-K^+\pi^+$) having a branching fraction of $4.6 \pm 1.2\%$ [2]. The corresponding invariant mass distribution is shown in Fig. 5, upper plots. This is the first measurement of charm production with a leading strange quark. The small peak left to the D_s^- are candidates of $D^- \to K^+K^-\pi^-$ having a branching fraction of $0.89 \pm 0.08\%$ [2]. The Λ_c^+ ($\bar{\Lambda}_c^-$) are reconstructed in the hadronic decay mode $pK^-\pi^+$ ($\bar{p}K^+\pi^-$) of which the branching fraction is $4.4 \pm 0.6\%$ [2]. The invariant mass distributions are shown in the lower plots of Fig. 5.

Compared to the D^-/D^+ case a significantly larger asymmetry is observed in the D_s^- and the Λ_c^+ mode. While it may be tempting to inteprete this observation as due to a leading particle effect of the production process, asymmetries may also result from simple phase space arguments. Note that in Σ^--Nucleon (N) collisions the energetically most favoured production of a D_s^- or a Λ_c^+ is the associate production in the $\Sigma^- N \to \Lambda_c^+ D_s^- \pi$ channel. This channel is by 740 and 1300 MeV lower in energy with respect to the energet-

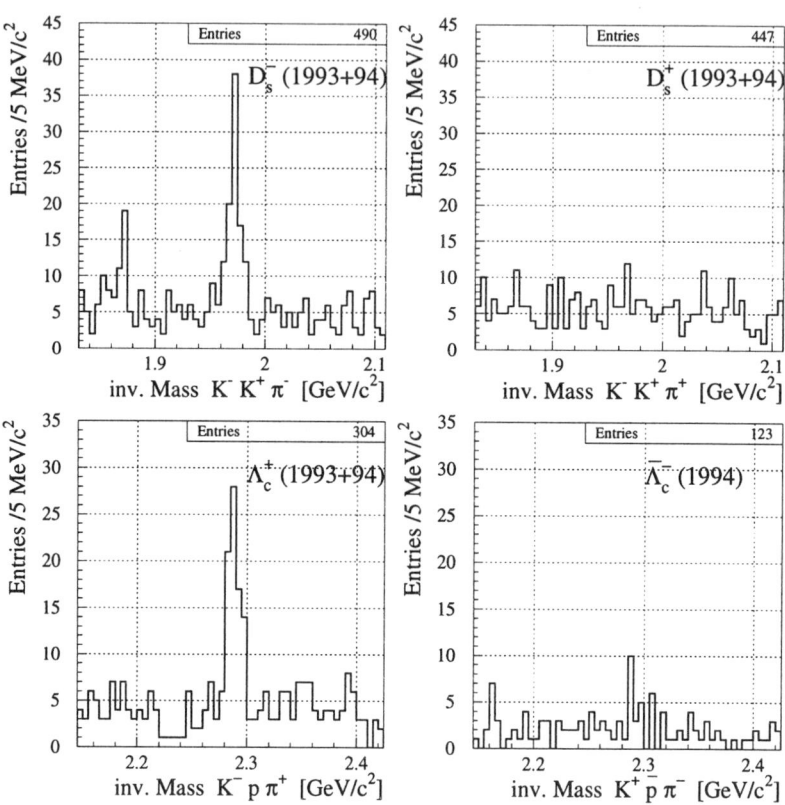

FIGURE 5. The upper plots show the invariant mass distributions of $K^-K^+\pi^+$ (left) and $K^+K^-\pi^-$-candidates (right), the lower plots show the invariant mass distribution of $pK^-\pi^+$ (left) and $\bar{p}K^+\pi^-$-candidates (right).

ically most favoured reactions $\Sigma^- N \to D_s^+ D_s^- \Sigma^-$ and $\Sigma^- N \to \bar{\Lambda}_c^- \Xi_c N$ to produce a D_s^+ or a $\bar{\Lambda}_c^-$, respectively. Detailed studies of the x_F distributions in future experiments may help to clarify the role of the two phenomena.

A-DEPENDENCE OF INCLUSIVE Ξ^- PRODUCTION

Using carbon and copper targets allows to measure the dependence of inclusive Ξ^--production in three different reaction modes, namely $\pi^- A \to \Xi^- X$, $n A \to \Xi^- X$ and $\Sigma^- A \to \Xi^- X$ as a function of the massnumber A of the target. The cross-section σ per nucleus is parametrised as

$$\sigma = \sigma_0 \cdot A^\alpha$$

with α being the attenuation factor and σ_0 the extrapolated cross-section for A=1. Considering that the nuclear density is about constant the nuclear radius r is proportional to $A^{1/3}$. The situation of no attenuation is expressed by $\alpha = 1$ and could thus be called volume-dependence, while a strong attenuation would be expressed by $\alpha = 2/3$ being referred to as surface-dependence. In general α reflects both the influences of the production mechanism and absorbsion effects by the nuclei as will be discussed later. Fig. 6 shows the attenuation factor α as a function of x_F of the Ξ^- cross section measured in reactions with carbon and copper nuclei for Σ^-, neutron and π^- induced reactions.

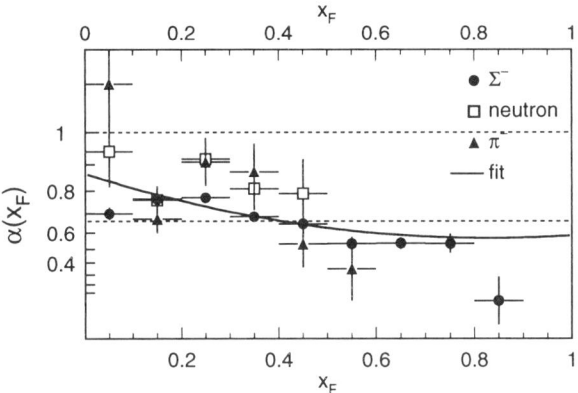

FIGURE 6. The attenuation factor α in case of an A^α dependence of the of the Ξ^- cross section measured in reactions with carbon and copper nuclei as a function of x_F (integrated over p_t) for Σ^-, neutron and π^- induced reactions. The solid line marks a polynomial fit to a compilation of target attenuation factors given in ref. [3].

The following table lists the x_F and p_T integrated attenuation factors together with σ_0 for the three reaction modes:

reaction mode	α	attenuation	σ_0 [mb] for $x_F > 0$ (extrapolated cross section for A=1)
$\pi^- A \to \Xi^- X$	0.931 ± 0.046	weak	0.086 ± 0.013
$n\ A \to \Xi^- X$	0.891 ± 0.034	weak	0.255 ± 0.029
$\Sigma^- A \to \Xi^- X$	0.679 ± 0.011	strong	0.999 ± 0.037

Qualitatively, the values of σ_0 can be understood considering the total cross section of proton-proton collisions is about $\sigma_{tot}(pp) \approx 40$ mb and that typically the production of one $s\bar{s}$ pair is suppressed by a factor $\approx 1/10$ (see Fig. 2 at large x_F). This would then refer to a total inclusive cross section $\sigma_{tot}^{s\bar{s}} \approx 4mb$ which represents an upper limit for the production cross section of the Ξ^- final state in Σ^- induced reaction. In the π^- and neutron induced Ξ^- production the cross section should be significantely lower because here two $s\bar{s}$ pairs have to be produced.

In case of the π and n induced reactions an attenuation factor α close to 1 are observed, while in the Σ^- induced reactions values of α closer to 2/3 are measured. In order to understand that trend, we note that in general the more scattering centres there are, the higher is the probability that several constituent quarks make an inelastic collisions within the target nuclei [5]. Being more specific the probability $P^{(1)}$ that *one* constituent quark of the incident projectile makes an inelastic collision with a target nucleus decreases monotonically with increasing mass number A, while the corresponding probabilities $P^{(2)}$, $P^{(3)}$ of *two* or *three* constituent quarks making inelastic collisions increase with increasing A.

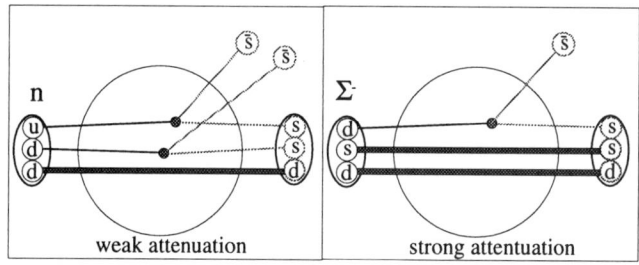

FIGURE 7. A sketch of the neutron and Σ^- induced Ξ^- production indicating that the production of two additional $s\bar{s}$ pairs is generally related to a weaker attenuation than the production of one $s\bar{s}$ pair.

Further one has to consider which are the dominant processes in the reaction modes studied (Fig. 7): In both the neutron and π^- induced reaction the dominant process is the production of *two* additional $s\bar{s}$ pairs which is governed by $P^{(2)}$ leading to a relatively weak attenuation. On the other hand in the Σ^- induced Ξ^- production the dominant process is the production of *one* additional $s\bar{s}$ pair being governed by $P^{(1)}$ thus leading to a relatively strong attenuation.

REFERENCES

1. M. I. Adamovich et al., Preprint CERN-PPE/97-23, submitted to Z.Phys.C
2. Review of Particles and Fields, Phys. Rev. **D 54** (1996) 1
3. Geist, W. M. *Nucl. Phys.* **A 525**, 149c (1991).
4. E687-Kollaboration (G. A. Alves *et al.*): Conference talk at the Workshop on '*Heavy Quarks at fixed Target*', St. Goar, Germany, Oct. 3-6, 1996
5. Takagi, F. *Phys. Rev.* **D 27**, 1461 (1983).

Physics Goals and Experimental Status of SELEX: Fermilab E781

Michael Procario[1]

Carnegie Mellon University
Pittsburgh, PA 15206

Abstract. SELEX is a fixed target experiment at Fermilab designed to do a systematic study of charm baryons. Data taking began in February, 1997, and preliminary charmed hadron signals have been observed.

INTRODUCTION

SELEX is a fixed target experiment at Fermilab designed to do a systematic study of charm baryons, with an emphasis on the study of charm strange baryons. The experiment is designed to optimize charm baryons rather than charm hadrons in general, since there are large charm meson samples already available from E687, E791, and CLEO and there will be a very large sample of charm mesons measured by FOCUS (E831) in the current fixed target run at Fermilab.

Charm baryons are less well understood than charm mesons. This is true both experimentally, where they are harder to produce and observe, as well as theoretically where the dynamics of their decay and spectroscopy are more complex. For these reasons, charm baryons provide an excellent testing ground for models used to explain charm meson physics. Our goals are:

- Measure the lifetimes of all weakly decaying charm baryons. The charm strange baryons are currently poorly measured.

- Study the excited states of all charm baryons. Many of the spin $\frac{1}{2}$ and spin $\frac{3}{2}$ states have been seen, but are not well measured.

- Study the hadronic decays of charm baryons, since all of the lifetime differences are due to hadronic decay processes.

- Study the semileptonic decays of charm baryons.

---

[1] representing the SELEX Collaboration

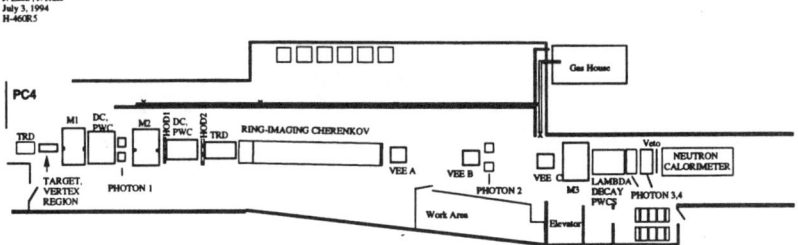

FIGURE 1. Experimental layout.

- Measure the production cross sections of charm baryons. Nonperturbative effects, such as the leading particle effect, could be quite important.

There will also be non-charm physics topics pursued. These will be either hyperon measurements, since our beam contains other hyperons than the Σ^-, or measurements that rely on the high resolution of the spectrometer for small angles with respect to the beam, such as Primakoff effect reactions. The small angle tracks are measured with silicon microstrips before and after the first two magnets, and have excellent momentum resolution.

EXPERIMENTAL DESIGN

The experiment is a three stage spectrometer with acceptance from $0.1 < x < 1.0$, which enhances baryon production relative to meson production. The spectrometer is instrumented with silicon microstrips before and after the target, and at small angles after the first and second magnets. There are also PWC's and drift chambers for measuring charged tracks. The are three lead glass calorimeters for photon detection and electron identification after each magnet. There is a transition radiation detector (TRD) that tags the type of beam particle, and a second TRD after the second magnet for electron identification. There is a ring imaging Cerenkov counter (RICH) for particle identification. The layout is shown in figure 1.

We use a Σ^- beam which also enhances baryon production relative to meson production. In addition, WA89 has shown that charm strange hadrons are enhanced relative to charm hadrons in a Σ^- beam. Charm strange baryons are the least studied of the charm baryons.

The experiment has a 20 plane silicon microstrip vertex detector for reconstructing the decays of short lived particles. The vertex resolution of the detector is enhanced by the forward acceptance of the spectrometer. Forward particles have higher momentum and therefore less multiple scattering and

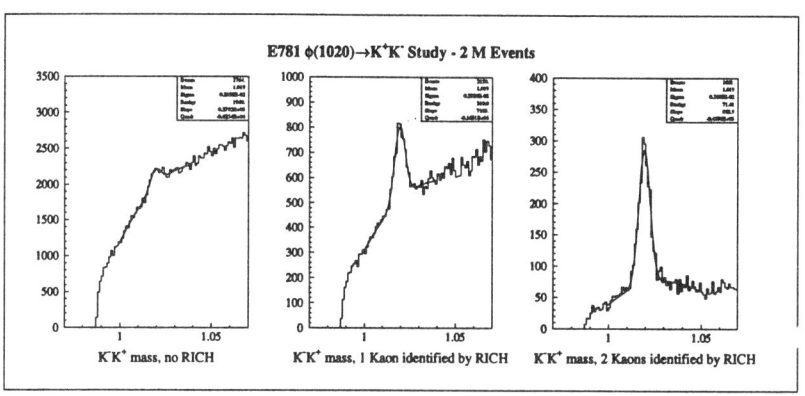

FIGURE 2. Example of RICH particle identification on the ϕ.

therefore better resolution. The very short lifetimes of the Ω_c and Ξ_c^0 will put a premium on vertex resolution.

In order to measure charm baryons it is necessary to identify protons, since most of the reconstructible final states will have a proton. In addition, kaon identification is critical for charm decays and doubly critical for charm strange decays. In order to achieve the required particle identification, we use a RICH. Because of the forward acceptance of the spectrometer, the solid angle covered by the RICH is small enough that cost of phototubes for the photon measurement is reasonable. Phototubes are simpler to operate, more robust and more efficient than other technologies used for RICH's. Figure 2 shows the benefit of the RICH in identifying $\phi \to K^+K^-$ decays.

The trigger is based on miss distance from the primary vertex. Since charm particles decay away from the primary vertex, and this property is used in almost all charm particle analyses, a trigger based on this property should be very efficient for charm and still give good rejection of light quark events. The trigger is done in software using tracks found after the second magnet, which has a momentum cutoff of 15 GeV/c. These tracks are extrapolated back to the vertex detector. The beam track is also measured in silicon microstrip detector before the target. If all these tracks are inconsistent with a single primary vertex, then the event is kept. We need ~ 10 milliseconds per event for this analysis on a 200 MHz SGI Challenge processor, and we use 20 such processors.

The Ω_c and Ξ_c^0 will be hard to trigger on because of their short lifetime. We can also trigger on events that are identified in the RICH. We trigger on events having either a proton and kaon, or an antiproton. These events represent only a small fraction of our triggered events.

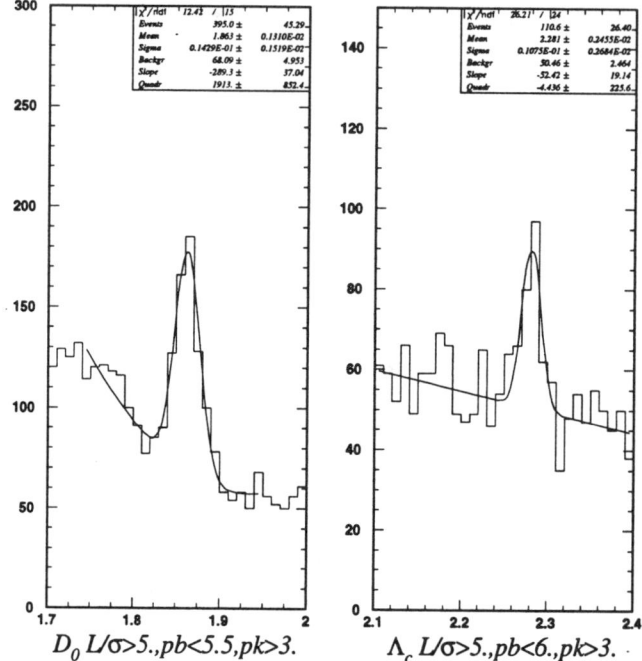

FIGURE 3. D^0 and Λ_c^+ signals from analysis of trigger detectors for 86 million events.

CURRENT STATUS

The run began in the summer of 1996 and continues until September 1997. SELEX is a new experiment and required extensive checkout, so data taking began in February 1997. All detectors are operational but only those used in the trigger are being fully analyzed. These consist of the beam silicon, beam transition radiation detector, vertex silicon, PWC chambers after the second magnet and the RICH. Signals will become larger when the rest of the detectors are included in the reconstruction.

Data from February has been analyzed to demonstrate our ability to observe charm. A total of 86 million events were searched for the decays $D^0 \to K^-\pi^+$ and $\Lambda_c^+ \to pK^-\pi^+$. Candidates were required to be separated from the primary vertex by at least 5σ; to point back to the primary with a $\chi^2 < 6$. Any kaon candidates were required to have have a RICH kaon probability more than 3 times greater then the RICH pion probability. A D^0 signal of 395 ± 46 events and a Λ_c signal of 110 ± 26 are observed. See figure 3.

These very preliminary results confirm that our charm baryon signals will be comparable to our charm meson signals as desired.

Recent Results from Experiment E835 at Fermilab

Peter Maas*
for the E835 Collaboration[†]

*Northwestern University
Evanston, IL

[†] E835 Collaboration includes:

M. Ambrogiani, S. Argiro, S. Bagnasco, W. Baldini, F. Bertini, D. Bettoni,
M. Bombonati, D. Bonsi, G. Borreani, A. Buzzo, R. Calabrese, M. Cardarelli,
A. Ceccucci, R. Cester, P. Dalpiaz, S. Frabetti, X. Fan, G. Garzoglio, K. E. Gollwitzer,
A. Hahn, S. Jin, J. Kasper, G. Lasio, M. Lovetere, E. Luppi, P. Maas, M. Macri,
M. Mandelkern, F. Marchetto, M. Marinelli, W. Marsh, M. Martini, R. McTaggart,
E. Menichetti, R. Mussa, M. Obertino, M. Pallavicini, N. Pastrone, C. Patrignani,
T. Pedlar, J. Peoples Jr., S. Pordes, E. Robutti, J. Rosen, L. Rossetto, P. Rumerio,
A. Santroni, M. Savrie, J. Schultz, K. K. Seth, J. Streets, G. Stancari, M. Thompson,
L. Tomassetti, S. Werkema, G. Zioulas

Abstract. E835 can measure the mass and width of resonant states produced in $\bar{p}p$ collisions at center of mass energies which span the spectrum of $c\bar{c}$ bound states (Charmonium). The experiment began taking data during Autumn, 1996. Results presented here include a measurement of the η_c.

Charmonium states are interesting because they provide an accessible means to study the strong force; both because production cross sections are not small and secondly because the charm quarks are non-relativistic. In an ideal case, one can imagine measuring the mass and widths of all the states and inverting the eigenvalue equation to determine the form of the potential [1,2].

Charmonium states were first observed in e^+e^- collisions as a resonance at $\sqrt{s}$ = 3097 MeV [3,4]. Following the discovery of the J/ψ, the ψ' was observed in e^+e^- collisions and the χ states were observed via the radiative processes $\psi' \to \chi + \gamma, \chi \to J/\psi + \gamma$. However other states (notably the $^1P_1$) could not be produced either in e^+e^- collisions or via radiative transitions. E704 [5] pioneered the use of internal gas jet targets to resonantly produce charmonium bound states in antiproton storage rings. The advantage of this technique is that all bound states which couple to a proton and an antiproton are capable of being formed and the resolution depends on the accuaracy of the knowledge of the antiproton beam momentum. The disadvantage is that one must reject copious hadronic backgrounds to have a sensitivity to the small but observable rate of events containing only photons and electrons due to resonant production of charmonium. Experiment E835 [6] is a continuation of experiment E760 [7] at Fermilab.

The goal of E835 is to improve on previous measurements of the η_c and $^1P_1$, find the η_c' and study the D states. The following components have been added to the E760 detector, both to increase the luminosity and enhance the ability of the detector at higher luminosity.

- Increase the density of the gas jet from 0.7×10^{14} atoms/cm^3 to 3.6×10^{14} atoms/cm^3

- Add a scintillating fiber detector to measure the polar angle of charged tracks

- Replace straw tubes to measure the azimuthal angle of charged tracks

- Add an additional layer of scintillator hodoscopes to improve the trigger

- Replace the mirrors in the Cerenkov detector

- Add pulse shaping and TDCs to the calorimeter signal processing chain to increase our sensitivity to events which overlap in time.

Figure 1 gives a schematic of the detector. Electrons are identified by signals in the hodoscopes, scintillating fibers, straw tubes, and the Cerenkov. Photons are identified by a signal in the calorimeter and no signal in either of the first two layers of the scintillator hodoscopes.

The efficiency of the electron identification is demonstrated in figure 2. The unshaded histogram shows the measured mass of the two charged tracks (with a Cerenkov tag) of all events which are consistent with having two charged tracks and either 0 or 2 additional clusters in the calorimeter at a center of mass energy of 3686 MeV. These events are classified according to a nearest neighbor algorithm [8] in an 11 dimensional space where each dimension represents a measured quantity. The shaded portion shows the mass distribution for events for whom 4 or 5 of the five nearest neighbors are electrons. The distribution is consistent with expectations for both exclusive and inclusive decays of the ψ' to e^+e^- pairs.

The ^1S states are identified via their decays to two photons. All events which have two large energy deposits in opposite hemispheres and no charged tracks (determined by the coincidence of the H1 and H2$'$ counters or the forward charged veto) are recorded to tape. Offline the events are required to fufill the following criteria:

- Exactly two clusters in the calorimeter classified as in time.

E835 EQUIPMENT LAYOUT

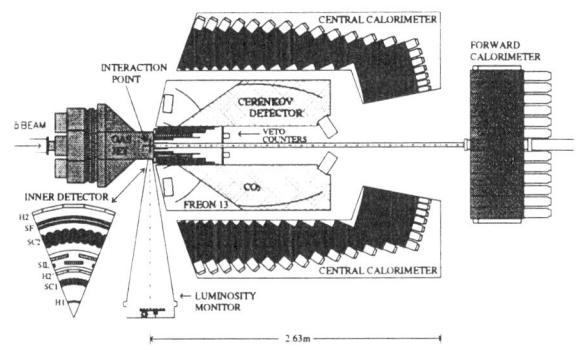

FIGURE 1. Schematic of the apparatus.

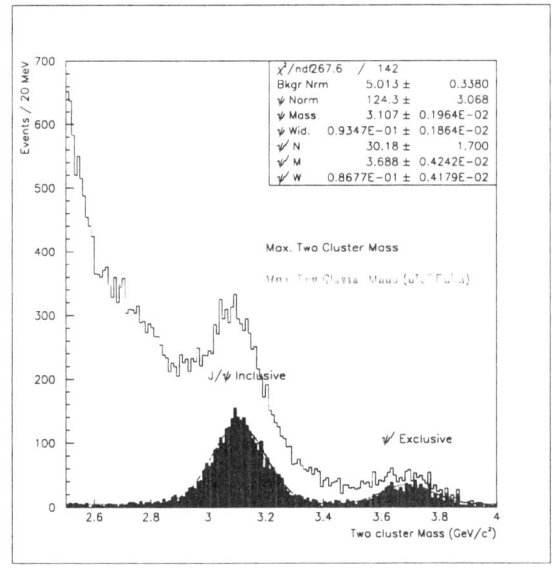

FIGURE 2. Invariant mass of two charged tracks recorded at $\sqrt{s} = 3686$ MeV.

- No out of time clusters

- No extra cluster with an undetermined time which pairs with one of the two clusters to form a π^0 mass

- Probability of kinematical fit greater than 10%.

- $|\cos\theta^*| < 0.2$

Figure 3 shows the number of diphoton events as a function of mass in the region of the η_c. While the experiment is not yet ready to quote values for the resonance parameters it is expected that the uncertainty on the width will be on the order of 30%.

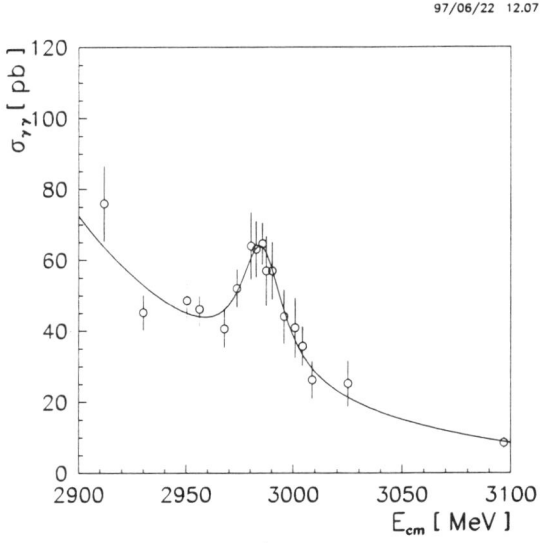

FIGURE 3. Diphoton invariant mass spectrum in the interval $2900 < \sqrt{s} < 3100$ MeV.

In conclusion, the experiment is functioning well and will produce the best measurement to date of the resonance parameters of the η_c. In addition, the experiment has submitted a letter of intent to continue measurements of charmonium states following the completion of the main injector project at FNAL.

This research has been supported in part by the U.S. DOE, the U.S. NSF, and the INFN, Italy. We wish to thank the Fermilab Accelerator Division, the Fermilab Computing Division and the Fermilab Physics Section and for their assistance in the construction, preparation and operation of E835.

REFERENCES

1. H. B. Thacker et al, *Phys. Rev.* **D18**, 274 (1978).
2. H. B. Thacker et al, *Phys. Rev.* **D18**, 287 (1978).
3. J. J. Aubert et al, *Phys. Rev. Lett.* **33**, 1404 (1974).
4. J. E. Augustin et al, *Phys. Rev. Lett.* **33**, 1406 (1974).
5. C. Baglin et al, *Nucl. Phys.* **B286**, 592 (1987).
6. T. A. Armstrong et al, **FERMILAB-PROPOSAL-P-835-REV**, (1992).
7. A. Ceccucci et al, *Nucl. Phys.* **A558**, 259 (1993).
8. M. Pallavicini, C. Patrignani, M. Pontil, A. Verri, The Nearest Neighbor Technique for Particle Identification, *Submitted to Nucl. Instr. and Meth.*

Recent Results from FNAL E791

A. J. Schwartz
(for the E791 Collaboration)

Physics Department, Princeton University, Princeton, NJ 08544

Abstract. We report limits on D^0-$\bar{D}^0$ mixing and CP violation, and measurements of the double-Cabibbo-suppressed decays $D^0 \to K^+\pi^-$ and $D^+ \to K^+\pi^-\pi^+$. We also report a measurement of the ratio $\Gamma(D^+ \to \rho^0 \ell^+ \nu)/\Gamma(D^+ \to \bar{K}^{*0}\ell^+\nu)$.

INTRODUCTION

E791 is a charm hadroproduction experiment which produced charm using a 500 GeV/c π^- beam incident on platinum and diamond target foils. A loose transverse energy trigger ($E_T \gtrsim 3$ GeV) was used to record a large sample of charm decays with minimal trigger bias. The experiment took data for six months (9/91–1/92) and recorded over 2×10^{10} events.

Downstream of the target foils was a silicon vertex detector followed by a two-magnet spectrometer, two threshold Čerenkov counters, an electromagnetic calorimeter, a hadronic calorimeter, and approximately 1 m of iron followed by scintillator to identify muons. The vertex detector consisted of 23 planes of silicon (6 upstream of the target and 17 downstream), and the spectrometer comprised 35 planes of multi-wire drift chambers. The multi-cell Čerenkov counters were used to discriminate among pions, kaons, and protons. The two counters used different gas mixtures and thus had different velocity thresholds; this allowed $\pi/K/p$ separation over a wide momentum range, 6 – 40 GeV/c. More details about the detector can be found in Ref. [1].

The offline analysis began by reconstructing tracks and fitting them to common vertices. Events with evidence of multiple vertices (indicative of charm decay) were kept for further analysis. In the analyses discussed here a number of variables were commonly used to reject backgrounds; these are listed in Table 1 along with typical cut values. The analyses also included cuts on the amount of light collected in the Čerenkov counters. For each reconstructed track, the Čerenkov light associated with it was compared to that expected for a π, K, or p of the given track momentum. The comparison resulted in a

TABLE 1. Variables used to reject backgrounds, and typical cut values.

Variable	Typical cut value
$\chi^2/$(d.o.f.) of the fits for production and decay vertices	< 3
Δz distance between production and decay vertices	$> 10\sigma_z$, where σ_z is the error on Δz
Δs distance between decay vertex and nearest target foil	$> 3\sigma_s$ (outside of foil)
Impact parameter of $\vec{p}_D$ with respect to production vertex	$< 40~\mu$m
p_T relative to D line-of-flight	< 250 MeV/c
Measured lifetime $(d \cdot m)/p$	< 5 ps

relative likelihood for each type of charged hadron, and these likelihoods were used to select specific hadrons.

The results presented here are recent and cover four areas: *1)* searches for D^0-$\bar{D}^0$ mixing; *2)* measurement of the rate of double-Cabibbo-suppressed decays; *3)* searches for CP violation; and *4)* measurement of the ratio $\Gamma(D^+ \to \rho^0 \ell^+ \nu)/\Gamma(D^+ \to \overline{K}^{*0} \ell^+ \nu)$. Here, and throughout this paper unless otherwise noted, charge-conjugate modes are assumed.

$D^0 - \bar{D}^0$ MIXING

In the Standard Model, D^0-$\bar{D}^0$ mixing proceeds via internal quark loops and long-distance effects; the predicted rate for the ratio $r_{\text{mix}} \equiv \Gamma(D^0 \to \bar{D}^0 \to \bar{f})/\Gamma(D^0 \to f)$ is less than $\sim 10^{-7}$. Measuring r_{mix} significantly larger than this would indicate new physics. In E791 we have searched for evidence of mixing by searching for $D^0 \to \bar{D}^0 \to K^+ \ell^- \nu$ semileptonic decays and $D^0 \to \bar{D}^0 \to K^+ \pi^-$, $K^+ \pi^- \pi^+ \pi^-$ hadronic decays.

Both types of decay are selected from a sample of $D^{*+} \to D^0 \pi^+$ decays in order to identify the flavor of the decaying D^0 at birth; the methods differ only in how the D^0 is identified when it decays. The semileptonic decay has the advantage that the final state is reached only via mixing, and hence observation of this final state would constitute unequivocal evidence for mixing. However, these decays have the disadvantage that the D^0 is not fully reconstructed due to the missing neutrino, and thus measurement of the D^0 mass and lifetime (used to reject background) is smeared. The hadronic decays have the advantage that they are fully reconstructed, but the final states can be reached via a double-Cabibbo suppressed (DCS) diagram, and thus observation of $D^0 \to K^+ \pi^-$, $K^+ \pi^- \pi^+ \pi^-$ is not easy to interpret. To separate the DCS and mixing contributions, one can study the lifetime distribution: DCS decays have an $e^{-\Gamma t}$ dependence, while decays due to mixing have a $t^2 e^{-\Gamma t}$ dependence. There also exists an interference term with a $t e^{-\Gamma t}$ dependence.

In the semileptonic decay, the component of the neutrino momentum transverse to the D^0 line-of-flight (determined from the positions of the production and decay vertices) is inferred by requiring that the vector sum of all transverse momentum (p_T) be zero. The neutrino's longitudinal momentum is then determined up to a two-fold ambiguity by requiring that $(P_\nu + P_K + P_\ell)^2 = m_D^2$. We resolve this ambiguity by choosing the solution which gives larger $\vec{p}_D$; this choice is found by Monte Carlo to be correct for the selected sample over half the time. The final event samples are shown in Fig. 1, which shows both right-sign (RS) events, in which the pion from $D^{*+} \to D^0 \pi^+$ and the kaon have opposite charge, and wrong-sign (WS) events, in which the pion and kaon have the same sign charge. A large D^0 peak is visible in the RS sample, while no peak is observed in the WS sample. We do an unbinned maximum likelihood (ML) fit for $Q \equiv m_{D^*} - m_{D^0} - m_\pi$ and the proper decay time t, in which the signal shapes are taken from the RS sample, the background shape in t is taken from the Q sideband, and the background shape in Q is taken from "mixed" events in which a D^0 from one event and a slow π^- from another event are combined. The fit results are: $r_{\text{mix}} = 0.16^{+0.42}_{-0.37}\%$ for the $Ke\nu$ channel, $0.06^{+0.44}_{-0.40}\%$ for the $K\mu\nu$ channel, and $0.11^{+0.30}_{-0.27}\%$ for both channels combined. This latter result implies $r_{\text{mix}} < 0.50\%$ at 90% C.L.

For the hadronic $D^0 \to K^+\pi^-$ and $D^0 \to K^+\pi^-\pi^+\pi^-$ decays, the final sample was selected using several neural networks in addition to standard selection criteria in order to maximize sensitivity. An unbinned ML fit was then done which included all potential contributions: mixed events, unmixed events, DCS decays, and several background processes. The fit results depend upon whether a DCS-mixing interference term is included, and whether CP violation is allowed. The results for three different assumptions are listed in Table 2. For the assumption of *no* mixing contribution – which is plausible

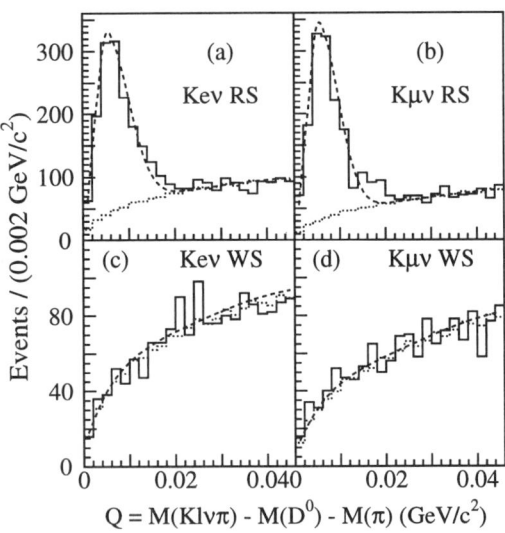

FIGURE 1. Right-sign and wrong-sign samples of $D^0 \to Ke\nu$ and $D^0 \to K\mu\nu$ decays.

TABLE 2. Mixing results from ML fit to $D^{*+} \to D^0\pi^+$, $D^0 \to K^+\pi^-, K^+\pi^-\pi^+\pi^-$ decays.

Fit assumptions	Result (%)
No assumptions (CP violation + interference)	$r_{\mathrm{mix}}(\bar{D}^0 \to D^0) = 0.18^{+0.43}_{-0.39} \pm 0.17$
	$r_{\mathrm{mix}}(D^0 \to \bar{D}^0) = 0.70^{+0.58}_{-0.53} \pm 0.18$
No CP violation, no interference	$r_{\mathrm{mix}} = 0.21 \pm 0.09 \pm 0.02$
No mixing	$r_{\mathrm{DCS}}(K\pi) = 0.68^{+0.34}_{-0.33} \pm 0.07$
	$r_{\mathrm{DCS}}(K\pi\pi) = 0.25^{+0.36}_{-0.34} \pm 0.03$

given it's very small amplitude relative to the DCS amplitude – the DCS contribution to $D^0 \to K^+\pi^-$ is inconsistent with zero, indicating a signal.

DOUBLE-CABIBBO-SUPPRESSED DECAYS

More explicit evidence for a DCS amplitude is obtained by searching for $D^+ \to K^+\pi^-\pi^+$ decays. Because the initial state is charged, there is no mixing contribution. The lowest-order (spectator) amplitude is double-Cabibbo-suppressed, and higher order amplitudes are believed to be negligible. From the spectator diagram we expect $\Gamma(D^+ \to K^+\pi^-\pi^+)/\Gamma(D^0 \to K^-\pi^+\pi^0) \approx \tan^4\theta_C$, and since $\Gamma(D^0 \to K^-\pi^+\pi^0) \simeq 3.2 \times \Gamma(D^+ \to K^-\pi^+\pi^+)$ (i.e., giving rise to the D^0/D^+ lifetime difference), we expect $\Gamma(D^+ \to K^+\pi^-\pi^+)/\Gamma(D^+ \to K^-\pi^+\pi^+) \approx 3.2\tan^4\theta_C$. This ratio is straightforward to measure as only charged particles are present in the final state.

There are three classes of backgrounds which contaminate the final sample: a) D^+ and D_s^+ multibody decays with missing neutral daughters; b) D^0 two-body decays which are inadvertently combined with a random track when fitted for the decay vertex; and c) D^+ and D_s^+ three-body decays with misidentified final state particles. This last class of background is especially troublesome as the misidentified decays can appear in the $K^+\pi^-\pi^+$ mass spectrum near m_{D^+}. To estimate their contribution, we plot the invariant mass of all events in the final sample as $K^+K^-\pi^+$ and $K^-\pi^+\pi^+$. For each distribution we fit for the number of events from $D_s^+, D^+ \to K^+K^-\pi^+$ and $D^+ \to K^-\pi^+\pi^+$, where the spectrum shapes (1 or 2 Gaussians and 2 or 1 "reflections") are obtained from data. The event yield for the mode(s) fit to a Gaussian is used in the other mass distribution fit, and we iterate until all yields converge. We independently estimate $D^+ \to \pi^-\pi^+\pi^+$ background from the $\pi^+\pi^-\pi^+$ mass distribution. Fixing these four background contributions, we fit the $K^+\pi^-\pi^+$ mass spectrum to Gaussian signals for D^+ and D_s^+ decay plus a smooth function to account for other backgrounds. The results are shown in Fig. 2a. The event yield for $D^+ \to K^+\pi^-\pi^+$ is 59±13 events, and normalizing to $D^+ \to K^-\pi^+\pi^+$ decays gives $\Gamma(D^+ \to K^+\pi^-\pi^+)/\Gamma(D^+ \to K^-\pi^+\pi^+) = (7.7\pm1.7\pm0.8)\times 10^{-3}$. This result equals $(3.0\pm0.8)\times\tan^4\theta_C$, which is consistent with expectations.

We subsequently study the Dalitz plot structure of these decays by selecting a subsample of well-identified events containing minimal background. The additional selection criteria are: *a)* the Čerenkov particle identification cut for the odd-charged pion is tightened, *b)* the proper lifetime is required to be greater than $2 \times \tau_{D^0}$, and *c)* events having $m_{K^-\pi^+\pi^+} \approx m_{D^+}$ are removed. These criteria leave 67 events, of which 42 ± 9 are estimated to be signal. The Dalitz plot of these events is shown in Fig. 2b. Doing an unbinned ML fit for the amplitudes of $D^+ \to \rho^0 K^+$, $D^+ \to \overline{K}^{*0}\pi^+$, and nonresonant decay gives the decay fractions $37 \pm 14 \pm 7\%$, $35 \pm 14 \pm 1\%$, and $36 \pm 14 \pm 7\%$, respectively.

CP VIOLATION

E791 has searched for CP violation in charged $D^+ \to f^+$ decays and neutral $D^0 \to f^0$ decays, where $f^+ = \{K^+K^-\pi^+, \phi\pi^+, \overline{K}^{*0}K^+, \pi^-\pi^+\pi^+\}$ and $f^0 = \{K^+K^-, \pi^+\pi^-\}$. For the charged decays, any CP asymmetry observed would constitute evidence for *direct* CP violation, as no mixing of neutral D's is involved. These modes are governed by singly-Cabibbo-suppressed amplitudes, where direct CP violating effects are believed most likely to appear. For charged and neutral decays the CP asymmetry parameter of interest is:

$$A_{CP} \equiv \frac{\Gamma(D \to f) - \Gamma(\bar{D} \to \bar{f})}{\Gamma(D \to f) + \Gamma(\bar{D} \to \bar{f})}. \quad (1)$$

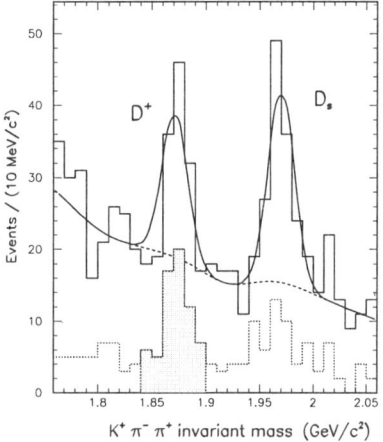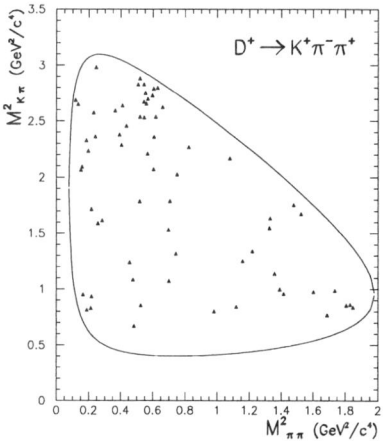

FIGURE 2. *a)* The $K^+\pi^-\pi^+$ invariant mass spectrum. The shaded events are selected for the Dalitz plot. *b)* The Dalitz plot of well-identified $D^+ \to K^+\pi^-\pi^+$ decays.

Since the incident beam for E791 is π^-, the production cross section for D^- is greater than that for D^+. To correct for this non-CP asymmetry, we normalize the yields of potentially CP-violating $D^\pm \to f^\pm$ decays to the yields of CP-conserving $D^\pm \to K^\mp \pi^\pm \pi^\pm$ decays. Defining $\eta(D^\pm) \equiv N(D^\pm \to f^\pm)/N(D^\pm \to K^\mp \pi^\pm \pi^\pm)$, we measure $[\eta(D^+)-\eta(D^-)]/[\eta(D^+)+\eta(D^-)]$. This ratio is equal to A_{CP} if the relationship $\varepsilon(D^+ \to f^+)/\varepsilon(D^+ \to K^-\pi^+\pi^+) = \varepsilon(D^- \to f^-)/\varepsilon(D^- \to K^+\pi^-\pi^-)$ holds, where ε represents trigger and reconstruction efficiencies. From Monte Carlo study we find that, for the E791 detector, this relationship is satisfied to very good precision.

We thus measure the asymmetry A_{CP} for four charged final states: $K^+K^-\pi^+$ (inclusive), $\phi\pi^+$, $\overline{K}^{*0}K^+$, and $\pi^-\pi^+\pi^+$, where $\phi \to K^+K^-$ and $\overline{K}^{*0} \to K^-\pi^+$. Similar analysis cuts are applied to all four modes as well as to the normalization mode $D^+ \to K^-\pi^+\pi^+$, with the exception that no Čerenkov particle identification cuts are applied to $K^-\pi^+\pi^+$. For the $\phi\pi^+$ sample, a cut $|m_{K^+K^-} - m_\phi| < 6$ MeV/c^2 is imposed; for the $\overline{K}^{*0}K^+$ sample, a cut $|m_{K^-\pi^+} - m_{K^*}| < 45$ MeV/c^2 is imposed. The final event samples are displayed in Figs. 3a–d. Doing a binned ML fit gives the following results: $A_{CP}(KK\pi, \text{inclusive}) = -0.014 \pm 0.029$; $A_{CP}(\phi\pi) = -0.028 \pm 0.036$; $A_{CP}(\overline{K}^{*0}K) = -0.010 \pm 0.050$; and $A_{CP}(\pi\pi\pi) = -0.017 \pm 0.042$. All measured values are consistent with zero.

For the neutral decays $D^0 \to K^+K^-$ and $D^0 \to \pi^+\pi^-$, a CP asymmetry can arise from either an asymmetry in D^0-$\bar{D}^0$ mixing or an asymmetry in the decay amplitudes. The branching fractions for $D^0 \to K^+K^-$, $\pi^+\pi^-$ are not understood: one naively expects $\Gamma(D^0 \to K^+K^-)/\Gamma(D^0 \to \pi^+\pi^-) \lesssim 1$, as both decays are singly-Cabibbo suppressed and the K^+K^- final state has less phase space. However, the ratio is measured to be ~ 3, indicating that higher-order diagrams and/or final state interactions are significant. In E791, $D^0 \to K^+K^-$ decays are selected from two-track vertices in which the tracks pass Čerenkov kaon identification cuts. Normalizing to $D^0 \to K^-\pi^+$ yields $\Gamma(D^0 \to K^+K^-)/\Gamma(D^0 \to K^-\pi^+) = 0.109 \pm 0.003 \pm 0.003$. To select $D^0 \to \pi^+\pi^-$ decays, we replace the Čerenkov kaon identification cut by a pion identification cut; this results in $\Gamma(D^0 \to \pi^+\pi^-)/\Gamma(D^0 \to K^-\pi^+) = 0.040 \pm 0.002 \pm 0.003$. The ratio $\Gamma(D^0 \to K^+K^-)/\Gamma(D^0 \to \pi^+\pi^-)$ is then $2.75 \pm 0.15 \pm 0.16$.

To search for CP violation, we tag the initial flavor of the D^0 using $D^{*+} \to D^0\pi^+$ decays; i.e., we require the presence of a slow pion and a Q near 5.8 MeV/c^2. The D^0 is then reconstructed in the K^+K^-, $\pi^+\pi^-$, and, for normalization, $K^-\pi^+$ decay channels. After a background subtraction (the spectrum of which is obtained from the D^0 mass sidebands), we obtain approximately 600 K^+K^- decays and 350 $\pi^+\pi^-$ decays. These yields give:

$$A_{CP}(K^+K^-) = -0.013 \pm 0.049 \pm 0.012 \quad (2)$$
$$A_{CP}(\pi^+\pi^-) = -0.049 \pm 0.078 \pm 0.030. \quad (3)$$

Both results are consistent with zero.

RATIO $\Gamma(D^+ \to \rho^0 \ell^+ \nu)/\Gamma(D^+ \to \overline{K}^{*0} \ell^+ \nu)$

The semileptonic decays $D^+ \to \rho^0 \ell^+ \nu$ and $D^+ \to \overline{K}^{*0} \ell^+ \nu$ proceed via spectator diagrams, and thus the decay rates are governed by form factors and Cabibbo-Kobayashi-Maskawa (CKM) matrix elements. For the $\rho^0 \ell^+ \nu$ final state, the CKM matrix element is V_{cd}; for the $\overline{K}^{*0} \ell^+ \nu$ final state, the matrix element is V_{cs}. A measurement of the partial widths thus yields either: (a) the CKM matrix elements given the form factors (i.e., calculated from theory), or (b) the form factors given the matrix elements. This latter possibility is interesting because the form factors for $D^+ \to \rho^0 \ell^+ \nu$ can be related to those for $B^+ \to \rho^0 \ell^+ \nu$ via Heavy Quark Effective Theory [2], and these latter form factors are needed to extract V_{ub} from a measurement of $B(B^+ \to \rho^0 \ell^+ \nu)$. Thus, measuring $D^+ \to \rho^0 \ell^+ \nu$ helps determine V_{ub}.

To select events, we require that the minimum kinematically allowed D^+ mass, $m_{\min} = p_T + \sqrt{m_{\text{vis}}^2 + p_T^2}$, be between 1.6 and 2.0 GeV/c^2. In this

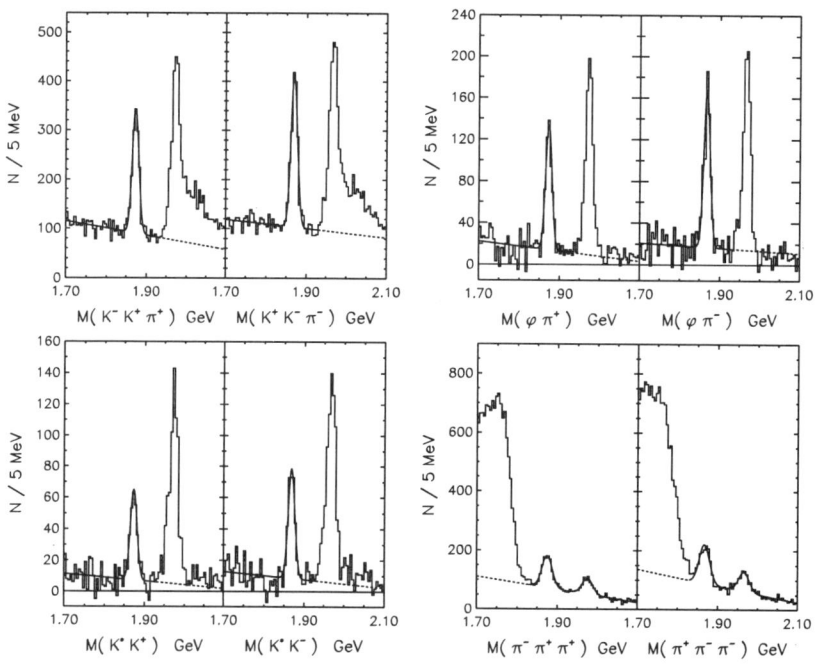

FIGURE 3. a) The $D^\pm \to K^+ K^- \pi^\pm$ inclusive sample; b) the $D^\pm \to \phi \pi^\pm$ sample; c) the $D^\pm \to \overline{K}^{*0} K^\pm$ sample; and d) the $D^\pm \to \pi^+ \pi^- \pi^\pm$ sample. The right-most peak appearing in all plots is due to $D_s^\pm$ decay.

expression, p_T is the momentum of the $\rho^0\ell$ or $\overline{K}^{*0}\ell$ system transverse to the D^+ line-of-flight, and $m_{\rm vis}^2 = (P_{\rho/K^*} + P_\ell)^2$. We reduce background from $D^+ \to K^-\pi^+\pi^+$, $D^+ \to \pi^-\pi^+\pi^+$, and $D_s^+ \to \phi\ell^+\nu \to K^+K^-\ell^+\nu$ decays by removing events with $\{m_{K^-\pi\pi}, m_{\pi\pi\pi}\} \approx m_{D^+}$ or $m_{KK} \approx m_\phi$. For the $\rho^0\ell\nu$ sample, we reduce non-charm background by requiring $-0.10 < m_{\rm miss}^2 < 0.15 \ ({\rm GeV}/c^2)^2$, where $m_{\rm miss}^2 \equiv m_D^2 + m_{\rm vis}^2 - 2m_D\sqrt{p_T^2 + m_{\rm vis}^2} \approx (P_D - P_{\rm vis})^2$. To reject $\overline{K}^{*0}\ell\nu$ decays – which are topologically similar to $\rho^0\ell\nu$ but have a rate 20 times higher – we apply more stringent Čerenkov hadron identification cuts and require $m_{\rm min}(\overline{K}^{*0}\ell) > 2.0 \ {\rm GeV}/c^2$ and $|m_{K\pi} - m_{K^*}| > 40 \ {\rm MeV}/c^2$.

The final $D^+ \to \rho^0\ell^+\nu$ sample is shown in Fig. 4a, and the final $D^+ \to \overline{K}^{*0}\ell^+\nu$ sample is shown in Fig. 4b. The top plots show the RS sample (hadrons having opposite charge) and the bottom plots show the WS sample (hadrons having the same sign charge). All RS plots exhibit clear signals while all WS plots show spectra with no structure, i.e., background. The signals are fit to Breit-Wigner shapes while the background spectra are fit to smooth functions $F(m) = N_0(m-m_0)^\alpha \exp[C_1(m-m_0) + C_2(m-m_0)^2]$. After background subtraction the event yields are 103 ± 25 $\rho^0\ell\nu$ decays and 1661 ± 75 $\overline{K}^{*0}\ell\nu$ decays. Correcting for acceptance and reconstruction efficiencies gives: $\Gamma(D^+ \to \rho^0\ell^+\nu)/\Gamma(D^+ \to \overline{K}^{*0}\ell^+\nu) = 4.7 \pm 1.3\%$.

SUMMARY

All E791 results presented here are summarized in Table 3. The table lists our measurements or limits and the previous results to which they can be

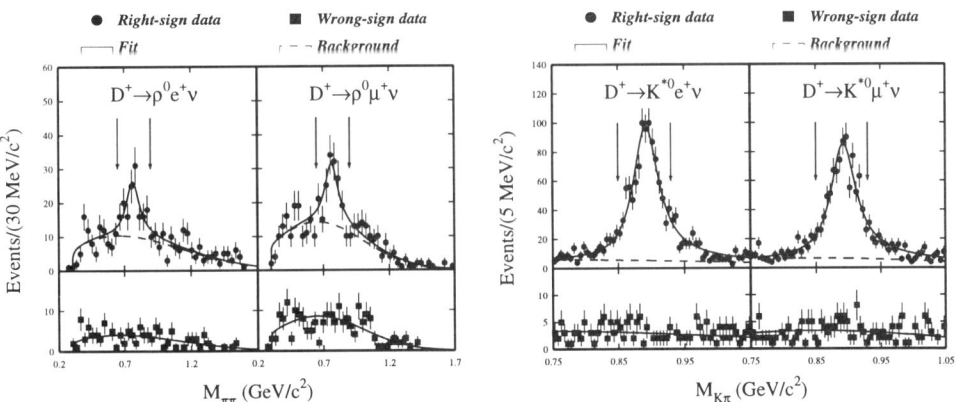

FIGURE 4. RS and WS samples of $D^+ \to \rho^0\ell^+\nu$ and $D^+ \to \overline{K}^{*0}\ell^+\nu$ decays.

TABLE 3. Summary of recent E791 results. All limits are at 90% C.L.

Physics topic	E791 Result	Previous result
Mixing in $K^{\pm}\ell^{\mp}\nu$	$r_{\text{mix}} < 0.50\%$	$r_{\text{mix}} < 0.56\%$ (FNAL E615 [3])
Mixing in $K^{\pm}\pi^{\mp}$, $K^{\pm}\pi^{\mp}\pi^{\pm}\pi^{\mp}$	$r_{\text{mix}} < 0.33\%$ (no interference term)	$r_{\text{mix}} < 0.37\%$ (FNAL E691 [4])
DCS $D^0 \to K^+\pi^-$	$r_{\text{DCS}} = 0.68^{+0.34}_{-0.33} \pm 0.07\%$	$r_{\text{DCS}} = 0.77 \pm 0.25 \pm 0.25\%$ (CLEO [5])
DCS $D^+ \to K^+\pi^-\pi^+$	$\Gamma(D^+ \to K^+\pi^-\pi^+)/\Gamma(D^+ \to K^-\pi^+\pi^+)$ $= (7.7 \pm 1.7 \pm 0.8) \times 10^{-3}$ $\begin{pmatrix} \rho^0 K^+ & 37 \pm 14 \pm 7\% \\ \overline{K}^{*0}\pi^+ & 35 \pm 14 \pm 1\% \\ \text{Nonresonant} & 36 \pm 14 \pm 7\% \end{pmatrix}$	$(7.2 \pm 2.3 \pm 1.7) \times 10^{-3}$ (FNAL E687 [6])
$\Gamma(D^0 \to K^+K^-)/\Gamma(D^0 \to \pi^+\pi^-)$	$2.75 \pm 0.15 \pm 0.16$	$2.53 \pm 0.46 \pm 0.19$ (FNAL E687 [7])
CP violation in:		
$D^+ \to K^-K^+\pi^+$	$A_{CP} = -0.014 \pm 0.029$	-0.031 ± 0.068 (FNAL E687 [8])
$D^+ \to \phi\pi^+$	-0.028 ± 0.036	0.066 ± 0.086 (FNAL E687 [8])
$D^+ \to \overline{K}^{*0}K^+$	-0.010 ± 0.050	-0.12 ± 0.13 (FNAL E687 [8])
$D^+ \to \pi^-\pi^+\pi^+$	-0.017 ± 0.042	None
$D^0 \to K^+K^-$	$-0.013 \pm 0.049 \pm 0.012$	$\begin{cases} 0.024 \pm 0.084 & \text{(FNAL E687 [8])} \\ 0.080 \pm 0.061 & \text{(CLEO [9])} \end{cases}$
$D^0 \to \pi^+\pi^-$	$-0.049 \pm 0.078 \pm 0.030$	None
$\Gamma(D^+ \to \rho^0\ell^+\nu)/\Gamma(D^+ \to \overline{K}^{*0}\ell^+\nu)$	$4.7 \pm 1.3\%$	$4.4^{+3.1}_{-2.5} \pm 1.4\%$ (FNAL E653 [10])

compared. The E791 results for mixing and the DCS decay $D^0 \to K^+\pi^-$ are similar to previous results, while the results for $D^+ \to K^+\pi^-\pi^+$, $\Gamma(D^0 \to K^+K^-)/\Gamma(D^0 \to \pi^+\pi^-)$, CP violation, and $\Gamma(D^+ \to \rho^0\ell^+\nu)/\Gamma(D^+ \to \overline{K}^{*0}\ell^+\nu)$ have significantly more precision than previous results.

REFERENCES

1. J. A. Appel, *Ann. Rev. Nucl. Part. Sci.* **42**, 367 (1992), and references therein; E. M. Aitala et al., *Phys. Rev. Lett.* **76**, 364 (1996).
2. N. Isgur and M. B. Wise, *Phys. Rev. D* **42**, 2388 (1990).
3. W. C. Louis et al., *Phys. Rev. Lett.* **56**, 1027 (1986).
4. J. C. Anjos et al., *Phys. Rev. Lett.* **60**, 1239 (1988).
5. D. Cinabro et al., *Phys. Rev. Lett.* **72**, 1406 (1994).
6. P. L. Frabetti et al., *Phys. Lett. B* **359**, 403 (1995).
7. P. L. Frabetti et al., *Phys. Lett. B* **321**, 295 (1994).
8. P. L. Frabetti et al., *Phys. Rev. D* **50**, R2953 (1994).
9. J. Bartelt et al., *Phys. Rev. D* **52**, 4860 (1995).
10. K. Kodama et al., *Phys. Lett. B* **316**, 455 (1993).

Recent Results from Fermilab E687 on Charm Spectroscopy

Paul Lebrun
for the Fermilab E687 collaboration

Fermi National Acclerator Laboratory
P.O. 500 Batavia Illinois, 60510, USA

Abstract. Recent analyses of charm spectroscopy from Fermilab fixed target experiment 687 [1] are summarized. Emphasis is placed on the phenomenology of Cabibbo suppression in the meson sector. Such transitions have been observed in the semileptonic modes and hadronic modes. While the former transitions give us an opportunity to observe the weak current and the CKM matrix, a systematic comparison of the latter transitions allow us to study strong interaction effects, and possibly, light quark spectroscopy.

INTRODUCTION

Fermilab Experiment 687 took its first beam about a decade ago. The data I will be presenting were obtained during the 1990-1991 run. Reconstructed data have been available for quite sometime, and numerous results have been published. However, thanks to meticulous efforts on the part of my collaborators, and results obtained by other experiments, we were able to substantially refine our charm reconstruction methods and study more difficult signals, such as doubly Cabibbo suppressed decays and some Cabibbo suppressed semileptonic decays. Rather then presenting the exhaustive list of measurements made by E687 in this very productive decade, I'll concentrate on most recent results in semileptonic decay of the D^+ and D^0 meson, and on the Dalitz analysis of the Cabibbo suppressed decay $D^+ \rightarrow \pi^+\pi^+\pi^-$ and the Cabibbo allowed decay $D_s \rightarrow \pi^+\pi^+\pi^-$.

In E687, charm particles were produced by photons with average tagged energies of approximately 200 GeV colliding a $\approx$ 4 cm. long Beryllium target and detected by a wide-acceptance, multi-purpose spectrometer which is described in detail elsewhere [2]. Charged particle tracking and momentum analysis was accomplished by a high resolution silicon microstrip detector, five stations of multi-wire proportional chambers and two large magnets operated with opposite polarities. A system of three multicell Čerenkov detectors working in threshold mode provided charged hadron identification ($\pi^{\pm}, K^{\pm}, p^{\pm}$) over a large momentum range. Two electromagnetic calorimeters, both composed of alternating layers of lead and scintillators, were used to detect electrons in complementary regions of the spectrometer: the inner electromagnetic calorimeter covered the central solid angle around the beam direction and detected particles passing through the fields of the two magnets; the outer electromagnetic calorimeter covered the outer angular anulus described by particles passing through the field of the first magnet, but not the second magnet. Muons were identified only in the central region of the spectrometer by the inner muon detector, composed of three scintillator planes and four proportional tube planes; shielding was provided by the upstream detectors (mainly the inner electromagnetic and the hadron calorimeter) and two blocks of steel. The hadron calorimeter was primarily used in the trigger.

All the analysis mentioned in this paper required at least two vertices, obtained either by the "candidate-driven" algorithm, valid for fully reconstructed final state, or by the "stand alone" method, where no apriori knowledge about the decay topology is assumed while forming vertices in the event. In order to obtain clean sample, in addition to conventional cuts such as those based on χ^2 from track or vertex fits, numerous vertex cuts had to be considered. For instance:

- isolation cuts: leftover tracks not found in the primary vertex were required to be inconsistent with emerging from the secondary vertex, and secondary tracks were required not to point towards the primary vertex.

- Point back cut: For fully reconstructed decays, the charm particle direction can be reconstructed and must points back to the primary vertex.

- Requiring that the secondary vertex be outside the beryllium target allows us to reject background due to secondary interaction.

In addition, particle identification played a crucial role in these analysis. for instance, the efficiencies of the Čerenkov system in presence of other tracks was carefully studied using $K_s^0 \to \pi^+\pi^-$, $\Lambda^0 \to p\pi^-$ and $\phi(1020) \to K^+K^-$ decays.

RESULTS

Cabibbo suppressed, semileptonic decays of the D meson

With the Cabibbo allowed semileptonic decays well established, experiments have begun turning their attention towards the more elusive Cabibbo suppressed, semileptonic decays ($D^0 \to \pi l\nu$ and $D^+ \to \rho l\nu$). These decays may be used to compare the functional dependence of form factors between Cabibbo favored and Cabibbo suppressed currents. In particular, E687 has observed the $D^0 \to \pi^- e^+ \nu$ and $D^0 \to \pi^- \mu^+ \nu$ (charge conjugate are always implied) [3]. Assuming the D^0 mass and using the direction of flight of the D^0 and the soft pion from the $D^{*+} \to D^0 \pi^+$ decay, it is possible to fully reconstruct the decay kinematics, resolving correctly the D^0 momentum twofold ambiguity approximately 80% of the time, and extract a signal. We obtained:

$$\frac{BR(D^0 \to \pi^- l^+ \nu_l)}{BR(D^0 \to K^- l^+ \nu_l)} = 0.101 \pm 0.020(stat) \pm 0.003(syst)$$

Assuming a single pole mass dependence for the form factors, we determined:

$$|\frac{V_{cd}}{V_{cs}}|^2 |\frac{f_+^\pi(0)}{f_+^K(0)}|^2 = 0.050 \pm 0.011 \pm 0.002$$

Finally, untarity constraints on the CKM matrix set a value for the ratio $|\frac{V_{cd}}{V_{cs}}|^2$ and we can compute:

$$|\frac{f_+^\pi(0)}{f_+^K(0)}|^2 = 1.00 \pm 0.11 \pm 0.02$$

More recently, we observed the first statistically significant signal for the vector meson Cabibbo suppressed decay $D^+ \to \rho^0 \mu^+ \nu$ [4]. This decay had to be reconstructed without the help of D^* trick. However, three charged tracks are in this final state which greatly ease the vertex reconstruction. In addition, the lifetime of D^+ is relatively large. The background to the $M(\pi^- \pi^+)$ invariant mass distribution is adequately described by three sources: other D^+ and D_s^+ semileptonic decays involving two pions, semileptonic decays of the D^0 produced in D^{*+} decays and charm hadronic decays. The $\pi^+ \pi^-$ invariant mass in this selected semileptonic decay sample is shown on figure 1. We measured the branching ratio of the decay mode $D^+ \to \rho^0 \mu^+ \nu$ (plus possible unobserved γ from $D^+ \to \eta' \mu^+ \nu$, $\eta' \to \gamma \rho^0$) with respect to the decay mode $D^+ \to \bar{K}^{*0} \mu^+ \nu$ to be

$$\frac{BR(D^+ \to \rho^0 \mu^+ \nu)}{BR(D^+ \to \bar{K}^{*0} \mu^+ \nu)} = 0.079 \pm 0.019 \pm 0.013$$

Hadronic decays

Amplitude analysis of non-leptonic Cabibbo favored [5] and suppressed [6] decays have been previously studied by E687. These analyses have emerged as an excellent tool for studying hadron dynamics. In particular, the D_s^+ decay into three pions is, in fact, the best candidate to proceed through an annihilation diagram, since annihilation of the two initial quarks is Cabibbo favored and not suppressed as in the D^+ decay. This annihilation amplitude seems also to manifest itself through markedly different final states: the $f_0(980)$ "oddball" -as stated in the discussions on hadron spectroscopy- appears in the D_s^+ decay is absent in the D^+ decay [7]. The Dalitz plots for these two decays are shown in figure 2. Doubly and singly Cabibbo suppression has also been studied in the $K^+K^-K^+$ channel [8] and in $K^+\pi^-\pi^+$ final state [9]

CONCLUSION

The observations of Cabibbo suppressed semileptonic D decays are consistent with the theoretical knowledge of the weak current in the quark sector. However, much remains to be done [10]: improved statistics are of course needed along with better experimental information on other semileptonic decays, such as $D^0 \to \rho^- l^+ \nu$ and $D^+ \to \eta' \mu^+ \nu$. It should be noted that these

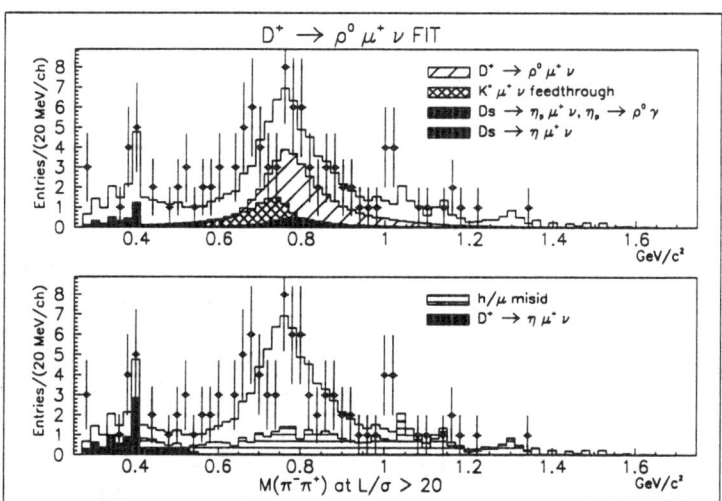

FIGURE 1. $M(\pi^+\pi^-)$ invariant mass reconstructed from $D^+ \to \rho^0 \mu^+ \nu$ candidates. The points are the data, the solid line is the total fit, the various fit components are represented with different hatching styles. The fit components are shown in two separate histograms for clarity.

interesting decay channels cannot be studied in isolation, one must also complete and improve the phenomenology of Charm. For instance, a study of charged hadronic five body decays of the D^+ and D_s^+ mesons has been recently published by E687 [11], showing that in all instances, resonant channel decay dominate. Finally, many topics, such as Charmed Baryon analysis [12] and the search for rare and forbidden decays [13] have been left out due to lack of time.

REFERENCES

1. E687 Collaboration coauthors:

   P.L.Frabetti (**Bologna**); H.W.K.Cheung, J.P.Cumalat, C.Dallapiccola, J.F.Ginkel, W.E.Johns, M.S.Nehring (**Colorado**); J.N.Butler, S.Cihangir, I.Gaines, P.H.Garbincius, L.Garren, S.A.Gourlay, D.J.Harding, P.Kasper, A.Kreymer, P.Lebrun, S.Shukla, M.Vittone (**Fermilab**); S.Bianco, F.L.Fabbri, S.Sarwar, A.Zallo (**Frascati**); R.Culbertson, R.Gardner, R.Greene, J.Wiss (**Illinois**); G.Alimonti, G.Bellini, M.Boschini, D.Brambilla, B.Caccianiga, L.Cinquini, M.Di Corato, M.Giammarchi, P.Inzani, F.Leveraro, S.Malvezzi, D.Menasce, E.Meroni, L.Moroni, D.Pedrini, L.Perasso, F.Prelz, A.Sala, S.Sala, D.Torretta (**Milano**); D.Buchholz, D.Claes, B.Gobbi, B.O'Reilly (**Northwestern**); J.M.Bishop, N.M.Cason, C.J.Kennedy, G.N.Kim, T.F.Lin, D.L.Puseljic, R.C.Ruchti, W.D.Shephard, J.A.Swiatek, Z.Y.Wu (**Notre Dame**); V.Arena, G.Boca, C.Castoldi, G.Gianini, S.P.Ratti,

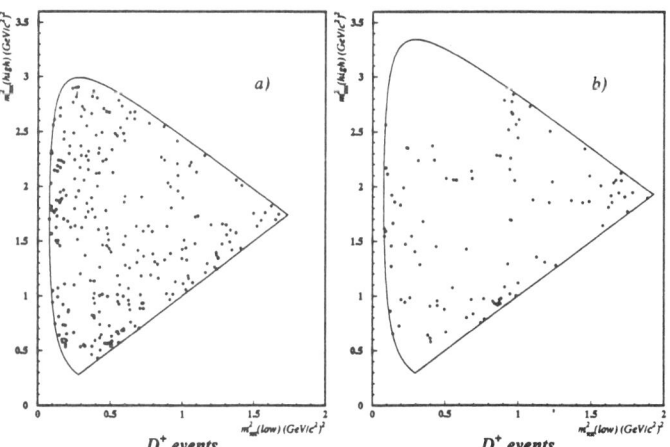

FIGURE 2. The D^+ (a) and D_s^+ (b) Dalitz plots for the $\pi^+\pi^-\pi^+$ channel, reconstructed using the "candidate-driven" method.

C.Riccardi L.Viola, P.Vitulo (**Pavia**); A.Lopez (**Puerto Rico**); G.P.Grim, V.S.Paolone, P.M.Yager (**Davis**); J.R.Wilson (**South Carolina**); P.D.Sheldon (**Vanderbilt**); F.Davenport (**North Carolina**); G.R.Blackett, K.Danyo, M.Pisharody, T.Handler (**Tennessee**); B.G.Cheon, J.S.Kang, K.Y.Kim (**Korea**)

2. P.L. Frabetti et al *Nucl. Instr. & Meth.* **A230** 519 (1992)
3. P.L. Frabetti et al *Phys. Lett.* **B382** 312 (1996)
4. P.L. Frabetti et al *Phys. Lett.* **B391** 235 (1997)
5. P.L. Frabetti et al *Phys. Lett.* **B331** 217 (1994)
6. P.L. Frabetti et al *Phys. Lett.* **B351** 591 (1995)
7. P.L. Frabetti et al *FERMILAB-PUB-97-045-E* (1997)
8. P.L. Frabetti et al *Phys. Lett.* **B363** 259 (1995)
9. P.L. Frabetti et al *Phys. Lett.* **B359** 403 (1995)
10. See next talk in this session, given Y. Zhang on E831, for anticipated improvements
11. P.L. Frabetti et al *Phys. Lett.* **B401** 131 (1997)
12. P.L. Frabetti et al *Phys. Lett.* **B365** 461 (1996)
13. P.L. Frabetti et al *Phys. Lett.* **B398** 239 (1997)

HYPERNUCLEAR SPECTROSCOPY AND WEAK DECAYS AT KEK

H. Noumi

High Energy Accelerator Organization (KEK), Tsukuba 305, JAPAN

1 Introduction

Λ hypernuclei provide unique opportunities to explore interior of nucleus since a Λ is free from the Pauli exclusion principle of the nucleon. Therefore, a Λ can be a good probe to investigate dynamics of baryonic many body system proceeded deeply inside nucleus.

In addition, there is an important role of studying Λ hypernuclei. Since the strangeness-involved two-baryon scattering experiments in free space are difficult, Λ hypernuclei provide invaluable information on the ΛN strong and weak interactions through their structure and weak decay mechanism.

There are a several experimental activities on hypernuclear physics at the K6 beam line of KEK-PS. They are proceeded with good pion beam at 1.06 GeV/c provided from K6 and high-resolution/wide-aperture kaon spectrometer, SKS [1]. Among them, in this article, two experimental programs are introduced. One is spectroscopic investigation of light Λ hypernuclei (E336) [2], and the other is asymmetry of nonmesonic weak decay of polarized $^5_\Lambda$He (E278) [3].

2 Spectroscopic Investigation of Light Λ hypernuclei

The E336 experiment aims at spectroscopic investigation of p-shell Λ hypernuclei via the (π^+, K^+) reaction with the energy resolution as good as 2 MeV. Precision spectroscopy of p-shell hypernuclear structure will reveal the effective ΛN interaction, since light hypernuclei have abundant information but reasonably simple structure.

Excitation energy spectra for $^7_\Lambda$Li, $^9_\Lambda$Be, $^{13}_\Lambda$C, and $^{16}_\Lambda$O were measured in E336. The (π^+, K^+) runs with the ^{12}C target (1.78g/cm^2) were occasionally inserted during the other target runs for the calibration of energy scale as well as for system check. The energy resolution is achieved to be $\sim$1.9 MeV (FWHM) for the $^{12}_\Lambda$C ground state in the present experiment, as demonstrated in Fig. 1. Two core excited states are also observed between two prominent peaks, which have been discovered in E140 for the first time [4]. Figs. 2-(a)$\sim$(d) correspond to the spectra for $^7_\Lambda$Li, $^9_\Lambda$Be, $^{13}_\Lambda$C, and $^{16}_\Lambda$O, respectively, which are based on a fraction of total data. As a guide, configurations of populated states predicted by theoretical calculations [5, 6, 7] are shown in the figures. It should be noted that the horizontal scale of the spectra has not been calibrated. Fine tuning has to be made so as to achieve the best resolution and efficiency.

The spectrum for $^7_\Lambda$Li is shown for the first time. Two peaks can be seen at the excitation energies (E_x) equal to 0 and 2 MeV. The spectrum also suggests yields at around the threshold energy of Λ-unbound region, where theory predicts the states of isospin equal to 1 [5]. Much higher resolution would be required to separate these states.

For $^9_\Lambda$Be, 4-peak structure can be seen in the region below $E_x=\sim$10 MeV. The structure has been predicted by the microscopic cluster model [6]. According to the model calculation, the 3rd peak should correspond to the lowest state of so-called the genuine hypernuclear (or supersymmetric) states [8].

The $^{13}_\Lambda$C spectrum is expected to reflect the structure of Λ-orbit and core ^{12}C state. The lowest two peaks correspond to the states a Λ sitting in s-orbit coupled to the core ground and 1st excited states, respectively. Excited states above $E_x=\sim$10 MeV can be seen. The configurations of them will be investigated with detail analysis of the position, width, and peak yield.

The $^{16}_\Lambda$O spectrum shows clearly separated four peaks. The particle-hole states corresponding to the peaks are indicated in the figure. The peak position is of our interest. Particularly, the position for the 3rd peak is related to the ΛN spin-orbit interaction. The (π^+, K^+) reaction populates the $p^\Lambda_{3/2}$ state predominantly because of its large momentum transfer. On the other hand, the recoilless (K^-, π^-) reaction populates almost only the $p^\Lambda_{1/2}$ state [9]. The spin-orbit splitting energy can be extracted comparing the position of these states, for which careful calibration of the horizontal scale is necessary.

3 Asymmetry of Nonmesonic Weak Decay of Polarized $^5_\Lambda$He

The hypernuclear nonmesonic decay can be understood as a process that the elementary two-body transition, $\Lambda N \to NN$, predominantly takes place in a hypernucleus. The $\Lambda N \to NN$ transition cannot be realized in free space. Therefore, the nonmesonic decay gives an important channel to investigate the two-body ΛN weak interaction. Since the process involves the large momentum transfer of 0.4 GeV/c, the degrees of freedom appear in short distance, such as heavier meson exchange potentials or quark dynamics, may play a significant role on the weak decay mechanism.

The E278 experiment aims at observing asymmetric weak decays of polarized $^5_\Lambda$He produced via the $(\pi^+, K^+ p)$ reaction on ^{6}Li. The angular distribution of nonmesonic decay protons from polarized $^5_\Lambda$He is described as $W(\theta) \propto 1 + A_1 P_\Lambda \cos\theta$ with respect to the Λ-spin polarization P_Λ, where the coefficient A_1 represents the asymmetry parameter of the nonmesonic decay. The asymmetry is a consequence of the interference of the parity-violating amplitude (PVA) with the parity-conserving one (PCA) in nonmesonic decay. The presence of the interference in p-shell hypernuclei has been first demonstrated in the previous E160 experiment at KEK [10], although the observed asymmetry has a large error of 40%. E278 has been thus carried out in order to improve the precision of the asymmetry parameter taking advantages of (a) determining the polarization of $^5_\Lambda$He employing the mesonic decay asymmetry, and (b) using the s-shell hypernucleus where the nuclear effects such as the final state interaction affect the asymmetry less.

Since SKS can accept kaons scatterred at angles ±15° simultaneously, the experimental technique to cancel the instrumental asymmetry in the first order by taking 2-fold ratio of up-down asymmetry measured by the decay counters placed above and below the target can be used. Monitoring $(\pi^+, \pi^{+\prime} X)$ which involves no asymmetry, the instrumental asymmetry is found to be as small as -0.011 ± 0.015 for $X=\pi^-$ and -0.008 ± 0.004 for $X=p$, where X represents the particles detected by the decay counters in coincidence with $(\pi^+, \pi^{+\prime})$ reactions. Fig. 3 shows the reconstructed excitation energy spectra for $^6_\Lambda$Li without any gate for decay events (upper), with gates for decay pions (middle), and decay protons (lower). The ground state of $^6_\Lambda$Li, which forms $^5_\Lambda$He immidiately, is clearly separated for each spectrum.

The preliminary results on the asymmetry of decay pions are shown as follows: $A_1=-0.114\pm0.046$ for $2°<|\theta_K|<7°$ (mean:4.4°) and $A_1=-0.201\pm0.060$ for $7°<|\theta_K|<15°$ (mean:9.8°), where only the statistical errors are taken into account. These values include the attenuation due to the finite solid angle of the decay counters. The attenuation is estimated to be ~ 0.82 by a Monte Calro simulation. The asymmetry parameters of the pion for $^5_\Lambda$He is expected to be almost same as that for free Λ [12]. If one takes -0.64 as the pion asymmetry parameter, the polarization is derived to be -0.38 at $\theta_K \sim 10°$. According to theoretical calculation, the polarization is 0.37 at $\theta_K=10°$ for the $^6_\Lambda$Li ground state [11]. The present result well agrees with the theoretical calculation.

In order to finalize the polarization and extract the nonmesonic-decay asymmetry parameter, Careful corrections on the observed asymmetry, due to the effects of finite solid angle and efficiency of the decay counters, contaminations in the gated spectra, and the attenuation by the final state interaction, have to be made.

4 Summary

The present article introduced two experiments, E336 and E278, among the hypernuclear experimental activities at K6 of KEK-PS. E336 demonstrates spectroscopic studies of p-shell hypernuclei with better quality, although the analyses are under way. Improvements on the resolution and statistics in the hypernuclear spectra will be made. E278 was carried out in order to improve the nonmesonic-decay asymmetry parameter, and the status of the data analysis for the asymmetry is close to the end. The measured polarization is consistent with the theoretical calculation. To finalize the results, fine corrections on the asymmetry due to the geometrical effects and the final state interaction must be made.

Acknowledgement

This presentation is based on the collaborations of E336 and E278. I should thank all the collaborators, in particular, H. Hotchi and S. Ajimura who are in charge of the present data analyses for E336 and E278, respectively.

References

[1] T. Fukuda et al., Nucl. Inst. and Meth. **A361**, 485(1995).

[2] O. Hashimoto et al., KEK-PS Proposal E336, (1994).

[3] T. Kishimoto et al., KEK-PS Proposal E278, (1992).

[4] T. Hasegawa et al., Phys. Rev. Lett. **74**, 224(1995).

[5] H. Bandō, T. Motoba, and J. Žofka, Int. J. Mod. Phys. **A5**, 4021(1990); T. Motoba, private communication (1996).

[6] T. Yamada et al., Phys. Rev. **C38**, 854(1988).

[7] K. Itonaga, T. Motoba, O. Richter, and M. Sotona, Phys. Rev. **C49**, 1045(1994).

[8] T. Motoba, H. Bandō, and K. Ikeda, Prog. Theor. Phys. **79**, 189(1983); R. H. Dalitz and A. Gal, Phys. Rev. Lett. **36**, 362(1976).

[9] W. Brückner et al., Phys. Lett., **79B**, 157(1978).

[10] S. Ajimura et al., Phys. Lett. **B282**, 293(1992).

[11] T. Motoba and K. Itonaga, Nucl. Phys., **A577**, 293c(1994).

[12] T. Motoba, H. Bandō, T. Fukuda, and J. Žofka, Nucl. Phys., **A534**, 597(1991).

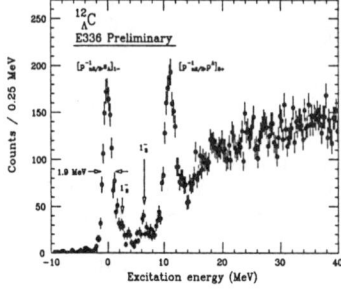

Fig. 1: Preliminary result of $^{12}_{\Lambda}$C spectrum produced via the (π^+, K^+) reaction shown as a function of the excitation energy.

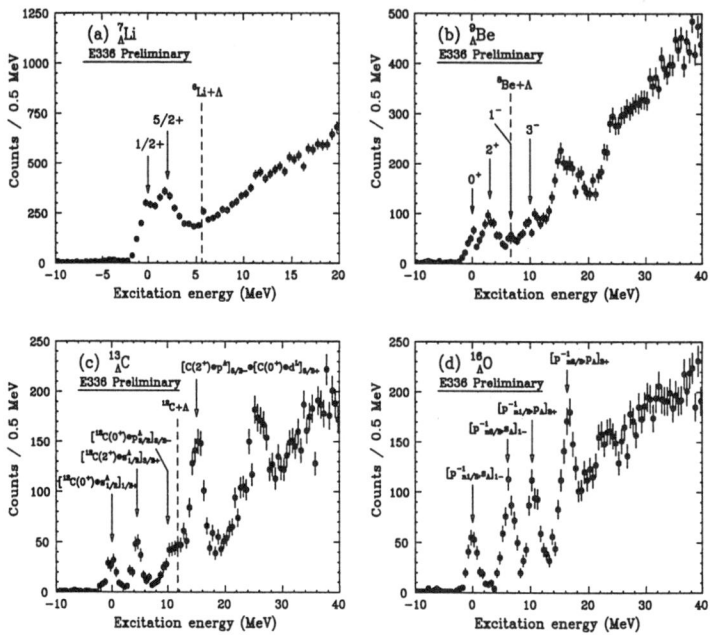

Fig. 2: Preliminary results of (a) $^{7}_{\Lambda}$Li, (b) $^{9}_{\Lambda}$Be, (c) $^{13}_{\Lambda}$C, and (d) $^{16}_{\Lambda}$O spectra produced via the (π^+, K^+) reactions shown as a function of excitation energy. They are based on 73%, 70%, 30%, and 21% of total data, respectively.

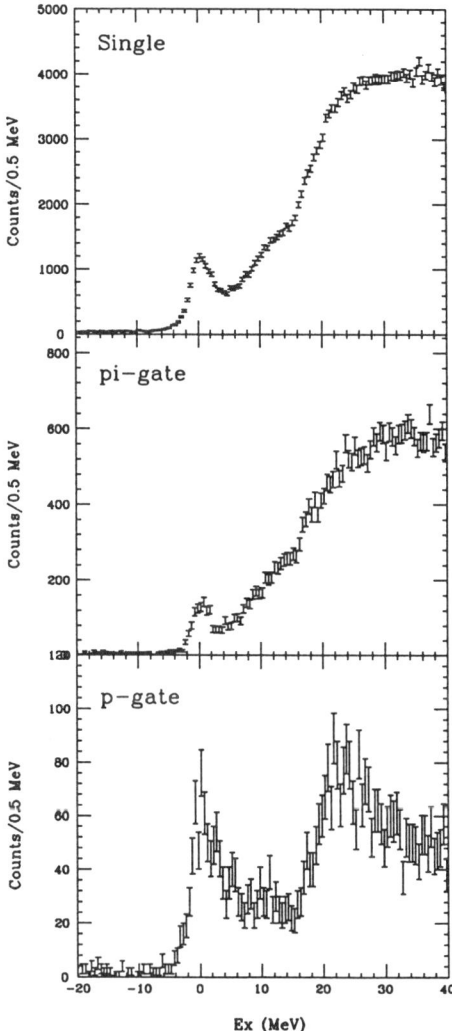

Fig. 3: Reconstructed excitation energy spectra for $^6_\Lambda$Li without any gate for decay events (upper), with gates for decay pions (middle), and decay protons (lower).

The Parity-Violating Asymmetry in the Weak Decay of Polarized Hypernuclei [1]

C. Bennhold[1], A. Parreño[2], and A. Ramos[2]

[1]*Center for Nuclear Studies, Department of Physics,*
The George Washington University, Washington, D. C. 20052
[2]*Departament d'Estructura i Constituents de la Matèria, Facultat de Física,*
Diagonal 647, E-08028 Barcelona, Spain

Abstract. The hypernuclear weak decay is examined in a shell model framework, using a complete strangeness-changing weak $\Lambda N \to NN$ transition potential that includes the exchange of π, K, η as well as ρ, ω, and K*. The total and partial rates as well as the parity-violating asymmetries are compared to experimental data for light s- and p-shell nuclei. The inclusion of heavier mesons strongly influences the asymmetry which is found to agree with new KEK data on $^5_\Lambda$He with much reduced uncertainties.

Parity violation in hadronic systems represents a unique tool to study aspects of the nonleptonic weak interaction between hadrons. In the $\Delta S=0$ sector the parity-violating part of the nuclear force has been studied for many years through nuclear transitions forbidden by the strong force and, more recently, in parity-violating NN scattering by measuring the asymmetry of longitudinally polarized protons. For the $\Delta S=1$ sector the analogous reaction $pn \to \Lambda p$ may be measured with new, high precision proton accelerators such as COSY. While measuring the $pn \to \Lambda p$ process has only recently been contemplated it has long been known that the time-inversed process $\Lambda N \to NN$ constitutes the dominant decay mode for all but the lightest hypernuclei. In this case the hypernucleus in its ground state provides the laboratory that allows the weak process to occur. On the other hand, this decay mode will probe this process at only one well-defined kinematics. Furthermore, one has to deal with the complications of hypernuclear structure. It is the latter aspect that

---

[1] This work was supported by US-DOE grant no. DE-FG02-95-ER40907 (CB), DGICYT contract no. PB92-0761 (Spain), the Generalitat de Catalunya grant no. GRQ94-1022 (AP and AR), a doctoral fellowship of the Ministerio de Educación y Ciencia (AP), and NATO Grant CRG 960132.

has recently been treated in as complete a framework as possible [1]. The initial hypernuclear and final nuclear shell model structure are taken into account through spectroscopic factors. All possible initial and final relative orbital angular momenta are included in the baryon-baryon system. Realistic initial and final short-range correlations obtained from ΛN and NN interactions based on the Nijmegen baryon-baryon potential are employed. The calculations were performed in a one-boson-exchange model that includes the long-ranged pion along with the η, K, ρ, ω and K* whose weak baryon-baryon-meson vertices were obtained using SU(6) and soft meson theorems for the PV vertices and the pole model for the PC vertices. By reducing nuclear structure uncertainties to a minimum this work provides a framework to test - or extract - these weak baryon-baryon-meson couplings.

The nonmesonic $\Lambda N \to NN$ decay rate is given by [1]

$$\Gamma_{nm} = \int \frac{d^3P}{(2\pi)^3} \int \frac{d^3k}{(2\pi)^3} \sum (2\pi)\delta(M_H - E_R - E_1 - E_2) \mid \mathcal{M} \mid^2 \quad (1)$$

where $\mathcal{M} = \langle \psi_R; \mathbf{P}\mathbf{k}\ S\ M_S\ T\ M_T | \hat{O}_{\Lambda N \to NN} |_{\Lambda} A \rangle$ is the transition amplitude.

Due to the interference between the parity conserving and parity violating amplitudes, the distribution of the emitted protons in the weak decay displays an angular asymmetry with respect to the polarization axis given by

$$\sigma(\chi) = \sigma_0(1 + P_y A_p(\chi)) \quad (2)$$

where P_y is the hypernuclear polarization, created in the production reaction, such as (π^+, K^+) at KEK and BNL or (γ, K^+) at TJNAF, and the expression for the asymmetry

$$A_p(\chi) = \frac{3}{J+1} \frac{Tr(\mathcal{M}\hat{S}_y\mathcal{M}^\dagger)}{Tr(\mathcal{M}\mathcal{M}^\dagger)} = \frac{3}{J+1} \frac{\sum_{M_i} \sigma(M_i) M_i}{\sum_{M_i} \sigma(M_i)} \cos\chi = A_p \cos\chi, \quad (3)$$

shown, for instance, in Ref. [2], defines the asymmetry parameter, A_p, characteristic of the hypernuclear weak decay process. The asymmetry in the distribution of protons is thus determined by the product $P_y A_p$. In the weak coupling scheme, simple angular momentum algebra relations relate the hypernuclear polarization to the Λ polarization. It is convenient to introduce the intrinsic Λ asymmetry parameter such that $P_y A_p = p_\Lambda a_\Lambda$, which is then characteristic of the elementary Λ decay process, $\tilde{\Lambda} N \to NN$, taking place in the nuclear medium.

The results for the usual observables are shown in Table 1 and demonstrate the significance of adding the heavier mesons. The rate is especially sensitive to the inclusion of the strange mesons. While including the ρ-meson has almost

TABLE 1. Weak decay observables for $^{12}_\Lambda$C

	Γ/Γ_Λ	Γ_n/Γ_p	a_Λ
π	0.885	0.104	-0.238
$+\rho$	0.859	0.095	-0.100
$+K$	0.497	0.030	-0.138
$+K^*$	0.760	0.049	-0.182
$+\eta$	0.683	0.058	-0.200
$+\omega$	0.753	0.068	-0.316

no effect the addition of kaon exchange reduces the total rate by almost 50%. The reduction is mostly compensated by the addition of the K^*, yielding a rate 15% below the pion-only decay rate. However, the results are also somewhat sensitive to the model used for the strong vertices (Nijmegen couplings are used here); improved YN potentials which narrow the range of the strong coupling constants are required to reduce this uncertainty.

The neutron- to proton-induced ratio is, as expected, quite sensitive to the isospin structure of the exchanged mesons. Including the K-exchange which interferes destructively with the pion amplitude in the neutron-induced channel leads to a reduction of the ratio by more than a factor of three. In the $T = 1$ PV channel, relevant for the n-induced rate, the K and K^* amplitudes have the same sign, whereas in both $T = 0$ channels the interference between the two strange mesons is destructive and, as a consequence, the p-induced rate is lowered with respect to the n-induced rate. The intrinsic asymmetry parameter, a_Λ, is also found to be very sensitive to the different mesons included in the model. This is the only observable which is changed dramatically by the inclusion of the ρ, reducing the pion-only value by more than a factor of two. Adding the other mesons increases a_Λ, leading to a result about 30% larger than for π-exchange alone.

TABLE 2. Weak decay observables for various hypernuclei

	$^5_\Lambda$He	$^{11}_\Lambda$B	$^{12}_\Lambda$C
Γ/Γ_Λ	0.414	0.611	0.753
EXP:	0.41 ± 0.14 [3]	$0.95 \pm 0.13 \pm 0.04$ [4]	$0.89 \pm 0.15 \pm 0.03$ [4]
Γ_n/Γ_p	0.073	0.084	0.068
EXP:	0.93 ± 0.55 [3]	$1.04^{+0.59}_{-0.48}$ [3]	$1.33^{+1.12}_{-0.81}$ [3]
		$2.16 \pm 0.58^{+0.45}_{-0.95}$ [4]	$1.87 \pm 0.59^{+0.32}_{-1.00}$ [4]
Γ_p/Γ_Λ	0.386	0.563	0.705
EXP:	0.21 ± 0.07 [3]	$0.30^{+0.15}_{-0.11}$ [4]	$0.31^{+0.18}_{-0.11}$ [4]
a_Λ	-0.273	-0.391	-0.316
$\mathcal{A}(0°)$		-0.120	-0.030
EXP:		-0.20 ± 0.10 [5]	-0.01 ± 0.10 [5]

The final ratio of proton- to neutron-induced rates, shown in Table 2, greatly underestimates the newer central experimental values, which still have very large experimental errors. The proton-induced rate which has errors of the same magnitude as the total rate is overpredicted by our calculations by up to a factor of two. It is somewhat surprising that while both individual rates appear in disagreement with the data their sum conspires to a total rate which reproduces the measurements.

Regarding the asymmetry parameter, comparison with experiment can only be made at the level of the measured proton asymmetry. As discussed above, this quantity is determined as a product of the asymmetry parameter A_p, characteristic of the weak decay, and the polarization of the hypernucleus, P_y, which must be determined theoretically. The previous experiment [5] measuring this asymmetry on $^{12}_\Lambda$C and $^{11}_\Lambda$B obtained only fairly small polarizations and, thus, small asymmetries with large experimental error bars, shown in Table 2.

In order to avoid the need for theoretical input and access A_p directly, a new experiment at KEK has measured the decay of polarized $^5_\Lambda$He, extracting both the pion asymmetry from the mesonic channel, $\mathcal{A}_{\pi^-}$, and the proton asymmetry from the nonmesonic decay, $\mathcal{A}$. Assuming that the asymmetry parameter a_{π^-} remains similar to that of the free Λ decay the hypernuclear polarization can now be obtained from the relation $P_y = \mathcal{A}_{\pi^-}/a_{\pi^-}$. This in turn can then be used as input, together with the measured value of $\mathcal{A}$, to determine the asymmetry parameter for the nonmesonic decay from the equality $A_p = \mathcal{A}/P_y$. Very preliminary results [6] yield an asymmetry parameter of -0.3 which is in good agreement with our theoretical result of -0.27. This result emphasizes the need to improve the individual proton- and neutron-induced rates. The newly finished FINUDA facility at DAΦNE [7] holds the potential to obtain these important observables with much improved accuracy.

REFERENCES

1. A. Parreño, A. Ramos, and C. Bennhold, Phys. Rev. C **56** (1997).
2. A. Ramos, E. van Meijgaard, C. Bennhold and B.K. Jennings, Nucl. Phys. **A544**, 703 (1992).
3. J.J. Szymanski et al., Phys. Rev. C **43**, 849 (1991).
4. H. Noumi et al., Phys. Rev. C **52**, 2936 (1995).
5. S. Ajimura et al., Phys. Lett. **B282**, 293 (1992).
6. H. Noumi et al., Proceedings for this conference.
7. A. Zenoni for the FINUDA collaboration, *Proceedings of the Second Workshop on Physics and Detectors for DAΦNE*, edited by R. Baldini, F. Bossi, G. Capon and G. Pancheri, (Frascati Physics Series vol. IV, 1995) 293.

Nonmesonic Weak Decays of Light Hypernuclei

T. Inoue

Department of Physics, University of Tokyo, Bunkyo, Tokyo 112 Japan

Abstract. We study the nonmesonic decay rates of light hypernuclei, $^4_\Lambda H, ^4_\Lambda He$ and $^5_\Lambda He$. The $\Lambda N \to NN$ transition is described by the one pion exchange mechanism and the direct quark mechanisms. We find that the superposed potential reproduce the nonmesonic decay rates of these hypernuclei.

I INTRODUCTION

The weak $\Lambda N \to NN$ transition plays a dominant role in the nonmesonic decay of hypernuclei. The direct quark (DQ) mechanism have been proposed for the transition [1], where a contact interaction between the constituent quarks of baryons causes the transition without exchanging mesons. The transition potential are calculated in the non-relativistic quark cluster model with the first order perturbation theory in $H_{eff}^{\Delta S=1}$.

We calculate the nonmesonic decay rates of light hypernuclei, $^4_\Lambda H, ^4_\Lambda He$ and $^5_\Lambda He$. We describe the decay of Λ inside a nucleus by a coherent sum of the DQ transition and the weak one pion exchanges (OPE) transition [2]. We employ a realistic wave function of the decaying Λ inside the hypernuclei [3].

II RESULTS AND DISCUSSION

Table 1 summarizes the our results for $^5_\Lambda$He. We list the partial decay rates, the proton induced decay rate (Γ_p), the neutron induced decay rate (Γ_n), the total nonmesonic decay rate ($\Gamma_{nm} = \Gamma_p + \Gamma_n$), and the n-p ratio ($R_{np} = \Gamma_n/\Gamma_p$). All the decay rates are written in the unit of Γ_Λ, the free Λ decay rate.

First we study the OPE mechanism. The channel c_p and d_p are dominant. This comes from the tensor part of the transition potential together with the tensor part of the final state interaction. One can see that the calculated Γ_p is in good agreement with experiment, while the calculated Γ_n is much smaller than experiment. The Γ_p is dominated by large contribution of the channels

TABLE 1. Calculated nonmesonic decay rates of $^5_\Lambda$He

	OPE only	DQ only	OPE + DQ	EXP [4]
a_p	0.0002	0.0167	0.0188	
b_p	0.0031	0.0113	0.0026	
c_p	0.1022	0.0548	0.2612	
d_p	0.0415	0	0.0415	
e_p	0.0346	0.0064	0.0207	
f_p	0.0093	0.0353	0.0763	
a_n	0.0003	0.0407	0.0356	
b_n	0.0063	0.0069	0.0264	
f_n	0.0185	0.0648	0.1437	
Γ_p	0.191	0.125	0.421	0.21 ± 0.07
Γ_n	0.025	0.112	0.206	0.20 ± 0.11
Γ_{nm}	0.216	0.237	0.627	0.41 ± 0.14
R_{np}	0.132	0.903	0.489	0.93 ± 0.55

c and d, which vanish in the neutron induced decay. Thus the calculated n-p ratio is much smaller than the experimental one. We turn to the DQ mechanism. One can see that the overall magnitudes of the decay rates are comparable to OPE. The components, however, are different. For the proton induced decays, OPE is dominated by the $I = 0$ final states, c_p, d_p, and e_p, while the DQ gives a large decay rates to the $I = 1$ final states, a_p, b_p, and f_p. The proton induced decay rate Γ_p in DQ is smaller than that of OPE. On the other hand, the neutron induced decay rate Γ_n in DQ is much larger than that of the OPE. This is due to large contribution of channel a_n and f_n. The total nonmesonic decay rate in DQ is roughly equal to that of OPE, while the n-p ratio in DQ is much larger than that of OPE and is about 0.9, which is closer to the experimental data.

The final results are given by the superposition of the OPE and DQ mechanism. The relative sign between the OPE potential and the DQ one is important when we consider the interference of these mechanisms. Namely, the sign of OPE potential must be chosen consistently with $H_{eff}^{\Delta S=1}$ and the quark model wave function of baryon in DQ potential. In order to determine the sign, we calculate the parity-violating part of the $\Lambda \to n\pi^0$ decay matrix element with the Hamiltonian and the wave function in the soft pion limit.

The final results are also listed in Table 1. We find that the neutron induced decay rate, Γ_n is enhanced from OPE much and becomes consistent with experimental data. On the other hand, the combined result overestimates the proton induced decay rate, Γ_p. Thus the total decay rate, Γ_{nm}, is slightly overestimated. The n-p ratio, R_{np}, is predicted in between the values for OPE and DQ, while the experimental data suggests a larger value.

Table 2 and Table 3 summarize the results for $^4_\Lambda$He and $^4_\Lambda$H respectively. One can see that, for $^4_\Lambda$He, our final results again provides a agreement to

TABLE 2. Calculated nonmesonic decay rates of $^4_\Lambda$He

	OPE only	DQ only	OPE+DQ	EXP [5]
Γ_p	0.145	0.060	0.214	0.15 ± 0.02
Γ_n	0.009	0.012	0.038	0.04 ± 0.02
Γ_{nm}	0.154	0.072	0.253	0.19 ± 0.04
R_{np}	0.061	0.202	0.178	0.27 ± 0.14

TABLE 3. Calculated nonmesonic decay rates of $^4_\Lambda$H

	OPE only	DQ only	OPE+DQ	EXP [6]
Γ_p	0.004	0.054	0.047	
Γ_n	0.017	0.057	0.126	
Γ_{nm}	0.022	0.110	0.174	0.15 ± 0.13
R_{np}	3.952	1.048	2.660	

the data except that $Gamma_p$ is again slightly overestimated. At present, the experimental data for the $^4_\Lambda$H are very limited, only the total nonmesonic decay rate is known. Our prediction agree with it. The result also suggest that the neutron induced decay is lager than the proton ne and thus n-p ratio, R_{np}, is larger than one. We anticipate new good data.

III CONCLUSION

We find that the OPE+DQ reproduce the present data for the nonmesonic decay of $^5_\Lambda$He, $^4_\Lambda$He and $^4_\Lambda$H. We made some predictions for the nonmesonic decay of $^4_\Lambda$H. We conclude that the DQ contribution is essential to understand the some features of the nonmesonic weak decays of light hypernuclei.

REFERENCES

1. T. Inoue, S. Takeuchi and M. Oka, Nucl. Phys. **A597**(1996)563
2. K. Takeuchi, H. Takaki and H. Bando, Prog. Theor. Phys. **73**(1985)841
3. T. Motoba and K. Itonaga, Prog. Theor. Phys. Suppl. **117**(1994)477
4. J.J. Szymanski *et al.* Phys. Rev. **C43**(1991)849
5. V.J. Zeps and G.B. Franklin for the E788 Collaboration, Proceedings of the 23rd INS International Symposium on Nuclear and Particle Physics with Meson Beams in 1 GeV/c Region, p.227
6. H. Outa, *et al.* Proceedings of the IV International Symposium on Weak and Electromagnetic Interactions in Nuclei, ed. by H. Ejiri, T. Kishimoto and T. Sato (World Scientific, 1995), p.532

Systems with Strangeness -2 at BNL

G. B. Franklin

Carnegie Mellon University
Pittsburgh, PA 15213

Abstract. The study of systems with strangeness -2 provides an opportunity to extend models of the baryon-baryon interactions which have been tested largely against data obtained in the non-strange sector. This paper presents motivations for examining the strangeness -2 sector and reviews the experimental difficulties responsible for the sparsity of data. The status of two ongoing experiments at the AGS, E885 and E906, are presented.

INTRODUCTION

Several ongoing AGS experiments have explored $S = -2$, two-baryon systems. There has been considerable effort aimed at finding evidence of the $S = -2$ H-Dibaryon, first predicted by Jaffe in 1977 [1]. Since the current status of these searches was covered in by B. Basselleck [2], this paper will discuss ΞN and $\Lambda\Lambda$ systems. In particular, I will review 1) the motivation for extending studies of baryon-baryon interactions to include systems with strange quarks, 2) the experimental difficulties responsible for the sparsity of existing data, and 3) the status AGS experiments E885 and E906.

The nucleon-nucleon interaction has been studied for decades. Our understanding of this interaction is largely based on a potential description which is, in turn, motivated from a OBE description for the long-range portion coupled with additional short-range ingredients. There are many parameters but the wealth of data introduce considerable constraints.

Studies in the $S = -1$ and $S = -2$ sector should be thought of as an extension of existing knowledge. The OBE terms fit to the NN data are transformed to systems with strangeness using SU(3) flavor symmetry. Although new parameters are introduced in the description of the short range terms, these parameters have only limited influence on the model's ability to describe existing data. The extension of NN models into the strange sector has been studied by the Nijmegen [3] and Jülich [4] groups.

It has long been recognized that competing descriptions which are essentially equivalent in the NN sector give quite differing predictions in the strangeness sector. For example, the well known Nijmegen D and F potentials are largely equivalent in the NN sector but don't even agree on the sign of the interaction for the ΞN case. [6] Thus a limited amount of reliable data in the strangeness sector can have a direct influence in our understanding of the NN interaction.

$\Lambda\Lambda$ AND Ξ HYPERNUCLEI

In the S=-1 sector, a limited amount of data on YN scattering exists. [5] Also, the binding energies of low and medium $\Lambda\Lambda$ hypernuclei have been measured with precision in emulsion and bubble chambers and the nuclear structure of a few hypernuclei have been studied using the (K^-,π^-) and (π^+,K^+) reactions. In contrast, the $S=-2$ sector data consists of a handful of kinematically reconstructed emulsion events. This includes seven events interpreted as the formation and decay of Ξ hypernuclei and three events interpreted as $\Lambda\Lambda$ hypernuclear formation and decay. An analysis of the Ξ hypernuclear events by Dover and Gal [7] concludes that the interpretation of these events are not "statistically unique, and in some cases the evidence is far from being compelling." However, they find that the data imply a Ξ-nucleus well depth around 21 to 24 MeV which in turn is close to the theoretical expectations of Nijmegen Model D.

Insight into the $\Lambda\Lambda$ interaction can be gained from the binding energies extracted from the three $\Lambda\Lambda$ hypernuclear events. Two of these events were reported in the 1960's, one by Danysz et al. [8] and one by Prowse [9]. Both these events appear to be statistical flukes since the probability that such hypernuclei would have been formed is quite low considering the limited integrated kaon flux. The event reported by Danysz et al., interpreted as $^{10}_{\Lambda\Lambda}\text{Be}$, has withstood further examination by Dalitz et al. [10]. Prowse's event, which he interpreted as the formation and decay of $^{6}_{\Lambda\Lambda}\text{He}$, requires a binding energy which appears to be inconsistent with the Danysz event and the details which would allow an independent study of the Prowse event are not available. The third event, found in a hybrid-emulsion experiment, was reported in 1991 by S. Aoiki et al. [11] Unfortunately, there are two possible interpretations of this event involving the formation of different hypernuclear species, either $^{10}_{\Lambda\Lambda}\text{Be}$ or $^{13}_{\Lambda\Lambda}\text{B}$. The interpretation favored by the experimentalists, $^{10}_{\Lambda\Lambda}\text{Be}$, requires a repulsive $\Lambda\Lambda$ interaction. The interpretation favored by several theorists [12], $^{13}_{\Lambda\Lambda}\text{B}$, allows for an attractive $\Lambda\Lambda$ interaction and is thus consistent with the earlier reported events. It also assumes a more probably formation channel. However, it requires the presence of an (unseen) neutron which must be almost completely collinear with other decay products to be consistent with the observed emulsion tracks.

It is clear that more experimental data are needed to clarify the $S=-2$ sector. The most straightforward means of exploration would appear to be the utilization of (K^-,K^+) reactions. For example, one could consider using a carbon target and creating $S = -2$ hypernuclei through the reactions

$$K^- + {}^{12}C \rightarrow \begin{cases} K^+ + {}^{12}_{\Xi}Be \\ K^+ + {}^{12}_{\Lambda\Lambda}Be. \end{cases}$$

The differential cross section for the formation of Ξ hypernuclei through (K^-, K^+) has been calculated to be around 300 nb/sr [13,14] and production of $\Lambda\Lambda$ hypernuclei could be two orders of magnitude smaller. The available kaon flux at the AGS yields around $10^{12} K^-$ mesons per running period. If one assumes a 2 g/cm^2 target, a 30 msr spectrometer, and typical decay and efficiency losses, one finds that a direct production experiment yields around 450 counts/year for Ξ hypernuclei and perhaps 4 to 40 counts/year for $\Lambda\Lambda$ hypernuclei; it is not possible to do systematic studies of $S = -2$ systems with these rates. To compensate, experiments must use thick targets combined with some type of "trick" to recover the necessary energy resolution. For example, the emulsion experiments obtain their resolution by measuring the range of the decay products.

AGS EXPERIMENTS IN PROGRESS

The E885 collaboration (BNL, Carnegie Mellon, Kyoto, Kyoto Sangyo, TRIUMF, Freiburg, Manitoba, and New Mexico) [15] used the AGS 2 GeV/c Kaon beam line and a carbon target to create $S = -2$ systems with the (K^-, K^+) reaction. The experiment ran during the summer of 1996. The data are being analyzed for several production channels:

$$K^- + {}^{12}C \rightarrow \begin{cases} K^+ + {}^{12}_{\Xi}Be \\ K^+ + \Xi^- + X \end{cases}$$

and for events with stopped Ξ^-

$$(\Xi^-, {}^{12}C)_{atom} \rightarrow \begin{cases} {}^{12}_{\Xi}B^* + n \\ H + X. \end{cases}$$

This experiment utilizes the (K^-, K^+) spectrometers used in the E813/E836 H Dibaryon searches, time-of-flight neutron detectors, and scintillating fiber range stacks located above and below a 5 $cm \times 7$ $cm \times 1$ cm carbon target. To maximize the number of stopped Ξ^- hyperons, the target was constructed from synthetic diamond to achieve a density of 3.2 g/cm^3. Although the (K^-, K^+) missing mass resolution (10 MeV/c^2 for Ξ^- production) is not sufficient to resolve individual hypernuclear states, calculated excitation curves such as those found in reference [13], can be compared with the data by folding them with the experimental resolution. Production channels involving an outgoing neutron can be measured to a few MeV/c^2. The scintillating fibers can be used to reject quasi-free Ξ production events and thus enhance the signal. They can also be used to search for a signature of H-Dibaryon production through its decay modes. The data reduction for E885 has been completed and we estimate the data includes approximately 20,000 stopped Ξ events. The first results should be available this fall.

The E906 collaboration (BNL, Carnegie Mellon, Kyoto, Kyoto Sangyo, TRI-UMF, Freiburg, New Mexico, and Tokyo) [16] will use the same beam line and in-beam instrumentation to search for $\Lambda\Lambda$ Hypernuclei by observing their characteristic π-mesonic decays. To facilitate detection of these decay products, the "Cylindrical Detector System" (CDS) was constructed and installed in the target area. This system features of 12 layers of cylindrical drift chambers and a time-of-flight hodoscope located in a solenoidal field. The experiment will use a ^9Be target and the CDS to identify pions from the reaction chain:

$$K^- +\,^9Be \to K^+ +_\Xi X$$
$$_\Xi X \to\,^5_{\Lambda\Lambda}H + X'$$
$$^5_{\Lambda\Lambda}H \to\,^5_\Lambda He + \pi^-$$
$$^5_\Lambda He \to\,^5Li + \pi^-.$$

The CDS system has been installed in the AGS 2 GeV/c separated beam line and an engineering run was completed in June. We expect to take production data in the Spring of 1998.

REFERENCES

1. R.L. Jaffe, *Phys. Rev. Let.* **38**, 195 (1977).
2. B. Basselleck, *these proceedings*.
3. M.M. Nagels, T.A. Ruken, and J.J. DeSwart, *Phys. Rev.* **D15** 2547 (1977).
4. A. Reuber et al., *AIP Conf. Proc.* **338**, 583 (1995). A. Reuber et al., *Nucl Phys.* **A570** 543 (1994).
5. G. Alexander et al., *Phys. Rev.* **173**, 1452 (1968); B. Sechi-Zorn et al., *Phys. Rev.* **175**, 1735 (1968); J.A. Kadyk et al., *Nucl Phys.* **B27**, 13 (1971).
6. C.B. Dover and A. Gal, *Progress in Particle and Nuclear Physics* **12** (1984).
7. C.B. Dover and A. Gal, *Annals of Phys.* **146** 309 (1983)
8. M. Danysz et al., *Nucl. Phys.* **49** 121 (1963).
9. D. Prowse, *Phys. Rev. Lett.* **17** 782 (1966).
10. R. Dalitz et al., *Proc. Royal Soc. Lond.* **A426** 1 (1989).
11. S. Aoiki et al., *Prog. Theor. Phys.* **85** 1287 (1991).
12. C.B. Dover et al., *Phys Rev.* **C44** 1905 (1991).
13. Yamamoto et al., *Prog. Theor. Phys.* **117** 281 (1994).
14. C.B. Dover et al., *Nucl. Phys.* **A572** 85 (1994).
15. M. May et al., *AGS Experiment E885* unpublished.
16. T. Fukuda et al., *AGS Experiment E906* unpublished.

Enhanced Production of $\Lambda\Lambda$ Pairs Near Threshold in (K^-, K^+) Reaction

KEK-PS E224 Collaboration

J.K. Ahn[*], S. Aoki[||], K.S.Chung[§§], M.S.Chung[#], H. En'yo[*], T.Fukuda[¶],
H. Funahashi[*], Y. Goto[*], A. Higashi[¶], M. Ieiri[§], T. Iijima[§], M. Iinuma[*], K. Imai[*],
Y. Itow[*], J.M. Lee[§§], S. Makino[*], A. Masaike[*], Y. Matsuda[*], Y. Matsuyama[¶],
S. Mihara[*], C. Nagoshi[¶], I. Nomura[††], I.S. Park[#], N.Saito[*], M. Sekimoto[¶],
Y.M. Shin[‡‡], K.S. Sim[#], R. Susukita[*], R. Takashima[†], F. Takeutchi[‡], P. Tlustý[§],
S. Weibe[‡‡], S. Yokkaichi[*], K. Yoshida[*], M. Yoshida[*], T. Yoshida[**], S. Yamashita[*]

[*]Kyoto Univ., [†]Kyoto Univ. of Education, [‡]Kyoto Sangyo Univ., [||]Kobe Univ.,
[¶]INS, [§]KEK, [**]Osaka City Univ., [††]Nat'l Inst. for Fusion Science,
[‡‡]Univ. of Saskatchewan, [#]Korea Univ., [§§]Yonsei Univ.

Presented by J.K. Ahn

Abstract. Enhanced production of $\Lambda\Lambda$ pairs are observed very near the threshold (around the mass of 2.24 GeV/c^2) beyond prediction of a two-step process model in (K^-, K^+) reaction at $p_{K^-} = 1.66$ GeV/c using the scintillating fiber target. The differential cross section for the $\Lambda\Lambda$ production averaged over $2.3° \leq \theta_{K^+} \leq 14.7°$ in the momentum region $0.95 \leq p_{K^+} \leq 1.3$ GeV/c was found to be $8.2 + 1.3$ μb/sr, and that for the enhancement approximately 3 μb/sr.

While the H dibaryon is predicted to be stable with respect to strong decays in many theoretical approaches, it has not yet been unambiguously observed experimentally. On the other hand, there has been little experimental effort to search for an unbound H particle, for example, as a $\Lambda\Lambda$ resonance near the threshold. Some decades ago a heavy liquid bubble chamber experiment [1] was performed using 2.1 GeV/c K^- to search for H dibaryon resonance in $\Lambda\Lambda$ invariant mass plot. There was, however, no conclusive observation for the H dibaryon resonance. It might be due to the composite target containing heavy nuclei and high momentum K^- particles, which cause a large nuclear medium effect and high energy Ξ^- hyperon produced. The $\Lambda\Lambda$ production in the (K^-, K^+) reaction originates mostly from the Ξ^- and intermediate meson-induced two-step processes, such as $K^- p \to \Xi^- K^+; \Xi^- p \to \Lambda\Lambda$ and $K^- p \to \pi^0 \Lambda; \pi^0 p \to \Lambda K^+$. Recently, a scintillating fiber (SCIFI) active target experiment has been performed to study the reaction mechanism of the $\Lambda\Lambda$ production in the (K^-, K^+) reaction, in parallel with the bound H particle search [2].

The experiment was carried out using a separated 1.66 GeV/c K^- beam at the KEK proton synchrotron. The momenta of outgoing particles were measured by a K^+ spectrometer, approximately 5 m long, consisting of a dipole magnet with a field integral of 1.08 T·m, twelve drift chamber planes, two Cherenkov detectors and three planes of hodoscopes for timing and triggering purpose. The heart of the experiment is a novel 4π detector consisting of scintillating fibers viewed by image intensifiers. The SCIFI target acts both an interaction target and track detector, and provides an image of charged particle tracks around the (K^-, K^+) reaction vertex. This SCIFI target has good sensitivity over as wide a range in H mass as possible. The SCIFI target was made of scintillating fibers of a 0.5 mm square shape, and has the effective volume of $8 \times 8 \times 10$ cm^3. The SCIFI target was viewed using two sets of the image intensifier tubes (IIT), which were arranged orthogonally along the X and Y directions. The intensified image was viewed with a CCD camera and digitized by a flash ADC. The details of the experimental setup are described elsewhere [2,3].

A total of eight thousand (K^-, K^+) events above $p_{K^+} \geq 0.95$ GeV/c were eye-scanned and categorized with respect to their event topologies, defined by the number of charged prongs, kinks and Λ particles [3]. Multiple eye-scanning for events of the two V topology yields 71 events corresponding to $^{12}C(K^-, K^+)\Lambda\Lambda X$ reaction. Out of 71 events 63 events have no additional charged prong while remaining 8 events have one stopping prong from the (K^-, K^+) primary vertex. The analysis is based on reconstructing events with two V shape particles which are the candidates of the decay of a $\Lambda\Lambda$ pair. The mass (m_V) of every V shape particle is reconstructed by using the track information from the SCIFI data. Λ candidates were selected by requiring that $| m_V - m_\Lambda |$ be less than 6 times the mass resolution of $\Lambda \to p\pi^-$ decays(15 MeV/c^2). Next a full kinematic fit (3C-fit) is performed to determine if the data agree with the $\Lambda\Lambda$ production at the K^- interaction vertex. Both Vs of a $\Lambda\Lambda$ event were required to fit the Λ hypothesis with a probability greater than that for K^0 hypothesis. The mean lifetime of Λ for both one and two Λ events(2.42 $\pm 0.28 \times 10^{-10}s$) is consistent with the world data, and it proves the identification of Λ is quite reliable. Finally only 37 events with all track lengths greater than 1 mm were retained, and used in a present analysis.

Fig.1 shows the $\Lambda\Lambda$ invariant mass spectrum. The invariant mass resolution is determined experimentally by plotting the normal distribution to the individual measurement with a standard deviation corresponding to the experimental error, which is found to be about 6.5 MeV/c^2 below 2.26 GeV/c^2. The detection efficiency of the two Λ decays in the SCIFI target was deduced from the Monte Carlo simulation based on the real data. The branching ratio for charged particle decay of the $\Lambda\Lambda$ pair was taken into account by using

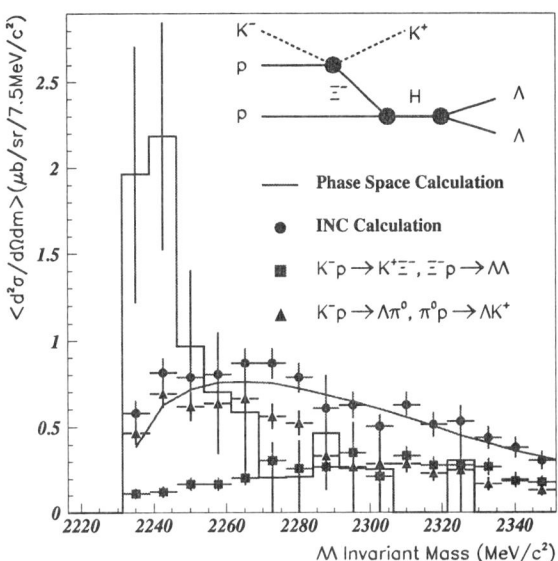

FIGURE 1. $\Lambda\Lambda$ invariant mass spectrum.

the factor $\mathrm{Br}^2(\Lambda \to p\pi^-) = 0.639^2$. The differential cross section integrated over the momentum region ($950 \leq p_{K^+} \leq 1300$ MeV/c) was then found to be $\langle d\sigma/d\Omega_L \rangle_{^{12}C(K^-,K^+)\Lambda\Lambda X} = 8.2 \pm 1.3$ μb/sr.

The $\Lambda\Lambda$ invariant mass spectrum represents a significant peak near the $\Lambda\Lambda$ threshold. We make our intranuclear cascade model calculation to speculate the origin of the peak. The total contribution of two-step processes are shown as closed circles in Fig. 1, and each contribution is represented with squares and triangles for Ξ^- and π^0-induced reactions, respectively. The total cross section for $\Xi^- p \to \Lambda\Lambda$ reaction is taken from the prediction of the Nijmegen D model (r_c=0.5 fm), which is approximately 2 mb at $p_{\Xi^-} \sim 0.5$ GeV/c, and that for $\pi^0 p \to \Lambda K^+$ reaction is taken from the prediction of the resonance model [4]. To prove a reliability of the intranuclear cascade model calculation, the shape of the invariant mass distribution is calculated by the approximate Kopylov-Komolova shape formula, which accounts for the phase space available for $K^{-12}C \to K^+ \Lambda\Lambda^{10}$Be 4-body reaction. We assume, for simplicity, that the remaining nucleus ^{10}Be should be in a ground state, and adopt the K^+ momentum distribution from the real data. Two results are well consistent with each other. However, none of two reproduces the peak near the threshold.

The enhancement of the invariant mass indicates a possible existence of $\Lambda\Lambda$ resonance which is referred to H-dibaryon resonance state. The differential cross section for the enhancement ($\sim 3\mu$b/sr) is approximately 6 times as large as the prediction of Aerts and Dover for the bound H production ($m_H \sim 2m_\Lambda$)

[2,5]. It is highly desirable to search for the $\Lambda\Lambda$ resonance state near the threshold. At BNL an (K^-, K^+) experiment (E906) [6] is now under data-taking with a cylindrical drift chamber system, and expected to achieve a factor of a few tens higher statistics for the $\Lambda\Lambda$ events. It is also probable that such studies could be carried out in the context of the strangeness physics program envisioned for the coming JHF in Japan.

REFERENCES

1. P. Beilliere et. al., *Phys. Lett.* **B39**, 671 (1972); G. Wilquet et. al., *Phys. Lett.* **B57**, 97 (1975).
2. J.K. Ahn et. al., *Phys. Lett.* **B378**, 53 (1996).
3. J.K. Ahn et. al., *KEK preprint* 96-158 (1997), *Nucl. Phys. A* in print.
4. C. Fuchs et. al., *e-print* nucl-th/9701065 (1997).
5. A.T.M. Aerts and C.B. Dover, *Phys. Rev.* **D28**, 450 (1983).
6. T. Fukuda and R.E. Chrien, *BNL AGS-E906* (1997).

$S = -1$ and $S = -2$ Few-Body Hypernuclei

B. F. Gibson

Theoretical Division, Los Alamos National Laboratory
Los Alamos, New Mexico 87545, U. S. A.

Abstract. The S=−1 and S=−2 few-body bound states are the focus of this discussion. Implications for our understanding of the baryon-baryon interaction are examined. Octet-octet coupling effects not found in conventional, non-strange nuclei are highlighted. TJNAF wave function tests for the S=−1 sector are noted. The need to identify S=−2 hypernuclei to explore the S=−2 strong interaction and to test model predictions is emphasized.

A strong impetus for investigating the structure and reactions of baryon systems is to understand the fundamental baryon-baryon interaction in the realm of nonperturbative QCD. Few-baryon systems play an important role, because we can calculate complete solutions to test a particular ansatz for the baryon-baryon force. Hypernuclei are crucial to this effort, because they permit us to probe models based upon our non strange sector experience outside the conventional world where they were developed. That is, we learn whether sophisticated models of the nucleon-nucleon (NN) interaction extrapolate successfully beyond the region in which the parameters were fitted or whether the models are only interpolative. For these reasons, the strange (S=−1 and S=−2) few-body hypernuclei are the focus of this brief discussion.

Pure one-boson-exchange (OBE) potential models provide both a quantitative fit to the extensive NN data base and a means to transform from S=0 into S=−1,−2. In particular, the Nijmegen models satisfy the criteria. [1] Other contemporary potential models yield similar few-nucleon system results. Predictions for low-energy properties such as rms radii, Coulomb energies, and asymptotic normalization constants are, indeed, robust. The addition of a small three-nucleon force to the Hamiltonian, in order to tune the binding energy to the experimental value of the triton, yields a good prediction of the alpha particle binding as well as the low-energy physical observables mentioned above. [2] We recognize that the OBE models are phenomenological, but the flexibility of the models is such that possible flaws, which would require introducing quark degrees of freedom, may only be observable in the S≠0 sector. This was addressed in the 1993 US/Japan seminar on the hyperon-nucleon interaction. [3]

Let me briefly reprise the triton results, which suggest an approach to Λ hypernuclei with a somewhat surprising outcome. The Argonne V_{14} potential [4] has been thoroughly studied and yields a triton binding energy of $\simeq 7.7$ MeV. The V_{14} model is particularly interesting because there also exists a V_{28} model, one which includes $NN - N\Delta$ coupling and is fitted to the same NN data set. Surprisingly, the triton binding energy is essentially unchanged.

Why is the octect-decuplet $(NN - N\Delta)$ coupling so well modeled implicitly by V_{14}? Can one extend this approach to the S=-1 octet-octet coupling? That is, can one represent the hyperon-nucleon coupled-channel $N\Lambda$ potential

$$\begin{pmatrix} V_{N\Lambda} & V_{NX} \\ V_{NX} & V_{N\Sigma} \end{pmatrix}$$

by an effective single-channel potential $\overline{V}_{N\Lambda}$?

In the non strange sector we observe that the ratio of neutron separation energies for neighboring s-shell nuclei is approximatly 3: $B_n(^3H)/B_n(^2H) \simeq 6/2 = 3$, and $B_n(^4He)/B_n(^3H) \simeq 20/6 \simeq 3$. If the physics of few-body systems is similar, then we might anticipate a factor of 3 in the ratio of Λ separation energies for neighboring Λ hypernuclei. Using $B_\Lambda(^4_\Lambda H) \simeq 2$ MeV as our basis, we would then predict $B_\Lambda(^5_\Lambda He) \simeq 3 \times B_\Lambda(^4_\Lambda H) \simeq 6$ MeV and $B_\Lambda(^3_\Lambda H) \simeq \frac{1}{3} \times B_\Lambda(^4_\Lambda H) \simeq \frac{2}{3}$ MeV. Simple, central force calculations using $\overline{V}_{N\Lambda}$ fitted to $B_\Lambda(^4_\Lambda H)$ plus low-energy scattering data confirm this simple analysis. [5]

However, the real world is more complex. Experimentally [6] we know that $B_\Lambda(^5_\Lambda He) \simeq 3.1$ MeV and $B_\Lambda(^3_\Lambda H) \simeq 0.13$ MeV. Our S=0 model experience does not extrapolate to S=-1. Explicit $N\Lambda - N\Sigma$ coupling is required. Moreover, there exist π^+ decay data that suggest the importance of explicit $N\Lambda - N\Sigma$ coupling in Λ hypernuclei. The open channels for Λ mesonic decay are $\Lambda \to \pi^- p$ and $\Lambda \to \pi^0 n$. However, experimentaly [7] there is observed a 5% braching ratio for $^4_\Lambda He \to \pi^+ + X$. Second order pion processes such as charge exchange $(\pi^- pp \to \pi^+ nn)$ are too small to explain more than 1%. [8] The virtual $p\Lambda \to n\Sigma^+$ transition followed by $\Sigma^+ N \to \pi^+ nN$ is the key.

Furthermore, explicit $N\Lambda - N\Sigma$ coupling was demonstrated in Ref. [9] to play a crucial role in driving the hypertriton Λ separation energy from 2/3 MeV toward 0.1 MeV. Gloeckle and coworkers [10] have since shown that the S=-1 Nijmegen soft-core potential yields a value for $B_\Lambda(^3_\Lambda H)$ which agrees with experiment. Additionally, a correct ordering of the A=4 isodoublet 0^+ and 1^+ states appears to require explicit $N\Lambda - N\Sigma$ coupling. [11] Finally, Monte Carlo calculations [12] have indicated that suppression of $\Lambda \otimes ^4He \leftrightarrow \Sigma \otimes ^4He^*$ coupling, because of the large excitation energy of the T=1 even parity $^4He^*$ states that result when the T=0 Λ converts to a T=1 Σ, can account for the anomalously low value of $B_\Lambda(^5_\Lambda He) = 3.1$ MeV. In the S=-1 sector we now need K^+ electroproduction to test the model wavefunctions that result from these calculations; ^{4}He $\to\,_\Lambda^4$H* is the logical first step at TJNAF.

Given that explicit octet-octet coupling plays a key role in S=-1 physics, we turn to the interesting puzzle that the single reported $_{\Lambda\Lambda}^6$He event [13] presents. Assuming that the $\Lambda\Lambda$ separation energy $B_{\Lambda\Lambda}(_{\Lambda\Lambda}^6\text{He}) = B(_{\Lambda\Lambda}^6\text{He}) - B(^4\text{He}) \simeq 10.9$ MeV is accurate (this interpretation is consistent with the two other accepted $\Lambda\Lambda$ events), we see that the matrix element $<V_{\Lambda\Lambda}>_{A=6}$ is weak: $- <V_{\Lambda\Lambda}> = B_{\Lambda\Lambda}(_{\Lambda\Lambda}^6\text{He}) - 2\times B_\Lambda(_\Lambda^5\text{He}) \simeq 10.9 - 2(3.1) = 4.7$ MeV. This value is comparable with that of the $N\Lambda$ interaction: $<V_{\Lambda\Lambda}>_{A=6} \simeq <V_{N\Lambda}>_{A=4}$. Both the $\Lambda\Lambda$ and $N\Lambda$ matrix elements are relatively small [14] compared with that of the nn interaction: $<V_{nn}> \simeq -7$ MeV. However, $\Lambda\Lambda$ and nn are analogs, belonging to the same $^1S_0$ multiplet. Why is $<V_{\Lambda\Lambda}>_{A=6}$ so small? In response to those who ask why $V_{\Lambda\Lambda}$ should be strong without OPE, we note that the Walecka model NN force has only ρ,ω and no OPE.

Can we measure $\Lambda\Lambda$ scattering? "Yes, indirectly." Two example Ξ^- capture reactions that could measure $a_{\Lambda\Lambda}$ are $\Xi^- d \to \Lambda\Lambda n$ and Ξ^{-7}Li $\to \Lambda\Lambda^6$He, where the spectator particle is detected in analogy with the a_{nn} measurements from $nd \to nnp$ and ^{3}H ^{3}H $\to nn^4$He.

Short of such data we ask about the constraints that an $\alpha\Lambda\Lambda$ model of $_{\Lambda\Lambda}^6$He can provide. In the analysis by Carr et al. [15], octet-octet coupling is essential. For an effective $\Lambda\Lambda$ potential whose strength is comparable to that of the NN force ($\overline{V}_{\Lambda\Lambda} \simeq V_{nn}$), we obtain overbinding of $_{\Lambda\Lambda}^6$He. In contrast, a coupled-channel ($\Lambda\Lambda - N\Xi$) potential of similar overall strength yields binding comparable to experiment, because of Pauli blocking. The α core saturates the $(1s)^4$ shell, forcing a 5th nucleon into a higher shell and significantly weakening the $N\Xi$ part of the force. In other words, by including $\Lambda\Lambda - N\Xi$ coupling explicitly, we can accomodate a relatively weak $<V_{\Lambda\Lambda}>_{A=6}$ even though the free space $\Lambda\Lambda - N\Xi$ potential is comparable in strength to the nn.

Alternatively, the S=-2 $_{\Lambda\Lambda}^4$H and $_{\Lambda\Lambda}^5$H hypernuclei should show evidence of enhanced binding. Whereas the ^{4}He core in $_{\Lambda\Lambda}^6$He must be excited by 40 MeV to permit the $\Lambda\Lambda \to N\Xi$ transition, the ^{2}H (^{3}H) core in $_{\Lambda\Lambda}^4$He ($_{\Lambda\Lambda}^5$H) is bound by an additional 6 (20) MeV following $\Lambda\Lambda \to N\Xi$ ($\Lambda\Lambda \to p\Xi^-$) conversion, to form a ^{3}H or ^{3}He (^{4}He) core.

Let us return to the question of why octet-deculpet coupling in the S=0 sector ($NN - N\Delta$) appears relatively unimportant, whereas octet-octet coupling in the S$\neq$0 sector is essential. There exist alternate possibilities: (i) the large Δ width, (ii) the large $N - \Delta$ mass difference, and (iii) the duality of particle physics. The first two are obvious; the third may not be. In the Maldelstam representation we can write the scattering amplitude M equivalently as $M(t,u)$ or $M(s,u)$. That is, we can use either t-channnel/u-channel variables or s-channel/u-channel variables. Does the t-channel meson-exchange picture of the OBE potential essentially subsume the s-channel resonance picture of the $NN - N\Delta$ conversion process, so that explicit octet-decuplet coupling is not required? The question is open.

Hypernuclear physics continues to be novel and exciting. New questions continue to arise. As a testing ground for S=0 based concepts, hypernuclei are unsurpassed.

The work of BFG was performed under the auspices of the U. S. Department of Energy. He gratefully acknowledges a Research Award for Senior Scientists by the Alexander von Humboldt Stiftung which made possible a stay with IKP-Theorie, Forschungszentrum Juelich where this was written.

REFERENCES

1. M. M. Nagels, T. A. Rijken, and J. J. de Swart, Phys. Rev. D **15**, 2547 (1977); **20**, 1633 (1979); P. M. M. Maessen, T. A. Rijken and J. J. de Swart, Phys. Rev. C **40**, 2226 (1989).
2. B. F. Gibson, Nucl. Phys. **A543**, 1c (1992).
3. J. J. de Swart, Proceedings of the *U.S. - Japan Seminar on the Hyperon-Nucleon Interaction* (World Scientific, Singapore, 1994), p. 37; M. Oka, *ibid.* p. 169.
4. R. B. Wiringa, R. A. Smith, and T. A. Ainsworth, Phys. Rev. C **29**, 1207 (1984)
5. R. C. Herndon and Y. C. Tang, Phys. Rev. **153**, 1091 (1967); **159**, 853 (1967); **165**, 1093 (1969); B. F. Gibson, A. Goldberg, and M. S. Weiss, Phys. Rev. C **6**, 741 (1972); *Few Particle Problems in Nuclear Interactions*, (North Holland, Amsterdam, 1972), p. 188; J. Dabrowski and E. Fedorynska, Nucl. Phys. **A210**, 509 (1973); B. F. Gibson and D. R. Lehman, Phys. Rev. C **11**, 29 (1975); A. Gal, Adv. Nucl. Physics **8**, 1 (1975).
6. M. Juric *et al.*, Nucl. Phys. **B52**, 1 (1973).
7. C. Mayeur *et al.*, Nuovo Cim. **44**, 698 (1966); G. Keyes *et al.*, Nuovo Cim. **31A**, 401 (1976).
8. A. Cieply and A. Gal, Phys. Rev. C **55**, 2715 (1996).
9. I. R. Afnan and B. F. Gibson, Phys. Rev. C **40**, R7 (1989); **41**, 2787 (1990).
10. K. Miyagawa *et al.*, Phys. Rev. C **51**, 2905 (1995).
11. B. F. Gibson and D. R. Lehman, Phys. Rev. C **37** 679 (1988).
12. J. A. Carlson, Proceedings of the *LAMPF Workshop on (π, K) Physics*, A. I. P. Conf. Proc. **224** (American Institute of Physics, New York, 1991), pp. 198.
13. D. J. Prowse, Phys. Rev. Lett. **17**, 782 (1966); see also R. H. Dalitz *et al.*, Proc. Roy. Soc. London **A426**, 1 (1989).
14. B. F. Gibson, A. Goldberg, and M. S. Weiss, Phys. Rev. **181**, 1486 (1969); A. R. Bodmer and Q. N. Usmani, Nucl. Phys. **A477**, 621 (1988); Phys. Rev. C **31**, 1400 (1985); A. R. Bodmer, Q. N. Usmani, and J. A. Carlson, Phys. Rev. C **29**, 684 (1984); Nucl. Phys. **A422**, 510 (1984); H. Bandō, Prog. Theor. Phys. **67**, 669 (1982).
15. S. B. Carr, I. R. Afnan, and B. F. Gibson, Nucl. Phys. **A** (submitted), "$^{6}_{\Lambda\Lambda}$He as a $\Lambda\Lambda$ Interaction Constraint".

Search for Strange Matter via Heavy Ion Activation

M.C. Perillo Isaac, Y.D. Chan, R. Clark, M.A. Deleplanque,
M.R. Dragowsky¶, P. Fallon, I.D. Goldman*, K. Nishiizumi†,
R-M. Larimer, I.Y. Lee, A.O. Macchiavelli, R.W. MacLeod,
E.B. Norman, L.S. Schroeder, F.S. Stephens

Lawrence Berkeley National Laboratory - Berkeley CA 94720
†*Space Sciences Laboratory, University of California, Berkeley CA 94720*
**University of Sao Paulo, Sao Paulo, Brazil*
¶*Oregon State University, Corvallis OR 97331*

Abstract.
In this paper we present the results of an experiment performed at LBNL's 88-Inch Cyclotron using GAMMASPHERE in which we searched for strange matter by applying the heavy ion activation technique to samples of meteorites. Our exploratory experiment improved existing experimental limits by 3 orders of magnitude.

INTRODUCTION

It is possible that strange quark matter with roughly equal numbers of up, down and strange quarks exists and that it is absolutely stable. Regarding strange matter as a three-flavor quark gas, the degree of freedom introduced by the strange flavor may compensate for the energy penalty associated with the strange quark mass. If this is the case, strange quark matter could be the true ground state of the strong interactions [1]. Theoretical calculations of the properties of bulk strange matter do not allow a firm conclusion as to whether strange matter is actually stable. If strange matter is stable and present in in the Galaxy, its existence can be probed through sensitive earth-based experiments.

The most stringent experimental limits on strange matter contents in normal matter are due to Brügger and collaborators [2]. Using Rutherford backscattering of heavy ions on natural samples they established upper limits for the abundance of strange nuggets with masses $10^2 < A < 10^7$ amu. Their limits were in the range $10^{-14} < N_{strange}/N_{nucleons} < 10^{-10}$, where $N_{strange}/N_{nucleons}$

TABLE 1. Expected signal from strange matter activation by 450 MeV ^{136}Xe. For $I = 5$ Mev the total energy release is expected to be of the order of 1 GeV, while if $I = 20$ MeV, the energy release is expected to be 3 GeV.

	$I = 5$ Mev			$I = 20$ MeV	
$A(amu)$	$T(keV)$	N_γ	$A(amu)$	$T(keV)$	N_γ
10^2	26170.14	43.17898	10^2	43832.52	72.32073
10^3	8275.725	136.5439	10^3	13861.06	228.6982
10^4	2617.014	431.7898	10^4	4383.252	723.2073
10^5	827.5724	1365.439	10^5	1386.106	2286.982
10^6	261.7014	4317.898	10^6	438.3252	7232.073
10^7	82.75724	13654.39	10^7	138.6106	22869.82
10^8	26.17014	43178.98	10^8	43.83252	72320.73

is the relative abundance of strange nuggets to the number of nucleons in normal matter.

Farhi and Jaffe [3] proposed an experiment to search for strange matter in different samples via heavy ion activation."Strangelets" would have a lower electric charge than normal matter of the same A, hence their Coulomb barrier would be lower. When a normal nucleus enters a strangelet the nucleons in the normal matter will "dissolve" into quarks inside the strange nugget. The extra binding energy and the beam kinetic energy will excite the strangelet and raise its temperature. This temperature T characterizes a spectrum of photons released by the de-excitation of the strange nugget. For an extensive review of the properties of strange matter, see [4].

Here we present the results of an experiment performed at the 88-Inch Cyclotron using GAMMASPHERE in which, by applying this technique, we were able to improve the existing experimental limit by up to five orders of magnitude for strange nuggets with masses between $A = 10^2$ and $A = 10^8$ amu.

EXPERIMENT

Strange matter presents a lower Coulomb barrier than ordinary nuclear matter with the same A. It is thus possible to excite strange matter in an experiment free of nuclear reaction backgrounds. In such a process the energy released will be $\Delta E = IA_B + K$ where I is the binding energy per nucleon of strange matter relative to the binding energy per nucleon of normal matter, A_B is the atomic mass of the beam nucleus and K is the kinetic energy of the beam. I is expected to be of the order of 5 to 20 MeV which means that a few GeV can be released in a single event. The expected signal depends on the strangelet mass, on I, and on the beam's characteristics. Table 1 shows the expected signal for strange nugget masses from 10^2 to 10^8 for two values of I, 5 and 20 MeV.

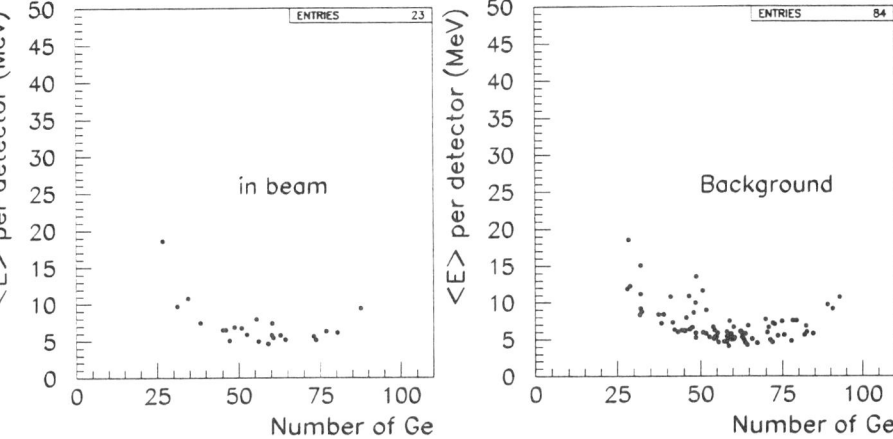

FIGURE 1. Comparison between in beam data (left) and background (right). In beam data corresponds to approximately 4 hours. Background corresponds to 15 hours. The same analysis conditions were applied in both cases, i.e., we consider only events which deposit at least 270 MeV in the Ge detectors

GAMMASPHERE is a gamma-ray detector array composed of 110 modules of high purity Ge detectors. Each module has a Ge detector surrounded by BGO crystals. The 4π solid angle coverage is shared by the Ge detector(45%) and the BGO crystals (55%). The threshold of each element, either BGO or Ge is of the order of 20 keV, when no absorbers are present.

Monte Carlo studies [5] using GEANT 3.21 show that the detection efficiency of any strange matter event is equal to 100%. The energy deposited in the Ge detectors depends on the value of I and on the strangelet mass. For a strangelet with mass A=100, the energy deposited in the Ge detectors in GAMMASPHERE is 25% of the total energy if I=5 MeV, i.e., 270 MeV.

We used a 250 enA beam of ^{136}Xe at 450 MeV, delivered by the 88-Inch Cyclotron at LBNL. A 0.8 g sample of the Allende meteorite was irradiated for approximately 4 hours. The rate of beam particles impinging on the target was 6.0×10^{10} per second and their range is 35.68 μm [6]. The number of events with more than 270 MeV observed in GAMMASPHERE's Ge detectors during this run was 23. High energy/high multiplicity events were also observed in the background and are due to cosmic ray showers. During a 15 hours background run, 84 events were observed with the same analysis conditions. We see no statistically significant excess of events above background. Figure 1 shows the raw data.

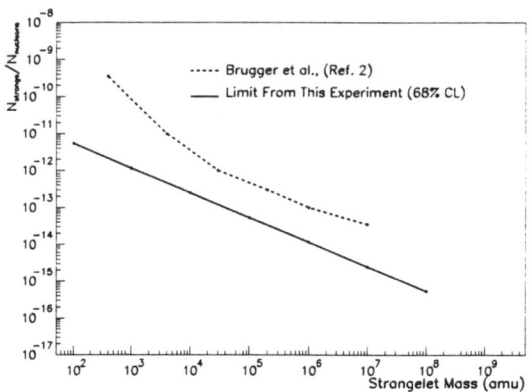

FIGURE 2. Upper limit for the abundance of strangelets $N_{strange}$ relative to the number of nucleons $N_{nucleons}$ as a function of the strangelet mass. The solid line represents the result of this experiment. The dashed line shows the limit obtained by Brügger and collaborators.

Our results are summarized in figure 2, where we compare our results with the experiment performed by Brügger [2]. Assuming a geometrical cross session for the interaction of a normal nucleus with a strangelet we are able to exclude the presence of strangelets relative to normal nucleons to the level of 10^{-12} for strangelets of mass 100, compared to levels of 10^{-9} from previous experiments. The use of this technique to probe strange matter contents in normal matter has effectively improved the existing limits by 3 orders of magnitude. We will be applying it to Moon soil samples in the near future.

REFERENCES

1. E. Witten, Phys. Rev. D **30**, 272, 1984.
2. M. Brügger et al., Nature **337**, 434, 1989.
3. E. Farhi and R. L. Jaffe, Phys Rev. D **32**, 2452, 1985.
4. For an extensive study of the properties of strange matter, see E. Farhi and R. L. Jaffe, Phys Rev. D **30**, 2379, 1984
5. K. Vetter, private communication
6. TRIM: Stopping and Range of Ions in Solids, J.F. Ziegler and J.P. Biersack, Pergamom Press, NY, 1985.

Strangeness Physics Sessions Summary

R. A. Schumacher

Department of Physics, Carnegie Mellon University, Pittsburgh, PA 15213

Abstract. Two sessions at this conference focused on strange particle physics. The first was a survey of strangeness production reactions, including photoproduciton, electroproduction, and production via $\bar{p}-p$ reactions. The other session concentrated on hypernuclear physics for strangeness $-1, -2$, and $-N$ systems. This summary outlines the papers that were presented.

STRANGENESS PRODUCTION

Recent Λ and Σ photoproduction results from the SAPHIR detector at Bonn were shown by Dr. Menze. New and detailed measurements of total and differential cross sections, as well as hyperon polarization data were presented. The total cross section data revealed a more complex resonance structure than was revealed in the past. None of the reactions models available today matched these new data in more than in a qualitative way. First-ever measurements of the Σ polarization in photoproduction were shown, and found to be opposite in sign to that of the Λ, in agreement with quark model predictions. Professor Li discussed the importance which accurate hyperon photoproduction data can play in understanding the structure of baryons. One example was the role of the second (and possible third) $S11$ resonances near 1650 MeV, which may have a quark structure which is mixed with nearby "molecular" structures of bound K's and Σ's or bound K's and Λ's. Precise data such as those from Bonn or from Jefferson Lab will make it possible to study this question. Professor Li also emphasized the interest in three-body final states $K - \pi - \Sigma$ and $K - \pi - \Lambda$ made in photoproduction, which have scarcely been studied, but which could reveal additional resonance structure, as hinted by sightings of states near 2 GeV by the Sphinx collaboration.

First results from strangeness electroproduction at Jefferson Lab were presented by Professor Baker, with preliminary looks at data from proton, deuteron, and carbon targets. The experiments used the small-aperture spectrometers in Hall C, with the eventual goal of separating the elements of the electroproduction cross sections for the first time. Particularly intriguing was an $(e, e'K)$ spectrum taken on ^{12}C leading to possible formation of the hypernucleus $^{11}_\Lambda B$. Since no one has ever succeeded in the electromagnetic production of hypernuclear states, the unresolved strength seen in the bound state region in this reaction was an indicator that perhaps Jefferson Lab will succeed at being the first to open this field of research. Theoretical considerations were discussed by Dr. Saghai, who described new off-shell treatments of spin 3/2 particles in his effective Lagrangian approach to electromagnetic strangeness production. He showed the significant effects produced by the off-shell treatments on the four components of the strangeness electroproduction cross section.

New data from the prolific PS185 experiment at LEAR for $\bar{p} + p \rightarrow \bar{Y} + Y$ were presented by Dr. Franz. It was shown that a small wiggle in the $\bar{\Lambda}\Lambda$ total cross section a mere 1 MeV above threshold, a wiggle which had been first seen by PS185, was in fact not present in subsequent data dedicated to map it out in detail. Sic transit gloria exotica. Data for the $\bar{p}+p \rightarrow \Sigma^- + \overline{\Sigma^-}$ reaction were presented; this reaction must go via either two meson exchange or via two quark annihilation/creation processes. In this way this reaction differs from other hyperon production channels which can go via a single exchange or annihilation/creation. The data showed a strong backward peak in the cross section, in contrast to all the other channels which display a "black disk" type of forward peaking, and thus provided a nice challenge to test hyperon production models. A longstanding goal of the PS185 measurements has been to find situations which distinguish between quark model and meson exchange model calculations of these reactions. New measurements were described for $\bar{p}$ on polarized protons; the amount of polarization predicted to be transferred to the created Λ's is large, but differs in sign between quark and meson exchange models. Professor Alberg discussed these differing predictions, pointing out that these results, when presented, may for the first time clearly show quark effects, or the lack of them, in hyperon production via anti-protons.

The status of the ϕ-factory at DAϕNE and the Finuda experiment which will exploit the 16 MeV kaons produced at this facility were presented by Professor Bressani. No results were available yet, but the facility appears to be within a year of starting operations.

HYPERNUCLEAR SYSTEMS

The hypernuclear spectroscopy session began with an overview of new results from the K6 beamline at KEK by Dr. Noumi. Remarkably good data, with about 2.5 MeV FWHM missing mass resolution, were shown for Λ hypernuclei of $^7Li, ^9Be, ^{13}C$, and ^{16}O. However, the goal of extracting the Λ-nucleus spin-orbit interaction strength remains as elusive as ever. On the other hand, the first-ever measurement of the weak decay asymmetry of the *non − mesonic* decay of a hypernuclear state was reported. It was for the case of $^5_\Lambda He$, for which the mesonic decay was used to estimate the degree of polarization of the produced hypernuclei. The experimental non-mesonic asymmetry was observed and the intrinsic decay asymmetry obtained by dividing out the measured polarization. The preliminary result of −0.3 (no error bar quoted) turned out to be in agreement with a calculation presented by Professor Bennhold, who obtained a value of −0.32 in a meson exchange model.

A discussion of the non-mesonic decays of hypernuclei was presented by Dr. Inoue, who compared the effects of one-pion exchange decays with a so-called direct-quark mechanism. The neutron and proton induced non-mesonic decay data could be better fitted with his model, which included pion exchange and substantial contributions from the $\Delta I = \frac{3}{2}$ amplitudes, than pure meson exchange models. The results were to be compared with those of Professor Bennhold, who also discussed these decays in a model where many mesons contribute to the decay process, but where there remain long-standing difficulties in fitting the separate neutron and proton induced rates.

Very interesting results on the formation of two-Λ systems were presented by Mr. Ahn for the KEK PS-E224 collaboration. From a sample of 8000 (K^-, K^+) events produced inside a scintillating fiber block, they identified 71 events with a topology consistent with a two-step process of Ξ^- production followed by $\Xi^- + p \to \Lambda + \Lambda$. The invariant mass distribution of the $\Lambda - \Lambda$ pairs showed a strong enhancement near or at the $\Lambda - \Lambda$ threshold. The enhancement was well in excess of what was expected based on an intranuclear cascade calculation, and hence was suggestive of either a significant $\Lambda - \Lambda$ attractive interaction near threshold, or perhaps a signal of the elusive H particle formed slightly above the threshold, or both. Experiments which can follow up on this finding were described by Professor Franklin; at BNL the E906 and E885 experiments are well suited to provide more statistics on the (K^-, K^+) reaction on light nuclei, and they hope to provide new results before the next conference in this series. Dr. Gibson discussed theoretical consideration in the formation of light $S = -2$ hypernuclei, especially the importance of the in-medium effects of $\Lambda - N \to \Sigma - N$ and $\Xi - N \to \Lambda - \Lambda$.

Finally, in the hunt for truly exotic systems, Ms. Perillo-Isaac discussed a search for pre-existing stable strangelets in meteorite and moon rock samples using the Berkeley cyclotron. Heavy-ion bombardment of these sample, should, according to a prediction by Farhi and Jaffe, produce a flood of photons as the heavy ions are absorbed into the strange matter. No such occurrences were seen.

Neutrino and Non-Accelerator Physics

Recent Double Beta Decay Experimental Results

C. Sean Sutton

Physics Department, Mount Holyoke College, South Hadley, Massachusetts 01075

NEMO Collaboration
CEN-Bordeaux-Gradignan, France; CFR-Gif/Yvette, France; CRN-Strasbourg, France; Department of Physics-Jyvaskyla, Finland; INEL-Idaho Falls, USA; INR-Kiev, Ukraine; ITEP-Moscow, Russia; JINR-Dubna, Russia; LAL-Orsay, France; LPC-Caen, France; MHC-South Hadley, USA.

Abstract. Double beta decay experiments continue to contribute illuminating and constraining factors in the fields of nuclear and particle physics. Recently, there have been a number of half-life measurements realized for two-neutrino double beta decay candidates, often with data samples of high statistical significance. Searches for neutrinoless double beta decay and for new Majoron emitting decay modes have resulted in half-life limits only. For the zero-neutrino mode, these limits have pushed the Majorana neutrino mass to less than 1 eV. Currently, the generation of small source mass experiments is coming to an end. In the spirit of the ^{76}Ge experiments, promising next generation research will need to study several kilograms of ultra-pure, isotopically-enriched double beta decay sources.

INTRODUCTION

Double beta decay, $\beta\beta$, is a second order, weak interaction first proposed by Geoppert-Mayer (1). A number of excellent review articles exist in the literature (2-4) on the topic for which this review should act as a supplement on more recent results. Presented here is a brief summary of the $\beta\beta$ modes and a discussion of the problems involved in isotope selection for successful $\beta\beta$ measurements. Next, sections dedicated to the most current two neutrino, zero neutrino, and zero neutrino with the emission of a Majoron are addressed. The review will conclude with a report on the status of experiments for the future.

The standard model allows $\beta\beta$ to occur with the emission of two electrons and two anti-electronic neutrinos, $(A,Z) \rightarrow (A,Z+2) + 2e^- + 2\nu_e$. This process, $\beta\beta 2\nu$, is currently marking its tenth anniversary of the first successful direct detection as well as a prosperous decade in which a total of seven isotopes have had $\beta\beta 2\nu$ half-life measurements. These measurements have provided interesting experimental nuclear physics, which constrain theoretical models. Somewhat more troublesome, but also of interest, are studies of decays to excited states of the daughter nuclei and those decays which involve double positron decay $(A,Z) \rightarrow (A,Z-2) + 2\beta^+ + 2\nu_e$, positron decay with electron capture and double electron capture.

A second class of $\beta\beta$ was proposed which fails to produce neutrinos. These decay modes deviate from the standard model and harbor the excitement of new frontiers for particle physics. One mode of this class is simply the neutrinoless $\beta\beta$ mode, $\beta\beta 0\nu$, which was first (5) proposed shortly after $\beta\beta 2\nu$ was predicted. The process violates lepton number $(A,Z) \rightarrow (A,Z+2) + 2e^-$ and the neutrino must be a massive Majorana particle and/or requires a presence of a right-handed lepton current. A second neutrinoless $\beta\beta$ involves the emission of a Majoron, $\beta\beta\chi$, $(A,Z) \rightarrow (A,Z+2) + 2e^- + \chi$. The spectrum of the summed electron energies for the different decay modes provides a clear signature for the different decay modes. A half-life measurement or limit for $\beta\beta 2\nu$, $\beta\beta 0\nu$, and $\beta\beta\chi$ for the different modes is dependent on kinematic factors and nuclear matrix elements (n.m.e). as given in equations 1-3 respectively.

$$[T_{1/2}(2\nu)]^{-1} = F_{2\nu}[M_{2\nu}]^2 \tag{1}$$

$$\begin{aligned}[T_{1/2}(0\nu)]^{-1} &= F_{0\nu}[M_{0\nu}]^2[<m_\nu>/m_e]^2 \\ &= C_{mm}[<m_\nu>/m_e]^2 + C_{\lambda\lambda}<\lambda>^2 + C_{\eta\eta}<\eta>^2 + C_{m\lambda}[<m_\nu>/m_e]<\lambda> + \\ &\quad C_{m\eta}[<m_\nu>/m_e]<\eta> + C_{\lambda\eta}<\lambda><\eta>\end{aligned} \tag{2}$$

$$[T_{1/2}(0\nu,\chi)]^{-1} = F_\chi[M_{0\nu}]^2[<g_{\nu,\chi}>]^2 \tag{3}$$

Here the F_i's and M_i's refer to the kinematic factors and n.m.e., the first of which are well determined while there is currently rich acitivity seeking accurate determination of n.m.e., both experimentally and with theoretical models. The term $<m_\nu>$ denotes the effective neutrino mass and $<\lambda>$ and $<\eta>$ address possible right-handed coupling strengths. The C_{jk}'s incorporate the relevent n.m.e and phase space integrals so that the quadratic polynomial of equation 2, (6) addresses proposed couplings. Finally, the term $<g_{\nu,\chi}>$ in equation 3 is the neutrino-Majoron coupling constant.

Indirect detection of $\beta\beta$ can be accomplished with geochemical and radiochemical (milking) methods. A geochemical method which provided the first convincing evidence of $\beta\beta$ was in a Te ore (7) and recent geochemical work (8) on ^{130}Te and ^{128}Te have resulted in some refinements of the half-life measurement of ^{128}Te, $(2.2\pm 0.3) \cdot 10^{24}$yr. Another recent geochemical investigation involving the $\beta^+\beta^+$ of ^{130}Ba (9) has resulted in a half-life limit of $> 4 \cdot 10^{21}$yr. However, the short coming of these indirect detection methods is that the decay mode is lost and the current interest in neutrinoless decays is not engaged.

Direct dectection of the 41 possible $\beta\beta$ and $\beta^+\beta^+$ candidate istopoes is restricted experimentally by the ubiquitous natural and anthropogenic radioactivity. The U and Th decay chains in particular have ^{214}Bi and ^{208}Tl that decay with gammas which can generate events which mimic $\beta\beta$ through double compton scattering and other processes. Thus, one elects to work with isotopes whose end point energy, $Q_{\beta\beta}$, is above 2.4 MeV, which leaves as $\beta\beta$ candidates ^{48}Ca, ^{82}Se, ^{96}Zr, ^{100}Mo, ^{116}Cd, ^{130}Te, ^{136}Xe, and ^{150}Nd. The isotope, ^{76}Ge, with a $Q_{\beta\beta}$ of 2.038 MeV is an exception to the rule, owing to the ultra-high purification techniques developed in semi-conductor technologies known as zone refinement. There are no $\beta^+\beta^+$ candidates with a $Q_{\beta\beta}$ above 1.1 MeV so $\beta^+\beta^+$ studies are not undertaken in direct detection experiments. Still the optimum choice of a $\beta\beta$-isotope depends on a number of factors, one of which is the n.m.e. that characterises the half-life decay. The n.m.e. have historically been determined theoretically, but $2\nu\beta\beta$ half-life measurements of the above isotopes are now focusing attention on the success of the various models. In what follows, the half-life limits will be given with the n.m.e. used in the original papers.

Two Neutrino $\beta\beta$ Measurements

The first successful measurement of a half-life for one of the above identified $\beta\beta$-isotopes was accomplished at the University of California Irvine (UCI) with a time projection chamber (TPC). The TPC measured the half-life of ^{82}Se (10) to be $(1.08^{+0.26}_{-0.06}) \cdot 10^{20}$yr which has recently been confirmed within errors by the NEMO-2 detector $(0.85\pm 0.1[\text{stat}]\pm 0.2[\text{syst}]) \cdot 10^{20}$ yr. The NEMO-2 detector is a tracking detector with a calorimeter operating in the Frejus Underground Laboratory and is described elsewhere (11).

A UCI-TPC has also measured a half-life for ^{48}Ca $(4.3^{+2.4}_{-1.1}[\text{stat}]\pm1.4[\text{syst}])\cdot10^{19}$ yr in the form of $CaCO_3$ (12). This result accords well with the n.m.e. calculated with the shell model, and the recent workshop in Prague has placed the shell model calculations in a favorable light over the Quasi Random Phase Approximation (QRPA) methods. Confirmation of the ^{48}Ca should come from the ELEGANT VI (13) and TGV (14) experiments.

Recent work with ^{150}Nd in the form of Nd_2O_3 with a TPC (15) at the Institute for Theoretical and Experimental Physics (ITEP) reports a half-life of $(1.88^{+0.66}_{-0.39}[\text{stat}]^{+0.19}_{-0.19}[\text{syst}])\cdot10^{19}$yr which confirms the earlier result of the UCI-TPC (16) but is somewhat higher than the revised UCI TPC analysis which reports $6.75\cdot10^{18}$yr (17).

Investigations of ^{100}Mo have been numerous and noteworthy of these works is the high statistical significance of the NEMO-2 measurement (18) shown in Fig. 1 The half-life reported here is $(0.95\pm0.04[\text{stat}]\pm0.09[\text{syst}])\cdot10^{19}$yr which fits within the errors of two earlier results (19, 20) and the recent and final result of the LBL-MHC-UNM-INEL collaboration (21) of $(0.76^{+0.22}_{-0.14})\cdot10^{19}$ yr. A 1995 reanalysis of the UCI-TPC (22) lowered the half-life and a second reanalysis just completed (16) reports a half-life of $0.75\cdot10^{19}$yr. This coupled with a reanalysis of NEMO-2 data (23) which yields a half-life $(0.751\pm0.028[\text{stat}]^{+0.053}_{-0.031}[\text{syst}])\cdot10^{19}$yr provides an extraordinary convergence with the LBL-MHC-UMN-INEL result. A remarkable shared result from vastly different detectors!

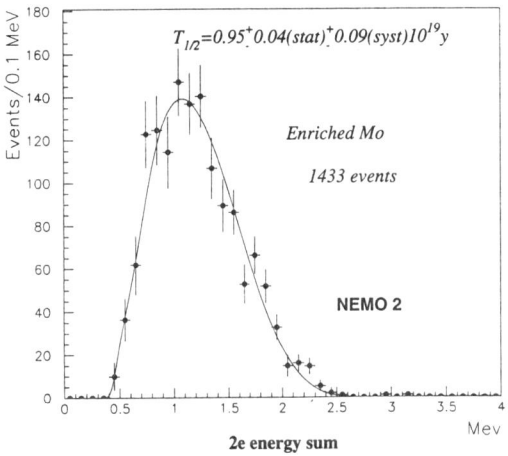

FIGURE 1. The $2\nu\beta\beta$ signal of ^{100}Mo in NEMO-2

The half-life for ^{116}Cd has also been measured by NEMO-2 $(3.75^{+0.35}_{-0.35}[\text{stat}]^{+0.21}_{-0.21}[\text{syst}])\cdot10^{19}$yr (24). This result is slightly higher than the ELEGANT V (25) measurement and a $^{116}CdWO_4$ crystal scintillators experiment result at the Institute for Nuclear Research (INR), Kiev (26). The INR, Kiev group cites possible ^{90}Sr contamination as an explanation.

Detectors for which the source is itself the detector have generated results for the three isotopes; ^{76}Ge, ^{130}Te, and ^{136}Xe. The ^{76}Ge studies have gone through a number of generations of studies. Currently, two experiments, the Heidelberg-Moscow (27) and IGEX (28) experiments, with kilograms of detectors enriched at the 86% level in ^{76}Ge, account for the active experimental investigations involving ^{76}Ge. Historically, the reported $2\nu\beta\beta$ half-life measurements of the Heidelberg-Moscow group $(1.77^{+0.01}_{-0.01}[\text{stat}]^{+0.13}_{-0.11}[\text{syst}])\cdot10^{21}$ yr and IGEX $(1.10\pm0.15)\cdot10^{21}$yr have differed. But an analysis (29) of the more recent data from IGEX causes this discrepancy to vanish and IGEX now reports a half-life of $(1.70\pm0.26)\cdot10^{21}$ yr. The change comes from a clearer understanding of the roles of ^{60}Co and ^{68}Ge in the earlier data.

In addition to studies performed with the ITEP-TPC, a xenon TPC operating in the Gotthard Underground Laboratory (30, 31) has carried out leading research on ^{136}Xe and reports a half-life measurement of $5.5\cdot10^{20}$yr. In this detector the xenon gas which fills the TPC at 5 atm is the source rather than a central anode.

Direct detection of 2νββ in ^{96}Zr and ^{130}Te has not been realized. Though ^{96}Zr has been studied with the NEMO-2 detector, the sample was too contaminated to provide a measurement and only a limit of $(1.5 \pm 0.7) \cdot 10^{21}$ yr is given. ^{130}Te has been investigated in the form of TeO$_2$ crystals at 10mK in the Gran Sasso Underground Laboratory. This thermal detector has excellent energy resolution but is best suited for 0νββ and reports no 2νββ half-life (32).

Experiments which study decays to excited states of daughter nuclei are numerous but only one half-life has been measured (33) and it is in ^{100}Mo to the first excited 0$^+$ state in ^{100}Ru. This result is being checked with a cleaner sample of ^{100}Mo.

Neutrinoless ββ Measurements

In the limits reported below there has been no attempt to use a consistent set of n.m.e. and cited limits are taken directly from the original sources. The reader is cautioned here that the reported limits select n.m.e. with more favorable outcomes and some of the advantages of one model for a particular isotope maybe contested by another model. For example, ^{100}Mo is found to be a very favorable nucleus in one study (34).

The study of neutrinoless decays has been lead by the two ^{76}Ge experiments mentioned above. The Heidelberg-Moscow experiment, which began earlier reported a limit on the half life of $>7.4 \cdot 10^{24}$ yr (90% CL) in January of 1997 (27). IGEX recently reported $>7.3 \cdot 10^{24}$ yr (90% CL) (29). The use of kilograms of enriched double beta decay isotopes makes these experiments the leaders in the research that will carry the field into the next millenium. Both programs have implemented pulse-shape discrimination to lower backgrounds by recognizing single-site events characteristic of ββ.

In Table 1 there is a listing of ββ0ν and ββχ results from the experiments introduced above with the most stringent limits identified in bold print. All limits are reported at the 90% CL with the exception of the ^{82}Se which is at the 68% CL. What is particularly interesting is that a result such as the ^{150}Nd limit on $<g_{v,\chi}>$ was achieved with only a few grams of material and given kinematical factors for this isotope it provides the most restrictive limit of all existing ββχ measurements.

TABLE 1: Listing of ββ0ν and ββχ Limits

Isotope	$T_{1/2}(0\nu)$ (yr)	$<m_\nu>$ (eV)	$<\eta>$	$<\lambda>$	$T_{1/2}(0\nu,\chi)$ (yr)	$<g_{v,\chi}>$	Ref.
^{82}Se	$2.7 \cdot 10^{22}$						10
^{76}Ge	**$7.4 \cdot 10^{24}$**	0.6	$6.4 \cdot 10^{-9}$	$1.1 \cdot 10^{-6}$	$7.9 \cdot 10^{21}$	$2.3 \cdot 10^{-4}$	27
^{100}Mo	$5.2 \cdot 10^{22}$	2.2	$2.5 \cdot 10^{-8}$	$3.7 \cdot 10^{-6}$	$5.4 \cdot 10^{21}$	$7.3 \cdot 10^{-5}$	19
^{116}Cd	$2.9 \cdot 10^{22}$	4.1	$5.9 \cdot 10^{-8}$	$5.3 \cdot 10^{-6}$	$1.2 \cdot 10^{21}$	$1.2 \cdot 10^{-4}$	26,24
^{130}Te	$1.8 \cdot 10^{22}$	3-6	$2.5 \cdot 10^{-8}$	$8 \cdot 10^{-6}$			32
^{136}Xe	$4.25 \cdot 10^{23}$	1.9-2.8			**$1.4 \cdot 10^{22}$**	$1.5 \cdot 10^{-4}$	31
^{150}Nd	$2.1 \cdot 10^{21}$	4			$5.3 \cdot 10^{20}$	$7 \cdot 10^{-5}$	16

The Future of ββ0ν and ββχ Experiments

The success of future experiments to probe the Majorana neutrino mass to 0.1eV will require kilograms of material characteristic of the ^{76}Ge experiments. The Heidelberg-Moscow experiment will certainly continue to be the front-runner. However, IGEX, if additional low background detectors are added to the experiment, shall prove to be very competative. At some point a result with a combined data set should be analyzed to take advantage of the large exposure.

The successor of the NEMO-2 experiment, NEMO-3, is funded and plans to come on line in 1998 with a capability of housing 10 kilograms of enriched material. This experiment is well suited to operate with any double beta decay isotope that can be formed into a thin foil. The experiment will initially focus on ^{100}Mo and small amounts of other as yet to be determined isotopes. A vote for the success of ELEGANT VI with ^{48}Ca comes from *SCIENCE* (35). The Milan group plans to increase the number of TeO$_2$ crystals to 20. The ITEP-TPC is to be upgraded to a 13m^3 atmospheric pressure TPC within the next couple of years and a liquid argon ionization chamber in Gran Saaso is coming on line.

The future of many of these proposed upgrades to experiments is uncertain as the facilities necessarily must scale up in size, radioactivity purities, and complexity of data acquisition/analysis to compete in this still very interesting pursuit of physics beyond the standard model.

REFERENCES

1. Geoppert-Mayer, M., *Phys.Rev.* **48**, 512 (1935).
2. Moe, M. and Vogel, P., *Annu. Rev.Nucl. Part. Sci.* **44**, 247 (1994).
3. Haxton, W.C., and Stephenson, G.S., *Prog. Part. Nucl.Phys.* **12**, 409 (1984).
4. Haxton, W.C., *Nucl. Phys. B* **31**, 88 (1993).
5. Furry, W.H., *Phys.Rev.* **59**, 1148 (1939).
6. Doi, M., et al., *Prog. of Theor. Phys.Suppl.* **83**, (1985).
7. Kirsten, T., et al., *Phys.Rev.Lett.* **20**, 1300 (1968).
8. Takaoka, N., et al., Phys.Rev. C. **53**, 1557 (1996).
9. Barabash, A.S., et al., *Physics of Atomic Nulcei*, **59**, 179 (1996).
10. Elliot. S.R. et al., *Phys. Rev. C.* **46**, 1535 (1992),. Elliott, S.R. et al., *Phys.Rev.Lett.* **59**, 2020 (1987).
11. Arnold, R., et al., *Nucl. Inst. Meth. A* **354**, 338 (1995).
12. Balysh, A., et al., *Phys. Rev. Lett.* **26**, 5186 (1996).
13. Hazama, R., et al., *Proc. of WEIN* **95**, 635 (1995).
14. Briancon, C., et al., *Nucl. Instr. Meth.* **A372**, 222 (1996).
15. Artemev, V., et al., *Phys. Lett B* **345**, 564 (1995).
16. Moe, M.K., et al., *Prog. Part. Nucl. Phys.* **32**, 247 (1994).
17. Moe, M.K., private comm.; details submitted to *Phys. Rev. C* (1997).
18. Dassie, D., et al;., *Phys. Rev. Lett.* **51** 2090 (1995).
19. Egiri, H. et al., *Nucl.Physics* **A611** 85 (1996).
20. Moe, M.K. et al., *Prog. Part. Nucl. Phys.* **32**, 247 (1994).
21. Alston-Garnjost, M., et al., *Phys. Rev. C.* **55** 474 (1997).
22. Nelson M.A., Ph.D. Thesis, University of California Irvine (1995).
23. Vareill, A., These de l'Universite de Bourdeaux I, numero d'ordre 1669 (1997).
24. Arnold, R., et al., *Z. Phys. C* **72**, 239 (1996).
25. Ejiri, N., et al., *J.Phys. Soc. of Japan* **64**, (1995).
26. Danevich, F.A., et al., *Phys.Lett. B* **344**, 72 (1995).
27. Gunther, M., et al., *Phys.Rev. D* **55**, 54 (1997).
28. Aalseth, L.E., et al., *Nucl. Phys. B, Proc Suppl.* **48**, 223 (1996).
29. Avignone, F.T., private comm. (1997).
30. Vuilleumier, J.-L., et al., *Nucl.Phys. B* **31**, 80 (1993).
31. Busto, J., private comm. (1997).
32. Alessandrello, A., et al., *Phys.Letts. B* **335**, 519 (1994).
33. Barabash,A.S., et al., *Phys.Letts B* **345**, 408 (1995).
34. Simkovic, F. et al., *Phys. Lett. B* **393** 267 (1997).
35. Normile, D., *SCIENCE* **276**, 1795 (1997).

Super–Kamiokande Solar Neutrino Analysis: The First 201.6 days' Results

Robert E. Sanford, Jr.[*] for the
Super–Kamiokande Collaboration

[*]*Louisiana State University
Baton Rouge, LA 70803*

Abstract. The solar neutrino measurement results of the first 201.6 days of analyzed data from Super–Kamiokande are presented.

INTRODUCTION

Previous measurements of solar neutrino fluxes have all reported deficits when compared to standard solar model predictions[1,2,3,4]. These deficits may indicate that solar neutrinos are oscillatings from one flavor (electron neutrino) to another flavor to which these experiments were largely or completely insensitive. Although these experiments were not able to definitively conclude whether neutrino oscillations are occurring or not, they were able to put constraints upon solar neutrino oscillation models. A leading model known as the Two Component Model with the Mikheyev–Smirnov–Wolfenstein[5] (MSW) effect has been constrained in its two dimensional parameter space to two areas often referred to as the "Large Mixing Angle Solutions" and the "Small Angle Mixing Solutions." The two solution types predict differing oscillation characteristics: small angle solutions are highly energy dependent and so would produce a non-standard energy flux spectrum, while large angle solutions are nearly energy independent and would generate an energy spectrum essentially similar to that assuming no oscillations occur. Further, large angle oscillations may also cause day and night flux differences due to the MSW effect of neutrino interactions with matter within the Earth. Distinguishing these characteristics required a new generation of neutrino detectors with high statistics and good energy and direction resolutions; Super–Kamiokande the first of such detectors became operational in April of 1996. This paper will briefly review the Super–Kamiokande experiment and the solar neutrino results of the first 201.6 days of data.

DETECTOR AND ANALYSIS DESCRIPTION

Super–Kamiokande or SuperK is a massive 50 kiloton water Cerenkov light detector, designed to detect solar Boron 8 neutrinos by neutrino/electron scattering. Equipped with 11146 inward-facing 50 cm photomultiplier tubes and 1885 outward-facing 20 cm anti-coincidence pmts, SuperK detects the Cerenkov light produced by electrons scattered by the neutrinos. By electronically recording the time-of-arrival information for each pmt, the scatter vertices can be reconstructed. Using the reconstructed vertex and the pattern of hit pmts in an event, the particles' directions of travel can also be determined.

Photon emitting sources comprised of Nickel and Californium provided calibration benchmarks for vertex reconstruction testing and energy scale determination by observing electron Compton scattered by the photons. The reconstructed vertex resolutions of the test sources in one dimension demonstrated one sigma distributions of less than 50 cm, better than the designed goal. Before energy scale determination could commence, detector geometry effects had to be reduced or eliminated so that test sources "looked" the same no matter where the sources were located in the detector. By adjusting for factors such as light attenuation, tube density, and light incidence angle onto each pmt, the energy of the test sources measured in effective charge units showed a less than 1% deviation among several locations measured within the fiducial volume. Monte Carlo simulations of the test sources provided the basis for absolute energy scale calibration.

To confirm and refine the absolute energy scale calibration, a second calibration system, a linear electron accelerator (LINAC), came on-line earlier this year to inject electrons of known energy directly into SuperK. Preliminary LINAC results coupled with the test source calibrations indicated that the energy scale used for this analysis was known to ±1.5%. More LINAC calibration tests and energy scale analysis work are still required to further reduce this uncertainty, since a 1% uncertainty in energy scale can translate into a several percent uncertainty in the energy flux calculation.

Flux determination is also highly sensitive to background levels. Cosmic ray muons entering the detector at a rate of about 3 hertz provide a significant background to the solar neutrino signal by breaking apart oxygen in the water, producing various unstable spallation nuclei. Optimization of cuts around muon tracks, both in time and space, to remove the maximum of spallation events while cutting a minimum of detector volume was achieved by vigorous study of muons of various energies over several time scales. Other backgrounds include gammas emitted from the surrounding rock and from the pmt glass itself. Spatial distributions of candidate neutrino signal events after spallation cuts show large populations at the walls of the detector due to these backgrounds. Such backgrounds were removed from the candidate set by making a 2 m fiducial cut from the surface tangent to the pmt faces, rejecting all events which lie outside of this 22.5 kiloton volume. With a cleaned data set and an absolute energy scale, analysis could then commence.

RESULTS

Events which pass spallation and fiducial volume cuts formed the final data set. The reconstructed particle directions for each event are projected onto the sun's direction at the time of the event (Figure 1), which reveal a large peak over background in the sun direction (defined as +1.0) and confirm that SuperK is detecting solar neutrino/electron scattering events. The flux rate over an assumed flat background is then calculated to be $2.65^{+0.09}_{-0.08}(stat.)^{+0.14}_{-0.10}(syst.)[*10^6/cm^2/s]$, only a factor of 0.400 of the flux predicted by the BP95 solar model. Indications of large angle oscillations are not found by comparing day and night fluxes (Figure 2); no statistically significant difference in flux levels are observed. Interpretation of the energy flux spectrum (Figure 3), which can distinguish between large and small angle solutions, is hampered by low statistics and the correlated systematic errors due to the uncertainties in the absolute energy scale. However, this preliminary analysis of only 201.6 days of data demonstrates the great capability of SuperK to measure the Boron 8 solar neutrino flux.

REFERENCES

[1] R. Davis Jr., *Phys. Rev*, 97, 766 (1955).
[2] K. S. Hirata et.al., *Phys Rev. D*, 44, 2241 (1991).
[3] A. Abazov et.al., *Phys. Rev. Lett.*, 67, 3332 (1991).
[4] P. Anselmann et.al., *Phys. Lett. B*, 285, 376 (1992).
[5] S. P. Mikheyev and A. Y. Smirnov, *Nuov. Cim.*, 9, 17 (1986).

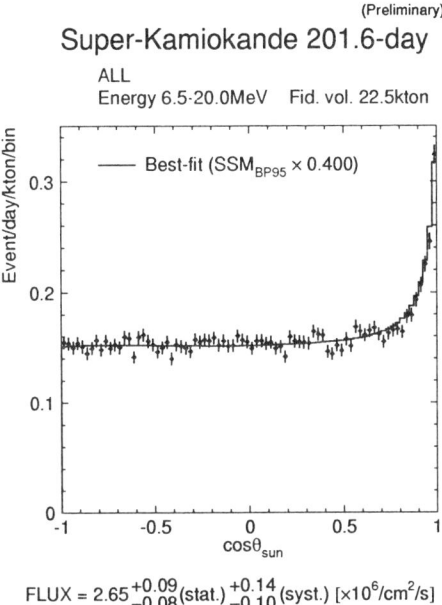

FIGURE 1. Direction to Sun Plot: The flux is extracted from these data by fitting this distribution to the sum of background model plus Monte Carlo of Boron 8 solar neutrinos.

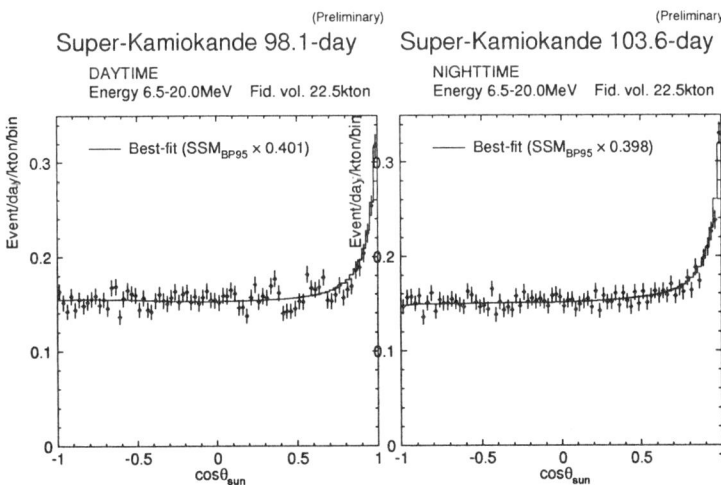

FIGURE 2. Day and Night Direction to Sun Flux Plots: These are statistically independent samples, the sum of which gives the plot in Figure 1.

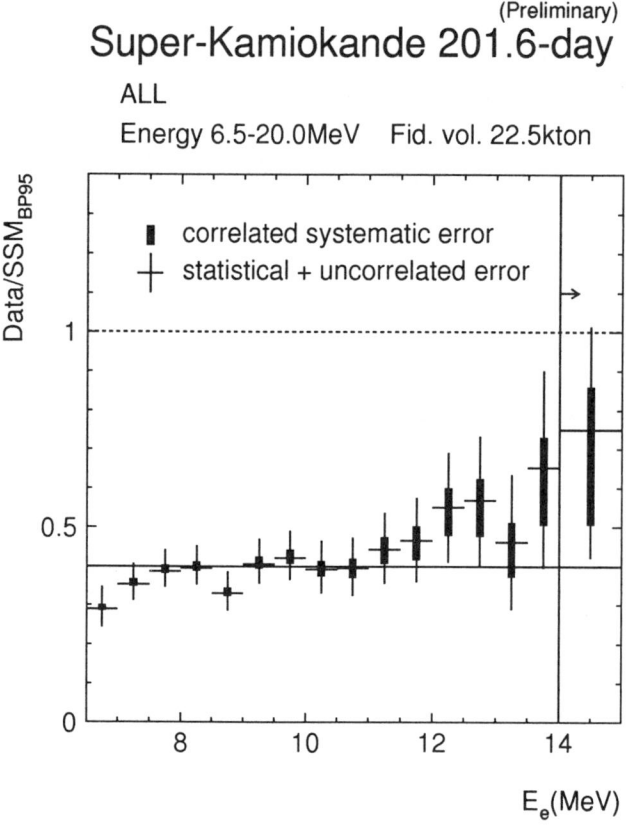

FIGURE 3. Energy Flux Spectrum: For each energy bin the same analysis is done as for the sample in Figure 1.

Toward Future Solar Neutrino Experiments

Robert E. Lanou, Jr.

Department of Physics, Brown University, Providence, RI 02912

Abstract. A new generation of solar neutrino experiments has recently begun with the initial operation of the SuperKamiokande detector and will soon be joined with the expected completion of SNO and Borexino. These detectors will provide the first direct tests of spectral deformation and possible flavor non-conservation for neutrinos from the Sun. The discoveries by these three experiments will no doubt define what direction the field will take; however, the outlines of several desirable capabilities for future detectors are clear and they present significant technical challenges. This paper[1] presents a review of the several efforts known to the author which aim to offer answers to these challenges.

Introduction

The basic guidelines for this talk assume that the "future" refers to a period beyond that of SuperKamiokande, SNO and Borexino and further, since I know of no projects presently fully funded beyond those, that the "toward" in the title is to emphasize there are a large number of activities in progress aiming to provide capability for solar neutrino experiments beyond the present ones. I will mention at the conclusion the reason for my belief that there will likely be a need for a next generation. In that connection I would like to emphasize that all solar neutrino experiments are hard and that the future ones may be more difficult than the present and require new techniques, new ideas and optimism. The projects which represent these activities constitute a continuous spectrum ranging from ideas on which work is just beginning, on through active R & D projects, and on to large prototypes — some of which will be large enough to actually see solar neutrinos. Among the later are ICARUS, Iodine and GNO (Gallium Neutrino Observatory). Among the former there are six other projects. All nine projects are summarized in the Figure as to their locations in the energy spectrum and general reaction type. While I will briefly discuss all of them, in the interest of time and because

---

[1] This work supported in part by DoE DE-FG02-88ER40452 & NSF PHY-9420744.

the three large prototypes may be more familiar to you, I will give a bit more detail on the less familiar R & D projects. Also, most represent an important new thrust into lower energies and, largely, in real-time. For all I will try to give some flavor of the goals, techniques and status.

The Specific Projects

HERON & HELLAZ: These projects are both based upon the use of helium as the target medium but they have radically different approaches and somewhat complementary goals. Both will utilize the elastic reaction, $\nu_{e,\mu,\tau} + e^- \rightarrow \nu_{e,\mu,\tau} + e^-$, for real-time detection in the energy region dominated by the p-p and Be^7 neutrinos. They will both measure the energy of the recoil electron and the overall rate. Turning now to the contrasting specifics.
HERON: This project uses He^4 in its superfluid state [1]. For purity from radioactivity, liquid helium is an ideal target. It has no long-lived isotopes and is self-cleaning in that all other atomic species freeze out. It has a relatively high density (0.14 g/cc), is inexpensive and standard commercial procedures exist for handling it in bulk. Recoil particle detection occurs by making use of the very high multiplicity ($\sim 10^8$ for an 80 keV e^-) of energy carriers (phonons/rotons) generated by the recoil. A full scale detector would utilize a fiducial mass of 10 tons as a single volume ($\sim$ 4 meter cube) but with segmented readout external to the liquid. The operating temperature would be 30mK. The expected event rate would be 18 p-p and 7 Be^7 events/day in the standard solar model (SSM). Approaches to background suppression include exclusion of any other material in the helium, use of event position, topology, differences in mean-free paths, energy and possibly track orientation. Neutron shielding would be water external to the detector. Among the physics and technology R&D issues to be addressed are the physics of an entirely new way of particle detection, the level of sensitivity and resolution possible, backgrounds from containment and engineering challenges. The R&D carried out so far [1] includes demonstration of the validity of the basic physics detection processes in a 3 liter prototype, a new finding of scintillation and directionality for α tracks (now being tested for with e^-'s) , Monte Carlo and material tests of backgrounds; work is also underway on improvements of sensitivity and the design of a larger prototype with multiple segmented read-out.
HELLAZ: This project [2] will utilize helium in gaseous form under high pressure (8 atmos. and 100 K) yielding a 0.003 g/cc density. Detection would be via a large time projection chamber (TPC) immersed in the helium thus the detected carriers are the charges resulting from ionization ($\sim 10^3$ for 90 keV). The TPC would use a gas mix of helium and methane. With the TPC, full track reconstruction is aimed for to provide a very powerful signature when referred to simultaneous solar position. The 7 ton fiducial volume (25 m long x 10 m diam.) would produce 11 p-p and 3 Be^7 per day (SSM). Approaches to

background suppression would include use of ultrapure materials in the TPC and specially purified methane, use of full track reconstruction (position, direction and energy); neutron shielding to be provided by a 77K CO_2 shield. Among the physics and technology R&D issues to be addressed are TPC operation in an unconventional pressure, temperature & mixture, drift dispersion over 10 m., backgrounds from components and engineering challenges. After some initial R&D at CERN a few years ago work is now resuming at FSU and WSU [2] on the construction of a 5 liter, high pressure, cold TPC where some of the issues related to TPC operation and drift dispersion will be addressed.

YBEX: This new effort [3] is intended to produce a real-time detector which would be flavor specific to ν_e's from the same portion of the spectrum sampled by Borexino — the line at $E_\nu = 862$keV; Borexino's events do not distinguishing among ν_e, ν_μ and ν_τ. (This function might conceivably be provided also to HERON or HELLAZ.) To achieve this, the scintillator would be loaded with Yb^{176} which is stable and has a 12% natural abundance. With a total Yb loading of $\sim 12\%$ wt., a 100 ton detector is estimated to detect yearly rates (SSM) for p-p : Be^7 : p-e-p : N,O of 20: 183: 26: 106 with backgrounds (after cuts) of 0.7: 4: 8: 50. The initiating reaction is by inverse beta decay from the 0^+ to 1^+ excited states of Lu: $\nu_e + Yb^{176} \rightarrow e^- + (Lu^{176})^*$. The produced e^- has $E_e = E_\nu - 301$ keV. In addition to the produced e^-, there are γ's resulting from two decay channels: in one, a γ of 71.5 keV in 50 nsec while in the second a prompt 144.4 keV γ is produced also tagging the energy level. In order to defeat the enormous background without having to purify the scintillator to 10^{-16} g/g and make use of these nearly unique event signatures it is proposed to segment the detector into 132 physical modules and those further into smaller segments by relative timing and to require e^- and γ's to be in the same logical cell. A particularly important question as to the feasibility of the experiment concerns the determination of the presently unmeasured cross sections although estimates have been made [3]. R&D has only just begun with some successful tests of scintillator loading with Yb (Se^{82} and Gd^{160} will also be tried) [3]; an experiment to do a low energy, 0° (p,n) cross section measurement has also been proposed.

Gallium Arsenide: This is perhaps the most ambitious project [4] and while it does not aim to be the "complete" solar neutrino detector it does give a measure of the task to realize a single detector more inclusive of reaction types and range of energies. Its principal aims are to make a model independent test of flavor non-conservation (including sterile ν's), to determine the precise energies and widths of the lines of Be^7(862keV), p-e-p(1.4MeV) and the end point of the p-p continuum using the three channels of elastic, charged current and neutral current scattering of ν's. Further, in the case of any observed flavor non-conservation, they wish to determine the MSW parameters precisely if that is the mechanism involved. To carry out such a program, a massive detector with ≤ 2 keV energy resolution is required. Their idea is to create a GaAs based, electronic, real-time detector. Gallium and arsenic are chosen because they have no long-lived isotopes or (n,γ) daughters and for the prospect of good resolution due to the high multiplicity of e-hole pairs and low noise when cooled. However, they estimate this would require a 125 tonne (60 t of Ga) in 40,000 hyperpure 3.2 kg segments. Among the primary reactions to be exploited are $\nu_{e,\mu,\tau} + e^- \to \nu_{e,\mu,\tau} + e^-$, $\nu_e + Ga^{71} \to e^- + Ge^{71}$ and $\nu_{e,\mu,\tau} + Ga^{71} \to \nu_{e,\mu,\tau} + (Ga^{71})^*$. Good resolution of the p-e-p and upper Be^7 line would be an important advance; however, a successful data analysis to achieve fully the above goals will be a very challenging one in that it relies upon use of the shape of the elastic differential cross-section to separate ν_e and $\nu_{\mu,\tau}$ as well as requiring a separation of the charged current p-p events. R&D on this project is just beginning. Extreme purity (from U,Th,K & C) of the GaAs as well as electronic performance for 3.2kg devices must be tested and assured. To this end, a 1kg boule of GaAs has been produced and preliminary tests of electrical properties appear promising.

Sodium Bromide : For completeness, mention should be made of a fully cryogenic project [5] which also aimed at exploiting the line spectra and upon which exploratory R&D has been carried out. The reaction was to have been $\nu_e + Br^{81} \to e^- + (Kr^{81})^*$ with delayed coicidence detection of the electron and de-excitation γ's. A precise determination of the event energy was to be achieved by phonon detection in a highly segmented crystal array at mK temperatures — a technique pioneered very successfully by this group for double-beta-decay experiments in other crystal types. The R&D on NaBr has not been promising; the thermal properties are much worse than expected. It is very difficult to grow large crystals and the signals are an order of magnitude too small. This work has ceased but is well summarized in [5].

Lithium: For some time, Li^7 has attracted interest as a radiochemical target with special relevance for the p-e-p and CNO neutrinos because the strength of the ground and first excited states can be accurately inferred from laboratory experiments [6]. The reaction is $\nu_e + Li^7 \to e^- + Be^7$ with a threshold of 862 keV. A particular challenge to the realization of a lithium-based detector has been the counting of the electron capture decays ($\tau_{1/2}$=53d) of the extracted Be^7 in which 90% go to the Li^7 ground state but produce an Auger electron of only 55 eV and is therefore not amenable to the usual proportional counter methods. Consequently, interest had centered on utilizing the decay to the first excited state by detecting the subsequent 474 keV gamma but, with only a 10% branching ratio, a 100 ton detector was required to achieve a rate of 0.5/day. Presently, a joint Russian-Italian project [7] is making interesting progress toward capitalizing on the use of the 90% channel thus obviating the need for so large a detector. In tests using cryogenic microcalorimeter techniques with accelerator produced samples of Be and BeO, they have detected Be^7 decays via the summed energy deposited by the Auger electron (55eV) and the recoiling nucleus (57eV) giving a well resolved peak at 112eV with a 24eV FWHM at ∼ 80% efficiency. The microcalorimeter consists of a 100x200x200 μm NTD thermistor at 45mK glued to the few μg sample. While further work along these lines is in progress, this result seems to establish feasibility of the method. Additionally, a prototype to hold 300 kg of liquid lithium has been constructed and they hope to fill it at Obninsk at the end of the year. Funds for this purpose have been applied for. The principal R&D work to be carried out with the prototype is study of the Be extraction process in forms most suitable for the microcalorimetry counting method.

GNO : The Gallium Neutrino Observatory [8] is designed to improve and extend the very successful gallium radiochemical technique. It is intended to do so in a phased approach during which the total mass in the target might eventually reach 100 tons. Among the principal goals are the long term ($\sim$ 11 year solar sunspot cycle) observation of the solar ν_e flux and significant improvement of the systemaic errors ($\leq 3\%$). The first phase, which could commence within the year with the original 30 tons and new counting electronics, is already fully approved. Proposals for $\sim$ 30 ton step-wise increments are expected later.

Iodine : This project [9] to build a 235 ton (100 tons of iodine from NaI in solution) prototype for a potentially much larger detector is already well underway. A radiochemical detector, it is similar to that of the pioneering experiment with chlorine but with important differences. Iodine is expected to have a significantly larger cross-section for ν_e, with different relative sensitivity to the Be^7 and B^8 fluxes and it is expected to be equipped with 12 hour cycling so that possible day-night flux variation could be tested for. The initiating reaction is $\nu_e + I^{127} \rightarrow e^- + (Xe^{127})^*$, Q = 662keV; however, transitions to the Xe ground state is forbidden but to the two, $3/2^+$ excited states at 125 and 322keV are allowed. The electron capture half-life of the extracted Xe^{127} is 36.4 days and proceeds 46% and 54% to excited levels in I^{127} yielding, respectively, γ's at 375 and 203keV. The counting signature would then be the post-electron capture Auger electron in coincidence with one or more of the γ's detected by additional NaI counters surrounding the proportional counter. This coincidence feature should provide an important control on backgrounds. Annual neutrino event rates are suggested to fall between 100 and 300 depending upon assumptions. Presently, 8 of the 10 tanks needed for the 100 ton experiment are in the mine at Homestake but are not yet fully installed nor filled. The processing and preparation of the NaI solution is projected to take six months. It is hoped that a major portion of the detector will be operating by year's end. Initially, the fast extraction system will not be implemented; however, it has been successfully tested in one module with a neutron source and found to give 99% recovery in 1 hour independent of source position. A major project to address the present lack of precise knowledge of the cross-sections to the excited states is under discussion with Russian colleagues. It is proposed to construct a 400kilo-Ci Ar^{40} calibration source to provide ν_e's of 814keV into a suitable test vessel; funds are being sought for this purpose.

ICARUS : As a step in the modular construction of a 6 kiloton proton decay detector, the first module (600 tons) to be installed in the Gran Sasso will, as part of its operational commissioning, be used in a solar neutrino experiment [10]. The target material is liquid argon (540 tons fiducial) in which the ionization products are drifted in a TPC and pattern recognition achieved by providing full track reconstruction of an event. Extensive R&D has been carried out with a 3 ton prototype. The primary goal is to search for direct evidence of ν oscillations in the B^8 flux by measuring the ratio of rates between the two channels: $\nu_{e,\mu,\tau} + e^- \to \nu_{e,\mu,\tau} + e^-$ and $\nu_e + Ar^{40} \to e^- + (K^{40})^*$, $Q = 1.504$ MeV. The threshold for the former is set by the minimum electron energy visible (expected to be $\simeq 4$ MeV) and for the latter by also reaching the Fermi level (F) at 4.384 MeV or by the various Gamow-Teller(G-T) levels below it. The signature for the $(K^{40})^*$ events will consist of an $E_e \geq 5$ MeV associated with one or more lower energy tracks (γ's) within a 50 cm sphere. Some discrimination can be made between the F and G-T transitions by the γ multiplicity. In a year, they expect a total of 290 elastic and 760 charge current (330 F and 430 G-T) events(SSM). Contamination from various other reactions appearing as the same event topologies are 18 "elastic" and 178 "charge current" (68 to F and 110 to G-T). Expected radioactive background is, respectively, 13 and 46 (3 and 43). A cone of $25°$ is expected to contain 65% of the elastic angular distribution and will be used with the Sun's position in background reduction. The planned for location of the module, and its neutron shield, is Hall C in Gran Sasso with cool-down by the end of 1998.

Conclusions

We might reasonably ask how likely is it that one or more future experiments will be needed and to what extent can we foresee just what shape it or they might take. I believe it is highly likely there will be a long run of new experiments before the physics is fully understood. It now seems inescapable that the results we currently have on ν's from the Sun are not due to any experimental artifacts. The legacy from the elegant and careful work by Homestake, Kamiokande and GALLEX/SAGE is the possibility of entirely new physics. The present group of SuperK, SNO and Borexino will directly test this proposition but are unlikely definitively to provide all answers. Three possible outcomes seem reasonable to think about: a) "Smoking gun" = "evidence for spectrum deformation and neutral current" in which case, what are the full parameters? Are there 2 or 3 ν's involved? Is it MSW vs. vacuum oscillations? Is there a magnetic moment? Are there sterile neutrinos? or b)"Loaded but not smoking" = "e.g., clear spectral deformation but not yet neutral current evidence", in which case, what is(are) the source(s) of the solar ν problem(s)? Is it still new physics? Is it astrophysics? or c)"Entirely new surprises", in which case, what relation do they have to the new terrestrial experiments now evolving? Is it still new physics or astrophysics? The discoveries of SuperK, SNO and Borexino will indeed shape the details of what

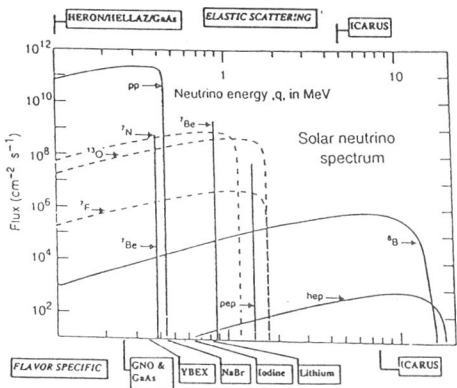

comes next. However, any detailed tests of the full spectrum, in real time and with reaction specificity await the capabilities sought by just such projects as those discussed here. They employ a rich, varied and imaginative range of new techniques and ideas to address hard problems. There is room for much optimism and for more new ideas as well.

REFERENCES

1. At Brown University. Bandler et al, *PRL* **74**,3169(1997); Adams et al, *PL* **B341**,431(1995); Bandler et al, *J. Low Temp. Phys.* **93**,785(1993); Bandler et al, *NIM* **A370**,578(1996).
2. At Florida State Univ., Wayne State Univ. & College de France. Arzarello et al, LPC-94-28; Seguinot et al LPC-92-31; Bonvicini *NP (Proc.Supp.)* **438**(1994); private communication H. Prosper and G. Bonvicini.
3. At Lucent Technology's Bell Laboratories. Raghavan, *PRL* **78**,3618(1997).
4. At Los Alamos and INS(Moscow). Private communication T.J. Bowles and V.N.Gavrin.
5. At INFN(Milano). Alessandrello et al,*Astropart. Phys.* **3**, 239(1995) and private communication E. Fiorini.
6. Bahcall, J.N. *Neutrino Astrophysics*, 372ff (and refs. therein),Cambridge University Press (1989).
7. At INFN(Genoa) & INR(Troitsk/Moscow). Galeazzi et al, *PL* **B398**,187(1997) and private communications S. Vitale and A. Kopylov.
8. At Gran Sasso Laboratory. Private communication W. Hampel; proposal available via http://kosmopc.mpi-hd.mpg.de/gallex/gallex.htm .
9. At University of Pennsylvania and Homestake Mine. Private communication K. Lande. Haxton, W.,*PRL* **60**,768(1988) and *PR* **C38**,2474(1988).
10. At Gran Sasso with a CERN-Italy-UCLA collaboration. Private communication D. Cline and www.aquila.infn.it/icarus/exp.html/; Bahcall et al,*PL* **B178**,324(1986).

The Atmospheric Neutrino Muon-like Fraction Above 1 GeV

R. Clark*,
For the IMB Collaboration

*Louisiana State University, Baton Rouge, Louisiana 70803, USA

Abstract. A total of 72 atmospheric neutrino events were found in a 2.1 kton-yr exposure of data from the Irvine-Michigan-Brookhaven detector. Each event had a vertex contained inside the fiducial volume and at least 0.95 GeV of visible Čerenkov energy. The ratio of ratios: $f = (\frac{muon-like}{total})_{Data}/(\frac{muon-like}{total})_{MC}$, was found to be $1.1^{+0.07}_{-0.12}(stat.) \pm 0.11(syst.)$. The zenith angle dependence of f is consistent with being flat. The regions of $sin^2(2\theta) > 0.5$, $\delta m^2 > 9.8 \times 10^{-3} eV^2$ and $sin^2(2\theta) > 0.7$, $\delta m^2 > 1.5 \times 10^{-2} eV^2$ have been excluded to the 90% confidence level for $\nu_\mu \to \nu_e$ and $\nu_\mu \to \nu_\tau$ oscillations respectively.

A large flux of pions (and a smaller flux of kaons) are produced by cosmic ray interactions in the upper atmosphere. These subsequently decay to produce muon and electron neutrinos which have been observed in underground detectors. The number and distribution of muon-like and electron-like neutrino interactions have been used to test the hypothesis of neutrino oscillations. It is favorable to use a ratio of the neutrino flavors since errors in the absolute flux largely cancel. Typically, the data is compared to the Monte Carlo prediction as a ratio of ratios $f = (\frac{muon-like}{total})_{Data}/(\frac{muon-like}{total})_{MC}$ which will be equal to one if the measured flavor ratio agrees with the predicted ratio. Since 1986, the Irvine-Michigan-Brookhaven (IMB) Detector and other experiments have reported values of f below one [1]. In 1997, the Kamiokande group reported a zenith angle dependence to f for their high energy contained events [2]. New results are presented here from a sample of high energy IMB contained neutrino events. Details of the IMB detector may be found elsewhere [3].

There were 236 live days in the IMB data used for this analysis which gives a total exposure of 2.1 kton-yr. Each event was required to have at least 0.95 GeV of visible Čerenkov energy and a vertex contained inside the fiducial volume. The events were all passed through data reduction software designed to remove events with entering tracks and with less than 1000 photoelectrons (p.e.s) of visible Čerenkov energy. Out of roughly 55 million events, 22,192 survived the software cuts. These were then visually scanned with a custom graphics display package to remove events with entering tracks missed by the data reduction software. The efficiency of the scanning process was estimated as 0.98 ± 0.02 for events with a vertex inside the fiducial volume. This breaks down as 0.98 ± 0.02 for $\nu_e + \bar{\nu}_e$ events and 0.99 ± 0.02 for $\nu_\mu + \bar{\nu}_\mu$ events. A total of 83 events were left after the initial scanning and 72 of these were fit with a vertex inside the fiducial volume.

A 29.1 kton-yr exposure of Monte Carlo events were generated in the total volume of the detector. Charged and neutral current interactions were generated for $\bar{\nu}_e$, ν_e, $\bar{\nu}_\mu$ and ν_μ. The fraction of these events with at least 1000 p.e.s of visible Čerenkov energy was 0.17. The atmospheric neutrino fluxes were based on a table by Agrawal et. al. [4] and ranged in energy from 0.8 GeV to 100 GeV. The fluxes from Honda et. al. [5] were also used. The results using both flux models are presented in Table 1. The median neutrino energy for all generated events was 1.6 GeV while the median energy for events surviving all cuts was 4.0 GeV.

The generated Monte Carlo events were treated the same as the data events. They were passed through the same data reduction software, visually scanned and hand fitted. Based on the Monte Carlo sample, the efficiency of the whole data reduction process for events with a vertex contained inside the fiducial volume and with at least 1000 p.e.s was calculated as 0.79 ± 0.02. Multiple track events comprised 61% of the sample with the rest being single track events. The fraction of neutral current events in the final sample was 9%.

A likelihood function $L = \prod_{i=1}^{n} P_i(x_i)$ was used to determine if the events were electron-like or muon-like. It was calculated by taking the product of n different probabilities P_i. These were determined from the ratio of distributions for the measured factor x_i based on both Monte Carlo and cosmic ray muon events. Different likelihood functions were applied to multiple and single track events [6]. The accuracy of L for single track events was calculated as 0.90 ± 0.03, while for multiple track events it was 0.78 ± 0.04 and the combined accuracy for all events was 0.85 ± 0.03. In the data sample, 47 events were identified as muon-like and 25 as electron-like. In the Monte Carlo sample, 57% were identified as muon-like and the rest were identified as electron-like. This is summarized in table 1.

TABLE 1. A summary of the data and Monte Carlo events showing the number of electron and muon-like events. See the text for an explanation of BGS and HKKM.

	Data sample	BGS sample	HKKM sample
Electron-like events	25	31.2	29.2
Muon-like events	47	41.9	40.4
All events	72	73.1	69.6

The ratio of ratios was found to be $f = 1.1^{+0.07}_{-0.12}(stat.) \pm 0.11(syst.)$. The zenith angle distribution of f is shown in figure 1 along with that from Kamiokande [2]. Consistency between the angular distributions of Kamiokande and this analysis were checked with a χ^2 test. Uncertainties from the two experiments were added in quadrature. From this χ^2 test, the probability that this result is a statistical fluctuation of the Kamiokande angular distribution is 5%. Using just the shape of the distributions alone, the probability is 23%.

This result has been used to exclude some of the parameter space for $\nu_\mu \to \nu_e$ and $\nu_\mu \to \nu_\tau$ oscillations. In general, the probability for neutrino oscillations is given by $P_{\nu_\mu \to \nu_x} = \sin^2(2\theta) \sin^2(\frac{1.27\delta m^2 D}{E_\nu})$, where θ is the mixing angle between the states, δm^2 is the difference of the squared neutrino masses, D is the distance from the neutrino creation point (km) and E_ν is the energy of the neutrino (GeV). To the 90% confidence level, the region excluded for $\nu_\mu \to \nu_e$ oscillations is $sin^2(2\theta) > 0.5$ and $\delta m^2 > 9.8x10^{-3}eV^2$ while the region excluded for $\nu_\mu \to \nu_\tau$ oscillations is $sin^2(2\theta) > 0.7$ and $\delta m^2 > 1.5x10^{-2}eV^2$. The shape of the angular distribution of f was not used in determining these regions. The excluded regions of this analysis are shown in figure 2 along with the allowed regions from Kamiokande [2]. Note that this analysis differs from

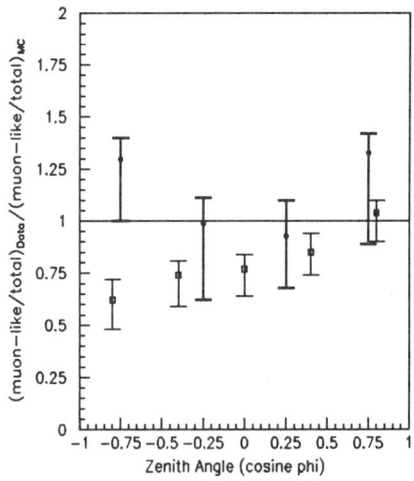

FIGURE 1. The zenith angle ($\cos\phi$) distribution of the ratio of ratios $(\frac{muon-like}{total})_{Data}/(\frac{muon-like}{total})_{MC}$. The filled circles are from this analysis while the open squares are taken from Kamiokande [2]. Upward going events are on the left side and the error bars are statistical only.

previous water Čerenkov results in that it does not rely heavily on separating out fully and partially contained events [2], or on rejecting multiple track events [1].

REFERENCES

1. K. S. Hirata et al., Phys. Lett. B **205**, 416 (1988); D. Casper et al., Phys. Rev. Lett. **66**, 2561 (1991); K. S. Hirata et al., Phys. Lett. B **280**, 146 (1992).
2. Y. Fukuda et al., Phys. Lett. B **335**, 237 (1994).
3. R. M. Bionta et al., Phys. Rev. Lett. **51**, 27 (1983); R. Claus et al., Nucl. Instrum. Methods Phys. Res., Sect. A **261**, 540 (1987); R. Becker-Szendy et al., Phys. Rev. D **42**, 2974 (1990); R. Becker-Szendy et al., Nucl. Instrum. Methods Phys. Res., Sect. A **352**, 629 (1995).
4. V. Agrawal et al., Phys. Rev. D **53**, 1314 (1996).
5. M. Honda et al., Phys. Rev. D **52**, 4985 (1995).
6. Details of the flavor identification likelihood function may be found in R. Clark, Ph. D. thesis, Louisiana State University (1997).

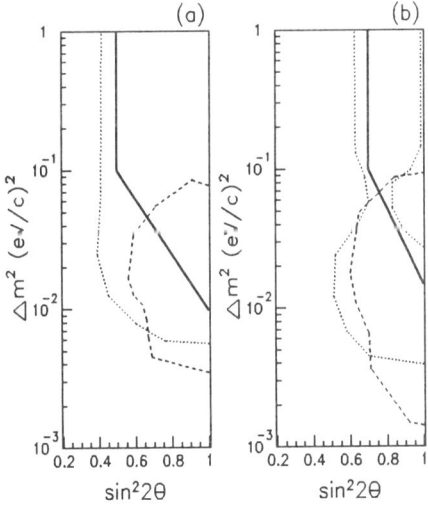

FIGURE 2. The parameter space regions for $\nu_\mu \to \nu_e$ (a) and $\nu_\mu \to \nu_\tau$ (b) oscillations excluded to the 90% confidence level by this work (solid lines). The allowed regions for the Kamiokande multi-GeV (dashed lines) and sub-GeV (dotted lines) sets [2] are also indicated. The sub-GeV set is similar to earlier IMB sub-GeV results.

The Atmospheric Neutrino Flavor Ratio in Soudan 2

Maury Goodman
For the Soudan 2 Collaboration

HEP362, Argonne National Lab, Argonne Ill. 60439, USA

Abstract. The Soudan 2 collaboration has measured the atmospheric neutrino flavor ratio with 2.63 kiloton years of exposure. Our measured flavor ratio is 0.67 ± 0.15(stat) +0.04-0.06(syst). The neutrino induced horizontal muon flux has been measured to be $\Phi_\mu = (4.12 \pm 1.1 \pm 0.58) \times 10^{-13} cm^{-2} sr^{-1} s^{-1}$.

I INTRODUCTION

The measurement of the atmospheric neutrino flavor ratio is of interest due to the apparent anomaly by some reported experiments [8,5,7,1,3] and the possible explanation of that anomaly in the context of neutrino oscillations. The double ratio

$$R \equiv \left(\frac{\nu_\mu}{\nu_e}\right)_{data} / \left(\frac{\nu_\mu}{\nu_e}\right)_{MC} \sim \left(\frac{tracks}{showers}\right)_{data} / \left(\frac{tracks}{showers}\right)_{MC}$$

allows a measurement which is independent of an absolute flux or exposure calculation.

Soudan 2 is an iron calorimeter with different experimental systematics from the water Cherenkov detectors and whose geometry and detection technique differ from the Frejus experiment. A large veto shield placed against the cavern wall allows the identification of particles entering the detector from the interactions of cosmic ray muons in the surrounding rock. We used these "rock" events to determine whether our low value of R could be due to contamination from such events.

II DETECTOR AND EXPOSURE

The 963 metric ton Soudan 2 experiment is located in the Soudan Underground Mine State Park, Minnesota, 710 meters underground. About 85% of the mass is provided by 1.6 mm thick sheets of corrugated steel. The sheets are stacked to form a hexagonal 'honeycomb' structure. Plastic drift tubes (1.0 m long and 15 mm in diameter) fill the spaces in the honeycomb. An 85% argon/15% CO_2 gas mixture is recirculated through the modules. Ionization deposited in the gas drifts toward the closer end of the tube in an 180 volt/cm electric field. The drift velocity is approximately 0.6 cm/μsec, which yields a maximum drift time of 83 μsec. The average density is 1.6 g/cc. Further details of module construction may be found in Reference [2].

The calorimeter is surrounded by a 1700 m^2 active shield designed to detect charged particles which enter or exit the detector cavern. The shield covers about 97% of the total solid angle. The basic element is an extruded aluminium manifold, consisting of eight hexagonal proportional tubes arranged in two layers of four. A two-layer coincidence is required to signal a high energy particle entering or leaving the cavern. The measured efficiency of a coincidence is 95%. Details of the shield performance can be found in Reference [11].

III DATA REDUCTION

We have analyzed data from 2.63 fiducial kton-year exposure taken between April 1989 and March 1995. During this period the detector was under construction, starting with a total mass of 275 tons and ending with the complete 963 tons. There were 75 million triggers taken. The goal of the data reduction is to obtain a sample of 'contained events', defined as those in which no primary particle in the event enters or leaves the fiducial volume of the detector. The fiducial volume is defined by a 20 cm depth cut.

The events are passed through a software filter to reject events with tracks entering or leaving the fiducial volume (mostly cosmic ray muons) or events which have the characteristics of radioactive background or electronic noise. Approximately 1 event per 1500 triggers passes this filter.

The selected events are then double scanned to check containment and to reject background events, using an interactive graphics program. The main backgrounds are residual radioactive and electronic noise, badly reconstructed cosmic ray muons, and events where muons pass down the gaps between individual modules, either finally entering a module and stopping or interacting in material in the gap and sending secondary tracks into the modules. Any event with a track which starts or ends on a gap, or which can be projected through a gap to the exterior of the detector is rejected. In addition, events with a vertex in the crack region are rejected. Differences between scanners are resolved by a second level scan. Approximately 1 event in 40 passed by the program filter is finally selected as contained. The average efficiency of individual scanners in selecting contained events was 93.5%. Further details of the event selection procedure can be found in Reference [9].

IV NEUTRINO MONTE CARLO

Monte Carlo events equivalent to 5.9 times the exposure of the real data were inserted randomly into the data stream and processed simultaneously with the data events, ensuring that they are treated identically. The neutrinos were generated using the BGS flux [4]. The variation of the ν intensity with the solar cycle was corrected using neutron monitor data [9,6].

At the low ν energies characteristic of the atmospheric flux the predominant interactions are quasi-elastic or resonance production. Full details of the event generation process and a detailed comparison with all available low energy data are given in Reference [9]. Nuclear physics effects were represented by the Fermi gas model. Rescattering of pions within the nucleus was applied using data obtained by comparison of bubble chamber ν interactions on deuterium and neon [10]. Particles produced in the neutrino interactions were tracked through the detector geometry using the EGS and GEISHA codes. The generated event was superimposed on a pulser trigger which reproduces noise and background in the detector as they vary with calendar time.

V EVENT CLASSIFICATION

The lepton flavour of each event is determined by the second level scanners who flag them as 'track', 'shower' or 'multiprong'. Single track events which have heavy ionization and are straight are separately classified as 'proton '. Proton recoils accompanying tracks and showers are an additional tag of quasi-elastic scattering and are ignored in the classification. Any second (non-proton) track or shower in the event results in a multiprong classification. Results are shown in Table 1. Events without (with) shield hits are labeled "gold" ("rock") events. As a test of the systematic uncertainties introduced by the classification process, all scanning was done independently by two groups prior to merging for the final results.

TABLE 1. Classifications for the contained events before corrections.

	Track	Shower	Multiprong	Proton
Data: gold	75	106	89	22
Data: rock	237	312	177	130
MC	461	445	432	48

The quality of the flavour assignment was measured using the Monte Carlo data. Table 2 shows the fraction of Monte Carlo events selected as contained which were classified in each category. It can be seen that 87% of events assigned as tracks have muon flavour and 96% of showers electron flavour. The ratio of accepted muon to electron charged current Monte Carlo events is approximately 1:1, different from the ratio of 2:1 for the $\pi \to \mu \to e$ decay chain. At these low energies threshold effects due to the difference in the muon and electron masses cause the generated event ratio to be approximately 1.5:1. Acceptance differences for high energy muons and electrons and the cuts required to remove background produced by cosmic ray muons passing down the gaps between modules further reduce the ratio.

TABLE 2. Monte Carlo identification matrix.

Generated	Assigned			
	Track	Shower	Multiprong	Proton
ν_μ cc	0.87	0.01	0.38	0.24
ν_e cc	0.05	0.96	0.44	0.04
Neutral current	0.08	0.03	0.18	0.72

VI BACKGROUND SUBTRACTION

The majority of the 1148 events classified as contained are due to the interactions of neutral particles (neutrons or photons) produced by muon interactions in the rock around the detector. Calculations show that only a few percent of such events will not have an accompanying charged track traversing our shield, which was placed as close to the cavern wall and as far away from the detector as possible to maximize the probability of detecting the accompanying charged particles. The efficiency of the shield has been measured using single cosmic ray muons detected in the main detector. It ranges from 81% during the early data runs before the geometrical coverage was complete to 93% at the end of this data period. Also, 8.9% of pulser events overlaid on Monte Carlo events had a random shield coincidence.

Our large sample of rock events enables us to investigate muon induced background by studying the depth distribution of the events in the detector. This allows us to simultaneously measure any backgrounds due to either shield inefficiency or contained events due to neutral particles entering the detector without being accompanied by charged particles in the shield. Neutrino events should be distributed uniformly throughout the detector, while background events are attenuated towards the center. We define a measure of the proximity of the event to the detector exterior by calculating the minimum perpendicular distance from the event vertex to the detector edge.

The depth distribution for tracks and showers from the gold, rock and neutrino Monte Carlo samples are shown in Figure 1. The points are the gold data, the shaded histogram is the neutrino Monte Carlo, normalized to the experiment exposure, and the unshaded histogram is the rock data, normalized to the same number of events as the data sample. It can be seen that the data more closely resembles the neutrino depth distribution than the rock background depth distribution. In the next section, we fit the shape of the gold data to the shapes of the neutrino Monte Carlo and rock samples to estimate the rock background. This produces a noticeable error in the flavor ratio obtained due to the statistics of the fit, but this is required to properly take into account the possibility of background contributions to the flavor ratio.

VII CALCULATION OF R

In using the depth distribution of the rock events to correct for background, we note that the measured flavor ratio as a function of shield multiplicity is observed to be a constant value of 0.76 ± 0.07. We then correct the track to shower ratio in the data by fitting the track and shower depth distribution to a sum of those in the rock events and Monte Carlo, constraining the flavor ratio of the rock events to its observed value. The result of the fit is 20.4 tracks and 26.9 showers in the gold sample are due to background, leading to a corrected neutrino induced rate of 54.6 tracks and 79.1 showers. From this we calculate $R = 0.67 \pm 0.15$, where the error includes the statistical error on the data, the statistical error on the Monte Carlo, and the error on the background subtraction from the fit.

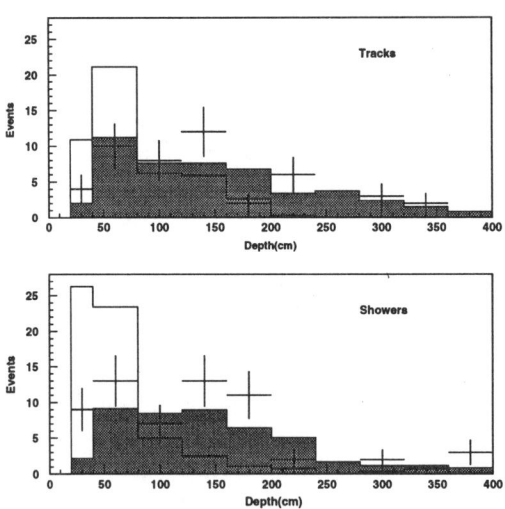

FIGURE 1. *The depth distributions for tracks (top) and showers (bottom)*

Systematic errors have been calculated based on errors on the flux models, neutrino cross sections, scanning, and the assumption that the flavor ratio of the background is independent of shield multiplicity. The latter systematic error, which can be measured by fitting the tracks and showers separately, leads to an asymmetric error. We calculate a total systematic error on R of +0.04 and - 0.06.

VIII HORIZONTAL NEUTRINO INDUCED MUON FLUX

Soudan 2 cannot resolve upward from downward going muons. But using horizontal muons, where the atmospheric muon contribution is significantly diminished, we can isolate neutrino induced throughgoing muons. The slant depth calculation uses U.S.G.S. digitized surface elevation data. Muon events which have a slant depth in excess of 14 km.w.e. are considered to be from the population of muons induced from atmospheric neutrinos. Acceptance to this population of muons varies slightly with zenith owing to topographical effects on the surface. The solid angle acceptance for this slant depth starts near 81 degrees, includes all of solid angle beyond 83 degrees, and totals 1.77 sr. The effective solid angle was calculated by Monte Carlo to be 84.86 m^2. Trigger and program efficiencies have been estimated to be 81%. The initial search has produced 14 events which pass the 175 cm track-length cut corresponding to a muon energy threshold of 600 MeV. The calculated flux is $\Phi_\mu = (4.12 \pm 1.1 \pm 0.58) \times 10^{-13} cm^{-2} sr^{-1} s^{-1}$. This matches the expected flux using our 600 MeV threshold.

IX CONCLUSIONS

We have measured the flavour ratio of ratios (R) in atmospheric neutrino interactions using a 1.52 kton-year exposure of Soudan 2. We find $R = 0.67 \pm 0.15^{+0.04}_{-0.06}$. This value is about 2σ from the expected value of 1.0 and is consistent with the anomalous ratios measured by the Kamiokande and IMB experiments. We note that since our acceptance matrix is different from those of the water Cherenkov experiments we would not expect to measure the same value of R, unless R=1.

REFERENCES

1. M. Aglietta et al., Europhys. Lett. **8**, 611 (1989).
2. W.W.M. Allison et al., NIM **A376**, 36 (1996).
3. W.W.M. Allison et al., Physics Lett. **B391**, 491 (1997).

4. G. Barr et al., G. Barr,T.K. Gaisser and T. Stanev, Phys Rev **D39**, 3532 (1989).
5. R. Becker-Szendy et al., Phys. Rev. **D46**, 3720 (1992).
6. b J. Beiber, Bartol Research Institute, private communication.
7. K. Daum et al., Z. Phys. **C66**, 417 (1995).
8. Y. Fukuda et al., Phys. Lett. **B335**, 237 (1994).
9. H. Gallagher, H. Gallagher, Neutrino Oscillation Searches with the Soudan 2 detector, PhD Thesis, University of Minnesota (1996).
10. R. Merenyi et al., Phys. Rev. **D45**,743 (1992).
11. W.P. Oliver et al., NIM **A276**, 371 (1989).

The 1000ton Liquid Scintillation Detector Project at Kamioka (Kam-LAND[1])

F.Suekane [2]

Faculty of Science
Tohoku University
Sendai, 980-77, JAPAN

Abstract. We are constructing 1,000ton liquid scintillation detector at the old Kamiokande cave in order to detect low energy (anti)neutrinos from various sources. The main physics target of this experiment is to measure the neutrino oscillation parameter; Δm^2 down to $10^{-5}eV^2$ by detecting reactor antineutrinos coming from 150 to 200 km away. An outline of this experiment is explained in this paper.

INTRODUCTION

Super Kamioka is the world's largest water Cherenkov detector and is presently observing various rare event phenomena with signals above 6MeV. After long deliberations of the future of the old Kamioka detector, it has been decided to convert it into a companion facility as the world's largest liquid scintillation detector specializing in rare events occurring at lower energies. Because liquid scintillator generally produces more photoelectrons than Cherenkov light per unit energy and has lower background in terms of radioactive contaminations, it is possible to detect low energy (anti)neutrino events down to few MeV level. Many interesting phenomena occur at these energies, such as antineutrinos from nuclear reactors, those from the Earth, ^7Be solar neutrinos, and so on. Delayed coincidence technique can be used to identify electron-type antineutrinos, reducing background events significantly. As this is the world largest liquid scintillation detector located in

---

[2] Email address: suekane@awa.tohoku.ac.jp
[1] The collaborators are; A.Suzuki, J.Shirai, F.Suekane, A.Hasegawa, T.Iwamoto, H.Ogawa, S.Enomoto, S.Kawakami, K.Mashiko, K.Oki, D.Takagi, T.Itoh, O.Tajima, T.Ujiie (Tohoku University), M.Higuchi(Tohoku Gakuin University), T.Chikamatsu(Miyagi Women's University)

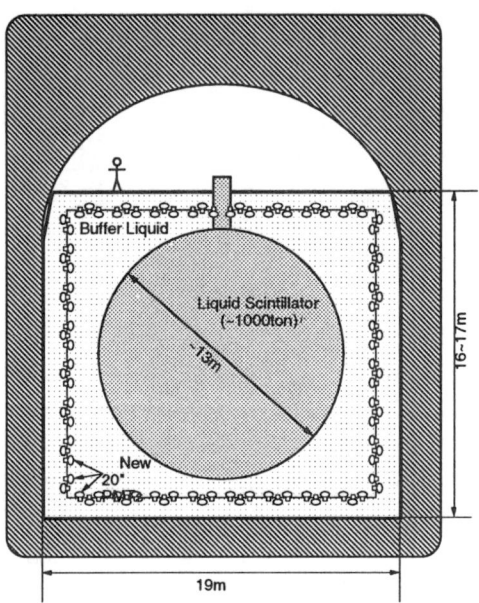

FIGURE 1. the Kam-LAND Detector

a very low background environment, we will be able to perform several low energy (anti)neutrino physics with the best sensitivities.

THE DETECTOR [1]

Fig.-1 shows a schematic view of the planned detector. The detector consists of $1,200m^3$ liquid scintillator surrounded by $3,300m^3$ buffer liquid. 1,500 new 20" photo multipliers look at the scintillator volume. The liquid scintillator will be a mineral oil base with the mineral oil concentration of 80% or more. Mineral oil has very good features such as, non-toxic, chemically stable, high flash point, transparent down to UV region, high H/C ratio, and very low radioactive contaminations. We are expecting to develop a scintillator with the light output of 30% Anthracene equivalent or more and light attenuation length of significantly longer than 7m. The liquid scintillator will be held in a large balloon with diameter of 13m. The buffer liquid is used as the shield for γ rays and neutrons come from detector walls and rocks, as well as cosmic ray veto counter. The material will be pure mineral oil to reduce the buoyant force of the scintillator balloon. A new 20" PMT is being developed to improve timing resolution to achieve a good position resolution. The PMTs will cover 25% of the effective area and 100 photoelectrons per 1MeV energy deposition in the scintillator are expected. Corresponding statistical energy resolution is 10% for 1MeV event. The tank will be made of stainless steel.

ANTINEUTRINO SIGNAL

Antineutrinos are detected using the delayed coincidence technique. That is, antineutrino produces a positron and a neutron through charged current interaction with a proton. The positron produces the prompt signal and the neutron produces the delayed signal when it is captured by a proton. The coincidence gate is a few hundred microsecond.

$$\bar{\nu}_e + p \to e^+ + n$$
$$\hookrightarrow n + p \to d + \gamma(2.2 MeV)$$

As an option, Gadolinium may be loaded in the liquid scintillator if accidental background rate turns out to be too high. In this case the neutron capture signal is 7.9MeV gamma rays and the coincidence gate can be set much shorter. The threshold of the charged current reaction is 1.8MeV. As the energy of the reactor antineutrino is very small compared to the proton mass, the antineutrino energy is directly related to the event energy; $E_{\bar{\nu}} = E_{obs} + 0.8 MeV$. This feature is very useful in studying deformation of the energy spectrum due to neutrino oscillation.

BACKGROUNDS

There expected to be several background sources which mimic the antineutrino signal. Uranium and Thorium contaminations in the liquid scintillator are expected to be less than 10^{-14} g/g and the potassium are expected to be less than 10^{-10} g/g. With this concentration, both the accidental and correlated background rate due to their decay chains and the fission are negligibly small. Rn has relatively short lifetime and its effect will become negligible a few months after the detector seal. The cosmic ray rate in the detector is roughly 0.3Hz and cosmic ray related backgrounds, such as prompt and delayed signal caused by spallations, will be reduced to a negligible level after 3s of dead time. The γ rays and neutrons from outside are shielded by 3m thick buffer liquid and outer scintillator shell region.

THE TEST BENCH

We are constructing an 1 ton scale liquid scintillator test bench detector. Basic studies of liquid scintillator properties, back ground properties, etc. are to be performed with it. After these studies, the detector will be brought to a reactor site in order to calibrate reactor antineutrino flux, event energy spectrum and detection efficiency using the same liquid scintillator used in the main detector.

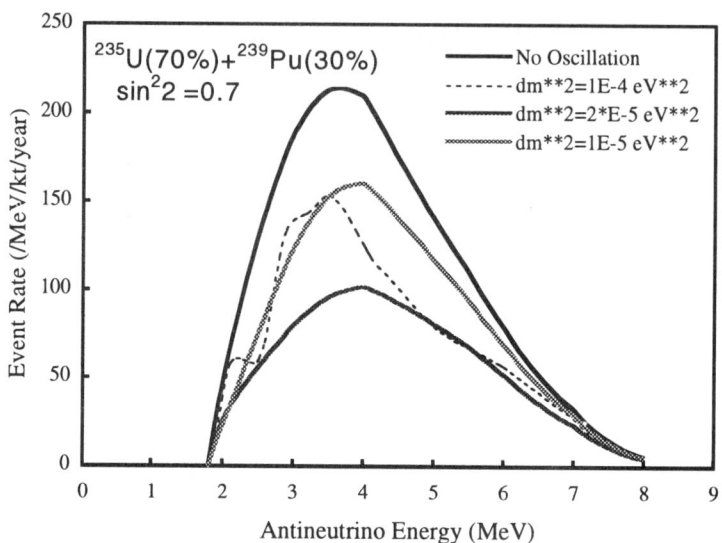

FIGURE 2. Event Energy Spectra with Oscillations

REACTOR ANTINEUTRINO OSCILLATION

In a nuclear reactor, antineutrinos are generated in beta decays of the fission products. A ^{235}U fission produces 6 antineutrinos accompanied by 200MeV of energy release in average. The energy spectra of reactor antineutrinos are shown in the references [2]. The typical energy of the antineutrino is a few MeV.

There are 17 commercial nuclear power plants in Japan, supplying one third (=120GW(thermal)) of the total electric power in the country [3]. At Kamioka, $3 \times 10^6/cm^2/s$ ($1 \times 10^6/cm^2/s$ for $E_{\bar{\nu}} \geq 1.8MeV$) of antineutrinos are coming from the reactors. Of them, 80% comes from the reactors located between 150 to 200km from Kamioka, that is, the flight distances are rather unique. The expected $\bar{\nu}p \rightarrow e^+n$ event rate is 700/kt/year for C_nH_{2n} target. The energy spectrum of the antineutrino event has the maximum at $E_{\bar{\nu}} = 4MeV$ as shown in the Fig.-2.

From these numbers, this experiment is sensitive at $\Delta m^2 > 10^{-5}eV^2$ and $\sin^2 2\theta > 0.2$ as shown in the Fig.-3. This covers both large angle MSW solution for the problem of solar neutrino deficit and $\nu_e - \nu_\mu$ mixing solution for atmospheric neutrino problem.

If neutrino oscillation exists, the event energy spectrum will be deformed as shown in the Fig.-2. As antineutrino energy can be measured, it is possible to measure this deformation to search for neutrino oscillation.

The reactor generating power has seasonal variation because energy demand is high in summers. There is at most 30% difference of antineutrino flux

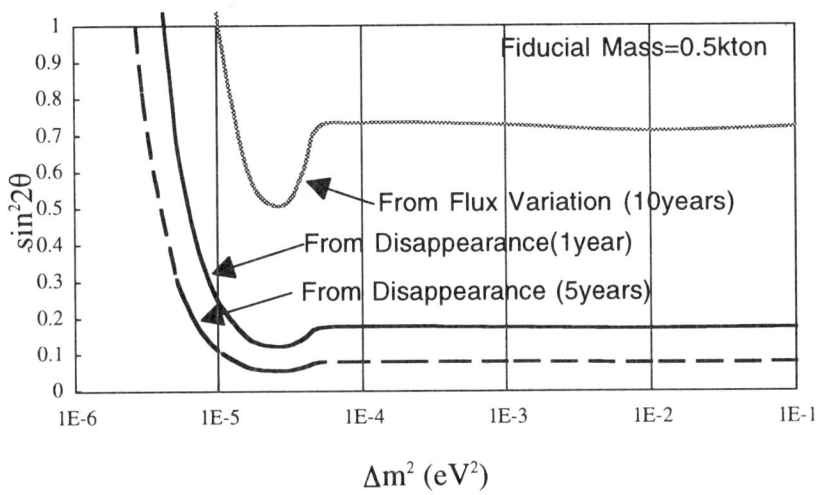

FIGURE 3. Sensitivity for the Neutrino Oscillation Parameters

between high flux month and low flux month. Making use of this difference, it is possible to perform background less-dependent search. The Fig.-3 also shows the sensitivities of this method.

OTHER PHYSICS POTENTIALS

Terrestrial Antineutrinos [4]

The earth is radiating 40TW of heat from the surface. Significant part of the energy is considered to come from decay energy of radioactive nuclei in the earth and there has been several discussions on possibilities of terrestrial neutrino detection [5]. According to some model, $2 \times 10^7/cm^2/s$ ($5 \times 10^5/cm^2/s$ for $E_{\bar{\nu}} \geq 1.8 MeV$) of effective antineutrinos flux come from the earth and we will see 60/kt/year of the terrestrial antineutrino events. Fig.-4 shows the terrestrial and reactor antineutrino event spectra at Kamioka. The observation of the terrestrial antineutrinos is important because they carry direct information of inner structure of the earth.

The Solar Neutrinos

There coming larger number of neutrinos from the sun than that of reactor antineutrinos. If the background is small enough to set the threshold energy at 0.25MeV, we will see 500/day/kt of ^7Be, pep, ^{13}N , ^{15}O neutrino events [6].

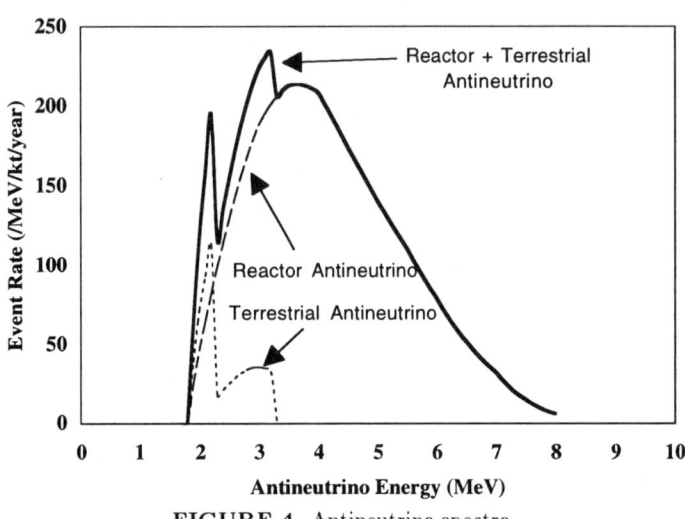

FIGURE 4. Antineutrino spectra

(Anti)Neutrinos from Supernova Burst

If there is a supernova burst at the distance of 10kpc, we will see 300 $\bar{\nu}_e p$ events within a few seconds [7].

The Double Beta Decays [8]

Liquid scintillators can dissolve Xe with relatively high concentration. By making use of the low background environment, it is possible to measure Majorana neutrino mass of $0.2 \sim 0.3 eV$ by loading hundred kilograms of ^{136}Xe in the liquid scintillator.

SCHEDULE

This project was approved in April, 1997. We will finish the construction by the year 2000 and will produce the first physics output from the year 2001.

SUMMARY

We are constructing 1000ton liquid scintillator detector at the old Kamiokande cave. The main physics target is to measure neutrino oscillation parameter down to $\Delta m^2 \sim 10^{-5} eV^2$ using reactor antineutrinos. There are several other physics topics. The construction will be finished by the year 2000 and we expect to produce the first physics output from the year 2001.

REFERENCES

1. A.Suzuki, talk at 'the Conference on Neutrino Physics at Miyako', Tohoku-HEP-Note-97-01, (1997)
2. For example, P.Vogel, et al., Phys. Rev. C, vol24, 1543(1981)
3. The Federation of Electric Power Companies/JAPAN, 'Status of The Electric Power Industry; 1996-1997'.
4. R.S.Raghavan, S.Achoenert, A.Suzuki, F.Suekane, J.Shirai, to be published.
5. G.Eder, Nucl.Phys., 78, 657 (1966), M.Kobayashi, et al., Geophys. Research Lett. vol.18, 633(1991), L.M.Krauss, et al. , NATURE vol.319, p191 (1984), G. Marx, Czech.J.Phys., B19, 1471(1969).
6. Borexino Proposal (INFN, Italy) 1992.
7. K. Hirata, et al., Phys. Rev. Lett., 58,1490(1987).
8. R.S.Raghavan, Phys. Rev. Lett., vol.72, 1411(1994).

Particle and Nuclear Astrophysics

Very High Energy Gamma-ray Astronomy

Michael Catanese

Dept. of Physics and Astronomy, Iowa State University, Ames, IA 50011, USA

Abstract. Very high energy (VHE) γ-ray astronomy (E>250 GeV) has, in recent years, established its place as an astrophysical discipline. The growing catalog of galactic and extragalactic sources, and the non-detection of prominent astrophysical objects, have impacted theories ranging from the environments of pulsars and active galactic nuclei (AGN) to the origins of cosmic rays. The current status of the field is reviewed, with particular emphasis on observations of AGN and their implications for AGN emission models.

INTRODUCTION

Very high energy γ-ray astronomy currently covers the energy range from approximately 300 GeV to 50 TeV. As such, it provides a complement to the observations between 30 MeV and 30 GeV covered by the Energetic Gamma-Ray Experiment Telescope (EGRET) on the Compton Gamma-Ray Observatory (CGRO). The low fluxes emitted above 300 GeV by cosmic sources require very large effective areas to be observed so VHE detectors are necessarily ground-based. The techniques employed at these energies detect the particles or Čerenkov light of air showers produced when γ-rays and cosmic rays interact in the atmosphere.

Results obtained with the imaging atmospheric Čerenkov technique (see below) have established the field of VHE γ-ray astronomy as an astrophysical discipline with very statistically significant detections of objects using standard analyses predicted to work by Monte Carlo simulations. With these detections, the lack of VHE emission from some other objects has become more constraining as well. The physical processes needed to produce γ-rays with $E \sim 1$ TeV are well studied in particle and nuclear physics so those results are employed by astrophysical theorists to explain the underlying environments of the celestial objects studied. In addition, the techniques for detecting γ-rays are derived from particle and nuclear physics experiments.

In this paper, we first give a very brief description of the imaging atmospheric Čerenkov technique and then discuss the observation status of the galactic sources of VHE γ-rays as well as some intriguing non-detections. Next we give some recent results on extragalactic sources and close by discussing what lies in the future for this field.

IMAGING ATMOSPHERIC ČERENKOV TECHNIQUE

In the imaging atmospheric Čerenkov technique [1], large optical reflectors focus the Čerenkov light flash produced by air showers onto an array of (~100-500) densely packed phototubes which view ~3° to 5° diameter regions of the sky. The Čerenkov flash is dim enough that the telescopes can only be operated at night when the Moon is below the horizon, at least without a filter. Because the showers develop several km up in the atmosphere, clear skies are also needed to detect the Čerenkov light. These telescopes have huge effective areas (~50,000 m^2) limited only by the size of the Čerenkov light pool. Thus, they are excellent monitors of short-term variations in γ-ray flux and are sensitive to very low fluxes of γ-rays. They have good angular resolution (FWHM $\lesssim$ 0°.3) for a γ-ray telescope so source location can be accurate to about 0°.05, reducing source-confusion problems which can plague lower energy γ-ray telescopes.

The Whipple Observatory γ-ray telescope [2] is still the standard in this field after 25 years of operation and the Whipple collaboration has led the effort in establishing the imaging technique. In recent years, several other groups have also built imaging Čerenkov telescopes (see [3] for a review) and they are beginning to push this field along as well.

GALACTIC SOURCE OBSERVATIONS

Plerions

Plerions are supernova remnants (SNRs) containing a central radio pulsar surrounded by a nebula. The energy source for these objects is the pulsar's rotational energy. The Crab is the most-studied of these objects and has become a standard candle for high energy telescopes. The plerions which have been observed to emit VHE γ-rays are listed in Table 1. This class of object seems to represent the brightest class of galactic VHE-emitting source. The VHE emission is apparently constant so they are good test sources for developing new techniques and detectors.

All of the sources listed in Table 1 are also detected in the 30 MeV - 30 GeV energy range by the EGRET experiment [7]. Interestingly, the emission

TABLE 1. Pulsars detected at VHE energies

Source	Flux ($E > 1$ TeV)	Ref.
Crab	$(2.2 \pm 0.5) \times 10^{-11}$cm^{-2}s^{-1}	[4]
PSR 1706-44	0.8×10^{-11}cm^{-2}s^{-1}	[5]
Vela		[6]

at EGRET energies is almost completely pulsed at the radio frequency of the central pulsar while the VHE emission has no detected pulsed component. The most common explanation for this is that the unpulsed VHE emission is produced by electrons upscattering low energy photons in the synchrotron nebula surrounding the pulsar [8]. The spectrum of the VHE emission can then be used to estimate the magnetic field in the nebula [9]. The origin of the pulsed emission is not agreed upon, but the detection of the end of pulsed spectrum would help decide between the competing models [10,11].

Shell-type Supernova Remnants

Shell-type SNRs do not generally have a central radio pulsar and the energy source for the emission from these objects is the supernova blast itself. SNRs are generally believed to be the sources of cosmic rays with energies below ~100-1000 TeV. For instance, supernova blast shocks are one of the few galactic sites capable of supplying the energy needed to maintain the cosmic ray population within the galaxy [12]. Also, the diffusive shock acceleration which occurs in the supernova shock wave naturally gives a power-law spectrum similar to the measured cosmic-ray spectrum after accounting for propagation effects [13].

If SNRs do produce cosmic rays, they also should emit γ-rays from the decay of π^0's produced from interactions between the cosmic rays and interstellar matter. Drury, Aharonian, and Völk [14] predict γ-ray fluxes which should be detectable by EGRET or a telescope with the sensitivity of the Whipple Observatory γ-ray detector. In fact, 3 SNRs are associated with EGRET unidentified sources [15]. The detection of VHE emission with the expected spectrum would be convincing evidence that SNRs are the source of cosmic rays in our galaxy.

However, initial observations of SNRs with the Whipple Observatory γ-ray telescope have not detected VHE emission [16]. Figure 1 shows the upper limits from the Whipple observations of 6 SNRs chosen as promising γ-ray candidates. Figure 1 indicates that the Whipple upper limits are below the extrapolation from the EGRET integral flux points and toward the low end of the predicted range of allowed fluxes. Also, EGRET's spectra are steeper than expected for the π^0 decay spectrum. Gaisser, Protheroe and Stanev [17] determine that these results imply that the source spectrum must be $\sim E^{-2.4}$

and that electron bremsstrahlung radiation dominates the low energy emission ($\lesssim 100$ MeV). A cutoff in the cosmic-ray spectrum below 10 TeV could also account for the non-detections by the Whipple Observatory, but that explanation and a source spectrum steeper than $\sim E^{-2}$ contradict expectations for the γ-ray emission from shock acceleration in the SNRs [16].

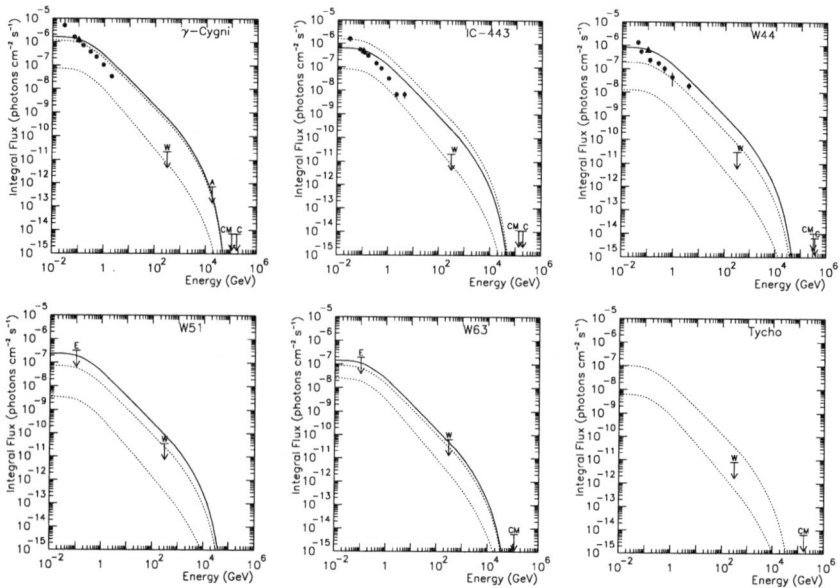

FIGURE 1. Whipple Observatory upper limits (W) shown with EGRET integral fluxes (E), and integral spectra. Extrapolations from the EGRET integral data points (solid curves), as well as an estimate of the allowed range of fluxes (dotted lines) are also shown. Upper limits from the CASA-MIA (CM) and Cygnus (C) air shower arrays, and the AIRO-BICC array (A) are shown as well. Figure adapted from [16].

EXTRAGALACTIC SOURCE OBSERVATIONS

One of the most surprising and exciting developments in γ-ray astronomy has been the detection of more than 50 extragalactic sources of γ-rays by the EGRET experiment [7]. With the exception of the Large Magellanic Cloud, these sources are all blazars. Blazars are AGN in which a relativistic jet is oriented nearly along our line of sight. The emission derives from this jet, so it is Doppler boosted in energy and luminosity. The Doppler boosting also reduces the apparent time scales of the variations in the emission.

The emission from blazars appears to consist of a low energy synchrotron component which extends from radio up to optical or even X-ray wavelengths [18], and a high energy component which in some cases extends to γ-ray wavelengths. The origin of the high energy emission is a matter of active debate but the main theories fall into two categories: electron models where the high energy emission is the result of inverse Compton scattering of low energy photons by the same electrons which produce the synchrotron emission [19,20] and proton models where the high energy emission is the result of a cascade initiated by protons interacting with the low energy synchrotron photons [21].

The only EGRET blazar detected above 300 GeV is the BL Lac object Mrk 421 [22]. The other unequivocally detected VHE-emitting blazar, Mrk 501 [23], is not detected by EGRET despite being quite similar to Mrk 421 at radio to soft X-ray wavelengths. One other blazar, 1ES 2344+514, has shown evidence of VHE emission [24] but requires confirmation. 1ES 2344+514 is also not an EGRET source.

These objects are all nearby ($z < 0.05$) so their VHE emission is not severely attenuated by pair-production with intergalactic infrared radiation fields [25,26]. This interaction almost certainly plays a role in the non-detection of most of the EGRET blazar sources which typically have redshifts >0.2. Also, the X-ray energy output for these objects is comparable to or larger than their optical energy output. In the electron models, this high X-ray emission implies a maximum electron energy which is higher than in other blazar types (like those typically seen by EGRET) which in turn implies higher maximum γ-ray energy [19]. Thus, it may be that the other types of blazar do not produce particles with enough energy to generate VHE γ-rays.

The VHE emission from these objects is characterized by rapid variability, a low mean level of emission relative to the Crab Nebula, and a spectrum which extends from 300 GeV to 5 TeV consistent with a single power law [27]. This latter feature constrains the density of the extragalactic IR emission at some wavelengths to well below that of direct IR measurement experiments [28] and eliminates some models of how the background is created [27].

The VHE emission of Mrk 421 appears to consist entirely of flaring activity with no steady emission at the sensitivity limit of existing telescopes [29]. The time scales of the variations of the VHE emission from Mrk 421 are very short. The most spectacular were observed in 1996 May (see Figure 2) [30]. In the first flare, on May 7, the flux doubled in ~1 hour and peaked at a flux 10 times that of the Crab Nebula and more than 5 times higher than had ever been observed before. The following night its flux had returned to its nominal flux level. In the second event, on May 15, the entire flare, which peaked at a flux 4.5 times the Crab Nebula flux, lasted *only 30 minutes*. The doubling time of this flare is ~10 minutes, implying an emission region, once Doppler boosting is accounted for, of only 100 minutes: about the same as the mean radius of the orbit of Uranus. This may imply that we are viewing processes very close to the base of the AGN jet.

The variation of the VHE emission from Mrk 501 is less rapid and of lower amplitude than that of Mrk 421, but in some respects is more extreme. As

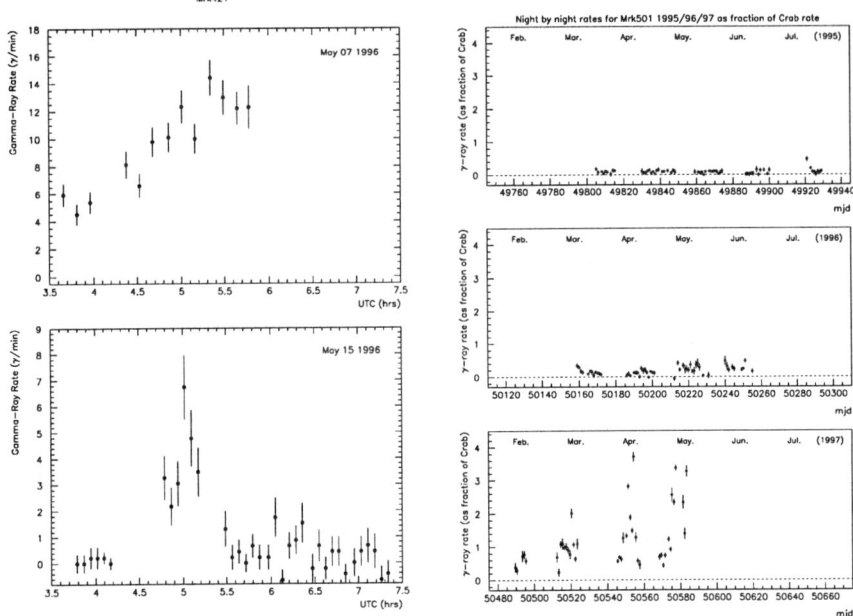

FIGURE 2. Left: Temporal histories of two γ-ray flares observed from Mrk 421 in 1996 by the Whipple Observatory. Top figure shows the flare observed on May 7 and the bottom shows the flare observed on May 15 (from [30]). Right: Temporal history of 3 years of VHE observations of Mrk 501 with the Whipple Observatory γ-ray telescope (from [31]).

shown in Figure 2, the first two years of observations of Mrk 501 at the Whipple Observatory showed that its emission was variable, but unlike Mrk 421, its emission never disappeared [31]. In 1997, Mrk 501 has been in an active high state which is still continuing after 5 months. During this time, the mean flux of Mrk 501 has been ∼1.5 times that of the Crab Nebula, making it the brightest object in the VHE sky. Curiously, despite this high activity, no hour-scale flares of the type seen in Mrk 421 have been observed.

As interesting as the VHE variability is in itself, the best tests of the emission models for these objects come from observations of flaring activity by several telescopes operating at different wavelengths. Figure 3 shows the two best multi-wavelength measurements of Mrk 421 [29] and Mrk 501 [32] obtained to date. The correlated variability of the VHE γ-rays and the other wavelength observations imply that they are produced in the same region of the jet. This information can then be used to estimate the Doppler factor of the jet and the magnetic field of the emission region [29,32]. Also, the detection of a very strong signal from Mrk 501 in the 50 keV to 150 keV range with the Oriented Scintillation Spectrometer Experiment (OSSE) of CGRO and the lack of a

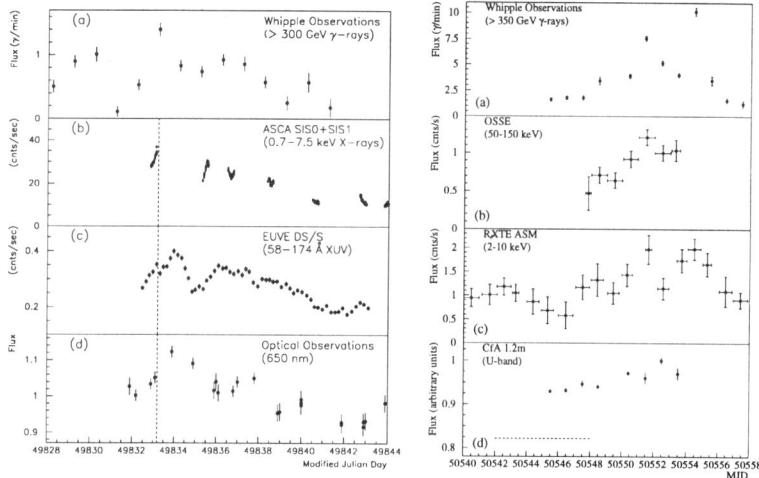

FIGURE 3. Multiwavelength temporal history of a flare from Mrk 421 (left) in 1995 April (from [29]) and one from Mrk 501 (right) in 1997 (from [32]).

detection for Mrk 421 by OSSE [33] may imply a higher maximum energy for γ-rays from Mrk 501 [19] if electrons are responsible for the γ-rays.

THE FUTURE

Very high energy γ-ray astronomy has begun to take its proper place in the field of high energy astrophysics, as evidenced by the results given in this paper. Still, there is much to be done in the field. For instance, the gap in energy between the satellite experiments and the ground-based telescopes must be closed in order to understand why the γ-ray sky has over 150 sources at $\sim$100 MeV while above 300 GeV there are only 5 so far. Also, larger fields of view need to be covered by VHE telescopes in order to enable all-sky surveys for unexpected sources of γ-rays. More consistent monitoring of the variable sources are also needed to get better pictures of their long-term behavior. The development of third generation VHE γ-ray telescopes is underway to overcome some of these deficiencies. Solar arrays such as STACEE [34] and CELESTE [35] and large single dish mirror telescopes like MAGIC [36] will attempt to close the energy gap. Arrays of telescopes like the proposed VERITAS experiment [37] will lower the energy threshold to 50-100 GeV, improve the sensitivity of the technique by an order of magnitude and also enable all-sky surveys and more consistent monitoring of variable sources. Finally, the MILAGRO experiment [38] will enable monitoring of the entire northern hemisphere sky above 1 TeV, 24 hours per day. These advancements make the future of VHE γ-ray astronomy seem very promising.

REFERENCES

1. Cawley, M. F., and Weekes, T. C., *Exp. Astron.* **6**, 7 (1995).
2. Cawley, M. F., et al., *Exp. Astron.* **1**, 173 (1990).
3. Weekes, T. C., *Space Sci. Rev.* **75**, 1 (1996).
4. Weekes, T. C., et al., *ApJ* **342**, 379 (1989).
5. Kifune, T., et al., *ApJ* **438**, L91 (1995).
6. Yoshikoshi, Ph.D. thesis, Univ. of Tokyo (1996).
7. Thompson, D. J., et al., *ApJS* **101**, 259 (1995).
8. deJager, O. C., and Harding, A. K., *ApJ* **396**, 161 (1992).
9. Hillas, A. M., et al., in preparation (1997).
10. Daugherty, J. K., and Harding, A. K., *ApJ* **458**, 278 (1996).
11. Cheng, K. S., Ho, C., and Ruderman, M. A., *ApJ* **300**, 500 (1984).
12. Drury, L. O'C., Markiewicz, W. J., and Völk, H. J., *A&A* **225**, 179 (1989).
13. Swordy, S. P., et al., *ApJ* **349**, 635 (1990).
14. Drury, L. O'C., Aharonian, F. A., and Völk, H. J., *A&A* **287**, 959 (1994).
15. Esposito, J. A., et al., *ApJ* **461**, 820 (1996).
16. Buckley, J. H., et al., *A&A*, submitted (1997).
17. Gaisser, T. K., Protheroe, R. J., Stanev, T., *ApJ*, submitted (1997).
18. Blandford, R. D., & Königl, A., *ApJ* **232**, 34 (1979).
19. Königl, A., *ApJ* **243**, 700 (1981).
20. Sikora, M., Begelman, M. C., and Rees, M. J., *ApJ* **421**, 153 (1994).
21. Mannheim, K., *A&A* **269**, 67 (1993).
22. Punch, M., et al., *Nature* **358**, 477 (1992).
23. Quinn, J. et al., *ApJ* **456**, L83 (1996).
24. Catanese, M., et al., in Proc. 25th ICRC, in press (1997).
25. Gould, J. R., & Schréder, G. P., *Phys. Rev.* **155**, 1408 (1967).
26. Stecker, F. W., de Jager, O. C., and Salamon, M., *ApJ* **415**, L71 (1993).
27. Zweerink, J., et al., *ApJ Letters*, submitted (1997).
28. Biller, S. D., et al., *ApJ* **445**, 227 (1995).
29. Buckley, J. H., et al., *ApJ* **472**, L9 (1996).
30. Gaidos, J. A., et al., *Nature* **383**, 319 (1996).
31. Quinn, J., et al., in *Proc. of the 25th ICRC* (Durban), in press (1997).
32. Catanese, M., et al., *ApJ Letters*, submitted (1997).
33. McNaron-Brown, K., et al., *ApJ* **451**, 575 (1995).
34. Quebert, J., et al., in *Towards a Major Atmospheric Cherenkov Detector - IV* (Padua), ed. M. Cresti, 248 (1995).
35. Ong, R. A., et al., *Nucl. Inst. Meth. A*, submitted (1997).
36. Mirzoyan, R., in *Proc. Int. Symp. on Extremely High Energy Cosmic Rays* (Tanashi, Japan), ed. M. Nagano, 205 (1996).
37. Weekes, T. C., et al., in *Proc. 25th ICRC* (Durban), in press (1997).
38. Yodh, G. B., *Space Sci. Rev.* **75**, 199 (1996).

The rp-process in X-ray bursts

H. Schatz*, L. Bildsten†, J. Görres*, M. Wiescher*,
F.-K. Thielemann‡

*University of Notre Dame, Dept. of Physics, Notre Dame, IN 46556
†University of California, Dept. of Physics and Dept. of Astronomy, Berkeley, CA 94720
‡Universität Basel, Institut für Physik, CH-4056 Basel, Switzerland

Abstract. The rp-process in X-ray bursts is investigated using a complete and updated nuclear reaction network including nuclei between H and Sn. This allows for the first time reliable calculations of the rp-process nucleosynthesis beyond ^{56}Ni. We find that for a 25 s burst the reaction flow reaches already Cd. The consequences for energy production, final composition of the ashes and fuel consumption are discussed.

I INTRODUCTION

Type I X-ray bursts are thermonuclear flashes on the surface of accreting neutron stars [1-3]. The accreted hydrogen and helium are ignited by a thermal fluctuation and burn explosively in a thermonuclear runaway. Under these conditions helium is burned via the 3α-reaction and the αp-process (a sequence of (α,p) and (p,γ) reactions, [4]), which provides seed nuclei for the hydrogen burning via the rp-process [5] (rapid proton capture and β-decays). For a long time the double magic nucleus ^{56}Ni was considered to be the endpoint of the rp-process in X-ray bursts, since a captured proton is so weakly bound that photodisintegration can remove it efficiently. In fact most of the previously used reaction networks ended at ^{56}Ni. It was argued, however, that the rp-process might well continue beyond ^{56}Ni and that this would have interesting consequences. Most important is the long standing problem of occasionally observed extremely short burst intervals of the order of several minutes. These burst intervals are too short to accrete enough fuel to power the second burst and can therefore not be explained by the simple thermonuclear flash model. Several solutions to this problem were offered assuming that the second burst is powered by fuel left over from the first burst [4,6]. [7] speculated however, that taking into account processing beyond ^{56}Ni would consume all hydrogen in a single burst and presented therefore an alternative explanation based on mixing of unburned fuel from outer layers into the burning zone. The rp-process beyond ^{56}Ni would also dramatically change the composition of the ashes of the nuclear burning and thus the composition of the outer crust of the neutron star, which would affect neutron star seismology calculations.

The development of nuclear reaction networks beyond ^{56}Ni was hampered in the past by the lack of experimental and theoretical information about the very unstable nuclei at the proton drip line. Nevertheless a few exploratory studies were performed by several authors that either ended at Se [8] or Y [5], or used only 16 nuclei between H and Cd [9]. Given the different results of these studies and the lack of detailed nucleosynthesis calculations beyond Y it seemed to be necessary to investigate the rp-process between Ni and Sn with an updated and complete reaction network.

II NETWORK CALCULATIONS

In a first step [10] calculated the rp-process nucleosynthesis between H and Kr for constant temperatures and densities. We extended their network up to ^{100}Sn [11] and added the most recent experimental information on the crucial isotopes ^{65}As, ^{69}Br, and ^{73}Rb. Charged particle reaction rates were calculated with the Hauser-Feshbach method and β-decay half-lives using the QRPA. We also included 2p-capture reactions that bridge proton unbound nuclei by proton capture on an equilibrium abundance of the unbound nucleus similar to the 3α-reaction. These reactions are typically slow, but they can well compete with the slow β-decays of ^{68}Se and ^{72}Kr thus accelerating the rp-process considerably. The reaction network was coupled to a 1 dimensional, 1 zone X-ray burst model that calculates temperature and density in the burning zone assuming constant pressure [12]. The raise of the burst is calculated selfconsistently from the generated nuclear energy, while during the burst decline the temperature curve is matched to typical observed luminositiy profiles. In all calculations we assumed an initial solar composition.

III RESULTS

A Reaction flow and energy production

Fig. 1 shows the energy production rate for a burst with a time constant of 25 s for the exponential decline of the luminosity (luminosity timescale). A striking feature are the pronounced structures that are a result of the fact that the energy production drops drastically each time a nucleus is reached, where further particle induced reactions are slow (waiting point). Therefore, each peak in the energy production rate corresponds to a waiting point (see Fig. 1). The reaction flow during the raise of the burst proceeds via the 3α-reaction to

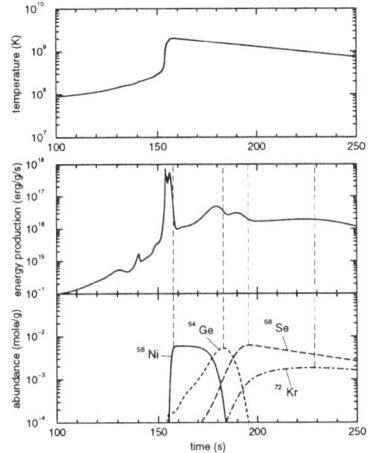
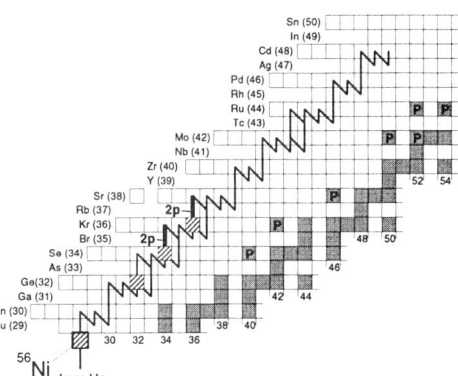

FIGURE 1. Temperature, energy production, and the abundances of the waiting points above $A = 56$ as a function of time during the burst.

FIGURE 2. The reaction flow during the burst decline. Only the strongest flows are shown. Hatched squares indicate waiting points, and filled squares stable isotopes.

^{12}C, via the αp-process to ^{41}Sc and then via the rp-process to ^{56}Ni. Waiting points along the path cause some structure in the energy production rate but are not discussed here. Processing beyond ^{56}Ni does not take place during the raise of the burst, since at the time when the reaction flow reaches ^{56}Ni the temperature is already so high, that photodisintegration inhibits any further net proton capture.

The rp-process continues beyond ^{56}Ni during the burst decline when the temperature has dropped below approximately 1.8 GK. Then, as can be seen in Fig. 1, ^{56}Ni is rapidly depleted and the rp-process proceeds into the Cd-region. The calculated reaction path is shown in Fig. 2. Important waiting points are ^{64}Ge, ^{68}Se, and to some extent ^{72}Kr. As can be seen in Fig 2. the previously neglected 2p-capture reactions on ^{68}Se and ^{72}Kr carry a significant fraction of the reaction flow. It is also interesting to note that 60% of the total burst energy are produced by processing beyond ^{56}Ni during the tail of the burst.

B Final composition

The final composition of the ashes can be found in Fig. 3. Obviously, the endpoint of the rp-process is neither the iron group nor a specific isotope at $A = 64$ or $A = 68$ as it had been suggested before, but a range of nuclei between $A = 68$ and $A = 100$ with a relatively flat distribution. Some material is already accumulated at the end of the network at $A = 100$.

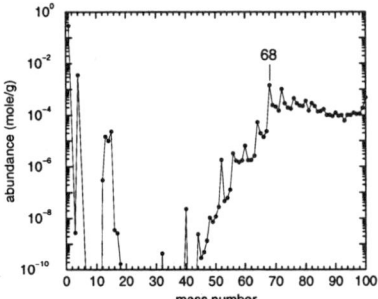

FIGURE 3. The final isotopic abundances summed in each mass chain as a function of mass number.

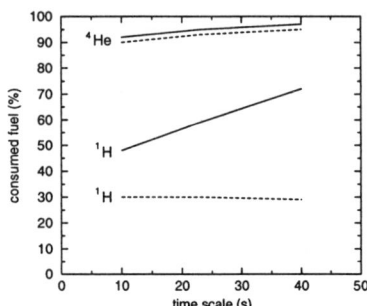

FIGURE 4. The fraction of consumed H and ^{4}He fuel. The dashed lines indicate results from a smaller network ending at ^{56}Ni.

C   Fuel consumption

In Fig. 4 we calculated the consumption of ^{1}H and ^{4}He was as a function of the luminosity timescale. Since the reaction flow reaches the end of the network already for a 25 s burst, the calculations at larger timescales underestimate the fuel comsumption somewhat. Shown are the results for our full reaction network as well as for a network ending at ^{56}Ni. It can be seen that ^{4}He is consumed almost completely, independent on burst duration and network size. This is a consequence of the fact that most (70%) of the ^{4}He is rapidly burned during the few seconds in the raise of the burst, and that all α-induced reactions occur below ^{56}Ni owing to the Coulomb barrier. On the other hand, Fig. 4 also shows that the use of the full reaction network is essential for calculating the H-consumption. However, even with a full reaction network significant amounts of H remain unburned (30-50%). This can be understood from the fact that the transition from the αp-process into the rp-process occurs approximately at ^{41}Sc. Therefore ^{41}Sc is the seed nucleus for the rp-process. Since 10 α-particles (and one proton) are consumed to produce one ^{41}Sc nucleus, and since the H/He number ratio is about 10 (solar value), the proton to seed ration is about 100. From this it follows that in order to consume all hydrogen, the rp-process would have to proceed on average up to $A = 141$. This is not possible, since the here considered burst timescales of 10-40 s are much shorter than the timescale of the rp-process owing to the slow β-decays along the reaction path. Therefore, significant amounts of hydrogen remain unburned and the H-consumption becomes roughly proportional to the burst timescale.

IV CONCLUSIONS

We developed the first complete and updated nuclear reaction network for rp-process calculations beyond ^{56}Ni up to ^{100}Sn. As a first step we used a simple X-ray burst model to investigate the influence of nuclear burning beyond ^{56}Ni. In contrast to [5] and [4] we find that the rp-process in typical X-ray bursts does not end at ^{64}Ge or ^{56}Ni but reaches at least nuclei around $A = 100$. The rp-process beyond ^{56}Ni produces 60% of the burst energy and waiting points cause pronounced structures in the energy production rate during the burst tail. It would be interesting to determine under what conditions these structures are observable. This would provide a unique opportunity to use X-ray burst luminosity profiles to constrain properties of very proton rich nuclei that are mostly not yet accessible by nuclear physics experiments. We also find that the ashes of the nuclear burning in X-ray bursts, and therefore the outer crust of an accreting neutron star does not consist of iron group nuclei as assumed previously, but of a mostly flat distribution of isotopes between $A = 68$ and $A = 100$. However, in contrast to speculations by [7] the synthesis of these heavier nuclei does not lead to a complete consumption of hydrogen. As a consequence, the models for the explanation of very short burst intervals that rely on the assumption that some hydrogen remains unburned are probably still viable. However, the amount of unburned hydrogen might be lower than assumed.

REFERENCES

1. S. E. Woosley and R. E. Taam, Nature **263**, 101 (1976).
2. L. Maraschi and A. Cavaliere, Highlights of Astronomy **4**, 127 (1977).
3. P. Joss, Nature **270**, 310 (1977).
4. S. E. Woosley and T. A. Weaver, in *High Energy Transients in Astrophysics*, Vol. 115 of *AIP Conference Proceedings*, edited by S. E. Woosley (American Institute of Physics, New York, 1984), p. 273.
5. R. K. Wallace and S. E. Woosley, *Ap. J. Suppl.* **45**, 389 (1981).
6. R. E. Taam *et al.*, *Ap. J.* **413**, 324 (1993).
7. M. Y. Fujimoto *et al.*, *Ap. J.* **319**, 902 (1987).
8. T. Hanawa, D. Sugimoto, and M.-A. Hashimoto, *Pub. Astr. Soc. Japan* **35**, 491 (1983).
9. R. K. Wallace and S. E. Woosley, in *High Energy Transients in Astrophysics*, Vol. 115 of *AIP Conference Proceedings*, ed. S. E. Woosley (American Institute of Physics, New York, 1984), p. 319.
10. L. Van Wormer *et al.*, *Ap. J.* **432**, 326 (1994).
11. H. Schatz *et al.*, *Phys. Rep.*, (1997), in press.
12. L. Bildsten, in *The Many Faces of Neutron Stars*, ed. A. Alpar, L. Buccheri, and J. Van Paradijs, (Dordrecht: Kluwer, 1997), in press

Neutrino Capture and r-Process Nucleosynthesis

Bradley S. Meyer

Department of Physics and Astronomy
Clemson University
Clemson, South Carolina 29634-1911

Abstract. Neutrino capture effects on free nucleons and heavy nuclei in r-process nucleosynthesis are discussed. Neutrino capture on heavy nuclei can "accelerate" the r-process; however, the dominant role neutrino capture plays is to set the neutron-to-proton ratio. In fact, network calculations indicate that neutrino capture on heavy nuclei is unlikely to be significant for r-process components that produce nuclei with mass number A near 195. The reason is that if the neutrino flux is sufficiently large to cause neutrino capture on heavy nuclei to be important, it was even larger in a earlier phase when ^{4}He and neutrons dominated the abundances and capture on the free neutrons drastically reduced the neutron-to-seed-nucleus ratio.

INTRODUCTION

The "r-process" is the name given to the nucleosynthesis process responsible for the production of roughly half of the naturally occurring isotopes in the solar system with mass number A greater than 70 [1]. The "r" is a mnemonic for "rapid", which indicates that the r-process occurs in an environment with a sufficiently high density of neutrons that neutron captures on heavy nuclei take place more *rapidly* than nuclear beta decay. In such an environment, nuclei considerably neutron rich of the beta-stable nuclei are created. These nuclei beta decay and then capture more neutrons, thereby increasing their mass. Once the neutrons disappear, the nuclei beta-decay back to stability, thereby giving rise to the abundance of certain neutron-rich stable species. For a review of the r-process, see [2,3].

The high densities and temperatures and short timescales inferred for the r-process strongly suggest it occurs in explosive events. The precise r-process site, however, remains a mystery. The most plausible site yet discussed is the neutrino-driven wind in core-collapse supernova explosions of massive stars [4–9]. In this scenario, neutrinos from a Kelvin-Helmholtz cooling nascent neutron star drive a tenuous wind from the star's surface. This wind may have the right combination of entropy, neutron richness, and expansion timescale to produce r-process isotopes. It "blows" for some tens of seconds after the collapse of the massive star's core.

A particularly fascinating aspect of this site is that it is probably the only place in the present universe where neutrinos completely dictate what happens. The neutrinos heat matter to high entropies, which in turn sets the pressure and determines the outflow velocity. They also set the neutron richness of the system via neutrino captures on free neutrons and protons. Because r-process yields are so sensitive to the above parameters set by the neutrinos, they may provide key diagnostics of the neutrinos and the explosion itself, if indeed this is the r-process site. The present brief paper explores what r-process yields might tell us about supernova dynamics via neutrino capture on nuclei.

KEY CONCEPTS

Two important mental pictures are important for discussing the neutron-driven wind scenario for the r-process. First is the wind itself. The wind begins at the surface of the nascent neutron star, which at this epoch has a radius of 10 − 100 km. The matter moves out radially at speeds of order 100−10,000 km/s, giving a mass-loss rate early ($\sim$ 1 second after core bounce) of order 10^{-2} $M_\odot$ s^{-1}. After several seconds, as the neutron star shrinks and falls deeper into its gravitational well, it becomes increasingly difficult to drive matter from the star, and the mass-loss rate drops to $\lesssim 10^{-5}$ $M_\odot$ s^{-1}, eventually declining to zero. The temperature near the neutron star surface is several ($\sim$ 5) MeV but drops rapidly in a wind element as it moves out from the star. Figure 1 shows the temperature $T_9 = T/10^9$ K as a function of a wind element's distance r from the center of the neutron star for a plausible wind trajectory. It was obtained for an adiabatic expansion of a wind element with an entropy per nucleon of 350 in units of Boltzmann's constant and traveling at 3×10^8 km/s. This trajectory is used throughout the rest of this paper. Notice that since the entropy is so high, the mass density $\rho \propto T^3$.

The other important picture is the basic nuclear dynamics in the expanding matter. Because the material is initially at such high temperature, it is comprised of free neutrons and protons. As the temperature falls, however, the free nucleons assemble into ^{4}He and some leftover neutrons since the matter is neutron rich. Some of the ^{4}He nuclei assemble into heavier nuclei with $A \geq 12$. Eventually, the matter consists of a few heavy nuclei and many neutrons and ^{4}He nuclei. Since ^{4}He does not capture neutrons, the many neutrons must capture onto the few heavy nuclei during the r-process phase of the expansion. The more neutrons per heavy seed nucleus, the more neutrons a given nucleus can capture during the r-process and the heavier the resultant nuclei can be.

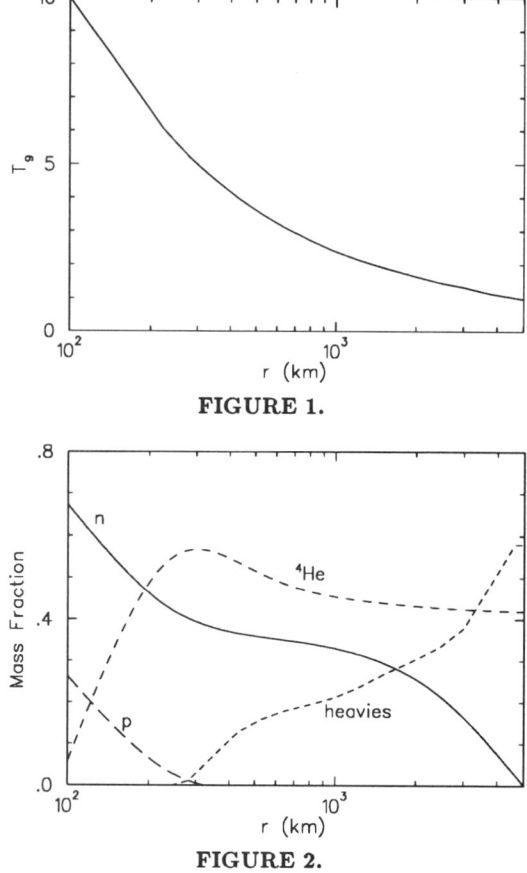

FIGURE 1.

FIGURE 2.

The key quantity then is R, the neutron-to-seed ratio, at the beginning of the r-process phase. It is sensitive to the entropy, neutron richness, and expansion timescale of the given wind element. Figure 2 shows the fraction of the mass in neutrons, protons, ^{4}He, and heavy nuclei in a wind element moving outward along the trajectory in figure 1. Notice how the mass fraction of heavy nuclei grows late in the expansion as neutrons capture onto heavy nuclei. The final r-process abundances as a function mass number A for this expansion are shown in figure 3. All calculations presented in this paper were performed with the Clemson nucleosynthesis code [10,11].

CALCULATIONS WITH NEUTRINOS

The reference r-process calculation presented above did not include the effect of ongoing neutrino capture on free nucleons and heavy nuclei. As a first

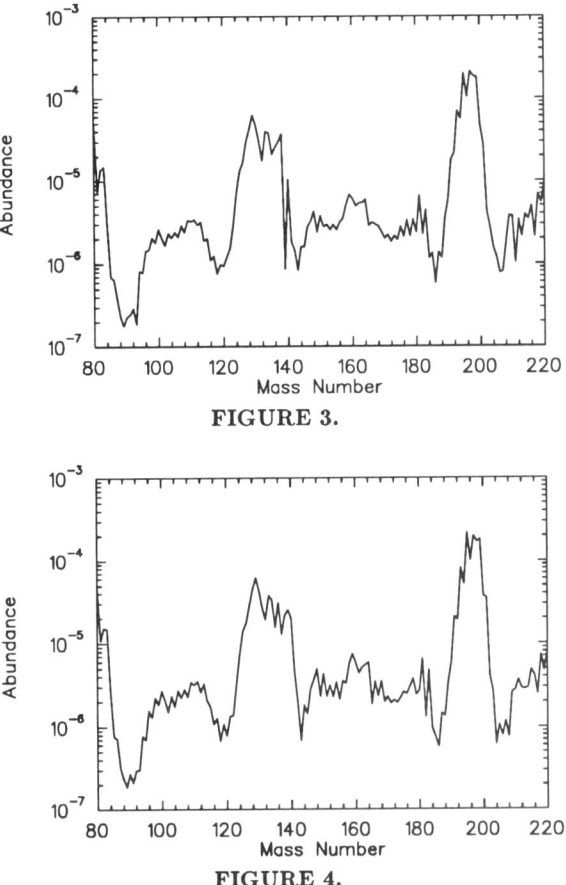

FIGURE 3.

FIGURE 4.

variation on the reference calculation, neutrino and anti-neutrino capture on heavy nuclei was included. The rates for neutrino captures on heavy nuclei were taken from reference [12]. The trajectory was again that shown in figure 1. The neutrino luminosities were taken to be 10^{52} ergs/s, and the neutrino spectra were taken to be blackbodies at temperatures 4 MeV and 7 MeV for the electron neutrinos and anti-neutrinos, respectively. The final r-process abundances for this calculation are shown in figure 4. The neutrino captures on heavy nuclei have subtly modified the final abundance distribution, for example near mass number 140; however, the final abundance distribution is not greatly different from that in figure 3.

The next calculation included neutrino and anti-neutrino capture on free nucleons in addition to capture on heavy nuclei. The rates on free nucleons were again taken from reference [12], and the neutrino temperatures and luminosities were as in the above calculation. The final r-process abundances

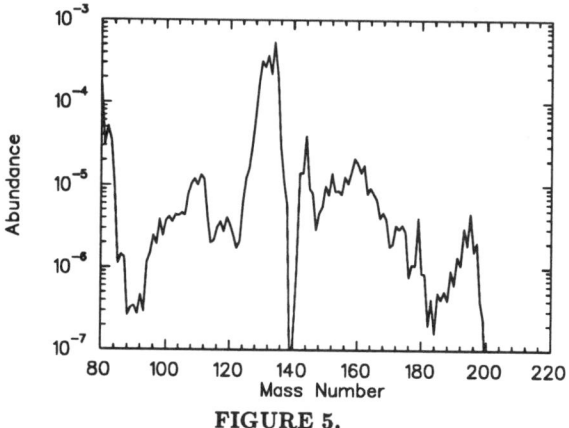
FIGURE 5.

for this calculation are shown in figure 5. This calculation shows the dramatic effect neutrino capture on free nucleons has on r-process yields.

The reason for this strong depletion of the yield of high-mass nuclei in this calculation is the so-called "alpha effect" [12]. Neutrino capture sets the neutron richness during the expansion of the wind from the neutron star. Anti-neutrino capture on protons drives the matter neutron rich, but neutrino capture on neutrons decreases the neutron richness of the material. As the wind moves outward, a steady state between neutrino and anti-neutrino capture is established. Because anti-neutrinos emerge from deeper within the neutron star, they have a higher temperature and capture more readily than do the neutrinos. Thus, the matter is neutron rich. Once the temperature in the wind element falls to the point that protons get locked up into ^{4}He (see figure 2), however, only neutrino capture on neutrons can occur since neutrino and anti-neutrino captures on ^{4}He are slow. This drastically reduces the abundance of free neutrons and, consequently, the neutron-to-seed ratio R, thereby resulting in much lower yields of high-mass r-process nuclei.

DISCUSSION

The drastic consequences of the alpha effect on r-process yields provides interesting constraints on supernova dynamics. In order to produce $A = 195$ nuclei in the r-process, it is necessary to have either sufficiently low neutrino fluxes that neutrino capture is ineffective or sufficiently high outflow velocities that neutrino captures do not have time to occur. It is worth noting in this context that the r-process model of Woosley et al. [8] produces $A = 195$ r-process nuclei $10 - 20$ seconds after core bounce, by which time the neutrino luminosity has fallen to less than 10^{51} ergs/s. In this scenario, then, neutrino capture during the r-process is not important because the rates are low.

These results also argue against a major role for neutrino-heavy nucleus interactions in wind expansions responsible for production of the $A = 195$ peak nuclei. If neutrino fluxes are sufficiently large to allow neutrino-nucleus interactions to be important during the r-process phase of the expansion, the alpha effect probably will be so strong that it will preclude synthesis of the $A = 195$ nuclei. On the other hand, if the alpha effect is not important, neutrino-nucleus interactions will be even less so since they will occur at even larger distance from the neutron star. A possible exception is the effect of neutrino spallation of protons from ^{4}He [10] which relies on the huge abundance of ^{4}He relative to that of the heavy nuclei to have any leverage. These considerations should serve as important caveats to those seeking to constrain supernova neutrinos via the r-process.

REFERENCES

1. Burbidge E. M., Burbidge G. R., Fowler W. A., and Hoyle F. *Rev. Mod. Phys.* **29**, 547 (1957).
2. Cowan J. C., Thielemann F.-K., and Truran J. W. *Phys. Rep.* **208**, 267 (1991).
3. Meyer B. S. *Ann. Rev. Astron. Astrophys.* **32**, 153 (1994).
4. Woosley S. E., and Hoffman R. D. *Astrophys. J.* **395**, 202 (1992).
5. Meyer B. S., Mathews G. J., Howard W. M., Woosley S. E., and Hoffman R. *Astrophys. J.* **399**, 656 (1992).
6. Howard W. M., Goriely S., Rayet M., and Arnould M. *Astrophys. J.* **417**, 713 (1993).
7. Takahashi K., Witti J., and Janka H.-T. *Astron. Astrophys. J.* **286**, 857 (1994).
8. Woosley S. E., Wilson J. R., Mathews G. J., Hoffman R. D., and Meyer B. *Astrophys. J.* **433**, 229 (1994).
9. Qian Y.-Z., and Woosley S. E. *Astrophys. J.* **471**, 331 (1996).
10. Meyer B. S. *Astrophys. J. Lett.* **449** 55 (1995).
11. Meyer B. S., Krishnan T. D., and Clayton D. D. *Astrophys. J.* **462**, 825 (1996).
12. Fuller G. M., and Meyer B. S. *Astrophys. J.* **453**, 792 (1995).

STRANGE STARS

Jes Madsen

Institute of Physics and Astronomy, University of Aarhus, DK-8000 Århus C, Denmark,
jesm@dfi.aau.dk

Abstract. The astrophysics of strange quark matter is reviewed, with special emphasis on properties of strange stars such as their formation and rotation.

INTRODUCTION

If three flavor quark matter (strange quark matter; SQM) is absolutely stable at zero pressure, "strange stars" [1–4] consisting completely of SQM (perhaps apart from a minor crust to be discussed below) may exist in nature. Such strange stars behave quite differently from neutron stars due to the unusual equation of state. For massless quarks the total energy density is given by $\rho = \rho_q + B$, and the total pressure by $P = P_q - B$, where B is the bag constant, ρ_q is the energy density of quarks, and the pressure of the quarks is $P_q = \rho_q/3$, since massless quarks are relativistic. The equation of state is thus given by $P = \frac{1}{3}(\rho - 4B)$. The exact equation of state taking into account $m_s \neq 0$ is very similar [3] since s-quarks are relativistic for low m_s and not present for high m_s. (I here assume that $\alpha_s = 0$; a non-zero α_s effectively corresponds to a reduction of B, and does not change the general picture).

The structure of a strange star is calculated from the Oppenheimer-Volkoff equation, describing the balance between gravity and pressure gradient, using the equation of state given above. The surface of the star corresponds to $P = 0$, a condition fulfilled for $\rho = 4B$, which for typical values of B is somewhat more than the density of ordinary nuclear matter! For stellar masses below $1M_\odot$ ($M_\odot$ is the solar mass) this density is almost constant throughout the star, so to a good approximation total mass and radius are related by $M \propto R^3$, a relation in striking contrast to ordinary neutron stars, where $M \propto R^{-3}$. This means that low-mass neutron stars and strange stars have widely different radii, possibly allowing observational distinction. However, stellar evolution normally leads to compact objects with masses near $1.4M_\odot$. For such a mass gravity rather than bag pressure plays the dominant stabilizing role, and there is no significant difference between neutron star and strange star radii. Also the maximum mass given by gravitational instability (the Chandrasekhar limit) is similar, of order $2M_\odot$. While ordinary neutron stars are unstable for masses below $0.1\,M_\odot$, strange stars have no minimum mass; the sequence continues smoothly to the domain of strangelets.

The only natural energy scale in the simple equation of state above is $B^{1/4}$. Thus there exists a homology transformation between strange star models for different values of B. In particular, the maximum mass of a strange star is given by

$$M_{\max} = 2.006 B_{145}^{-1/2} M_\odot, \qquad (1)$$

where $B_{145} \equiv B/(145\text{MeV})^4$. The corresponding minimal radius, maximal moment of inertia, maximal central density, and surface density, are given by $R_{\min} = 10.94 B_{145}^{-1/2}$km, $I_{\max} = 2.256 \times 10^{45} B_{145}^{-3/2}$g cm^2, $\rho_{\max} = 1.97 \times 10^{15} B_{145}$g cm^{-3}, $\rho_{\text{surf}} = 4.102 \times 10^{14} B_{145}$g cm^{-3}, respectively. The minimal rotation period (the so-called Kepler period corresponding to mass-shedding at the equator), is given by

$$P_{\min} = 0.66 B_{145}^{-1/2}\text{ms}. \qquad (2)$$

Bare strange stars (strange stars with quark matter all the way to the surface) have quite unusual properties. The density abruptly jumps from 0 to ρ_{surf} and stays almost constant through the interior (except when the mass is close to $M_{\max}$). The plasma frequency of the star is huge, meaning that photons with energies below 20MeV are reflected from the surface, whereas the star itself can only emit photons with higher energies [3,5]. Even more important, because of the strong interaction binding of the surface material, the star is not subject to the "Eddington limit", which for ordinary neutron stars limits the luminosity to be below 10^{38} erg/s (for higher luminosities the radiation pressure would exceed the gravitational attraction and expel the surface layers). As discussed below, this could lead to important "applications" of strange stars.

The description so far is oversimplified because real strange stars may have surfaces more like ordinary neutron stars. In particular, a solid crust of ordinary material may form from accretion by the strange star after formation, or from material that was not converted during neutron star burning. Such a crust may be held up by the extreme, outward directed electrostatic potential of 10^{17}–10^{18}V/cm, created by the electron atmosphere with a thickness of a few hundred Fermi. This atmosphere merely expresses that the electrostatic binding of electrons is weaker than the strong binding of quarks; therefore the electron distribution does not end abruptly like that of quarks (the detailed structure was found from a Thomas-Fermi calculation by Alcock, Farhi and Olinto [3]; see also [6]).

The electrostatic potential can sustain a significant crust of ordinary neutron star material. The limit is given by the neutron drip density (4×10^{11} gcm^{-3}), above which neutrons drip out of nuclei and would be swallowed by the quark phase. This crust may be decisive for interpretation of pulsar behavior.

As emphasized by Glendenning, Kettner and Weber [7,8], the existence of crusts not only changes the mass-radius relation for strange stars, but also opens a rich plethora of new stellar configurations. In particular, one may have a sequence of "strange dwarfs", much like white dwarfs except for an SQM core. At present there is no well-studied model for formation of such strange dwarfs.

Studies of strange stars have not been pursued to the degree of detail known for ordinary neutron stars, and it is premature to draw any detailed conclusions. However, in the following, I shall look at some of the properties expected and emphasize the possible observable differences between neutron stars and strange stars.

A distinction between strange stars and neutron stars was for a long time believed to be a much more rapid cooling of SQM due to neutrino emitting weak interactions involving the quarks [3]. Thus a strange star was presumed to be much colder than a neutron star of similar age, a signature potentially observable from x-ray satellites. Only a few other mechanisms, such as kaon condensates may mimic the speed of quark matter neutrino cooling. Recently the story has been complicated by the finding that ordinary neutron β-decay may be energetically allowed in nuclear matter [9], so that the cooling rate can be comparable to that of SQM. For this reason I shall not discuss the issue here, but refer the reader to an excellent review of neutron star cooling in [10], and to the reinvestigation of strange star cooling in [11].

PULSAR GLITCHES

One important feature seems to distinguish strange stars from neutron stars in a manner with observable consequences, and that is the distribution of the moment of inertia inside the star. Ordinary neutron stars older than a few months have a crust made of a crystal lattice or an ordered inhomogeneous medium reaching from the surface down to regions with density 2×10^{14} g cm^{-3}. This crust contains about 1% of the total moment of inertia. Strange stars in contrast can only support a crust with density below the neutron drip density (4.3×10^{11} g cm^{-3}). This is because free neutrons would be absorbed and converted by the strange matter. Such a strange star crust contains at most a few times 10^{-5} of the total moment of inertia. This is an upper bound, since the strange star may have no crust at all, depending on its prior evolution.

The difference in the moment of inertia stored in the crust of neutron stars and strange stars seems to pose significant difficulties for explaining the glitch-phenomenon observed in radio pulsars with models based on strange stars

[2,3,12]. Glitches are observed as a sudden speed-up in the rotation rate of pulsars. The fractional change in rotation rate Ω is $\Delta\Omega/\Omega \approx 10^{-6}$—$10^{-9}$, and the corresponding fractional change in the spin-down rate $\dot{\Omega}$ is of order $\Delta\dot{\Omega}/\dot{\Omega} \approx 10^{-2}$—$10^{-3}$. Regardless of the detailed model for the glitch phenomenon these jumps must involve the decoupling and recoupling of a component in the star containing a significant fraction of the total moment of inertia. This role is played by the inner crust of an ordinary neutron star, but the crust around a strange star is smaller; less than a few times $10^{-5} M_\odot$, with I_{crust}/I around a few times 10^{-5} for ordinary neutron star masses of $1.4 M_\odot$ (higher for less massive stars). These numbers are based on models [13] assuming a maximum mass crust, i.e. a crust reaching neutron drip density at the base. Strange stars may have sufficiently massive crusts to account for glitches, but only for a quite limited range of parameters.

Other possibilities for glitches in strange stars could involve a crust composed of strangelets (cf. the "quark-alpha" scenario in [14]), not to mention the possibility of a quark-hadron mixed phase [15-17]. There is still a lack of any detailed model for how the magnetic field structure and other crucial aspects of a pulsar can be modeled for strange stars. Presumably a strange star cannot do the job without significant structure, such as a crust and/or superfluidity/superconductivity in certain regions. These issues have only been very superficially studied and need further consideration. The present lack of such models should not be used to dismiss the possibility of strange stars.

STRANGE STAR OSCILLATION AND ROTATION

Perhaps the most interesting difference between neutron stars and strange stars is related to the damping of instabilities. Strange stars may have radial oscillations with a fundamental period of 0.06–0.3 ms [18], but these are characterized by rapid damping in a matter of seconds [19-22]. This is due to the extremely high viscosity of SQM.

The large viscosity also plays a role in setting the maximum rotation limit for strange pulsars (or hybrid stars with SQM cores). The ultimate rotation limit corresponds to mass-shedding from the equator of the star (this is called the Kepler limit and is of order 0.6 msec for a strange star, Eq. (2); see Zdunik [23] for a review). But before reaching such rotation rates, the pulsars become unstable to non-radial deformations and are slowed down by emission of gravitational radiation. Shear and bulk viscosities tend to stabilize the star against these instabilities [20,24], and the high value for the bulk viscosity may mean that strange pulsars in contrast to ordinary pulsars can reach submillisecond periods [22]. Thus the discovery of very fast pulsars may be an indication favoring the existence of strange stars.

The bulk viscosity of strange quark matter depends on the rate of the non-leptonic interaction [25,26]

$$u + d \leftrightarrow s + u. \tag{3}$$

This reaction changes the concentrations of down and strange quarks in response to the density changes involved in vibration or rotational instabilities, thereby causing dissipation. This dissipation is most efficient if the rate of reaction (3) is comparable to the frequency of the density change. If the weak rate is very small, the quark concentrations keep their original values in spite of a periodic density fluctuation, whereas a very high weak rate means that the matter immediately adjusts to follow the true equilibrium values reversibly. But in the intermediate range dissipation due to PdV-work is important.

The importance of dissipation due to Eq. (3) was first stressed by Wang and Lu [19] in the case of neutron stars with quark cores. These authors made a numerical study of the evolution of the vibrational energy of a neutron star with an $0.2 M_\odot$ quark core, governed by the energy dissipation due to Eq. (3). Sawyer [20] expressed the damping in terms of the bulk viscosity, a function of temperature and oscillation frequency, which he tabulated for a range of densities and strange quark masses. Sawyer's tabulation has later been used in studies of quark star vibration [21], and of the gravitational radiation reaction instability determining the maximum rotation rate of pulsars [24]. The latter study concluded, that the bulk viscosity is large enough to be important for temperatures exceeding 0.01 MeV, but that it should be a few orders of magnitude larger to generally dominate the stability properties.

However, as has been pointed out in [22], the bulk viscosities in [20] depend on the assumption, that the rate of Eq. (3) can be expanded to first order in $\delta\mu = \mu_s - \mu_d$, where $\mu_i \approx 300$MeV are the quark chemical potentials. This assumption is not correct at low temperatures ($2\pi T \ll \delta\mu$), where the dominating term in the rate is proportional to $\delta\mu^3$. Furthermore, Sawyer's rate is too small by an overall factor of 3, and a discrepancy of 2–3 orders of magnitude, perhaps due to unit conversions, appears as well. Taken together, these effects lead to an upward correction of the bulk viscosity by several orders of magnitude, and thereby increases the importance for the astrophysical applications. The non-linearity of the rate also means, that the bulk viscosity is no longer independent of the amplitude of the density variations.

Even at very low temperatures, high amplitude oscillations are damped in fractions of a second, and those of low amplitude in a matter of minutes, if one takes into account, that the heat released by viscous dissipation can speed up the damping of vibrations [22].

Investigations by Colpi and Miller [24] based on older viscosities indicated, that the minimal rotation period of strange stars might be set by the gravitational radiation reaction instability of $m = 2$ or $m = 3$ modes at or just below 1 millisecond. With the new, much larger, viscosities, the non-axisymmetric instabilities will be suppressed, and it is not unreasonable to expect, that the maximum rotation frequency of strange stars will be close to the Keplerian limit, Eq. (2). Detailed numerical calculations including the new viscosities and effects of dissipative heating, are required to settle the issue, but they are complicated by the non-linear behavior of the new bulk viscosity.

GAMMA-RAY BURSTERS

Strange stars because of their high surface density, strong binding (making it possible to circumvent the Eddington limit), and special emission properties have been suggested as explanations for some of the more mysterious cosmic events, namely γ-ray bursters. These are bursts of γ-rays of a few seconds duration, coming from unidentified sources which are presumably at extragalactic distances.

No consensus exists concerning the nature of these bursts, but Alcock, Farhi and Olinto [27] suggested a detailed model for the most prominent of the bursters, the one on 5 March 1979. Their model is based on an impact of a $10^{-8} M_\odot$ lump of SQM on a rotating strange star, and the authors are able to explain most of the observations concerning energetics and time-scales under the assumption that the burster is located in a supernova remnant in the Large Magellanic Cloud, as position measurements seem to indicate.

γ-ray bursters at truly cosmological distances could be due to collisions of two strange stars in binary systems [28], each collision releasing 10^{50} ergs in the form of gamma rays over a time-scale of 0.2 s.

There are, however, literally hundreds of different models for γ-bursts, and in spite of improved observational data the interpretation is at present unclear.

A recent identification of the x-ray source Her X-1 as a strange star [29] was based on incorrect use of the bag model [30].

ALL PULSARS OR NO PULSARS ARE STRANGE

If strange quark matter is stable, strange stars may be formed during supernova-explosions, and neutron stars can be converted to strange stars by a number of different mechanisms, such as pressure-induced transformation to uds-quark matter via ud-quark matter, sparking by high-energy neutrinos, or triggering due to the intrusion of a quark nugget. These and other possibilities were described by Alcock, Farhi, and Olinto [3].

As soon as a lump of strange matter comes in contact with free neutrons it starts converting them into strange matter. The burning of a neutron star into a strange star was discussed by Baym *et al.* [31] and Olinto [32], and it was shown that the star would be converted on a rather small time-scale set by quark diffusion and flavor-changing weak interactions. Later studies [33,34] found burning times in the range of $1-10^3$ seconds under various parameter assumptions (see also [35] for a review). For the fastest burning times, the energy liberated may be important for the supernova mechanism and supernova neutrino bursts. Horvath and Benvenuto [36] have questioned the stability of "slow" neutron combustion and suggested that the conversion takes place much faster as a detonation. So far, the investigations of neutron star burning have been rather crude, neglecting many aspects of transport theory, heat conduction etc. A detailed study of this phenomenon would be interesting.

Perhaps the most likely mechanisms for initiating the formation of a strange star involves either a seed of SQM in the star, or thermal formation of quark matter bubbles. The thermal formation mechanism was studied in [37], where it was concluded, that quark matter bubbles nucleate (possibly followed by burning of the star into SQM) in neutron stars/supernovae if the bag constant is low, and if the temperature exceeds a few MeV (thus the process is most likely during the supernova explosion itself). Should thermal nucleation not take place, one of the other mechanisms mentioned above must be relied on. Apart from seed-induced burning, all of these are likely to be much less efficient than thermal nucleation.

Some pulsars are known to exist in binaries with orbits decaying because of emission of gravitational waves. If such a system contains a strange star, some pollution of strange quark matter will spread in the Galaxy and be absorbed by the main sequence progenitors of new generations of supernovae, which in other words act as huge surface area, long integration time detectors. If just a single lump of SQM is absorbed, it is sufficient to transform the new pulsar into a strange star. Thus, if SQM is absolutely stable, ALL pulsars are strange stars. And if the argument is reversed: should it somehow be possible to prove that non-strange pulsars exist, then SQM cannot be absolutely stable! A simple but very powerful argument (see Madsen [38] for details).

ACKNOWLEDGMENTS

This work was supported in part by the Theoretical Astrophysics Center under the Danish National Research Foundation.

REFERENCES

1. Witten, E., *Phys. Rev. D* **30**, 272 (1984).
2. Haensel, P., Zdunik, J. L., and Schaeffer, R., *Astron. Astrophys.* **160**, 121 (1986).
3. Alcock, C., Farhi, E., and Olinto, A., *Astrophys. J.* **310**, 261 (1986).
4. Alcock, C., in *Strange Quark Matter in Physics and Astrophysics*, edited by J. Madsen and P. Haensel (Nucl. Phys. B (Proc. Suppl.), **24B**, 1991), pp. 93–102.
5. Chmaj, T., Haensel, P., and Słomiński, W., in *Strange Quark Matter in Physics and Astrophysics*, edited by J. Madsen and P. Haensel (Nucl. Phys. B (Proc. Suppl.), **24B**, 1991), pp. 40–44.
6. Kettner, C., Weber, F., Weigel, M. K., and Glendenning, N. K., *Phys. Rev. D* **51**, 1440 (1995).

7. Glendenning, N. K., Kettner, C., and Weber, F., *Phys. Rev. Lett.* **74**, 3519 (1995).
8. Glendenning, N. K., Kettner, C., and Weber, F., *Astrophys. J.* **450**, 253 (1995).
9. Lattimer, J. M., Pethick, C. J., Prakash, M., and Haensel, P., *Phys. Rev. Lett.* **66**, 2701 (1991).
10. Pethick, C. J., *Rev. Mod. Phys.* **64**, 1133 (1992).
11. Schaab, C., Hermann, B., Weber, F., and Weigel, M. K., *Astrophys. J.* **480**, L111 (1997).
12. Alpar, M. A., *Phys. Rev. Lett.* **58**, 2152 (1987).
13. Glendenning, N. K., and Weber, F., *Astrophys. J.* **400**, 647 (1992).
14. Benvenuto, O. G., and Horvath, J. E., *Mon. Not. R. Ast. Soc.* **247**, 584 (1990).
15. Glendenning, N. K., *Phys. Rev. D* **46**, 1274 (1992).
16. Heiselberg, H., Pethick, C. J., and Staubo, E. F., *Phys. Rev. Lett.* **70**, 1355 (1993).
17. Glendenning, N. K., and Pei, S., *Phys. Rev. C* **52**, 2250 (1995).
18. Datta, B., Sahu, P. K., Anand, J. D., and Goyal, A., *Phys. Lett. B* **283**, 313 (1992).
19. Wang, Q. D., and Lu, T., *Phys. Lett. B* **148**, 211 (1984).
20. Sawyer, R. F., *Phys. Lett. B* **233**, 412 (1989).
21. Cutler, C., Lindblom, L., and Splinter, R. J., *Astrophys. J.* **363**, 603 (1990).
22. Madsen, J., *Phys. Rev. D* **46**, 3290 (1992).
23. Zdunik, J. L., in *Strange Quark Matter in Physics and Astrophysics*, edited by J. Madsen and P. Haensel (Nucl. Phys. B (Proc. Suppl.), **24B**, 1991), pp. 119–124.
24. Colpi, M., and Miller, J. C., *Astrophys. J.* **388**, 513 (1992).
25. Madsen, J., *Phys. Rev. D* **47**, 325 (1993).
26. Heiselberg, H., *Phys. Scr.* **46**, 485 (1992).
27. Alcock, C., Farhi, E., and Olinto, A., *Phys. Rev. Lett.* **57**, 2088 (1986).
28. Haensel, P., Paczyński, B., and Amsterdamski, P., *Astrophys. J.* **375**, 209 (1991).
29. Li, X.-D., Dai, Z.-G., and Wang, Z.-R., *Astron. Astrophys.* **303**, L1 (1995).
30. Madsen, J., *Astron. Astrophys.* **318**, 466 (1997).
31. Baym G., et al., *Phys. Lett. B* **160**, 181 (1985).
32. Olinto, A. V., *Phys. Lett. B* **192**, 71 (1987).
33. Heiselberg, H., Baym, G., and Pethick, C. J., in *Strange Quark Matter in Physics and Astrophysics*, edited by J. Madsen and P. Haensel (Nucl. Phys. B (Proc. Suppl.), **24B**, 1991), pp. 144–147.
34. Olesen, M. L., and Madsen, J., in *Strange Quark Matter in Physics and Astrophysics*, edited by J. Madsen and P. Haensel (Nucl. Phys. B (Proc. Suppl.), **24B**, 1991), pp. 170–174.
35. Olinto, A., in *Strange Quark Matter in Physics and Astrophysics*, edited by J. Madsen and P. Haensel (Nucl. Phys. B (Proc. Suppl.), **24B**, 1991), pp. 103–109.
36. Horvath, J. E., and Benvenuto, O. G., *Phys. Lett. B* **213**, 516 (1988).
37. Olesen, M. L., and Madsen, J., *Phys. Rev. D* **49**, 2698 (1994).
38. Madsen, J., *Phys. Rev. Lett.* **61**, 2909 (1988).

The Equation of State in Nucleon and Strange Stars

Madappa Prakash

Department of Physics and Astronomy
State University of New York at Stony Brook
Stony Brook, New York 11794

Abstract. The equation of state (EOS) is an essential physical ingredient in simulations of gravitational collapse, supernovae, old and newly born neutron stars, and binary mergers of compact stars. In this talk, I will examine (1) the extent to which stringent constraints may be placed on the EOS by a comparison of calculations with the available data on some basic neutron star properties; and (2) some astrophysical consequences of the possible presence of strangeness, in the form of baryons, notably the Λ and Σ^-, or as a Bose condensate, such as a K^- meson condensate, or in the form of strange quarks.

INTRODUCTION

Observed neutron star masses [1,2] are shown in Fig. 1. The smallest range that is consistent with all of the data has an upper limit of $1.44 M_\odot$ from PSR1913+16 and a lower limit of $1.36 M_\odot$ from the precisely measured total mass of PSR2127+11C and its companion. The upper limit, $M = 1.44 M_\odot$, provides constraints on the neutron star EOS. A conservative estimate of this upper limit is shown by the dashed line at $M = 1.5 M_\odot$. Any neutron star EOS has to support a maximum mass of at least this value.

The fact that all the measured neutron star masses consistently lie within a narrow range around $1.4 M_\odot$ is intriguing. Since neutron stars are formed in the gravitational core collapse of massive stars, their masses may depend on the structure of the progenitor star. The cores of stars which evolve into neutron stars have precollapse masses of about $1.4 M_\odot$, which introduces a natural mass scale for possible neutron star masses. The final neutron star mass may, however, depend on the amount of accretion at times subsequent to a neutron star's birth. Thus, rigorous arguments for the happenstance of observations are not yet available. Later we will return to the question of whether or not it is the nature of strong interactions which restricts these stars to this mass range.

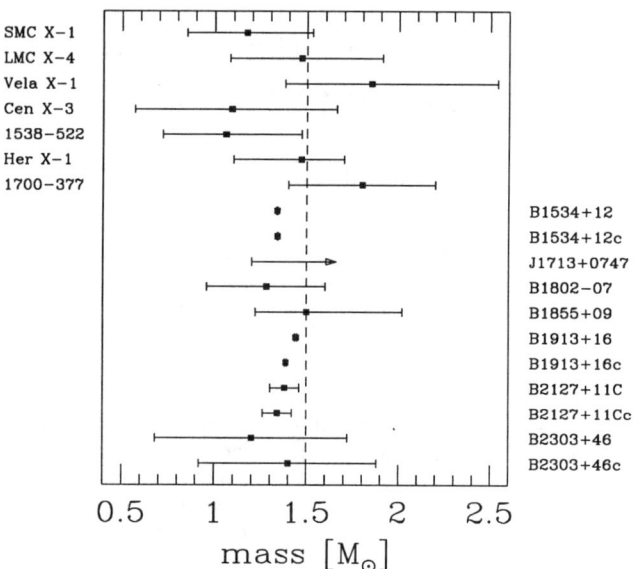

FIGURE 1. Measured neutron star masses. Error bars indicate 95% confidence limits.

NEUTRON STAR STRUCTURE

In hydrostatic equilibrium, the structure of a spherically symmetric neutron star is determined by the Tolman-Oppenheimer-Volkov (TOV) equations [3]:

$$\frac{dMc^2}{dr} = 4\pi r^2 \epsilon, \quad \frac{dP}{dr} = -\frac{GM\epsilon}{c^2 r^2}\left[1 + \frac{P}{\epsilon}\right]\left[1 + \frac{4\pi r^3 P}{Mc^2}\right]\left[1 - \frac{2GM}{c^2 r}\right]^{-1}. \quad (1)$$

Above, G is the gravitational constant, P is the pressure, ϵ is the energy density inclusive of the rest mass density, and M is the enclosed gravitational mass. The quantity $R_s = 2GM/c^2$ is known as the Schwarzschild radius. The gravitational and baryon masses of the star are defined by

$$M_G c^2 = \int_0^R dr\, 4\pi r^2\, \epsilon, \quad M_A c^2 = m_A \int_0^R dr\, 4\pi r^2\, n\left[1 - \frac{2GM}{c^2 r}\right]^{-1/2}, \quad (2)$$

where m_A is the baryonic mass and n is the baryon number density. The binding energy of the star is then $B.E. = (M_A - M_G)c^2$.

By specifying the EOS of enclosed matter, $P = P(\epsilon)$, the structure of the star is determined by choosing a central pressure $P_c = P(\epsilon_c)$ at $r = 0$ and integrating the coupled differential equations in Eq. (1) out to the star surface at $r = R$ determined by the condition $P(r = R) = 0$. The significance of general relativity (GR) may be gauged by the magnitude of R_s/R. When $R_s/R \ll 1$, the structure is essentially determined by Newtonian gravity. The surface approaches the event horizon as $R_s/R \to 1$; larger values result in black hole configurations.

Constraints on the equation of state

Stringent constraints may be placed on the EOS if measurements of stars' masses, radii, pulsar frequencies, moments of inertia, *etc.* are available. To see how this works, consider the dependence of some of these properties on the mass and radius in a so-called $M - R$ diagram.

1. Eq. (1) requires that the radius $R > R_s = 2GM/c^2$. This yields $M/M_\odot \leq R/R_s(\odot)$, where $R_s(\odot) = 2GM_\odot/c^2 \cong 2.95$ km.

2. Since the pressure in the center is finite, $P_c < \infty$, $R > (9/8)R_s$ [4]. This translates to $M/M_\odot \leq (8/9)R/R_s(\odot)$.

3. Since the adiabtic sound speed $c_s = (dP/d\epsilon)^{1/2} \leq c$, where c is the speed of light, one has $R > 1.39 R_s$, giving $M/M_\odot \leq R/(1.39 \times R_s(\odot))$.

4. If, instead of employing the causal EOS $P = \epsilon$ at all densities, one requires it to hold above a fiducial density $n_t \cong 2n_0$ (below which the EOS is presumed to be known), the limit $R > 1.52 R_s$ is obtained [5], yielding $M/M_\odot \leq R/(1.52 \times R_s(\odot))$.

In practice, however, these restrictions allow a large class of EOSs to be consistent with observations. One is thus forced to utilize additional constraints that are based on observations. In the hope that continued measurements will prescribe stringent limits, we will note the constraints employed currently.

1. The limiting or maximum mass M_{max} should exceed the largest of the observed neutron star masses. Currently, this condition is taken to imply $M_{max} \geq 1.44 M_\odot$, the most accurately measured neutron star mass [1,2].

2. Nearly all (up to 99%) of the binding energy is released in the form of neutrinos during the birth of a neutron star, after the gravitational core collapse of a massive ($\geq 8 M_\odot$) star results in a type II supernova. Estimates [6] of the energy released in neutrinos from the SN1987A explosion lie in the range $(2 - 4) \times 10^{53}$ ergs. This places a restriction on the EOS that the B.E $\geq (2 - 4) \times 10^{53}$ ergs.

3. The Keplerian frequency of the star (this is the rotational frequency Ω_K beyond which the star will begin to shed mass at the equator) should exceed the spin period of the fastest spinning pulsar, namely that of

PSR1957+20. This translates to $P_K \geq 1.56$ ms. (In reality, a star may spin at a frequency lower than Ω_K due to viscous effects. Choosing the larger Keplerian frequency thus gives an upper limit.) GR calculations of rapidly rotating stars give [5]

$$\Omega_K \cong 7.7 \times 10^3 (M_{max}/M_\odot)^{1/2} (R_{max}/10 \text{ km})^{-3/2} \text{ s}^{-1}, \qquad (3)$$

where M_{max} and R_{max} refer to the mass and radius of the non-rotating spherical configuration. It is worthwhile to note that the discovery of a sub-millisecond pulsar (say of 0.5 ms), as was purported [7] to be the case in the wake of the SN1987A explosion and later retracted [8], would place rather severe limits on the EOS.

4. Another limit [9,10] employs the maximum moment of inertia, expressed in terms of M_{max} and R_{max} of the non-rotating configurations:

$$I_{max} = 0.6 \times 10^{45} \frac{(M_{max}/M_\odot)(R_{max}/10 \text{ km})^2}{1 - 0.295(M_{max}/M_\odot)/(R_{max}/10 \text{ km})} \text{ g cm}^2. \qquad (4)$$

Limits on I from a study of glitches would severely limit the allowed region in the $M - R$ plane.

In Fig. 2, the implications of the various restrictions mentioned above are considered. Besides the theoretical constraints, observational constraints imposed by mass, moments of inertia, and pulsar periods are illustrated. Also shown are the $M-R$ relationships of two representative EOSs. To date, data are consistent with a wide variety of EOSs, which highlights the need for continuing observations.

The Radius of the Neutron Star RXJ185635-3754

Although accurate masses of several neutron stars are beginning to be available [1,2], a precise measurement of the radius does not yet exist. Recently, Walter et al. [11] have identified a star, RXJ185635-3754, in both soft X-rays and optical light which appears to be an isolated neutron star. The observed X-rays are consistent with black body emission with a temperature of about 57 eV and very little extinction. In addition, the fortuitous location of the star in the foreground of the R CrA molecular cloud limits the distance to $D < 120$ pc. The fact that the source is not observable in radio and its lack of variability in X-rays implies that it is not a pulsar unlike other identified radio-silent isolated neutron stars [12]. It may give us the clearest view so far of the surface of a neutron star without complications from non-thermal emission in a magnetosphere. Coupled with an accurate measurement of its distance, its radius and the EOS of matter above nuclear density could be constrained.

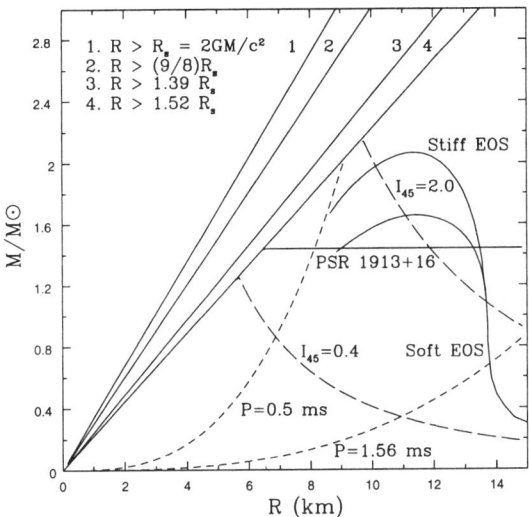

FIGURE 2. Valid EOSs must produce neutron stars which lie to the right of curve labeled 4. Current observational constraints include $M > M_{PSR1913+16}$, $I > 0.4 \times 10^{45}$ g cm² (Crab pulsar), and $P_{min} < 1.6$ ms (PSR1957+20). The promise of future observations is illustrated with curves of other values of I and P.

Assuming blackbody emission from the entire surface of a neutron star, an observed flux f_∞ and temperature T_∞ (the subscript ∞ denotes observations made at a very large distance from the star) implies that the local stellar radius R and mass M are related by [13]

$$R_\infty \equiv D\sqrt{f_\infty/(\sigma T_\infty^4)} = R/\sqrt{1 - R_s/R}, \tag{5}$$

where D is the distance to the star, σ is the Stefan-Boltzmann constant and the Schwarzschild radius $R_s = 2GM/c^2 = 2.954(M/M_\odot)$ km. General relativity (GR) constrains the stellar radius to be larger than R_s and additionally imposing causality restricts [5,14] the radius further: $R \geq 1.52 R_s$. Coincidentally, the largest permitted mass, $R_\infty c^2/(3^{3/2}G)$, occurs when $R = 1.5R_s$. These relations are illustrated in Fig. 3, and four values of R_∞ are shown. An upper limit to R_∞ would constrain the mass and radius of the star to lie under that R_∞ curve and to the right of the line $R = 1.52R_s$.

X-ray observations of RXJ185635-3754 alone give $R_\infty \approx 7.3(D/120 \text{ pc})$ km for a best-fit blackbody. This, however, is incompatible with the knowledge gained from supernova modeling that neutron stars cannot be formed with less than about 1 $M_\odot$ which requires $R_\infty > 7.6$ km. The optical flux is about a factor of 2.5 brighter than what is predicted for the X-ray blackbody, and the probable consequence of an Fe-rich neutron star atmosphere would be to decrease the predicted optical flux further [15]. This discrepancy can be partially reduced if the neutron star has a temperature differential across its surface [16]. For example, a model which considers possible effects of magnetic fields upon the surface conductivities [17], can yield satisfactory fits to the X-ray data with an increase in the predicted optical flux up to 50% and predicted radii $R_\infty \lesssim 11(D/120 \text{ pc})$ km. This represents a nearly 50% increase in the inferred radius. Another class of models, which has a uniform hot region surrounded by a region with a uniformly lower temperature, can yield acceptable fits with radii $R_\infty \lesssim 13(D/120 \text{ pc})$ km, with increases in the predicted optical flux up to a factor of 2.2 [16]. Both of these models must be constrained by the lack of variability observed in this star. And if a hot region is to be produced by accretion, the relatively high flux implies a large accretion rate and thus a small stellar velocity $\lesssim 80$ km/s [16] if the Bondi-Hoyle accretion model is adopted. This is less than 20% of the mean observed pulsar velocities [18]. It is possible that there may be spectral features in the optical similar to those inferred for Geminga [19]. Optical spectra will be necessary to clarify this situation. Notwithstanding these caveats, it appears that this star has a very small radius, $R \lesssim 11$ km.

Small values for R could severely restrict the high-density EOS. The composition of a neutron star chiefly depends on the nature of strong interactions, which are not well understood in dense matter. We have therefore investigated many of the possible models [20], and typical $M - R$ curves are shown in Figs. 3 through 5. Different EOSs are distinguished by the different symbols used to denote their respective maximum masses. The EOSs of Refs. [21,22] are based on a potential model description. $M - R$ curves for three choices of the Hamiltonian are shown in Fig. 3 (solid squares), the differences reflecting uncertainties in the three-body interaction at high density. Such uncertainties at high density are further illustrated by the parametrized EOSs of Prakash et al. [23] (filled circles), and for recent field-theoretical EOSs [24] (asterisks). A few general features emerge from these results: (1) the minimum radius occurs for the maximum mass star, and (2) the range of possible radii in the mass window $1.0 M_\odot < M < M_{max}$ is significantly smaller for the stiffer EOSs (e.g., those of Refs. [21,22] and Ref. [24]) than for relatively softer EOSs (e.g., those of Ref. [23]). Available results of Dirac-Brueckner calculations [25] (not shown in the figure) exhibit similar features.

In Fig. 4, we show $M - R$ curves for cases in which strangeness-bearing components such as hyperons, a kaon condensate, or strange quarks may be present in addition to nucleons. Depending on the interactions, these additional components can appear separately or in combination with one another. In either case, the EOS is softened in its high-density behavior relative to the case in which only nucleons are present. The results [27] for matter with hyperons (solid squares), are based on a field-theoretical model [28], in which the hyperon interactions are constrained at nuclear density by hypernuclear

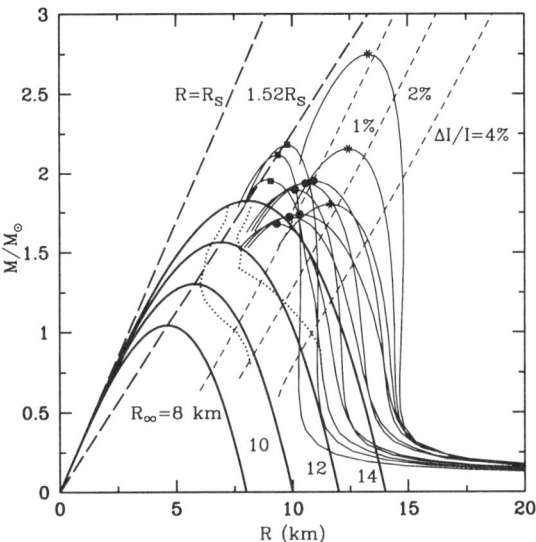

FIGURE 3. $M - R$ curves for several recent EOSs which permit only nucleons and leptons to be present. The curves labeled $R = R_s$ and $R = 1.52 R_s$ show the limits imposed by GR and GR + causality, respectively. Also displayed are contours of $R_\infty = 8, 10, 12$, and 14 km, and $\Delta I/I = 1, 2$, and 4 % (see text). The dotted lines indicate the allowed region inferred from analyses of X-ray bursts (see Ref. 26).

data. There is, however, considerable uncertainty in the high density behavior of these EOSs. Also shown, as asterisks, are cases in which $u, d,$ and s quarks are present together with strange baryons. In both situations, the softening induced by additional fermionic species significantly reduces both the maximum mass and the minimum radius.

It is a generic feature of the GR structure equations that especially small radii are obtained if the pressure varies relatively slowly with density in the vicinity of ordinary nuclear density and rapidly stiffens at higher densities [5]. Such conditions are optimally met if a Bose condensate appears around $3 - 4$ times the nuclear density. The resulting phase transition delays a pressure increase until a density of, say, 6–7 times the nuclear density is reached. The suggestion [29] that a kaon condensate could occur in compact stars is supported by in-medium effects observed in kaonic atoms and in subthreshold kaon production in heavy-ion collisions [30]. Thorsson et $al.$ [31] have explored such models in detail, and representative calculations are shown in Fig. 5 as solid circles. Depending on the overall strength of the condensation, which is determined by the strangeness content of the proton, radii as small as 7 km are obtained for stars with maximum masses in the range $1.45 - 1.5 M_\odot$.

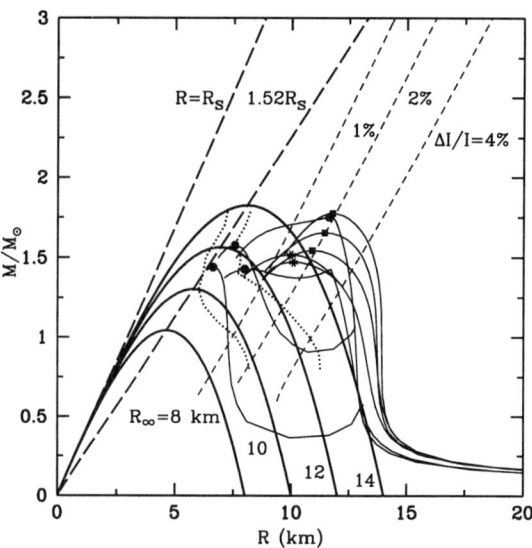

FIGURE 4. Same as Fig. 3, but for EOSs which permit strangeness-bearing components to be present in addition to nucleons and leptons (see text).

Stars with larger masses, however, have larger radii. It appears that with the constraints $M > 1$ $M_\odot$ and $M_{max} > 1.44$ $M_\odot$, stars with kaon condensates can have R_∞ as small as 9 km. With the same constraints, only EOS parametrizations with broad phase transitions, such as those induced by pion or kaon condensates, can yield such small values of R_∞ [5]. We note that in stars with multiple strange components, only those in which a kaon condensate appears first will have relatively compact configurations. However, it is currently unknown which component might appear first [20].

Another class of EOS exists in which relatively small radii can be obtained, namely those resulting from the conjecture [32] that strange quark matter is the ultimate ground state of matter. Use of perturbative QCD and an MIT-type bag model for the EOS results in self-bound strange quark matter stars. Typical $M - R$ curves are shown in Fig. 5. Unlike normal neutron stars, maximum mass stars have nearly the largest radii possible for a given EOS. If the strange quark mass $m_s = 0$ and interactions are neglected ($\alpha_c = 0$), the maximum mass is related to the bag constant B by $M_{max} = 2.033\sqrt{\frac{56 \text{ MeV fm}^{-3}}{B}}$ $M_\odot$. The existence of an energy ceiling of 939 MeV for zero pressure matter requires that $B < 95$ MeV fm^{-3} [32]. The constraint that $M_{max} > 1.44$ $M_\odot$ is thus automatically satisfied. In addition, the locus of maximum masses is given simply by $R = 1.85 R_s$ [5] and is shown in Fig. 5. The addition of a finite strange quark mass and/or interactions produces maximum mass configura-

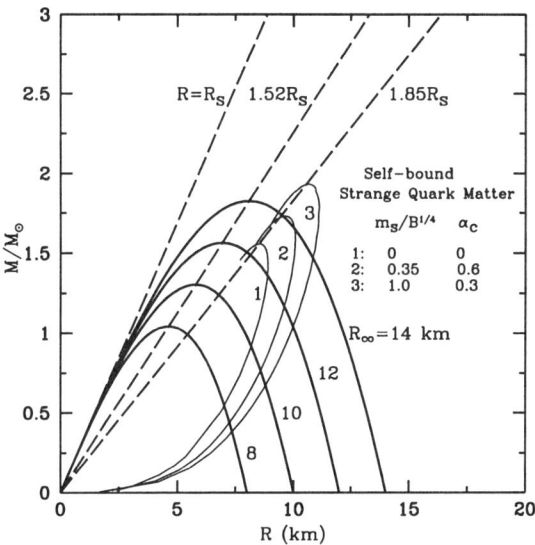

FIGURE 5. Same as Fig. 3, but for EOSs describing self-bound strange quark matter. The additional curve $R = 1.85 R_s$ represents the loci of maximum masses for $m_s = 0$ and $\alpha_c = 0$. For curves labelled 1, 2, and 3, values of B are 94.9, 57.4, and 64.2 MeV fm^{-3}, respectively.

tions which have only marginally larger radii than this relation implies; also, such stars have larger radii for every mass than do strange quark stars with $m_s = 0$ and $\alpha_c = 0$. Strange quark stars with crusts [33] have larger radii than those with bare surfaces. Thus, the non-interacting case shown in Fig. 5 with $m_s = 0$ and $B = 95$ MeV fm^{-3} gives the most compact self-bound configuration possible. Coupled with the additional constraint $M > 1$ M$_\odot$ from supernova models, strange quark stars cannot have $R < 8$ km or $R_\infty < 10.5$ km, considerably larger than the limits for a kaon condensate EOS.

A small radius for RXJ185635-3754 might imply an EOS inconsistent with glitch models [34] in which the inferred change in the star's spindown rate during a glitch is assumed to be related to the fractional amount of moment of inertia $\Delta I/I$ contained in the star's superfluid crust. Ravenhall and Pethick [35] have noted that $\Delta I/I$ is a nearly unique function of M and R with only a slight sensitivity to the EOS for a given value of the density demarking the crust-core interface. For a large number of EOSs, we have explicitly determined the location of $\Delta I/I$ contours, which are shown as thin dashed lines in the figures, assuming that the interface occurs at a density of $n_{cc} = 0.074$ fm^{-3} and an enthalpy of $\mathcal{H}_{cc} = 24.1$ MeV. The dependence on the interface parameters is

$$\Delta I/I \propto n_{cc}\Delta R = n_{cc}R[e^h - 1]/[\Lambda e^h - 1], \tag{6}$$

where $\Lambda = (1 - 2GM/Rc^2)^{-1}$, $h = 2\mathcal{H}_{cc}/m_n c^2$ and m_n is the neutron mass. For matter in beta equilibrium, $\mathcal{H}_{cc} = \mu_n(n_{cc}) - \mu_n(0)$, where μ_n is the neutron chemical potential. Thus, the location of a $\Delta I/I$ contour is not sensitive to the high-density EOS but only to the EOS below nuclear density and the density of the crust-core interface; decreasing the density n_{cc} moves the loci to the right. The largest glitches observed imply $\Delta I/I > 0.01$, which together with $M > 1 M_\odot$ implies $R_\infty > 9.5$ km. A limit of $\Delta I/I > 0.04$ implies $R_\infty > 13$ km. Note that non kaon condensate-containing stars with $M > 1 M_\odot$ must have $R_\infty > 12.5$ km.

Recently, Haberl [36] *et al.* have reported the discovery of another isolated old neutron star RXJ0720.4-3125, which is a soft X-ray source with 8.39 s pulsations. The best fit blackbody temperature is 79 ± 4 eV and the estimated surface magnetic field is less than 10^{10} G. Their quoted flux and temperature imply that $R_\infty = 6.82(D/300 \text{ pc})$ km. However, the lack of any distance determination for this star makes a radius estimation uncertain at this time.

More precise distance and spectral measurements for RXJ185635-3754 are needed to fix the radius of this star more accurately. Planned HST measurements of both the star's parallax and proper motion should yield the star's distance and velocity to about 10% accuracy. It will also determine if accretion is a plausible heating source for this star and may even allow the identification of the star's birthplace and age. In addition, the next HST measurements will be at different optical wavelengths than those obtained to date, possibly permitting more detailed atmospheric modeling of this star than the blackbody analyses performed so far.

THE FATE OF A NEWBORN NEUTRON STAR

After a supernova explosion, the gravitational mass of the remnant is less than 1 $M_\odot$. It is lepton rich and has an entropy per baryon of $S \simeq 1$ (in units of Boltzmann's constant k_B). The leptons include both electrons and neutrinos, the latter being trapped in the star because their mean free paths in the dense matter are of order 1 cm, whereas the stellar radius is about 15 km. Accretion onto the neutron star increases its mass to the 1.3–1.5 $M_\odot$ range, and should mostly cease after a second. It then takes about 10–15 s [6] for the trapped neutrinos to diffuse out, and in the diffusion process they leave behind most of their energy, heating the protoneutron star to fairly uniform entropy values of about $S = 2$. Cooling continues as thermally-produced neutrinos diffuse out and are emitted. After about 50 s, the star becomes completely transparent to neutrinos, and the neutrino luminosity drops precipitously [37].

Denoting the maximum mass of a cold, catalyzed neutron star by M_{max} and the maximum mass of the protoneutron star with abundant trapped leptons by M_{max}^L, there are two possible ways that a black hole could form after a supernova explosion. First, accretion of sufficient material could increase the remnant's mass to a value greater than either M_{max} or M_{max}^L and produce a black hole, which then appears on the accretion time scale [38]. Second, if exotic matter plays a role and if accretion is insignificant after a few seconds, then for $M_{max}^L > M > M_{max}$, where M is the final remnant mass, a black hole will form as the neutrinos diffuse out [20,27,31,39,40] on the deleptonization time scale of 10–15 s.

The existence of metastable neutron stars has some interesting implications. First, it could explain why no neutron star is readily apparent in the remnant of SN1987A despite our knowledge that one existed until at least 12 s after the supernova's explosion. Second, it would suggest that a significant population of relatively low mass black holes exists [41], one of which could be the compact object in the X-ray binary 4U1700-37 [2].

Neutrino-poor versus neutrino-rich stars

A detailed discussion of the composition and structure of protoneutron stars may be found in Refs. [20,42]. The main findings were that the structure depends more sensitively on the composition of the star than its entropy and that the trapped neutrinos play an important role in determining the composition. Since the structure is chiefly determined by the pressure of the strongly interacting constituents and the nature of the strong interactions is poorly understood at high density, several models of dense matter, including matter with strangeness-rich hyperons, a kaon condensate and quark matter were studied there.

Evolutionary calculations [6,39] without accretion show that it takes on the order of 10–15 s for the trapped neutrino fraction to vanish for a nucleons-only EOS. To see qualitatively what might transpire during the early evolution, we show in Fig. 6 the dependence of the maximum stellar mass upon the trapped neutrino fraction Y_{ν_e}, which decreases during the evolution. When the only hadrons are nucleons (np), the maximum mass increases with decreasing Y_{ν_e}, whereas when hyperons (npH) or kaons (npK) are also present, it decreases. Further, the rate of decrease accelerates for rather small values of Y_{ν_e}. Coupled with this is the fact that the central density of stars will tend to increase during deleptonization. The implication is clear. *If hyperons, kaons, or other negatively-charged hadronic species are present, an initially stable star can change into a black hole after most of the trapped neutrinos have left, and this takes 10 − 15 s. This happens only if the remnant mass M satisfies $M_{max}^L > M > M_{max}$.*

It must be emphasized that the maximum mass of the cold catalyzed star still remains uncertain due to the uncertainty in strong interactions at high density. At present, all nuclear models can only be effectively constrained

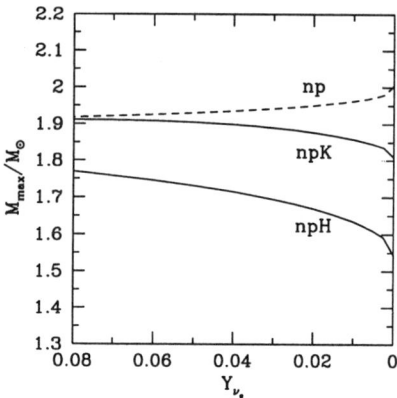

FIGURE 6. Maximum neutron star mass as a function of Y_{ν_e} for hadronic matter with only nucleons (np) or with nucleons and hyperons (npH) or kaons (npK).

at nuclear density and by the condition of causality at high density. The resulting uncertainty is evident from the range of possible maximum masses predicted by the different models. Notwithstanding this uncertainty, our findings concerning the effects of neutrino trapping offer intriguing possibilities for distinguishing between the different physical states of matter. These possibilities include both black hole formation in supernovae and the signature of neutrinos to be expected from supernovae.

Supernova SN1987A

On February 23 of 1987, neutrinos were observed [43] from the explosion of supernova SN1987A, indicating that a neutron star, not a black hole, was initially present. (The appearance of a black hole would have caused an abrupt cessation of any neutrino signal [37].) The neutrino signal was observed for a period of at least 12 s, after which counting statistics fell below measurable limits. From the handful of events observed, only the average neutrino energy, ~ 10 MeV, and the total binding energy release of $\sim (0.1 - 0.2)M_\odot$ could be estimated.

These estimates, however, do not shed much light on the composition of the neutron star. This is because, to lowest order, the average neutrino energy is fixed by the neutrino mean free path in the outer regions of the protoneutron star. Further, the binding energy exhibits a universal relationship [20] for a wide class of EOSs, including those with strangeness bearing components, namely

$$B.E. = (0.065 \pm 0.01)(M_B/M_\odot)^2 M_\odot, \qquad (7)$$

where M_B is the baryonic mass. This allows us only to determine a remnant gravitational mass of $(1.14 - 1.55)M_\odot$, but not the composition.

The ever-decreasing optical luminosity (light curve) [44] of the remnant of SN1987A suggests two arguments against the continued presence of a neutron star. First, accretion onto a neutron star at the Eddington limit is already ruled out for the usual hydrogen-dominated Thomson electron scattering opacity. (However, if the atmosphere surrounding the remnant contains a sufficient amount of iron-like elements [45] the appropriate Eddington limit is much lower.) Second, a Crab-like pulsar cannot exist in SN1987A, since the emitted magnetic dipole radiation would be observed in the light curve. Either the magnetic field or the spin rate of the neutron star remnant would have to be much less than in the case of the Crab and what is inferred from other young neutron stars. The spin rate of a newly formed neutron star is expected to be high; however, the time scale for the generation of a significant magnetic field is not well known and could be greater than 10 years.

Although most of the binding energy is released during the initial accretion and collapse stage in about a second after bounce, the neutrino signal continued for a period of at least 12 s. The compositionally-induced changes in the structure of the star occur on the deleptonization time scale, which we have estimated to be of order 10–15 s [20], not on the binding energy release time scale. Thus, the duration of the neutrino signal from SN1987A was comparable to the time required for the neutrinos initially trapped in the star to leave. However, counting statistics prevented measurement of a longer duration, and this unfortunate happenstance prevents one from distinguishing a model in which negatively-charged matter appears and a black hole forms from a less exotic model, in which a neutron star still exists. As we have pointed out, the maximum stable mass drops by as much as $0.2 M_\odot$ when the trapped neutrinos depart if negatively charged particles are present, which could be enough to cause collapse to a black hole.

Observed neutron stars lie in a very small range of gravitational masses (see Fig. 1). Estimates [46,47] of the gravitational mass of the remnant of SN1987A, based on the observed amounts of ejected ^{56}Ni and/or the total explosion energy, lie in the range $(1.40 - 1.56)M_\odot$. This range extends above the largest accurately known value for a neutron star mass, 1.44 $M_\odot$, so the possibility exists that the neutron star initially produced in SN1987A could be unstable in the cold, deleptonized state. In this case, SN1987A would have become a black hole once it had deleptonized, and no further signal would be expected. Should this scenario be observationally verified, it would provide strong evidence for the appearance of strange matter.

Future Directions

In an optimistic scenario, several thousand neutrinos from a typical galactic supernova might be seen in upgraded neutrino detectors (for rough characteristics of present and future neutrino detectors, see Ref. [48].) Among the interesting features that could be sought are:

1. possible cessation of a neutrino signal, due to black hole formation;

2. possible burst or light curve feature associated with the onset of negatively-charged, strongly interacting matter near the end of deleptonization, whether or not a black hole is formed;

3. identification of the deleptonization/cooling epochs by changes in luminosity evolution or neutrino flavor distribution;

4. determination of a radius-mean free path correlation from the luminosity decay time or the onset of neutrino transparency; and

5. determination of the neutron star mass from the universal binding energy-mass relation.

Acknowledgement

I gratefully acknowledge collaborations with G.E. Brown, J.M. Lattimer, P.J. Ellis, I. Bombaci, Manju Prakash, R. Knorren, and J.R. Cooke. This work was supported by the U. S. Dept. of Energy under grants DOE/DE/FG02-87ER-40388 and by the NASA grant NAG52863.

REFERENCES

1. S.E. Thorsett et al., Astrophys. J. **405** (1994) L29; M. H. van Kerkwijk et al., Astron. & Astrophys. **303** (1995) 497.
2. G.E. Brown et al., Astrophys. J. **463** (1996) 297.
3. R.C. Tolman, Proc. Nat. Acad. Sci. USA **20** (1934) 3; J.R. Oppenheimer and G.M. Volkoff, Phys. Rev. **55** (1939) 374.
4. S. Weinberg, Gravitation and Cosmology: Principles and Applications of the General Theory of Relativity (New York: Wiley) 1972.
5. J.M. Lattimer et al., Astrophys. J. **355** (1990) 241.
6. A. Burrows and J.M. Lattimer, Astrophys. J. **307** (1986) 178; A. Burrows, Ann. Rev. Nucl. Sci. **40** (1990) 181.
7. C. Kristian et al., Nature **338** (1989) 234.
8. C. Pennypacker et al., private communication (1990).
9. P. Haensel, Copernicus Astronomical Center Preprint (1990).
10. C.J. Pethick et al., Nucl. Phys. **A584** (1995) 675.
11. F.M. Walter et al., Nature, **379** (1996) 233; Nature, (1997), in press.

12. P.A. Caraveo et al., *Astron. Astrophys. Rev.*, **7** (1996) 209.
13. W.H.G Lewin et al., *Space. Sci. Rev.*, **62** (1993) 223, and references therein.
14. N.K. Glendenning, *Phys. Rev.* **D46** (1992) 4161.
15. G. G. Pavlov et al., *Astrophys. J.* **472** (1996) L33.
16. P. An et al., in preparation.
17. G. Greenstein and G.J. Hartke, *Astrophys. J.* **271** (1983) 283.
18. A.G. Lyne and D.R. Lorimer, *Nature*, **369** (1994) 127.
19. G.F. Bignami et al., *Astrophys. J. Lett.* **456** (1996) L111.
20. M. Prakash et al., *Phys. Rep.* **280** (1997) 1.
21. B. Friedman and V.R. Pandharipande, *Nucl. Phys.* **A361** (1981) 502.
22. R.B. Wiringa et al., *Phys. Rev.* **C38** (1988) 1010.
23. M. Prakash et al., *Phys. Rev. Lett.* **61** (1988) 2518.
24. H. Müller and B.D. Serot, *Nucl. Phys.* **606** (1996) 508.
25. H. Müther et al., *Phys. Lett.* **199** (1987) 469; L. Engvik et al., *Phys. Rev. Lett.* **73** (1994) 2650.
26. L. Titarchuk, *Astrophys. J.* **429** (1994) 340; F. Haberl and L. Titarchuk, *Astron. & Astrophys.* **299** (1995) 414; I. Bombaci, *Phys. Rev.* **C55** (1997) 1587.
27. M. Prakash et al., *Phys. Rev.* **D52** (1994) 661.
28. N.K. Glendenning and S.A. Moszkowski, *Phys. Rev. Lett.* **67** (1991) 2414; N.K. Glendenning, *Phys. Rev.* **D46** (1992) 1274.
29. D.B. Kaplan and A.E. Nelson, *Phys. Lett.* **B175** (1986) 57
30. E. Friedman et al., *Nucl. Phys.* **A579** (1995) 518; A. Schröter et al., *Z. Phys.* **A350** (1994) 101; R. Barth et al., *Phys. Rev. Lett.* **78** (1997) 4007.
31. V. Thorsson et al., *Nucl. Phys.* **A572** (1994) 693.
32. E. Witten, *Phys. Rev.* **D30** (1984) 272; C. Alcock and A. Olinto, *Ann. Rev. Nucl. Sci.*, **38** (1988) 161; Manju Prakash et al., *Phys. Lett.* **B243** (1990) 175.
33. N.K. Glendenning and F. Weber, *Astrophys. J.* **400** (1992) 672.
34. B. Link et al., *Nature* **359** (1992) 616; B. Datta and M. A. Alpar, *Astron. & Astrophys.* **275** (1993) 210.
35. D.G. Ravenhall and C.J. Pethick, *Astrophys. J.* **424** (1994) 846 9; see also C.P. Lorenz et al., *Phys. Rev. Lett.* **70** (1993) 37
36. F. Haberl et al., *Astron. & Astrophys.* in press.
37. A. Burrows, *Astrophys. J.* **334** (1988) 891.
38. G.E. Brown et al., *Comments Astrophys.* **16** (1992) 153.
39. W. Keil and H.T. Janka, *Astron. & Astrophys.* **296** (1994) 145.
40. N.K. Glendenning, *Astrophys. J.* **448** (1995) 797.
41. G.E. Brown and H.A. Bethe, *Astrophys. J.* **423** (1994) 659.
42. P.J. Ellis et al., *Comments on Nucl. and Part. Phys.* **22** (1996) 63.
43. K. Hirata et al., *Phys. Rev. Lett.* **58** (1987) 1490; R.M. Bionta et al., *Phys. Rev. Lett.* **58** (1987) 1494.
44. S. Kumagai et al., *Astron. & Astrophys.* **243** (1991) L13.
45. K. Chen and S.A. Colgate, Los Alamos preprint LA-UR-95-2972 (1995).
46. F.-K. Thielemann et al., *Astrophys. J.* **349** (1990) 222.
47. H.A. Bethe and G.E. Brown, *Astrophys. J.* **445** (1995) L129.
48. A. Burrows et al., *Phys. Rev.* **D45** (1992) 3361.

Composition and Energy Spectra of Cosmic Rays – Implications for Cosmic Ray Origins

Michael L. Cherry

Dept. of Physics and Astronomy, Louisiana State Univ., Baton Rouge, LA 70803

Abstract. A brief review is presented of the energy spectrum and composition of the cosmic rays up to the knee region near 10^{15} eV. The measurements suggest a picture based on acceleration in supernova shocks, including shocks produced by interaction with the winds of massive pre-supernova stars, coupled with a leaky box model of propagation through the galaxy.

INTRODUCTION

The initial observational fact to be recognized when discussing the nature and origin of the high energy cosmic rays is the smooth, almost featureless power law spectrum extending from approximately 10 GeV (where the effects of the solar wind and the earth's geomagnetic field become small) to above 10^{20} eV (Fig. 1). Over a range of ten orders of magnitude in energy the flux decreases by more than 26 orders of magnitude. At energies below ~ 1 TeV/nucleon, where balloon and spacecraft measurements can be made of energy and charge on a direct particle-by-particle basis, the individual elemental spectra can be seen to follow the expected power law (Fig. 2). It can already be seen, however, in Fig. 2 that the spectral slope is somewhat flatter for iron than for the H, He, and C.

If these individual elemental power law spectra are extended up in energy, then in the region $10^{15} - 10^{16}$ eV/nucleus, the iron flux becomes comparable to the flux of H and He. This region of the spectrum is shown in more detail in Fig. 3. It becomes clear here that the spectrum of Fig. 1 is in fact approximated by two power laws with a break or "knee" near $10^{15} - 10^{16}$ eV/nucleus. This region is interesting for several reasons:

1) The knee marks a demarcation point between different experimental techniques. Below 10^{15} eV, direct measurements can be made with experiments carried into space or to the top of the atmosphere on satellites or high altitude balloons. Above $\sim 10^{15}$ eV, measurements must be performed in a more

indirect fashion by using large ground-based air shower arrays. One therefore has a change of experimental techniques right where the spectrum appears to bend.

2) Accelerator data on proton-proton collisions at $\sqrt{s} = 2$ TeV correspond to a fixed target energy of 2×10^{15} eV. There is no reason to expect any change in the nature of nucleon-nucleon collisions at such an energy, but the situation is not so clear for nucleus-nucleus interactions. A sizable fraction of the cosmic rays near the knee are relatively heavy nuclei (C-Fe) interacting on air (N). Our knowledge about nucleus-nucleus collisions extends only up to SPS energies (200 GeV/nucleon, corresponding to 10^{13} eV/nucleus for cosmic ray iron). Although the superposition model of nucleus-nucleus interactions appears to work reasonably well [1], even at SPS energies there are indications of discrepancies between the measurements and the predictions [2]. From an experimental point of view, our knowledge of nucleus-nucleus collisions near 10^{15} eV is not particularly well-founded.

3) A supernova remnant expanding freely into the surrounding interstellar medium begins to slow down at a radius of a few parsecs, as the mass of the swept-up interstellar material approaches the mass of the outward-flowing ejecta. At a scale size ~ 1 pc, protons with energies in excess of about 10^{15} eV are no longer contained by the expanding remnant's magnetic fields and thus escape. This loss of particles implies that cosmic rays cannot be accelerated to energies beyond the knee by the standard mechanism of acceleration by supernova shocks [3,4].

For several reasons, then, having to do with the systematics of the experimental techniques, the nucleus-nucleus interaction physics, and the astrophysics of the source acceleration mechanism, the region near the knee in the spectrum is an interesting one. There is a great need to push the balloon and satellite measurements up in energy as far as possible, and simultaneously to extend the air shower measurements down in energy as far as possible so as to obtain overlap and check for consistency between the different techniques.

In this paper, I will sketch the picture of acceleration of the cosmic rays in galactic supernova shocks, together with propagation through a galaxy modelled by a leaky box. I will emphasize the importance of the high energy measurements near the knee region, and use the high energy spectrum and lower energy composition measurements to argue for the contribution of massive Wolf-Rayet stars as acceleration sites for the cosmic rays.

SUPERNOVA SHOCK ACCELERATION PLUS LEAKY BOX MODEL

The cosmic rays are a hot gas filling the galactic disk and exerting an outward pressure. The outward expansion is balanced by the gravitational force on the interstellar matter to which the galaxy's magnetic field is anchored.

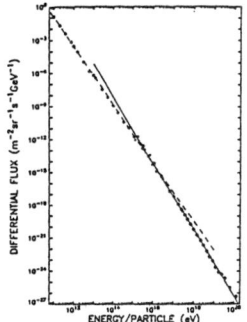

Fig. 1 The differential energy spectrum of all cosmic rays at the Earth.

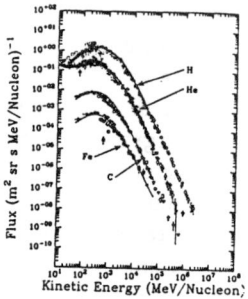

Fig. 2. Individual spectra of H, He, C, Fe.

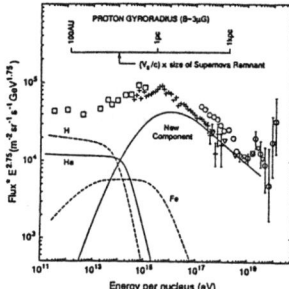

Fig. 3. Measured "all particle" spectrum together with expected spectra of H, He, Fe extrapolated from low energies [11].

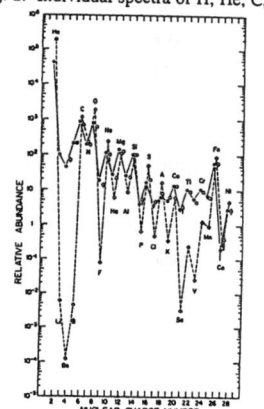

Fig. 4. Measured cosmic ray elemental composition (closed symbols) and solar system abundances (open symbols) [8].

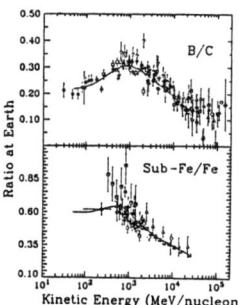

Fig. 5. Measured secondary-to-primary ratios [8].

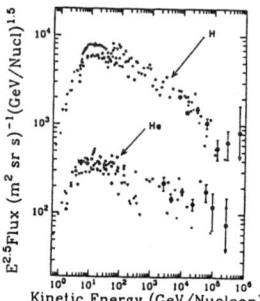

Fig. 6. H and He spectra. Large solid points are newest JACEE results [19].

The cosmic ray pressure inflates loops in the magnetic field that extend out beyond the disk, leading to losses of relativistic particles from the galaxy, perhaps through a galactic wind. The cosmic ray gas therefore requires constant replenishment. If V_{disk} is the volume of the galactic disk, $\rho_{cr} \sim 1$ eV/cm^3 is the cosmic ray energy density, and $\tau \sim 6 \times 10^6$ yrs is the cosmic ray age, the power required to supply the galactic cosmic rays is [5] $L_{cr} = V_{disk}\, \rho_{cr}/\tau \sim 5 \times 10^{40}$ erg/sec. The cosmic ray luminosity can be supplied by supernova explosions into the interstellar medium: If a type II supernova ejects $10 M_\odot$ into the interstellar medium with velocity 5000 km/sec once every 100 years, the power provided by supernovae is $L_{SN} \sim 10^{42}$ erg/sec. If the efficiency for producing cosmic rays is several percent, the required cosmic ray power can be supplied by galactic supernovae.

X-ray and radio observations give direct evidence for the presence of strong shocks and the production of relativistic particles with power law spectra in expanding supernova remnants [6]. The well-developed theory of shock acceleration gives a natural explanation for the observed power law spectrum, and is observed to operate in locales where detailed measurements are possible: e.g., in the earth's magnetosphere and in solar flares. The maximum energy to which a shock can accelerate particles depends on the the field strength B and the size of the acceleration region R. Very generally,

$$E_{max} \sim \frac{u}{c} q B R \quad ,$$

where q is the charge of the particle, u is the shock velocity, and c is the speed of light. This expression holds true for observed shocks on scales from the earth's bow shock ($E_{max} \sim 30$ keV) to solar flares (E_{max} up to 10 GeV), and should be applicable all the way up to the case of supernova remnants at $E_{max} \sim 10^{15}$ eV.

By comparing the measured composition of the cosmic rays near 1 GeV with terrestrial and solar abundances (Fig. 4), one finds that the cosmic rays are enhanced in the Li-Be-B and the sub-iron elements due to spallation of the primary cosmic rays during their propagation through the interstellar medium [7,8]. Above ~ 1 GeV, where solar modulation and magnetospheric effects become small, the ratio of secondary to primary cosmic rays is observed to decrease with increasing energy (Fig. 5), suggesting that the cosmic ray path length decreases with increasing energy as $\sim E^{-0.6}$.

One can write a cosmic ray transport equation incorporating these results [5]. A solution can be obtained by assuming a simple "leaky box" model of propagation [9,10]: If particles of charge Z propagate freely through the galaxy with some escape mean free path $\lambda_{esc} = 10.8(4Z/E)^{0.6}$ g/cm^2, where E is in GeV, and one ignores energy gains and losses and convection, then for a primary nucleus, where the feed-down from fragmentation of heavier nuclei is unimportant,

$$N(E) = \frac{Q(E)\,\lambda_{esc}}{1+\lambda_{esc}/\lambda_{int}} \sim \frac{E^{-\alpha-.6}}{1+\lambda_{esc}/\lambda_{int}} \quad.$$

Here $Q(E) = aE^{-\alpha}$ is the power-law source term and λ_{int} is the interaction mean free path. In the proton case, where $\lambda_{esc} \ll \lambda_{int}$,

$$protons \implies N(E) \sim E^{-\alpha-.6} \quad.$$

The observed $E^{-2.7}$ spectrum at earth implies a source spectrum of the form $E^{-2.1}$. At the maximum energy E_{max}, the source spectrum and the corresponding spectrum observed at earth then presumably steepen or cut off.

In the iron case, the situation is different. Here the interaction mean free path is much smaller: $\lambda_{int} \sim 2$ g/cm^2. At low energies,

$$iron: \quad \lambda_{int} < \lambda_{esc} \implies N(E) \sim Q(E) \sim E^{-2.1} \quad.$$

At higher energies the spectrum steepens:

$$iron: \quad \lambda_{esc} < \lambda_{int} \implies N(E) \sim Q\lambda_{esc} \sim E^{-2.7} \quad.$$

The iron will steepen at an energy 26 times higher than that of the protons due to the maximum accelerator energy, but an additional high energy steepening will come from the propagation effects.

The resulting picture is shown in Fig. 3, where the measured all-particle spectrum is shown. Fig. 3 also shows the expected elemental spectra extrapolated up to the knee region [11]. At low energies, the H, He, and Fe spectra are shown with the slopes measured by balloon and spacecraft experiments [12–14]. Based on the JACEE balloon data of ref. [14], a cutoff is assumed for the hydrogen at 40 TeV, and similarly at 40 Z TeV for the heavier species. In order for the sum of the elemental spectra to account for the total observed flux, a new component must be assumed of unknown origin [11].

This picture has two main problems: First, the maximum energy for supernova acceleration is typically [4]

$$E_{max} \sim 10^{14} Z \left(\frac{B}{1\mu G}\right) eV \quad.$$

This upper limit to the spectrum can be extended upward [15–18] by invoking reacceleration from multiple sources, perpendicular shocks, stronger fields, or larger sources (e.g., AGNs or a galactic wind). Nevertheless, the existence of particles at energies significantly higher than 10^{14} eV poses a problem for the standard supernova shock/leaky box model. Second, a fairly accurate (and therefore perhaps implausible?) tuning is required to match the normalization at the knee: The low energy component must be cut off and the new component must turn on at about the same point with just the right normalization in order to result in a remarkably smooth transition at the knee.

The motivation for the 40 TeV proton cutoff in Fig. 3 came from the preliminary results of the JACEE balloon experiment [14], which reported a possible bend in the hydrogen spectrum at 40 TeV but no corresponding feature for the helium. Nilsen [19] reports at this meeting, however, that with significantly better statistics and with improvements in the background subtraction procedures, JACEE now sees a proton spectrum consistent with a single power law up to 800 TeV (Fig. 6).

A break in the spectrum must occur at some point near the knee. Otherwise the individual element fluxes extrapolated as continuing power laws up to high energy will exceed the total all-particle flux measured by the air shower experiments. If one assumes, then, that the proton spectrum steepens at 10^{15} eV (just above the JACEE range), and the heavier elements correspondingly steepen at 10^{15} Z eV, then the result is shown in Fig. 7. The solid line is the sum of the individual elemental spectra. The summed spectrum agrees very well with the observations up to above the knee. Fig. 7 suggests that there is no unknown new component. Rather, the spectra do not break abruptly at some E_{max}, but rather only steepen and then continue upward in energy to well beyond the nominal E_{max}.

WOLF-RAYET STARS AND COSMIC RAY ORIGIN

The sun's corona is sufficiently hot that some of its hydrogen and helium can escape and stream out as an ionized plasma into the interplanetary medium. The sun's mass loss rate is $\dot{M} \sim 10^{-12} M_\odot$/yr. The solar wind moves past the earth at 400 km/sec, and the bow shock it creates is observed to be an efficient accelerator of energetic particles. The magnetized plasma moving out from the rotating sun sets up a Parker spiral configuration of the interplanetary field [20] such that, at sufficiently great radial distances from the sun, the field lines are largely perpendicular to the outward-flowing wind.

The Lagage-Cesarsky supernova shock acceleration model [4] was based on the assumption of quasi-parallel shocks. In a perpendicular shock [15,21], drift along the magnetic field moves the particles parallel to the shock boundary, enabling the particles to cross the shock boundary multiple times and gain considerably more energy than in the case of a parallel shock. Acceleration at perpendicular shocks is observed in the solar system in the case of corotating interaction regions where rapidly-moving plasma ejected from recent solar flare sites overtakes slowly moving plasma ejected previously.

The solar wind involves modest parameters compared to those in Wolf-Rayet systems [22]. Wolf-Rayet stars are massive stars ($M > 20 M_\odot$) whose coronas are sufficiently hot and sufficiently extended that their mass loss rates and wind velocities are extremely high: $\dot{M} \sim 10^{-5} M_\odot/yr$ and $v_{wind} \sim 2000$ km/sec. Wolf-Rayet magnetic fields are $\sim 10^3-10^4$ G at the stellar surface and 3 G at 10^{14} cm [23]. In a case where a Wolf-Rayet star produces a supernova, the supernova ejecta expand into the Wolf-Rayet wind, and the maximum energy (noting that Br is a constant in a Parker spiral) is expected to be [24]

$$E_{max} \sim 10^{17} Z \left(\frac{Br}{3 \times 10^{14} \text{ G} - \text{cm}}\right) eV \ .$$

Our galaxy contains 300-1000 Wolf-Rayet stars with an average lifetime of 10^5 yrs. The Wolf-Rayet supernova rate is therefore $\sim 1/100$-300 yrs compared to an overall galactic supernova rate of 1/30 yrs. In other words, 10-30% of supernovae may be due to Wolf-Rayet stars, so that acceleration into Wolf-Rayet winds provides a mechanism for producing significant numbers of particles at energies well above the standard Lagage-Cesarsky limit [25].

The cosmic ray source composition can be derived by correcting the measured composition for the effects of propagation through the galaxy [7,8]. The resulting source composition is shown in Fig. 8 as a function of first ionization potential and mass [26–28]. Refractory elements should be preferentially bound in grains. Assuming then that only gas ions (and not grains) are accelerated, one expects that the refractory elements should be depleted in the cosmic rays. It is clear from Fig. 8 that they are not. The implication is that standard interstellar gas is not the source material for the observed cosmic rays.

From the point of view of nucleosynthesis, it should be noted that ^{20}Ne, Mg, Al, and Na are all produced by carbon burning. Yet Fig. 8 shows that the ^{20}Ne is suppressed by a factor ~ 8. Similarly, S, Ar, Si, and Ca are all produced by oxygen and silicon burning, and yet the S and Ar are low by a factor of 4. The ^{22}Ne/^{20}Ne cosmic ray source ratio is 4.4 times the solar and interstellar medium ratio, and the C/O ratio is 1.7 times the solar/ISM value. The galactic cosmic ray source abundances appear to be inconsistent with the values expected from nucleosynthesis, and suggest a component enriched by the presence of helium-burning material [29,30]. In massive stars, the CNO cycle converts the initial CNO into ^{14}N, ^{18}O, and ^{22}Ne, and transforms ^{4}He into ^{12}C. Numerous authors have pointed out that high-mass Wolf-Rayet stars have had their outer layers stripped off by their strong winds, to the point where the stars' helium-burning layers are exposed. The wind material is then highly enriched in the products of helium burning [24].

Based on the recent H and He results from the JACEE balloon experiment [19], the hydrogen has a slightly steeper spectrum than the helium in the region below the knee:

$$\frac{dn}{dE} \sim \begin{cases} E^{-2.80 \pm 0.04} & hydrogen \\ E^{-2.68 \pm 0.06} & helium \end{cases}$$

The spectral index depends on the compression ratio across the shock, so a steeper spectrum for H than for He (and the slightly different spectral indices seen for some of the heavier elements [14]) suggest different sources. Biermann and collaborators [24] have argued strongly for a model in which hydrogen is accelerated by supernova shock waves in the interstellar medium, and the helium and heavies are produced by supernova shocks expanding into Wolf-Rayet winds.

CONCLUSION

The combination of composition and spectrum measurements is providing direct particle-by-particle results nearly up to the knee region. The most recent spectral results are consistent with a steepening in the proton spectrum near 10^{15} eV, and with modest additional exposure on either the Space Station or long duration balloons, it should be possible to resolve the question of what is happening at the knee. The Wolf-Rayet/multiple supernova source picture seems promising, but additional theoretical work is needed to fully explain the break in the spectrum and the acceleration mechanism (and its cutoff behavior) above the knee.

I thank the conference organizers for their hospitality at this meeting, and acknowledge the support of the Louisiana State Board of Regents, NASA EPSCoR, and NSF Particle Physics.

REFERENCES

1. G. Schatz et al., J. Phys. G. **20**, 1267 (1994).
2. P.V. Deines-Jones et al., Phys. Rev. **C53**, 3044 (1996); M.L. Cherry et al., these proceedings.
3. I. Axford, Proc. 17th Intl. Cosmic Ray Conf., Paris **12**, 155 (1981).
4. P.O. Lagage and C.J. Cesarsky, Astron. Ap. **125**, 249 (1983).
5. T.K. Gaisser, *Cosmic Rays and Particle Physics*, Cambridge Univ. Press, Cambridge (1990).
6. K. Koyama et al., Nature **378**, 255 (1995).
7. M. Garcia-Munoz et al., Ap. J. Suppl. **64**, 269 (1987).
8. J.P. Wefel, in *Cosmic Rays, Supernovae, and the Interstellar Medium*, ed. by M.M. Shapiro, R. Silberberg, and J.P. Wefel, Kluwer, Dordrecht, p. 29 (1991).
9. J.F. Ormes and P.S. Freier, Ap. J. **222**, 471 (1978).
10. C.J. Cesarsky, Ann. Rev. Astron. Ap. **18**, 289 (1980).
11. J. Waddington et al., NASA Cosmic Ray Program Working Group report (1992).
12. S.P. Swordy et al., Proc. 23rd Intl. Cosmic Ray Conf., Calgary **5**, 243 (1993).
13. J.M. Grunsfeld et al., Ap. J. Lett. **327**, L31 (1988); D. Müller et al., Ap.J. **374**, 356 (1991); S.P. Swordy et al., Ap. J. **349**, 625 (1990).

14. K. Asakimori et al., Proc. 23rd Intl. Cosmic Ray Conf., Calgary **2**, 21 and 25 (1995); Y. Takahashi, in *The Sun and Beyond*, ed. by Tran Thanh Van, World Scientific (1996).
15. J.R. Jokipii, Ap. J. **313**, 842 (1987).
16. H.J. Volk and P.J. Biermann, Ap.J. Lett. **333**, L65 (1988).
17. Auger Project Design Report, Fermilab (1995).
18. J.R. Jokipii and G. Morfill, Ap. J. **312**, 170 (1987).
19. B. Nilsen et al., these proceedings; M.L. Cherry et al., Proc. 25th Intl. Cosmic Ray Conf., Durban, to be published (1997).
20. E.N. Parker, Ap.J. **128**, 664 (1958).
21. D.C. Ellison et al., Publ. Astron. Soc. Pac. **106**, 780 (1994).
22. K.A. Van der Hucht and B. Hidayat, eds., *Wolf-Rayet Stars and Interrelations with Other Massive Stars in the Galaxy*, IAU Symp. **43**, Kluwer, Dordrecht, (1991); K.A. Van der Hucht and P.M. Williams, eds, *Wolf-Rayet Stars: Binaries, Colliding Winds, Evolution*, IAU Symp. **163** (1995).
23. M. Maheswaran and J.P. Casinelli, Ap. J. **386**, 695 (1992).
24. P.L. Biermann, Astron. Ap. **271**, 649 (1993); P.L. Biermann and R.G. Strom, Astron. Ap. **275**, 659 (1993); P.L. Biermann, T.K. Gaisser, and T. Stanev, Phys. Rev. **D51**, 3450 (1995).
25. P.L. Biermann and J.P. Cassinelli, Astron. Ap. **277**, 691 (1993).
26. J.P. Meyer, Ap. J. Suppl. **57**, 173 (1985).
27. J.P. Meyer, L. O'C. Drury, and D.C. Ellison, Ap. J., to be published (1997).
28. D.C. Ellison, L. O'C. Drury, and J.P. Meyer, Ap. J., to be published (1997).
29. M. Cassé and J.A. Paul, Ap.J. **258**, 860 (1982).
30. R. Silberberg et al., Ap. J. **363**, 265 (1990).

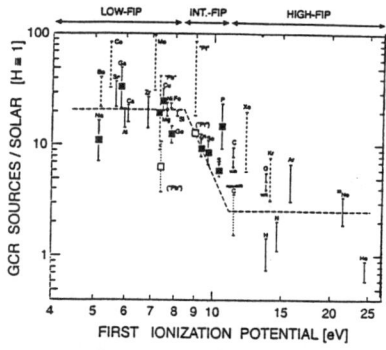

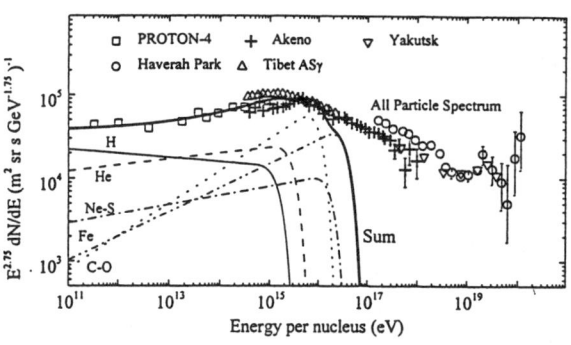

Fig. 7. All particle spectrum and summed JACEE spectra.

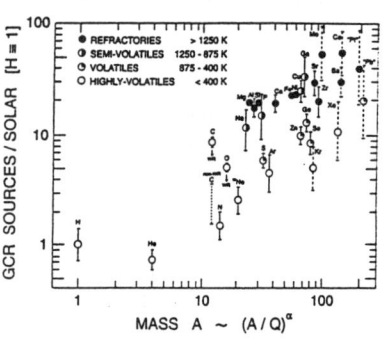

Fig. 8. Cosmic ray source abundances [27, 28].

Cosmic Ray H and He Spectra from 2 to 800 TeV/nucleon from the JACEE Experiments

B.S. Nilsen[1], K. Asakimori[2], T.H. Burnett[3], M.L. Cherry[1],
K. Chevli[4], M.J. Christl[5], S. Dake[6], J.H. Derrickson[5],
W.F. Fountain[5], M. Fuki[7], J.C. Gregory[4], T. Hayashi[8],
A. Iyono[9], J. Iwai[3], J. Johnson[4], M. Kobayashi[10], J. Lord[3],
O. Miyamura[11], K.H. Moon[5a], H. Oda[6], T. Ogata[12],
E.D. Olson[3b], T.A. Parnell[5], F.E. Roberts[5], K. Sengupta[1c],
T. Shiina[4], S.C. Strausz[3], T. Sugitate[11], Y. Takahashi[4],
T. Tominaga[11], J.W. Watts[5], J.P. Wefel[1], B. Wilczynska[13],
H. Wilczynski[13], R.J. Wilkes[3], W. Wolter[13], H. Yokomi[14],
and E. Zager[3].

[1] *Dept. of Physics and Astronomy, Louisiana State Univ., Baton Rouge, LA 70803*
[2] *Kobe Women's Junior College, Kobe, Japan*
[3] *Dept. of Physics, Univ. of Washington, Seattle, WA 98195*
[4] *Dept. of Physics, Univ. of Alabama, Huntsville, AL 35899*
[5] *NASA Marshall Space Flight Center, Huntsville, AL 35812*
[6] *Kobe Univ., Kobe Japan*
[7] *Kochi Univ., Kochi, Japan*
[8] *Waseda Univ., Tokyo, Japan*
[9] *Okayama Univ. of Science, Okuyama, Japan*
[10] *KEK, Tsukuba, Japan*
[11] *Hiroshima Univ., Hiroshima, Japan*
[12] *Inst. for Cosmic Ray Research, Tokyo, Japan*
[13] *Inst. for Nuclear Physics, Krakow, Poland*
[14] *Tezukayama Univ., Nara, Japan.*

Abstract. Results for the cosmic ray hydrogen and helium spectra up to 800 TeV, near the "knee" region, are presented. There is no sign of a break in either the hydrogen or helium spectra. The differential power law slopes are 2.80 ± 0.04 for hydrogen and 2.68 ± 0.06 for helium. With these new H and He measurements, together with earlier reported results for the heavier elements, the sum of the spectra give an all-particle spectrum that is in good agreement with the all-particle spectrum measured using extensive air showers.

By making cosmic ray composition measurements through the all-particle spectrum's "knee" region (10^{14}–10^{16} eV), we hope to learn why the all-particle spectrum changes slope from ~ 2.6 to ~ 3.0 [1]. Below the knee, instruments on balloons and satellites make direct measurements of the charge of individual cosmic rays; but near the knee and above, the all-particle measurements rely on indirect extensive air shower techniques. The JACEE (Japanese-American Cooperative Emulsion Experiment) collaboration has now measured the hydrogen and helium cosmic ray spectra up to 800 TeV and 400 TeV/n respectively.

JACEE uses electron-sensitive nuclear emulsion and x-ray film as the sensitive elements in a large-area balloon-borne thin electromagnetic sampling calorimeter. (See reference [2] and the references therein for a complete discussion of the techniques used.) JACEE has now completed fifteen flights, including two long duration flights (>120 hrs) from Australia to South America and four flights (>200 hrs) in Antarctica. We have now completed the analysis of the hydrogen and helium cosmic ray spectra through JACEE flight 12, thereby nearly doubling the total exposure reported previously [3]. When we finish the analysis of flights 13 and 14, the exposure will double again.

The hydrogen and helium differential spectra are shown in Fig. 1. The JACEE results agree well with earlier measurements. There are now 25 hy-

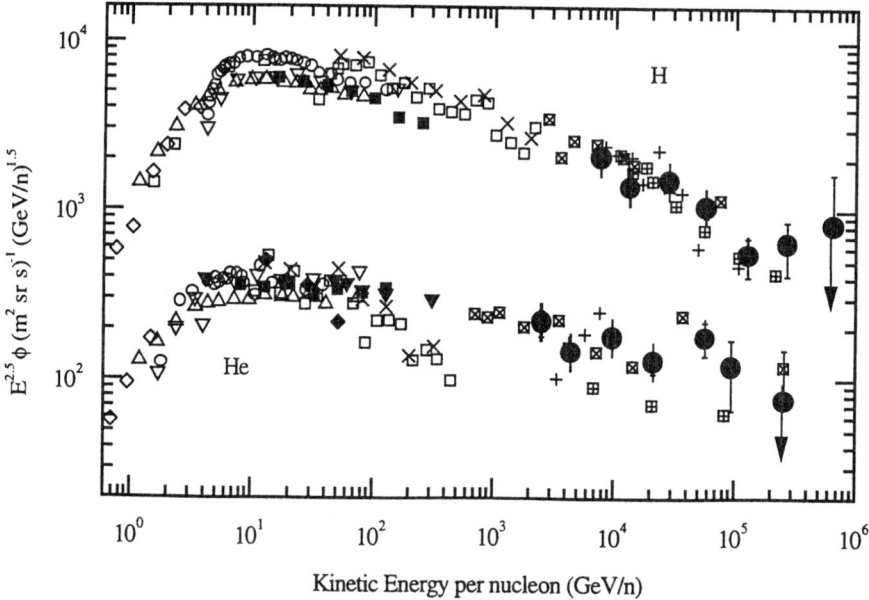

FIGURE 1. Differential spectra for hydrogen and helium (•). Also shown for comparison are other measurements, see references in [2].

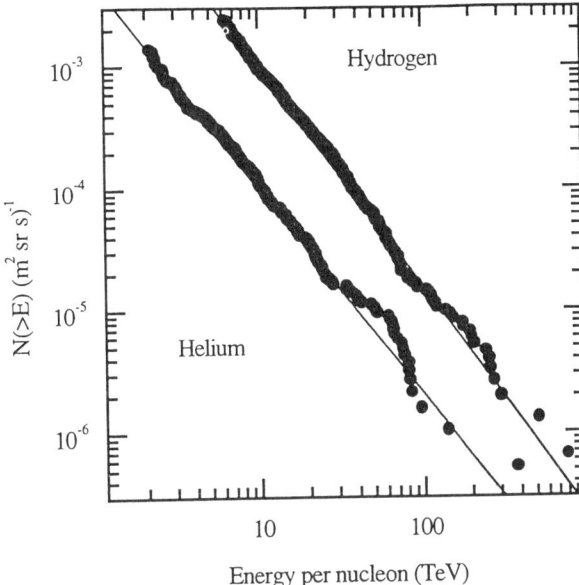

FIGURE 2. Integral spectra for hydrogen and helium. Each point corresponds to one more event than the point to the right. At high energies the correlated nature of an integral spectrum and the low statistics are responsible for the characteristic "waviness".

drogen events above 90 TeV and 37 helium events above 25 TeV/n. The differential spectra are, however, sensitive to the binning used. Therefore, we fit to the integral spectra shown in Fig. 2.

There is no sign of a break in either distribution shown in Fig. 2. This is contrary to earlier results [3] which suggested a break in the hydrogen spectrum at about 40 TeV but no comparable break in the helium spectrum. Fitting the spectra to a single power law gives $N(>E)|_H = (6.2^{+0.5}_{-0.3}) \times 10^{-2} (\frac{E}{1TeV})^{-1.80\pm0.04}$ (m² sr s)⁻¹ and $N(>E)|_{He} = (4.7\pm0.1) \times 10^{-3} (\frac{E}{1TeV/n})^{-1.68\pm0.06}$ (m² sr s)⁻¹.

In 1992 the Cosmic Ray Program Working Group issued a report advocating composition measurements through the knee region [4]. The JACEE results now make possible a direct comparison of the balloon-borne and air shower results at these energies. Using these new hydrogen and helium spectra, along with fits to preliminary JACEE CNO, Ne-S, and Fe group data [5], an estimate of the all-particle spectrum can be made (Fig. 3). The cutoff in the spectra are chosen here to be at Z×1000 TeV/nucleus, where Z is the particle charge, as suggested by supernova shock wave acceleration models. We assume no flattening in the Ne-S and Fe group spectra (as might be expected when the rigidity-dependent escape mean free path begins to dominate the interaction mean free path [6]). The all-particle spectrum obtained from summing

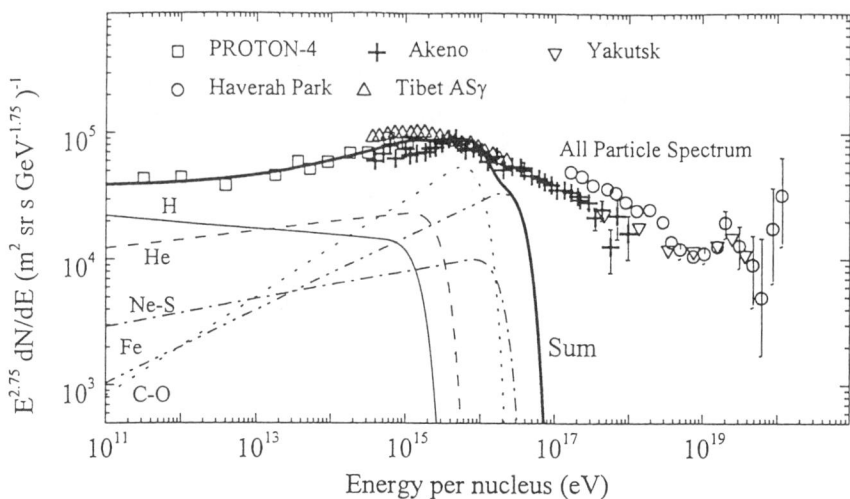

FIGURE 3. The all-particle spectrum plus the elemental spectra measured by JACEE. The thick solid line is the sum of the elemental spectra.

these spectra (assuming power laws with no flattening and a cutoff at $Z \times 1000$ TeV/nucleus) is remarkably close to the observed all-particle spectrum.

With these new results, JACEE is beginning to determine the composition approaching the all-particle knee region. With the analysis of JACEE flights 13 and 14 we will either find a knee in the hydrogen and/or helium spectra or push up the position of the knee to still higher energies.

REFERENCES

[a] Deceased
[b] Present Address: WRQ Inc., 1500 Dexter Ave. N., Seattle, WA 9809
[c] Present Address: Horizon Computer Corp., 5 Lincoln Hwy., Edison, N.J. 08820

1. Amenomori, M. et al., Ap. J. **461**, 408 (1996)
2. Asakimori, K. et al., submitted to Ap. J. (1997)
3. Asakimori, K. et al., 23rd Intl. Cosmic Ray Conf. (Calgary) **2**, 21 & **2**, 25 (1993)
4. Waddington, C. J. et al., "Galactic Origin and the Acceleration Limit", NASA report, September 1992
5. Takahashi, Y., Proc. IInd Recontres du Vietnam, "The Sun and Beyond", Ho Chi-Minh City, ed. Tran Thanh Van (World Scientific 1996)
6. Esposito, J. A. et al., Ap. J. **351**, 459 (1990)

AUTHOR INDEX

A

Abbott, D., 561
Abbott, D. J., 213
Abela, R., 429
Abreu, M. C., 233
Adams, J. M., 399
Adler, R., 372
Ahmidouch, A., 213, 561
Ahn, J. K., 923
Ajaka, J., 567
Alberg, M. A., 862
Alessandro, B., 233
Alexa, C., 233
Alexopoulos, T., 812
Alkofer, R., 746
Amann, J. F., 826
Amatuni, T. S., 561
Ambrogiani, M., 887
Amsbaugh, J. F., 416
Anaya, J. M., 399
Andrić, D., 717
Angelopoulos, A., 372
Aoki, S., 923
Apostolakis, A., 372
Argiro, S., 887
Armstrong, C., 561
Armstrong, C. S., 213
Armstrong, T., 419
Arrington, J., 213, 561
Asakimori, K., 1031
Ashery for the E791 Collaboration, D., 451
Aslanides, E., 372
Assamagan, K. A., 213, 561
Astafyeva, N. M., 269
Astruc, J., 233
Awes, T. C., 643

B

Backenstoss, G., 372, 717
Bagaturia, J., 429
Baglin, C., 233
Bagnasco, S., 887
Baker, F. T., 342
Baker, O. K., 213, 561
Baldini, W., 887
Baldit, A., 233
Bargassa, P., 372
Barkov, L. M., 792
Barnett, M. A., 558
Barrow, S., 561
Barrow, S. P., 213
Barth, J., 849
Bassalleck, B., 457
Bavink, S., 704
Beard, K., 561
Beatty, D., 561
Beatty, D. P., 213
Beck, D. H., 213
Beck, M., 416
Beddo, M. E., 643
Bedjidian, M., 233
Bee, C. P., 372
Beedoe, S., 561
Beedoe, S. Y., 213
Behnke, O., 372
Beise, E., 561
Beise, E. J., 213
Bellaich, F., 233
Bellemann, F., 704
Belz, E., 561
Belz, J., 763
Belz, J. E., 213
Benante, J., 792
Benelli, A., 372
Bennhold, C., 849, 912
Beole, S., 233
Berdoz, A. R., 328, 351
Berg, A., 704
Bergman, D., 812
Bertin, V., 372
Bertini, F., 887
Bertl, W., 429
Bettoni, D., 887
Bhat, C., 812
Bhattacharya, T., 736
Bijker, R., 519
Bildsten, L., 987
Bimbot, L., 342
Birchall, J., 328, 351
Bisplinghoff, J., 704

1035

Blanc, F., 372
Blanpied, G., 547
Blaylock, G., 101
Blazey, G. C., 18
Blecher, M., 547
Bloch, P., 372
Blotz, A., 670
Bochna, C., 561
Bochna, C. W., 213
Bockhorst, M., 849
Bohlscheid, G., 704
Boldea, V., 233
Bolton, R. D., 826
Bombonati, M., 887
Bonazzola, G., 233
Bonsi, D., 887
Bordalo, P., 233
Borhani, A., 233
Borreani, G., 887
Bosnar, D., 717
Bosted, P., 588
Bosted, P. E., 213
Bowles, T. J., 399
Bowman, J. D., 328, 351
Brash, E. J., 213, 342
Braun, W., 849
Breuer, H., 213, 561, 717
Brooks, M., 34
Brown, B. A., 408
Brown, C. N., 643
Brown, D. H., 792
Brown, H. N., 792
Bruins, E., 561
Buchalla, G., 49
Buchanan, C., 419
Buchmann, A. J., 685
Bugg, B., 141
Bunce, G., 792
Burgwinkler, R., 849
Burnett, T. H., 1031
Bush, J. D., 643
Bussère, A., 233
Buzzo, A., 887

C

Cadman, R. V., 213
Calabrese, R., 887
Campbell, J. R., 328, 351

Camps, J., 408
Capony, V., 233
Caracappa, A., 547
Cardarelli, M., 887
Cardman, L., 213
Carey, R. M., 792
Carey, T. A., 643
Carlini, R., 561
Carlini, R. D., 213
Carlson, P., 372
Carroll, M., 372
Carvalho, J., 372
Cason for the E852 Collaboration, N. M., 471
Castor, J., 233
Catanese, M., 979
Cawley, E., 372
Ceccucci, A., 887
Cerú, M., 233
Cester, R., 887
Cha, J., 213, 561
Chambon, T., 233
Chan, Y. D., 931
Chang, T. H., 643
Chang for the E877 Collaboration, W., 247
Chant, N., 561
Chant, N. S., 213
Charalambous, S., 372
Chardin, G., 372
Chatterjee, L., 141
Chaurand, B., 233
Chen, Y., 34, 826
Cherli, K., 1031
Cherry, M. L., 253, 1022, 1031
Chertok, M. B., 372
Chertovskikh, A., 792
Chevrot, I., 233
Cheynis, B., 233
Chiavassa, E., 233
Chrisl, M. J., 1031
Chung, K. S., 923
Chung, M. S., 923
Chupp, T. E., 399
Church, E. D., 570
Cicalo, C., 233
Clark, R., 931
Clark for the IMB Collaboration, R., 958
Cody, A., 372
Cohn, N., 141

Collins, G., 213
Constantinescu, S., 233
Conzett, H. E., 403
Coon, S. A., 368
Cooper, M., 34
Cooper, M. D., 826
Cooper, P. S., 34, 826
Cooper, W. E., 643
Corbin, B., 419
Corcoran, M. D., 628
Cotanch, S. R., 515, 733
Cothran, C., 213, 561
Coulter, K. P., 399
Cullen, J., 792
Cummings, W. J., 213, 561
Cushman, P., 792

D

Dabrowska, A., 253
Dabrowski, W., 233
Dalpiaz, P., 887
Danagoulian, S., 213, 561
Danby, G. T., 792
Danielsson, M., 372
Dannbam, K., 743
Davis, C. A., 328, 351
Day, D., 561
Debevec, P. T., 792
DeBraeckeler, L., 416
De Falco, A., 233
Deines-Jones, P., 253
Dejardin, M., 372
Deleplanque, M. A., 931
Dellacasa, G., 233
De Marco, N., 233
DeMoor, P., 408
Deng, H., 792
Deninger, W., 792
Derre, J., 372
Derrickson, J. H., 1031
DeSchepper, D., 561
de Simone for the KLOE Collaboration, P., 819
Deutsch, J., 408
Devaux, A., 233
Devlin, T. J., 812
Dewey, M. S., 399
Dhawan, S. K., 792

Disco, A., 792
Dita, S., 233
Djalali, C., 547
Döbbeling, H., 717
Dobrotin, N. A., 269
Dooling, T., 717
Doornbos, J., 812
Doyle, J., 387
Dragowsky, M. R., 931
Drake, S., 1031
Drapier, O., 233
Dremin, I. M., 269
Druzhinin, V. P., 792
Ducret, J.-E., 561
Dugas, J., 253
Duncan, F., 561
Duncan, F. A., 213
Dunlop for the E866 and E917 Collaborations, J. C., 259
Dunne, J., 561
Dunne, J. A., 213
Duong, L., 792
Dutta, D., 213, 561
Dutto, G., 322
Dzemidzic, M., 34, 826

E

Ealet, A., 372
Earle, W., 792
Eckart, B., 372
Eden, T., 213, 561
Efemenko, Y., 141
Efstathiadis, E., 792
Eichenbaum, A., 812
Eilerts for the E865 Collaboration, S., 774
Eleftheriadis, C., 372
Elliot, S. R., 399
Empl, A., 34
Endo, K., 792
Engler, R., 429
Ent, R., 213, 561
En'yo, H., 923
Ernst, J., 704, 849
Erwin, A., 812
Espagnon, B., 233
Evangelou, I., 372

F

Faessler, A., 685
Fallon, P., 931
Fan, X., 887
Faravel, L., 372
Fargeix, J., 233
Farley, F. J. M., 792
Fassnacht, P., 372
Fayard, C., 855
Fazely, A., 141
Fedchak, J., 561
Fedotovich, G. V., 792
Fei, X., 792
Feinberg, E. L., 269
Felder, C., 372
Fellbaum for the E143 Collaboration, J., 588
Ferreira-Marques, R., 372
Fetscher, W., 372
Feurstein, M., 756
Fidecaro, M., 372
Filipčič, A., 372
Filippone, B. W., 213
Fleuret, F., 233
Force, P., 233
Foreman, W., 826
Forest, T. A., 213
Fortune, H. T., 213, 561
Fountain, W. F., 1031
Frabetti, S., 887
Francis, D., 372
Frank, M., 753
Franklin, G. B., 919
Franz for the PS 185 Collaboration, J., 858
Freedman, S. J., 399
Freedman (No manuscript provided by speaker.), S., 110
Frolov, V., 561
Frolov, V. V., 213
Fry, J., 372
Fujikawa, B. K., 399
Fujino, D. H., 316
Fujino for the CLEO Collaboration, D. H., 865
Fuki, M., 1031
Fukuda, T., 923
Fuller, G., 160
Funahashi, H., 923

Furić, M., 717

G

Gabathuler, E., 372
Gabriel, T., 141
Gagliardi, C., 34
Gagliardi, C. A., 643, 826
Galik for the CLEO Collaboration, R. S., 489
Gallio, M., 233
Gamet, R., 372
Ganberg, L., 664
Gao, H., 213, 561
Garcia, A., 399
Gardner, R. W., 654, 786
Gardner, S., 383
Garreta, D., 372
Garvey, G. T., 643
Garzoglio, G., 887
Gavrilov, Y. K., 233
Geer, S., 419
Geesaman, D. F., 213, 561, 643
Geller, J., 792
Gemme, C., 659
Gerber, H.-J., 372
Gerhaeuser, J., 792
Gerschel, C., 233
Gerstner, G. M., 558
Gibson, B. F., 927
Gilman, R., 213, 342, 561
Ginther, G., 632
Giordano, G., 547
Giron, S., 792
Giubellino, P., 233
Glander, K. H., 849
Glashausser, C., 342
Glozman, L. Y., 743
Go, A., 372
Goers, S., 849
Goldman, I. D., 931
Gollwitzer, K. E., 887
Golub, R., 387
Golubev, V. B., 792
Golybeva, M. B., 233
Gonchanova, L. A., 269
Gonin, M., 233
Goodman for the MINOS Collaboration, M., 300

Goodman for the Soudan 2 Collaboration, M., 962
Gorodetzky, P., 233
Görres, J., 987
Goto, Y., 923
Govaerts, J., 408
Gram, P. A. M., 717
Green, A. A., 328, 351
Green, G. L., 399
Green, P. W., 328, 351
Gregory, J. C., 1031
Gregory, N. K., 717
Grigorev, D. N., 792
Grosse Perdekamp, M., 792
Grossiord, J. Y., 233
Grossmann, A., 429, 792
Guaita, P., 233
Guber, F. F., 233
Gueye, P., 561
Gueye, P. L. J., 213
Guichard, A., 233
Gunter, J., 447
Gupta, R., 736
Gustafson, R., 419
Gustafsson, K. K., 213
Guyot, C., 372

H

Haeberlen, U., 792
Hahn, A., 887
Haidenbauer, J., 725
Haim, D., 826
Hall, L. J., 197
Hamian, A. A., 351
Hanien, A. A., 328
Hannappel, J., 849
Hansen, J. O., 561
Hansen, J-O., 213
Haroutunian, R., 233
Harrison, R., 826
Hart, G., 826
Harvey, M., 213
Haselden, A., 372
Hasinoff for the KEK-246 Collaboration, M. D., 842
Hawker, E. A., 643
Hayashi, T., 1031
Hayman, P. J., 372

Hazen, E. S., 792
He, X. C., 643
Healey, D. C., 328, 351
Helmer, R., 328, 351
Helminen, C., 743
Hemmick, T. K., 67
Henley, E., 365
Henley, E. M., 862
Henrich, C., 704
Henry-Couannier, F., 372
Hertzog, D. W., 481, 792
Hicks, K., 547
Higashi, A., 923
Hill for the E864 Collaboration, J. C., 465
Hinterberger, F., 704
Hinterberger for the EDDA Collaboration, F., 713
Hinton, W., 213, 561
Hirabayashi, H., 792
Hoblit, S., 547
Hoffart, A., 717
Hoffman-Rothe, P., 567
Hogan, G. E., 34, 826
Hollander, R. W., 372
Holstein, B., 408
Holstein, B. R., 438
Holt, R., 561
Holt, R. J., 213
Holynski, R., 253
Hseuh, H., 792
Hu, M., 419
Hubert, E., 372
Hughes, B. J., 792
Hughes, V. W., 429, 792
Hungerford, E. V., 826
Hungerford III, E. V., 34
Hwang, S.-R., 399

I

Ibald, R., 704
Ichii, S., 792
Idzik, M., 233
Ieiri, M., 923
Iijima, T., 923
Iinuma, M., 923
Imai, K., 923
Ingram, C. H. Q., 717

Inoue, T., 916
Isenhower, L. D., 643
Ishida, K., 792
Itow, Y., 923
Ivanov, Y. M., 722
Iwai, J., 1031
Iyono, A., 1031

J

Jackson, C., 561
Jackson, H. E., 213, 561
Jackson, J. W., 792
Jahn, R., 704
Jarczyk, L., 704
Jeon, S., 264
Ji, C.-R., 515
Jia, L., 792
Jin, S., 887
Johnson, J., 1031
Jon-And, K., 372
Jones, C., 561
Jones, G. L., 399
Jones, M. K., 342
Jones, W. V., 253
Jönsson, L., 605
Joosten, R., 704
Jöper, N., 849
Jouan, D., 233
Jui, C. C. H., 826
Jungmann, K., 429, 792
Jüngst, H. J., 849

K

Kadantser, S., 328
Kadantsev, S., 351
Kaidalov, A. B., 720
Kajita, T., 146
Kalinowsky, H., 849
Kamyshkov, Y., 141
Kamyshkov, Y. A., 335
Kaplan, D. M., 643
Karavitcheva, T. L., 233
Karpuchin, V., 429
Kasper, J., 887
Kaufman, S., 561
Kaufman, S. B., 643

Kawall, D., 792
Kelly, J. J., 561
Keppel, C., 561, 585
Keppel, C. E., 213
Kettle, P.-R., 372
Khandaker, M., 547, 561
Khandaker, M. A., 213
Khazin, B. I., 792
Kim, W., 561
Kindem, J., 792
Kinney, E., 561
Kinney, E. R., 213
Kinoshita, T., 792
Kirch, U., 849
Kirchner, R., 408
Kirk, P. N., 643
Kisel, I., 429
Kistner, O. C., 547
Klein, A., 213, 561, 717
Klein, F., 849
Klein, F. J., 849
Klein, S., 274
Klempt, E., 849
Kluberg, L., 233
Knott, J. E., 826
Kobayashi, M., 1031
Koch, K., 717
Koch, V., 227
Kochowski, C., 372
Koetke, D. D., 34, 643, 826
Köhler, J., 717
Kokkas, P., 372
Koltenuk, D., 561
Koltenuk, D. M., 213
Kopeliovich, V. B., 524
Korenchenko, A., 429
Korkmaz, E., 359
Kossakowski, R., 233
Kotelnikov, K. A., 269
Kotliński, B., 717
Kozela, A., 704
Kozlowski, T., 826
Kramer, L., 561
Krause, R., 387
Kravchuk, N., 429
Kreuger, R., 372
Kriener, F., 792
Kroedel, M., 717
Kroupa, M. A., 826
Krutenkova, A. P., 720

Kuchinsky, N., 429
Kuczewski, A., 547
Kudzia, D., 253
Kumbartzki, G., 213, 342
Kunz, P. D., 862
Kurepin, A. B., 233
Kurokawa, S., 792
Kuze for the H1 and ZEUS Collaboration, M., 612
Kuznetsov, Y., 328, 351
Kyle, G., 643, 717

L

Lajoie for the E864 Collaboration, J. G., 241
Lamot, G. H., 855
Lan, K., 34, 826
Landaud, G., 233
Lanou, R. E., 951
Laptev, V. D., 34
Larimer, R.-M., 931
Larson, R., 792
Lasio, G., 887
Laxdal, R., 328, 351
Le Bornec, Y., 233
Lebrun for the E867 Collaboration, P., 901
Lee, D. M., 643
Lee, F. S., 826
Lee, I. Y., 931
Lee, J. M., 923
Lee, L., 328, 351
Lee, W. M., 643
Lee, Y. Y., 792
Le Gac, R., 372
Lehmann, A., 717
Leimgruber, F., 372
Leitch, M. J., 643
Leviatan, A., 519
Levy, C. D. P., 322, 328, 351
Li, Z., 510, 852
Lichtenberg, D. B., 689
Lindgren, M., 419
Link, J., 849
Liolios, A., 372
Lipkin, H. J., 504
Lising, L. J., 399
Liu, F., 826

Logashenko, I., 792
Lopatin, I. V., 499
Lord, J., 1031
Lorenzon, W., 561
Lourenco, C., 233
Lovetere, M., 887
Lowry, M., 547
Lucas, M., 547
Lung, A., 561
Lung, A. F., 213
Luppi, E., 887
Luquin, L., 233

M

Maas, P., 887
Macchiavelli, A. O., 931
Macciotta, P., 233
Machado, E., 372
Machleidt, R., 707
Machner, H., 704
Mack, D., 561
Mack, D. J., 213
MacLead, R. W., 931
Macri, M., 887
Madey, R., 213, 561
Madsen, J., 999
Magiera, A., 704
Makino, S., 923
Makins, N., 643
Maltman, K., 736
Mandelkern, M., 887
Mandić, I., 372
Manley, D. M., 494
Manthos, N., 372
Manweiler, R., 34, 826
Mapes, M., 792
Marchetto, F., 887
Mardor, I., 640
Mardor, Y., 637
Marel, G., 372
Marinelli, M., 887
Markowitz, P., 213, 561
Markum, H., 756
Marriner, J., 419
Marsh, W., 887
Martens, M., 419
Martin, J., 561
Martin, P., 812

Martini, M., 887
Martynov, A. G., 269
Marzari-Chiesa, A., 233
Masaike, A., 923
Maschuw, R., 704
Masera, M., 233
Masoni, A., 233
Mateos, A., 561
Mateos, A. O., 717
Matone, G., 547
Matsuda, Y., 923
Matsuyama, Y., 923
Mayer-Kuckuk, T., 704
Mayes, B. W., 826
McFarlane, K., 561
McFarlane, K. W., 213
McGaughey, P. L., 643
McIntyre, J., 342
McKellar, B. H. J., 368
McKeown, R. D., 213, 423
McNabb, R., 792
McTaggart, R., 887
Mecke, A., 746
Meekins, D., 561
Meekins, D. G., 213
Meissner, T., 365, 753
Meissner, U. G., 730
Meng, W., 792
Menichetti, E., 887
Menze, D., 849
Merkel, J., 429
Mertler, G., 704
Merzliakov, Y., 792
Meyer, B. S., 992
Meyer, C. A., 91
Meyer, V., 429
Meziani, Z-E., 213
Miceli, L., 547
Michaelian, K., 717
Michaels for the Hall A Collaboration, R., 307
Mihara, S., 923
Mikuž, M., 372
Miller, D., 792
Miller, G. A., 621
Miller, J., 372
Miller, J. P., 792
Miller, M., 561
Miller, M. A., 213
Milner, R., 561

Mintz, S. L., 558
Mischke, R. E., 34, 328, 351, 826
Mitchell, J., 561
Mitchell, J. H., 213
Miyamura, O., 1031
Mizumachi, Y., 792
Mizutani, T., 855
Mkrtchyan, H., 561
Mkrtchyan, H. G., 213
Mohring, R., 561
Mohring, R. M., 213
Moiseenko, A., 429
Monich, V., 792
Montanet, F., 372
Moon, K. H., 1031
Morse, W. M., 792
Moss, J. M., 643
Mourgues, S., 233
Mull, V., 730
Muller, A., 372
Müller, B., 82
Müller, T., 419
Munkel, J., 704
Mussa, R., 887
Musso, A., 233
Mzavia, D., 429
Mzt, T., 849

N

Nagoshi, C., 923
Nakada, T., 372
Napolitano, J., 213, 530
Nathan, A. M., 213
Naviliat-Cunic, O., 408
Negele, J., 3
Nemoto, Y., 750
Neuerburg, W., 849
Neumann-Cosel, P. V., 704
Nico, J. S., 399
Niculescu, G., 213, 561
Niculescu, I., 213, 561
Nilsen, B. S., 253, 1031
Nishiizumi, K., 931
Nodland, B., 432
Nomura, I., 923
Nord, P. M., 643
Norman, E. B., 931
Noumi, H., 907

Numao, T., 769, 807

O

Obertino, M., 887
O'Connell, H. B., 383
Oda, H., 1031
Ogata, T., 1031
Ohlsson-Malek, F., 233
Oka, M., 750
Olson, E. D., 1031
Olszewski, A., 253
O'Neill, T. G., 213, 561
Orlov, Y., 792
Ostrick for the A3 Collaboration, M., 541
Otto, T., 408
Ouyang, J., 792
Owen, B. R., 213

P

Page, S. A., 328, 351, 438
Pagels, B., 372
Pai, C., 792
Pallavicini, M., 887
Papadopoulos, I., 372
Papavassiliou, V., 643
Park, B. K., 643
Park, I. S., 923
Parnell, T. A., 1031
Parreño, A., 912
Pastrone, N., 887
Pate, S., 213
Patrignani, C., 887
Paul, E., 849
Pavlopoulos, P., 372
Pearson, C., 792
Pedlar, T., 887
Pee, H. V., 849
Peng, J. C., 643
Peoples Jr., J., 887
Perdrisat, C. F., 342
Perillo Isaac, M. C., 931
Petiau, P., 233
Petitt, G., 643
Petković, T., 717
Petrunin, A. A., 722

Piccotti, A., 233
Piekarz, H., 694
Piilonen, L., 826
Piilonen, L. E., 34
Pillai, C., 826
Pinsky, L., 826
Pinto da Cunha, J., 372
Pizzi, J. R., 233
Planinić, M., 717
Plasil, F., 141
Plötzke, R., 849
Policarpo, A., 372
Polivka, G., 372
Polk, I., 792
Polly, C., 792
Polukhina, N. G., 269
Pomianowski for the CLEO Collaboration, P. A., 801
Pordes, S., 887
Potterveld, D., 561
Potterveld, D. H., 213
Pourkaviani, M., 558
Prado Da Silva, W. L., 233
Prakash, M., 1007
Preedom, B., 547
Price, J. W., 213, 561
Prieels, R., 408
Prige, R., 792
Procario for the SELEX Collaboration, M., 883
Puddu, G., 233
Punjabi, V., 342

Q

Quéméner, G., 342
Quin, P. A., 408

R

Racca, C., 233
Rakness, G. L., 213
Ralston, J. P., 432
Ramello, L., 233
Ramos, A., 912
Ramos, S., 233
Ramsay, W. D., 328, 351
Rankowitz, S., 792

Ransome, R., 213, 342
Rato-Mendes, P., 233
Rawlinson, A. A., 368
Ray, R., 419
Rebreyend, D., 547
Redin, S. I., 792
Redwine, R. P., 717
Reimer, P. E., 643
Reinhard, I., 429
Reinhardt, H., 664
Reinhold, J., 213, 561
Reitzner, S. D., 328, 351
Renker, D., 429
Riccati, L., 233
Rickenbach, R., 372
Rind, O., 792
Riska, D. O., 743
Roberts, B. L., 372, 792
Roberts, F. E., 1031
Robertson, H. G. R., 399
Robutti, E., 887
Rock for the E154 Collaboration, S., 579
Romana, A., 233
Rosen, J., 887
Rosendaal, D., 704
Rossen, P. V., 704
Rossetto, L., 887
Rowntree, D., 717
Roy, G., 328, 351
Ruf, T., 372
Rumerio, P., 887
Rutt, P. M., 213, 342
Ryskulov, N., 792

S

Sadler, M. E., 643
Saghai, B., 567
Saghi, B., 855
Saito, N., 923
Sakeliou, L., 372
Sakhelashvili, T., 429
Salgado, C., 561
Sammarruca, F., 707, 708
Sandacz, A., 573
Sandberg, J., 792
Sanders, P., 372
Sandorfi, A. M., 547

Sanford for the Super-Kamiokande Collaboration, R. E., 946
Santoni, C., 372
Santoro for the D0 Collaboration, A., 699
Santroni, A., 887
Sartori, S., 233
Sato, T., 792
Saturnini, P., 233
Savage, G., 213
Savrie, M., 887
Scadron, M. D., 673
Scannapieco, E., 274
Schaerf, C., 547
Schäfer, M., 372
Schaller, L. A., 372
Schatz, H., 987
Schietinger, T., 372
Schiffer, J. P., 561
Schilling, S., 826
Schmidt, P. V., 429
Schmor, P., 351
Schmor, P. W., 322, 328
Schnetzer, S. R., 812
Schnitker, H., 704
Scho, K., 704
Schopper, A., 372
Schroeder, L. S., 931
Schultz, J., 887
Schumacher, M., 849
Schumacher, R. A., 935
Schune, P., 372
Schütz, C., 725
Schuurmans, P., 408
Schwartz, A., 530
Schwartz for the E791 Collaboration, A. J., 892
Schwille, W. J., 849
Scomparin, E., 233
Sealock, R. M., 547
Sedykh, S., 792
Segel, R. E., 213, 561
Sekimoto, M., 923
Sekulovich, A. M., 351
Selden, J., 643
Semertzidis, Y. K., 792
Sengupta, K., 1031
Sennhauser, U., 717
Serci, S., 233
Serednyakov, S., 792

Seth, K. K., 887
Severijns, N., 408
Shaoian, R., 233
Shatunov, Y. M., 792
Shiina, T., 1031
Shin, Y. M., 923
Shuryak, E., 670
Shutt, R., 792
Silva, S., 233
Sim, K. S., 923
Simicevic, N., 213
Šimičević, N., 717
Slaughter, J., 648
Smend, F., 849
Smyrski, J., 704
Snow, G., 419
Snydstrup, L., 792
Soares, A., 372
Sobie, R. J., 832
Solodov, E., 792
Somalwar, S. V., 812
Sonderegger, P., 233
Sondheim, W. E., 643
Soukas, A., 792
Soukup, J., 328, 351
Speth, J., 725, 730
Stancari, G., 887
Stanislaus, S., 34
Stanislaus, T. D. S., 826
Stankus, P. W., 643
Stantz, K. M., 826
Stanz, K., 34
Steiger, T. D., 399
Stephens, F. S., 931
Stephenson, E. J., 708
Stern, B. E., 524
Stillman, A., 792
Stinson, G. M., 328, 351
Stocki, T. J., 351
Stocki, T. S., 328
Stoler, P., 213, 561
Stoler for the E94-014 Collaboration, P., 552
Stone, R., 812
Storm, D. W., 416
Straumann, U.
Strausz, S. C., 1031
Streets, J., 419, 887
Ströher, H., 547
Strzalkowski, A., 704

Suekane, F., 969
Sugitate, T., 1031
Sulak, L. R., 792
Suleiman, R., 213, 561
Sum, V., 351
Sun, V., 328
Susukita, R., 923
Sutton for the NEMO Collaboration, C. S., 941
Svoboda, R., 141
Swanson, E., 416
Swanson, E. S., 515
Swartz, K. B., 416
Szarska, M., 253
Szczepaniak, A., 515
Szymanski, J., 34
Szymanski, J. J., 826

T

Tabakin, F., 567
Taderosyan, V., 561
Takahashi, Y., 1031
Takashima, R., 923
Takeutchi, F., 923
Takizawa, M., 750
Talamonti, C., 779
Tallerico, T., 792
Tanaka, M., 297, 792
Tang, L., 213, 561
Tarrago, X., 233
Tauscher, L., 372
Teasdale, W. A., 399
Temnikov, P., 233
Terburg, B., 561
Terburg, B. P., 213
Thibault, C., 372
Thielemann, F.-K., 987
Thomas, A. W., 383
Thomas, E., 408
Thompson, A. K., 399
Thompson, M., 812, 887
Thompson, T. N., 643
Thomson, G. B., 812
Thorn, C. E., 547
Thornton, S. T., 547
Thurner, S., 756
Timmermans, C., 792
Titov, N., 351

Titov, N. A., 328
Tlustý, P., 923
Todenhagen for the H1 and ZEUS Collaborations, R., 591
Toldo, F., 792
Tölle, R., 704
Tomassetti, L., 887
Tominaga, T., 1031
Tonnison, J., 547
Topilskaya, N. S., 233
Touchard, F., 372
Touramanis, C., 372
Towell, R. S., 643
Träger, K., 429
Tran, M.-Q., 849
Trezeciak, R., 717
Triantis, F., 372
Tribble, R. E., 34, 643, 826
Trofimov, A., 792
Trzupek, A., 253
Tu, X. L., 826
Turchinetz, W., 561

U

Ullrich, H., 717
Ullrich, T. S., 279
Urner, D., 792
Usai, G., 233

V

Van Ausdeln, L. A., 826
Van Beveren, E., 372
Van Eijk, C. W. E., 372
Van Geert, A., 408
van Kolck, U.
Vanneste, L., 408
van Oers, W. T. H., 322, 328, 351
Van Orden, J. W., 730
van Schager, J. P. S., 416
van Westrum, D., 213, 561
Vasiliev, M. A., 643
Vercellin, E., 233
Vereecke, B., 408
Vlachos, S., 372
Vogt (No manuscript provided by speaker.), A., 125

von Smekal, L., 746
von Walter, P., 792
von Witsch, W., 34, 826
Vulcan, W. F., 213

W

Waddington, C. J., 253
Wagner, G., 685
Wagner, G. J., 679
Walet, N. R., 739
Walter, H. K., 429
Wang, M., 717
Wang, M. H., 717
Wang, Y. C., 643
Wang, Z. F., 643
Wang for the H1 and ZEUS Collaborations, S. M., 599
Wasserman, E. G., 399
Watts, J. W., 1031
Webb, J. C., 643
Weber, F., 181
Weber, P., 372
Wedemeyer, R., 849
Wefel, J. P., 253
Wehnes, F., 849
Weibe, S., 923
Weigel, H., 664
Welch, P., 561
Werkema, S., 887
Wester, W., 419
Wetel, J. P., 1031
Weyer, H. J., 717
Whisnant, C. S., 547
White, H., 812
Whitehouse, D., 826
Wiegers, B., 849
Wieland, F. W., 849
Wiescher, M., 987
Wietfeldt, F. E., 399
Wigger, O., 372
Wight, G. W., 322, 328
Wijesooriya, K., 342
Wilczynska, B., 253, 1031
Wilczynski, H., 253, 1031
Wildi, M., 717
Wilets, L., 862
Wilkerson, J. F., 399
Wilkes, R. J., 1031

Wilkes for the K2K Collaboration, R. J., 311
Williams, R. A., 733
Williamson, C., 561
Williamson, S. E., 213
Willis, J. L., 643
Willis, N., 233
Willmann, L., 429
Wilson, K. E., 717
Winn, D., 792
Wirtz, H. P., 429
Wise, D., 643
Wisskircher, J., 849
Witkowski, M. T., 213
Wittmann for the WA89 Collaboration, E. B., 875
Wolter, M., 372
Wolter, W., 253, 1031
Woo, R. J., 351
Wood, S., 561
Wood, S. A., 213
Woodle, K., 792
Worstell, W. A., 792
Wosiek, B., 253
Wozniak, K., 253
Wright, B. K., 826
Wright, D. C., 416
Wright, S. C., 826
Wright for the E614 Collaboration, D. H., 838
Wurzinger, R., 704

Y

Yamamoto, A., 792
Yamashita, S., 923
Yan, C., 213, 561
Yang, J.-C., 561
Yeche, C., 372
Yokkaichi, S., 923
Yokomi, H., 1031
Yoshida, K., 923
Yoshida, M., 923
Yoshida, T., 923
Young, G. R., 643
Yu, J., 561

Z

Zager, E., 1031
Zainea, G. D., 342
Zavrtanik, D., 372
Zeidman, B., 213, 561, 643
Zelenski, A., 351
Zelenski, A. N., 322, 328
Zhang, H., 547
Zhang, Y., 826
Zhao, W., 561
Zhao, X., 547
Zhao, Z., 416
Zhu, Y. C., 476
Zihlmann, B., 561
Zimmerman, D., 372, 792
Ziock, K. O. H., 34, 826
Zioulas, G., 887
zu Putlitz, G., 429, 792

AIP Conference Proceedings

	Title	L.C. Number	ISBN
No. 325	Conference on NASA Centers for Commercial Development of Space (Albuquerque, NM 1995)	94-73604	1-56396-431-7
No. 326	Accelerator Physics at the Superconducting Super Collider (Dallas, TX 1992-1993)	94-73609	1-56396-354-X
No. 327	Nuclei in the Cosmos III Third International Symposium on Nuclear Astrophysics (Assergi, Italy 1994)	95-75492	1-56396-436-8
No. 328	Spectral Line Shapes, Volume 8 12th ICSLS (Toronto, Canada 1994)	94-74309	1-56396-326-4
No. 329	Resonance Ionization Spectroscopy 1994 Seventh International Symposium (Bernkastel-Kues, Germany 1994)	95-75077	1-56396-437-6
No. 330	E.C.C.C. 1 Computational Chemistry F.E.C.S. Conference (Nancy, France 1994)	95-75843	1-56396-457-0
No. 331	Non-Neutral Plasma Physics II (Berkeley, CA 1994)	95-79630	1-56396-441-4
No. 332	X-Ray Lasers 1994 Fourth International Colloquium (Williamsburg, VA 1994)	95-76067	1-56396-375-2
No. 333	Beam Instrumentation Workshop (Vancouver, B. C., Canada 1994)	95-79635	1-56396-352-3
No. 334	Few-Body Problems in Physics (Williamsburg, VA 1994)	95-76481	1-56396-325-6
No. 335	Advanced Accelerator Concepts (Fontana, WI 1994)	95-78225	1-56396-476-7 (set) 1-56396-474-0 (Book) 1-56396-475-9 (CD-Rom)
No. 336	Dark Matter (College Park, MD 1994)	95-76538	1-56396-438-4
No. 337	Pulsed RF Sources for Linear Colliders (Montauk, NY 1994)	95-76814	1-56396-408-2
No. 338	Intersections Between Particle and Nuclear Physics 5th Conference (St. Petersburg, FL 1994)	95-77076	1-56396-335-3

	Title	L.C. Number	ISBN
No. 339	Polarization Phenomena in Nuclear Physics Eighth International Symposium (Bloomington, IN 1994)	95-77216	1-56396-482-1
No. 340	Strangeness in Hadronic Matter (Tucson, AZ 1995)	95-77477	1-56396-489-9
No. 341	Volatiles in the Earth and Solar System (Pasadena, CA 1994)	95-77911	1-56396-409-0
No. 342	CAM -94 Physics Meeting (Cacun, Mexico 1994)	95-77851	1-56396-491-0
No. 343	High Energy Spin Physics Eleventh International Symposium (Bloomington, IN 1994)	95-78431	1-56396-374-4
No. 344	Nonlinear Dynamics in Particle Accelerators: Theory and Experiments (Arcidosso, Italy 1994)	95-78135	1-56396-446-5
No. 345	International Conference on Plasma Physics ICPP 1994 (Foz do Iguaçu, Brazil 1994)	95-78438	1-56396-496-1
No. 346	International Conference on Accelerator-Driven Transmutation Technologies and Applications (Las Vegas, NV 1994)	95-78691	1-56396-505-4
No. 347	Atomic Collisions: A Symposium in Honor of Christopher Bottcher (1945-1993) (Oak Ridge, TN 1994)	95-78689	1-56396-322-1
No. 348	Unveiling the Cosmic Infrared Background (College Park, MD, 1995)	95-83477	1-56396-508-9
No. 349	Workshop on the Tau/Charm Factory (Argonne, IL, 1995)	95-81467	1-56396-523-2
No. 350	International Symposium on Vector Boson Self-Interactions (Los Angeles, CA 1995)	95-79865	1-56396-520-8
No. 351	The Physics of Beams Andrew Sessler Symposium (Los Angeles, CA 1993)	95-80479	1-56396-376-0
No. 352	Physics Potential and Development of $m^+ m^-$ Colliders: Second Workshop (Sausalito, CA 1994)	95-81413	1-56396-506-2
No. 353	13th NREL Photovoltaic Program Review (Lakewood, CO 1995)	95-80662	1-56396-510-0
No. 354	Organic Coatings (Paris, France, 1995)	96-83019	1-56396-535-6
No. 355	Eleventh Topical Conference on Radio Frequency Power in Plasmas (Palm Springs, CA 1995)	95-80867	1-56396-536-4

	Title	L.C. Number	ISBN
No. 356	The Future of Accelerator Physics (Austin, TX 1994)	96-83292	1-56396-541-0
No. 357	10th Topical Workshop on Proton-Antiproton Collider Physics (Batavia, IL 1995)	95-83078	1-56396-543-7
No. 358	The Second NREL Conference on Thermophotovoltaic Generation of Electricity	95-83335	1-56396-509-7
No. 359	Workshops and Particles and Fields and Phenomenology of Fundamental Interactions (Puebla, Mexico 1995)	96-85996	1-56396-548-8
No. 360	The Physics of Electronic and Atomic Collisions XIX International Conference (Whistler, Canada, 1995)	95-83671	1-56396-440-6
No. 361	Space Technology and Applications International Forum (Albuquerque, NM 1996)	95-83440	1-56396-568-2
No. 362	Two-Center Effects in Ion-Atom Collisions (Lincoln, NE 1994)	96-83379	1-56396-342-6
No. 363	Phenomena in Ionized Gases XXII ICPIG (Hoboken, NJ, 1995)	96-83294	1-56396-550-X
No. 364	Fast Elementary Processes in Chemical and Biological Systems (Villeneuve d'Ascq, France, 1995)	96-83624	1-56396-564-X
No. 365	Latin-American School of Physics XXX ELAF Group Theory and Its Applications (México City, México, 1995)	96-83489	1-56396-567-4
No. 366	High Velocity Neutron Stars and Gamma-Ray Bursts (La Jolla, CA 1995)	96-84067	1-56396-593-3
No. 367	Micro Bunches Workshop (Upton, NY, 1995)	96-83482	1-56396-555-0
No. 368	Acoustic Particle Velocity Sensors: Design, Performance and Applications (Mystic, CT, 1995)	96-83548	1-56396-549-6
No. 369	Laser Interaction and Related Plasma Phenomena (Osaka, Japan 1995)	96-85009	1-56396-445-7
No. 370	Shock Compression of Condensed Matter-1995 (Seattle, WA 1995)	96-84595	1-56396-566-6
No. 371	Sixth Quantum 1/f Noise and Other Low Frequency Fluctuations in Electronic Devices Symposium (St. Louis, MO, 1994)	96-84200	1-56396-410-4

Title	L.C. Number	ISBN
No. 372 Beam Dynamics and Technology Issues for + - Colliders 9th Advanced ICFA Beam Dynamics Workshop (Montauk, NY, 1995)	96-84189	1-56396-554-2
No. 373 Stress-Induced Phenomena in Metallization (Palo Alto, CA 1995)	96-84949	1-56396-439-2
No. 374 High Energy Solar Physics (Greenbelt, MD 1995)	96-84513	1-56396-542-9
No. 375 Chaotic, Fractal, and Nonlinear Signal Processing (Mystic, CT 1995)	96-85356	1-56396-443-0
No. 376 Chaos and the Changing Nature of Science and Medicine: An Introduction (Mobile, AL 1995)	96-85220	1-56396-442-2
No. 377 Space Charge Dominated Beams and Applications of High Brightness Beams (Bloomington, IN 1995)	96-85165	1-56396-625-7
No. 378 Surfaces, Vacuum, and Their Applications (Cancun, Mexico 1994)	96-85594	1-56396-418-X
No. 379 Physical Origin of Homochirality in Life (Santa Monica, CA 1995)	96-86631	1-56396-507-0
No. 380 Production and Neutralization of Negative Ions and Beams / Production and Application of Light Negative Ions (Upton, NY 1995)	96-86435	1-56396-565-8
No. 381 Atomic Processes in Plasmas (San Francisco, CA 1996)	96-86304	1-56396-552-6
No. 382 Solar Wind Eight (Dana Point, CA 1995)	96-86447	1-56396-551-8
No. 383 Workshop on the Earth's Trapped Particle Environment (Taos, NM 1994)	96-86619	1-56396-540-2
No. 384 Gamma-Ray Bursts (Huntsville, AL 1995)	96-79458	1-56396-685-9
No. 385 Robotic Exploration Close to the Sun: Scientific Basis (Marlboro, MA 1996)	96-79560	1-56396-618-2
No. 386 Spectral Line Shapes, Volume 9 13th ICSLS (Firenze, Italy 1996)		1-56396-656-5
No. 387 Space Technology and Applications International Forum (Albuquerque, NM 1997)	96-80254	1-56396-679-4 (Case set) 1-56396-691-3 (Paper set)
No. 388 Resonance Ionization Spectroscopy 1996 Eighth International Symposium (State College, PA 1996)	96-80324	1-56396-611-5

Title	L.C. Number	ISBN
No. 389 X-Ray and Inner-Shell Processes 17th International Conference (Hamburg, Germany 1996)	96-80388	1-56396-563-1
No. 390 Beam Instrumentation Proceedings of the Seventh Workshop (Argonne, IL 1996)	97-70568	1-56396-612-3
No. 391 Computational Accelerator Physics (Williamsburg, VA 1996)	97-70181	1-56396-671-9
No. 392 Applications of Accelerators in Research and Industry: Proceedings of the Fourteenth International Conference (Denton, TX 1996)	97-71846	1-56396-652-2
No. 393 Star Formation Near and Far Seventh Astrophysics Conference (College Park, MD 1996)	97-71978	1-56396-678-6
No. 394 NREL/SNL Photovoltaics Program Review Proceedings of the 14th Conference— A Joint Meeting (Lakewood, CO 1996)	97-72645	1-56396-687-5
No. 395 Nonlinear and Collective Phenomena in Beam Physics (Arcidosso, Italy 1996)	97-72970	1-56396-668-9
No. 396 New Modes of Particle Acceleration— Techniques and Sources (Santa Barbara, CA 1996)	97-72977	1-56396-728-6
No. 397 Future High Energy Colliders (Santa Barbara, CA 1997)	97-73333	1-56396-729-4
No. 398 Advanced Accelerator Colliders Seventh Workshop (Lake Tahoe, CA 1996)	97-72788	1-56396-697-2 (set) 1-56396-727-8 (cloth) 1-56396-726-X (CD-Rom)
No. 399 The Changing Role of Physics Departments: Proceedings of International Conference on Undergraduate Physics Education (College Park, MD 1996)	97-74866	1-56396-698-0
No. 400 High Energy Physics First Latin Symposium (Yucatan, México 1996)	97-73971	1-56396-686-7
No. 402 Astrophysical Implications of the Laboratory Study of Presolar Materials (St. Louis, MO 1996)	97-74679	1-56396-664-6
No. 401 Thermophotovoltaic Generation of Electricity Third NREL Conference (Colorado Springs, CO 1997)	97-74374	1-56396-734-0

Title	L.C. Number	ISBN
No. 403 Radio Frequency Power in Plasmas 12th Topical Conference (Savannah, GA 1997)	97-74472	1-56396-709-X
No. 404 Future Generations Photovoltaic Technologies First NREL Conference (Denver, CO 1997)	97-74386	1-56396-704-9
No. 405 Beam Stability and Nonlinear Dynamics (Santa Barbara, CA 1996)	97-74676	1-56396-731-6
No. 406 Laser Interaction and Related Plasma Phenomena 13th International Conference (Monterey, CA 1997)	97-76763	1-56396-696-4
No. 407 Deep Inelastic Scattering and QCD 5th International Workshop (Chicago, IL 1997)	97-74677	1-56396-716-2
No. 408 The Ultraviolet Universe at Low and High Redshift: Probing the Progress of Galaxy Evolution (College Park, MD 1997)	97-76762	1-56396-708-1
No. 409 Dense 2-Pinches 4th International Conference (Vancouver, Canada 1997)	97-76959	1-56396-610-7
No. 410 Proceedings of the 4th Compton Symposium (Williamsburg, VA 1997)	97-77179	1-56396-659-X
No. 411 Applied Non-Linear Dynamics Near the Millenium (San Diego, CA 1997)	97-77035	1-56396-736-7
No. 412 Intersections Between Particle and Nuclear Physics 6th Conference (Big Sky, MT 1997)	97-0564	1-56396-712-X
No. 413 Towards X-Ray Free Electron Lasers Workshop on Single Pass, High Gain FELs Starting from Noise, Aiming at Coherent X-Rays (Garda Lake, Italy 1997)	97-06161	1-56396-744-8
No. 414 Two-Dimensional Turbulence in Plasmas and Fluids Research Workshop (Canberra, Australia 1997)	97-06162	1-56396-764-2
No. 415 Beyond the Standard Model V Fifth Conference (Balholm, Norway 1997)	97-77246	1-56396-735-9
No. 416 Similarities and Differences Between Atomic Nuclei and Clusters: Toward a Unified Development of Cluster Science (Tsukuba, Japan 1997)		1-56396-714-6
No. 417 Synchrotron Radiation Instrumentation Tenth US National Conference (Ithaca, NY 1997)97-77402		1-56396-742-1